THE
TRANSFORMS
AND
APPLICATIONS
HANDBOOK

The Electrical Engineering Handbook Series

Series Editor

RICHARD C. DORF
University of California, Davis

Titles Included in the Series

The Biomedical Engineering Handbook, *Joseph D. Bronzino*
The Circuits and Filters Handbook, *Wai-Kai Chen*
The Transforms and Applications Handbook, *Alexander D. Poularikas*
The Control Handbook, *William S. Levine*

THE
TRANSFORMS
AND
APPLICATIONS
HANDBOOK

Editor-in-Chief

ALEXANDER D. POULARIKAS
University of Alabama
Huntsville, Alabama

CRC PRESS

 **IEEE PRESS**

A CRC Handbook Published in Cooperation with IEEE Press

Library of Congress Cataloging-in-Publication Data

The transforms and applications handbook / editor-in-chief, Alexander Poularikas.
      p.   cm. -- (The electrical engineering handbook series)
   Includes bibliographical references (p. -) and index.
   ISBN 0-8493-8342-0
   1. Transformations (Mathematics)--Handbooks, manuals, etc. I. Poularikas, Alexander
D., 1933– . II. Series.
   QA601.T73 1995
   515'.732--dc20
                            95-2513
                            CIP

© 1996 by CRC Press, Inc.

No claim to original U, S. Government works
International Standard Book Number 0-8493-8342-0
Library of Congress Card Number 95-2513
Printed in the United States of America 1 2 3 4 5 6 7 8 9 0
Printed on acid-free paper

Contents

Preface

The purpose of *The Transforms and Applications Handbook* is to include in a single volume the most important mathematical transforms frequently used by engineers and scientists. The book was also written with the advanced undergraduate and graduate students in mind. Each chapter covers one of the transforms, accompanied by a number of examples that are included to elucidate the use of the transform and its properties. Applications to different areas are also included in each chapter. This inclusion gives readers of different backgrounds the opportunity to get familiar with the wide spectrum of applications of these transforms. We believe that having all these useful transforms included in one book will be of great value to scientists, engineers, and students.

The information is organized into twelve chapters. Each chapter covers one of the transforms, except the first chapter, which contains material useful to the rest of the book. In addition, the first chapter enhances some topics that are treated less extensively in the other chapters. For example, numerous properties of the delta function and some of the classical orthogonal functions are included. The remaining chapters cover the following transforms: Fourier, cosine and sine, Hartley, Laplace, Z-, Hilbert, Radon and Abel, time-frequency, wavelet, Hankel, and Mellin. The appendices include *Complex Variables*, *Definite Integrals*, and *Matrices*.

A number of graphs in the text and specifically all the graphs in Chapter 1 have been produced using MATLAB® numeric computation and visualization software. MATLAB is developed by The MathWorks, Inc., which is located at 24 Prime Park Way, Natick, MA 01760 (phone: 508-653-1415, fax: 508-653-2997, email: info@mathworks.com, http://www.mathworks.com).

The Editor and CRC Press would be extremely grateful if the readers send their opinion about *The Handbook*, any errors they may detect, suggestions of additional material for new editions, and suggestions for deleting material.

The Handbook is the testimony of the efforts of our colleagues, whose contributions were invaluable, Joel Claypool of CRC Press, Publisher, the commitment of the Editor-in-Chief of this series, Dr. Richard Dorf, and others.

Alexander D. Poularikas
Editor

MATLAB is a registered trademark of The MathWorks, Inc.

Contributors

Fay Boudreaux-Bartels   Department of Electrical Engineering, University of Rhode Island, Kingston, Rhode Island

Jacqueline Bertrand   CNRS-LPTM, Universite de Paris VII, Paris, France

Pierre Bertrand   ONERA/DES, Chatillon, France

Stanley Deans   Department of Physics, University of Central Florida, Tampa, Florida

Stefan Hahn   Warsaw University of Technology, Institute of Radioelectronics, Warsaw, Poland

Kenneth Howell   Mathematics Department, University of Alabama in Huntsville, Huntsville, Alabama

Kraig J. Olejniczak   Department of Electrical Engineering, University of Arkansas, Fayetteville, Arkansas

Jean-Philippe Ovarlez   ONERA/DES, Palaiseau, France

Robert Piessens   Katholieke Universieit Leuven, Department of Computer Science, Heverlee, Belgium

Alexander D. Poularikas   Electrical and Computer Engineering, University of Alabama in Huntsville, Huntsville, Alabama

Samuel Seely   Westbrook, Connecticut

Yunlong Sheng   Department of Physics, Laval University, Quebec, Canada

Pat Yip   Communications Research Laboratories, McMaster University, Ontario, Canada

1

Signals and Systems

Alexander D. Poularikas

CONTENTS

1.1   Introduction to signals

A knowledge of a broad range of signals is of practical importance in describing human experience. In engineering systems, signals may carry information or energy. The signals with which we are concerned may be the cause of an event or the consequence of an action.

The characteristics of a signal may be of a broad range of shapes, amplitudes, time duration, and perhaps other physical properties. In many cases, the signal will be expressed in analytic form; in other cases, the signal may be given only in graphical form.

It is the purpose of this chapter to introduce the mathematical representations of signals, their properties, and some of their applications. These representations are in different formats depending on whether the signals are periodic or truncated, or whether they are deduced from graphical representations.

Signals may be classified as follows:

1. Phenomenological classification is based on the evolution type of signal, that is, a perfectly predictable evolution defines a deterministic signal and a signal with unpredictable behavior is called a **random signal**.

2. Energy classification separates signals into **energy signals**, those having finite energy, and **power signals**, those with a finite average power and infinite energy.

3. Morphological classification is based on whether signals are continuous, quantitized, sampled, or digital signals.

4. Dimensional classification is based on the number of independent variables.

5. Spectral classification is based on the shape of the frequency distribution of the signal spectrum.

1.1.1 *Functions (signals), variables, and point sets*

The **rule of correspondence** from a set S_x of real or complex numbers x to a real or complex number

$$y = f(x) \tag{1.1.1}$$

is called a function of the argument x. Equation (1.1.1) specifies a value (or values) y of the variable y (set of values in Y) corresponding to each suitable value of x in X. In (1.1.1) x is the **independent** variable and y is the **dependent** variable.

A function of n variables $x_1, x_2, \ldots, x_n$ associates values

$$y = f(x_1, x_2, \ldots, x_n) \tag{1.1.2}$$

of a dependent variable y with ordered sets of values of the independent variables $x_1, x_2, \ldots, x_n$.

The set S_x of the values of x (or sets of values of $x_1, x_2, \ldots, x_n$) for which the relationships (1.1.1) and (1.1.2) are defined constitutes the **domain** of the function. The corresponding set S_y of values of y is the **range** of the function.

A **single-valued** function produces a single value of the dependent variable for each value of the argument. A **multiple-valued** function attains two or more values for each value of the argument.

The function $y(x)$ has an **inverse** function $x(y)$ if $y = y(x)$ implies $x = x(y)$.

A function $y = f(x)$ is **algebraic** of x if and only if x and y satisfy a relation of the form $F(x, y) = 0$, where $F(x, y)$ is a polynomial in x and y. The function $y = f(x)$ is **rational** if $f(x)$ is a polynomial or is a quotient of two polynomials.

A real or complex function $y = f(x)$ is **bounded** on a set S_x if and only if the corresponding set S_y of values y is bounded. Furthermore, a real function $y = f(x)$ has an **upper bound**, **least upper bound**, **lower bound**, **greatest lower bound**, **maximum**, or **minimum** on S_x if this is also true for the corresponding set S_y.

Neighborhood

Given any finite real number a, an open neighborhood of the point a is the set of all points $\{x\}$ such that $|x - a| < \delta$ for any positive real number δ.

An open neighborhood of the point $(a_1, a_2, \ldots, a_n)$, where all a_i are finite, is the set of all points $(x_1, x_2, \ldots, x_n)$ such that $|x_1 - a_1| < \delta, |x_2 - a_2| < \delta, \ldots$ and $|x_n - a_n| < \delta$ for some positive real number δ.

Open and closed sets

A point P is a **limit** point (accumulation point) of the point set S if and only if every neighborhood of P has a neighborhood contained entirely in S, other than P itself.

A limit point P is an interior point of S if and only if P has a neighborhood contained entirely in S. Otherwise P is a **boundary** point.

A point P is an **isolated** point of S if and only if P has a neighborhood in which P is the only point belonging to S.

A point set is **open** if and only if it contains only interior points.

A point set is **closed** if and only if it contains all its limit points; a finite set is closed.

TABLE 1.2.1
Operations with limits.

$$\lim_{x \to a}[f(x) + g(x)] = \lim_{x \to a} f(x) + \lim_{x \to a} g(x)$$

$$\lim_{x \to a}[b\,f(x)] = b \lim_{x \to a} f(x)$$

$$\lim_{x \to a}[f(x)g(x)] = \lim_{x \to a} f(x) \lim_{x \to a} g(x)$$

$$\lim_{x \to a} \frac{f(x)}{g(x)} = \frac{\lim_{x \to a} f(x)}{\lim_{x \to a} g(x)} \qquad \left(\lim_{x \to a} g(x) \neq 0 \right)$$

$a =$ may be finite or infinite

1.1.2 *Limits and continuous functions*

1. A single-valued function $f(x)$ has a **limit**

$$\lim_{x \to a} f(x) = L, \qquad L = \text{finite}$$

as $x \to a\{f(x) \to L$ as $x \to a\}$ if and only if for each positive real number ε there exists a real number δ such that $0 < |x - a| < \delta$ implies that $f(x)$ is defined and $|f(x) - L| < \varepsilon$.

2. A single-valued function $f(x)$ has a limit

$$\lim_{x \to \infty} f(x) = L, \qquad L = \text{finite}$$

as $x \to \infty$ if and only if for each positive real number ε there exists a real number N such that $x > N$ implies that $f(x)$ is defined and $|f(x) - L| < \varepsilon$.

Operations with limits

If limits exist, Table 1.2.1 gives the limit operations.

Asymptotic relations between two functions

Given two real or complex functions $f(x), g(x)$ of a real or complex variable x, we write

1. $f(x) = 0[g(x)]$; $f(x)$ is **of the order** $g(x)$ as $x \to a$ if and only if there is a neighborhood of $x = a$ such that $|f(x)/g(x)|$ is bounded.

2. $f(x) \sim g(x)$; $f(x)$ is **asymptotically proportional** to $g(x)$ as $x \to a$ if and only if $\lim_{x \to a}[f(x)/g(x)]$ exists and it is not zero.

3. $f(x) \cong g(x)$; $f(x)$ is **asymptotically equal** to $g(x)$ as $x \to a$ if and only if

$$\lim_{x \to a}[f(x)/g(x)] = 1.$$

4. $f(x) = o[g(x)]$; $f(x)$ becomes negligible compared with $g(x)$ if and only if

$$\lim_{x \to a}[f(x)/g(x)] = 0.$$

5. $f(x) = \varphi(x) + O[g(x)]$ if $f(x) - \varphi(x) = O[g(x)]$
$f(x) = \varphi(x) + o[g(x)]$ if $f(x) - \varphi(x) = o[g(x)]$

Uniform convergence

1. A single-valued function $f(x_1, x_2)$ **converges uniformly** on a set S of values of x_2, $\lim_{x_1 \to a} f(x_1, x_2) = L(x_2)$ if and only if for each positive real number ε there exists a real number δ such that $0 < |x_1 - a| < \delta$ implies that $f(x_1, x_2)$ is defined and $|f(x_1, x_2) - L(x_2)| < \varepsilon$ for all x_2 in S (δ is independent of x_2).

2. A single-valued function $f(x_1, x_2)$ **converges uniformly** on a set S of values of x_2, $\lim_{x_1 \to \infty} f(x_1, x_2) = L(x_2)$ if and only if for each positive real number ε there exists a real number N such that for $x_1 > N$ implies that $f(x_1, x_2)$ is defined and $|f(x_1, x_2) - L(x_2)| < \varepsilon$ for all x_2 in S.

3. A **sequence** of functions $f_1(x), f_2(x), \ldots$ **converges uniformly** in a set S of values of x to a finite and unique function

$$\lim_{n \to \infty} f_n(x) = f(x)$$

if and only if for each positive real number ε there exists a real integer N such that for $n > N$ implies that $|f_n(x) - f(x)| < \varepsilon$ for all n in S.

Continuous functions

1. A single-valued function $f(x)$ defined in the neighborhood of $x = a$ is **continuous** at $x = a$ if and only if for every positive real number ε there exists a real number δ such that $|x - a| < \delta$ implies $|f(x) - f(x)| < \varepsilon$.

2. A function is **continuous on a series of points** (interval or region) if and only if it is continuous at each point of the set.

3. A real function continuous on a bounded closed interval $[a, b]$ is bounded on $[a, b]$ and assumes every value between and including its g.l.b. (greatest lower bound) and its l.u.b. (least upper bound) at least once on $[a, b]$.

4. A function $f(x)$ is **uniformly continuous** on a set S if and only if for each positive real number ε there exists a real number δ such that $|x - X| < \delta$ implies $|f(x) - f(X)| < \varepsilon$ for all X in S.

If a function is continuous in a bounded closed interval $[a, b]$, it is uniformly continuous on $[a, b]$.

If $f(x)$ and $g(x)$ are continuous at a point, so are the functions $f(x) + g(x)$ and $f(x)g(x)$.

Limits

1. A function $f(x)$ of a real variable x has the **right-hand** limit $\lim_{x \to a+} f(x) = f(a+) = L_+$ at $x = a$ if and only if for each positive real number ε there exists a real number δ such that $0 < x - a < \delta$ implies that $f(x)$ is defined and $|f(x) - L_+| < \varepsilon$.

2. A function $f(x)$ of a real variable x has the **left-hand** limit $\lim_{x \to a-} f(x) = f(a-) = L_-$ at $x = a$ if and only if for each positive real number ε there exists a real number δ such that $0 < a_- < \delta$ implies that $f(x)$ is defined and $|f(x) - L_-| < \varepsilon$.

3. If $\lim_{x \to a} f(x)$ exists, then $\lim_{x \to a+} f(x) = \lim_{x \to a-} f(x) = \lim_{x \to a} f(x)$. Conversely, $\lim_{x \to a-} f(x) = \lim_{x \to a+} f(x)$ implies the existence of $\lim_{x \to a} f(x)$.

4. The function $f(x)$ is **right continuous** at $x = a$ if $f(a+) = f(a)$.

5. The function $f(x)$ is **left continuous** at $x = a$ if $f(a-) = f(a)$.

6. A real function $f(x)$ has a **discontinuity of the first kind** at point $x = a$ if $f(a+)$ and $f(a-)$ exist. The greatest difference between two of these numbers $f(a)$, $f(a+)$,

$f(a-)$ is the **saltus** of $f(x)$ at the discontinuity. The discontinuities of the first kind of $f(x)$ constitute a discrete and countable set.

7. A real function $f(x)$ is **piecewise continuous** in an interval I if and only if $f(x)$ is continuous throughout I except for a finite number of discontinuities of the first kind.

Monotonicity

1. A real function $f(x)$ of a real variable x is **strongly monotonic** in the open interval (a, b) if $f(x)$ increases as x increases in (a, b), or if $f(x)$ decreases as x decreases in (a, b).

2. A function $f(x)$ is **weakly monotonic** in (a, b) if $f(x)$ does not decrease, or if $f(x)$ does not increase in (a, b). Analogous definitions apply to monotonic sequences.

3. A real function of a real variable x is of **bounded variation** in the interval (a, b) if and only if there exists a real number M such that

$$\sum_{i=1}^{m} |f(x_i) - f(x_{i-1})| < M \qquad \text{for all partitions}$$

$$a = x_0 < x_1 < x_2 < \cdots < x_m = b$$

of the interval (a, b). If $f(x)$ and $g(x)$ are of bounded variation in (a, b), then $f(x)+g(x)$ and $f(x)g(x)$ are of bounded variation also. The function $f(x)$ is of bounded variation in every finite open interval where $f(x)$ is bounded and has a finite number of relative maxima and minima and discontinuities (Dirichlet conditions).

A function of bounded variation in (a, b) is bounded in (a, b), and its discontinuities are only of the first kind.

Table 1.2.2 presents some useful mathematical functions.

1.1.3   *Energy and power signals*

Energy signals

If we consider any signal $f(t)$ as denoting a voltage that exists across a 1-ohm resistor, then

$$\frac{f^2(t)}{1} = f(t)\frac{f(t)}{1} = f(t)i(t) = \text{power } VA$$

Therefore, the integral

$$E = \int_a^b f^2(t)\, dt \qquad \text{joule} \tag{1.3.1}$$

representing the energy dissipated in the resistor during the time interval (a, b). A signal is called **energy signal** if

$$\int_{-\infty}^{\infty} f^2(t)\, dt < \infty \tag{1.3.2}$$

Power signals

Power signals are defined by the relation

$$0 \le \lim_{T \to \infty} \frac{1}{2T} \int_{-T}^{T} f^2(t)\, dt < \infty \tag{1.3.3}$$

TABLE 1.2.2
Some useful mathematical functions.

---

1. Signum Function

$$\text{sgn}(t) = \begin{cases} 1 & t > 0 \\ 0 & t = 0 \\ -1 & t < 1 \end{cases}$$

2. Step Function

$$u(t) = \frac{1}{2} + \frac{1}{2}\,\text{sgn}(t) = \begin{cases} 1 & t > 0 \\ 0 & t < 0 \end{cases}$$

3. Ramp Function

$$r(t) = \int_{-\infty}^{t} u(\tau)\,d\tau = tu(t)$$

4. Pulse Function

$$p_a(t) = u(t+a) - u(t-a) = \begin{cases} 1 & |t| < a \\ 0 & |t| > a \end{cases}$$

5. Triangular Pulse

$$\Lambda_a(t) = \begin{cases} 1 - \frac{|t|}{a} & |t| < a \\ 0 & |t| > a \end{cases}$$

6. Sinc Function

$$\sin c_a(t) = \frac{\sin at}{t}, \qquad -\infty < t < \infty$$

7. Gaussian Function

$$g_a(t) = e^{-at^2}, \qquad -\infty < t < \infty$$

8. Error Function

$$\text{erf}(t) = \frac{2}{\sqrt{\pi}} \int_0^t e^{-\tau^2}\,d\tau = \frac{2}{\sqrt{\pi}} \sum_{n=0}^{\infty} \frac{(-1)^n t^{2n+1}}{n!(2n+1)}$$

Properties:

$$\text{erf}(\infty) = 1,\ \text{erf}(0) = 0,\ \text{erf}(-t) = -\text{erf}(t)$$

$$\text{erfc}(t) = \text{complementary error function} = 1 - \text{erf}(t) = \frac{2}{\sqrt{\pi}} \int_t^{\infty} e^{-\tau^2}\,d\tau$$

9. Exponential Function

$$f(t) = e^{-at}u(t), \qquad t \geq 0$$

10. Double Exponential

$$f(t) = e^{-a|t|}, \qquad -\infty < t < \infty$$

11. Lognormal Function

$$f(t) = \frac{1}{t} e^{-\ln^2 t/2}, \qquad 0 < t < \infty$$

12. Rayleigh Function

$$f(t) = te^{-t^2/2}, \qquad 0 < t < \infty$$

For complex-valued signals, we must introduce $|f(t)|^2$ instead of $f^2(t)$.

We may represent the energy in a finite interval in terms of the coefficients of the basis function φ_i; that is, we write the energy integral in the form

$$E = \int_a^b f^2(t)\,dt = \int_a^b f(t) \sum_{n=0}^{\infty} c_n \varphi_n(t)\,dt = \sum_{n=0}^{\infty} c_n \int_a^b f(t)\varphi_n(t)\,dt = \sum_{n=0}^{\infty} c_n^2 \|\varphi_n(t)\|^2$$

$$(1.3.4)$$

where

$$\int_a^b f(t)\varphi_n(t)\, dt = c_n \int_a^b \varphi_n^2(t)\, dt \doteq c_n \|\varphi_n(t)\|^2$$

Because the square of the norm $\|\varphi_n(t)\|^2$ is the energy associated with the nth orthogonal function, (1.3.4) shows that the energy of the signal is the sum of the energies of its individual orthogonal components weighted by c_n. Note that this is the Parseval theorem. This equation shows that the set $\{\varphi_n(t)\}$ forms an orthogonal (complete) set, and the signal energy can be calculated from this representation.

Example

(a) $\int_0^\infty u^2(t)\, dt = \int_0^\infty dt = \infty$; $\lim_{T\to\infty} \frac{1}{2T} \int_{-T}^T u^2(t)\, dt = \lim_{T\to\infty} \frac{1}{2T} \int_0^T dt =$ $\lim_{T\to\infty} \frac{1}{2T} \left(t|_0^T\right) = \frac{1}{2} < \infty$. This implies that $u(t)$ is a power signal.

(b) The signal $e^{-at}u(t), a > 0$ is an energy signal. ∎

1.2 Distributions, delta function

1.2.1 Introduction

The **delta** function $\delta(t)$ often called the **impulse** or **Dirac delta** function, occupies a central place in signal analysis. Many physical phenomena such as point sources, point charges, concentrated loads on structures, and voltage or current sources, acting for very short times, can be modeled as delta functions.

Strictly speaking, delta functions are not functions in the accepted mathematical sense, and they cannot be treated with rigor within the framework of classical analysis. However, if distributions are introduced, then the concept of a delta function and operations on delta functions can be given a precise meaning.

1.2.2 Testing functions

A **distribution** is a generalization of a function. Within the framework of distributions, any function encountered in applications, such as unit-step functions and pulses, may be differentiated as many times as we desire, and any convergent series of functions may be differentiated term by term.

A **testing function** $\varphi(t)$ is a real-valued function of the real variable that can be differentiated an arbitrary number of times, and which is identical to zero outside a finite interval.

Example
Testing function

$$\varphi(t, a) = \begin{cases} e^{-\frac{a^2}{a^2-t^2}} & |t| < a \\ 0 & |t| \geq a \end{cases} \qquad (2.2.1)$$

∎

Properties

 1. If $f(t)$ can be differentiated arbitrarily often

$$\psi(t) = f(t)\varphi(t) = \text{testing function}$$

2. If $f(t)$ is zero outside a finite interval

$$\psi(t) = \int_{-\infty}^{\infty} f(\tau)\varphi(t-\tau)\,d\tau, \qquad -\infty < t < \infty = \text{testing function}$$

3. A sequence of testing functions, $\{\varphi_n\}$ $1 \leq n < \infty$, converges to zero if all φ_n are identically zero outside some interval independent of n and each φ_n, as well as all of its derivatives, tends uniformly to zero.

4. Testing functions belong to a set D, where D is a linear vector space, and if $\varphi_1 \in D$ and $\varphi_2 \in D$, then $\varphi_1 + \varphi_2 \in D$ and $a\varphi_1 \in D$ for any number a.

1.2.3 *Definition of distributions*

A **distribution** (or **generalized** function) $g(t)$ is a process of assigning to an arbitrary test function $\varphi(t)$ a **number** $N_g[\varphi(t)]$. A distribution is also a functional.

Example
An ordinary function $f(t)$ is a distribution if

$$\int_{-\infty}^{\infty} f(t)\varphi(t)\,dt = N_f[\varphi(t)] \tag{2.3.1}$$

exists for every test function $\varphi(t)$ in the set. For example, if $f(t) = u(t)$ then

$$\int_{-\infty}^{\infty} u(t)\varphi(t)\,dt = \int_{0}^{\infty} \varphi(t)\,dt \tag{2.3.2}$$

The function $u(t)$ is a distribution that assigns to $\varphi(t)$ a number equal to its area from zero to infinity. ∎

Properties of distributions

1. Linearity–Homogeneity

$$\int_{-\infty}^{\infty} g(t)[a_1\varphi_1(t) + a_2\varphi_2(t)]\,dt = a_1 \int_{-\infty}^{\infty} g(t)\varphi_1(t)\,dt + a_2 \int_{-\infty}^{\infty} g(t)\varphi_2(t)\,dt \tag{2.3.3}$$

   for all test functions and all numbers a_i.

2. Summation

$$\int_{-\infty}^{\infty} [g_1(t) + g_2(t)]\varphi(t)\,dt = \int_{-\infty}^{\infty} g_1(t)\varphi(t)\,dt + \int_{-\infty}^{\infty} g_2(t)\varphi(t)\,dt \tag{2.3.4}$$

3. Shifting

$$\int_{-\infty}^{\infty} g(t-t_0)\varphi(t)\,dt = \int_{-\infty}^{\infty} g(t)\varphi(t+t_0)\,dt \tag{2.3.5}$$

4. Scaling

$$\int_{-\infty}^{\infty} g(at)\varphi(t)\,dt = \frac{1}{|a|} \int_{-\infty}^{\infty} g(t)\varphi\left(\frac{t}{a}\right) dt \tag{2.3.6}$$

5. Even Distribution

$$\int_{-\infty}^{\infty} g(t)\varphi(t)\,dt = 0, \qquad \varphi(t) = \text{odd} \tag{2.3.7}$$

6. Odd Distribution

$$\int_{-\infty}^{\infty} g(t)\varphi(t)\,dt = 0, \qquad \varphi(t) = \text{even} \tag{2.3.8}$$

7. Derivative

$$\int_{-\infty}^{\infty} \frac{dg(t)}{dt}\varphi(t)\,dt = g(t)\varphi(t)|_{-\infty}^{\infty} - \int_{-\infty}^{\infty} g(t)\frac{d\varphi(t)}{dt}\,dt$$

$$= -\int_{-\infty}^{\infty} g(t)\frac{d\varphi(t)}{dt}\,dt \tag{2.3.9}$$

where the integrated term is equal to zero in view of the properties of testing functions.

8. The nth Derivative

$$\int_{-\infty}^{\infty} \frac{d^n g(t)}{dt^n}\varphi(t)\,dt = (-1)^n \int_{-\infty}^{\infty} g(t)\frac{d^n \varphi(t)}{dt^n}\,dt \tag{2.3.10}$$

9. Product with Ordinary Function

$$\int_{-\infty}^{\infty} [g(t)f(t)]\varphi(t)\,dt = \int_{-\infty}^{\infty} g(t)[f(t)\varphi(t)]\,dt \tag{2.3.11}$$

provided that $f(t)\varphi(t)$ belongs to the set of test functions.

10. Convolution

$$\int_{-\infty}^{\infty} \left[\int_{-\infty}^{\infty} g_1(\tau)g_2(t-\tau)\,d\tau\right]\varphi(t)\,dt$$

$$= \int_{-\infty}^{\infty} g_1(\tau)\left[\int_{-\infty}^{\infty} g_2(t-\tau)\varphi(t)\,dt\right]d\tau \tag{2.3.12}$$

by formal change of the order of integration.

DEFINITION
A sequence of distributions $\{g_n(t)\}_1^\infty$ is said to converge to the distribution $g(t)$ if

$$\lim_{n\to\infty}\int_{-\infty}^{\infty} g_n(t)\varphi(t)\,dt = \int_{-\infty}^{\infty} g(t)\varphi(t)\,dt \tag{2.3.13}$$

for all φ belonging to the set of test functions.

11. Every distribution is the limit, in the sense of distributions, of a sequence of infinitely differentiable functions.

12. If $g_n(t) \to g(t)$ and $r_n(t) \to r(t)$ (r is a distribution), and the numbers $a_n \to a$, then

$$\frac{d}{dt}g_n(t) \to \frac{dg(t)}{dt}, \qquad g_n(t) + r_n(t) \to g(t) + r(t), \qquad a_n g_n(t) \to a g(t) \tag{2.3.14}$$

13. Any distribution $g(t)$ may be differentiated as many times as desired. That is, the derivative of any distribution always exists and it is a distribution.

1.2.4   *The delta function*

Properties

Based on the distribution properties, the properties of the delta function are given below.

1. The delta function is a distribution assigning to the function $\varphi(t)$ the number $\varphi(0)$; thus

$$\int_{-\infty}^{\infty} \delta(t)\varphi(t)\, dt = \varphi(0) \tag{2.4.1}$$

2. Shifted

$$\int_{-\infty}^{\infty} \delta(t - t_0)\varphi(t)\, dt = \varphi(t_0) \tag{2.4.2}$$

3. Scaled

$$\int_{-\infty}^{\infty} \delta(at)\varphi(t)\, dt = \frac{1}{|a|} \int_{-\infty}^{\infty} \delta(t)\varphi\left(\frac{t}{a}\right) dt = \frac{1}{|a|}\varphi(0)$$

From (2.4.1) we have the identity

$$\delta(at) = \frac{1}{|a|}\delta(t)$$

and hence $(a = -1)$

$$\delta(-t) = \delta(t) = \text{even} \tag{2.4.3}$$

4. Multiplication by Continuous Function

$$\int_{-\infty}^{\infty} [\delta(t)f(t)]\varphi(t)\, dt = \int_{-\infty}^{\infty} \delta(t)[f(t)\varphi(t)]\, dt = f(0)\varphi(0)$$

If $f(t)$ is continuous at 0, then

$$f(t)\delta(t) = f(0)\delta(t) \tag{2.4.4}$$

and

$$t\delta(t) = 0 \tag{2.4.5}$$

5. Derivatives

$$\int_{-\infty}^{\infty} \frac{d\delta(t)}{dt}\varphi(t)\, dt = -\frac{d\varphi(0)}{dt}$$

$$\int_{-\infty}^{\infty} \frac{d\delta(t - t_0)}{dt}\varphi(t) = -\frac{d\varphi(t_0)}{dt} \tag{2.4.6}$$

$$\int_{-\infty}^{\infty} \frac{d^n\delta(t)}{dt^n}\varphi(t)\, dt = (-1)^n \frac{d^n\varphi(0)}{dt^n} \tag{2.4.7}$$

$$\int_{-\infty}^{\infty} \frac{d\delta(t)}{dt}f(t)\varphi(t)\, dt = -\int_{-\infty}^{\infty} \delta(t)\frac{d[f(t)\varphi(t)]}{dt}\, dt$$

$$= -f(0)\frac{d\varphi(0)}{dt} - \frac{df(0)}{dt}\varphi(0) \tag{2.4.8}$$

$$f(t)\frac{d\delta(t)}{dt} = -\frac{df(0)}{dt}\delta(t) + f(0)\frac{d\delta(t)}{dt} \tag{2.4.9}$$

$$t\frac{d\delta(t)}{dt} = -\delta(t) \tag{2.4.10}$$

Set $f(t) = \varphi(t) = 1$ in (2.4.8) to find the relation

$$\int_{-\infty}^{\infty} \frac{d\delta(t)}{dt}\, dt = 0 \qquad \left[\frac{d\delta(t)}{dt}\ \text{is an odd function}\right] \tag{2.4.11}$$

$$f(t)\frac{d^n\delta(t)}{dt^n} = \sum_{k=0}^{n}(-1)^k \frac{n!}{k!(n-k)!}\frac{d^k f(0)}{dt^k}\frac{d^{n-k}\delta(t)}{dt^{n-k}} \tag{2.4.12}$$

From

$$\int_{-\infty}^{\infty} \frac{du(t)}{dt}\varphi(t)\, dt = u(t)\varphi(t)\Big|_{-\infty}^{\infty} - \int_{-\infty}^{\infty} u(t)\frac{d\varphi(t)}{dt}\, dt$$

$$= -\int_{0}^{\infty} \frac{d\varphi(t)}{dt}\, dt = -\varphi(t)\int_{0}^{\infty} = \varphi(0)$$

and comparing with (2.4.1) we find that

$$\delta(t) = \frac{du(t)}{dt} \tag{2.4.13}$$

Therefore the generalized derivatives of discontinuous function contain impulses. A_n is the jump at the discontinuity point $t = t_n$ of the expression $A_n\delta(t - t_n)$. Also

$$\frac{d\delta(t)}{dt} = \frac{d^2 u(t)}{dt^2} \qquad \text{or} \qquad u(t) + u(-t) = 1$$

hence

$$\frac{du(-t)}{dt} = -\delta(t) \tag{2.4.14}$$

$$\delta(t - t_0) = \frac{du(t - t_0)}{dt} \tag{2.4.15}$$

If $r(t)$ has a finite or countably infinite number of zeros at t_n on the entire t axis and these points $r(t)$ have a continuous derivative $dr(t_n)/dt \neq 0$, then

$$\delta[r(t)] = \sum_{n} \frac{\delta(t - t_n)}{\left|\frac{dr(t_n)}{dt}\right|} \tag{2.4.16}$$

Hence, we obtain

$$\delta(t^2 - 1) = \frac{1}{2}\delta(t - 1) + \frac{1}{2}\delta(t + 1) \tag{2.4.17}$$

$$\delta(\sin t) = \sum_{n=-\infty}^{\infty} \delta(t - n\pi) \tag{2.4.18}$$

In addition, the following relation is also true:

$$\frac{d\delta[r(t)]}{dt} = \sum_{n} \frac{\frac{d\delta(t - t_n)}{dt}}{\frac{dr(t)}{dt}\left|\frac{dr(t_n)}{dt}\right|} \tag{2.4.19}$$

6. Integrals

$$\int_{-\infty}^{\infty} A\delta(t - t_0)\, dt = A \tag{2.4.20}$$

for all t_0

$$\delta(t - t_1) * \delta(t - t_2) = \text{convolution}$$

$$= \int_{-\infty}^{\infty} \delta(\tau - t_1)\delta(t - \tau - t_2)\,d\tau = \delta[t - (t_1 + t_2)] \quad (2.4.21)$$

$$f(t) * \delta(t) = \int_{-\infty}^{\infty} f(t - \tau)\delta(\tau)\,d\tau = f(t - 0) = f(t) \quad (2.4.22)$$

Distributions as generalized limits

We can define a distribution as a generalized limit of a sequence $f_n(t)$ of ordinary function. If there exists a sequence $f_n(t)$ such that the limit

$$\lim_{n \to \infty} \int_{-\infty}^{\infty} f_n(t)\varphi(t)\,dt \quad (2.4.23)$$

exists for every test function in the set, then the result is a number depending on $\varphi(t)$. Hence we may define a distribution $g(t)$ as

$$g(t) = \lim f_n(t) \quad (2.4.24)$$

and, therefore, equivalently

$$\delta(t) = \lim f_n(t) \quad (2.4.25)$$

Consider the two sequences shown in Figures 2.4.1a and 2.4.2b. The rectangular pulse sequence is given by

$$p_\varepsilon(t) = \frac{u(t) - u(t - \varepsilon)}{\varepsilon}$$

and has area unity whatever the value of ε. Because $\varphi(t)$ is continuous, it follows that

$$\lim_{\varepsilon \to 0} \int_{-\infty}^{\infty} p_\varepsilon(t)\varphi(t)\,dt = \lim_{\varepsilon \to 0} \frac{1}{\varepsilon} \int_0^\varepsilon \varphi(t)\,dt = \lim_{\varepsilon \to 0} \varphi(0)\frac{1}{\varepsilon} \int_0^\varepsilon dt = \varphi(0)$$

and therefore

$$\delta(t) = \lim_{\varepsilon \to 0} p_\varepsilon(t) \quad (2.4.26)$$

Similarly, from

$$\lim_{\varepsilon \to 0} \frac{1}{\sqrt{\varepsilon\pi}} \int_{-\infty}^{\infty} e^{-t^2/\varepsilon}\varphi(t)\,dt \cong \frac{\varphi(0)}{\sqrt{\varepsilon\pi}} \int_{-\infty}^{\infty} e^{-t^2/\varepsilon}\,dt = \varphi(0)$$

it follows that

$$\delta(t) = \lim_{\varepsilon \to 0} \frac{e^{-t^2/\varepsilon}}{\sqrt{\varepsilon\pi}} \quad (2.4.27)$$

If we use the sequence

$$\delta(t) = \lim_{\omega \to \infty} \frac{\sin \omega t}{\pi t}$$

we find that

$$\delta(t) = \lim_{a \to \infty} \frac{1}{2\pi} \int_{-a}^{a} e^{+j\omega t}\,d\omega = \lim_{a \to \infty} \frac{\sin at}{\pi t} = \frac{1}{2\pi} \int_{-\infty}^{\infty} e^{+j\omega t}\,d\omega \quad (2.4.28)$$

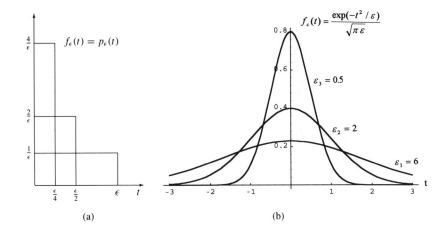

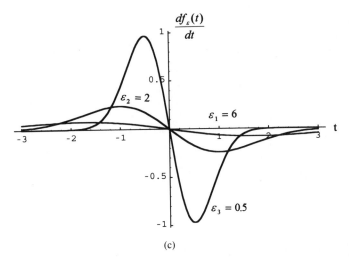

(c)

FIGURE 2.4.1

Also

$$\delta(t - t_0) = \frac{1}{2\pi} \int_{-\infty}^{\infty} e^{-j\omega(t-t_0)} \, d\omega \tag{2.4.29}$$

Further

$$\int_{-\infty}^{\infty} \cos \omega t \, d\omega = \lim_{\Omega \to \infty} \int_{-\Omega}^{\Omega} \cos \omega t \, d\omega$$

$$= \lim_{\Omega \to \infty} \frac{2 \sin \Omega t}{t}$$

$$= \lim_{\Omega \to \infty} 2\pi \frac{\sin \Omega t}{\pi t} = 2\pi \delta(t) \tag{2.4.30}$$

Figure 2.4.1c shows the derivatives of the sequence (2.4.27).   The following examples will elucidate some of the delta properties and the use of the delta function in Table 2.4.1.

TABLE 2.4.1
Delta function properties.

1. $\delta(at) = \dfrac{1}{|a|}\delta(t)$

2. $\delta\left(\dfrac{t - t_0}{a}\right) = |a|\delta(t - t_0)$

3. $\delta(at - t_0) = \dfrac{1}{|a|}\delta\left(t - \dfrac{t_0}{a}\right)$

4. $\delta(-t + t_0) = \delta(t - t_0)$

5. $\delta(-t) = \delta(t); \qquad \delta(t) = \text{even function}$

6. $\displaystyle\int_{-\infty}^{\infty} \delta(t)f(t)\,dt = f(0)$

7. $\displaystyle\int_{-\infty}^{\infty} \delta(t - t_0)f(t) = f(t_0)$

8. $f(t)\delta(t) = f(0)\delta(t)$

9. $f(t)\delta(t - t_0) = f(t_0)\delta(t - t_0)$

10. $t\delta(t) = 0$

11. $\displaystyle\int_{-\infty}^{\infty} A\delta(t)\,dt = \int_{-\infty}^{\infty} A\delta(t - t_0)\,dt = A$

12. $f(t) * \delta(t) = \text{convolution} = \displaystyle\int_{-\infty}^{\infty} f(t - \tau)\delta(\tau)\,d\tau = f(t)$

13. $\delta(t - t_1) * \delta(t - t_2) = \displaystyle\int_{-\infty}^{\infty} \delta(\tau - t_1)\delta(t - \tau - t_2)\,d\tau = \delta[t - (t_1 + t_2)]$

14. $\displaystyle\sum_{n=-N}^{N} \delta(t - nT) * \sum_{n=-N}^{N} \delta(t - nT) = \sum_{n=-2N}^{2N} (2N + 1 - |n|)\delta(t - nT)$

15. $\displaystyle\int_{-\infty}^{\infty} \dfrac{d\delta(t)}{dt} f(t)\,dt = -\dfrac{df(0)}{dt}$

16. $\displaystyle\int_{-\infty}^{\infty} \dfrac{d\delta(t - t_0)}{dt} f(t)\,dt = -\dfrac{df(t_0)}{dt}$

17. $\displaystyle\int_{-\infty}^{\infty} \dfrac{d^n\delta(t)}{dt^n} f(t)\,dt = (-1)^n \dfrac{d^n f(0)}{dt^n}$

18. $f(t)\dfrac{d\delta(t)}{dt} = -\dfrac{df(0)}{dt}\delta(t) + f(0)\dfrac{d\delta(t)}{dt}$

19. $t\dfrac{d\delta(t)}{dt} = -\delta(t)$

20. $t^n \dfrac{d^m \delta(t)}{dt^m} = \begin{cases} (-1)^n n!\delta(t), & m = n \\[2mm] (-1)^n \dfrac{m!}{m-n!} \dfrac{d^{m-n}\delta(t)}{dt^{m-n}}, & m > n \\[2mm] 0, & m < n \end{cases}$

TABLE 2.4.1
(continued)

21. $\displaystyle\int_{-\infty}^{\infty} \frac{d\delta(t)}{dt} = 0, \qquad \frac{d\delta(t)}{dt} = \text{odd function}$

22. $\displaystyle f(t) * \frac{d\delta(t)}{dt} = \frac{df(t)}{dt}$

23. $\displaystyle f(t)\frac{d^n\delta(t)}{dt^n} = \sum_{k=0}^{n}(-1)^k \frac{n!}{k!(n-k)!} \frac{d^k f(0)}{dt^k} \frac{d^{n-k}\delta(t)}{dt^{n-k}}$

24. $\displaystyle \frac{\partial\delta(yt)}{\partial y} = -\frac{1}{y^2}\delta(t)$

25. $\displaystyle \delta(t) = \frac{du(t)}{dt}$

26. $\displaystyle \frac{d^n\delta(-t)}{dt^n} = (-1)^n \frac{d^n\delta(t)}{dt^n}, \; \left\{\frac{d^n\delta(t)}{dt^n} \text{ is even if } n \text{ is even, and odd if } n \text{ is odd.}\right\}$

27. $\displaystyle (\sin at)\frac{d\delta(t)}{dt} = -a\delta(t)$

28. $\displaystyle \frac{d\delta(t)}{dt} = \frac{d^2u(t)}{dt^2}$

29. $\displaystyle -\delta(t) = \frac{du(-t)}{dt}$

30. $\displaystyle \delta(t - t_0) = \frac{du(t - t_0)}{dt}$

31. $\displaystyle \frac{d\,\text{sgn}(t)}{dt} = 2\delta(t)$

32. $\displaystyle \delta[r(t)] = \sum_n \frac{\delta(t - t_n)}{\left|\frac{dr(t_n)}{dt}\right|}, \qquad t_n = \text{zeros of } r(t), \quad \frac{dr(t_n)}{dt} \neq 0$

33. $\displaystyle \frac{d\delta[r(t)]}{dt} = \sum_n \frac{\frac{d\delta(t-t_n)}{dt}}{\frac{dr(t)}{dt}\left|\frac{dr(t_n)}{dt}\right|}, \qquad t_n = \text{zeros of } r(t), \quad \frac{dr(t_n)}{dt} \neq 0, \; \frac{dr(t)}{dt} \neq 0$

34. $\displaystyle \delta(\sin t) = \sum_{n=-\infty}^{\infty} \delta(t - n\pi)$

35. $\displaystyle \delta(t^2 - 1) = \frac{1}{2}\delta(t - 1) + \frac{1}{2}\delta(t + 1)$

36. $\displaystyle \delta(t^2 - a^2) = \frac{1}{2a}[\delta(t + a) + \delta(t - a)]$

37. $\displaystyle \delta(t) = \lim_{\varepsilon \to 0} \frac{e^{-t^2/\varepsilon}}{\sqrt{\varepsilon\pi}}$

38. $\displaystyle \delta(t) = \lim_{\omega \to \infty} \frac{\sin \omega t}{\pi t}$

39. $\displaystyle \delta(t) = \lim_{\varepsilon \to 0} \frac{1}{\pi} \frac{\varepsilon}{t^2 + \varepsilon^2}$

40. $\displaystyle \delta(t) = \frac{1}{2\pi}\int_{-\infty}^{\infty} \cos \omega t \, d\omega$

TABLE 2.4.1
(continued)

41. $\dfrac{df(t)}{dt} = \dfrac{d}{dt}[tu(t) - (t-1)u(t-1) - u(t-1)]$

$= t\delta(t) + u(t) - (t-1)\delta(t-1) - u(t-1) - \delta(t-1)$

42. $\text{comb}_T(t) = \displaystyle\sum_{n=-\infty}^{\infty} \delta(t - nT), \qquad f(t)\,\text{comb}_T(t) = \sum_{n=-\infty}^{\infty} f(nT)\delta(t - nT)$

$\text{COMB}_{\omega_0}(\omega) = \mathcal{F}\{\text{comb}_T(t)\} = \omega_0 \displaystyle\sum_{n=-\infty}^{\infty} \delta(t - n\omega_0), \qquad \omega_0 = \dfrac{2\pi}{T}$

Example
Equivalence of expressions involving the delta functions:

(a) $(\cos t + \sin t)\delta(t) = \delta(t)$

(b) $\cos 2t + \sin t\,\delta(t) = \cos 2t$

(c) $1 + 2e^{-t}\delta(t-1) = 1 + 2e^{-1}\delta(t-1)$ ∎

Example
The values of the following integrals are

$$\int_{-\infty}^{\infty}(t^2 + 4t + 5)\delta(t)\,dt = 0^2 + 4\cdot 0 + 5 = 5, \qquad \int_{-\infty}^{\infty}\frac{(1 + \cos t)\delta(t)}{1 + 2e^t}\,dt = \frac{2}{1+2}$$

$$\int_{-\infty}^{\infty} t^2 \sum_{k=1}^{n}\delta(t-k)\,dt = \sum_{k=1}^{n} k^2 = \frac{1}{6}[n(n+1)(2n+1)] \qquad ∎$$

Example
The first derivative of the functions is

$$\frac{d}{dt}(2u(t+1) + u(1-t)) = \frac{d}{dt}(2u(t+1) + u[-(t-1)]) = 2\delta(t+1) - \delta(t-1)$$

$$\frac{d}{dt}([2 - u(t)]\cos t) = \frac{d}{dt}(2\cos t - u(t)\cos t)$$
$$= 2\sin t - \delta(t)\cos t + u(t)\sin t$$
$$= (u(t) - 2)\sin t - \delta(t)$$

$$\frac{d}{dt}\left(\left[u\left(t - \frac{\pi}{2}\right) - u(t-\pi)\right]\sin t\right) = \left[\delta\left(t - \frac{\pi}{2}\right) - \delta(t-\pi)\right]\sin t$$
$$+ \left[u\left(t - \frac{\pi}{2}\right) - u(t-\pi)\right]\cos t$$
$$= \delta\left(t - \frac{\pi}{2}\right)$$
$$+ \left[u\left(t - \frac{\pi}{2}\right) - u(t-\pi)\right]\cos t \qquad ∎$$

Example

The values of the following integrals are

$$\int_{-\infty}^{\infty} e^{2t} \sin 4t \frac{d^2\delta(t)}{dt^2}\, dt = (-1)^2 \frac{d^2}{dt^2}[e^{2t}\sin 4t]\bigg|_{t=0} = 2 \times 2 \times 4 = 16$$

$$\int_{-\infty}^{\infty} (t^3 + 2t + 3)\left(\frac{d\delta(t-1)}{dt} + 2\frac{d^2\delta(t-2)}{dt^2}\right)dt = \int_{-\infty}^{\infty}(t^3+2t+3)\frac{d\delta(t-1)}{dt}\,dt$$

$$+ 2\int_{-\infty}^{\infty}(t^3+2t+3)\frac{d^2\delta(t-2)}{dt^2}\,dt$$

$$= (-1)(3t^2+2)\big|_{t=1} + (-1)^2 2(6t)\big|_{t=2}$$

$$= -5 + 24 = 19 \qquad\blacksquare$$

Example

The values of the following integrals are

$$\int_0^4 e^{4t}\delta(2t-3)\,dt = \int_0^4 e^{4t}\delta\left[2\left(t-\frac{3}{2}\right)\right] = \frac{1}{2}\int_0^4 e^{4t}\delta\left(t-\frac{3}{2}\right)dt = \frac{1}{2}e^{4\frac{3}{2}} = \frac{1}{2}e^6$$

$$\int_0^4 e^{4t}\delta(3-2t)\,dt = \int_0^4 e^{4t}\delta[-(2t-3)]\,dt = \frac{1}{2}\int_0^4 e^{4t}\delta(2t-3)\,dt = \frac{1}{2}e^6$$

$$\int_{-\infty}^{\infty} e^{at}\delta(\sin t)\,dt = \int_{-\infty}^{\infty} e^{at}\sum_{n=-\infty}^{\infty}\frac{\delta(t-n\pi)}{(-1)^n}\,dt$$

$$= \sum_{n=-\infty}^{\infty}\frac{1}{(-1)^n}\int_{-\infty}^{\infty}e^{at}\delta(t-n\pi)\,dt$$

$$= \sum_{n=-\infty}^{\infty}\frac{1}{(-1)^n}e^{an\pi} \qquad\blacksquare$$

Example

The values of the following integrals are

$$\int_{-2\pi}^{2\pi} e^{at}\delta(t^2-\pi^2)\,dt = \int_{-2\pi}^{2\pi} e^{at}\frac{1}{2\pi}[\delta(t-\pi)+\delta(t+\pi)]\,dt$$

$$= \frac{1}{2\pi}[e^{a\pi}+e^{-a\pi}]$$

$$= \frac{\cosh a\pi}{\pi}$$

$$\int_{-\pi}^{\pi} \cosh\theta\,\delta(\cos\theta)\,d\theta = \int_{-\pi}^{\pi} \cosh\theta\left[\frac{\delta\left(\theta+\frac{\pi}{2}\right)}{\left|\sin\left(-\frac{\pi}{2}\right)\right|} + \frac{\delta\left(\theta-\frac{\pi}{2}\right)}{\left|\sin\frac{\pi}{2}\right|}\right]d\theta$$

$$= \cosh\left(-\frac{\pi}{2}\right) + \cosh\frac{\pi}{2}$$

$$= 2\cosh\frac{\pi}{2} \qquad\blacksquare$$

1.2.5   *The gamma and beta functions*

The gamma function is defined by the formula

$$\Gamma(z) = \int_0^\infty e^{-t} t^{z-1}\, dt, \qquad \text{Re}\{z\} > 0 \tag{2.5.1}$$

We shall mainly concentrate on the positive values of z and we shall take the following relationship as the basic definition of the **gamma function**:

$$\Gamma(x) = \int_0^\infty e^{-t} t^{x-1}\, dt, \qquad x > 0 \tag{2.5.2}$$

The gamma function converges for all positive values of x.

   The **incomplete gamma function** is given by

$$\gamma(x, \tau) = \int_0^\tau t^{x-1} e^{-t}\, dt, \qquad x > 0,\ \tau > 0 \tag{2.5.3}$$

   The **beta function** is a function of two arguments and is given by

$$B(x, y) = \int_0^1 t^{x-1}(1 - t)^{y-1}\, dt, \qquad x > 0,\ y > 0 \tag{2.5.4}$$

The beta function is related to the gamma function as follows:

$$B(x, y) = \frac{\Gamma(x)\Gamma(y)}{\Gamma(x + y)} \tag{2.5.5}$$

Integral expressions of $\Gamma(x)$

If we set $u = e^{-t}$ in (2.5.3), then $1/u = e^t$, $\log_e(1/u) = t$, $-(1/u)du = dt$, and $[\log_e(1/u)]^{x-1} = t^{x-1}$, for the limits $t = 0\ u = 1$, and $t = \infty\ u = 0$. Hence

$$\Gamma(x) = \int_0^\infty t^{x-1} e^{-t}\, dt = -\int_1^0 \left[\log_e\left(\frac{1}{u}\right)\right]^{x-1} u\frac{1}{u}\, du = \int_0^1 \left[\log_e\left(\frac{1}{u}\right)\right]^{x-1} du \tag{2.5.6}$$

   Starting from the definitions and setting $t = m^2$ ($dt = 2m\, dm$) we obtain (limits are the same)

$$\Gamma(x) = \int_0^\infty t^{x-1} e^{-t}\, dt = \int_0^\infty m^{2(x-1)} e^{-m^2} 2m\, dm = 2\int_0^\infty m^{2x-1} e^{-m^2}\, dm \tag{2.5.7}$$

Properties and specific evaluations of $\Gamma(x)$

Setting $x + 1$ in place of x we obtain

$$\Gamma(x + 1) = \int_0^\infty t^{x+1-1} e^{-t}\, dt = \int_0^\infty t^x e^{-t}\, dt$$

$$= -\int_0^\infty t^x\, d(e^{-t}) = -t^x e^{-t}\Big|_0^\infty + \int_0^\infty x t^{x-1} e^{-t}\, dt$$

$$= x\Gamma(x) \tag{2.5.8}$$

From the above relation we also obtain

$$\Gamma(x) = \frac{\Gamma(x + 1)}{x} \tag{2.5.9}$$

$$\Gamma(x) = (x - 1)\Gamma(x - 1) \tag{2.5.10}$$

$$\Gamma(-x) = \frac{\Gamma(x-1)}{-x}, \qquad x \neq 0, 1, 2, \ldots \tag{2.5.11}$$

From (2.5.2) with $x = 1$ we find that $\Gamma(1) = 1$. Using (2.5.8) we obtain

$$\Gamma(2) = \Gamma(1+1) = 1\Gamma(1) = 1 \cdot 1 = 1,$$

$$\Gamma(3) = \Gamma(2+1) = 2\Gamma(2) = 2 \cdot 1,$$

$$\Gamma(4) = \Gamma(3+1) = 3\Gamma(3) = 3 \cdot 2 \cdot 1.$$

Hence we obtain

$$\Gamma(n+1) = n\Gamma(n) = n(n-1)! = n!, \qquad n = 0, 1, 2, \ldots \tag{2.5.12}$$

$$\Gamma(n) = (n-1)!, \qquad n = 0, 1, 2, \ldots \tag{2.5.13}$$

To find $\Gamma(\frac{1}{2})$ we first set $t = u^2$

$$\Gamma\left(\frac{1}{2}\right) = \int_0^\infty t^{-1/2} e^{-t}\, dt = \int_0^\infty 2e^{-u^2}\, du, \qquad (t = u^2)$$

Hence its square value is

$$\Gamma^2\left(\frac{1}{2}\right) = \left[\int_0^\infty 2e^{-x^2}\, dx\right]\left[\int_0^\infty 2e^{-y^2}\, dy\right]$$

$$= 4\int_0^\infty \left[\int_0^\infty e^{-y^2}\, dy\right]e^{-x^2}\, dx = 4\int_0^{\pi/2}\left[\int_0^\infty e^{-r^2} r\, dr\right]d\theta$$

$$= 4\frac{\pi}{2} \cdot \frac{1}{2} = \pi$$

and thus

$$\Gamma\left(\frac{1}{2}\right) = \sqrt{\pi} \tag{2.5.14}$$

Next let us find the expression for $\Gamma(n + \frac{1}{2})$ for integer positive value of n. From (2.5.10) we obtain

$$\Gamma\left(n+\frac{1}{2}\right) = \Gamma\left(\frac{2n+1}{2}\right) = \left(\frac{2n+1}{2} - 1\right)\Gamma\left(\frac{2n+1}{2} - 1\right) = \frac{2n-1}{2}\Gamma\left(\frac{2n-1}{2}\right)$$

$$= \left(\frac{2n-1}{2}\right)\left(\frac{2n-3}{2}\right)\Gamma\left(\frac{2n-3}{2}\right)$$

If we proceed to apply (2.5.10) we finally obtain

$$\Gamma\left(n+\frac{1}{2}\right) = \frac{(2n-1)(2n-3)(2n-5)\cdots(3)(1)\sqrt{\pi}}{2^n} \tag{2.5.15}$$

Similarly we obtain

$$\Gamma\left(n+\frac{3}{2}\right) = \frac{(2n+1)(2n-1)(2n-3)\cdots(3)(1)\sqrt{\pi}}{2^{n+1}} \tag{2.5.16}$$

$$\Gamma\left(n-\frac{1}{2}\right) = \frac{(2n-3)(2n-5)\cdots(3)(1)\sqrt{\pi}}{2^{n-1}} \tag{2.5.17}$$

Example

To find the ratio $\Gamma(x+n)/\Gamma(x-n)$ where n is a positive integer and $x - n \neq 0, -1, -2, \ldots$ we proceed as follows (see [2.5.10]):

$$\frac{\Gamma(x+n)}{\Gamma(x-n)} = \frac{(x+n-1)\Gamma(x+n-1)}{\Gamma(x-n)} = \frac{(x+n-1)(x+n-2)\Gamma(x+n-2)}{\Gamma(x-n)} = \cdots$$

$$= \frac{(x+n-1)(x+n-2)(x+n-3)\cdots(x+n-2n)\Gamma(x+n-2n)}{\Gamma(x-n)}$$

$$= (x+n-1)(x+n-2)\cdots(x-n) \qquad (2.5.18)$$

∎

Example

Applying (2.5.10) we find

$$2^n\Gamma(n+1) = 2^n n\Gamma(n) = 2^n n(n-1)\Gamma(n-1) = \cdots = 2^n n(n-1)(n-2)\cdots 2 \cdot 1$$

$$= 2^n n! = (2 \cdot 1)(2 \cdot 2)(2 \cdot 3)\cdots(2 \cdot n) = 2 \cdot 4 \cdot 6 \cdots 2n \qquad (2.5.19)$$

If $n - 1$ is substituted in place of n, we obtain

$$2 \cdot 4 \cdot 6 \cdots (2n-2) = 2^{n-1}\Gamma(n) \qquad (2.5.20)$$

∎

Example

Based on the Legendre duplication formula

$$\frac{\Gamma(2n)}{\Gamma(n)} = \frac{\Gamma\left(n+\frac{1}{2}\right)}{\sqrt{\pi}2^{1-2n}} \qquad (2.5.21)$$

we can find the ratio $\Gamma(n+\frac{1}{2})/(\sqrt{\pi}\Gamma(n+1))$ as follows:

$$\frac{\Gamma\left(n+\frac{1}{2}\right)}{\sqrt{\pi}\Gamma(n+1)} = \frac{\Gamma(2n)2^{1-2n}}{\Gamma(n)\Gamma(n+1)} = \frac{\Gamma(2n)2^{1-2n}2^n}{\Gamma(n)2^n\Gamma(n+1)} = \frac{\Gamma(2n)2^{1-n}}{\Gamma(n)2 \cdot 4 \cdot 6 \cdots 2n}$$

(see previous example). But

$$1 \cdot 3 \cdot 5 \cdots (2n-1) = \frac{1 \cdot 2 \cdot 3 \cdot 4 \cdot 5 \cdots (2n-2)(2n-1)}{2 \cdot 4 \cdots (2n-2)} = \frac{\Gamma(2n)}{2^{n-1}\Gamma(n)} \qquad (2.5.22)$$

and hence

$$\frac{\Gamma\left(n+\frac{1}{2}\right)}{\sqrt{\pi}\Gamma(n+1)} = \frac{1 \cdot 3 \cdot 5 \cdots (2n-1)}{2 \cdot 4 \cdot 6 \cdots 2n} \qquad (2.5.23)$$

∎

Remarks on gamma function

1. The gamma function is continuous at every x except 0 and the negative integers.
2. The second derivative is positive for every $x > 0$, and this indicates that the curve $y = \Gamma(x)$ is concave upward for all $x > 0$.

3. $\Gamma(x) \to +\infty$ as $x \to 0+$ through positive values and as $x \to +\infty$.

4. $\Gamma(x)$ becomes, alternatively, negatively infinite and positively infinite at negative integers.

5. $\Gamma(x)$ attains a single minimum for $0 < x < \infty$ and is located between $x = 1$ and $x = 2$.

The **beta function** is defined by

$$B(x, y) = \int_0^1 t^{x-1}(1 - t)^{y-1} \, dt, \qquad x > 0, y > 0 \tag{2.5.24}$$

From the above definition we write

$$B(y, x) = \int_0^1 t^{y-1}(1 - t)^{x-1} \, dt = -\int_1^0 (1 - s)^{y-1} s^{x-1} \, ds = \int_0^1 s^{x-1}(1 - s)^{y-1} \, ds$$

$$= B(x, y) \tag{2.5.25}$$

where we set $1 - t = s$.

If we set $t = \sin^2 \theta$, $dt = 2 \sin \theta \cos \theta d\theta$ and the limits of θ 0 and $\pi/2$, then

$$B(x, y) = \int_0^{\pi/2} 2 \sin^{2x-1} \theta \cos^{2y-1} \theta \, d\theta \tag{2.5.26}$$

The integral representation of the beta function is given by

$$B(x, y) = \int_0^\infty \frac{u^{x-1} du}{(u + 1)^{x+y}}, \qquad x > 0, y > 0 \tag{2.5.27}$$

Set $t = pt$ in (2.5.1) and find the relation

$$\int_0^\infty e^{-pt} t^{z-1} \, dt = \frac{\Gamma(z)}{p(z)}, \qquad \text{Re}\{p\} > 0 \tag{2.5.28}$$

Next set $p = 1 + u$ and $z = x + y$ in the above equation to find that

$$\frac{1}{(1 + y)^{x+y}} = \frac{1}{\Gamma(x + y)} \int_0^\infty e^{-(1+u)t} t^{x+y-1} \, dt \tag{2.5.29}$$

Substituting (2.5.29) in (2.5.27), we obtain

$$B(x, y) = \frac{1}{\Gamma(x + y)} \int_0^\infty e^{-t} t^{x+y-1} \, dt \int_0^\infty e^{-ut} u^{x-1} \, du$$

$$= \frac{\Gamma(x)}{\Gamma(x + y)} \int_0^\infty e^{-t} t^{y-1} \, dt = \frac{\Gamma(x)\Gamma(y)}{\Gamma(x + y)} \tag{2.5.30}$$

It can be shown that

$$B(p, 1 - p) = \frac{\pi}{\sin p\pi}, \qquad 0 < p < 1 \tag{2.5.31}$$

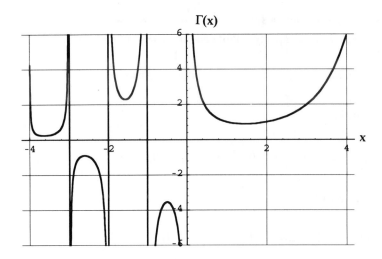

$\Gamma(\mathbf{x})$

FIGURE 2.5.1
The gamma function.

From the identities $\Gamma(x+1) = x\Gamma(x)$, $\Gamma(-x) = \Gamma(1-x)/(-x)$, $B(x, y) = \Gamma(x)\Gamma(y)/\Gamma(x+y)$ together with (2.5.31), we obtain

$$\Gamma(p)\Gamma(1-p) = \frac{\pi}{\sin p\pi}, \qquad p \text{ is nonintegral} \tag{2.5.32}$$

Example
To show that

$$\int_0^\infty t^{n-1}e^{-(a+1)t}\, dt = \frac{\Gamma(n)}{(a+1)^n}, \qquad n > 0, a > -1$$

we set $t = (a+1)^{-1}y$. Hence

$$\int_0^\infty t^{n-1}e^{-(a+1)t}\, dt = \int_0^\infty \left(\frac{y}{a+1}\right)^{n-1} e^{-y}\frac{dy}{a+1} = (a+1)^{-n}\int_0^\infty y^{n-1}e^{-y}\, dy$$

$$= \frac{\Gamma(n)}{(a+1)^n} \qquad\qquad\qquad ∎$$

Example
To evaluate the integral $\int_0^\infty e^{-x^2}\, dx$ we write it in the form

$$\int_0^\infty x^0 e^{-x^2}\, dx$$

which, if compared with the integral in Table 2.5.1, we have the correspondence $a = 0$, $b = 1$, $c = 2$. Hence we obtain

$$\int_0^\infty e^{-x^2}\, dx = \frac{\Gamma\left(\frac{a+1}{c}\right)}{cb^{(a+1)/c}} = \frac{\Gamma\left(\frac{0+1}{2}\right)}{2 \cdot 1^{1/2}} = \frac{\sqrt{\pi}}{2} \qquad\qquad ∎$$

TABLE 2.5.1
Gamma and beta function relations.

$$\Gamma(x) = \int_0^\infty e^{-t} t^{x-1}\, dt \qquad\qquad x > 0$$

$$\Gamma(x) = \int_0^\infty 2u^{2x-1} e^{-u^2}\, du \qquad\qquad x > 0$$

$$\Gamma(x) = \int_0^1 \left[\log\left(\frac{1}{r}\right) \right]^{x-1} dr \qquad\qquad x > 0$$

$$\Gamma(x) = \frac{\Gamma(x+1)}{x} \qquad\qquad x \neq 0, -1, -2, \ldots$$

$$\Gamma(x) = (x-1)\Gamma(x-1) \qquad\qquad x \neq 0, -1, -2, \ldots$$

$$\Gamma(-x) = \frac{\Gamma(1-x)}{-x} \qquad\qquad x \neq 0, 1, 2, \ldots$$

$$\Gamma(n) = (n-1)! \qquad\qquad n = 1, 2, 3, \ldots, \qquad 0! = 1$$

$$\Gamma\left(\frac{1}{2}\right) = \sqrt{\pi}$$

$$\Gamma\left(n + \frac{1}{2}\right) = \frac{1 \cdot 3 \cdot 5 \cdots (2n-1)\sqrt{\pi}}{2^n} \qquad\qquad n = 1, 2, \ldots$$

$$\Gamma\left(n + \frac{3}{2}\right) = \frac{(2n+1)(2n-1)(2n-3)\cdots(3)(1)\sqrt{\pi}}{2^{n+1}} \qquad\qquad n = 1, 2, \ldots$$

$$\Gamma\left(n - \frac{1}{2}\right) = \frac{(2n-3)(2n-5)\cdots(3)(1)\sqrt{\pi}}{2^{n-1}} \qquad\qquad n = 1, 2, \ldots$$

$$\Gamma(n+1) = \frac{2 \cdot 4 \cdot 6 \cdots 2n}{2^n} \qquad\qquad n = 1, 2, \ldots$$

$$\Gamma(2n) = 1 \cdot 3 \cdot 5 \cdots (2n-1)\Gamma(n)2^{1-n} \qquad\qquad n = 1, 2, \ldots$$

$$\frac{\Gamma(2n)}{\Gamma(n)} = \frac{\Gamma\left(n + \frac{1}{2}\right)}{\sqrt{\pi}2^{1-2n}} \qquad\qquad n = 1, 2, \ldots$$

$$\Gamma(x)\Gamma(1-x) = \frac{\pi}{\sin x\pi} \qquad\qquad x \neq 0, \pm 1, \pm 2, \ldots$$

$$n! = \left(\frac{n}{e}\right)^n \sqrt{2\pi n} + h \qquad\qquad n = 1, 2, \ldots, \quad 0 < \frac{h}{n!} < \frac{1}{12n}$$

$$\int_0^\infty t^a e^{-bt^c}\, dt = \frac{\Gamma\left(\frac{a+1}{c}\right)}{cb^{(a+1)/c}} \qquad\qquad a > -1,\ b > 0,\ c > 0$$

$$B(x, y) = \int_0^1 t^{x-1}(1-t)^{y-1}\, dt \qquad\qquad x > 0,\ y > 0$$

$$B(x, y) = \int_0^{\pi/2} 2\sin^{2x-1}\theta \cos^{2y-1}\theta\, d\theta \qquad\qquad x > 0,\ y > 0$$

$$B(x, y) = \int_0^\infty \frac{u^{x-1}}{(u+1)^{x+y}}\, du \qquad\qquad x > 0,\ y > 0$$

$$B(x, y) = \frac{\Gamma(x)\Gamma(y)}{\Gamma(x+y)}$$

TABLE 2.5.1
(continued)

$B(x, y) = B(y, x)$

$B(x, 1 - x) = \dfrac{\pi}{\sin x\pi}$ $0 < x < 1$

$B(x, y) = B(x + 1, y) + B(x, y + 1)$ $x > 0, \ y > 0$

$B(x, n + 1) = \dfrac{1 \cdot 2 \cdots n}{x(x + 1) \cdots (x + n)}$ $x > 0$

TABLE 2.5.2
$\Gamma(x)$, $1 \le x \le 1.99$.

x	0	1	2	3	4	5	6	7	8	9
1.0	1.0000	.9943	.9888	.9835	.9784	.9735	.9687	.9642	.9597	.9555
.1	.9514	.9474	.9436	.9399	.9364	.9330	.9298	.9267	.9237	.9209
.2	.9182	.9156	.9131	.9108	.9085	.9064	.9044	.9025	.9007	.8990
.3	.8975	.8960	.8946	.8934	.8922	.8912	.8902	.8893	.8885	.8879
.4	.8873	.8868	.8864	.8860	.8858	.8857	.8856	.8856	.8857	.8859
.5	.8862	.8866	.8870	.8876	.8882	.8889	.8896	.8905	.8914	.8924
.6	.8935	.8947	.8959	.8972	.8986	.9001	.9017	.9033	.9050	.9068
.7	.9086	.9106	.9126	.9147	.9168	.9191	.9214	.9238	.9262	.9288
.8	.9314	.9341	.9368	.9397	.9426	.9456	.9487	.9518	.9551	.9584
.9	.9618	.9652	.9688	.9724	.9761	.9799	.9837	.9877	.9917	.9958

1.3   Convolution and correlation

1.3.1   *Convolution*

Convolution of functions, although a mathematical relation, is extremely important to engineers. If the impulse response of a system is known, that is, the response of the system to a delta function input, the output of the system is the convolution of the input and its impulse response. The convolution of two functions is given by

$$g(t) \doteq f(t) * h(t) = \int_{-\infty}^{\infty} f(\tau)h(t - \tau)\, d\tau \tag{3.1.1}$$

PROOF   Let $f(t)$ be written as a sum of elementary functions $f_i(t)$. The output $g(t)$ is also given by the sum of the outputs $g_i(t)$ due to each elementary function $f_i(t)$. Hence

$$f(t) = \sum_i f_i(t), \qquad g(t) = \sum_i g_i(t) \tag{3.1.2}$$

If $\Delta\tau$ is sufficiently small, the area of $f_i(t)$ equals $f(\tau_i)\Delta\tau$ (see Figure 3.1.1). Hence, the output is approximately $f(\tau_i)\Delta\tau h(t - \tau_i)$ because $f_i(t)$ is concentrated near the point τ_i. As $\Delta\tau \to 0$, we thus conclude that

$$\sum_i g_i(t) \cong \sum_i f(\tau_i)h(t - \tau_i)\Delta\tau \to \int_{-\infty}^{\infty} f(\tau)h(t - \tau)\, d\tau \qquad\blacksquare$$

For causal systems, the impulse response is

$$h(t) = 0, \qquad t < 0 \tag{3.1.3}$$

and, therefore, the output of the system becomes

$$g(t) = \int_{-\infty}^{t} f(\tau)h(t - \tau)\,d\tau = \int_{0}^{\infty} f(t - \tau)h(\tau)\,d\tau \tag{3.1.4}$$

If, also, $f(t) = 0$ for $t < 0$, then $g(t) = 0$ for $t < 0$; for $t > 0$ we obtain

$$g(t) = \int_{0}^{t} f(\tau)h(t - \tau)\,d\tau = \int_{0}^{t} f(t - \tau)h(\tau)\,d\tau \tag{3.1.5}$$

The convolution does not exist for all functions. The sufficient conditions are

1. Both $f(t)$ and $h(t)$ must be absolutely integrable in the interval $(-\infty, 0]$.
2. Both $f(t)$ and $h(t)$ must be absolutely integrable in the interval $[0, \infty)$.
3. Either $f(t)$ or $h(t)$ (or both) must be absolutely integrable in the interval $(-\infty, \infty)$.

For example, the convolution $\cos \omega_0 t * \cos \omega_0 t$ does not exist.

Example
If the functions to be convoluted are

$$f(t) = 1, \qquad 0 < t < 1, \qquad h(t) = e^{-t}u(t)$$

then the output is given by

$$g(t) = \int_{-\infty}^{\infty} f(\tau)h(t - \tau)\,d\tau$$

The ranges are

1. $-\infty < t < 0$. No overlap of $f(t)$ and $h(t)$ takes place. Hence, $g(t) = 0$.
2. $0 < t < 1$. Overlap occurs from 0 to t. Hence

$$g(t) = \int_{0}^{t} 1 \cdot e^{-(t-\tau)}\,d\tau = e^{-t}\int_{0}^{t} e^{\tau}\,d\tau = 1 - e^{-t}$$

3. $1 < t < \infty$. Overlap occurs from 0 to 1. Hence

$$g(t) = \int_{0}^{1} e^{-(t-\tau)}\,d\tau = e^{-t}(e - 1) \qquad \blacksquare$$

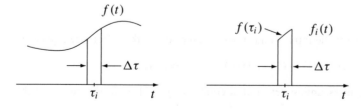

FIGURE 3.1.1

DEFINITION   Convolution Systems

The convolution of any continuous and discrete system is given respectively by

$$y(t) = \int_{-\infty}^{\infty} h(t, \tau)x(\tau)\, d\tau \tag{3.1.6}$$

$$y(n) = \sum_{m=-\infty}^{\infty} h(n, m)x(m) \tag{3.1.7}$$

If the systems are time invariant the kernels $h(\cdot)$ are functions of the difference of their argument. Hence

$$h(n, m) = h(n - m), \qquad h(t, \tau) = h(t - \tau)$$

and therefore

$$y(t) = \int_{-\infty}^{\infty} x(\tau)h(t - \tau)\, d\tau \tag{3.1.8}$$

$$y(n) = \sum_{m=-\infty}^{\infty} x(m)h(n - m) \tag{3.1.9}$$

DEFINITION   Impulse response

The impulse response $h(t)$ of a system is the result of a delta function input to the system. Its value at t is the response to a delta function at $t = 0$.

Example

The voltage $v_c(t)$ across the capacitor of an RC circuit in series with an input voltage source $v(t)$ is given by

$$\frac{dv_c(t)}{dt} + \frac{1}{RC}v_c(t) = \frac{1}{RC}v(t)$$

For a given initial condition $v_c(t_0)$ at time $t = t_0$ the solution is

$$v_c(t) = e^{(-t-t_0)/RC}v_c(t_0) + \frac{1}{RC}\int_{t_0}^{t} e^{-(t-\tau)/RC}v(\tau)\, d\tau, \qquad t \geq t_0$$

For a finite initial condition and $t_0 \to -\infty$ the above equation is written in the form

$$v_c(t) = \frac{1}{RC}\int_{-\infty}^{\infty} e^{-(t-\tau)/RC}u(t - \tau)v(\tau)\, d\tau = \left(\frac{1}{RC}e^{-t/RC}\right) * v(t)$$

Therefore, the impulse response of this system is

$$h(t) = \frac{1}{RC}e^{-t/RC}u(t) \qquad\qquad \blacksquare$$

Example

A discrete system that smooths the input signal $x(n)$ is described by the difference equation

$$y(n) = ay(n - 1) + (1 - a)x(n), \qquad n = 0, 1, 2, \ldots$$

By repeated substitution and assuming zero initial condition $y(-1) = 0$, the output of the system is given by

$$y(n) = (1 - a)\sum_{m=0}^{n} a^{n-m}x(m), \qquad n = 0, 1, 2, \ldots \tag{3.1.10}$$

If we define the impulse response of the system by

$$h(n) = (1 - a)a^n, \qquad n = 0, 1, 2, \ldots$$

the system has an input-output relation

$$y(n) = \sum_{m=-\infty}^{\infty} h(n - m)x(m)$$

which indicates that the system is a convolution one.   ∎

Example

A **pure delay** system is defined by

$$y(t) = \int_{-\infty}^{\infty} \delta(t - t_0 - \tau)x(\tau)\, d\tau = x(t - t_0) \qquad (3.1.11)$$

which shows that its impulse response is $h(t) = \delta(t - t_0)$.   ∎

DEFINITION   Nonanticipative Convolution System

A system, discrete or continuous, is nonanticipative if and only if its impulse response is

$$h(t) = 0, \qquad t < 0$$

with t ranging over the range in which the system is defined.

   If the delay t_0 of a pure delay system is positive, then the system is nonanticipative; and if it is negative, the system is anticipative.

1.3.2   Convolution properties

Commutative

$$y(t) = \int_{-\infty}^{\infty} f(\tau)h(t - \tau)\, d\tau = \int_{-\infty}^{\infty} f(t - \tau)h(\tau)\, d\tau$$

Set $t - \tau = \tau'$ in the first integral, and then rename the dummy variable τ' to τ.

Distributive

$$g(t) = f(t) * [h_1(t) + h_2(t)] = f(t) * h_1(t) + f(t) * h_2(t)$$

This property follows directly as a result of the linear property of integration.

Associative

$$[f(t) * h_1(t)] * h_2(t)] = f(t) * [h_1(t) * h_2(t)]$$

Shift invariance

If $g(t) = f(t) * h(t)$, then

$$g(t - t_0) = f(t - t_0) * h(t) = \int_{-\infty}^{\infty} f(\tau - t_0)h(t - \tau)\, d\tau$$

Write $g(t)$ in its integral form, substitute $t - t_0$ for t, set $\tau + t_0 = \tau'$, and then rename the dummy variable.

Area property

$$A_f = \int_{-\infty}^{\infty} f(t)\, dt = \text{area}$$

$$m_f = \int_{-\infty}^{\infty} tf(t)\, dt = \text{first moment}$$

$$K_f = \frac{m_f}{A_f} = \text{center of gravity}$$

The convolution $g(t) = f(t) * h(t)$ leads to

$$A_g = A_f A_h$$

$$K_g = K_f + K_h$$

PROOF

$$m_g = \int_{-\infty}^{\infty} tg(t)\, dt = \int_{-\infty}^{\infty} t \left[\int_{-\infty}^{\infty} f(\tau)h(t-\tau)\, dt \right] dt$$

$$= \int_{-\infty}^{\infty} f(\tau) \left[\int_{-\infty}^{\infty} th(t-\tau)\, dt \right] d\tau$$

$$= \int_{-\infty}^{\infty} f(\tau) \left[\int_{-\infty}^{\infty} (\lambda + \tau)h(\lambda)\, d\lambda \right] d\tau, \qquad t - \tau = \lambda$$

$$= \int_{-\infty}^{\infty} f(\tau)\, d\tau \int_{-\infty}^{\infty} \lambda h(\lambda)\, d\lambda + \int_{-\infty}^{\infty} \tau f(\tau)\, d\tau \int_{-\infty}^{\infty} h(\lambda)\, d\lambda = A_f m_h + m_f A_h$$

$$\frac{m_g}{A_f A_h} = \frac{m_g}{A_g} \doteq K_g = \frac{A_f m_h + m_f A_h}{A_f A_h} = K_h + K_f \qquad \blacksquare$$

Scaline property

If $g(t) = f(t) * h(t)$ then $f(\frac{t}{a}) * h(\frac{t}{a}) = |a|g(\frac{t}{a})$.

PROOF

$$\int_{-\infty}^{\infty} f\left(\frac{\tau}{a}\right) h\left(\frac{t-\tau}{a}\right) d\tau = \int_{-\infty}^{\infty} f\left(\frac{\tau}{a}\right) h\left(\frac{t}{a} - \frac{\tau}{a}\right) d\tau$$

$$= |a| \int_{-\infty}^{\infty} f(r)h\left(\frac{t}{a} - r\right) dr = |a|g\left(\frac{t}{a}\right) \qquad \blacksquare$$

Complex-valued functions

$$g(t) = f(t) * h(t) = [f_r(t) + jf_i(t)] * [h_r(t) + jh_i(t)]$$

$$= [f_r(t) * h_r(t) - f_i(t) * h_i(t)] + j[f_r(t) * h_i(t) + f_i(t) * h_r(t)]$$

Derivative of delta function

$$g(t) = f(t) * \frac{d\delta(t)}{dt} = \int_{-\infty}^{\infty} f(\tau)\frac{d}{dt}\delta(t-\tau)\, d\tau = \frac{d}{dt}\int_{-\infty}^{\infty} f(\tau)\delta(t-\tau)\, d\tau = \frac{df(t)}{dt}$$

Moment expansion

Expand $f(t - \tau)$ in Taylor series about the point $\tau = 0$

$$f(t - \tau) = f(t) - \tau f^{(1)}(t) + \frac{\tau^2}{2!} f^{(2)}(t) + \cdots + \frac{(-\tau)^{n-1}}{(n-1)!} f^{(n-1)}(t) + e_n$$

Insert into convolution integral

$$g(t) = f(t) \int_{-\infty}^{\infty} h(\tau)\, d\tau - f^{(1)} \int_{-\infty}^{\infty} \tau h(\tau)\, d\tau + \frac{f^{(2)}(t)}{2!} \int_{-\infty}^{\infty} \tau^2 h(\tau)\, d\tau$$

$$+ \cdots + \frac{f^{(n-1)}(t)}{(n-1)!} (-1)^{n-1} \int_{-\infty}^{\infty} \tau^{n-1} h(\tau)\, d\tau + E_n$$

$$= m_{h0} f(t) - m_{h1} f^{(1)}(t) + \frac{m_{h2}}{2!} f^{(2)}(t) + \cdots + \frac{(-1)^{n-1}}{(n-1)!} m_{h(n-1)} f^{(n-1)}(t) + E_n$$

where bracketed numbers in exponents indicate order of differentiation.

Truncation error

Because

$$e_n = \frac{(-\tau)^n}{n!} f^{(n)}(t - \tau_1), \qquad 0 \le \tau_1 \le \tau$$

$$E_n = \frac{1}{n!} \int_{-\infty}^{\infty} (-\tau)^n f^{(n)}(t - \tau_1) h(\tau)\, d\tau$$

Because τ_1 depends on τ, the function $f^{(n)}(t - \tau_1)$ cannot be taken outside the integral. However, if $f^{(n)}(t)$ is continuous and $t^n h(t) \ge 0$, then

$$E_n = \frac{1}{n!} f^{(n)}(t - \tau_0) \int_{-\infty}^{\infty} (-\tau)^n h(\tau)\, d\tau = \frac{(-1)^n m_{hn}}{n!} f^{(n)}(t - \tau_0)$$

where τ_0 is some constant in the interval of integration.

Fourier transform

$$\mathcal{F}\{f(t) * h(t)\} = F(\omega) H(\omega)$$

PROOF

$$\int_{-\infty}^{\infty} \left[\int_{-\infty}^{\infty} f(\tau) h(t - \tau)\, d\tau \right] e^{-j\omega t}\, dt = \int_{-\infty}^{\infty} f(\tau) \int_{-\infty}^{\infty} h(t - \tau) e^{-j\omega t}\, dt\, d\tau$$

$$\int_{-\infty}^{\infty} f(\tau) e^{-j\omega \tau}\, d\tau \int_{-\infty}^{\infty} h(r) e^{-j\omega r}\, dr, \qquad t - \tau = r \qquad\qquad \blacksquare$$

Inverse Fourier transform

$$\frac{1}{2\pi} \int_{-\infty}^{\infty} F(\omega) H(\omega) e^{-j\omega t}\, d\omega = \int_{-\infty}^{\infty} f(\tau) h(t - \tau)\, d\tau$$

Band-limited function

If $f(t)$ is σ-band limited, then the output of a system is

$$g(t) = \int_{-\infty}^{\infty} f(\tau) h(t - \tau) \, d\tau = \sum_{n=-\infty}^{\infty} T f(nT) h_\sigma(t - nT)$$

where

$$h_\sigma(t) = \frac{1}{2\pi} \int_{-\sigma}^{\sigma} H(\omega) e^{j\omega t} \, d\omega$$

PROOF

$$H_\sigma(\omega) = p_\sigma(\omega) H(\omega),$$

hence

$$G(\omega) = F(\omega) H(\omega)$$

$$= \bar{F}(\omega) p_\sigma(\omega) H(\omega)$$

$$= \bar{F}(\omega) H_\sigma(\omega), \qquad \bar{F}(\omega) = F(\omega) \qquad \text{for} -\sigma < \omega < \sigma$$

$$g(t) = \bar{f}(t) * h_\sigma(t) = \left[\sum_{n=-\infty}^{\infty} T f(nT) \delta(t - nT) \right] * h_\sigma(t) = \sum_{n=-\infty}^{\infty} T f(nT) h_\sigma(t - nT) \qquad \blacksquare$$

The convolution properties are given in Table 3.2.1.

Stability of convolution systems

DEFINITION   *Bounded Input Bounded Output (BIBO) Stability*

A discrete or continuous convolution system with impulse response h is BIBO stable if and only if the impulse satisfies the inequality, $\sum_n |h| < \infty$ or $\int_R |h(t)| \, dt < \infty$. If the system is BIBO stable then

$$\sup |y(n)| \le \sum_n |h(n)| \sup |x(n)|, \qquad \sup |y(t)| \le \int_R |h(t)| \, dt \sup |x(t)|, \qquad t \in R$$

for every finite amplitude input $x(t)$ (y is the input of the system).

Example

If the impulse response of a discrete system is $h(n) = ab^n$, $n = 0, 1, 2, \ldots$, then

$$\sum_{n=0}^{\infty} |h(n)| = \sum_{n=0}^{\infty} |a| \, |b|^n = \begin{cases} |a| \frac{1}{1-|b|} & |b| < 1 \\ \infty & |b| \ge 1 \end{cases}$$

The above indicates that for $|b| < 1$ the system is BIBO and for $|b| \ge 1$ the system is unstable.

$\blacksquare$

Example

If $h(t) = u(t)$ then $|h(t)| = \int_0^{\infty} |u(t)| \, dt = \infty$, which indicates the system is not BIBO stable.

$\blacksquare$

TABLE 3.2.1
Convolution properties.

---

1. Commutative

$$g(t) = \int_{-\infty}^{\infty} f(\tau)h(t-\tau)\,d\tau = \int_{-\infty}^{\infty} f(t-\tau)h(\tau)\,d\tau$$

2. Distributive

$$g(t) = f(t) * [h_1(t) + h_2(t)] = f(t) * h_1(t) + f(t) * h_2(t)$$

3. Associative

$$[f(t) * h_1(t)] * h_2(t) = f(t) * [h_1(t) * h_2(t)]$$

4. Shift Invariance

$$g(t) = f(t) * h(t)$$

$$g(t-t_0) = f(t-t_0) * h(t) = \int_{-\infty}^{\infty} f(\tau-t_0)h(t-\tau)\,d\tau$$

5. Area Property

$$A_f = \text{area of } f(t),$$

$$m_f = \int_{-\infty}^{\infty} tf(t)\,dt = \text{first moment}$$

$$K_f = \frac{m_f}{A_f} = \text{center of gravity}$$

$$A_g = A_f A_h, \qquad K_g = K_f + K_h$$

6. Scaling

$$g(t) = f(t) * h(t)$$

$$f\left(\frac{t}{a}\right) * h\left(\frac{t}{a}\right) = |a|g\left(\frac{t}{a}\right)$$

7. Complex Valued Functions

$$g(t) = f(t) * h(t) = [f_r(t) * h_r(t) - f_i(t) * h_i(t)] + j[f_r(t) * h_i(t) + f_i(t) * h_r(t)]$$

8. Derivative

$$g(t) = f(t) * \frac{d\delta(t)}{dt} = \frac{df(t)}{dt}$$

9. Moment Expansion

$$g(t) = m_{h0}f(t) - m_{h1}f^{(1)}(t) + \frac{m_{h2}}{2!}f^{(1)}(t) + \cdots + \frac{(-1)^{n-1}}{n-1!}m_{h(n-1)}f^{(n-1)}(t) + E_n$$

$$m_{hk} = \int_{-\infty}^{\infty} \tau^k h(\tau)\,d\tau$$

$$E_n = \frac{(-1)^n m_{hn}}{n!}f^{(n)}(t-\tau_0), \qquad \tau_0 = \text{constant in the interval of integration}$$

10. Fourier Transform

$$\mathcal{F}\{f(t) * h(t)\} = F(\omega)H(\omega)$$

11. Inverse Fourier Transform

$$\frac{1}{2\pi}\int_{-\infty}^{\infty} F(\omega)H(\omega)e^{j\omega t}\,d\omega = \int_{-\infty}^{\infty} f(\tau)h(t-\tau)\,d\tau$$

12. Band-limited Function

$$g(t) = \int_{-\infty}^{\infty} f(\tau)h(t-\tau)\,d\tau = \sum_{n=-\infty}^{\infty} Tf(nT)h_\sigma(t-nT)$$

$$h_\sigma(t) = \frac{1}{2\pi}\int_{-\sigma}^{\sigma} H(\omega)e^{j\omega t}\,d\omega, \qquad f(t) = \sigma - \text{band limited} = 0, \qquad |t| > \sigma$$

TABLE 3.2.1
(continued)

13. Cyclical Convolution

$$x(n) \otimes y(n) = \sum_{m=0}^{N-1} x((n-m) \bmod N) y(m)$$

14. Discrete-Time

$$x(n) * y(n) = \sum_{m=-\infty}^{\infty} x(n-m) y(m)$$

15. Sampled

$$x(nT) * y(nT) = T \sum_{m=-\infty}^{\infty} x(nT - mT) y(mT)$$

Harmonic inputs

If the input function is of complex exponential order $e^{j\omega t}$ then its output is

$$y(t) = \int_{-\infty}^{\infty} h(\tau) e^{j\omega(t-\tau)} \, d\tau = e^{j\omega t} \int_{-\infty}^{\infty} h(\tau) e^{-j\omega \tau} \, d\tau = H(\omega) e^{j\omega t}$$

The above equation indicates that the output is the same as the input $e^{j\omega t}$ with its amplitude modified by $|H(\omega)|$ and its phase by $\tan^{-1}(H_i(\omega)/H_r(\omega))$ where $H_r(\omega) = \mathrm{Re}\{H(\omega)\}$ and $H_i(\omega) = \mathrm{Im}\{H(\omega)\}$.

For the discrete case we have the relation

$$y(n) = e^{j\omega n} H(e^{j\omega})$$

where

$$H(e^{j\omega}) = \sum_{n=-\infty}^{\infty} h(n) e^{-j\omega n}$$

1.4 Correlation

The **cross-correlation** of two different functions is defined by the relation

$$R_{fh}(t) \doteq f(t) \diamond h(t) = \int_{-\infty}^{\infty} f(\tau) h(\tau - t) \, d\tau = \int_{-\infty}^{\infty} f(\tau + t) h(\tau) \, d\tau \qquad (4.1)$$

When $f(t) = h(t)$ the correlation operation is called **autocorrelation**.

$$R_{ff}(t) \doteq f(t) \diamond f(t) = \int_{-\infty}^{\infty} f(\tau) f(\tau - t) \, d\tau = \int_{-\infty}^{\infty} f(\tau + t) f(\tau) \, d\tau \qquad (4.2)$$

For complex functions the correlation operations are given by

$$R_{fh}(t) \doteq f(t) \diamond h^*(t) = \int_{-\infty}^{\infty} f(\tau) h^*(\tau - t) \, d\tau \tag{4.3}$$

$$R_{ff}(t) \doteq f(t) \diamond f^*(t) = \int_{-\infty}^{\infty} f(\tau) f^*(\tau - t) \, d\tau \tag{4.4}$$

The two basic properties of correlation are

$$f(t) \diamond h(t) \neq h(t) \diamond f(t) \tag{4.5}$$

$$|R_{ff}(t)| \doteq |f(t) \diamond f^*(t)| = \left| \int_{-\infty}^{\infty} f(\tau) f^*(\tau - t) \, d\tau \right|$$

$$\leq \left[\int_{-\infty}^{\infty} |f(\tau)|^2 \, d\tau \right]^{1/2} \left[\int_{-\infty}^{\infty} |f(\tau - t)|^2 \, d\tau \right]^{1/2}$$

$$= \int_{-\infty}^{\infty} |f(\tau)|^2 \, d\tau \leq R_{ff}(0) \tag{4.6}$$

Example
The cross-correlation of the following two functions, $f(t) = p(t)$ and $h(t) = e^{-(t-3)} u(t-3)$, is given by

$$R_{fh}(t) = \int_{-\infty}^{\infty} p(\tau) e^{-(\tau - t - 3)} u(\tau - t - 3) \, d\tau$$

The ranges of t are

1. $t > -2 : R_{fh}(t) = 0$ (no overlap of function)
2. $-4 < t < -2 : R_{fh}(t) = \int_{3+t}^{1} e^{-(\tau - t - 3)} \, d\tau = 1 - e^2 e^t$
3. $-\infty < t < -4 : R_{fh}(t) = \int_{-1}^{1} e^{-(\tau - t - 3)} \, d\tau = e^t e^2 (e^2 - 1)$ ∎

The discrete form of correlation is given by

$$x(n) \diamond y(n) = \sum_{m=-\infty}^{\infty} x(m - n) y^*(m) \equiv \text{crosscorrelation} \tag{4.7}$$

$$x(n) \diamond x(n) = \sum_{m=-\infty}^{\infty} x(m - n) x^*(m) \equiv \text{autocorrelation} \tag{4.8}$$

$$x(n) \diamond y(nT) = T \sum_{m=-\infty}^{\infty} x(mT - nT) y^*(mT) \equiv \text{sampled cross-correlation} \tag{4.9}$$

1.5   Orthogonality of signals

1.5.1   *Introduction*

Modern analysis regards some classes of functions as multidimensional vectors introducing the definition of inner products and expansion in term of orthogonal functions (base functions).

In this section functions $\Phi(t)$, $f(t)$, $F(x)$, ... symbolize either functions of one independent variable t, or, for brevity, a function of a set n independent variables $t^1, t^2, \ldots, t^n$. Hence $dt = dt^1 dt^2 \ldots dt^n$.

A real or complex function $f(t)$ defined on the measurable set E of elements $\{t\}$ is **quadratically integrable** on E if and only if

$$\int_E |f(\tau)|^2 \, d\tau$$

exists in the sense of Lebesque. The class L_2 of all real or complex functions is quadratically integrable on a given interval if one regards the functions $f(t)$, $h(t)$, ... as vectors and defines

> Vector sum of $f(t)$ and $h(t)$ as $f(t) + h(t)$
>
> Product of $f(t)$ by a scalar α as $\alpha f(t)$

The **inner product** of $f(t)$ and $h(t)$ is defined as

$$\langle f, h \rangle \doteq \int_I \gamma(\tau) f^*(\tau) h(\tau) \, d\tau \tag{5.1.1}$$

where $\gamma(\tau)$ is a real nonnative function (**weighing function**) quadratically integrable on I.

Norm

The norm in L_2 is the quantity

$$\|f\| = [\langle f, f \rangle]^{1/2} \doteq \left[\int_I \gamma(\tau) |f(\tau)|^2 \, d\tau \right]^{1/2} \tag{5.1.2}$$

If $\|f\|$ exists and is different from zero, the function is normalizable.

Normalization

$$\frac{f(t)}{\|f\|} = \text{unit norm}$$

Inequalities

If $f(t)$, $h(t)$ and the nonnegative weighting function $\gamma(t)$ are quadratically integrable on I then

Cauchy-Schwarz inequality

$$|\langle f(t), h(t) \rangle| \doteq \left| \int_I \gamma(\tau) f^* h \, d\tau \right|^2 \leq \int_I \gamma |f|^2 \, d\tau \int_I \gamma |h|^2 \, d\tau \doteq \langle f, f \rangle \langle h, h \rangle \tag{5.1.3}$$

Minkowski inequality

$$\|f + h\| \doteq \left(\int_I \gamma |f + h|^2 \, d\tau \right)^{1/2}$$

$$\leq \left(\int_I \gamma |f|^2 \, d\tau \right)^{1/2} + \left(\int_I \gamma |h|^2 \, d\tau \right)^{1/2}$$

$$= \|f\| + \|h\| \tag{5.1.4}$$

Convergence in mean

The space L_2 admits the **distance function** (matric)

$$d\langle f, h \rangle \doteq \| f - h \| \doteq \left[\int_I \gamma(\tau) |f(\tau) - h(\tau)|^2 \, d\tau \right]^{1/2} \tag{5.1.5}$$

The root-mean-square difference of the above equation between the two functions $f(t)$ and $h(t)$ is equal to zero if and only if $f(t) = h(t)$ for almost all t in I.

Every sequence in I of functions $r_0(t), r_1(t), r_2(t), \ldots$ **converges in the mean** to the limit $r(t)$ if and only if

$$d^2 \langle r_n, r \rangle \doteq \| r_n - r \|^2 \doteq \int_I \gamma(\tau) |r_n(\tau) - r(\tau)|^2 \, d\tau \to 0 \quad \text{as } n \to \infty \tag{5.1.6}$$

Therefore we define limit in the mean

$$\underset{n \to \infty}{\text{l.i.m}} \, r_n(t) = r(t) \tag{5.1.7}$$

Convergence in the mean does not necessarily imply convergence of the sequence at every point, nor does convergence of a sequence at all points on I imply convergence in the mean.

Riess-Fischer theorem

The L_2 space with a given interval I is **complete**; every sequence of quadratically integrable functions $r_0(t), r_1(t), r_2(t), \ldots$ such that $\text{l.i.m}_{m \to \infty, n \to \infty} |r_m - r_n| = 0$ (**Cauchy sequence**), converges in the mean to a quadratically integrable function $r(t)$ and defines $r(t)$ uniquely for almost all t in I.

Orthogonality

Two quadratically integrable functions $f(t), h(t)$ are **orthogonal** on I if and only if

$$\langle f, h \rangle = \int_I \gamma(\tau) f^*(\tau) h(\tau) \, d\tau = 0 \tag{5.1.8}$$

Orthonormal

A set of function $r_i(t), i = 1, 2, \ldots$ is an **orthonormal** set if and only if

$$\langle r_i, r_j \rangle \doteq \int_I \gamma(\tau) r_i^*(\tau) r_j(\tau) \, d\tau = \delta_{ij} = \begin{cases} 0 \text{ if } i \neq j \\ 1 \text{ if } i = j \end{cases} \quad (i, j = 1, 2, \ldots) \tag{5.1.9}$$

Every set of normalizable mutually orthogonal functions is linearly independent.

Bessel's inequalities

Given a finite or infinite orthonormal set $\varphi_1(t), \varphi_2(t), \varphi_3(t), \ldots$ and any function $f(t)$ quadratically integrable over I

$$\sum_i |\langle \varphi_i, f \rangle|^2 \leq \langle f, f \rangle \tag{5.1.10}$$

The equal sign applies if and only if $f(t)$ belongs to the space spanned by all $\varphi_i(t)$.

Complete orthonormal set of functions (orthonormal bases)

A set of functions $\{\varphi_i(t)\}, i = 1, 2, \ldots,$ in L_2 is a complete orthonormal set if and only if the set satisfies the following conditions:

1. Every quadratically integrable function $f(t)$ can be expanded in the form

$$f(t) = \langle f, \varphi_1 \rangle \varphi_1 + \langle f, \varphi_2 \rangle \varphi_2 + \cdots + \langle f, \varphi_i \rangle \varphi_i + \cdots, \qquad i = 1, 2, \ldots$$

2. If (1) above is true, then

$$\langle f, f \rangle = |\langle f, \varphi_1 \rangle|^2 + |\langle f, \varphi_2 \rangle|^2 + \cdots$$

which is the completeness relation (Parseval's identity).

3. For any pair of functions $f(t)$ and $h(t)$ in L_2, the relation holds

$$\langle f, h \rangle = \langle f, \varphi_1 \rangle \langle h, \varphi_1 \rangle + \langle f, \varphi_2 \rangle \langle h, \varphi_2 \rangle + \cdots$$

4. The orthonormal set $\varphi_1(t), \varphi_2(t), \varphi_3(t), \ldots$ is not contained in any other orthonormal set in L_2.

The above conditions imply the following: Given a complete orthonormal set $\{\varphi_i(t)\}, i = 1, 2, \ldots$ in L_2 and a set of complex numbers $\langle f, \varphi_1 \rangle, \langle f, \varphi_2 \rangle + \cdots$ such that $\sum_{i=1}^{\infty} |\langle f, \varphi_i \rangle|^2 < \infty$, there exists a quadratically integrable function $f(t)$ such that $\langle f, \varphi_1 \rangle \varphi_1 + \langle f, \varphi_2 \rangle \varphi_2 + \cdots$ converges in the mean to $f(t)$.

Gram-Schmidt orthonormalization process

Given any countable (finite or infinite) set of linear independent functions $r_1(t), r_2(t), \ldots$ normalizable in I, there exists an orthogonal set $\varphi_1(t), \varphi_2(t), \ldots$ spanning the same space of functions. Hence

$$\varphi_1 = r_1, \qquad \varphi_2 = r_2 - \frac{\int_I \varphi_1 r_2 \, dt}{\int_I \varphi_1^2 \, dt}, \qquad \varphi_3 = r_3 - \frac{\int_I \varphi_1 r_3 \, dt}{\int_I \varphi_1^2 \, dt} \varphi_1 - \frac{\int_I \varphi_2 r_3 \, dt}{\int_I \varphi_2^2 \, dt} \varphi_2, \text{ etc. (5.1.11)}$$

For creating an orthonormal set we proceed as follows:

$$\varphi_i(t) = \frac{v_i(t)}{\|v_i(t)\|} = \frac{v_i(t)}{+\sqrt{\langle v_i, v_i \rangle}}$$

$$v_1(t) = r_1(t), \qquad v_{i+1}(t) = r_{i+1}(t) - \sum_{k=1}^{i} \langle \varphi_k, r_{i+1} \rangle \varphi_k(t), \qquad i = 1, 2, \ldots \qquad (5.1.12)$$

Series approximation

If $f(t)$ is a quadratically integrable function, then

$$\int_I |f_n(t) - f(t)|^2 \, dt$$

yields the **least mean square error**. The set $\{\varphi_i(t)\}, i = 1, 2, \ldots$ is orthonormal and the approximation to $f(t)$ is

$$f_n(t) = a_1 \varphi_1(t) + a_2 \varphi_2(t) + \cdots + a_n \varphi_n(t), \qquad n = 1, 2, \ldots \qquad (5.1.13)$$

1.5.2 *Legendre polynomials*

1.5.2.1 Relations of Legendre polynomials

Legendre polynomials are closely associated with physical phenomena for which spherical geometry is important. The polynomials $P_n(t)$ are called Legendre polynomials in honor of their discoverer, and they are given by

$$P_n(t) = \sum_{k=0}^{[n/2]} \frac{(-1)^k (2n-2k)! \, t^{n-2k}}{2^n k! (n-k)! (n-2k)!} \tag{5.2.1}$$

$$[n/2] = \begin{cases} n/2 & n \text{ even} \\ (n-1)/2 & n \text{ odd} \end{cases}$$

$$\frac{1}{\sqrt{1-2st+s^2}} = \begin{cases} \sum_{n=0}^{\infty} P_n(t)s^n & |s| < 1 \\ \sum_{n=0}^{\infty} P_n(t)s^{-n-1} & |s| > 1 \text{ generating function} \end{cases} \tag{5.2.1a}$$

Table 5.2.1 gives the first eight Legendre polynomials. Figure 5.2.1 shows the first six Legendre polynomials.

Rodrigues formula

$$P_n(t) = \frac{1}{2^n n!} \frac{d^n}{dt^n} (t^2 - 1)^n, \qquad n = 0, 1, 2, \dots \tag{5.2.2}$$

Recursive formulas

$$(n+1) P_{n+1}(t) - (2n+1)t P_n(t) + n P_{n-1}(t) = 0 \tag{5.2.3}$$

$$P'_{n+1}(t) - t P'_n(t) = (n+1) P_n(t), \qquad (P'(t) \doteq \text{ derivative of } P(t)) \tag{5.2.4}$$

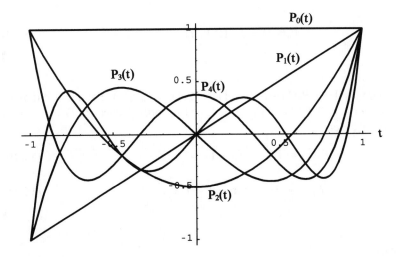

FIGURE 5.2.1

TABLE 5.2.1
Legendre polynomials.

$P_0 = 1$

$P_1 = t$

$P_2 = \dfrac{3}{2}t^2 - \dfrac{1}{2}$

$P_3 = \dfrac{5}{2}t^3 - \dfrac{3}{2}t$

$P_4 = \dfrac{35}{8}t^4 - \dfrac{30}{8}t^2 + \dfrac{3}{8}$

$P_5 = \dfrac{63}{8}t^5 - \dfrac{70}{8}t^3 + \dfrac{15}{8}t$

$P_6 = \dfrac{231}{16}t^6 - \dfrac{315}{16}t^4 + \dfrac{105}{16}t^2 - \dfrac{5}{16}$

$P_7 = \dfrac{429}{16}t^7 - \dfrac{693}{16}t^5 + \dfrac{315}{16}t^3 - \dfrac{35}{16}t$

$$t P_n'(t) - P_{n-1}'(t) = n P_n(t) \tag{5.2.5}$$

$$P_{n+1}'(t) - P_{n-1}'(t) = (2n + 1) P_n(t) \tag{5.2.6}$$

$$(t^2 - 1) P_n'(t) = nt P_n(t) - n P_{n-1}(t) \tag{5.2.7}$$

$$P_0(t) = 1, \qquad P_1(t) = t \tag{5.2.8}$$

Example
From (5.2.1) when n is even implies $P_n(-t) = P_n(t)$ and when n is odd $P_n(-t) = -P_n(t)$. Therefore

$$P_n(-t) = (-1)^n P_n(t) \tag{5.2.9}$$

∎

Example
From (5.2.7) $t = 1$ implies $0 = n P_n(1) - n P_{n-1}(1)$ or $P_n(1) = P_{n-1}(1)$. For $n = 1$ it implies $P_1(1) = P_0(1) = 1$. For $n = 2$ $P_2(1) = P_1(1) = 1$, and so forth. Hence $P_n(1) = 1$. From (5.2.9) $P_n(-1) = (-1)^n$. Hence

$$P_n(1) = 1, \qquad P_n(-1) = (-1)^n \tag{5.2.10}$$

$$P_n(t) < 1 \qquad \text{for } -1 < t < 1 \tag{5.2.11}$$

∎

Example
From (5.2.7) we get

$$\frac{d}{dt}[(1 - t^2)] P_n'(t)] = n P_{n-1}'(t) - n P_n(t) - nt P_n'(t)$$

Use (5.2.5) to find

$$\frac{d}{dt}[(1-t^2)P_n'(t)] + n(n+1)P_n(t) = 0$$

or

$$(1-t^2)P_n''(t) - 2t P_n'(t) + n(n+1)P_n(t) = 0 \qquad (5.2.12)$$

We have deduced the Legendre polynomials $y = P_n(t)$ ($n = 0, 1, 2, \ldots$) as the solution of the linear second-order ordinary differential equation

$$(1-t^2)y''(t) - 2ty'(t) + n(n+1)y(t) = 0 \qquad (5.2.12a)$$

called the **Legendre differential equation.**

If we let $x = \cos\varphi$ then the above equation transforms to the trigonometric form

$$y'' + (\cot\varphi)y' + n(n+1)y = 0 \qquad (5.2.12b)$$

It can be shown that (5.2.12a) has solutions of a first kind

$$y = C_0 \left[1 - \frac{n(n+1)}{2!}t^2 + \frac{n(n+1)(n-2)(n+3)}{4!}t^4 - \cdots \right]$$

$$+ C_1 \left[t - \frac{(n-1)(n+2)}{3!}t^3 + \frac{(n-1)(n+2)(n-3)(n+4)}{5!}t^5 - \cdots \right] \qquad (5.2.12c)$$

valid for $|t| < 1$, C_0 and C_1 being arbitrary constants. ∎

Schläfli's integral formula

$$P_n(t) = \frac{1}{2\pi j} \int_C \frac{(z^2-1)^n}{2^n(z-t)^{n+1}} \, dz \qquad (5.2.13)$$

where C is any regular, simple, closed curve surrounding t.

1.5.2.2 Complete orthonormal system, $\left\{ \left[\frac{1}{2}(2n+1) \right]^{1/2} P_n(t) \right\}$

The Legendre polynomials are orthogonal in $[-1, 1]$

$$\int_{-1}^{1} P_n(t) P_m(t) \, dt = 0 \qquad (5.2.14)$$

$$\int_{-1}^{1} [P_n(t)]^2 \, dt = \frac{2}{2n+1} \qquad (5.2.15)$$

and therefore the set

$$\varphi_n(t) = \sqrt{\frac{2n+1}{2}} P_n(t) \qquad (5.2.16)$$

is orthonormal.

Series expansion

If $f(t)$ is integrable in $[-1, 1]$ then

$$f(t) = \sum_{n=0}^{\infty} a_n P_n(t) \qquad (5.2.16a)$$

$$a_n = \frac{2n+1}{2} \int_{-1}^{1} f(t) P_n(t)\, dt \qquad (5.2.16b)$$

For even $f(t)$, the series will contain term $P_n(t)$ of even index; if $f(t)$ is odd, the term of odd index only.

If the real function $f(t)$ is piecewise smooth in $(-1, 1)$ and if it is square integrable in $(-1, 1)$, then the series (5.2.16a) converges to $f(t)$ at every continuity point of $f(t)$.

Change of range

If a function $f(t)$ is defined in $[a, b]$, it is sometimes necessary in the applications to expand the function in a series of orthogonal polynomials in this interval. Clearly the substitution

$$t = \frac{2}{b-a}\left[x - \frac{b+a}{2}\right], \qquad a < b, \qquad \left[x = \frac{b-a}{2}t + \frac{b+a}{2}\right] \qquad (5.2.17)$$

transform the interval $[a, b]$ of the x-axis into the interval $[-1, 1]$ of the t-axis. It is, therefore, sufficient to consider the expansion in series of Legendre polynomials of

$$f\left[\frac{b-a}{2}t + \frac{b+a}{2}\right] = \sum_{n=0}^{\infty} a_n P_n(t) \qquad (5.2.18a)$$

$$a_n = \frac{2n+1}{2} \int_{-1}^{1} f\left[\frac{b-a}{2}t + \frac{b+a}{2}\right] P_n(t)\, dt \qquad (5.2.18b)$$

The above equation can also be accomplished as follows:

$$f(t) = \sum_{n=0}^{\infty} a_n X_n(t) \qquad (5.2.19a)$$

$$X_n(t) = \frac{1}{n!(b-a)^n} \frac{d^n (t-a)^n (t-b)^n}{dt^n} \qquad (5.2.19b)$$

$$a_n = \frac{2n+1}{b-a} \int_{b}^{a} f(t) X_n(t)\, dt \qquad (5.2.19c)$$

Example
Suppose $f(t)$ is given by

$$f(t) = \begin{cases} 0 & -1 \leq t < a \\ 1 & a < t \leq 1 \end{cases}$$

Then from (5.2.16b)

$$a_n = \frac{2n+1}{2} \int_{a}^{1} P_n(t)\, dt$$

Using (5.2.6), and noting that $P_n(1) = 1$, we obtain

$$a_n = -\frac{1}{2}[P_{n+1}(a) - P_{n-1}(a)], \qquad a_0 = \frac{1}{2}(1 - a)$$

which leads to the expansion

$$f(t) \cong \frac{1}{2}(1 - a) - \frac{1}{2}\sum_{n=1}^{\infty}[P_{n+1}(a) - P_{n-1}(a)]P_n(t), \qquad -1 < t < 1 \qquad \blacksquare$$

Example

Suppose $f(t)$ is given by

$$f(t) = \begin{cases} -1 & -1 \leq t < 0 \\ 1 & 0 < t \leq 1 \end{cases}$$

The function is an odd function and, therefore, $f(t)P_n(t)$ is an odd function of $P_n(t)$ with even index. Hence a_n are zero for $n = 0, 2, 4, \ldots$. For odd index n, the product $f(t)P_n(t)$ is even and hence

$$a_n = \left(n + \frac{1}{2}\right) \int_{-1}^{1} f(t)P_n(t)\, dt = 2\left(n + \frac{1}{2}\right) \int_{0}^{1} P_n(t)\, dt, \qquad n = 1, 3, 5, \ldots$$

Using (5.2.6) and setting $n = 2k + 1$, $k = 0, 1, 2, \ldots$ we obtain

$$a_{2k+1} = (4k + 3) \int_{0}^{1} P_{2k+1}(t)\, dt = \int_{0}^{1} [P'_{2k+2}(t) - P'_{2k}(t)]\, dt$$

$$= [P_{2k+2}(t) - P_{2k}(t)]\big|_{0}^{1} = P_{2k}(0) - P_{2k+2}(0)$$

where we have used the property $P_n(1) = 1$ for all n. But

$$P_{2n}(0) = \begin{pmatrix} -\frac{1}{2} \\ n \end{pmatrix} = \frac{(-1)^n (2n)!}{2^{2n}(n!)^2} \tag{5.2.20}$$

and, thus, we have

$$a_{2k+1} = \frac{(-1)^k (2k)!}{2^{2k}(k!)^2} - \frac{(-1)^{k+1}(2k+2)!}{2^{2k+2}[(k+1)!]^2} = \frac{(-1)^k (2k)!}{2^{2k}(k!)^2}\left[1 + \frac{2k+1}{2k+2}\right]$$

$$= \frac{(-1)^k (2k)!(4k+3)}{2^{2k+1}k!(k+1)!}$$

The expansion is

$$f(t) = \sum_{k=0}^{\infty} \frac{(-1)^k (2k)!(4k+3)}{2^{2k+1}k!(k+1)!} P_{2k+1}(t), \qquad -1 \leq t \leq 1 \tag{5.2.21}$$

∎

1.5.2.3 Associated Legendre polynomials

If m is a positive integer and $-1 \leq t \leq 1$, then

$$P_n^m(t) = (1 - t^2)^{m/2} \frac{d^m P_n(t)}{dt^m}, \qquad m = 1, 2, \ldots, n \tag{5.2.22}$$

where $P_n^m(t)$ is known as the **associated Legendre function** or **Ferrer's functions**.

Rodrigues formula

$$P_n^m(t) = \frac{(1 - t^2)^{m/2}}{2^n n!} \frac{d^{n+m}}{dt^{n+m}}(t^2 - 1)^n, \qquad m = 1, 2, \ldots, n; \ n + m \geq 0 \tag{5.2.23}$$

Properties

$$P_n^{-m}(t) = (-1)^m \frac{(n-m)!}{(n+m)!} P_n^m(t) \tag{5.2.24}$$

$$P_n^0(t) = P_n(t) \tag{5.2.25}$$

$$(n-m+1)P_{n+1}^m(t) - (2n+1)tP_n^m(t) + (n+m)P_{n-1}^m(t) = 0 \tag{5.2.26}$$

$$(1-t^2)^{1/2}P_n^m(t) = \frac{1}{2n+1}\left[P_{n+1}^{m+1}(t) - P_{n-1}^{m+1}(t)\right] \tag{5.2.27}$$

$$(1-t^2)^{1/2}P_n^m(t) = \frac{1}{2n+1}[(n+m)(n+m-1)P_{n-1}^{m-1}(t)$$

$$- (n-m+1)(n-m+2)P_{n+1}^{m-1}(t)] \tag{5.2.28}$$

$$P_n^m(t) = 2mt(1-t^2)^{-1/2}P_n^m(t) - [n(n+1)-m(m-1)]P_n^{m-1}(t) \tag{5.2.29}$$

$$\int_{-1}^1 P_n^m(t)P_k^m(t)\,dt = 0, \qquad k \neq n \tag{5.2.30}$$

$$\int_{-1}^1 [P_n^m(t)]^2\,dt = \frac{2(n+m)!}{(2n+1)(n-m)!} \tag{5.2.31}$$

Example

To evaluate the integral $\int_{-1}^1 t^m P_n(t)\,dt$ we use the Rodrigues formula and proceed as follows:

$$\int_{-1}^1 t^m P_n(t)\,dt = \frac{1}{2^n n!} \int_{-1}^1 t^m D^n[(t^2-1)^n]\,dt, \qquad \left(D^n = \frac{d^n}{dt^n}\right)$$

$$= \frac{1}{2^n n!} \left[[t^m D^{n-1}(t^2-1)^n]\big|_{t=-1}^{t=1} - m\int_{-1}^1 t^{m-1}D^{n-1}[(t^2-1)^n]\,dt\right]$$

where integration by parts was used. The left expression is zero because of the presence of the expression $(t^2-1)^n$.

(a) For $m < n$ and after m integrations by parts we obtain

$$\int_{-1}^1 t^m P_n(t)\,dt = \frac{(-1)^m m!}{2^n n!} \int_{-1}^1 D^{n-m}[(t^2-1)^n]\,dt$$

$$= \frac{(-1)^m m!}{2^n n!} [D^{n-m-1}(t^2-1)^n]\big|_{t=-1}^{t=1} = 0, \qquad m < n$$

(b) $m \geq n$. Integrate n times by parts to find the following expression:

$$\int_{-1}^1 t^m P_n(t)\,dt = C_{mn} \int_{-1}^1 t^{m-n}(t^2-1)^n\,dt$$

where

$$C_{mn} = \frac{(-1)^m m(m-1)(m-2)\cdots(m-[n-1])}{2^n n!}$$

Multiplying numerator and denomenator by $(m-n)!$ and incorporating the $(-1)^n$ in the integrand, we obtain

$$\int_{-1}^{1} t^m P_n(t)\, dt = \frac{m!}{2^n n!(m-n)!} \int_{-1}^{1} t^{m-n}(1-t^2)^n\, dt, \qquad m \geq n$$

If $m-n$ is odd the integrand is an odd function and hence is equal to zero. If $m-n$ is even then the integrand is even and hence

$$\int_{-1}^{1} t^m P_n(t)\, dt = \frac{m!2}{2^n n!(m-n)!} \int_0^1 t^{m-n}(1-t^2)^n\, dt$$

$$= \frac{m!\Gamma\left(\frac{m-n+1}{2}\right)}{2^{n-1}(m-n)!(m+n+1)\Gamma\left(\frac{m+n+1}{2}\right)}, \qquad m \geq n, \; m-n \text{ is even}$$

If $m = n$

$$\int_{-1}^{1} t^m P_n(t)\, dt = \frac{n!\Gamma\left(\frac{1}{2}\right)}{2^{n-1}(2n+1)\left(\frac{2n-1}{2}\right)\left(\frac{2n-3}{2}\right)\cdots\left(\frac{3}{2}\right)\left(\frac{1}{2}\right)\Gamma\left(\frac{1}{2}\right)}$$

$$= \frac{n!2^n}{2^{n-1}(2n+1)(2n-1)(2n-3)\cdots(3)(1)}$$

$$= \frac{n!2^n 2^n n!}{2^{n-1}(2n+1)(2n)(2n-1)(2n-2)(2n-3)\cdots(3)(2)(1)}$$

$$= \frac{2^{n+1}(n!)^2}{(2n+1)!}$$

Hence

$$\int_{-1}^{1} t^m P_n(t)\, dt = \begin{cases} 0 & m < n \\ 0 & m \geq n, \; m-n \text{ is odd} \\ \frac{m!\Gamma\left(\frac{m-n+1}{2}\right)}{2^{n-1}(m-n)!(m+n+1)\Gamma\left(\frac{m+n+1}{2}\right)} & m > n, \; m-n \text{ is even} \\ \frac{2^{n+1}(n!)^2}{(2n+1)!} & m = n \end{cases}$$

∎

Example

To find $P_{2n}(0)$ we use the summation

$$P_{2n}(t) = \frac{(-1)^n}{2^{2n-1}} \sum_{k=0}^{n} \frac{(-1)^k (2n+2k-1)!}{(2k)!(n+k-1)!(n-k)!} t^{2k}$$

with $k = 0$. Hence

$$P_{2n}(0) = \frac{(-1)^n(2n-1)!}{2^{2n-1}(n-1)!n!} = \frac{(-1)^n 2n[(2n-1)!]}{2^{2n}n[(n-1)!]n!} = \frac{(-1)^n(2n)!}{2^{2n}(n!)^2}$$

∎

Example

To evaluate $\int_0^1 P_m(t)\, dt$ for $m \neq 0$ we must consider the two cases: m being odd and m being even.

(a) m is even and $m \neq 0$

$$\int_0^1 P_m(t)\, dt = \frac{1}{2}\int_{-1}^{1} P_m(t)\, dt = \frac{1}{2}\int_{-1}^{1} P_m(t)\cdot 1\, dt = \frac{1}{2}\int_{-1}^{1} P_m(t)P_0(t)\, dt = 0$$

The result is due to the orthogonality principle.

(b) m is odd and $m \neq 0$. From the relation (see Table 5.2.2)

$$\int_t^1 P_m(t)\,dt = \frac{1}{2m+1}[P_{m-1}(t) - P_{m+1}(t)]$$

with $t = 0$ we obtain

$$\int_0^1 P_m(t)\,dt = \frac{1}{2m+1}[P_{m-1}(0) - P_{m+1}(0)]$$

Using the results of the previous example, we obtain

$$\int_0^1 P_m(t)\,dt = \frac{1}{2m+1}\left[\frac{(-1)^{\frac{m-1}{2}}(m-1)!}{2^{m-1}\left[\left(\frac{m-1}{2}\right)!\right]^2} - \frac{(-1)^{\frac{m+1}{2}}(m+1)!}{2^{m+1}\left[\left(\frac{m+1}{2}\right)!\right]^2}\right]$$

$$= \frac{(-1)^{\frac{m-1}{2}}(m-1)!(2m+1)(m+1)}{(2m+1)2^{m+1}\left(\frac{m+1}{2}\right)!\left(\frac{m+1}{2}\right)\left(\frac{m-1}{2}\right)!} = \frac{(-1)^{\frac{m-1}{2}}(m-1)!}{2^m\left(\frac{m+1}{2}\right)!\left(\frac{m-1}{2}\right)!},$$

$$m \text{ is odd} \qquad\qquad\qquad \blacksquare$$

Example

One hemisphere of a homogeneous spherical solid is maintained at 300°C while the other half is kept at 75°C. To find the temperature distribution we must use the equation for heat conduction

$$\frac{\partial T}{\partial t} = \frac{k}{\rho c}\left(\nabla^2 T + \frac{\partial Q}{\partial t}\right)$$

where T is temperature, t is time, k is the thermal conductivity, ρ is the density, c is specific heat, and $\partial Q/\partial t$ is the rate of heat generation. Because of the steady-state condition of the problem, $\partial T/\partial t = \partial Q/\partial t = 0$. Hence, the equation becomes

$$\nabla^2 T = \frac{\partial^2 T}{\partial x^2} + \frac{\partial^2 T}{\partial y^2} + \frac{\partial^2 T}{\partial z^2} = \frac{\partial}{\partial r}\left(r^2\frac{\partial T}{\partial r}\right) + \frac{1}{\sin\varphi}\frac{\partial}{\partial\varphi}\left(\sin\varphi\frac{\partial T}{\partial\varphi}\right) = 0$$

where T is independent of θ.

Assuming a solution of the form

$$T = FG = f(r)g(\varphi)$$

we obtain

$$\frac{\partial T}{\partial r} = G\frac{dF}{dr}, \qquad \frac{\partial^2 T}{\partial r^2} = G\frac{d^2 F}{dr^2}$$

Similarly, we obtain

$$\frac{\partial T}{\partial\varphi} = F\frac{dG}{d\varphi}, \qquad \frac{\partial^2 T}{\partial\varphi^2} = F\frac{d^2 G}{d\varphi^2}$$

Introducing these relations in the Laplacian we obtain

$$2rG\frac{dF}{dr} + r^2 G\frac{d^2 F}{dr^2} + F\frac{dG}{d\varphi}\cot\varphi + F\frac{d^2 G}{d\varphi^2} = 0$$

or

$$\frac{2r\frac{dF}{dr} + r^2\frac{d^2 F}{dr^2}}{F} = -\frac{\frac{dG}{d\varphi}\cot\varphi + \frac{d^2 G}{dr^2}}{G}$$

Setting the above ratios equal to positive constant k^2, $k \neq 0$ we obtain

$$r^2 \frac{d^2 F}{dr^2} + 2r \frac{dF}{dr} - k^2 F = 0$$

$$\frac{d^2 G}{d\varphi^2} + (\cot \varphi) \frac{dG}{d\varphi} + k^2 G = 0$$

For $k^2 = n(n + 1)$, we recognize that the above equation is the Legendre equation with G playing the role of y. Thus a particular solution is

$$G = C_n P_n(\cos \varphi)$$

where C_n is an arbitrary constant. With $k^2 = n(n + 1)$ the general solution for F is given by

$$F = S_n r^n + \frac{B_n}{r^{n+1}}$$

where S_n and B_n are arbitrary constants. Because for $r = 0$ the second term becomes infinity, we set $B_n = 0$. Hence the product solution is

$$T = FG = S_n C_n r^n P_n(\cos \varphi) = D_n r^n P_n(\cos \varphi)$$

Because Legendre polynomials are continuous we must create a procedure to alleviate this problem. We denote the excess of the temperature T on the upper half of the surface over that of T on the lower half. On the bounding great circle between these halves, we arbitrarily set it equal to $(300 - 75)/2$. We then have

$$T_E(\varphi) = \begin{cases} 225 & 0 \leq \varphi < \pi/2 \\ 0 & \pi/2 < \varphi \leq \pi \\ 225/2 & \varphi = \pi/2 \end{cases}$$

If we let $x = \cos \varphi$ then $T_E(\varphi)$ becomes $f(x)$

$$f(x) = \begin{cases} 225 & 0 < x \leq 1 \\ 0 & -1 \leq x < 0 \\ 225/2 & x = 0 \end{cases}$$

Next we expand $f(x)$ in the form

$$f(x) = \sum_{n=0}^{\infty} a_n P_n(x), \qquad a_n = \frac{2n + 1}{2} \int_0^1 f(x) P_n(x)\, dx$$

$$= 225 \left[\frac{1}{2} + \frac{3}{4} P_1(x) - \frac{7}{16} P_3(x) + \frac{11}{32} P_5(x) - \cdots \right]$$

Setting $D_n = a_n / R^n$, where a_n is the coefficient of $P_n(x)$ and R is the radius of the solid, the solution is given by

$$T(r, \varphi) = 75 + \sum_{n=0}^{\infty} a_n \left(\frac{r}{R} \right)^n P_n(\cos \varphi)$$

$$= 75 + 225 \left[\frac{1}{2} + \frac{3}{4} \left(\frac{r}{R} \right) P_1(\cos \varphi) - \frac{7}{16} \left(\frac{r}{R} \right)^2 P_3(\cos \varphi) \right.$$

$$\left. + \frac{11}{32} \left(\frac{r}{R} \right)^5 P_5(\cos \varphi) - \cdots \right] \qquad \blacksquare$$

Table 5.2.2 gives relationships of Legendre and associated Legendre functions.

TABLE 5.2.2
Properties of Legendre and associate Legendre functions.

1. $\dfrac{1}{\sqrt{1 - 2tx + x^2}} = \displaystyle\sum_{n=0}^{\infty} P_n(t) x^n, \qquad |t| \le 1, \; |x| < 1$

2. $P_n(t) = \displaystyle\sum_{k=0}^{[n/2]} \dfrac{(-1)^k (2n - 2k)! \, t^{n-2k}}{2^n k! (n-k)! (n-2k)!}, \qquad [n/2] = \dfrac{n}{2}, n \text{ is even}; [n/2] = (n-1)/2, n \text{ is odd}$

3. $P_0(t) = 1$

4. $P_{2n}(0) = \begin{pmatrix} -\frac{1}{2} \\ n \end{pmatrix} = \dfrac{(-1)^n (2n)!}{2^{2n} (n!)^2}, \qquad\qquad n = 0, 1, 2, \ldots$

5. $P_{2n+1}(0) = 0, \qquad\qquad n = 0, 1, 2, \ldots$

6. $P_{2n}(-t) = P_{2n}(t), \qquad P_{2n+1}(-t) = -P_{2n+1}(t), \qquad n = 0, 1, 2, \ldots$

7. $P_n(-t) = (-1)^n P_n(t), \qquad\qquad n = 0, 1, 2, \ldots$

8. $P_n(1) = (1), \qquad\qquad n = 0, 1, 2, \ldots;$

   $P_n(-1) = (-1)^n, \qquad\qquad n = 0, 1, 2, \ldots$

9. $P_n(t) = \dfrac{1}{2^n n!} \dfrac{d^n}{dt^n} (t^2 - 1)^n = \text{Rodrigues formula}, \qquad n = 0, 1, 2, \ldots$

10. $(n+1) P_{n+1}(t) - (2n+1) t P_n(t) + n P_{n-1}(t) = 0, \qquad n = 1, 2, \ldots$

11. $P'_{n+1}(t) - 2t P'_n(t) + P'_{n-1}(t) - P_n(t) = 0, \qquad n = 1, 2, \ldots$

12. $P'_{n-1}(t) = P_n(t) + 2t P'_n(t) - P'_{n+1}(t)$

13. $P'_{n+1}(t) = P_n(t) + 2t P'_n(t) - P'_{n-1}(t)$

14. $P'_{n+1}(t) - t P'_n(t) = (n+1) P_n(t)$

15. $t P'_n(t) - P'_{n-1}(t) = n P_n(t)$

16. $P'_{n+1}(t) - P'_{n-1}(t) = (2n+1) P_n(t)$

17. $(1 - t^2) P'_n(t) = n P_{n-1}(t) - nt P_n(t)$

18. $|P_n(t)| < 1, \qquad\qquad -1 < t < 1$

19. $P_{2n}(t) = \dfrac{(-1)^n}{2^{2n-1}} \displaystyle\sum_{k=0}^{n} \dfrac{(-1)^k (2n + 2k - 1)!}{(2k)! (n+k-1)! (n-k)!} t^{2k}, \qquad n = 0, 1, 2, \ldots$

20. $(1 - t^2) P'_n(t) = (n+1)[t P_n(t) - P_{n+1}(t)], \qquad n = 0, 1, 2, \ldots$

21. $\displaystyle\int_{-1}^{1} P_n(t) \, dt = 0, \qquad\qquad n = 1, 2, \ldots$

22. $|P_n(t)| \le 1, \qquad\qquad |t| \le 1$

23. $\displaystyle\int_{-1}^{1} P_n(t) P_m(t) \, dt = 0, \qquad\qquad n \neq m$

TABLE 5.2.2
(continued)

24. $$\int_{-1}^{1} [P_n(t)]^2 \, dt = \frac{2}{2n+1},$$ $n = 0, 1, 2, \ldots$

25. $$\frac{1}{2} \int_{-1}^{1} t^m P_s(t) \, dt = \frac{m(m-2)\cdots(m-s+2)}{(m+s+1)(m+s-1)\cdots(m+1)},$$ m, s are even

26. $$\frac{1}{2} \int_{-1}^{1} t^m P_s(t) \, dt = \frac{(m-1)(m-3)\cdots(m-s+2)}{(m+s+1)(m+s-1)\cdots(m+2)},$$ m, s are odd

27. $$\int_{-1}^{1} t P_n(t) P_{n-1}(t) \, dt = \frac{2n}{4n^2-1},$$ $n = 1, 2, \ldots$

28. $$\int_{-1}^{1} P_n(t) P_{n+1}'(t) \, dt = 2,$$ $n = 0, 1, 2, \ldots$

29. $$\int_{-1}^{1} t P_n'(t) P_n(t) \, dt = \frac{2n}{2n+1},$$ $n = 0, 1, 2, \ldots$

30. $$\int_{-1}^{1} (1-t^2) P_n'(t) P_k'(t) \, dt = 0,$$ $k \neq n$

31. $$\int_{-1}^{1} (1-t)^{-1/2} P_n(t) \, dt = \frac{2\sqrt{2}}{2n+1},$$ $n = 0, 1, 2, \ldots$

32. $$\int_{-1}^{1} t^2 P_{n+1}(t) P_{n-1}(t) \, dt = \frac{2n(n+1)}{(4n^2-1)(2n+3)},$$ $n = 1, 2, \ldots$

33. $$\int_{-1}^{1} (t^2-1) P_{n+1}(t) P_n'(t) \, dt = \frac{2n(n+1)}{(2n+1)(2n+3)},$$ $n = 1, 2, \ldots$

34. $$\int_{-1}^{1} t^n P_n(t) \, dt = \frac{2^{n+1}(n!)^2}{(2n+1)!},$$ $n = 0, 1, 2, \ldots$

35. $$\int_{-1}^{1} t^2 [P_n(t)]^2 \, dt = \frac{2}{(2n+1)^2} \left[\frac{(n+1)^2}{2n+3} + \frac{n^2}{2n-1} \right]$$

36. $$P_n^m(t) = (1-t^2)^{m/2} \frac{d^m}{dt^m} P_n(t),$$ $m \geq 0$

37. $$P_n^m(t) = \frac{1}{2^n n!} (1-t^2)^{m/2} \frac{d^{n+m}}{dt^{n+m}} [(t^2-1)^n],$$ $m+n \geq 0$

38. $$P_n^{-m}(t) = (-1)^m \frac{(n-m)!}{(n+m)!} P_n^m(t)$$

39. $$P_n^0(t) = P_n(t)$$

40. $$(n-m+1) P_{n+1}^m(t) - (2n+1)t P_n^m(t) + (n+m) P_{n-1}^m(t) = 0$$

41. $$(1-t^2)^{1/2} P_n^m(t) = \frac{1}{2n+1} \left[P_{n+1}^{m+1}(t) - P_{n-1}^{m+1}(t) \right]$$

42. $$(1-t^2)^{1/2} P_n^m(t) = \frac{1}{2n+1} \left[(n+m)(n+m-1) P_{n-1}^{m-1}(t) \right.$$

    $$\left. -(n-m+1)(n-m+2) P_{n+1}^{m-1}(t) \right]$$

43. $$P_n^{m+1}(t) = 2mt(1-t^2)^{-1/2} P_n^m(t) - [n(n+1) - m(m-1)] P_n^{m-1}(t)$$

TABLE 5.2.2
(continued)

44. $\displaystyle\int_{-1}^{1} P_n^m(t) P_k^m(t)\, dt = 0,$ $\hspace{4cm} k \neq n$

45. $\displaystyle\int_{-1}^{1} [P_n^m(t)]^2\, dt = \frac{2(n+m)!}{(2n+1)(n-m)!}$

46. $P_n^m(-t) = (-1)^{n+m} P_n^m(t)$

47. $P_n^m(\pm 1) = 0,$ $\hspace{4.5cm} m > 0$

48. $P_{2n}^1(0) = 0, \qquad P_{2n+1}^1(0) = \dfrac{(-1)^n (2n+1)!}{2^{2n}(n!)^2}$

49. $P_n^m(0) = 0,$ $\hspace{4.5cm} n + m$ is odd

$P_n^m(0) = (-1)^{(n-m)/2} \dfrac{(n+m)!}{2^n [(n-m)/2]! [(n+m)/2]!},$ $\hspace{1cm} n + m$ is even

50. $\displaystyle\int_{-1}^{1} P_n^m(t) P_n^k(t)(1-t^2)^{-1}\, dt = 0,$ $\hspace{2.5cm} k \neq 0$

51. $\displaystyle\int_{-1}^{1} (1-t^2)^{-1/2} P_{2m}(t)\, dt = \left[\frac{\Gamma\left(\frac{1}{2}+m\right)}{m!} \right]^2$

52. $\displaystyle\int_{-1}^{1} t(1-t^2)^{-1/2} P_{2m+1}(t)\, dt = \frac{\Gamma\left(\frac{1}{2}+m\right)\Gamma\left(\frac{3}{2}+m\right)}{m!(m+1)!}$

53. $\displaystyle\int_{t}^{1} P_n(t)\, dt = \frac{1}{2n+1}[P_{n-1}(t) - P_{n+1}(t)]$

54. $\displaystyle\int_{0}^{1} t^q P_n(t)\, dt = \Gamma(q+1) \sum_{k=0}^{n} \frac{(-1)^k \Gamma(n+k+1)}{2^k k! \Gamma(n-k+1)\Gamma(q+k+2)},$ $\hspace{0.5cm} q > -1$

55. $\displaystyle\int_{0}^{1} t^{-1/2} P_n(t)\, dt = \begin{cases} \dfrac{2(-1)^{n/2}}{2n+1} & n \text{ is even} \\[2ex] \dfrac{2(-1)^{(n-1)/2}}{2n+1} & n \text{ is odd} \end{cases}$

56. $\displaystyle\int_{0}^{1} t^{1/2} P_n(t)\, dt = \begin{cases} \dfrac{2(-1)^{(n+2)/2}}{(2n-1)(2n+3)} & n \text{ is even} \\[2ex] \dfrac{2(-1)^{(n+3)/2}}{(2n-1)(2n+3)} & n \text{ is odd} \end{cases}$

1.5.3 *Hermite polynomials*

1.5.3.1 Generating function

If we define the Hermite polynomial by the Rodrigues formula

$$H_n(t) = (-1)^n e^{t^2} \frac{d^n e^{-t^2}}{dt^n}, \qquad n = 0, 1, 2, \ldots, \quad -\infty < t < \infty \qquad (5.3.1)$$

The first few Hermite polynomials are

$$H_0(t) = 1,$$

$$H_1(t) = 2t,$$

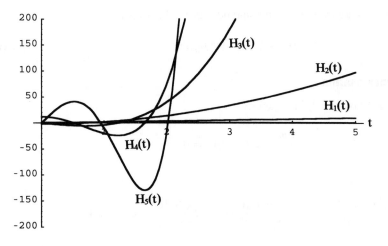

FIGURE 5.3.1

$$H_2(t) = 4t^2 - 2,$$

$$H_3(t) = 8t^3 - 12t,$$

$$H_4(t) = 16t^4 - 48t + 12,$$

$$H_5(t) = 32t^5 - 160t^3 + 120t$$

and therefore

$$H_n(t) = \sum_{k=0}^{[n/2]} \frac{(-1)^k n!}{k!(n-2k)!} (2t)^{n-2k} \tag{5.3.2}$$

$$[n/2] \equiv \text{largest integer} \le n/2$$

The Hermite polynomials are orthogonal with weight $\gamma(t) = e^{-t^2}$ on the interval $(-\infty, \infty)$.
The relation between Hermite polynomial and the generating function is

$$w(t, x) = e^{2tx - x^2} = \sum_{n=0}^{\infty} \frac{H_n(t)}{n!} x^n, \qquad |x| < \infty \tag{5.3.3}$$

Because $w(t, x)$ is the entire function in x it can be expanded in Taylor's series at $x = 0$ with
$|x| < \infty$. Hence the derivatives of the expansion are

$$\left(\frac{\partial^n w}{\partial x^n} \right)\Big|_{x=0} = e^{t^2} \left[\frac{\partial^n}{\partial x^n} e^{-(t-x)^2} \right]_{x=0} = (-1)^n e^{t^2} \left[\frac{d^n e^{-u^2}}{du^n} \right]_{u=t} \doteq H_n(t)$$

Figure 5.3.1 shows several Hermite polynomials.

Example
Let $t = 0$ in (5.3.3) and expand e^{-x^2} in power series. Comparing equal powers of both sides
we find that

$$H_{2n}(0) = (-1)^n \frac{(2n)!}{n!} \qquad \blacksquare$$

Hermite polynomials are even for even n and odd for n odd. Hence

$$H_n(-t) = (-1)^n H_n(t) \tag{5.3.4}$$

1.5.3.2 Recurrence relation

If we substitute $w(t, x)$ of (5.3.3) into identity

$$\frac{\partial w}{\partial x} - 2(t - x)w = 0$$

we obtain

$$\sum_{n=0}^{\infty} \frac{H_{n+1}(t)}{n!} x^n - 2t \sum_{n=0}^{\infty} \frac{H_n(t)}{n!} x^n + 2 \sum_{n=0}^{\infty} \frac{H_n(t)}{n!} x^{n+1} = 0$$

or

$$\sum_{n=1}^{\infty} [H_{n+1}(t) - 2t H_n(t) + 2n H_{n-1}(t)] \frac{x^n}{n!} + H_1(t) - 2t H_0(t) = 0$$

But $H_1(t) - 2t H_0(t) = 0$ and hence

$$H_{n+1}(t) - 2t H_n(t) - 2n H_{n-1}(t) = 0, \qquad n = 1, 2, \ldots \tag{5.3.5}$$

If we use

$$\frac{\partial w}{\partial x} - 2xw = 0$$

we obtain

$$H_n'(t) = 2n H_{n-1}(t), \qquad n = 1, 2, \ldots \tag{5.3.6}$$

Eliminating $H_{n-1}(t)$ from (5.3.6) and (5.3.5), we obtain

$$H_{n+1}(t) - 2t H_n(t) + H_n'(t) = 0, \qquad n = 0, 1, 2, \ldots \tag{5.3.7}$$

Differentiate (5.3.6), combine with (5.3.5), and use the relation $H_{n+1}' = 2(n + 1)H_{(n+1)-1}$; we obtain

$$H_n'' - 2t H_n'(t) + 2n H_n(t) = 0, \qquad n = 0, 1, 2, \ldots \tag{5.3.8}$$

From the above equation, with $y = H_n(t)$ ($n = 0, 1, 2, \ldots$), we observe that the Hermite polynomials are the solution to the second-order ordinary differential equation known as the **Hermite equation**

$$y'' - 2ty' + 2ny = 0 \tag{5.3.9}$$

1.5.3.3 Integral representation and integral equation

The integral representation of Hermite polynomials is given by

$$H_n(t) = \frac{(-1)^n 2^n e^{t^2}}{\sqrt{\pi}} \int_{-\infty}^{\infty} e^{-x^2 + j2tx} x^n \, dx, \qquad n = 0, 1, 2, \ldots \tag{5.3.10}$$

The integral equation satisfied by the Hermite polynomials is

$$e^{-t^2/2} H_n(t) = \frac{1}{j^n \sqrt{2\pi}} \int_{-\infty}^{\infty} e^{jty} e^{-y^2/2} H_n(y)\, dy, \qquad n = 0, 1, 2, \ldots \qquad (5.3.10a)$$

Also, because $H_{2m}(t)$ is an even function and $H_{2m+1}(t)$ is an odd function, then the above equation implies the following two integrals:

$$e^{-t^2/2} H_{2m}(t) = (-1)^m \sqrt{\frac{2}{\pi}} \int_0^{\infty} e^{-y^2/2} H_{2m}(y) \cos ty\, dy$$

$$e^{-t^2/2} H_{2m+1}(t) = (-1)^m \sqrt{\frac{2}{\pi}} \int_0^{\infty} e^{-y^2/2} H_{2m+1}(y) \sin ty\, dy, \qquad m = 0, 1, 2, \ldots \quad (5.3.11)$$

1.5.3.4   Orthogonality relation: Hermite series

The **orthogonality property** of the Hermite polynomials is given by

$$\int_{-\infty}^{\infty} e^{-t^2} H_m(t) H_n(t)\, dt = 0 \qquad \text{if } m \neq n \qquad (5.3.12)$$

and

$$\int_{-\infty}^{\infty} e^{-t^2} H_n^2(t)\, dt = 2^n n! \sqrt{\pi}, \qquad n = 0, 1, 2, \ldots \qquad (5.3.13)$$

Therefore, the **orthonormal** Hermite polynomials are

$$\varphi_n(t) = (2^n n! \sqrt{\pi})^{-1/2} e^{-t^2/2} H_n(t), \qquad n = 0, 1, 2, \ldots, \quad -\infty < t < \infty \qquad (5.3.14)$$

THEOREM 5.3.1
If $f(t)$ is piecewise smooth in every finite interval $[-a, a]$ and

$$\int_{-\infty}^{\infty} e^{-t^2} f^2(t)\, dt < \infty$$

then the Hermite series

$$f(t) = \sum_{n=0}^{\infty} C_n H_n(t), \qquad -\infty < t < \infty \qquad (5.3.15)$$

$$C_n = \frac{1}{2^n n! \sqrt{\pi}} \int_{-\infty}^{\infty} e^{-t^2} f(t) H_n(t)\, dt \qquad (5.3.16)$$

converges pointwise to $f(t)$ at every continuity point and converges at $[f(t+) - f(t-)]/2$ at points of discontinuity.

Example
The function $f(t) = t^{2p}$, $p = 0, 1, 2, \ldots$ satisfies Theorem 5.3.1 and it is even. Hence,

$$t^{2p} = \sum_{n=0}^{p} C_{2n} H_{2n}(t)$$

where

$$C_{2n} = \frac{1}{2^{2n}(2n)!\sqrt{\pi}} \int_{-\infty}^{\infty} e^{-t^2} t^{2p} H_{2n}(t)\, dt$$

$$= \frac{1}{2^{2n}(2n)!\sqrt{\pi}} \int_{-\infty}^{\infty} t^{2p} \frac{d^{2n}}{dt^{2n}} (e^{-t^2})\, dt = \frac{1}{2^{2n}(2n)!\sqrt{\pi}} \frac{(2p)!}{(2p-2n)!} \int_{-\infty}^{\infty} e^{-t^2} t^{2p-2n}\, dt$$

$$= \frac{1}{2^{2n}(2n)!\sqrt{\pi}} \frac{(2p)!}{(2p-2n)!} \Gamma\left(p - n + \frac{1}{2}\right)$$

to find C_{2n}, integration by parts was performed n times.   ∎

Example
The function e^{at}, where a is an arbitrary number, satisfies Theorem 5.3.1. Hence

$$e^{at} = \sum_{n=0}^{\infty} C_n H_n(t)$$

where

$$C_n = \frac{1}{2^n n!\sqrt{\pi}} \int_{-\infty}^{\infty} e^{at} e^{-t^2} H_n(t)\, dt = \frac{(-1)^n}{2^n n!\sqrt{\pi}} \int_{-\infty}^{\infty} e^{at} \frac{d^n}{dt^n} (e^{-t^2})\, dt$$

$$= \frac{a^n}{2^n n!\sqrt{\pi}} \int_{-\infty}^{\infty} e^{at-t^2}\, dt = \frac{a^n}{2^n n!} e^{-a^2/4}$$ ∎

Example
The $\mathrm{sgn}(t)$ function is odd and hence its expansion takes the form

$$\mathrm{sgn}(t) = \sum_{n=0}^{\infty} C_{2n+1} H_{2n+1}(t)$$

where

$$C_{2n+1} = \frac{1}{2^{2n+1}(2n+1)!\sqrt{\pi}} \int_{-\infty}^{\infty} e^{-t^2} H_{2n+1}(t)\, \mathrm{sgn}(t)\, dt$$

$$= \frac{1}{2^{2n}(2n+1)!\sqrt{\pi}} \int_{0}^{\infty} e^{-t^2} H_{2n+1}(t)\, dt$$

Use the identity

$$e^{-t^2} H_n(t) = -\frac{d}{dt}[e^{-t^2} H_{n-1}(t)]$$

which results from (5.3.5) and (5.3.6), to find that

$$C_{2n+1} = \frac{H_{2n}(0)}{2^{2n}(2n+1)!\sqrt{\pi}} = \frac{(-1)^n}{2^{2n}(2n+1)!n!\sqrt{\pi}}$$ ∎

Table 5.3.1 gives the Hermite relationships.

TABLE 5.3.1
Properties of the Hermite polynomials.

1. $H_n(t) = (-1)^n e^{t^2} \dfrac{d^n e^{-t^2}}{dt^n}$

2. $H_n(t) = \displaystyle\sum_{k=0}^{[n/2]} \dfrac{(-1)^k n!}{k!(n-2k)!} (2t)^{n-2k}$

    $[n/2] = $ largest integer $\leq n/2$

3. $e^{2tx-x^2} = \displaystyle\sum_{n=0}^{\infty} H_n(t) \dfrac{x^n}{n!}$

4. $H_{2n}(0) = (-1)^n \dfrac{(2n)!}{n!}$

5. $H_{2n+1}(0) = 0, \; H'_{2n}(0) = 0, \; H'_{2n+1}(0) = (-1)^n \dfrac{(2n+2)!}{(n+1)!}$

6. $H_n(-t) = (-1)^n H_n(t)$

7. $H_{2n}(t)$ are even functions, $\qquad H_{2n+1}(t)$ are odd functions

8. $H_{n+1}(t) - 2t H_n(t) + 2n H_{n-1}(t) = 0, \qquad n = 1, 2, \ldots$

9. $H'_n(t) = 2n H_{n-1}(t), \qquad\qquad n = 1, 2, \ldots$

10. $H_{n+1}(t) - 2t H_n(t) + H'_n(t) = 0$

11. $H''_n(t) - 2t H'_n(t) + 2n H_n(t) = 0$

12. $H_n(t) = \dfrac{(-1)^n 2^n e^{t^2}}{\sqrt{\pi}} \displaystyle\int_{-\infty}^{\infty} e^{-x^2 + j2tx} x^n \, dx$

13. $e^{-t^2/2} H_n(t) = \dfrac{1}{j^n \sqrt{2\pi}} \displaystyle\int_{-\infty}^{\infty} e^{jty} e^{-y^2/2} H_n(y) \, dy = $ integral equation

14. $e^{-t^2/2} H_{2m}(t) = (-1)^m \sqrt{\dfrac{2}{\pi}} \displaystyle\int_{0}^{\infty} e^{-y^2/2} H_{2m}(y) \cos ty \, dy$

15. $e^{-t^2/2} H_{2m+1}(t) = (-1)^m \sqrt{\dfrac{2}{\pi}} \displaystyle\int_{0}^{\infty} e^{-y^2/2} H_{2m+1}(y) \sin ty \, dy$

16. $\displaystyle\int_{-\infty}^{\infty} e^{-t^2} H_m(t) H_n(t) \, dt = 0, \qquad$ if $m \neq n$

17. $\displaystyle\int_{-\infty}^{\infty} e^{-t^2} H_n^2(t) \, dt = 2^n n! \sqrt{\pi}$

18. $f(t) = \displaystyle\sum_{n=0}^{\infty} C_n H_n(t)$

    $C_n = \dfrac{1}{2^n n! \sqrt{\pi}} \displaystyle\int_{-\infty}^{\infty} e^{-t^2} f(t) H_n(t) \, dt$

19. $\displaystyle\int_{-\infty}^{\infty} t^k e^{-t^2} H_n(t) \, dt = 0, \qquad k = 0, 1, 2, \ldots, n-1$

TABLE 5.3.1
(continued)

20. $\displaystyle\int_{-\infty}^{\infty} t^2 e^{-t^2} H_n^2(t)\, dt = \sqrt{\pi}\, 2^n n!\left(n + \frac{1}{2}\right)$

21. $\displaystyle\int_{-\infty}^{\infty} x^n e^{-x^2} H_n(tx)\, dx = \frac{\sqrt{\pi}\, n!}{2} P_n(t)$

22. $\displaystyle\int_{-\infty}^{\infty} e^{-2t^2} H_n^2(t)\, dt = 2^{n-\frac{1}{2}} \Gamma\left(n + \frac{1}{2}\right)$

23. $\dfrac{d^m H_n(t)}{dt^m} = \dfrac{2^m n!}{(n-m)!} H_{n-m}(t), \qquad m < n$

24. $\displaystyle\int_{-\infty}^{\infty} e^{-a^2 t^2} H_{2n}(t)\, dt = \frac{(2n)!}{n!} \frac{\sqrt{\pi}}{a} \left(\frac{1-a^2}{a^2}\right)^n, \qquad a > 0$

TABLE 5.4.1
Laguerre polynomials.

$L_0(t) = 1$

$L_1(t) = -t + 1$

$L_2(t) = \dfrac{1}{2!}(t^2 - 4t + 2)$

$L_3(t) = \dfrac{1}{3!}(-t^3 + 9t^2 - 18t + 6)$

$L_4(t) = \dfrac{1}{4!}(t^4 - 16t^3 + 72t^2 - 96t + 24)$

1.5.4 *Laguerre polynomials*

Generating function and Rodrigues formula

The generating function for the Laguerre polynomials is given by

$$w(t, x) = (1 - x)^{-1} \exp\left[-\frac{tx}{1-x}\right] = \sum_{n=0}^{\infty} L_n(t)x^n, \qquad |x| < 1, \qquad 0 \le t < \infty \quad (5.4.1)$$

By expressing the exponential function in a series, realizing that

$$\binom{-k-1}{m} = (-1)^m \binom{k+m}{m}$$

and finally making the change of index $m = n - k$, (5.4.1) leads to

$$L_n(t) = \sum_{k=0}^{n} \frac{(-1)^k n!\, t^k}{(k!)^2 (n-k)!} \tag{5.4.2}$$

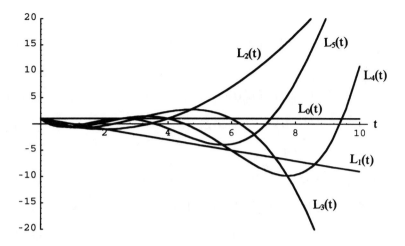

FIGURE 5.4.1

The Rodrigues formula for creating Laguerre polynomials is given by

$$L_n(t) = \frac{e^t}{n!} \frac{d^n}{dt^n}(t^n e^{-t}), \qquad n = 0, 1, 2, \ldots \tag{5.4.3}$$

which can be verified by application of the Leibniz formula

$$\frac{d^n}{dt^n}(fg) = \sum_{k=0}^{n} \binom{n}{k} \frac{d^{n-k}}{dt^{n-k}} \frac{d^k g}{dt^k}, \qquad n = 1, 2, \ldots \tag{5.4.4}$$

For a real $a > -1$ the general Laguerre polynomials are defined by the formula

$$L_n^a(t) = e^t \frac{t^{-a}}{n!} \frac{d^n}{dt^n}(e^{-t} t^{n+a}), \qquad n = 0, 1, 2, \ldots \tag{5.4.5a}$$

Using Leibniz's formula

$$L_n^a(t) = \sum_{k=0}^{n} \frac{\Gamma(n+a+1)}{\Gamma(k+a+1)} \frac{(-t)^k}{k!(n-k)!} \tag{5.4.5b}$$

Table 5.4.1 gives a few Laguerre polynomials. Figure 5.4.1 shows several Laguerre polynomials.

Recurrence relations

The generating function $w(t, x)$, (5.4.1) satisfies the identity

$$(1 - x^2)\frac{\partial w}{\partial x} + (t - 1 + x)w = 0 \tag{5.4.6}$$

Substituting (5.4.1) in (5.4.6) and equating the coefficients of x^n to zero, we obtain

$$(n + 1)L_{n+1}(t) + (t - 1 - 2n)L_n(t) + nL_{n-1}(t) = 0, \qquad n = 1, 2, \ldots \tag{5.4.7}$$

Similarly substituting (5.4.1) into

$$(1 - x)\frac{\partial w}{\partial t} + xw = 0 \tag{5.4.8}$$

we obtain the relation

$$L'_n(t) - L'_{n-1}(t) + L_{n-1}(t) = 0, \qquad n = 1, 2, \ldots \tag{5.4.9}$$

From this we obtain

$$L'_{n+1}(t) = L'_n(t) - L_n(t) \tag{5.4.10}$$

$$L'_{n-1}(t) = L'_n(t) + L_{n-1}(t) \tag{5.4.11}$$

From (5.4.7) by differentiation we find

$$(n + 1)L'_{n+1}(t) + (t - 1 - 2n)L'_n(t) + L_n(t) + nL'_{n-1}(t) = 0 \tag{5.4.12}$$

Eliminating $L'_{n+1}(t)$ and $L'_{n-1}(t)$ by using (5.4.10), (5.4.11), and (5.4.12), we obtain

$$tL'_n(t) = nL_n(t) - nL_{n-1}(t) \tag{5.4.13}$$

By differentiating (5.4.13) and using (5.4.9), we obtain

$$tL''_n(t) + L'_n(t) = -nL_{n-1}(t)$$

Next, eliminating $L_{n-1}(t)$ using (5.4.13) we obtain

$$tL''_n(t) + (1 - t)L'_n(t) + nL_n(t) = 0 \tag{5.4.14}$$

Setting $y = L_n(t)$ ($n = 0, 1, 2, \ldots$), we conclude that all $L_n(t)$ are the solution to the Laguerre equation

$$ty'' + (1 - t)y' + ny = 0 \tag{5.4.15}$$

Orthogonality, Laguerre series

The orthogonality relations for Laguerre polynomials are

$$\int_0^\infty e^{-t} L_n(t)L_m(t)\, dt = 0, \qquad n \neq m \tag{5.4.16}$$

$$\int_0^\infty e^{-t} [L_n(t)]^2\, dt = \frac{\Gamma(n + 1)}{n!} = 1, \qquad n = 0, 1, 2, \ldots \tag{5.4.17}$$

For the generalized Laguerre polynomials, the orthogonality relations

$$\int_0^\infty e^{-t} t^a L_m^a(t)L_n^a(t)\, dt = 0, \qquad n \neq m,\ a > -1$$

$$\int_0^\infty e^{-t} t^a [L_n^a(t)]^2\, dt = \frac{\Gamma(n + a + 1)}{n!}, \qquad a > -1,\ n = 0, 1, 2, \ldots \tag{5.4.18}$$

The orthogonal system for the generalized polynomials on the interval $0 \leq t < \infty$ is

$$\varphi_n^a(t) = \left[\frac{n!}{\Gamma(n + a + 1)} \right]^{1/2} e^{-t/2} t^{a/2} L_n^a(t), \qquad n = 0, 1, 2, \ldots \tag{5.4.19}$$

The Laguerre series is given by

$$f(t) = \sum_{n=0}^\infty C_n L_n(t), \qquad 0 \leq t < \infty \tag{5.4.20}$$

where

$$C_n = \int_0^\infty e^{-t} f(t) L_n(t)\, dt, \qquad n = 0, 1, 2, \dots \tag{5.4.21}$$

THEOREM 5.4.1
If $f(t)$ is piecewise smooth in every finite interval $t_1 \le t \le t_2$, $0 < t_1 < t_2 < \infty$ and

$$\int_0^\infty e^{-t} f^2(t)\, dt < \infty$$

then the Laguerre series converges pointwise to $f(t)$ at every continuity point of $f(t)$, and at the points of discontinuity the series converges to $[f(t) - f(t-)]/2$.
  If we set $a = m =$ integer $(m = 0, 1, 2, \dots)$, then (5.4.5b) becomes

$$L_n^m(t) = \sum_{k=0}^n \frac{(-1)^k (n+m)!\, t^k}{(n-k)!(m+k)!k!}, \qquad m = 0, 1, 2, \dots \tag{5.4.22}$$

The Rodrigues formula is

$$L_n^m(t) = \frac{1}{n!} e^t t^{-m} \frac{d^n}{dt^n}(e^{-t} t^{n+m}) \tag{5.4.23}$$

Example
The function t^b can be expanded in series

$$t^b = \sum_{n=0}^\infty C_n L_n^a(t), \qquad b > -\frac{1}{2}(a+1)$$

$$
\begin{aligned}
C_n &= \frac{n!}{\Gamma(n+a+1)} \int_0^\infty t^{b+a} e^{-t} L_n(t)\, dt \\[2mm]
&= \frac{n!}{\Gamma(n+a+1)} \int_0^\infty e^{-t} t^{b+a} \frac{e^t t^{-a}}{n!} \frac{d^n}{dt^n}(t^{n+a} e^{-t})\, dt \\[2mm]
&= \frac{1}{\Gamma(n+a+1)} \int_0^\infty t^b \frac{d^n}{dt^n}(t^{n+a} e^{-t})\, dt \\[2mm]
&= \frac{(-1)^n b(b-1)\cdots(b-n+1)}{\Gamma(n+a+1)} \int_0^\infty e^{-t} t^{b+a}\, dt \\[2mm]
&= (-1)^n \frac{\Gamma(b+1)}{\Gamma(n+b+1)\Gamma(b-n+1)} \int_0^\infty e^{-t} t^{(b+a+1)-1}\, dt \\[2mm]
&= (-1)^n \frac{\Gamma(b+1)\Gamma(b+a+1)}{\Gamma(n+b+1)\Gamma(b-n+1)}
\end{aligned}
$$

The steps to find C_n were: a) substitution of (5.4.5), b) integration by parts n times, and c) multiplication of numerator and denominator by $\Gamma(b-n+1)$. In particular if $b = m =$ positive integer

$$t^m = \Gamma(m+a+1) m! \sum_{n=0}^m \frac{(-1)^n L_n^a(t)}{\Gamma(n+a+1)(m-n)!},$$

$$0 < t < \infty,\ a > -1,\ \text{and } m = 0, 1, 2, \dots$$

If $a = 0$ we obtain the expansion

$$t^m = \Gamma(m+1)m! \sum_{n=0}^{m} \frac{(-1)^m L_n(t)}{n!(m-n)!}$$ ∎

Example

The function $f(t) = e^{-bt}$, with $b > -1/2$ and $t > 0$, is expanded as follows

$$C_n = \frac{n!}{\Gamma(n+a+1)} \int_0^\infty e^{-(b+1)t} t^a L_n^a(t)\, dt = \frac{1}{\Gamma(n+a+1)} \int_0^\infty e^{-bt} \frac{d^n}{dt^n}(e^{-t} t^{n+a})\, dt$$

$$= \frac{b^n}{\Gamma(n+a+1)} \int_0^\infty e^{-(b+1)t} t^{n+a}\, dt = \frac{b^n}{(b+1)^{n+a+1}}, \qquad n = 0, 1, 2, \ldots$$

and thus

$$e^{-bt} = (b+1)^{-a-1} \sum_{n=0}^{\infty} \left(\frac{b}{b+1}\right)^n L_n^a(t), \qquad 0 \le t < \infty$$

For $a = 0$

$$e^{-bt} = (b+1)^{-1} \sum_{n=0}^{\infty} \left(\frac{b}{b+1}\right)^n L_n(t), \qquad 0 \le t < \infty$$ ∎

Table 5.4.2 gives relationships of Laguerre polynomials.

1.5.5 *Chebyshev polynomials*

The Chebyshev polynomials can be derived from the Gegenbauer polynomials, and are given by

$$T_n(t) = \frac{n}{2} \sum_{k=0}^{[n/2]} \frac{(-1)^k (n-k-1)!}{k!(n-2k)!}(2t)^{n-2k}, \qquad -1 < t < 1 \tag{5.5.1}$$

The Chebyshev polynomials of the second kind are simply

$$U_n(t) = C_n^1(t), \qquad n = 0, 1, 2, \ldots \tag{5.5.2a}$$

where $C_n^1(t)$ is the Gegenbauer polynomial with $\lambda = 1$

$$C_n^\lambda(t) = (-1)^n \sum_{k=0}^{[n/2]} \binom{-\lambda}{n-k} \binom{n-k}{k} (2t)^{n-2k} \tag{5.5.2b}$$

Hence the second kind Chebyshev polynomials are

$$U_n(t) = \sum_{k=0}^{[n/2]} \binom{n-k}{k} (-1)^k (2t)^{n-2k} \tag{5.5.3}$$

The recurrence relations are

$$T_{n+1}(t) - 2t T_n(t) + T_{n-1}(t) = 0 \tag{5.5.4}$$

$$U_{n+1}(t) - 2t U_n(t) + U_{n-1}(t) = 0 \tag{5.5.5}$$

TABLE 5.4.2
Properties of the Laguerre polynomials.

1. $L_n(t) = \sum_{k=0}^{n} \frac{(-1)^k n! t^k}{(k!)^2 (n-k)!} = \sum_{k=0}^{n} (-1)^k \frac{1}{k!} \begin{pmatrix} n \\ k \end{pmatrix} t^k$

2. $L_n(t) = \frac{e^t}{n!} \frac{d^n}{dt^n} (t^n e^{-t})$

3. $(n+1)L_{n+1}(t) + (t - 1 - 2n)L_n(t) + nL_{n-1}(t) = 0$

4. $L_n'(t) - L_{n-1}'(t) + L_{n-1}(t) = 0,$ \hfill $n = 1, 2, 3, \ldots$

5. $(n+1)L_{n+1}'(t) + (t - 1 - 2n)L_n'(t) + L_n(t) + nL_{n-1}'(t) = 0,$ \hfill $n = 1, 2, 3, \ldots$

6. $L_{n+1}'(t) = L_n'(t) - L_n(t)$

7. $tL_n'(t) = nL_n(t) - nL_{n-1}(t),$ \hfill $n = 1, 2, 3, \ldots$

8. $tL_n''(t) + (1-t)L_n'(t) + nL_n(t) = 0,$ Laguerre differential equation

9. $L_n'(t) - L_{n-1}'(t) + L_{n-1}(t) = 0$

10. $\int_0^{\infty} e^{-t} L_n(t) L_k(t)\, dt = 0,$ \hfill $k \neq n$

11. $\int_0^{\infty} e^{-t} [L_n(t)]^2\, dt = 1$

12. $f(t) = \sum_{n=0}^{\infty} C_n L_n(t),$ \hfill $0 \leq t < \infty$

$C_n = \int_0^{\infty} e^{-t} f(t) L_n(t)\, dt$

13. $L_n(0) = 1,\ L_n'(0) = -n,\ L_n''(0) = \frac{1}{2} n(n-1)$

14. $L_n^m(t) = (-1)^m \frac{d^m}{dt^m} [L_{n+m}(t)],$ \hfill $m = 0, 1, 2, \ldots$

15. $L_n^m(t) = \sum_{k=0}^{n} \frac{(-1)^k (n+m)! t^k}{(n-k)!(m+k)! k!},$ \hfill $m = 0, 1, 2, \ldots$

16. $(n+1)L_{n+1}^m(t) + (t - 1 - 2n - m)L_n^m(t) + (n+m)L_{n-1}^m(t) = 0$

17. $tL_n^{m\prime}(t) - nL_n^m(t) + (n+m)L_{n-1}^m(t) = 0$

18. $L_n^m(t) = \frac{1}{n!} e^t t^{-m} \frac{d^n}{dt^n} (e^{-t} t^{n+m}) =$ Rodrigues formula

19. $L_{n-1}^m(t) + L_n^{m-1}(t) - L_n^n(t) = 0$

20. $L_n^{m\prime}(t) = -L_{n-1}^{m+1}(t)$

21. $L_n^m(0) = \frac{(n+m)!}{n! m!}$

TABLE 5.4.2
(continued)

22. $\displaystyle\int_0^\infty e^{-t}t^k L_n(t)\,dt = \begin{cases} 0 & k < n \\ (-1)^n n! & k = n \end{cases}$

23. $\displaystyle\int_0^t L_k(x)L_n(t-x)\,dx = \int_0^t L_{n+k}(x)\,dx = L_{n+k}(t) - L_{n+k+1}(t)$

24. $\displaystyle\int_t^\infty e^{-x}L_n^m(x)\,dx = e^{-t}[L_n^m(t) - L_{n-1}^m(t)],$ $\qquad\qquad m = 0, 1, 2, \ldots$

25. $\displaystyle\int_0^t (t-x)^m L_n(x)\,dx = \frac{m!n!}{(m+n+1)!}t^{m+1}L_n^{m+1}(t),$ $\qquad m = 0, 1, 2, \ldots$

26. $\displaystyle\int_0^1 x^a(1-x)^{b-1}L_n^a(tx)\,dx = \frac{\Gamma(b)\Gamma(n+a+1)}{\Gamma(n+a+b+1)}L_n^{a+b}(t),$ $\qquad a > -1,\ b > 0$

27. $\displaystyle\int_0^\infty e^{-t}t^a L_n^a(t)L_k^a(t)\,dt = 0,$ $\qquad\qquad k \neq n, a > -1$

28. $\displaystyle\int_0^\infty e^{-t}t^a[L_n^a(t)]^2\,dt = \frac{\Gamma(n+a+1)}{n!},$ $\qquad\qquad a > -1$

29. $\displaystyle\int_0^\infty e^{-t}t^{a+1}[L_n^a(t)]^2\,dt = \frac{\Gamma(n+a+1)}{n!}(2n+a+1),$ $\qquad a > -1$

30. $\displaystyle L_n^{-1/2}(t) = \frac{(-1)^n}{2^{2n}n!}H_{2n}(\sqrt{t})$

31. $\displaystyle L_n^{1/2}(t) = \frac{(-1)^n}{2^{2n+1}n!}\frac{H_{2n+1}(\sqrt{t})}{\sqrt{t}}$

32. $\displaystyle f(t) = \sum_{n=0}^\infty C_n L_n^m(t)$

$\displaystyle C_n = \frac{n!}{\Gamma(n+m+1)}\int_0^\infty e^{-t}t^m f(t)L_n^m(t)\,dt$

33. $\displaystyle t^p = p!\sum_{n=0}^p \binom{p}{n}(-1)^n L_n(t)$

34. $\displaystyle e^{-at} = (a+1)^{-1}\sum_{n=0}^\infty \left(\frac{a}{a+1}\right)^n L_n(t),$ $\qquad\qquad a > -\frac{1}{2}$

35. $\displaystyle\int_0^\infty \frac{e^{-tx}}{x+1}\,dx = \sum_{n=0}^\infty \frac{L_n(t)}{n+1}$

The orthogonality properties are

$$\int_{-1}^1 (1-t^2)^{-1/2}T_n(t)T_k(t)\,dt = 0, \qquad k \neq n \qquad\qquad (5.5.6)$$

$$\int_{-1}^1 (1-t^2)^{-1/2}U_n(t)U_k(t)\,dt = 0, \qquad k \neq n \qquad\qquad (5.5.7)$$

TABLE 5.5.1
Properties of the Chebyshev polynomials.

---

1. $(1 - t^2)\dfrac{d^2 y}{dt^2} - t\dfrac{dy}{dt} + n^2 y = 0;\; y(t) = T_n(t)$

2. $T_n(t) = \dfrac{n}{2}\displaystyle\sum_{k=0}^{[n/2]} \dfrac{(-1)^k (n - k - 1)!}{k!(n - 2k)!}(2t)^{n-2k}, \qquad n = 1, 2, \ldots, \; [n/2] = \text{largest integer} \le n/2$

3. $T_n(t) = \dfrac{(-2)^n n!}{(2n)!}\sqrt{1 - t^2}\dfrac{d^n}{dt^n}(1 - t^2)^{n - \frac{1}{2}}$, Rodrigues formula

4. $T_n(t) = \cos(n\cos^{-1} t)$

5. $\dfrac{1 - st}{1 - 2st + s^2} = \displaystyle\sum_{n=0}^{\infty} T_n(t)s^n$, generating function

6. $T_{n+1}(t) = 2t T_n(t) - T_{n-1}(t)$

7. $\displaystyle\int_{-1}^{1} \dfrac{T_n(t)T_m(t)}{\sqrt{(1 - t^2)}}\,dt = \begin{cases} 0 & n \ne m \\[2mm] \pi/2 & n = m \ne 0 \\[2mm] \pi & n = m = 0 \end{cases}$

8. $T_n(1) = 1,\; T_n(-1) = (-1)^n,\; T_{2n}(0) = (-1)^n,\; T_{2n+1}(0) = 0$

The governing differential equations for $T_n(t)$ and $U_n(t)$ are, respectively,

$$(1 - t^2)y'' - ty' + n^2 y = 0 \tag{5.5.8}$$

$$(1 - t^2)y'' - 3ty' + n(n + 2)y = 0 \tag{5.5.9}$$

The following are relationships between the two Chebyshev types:

$$T_n(t) = U_n(t) - tU_{n-1}(t) \tag{5.5.10}$$

$$(1 - t^2)U_n(t) = tT_n(t) - T_{n+1}(t) \tag{5.5.11}$$

Table 5.5.1 gives relationships for the Chebyshev polynomials.

If we set $t = \cos\theta$ in (5.5.8), we find that it reduces to

$$\frac{d^2 y}{d\theta^2} + n^2 y = 0$$

with solution $\cos n\theta$ and $\sin n\theta$. Therefore, if we set $T_n(\cos\theta) = C_n \cos n\theta$, we find that $C_n = 1$ for all n because $T_n(1) = 1$ for all n. Hence

$$T_n(t) = \cos n\theta = \cos(n\cos^{-1} t) \tag{5.5.12}$$

Similarly

$$U_n(t) = \frac{\sin[(n + 1)\cos^{-1} t]}{\sqrt{1 - t^2}} \tag{5.5.13}$$

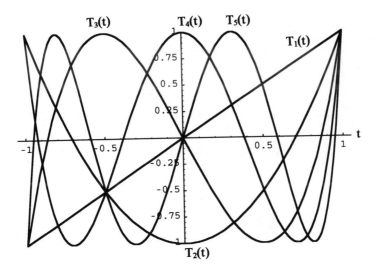

FIGURE 5.5.1

The generating function for the Chebyshev polynomial is

$$\frac{1 - st}{1 - 2st + s^2} = \sum_{n=0}^{\infty} T_n(t)s^n \tag{5.5.14}$$

The generalized Rodrigues formula is

$$T_n(t) = \frac{(-2)^n n!}{(2n)!} \sqrt{1 - t^2} \frac{d^n}{dt^n} (1 - t^2)^{n - \frac{1}{2}} \tag{5.5.15}$$

Figure 5.5.1 shows several Chevyshev polynomials.

1.5.6 *Bessel functions*

Bessel functions of the first kind

General relations   The solution of Bessel's equation

$$y'' + \frac{1}{t}y' + \left(1 - \frac{n^2}{t^2}\right)y = 0, \qquad n = 0, 1, 2, \ldots \tag{5.6.1}$$

is the function $y = J_n(t)$, known as the **Bessel function of the first kind and order** n. The Bessel function is defined by the series

$$J_n(t) = \sum_{k=0}^{\infty} \frac{(-1)^k (t/2)^{n+2k}}{k!(n+k)!}, \qquad -\infty < t < \infty \tag{5.6.2}$$

We can find (5.6.2) by expanding the function $w(t, x)$ in series of the two exponentials $\exp(tx/2)$ and $\exp(-t/2x)$ in the form

$$w(t, x) \doteq e^{\frac{1}{2}t(x - \frac{1}{x})} = \sum_{n=-\infty}^{\infty} J_n(t)x^n, \qquad x \neq 0 \tag{5.6.3}$$

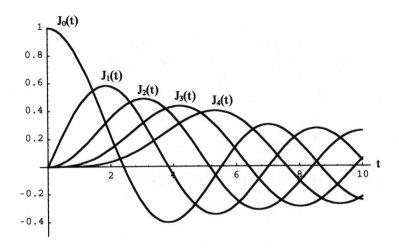

FIGURE 5.6.1

By setting $m = -n$ in (5.6.2) we obtain

$$J_{-n}(t) = \sum_{k=0}^{\infty} \frac{(-1)^k (t/2)^{2k-n}}{k!(k-n)!} = \sum_{k=n}^{\infty} \frac{(-1)^k (t/2)^{2k-n}}{k!(k-n)!}$$

because $1/[(k-n)!] = 0$ for $k = 0, 1, 2, \ldots, n-1$ ($\Gamma(n) = \infty$ for negative n). Setting $k = m + n$, we obtain

$$J_{-n}(t) = \sum_{m=0}^{\infty} \frac{(-1)^{m+n} (t/2)^{2m+n}}{m!(m+n)!} \tag{5.6.4}$$

from which it follows that

$$J_{-n}(t) = (-1)^n J_n(t), \qquad n = 0, 1, 2, \ldots \tag{5.6.5}$$

Equating like terms in the expanded form of (5.6.3), we obtain

$$J_0(0) = 1, \qquad J_n(0) = 0, \qquad n \neq 0 \tag{5.6.6}$$

Figure 5.6.1 shows several Bessel functions of the first kind and zero order.

Bessel functions of nonintegral order

The Bessel functions of a noninteger number are given by

$$J_\nu(t) = \sum_{k=0}^{\infty} \frac{(-1)^k (t/2)^{2k+\nu}}{k!\Gamma(k+\nu+1)}, \qquad \nu \geq 0 \tag{5.6.7a}$$

$$J_{-\nu}(t) = \sum_{k=0}^{\infty} \frac{(-1)^k (t/2)^{2k-\nu}}{k!\Gamma(k-\nu+1)}, \qquad \nu \geq 0 \tag{5.6.7b}$$

The two functions $J_{-\nu}(t)$ and $J_\nu(t)$ are linear independent for noninteger values of ν and they do not satisfy any generating-function relation. The functions $J_{-\nu}(0) = \infty$ and $J_\nu(0)$ remain finite. Both share most of the properties of $J_n(t)$ and $J_{-n}(t)$.

Recurrence Relation

$$\frac{d}{dt}[t^\nu J_\nu(t)] = \frac{d}{dt}\sum_{k=0}^{\infty}\frac{(-1)^k(t)^{2k+2\nu}}{2^{2k+\nu}k!\Gamma(k+\nu+1)} = t^\nu\sum_{k=0}^{\infty}\frac{(-1)^k(t/2)^{2k+(\nu-1)}}{k!\Gamma(k+\nu)}$$

$$= t^\nu J_{\nu-1}(t) \tag{5.6.8}$$

Similarly

$$\frac{d}{dt}[t^{-\nu}J_\nu(t)] = -t^{-\nu}J_{\nu+1}(t) \tag{5.6.9}$$

Differentiate (5.6.8) and (5.6.9) and dividing by t^ν and $t^{-\nu}$, respectively, we find

$$J_\nu'(t) + \frac{\nu}{t}J_\nu(t) = J_{\nu-1}(t) \tag{5.6.10}$$

$$J_\nu'(t) - \frac{\nu}{t}J_\nu(t) = -J_{\nu+1}(t) \tag{5.6.11}$$

Set $\nu = 0$ in (5.6.11) to obtain

$$J_0'(t) = -J_1(t) \tag{5.6.12}$$

Add and subtract (5.6.10) and (5.6.11) to find, respectively, the relations

$$2J_\nu'(t) = J_{\nu-1}(t) - J_{\nu+1}(t) \tag{5.6.13}$$

$$\frac{2\nu}{t}J_\nu(t) = J_{\nu-1}(t) + J_{\nu+1}(t) \tag{5.6.14}$$

The last relation is known as the **three-term recurrence formula**. Repeated operations result in

$$\left(\frac{d}{tdt}\right)^m[t^\nu J_\nu(t)] = t^{\nu-m}J_{\nu-m}(t) \tag{5.6.15}$$

$$\left(\frac{d}{tdt}\right)^m[t^{-\nu}J_\nu(t)] = (-1)^m t^{-\nu-m}J_{\nu+m}(t) \tag{5.6.16}$$

Example

We proceed to find the following derivative

$$\frac{d}{dt}[t^\nu J_\nu(at)] = \frac{d}{dt}\left[\left(\frac{u}{a}\right)^\nu J_\nu(u)\right] = \frac{d}{du}\left[\frac{u^\nu}{a^\nu}J_\nu(u)\right]\frac{du}{dt}$$

$$= a^{-\nu}\frac{d}{du}[u^\nu J_\nu(u)]a = a^{1-\nu}[u^\nu J_{\nu-1}(u)]$$

$$= a^{1-\nu}[(at)^\nu J_{\nu-1}(at)] = at^\nu J_{\nu-1}(at)$$

where (5.6.8) was used. ∎

Example

Differentiate (5.6.13) to find

$$\frac{d^2 J_\nu(t)}{dt^2} = \frac{1}{2}\left(\frac{dJ_{\nu-1}(t)}{dt} - \frac{dJ_{\nu+1}(t)}{dt}\right)$$

Then apply the same equation to each derivative on the right side to find

$$\frac{d^2 J_\nu(t)}{dt^2} = \frac{1}{2}\left[\frac{1}{2}[J_{\nu-2}(t) - J_\nu(t)] - \frac{1}{2}[J_\nu(t) - J_{\nu+2}(t)]\right]$$

$$= \frac{1}{2^2}[J_{\nu-2}(t) - 2J_\nu(t) + J_{\nu+2}(t)]$$

Similarly we find

$$\frac{d^3 J_\nu(t)}{dt^3} = \frac{1}{2^3}[J_{\nu-3}(t) - 3J_{\nu-1}(t) + 3J_{\nu+1}(t) - J_{\nu+3}(t)] \qquad \blacksquare$$

Integral representation

Set $x = \exp(-j\varphi)$ in (5.6.3), multiply both sides by $\exp(jn\varphi)$, and integrate the results from 0 to π. Hence

$$\int_0^\pi e^{j(n\varphi - t\sin\varphi)}\,d\varphi = \sum_{k=-\infty}^\infty J_k(t)\int_0^\pi e^{j(n-k)\varphi}\,d\varphi \qquad (5.6.17)$$

Expand on both sides the exponentials in Eauler's formula; equate the real and imaginary parts and use the relation

$$\int_0^\pi \cos(n - k)\varphi\,d\varphi = \begin{cases} 0 & k \neq 0 \\ \pi & k = n \end{cases}$$

to find that all terms of the infinite sum vanish except for $k = n$. Hence we obtain

$$J_n(t) = \frac{1}{\pi}\int_0^\pi \cos(n\varphi - t\sin\varphi)\,d\varphi, \qquad n = 0, 1, 2, \ldots \qquad (5.6.18)$$

When $n = 0$, we find

$$J_0(t) = \frac{1}{\pi}\int_0^\pi \cos(t\sin\varphi)\,d\varphi \qquad (5.6.19)$$

For a Bessel function with nonintegral order, the Poisson formula is

$$J_\nu(t) = \frac{(t/2)^\nu}{\sqrt{\pi}\,\Gamma\left(\nu + \frac{1}{2}\right)}\int_{-1}^1 (1 - x^2)^{\nu - \frac{1}{2}} e^{jtx}\,dx, \qquad \nu > -\frac{1}{2},\ t > 0 \qquad (5.6.20)$$

Set $x = \cos\theta$ to obtain

$$J_\nu(t) = \frac{(t/2)^\nu}{\sqrt{\pi}\,\Gamma\left(\nu + \frac{1}{2}\right)}\int_0^\pi \cos(t\cos\theta)\sin^{2\nu}\theta\,d\theta, \qquad \nu > -\frac{1}{2},\ t > 0 \qquad (5.6.21)$$

Integrals involving Bessel functions

Start with the identities

$$\frac{d}{dt}[t^\nu J_\nu(t)] = t^\nu J_{\nu-1}(t) \qquad (5.6.22)$$

$$\frac{d}{dt}[t^{-\nu} J_\nu(t)] = -t^{-\nu} J_{\nu+1}(t) \qquad (5.6.23)$$

and directly integrate to find

$$\int t^\nu J_{\nu-1}(t)\,dt = t^\nu J_\nu(t) + C \qquad (5.6.24)$$

$$\int t^{-\nu} J_{\nu+1}(t)\, dt = -t^{-\nu} J_\nu(t) + C \qquad (5.6.25)$$

where C is the constant of integration.

Example
We apply the integration procedure to find

$$\int t^2 J_2(t)\, dt = \int t^3 [t^{-1} J_2(t)]\, dt = -\int t^3 \frac{d}{dt}[t^{-1} J_1(t)]\, dt$$

$$= -t^2 J_1(t) + 3\int t J_1(t)\, dt = -t^2 J_1(t) - 3\int t[-J_1(t)]\, dt$$

$$= -t^2 J_1(t) - 3\int t\left[\frac{d}{dt} J_0(t)\right] dt = -t^2 J_1(t) - 3t J_0(t) + 3\int J_0(t)\, dt$$

The last integral has no closed solution.   ∎

Example
If $a > 0$ and $b > 0$, then (see [5.6.19])

$$\int_0^\infty e^{-at} J_0(bt)\, dt = \int_0^\infty e^{-at}\, dt\, \frac{2}{\pi} \int_0^{\pi/2} \cos(bt \sin\varphi)\, d\varphi$$

$$= \frac{2}{\pi} \int_0^{\pi/2} d\varphi \int_0^\infty e^{-at} \cos(bt \sin\varphi)\, dt$$

$$= \frac{2}{\pi} \int_0^{\pi/2} \frac{a\, d\varphi}{a^2 + b^2 \sin^2\varphi} = \frac{1}{\sqrt{a^2 + b^2}} \qquad ∎$$

Example
For $a > 0$, $b > 0$, and $\nu > -1$ (ν is real), then

$$\int_0^\infty e^{-a^2 t^2} J_\nu(bt) t^{\nu+1}\, dt = \int_0^\infty e^{a^2 t^2} t^{\nu+1}\, dt \sum_{k=0}^\infty \frac{(-1)^k (bt/2)^{\nu+2k}}{k!\,\Gamma(k+\nu+1)}$$

$$= \sum_{k=0}^\infty \frac{(-1)^k}{k!\,\Gamma(k+\nu+1)} \left(\frac{b}{2}\right)^{\nu+2k} \int_0^\infty e^{-a^2 t^2} t^{2\nu+2k+1}\, dt$$

$$= \sum_{k=0}^\infty \frac{(-1)^k}{k!\,\Gamma(k+\nu+1)} \left(\frac{b}{2}\right)^{\nu+2k} \frac{1}{2a^{2\nu+2k+2}} \int_0^\infty e^{-r} r^{\nu+k}\, dr$$

$$= \frac{b^\nu}{(2a^2)^{\nu+1}} \sum_{k=0}^\infty \frac{\left(-\frac{b^2}{4a^2}\right)^k}{k!} = \frac{b^\nu}{(2a^2)^{\nu+1}} e^{-b^2/4a^2} \qquad (5.6.26)$$

where the last integral is the gamma function and the summation is the exponential expression.   ∎

The usual method to find definite integrals involving Bessel functions is to replace the Bessel function by its series representation. To illustrate the technique, let us find the value of the integral

$$I = \int_0^\infty e^{-at} t^p J_p(bt) dt, \qquad p > -\frac{1}{2}, \quad a > 0, \ b > 0$$

$$= \sum_{k=0}^\infty \frac{(-1)^k (b/2)^{2k+p}}{k! \Gamma(k+p+1)} \int_0^\infty e^{-at} t^{2k+2p} \, dt$$

$$= b^p \sum_{k=0}^\infty \frac{(-1)^k \Gamma(2k+2p+1)}{2^{2k+p} k! \Gamma(k+p+1)} (a^2)^{-(p+\frac{1}{2})-k} (b^2)^k \qquad (5.6.27)$$

where the last integral is in the form of a gamma function. But we know that

$$\binom{-r}{k} = (-1)^k \binom{r+k-1}{k}, \qquad \binom{n}{k} = \binom{n}{n-k}$$

$$\binom{n+1}{k+1} = \binom{n}{k+1} + \binom{n}{k}, \qquad 0 \le k \le n-1$$

and thus we obtain

$$\frac{(-1)^k \Gamma(2k+2p+1)}{2^{2k+p} k! \Gamma(k+p+1)} = \frac{(-1)^k 2^p \Gamma\left(p+k+\frac{1}{2}\right)}{\sqrt{\pi} k!}$$

$$= \frac{(-1)^k}{\sqrt{\pi}} 2^p \Gamma\left(p+\frac{1}{2}\right) \binom{p+k-\frac{1}{2}}{k}$$

$$= \frac{2^p \Gamma\left(p+\frac{1}{2}\right)}{\sqrt{\pi}} \binom{-\left(p+\frac{1}{2}\right)}{k} \qquad (5.6.28)$$

Therefore, (5.6.27) becomes

$$I = \int_0^\infty e^{-at} t^p J_p(bt) \, dt$$

$$= \frac{(2b)^p \Gamma\left(p+\frac{1}{2}\right)}{\sqrt{\pi}} \sum_{k=0}^\infty \binom{-\left(p+\frac{1}{2}\right)}{k} (a^2)^{-(p+(1/2))-k} (b^2)^k$$

$$= \frac{(2b)^p \Gamma\left(p+\frac{1}{2}\right)}{\sqrt{\pi} (a^2+b^2)^{p+\frac{1}{2}}}, \qquad p > -\frac{1}{2}, \ a > 0, \ b > 0 \qquad (5.6.29)$$

Setting $p = 0$ in this equation we find

$$\int_0^\infty e^{-at} J_0(bt) = \frac{1}{[a^2+b^2]^{1/2}}, \qquad a > 0, \ b > 0 \qquad (5.6.30)$$

Set $a = 0+$ in this equation to obtain

$$\int_0^\infty J_0(bt) \, dt = \frac{1}{b}, \qquad b > 0 \qquad (5.6.31)$$

By assuming the real approaches zero and writing a as pure imaginary, (5.6.30) becomes

$$\int_0^\infty e^{-jat} J_0(bt) \, dt = \begin{cases} \frac{1}{(b^2-a^2)^{1/2}} & b > a \\ \frac{-j}{(a^2-b^2)^{1/2}} & b < a \end{cases} \qquad (5.6.32)$$

The above integral, by equating real and imaginary parts, becomes

$$\int_0^\infty \cos(at)\,J_0(bt)\,dt = \frac{1}{(b^2 - a^2)^{1/2}}, \qquad b > a \tag{5.6.33}$$

$$\int_0^\infty \sin(at)\,J_0(bt)\,dt = \frac{1}{(a^2 - b^2)^{1/2}}, \qquad b < a \tag{5.6.34}$$

Example

To evaluate the integral $\int_0^b t\,J_0(at)\,dt$, we proceed as follows:

$$\int_0^\infty t\,J_0(at)\,dt = \int_0^\infty \frac{1}{a}\frac{d}{dt}[t\,J_1(at)]\,dt$$

$$= \frac{1}{a}[t\,J_1(at)]\Big|_{t=0}^b = \frac{b}{a}J_1(ab), \qquad a \neq 0 \tag{5.6.35}$$

where (5.6.8) with $\nu = 1$ was used. ∎

Example

To evaluate the integral $I = \int_0^b t^2\,J_0(at)\,dt$, where a is a constant and nonequal to zero, we proceed as follows (set $at = r$):

$$I = \frac{1}{a^3}\int_0^{ab} r^2\,J_0(r)\,dr = \frac{1}{a^3}\int_0^{ab} r\,r\,J_0(r)\,dr$$

$$= \frac{1}{a^3}\int_0^{ab} r\frac{d}{dr}[r\,J_1(r)]\,dr = \frac{1}{a^3}\left[a^2 b^2 J_1(ab) - \int_0^{ab} r\,J_1(r)\,dr\right]$$

But (see [5.6.23])

$$\int_0^{ab} r\,J_1(r)\,dr = -\int_0^{ab} r\frac{d}{dr}[J_0(r)]\,dr = -r\,J_0(r)|_{r=0}^{ab}$$

$$+ \int_0^{ab} J_0(r)\,dr = -ab\,J_0(ab) + \int_0^{ab} J_0(r)\,dr$$

and therefore

$$I = \frac{1}{a^3}\left[a^2 b^2 J_1(ab) + ab\,J_0(ab) - \int_0^{ab} J_0(r)\,dr\right] \tag{5.6.36}$$

The integral can be approximately evaluated with any desired accuracy by termwise integration of the series of $J_0(t)$. Hence we write

$$\int_0^{ab} J_0(t)\,dt = ab - \frac{a^3 b^3}{3 \cdot 2^2} + \frac{a^5 b^5}{5 \cdot 2^4 \cdot (2!)^2} - \frac{a^7 b^7}{7 \cdot 2^6 \cdot (3!)^2} + \cdots \qquad ∎$$

Fourier Bessel series

A Bessel series is a member of the class of generalized Fourier series. It is defined by

$$f(t) = \sum_{n=1}^\infty c_n\,J_\nu(t_n t), \qquad 0 < t < a, \quad \nu > -\frac{1}{2} \tag{5.6.37}$$

where c's are the expansion coefficient constants and t_n's $(n = 1, 2, 3, \ldots)$ are the zeros (positive roots) of the function

$$J_\nu(t_n t), \qquad n = 1, 2, 3, \ldots \tag{5.6.38}$$

The orthogonality property is defined as follows $(\nu > -1)$:

$$\int_0^a t J_\nu(t_m t) J_\nu(t_n t) \, dt = 0, \qquad m \neq n \tag{5.6.39}$$

with weight t. It can also be shown that

$$\int_0^a t [J_\nu(t_n t)]^2 \, dt = \frac{a^2}{2} [J_{\nu+1}(t_n a)]^2 \tag{5.6.40}$$

THEOREM 5.6.1
If a real function $f(t)$ is piecewise continuous on $(0, a)$ and is of bounded variation in every subinterval $[t_1, t_2]$ where $0 < t_1 < t_2 < a$, then if the integral

$$\int_0^a \sqrt{t} |f(t)| \, dt$$

is finite, the Fourier-Bessel series converges to $f(t)$ at every continuity point of $f(t)$ and to $[f(t+) - f(t-)]/2$ at every discontinuity point.

To begin, multiply (5.6.37) by $t J_\nu(t_m t)$ and integrate from 0 to a. Assuming that termwise integration is permitted, we obtain

$$\int_0^a t f(t) J_\nu(t_m t) \, dt = \sum_{n=1}^\infty c_n \int_0^a t J_\nu(t_m t) J_\nu(t_n t) \, dt$$

$$= c_m \int_0^a [J_\nu(t_m t)]^2 \, dt \tag{5.6.41}$$

because the integral is zero if $n \neq m$ (see [5.6.39]). Hence from this equation we obtain

$$c_n = \frac{2}{a^2 [J_{\nu+1}(t_n a)]^2} \int_0^a t f(t) J_\nu(t_n t) \, dt, \qquad n = 1, 2, 3, \ldots \tag{5.6.42}$$

Example
Find the Fourier-Bessel series for the function

$$f(t) = \begin{cases} t & 0 < t < 1 \\ 0 & 1 < t < 2 \end{cases}$$

corresponding to the set of functions $\{J_1(t_n t)\}$ where t_n satisfies $J_1(2t_n) = 0$ $(n = 1, 2, 3, \ldots)$. ∎

Solution   We write the solution

$$f(t) = \sum_{n=1}^\infty c_n J_1(t_n t), \qquad 0 < t < 2$$

where

$$c_n = \frac{1}{2[J_2(2t_n)]^2} \int_0^2 t f(t) J_1(t_n t)\, dt$$

$$= (\cdot) \int_0^1 t^2 J_1(t_n t)\, dt \qquad (\text{let } r = t_n t)$$

$$= (\cdot) \frac{1}{t_n^3} \int_0^{t_n} r^2 J_1(r)\, dr \qquad (\text{apply } [5.6.22])$$

$$= (\cdot) \frac{1}{t_n^3} \int_0^{t_n} \frac{d}{dt}[r^2 J_2(r)]\, dr$$

$$= \frac{1}{2[J_2(2t_n)]^2 t^3} t_n^2 J_2(t_n) = \frac{J_2(t_n)}{2t_n[J_2(2t_n)]^2}, \qquad n = 1, 2, 3, \ldots \qquad \blacksquare$$

Example
To express the function $f(t) = 1$ on the open interval $0 < t < a$ as an infinite series of Bessel functions of zero order, we proceed as follows (see [5.6.42]):

$$c_n = \frac{2}{a^2[J_1(t_n a)]^2} \int_0^a t \cdot 1 \cdot J_0(t_n t)\, dt$$

$$= \frac{2}{a^2[J_1(t_n a)]^2} \int_0^a \frac{1}{t_n} \frac{d}{dt}[t J_1(t_n t)]\, dt \qquad (\text{see } [5.6.22])$$

$$= \frac{2}{a^2 t_n[J_1(t_n a)]^2} [t J_1(t_n t)]\Big|_{t=0}^{a} = \frac{2}{a t_n J_1(t_n a)}$$

Hence the expression is

$$1 = 2 \sum_{n=1}^{\infty} \frac{J_n(t_n t)}{t_n J_1(t_n a)}, \qquad 0 < t < a \qquad \blacksquare$$

Example
Let us expand the function $f(t) = t^2$, $0 \le t \le 1$, in a series of the form

$$c_1 J_0(t_1 t) + c_2 J_0(t_2 t) + \cdots$$

where t_n denotes the nth positive zero of $J_0(t)$. From (5.6.42) we obtain $(a = 1)$

$$c_n = \frac{2}{[J_1(t_n)]^2} \int_0^1 t^3 J_0(t_n t)\, dt$$

$$= \frac{2}{[J_1(t_n)]^2} \int_0^{t_n} \frac{r^3}{t_n^3} J_0(r) \frac{dr}{t_n} = \frac{2}{t_n^4[J_1(t_n)]^2} \int_0^{t_n} r^2 \frac{d}{dr}[r J_1(r)]$$

$$= \frac{2}{t_n^4[J_1(t_n)]^2} \left[r^3 J_1(r)\Big|_{r=0}^{t_n} - 2 \int_0^{t_n} r^2 J_1(r)\, dr \right]$$

$$= \frac{2}{t_n^4[J_1(t_n)]^2} \left[t_n^3 J_1(t_n) - 2 \int_0^{t_n} \frac{d}{dr}[r^2 J_2(r)]\, dr \right]$$

$$= \frac{2}{t_n^4[J_1(t_n)]^2} [t_n^3 J_1(t_n) - 2t_n^2 J_2(t_n)] \qquad \blacksquare$$

Table 5.6.1 gives Bessel function relationships. Tables 5.6.2 and 5.6.3 give numerical values for Bessel functions and Table 5.6.4 gives the zeros of several Bessel functions.

TABLE 5.6.1
Properties of Bessel functions of the first kind.

---

1. $J_n(t) = \sum_{k=0}^{\infty} \dfrac{(-1)^k (t/2)^{n+2k}}{k!(n+k)!}$, $\qquad\qquad -\infty < t < \infty, \; n = 0, 1, 2, 3, \ldots$

2. $J_{-n}(t) = \sum_{m=0}^{\infty} \dfrac{(-1)^{m+n} (t/2)^{2m+n}}{m!(m+n)!}$, $\qquad -\infty < t < \infty, \; n = 0, 1, 2, 3, \ldots$

3. $J_{-n}(t) = (-1)^n J_n(t)$, $\qquad\qquad\qquad\qquad n = 0, 1, 2, 3, \ldots$

4. $J_0(0) = 1, \qquad J_n(0) = 0$, $\qquad\qquad\qquad n \neq 0$

5. $J_\nu(t) = \sum_{k=0}^{\infty} \dfrac{(-1)^k (t/2)^{2k+\nu}}{k! \Gamma(k+\nu+1)}$, $\qquad\qquad \nu \geq 0, \; \nu \text{ is noninteger}$

6. $J_{-\nu}(t) = \sum_{k=0}^{\infty} \dfrac{(-1)^k (t/2)^{2k-\nu}}{k! \Gamma(k-\nu+1)}$, $\qquad\qquad \nu \geq 0, \; \nu \text{ is noninteger}$

7. $\dfrac{d}{dt} [t^\nu J_\nu(t)] = t^\nu J_{\nu-1}(t)$

8. $\dfrac{d}{dt} [t^\nu J_\nu(at)] = at^\nu J_{\nu-1}(at)$

9. $\dfrac{d}{dt} [t^{-\nu} J_\nu(t)] = -t^{-\nu} J_{\nu+1}(t)$

10. $\dfrac{d^2 J_\nu(t)}{dt^2} = \dfrac{1}{2^2} [J_{\nu-2}(t) - 2J_\nu(t) + J_{\nu+2}(t)]$

11. $\dfrac{d^3 J_\nu(t)}{dt^3} = \dfrac{1}{2^3} [J_{\nu-3}(t) - 3J_{\nu-1}(t) + 3J_{\nu+1}(t) - J_{\nu+3}(t)]$

12. $J_\nu'(t) + \dfrac{\nu}{t} J_\nu(t) = J_{\nu-1}(t)$

13. $J_\nu'(t) - \dfrac{\nu}{t} J_\nu(t) = -J_{\nu+1}(t)$

14. $J_0'(t) = -J_1(t)$

15. $2J_\nu'(t) = J_{\nu-1}(t) - J_{\nu+1}(t)$

16. $\dfrac{2\nu}{t} J_\nu(t) = J_{\nu-1}(t) + J_{\nu+1}(t)$

17. $\left(\dfrac{d}{t\,dt}\right)^m [t^\nu J_\nu(t)] = t^{\nu-m} J_{\nu-m}(t)$

18. $\left(\dfrac{d}{t\,dt}\right)^m [t^{-\nu} J_\nu(t)] = (-1)^m t^{-\nu-m} J_{\nu+m}(t)$

19. $J_1'(0) = \dfrac{1}{2}, \qquad J_n'(0) = 0$, $\qquad\qquad\qquad n > 1$

20. $J_n(t+r) = \sum_{k=-\infty}^{\infty} J_k(t) J_{n-k}(r)$

21. $J_0(2t) = [J_0(t)]^2 + 2 \sum_{k=1}^{\infty} (-1)^k [J_k(t)]^2$

TABLE 5.6.1
(continued)

22. $|J_0(t)| \leq 1, \qquad |J_n(t)| \leq \dfrac{1}{\sqrt{2}},$ $\qquad\qquad$ $n = 1, 2, 3, \ldots$

23. $e^{jt \sin \theta} = \displaystyle\sum_{n=-\infty}^{\infty} J_n(t) e^{jn\theta}$

24. $\cos(t \sin \theta) = J_0(t) + 2 \displaystyle\sum_{n=1}^{\infty} J_{2n}(t) \cos(2n\theta)$

25. $\cos(t \cos \theta) = J_0(t) + 2 \displaystyle\sum_{n=1}^{\infty} (-1)^n J_{2n}(t) \cos(2n\theta)$

26. $\sin(t \sin \theta) = 2 \displaystyle\sum_{n=1}^{\infty} J_{2n-1}(t) \sin[(2n-1)\theta]$

27. $\sin(t \cos \theta) = 2 \displaystyle\sum_{n=0}^{\infty} (-1)^n J_{2n+1}(t) \cos[(2n+1)\theta]$

28. $\cos t = J_0(t) + 2 \displaystyle\sum_{n=1}^{\infty} (-1)^n J_{2n}(t)$

29. $\sin t = 2 \displaystyle\sum_{n=1}^{\infty} (-1)^n J_{2n-1}(t)$

30. $J_\nu(t) J_{1-\nu}(t) + J_{-\nu}(t) J_{\nu-1}(t) = \dfrac{2 \sin \nu\pi}{\pi t}$ $\qquad\qquad$ Lommel's formula

31. $\dfrac{d}{dt}[t J_\nu(t) J_{\nu+1}(t)] = t[[J_\nu(t)]^2 - [J_{\nu+1}(t)]^2]$

32. $\dfrac{d}{dt}[t^2 J_{\nu-1}(t) J_{\nu+1}(t)] = 2t^2 J_\nu(t) J_\nu'(t)$

33. $J_{1/2}(t) = \sqrt{\dfrac{2}{\pi t}} \sin t, \qquad J_{-1/2}(t) = \sqrt{\dfrac{2}{\pi t}} \cos t$

34. $J_{1/2}(t) J_{-1/2}(t) = \dfrac{\sin 2t}{\pi t}, \qquad [J_{1/2}(t)]^2 + [J_{-1/2}(t)]^2 = \dfrac{2}{\pi t}$

35. $[J_0(t)]^2 = \displaystyle\sum_{n=0}^{\infty} \dfrac{(-1)^n (2n)!}{(n!)^4} \left(\dfrac{t}{2}\right)^{2n}$ $\qquad\qquad$ Cauchy product

36. $J_n(t) = \dfrac{1}{\pi} \displaystyle\int_0^\pi \cos(n\varphi - t \sin \varphi) \, d\varphi$

37. $J_0(t) = \dfrac{1}{\pi} \displaystyle\int_0^\pi \cos(t \sin \varphi) \, d\varphi$

38. $J_\nu(t) = \dfrac{(t/2)^\nu}{\sqrt{\pi} \, \Gamma\left(\nu + \frac{1}{2}\right)} \displaystyle\int_{-1}^1 (1 - x^2)^{\nu-\frac{1}{2}} e^{jtx} \, dx,$ $\qquad \nu > -\frac{1}{2}, \; t > 0$

39. $J_\nu(t) = \dfrac{(t/2)^\nu}{\sqrt{\pi} \, \Gamma\left(\nu + \frac{1}{2}\right)} \displaystyle\int_0^\pi \cos(t \cos \theta) \sin^{2\nu} \theta \, d\theta,$ $\qquad \nu > -\frac{1}{2}, \; t > 0$

40. $\displaystyle\int t^\nu J_{\nu-1}(t) \, dt = t^\nu J_\nu(t) + C,$ $\qquad\qquad$ $C = \text{constant}$

41. $\displaystyle\int t^{-\nu} J_{\nu+1}(t) \, dt = -t^{-\nu} J_\nu(t) + C,$ $\qquad\qquad$ $C = \text{constant}$

TABLE 5.6.1
(continued)

---

42. $[1 + (-1)^n]J_n(t) = \dfrac{2}{\pi}\displaystyle\int_0^\pi \cos n\varphi \cos(t \sin \varphi)\, d\varphi,$ $n = 0, 1, 2, \ldots$

43. $J_{2k}(t) = \dfrac{1}{\pi}\displaystyle\int_0^\pi \cos 2k\varphi \cos(t \sin \varphi)\, d\varphi,$ $k = 0, 1, 2, \ldots$

44. $J_{2k+1}(t) = \dfrac{1}{\pi}\displaystyle\int_0^\infty \sin[(2k+1)\varphi]\sin(t \sin \varphi)\, d\varphi,$ $k = 0, 1, 2, \ldots$

45. $\displaystyle\int_0^\pi \cos[(2k+1)\varphi]\cos(t \sin \varphi)\, d\varphi,$ $k = 0, 1, 2, \ldots$

45. $\displaystyle\int_0^\pi \sin 2k\varphi \sin(t \sin \varphi)\, d\varphi = 0,$ $k = 0, 1, 2, \ldots$

47. $J_0(t) = \dfrac{2}{\pi}\displaystyle\int_0^1 \dfrac{\cos tx}{\sqrt{1 - x^2}}\, dx$

48. $\dfrac{2 \sin t}{t} = \sqrt{\dfrac{2\pi}{t}}\, J_{1/2}(t)$

49. $\displaystyle\int t J_0(t)\, dt = t J_1(t) + C$

50. $\displaystyle\int t^2 J_0(t)\, dt = t^2 J_1(t) + t J_0(t) - \int J_0(t)\, dt + C$

51. $\displaystyle\int t^3 J_0(t)\, dt = (t^3 - 4t)J_1(t) + 2t^2 J_0(t) + C$

52. $\displaystyle\int J_1(t)\, dt = -J_0(t) + C$

53. $\displaystyle\int t J_1(t)\, dt = -t J_0(t) + \int J_0(t)\, dt + C$

54. $\displaystyle\int t^2 J_1(t)\, dt = 2t J_1(t) - t^2 J_0(t) + C$

55. $\displaystyle\int t^3 J_1(t)\, dt = 3t^2 J_1(t) - (t^3 - 3t)J_0(t) - 3\int J_0(t)\, dt + C$

56. $\displaystyle\int J_3(t)\, dt = -J_2(t) - 2t^{-1}J_1(t) + C$

57. $\displaystyle\int t^{-1}J_1(t)\, dt = -J_1(t) + \int J_0(t)\, dt + C$

58. $\displaystyle\int t^{-2}J_2(t)\, dt = -\dfrac{2}{3t^2}J_1(t) - \dfrac{1}{3}J_1(t)$
$$+ \dfrac{1}{3t}J_0(t) + \dfrac{1}{3}\int J_0(t)\, dt + C$$

59. $\displaystyle\int J_0(t)\cos t\, dt = t J_0(t)\cos t + t J_1(t)\sin t + C$

60. $\displaystyle\int J_0(t)\sin t\, dt = t J_0(t)\sin t - t J_1(t)\cos t + C$

61. $\displaystyle\int_0^\infty e^{-at}t^p J_p(bt)\, dt = \dfrac{(2b)^p \Gamma\left(p + \frac{1}{2}\right)}{\sqrt{\pi}(a^2 + b^2)^{p+\frac{1}{2}}},$ $p > -\frac{1}{2},\ a > 0,\ b > 0$

TABLE 5.6.1
(continued)

62. $\displaystyle\int_0^\infty e^{-at} J_0(bt)\, dt = \frac{1}{(a^2 + b^2)^{1/2}},$ $a > 0,\ b > 0$

63. $\displaystyle\int_0^\infty J_0(bt)\, dt = \frac{1}{b},$ $b > 0$

64. $\displaystyle\int_0^\infty J_{n+1}(t) = \int_0^\infty J_{n-1}(t)\, dt,$ $n = 1, 2, \ldots$

65. $\displaystyle\int_0^\infty J_n(at)\, dt = \frac{1}{a}$

66. $\displaystyle\int_0^\infty t^{-1} J_n(t)\, dt = \frac{1}{n},$ $n = 1, 2, \ldots$

67. $\displaystyle\int_0^\infty e^{-at} t^{p+1} J_p(bt)\, dt = \frac{2^{p+1}\Gamma\left(p + \frac{3}{2}\right)}{\sqrt{\pi}}\, \frac{ab^p}{(a^2 + b^2)^{p+\frac{3}{2}}},$ $p > -1,\ a > 0,\ b > 0$

68. $\displaystyle\int_0^\infty t^2 e^{-at} J_0(bt)\, dt = \frac{2a^2 - b^2}{(a^2 + b^2)^{5/2}},$ $a > 0,\ b > 0$

69. $\displaystyle\int_0^\infty e^{-at^2} t^{p+1} J_p(bt)\, dt = \frac{b^p e^{-b^2/4a}}{(2a)^{p+1}},$ $p > -1,\ a > 0,\ b > 0$

70. $\displaystyle\int_0^\infty e^{-at^2} t^{p+3} J_p(bt)\, dt = \frac{b^p}{2^{p+1}a^{p+2}}\left(p + 1 - \frac{b^2}{4a}\right) e^{-b^2/4a},$ $p > -1,\ a > 0,\ b > 0$

71. $\displaystyle\int_0^\infty t^{-1} \sin t\, J_0(bt)\, dt = \arcsin\left(\frac{1}{b}\right),$ $b > 1$

72. $\displaystyle\int_0^{\pi/2} J_0(t \cos\varphi) \cos\varphi\, d\varphi = \frac{\sin t}{t}$

73. $\displaystyle\int_0^{\pi/2} J_1(t \cos\varphi)\, d\varphi = \frac{1 - \cos t}{t}$

74. $\displaystyle\int_0^\infty e^{-t\cos\varphi} J_0(t \sin\varphi) t^n dt = n! P_n(\cos\varphi),$ $0 \le \varphi < \pi$

$P_n(t) = n$th Legendre polynomial

75. $\displaystyle\int_0^\infty t(t^2 + a^2)^{-1/2} J_0(bt)\, dt = \frac{e^{-ab}}{b},$ $a \ge 0,\ b > 0$

76. $\displaystyle\int_0^\infty \frac{J_p(t)}{t^m}\, dt = \frac{\Gamma((p + 1 - m)/2)}{2^m \Gamma((p + 1 + m)/2)},$ $m > \frac{1}{2},\ p - m > -1$

77. $\displaystyle\frac{1}{8}(1 - t^2) = \sum_{n=1}^\infty \frac{J_0(k_n t)}{k_n^3 J_1(k_n)},$ $0 \le t \le 1,\ J_0(k_n) = 0,$

$n = 1, 2, \ldots$

78. $\displaystyle t^p = 2\sum_{n=1}^\infty \frac{J_p(k_n t)}{k_n J_{p+1}(k_n)},$ $0 < t < 1,\ J_p(k_n) = 0,$

$n = 1, 2, \ldots$

79. $\displaystyle t^{p+1} = 2^2(p + 1)\sum_{n=1}^\infty \frac{J_{p+1}(k_n t)}{k_n^2 J_{p+1}(k_n)},$ $0 < t < 0,\ p > -1/2,$

80. $J_p(k_n) = 0,\ n = 1, 2, \ldots$

TABLE 5.6.2

					$J_0(x)$					
x	0	.1	.2	.3	.4	.5	.6	.7	.8	.9
0	1.0000	.9975	.9900	.9776	.9604	.9385	.9120	.8812	.8463	.8075
1	.7652	.7196	.6711	.6201	.5669	.5118	.4554	.3980	.3400	.2818
2	.2239	.1666	.1104	.0555	.0025	−.0484	−.0968	−.1424	−.1850	−.2243
3	−.2601	−.2921	−.3202	−.3443	−.3643	−.3801	−.3918	−.3992	−.4026	−.4018
4	−.3971	−.3887	−.3766	−.3610	−.3423	−.3205	−.2961	−.2693	−.2404	−.2097
5	−.1776	−.1443	−.1103	−.0758	−.0412	−.0068	.0270	.0599	.0917	.1220
6	.1506	.1773	.2017	.2238	.2433	.2601	.2740	.2851	.2931	.2981
7	.3001	.2991	.2951	.2882	.2786	.2663	.2516	.2346	.2154	.1944
8	.1717	.1475	.1222	.0960	.0692	.0419	.0146	−.0125	−.0392	−.0653
9	−.0903	−.1142	−.1367	−.1577	−.1768	−.1939	−.2090	−.2218	−.2323	−.2403
10	−.2459	−.2490	−.2496	−.2477	−.2434	−.2366	−.2276	−.2164	−.2032	−.1881
11	−.1712	−.1528	−.1330	−.1121	−.0902	−.0677	−.0446	−.0213	.0020	.0250
12	.0477	.0697	.0908	.1108	.1296	.1469	.1626	.1766	.1887	.1988
13	.2069	.2129	.2167	.2183	.2177	.2150	.2101	.2032	.1943	.1836
14	.1711	.1570	.1414	.1245	.1065	.0875	.0679	.0476	.0271	.0064
15	−.0142	−.0346	−.0544	−.0736	−.0919	−.1092	−.1253	−.1401	−.1533	−.1650

When $x > 15.9$,

$$J_0(x) \simeq \sqrt{\left(\frac{2}{\pi x}\right)} \left\{ \sin\left(x + \frac{1}{4}\pi\right) + \frac{1}{8x} \sin\left(x - \frac{1}{4}\pi\right) \right\}$$

$$\simeq \frac{.7979}{\sqrt{x}} \left\{ \sin(57.296x + 45)° + \frac{1}{8x} \sin(57.296x - 45)° \right\}$$

					$J_1(x)$					
x	0	.1	.2	.3	.4	.5	.6	.7	.8	.9
0	.0000	.0499	.0995	.1483	.1960	.2423	.2867	.3290	.3688	.4059
1	.4401	.4709	.4983	.5220	.5419	.5579	.5699	.5778	.5815	.5812
2	.5767	.5683	.5560	.5399	.5202	.4971	.4708	.4416	.4097	.3754
3	.3391	.3009	.2613	.2207	.1792	.1374	.0955	.0538	.0128	−.0272
4	−.0660	−.1033	−.1386	−.1719	−.2028	−.2311	−.2566	−.2791	−.2985	−.3147
5	−.3276	−.3371	−.3432	−.3460	−.3453	−.3414	−.3343	−.3241	−.3110	−.2951
6	−.2767	−.2559	−.2329	−.2081	−.1816	−.1538	−.1250	−.0953	−.0652	−.0349
7	−.0047	.0252	.0543	.0826	.1096	.1352	.1592	.1813	.2014	.2192
8	.2346	.2476	.2580	.2657	.2708	.2731	.2728	.2697	.2641	.2559
9	.2453	.2324	.2174	.2004	.1816	.1613	.1395	.1166	.0928	.0684
10	.0435	.0184	−.0066	−.0313	−.0555	−.0789	−.1012	−.1224	−.1422	−.1603
11	−.1768	−.1913	−.2039	−.2143	−.2225	−.2284	−.2320	−.2333	−.2323	−.2290
12	−.2234	−.2157	−.2060	−.1943	−.1807	−.1655	−.1487	−.1307	−.1114	−.0912
13	−.0703	−.0489	−.0271	−.0052	.0166	.0380	.0590	.0791	.0984	.1165
14	.1334	.1488	.1626	.1747	.1850	.1934	.1999	.2043	.2066	.2069
15	.2051	.2013	.1955	.1879	.1784	.1672	.1544	.1402	.1247	.1080

When $x > 15.9$,

$$J_1(x) \simeq \sqrt{\left(\frac{2}{\pi x}\right)} \left\{ \sin\left(x - \frac{1}{4}\pi\right) + \frac{3}{8x} \sin\left(x + \frac{1}{4}\pi\right) \right\}$$

$$\simeq \frac{.7979}{\sqrt{x}} \left\{ \sin(57.296x - 45)° + \frac{3}{8x} \sin(57.296x + 45)° \right\}$$

TABLE 5.6.3

					$J_2(x)$					
x	0	.1	.2	.3	.4	.5	.6	.7	.8	.9
0	.0000	.0012	.0050	.0112	.0197	.0306	.0437	.0588	.0758	.0946
1	.1149	.1366	.1593	.1830	.2074	.2321	.2570	.2817	.3061	.3299
2	.3528	.3746	.3951	.4139	.4310	.4461	.4590	.4696	.4777	.4832
3	.4861	.4862	.4835	.4780	.4697	.4586	.4448	.4283	.4093	.3879
4	.3641	.3383	.3105	.2811	.2501	.2178	.1846	.1506	.1161	.0813

When $0 \leq x < 1$, $J_2(x) \simeq \frac{x^2}{8}\left(1 - \frac{x^2}{12}\right)$.

					$J_3(x)$					
x	0	.1	.2	.3	.4	.5	.6	.7	.8	.9
0	.0000	.0000	.0002	.0006	.0013	.0026	.0044	.0069	.0102	.0144
1	.0196	.0257	.0329	.0411	.0505	.0610	.0725	.0851	.0988	.1134
2	.1289	.1453	.1623	.1800	.1981	.2166	.2353	.2540	.2727	.2911
3	.3091	.3264	.3431	.3588	.3734	.3868	.3988	.4092	.4180	.4250
4	.4302	.4333	.4344	.4333	.4301	.4247	.4171	.4072	.3952	.3811

When $0 \leq x < 1$, $J_3(x) \simeq \frac{x^3}{48}\left(1 - \frac{x^2}{16}\right)$.

					$J_4(x)$					
x	0	.1	.2	.3	.4	.5	.6	.7	.8	.9
0	.0000	.0000	.0000	.0000	.0001	.0002	.0003	.0006	.0010	.0016
1	.0025	.0036	.0050	.0068	.0091	.0118	.0150	.0188	.0232	.0283
2	.0340	.0405	.0476	.0556	.0643	.0738	.0840	.0950	.1067	.1190
3	.1320	.1456	.1597	.1743	.1891	.2044	.2198	.2353	.2507	.2661
4	.2811	.2958	.3100	.3236	.3365	.3484	.3594	.3693	.3780	.3853

When $0 \leq x < 1$, $J_4(x) \simeq \frac{x^4}{384}\left(1 - \frac{x^2}{20}\right)$.

TABLE 5.6.4
Zeros of $J_0(x)$, $J_1(x)$, $J_2(x)$, $J_3(x)$, $J_4(x)$, $J_5(x)$.

m	$j_{0,m}$	$j_{1,m}$	$j_{2,m}$	$j_{3,m}$	$j_{4,m}$	$j_{5,m}$
1	2.4048	3.8317	5.1356	6.3802	7.5883	8.7715
2	5.5201	7.0156	8.4172	9.7610	11.0647	12.3386
3	8.6537	10.1735	11.6198	13.0152	14.3725	15.7002
4	11.7915	13.3237	14.7960	16.2235	17.6160	18.9801
5	14.9309	16.4706	17.9598	19.4094	20.8269	22.2178
6	18.0711	19.6159	21.1170	22.5827	24.0190	25.4303
7	21.2116	22.7601	24.2701	25.7482	27.1991	28.6266
8	24.3525	25.9037	27.4206	28.9084	30.3710	31.8117
9	27.4935	29.0468	30.5692	32.0649	33.5371	34.9888
10	30.6346	32.1897	33.7165	35.2187	36.6990	38.1599

1.5.7 Zernike polynomials

Zernike polynomials are a set of complex exponentials that form a complete orthogonal set over the interior of the unit circle. Polynomial representation of optical wave fronts is essential in the analysis of interferometric test data, for example, to assess optical system performance. One such set, which is attractive for its simple rotational properties, is the **circle polynomials or Zernike polynomials**. The set of these polynomials is denoted by

$$V_{nl}(x, y) = V_{nl}(r\cos\theta, r\sin\theta) = V_{nl}(r, \theta) = R_{nl}(r)e^{jl\theta} \tag{5.7.1}$$

where

n is a nonnegative integer, $n \geq 0$

l is an integer subject to constraints: $n - |l|$ is even and $|l| \leq n$

r is the length of vector from origin to (x, y) point

θ is the angle between r- and x-axis in the counterclockwise direction

The orthogonality property is expressed by the formula

$$\iint\limits_{x^2+y^2\leq 1} V_{nl}^*(r, \theta)V_{mk}(r, \theta)r\,dr\,d\theta = \frac{\pi}{n+1}\delta_{mn}\delta_{k\ell} \tag{5.7.2}$$

where δ_{ij} is the Kronecker symbol. The real-valued radial polynomials satisfy the orthogonality relation

$$\int_0^1 R_{nl}(r)R_{ml}(r)r\,dr = \frac{1}{2(n+1)}\delta_{mn} \tag{5.7.3}$$

The radial polynomials are given by

$$R_{n\pm|l|}(r) = \frac{1}{\left(\frac{n-|l|}{2}\right)! r^m}\left[\frac{d}{d(r^2)}\right]^{\frac{n-|l|}{2}}\left[(r^2)^{\frac{n+|l|}{2}}(r^2-1)^{\frac{n-|l|}{2}}\right]$$

$$= \sum_{s=0}^{\frac{n-|l|}{2}}(-1)^s \frac{(n-s)!}{s!\left(\frac{n+|l|}{2}-s\right)!\left(\frac{n-|l|}{2}-s\right)!}r^{n-2s} \tag{5.7.4}$$

For all permissible values of n and $|l|$

$$R_{n\pm|l|} = 1, \qquad R_{n|l|}(r) = R_{n(-|l|)}(r) \tag{5.7.5}$$

Table 5.7.1 gives the explicit form of the function $R_{n|l|}(r)$.

A relation between radial Zernike polynomials and Bessel functions of the first kind is given by

$$\int_0^1 R_{n|l|}(r)J_n(vr)r\,dr = (-1)^{\frac{n-|l|}{2}}\frac{J_{n+1}(v)}{v} \tag{5.7.6}$$

From (5.7.1) we obtain the following real Zernike polynomials:

$$U_{nl} = \frac{1}{2}[V_{nl} + V_{n(-l)}] = R_{nl}(r)\cos l\theta, \qquad l \neq 0$$

$$U_{n(-l)} = \frac{1}{2j}[V_{nl} - V_{n(-l)}] = R_{nl}(r)\sin l\theta, \qquad l \neq 0 \tag{5.7.7}$$

$$V_{n0} = R_{n0}(r)$$

Figure 5.7.1 shows the function U_{nl} for a few radial modes.

TABLE 5.7.1
The radial polynomials $R_{n|l|}(r)$ for $|l| \leq 8, n \leq 8$.

$\dfrac{n}{|l|}$	0	1	2	3	4	5	6	7	8
0	1		$2r^2 - 1$		$6r^4 - 6r^2 + 1$		$20r^6 - 30r^4 + 12r^2 - 1$		$70r^8 - 140r^6 + 90r^4 - 20r^2 + 1$
1		r		$3r^3 - 2r$		$10r^5 - 12r^3 + 3r$		$35r^7 - 60r^5 + 30r^3 - 4r$	
2			r^2		$4r^4 - 3r^2$		$15r^6 - 20r^4 + 6r^2$		$56r^8 - 105r^6 + 60r^4 - 10r^2$
3				r^3		$5r^5 - 4r^3$		$21r^7 - 30r^5 + 10r^3$	
4					r^4		$6r^6 - 5r^4$		$28r^8 - 42r^6 + 15r^4$
5						r^5		$7r^7 - 6r^5$	
6							r^6		$8r^8 - 7r^6$
7								r^7	
8									r^8

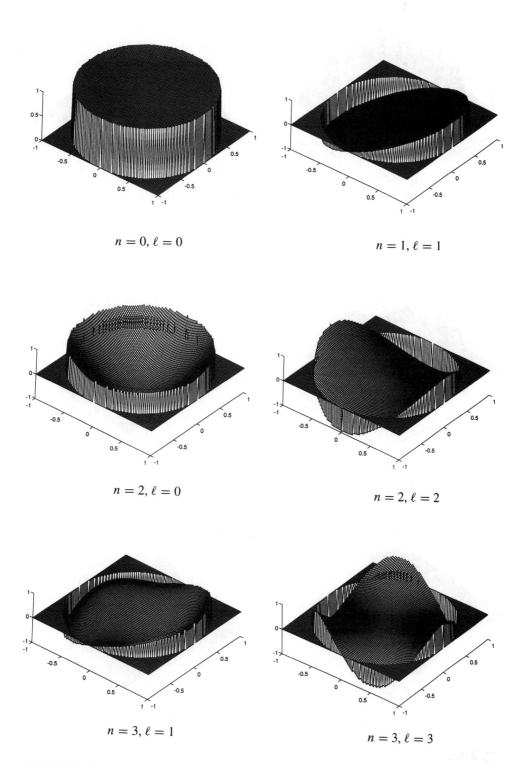

$n = 0, \ell = 0$

$n = 1, \ell = 1$

$n = 2, \ell = 0$

$n = 2, \ell = 2$

$n = 3, \ell = 1$

$n = 3, \ell = 3$

FIGURE 5.7.1

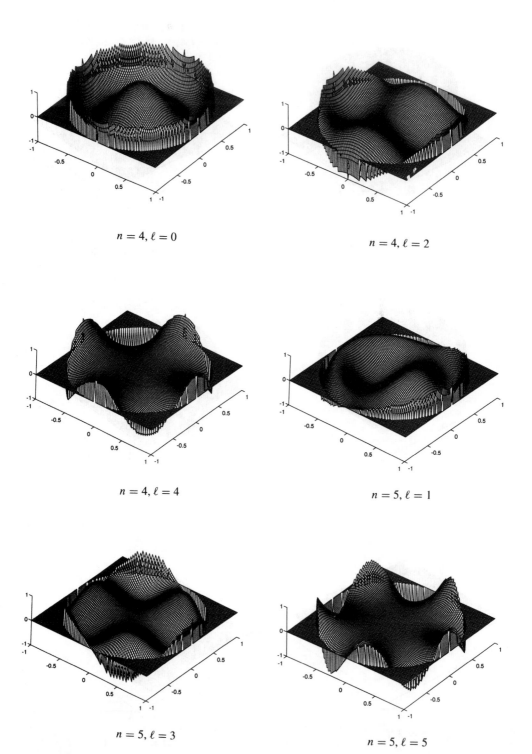

$n = 4, \ell = 0$

$n = 4, \ell = 2$

$n = 4, \ell = 4$

$n = 5, \ell = 1$

$n = 5, \ell = 3$

$n = 5, \ell = 5$

FIGURE 5.7.1
(continued)

Expansion in Zernike polynomials

If $f(x, y)$ is a piecewise continuous function, we can expand this function in Zernike polynomials in the form

$$f(x, y) = \sum_{n=0}^{\infty} \sum_{l=-\infty}^{\infty} A_{nl} V_{nl}(x, y), \qquad n - |l| \text{ is even, } |l| \leq n \qquad (5.7.8)$$

Multiplying by $V_{nl}^*(x, y)$, integrating over the unit circle, and taking into consideration the orthogonality property we obtain

$$A_{nl} = \frac{n+1}{\pi} \int_0^{2\pi} \int_0^1 V_{nl}^*(r, \theta) f(r \cos \theta, r \sin \theta) r \, dr \, d\theta$$

$$= \frac{n+1}{\pi} \iint_{x^2+y^2 \leq 1} V_{nl}^*(x, y) f(x, y) \, dx \, dy = A_{n(-l)}^* \qquad (5.7.9)$$

with restrictions of the values of n and l as shown above. A_{nl}'s are also known as **Zernike moments**.

Example
Expand the function $f(x, y) = x$ in Zernike polynomials.

Solution   We write $f(r \cos \theta, r \sin \theta) = r \cos \theta$ and observe that r has exponent (degree) one. Therefore, the values of n will be 0, 1 and because $n - |l|$ must be even, l will take 0, 1 and -1 values. We then write

$$f(x, y) = \sum_{n=0}^{\infty} \sum_{l=-\infty}^{\infty} A_{nl} R_{nl}(r) e^{jl\theta}$$

$$= \sum_{n=0}^{l} (A_{n(-1)} R_{n(-1)}(r) e^{-j\theta} + A_{n0} R_{n0}(r) + A_{n1} R_{n1}(r) e^{j\theta})$$

$$= A_{00} R_{00}(r) + A_{1(-1)} R_{1(-1)}(r) e^{-j\theta} + A_{11} R_{11}(r) e^{j\theta} \qquad (5.7.10)$$

where three terms were dropped because they did not obey the condition that $n - |l|$ is even. From (5.7.5) $R_{1(-1)}(r) = R_{11}(r)$ and hence we obtain

$$A_{00} = \frac{1}{\pi} \int_0^{2\pi} \int_0^1 R_{00}(r) r \cos \theta r \, dr \, d\theta = 0$$

$$A_{1(-1)} = \frac{2}{\pi} \int_0^{2\pi} \int_0^1 R_{11}(r) r \cos \theta e^{-j\theta} r \, dr \, d\theta = \frac{1}{2}$$

$$A_{11} = \frac{2}{\pi} \int_0^{2\pi} \int_0^1 R_{11}(r) r \cos \theta e^{j\theta} r \, dr \, d\theta = \frac{1}{2}$$

Therefore, the expansion becomes

$$f(x, y) = \frac{1}{2} r e^{j\theta} + \frac{1}{2} r e^{-j\theta} = r \cos \theta = R_{11}(r) \cos \theta = x$$

as was expected.   ∎

The radial polynomials $R_{nl}(r)$ are real valued and if $f(x, y)$ is real, that is, image intensity, it is often convenient to expand in real-valued series. The real expansion corresponding to (5.7.8) is

$$f(x, y) = \sum_{n=0}^{\infty} \sum_{l=0}^{\infty} (C_{nl} \cos l\theta + S_{nl} \sin l\theta) R_{nl}(r) \tag{5.7.11}$$

where $n - l$ is even and $l < n$. Observe that l takes only positive value. The unknown constants are found from

$$\begin{bmatrix} C_{nl} \\ S_{nl} \end{bmatrix} = \frac{2n + 2}{\pi} \int_0^1 \int_0^{2\pi} r\, dr\, d\theta f(r \cos \theta, r \sin \theta) R_{nl}(r) \begin{bmatrix} \cos l\theta \\ \sin l\theta \end{bmatrix},$$
$$l \neq 0 \tag{5.7.12}$$

$$C_{n0} = A_{n0} = \frac{1}{\pi} \int_0^1 \int_0^{2\pi} r\, dr\, d\theta f(r \cos \theta, r \sin \theta) R_{nl}(r), \qquad l = 0 \tag{5.7.13a}$$

$$S_{n0} = 0, \qquad l = 0 \tag{5.7.13b}$$

If the function is axially symmetric only the cosine terms are needed. The connection between real and complex Zernike coefficients are

$$C_{nl} = 2 \operatorname{Re}\{A_{nl}\} \tag{5.7.14a}$$

$$S_{nl} = -2 \operatorname{Im}\{A_{nl}\} \tag{5.7.14b}$$

$$A_{nl} = (C_{nl} - j S_{nl})/2 = (A_{n(-l)})^* \tag{5.7.14c}$$

Figure 5.7.2 shows the reconstruction of the letter Z using different orders of Zernike moments.

1.6 Sampling of signals

Two critical questions in signal sampling are: First, do the sampled values of a function adequately represent the system? Second, what must the sampling interval be in order that an optimum recovery of the signal can be accomplished from the sampled values?

The value of the function at the sampling points is the **sampled value**, the time that separates the sampling points is the **sampling interval**, and the reciprocal of the sampling interval is the **sampling frequency** or **sampling rate**.

If the sampling interval T_s is chosen to be constant, and $n = 0 \pm 1, \pm 2, \ldots$, the sampled signal is

$$f_s(t) = f(t) \sum_{n=-\infty}^{\infty} \delta(t - nT_s) = \sum_{n=-\infty}^{\infty} f(nT_s)\delta(t - nT_s) \tag{6.1}$$

Its Fourier transform is

$$F_s(\omega) \doteq \mathcal{F}\{f_s(t)\} = \sum_{n=-\infty}^{\infty} f(nT_s)\mathcal{F}\{\delta(t - nT_s)\} = \sum_{n=-\infty}^{\infty} f(nT_s)e^{-jn\omega T_s} \tag{6.2}$$

Original

n up to 5

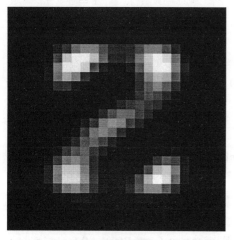

n up to 10

n up to 15

n up to 20

FIGURE 5.7.2

We can also represent the Fourier transform of a sampled function as follows:

$$F_s(\omega) \doteq \mathcal{F}\left\{ f(t) \sum_{n=-\infty}^{\infty} \delta(t - nT_s) \right\} = \frac{1}{2\pi} \mathcal{F}\{f(t)\} * \mathcal{F}\left\{ \sum_{n=-\infty}^{\infty} \delta(t - nT_s) \right\}$$

$$= \frac{1}{2\pi} F(\omega) * \left[\frac{2\pi}{T_s} \sum_{n=-\infty}^{\infty} \delta(\omega - n\omega_s) \right]$$

$$= \frac{1}{T_s} \sum_{n=-\infty}^{\infty} \int_{-\infty}^{\infty} F(x)\delta(\omega - n\omega_s - x)\,dx = \frac{1}{T_s} \sum_{n=-\infty}^{\infty} F(\omega - n\omega_s)$$

$$= \frac{1}{T_s} \sum_{n=-\infty}^{\infty} F(\omega + n\omega_s), \qquad \omega_s = \frac{2\pi}{T_s} \tag{6.3}$$

$F_s(\omega)$ is periodic with period ω_s in the frequency domain.

Example

$$\mathcal{F}\left\{ e^{-|t|} \sum_{n=-\infty}^{\infty} \delta(t - nT_s) \right\} \doteq \mathcal{F}_s(\omega) = \frac{1}{T_s} \sum_{n=-\infty}^{\infty} \frac{2}{1 + (\omega - n\omega_s)^2}$$ ∎

1.6.1   *The sampling theorem*

It can be shown that it is possible for a **band-limited** signal $f(t)$ to be exactly specified by its sampled values provided that the time distance between sample values does not exceed a critical sampling interval.

THEOREM 6.1.1
A finite energy function f(t) having a band-limited Fourier transform, $F(\omega) = 0$ for $|\omega| \geq \omega_N$, can be completely reconstructed from its sampled values $f(nT_s)$ (see Figure 6.1.1), with

$$f(t) = \sum_{n=-\infty}^{\infty} T_s f(nT_s) \left\{ \frac{\sin\left[\frac{\omega_s(t - nT_s)}{2} \right]}{\pi(t - nT_s)} \right\}, \qquad \omega_s = \frac{2\pi}{T_s} \tag{6.1.1}$$

provided that

$$\frac{2\pi}{\omega_s} = T_s \leq \frac{\pi}{\omega_N} = \frac{1}{2f_N} = \frac{T_N}{2}$$

The function within the braces, which is the sinc function, is often called the interpolation function to indicate that it allows an interpolation between the sampled values to find $f(t)$ for all t.

PROOF   Employ (6.3) and Figure 6.1.1c to write

$$F(\omega) = p_{\omega_s/2}(\omega) T_s F_s(\omega) \tag{6.1.2}$$

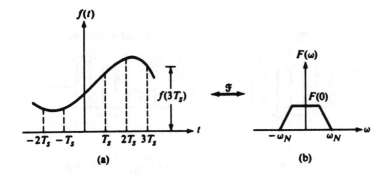

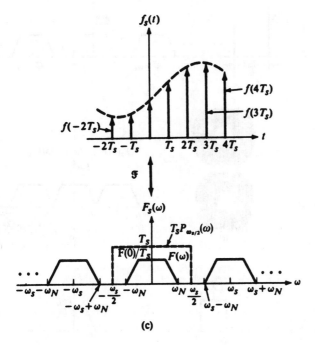

FIGURE 6.1.1

By (6.1.2), the above equation becomes

$$f(t) = \mathcal{F}^{-1}\{F(\omega)\} = \mathcal{F}^{-1}\left\{ p_{\omega_s/2}(\omega) T_s \sum_{n=-\infty}^{\infty} f(nT_s) e^{-jn\omega T_s} \right\}$$

$$= T_s \sum_{n=-\infty}^{\infty} f(nT_s) \mathcal{F}^{-1}\{p_{\omega_s/2}(\omega) e^{-jn\omega T_s}\}$$

By application of the frequency-shift property of the Fourier transform, this equation proves the theorem. ∎

The sampling time

$$T_s = \frac{T_N}{2} = \frac{1}{2f_N} \tag{6.1.3}$$

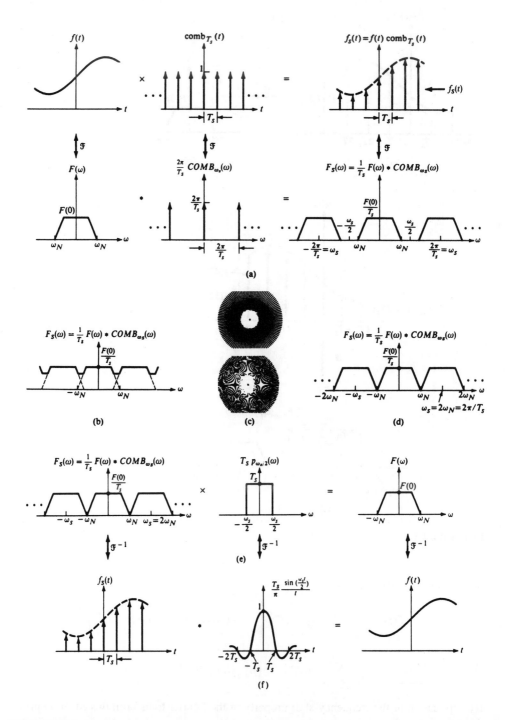

FIGURE 6.1.2
The effect of undersampling is shown in (c).

is related to the **Nyquist interval**. It is the largest time interval that can be used for sampling of a band-limited signal and still allows recovering of the signal without distortion. If, however, the sampling time is larger than the Nyquist interval, overlap of spectra takes place, known as **aliasing**, and no perfect reconstruction of the band-limited signal is possible. Figure 6.1.2 shows the delta sampling representation and recovery of a band-limited signal. The following definitions have been used in the figure:

$$\text{comb}_{T_s}(t) = \sum_{n=-\infty}^{\infty} \delta(t - nT_s) \tag{6.1.4}$$

$$\text{COMB}_{\omega_s}(\omega) = \sum_{n=-\infty}^{\infty} \delta(\omega - \omega_s) \tag{6.1.5}$$

Frequency sampling

Analogous to the time-sampling theorem, a frequency-sampling equivalent also exists.

THEOREM 6.1.2
A time function $f(t)$ *that is time limited so that*

$$f(t) = 0, \qquad |t| > T_N \tag{6.1.6}$$

possesses a Fourier transform that can be uniquely determined from its samples at distances $n\pi / T_N$, *and is given by*

$$F(\omega) = \sum_{n=-\infty}^{\infty} F\left(n\frac{\pi}{T_N}\right) \frac{\sin(\omega T_N - n\pi)}{\omega T_N - n\pi} \tag{6.1.7}$$

where the sampling is at the Nyquist rate.

Sampling with a train of rectangular pulses

The Fourier transform of a band-limited function sampled with periodic pulses is given by (see Figure 6.1.3)

$$F_s(\omega) = \text{F}\{f(t)f_p(t)\} = \frac{1}{2\pi}F(\omega) * F_p(\omega)$$

$$= \frac{1}{2\pi}F(\omega) * \left\{ \sum_{n=-\infty}^{\infty} 2\pi \frac{\sin\left(\frac{n\omega_s \tau}{2}\right)}{\frac{n\omega_s \tau}{2}} \delta(\omega - n\omega_s) \right\}$$

$$= \sum_{n=-\infty}^{\infty} \frac{\sin\left(\frac{n\omega_s \tau}{2}\right)}{\frac{n\omega_s \tau}{2}} \int_{-\infty}^{\infty} \delta(x - n\omega_s)F(\omega - x)\,dx$$

$$= \sum_{n=-\infty}^{\infty} \frac{\sin\left(\frac{n\omega_s \tau}{2}\right)}{\frac{n\omega_s \tau}{2}} F(\omega - n\omega_s) \tag{6.1.8}$$

where τ is the width of the pulse. The above expression indicates that as long as $\omega_s > 2\omega_N$, the spectrum of the sampled signal contains no overlapping spectra of $f(t)$ and can be recovered using a low-pass filter.

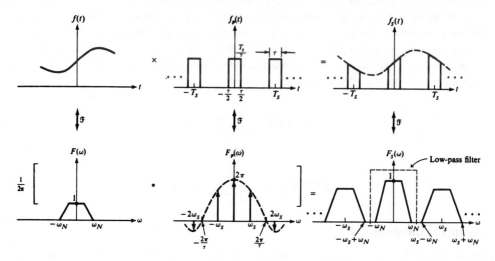

FIGURE 6.1.3

1.6.2 *Extensions of the sampling theorem*

The sampling theorem of a band-limited function of n variables is given by the following theorem:

THEOREM 6.2.1

Let $f(t_1, t_2, \ldots, t_n)$ *be a function of n real variables, whose n-dimensional Fourier integral exists and is identically zero outside an n-dimensional rectangle and is symmetrical about the origin; that is,*

$$g(y_1, y_2, \ldots, y_n) = 0, \qquad |y_k| > |\omega_k|, \qquad k = 1, 2, \ldots, n \qquad (6.2.1)$$

Then

$$f(t_1, t_2, \ldots, t_n) = \sum_{m_1=-\infty}^{\infty} \cdots \sum_{m_n=-\infty}^{\infty} f\left(\frac{\pi m_1}{\omega_1}, \ldots, \frac{\pi m_n}{\omega_n}\right)$$

$$\times \frac{\sin(\omega_1 t_1 - m_1\pi)}{\omega_1 t_1 - m_1\pi} \cdots \frac{\sin(\omega_n t_n - m_n\pi)}{\omega_n t_n - m_n\pi} \qquad (6.2.2)$$

An additional theorem on the sampling of band-limited signals follows.

THEOREM 6.2.2

Let $f(t)$ *be a continuous function with finite Fourier transform* $F(\omega)[F(\omega) = 0$ *for* $|\omega| > 2\pi f_N]$. *Then*

$$f(t) = \sum_{k=-\infty}^{\infty} \left[\xi(kh) + (t - kh)\xi^{(1)}(kh) + \cdots + \frac{(t - kh)^R}{R!}\xi^{(R)}(kh)\right]$$

$$\times \left[\frac{\sin\frac{\pi}{h}(t - kh)}{\frac{\pi}{h}(t - kh)}\right]^{R+1} \qquad (6.2.3)$$

where:

R is the highest derivative order

$h = (R + 1)/(2 f_N)$

$\xi^{(R)}(kh)$ is the Rth derivative of the function $\xi(.)$

$\xi^{(j)}(kh) = \sum_{i=0}^{j} \binom{j}{i} \left(\frac{\pi}{h}\right)^{j-1} \Gamma_{R+1}^{(j-1)} f^{(i)}(kh)$

$\Gamma_a^{(\beta)} = \frac{d^\beta}{dt^\beta} \left[\left(\frac{t}{\sin t}\right)^\alpha \right]_{t=0}$

$\Gamma_\alpha^{(0)} = 1,\ \Gamma_\alpha^{(2)} = \frac{\alpha}{3},\ \Gamma_\alpha^{(4)} = \frac{\alpha(5\alpha+2)}{15},\ \Gamma_\alpha^{(6)} = \frac{\alpha(35\alpha^2+42\alpha+16)}{63}, \dots, \Gamma_\alpha^{(\beta)} = 0$ *for odd β*

Papoulis extensions

The band-limited signal

$$f(t) = \frac{1}{2\pi} \int_{-W_1}^{W_1} F(\omega) e^{j\omega t}\, d\omega \tag{6.2.4}$$

can be represented by

$$f(t) = \sum_{n=-\infty}^{\infty} f(nT) \frac{\sin w_0(t - nT)}{w_2(t - nT)} \tag{6.2.5}$$

where

$$w_2 = \frac{\pi}{T} \geq w_1, \qquad w_1 \leq w_0 \leq 2w_2 - w_1$$

THEOREM 6.2.3

Given an arbitrary sequence of numbers $\{a_n\}$, if we form the sum

$$x(t) = \sum_{n=-\infty}^{\infty} a_n \frac{\sin w_0(t - nT)}{w_2(t - nT)} \tag{6.2.6}$$

then $x(t)$ is band limited by w_0.

The sampling expansion of $f^2(t)$ is given by

$$f^2(t) = \sum_{n=-\infty}^{\infty} f^2(nT) \frac{\sin w_0(t - nT)}{w_2(t - nT)} \tag{6.2.7}$$

where

$$w_2 = \frac{\pi}{T},\ w_2 \geq 2w_1,\ 2w_1 \leq w_0 \leq 2w_2 - 2w_1,\ T \leq \frac{\pi}{2w_1}$$

The band-limited signal given in (6.2.4) can be expressed in terms of the sample values $g(nT)$ of the output

$$g(t) = \frac{1}{2\pi} \int_{-w_1}^{w_1} F(\omega) H(\omega) e^{j\omega t}\, d\omega \tag{6.2.8}$$

of a system with transfer function $H(\omega)$ driven by $f(t)$. The sampling expansion of $f(t)$ is

$$f(t) = \sum_{n=-\infty}^{\infty} g(nT) y(t - nT) \tag{6.2.9}$$

where

$$y(t) = \frac{1}{2w_1} \int_{-w_1}^{w_1} \frac{e^{j\omega t}}{H(\omega)}\, d\omega \tag{6.2.10}$$

1.7 Asymptotic series

Functions such as $f(z)$ and $\varphi(z)$ are defined on a set R in the complex plane. By a neighborhood of z_0 we mean an open disc $|z - z_0| < \delta$ if z_0 is at a finite distance, and a region $|z| > \delta$ if z_0 is the point at infinity.

$f = O(\varphi)$ and $f = o(\varphi)$ notation

We write $f = O(\varphi)$ if there exists a constant A such that $|f| \leq A|\varphi|$ for all z in R.

We also write $f = O(\varphi)$ as $z \to z_0$ if there exists a constant A and a neighborhood U of z_0 such that $|f| \leq A|\varphi|$ for all points in the intersection of U and R.

We write $f = o(\varphi)$ as $z \to z_0$ if, for any positive number ε, there exists a neighborhood U of z_0 such that $|f| \leq \varepsilon|\varphi|$ for all points z of the intersection of U and R.

More simply, if φ does not vanish on R, $f = O(\varphi)$ means that f/φ is bounded, $f = o(\varphi)$ means that f/φ tends to zero as $z \to z_0$.

Asymptotic sequence

A sequence of functions $\{\varphi_n(z)\}$ is called an **asymptotic sequence** as $z \to z_0$ if there is a neighborhood of z_0 in which none of the functions vanish (except the point z_0) and if for all n

$$\varphi_{n+1} = o(\varphi_n) \quad \text{as } z \to z_0$$

For example, if z_0 is finite $\{(z - z_0)^n\}$ is an asymptotic sequence as $z \to z_0$, and $\{z^{-n}\}$ is as $z \to \infty$.

Poincaré sense asymptotic sequence

The formal series

$$f(z) \cong \sum_{n=0}^{\infty} a_n \varphi_n(z) \tag{7.1}$$

which is not necessarily convergent, is an asymptotic expansion of $f(z)$ in the Poincaré sense with respect to the asymptotic sequence $\{\varphi_n(z)\}$, if for every value of m,

$$f(z) - \sum_{n=0}^{\infty} a_n \varphi_n(z) = o(\varphi_m(z)) \tag{7.2}$$

as $z \to z_0$.

Because

$$f(z) - \sum_{n=0}^{m-1} a_n \varphi_n(z) = a_m \varphi_m(z) + o(\varphi_m(z)) \tag{7.3}$$

the partial sum

$$\sum_{n=0}^{m-1} a_n \varphi_n(z) \tag{7.4}$$

is an approximation to $f(z)$ with an error $O(\varphi_m)$ as $z \to z_0$; this error is of the same order of magnitude as the first term omitted. If such an asymptotic expansion exists, it is unique, and the coefficients are given successively by

$$a_m = \frac{\lim_{z \to z_0} \left\{ f(z) - \sum_{n=0}^{m-1} a_n \varphi_n(z) \right\}}{\varphi_m(z)} \tag{7.5}$$

Hence for a function $f(z)$ we write

$$f(z) \cong \sum_{n=0}^{\infty} a_n \varphi_n(z) \tag{7.6}$$

Asymptotic approximation

A partial sum of (7.6) is called an **asymptotic approximation** to $f(z)$. The first term is called the **dominant term**.

The above definition applies equally well for a real variable x.

Asymptotic power series

We shall assume that the transformation $z' = 1/(z - z_0)$ has been done for limit points z_0 located at a finite distance. Hence we can always consider expansions as z approaches infinity in a sector $a < ph\ z < \beta$; or, for real value x, as x approaches infinity or as x approaches negative infinity.

The divergence series

$$f(z) = \sum_{n=0}^{\infty} \frac{a_n}{z_n} = a_0 + \frac{a_1}{z} + \frac{a_2}{z^2} + \cdots + \frac{a_n}{z^n} + \cdots \tag{7.7}$$

in which the sum of the first $(n + 1)$ terms is $S_n(z)$, is said to be an **asymptotic expansion** of a function $f(z)$ for a given range of values of arg z, if the expansion $R_n(z) = z^n \{ f(z) - S_n(z) \}$ satisfies the condition

$$\lim_{|z| \to \infty} R_n(z) = 0 \qquad (n \text{ is fixed}) \tag{7.8}$$

even though

$$\lim_{n \to \infty} |R_n(z)| = \infty \qquad (z \text{ is fixed})$$

When this is true, we can make

$$|z^n \{ f(z) - S_n(z) \}| < \varepsilon \tag{7.9}$$

where ε is arbitrarily small, by taking $|z|$ sufficiently large. This definition is due to Poincaré.

Example

For real x, integration on the real axis and repeated integration by parts, we obtain

$$f(x) = \int_x^\infty t^{-1} e^{x-t}\, dt = \frac{1}{x} - \frac{1}{x^2} + \frac{2!}{x^3} - \cdots + \frac{(-1)^{n-1}(n-1)!}{x^n} + (-1)^n n! \int_x^\infty \frac{e^{x-t} dt}{t^{n+1}}$$

If we consider the expansion

$$u_{n-1} = \frac{(-1)^{n-1}(n-1)!}{x^n}$$

we can write

$$\sum_{m=0}^{n} u_m = \frac{1}{x} - \frac{1}{x^2} + \frac{2!}{x^3} - \cdots + \frac{(-1)^n n!}{x^{n+1}} = S_n(x)$$

But $|u_m/u_{m-1}| = mx^{-1} \to \infty$ as $m \to \infty$. The series $\sum u_m$ is divergent for all values of x. However, the series can be used to calculate $f(x)$.

For a fixed n, we can calculate S_n from the relation

$$f(x) - S_n(x) = (-1)^{n+1}(n+1)! \int_x^\infty \frac{e^{x-t}dt}{t^{n+2}}$$

Because $\exp(x - t) \le 1$,

$$|f(x) - S_n(x)| = (n+1)! \int_x^\infty \frac{e^{x-t}dt}{t^{n+2}} < (n+1)! \int_x^\infty \frac{dt}{t^{n+2}} = \frac{n!}{x^{n+1}}$$

For large values of x the right-hand member of the above relation is very small. This shows that the value of $f(x)$ can be calculated with great accuracy for large values of x, by taking the sum of a suitable number of terms of the series $\sum u_m$. From the last relation we obtain

$$\left|x^n\{f(x) - S_n(x)\}\right| < n!x^{-1} \to 0 \qquad \text{as } x \to \infty$$

which satisfies the asymptotic expansion condition. ∎

Operation of asymptotic power series

Let the following two functions possess asymptotic expansions:

$$f(x) \approx \sum_{n=0}^\infty \frac{a_n}{x^n}, \qquad g(x) \approx \sum_{n=0}^\infty \frac{b_n}{x^n} \qquad \text{as } x \to \infty$$

on the real axis.

(a) If A is constant

$$Af(x) \approx \sum_{n=0}^\infty \frac{Aa_n}{x^n} \tag{7.10}$$

(b)

$$f(x) + g(x) \approx \sum_{n=0}^\infty \frac{a_n + b_n}{x^n} \tag{7.11}$$

(c)

$$f(x)g(x) \approx \sum_{n=0}^\infty \frac{c_n}{x^n}$$

$$c_n = a_0 b_n + a_1 b_{n-1} + \cdots + a_{n-1}b_1 + a_n b_0 \tag{7.12}$$

(d) If $a_0 \ne 0$, then

$$\frac{1}{f(x)} \approx \frac{1}{a_0} + \sum_{n=1}^\infty \frac{d_n}{x^n}, \qquad x \to \infty \tag{7.13}$$

The function $1/f(x)$ tends to a finite limit $1/a_0$ as x approaches infinity. Hence,

$$\left(\frac{1}{f(x)} - \frac{1}{a_0}\right) / (1/x) = x\left(\frac{1}{a_0 + (a_1/x) + O(1/x^2)} - \frac{1}{a_0}\right)$$

$$= \frac{-a_1 + O\left(\frac{1}{x}\right)}{a_0[a_0 + (a_1/x) + O(1/x^2)]} \to -\frac{a_1}{a_0^2} = d_1$$

Similarly we obtain

$$\left(\frac{1}{f(x)} - \frac{1}{a_0} + \frac{a_1}{a_0^2 x}\right) \bigg/ \left(\frac{1}{x^2}\right) \to \frac{a_1^2 - a_0 a_2}{a_0^3} = d_2$$

and so on.

In general, any rational function of $f(x)$ has an asymptotic power series expansion provided that the denominator does not tend to zero as x approaches infinity.

(e) If $f(x)$ is continuous for $x > a > 0$ and if $x > a$, then

$$F(x) = \int_x^\infty \left(f(t) - a_0 - \frac{a_1}{t}\right) dt \approx \frac{a_2}{x} + \frac{a_3}{2x^2} + \cdots + \frac{a_{n+1}}{nx^n} + \cdots \qquad (7.14)$$

(f) If $f(x)$ has a continuous derivative $f'(x)$, and if $f'(x)$ possesses an analytic power series expansion as x approaches infinity, the expression is

$$f'(x) \approx - \sum_{n=2}^\infty \frac{(n-1)a_{n-1}}{x^n} \qquad (7.15)$$

(g) It is permissible to integrate an asymptotic expansion term-by-term. The resulting series is the expansion of the integral of the function represented by the original series.

Let

$$f(x) \approx \sum_{m=2}^\infty a_m x^{-m} \qquad \text{and} \qquad S_n = \sum_{m=2}^n a_m x^{-m}$$

Then, given any positive number ε, we can find x_0 such that

$$|f(x) - S_n(x)| < \varepsilon |x|^{-n} \qquad \text{for } x > x_0$$

Hence

$$\left| \int_x^\infty f(x)\, dx - \int_x^\infty S_n(x)\, dx \right| \le \int_x^\infty |f(x) - S_n(x)|\, dx < \frac{\varepsilon}{(n-1)x^{n-1}}$$

However,

$$\int_x^\infty S_n(x)\, dx = \frac{a_2}{x} + \frac{a_3}{2x^2} + \cdots + \frac{a_n}{(n-1)x^{n-1}}$$

and therefore

$$\int_x^\infty f(x)\, dx \approx \sum_{m=2}^\infty \frac{a_m}{(m-1)x^{m-1}}$$

Example
The Fresnel integrals

$$\int_u^\infty \cos(\theta^2)\, d\theta, \qquad \int_u^\infty \sin(\theta^2)\, d\theta \qquad (7.16)$$

can be written in the form

$$\int_{u^2}^\infty \frac{\cos t}{\sqrt{t}}\, dt, \qquad \int_{u^2}^\infty \frac{\sin t}{\sqrt{t}}\, dt$$

These are particular cases of the real and imaginary parts of the integral

$$f(x, a) = \int_x^\infty \frac{e^{jt}}{t^a}\, dt \qquad (7.17)$$

Integrating by parts we obtain

$$f(x, a) = \frac{je^{jx}}{x^a} - jaf(x, a+1) = \frac{je^{jx}}{x^a} \sum_{r=0}^{n} \frac{\Gamma(a+r)}{\Gamma(a)(jx)^r}$$

$$+ \frac{1}{j^{n+1}} \frac{\Gamma(a+n+1)}{\Gamma(a)} f(x, a+n+1) \qquad (7.18)$$

Hence

$$f(x, a) \approx \frac{je^{jx}}{x^a} \sum_{r=0}^{\infty} \frac{\Gamma(a+r)}{\Gamma(a)(jx)^r} \qquad (7.19)$$

as x approaches infinity. The absolute value of the remainder after $n + 1$ terms is

$$\frac{\Gamma(a+n+1)}{\Gamma(a)} \left| \int_x^{\infty} \frac{e^{jt}}{t^{a+n+1}} dt \right| \leq \frac{\Gamma(a+n+1)}{\Gamma(a)} \int_x^{\infty} \frac{dt}{t^{a+n+1}} = \frac{\Gamma(a+n)}{\Gamma(a)x^{a+n}}$$

Hence the remainder after n terms does not exceed in absolute value the absolute value of the $(n + 1)$th term, which proves the result. ∎

References

[1] Abdul, J. Jerry. 1977. "The Shannon Sampling Theorem—Its Various Extensions and Applications: A Tutorial Review." *Proc. IEEE*, 65:1565–1596.

[2] Andrews, L. C. 1985. *Special Functions for Engineers and Applied Mathematicians*. New York, NY: Macmillan.

[3] Copson, E. T. 1965. *Asymptotic Expansions*. New York, NY: Cambridge Unversity Press.

[4] Erdélyi, A. 1956. *Asymptotic Expansions*. New York, NY: Dover Publications.

[5] Gel'fand, I. M. and Shilov, G. E. 1964. *Generalized Functions*. New York: Academic Press.

[6] Hoskins, R. F. 1979. *Generalized Functions*. Chichester, England: Ellis Horwood Limited.

[7] Lebedev, N. N. 1972. *Special Functions and Their Applications*. New York, NY: Dover Publications.

[8] Lighthill, M. J. 1964. *Introduction to Fourier Analysis and Generalized Functions*. London, England: Cambridge at the University Press.

[9] Papoulis, A. 1977. *Signal Analysis*. New York, NY: McGraw-Hill.

[10] Papoulis, A. 1968. *Systems and Transforms with Applications in Optics*. New York, NY: McGraw-Hill.

[11] Sansone, G. 1959. *Orthogonal Functions*. New York, NY: Interscience Publishers.

2

Fourier Transforms

Kenneth B. Howell

CONTENTS

2.1 Introduction and basic definitions

The Fourier transform is certainly one of the best known of the integral transforms and vies with the Laplace transform as being the most generally useful. Since its introduction by Fourier in the early 1800s, it has found use in innumerable applications and has, itself, led to the development of other transforms. Today the Fourier transform is a fundamental tool in engineering science. Its importance has been enhanced by the development in the twentieth century of generalizations extending the set of functions that can be Fourier transformed and by the development of efficient algorithms for computing the discrete version of the Fourier transform.

There are two parts to this article on the Fourier transform. The first (Sections 2.1 through 2.4) contains the fundamental theory necessary for the intelligent use of the Fourier transform in practical problems arising in engineering. The second part (Sections 2.5 through 2.8) is devoted to applications in which the Fourier transform plays a significant role. This part contains both fairly detailed descriptions of specific applications and fairly broad overviews of classes of applications.

This particular section deals with the basic definition of the Fourier transform and some of the integrals used to compute Fourier transforms. Two definitions for the transform are given. First, the classical definition is given in Subsection 2.1.1. This is the integral formula for directly computing transforms generally found in elementary texts. From this formula many of the basic formulas and identities involving the Fourier transform can be derived. Inherent in the classical definition, however, are integrability conditions that cannot be satisfied by many functions routinely arising in applications. For this reason, more general definitions of the Fourier transform are briefly discussed in Subsections 2.1.3 and 2.1.4. These more general definitions will also help clarify the role of generalized functions in Fourier analysis.

The computation of Fourier transforms often involves the evaluation of integrals, many of which cannot be evaluated by the elementary methods of calculus. For this reason, this section also contains a brief discussion illustrating the use of the residue theorem in computing certain integrals as well as a brief discussion of how to deal with certain integrals containing singularities in the integrand.

2.1.1 *Basic definition, notation, and terminology*

If $\phi(s)$ is an absolutely integrable function on $(-\infty, \infty)$ (i.e., $\int_{-\infty}^{\infty} |\phi(s)| \, ds < \infty$), then the (direct) Fourier transform of $\phi(s)$, $\mathcal{F}[\phi]$, and the Fourier inverse transform of $\phi(s)$, $\mathcal{F}^{-1}[\phi]$, are the functions given by

$$\mathcal{F}[\phi]|_x = \int_{-\infty}^{\infty} \phi(s) e^{-jxs} \, ds, \tag{2.1.1.1}$$

and

$$\mathcal{F}^{-1}[\phi]\big|_x = \frac{1}{2\pi} \int_{-\infty}^{\infty} \phi(s) e^{jxs} \, ds. \tag{2.1.1.2}$$

Example 2.1.1.1
If $\phi(s) = e^{-s} u(s)$, then

$$\mathcal{F}[\phi]|_x = \int_{-\infty}^{\infty} e^{-s} u(s) e^{-jxs} \, ds = \int_0^{\infty} e^{-(1+jx)s} \, ds = \frac{1}{1 + jx}$$

and

$$\mathcal{F}^{-1}[\phi]\big|_x = \frac{1}{2\pi} \int_{-\infty}^{\infty} e^{-s} u(s) e^{jxs} \, ds = \frac{1}{2\pi} \int_0^{\infty} e^{-(1-jx)s} \, ds = \frac{1}{2\pi - j2\pi x}. \qquad \blacksquare$$

Example 2.1.1.2
For $\alpha > 0$, the transform of the corresponding pulse function,

$$p_\alpha(s) = \begin{cases} 1, & \text{if } |s| < \alpha \\ 0, & \text{if } \alpha < |s| \end{cases},$$

is

$$\mathcal{F}[p_\alpha]|_x = \int_{-\alpha}^{\alpha} e^{-jxs} \, ds = \frac{e^{j\alpha x} - e^{-j\alpha x}}{jx} = \frac{2}{x} \sin(\alpha x). \qquad \blacksquare$$

A function, ψ, is said to be "classically transformable" if either

1. ψ is absolutely integrable on the real line, or

2. ψ is the Fourier transform (or Fourier inverse transform) of an absolutely integrable function, or

3. ψ is a linear combination of an absolutely integrable function and a Fourier transform (or Fourier inverse transform) of an absolutely integrable function.

If ϕ is classically transformable but not absolutely integrable, then it can be shown that formulas (2.1.1.1) and (2.1.1.2) can still be used to define $\mathcal{F}[\phi]$ and $\mathcal{F}^{-1}[\phi]$ provided the limits are taken symmetrically, that is

$$\mathcal{F}[\phi]|_x = \lim_{a\to\infty} \int_{-a}^{a} \phi(s)e^{-jxs}\, ds$$

and

$$\mathcal{F}^{-1}[\phi]\Big|_x = \frac{1}{2\pi} \lim_{a\to\infty} \int_{-a}^{a} \phi(s)e^{jxs}\, ds.$$

In most applications involving Fourier transforms, the functions of time, t, or position, x, are denoted using lower case letters—for example: f and g. The Fourier transforms of these functions are denoted using the corresponding upper case letters—for example: $F = \mathcal{F}[f]$ and $G = \mathcal{F}[g]$. The transformed functions can be viewed as functions of angular frequency, ω. Along these same lines it is standard practice to view a signal as a pair of functions, $f(t)$ and $F(\omega)$, with $f(t)$ being the "time domain representation of the signal" and $F(\omega)$ being the "frequency domain representation of the signal."

2.1.2   Alternate definitions

Pairs of formulas other than formulas (2.1.1.1) and (2.1.1.2) are often used to define $\mathcal{F}[\phi]$ and $\mathcal{F}^{-1}[\phi]$. Some of the other formula pairs commonly used are:

$$\mathcal{F}[\phi]|_x = \int_{-\infty}^{\infty} \phi(s)e^{-j2\pi xs}\, ds, \qquad \mathcal{F}^{-1}[\phi]\Big|_x = \int_{-\infty}^{\infty} \phi(s)e^{j2\pi xs}\, ds \qquad (2.1.2.1)$$

and

$$\mathcal{F}[\phi]|_x = \frac{1}{\sqrt{2\pi}} \int_{-\infty}^{\infty} \phi(s)e^{-jxs}\, ds, \qquad \mathcal{F}^{-1}[\phi]\Big|_x = \frac{1}{\sqrt{2\pi}} \int_{-\infty}^{\infty} \phi(s)e^{jxs}\, ds. \qquad (2.1.2.2)$$

Equivalent analysis can be performed using the theory arising from any of these pairs; however, the resulting formulas and equations will depend on which pair is used. For this reason care must be taken to insure that, in any particular application, all the Fourier analysis formulas and equations used are derived from the same defining pair of formulas.

Example 2.1.2.1
Let $\phi(t) = e^{-t}u(t)$ and let Ψ_1, Ψ_2, and Ψ_3 be the Fourier transforms of ϕ as defined, respectively, by formulas (2.1.1.1), (2.1.2.1), and (2.1.2.2). Then,

$$\Psi_1(\omega) = \int_{-\infty}^{\infty} e^{-t}u(t)e^{-jt\omega}\, dt = \frac{1}{1+j\omega},$$

$$\Psi_2(\omega) = \int_{-\infty}^{\infty} e^{-t}u(t)e^{-j2\pi t\omega}\, dt = \frac{1}{1+j2\pi\omega},$$

and

$$\Psi_3(\omega) = \frac{1}{\sqrt{2\pi}} \int_{-\infty}^{\infty} e^{-t}u(t)e^{-jt\omega}\, dt = \frac{1}{\sqrt{2\pi}} \cdot \frac{1}{1+j\omega}. \qquad \blacksquare$$

2.1.3 *The generalized transforms*

Many functions and generalized functions[1] arising in applications are not sufficiently integrable to apply the definitions given in subsection 2.1.1 directly. For such functions it is necessary to employ a generalized definition of the Fourier transform constructed using the set of "rapidly decreasing test functions" and a version of Parseval's equation (see subsection 2.2.14).

A function, ϕ, is a "rapidly decreasing test function" if

1. every derivative of ϕ exists and is a continuous function on $(-\infty, \infty)$, and
2. for every pair of nonnegative integers, n and p,

$$|\phi^{(n)}(s)| = O(|s|^{-p}) \text{ as } |s| \to \infty.$$

The set of all rapidly decreasing test functions is denoted by $\mathcal{S}$ and includes the Gaussian functions as well as all test functions that vanish outside of some finite interval (such as those discussed in the first chapter of this handbook). If ϕ is a rapidly decreasing test function then it is easily verified that ϕ is classically transformable and that both $\mathcal{F}[\phi]$ and $\mathcal{F}^{-1}[\phi]$ are also rapidly decreasing test functions. It can also be shown that $\mathcal{F}^{-1}[\mathcal{F}[\phi]] = \phi$. Moreover, if f and G are classically transformable, then

$$\int_{-\infty}^{\infty} \mathcal{F}[f]|_x \, \phi(x) \, dx = \int_{-\infty}^{\infty} f(y) \, \mathcal{F}[\phi]|_y \, dy \tag{2.1.3.1}$$

and

$$\int_{-\infty}^{\infty} \mathcal{F}^{-1}[G]|_x \, \phi(x) \, dx = \int_{-\infty}^{\infty} G(y) \, \mathcal{F}^{-1}[\phi]\Big|_y \, dy. \tag{2.1.3.2}$$

If f is a function or a generalized function for which the right-hand side of equation (2.1.3.1) is well defined for every rapidly decreasing test function, ϕ, then the generalized Fourier transform of f, $\mathcal{F}[f]$, is that generalized function satisfying (2.1.3.1) for every ϕ in $\mathcal{S}$. Likewise, if G is a function or generalized function for which the right-hand side of (2.1.3.2) is well defined for every rapidly decreasing test function, ϕ, then the generalized inverse Fourier transform of G, $\mathcal{F}^{-1}[G]$, is that generalized function satisfying equation (2.1.3.2) for every ϕ in $\mathcal{S}$.

Example 2.1.3.1
Let α be any real number. Then, for every rapidly decreasing test function ϕ,

$$\int_{-\infty}^{\infty} \mathcal{F}[e^{j\alpha y}]\Big|_x \, \phi(x) \, dx = \int_{-\infty}^{\infty} e^{j\alpha y} \mathcal{F}[\phi]|_y \, dy$$

$$= 2\pi \left[\frac{1}{2\pi} \int_{-\infty}^{\infty} \mathcal{F}[\phi]|_y \, e^{j\alpha y} \, dy \right]$$

$$= 2\pi \, \mathcal{F}^{-1}[\mathcal{F}[\phi]]\Big|_{\alpha}$$

$$= 2\pi \phi(\alpha)$$

$$= \int_{-\infty}^{\infty} 2\pi \delta(x - \alpha)\phi(x) \, dx$$

---

[1] For a detailed discussion of generalized functions, see the first chapter in this handbook.

where $\delta(x)$ is the delta function. This shows that, for every ϕ in S,

$$\int_{-\infty}^{\infty} 2\pi \delta(x - \alpha)\phi(x)\, dx = \int_{-\infty}^{\infty} e^{j\alpha y} \mathcal{F}[\phi]\big|_y \, dy$$

and, thus,

$$\mathcal{F}[e^{j\alpha y}]\big|_x = 2\pi\delta(x - \alpha).$$ ∎

Any (generalized) function whose Fourier transform can be computed via the above generalized definition is called "transformable." The set of all such functions is sometimes called the set of "tempered generalized functions" or the set of "tempered distributions." This set includes any piecewise continuous function, f, which is also polynomially bounded, that is, which satisfies

$$|f(s)| = O(|s|^p) \qquad \text{as } |s| \to \infty$$

for some $p < \infty$. Finally, it should also be noted that if f is classically transformable, then it is transformable, and the generalized definition of $\mathcal{F}[f]$ yields exactly the same function as the classical definition.

2.1.4 *Further generalization of the generalized transforms*

Unfortunately, even with the theory discussed in the previous subsection, it is not possible to define or discuss the Fourier transform of the real exponential, e^t. It may be of interest to note, however, that a further generalization that does permit all exponentially bounded functions to be considered "Fourier transformable" is currently being developed using a recently discovered alternate set of test functions. This alternate set, denoted by $\mathcal{G}$, is the subset of the rapidly decreasing test functions that satisfy the following two additional properties:

1. Each test function is an analytic test function on the entire complex plane.

2. Each test function, $\phi(x + jy)$, satisfies

$$\phi(x + jy) = O\left(e^{-\gamma|x|}\right) \qquad \text{as } x \to \pm\infty$$

for every real value of y and γ.

The second additional property of these test functions insures that all exponentially bounded functions are covered by this theory. The very same computations given in example 2.1.3.1 can be used to show that, for any **complex** value, $\alpha + j\beta$,

$$\mathcal{F}\left[e^{j(\alpha+j\beta)t}\right]\big|_\omega = 2\pi\delta_{\alpha+j\beta}(\omega),$$

where $\delta_{\alpha+j\beta}(t)$ is "the delta function at $\alpha + j\beta$." This delta function, $\delta_{\alpha+j\beta}(t)$, is the generalized function satisfying

$$\int_{-\infty}^{\infty} \delta_{\alpha+j\beta}(t)\phi(t)\, dt = \phi(\alpha + j\beta)$$

for every test function, $\phi(t)$, in $\mathcal{G}$. In particular, letting $\alpha = -j$,

$$\mathcal{F}[e^t]\big|_\omega = 2\pi\delta_{-j}(\omega)$$

and

$$\mathcal{F}\left[\delta_j(t)\right]\big|_\omega = e^t.$$

In addition to allowing delta functions to be defined at complex points, the analyticity of the test functions allows a generalization of translation. Let $\alpha + j\beta$ be any complex number and $f(t)$ any (exponentially bounded) (generalized) function. The "generalized translation of $f(t)$ by $\alpha + j\beta$," denoted by $T_{\alpha+j\beta} f(t)$, is that generalized function satisfying

$$\int_{-\infty}^{\infty} T_{\alpha+j\beta} f(t)\phi(t)\, dt = \int_{-\infty}^{\infty} f(t)\phi(t + (\alpha + j\beta))\, dt \qquad (2.1.4.1)$$

for every test function, $\phi(t)$, in $\mathcal{G}$. So long as $\beta = 0$ or $f(t)$ is, itself, an analytic function on the entire complex plane, then the generalized translation is exactly the same as the classical translation,

$$T_{\alpha+j\beta} f(t) = f(t - (\alpha + j\beta)).$$

It may be observed, however, that equation (2.1.4.1) defines the generalized function $T_{\alpha+j\beta} f$ even when $f(z)$ is not defined for nonreal values of z.

2.1.5 Use of the residue theorem

Often a Fourier transform or inverse transform can be described as an integral of a function that either is analytic on the entire complex plane, or else has a few isolated poles in the complex plane. Such integrals can often be evaluated through intelligent use of the residue theorem from complex analysis (see Appendix 1). Two examples illustrating such use of the residue theorem will be given in this subsection. The first example illustrates its use when the function is analytic throughout the complex plane, while the second example illustrates its use when the function has poles off the real axis. The use of the residue theorem to compute transforms when the function has poles on the real axis will be discussed in the next subsection.

Example 2.1.5.1   Transform of an analytic function
Consider computing the Fourier transform of $g(t) = e^{-t^2}$,

$$G(\omega) = \mathcal{F}\left[g(t)\right]\big|_\omega = \int_{-\infty}^{\infty} e^{-t^2} e^{-j\omega t}\, dt.$$

Because

$$t^2 + j\omega t = \left(t + j\frac{\omega}{2}\right)^2 + \frac{\omega^2}{4},$$

it follows that

$$G(\omega) = e^{-\frac{1}{4}\omega^2} \int_{-\infty}^{\infty} \exp\left[-\left(t + j\frac{\omega}{2}\right)^2\right] dt$$

$$= e^{-\frac{1}{4}\omega^2} \int_{-\infty+j\frac{\omega}{2}}^{\infty+j\frac{\omega}{2}} e^{-z^2}\, dz. \qquad (2.1.5.1)$$

Consider, now, the integral of e^{-z^2} over the contour C_γ, where, for each $\gamma > 0$, $C_\gamma = C_{1,\gamma} + C_{2,\gamma} + C_{3,\gamma} + C_{4,\gamma}$ is the contour in Figure 2.1. Because e^{-z^2} is analytic everywhere on the complex plane, the residue theorem states that

$$0 = \int_{C_\gamma} e^{-z^2}\, dz$$

$$= \int_{C_{1,\gamma}} e^{-z^2}\, dz + \int_{C_{2,\gamma}} e^{-z^2}\, dz + \int_{C_{3,\gamma}} e^{-z^2}\, dz + \int_{C_{4,\gamma}} e^{-z^2}\, dz.$$

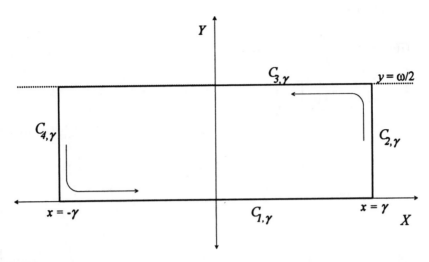

FIGURE 2.1
Contour for computing $\mathcal{F}[e^{-t^2}]$.

Thus,

$$-\int_{C_{3,\gamma}} e^{-z^2}\,dz = \int_{C_{1,\gamma}} e^{-z^2}\,dz + \int_{C_{2,\gamma}} e^{-z^2}\,dz + \int_{C_{4,\gamma}} e^{-z^2}\,dz. \qquad (2.1.5.2)$$

Now,

$$\lim_{\gamma\to\infty}\int_{C_{2,\gamma}} e^{-z^2}\,dz = \lim_{\gamma\to\infty}\int_{y=0}^{\omega/2} e^{-(\gamma+jy)^2}\,dy$$

$$= \lim_{\gamma\to\infty} e^{-\gamma^2}\int_{y=0}^{\omega/2} e^{y^2-j2\gamma y}\,dy$$

$$= 0.$$

Likewise,

$$\lim_{\gamma\to\infty}\int_{C_{4,\gamma}} e^{-z^2}\,dz = 0,$$

while

$$\lim_{\gamma\to\infty}\int_{C_{3,\gamma}} e^{-z^2}\,dz = \lim_{\gamma\to\infty}\int_{\gamma+j\frac{\omega}{2}}^{-\gamma+j\frac{\omega}{2}} e^{-z^2}\,dz = -\int_{-\infty+j\frac{\omega}{2}}^{\infty+j\frac{\omega}{2}} e^{-z^2}\,dz$$

and

$$\lim_{\gamma\to\infty}\int_{C_{1,\gamma}} e^{-z^2}\,dz = \lim_{\gamma\to\infty}\int_{x=-\gamma}^{\gamma} e^{-x^2}\,dx = \int_{-\infty}^{\infty} e^{-x^2}\,dx.$$

The last integral is well known and equals $\sqrt{\pi}$. Combining equations (2.1.5.1) and (2.1.5.2) with the above limits yields

$$G(\omega) = e^{-\frac{1}{4}\omega^2} \int_{-\infty+j\frac{\omega}{2}}^{\infty+j\frac{\omega}{2}} e^{-z^2} \, dz$$

$$= e^{-\frac{1}{4}\omega^2} \lim_{\gamma\to\infty} \left[-\int_{C_{3,\gamma}} e^{-z^2} \, dz \right]$$

$$= e^{-\frac{1}{4}\omega^2} \lim_{\gamma\to\infty} \left[\int_{C_{1,\gamma}} e^{-z^2} \, dz + \int_{C_{2,\gamma}} e^{-z^2} \, dz + \int_{C_{4,\gamma}} e^{-z^2} \, dz \right]$$

$$= e^{-\frac{1}{4}\omega^2} \sqrt{\pi} .$$

So,

$$\mathcal{F}[e^{-t^2}]\Big|_{\omega} = G(\omega) = \sqrt{\pi} e^{-\frac{1}{4}\omega^2}. \qquad \blacksquare$$

Example 2.1.5.2   *Transform of a function with a pole off the real axis*
Consider computing the Fourier inverse transform of $F(\omega) = \frac{1}{1+\omega^2}$

$$f(t) = \mathcal{F}^{-1}[F(\omega)]\Big|_t = \frac{1}{2\pi} \int_{-\infty}^{\infty} \frac{e^{jt\omega}}{1+\omega^2} \, d\omega. \qquad (2.1.5.3)$$

For $t = 0$,

$$f(0) = \frac{1}{2\pi} \int_{-\infty}^{\infty} \frac{1}{1+\omega^2} \, d\omega = \frac{1}{2\pi} \arctan \omega \Big|_{-\infty}^{\infty} = \frac{1}{2}. \qquad (2.1.5.4)$$

To evaluate $f(t)$ when $t \neq 0$, first observe that the integrand in formula (2.1.5.1), viewed as a function of the complex variable,

$$\Phi(z) = \frac{e^{jtz}}{1+z^2},$$

has simple poles at $z = \pm j$. The residue at $z = j$ is

$$\text{Res}_j[\Phi] = \lim_{z\to j}(z-j)\Phi(z) = \lim_{z\to j}(z-j)\left[\frac{e^{jtz}}{(z-j)(z+j)} \right] = \frac{1}{2j} e^{-t},$$

while the residue at $z = -j$ is

$$\text{Res}_{-j}[\Phi] = \lim_{z\to -j}(z+j)\Phi(z) = -\frac{1}{2j} e^t.$$

For each $\gamma > 1$, let C_γ, $C_{+,\gamma}$, and $C_{-,\gamma}$ be the curves sketched in Figure 2.2. By the residue theorem:

$$\int_{C_\gamma} \frac{e^{jtz}}{1+z^2} \, dz + \int_{C_{+,\gamma}} \frac{e^{jtz}}{1+z^2} \, dz = 2\pi j \, \text{Res}_j[\Phi] = \pi e^{-t}$$

and

$$-\int_{C_\gamma} \frac{e^{jtz}}{1+z^2} \, dz + \int_{C_{-,\gamma}} \frac{e^{jtz}}{1+z^2} \, dz = 2\pi j \, \text{Res}_{-j}[\Phi] = -\pi e^t.$$

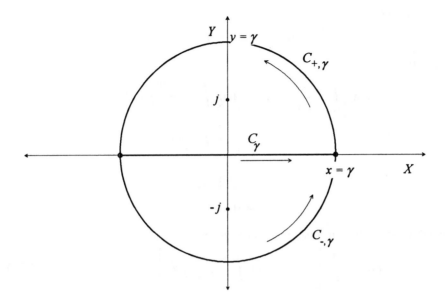

FIGURE 2.2
Contours for computing $\mathcal{F}^{-1}[1/(1+\omega^2)]$.

Combining these calculations with equation (2.1.5.3) yields

$$f(t) = \frac{1}{2\pi} \int_{-\infty}^{\infty} \frac{e^{jt\omega}}{1+\omega^2} \, d\omega$$

$$= \frac{1}{2\pi} \lim_{\gamma \to \infty} \int_{C_\gamma} \frac{e^{jtz}}{1+z^2} \, dz$$

$$= \frac{1}{2\pi} \left[\pi e^{-t} - \lim_{\gamma \to \infty} \int_{C_{+,\gamma}} \frac{e^{jtz}}{1+z^2} \, dz \right] \qquad (2.1.5.5)$$

and

$$f(t) = \frac{1}{2\pi} \int_{-\infty}^{\infty} \frac{e^{jt\omega}}{1+\omega^2} \, d\omega$$

$$= \frac{1}{2\pi} \lim_{\gamma \to \infty} \int_{C_\gamma} \frac{e^{jtz}}{1+z^2} \, dz$$

$$= \frac{1}{2\pi} \left[\pi e^{t} + \lim_{\gamma \to \infty} \int_{C_{-,\gamma}} \frac{e^{jtz}}{1+z^2} \, dz \right]. \qquad (2.1.5.6)$$

Now,

$$\left| \int_{C_{+,\gamma}} \frac{e^{jtz}}{1+z^2} \, dz \right| = \left| \int_0^\pi \frac{e^{jt\gamma(\cos\theta + j\sin\theta)}}{1+\gamma^2 e^{j2\theta}} \gamma e^{j\theta} \, d\theta \right|$$

$$< \int_0^\pi \left| \frac{e^{jt\gamma(\cos\theta + j\sin\theta)}}{1+\gamma^2 e^{j2\theta}} \gamma e^{j\theta} \right| \, d\theta$$

$$< \int_0^\pi \frac{e^{-t\gamma\sin\theta}}{\gamma^2 - 1} \gamma \, d\theta.$$

So long as $t > 0$ and $0 \leq \theta \leq \pi$,

$$0 \leq e^{-t\gamma \sin\theta} \leq 1.$$

Thus, for $t > 0$,

$$\lim_{\gamma \to \infty} \left| \int_{C_{+,\gamma}} \frac{e^{jtz}}{1+z^2} \, dz \right| < \lim_{\gamma \to \infty} \int_0^\pi \frac{e^{-t\gamma \sin\theta}}{\gamma^2 - 1} \gamma \, d\theta$$

$$< \lim_{\gamma \to \infty} \int_0^\pi \frac{\gamma}{\gamma^2 - 1} \, d\theta$$

$$< \lim_{\gamma \to \infty} \frac{2\pi\gamma}{\gamma^2 - 1}$$

$$= 0.$$

Combining this last result with equation (2.1.5.5) gives

$$f(t) = \frac{1}{2\pi} \left[\pi e^{-t} - \lim_{\gamma \to \infty} \int_{C_{+,\gamma}} \frac{e^{jtz}}{1+z^2} \, dz \right] = \frac{1}{2} e^{-\pi} \qquad (2.1.5.7)$$

whenever $t > 0$.

In a similar fashion, it is easy to show that if $t < 0$,

$$\lim_{\gamma \to \infty} \left| \int_{C_{-,\gamma}} \frac{e^{jtz}}{1+z^2} \, dz \right| < \lim_{\gamma \to \infty} \int_\pi^{2\pi} \frac{e^{-t\gamma \sin\theta}}{\gamma^2 - 1} \gamma \, d\theta$$

$$< \lim_{\gamma \to \infty} \int_\pi^{2\pi} \frac{\gamma}{\gamma^2 - 1} \, d\theta$$

$$= 0,$$

which, combined with equation (2.1.5.6), yields

$$f(t) = \frac{1}{2\pi} \left[\pi e^t + \lim_{\gamma \to \infty} \int_{C_{-,\gamma}} \frac{e^{jtz}}{1+z^2} \, dz \right] = \frac{1}{2} e^t \qquad (2.1.5.8)$$

whenever $t < 0$.

Finally, it should be noted that formulas (2.1.5.4), (2.1.5.7), and (2.1.5.8) can be written more concisely as

$$f(t) = \frac{1}{2} e^{-|t|}. \qquad \blacksquare$$

2.1.6 *Cauchy principle values*

The Cauchy principle value (CPV) at $x = x_0$ of an integral, $\int_{-\infty}^\infty \phi(x) \, dx$, is

$$\text{CPV} \int_{-\infty}^\infty \phi(x) \, dx = \lim_{\epsilon \to 0^+} \left[\int_{-\infty}^{x_0 - \epsilon} \phi(x) \, dx + \int_{x_0 + \epsilon}^\infty \phi(x) \, dx \right]$$

provided the limit exists. So long as ϕ is an integrable function, it should be clear that

$$\text{CPV} \int_{-\infty}^\infty \phi(x) \, dx = \int_{-\infty}^\infty \phi(x) \, dx.$$

It is when ϕ is not an integrable function that the Cauchy principle value is useful. In particular, the Fourier transform and Fourier inverse transform of any function with a singularity of the

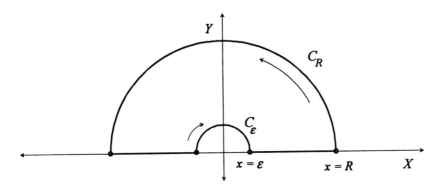

FIGURE 2.3
Contour for computing $\mathcal{F}^{-1}[1/\omega]$.

form $(x - x_0)^{-1}$ can be evaluated as the Cauchy principle values at $x = x_0$ of the integrals in formulas (2.1.1.1) and (2.1.1.2).

Example 2.1.6.1
Consider evaluating the inverse transform of $F(\omega) = \frac{1}{\omega}$. Because of the $\frac{1}{\omega}$ singularity, $f = \mathcal{F}^{-1}[F]$ is given by

$$f(t) = \frac{1}{2\pi} \text{CPV} \int_{-\infty}^{\infty} \frac{1}{\omega} e^{j\omega t} \, d\omega$$

or, equivalently, by

$$f(t) = \frac{1}{2\pi} \lim_{\substack{\epsilon \to 0^+ \\ R \to +\infty}} \left[\int_{-R}^{-\epsilon} \frac{1}{z} e^{jtz} \, dz + \int_{\epsilon}^{R} \frac{1}{z} e^{jtz} \, dz \right]. \tag{2.1.6.1}$$

Because $\frac{1}{\omega}$ is an odd function, $f(0)$ is easily evaluated,

$$f(0) = \frac{1}{2\pi} \lim_{\substack{\epsilon \to 0^+ \\ R \to +\infty}} \left[\int_{-R}^{-\epsilon} \frac{1}{\omega} \, d\omega + \int_{\epsilon}^{R} \frac{1}{\omega} \, d\omega \right] = 0. \tag{2.1.6.2}$$

To evaluate $f(t)$ when $t > 0$, first observe that the only pole of the integrand in formula (2.1.6.1),

$$\Phi(z) = \frac{1}{z} e^{jtz},$$

is at $z = 0$. For each $0 < \epsilon < R$, let C_ϵ and C_R be the semicircles indicated in Figure 2.3. By the residue theorem,

$$\int_{-R}^{-\epsilon} \frac{1}{z} e^{jtz} \, dz + \int_{\epsilon}^{R} \frac{1}{z} e^{jtz} \, dz + \int_{C_\epsilon} \frac{1}{z} e^{jtz} \, dz + \int_{C_R} \frac{1}{z} e^{jtz} \, dz = 0.$$

This, combined with equation (2.1.6.1), yields

$$f(t) = -\frac{1}{2\pi} \left[\lim_{\epsilon \to 0^+} \int_{C_\epsilon} \frac{1}{z} e^{jtz} \, dz + \lim_{R \to \infty} \int_{C_R} \frac{1}{z} e^{jtz} \, dz \right], \tag{2.1.6.3}$$

provided the limits exist. Now,

$$\lim_{\epsilon \to 0^+} \int_{C_\epsilon} \frac{1}{z} e^{jtz} \, dz = \lim_{\epsilon \to 0^+} \int_\pi^0 \frac{1}{\epsilon e^{j\theta}} e^{jt\epsilon(\cos\theta + j\sin\theta)} j\epsilon e^{j\theta} \, d\theta$$

$$= j \lim_{\epsilon \to 0^+} \int_\pi^0 e^{-\epsilon t(\sin\theta + j\cos\theta)} \, d\theta$$

$$= j \int_\pi^0 e^0 \, d\theta$$

$$= -j\pi. \tag{2.1.6.4}$$

Similarly,

$$\int_{C_R} \frac{1}{z} e^{jtz} \, dz = j \int_0^\pi e^{-Rt(\sin\theta + j\cos\theta)} \, d\theta.$$

Here, because $t > 0$, the integrand is uniformly bounded and vanishes as $R \to \infty$. Thus,

$$\lim_{R \to \infty} \int_{C_R} \frac{1}{z} e^{jtz} \, dz = 0. \tag{2.1.6.5}$$

With equations (2.1.6.4) and (2.1.6.5), equation (2.1.6.3) becomes

$$f(t) = -\frac{1}{2\pi} \left[\lim_{\epsilon \to 0^+} \int_{C_\epsilon} \frac{1}{z} e^{jtz} \, dz + \lim_{R \to \infty} \int_{C_R} \frac{1}{z} e^{jtz} \, dz \right] = \frac{j}{2}. \tag{2.1.6.6}$$

By replacing C_ϵ and C_R with corresponding semicircles in the lower half-plane, the approach used to evaluate $f(t)$ when $0 < t$, can be used to evaluate $f(t)$ when $t < 0$. The computations are virtually identical, except for a reversal of the orientation of the contour of integration, and yield

$$f(t) = -\frac{j}{2}, \tag{2.1.6.7}$$

when $t < 0$.

Finally, it should be noted that formulas (2.1.6.2), (2.1.6.6), and (2.1.6.7) can be written more concisely as

$$\mathcal{F}^{-1}\left[\frac{1}{\omega} \right]\bigg|_t = f(t) = \frac{j}{2} \operatorname{sgn}(t). \qquad\blacksquare$$

2.2 General identities and relations

Some of the more general identities commonly used in computing and manipulating Fourier transforms and inverse transforms are described here. Brief (nonrigorous) derivations of some are presented, usually employing the classical transforms (formulas [2.1.1.1] and [2.1.1.2]). Unless otherwise stated, however, each identity may be assumed to hold for the generalized transforms as well.

2.2.1 Invertibility

The Fourier transform and the Fourier inverse transform, $\mathcal{F}$ and $\mathcal{F}^{-1}$, are operational inverses, that is,

$$\Psi = \mathcal{F}[\phi] \Leftrightarrow \mathcal{F}^{-1}[\Psi] = \phi.$$

Equivalently,

$$\mathcal{F}^{-1}[\mathcal{F}[f]] = f \qquad \text{and} \qquad \mathcal{F}[\mathcal{F}^{-1}[F]] = F.$$

Example 2.2.1.1
Because $\mathcal{F}[e^{-t}u(t)]\big|_{\omega} = \frac{1}{1+j\omega}$ (see Example 2.1.1.1),

$$\mathcal{F}^{-1}\left[\frac{1}{1+j\omega}\right]\bigg|_{t} = e^{-t}u(t). \qquad \blacksquare$$

2.2.2 Near-equivalence (symmetry of the transforms)

Computationally, the classical formulas for $\mathcal{F}[\phi(s)]\big|_x$ and $\mathcal{F}[\phi(s)]\big|_x$ (formulas [2.1.1.1] and [2.1.1.2]) are virtually the same, differing only by the sign in the exponential and the factor of $(2\pi)^{-1}$ in (2.1.1.2). Observing that

$$\int_{-\infty}^{\infty} \phi(s)e^{-jxs}\,ds = 2\pi\left[\frac{1}{2\pi}\int_{-\infty}^{\infty}\phi(s)e^{j(-x)s}\,ds\right] = 2\pi\left[\frac{1}{2\pi}\int_{-\infty}^{\infty}\phi(s)e^{jx(-s)}\,ds\right]$$

leads to the "near equivalence" identity,

$$\mathcal{F}[\phi(s)]\big|_x = 2\pi\mathcal{F}^{-1}[\phi(s)]\big|_{-x} = 2\pi\mathcal{F}^{-1}[\phi(-s)]\big|_x. \qquad (2.2.2.1)$$

Likewise,

$$\mathcal{F}^{-1}[\phi(s)]\big|_x = \frac{1}{2\pi}\mathcal{F}[\phi(s)]\big|_{-x} = \frac{1}{2\pi}\mathcal{F}[\phi(-s)]\big|_x. \qquad (2.2.2.2)$$

Example 2.2.2.1
Using near-equivalence and results of example 2.1.1.1,

$$\mathcal{F}[e^{s}u(-s)]\big|_x = 2\pi\mathcal{F}^{-1}[e^{-s}u(s)]\big|_x = 2\pi\left[\frac{1}{2\pi - j2\pi x}\right] = \frac{1}{1-jx}. \qquad \blacksquare$$

2.2.3 Conjugation of transforms

Using z^* to denote the complex conjugate of any complex quantity, z, it can be observed that

$$\left(\int_{-\infty}^{\infty} f(t)e^{-j\omega t}\,dt\right)^{*} = \int_{-\infty}^{\infty} f^{*}(t)e^{j\omega t}\,dt.$$

Thus,

$$\mathcal{F}[f]^{*} = 2\pi\mathcal{F}^{-1}[f^{*}]. \qquad (2.2.3.1)$$

Likewise

$$\left(\mathcal{F}^{-1}[f]\right)^{*} = \frac{1}{2\pi}\mathcal{F}[f^{*}]. \qquad (2.2.3.2)$$

2.2.4   Linearity

If α and β are any two scalar constants, then it follows from the linearity of the integral that

$$\mathcal{F}[\alpha f + \beta g] = \alpha \mathcal{F}[f] + \beta \mathcal{F}[g]$$

and

$$\mathcal{F}^{-1}[\alpha F + \beta G] = \alpha \mathcal{F}^{-1}[F] + \beta \mathcal{F}^{-1}[G].$$

Example 2.2.4.1

Using linearity and the transforms computed in Examples 2.1.1.1 and 2.2.2.1,

$$\mathcal{F}\left[e^{-|t|}\right]\Big|_x = \mathcal{F}\left[e^{-t}u(t) + e^t u(-t)\right]\Big|_\omega = \frac{1}{1+j\omega} + \frac{1}{1-j\omega} = \frac{2}{1+\omega^2}$$

and

$$\mathcal{F}\left[\operatorname{sgn}(t)e^{-|t|}\right]\Big|_x = \mathcal{F}\left[e^{-t}u(t) - e^t u(-t)\right]\Big|_\omega = \frac{1}{1+j\omega} - \frac{1}{1-j\omega} = \frac{-2\omega j}{1+\omega^2}. \qquad ∎$$

2.2.5   Scaling

If α is any nonzero real number, then, using the substitution $\tau = \alpha t$,

$$\int_{-\infty}^{\infty} f(\alpha t) e^{-jt\omega}\, dt = \frac{1}{|\alpha|} \int_{-\infty}^{\infty} f(\tau) e^{-j\frac{\tau\omega}{\alpha}}\, d\tau.$$

Letting $F(\omega) = \mathcal{F}[f(t)]\big|_\omega$, this can be rewritten as

$$\mathcal{F}\left[f(\alpha t)\right]\big|_\omega = \frac{1}{|\alpha|} F\left(\frac{\omega}{\alpha}\right). \tag{2.2.5.1}$$

Likewise,

$$\mathcal{F}^{-1}\left[F(\alpha\omega)\right]\big|_t = \frac{1}{|\alpha|} f\left(\frac{t}{\alpha}\right). \tag{2.2.5.2}$$

Example 2.2.5.1

Using identity (2.2.5.1) and the results from example 2.2.4.1:

$$\mathcal{F}\left[e^{-|\alpha t|}\right]\big|_\omega = \frac{1}{|\alpha|} \cdot \frac{2}{1+\left(\frac{\omega}{\alpha}\right)^2} = \frac{2|\alpha|}{\alpha^2 + \omega^2}. \qquad ∎$$

2.2.6   Translation and multiplication by exponentials

If $F(\omega) = \mathcal{F}[f(t)]\big|_\omega$ and α is any real number, then

$$\mathcal{F}\left[f(t-\alpha)\right]\big|_\omega = e^{-j\alpha\omega} F(\omega), \tag{2.2.6.1}$$

$$\mathcal{F}\left[e^{j\alpha t} f(t)\right]\big|_\omega = F(\omega - \alpha), \tag{2.2.6.2}$$

$$\mathcal{F}^{-1}\left[F(\omega - \alpha)\right]\big|_t = e^{j\alpha t} f(t), \tag{2.2.6.3}$$

and

$$\mathcal{F}^{-1}\left[e^{j\alpha\omega} F(\omega)\right]\big|_t = f(t+\alpha). \tag{2.2.6.4}$$

These formulas are easily derived from the classical definitions. Identity (2.2.6.2), for example, comes directly from the observation that

$$\int_{-\infty}^{\infty} e^{j\alpha t} f(t) e^{-j\omega t}\, dt = \int_{-\infty}^{\infty} f(t) e^{-j(\omega-\alpha)t}\, dt.$$

In general, identities (2.2.6.1) through (2.2.6.4) are not valid when α is not a real number. An exception to this occurs when f is an analytic function on the entire complex plane. Then identities (2.2.6.1) and (2.2.6.4) do hold for all complex values of α. Likewise, identities (2.2.6.2) and (2.2.6.3) may be used whenever α is complex provided F is an analytic function on the entire complex plane.

Example 2.2.6.1
Let $g(t) = e^{-t^2}$. It can be shown that $g(t)$ is analytic on the entire complex plane and that its Fourier transform is

$$G(\omega) = \sqrt{\pi} \exp\left[-\frac{1}{4}\omega^2\right]$$

(see Example 2.1.5.1 or Example 2.2.11.2). If β is any real value, then

$$\mathcal{F}\left[e^{-t^2+2\beta t}\right]\Big|_{\omega} = \mathcal{F}\left[e^{j(-j2\beta)t}e^{-t^2}\right]\Big|_{\omega}$$

$$= \sqrt{\pi} \exp\left[-\frac{1}{4}(\omega - (-j2\beta))^2\right]$$

$$= \sqrt{\pi} e^{\beta^2} \exp\left[-\frac{1}{4}\omega^2 + j\beta\omega\right]. \qquad \blacksquare$$

2.2.7 Complex translation and multiplication by real exponentials

Using the "generalized" notion of translation discussed in subsection 2.1.4, it can be shown that for any complex value, $\alpha + j\beta$,

$$\mathcal{F}\left[T_{\alpha+j\beta} f(t)\right]\Big|_{\omega} = e^{-j(\alpha+j\beta)\omega} F(\omega),$$

$$\mathcal{F}\left[e^{j(\alpha+j\beta)t} f(t)\right]\Big|_{\omega} = T_{\alpha+j\beta} F(\omega),$$

$$\mathcal{F}^{-1}\left[T_{\alpha+j\beta} F(\omega)\right]\Big|_{t} = e^{j(\alpha+j\beta)t} f(t),$$

and

$$\mathcal{F}^{-1}[e^{j(\alpha+j\beta)\omega} F(\omega)]\Big|_{t} = T_{-(\alpha+j\beta)} f(t).$$

Letting $\alpha = 0$ and $\beta = -\gamma$, these identities become

$$\mathcal{F}[T_{-j\gamma} f(t)]\Big|_{\omega} = e^{-\gamma\omega} F(\omega),$$

$$\mathcal{F}\left[e^{\gamma t} f(t)\right]\Big|_{\omega} = T_{-j\gamma} F(\omega),$$

$$\mathcal{F}^{-1}\left[T_{-j\gamma} F(\omega)\right]\Big|_{t} = e^{\gamma t} f(t),$$

and

$$\mathcal{F}^{-1}[e^{\gamma\omega} F(\omega)]\big|_{t} = T_{j\gamma} f(t).$$

Caution must be exercised in the use of these formulas. It is true that $T_{\alpha+j\beta} f(t) = f(t - (\alpha + j\beta))$ whenever $\beta = 0$ or $f(z)$ is analytic on the entire complex plane. However, if $f(z)$ is not analytic and $\beta \neq 0$, then it is quite possible that $T_{\alpha+j\beta} f(t) \neq f(t - (\alpha + j\beta))$, even if $f(t - (\alpha + j\beta))$ is well defined. In these cases $T_{\alpha+j\beta} f(t)$ should be treated formally.

Example 2.2.7.1
By the above

$$\mathcal{F}\left[e^t u(t)\right]\Big|_\omega = \mathcal{F}\left[e^{2t} e^{-t} u(t)\right]\Big|_\omega = T_{-2j}\left[\frac{1}{1+j\omega}\right].$$

Note, however, that

$$\mathcal{F}\left[-e^t u(-t)\right]\Big|_\omega = \frac{-1}{1-j\omega} = \frac{1}{1+j(\omega-(-2j))}.$$

Because $e^t u(t)$ and $-e^t u(-t)$ certainly are not equal, it follows that their transforms are not equal,

$$T_{-2j}\left[\frac{1}{1+j\omega}\right] \neq \frac{1}{1+j(\omega-(-2j))}. \qquad \blacksquare$$

2.2.8 Modulation

The "modulation formulas,"

$$\mathcal{F}\left[\cos(\omega_0 t) f(t)\right]\big|_\omega = \frac{1}{2}\left[F(\omega-\omega_0) + F(\omega+\omega_0)\right] \tag{2.2.8.1}$$

and

$$\mathcal{F}\left[\sin(\omega_0 t) f(t)\right]\big|_\omega = \frac{1}{2j}\left[F(\omega-\omega_0) - F(\omega+\omega_0)\right] \tag{2.2.8.2}$$

are easily derived from identity (2.2.6.2) using the well-known formulas

$$\cos(\omega_0 t) = \frac{1}{2}\left[e^{j\omega_0 t} + e^{-j\omega_0 t}\right]$$

and

$$\sin(\omega_0 t) = \frac{1}{2j}\left[e^{j\omega_0 t} - e^{-j\omega_0 t}\right].$$

Example 2.2.8.1
For $\alpha > 0$, the function

$$f(t) = \begin{cases} \cos\left(\frac{\pi}{2\alpha}t\right), & \text{if } -\alpha \leq t \leq \alpha \\ 0, & \text{otherwise} \end{cases}$$

can be written as

$$f(t) = \cos\left(\frac{\pi}{2\alpha}t\right) p_\alpha(t).$$

Thus, using identity (2.2.8.1) and the results of example 2.1.1.2,

$$F(\omega) = \mathcal{F}\left[\cos\left(\frac{\pi}{2\alpha}t\right) p_\alpha(t)\right]\Big|_\omega$$

$$= \frac{1}{2}\left[\frac{2}{\omega - \frac{\pi}{2\alpha}}\sin\left(\alpha\left[\omega - \frac{\pi}{2\alpha}\right]\right) + \frac{2}{\omega + \frac{\pi}{2\alpha}}\sin\left(\alpha\left[\omega + \frac{\pi}{2\alpha}\right]\right)\right]$$

$$= \frac{4\alpha\pi}{\pi^2 - 4\alpha^2\omega^2}\cos(\alpha\omega). \qquad \blacksquare$$

2.2.9   *Products and convolution*

If $F = \mathcal{F}[f]$ and $G = \mathcal{F}[g]$, then the corresponding transforms of the products, fg and FG, can be computed using the identities

$$\mathcal{F}[fg] = \frac{1}{2\pi} F * G \tag{2.2.9.1}$$

and

$$\mathcal{F}^{-1}[FG] = f * g, \tag{2.2.9.2}$$

provided the convolutions, $F * G$ and $f * g$, exist. Conversely, as long as the convolutions exist,

$$\mathcal{F}[f * g] = FG \tag{2.2.9.3}$$

and

$$\mathcal{F}^{-1}[F * G] = 2\pi fg. \tag{2.2.9.4}$$

Identity (2.2.9.1) can be derived as follows:

$$\int_{-\infty}^{\infty} f(t)g(t)e^{-j\omega t}\, dt = \int_{-\infty}^{\infty} \left(\frac{1}{2\pi} \int_{-\infty}^{\infty} F(s)e^{jst}\, ds \right) g(t)e^{-j\omega t}\, dt$$

$$= \frac{1}{2\pi} \int_{-\infty}^{\infty} F(s) \int_{-\infty}^{\infty} g(t)e^{-j(\omega-s)t}\, dt\, ds$$

$$= \frac{1}{2\pi} \int_{-\infty}^{\infty} F(s)G(\omega - s)\, ds.$$

The other identities can be derived in a similar fashion.

Example 2.2.9.1
From direct computation, if $\beta > 0$, then

$$\mathcal{F}^{-1}[e^{-\beta\omega}u(\omega)]\Big|_t = \frac{1}{2\pi} \int_0^{\infty} e^{(jt-\beta)\omega}\, d\omega = \frac{1}{2\pi} \cdot \frac{1}{\beta - jt}.$$

And so,

$$\mathcal{F}\left[\frac{1}{\beta - jt} \right]\Big|_\omega = 2\pi e^{-\beta\omega}u(\omega).$$

Applying identity (2.2.9.1),

$$\mathcal{F}\left[\frac{1}{10 - 7tj - t^2} \right]\Big|_\omega = \mathcal{F}\left[\frac{1}{2 - jt} \cdot \frac{1}{5 - jt} \right]\Big|_\omega$$

$$= \frac{1}{2\pi} \left[2\pi e^{-2\omega}u(\omega) \right] * \left[2\pi e^{-5\omega}u(\omega) \right]$$

$$= 2\pi \int_{-\infty}^{\infty} e^{-2s}u(s)e^{-5(\omega-s)}u(\omega - s)\, ds$$

$$= \begin{cases} 0, & \text{if } \omega < 0 \\ \frac{2\pi}{3}\left[e^{2\omega} - e^{-5\omega} \right], & \text{if } 0 < \omega \end{cases}$$

∎

Example 2.2.9.2

By straightforward computations it is easily verified that for $\alpha > 0$,

$$\mathcal{F}\left[p_{\alpha/2}(t)\right]\big|_{\omega} = \frac{2}{\omega} \sin\left(\frac{\alpha}{2}\omega\right)$$

and

$$p_{\alpha/2}(t) * p_{\alpha/2}(t) = \alpha \Lambda_{\alpha}(t),$$

where $p_{\alpha/2}(t)$ is the pulse function,

$$p_{\alpha/2}(t) = \begin{cases} 1, & \text{if } |t| < \frac{\alpha}{2} \\ 0, & \text{if } \frac{\alpha}{2} < |t| \end{cases},$$

and $\Lambda_{\alpha}(t)$ is the triangle function,

$$\Lambda_{\alpha}(t) = \begin{cases} 1 - \frac{|t|}{\alpha}, & \text{if } |t| < \alpha \\ 0, & \text{if } \alpha < |t| \end{cases}.$$

Using identity (2.2.9.3)

$$\mathcal{F}\left[\Lambda_{\alpha}(t)\right]\big|_{\omega} = \frac{1}{\alpha}\,\mathcal{F}[p_{\alpha/2}(t) * p_{\alpha/2}(t)]\big|_{\omega}$$

$$= \frac{1}{\alpha}\left(\frac{2}{\omega}\sin\left(\frac{\alpha}{2}\omega\right)\right)\left(\frac{2}{\omega}\sin\left(\frac{\alpha}{2}\omega\right)\right)$$

$$= \frac{4}{\alpha\omega^2}\sin^2\left(\frac{\alpha}{2}\omega\right). \qquad \blacksquare$$

2.2.10 Correlation

The cross-correlation of two functions, $f(t)$ and $g(t)$, is another function, denoted by $f(t) \star g(t)$, given by

$$f(t) \star g(t) = \int_{-\infty}^{\infty} f^*(s)g(t+s)\,ds, \qquad (2.2.10.1)$$

where $f^*(s)$ denotes the complex conjugate of $f(s)$. The notation $\rho_{fg}(t)$ is often used instead of $f(t) \star g(t)$. The Wiener–Khintchine theorem states that, provided the correlations exist,

$$\mathcal{F}[f(t) \star g(t)]\big|_{\omega} = F^*(\omega)G(\omega) \qquad (2.2.10.2)$$

and

$$\mathcal{F}[f^*(t)g(t)]\big|_{\omega} = \frac{1}{2\pi}F(\omega) \star G(\omega), \qquad (2.2.10.3)$$

where $F = \mathcal{F}[f]$ and $G = \mathcal{F}[g]$. Derivations of these formulas are similar to the analogous identities involving convolution.

For a given function, $f(t)$, the corresponding autocorrelation function is simply the cross-correlation of $f(t)$ with itself,

$$f(t) \star f(t) = \int_{-\infty}^{\infty} f^*(s)f(t+s)\,ds. \qquad (2.2.10.4)$$

Often autocorrelation is denoted by $\rho_f(t)$ instead of $f(t) \star f(t)$. For autocorrelation, formulas (2.2.10.2) and (2.2.10.3) simplify to

$$\mathcal{F}[f(t) \star f(t)]\big|_{\omega} = |F(\omega)|^2 \qquad (2.2.10.5)$$

and

$$\mathcal{F}\left[|f(t)|^2\right]\Big|_\omega = \frac{1}{2\pi}F(\omega) \star F(\omega). \tag{2.2.10.6}$$

2.2.11   *Differentiation and multiplication by polynomials*

If $f(t)$ is differentiable for all t and vanishes as $t \to \pm\infty$, then the Fourier transform of the derivative of the function can be related to the transform of the undifferentiated function through the use of integration by parts,

$$\int_{-\infty}^{\infty} f'(t)e^{-j\omega t}\,dt = f(t)e^{-j\omega t}\Big|_{-\infty}^{\infty} + j\omega\int_{-\infty}^{\infty} f(t)e^{-j\omega t}\,dt$$

$$= j\omega\int_{-\infty}^{\infty} f(t)e^{-j\omega t}\,dt.$$

In more concise form this can be written

$$\mathcal{F}\left[f'(t)\right]\Big|_\omega = j\omega F(\omega), \tag{2.2.11.1}$$

where $F = \mathcal{F}[f]$. By near equivalence, if $G(\omega)$ is differentiable for all ω and vanishes as $\omega \to \pm\infty$, then

$$\mathcal{F}^{-1}\left[G'(\omega)\right]\Big|_t = -jtg(t), \tag{2.2.11.2}$$

where $g = \mathcal{F}^{-1}[G]$. Similar derivations yield

$$\mathcal{F}\left[tf(t)\right]\Big|_\omega = jF'(\omega) \tag{2.2.11.3}$$

and

$$\mathcal{F}^{-1}[\omega G(\omega)]\Big|_t = -jg'(t), \tag{2.2.11.4}$$

provided $tf(t)$ and $\omega G(\omega)$ are suitably integrable.

Example 2.2.11.1
Using identity (2.2.11.1),

$$\mathcal{F}\left[\frac{j}{(1-jt)^2}\right]\Big|_\omega = \mathcal{F}\left[\frac{d}{dt}\left(\frac{1}{1-jt}\right)\right]\Big|_\omega$$

$$= j\omega\mathcal{F}\left[\frac{1}{1-jt}\right]\Big|_\omega$$

$$= j\omega 2\pi e^{-\omega}u(\omega). \qquad\blacksquare$$

Example 2.2.11.2
Let $\alpha > 0$ and $g(t) = e^{-\alpha t^2}$. It is easily verified that

$$\frac{dg}{dt} = -2\alpha t g(t). \tag{2.2.11.5}$$

Taking the Fourier transform of each side and using identities (2.2.11.1) and (2.2.11.3) yields

$$j\omega G(\omega) = -2\alpha j\frac{dG}{d\omega}.$$

The solution to this first-order differential equation is easily computed. It is

$$G(\omega) = A \exp\left[-\frac{1}{4\alpha}\omega^2\right].$$

The value of the constant of integration, A, can be determined[2] by noting that

$$A = G(0) = \int_{-\infty}^{\infty} e^{-\alpha t^2}\, dt.$$

The value of this last integral is well known to be $\sqrt{\frac{\pi}{\alpha}}$. Thus,

$$G(\omega) = \sqrt{\frac{\pi}{\alpha}} \exp\left[-\frac{1}{4\alpha}\omega^2\right].$$ ∎

It should be noted that if f' and F' are assumed to be the classical derivatives of f and F, that is

$$f'(t) = \lim_{\Delta t \to 0} \frac{f(t + \Delta t) - f(t)}{\Delta t}$$

and

$$F'(\omega) = \lim_{\Delta\omega \to 0} \frac{F(\omega + \Delta\omega) - F(\omega)}{\Delta\omega},$$

then application of the above identities is limited by requirements that the functions involved be suitably smooth and that they vanish at infinity. These limitations can be eliminated, however, by interpreting f' and F' in a more generalized sense. In this more generalized interpretation, f' and F' are defined to be the (generalized) functions satisfying the "generalized" integration by parts formulas,

$$\int_{-\infty}^{\infty} f'(t)\phi(t)\, dt = -\int_{-\infty}^{\infty} f(t)\phi'(t)\, dt$$

and

$$\int_{-\infty}^{\infty} F'(\omega)\phi(\omega)\, d\omega = -\int_{-\infty}^{\infty} F(\omega)\phi'(\omega)\, d\omega,$$

for every test function, ϕ (with ϕ' denoting the classical derivative of ϕ). As long as the function being differentiated is piecewise smooth and continuous, then there is no difference between the classical and the generalized derivative. If, however, the function, $f(x)$, has jump discontinuities at $x = x_1, x_2, \ldots, x_N$, then

$$f'_{\text{generalized}} = f'_{\text{classical}} + \sum_k J_k \delta_{x_k},$$

where J_k denotes the "jump" in f at $x = x_k$,

$$J_k = \lim_{\Delta x \to 0^+} f(x_k + \Delta x) - f(x_k - \Delta x).$$

It is not difficult to show that the product rule, $(fg)' = f'g + fg'$, holds for the generalized derivative as well as the classical derivative.

Example 2.2.11.3

Consider the step function, $u(t)$. The classical derivative of u is clearly 0, because the graph of u consists of two horizontal half-lines (with slope zero). Using the generalized integration

---

[2] A method for determining A using Bessel's equality is described is subsection 2.2.15.

by parts formula, however,

$$\int_{-\infty}^{\infty} u'(t)\phi(t)\, dt = -\int_{-\infty}^{\infty} u(t)\phi'(t)\, dt$$

$$= -\int_{0}^{\infty} \phi'(t)\, dt$$

$$= \phi(0)$$

$$= \int_{-\infty}^{\infty} \delta(t)\phi(t)\, dt,$$

showing that $\delta(t)$ is the generalized derivative of $u(t)$.   ∎

Example 2.2.11.4
Using the generalized derivative and identity (2.2.11.3),

$$\mathcal{F}\left[\frac{t}{1-jt}\right]\bigg|_{\omega} = j\frac{d}{d\omega}\left(\mathcal{F}\left[\frac{1}{1-jt}\right]\bigg|_{\omega}\right)$$

$$= j\frac{d}{d\omega}(2\pi e^{-\omega}u(\omega))$$

$$= 2\pi j\left[\frac{de^{-\omega}}{d\omega}u(\omega) + e^{-\omega}u'(\omega)\right]$$

$$= 2\pi j\left[-e^{-\omega}u(\omega) + \delta(\omega)\right].$$ ∎

The extension of formulas (2.2.11.1) through (2.2.11.4) to the corresponding identities involving higher-order derivatives is straightforward. If n is any positive integer, then

$$\mathcal{F}\left[f^{(n)}(t)\right]\big|_{\omega} = (j\omega)^n F(\omega), \tag{2.2.11.6}$$

$$\mathcal{F}^{-1}\left[F^{(n)}(\omega)\right]\big|_{t} = (-jt)^n f(t), \tag{2.2.11.7}$$

$$\mathcal{F}\left[t^n f(t)\right]\big|_{\omega} = j^n F^{(n)}(\omega), \tag{2.2.11.8}$$

and

$$\mathcal{F}^{-1}\left[\omega^n F(\omega)\right]\big|_{t} = (-j)^n f^{(n)}(t). \tag{2.2.11.9}$$

Again, these identities hold for all transformable functions as long as the derivatives are interpreted in the generalized sense.

2.2.12 *Moments*

For any suitably integrable function, $f(t)$, and nonnegative integer, n, the "nth moment of f" is the quantity

$$m_n(f) = \int_{-\infty}^{\infty} t^n f(t)\, dt.$$

Because

$$\int_{-\infty}^{\infty} t^n f(t)\, dt = \mathcal{F}\left[t^n f(t)\right]\big|_{0},$$

it is clear from identity (2.2.11.8) that

$$m_n(f) = j^n F^{(n)}(0).$$

2.2.13   *Integration*

If $F(\omega)$ and $G(\omega)$ are the Fourier transforms of $f(t)$ and $g(t)$, and $g(t) = \frac{f(t)}{t}$, then $tg(t) = f(t)$ and, by identity (2.2.11.3), $jG'(\omega) = F(\omega)$. Integrating this gives

$$G(\omega) - G(\alpha) = -j \int_\alpha^\omega F(s)\, ds,$$

where α can be any real number. This can be written

$$\mathcal{F}\left[\frac{f(t)}{t}\right]\bigg|_\omega = -j \int_\alpha^\omega F(s)\, ds + c_\alpha \tag{2.2.13.1}$$

where $c_\alpha = G(\alpha)$. For certain general types of functions and choices of α, the value of c_α is easily determined. For example, if $f(t)$ is also absolutely integrable, then

$$\mathcal{F}\left[\frac{f(t)}{t}\right]\bigg|_\omega = -j \int_{-\infty}^\omega F(s)\, ds, \tag{2.2.13.2}$$

while if $f(t)$ is an even function

$$\mathcal{F}\left[\frac{f(t)}{t}\right]\bigg|_\omega = -j \int_0^\omega F(s)\, ds, \tag{2.2.13.3}$$

provided the integrals are well defined.

It can also be shown that as long as the limit of $\frac{F(\omega)}{\omega}$ exists as $\omega \to 0$, then for each real value of α there is a constant, c_α, such that

$$\mathcal{F}\left[\int_\alpha^t f(s)\, ds\right]\bigg|_\omega = -j \frac{F(\omega)}{\omega} + c_\alpha \delta(\omega). \tag{2.2.13.4}$$

If $f(t)$ is an even function, then

$$\mathcal{F}\left[\int_0^t f(s)\, ds\right]\bigg|_\omega = -j \frac{F(\omega)}{\omega}, \tag{2.2.13.5}$$

while if $f(t)$ and $\int_{-\alpha}^t f(s)\, ds$ are absolutely integrable, then

$$\mathcal{F}\left[\int_{-\infty}^t f(s)\, ds\right]\bigg|_\omega = -j \frac{F(\omega)}{\omega}. \tag{2.2.13.6}$$

Example 2.2.13.1
Let α and β be positive,

$$f(t) = e^{-\alpha|t|} - e^{-\beta|t|},$$

and

$$g(t) = \frac{f(t)}{t} = \frac{e^{-\alpha|t|} - e^{-\beta|t|}}{t}.$$

Both functions are easily verified to be transformable with

$$F(\omega) = \mathcal{F}[e^{-\alpha|t|} - e^{-\beta|t|}]\big|_\omega = \frac{2\alpha}{\alpha^2 + \omega^2} - \frac{2\beta}{\beta^2 + \omega^2}.$$

Because $f(t)$ is even, formula (2.2.13.3) applies, and

$$G(\omega) = \mathcal{F}\left[\frac{e^{-\alpha|t|} - e^{-\beta|t|}}{t}\right]\Bigg|_{\omega}$$

$$= -j\int_0^{\omega} F(s)\,ds$$

$$= -j\int_0^{\omega}\left(\frac{2\alpha}{\alpha^2 + s^2} - \frac{2\beta}{\beta^2 + s^2}\right) ds$$

$$= -2j\left(\arctan\left(\frac{\omega}{\alpha}\right) - \arctan\left(\frac{\omega}{\beta}\right)\right). \qquad (2.2.13.7)$$

∎

Example 2.2.13.2
Applying the same analysis done in the previous example but using

$$f(t) = 1 - e^{-\beta|t|}$$

leads, formally, to

$$\mathcal{F}\left[\frac{1 - e^{-\beta|t|}}{t}\right]\Bigg|_{\omega} = -j\int_0^{\omega}\left(2\pi\delta(s) - \frac{2\beta}{\beta^2 + s^2}\right) ds$$

$$= -2\pi j\int_0^{\omega}\delta(s)\,ds + 2j\arctan\left(\frac{\omega}{\beta}\right).$$

Unfortunately, this is of little value because

$$\int_\alpha^{\omega}\delta(s)\,ds$$

is not well defined if $\alpha = 0$. However, because

$$\lim_{\alpha\to 0^+} e^{-\alpha|t|} = 1,$$

and

$$\lim_{\alpha\to 0^+}\arctan\left(\frac{\omega}{\alpha}\right) = \begin{cases} \frac{\pi}{2}, & \text{if } 0 < \omega \\ -\frac{\pi}{2}, & \text{if } \omega < 0 \end{cases}$$

$$= \frac{\pi}{2}\,\mathrm{sgn}(\omega),$$

it can be argued, using equation (2.2.13.7), that

$$\mathcal{F}\left[\frac{1 - e^{-\beta|t|}}{t}\right]\Bigg|_{\omega} = \lim_{\alpha\to 0^+}\mathcal{F}\left[\frac{e^{-\alpha|t|} - e^{-\beta|t|}}{t}\right]\Bigg|_{\omega}$$

$$= \lim_{\alpha\to 0^+} -2j\left(\arctan\left(\frac{\omega}{\alpha}\right) - \arctan\left(\frac{\omega}{\beta}\right)\right)$$

$$= -j\pi\,\mathrm{sgn}(\omega) + 2j\arctan\left(\frac{\omega}{\beta}\right). \qquad (2.2.13.8)$$

∎

2.2.14 *Parseval's equality*

Parseval's equality is

$$\int_{-\infty}^{\infty} f(t)g^*(t)\,dt = \frac{1}{2\pi}\int_{-\infty}^{\infty} F(\omega)G^*(\omega)\,d\omega \tag{2.2.14.1}$$

and is valid whenever the integrals make sense. Closely related to Parseval's equality are the two "fundamental identities,"

$$\int_{-\infty}^{\infty} f(x)\,\mathcal{F}[h]|_x\,dx = \int_{-\infty}^{\infty} \mathcal{F}[f]|_y\,h(y)\,dy \tag{2.2.14.2}$$

and

$$\int_{-\infty}^{\infty} F(y)\,\mathcal{F}^{-1}[H]\big|_y\,dy = \int_{-\infty}^{\infty} \mathcal{F}^{-1}[F]\big|_x\,H(x)\,dx. \tag{2.2.14.3}$$

Derivations of these identities are straightforward. Identity (2.2.14.2), for example, follows immediately from

$$\int_{-\infty}^{\infty} f(x)\left(\int_{-\infty}^{\infty} h(y)e^{-jxy}\,dy\right)dx = \int_{-\infty}^{\infty}\int_{-\infty}^{\infty} f(x)h(y)e^{-jxy}\,dy\,dx$$

$$= \int_{-\infty}^{\infty}\left(\int_{-\infty}^{\infty} f(x)e^{-jxy}\,dx\right)h(y)\,dy.$$

Parseval's equality can then, in turn, be derived from identity (2.2.14.2) and the observation that

$$g^*(t) = \left(\mathcal{F}^{-1}[G]\big|_t\right)^* = \frac{1}{2\pi}\int_{-\infty}^{\infty} G^*(\omega)e^{-j\omega t}\,d\omega = \frac{1}{2\pi}\,\mathcal{F}[G^*]\big|_t\,.$$

2.2.15 *Bessel's equality*

Bessel's equality,

$$\int_{-\infty}^{\infty} |f(t)|^2\,dt = \frac{1}{2\pi}\int_{-\infty}^{\infty} |F(\omega)|^2\,d\omega, \tag{2.2.15.1}$$

is obtained directly from Parseval's equality by letting $g = f$.

Example 2.2.15.1

Let $\alpha > 0$ and $f(t) = p_\alpha(t)$, where $p_\alpha(t)$ is the pulse function. It is easily verified that

$$F(\omega) = \mathcal{F}[p_\alpha(t)]|_\omega = \int_{-\alpha}^{\alpha} e^{-j\omega t}\,dt = \frac{2}{\omega}\sin(\alpha\omega).$$

So, using Bessel's equality,

$$\int_{-\infty}^{\infty}\left|\frac{2}{\omega}\sin(\alpha\omega)\right|^2\,d\omega = 2\pi\int_{-\infty}^{\infty}\left|\frac{1}{2\alpha}p_\alpha(t)\right|^2\,dt$$

$$= \frac{2\pi}{4\alpha^2}\int_{-\alpha}^{\alpha} dt$$

$$= \frac{\pi}{\alpha}.$$

Example 2.2.15.2

Let $\alpha > 0$. In Example 2.2.11.2 it was shown that the Fourier transform of $g(t) = e^{-\alpha t^2}$ is $G(\omega) = A \exp\left[-\frac{1}{4\alpha}\omega^2\right]$. The positive constant A can be determined by noting that, by Bessel's equality,

$$\int_{-\infty}^{\infty} \left|e^{-\alpha t^2}\right|^2 dt = \frac{1}{2\pi} \int_{-\infty}^{\infty} \left|A \exp\left[-\frac{1}{4\alpha}\omega^2\right]\right|^2 d\omega.$$

Letting $\omega = 2\alpha\tau$ this becomes, after a little simplification,

$$\int_{-\infty}^{\infty} e^{-2\alpha t^2} dt = \frac{\alpha}{\pi} A^2 \int_{-\infty}^{\infty} e^{-2\alpha\tau^2} d\tau.$$

Dividing out the integrals and solving for A yields

$$A = \sqrt{\frac{\pi}{\alpha}},$$

where the positive square root is taken because

$$A = G(0) = \int_{-\infty}^{\infty} e^{-\alpha t^2} dt > 0. \qquad \blacksquare$$

2.2.16   The bandwidth theorem

If $f(t)$ is a function whose value may be considered as "negligible" outside of some interval, (t_1, t_2), then the length of that interval, $\Delta t = t_2 - t_1$, is the effective duration of $f(t)$. Likewise, if $F(\omega)$ is the Fourier transform of $f(t)$, and $F(\omega)$ can be considered as "negligible" outside of some interval, (ω_1, ω_2), then $\Delta\omega = \omega_2 - \omega_1$ is the effective bandwidth of $f(t)$.

The essence of the bandwidth theorem is that there is a universal positive constant, γ, such that the effective duration, Δt, and effective bandwidth, $\Delta\omega$, of any function (with finite Δt or finite $\Delta\omega$) satisfies

$$\Delta t \, \Delta\omega \geq \gamma.$$

Thus, it is not possible to find a function whose effective bandwidth and effective duration are both arbitrarily small.

There are, in fact, several versions of the bandwidth theorem, each applicable to a particular class of functions. The two most important versions involve absolutely integrable functions and finite energy functions. They are described in greater detail in Subsections 2.3.3 and 2.3.5, respectively. Also in these subsections are appropriate precise definitions of effective duration and effective bandwidth.

Because it is the basis of the Heisenberg uncertainty principle of quantum mechanics, the bandwidth theorem is often, itself, referred to as the uncertainty principle of Fourier analysis.

2.3   Transforms of specific classes of functions

In many applications one encounters specific classes of functions in which either the functions or their transforms satisfy certain particular properties. Several such classes of functions are discussed below.

2.3.1 *Real/imaginary valued even/odd functions*

Let $F(\omega)$ be the Fourier transform of $f(t)$. Then, assuming $f(t)$ is integrable,

$$F(\omega) = \int_{-\infty}^{\infty} f(t)e^{-j\omega t}\, dt$$

$$= \int_{-\infty}^{\infty} f(t)[\cos(\omega t) - j\sin(\omega t)]\, dt$$

$$= \int_{-\infty}^{\infty} f(t)\cos(\omega t)\, dt - j \int_{-\infty}^{\infty} f(t)\sin(\omega t)\, dt. \qquad (2.3.1.1)$$

If $f(t)$ is an even function, then

$$\int_{-\infty}^{\infty} f(t)\sin(\omega t)\, dt = 0$$

and equation (2.3.1.1) becomes

$$F(\omega) = \int_{-\infty}^{\infty} f(t)\cos(\omega t)\, dt = 2 \int_{0}^{\infty} f(t)\cos(\omega t)\, dt,$$

which is clearly an even function of ω and is real valued whenever f is real valued. Likewise, if $f(t)$ is an odd function, then

$$\int_{-\infty}^{\infty} f(t)\cos(\omega t)\, dt = 0,$$

and equation (3.1.1) reduces to

$$F(\omega) = -j \int_{-\infty}^{\infty} f(t)\sin(\omega t)\, dt = -2j \int_{0}^{\infty} f(t)\sin(\omega t)\, dt,$$

which is clearly an odd function of ω and is imaginary valued as long as f is real valued.

These and related relations are summarized in Table 2.1.

On occasion it is convenient to decompose a function, $f(t)$, into its even and odd components, $f_e(t)$ and $f_o(t)$,

$$f(t) = f_e(t) + f_o(t),$$

where

$$f_e(t) = \frac{1}{2}[f(t) + f(-t)] \qquad \text{and} \qquad f_o(t) = \frac{1}{2}[f(t) - f(-t)].$$

If $f(t)$ is a real-valued function with Fourier transform

$$F(\omega) = R(\omega) + jI(\omega),$$

TABLE 2.1
$(F = \mathcal{F}[f])$.

$f(t)$ is even	$\Leftrightarrow$	$F(\omega)$ is even
$f(t)$ is real and even	$\Leftrightarrow$	$F(\omega)$ is real and even
$f(t)$ is imaginary and even	$\Leftrightarrow$	$F(\omega)$ is imaginary and even
$f(t)$ is odd	$\Leftrightarrow$	$F(\omega)$ is odd
$f(t)$ is real and odd	$\Leftrightarrow$	$F(\omega)$ is imaginary and odd
$f(t)$ is imaginary and odd	$\Leftrightarrow$	$F(\omega)$ is real and odd

where $R(\omega)$ and $I(\omega)$ denote, respectively, the real and imaginary parts of $F(\omega)$, then, by the above discussion it follows that

$$F_e(\omega) = R(\omega) = \mathcal{F}[f_e(t)]|_\omega, \tag{2.3.1.2}$$

$$F_o(\omega) = jI(\omega) = \mathcal{F}[f_o(t)]|_\omega, \tag{2.3.1.3}$$

$$f_e(t) = \mathcal{F}^{-1}[F_e(\omega)]|_t = \frac{1}{\pi} \int_0^\infty R(\omega) \cos(\omega t)\, d\omega, \tag{2.3.1.4}$$

and

$$f_o(t) = \mathcal{F}^{-1}[F_o(\omega)]|_t = -\frac{1}{\pi} \int_0^\infty I(\omega) \sin(\omega t)\, d\omega. \tag{2.3.1.5}$$

Rewriting $F(\omega)$ in terms of its amplitude, $A(\omega) = |F(\omega)|$, and phase, $\phi(\omega)$,

$$F(\omega) = A(\omega)e^{j\phi(\omega)},$$

it is easily seen that

$$R(\omega) \cos(\omega t) - I(\omega) \sin(\omega t) = A(\omega)[\cos\phi(\omega) \cos(\omega t) - \sin\phi(\omega) \sin(\omega t)]$$

$$= A(\omega) \cos(\omega t + \phi(\omega)).$$

Thus, by equations (2.3.1.4) and (2.3.1.5), if $f(t)$ is real, then

$$f(t) = f_e(t) + f_o(t) = \frac{1}{\pi} \int_0^\infty A(\omega) \cos(\omega t + \phi(\omega))\, d\omega. \tag{2.3.1.6}$$

2.3.2   Absolutely integrable functions

If $f(t)$ is absolutely integrable (i.e., $\int_{-\infty}^\infty |f(t)|\, dt < \infty$) then the integral defining $F(\omega)$,

$$F(\omega) = \mathcal{F}[f(t)]|_\omega = \int_{-\infty}^\infty f(t)e^{-j\omega t}\, dt$$

is well defined and well behaved. As a consequence, $F(\omega)$ is well defined for every ω and is a reasonably well behaved function on $(-\infty, \infty)$. One immediate observation is that for such functions,

$$F(0) = \int_{-\infty}^\infty f(t)\, dt.$$

It is also worth noting that for any ω,

$$|F(\omega)| = \left| \int_{-\infty}^\infty f(t)e^{-j\omega t}\, dt \right| \le \int_{-\infty}^\infty \left| f(t)e^{-j\omega t} \right|\, dt = \int_{-\infty}^\infty |f(t)|\, dt.$$

The following can also be shown:

1. $F(\omega)$ is a continuous function of ω and for each $-\infty < \omega_0 < \infty$,

$$\lim_{\omega \to \omega_0} F(\omega) = \int_{-\infty}^\infty f(t)e^{-j\omega_0 t}\, dt$$

2. (The Riemann-Lebesgue lemma)

$$\lim_{\omega \to \pm\infty} F(\omega) = 0.$$

As shown in the next example, care must be exercised not to assume these facts when $f(t)$ is not absolutely integrable.

Example 2.3.2.1
Consider the transform, $F(\omega)$, of $f(t) = \text{sinc}(t) = \frac{1}{t}\sin(t)$. The function $f(t)$ is not absolutely integrable. Because

$$\mathcal{F}^{-1}\left[\pi p_1(t)\right]\big|_\omega = \frac{1}{2\pi}\int_{-\infty}^{\infty} \pi p_1(t)e^{j\omega t}\, dt = \text{sinc}(\omega)$$

it follows that

$$F(\omega) = \mathcal{F}[\text{sinc}(t)]\big|_\omega = \pi p_1(\omega).$$

Clearly

$$\lim_{\omega \to 1^+} F(\omega) = 0 \qquad \text{and} \qquad \lim_{\omega \to 1^-} F(\omega) = \pi,$$

while, using the residue theorem, it is easily shown that

$$F(1) = \mathcal{F}[\text{sinc}(t)]\big|_{\omega=1} = \int_{-\infty}^{\infty} \text{sinc}(t)e^{jt}\, dt = \frac{\pi}{2}.$$

Thus, $F(\omega)$ is not continuous.   ∎

Analogous results hold when taking inverse transforms of absolutely integrable functions. If $F(\omega)$ is absolutely integrable and $f = \mathcal{F}^{-1}[F]$, then

$$f(0) = \frac{1}{2\pi}\int_{-\infty}^{\infty} F(\omega)\, d\omega,$$

and, for all real t,

$$|f(t)| \le \frac{1}{2\pi}\int_{-\infty}^{\infty} |F(\omega)|\, d\omega.$$

Furthermore,

1′. $f(t)$ is a continuous function of t and for each $-\infty < t_0 < \infty$,

$$\lim_{t \to t_0} f(t) = \frac{1}{2\pi}\int_{-\infty}^{\infty} F(\omega)e^{j\omega t_0}\, d\omega.$$

2′. (The Riemann-Lebesgue lemma)

$$\lim_{t \to \pm\infty} f(t) = 0.$$

2.3.3   *The bandwidth theorem for absolutely integrable functions*

Assume that both $f(t)$ and its Fourier transform, $F(\omega)$, are absolutely integrable. Let $\bar{t}$ and $\bar{\omega}$ be any two fixed values for t and ω such that $f(\bar{t}) \neq 0$ and and $F(\bar{\omega}) \neq 0$. The corresponding effective duration, Δt, and the corresponding effective bandwidth, $\Delta\omega$, are the values satisfying

$$\int_{-\infty}^{\infty} |f(t)|\, dt = |f(\bar{t})|\Delta t$$

and

$$\int_{-\infty}^{\infty} |F(\omega)|\, d\omega = |F(\bar{\omega})|\Delta\omega.$$

The bandwidth theorem for absolutely integrable functions states that

$$\Delta t\, \Delta\omega \geq 2\pi.$$

Moreover, using $\bar{t} = \bar{\omega} = 0$,

$$\Delta t\, \Delta\omega = 2\pi$$

whenever $f(t)$ and $F(\omega)$ are both real nonnegative functions (or real nonpositive functions) and neither $f(0)$ nor $F(0)$ vanishes.

The choice of the values for $\bar{t}$ and $\bar{\omega}$ depends on the use to be made of the bandwidth theorem. One standard choice for $\bar{t}$ and $\bar{\omega}$ is as the centroids of $|f(t)|$ and $|F(\omega)|$,

$$\bar{t} = \frac{\int_{-\infty}^{\infty} t\,|f(t)|\, dt}{\int_{-\infty}^{\infty} |f(t)|\, dt} \qquad \text{and} \qquad \bar{\omega} = \frac{\int_{-\infty}^{\infty} \omega\,|F(\omega)|\, d\omega}{\int_{-\infty}^{\infty} |F(\omega)|\, d\omega}.$$

Alternatively, to minimize the values used for the effective duration and effective bandwidth, $\bar{t}$ and $\bar{\omega}$ can be chosen to maximize the values of $|f(\bar{t})|$ and $|F(\bar{\omega})|$. Clearly, choosing $\bar{t} = 0$ and $\bar{\omega} = 0$ is especially appropriate if both $f(t)$ and $F(\omega)$ are real valued even functions with maximums at the origin.

The above version of the bandwidth theorem is very easily derived. Because $f(t)$ and $F(\omega)$ are both absolutely integrable,

$$|F(\bar{\omega})| \leq \int_{-\infty}^{\infty} |f(t)|\, dt = |f(\bar{t})|\Delta t$$

and

$$|f(\bar{t})| \leq \frac{1}{2\pi} \int_{-\infty}^{\infty} |F(\omega)|\, d\omega = \frac{1}{2\pi}|F(\bar{\omega})|\Delta\omega.$$

Thus,

$$\Delta t\, \Delta\omega \geq \frac{|F(\bar{\omega})|}{|f(\bar{t})|} \times \frac{2\pi |f(\bar{t})|}{|F(\bar{\omega})|} = 2\pi.$$

Clearly, if both $f(t)$ and $F(\omega)$ are real and nonnegative and neither $f(0)$ or $F(0)$ vanish, then the above inequalities can be replaced with

$$F(0) = \int_{-\infty}^{\infty} f(t)\, dt = f(0)\Delta t,$$

$$f(0) = \frac{1}{2\pi} \int_{-\infty}^{\infty} F(\omega)\, d\omega = F(0)\Delta\omega,$$

and

$$\Delta t\, \Delta\omega = \frac{F(0)}{f(0)} \times \frac{2\pi f(0)}{F(0)} = 2\pi.$$

Example 2.3.3.1

Let $\alpha > 0$ and $f(t) = e^{-\alpha|t|}$. The transform of $f(t)$ is

$$F(\omega) = \frac{2\alpha}{\alpha^2 + \omega^2}.$$

Observe that both $f(t)$ and $F(\omega)$ are even functions with maximums at the origin. It is therefore appropriate to use $\bar{t} = 0$ and $\bar{\omega} = 0$ to compute the effective duration and effective bandwidth,

$$\Delta t = \frac{1}{|f(0)|} \int_{-\infty}^{\infty} |f(t)| \, dt = \int_{-\infty}^{\infty} e^{-\alpha|t|} \, dt = 2 \int_{0}^{\infty} e^{-\alpha t} \, dt = \frac{2}{\alpha}$$

and

$$\Delta\omega = \frac{1}{|F(0)|} \int_{-\infty}^{\infty} |F(\omega)| \, d\omega = \frac{\alpha}{2} \int_{-\infty}^{\infty} \frac{2\alpha}{\alpha^2 + \omega^2} \, d\omega = \alpha\pi.$$

The products of these measures of effective bandwidth and duration are

$$\Delta t \, \Delta\omega = \left(\frac{2}{\alpha}\right)(\alpha\pi) = 2\pi,$$

as predicted by the bandwidth theorem. ∎

2.3.4   *Square integrable ("finite energy") functions*

A function, $f(t)$, is square integrable if

$$\int_{-\infty}^{\infty} |f(t)|^2 \, dt < \infty.$$

For many applications, it is natural to define the energy, E, in a function (or signal), $f(t)$, by

$$E = E[f] = \int_{-\infty}^{\infty} |f(t)|^2 \, dt.$$

For this reason square integrable functions are also called finite energy functions. By Bessel's equality,

$$E[f] = \int_{-\infty}^{\infty} |f(t)|^2 \, dt = \frac{1}{2\pi} \int_{-\infty}^{\infty} |F(\omega)|^2 \, d\omega, \qquad (2.3.4.1)$$

where $F(\omega)$ is the Fourier transform of $f(t)$. This shows that a function is square integrable if and only if its transform is also square integrable. It also indicates why $|F(\omega)|^2$ is often referred to as either the "energy spectrum" or the "energy spectral density" of $f(t)$.

2.3.5   *The bandwidth theorem for finite energy functions*

Assume that $f(t)$ and its Fourier transform, $F(\omega)$, are finite energy functions, and let the effective duration, Δt, and the effective bandwidth, $\Delta\omega$, be given by the "standard deviations,"

$$(\Delta t)^2 = \frac{\int_{-\infty}^{\infty} (t - \bar{t})^2 |f(t)|^2 \, dt}{\int_{-\infty}^{\infty} |f(t)|^2 \, dt}$$

and

$$(\Delta\omega)^2 = \frac{\int_{-\infty}^{\infty} (\omega - \bar{\omega})^2 |F(\omega)|^2 \, d\omega}{\int_{-\infty}^{\infty} |F(\omega)|^2 \, d\omega},$$

where $\bar{t}$ and $\bar{\omega}$ are the mean values of t and ω,

$$\bar{t} = \frac{\int_{-\infty}^{\infty} t |f(t)|^2 \, dt}{\int_{-\infty}^{\infty} |f(t)|^2 \, dt} \qquad \text{and} \qquad \bar{\omega} = \frac{\int_{-\infty}^{\infty} \omega |F(\omega)|^2 \, d\omega}{\int_{-\infty}^{\infty} |F(\omega)|^2 \, d\omega}.$$

Using the energy of $f(t)$,

$$E = \int_{-\infty}^{\infty} |f(t)|^2 \, dt = \frac{1}{2\pi} \int_{-\infty}^{\infty} |F(\omega)|^2 \, d\omega,$$

the effective duration and effective bandwidth can be written more consisely as

$$\Delta t = \sqrt{\frac{1}{E} \int_{-\infty}^{\infty} (t - \bar{t})^2 |f(t)|^2 \, dt}$$

and

$$\Delta \omega = \sqrt{\frac{1}{2\pi E} \int_{-\infty}^{\infty} (\omega - \bar{\omega})^2 |F(\omega)|^2 \, d\omega}.$$

The bandwidth theorem for finite energy functions states that, if the above quantities are well defined (and finite) and

$$\lim_{t \to \pm\infty} t |f(t)|^2 = 0,$$

then

$$\Delta t \, \Delta \omega \geq \frac{1}{2}.$$

Moreover,

$$\Delta t \, \Delta \omega = \frac{1}{2}$$

if and only if $f(t)$ is a Gaussian,

$$f(t) = A e^{-\alpha t^2},$$

for some $\alpha > 0$.

The reader should be aware that the effective duration and effective bandwidth defined in this subsection are not the same as the effective duration and effective bandwidth previously defined in Subsection 2.3.3. Nor do these definitions necessarily agree with the definitions given for the analogous quantities defined later in the subsections on reconstructing sampled functions.

Example 2.3.5.1

Let $\alpha > 0$ and $f(t) = e^{-\alpha|t|}$. The transform of $f(t)$ is

$$F(\omega) = \frac{2\alpha}{\alpha^2 + \omega^2}.$$

Because $tf(t)$ and $\omega F(\omega)$ are both odd functions, it is clear that $\bar{t} = 0$ and $\bar{\omega} = 0$. The energy is

$$E = \int_{-\infty}^{\infty} |e^{-\alpha|t|}|^2 \, dt = 2 \int_{0}^{\infty} e^{-2\alpha t} \, dt = \frac{1}{\alpha}.$$

Using integration by parts, the corresponding effective duration and effective bandwidth are easily computed,

$$\Delta t = \sqrt{\frac{1}{E} \int_{-\infty}^{\infty} (t - \bar{t})^2 |f(t)|^2 \, dt}$$

$$= \sqrt{2\alpha \int_{0}^{\infty} t^2 e^{-2\alpha t} \, dt}$$

$$= \frac{\sqrt{2}}{2\alpha}$$

and

$$\Delta\omega = \sqrt{\frac{1}{2\pi E} \int_{-\infty}^{\infty} (\omega - \bar{\omega})^2 |F(\omega)|^2 \, d\omega}$$

$$= \sqrt{\frac{\alpha}{2\pi} \int_{-\infty}^{\infty} \omega^2 \left(\frac{2\alpha}{\alpha^2 + \omega^2} \right)^2 \, d\omega}$$

$$= \sqrt{\frac{\alpha^3}{\pi} \int_{-\infty}^{\infty} \omega \frac{2\omega}{(\alpha^2 + \omega^2)^2} \, d\omega}$$

$$= \alpha.$$

(By comparison, treating $f(t)$ and $F(\omega)$ as absolutely integrable functions [Example 2.3.3.1] led to an effective duration of $\frac{2}{\alpha}$ and an effective bandwidth of $\alpha\pi$.)

The products of these measures of bandwidth and duration computed here are

$$\Delta t \Delta\omega = \frac{\sqrt{2}}{2\alpha}\alpha = \frac{\sqrt{2}}{2} > \frac{1}{2},$$

as predicted by the bandwidth theorem for finite energy functions. ▌

2.3.6   *Functions with finite duration*

A function, $f(t)$, has finite duration (with duration $2T$) if there is a $0 < T < \infty$ such that

$$f(t) = 0 \qquad \text{whenever } T < |t|.$$

The transform, $F(\omega)$, of such a function is given by a proper integral over a finite interval,

$$F(\omega) = \int_{-T}^{T} f(t)e^{-j\omega t} \, dt. \tag{2.3.6.1}$$

Any piecewise continuous function with finite duration is automatically absolutely integrable and automatically has finite energy, and, so, the discussions in the previous subsections apply to such functions. In addition, if $f(t)$ is a piecewise continuous function of finite duration (with duration $2T$), then, for every nonnegative integer, n, $t^n f(t)$ is also a piecewise continuous finite duration function with duration $2T$, and using identity (2.2.11.8),

$$F^{(n)}(\omega) = \mathcal{F}[(-jt)^n f(t)]\big|_{\omega} = \int_{-T}^{T} (-jt)^n f(t)e^{-j\omega t} \, dt.$$

From the discussion in Subsection 2.3.2, it is apparent that the transform of a piecewise continuous function with finite duration must be classically differentiable up to any order, and that every derivative is continuous.

It should be noted that the integral defining $F(\omega)$ in formula (2.3.6.1) is, in fact, well defined for every complex $\omega = x + jy$. It is not difficult to show that the real and imaginary parts of $F(x + jy)$ satisfy the Cauchy–Riemann equations of complex analysis (see Chapter 1). Thus, $F(\omega)$ is an analytic function on both the real line and the complex plane. As a consequence, it follows that the transform of a finite duration function cannot vanish (or be any constant value) over any nontrivial subinterval of the real line. In particular, no function of finite duration can also be band limited (see Subsection 2.3.7).

Another important feature of finite duration functions is that their transforms can be reconstructed using a discrete sampling of the transforms. This is discussed more fully in Section 2.5.

2.3.7 *Band–limited functions*

Let $f(t)$ be a function with Fourier transform $F(\omega)$. The function, $f(t)$, is said to be band limited if there is a $0 < \Omega < \infty$ such that

$$F(\omega) = 0 \qquad \text{whenever } \Omega < |\omega|.$$

The quantity 2Ω is called the bandwidth of $f(t)$.

By the near equivalence of the Fourier and inverse Fourier transforms, it should be clear that $f(t)$ satisfies properties analogous to those satisfied by the transforms of finite duration functions. In particular

$$f(t) = \frac{1}{2\pi} \int_{-\Omega}^{\Omega} F(\omega) e^{j\omega t} \, d\omega, \tag{2.3.7.1}$$

and, for any nonnegative integer, n, $f^{(n)}(t)$ is a well-defined continuous function given by

$$f^{(n)}(t) = \frac{1}{2\pi} \int_{-\Omega}^{\Omega} (j\omega)^n F(\omega) e^{j\omega t} \, d\omega.$$

Letting $t = x + jy$ in equation (2.3.7.1), it is easily verified that $f(x + jy)$ is a well-defined analytic function on both the real line and on the entire complex plane. From this it follows that if $f(t)$ is band limited, then $f(t)$ cannot vanish (or be any constant value) over any nontrivial subinterval of the real line. Thus, no band–limited function can also be of finite duration. This fact must be considered in many practical applications where it would be desirable (but, as just noted, impossible) to assume that the functions of interest are both band–limited and of finite duration.

Another most important feature of band–limited functions is that they can be reconstructed using a discrete sampling of their values. This is discussed more thoroughly in Section 2.5.

2.3.8 *Finite power functions*

For a given function, $f(t)$, the average autocorrelation function, $\bar{\rho}_f(t)$, is defined by

$$\bar{\rho}_f(t) = \lim_{T \to \infty} \frac{1}{2T} \int_{-T}^{T} f^*(s) f(t + s) \, ds, \tag{2.3.8.1}$$

or, equivalently, by

$$\bar{\rho}_f(t) = \lim_{T \to \infty} \frac{1}{2T} f_T(t) \star f_T(t) \tag{2.3.8.2}$$

where the $\star$ denotes correlation (see Subsection 2.2.10), and $f_T(t)$ is the truncation of $f(t)$ at $t = \pm T$,

$$f_T(t) = f(t)p_T(t) = \begin{cases} f(t), & \text{if } -T \leq t \leq T \\ 0, & \text{otherwise} \end{cases}. \tag{2.3.8.3}$$

If $\bar{\rho}_f(t)$ is a well-defined function (or generalized function), then $f(t)$ is called a finite power function.

The power spectrum or power spectral density, $P(\omega)$, of a finite power function, $f(t)$, is defined to be the Fourier transform of its average autocorrelation,

$$P(\omega) = \mathcal{F}[\bar{\rho}_f(t)]\big|_\omega = \int_{-\infty}^{\infty} \bar{\rho}_f(t)e^{-j\omega t}\, dt. \tag{2.3.8.4}$$

Using formula (2.3.8.2) for $\bar{\rho}_f(t)$ and recalling the Wiener–Khintchine theorem (Subsection 2.2.10),

$$\mathcal{F}[\bar{\rho}_f(t)]\big|_\omega = \lim_{T \to \infty} \frac{1}{2T}\mathcal{F}[f_T(t) \star f_T(t)]\big|_\omega = \lim_{T \to \infty} \frac{1}{2T}|F_T(\omega)|^2$$

where $F_T(\omega)$ is the Fourier transform of $f_T(t)$,

$$F_T(\omega) = \int_{-\infty}^{\infty} f(t)p_T(t)e^{-j\omega t}\, dt = \int_{-T}^{T} f(t)e^{-j\omega t}\, dt.$$

Thus, an alternate formula for the power spectrum is

$$P(\omega) = \lim_{T \to \infty} \frac{1}{2T}\left| \int_{-T}^{T} f(t)e^{-j\omega t}\, dt \right|^2. \tag{2.3.8.5}$$

The average power in $f(t)$ is defined to be

$$\bar{\rho}_f(0) = \lim_{T \to \infty} \frac{1}{2T}\int_{-T}^{T} |f(s)|^2\, ds. \tag{2.3.8.6}$$

Because $P(\omega) = \mathcal{F}[\bar{\rho}_f(t)]\big|_\omega$, this is equivalent to

$$\bar{\rho}_f(0) = \mathcal{F}^{-1}[P(\omega)]\big|_0 = \frac{1}{2\pi}\int_{-\infty}^{\infty} P(\omega)\, d\omega.$$

A number of properties of the average autocorrelation should be noted. They are:

1. $\bar{\rho}_f(t)$ is invariant under a shift in $f(t)$, that is, if $g(t) = f(t - t_0)$, then $\bar{\rho}_g(t) = \bar{\rho}_f(t)$.

2. $\bar{\rho}_f(t)$ and $|\bar{\rho}_f(t)|$ each has a maximum value at $t = 0$.

3. $(\bar{\rho}_f(t))^* = \bar{\rho}_f(-t)$. Thus, as is often the case, if $f(t)$ is a real-valued function, then $\bar{\rho}_f(t)$ is an even real-valued function.

As a consequence of the second property above, any function, $f(t)$, satisfying

$$\lim_{T \to \infty} \frac{1}{2T}\int_{-T}^{T} |f(s)|^2\, ds < \infty$$

is a finite power function.

The three properties listed above are easily derived. For the first,

$$\bar{\rho}_g(t) = \lim_{T \to \infty} \frac{1}{2T} \int_{-T}^{T} f^*(s - t_0) f(s - t_0 + t) \, ds$$

$$= \lim_{T \to \infty} \frac{1}{2T} \int_{-T-t_0}^{T-t_0} f^*(\sigma) f(\sigma + t) \, d\sigma$$

$$= \lim_{T \to \infty} \frac{1}{2T} \int_{-T}^{T} f^*(\sigma) f(\sigma + t) \, d\sigma$$

$$+ \lim_{T \to \infty} \frac{1}{2T} \int_{T}^{T-t_0} f^*(\sigma) f(\sigma + t) \, d\sigma$$

$$+ \lim_{T \to \infty} \frac{1}{2T} \int_{-T-t_0}^{-T} f^*(\sigma) f(\sigma + t) \, d\sigma.$$

The first limit in the last line above equals $\bar{\rho}_f(t)$ while the other limits, involving integrals over intervals of fixed bounded length, must vanish.

From an application of the Schwarz inequality,

$$\left| \int_{-T}^{T} f^*(s) f(s + t) \, ds \right|^2 \le \int_{-T}^{T} |f^*(s)|^2 \, ds \int_{-T}^{T} |f(s + t)|^2 \, ds,$$

it follows, after taking the limit, that

$$|\bar{\rho}_f(t)|^2 \le |\bar{\rho}_f(0)|^2.$$

Hence, at $t = 0$, $|\bar{\rho}_f(t)|$ has a maximum (as does $\bar{\rho}_f(t)$, because $\bar{\rho}_f(0) = |\bar{\rho}_f(0)|$).

Finally, using the substitution $\sigma = s + t$,

$$\left(\bar{\rho}_f(t)\right)^* = \left(\lim_{T \to \infty} \frac{1}{2T} \int_{-T}^{T} f^*(s) f(t + s) \, ds \right)^*$$

$$= \lim_{T \to \infty} \frac{1}{2T} \int_{-T}^{T} f(s) f^*(t + s) \, ds$$

$$= \lim_{T \to \infty} \frac{1}{2T} \int_{-T}^{T} f(\sigma - t) f^*(\sigma) \, d\sigma$$

$$= \bar{\rho}_f(-t).$$

If $f(t)$ is a finite energy function, then, trivially, it is also a finite power function (with zero average power). Nontrivial examples of finite power functions include periodic functions, nearly periodic functions, constants, and step functions. Finite energy functions also play a significant role in signal-processing problems dealing with noise.

Example 2.3.8.1

Consider the step function,

$$u(t) = \begin{cases} 0, & \text{if } t < 0 \\ 1, & \text{if } 0 < t \end{cases}.$$

For $0 \leq t$,

$$\bar{\rho}_u(t) = \lim_{T \to \infty} \frac{1}{2T} \int_{-T}^{T} u(s)u(s+t)\,ds$$

$$= \lim_{T \to \infty} \frac{1}{2T} \int_{0}^{T} ds$$

$$= \frac{1}{2}.$$

Because the step function is a real function, its average autocorrelation function must be an even function. Thus, for all t,

$$\bar{\rho}_u(t) = \frac{1}{2},$$

showing that the step function is a finite power function. Its average power, $\bar{\rho}_u(0)$, is equal to $1/2$, and its power spectrum is

$$P(\omega) = \mathcal{F}\left[\frac{1}{2}\right]\Big|_{\omega} = \pi \delta(\omega). \qquad \blacksquare$$

Example 2.3.8.2
Consider now the function

$$f(t) = \begin{cases} 0, & \text{if } t \leq 0 \\ \sin t, & \text{if } 0 \leq t \end{cases}.$$

For $0 \leq t$,

$$\bar{\rho}_f(t) = \lim_{T \to \infty} \frac{1}{2T} \int_{-T}^{T} f(s)f(s+t)\,ds$$

$$= \lim_{T \to \infty} \frac{1}{2T} \int_{0}^{T} \sin(s)\sin(s+t)\,ds$$

$$= \lim_{T \to \infty} \frac{1}{2T} \int_{0}^{T} \sin(s)[\sin(s)\cos(t) + \cos(s)\sin(t)]\,ds$$

$$= \lim_{T \to \infty} \frac{1}{2T} \left[\cos(t) \int_{0}^{T} \sin^2(s)\,ds + \sin(t) \int_{0}^{T} \sin(s)\cos(s)\,ds \right]$$

$$= \lim_{T \to \infty} \frac{1}{2T} \left[\cos(t) \left(\frac{T}{2} - \frac{\sin(2T)}{4} \right) - \sin(t) \frac{\sin^2(T)}{2} \right]$$

$$= \frac{1}{4}\cos(t).$$

Because $\bar{\rho}_f(t)$ is even,

$$\bar{\rho}_f(t) = \frac{1}{4}\cos(t)$$

for all t. The average power is

$$\bar{\rho}_f(0) = \frac{1}{4},$$

and the power spectrum is

$$P(\omega) = \mathcal{F}\left[\frac{1}{4}\cos(t)\right]\bigg|_{\omega} = \frac{\pi}{4}[\delta(\omega - 1) + \delta(\omega + 1)].$$ ∎

2.3.9   Periodic functions

Let $0 < p < \infty$. A function, $f(t)$, is periodic (with period p) if

$$f(t + p) = f(t)$$

for every real value of t. The Fourier series, $FS[f]$, for such a function is given by

$$FS[f]|_t = \sum_{n=-\infty}^{\infty} c_n e^{jn\Delta\omega t},$$ (2.3.9.1)

where

$$\Delta\omega = \frac{2\pi}{p}$$

and, for each n,

$$c_n = \frac{1}{p} \int_{\text{period}} f(t) e^{-jn\Delta\omega t} \, dt.$$ (2.3.9.2)

(Because of the periodicity of the integrand, the integral in formula (2.3.9.2) can be evaluated over any interval of length p.)

As long as $f(t)$ is at least piecewise smooth, its Fourier series will converge, and at every value of t at which $f(t)$ is continuous,

$$f(t) = \sum_{n=-\infty}^{\infty} c_n e^{jn\Delta\omega t}.$$

At points where $f(t)$ has a "jump" discontinuity, the Fourier series converges to the midpoint of the jump. In any immediate neighborhood of a jump discontinuity any finite partial sum of the Fourier series,

$$\sum_{n=-N}^{N} c_n e^{jn\Delta\omega t},$$

will oscillate wildly and will, at points, significantly over– and undershoot the actual value of $f(t)$ ("Ringing" or Gibbs phenomena).

Because periodic functions are not at all integrable over the entire real line, the standard integral formula, formula (2.1.1.1), cannot be used to find the Fourier transform of $f(t)$. Using the generalized theory, however, it can be shown that as generalized functions,

$$f(t) = \sum_{n=-\infty}^{\infty} c_n e^{jn\Delta\omega t}$$ (2.3.9.3)

and that the Fourier transform of $f(t)$ is given by

$$F(\omega) = \mathcal{F}\left[\sum_{n=-\infty}^{\infty} c_n e^{jn\Delta\omega t}\right]\bigg|_{\omega}$$

$$= \sum_{n=-\infty}^{\infty} c_n \mathcal{F}\left[e^{jn\Delta\omega t}\right]\bigg|_{\omega}$$

$$= \sum_{n=-\infty}^{\infty} c_n 2\pi \delta(\omega - n\Delta\omega).$$

It should be noted that $F(\omega)$ is a regular array of delta functions with spacing inversely proportional to the period of $f(t)$ (see Subsection 2.3.10).

If $f(t)$ is periodic (with period p), then $f(t)$ is a finite power function (but is not, unless $f(t)$ is the zero function, a finite energy function). The average autocorrelation, $\bar{\rho}_f(t)$, will also be periodic and have period p. Formula (2.3.8.1) reduces to

$$\bar{\rho}_f(t) = \frac{1}{p}\int_{\text{period}} f^*(s)f(s+t)\,ds. \tag{2.3.9.4}$$

Because $\bar{\rho}_f(t)$ is periodic, it can also be expanded as a Fourier series,

$$\bar{\rho}_f(t) = \sum_{n=-\infty}^{\infty} a_n e^{jn\Delta\omega t}, \tag{2.3.9.5}$$

and the power spectrum is the regular array of delta functions,

$$P(\omega) = \sum_{n=-\infty}^{\infty} a_n 2\pi \delta(\omega - n\Delta\omega).$$

A useful relation between the Fourier coefficients of $\bar{\rho}_f(t)$,

$$a_n = \frac{1}{p}\int_{\text{period}} \bar{\rho}_f(t)e^{-jn\Delta\omega t}\,dt, \tag{2.3.9.6}$$

and the Fourier coefficients of $f(t)$,

$$c_n = \frac{1}{p}\int_{\text{period}} f(t)e^{-jn\Delta\omega t}\,dt, \tag{2.3.9.7}$$

is easily derived. Inserting formula (2.3.9.4) for $\bar{\rho}_f(t)$ into formula (2.3.9.6), rearranging, and using the substitution $\tau = s + t$,

$$a_n = \frac{1}{p}\int_{\text{period}}\left[\frac{1}{p}\int_{\text{period}} f^*(s)f(s+t)\,ds\right]e^{-jn\Delta\omega t}\,dt$$

$$= \frac{1}{p}\int_{\text{period}}\frac{1}{p}f^*(s)\left[\int_{\text{period}} f(s+t)e^{-jn\Delta\omega t}\,dt\right]ds$$

$$= \frac{1}{p}\int_{\text{period}}\frac{1}{p}f^*(s)\left[\int_{\text{period}} f(\tau)e^{-jn\Delta\omega(\tau-s)}\,d\tau\right]ds$$

$$= \left[\frac{1}{p}\int_{\text{period}} f^*(s)e^{jn\Delta\omega s}\,ds\right]\left[\frac{1}{p}\int_{\text{period}} f(\tau)e^{-jn\Delta\omega \tau}\,d\tau\right]$$

$$= c_n^* c_n.$$

Thus, $a_n = |c_n|^2$.

In summary, if $f(t)$ is periodic with period p, then so is its average autocorrelation function, $\bar{\rho}_f(t)$. Moreover (as generalized functions)

$$f(t) = \sum_{n=-\infty}^{\infty} c_n e^{jn\Delta\omega t}, \tag{2.3.9.8}$$

$$F(\omega) = 2\pi \sum_{n=-\infty}^{\infty} c_n \delta(\omega - n\Delta\omega), \tag{2.3.9.9}$$

$$\bar{\rho}_f(t) = \sum_{n=-\infty}^{\infty} |c_n|^2 e^{jn\Delta\omega t}, \tag{2.3.9.10}$$

and

$$P(\omega) = 2\pi \sum_{n=-\infty}^{\infty} |c_n|^2 \delta(\omega - n\Delta\omega), \tag{2.3.9.11}$$

where $F(\omega)$ is the Fourier transform of $f(t)$, $P(\omega)$ is the power spectrum of $f(t)$,

$$\Delta\omega = \frac{2\pi}{p}, \tag{2.3.9.12}$$

and, for each n,

$$c_n = \frac{1}{p} \int_{\text{period}} f(t) e^{-jn\Delta\omega t}\, dt. \tag{2.3.9.13}$$

Analogous formulas are valid if $G(\omega)$ is a periodic function with period P. In particular, its inverse transform is

$$g(t) = \sum_{k=-\infty}^{\infty} C_k \delta(t - k\Delta t), \tag{2.3.9.14}$$

where

$$\Delta t = \frac{2\pi}{P}$$

and, for each k,

$$C_k = \frac{1}{P} \int_{\text{period}} G(\omega) e^{jk\Delta t\omega}\, d\omega.$$

Again, because of periodicity, the integral can be evaluated over any interval of length P.

Example 2.3.9.1 *Fourier series and transform of a periodic function*

Consider the "saw" function,

$$\text{saw}(t) = \begin{cases} t, & \text{if } -1 \le t < 1 \\ \text{saw}(t+2), & \text{for all } t \end{cases}.$$

The graph of this saw function is sketched in Figure 2.4. Here, because the period is $p = 2$, formula (2.3.9.12) becomes

$$\Delta\omega = \frac{2\pi}{p} = \pi,$$

and formula (2.3.9.13) becomes

$$c_n = \frac{1}{2} \int_{-1}^{1} t e^{-jn\pi t}\, dt = \begin{cases} 0, & \text{if } n = 0 \\ (-1)^n \frac{j}{n\pi}, & \text{if } n = \pm 1, \pm 2, \pm 3, \ldots \end{cases}.$$

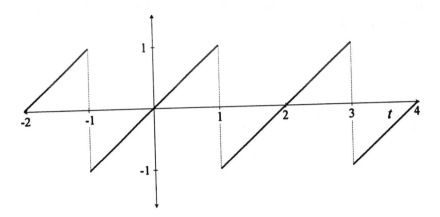

FIGURE 2.4
The saw function.

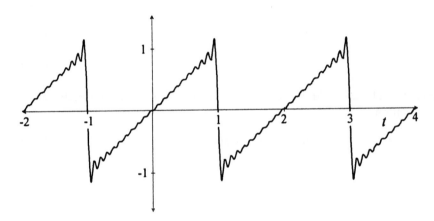

FIGURE 2.5
Partial sum of the saw function's Fourier series.

Using equations (2.3.9.8) and (2.3.9.9),

$$\text{saw}(t) = \sum_{\substack{n=-\infty \\ n \neq 0}}^{\infty} (-1)^n \frac{j}{n\pi} e^{jn\pi t}$$

and

$$\mathcal{F}[\text{saw}(t)]|_\omega = j \sum_{\substack{n=-\infty \\ n \neq 0}}^{\infty} (-1)^n \frac{2}{n} \delta(\omega - n\pi).$$

The graph of the Nth partial sum approximation to saw(t),

$$\sum_{\substack{n=-N \\ n \neq 0}}^{N} (-1)^n \frac{j}{n\pi} e^{jn\pi t},$$

is sketched in Figure 2.5 (with $N = 20$), and the graph of the imaginary part of $\mathcal{F}[\text{saw}(t)]|_\omega$ is

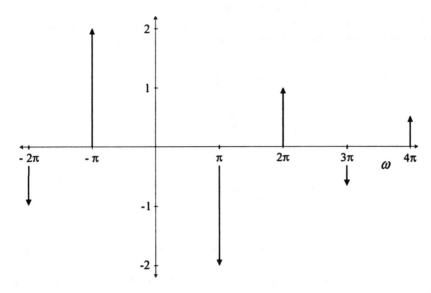

FIGURE 2.6
Fourier transform of the saw function (imaginary part).

sketched in Figure 2.6. The Gibbs phenomenon is evident in Figure 2.5. Formulas (2.3.9.10) and (2.3.9.11) for the autocorrelation function, $\bar{\rho}_{\text{saw}}(t)$, and the power spectrum, $P(\omega)$, yield

$$\bar{\rho}_{\text{saw}}(t) = \frac{1}{\pi^2} \sum_{\substack{n=-\infty \\ n \neq 0}}^{\infty} \frac{1}{n^2} e^{jn\pi t}$$

and

$$P(\omega) = \frac{2}{\pi} \sum_{\substack{n=-\infty \\ n \neq 0}}^{\infty} \frac{1}{n^2} \delta(\omega - n\pi). \qquad \blacksquare$$

2.3.10   *Regular arrays of delta functions*

Let $\Delta x > 0$. The array $\phi(x)$ is called a regular array of delta functions (with spacing Δx) if

$$\phi(x) = \sum_{n=-\infty}^{\infty} \phi_n \delta(x - n\Delta x),$$

where the ϕ_n's denote fixed values. Such arrays arise in sampling and as transforms of periodic functions. They are also useful in describing discrete probability distributions (see Examples 2.3.10.2 and 2.3.10.3 below).

Example 2.3.10.1
The transform of the saw function from Example 2.3.9.1,

$$\mathcal{F}\left[\text{saw}(t)\right]\big|_{\omega} = j \sum_{\substack{n=-\infty \\ n \neq 0}}^{\infty} (-1)^n \frac{2}{n} \delta(\omega - n\pi),$$

is a regular array of delta functions with spacing $\Delta\omega = \pi$. $\blacksquare$

Let $f(t)$ be a function with Fourier transform $F(\omega)$. A straightforward extension and restatement of the results in the previous subsection is that $f(t)$ is periodic if and only if $F(\omega)$ is a regular array of delta functions. The period, p, of $f(t)$, and the spacing, $\Delta\omega$, of $F(\omega)$ are related by

$$p\Delta\omega = 2\pi.$$

Moreover,

$$f(t) = \frac{1}{2\pi} \sum_{n=-\infty}^{\infty} F_n e^{jn\Delta\omega t}$$

and

$$F(\omega) = \sum_{n=-\infty}^{\infty} F_n \delta(\omega - n\Delta\omega),$$

where, for each n,

$$F_n = \frac{2\pi}{p} \int_{\text{period}} f(t) e^{-jn\Delta\omega t}\, dt. \tag{2.3.10.1}$$

Conversely, if $g(t)$ is a function with Fourier transform $G(\omega)$, then $g(t)$ is a regular array of delta functions if and only if $G(\omega)$ is periodic. The spacing of $g(t)$, Δt, and the period of $G(\omega)$, P, are related by

$$P\Delta t = 2\pi.$$

Moreover,

$$g(t) = \sum_{k=-\infty}^{\infty} g_k \delta(t - k\Delta t)$$

and

$$G(\omega) = \sum_{k=-\infty}^{\infty} g_{-k} e^{jk\Delta t\omega},$$

where, for each k,

$$g_k = \frac{1}{P} \int_{\text{period}} G(\omega) e^{jk\Delta t\omega}\, d\omega.$$

Example 2.3.10.2

For any $\lambda > 0$, the corresponding Poisson probability distribution is given by

$$\phi_\lambda(t) = e^{-\lambda} \sum_{n=0}^{\infty} \frac{\lambda^n}{n!} \delta(t - n).$$

Its Fourier transform, $\Psi_\lambda(\omega)$, is given by

$$\Psi_\lambda(\omega) = e^{-\lambda} \sum_{n=0}^{\infty} \frac{\lambda^n}{n!} e^{-jn\omega}.$$

Recalling the Taylor series for the exponential,

$$\Psi_\lambda(\omega) = e^{-\lambda} \sum_{n=0}^{\infty} \frac{1}{n!} \left(\lambda e^{-j\omega}\right)^n$$

$$= e^{-\lambda} e^{\lambda e^{-j\omega}}$$

$$= e^{-\lambda(1 - \cos\omega + j\sin\omega)},$$

which is clearly a periodic function with period $P = 2\pi$. It can also be seen that the amplitude, $A(\omega)$, and the phase, $\Theta(\omega)$, of $\Psi_\lambda(\omega)$ are given by

$$A(\omega) = e^{-\lambda(1-\cos\omega)} \qquad \text{and} \qquad \Theta(\omega) = -\lambda\sin\omega. \qquad\blacksquare$$

Example 2.3.10.3

For any nonnegative integer, n, and $0 \le p \le 1$, the corresponding binomial probability distribution is given by

$$b_{n,p}(t) = \sum_{k=0}^{n} \binom{n}{k} p^k q^{n-k} \delta(t-k)$$

where $q = 1 - p$. The Fourier transform of $b_{n,p}$ is given by

$$B_{n,p}(\omega) = \sum_{k=0}^{n} \binom{n}{k} p^k q^{n-k} e^{-jk\omega} = \sum_{k=0}^{n} \binom{n}{k} (pe^{-j\omega})^k q^{n-k}.$$

By the binomial theorem this can be rewritten as

$$B_{n,p}(\omega) = (pe^{-j\omega} + q)^n,$$

which is clearly periodic with period $P = 2\pi$. $\blacksquare$

A regular array of delta functions,

$$g(t) = \sum_{k=-\infty}^{\infty} g_k \delta(t - k\Delta t),$$

cannot be a finite energy function (unless all the g_k's vanish), but, if the g_k's are bounded, can be treated as a finite power function with average autocorrelation function, $\bar{\rho}_g(t)$, and power spectrum, $P(\omega)$, given by

$$\bar{\rho}_g(t) = \sum_{k=-\infty}^{\infty} A_k \delta(t - k\Delta t)$$

and

$$P(\omega) = \sum_{k=-\infty}^{\infty} A_k e^{-jk\Delta t\omega},$$

where

$$A_k = \lim_{M\to\infty} \frac{1}{2M\Delta t} \sum_{m=-M}^{M} g_m^* g_{m+k}.$$

It should be noted, however, that if

$$\sum_{m=-\infty}^{\infty} |g_m|^2 < \infty,$$

then the A_k's will all be zero.

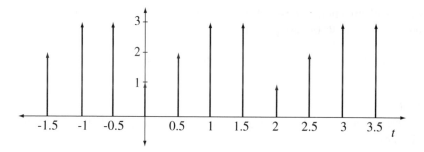

FIGURE 2.7
A regular periodic array of delta functions.

2.3.11 *Periodic arrays of delta functions*

Regular periodic arrays of delta functions are of considerable importance because the formulas for the discrete Fourier transforms can be based directly on formulas derived in computing transforms of regular arrays that are also periodic. For an array with spacing Δx,

$$\phi(x) = \sum_{k=-\infty}^{\infty} \phi_k \delta(x - k\Delta x),$$

to also be periodic with period p,

$$\phi(x + p) = \phi(x),$$

it is necessary that there be a positive integer, N, called the index period, such that

$$\phi_{k+N} = \phi_k \qquad \text{for all } k.$$

The index period, spacing, and period of $\phi(x)$ are related by

period of $\phi(x)$ = (index period of $\phi(x)$) × (spacing of $\phi(x)$).

Example 2.3.11.1
The regular periodic array,

$$f(t) = \sum_{k=-\infty}^{\infty} f_k \delta(t - k\Delta t),$$

with spacing $\Delta t = \frac{1}{2}$, index period $N = 4$, and $(f_0, f_1, f_2, f_3) = (1, 2, 3, 3)$, is sketched in Figure 2.7. Note that $f_4 = f_0$, $f_5 = f_1, \ldots$, and that the period of $f(t)$ is $4\Delta t = 2$. ∎

Let

$$f(t) = \sum_{k=-\infty}^{\infty} f_k \delta(t - k\Delta t)$$

be a regular periodic array with spacing Δt, index period N, and period $p = N\Delta t$. From the discussion in the subsection on regular arrays, it is evident that the Fourier transform of $f(t)$

is also a regular periodic array of delta functions,

$$F(\omega) = \sum_{n=-\infty}^{\infty} F_n \delta(\omega - n\Delta\omega). \tag{2.3.11.1}$$

Also, $f(t)$ can be expressed as a corresponding Fourier series,

$$f(t) = \frac{1}{2\pi} \sum_{n=-\infty}^{\infty} F_n e^{jn\Delta\omega t}. \tag{2.3.11.2}$$

The spacing, $\Delta\omega$, and period, P, of $F(\omega)$ are related to the spacing, Δt, and period, p, of $f(t)$ by

$$\Delta\omega = \frac{2\pi}{p} \qquad \text{and} \qquad P = \frac{2\pi}{\Delta t}.$$

The index period, M, of $F(\omega)$ is given by

$$M = \frac{P}{\Delta\omega} = \frac{(2\pi/\Delta t)}{(2\pi/p)} = \frac{p}{\Delta t} = N.$$

Using equation (2.3.10.1),

$$F_n = \frac{2\pi}{p} \int_{t=-\frac{\Delta t}{2}}^{p-\frac{\Delta t}{2}} \left(\sum_{k=-\infty}^{\infty} f_k \delta(t - k\Delta t) \right) e^{-jn\Delta\omega t} \, dt. \tag{2.3.11.3}$$

But, as is easily verified,

$$\int_{t=-\frac{\Delta t}{2}}^{p-\frac{\Delta t}{2}} \delta(t - k\Delta t) e^{-jn\Delta\omega t} \, dt = \begin{cases} e^{-jnk\Delta\omega\Delta t}, & \text{if } 0 \le k \le N-1 \\ 0, & \text{otherwise} \end{cases},$$

and

$$\Delta\omega\Delta t = \frac{2\pi\Delta t}{p} = \frac{2\pi}{N}.$$

Thus, equation (2.3.11.3) reduces to

$$F_n = \frac{2\pi}{N\Delta t} \sum_{k=0}^{N-1} f_k e^{-j\frac{2\pi}{N}nk}. \tag{2.3.11.4}$$

A similar set of calculations yields the inverse relation,

$$f_k = \frac{1}{N\Delta\omega} \sum_{n=0}^{N-1} F_n e^{j\frac{2\pi}{N}kn}. \tag{2.3.11.5}$$

Formulas for the autocorrelation function, $\bar{\rho}_f(t)$, and the power spectrum, $P(\omega)$, follow immediately from the above and the discussion in Subsections 2.3.9 and 2.3.10. They are

$$\bar{\rho}_f(t) = \sum_{k=-\infty}^{\infty} A_k \delta(t - k\Delta t), \tag{2.3.11.6}$$

where

$$A_k = \frac{1}{N\Delta t} \sum_{m=0}^{N-1} f_m^* f_{m+k}, \tag{2.3.11.7}$$

and

$$P(\omega) = \frac{1}{2\pi} \sum_{n=-\infty}^{\infty} |F_n|^2 \delta(\omega - n\Delta\omega). \tag{2.3.11.8}$$

Example 2.3.11.2   The comb function

For each $\Delta x > 0$, the corresponding comb function is

$$\text{comb}_{\Delta x}(x) = \sum_{k=-\infty}^{\infty} \delta(x - k\Delta x).$$

With index period $N = 1$ and with the spacing equal to the period, the comb function is the simplest possible nonzero regular periodic array. By the above discussion,

$$F(\omega) = \mathcal{F}\left[\text{comb}_{\Delta t}(t)\right]\big|_{\omega}$$

must also be a regular periodic array,

$$F(\omega) = \sum_{n=-\infty}^{\infty} F_n \delta(\omega - n\Delta\omega),$$

where

$$\Delta\omega = \frac{2\pi}{\Delta t}.$$

Because the index period of $F(\omega)$ must also be $N = 1$,

$$F_n = F_0 = \frac{2\pi}{\Delta t} \sum_{k=0}^{0} f_k e^{-j\frac{2\pi}{N} 0 \cdot k} = \Delta\omega,$$

for all n. Combining the last few equations gives

$$\mathcal{F}\left[\text{comb}_{\Delta t}(t)\right]\big|_{\omega} = \sum_{n=-\infty}^{\infty} \Delta\omega \delta(\omega - n\Delta\omega) = \Delta\omega\, \text{comb}_{\Delta\omega}(\omega),$$

where

$$\Delta\omega = \frac{2\pi}{\Delta t}.$$

From formulas (2.3.11.6), (2.3.11.7), and (2.3.11.8), the average correlation function and the power spectrum for $\text{comb}_{\Delta t}(t)$ are given by

$$\bar{\rho}(t) = \frac{1}{\Delta t} \sum_{k=-\infty}^{\infty} \delta(t - k\Delta t) = \frac{1}{\Delta t} \text{comb}_{\Delta t}(t)$$

and

$$P(\omega) = \frac{\Delta\omega}{\Delta t} \sum_{n=-\infty}^{\infty} \delta(\omega - n\Delta\omega) = \frac{\Delta\omega}{\Delta t} \text{comb}_{\Delta\omega}(\omega).$$

In addition, using equation (2.3.11.2), the comb function can be expressed as a Fourier series,

$$\text{comb}_{\Delta t}(t) = \frac{\Delta\omega}{2\pi} \sum_{n=-\infty}^{\infty} e^{jn\Delta\omega t} = \frac{1}{\Delta t} \sum_{n=-\infty}^{\infty} e^{jn\Delta\omega t}. \qquad \blacksquare$$

2.3.12 *Powers of variables and derivatives of delta functions*

In Example 2.1.3.1 it was shown that, for any real value of α,

$$\mathcal{F}\left[e^{j\alpha t}\right]\Big|_\omega = 2\pi\delta(\omega - \alpha).$$

Letting $\alpha = 0$, this gives

$$\mathcal{F}\left[1\right]\Big|_\omega = 2\pi\delta(\omega),$$

and, by symmetry or near equivalence,

$$\mathcal{F}\left[\delta(t)\right]\Big|_\omega = 1.$$

Now, let n be any nonnegative integer. Because, trivially, $x^n = x^n \cdot 1$, it immediately follows from an application of identities 2.2.11.6 through 2.2.11.9 that

$$\mathcal{F}\left[t^n\right]\Big|_\omega = j^n 2\pi\delta^{(n)}(\omega), \tag{2.3.12.1}$$

$$\mathcal{F}^{-1}\left[\omega^n\right]\Big|_t = (-j)^n \delta^{(n)}(t), \tag{2.3.12.2}$$

$$\mathcal{F}\left[\delta^{(n)}(t)\right]\Big|_\omega = (j\omega)^n, \tag{2.3.12.3}$$

and

$$\mathcal{F}^{-1}\left[\delta^{(n)}(\omega)\right]\Big|_t = \frac{(-jt)^n}{2\pi}, \tag{2.3.12.4}$$

where $\delta^{(n)}(x)$ is the nth (generalized) derivative of the delta function.

2.3.13 *Negative powers and step functions*

The basic relation between step functions and negative powers is

$$\mathcal{F}\left[\mathrm{sgn}(t)\right]\Big|_\omega = -j\frac{2}{\omega}, \tag{2.3.13.1}$$

where $\mathrm{sgn}(t)$ is the signum function,

$$\mathrm{sgn}(t) = \begin{cases} -1, & \text{if } t < 0 \\ +1, & \text{if } 0 < t \end{cases}.$$

Because the step function,

$$u(t) = \begin{cases} 0, & \text{if } t < 0 \\ 1, & \text{if } 0 < t \end{cases},$$

can be written in terms of the signum function,

$$u(t) = \frac{1}{2}[\mathrm{sgn}(t) + 1],$$

formula (2.3.13.1) is equivalent to

$$\mathcal{F}\left[u(t)\right]\Big|_\omega = \pi\delta(\omega) - j\frac{1}{\omega}. \tag{2.3.13.2}$$

A number of useful formulas can be easily derived from equations (2.3.13.1) and (2.3.13.2) with the aid of various identities from the identities in Subsection 2.2. Some of these formulas are

$$\mathcal{F}\left[\frac{1}{t}\right]\bigg|_\omega = -j\pi \, \text{sgn}(\omega),\tag{2.3.13.3}$$

$$\mathcal{F}\left[t^{-n}\right]\big|_\omega = -j\pi \frac{(-j\omega)^{n-1}}{(n-1)!}\,\text{sgn}(\omega),\tag{2.3.13.4}$$

$$\mathcal{F}\left[|t|\right]\big|_\omega = -\frac{2}{\omega^2},\tag{2.3.13.5}$$

$$\mathcal{F}\left[t^n \, \text{sgn}(t)\right]\big|_\omega = (-j)^{n+1}\frac{2n!}{\omega^{n+1}},\tag{2.3.13.6}$$

$$\mathcal{F}\left[\text{ramp}(t)\right]\big|_\omega = j\pi\delta'(\omega) - \frac{1}{\omega^2},\tag{2.3.13.7}$$

and

$$\mathcal{F}\left[t^n u(t)\right]\big|_\omega = j^n \pi \delta^{(n)}(\omega) + n!\left(\frac{-j}{\omega}\right)^{n+1}.\tag{2.3.13.8}$$

In these formulas n denotes an arbitrary positive integer.

Derivations of formulas (2.3.13.1) and (2.3.13.2) are easily obtained. One derivation starts with the observation that, for any $\alpha < 0$,

$$u(t) = \int_\alpha^t \delta(s)\,ds.$$

By identity (2.2.13.4), with $f(t) = \delta(t)$ and $F(\omega) = \mathcal{F}[\delta(t)]|_\omega = 1$,

$$\mathcal{F}\left[u(t)\right]\big|_\omega = \mathcal{F}\left[\int_\alpha^t f(s)\,ds\right]\bigg|_\omega$$

$$= -j\frac{F(\omega)}{\omega} + c\delta(\omega)$$

$$= -j\frac{1}{\omega} + c\delta(\omega),\tag{2.3.13.9}$$

where c is some constant. From this,

$$\mathcal{F}\left[\text{sgn}(t)\right]\big|_\omega = \mathcal{F}\left[2u(t) - 1\right]\big|_\omega$$

$$= 2\left[-j\frac{1}{\omega} + c\delta(\omega)\right] - 2\pi\delta(\omega)$$

$$= -j\frac{2}{\omega} + 2(c - \pi)\delta(\omega).\tag{2.3.13.10}$$

Because $\text{sgn}(t)$ is an odd function, so is $\mathcal{F}\left[\text{sgn}(t)\right]|_\omega$ and, hence, so is the right-hand side of equation (2.3.13.10). But, because the delta function is even, this is possible only if $c = \pi$. Plugging this only possible choice for c into equations (2.3.13.9) and (2.3.13.10) gives formulas (2.3.13.1) and (2.3.13.2).

Example 2.3.13.1 Derivation of formulas (2.3.13.6) and (2.3.13.5)
Using identity (2.2.11.8),

$$\mathscr{F}\left[t^n \operatorname{sgn}(t)\right]\Big|_\omega = j^n \frac{d^n}{d\omega^n} \mathscr{F}\left[\operatorname{sgn}(t)\right]\Big|_\omega$$

$$= j^n \frac{d^n}{d\omega^n}\left(-j\frac{2}{\omega}\right)$$

$$= (-j)^{n+1} \frac{2n!}{\omega^{n+1}}.$$

Using this and the observation that

$$|t| = t \operatorname{sgn}(t),$$

it immediately follows that

$$\mathscr{F}\left[|t|\right]\Big|_\omega = \mathscr{F}\left[t \operatorname{sgn}(t)\right]\Big|_\omega = (-j)^{1+1} \frac{2(1!)}{\omega^{1+1}} = -\frac{2}{\omega^2}. \qquad \blacksquare$$

One technical flaw in the above discussion should be noted. If $\phi(x)$ is any function continuous at $x = 0$, and $n \geq 1$, then, from a strict mathematical point of view, the function $x^{-n}\phi(x)$ is not integrable over any interval containing $x = 0$. Because of this, it is not possible to define $\mathscr{F}\left[t^{-n}\right]\Big|_\omega$ or $\mathscr{F}^{-1}\left[\omega^{-n}\right]\Big|_t$ via the classical integral formulas. Neither is it possible for the function x^{-n} to be treated as a generalized function. However, the function $\ln|x|$ is integrable over any finite interval and can be treated as a legitimate generalized function, as can any of its generalized derivatives (as defined in Subsection 2.2.11). It is possible to justify rigorously the formulas given in this subsection, as well as any other standard use of x^{-n}, by agreeing that $\frac{1}{x}$ is actually a symbol for the generalized derivative of $\ln|x|$, and that, more generally, for any positive integer n, x^{-n} is a symbol for

$$\frac{(-1)^{n-1}}{(n-1)!}\frac{d^n}{dx^n}\ln|x|$$

where the derivatives are taken in the generalized sense as described in Subsection 2.2.11.

2.3.14 Rational functions

Rational functions often turn out to be the transforms of functions of interest. The simplest nontrivial rational function is given by

$$F(\omega) = \frac{1}{(\omega - \lambda)^m},$$

where m is a positive integer and λ is some complex constant. Using the elementary identities and the material from the previous subsection, it can be directly verified that

$$\mathscr{F}^{-1}\left[\frac{1}{(\omega-\lambda)^m}\right]\Big|_\omega = j\frac{(jt)^{m-1}}{(m-1)!}e^{j\lambda t}\Gamma_\alpha(t) \qquad (2.3.14.1)$$

where α is the imaginary part of λ and

$$\Gamma_\alpha(t) = \begin{cases} u(t), & \text{if } 0 < \alpha \\ \frac{1}{2}\operatorname{sgn}(t), & \text{if } \alpha = 0 \\ -u(-t), & \text{if } \alpha < 0 \end{cases}.$$

More generally, if $F(\omega)$ is any rational function, then $F(\omega)$ can be written

$$F(\omega) = P(\omega) + R(\omega),$$

where $P(\omega)$ is a polynomial,

$$P(\omega) = \sum_{n=0}^{N} c_n \omega^n,$$

and $R(\omega)$ is the quotient of two polynomials,

$$R(\omega) = \frac{N(\omega)}{D(\omega)},$$

in which the degree of the numerator is strictly less than the degree of the denominator. According to formula (2.3.12.1) the inverse transform of $P(\omega)$ is simply a linear combination of derivatives of delta functions,

$$\mathcal{F}^{-1}\left[P(\omega)\right]\big|_t = \sum_{n=0}^{N} (-j)^n c_n \delta^{(n)}(t).$$

Letting $\lambda_1, \lambda_2, \ldots, \lambda_K$ be the distinct roots of $D(\omega)$ and $M_1, M_2, \ldots, M_K$ the corresponding multiplicities of the roots, $R(\omega)$ can be written in the partial fraction expansion,

$$R(\omega) = \sum_{k=1}^{K} \sum_{m=1}^{M_k} \frac{a_{k,m}}{(\omega - \lambda_k)^m}.$$

Thus, applying formula (2.3.14.1),

$$\mathcal{F}^{-1}\left[R(\omega)\right]\big|_t = j \sum_{k=1}^{K} e^{j\lambda_k t} \Gamma_{\alpha_k}(t) \sum_{m=1}^{M_k} a_{k,m} \frac{(jt)^{m-1}}{(m-1)!}, \tag{2.3.14.2}$$

where, for each k, α_k is the imaginary part of λ_k.

Fourier transforms of rational functions can be computed using the same approach as just described for inverse transforms of rational functions.

Example 2.3.14.1
Let

$$F(\omega) = \frac{N(\omega)}{D(\omega)} = \frac{5\omega + 9 - 10j}{\omega^2 - 4j\omega - 13}.$$

Using the quadratic formula, the roots of $D(\omega)$ are found to be

$$\lambda = \frac{4j \pm \sqrt{(4j)^2 + 4(13)}}{2} = \pm 3 + 2j.$$

$F(\omega)$ can then be expanded

$$F(\omega) = \frac{5\omega + 9 - 10j}{\omega^2 - 4j\omega - 13} = \frac{A}{\omega - (3 + 2j)} + \frac{B}{\omega - (-3 + 2j)}.$$

Solving for A and B gives

$$F(\omega) = \frac{4}{\omega - (3 + 2j)} + \frac{1}{\omega - (-3 + 2j)},$$

whose inverse transform can be computed directly from formula (2.3.14.2),

$$f(t) = j \left[4e^{j(3+2j)t} \Gamma_2(t) + e^{j(-3+2j)t} \Gamma_2(t) \right]$$

$$= 4je^{(-2+3j)t} u(t) + je^{(-2-3j)t} u(t)$$

$$= j \left[4e^{j3t} + e^{-j3t} \right] e^{-2t} u(t). \qquad \blacksquare$$

2.3.15   Causal functions

A function, $f(t)$, is said to be "causal" if

$$f(t) = 0 \qquad \text{whenever } t < 0.$$

Such functions arise in the study of causal systems and are of obvious importance in describing phenomena that have well-defined "starting points."

Let $f(t)$ be a real causal function with Fourier transform $F(\omega)$, and let $R(\omega)$ and $I(\omega)$ be the real and imaginary parts of $F(\omega)$,

$$F(\omega) = R(\omega) + jI(\omega).$$

Then $R(\omega)$ is even, $I(\omega)$ is odd, and, provided the integrals are suitably well defined,

$$f(t) = \frac{2}{\pi} \int_0^\infty R(\omega) \cos(\omega t) \, d\omega \qquad \text{for } 0 < t, \qquad (2.3.15.1)$$

$$f(t) = -\frac{2}{\pi} \int_0^\infty I(\omega) \sin(\omega t) \, d\omega \qquad \text{for } 0 < t, \qquad (2.3.15.2)$$

$$\int_0^\infty |f(t)|^2 \, dt = \frac{1}{\pi} \int_{-\infty}^\infty |R(\omega)|^2 \, d\omega, \qquad (2.3.15.3)$$

and

$$\int_0^\infty |f(t)|^2 \, dt = \frac{1}{\pi} \int_{-\infty}^\infty |I(\omega)|^2 \, d\omega. \qquad (2.3.15.4)$$

In addition, if $f(t)$ is bounded at the origin, then provided the integrals exist,

$$R(\omega) = \frac{1}{\pi} \int_{-\infty}^\infty \frac{I(s)}{\omega - s} \, ds \qquad (2.3.15.5)$$

and

$$I(\omega) = -\frac{1}{\pi} \int_{-\infty}^\infty \frac{R(s)}{\omega - s} \, ds. \qquad (2.3.15.6)$$

The last two integrals are Hilbert transforms and may be defined using Cauchy principle values (see Subsection 2.1.6).

Conversely, it can be shown that if $R(\omega)$ and $I(\omega)$ are real-valued functions (with $R(\omega)$ even and $I(\omega)$ odd) satisfying either (2.3.15.1) or (2.3.15.6), then

$$f(t) = \mathcal{F}^{-1} \left[R(\omega) + jI(\omega) \right]|_t$$

must be a causal function.

Derivations of equations (2.3.15.1) through (2.3.15.6) are quite straightforward. First, observe that because $f(t)$ vanishes for negative values of t, then

$$f(t) = 2f_e(t) = 2f_o(t) \qquad \text{for } 0 < t,$$

where $f_e(t)$ and $f_o(t)$ are the even and odd components of $f(t)$. Equations (2.3.15.1) and (2.3.15.2) then follow immediately from equations (2.3.1.4) and (2.3.1.5), while equations (2.3.15.3) and (2.3.15.4) are simply Bessel's equality combined with equations from Subsection 2.3.1 and the subsequent observation that

$$\int_0^\infty |f(t)|^2 \, dt = 4 \int_0^\infty |f_e(t)|^2 \, dt = 2 \int_\infty^\infty |f_e(t)|^2 \, dt = 2 \int_\infty^\infty |f_o(t)|^2 \, dt.$$

Finally, for equation (2.3.15.5) observe that

$$f_e(t) = f_o(t) \operatorname{sgn}(t) \qquad \text{and} \qquad f_o = f_e(t) \operatorname{sgn}(t).$$

Thus, using results from Subsections 2.3.1, 2.2.9, and 2.3.13,

$$R(\omega) = \mathcal{F}\left[f_e(t)\right]|_\omega$$

$$= \mathcal{F}\left[f_o(t) \operatorname{sgn}(t)\right]|_\omega$$

$$= \frac{1}{2\pi}\mathcal{F}\left[f_o(t)\right]|_\omega * \mathcal{F}\left[\operatorname{sgn}(t)\right]|_\omega$$

$$= \frac{1}{2\pi} jI(\omega) * \left(-j\frac{2}{\omega}\right)$$

$$= \frac{1}{\pi}\int_{-\infty}^\infty \frac{I(s)}{\omega - s} \, ds,$$

which is equation (2.3.15.5). Similar computations yield (2.3.15.6).

Example 2.3.15.1

Assume $f(t)$ is a causal function whose transform, $F(\omega)$, has real part

$$R(\omega) = \delta(\omega - \alpha) + \delta(\omega + \alpha),$$

for some $\alpha > 0$. Then, according to formula (2.3.15.1), for $t > 0$

$$f(t) = \frac{2}{\pi}\int_0^\infty [\delta(\omega - \alpha) + \delta(\omega + \alpha)] \cos(\omega t) \, d\omega = \frac{2}{\pi}\cos(\alpha t),$$

and by formula (2.3.15.6),

$$I(\omega) = -\frac{1}{\pi}\int_{-\infty}^\infty \frac{[\delta(s - \alpha) + \delta(s + \alpha)]}{\omega - s} \, ds$$

$$= -\left[\frac{1}{\pi(\omega - \alpha)} + \frac{1}{\pi(\omega + \alpha)}\right]$$

$$= \frac{2\omega}{\pi(\alpha^2 - \omega^2)}.$$

Thus,

$$f(t) = \frac{2}{\pi}\cos(\alpha t)u(t)$$

and

$$F(\omega) = \delta(\omega - \alpha) + \delta(\omega + \alpha) + j\frac{2\omega}{\pi(\alpha^2 - \omega^2)}. \qquad \blacksquare$$

2.3.16   *Functions on the half-line*

Strictly speaking, functions defined only on the half-line, $0 < t < \infty$, do not have Fourier transforms. Fourier analysis in problems involving such functions can be done by first extending the functions (i.e., systematically defining the values of the functions at negative values of t), and then taking the Fourier transforms of the extensions. The choice of extension will depend on the problem at hand and the preferences of the individual. Three of the most commonly used extensions are the null extension, the even extension, and the odd extension. Given a function, $f(t)$, defined only for $0 < t$, the null extension is

$$f_{\text{null}}(t) = \begin{cases} f(t), & \text{if } 0 < t \\ 0, & \text{if } t < 0 \end{cases},$$

the even extension is

$$f_{\text{even}}(t) = \begin{cases} f(t), & \text{if } 0 < t \\ f(-t), & \text{if } t < 0 \end{cases},$$

and the odd extension is

$$f_{\text{odd}}(t) = \begin{cases} f(t), & \text{if } 0 < t \\ -f(-t), & \text{if } t < 0 \end{cases}.$$

If $f(t)$ is reasonably well behaved (say, continuous and differentiable) on $0 < t$, then any of the above extensions will be similarly well behaved on both $0 < t$ and $t < 0$. At $t = 0$, however, the extended function is likely to have singularities that must be taken into account, especially if transforms of the derivatives are to be taken. It is recommended that the generalized derivative be explicitly used. Assume, for example, that $f(t)$ and its first two derivatives are continuous on $0 < t$, and that the limits

$$f(0) = \lim_{t \to 0^+} f(t) \qquad \text{and} \qquad f'(0) = \lim_{t \to 0^+} f'(t)$$

exist. Let $\hat{f}(t)$ be any of the above extensions of $f(t)$, and, for convenience, let $d\hat{f}/dt$ and $D\hat{f}$ denote, respectively, the classical derivative of $\hat{f}(t)$ and the generalized derivative of $\hat{f}(t)$. Recalling the relation between the classical and generalized derivatives (see Subsection 2.2.11),

$$D\hat{f} = \frac{d\hat{f}}{dt} + J_0 \delta(t)$$

and

$$D^2 \hat{f} = \frac{d^2 \hat{f}}{dt^2} + J_0 \delta'(t) + J_1 \delta(t),$$

where J_0 and J_1 are the "jumps" in $\hat{f}(t)$ and $\hat{f}'(t)$ at $t = 0$,

$$J_0 = \lim_{t \to 0^+} \left[\hat{f}(t) - \hat{f}(-t) \right]$$

and

$$J_1 = \lim_{t \to 0^+} \left[\hat{f}'(t) - \hat{f}'(-t) \right].$$

Computing these jumps for the above extensions yield the following:

$$D f_{\text{null}} = \frac{d f_{\text{null}}}{dt} + f(0)\delta(t), \qquad (2.3.16.1)$$

$$D^2 f_{\text{null}} = \frac{d^2 f_{\text{null}}}{dt^2} + f(0)\delta'(t) + f'(0)\delta(t), \qquad (2.3.16.2)$$

$$D f_{\text{even}} = \frac{d f_{\text{even}}}{dt}, \qquad (2.3.16.3)$$

$$D^2 f_{\text{even}} = \frac{d^2 f_{\text{even}}}{dt^2} + 2 f'(0)\delta(t), \qquad (2.3.16.4)$$

$$D f_{\text{odd}} = \frac{d f_{\text{odd}}}{dt} + 2 f(0)\delta(t), \qquad (2.3.16.5)$$

and

$$D^2 f_{\text{odd}} = \frac{d^2 f_{\text{odd}}}{dt^2} + 2 f(0)\delta'(t). \qquad (2.3.16.6)$$

An example of the use of Fourier transforms in problems on the half-line is given in Subsection 2.8.4. This example also illustrates how the data in the problem determine the appropriate extension for the problem.

2.3.17 *Functions on finite intervals*

If a function, $f(t)$, is defined only on a finite interval, $0 < t < L$, then it can be expanded into any of a number of "Fourier series" over the interval. These series equal $f(t)$ over the interval but are defined on the entire real line. Thus, each series corresponds to a particular extension of $f(t)$ to a function defined for all real values of t, and, with care, Fourier analysis can be done using the series in place of the original functions. Among the best known "Fourier series" for such functions are the sine series and the cosine series.

The sine series for $f(t)$ over $0 < t < L$ is

$$S[f]\big|_t = \sum_{k=1}^{\infty} b_k \sin\left(\frac{k\pi t}{L}\right),$$

where

$$b_k = \frac{2}{L} \int_0^L f(t) \sin\left(\frac{k\pi t}{L}\right) dt.$$

This series can be viewed as an odd periodic extension of $f(t)$. The Fourier transform of the sine series is

$$\mathcal{F}\left[S[f]\big|_t \right]\Big|_\omega = j\pi \sum_{k=1}^{\infty} b_k \left[\delta\left(\omega + \frac{k\pi}{L}\right) - \delta\left(\omega - \frac{k\pi}{L}\right) \right]$$

$$= \sum_{k=-\infty}^{\infty} B_k \delta\left(\omega - \frac{k\pi}{L}\right),$$

where

$$B_k = \begin{cases} -j\pi b_k, & \text{if } 0 < k \\ 0, & \text{if } k = 0 \\ j\pi b_{-k}, & \text{if } k < 0 \end{cases}.$$

The cosine series for $f(t)$ over $0 < t < L$ is

$$C[f]\big|_t = a_0 + \sum_{k=1}^{\infty} a_k \cos\left(\frac{k\pi t}{L}\right),$$

where

$$a_0 = \frac{1}{L} \int_0^L f(t)\, dt$$

and, for $k \neq 0$,

$$a_k = \frac{2}{L} \int_0^L f(t) \cos\left(\frac{k\pi t}{L}\right) dt.$$

This series can be viewed as an even periodic extension of $f(t)$. The Fourier transform of the cosine series is

$$\mathcal{F}\left[C[f]|_t \right]\Big|_\omega = 2\pi a_0 \delta(\omega) + \pi \sum_{k=1}^{\infty} a_k \left[\delta\left(\omega - \frac{k\pi}{L}\right) + \delta\left(\omega + \frac{k\pi}{L}\right) \right]$$

$$= \sum_{k=-\infty}^{\infty} A_k \delta\left(\omega - \frac{k\pi}{L}\right),$$

where

$$A_k = \begin{cases} \pi a_k, & \text{if } 0 < k \\ 2\pi a_0, & \text{if } k = 0 \\ \pi a_{-k}, & \text{if } k < 0 \end{cases}.$$

The choice of which series to use depends strongly on the actual problem at hand. For example, because the sine functions in the sine series expansion vanish at $t = 0$ and $t = L$, sine series expansions tend to be most useful when the functions of interest are to vanish at both of the end points of the interval. For problems in which the first derivatives are expected to vanish at both end points, the cosine series tends to be a better choice. Other boundary conditions suggest other choices for the appropriate Fourier series. In addition, the equations to be satisfied must be considered in choosing the series to be used. Unfortunately, the development of a reasonably complete criteria for choosing the appropriate "Fourier series" in general goes beyond the scope of this chapter. It is recommended that texts covering eigenfunction expansions and Sturm-Liouville problems be consulted.[3]

2.4 Extensions of the Fourier transform and other closely related transforms

A number of applications call for transforms that are closely related to the Fourier transform. This section presents a brief survey and development of some of the transforms having a particularly close relation to the Fourier transform. Many of them, in fact, can be viewed as natural modifications or direct extensions of the transforms defined and developed in the previous sections, or else are special cases of these modifications and extensions.

2.4.1 *Multidimensional Fourier transforms*

The extension of Fourier analysis to handle functions of more than one variable is quite straightforward. Assuming the functions are suitably integrable, the Fourier transform of

---

[3]See, for example, *Elementary Differential Equations and Boundary Value Problems* by Boyce and DiPrima, *Applied Analysis by the Hilbert Space Method* by Holland or *Partial Differential Equations and Boundary-Value Problems with Applications* by Pinsky.

$f(x, y)$ is

$$F(\omega, v) = \int_{-\infty}^{\infty} \int_{-\infty}^{\infty} f(x, y) e^{-j(\omega x + v y)} \, dx \, dy$$

and the Fourier transform of $f(x, y, z)$ is

$$F(\omega, v, \mu) = \int_{-\infty}^{\infty} \int_{-\infty}^{\infty} \int_{-\infty}^{\infty} f(x, y, z) e^{-j(\omega x + v y + \mu z)} \, dx \, dy \, dz.$$

More generally, using vector notation with $\mathbf{t} = (t_1, t_2, \ldots, t_n)$ and $\boldsymbol{\omega} = (\omega_1, \omega_2, \ldots, \omega_n)$, the "$n$-dimensional Fourier transform" is defined by

$$\mathcal{F}\left[f(\mathbf{t})\right]|_{\omega} = \int_{-\infty}^{\infty} \int_{-\infty}^{\infty} \cdots \int_{-\infty}^{\infty} f(\mathbf{t}) e^{-j\boldsymbol{\omega} \cdot \mathbf{t}} \, dt_1 \, dt_2 \cdots dt_n, \qquad (2.4.1.1)$$

assuming $f(\mathbf{t})$ is sufficiently integrable. The inverse n-dimensional Fourier inverse transform is given by

$$\mathcal{F}^{-1}\left[F(\omega)\right]|_{\mathbf{t}} = (2\pi)^{-n} \int_{-\infty}^{\infty} \int_{-\infty}^{\infty} \cdots \int_{-\infty}^{\infty} F(\omega) e^{-j\boldsymbol{\omega} \cdot \mathbf{t}} \, d\omega_1 \, d\omega_2 \cdots d\omega_n, \qquad (2.4.1.2)$$

provided $F(\omega)$ is suitably integrable.

For many functions of $\mathbf{t}$ and $\boldsymbol{\omega}$ that are not suitably integrable, the generalized n-dimensional Fourier and inverse Fourier transforms can be defined using the n-dimensional analogs of the rapidly decreasing test functions described in Subsections 2.1.3 and 2.1.4.

Analogs to the identities discussed in Section 2.2 can be easily derived for the n-dimensional transforms. In particular, $\mathcal{F}$ and $\mathcal{F}^{-1}$ are inverses of each other, that is,

$$F(\omega) = \mathcal{F}\left[f(\mathbf{t})\right]|_{\omega} \Leftrightarrow \mathcal{F}^{-1}\left[F(\omega)\right]|_{\mathbf{t}} = f(\mathbf{t}).$$

The near equivalence (or symmetry) relations for the n-dimensional transforms are

$$\mathcal{F}^{-1}\left[\phi(\mathbf{x})\right]|_{\mathbf{y}} = (2\pi)^{-n} \mathcal{F}\left[\phi(-\mathbf{x})\right]|_{\mathbf{y}} = (2\pi)^{-n} \mathcal{F}\left[\phi(\mathbf{x})\right]|_{-\mathbf{y}}$$

and

$$\mathcal{F}\left[\phi(\mathbf{x})\right]|_{\mathbf{y}} = (2\pi)^{n} \mathcal{F}^{-1}\left[\phi(-\mathbf{x})\right]|_{\mathbf{y}} = (2\pi)^{n} \mathcal{F}^{-1}\left[\phi(\mathbf{x})\right]|_{-\mathbf{y}}.$$

An abbreviated listing of identities for the n-dimensional transforms are given in Table 2.2. In this table, $\phi(\mathbf{x}) * \psi(\mathbf{x})$ and $\phi(\mathbf{x}) \star \psi(\mathbf{x})$ denote the n-dimensional convolution and correlation,

$$\phi(\mathbf{x}) * \psi(\mathbf{x}) = \int_{-\infty}^{\infty} \int_{-\infty}^{\infty} \cdots \int_{-\infty}^{\infty} \phi(\mathbf{s}) \psi(\mathbf{x} - \mathbf{s}) \, ds_1 \, ds_2 \cdots ds_n$$

and

$$\phi(\mathbf{x}) \star \psi(\mathbf{x}) = \int_{-\infty}^{\infty} \int_{-\infty}^{\infty} \cdots \int_{-\infty}^{\infty} \phi^*(\mathbf{s}) \psi(\mathbf{x} + \mathbf{s}) \, ds_1 \, ds_2 \cdots ds_n.$$

There is one particularly useful n-dimensional identity that does not have a direct analog to the identities given in Section 2.2 (though it can be viewed as a generalization of the scaling formula). If $\mathbf{A}$ is a real, invertible, $n \times n$ matrix and $F(\omega) = \mathcal{F}[f(\mathbf{t})]|_{\omega}$, then

$$\mathcal{F}\left[f(\mathbf{t}\mathbf{A})\right]|_{\omega} = \frac{1}{|\mathbf{A}|} F(\omega \mathbf{A}^{-\mathrm{T}}) \qquad (2.4.1.3)$$

where $|\mathbf{A}|$ is the determinant of $\mathbf{A}$ and $\mathbf{A}^{-\mathrm{T}}$ is the inverse of the transpose of $\mathbf{A}$ (equivalently $\mathbf{A}^{-\mathrm{T}}$ is the transpose of the inverse of $\mathbf{A}$),

$$\mathbf{A}^{-\mathrm{T}} = (\mathbf{A}^{\mathrm{T}})^{-1} = (\mathbf{A}^{-1})^{\mathrm{T}}.$$

TABLE 2.2
Identities for multi-dimensional transforms (α is
any nonzero real number, $\mathbf{t}_0$ and $\boldsymbol{\omega}_0$ are fixed
n-dimensional points, and $F(\boldsymbol{\omega}) = \mathcal{F}[f(\mathbf{t})]|_\omega$
and $G(\boldsymbol{\omega}) = \mathcal{F}[g(\mathbf{t})]|_\omega$).

$h(\mathbf{t})$	$H(\boldsymbol{\omega}) = \mathcal{F}[h(\mathbf{t})]|_\omega$
$f(\alpha\mathbf{t})$	$\dfrac{1}{|\alpha|} F\left(\dfrac{\omega}{\alpha}\right)$
$f(\mathbf{tA})$	$\dfrac{1}{|\mathbf{A}|} F(\boldsymbol{\omega}\mathbf{A}^{-\mathrm{T}})$
$f(\mathbf{t} - \mathbf{t}_0)$	$e^{-j\mathbf{t}_0 \cdot \boldsymbol{\omega}} F(\boldsymbol{\omega})$
$e^{j\boldsymbol{\omega}_0 \cdot \mathbf{t}} f(\mathbf{t})$	$F(\boldsymbol{\omega} - \boldsymbol{\omega}_0)$
$\dfrac{\partial f}{\partial t_k}$	$j\omega_k F(\omega)$
$t_k f(\mathbf{t})$	$j \dfrac{\partial F}{\partial \omega_k}$
$f(\mathbf{t})g(\mathbf{t})$	$\dfrac{1}{2\pi} F(\boldsymbol{\omega}) * G(\boldsymbol{\omega})$
$f(\mathbf{t}) * g(\mathbf{t})$	$F(\boldsymbol{\omega})G(\boldsymbol{\omega})$
$f(\mathbf{t}) \star g(\mathbf{t})$	$F^*(\boldsymbol{\omega})G(\boldsymbol{\omega})$
$f^*(\mathbf{t})g(\mathbf{t})$	$\dfrac{1}{2\pi} F(\boldsymbol{\omega}) \star G(\boldsymbol{\omega})$

Likewise if $f(\mathbf{t}) = \mathcal{F}^{-1}[F(\boldsymbol{\omega})]|_\mathbf{t}$, then

$$\mathcal{F}^{-1}[F(\boldsymbol{\omega}\mathbf{A})]|_\mathbf{t} = \frac{1}{|\mathbf{A}|} f(\mathbf{t}\mathbf{A}^{-\mathrm{T}}). \tag{2.4.1.4}$$

The derivation of either of these identities is relatively simple. Letting $\mathbf{s} = \mathbf{tA}$ and recalling the change of variables formula for multiple integrals,

$$\mathcal{F}[f(\mathbf{tA})]|_\omega = \int_{-\infty}^{\infty} \int_{-\infty}^{\infty} \cdots \int_{-\infty}^{\infty} f(\mathbf{tA})e^{-j\boldsymbol{\omega} \cdot \mathbf{t}}\, dt_1\, dt_2 \cdots dt_n$$

$$= \int_{-\infty}^{\infty} \int_{-\infty}^{\infty} \cdots \int_{-\infty}^{\infty} f(\mathbf{s})e^{-j\boldsymbol{\omega} \cdot (\mathbf{sA}^{-1})} \left| \frac{\partial(t_1, t_2, \ldots, t_n)}{\partial(s_1, s_2, \ldots, s_n)} \right| ds_1\, ds_2 \ldots ds_n. \tag{2.4.1.5}$$

Now

$$\frac{\partial s_i}{\partial t_j} = A_{j,i}$$

and, so, the Jacobian in (2.4.1.5) is

$$\left| \frac{\partial(t_1, t_2, \ldots, t_n)}{\partial(s_1, s_2, \ldots, s_n)} \right| = \left| \frac{\partial(s_1, s_2, \ldots, s_n)}{\partial(t_1, t_2, \ldots, t_n)} \right|^{-1} = \frac{1}{|\mathbf{A}|}.$$

From linear algebra and the definition of the transpose

$$\omega \cdot (\mathbf{s}\mathbf{A}^{-1}) = \left(\omega(\mathbf{A}^{-1})^{\mathrm{T}}\right) \cdot \mathbf{s} = (\omega\mathbf{A}^{-\mathrm{T}}) \cdot \mathbf{s}.$$

Thus equation (2.4.1.5) can be written

$$\mathcal{F}\left[f(\mathbf{t}\mathbf{A})\right]\big|_\omega = \int_{-\infty}^\infty \int_{-\infty}^\infty \cdots \int_{-\infty}^\infty f(\mathbf{s})e^{-j(\omega\mathbf{A}^{-\mathrm{T}})\cdot\mathbf{s}}\frac{1}{|\mathbf{A}|}\,ds_1\,ds_2\cdots ds_n$$

$$= \frac{1}{|\mathbf{A}|}F(\omega\mathbf{A}^{-\mathrm{T}}).$$

An example of how (2.4.1.3) can be used to compute transforms will be given in Section 2.4.2.

2.4.2   *Multidimensional transforms of separable functions*

A function of two variables, $f(x, y)$, is separable if it can be written as the product of two single variable functions,

$$f(x, y) = f_1(x)f_2(y). \tag{2.4.2.1}$$

The transform of such a function is easily computed provided $F_1(\omega) = \mathcal{F}[f_1(x)]\big|_\omega$ and $F_2(v) = \mathcal{F}\left[f_2(y)\right]\big|_v$ are known. Then

$$F(\omega, v) = \mathcal{F}\left[f(x, y)\right]\big|_{(\omega,v)}$$

$$= \int_{-\infty}^\infty \int_{-\infty}^\infty f_1(x)f_2(y)e^{-j(\omega x+vy)}\,dx\,dy$$

$$= \int_{-\infty}^\infty f_1(x)e^{-j\omega x}\,dx \int_{-\infty}^\infty f_2(y)e^{-jvy}\,dy$$

$$= F_1(\omega)F_2(v).$$

More generally, $f(\mathbf{t})$ is said to be separable if there are n functions of a single variable, $f_1(t_1), f_2(t_2), \ldots, f_n(t_n)$, such that

$$f(t_1, t_2, \ldots, t_n) = f_1(t_1)f_2(t_2)\cdots f_n(t_n). \tag{2.4.2.2}$$

The Fourier transform of such a function is another separable function

$$F(\omega_1, \omega_2, \ldots, \omega_n) = F_1(\omega_1)F_2(\omega_2)\cdots F_n(\omega_n),$$

where, for each k, $F_k(\omega_k)$ is the one-dimensional Fourier transform of $f_k(t_k)$.

Likewise, if

$$F(\omega_1, \omega_2, \ldots, \omega_n) = F_1(\omega_1) F_2(\omega_2) \cdots F_n(\omega_n),$$

then the n-dimensional inverse Fourier transform is

$$f(t_1, t_2, \ldots, t_n) = f_1(t_1) f_2(t_2) \cdots f_n(t_n)$$

where, for each k, $f_k(t_k)$ is the one-dimensional inverse Fourier transform of $F_k(\omega_k)$.

Example 2.4.2.1

The two-dimensional rectangular aperture function (with half-widths α and β) is

$$\eta_{\alpha, \beta}(x, y) = \begin{cases} 1, & \text{if } |x| < \alpha \text{ and } |y| < \beta \\ 0, & \text{if } \alpha < |x| \text{ or } \beta < |y| \end{cases}$$

or, equivalently,

$$\eta_{\alpha, \beta}(x, y) = p_\alpha(x) p_\beta(y).$$

Its Fourier transform is

$$
\begin{aligned}
N_{\alpha, \beta}(\omega, v) &= \int_{-\infty}^{\infty} \int_{-\infty}^{\infty} \eta_{\alpha, \beta}(x, y) e^{-j(\omega x + v y)} \, dx \, dy \\
&= \int_{-\infty}^{\infty} p_\alpha(x) e^{-j\omega x} \, dx \int_{-\infty}^{\infty} p_\beta(y) e^{-j v y} \, dy \\
&= \left(\frac{2 \sin(\alpha \omega)}{\omega} \right) \left(\frac{2 \sin(\beta v)}{v} \right) \\
&= \frac{4}{\omega v} \sin(\alpha \omega) \sin(\beta v).
\end{aligned}
$$

Example 2.4.2.2

The three-dimensional delta function, $\delta(x, y, z)$, is defined as the generalized function such that, if $\phi(x, y, z)$ is any function of three variables continuous at the origin,

$$\int_{-\infty}^{\infty} \int_{-\infty}^{\infty} \int_{-\infty}^{\infty} \delta(x, y, z) \phi(x, y, z) \, dx \, dy \, dz = \phi(0, 0, 0).$$

Because

$$
\begin{aligned}
\int_{-\infty}^{\infty} &\int_{-\infty}^{\infty} \int_{-\infty}^{\infty} \delta(x) \delta(y) \delta(y) \phi(x, y, z) \, dx \, dy \, dz \\
&= \int_{-\infty}^{\infty} \delta(z) \int_{-\infty}^{\infty} \delta(y) \int_{-\infty}^{\infty} \delta(x) \phi(x, y, z) \, dx \, dy \, dz \\
&= \int_{-\infty}^{\infty} \delta(z) \int_{-\infty}^{\infty} \delta(y) \phi(0, y, z) \, dy \, dz \\
&= \int_{-\infty}^{\infty} \delta(z) \phi(0, 0, z) \, dz \\
&= \phi(0, 0, 0),
\end{aligned}
$$

it is clear that

$$\delta(x, y, z) = \delta(x)\delta(y)\delta(y)$$

and

$$\mathcal{F}\left[\delta(x, y, z)\right]|_{(\omega,\nu,\mu)} = \mathcal{F}[\delta(x)]|_\omega \,\mathcal{F}\left[\delta(y)\right]|_\nu \,\mathcal{F}\left[\delta(z)\right]|_\mu = 1 \cdot 1 \cdot 1 = 1. \qquad ∎$$

In using formulas (2.4.2.1) or (2.4.2.2) care must be taken to account for all the variables especially if the function depends explicitly on only a small subset of the variables. This can be done by including on the right-hand side of (2.4.2.1) or (2.4.2.2) the unit constant function,

$$1(s) = 1 \qquad \text{for all } s,$$

for each variable, s, not explicitly involved in the computation of the function.

Example 2.4.2.3

The vertical slit aperture of half width α is the function of two variables given by

$$\text{vslit}_\alpha(x, y) = \begin{cases} 1, & \text{if } |x| < \alpha \\ 0, & \text{if } \alpha < |x| \end{cases}$$

or, equivalently,

$$\text{vslit}_\alpha(x, y) = p_\alpha(x) = p_\alpha(x)1(y).$$

Its Fourier transform is given by

$$\mathcal{F}\left[\text{vslit}_\alpha(x, y)\right]|_{(\omega,\nu)} = \mathcal{F}\left[p_\alpha(x)\right]|_\omega \,\mathcal{F}[1]|_\nu = \frac{2}{\omega}\sin(\alpha\omega) \cdot 2\pi\delta(\nu)$$

and not by

$$\mathcal{F}\left[\text{vslit}_\alpha(x, y)\right]|_{(\omega,\nu)} = \mathcal{F}\left[p_\alpha(x)\right]|_\omega = \frac{2}{\omega}\sin(\alpha\omega). \qquad ∎$$

Example 2.4.2.4

The three-dimensional vertical line source function is

$$l(x, y, z) = \delta(z).$$

Its Fourier transform is

$$\mathcal{F}\left[l(x, y, z)\right]|_{(\omega,\nu,\mu)} = \mathcal{F}\left[1(x)1(y)\delta(z)\right]|_{(\omega,\nu,\mu)} = 2\pi\delta(\omega)2\pi\delta(\nu) \cdot 1 = 4\pi^2\delta(\omega)\delta(\nu). \qquad ∎$$

Often, functions that are not separable in one set of coordinates are separable in another set of coordinates. In such cases one of the generalized scaling identities, identities (2.4.1.3) and (2.4.1.4), can simplify the computations.

Example 2.4.2.5

Let $\mathcal{P}$ be the parallelogram bounded by the lines $y = \pm 1$ and $x = y \pm 1$, and consider the two-dimensional aperture function

$$n_{\mathcal{P}}(x, y) = \begin{cases} 1, & \text{if } (x, y) \text{ is in } \mathcal{P} \\ 0, & \text{otherwise} \end{cases}.$$

Note that $\eta_{\mathcal{P}}(x, y) = 1$ if and only if

$$-1 < y < 1 \qquad \text{and} \qquad -1 < x - y < 1. \qquad (2.4.2.3)$$

Let

$$\mathbf{A} = \begin{bmatrix} 1 & 0 \\ -1 & 1 \end{bmatrix}.$$

$\mathbf{A}^T$ and the determinant of $\mathbf{A}$ are easily computed,

$$|\mathbf{A}| = 1 \qquad \text{and} \qquad \mathbf{A}^{-T} = \begin{bmatrix} 1 & -1 \\ 0 & 1 \end{bmatrix}^{-1} = \begin{bmatrix} 1 & 1 \\ 0 & 1 \end{bmatrix}.$$

For each $\mathbf{x} = (x, y)$ and $\boldsymbol{\omega} = (\omega, v)$ let

$$\hat{\mathbf{x}} = (\hat{x}, \hat{y}) = \mathbf{xA} = (x, y) \begin{bmatrix} 1 & 0 \\ -1 & 1 \end{bmatrix} = (x - y, y)$$

and

$$\hat{\boldsymbol{\omega}} = (\hat{\omega}, \hat{v}) = \boldsymbol{\omega}\mathbf{A}^{-T} = (\omega, v) \begin{bmatrix} 1 & 1 \\ 0 & 1 \end{bmatrix} = (\omega, \omega + v).$$

It is easily verified that conditions (2.4.2.3) are equivalent to

$$-1 < \hat{y} < 1 \qquad \text{and} \qquad -1 < \hat{x} < 1.$$

Thus,

$$\eta_{\mathcal{P}}(x, y) = \eta_{1,1}(\hat{x}, \hat{y}) = \eta_{1,1}(\mathbf{xA})$$

where $\eta_{1,1}(\hat{x}, \hat{y})$ is the rectangular aperture function of example 2.4.2.1. Using these results and the generalized scaling identity, (2.4.1.3),

$$\mathcal{F}\left[\eta_{\mathcal{P}}(x, y)\right]\big|_{(\omega, v)} = \mathcal{F}\left[\eta_{1,1}(\mathbf{xA})\right]\big|_{\boldsymbol{\omega}}$$

$$= \frac{1}{|\mathbf{A}|} N_{1,1}(\boldsymbol{\omega}\mathbf{A}^{-T})$$

$$= N_{1,1}(\omega, \omega + v)$$

$$= \frac{4}{\omega(\omega + v)} \sin(\omega) \sin(\omega + v). \qquad \blacksquare$$

2.4.3 *Transforms of circularly symmetric functions and the Hankel transform*

Replacing (x, y) and (ω, v) with their polar equivalents,

$$(x, y) = (r \cos\theta, r \sin\theta)$$

and

$$(\omega, v) = (\rho \cos\phi, \rho \sin\phi),$$

and using a well-known trigonometric identity, the formula for the two-dimensional Fourier transform, $F(\omega, \nu) = \mathcal{F}[f(x, y)]|_{(\omega,\nu)}$, becomes

$$F(\rho \cos \phi, \rho \sin \phi) = \int_0^\infty \int_{-\pi}^\pi f(r \cos \theta, r \sin \theta) e^{-jr\rho(\cos \theta \cos \phi + \sin \theta \sin \phi)} r \, d\theta \, dr$$

$$= \int_0^\infty \int_{-\pi}^\pi f(r \cos \theta, r \sin \theta) e^{-jr\rho \cos(\theta - \phi)} r \, d\theta, dr. \qquad (2.4.3.1)$$

Likewise in polar coordinates the formula for the two-dimensional inverse Fourier transform, $f(x, y) = \mathcal{F}[F(\omega, \nu)]|_{(x,y)}$, is

$$f(r \cos \theta, r \sin \theta) = \frac{1}{4\pi^2} \int_0^\infty \int_{-\pi}^\pi F(\rho \cos \phi, \rho \sin \phi) e^{jr\rho \cos(\theta - \phi)} \rho \, d\phi \, d\rho. \qquad (2.4.3.2)$$

If $f(r \cos \theta, r \sin \theta)$ is separable with respect to r and θ,

$$f(r \cos \theta, r \sin \theta) = f_r(r) f_\theta(\theta),$$

then (2.4.3.1) becomes

$$F(\rho \cos \phi, \rho \sin \phi) = \int_0^\infty f_r(r) r K^-(r\rho, \phi) \, dr, \qquad (2.4.3.3)$$

where

$$K^-(z, \phi) = \int_{-\pi}^\pi f_\theta(\theta) e^{-jz \cos(\theta - \phi)} \, d\theta.$$

Observe that the integrand for $K^-(r, \phi)$ must be periodic with period 2π. Thus, letting $\theta' = \theta - \phi$,

$$K^-(z, \phi) = \int_{-\pi}^\pi f_\theta(\theta' + \phi) e^{-jz \cos(\theta')} \, d\theta'. \qquad (2.4.3.4)$$

Likewise, if $F(\rho \cos \phi, \rho \sin \phi)$ is separable with respect to ρ and ϕ,

$$F(\rho \cos \phi, \rho \sin \phi) = F_\rho(\rho) F_\phi(\phi),$$

then formula (2.4.3.2) becomes

$$f(r \cos \theta, r \sin \theta) = \frac{1}{4\pi^2} \int_0^\infty F_\rho(\rho) \rho K^+(r\rho, \theta) \, d\rho, \qquad (2.4.3.5)$$

where

$$K^+(z, \theta) = \int_{-\pi}^\pi F_\phi(\phi' + \theta) e^{jz \cos(\phi')} \, d\phi'. \qquad (2.4.3.6)$$

The above formulas simplify considerably when circular symmetry can be assumed for either $f(x, y)$ or $F(\omega, \nu)$. If follows immediately from (2.4.3.3) through (2.4.3.6) that if either $f(x, y)$ or $F(\omega, \nu)$ is circularly symmetric, that is,

$$f(r \cos \theta, r \sin \theta) = f_r(r) \qquad \text{or} \qquad F(\rho \cos \phi, \rho \sin \phi) = F_\rho(\rho),$$

then, in fact, both $f(x, y)$ and $F(\omega, \nu)$ are circularly symmetric and can be written

$$f(r \cos \theta, r \sin \theta) = f_r(r) \qquad \text{and} \qquad F(\rho \cos \phi, \rho \sin \phi) = F_\rho(\rho).$$

In such cases it is convenient to use the Bessel function identity

$$2\pi J_0(z) = \int_{-\pi}^\pi \cos(z \cos w) \, dw,$$

where $J_0(z)$ is the 0th order Bessel function of the first kind.[4] It is easily verified that

$$\int_{-\pi}^{\pi} \sin(r\rho \cos w) dw = 0$$

and so

$$K^{\pm}(r\rho, w) = \int_{-\pi}^{\pi} e^{\pm jr\rho \cos w} dw = \int_{-\pi}^{\pi} \cos(r\rho \cos w) dw = 2\pi J_0(r\rho)$$

and equations (2.4.3.3) and (2.4.3.5) reduce to

$$F_\rho(\rho) = 2\pi \int_0^\infty f_r(r) J_0(r\rho) r\, dr \tag{2.4.3.7}$$

and

$$f_r(r) = \frac{1}{2\pi} \int_0^\infty F_\rho(\rho) J_0(r\rho) \rho\, d\rho. \tag{2.4.3.8}$$

The 0th order Hankel transform of $g(r)$ is defined to be

$$\hat{g}(\rho) = \int_0^\infty g(r) J_0(r\rho) r\, dr.$$

Such transforms are the topic of Chapter 9 of this handbook. It should be noted that (2.4.3.7) and (2.4.3.8) can be expressed in terms of 0th order Hankel transforms,

$$F_\rho(\rho) = 2\pi \widehat{f_r}(\rho) \qquad \text{and} \qquad f_r(r) = \frac{1}{2\pi} \widehat{F_\rho}(r). \tag{2.4.3.9}$$

From this it should be clear that 0th order Hankel transforms can be viewed as two-dimensional Fourier transforms of circularly symmetric functions. This allows fairly straightforward derivation of many of the properties of these Hankel transforms from corresponding properties of the Fourier transform. For example, letting $g(r) = 2\pi f(r)$ in (2.4.3.7) and (2.4.3.8) leads immediately to the inversion formula for the 0th order Hankel transform,

$$g(r) = \int_0^\infty \hat{g}(\rho) J_0(r\rho) \rho\, d\rho.$$

(For further discussion of the Hankel transforms, see Chapter 9 of this handbook.)

Example 2.4.3.1
Let $a > 0$ and let $f(x, y)$ be the corresponding circular aperture function,

$$f(x, y) = \begin{cases} 1, & \text{if } x^2 + y^2 < a^2 \\ 0, & \text{otherwise} \end{cases}$$

This function is circularly symmetric with

$$f(x, y) = f_r(r) = \begin{cases} 1, & \text{if } 0 \leq r < a \\ 0, & \text{otherwise} \end{cases}.$$

Its Fourier transform must also be circularly symmetric and, using (2.4.3.7), is given by

$$F(\omega, \nu) = F_\rho(\rho) = 2\pi \int_0^a J_0(r\rho) r\, dr. \tag{2.4.3.10}$$

---

[4]See Chapter 1 for additional information on Bessel functions.

Letting $z = r\rho$ and using the Bessel function identity

$$\frac{d}{dz}[zJ_1(z)] = zJ_0(z),$$

where $J_1(z)$ is the first-order Bessel function of the first kind, the computation of this transform is easily completed,

$$F_\rho(\rho) = 2\pi\rho^{-2} \int_0^{a\rho} J_0(z)z\,dz$$

$$= 2\pi\rho^{-2} \int_0^{a\rho} \frac{d}{dz}[zJ_1(z)]\,dz$$

$$= 2\pi\rho^{-2}[a\rho J_1(a\rho)]$$

$$= \frac{2\pi a}{\rho}J_1(a\rho).$$ ∎

2.4.4 *Half-line sine and cosine transforms*

Half-line sine and cosine transforms are usually taken only of functions defined on just the half-line $0 < t < \infty$. For such a function, $f(t)$, the corresponding (half-line) sine transform is

$$F_S(\omega) = S\,[f(t)]|_\omega = \int_0^\infty f(t)\sin(\omega t)\,dt \qquad (2.4.4.1)$$

and the corresponding (half-line) cosine transform is

$$F_C(\omega) = C\,[f(t)]|_\omega = \int_0^\infty f(t)\cos(\omega t)\,dt. \qquad (2.4.4.2)$$

These formulas define $F_S(\omega)$ and $F_C(\omega)$ for all real values of ω, with $F_S(\omega)$ being an odd function of ω, and $F_C(\omega)$ being an even function of ω.

The half-line sine and cosine transforms are directly related to the standard Fourier transforms of the odd and even extensions of $f(t)$,

$$f_{\text{odd}}(t) = \begin{cases} f(t), & \text{if } 0 < t \\ -f(-t), & \text{if } t < 0 \end{cases}$$

and

$$f_{\text{even}}(t) = \begin{cases} f(t), & \text{if } 0 < t \\ f(-t), & \text{if } t < 0 \end{cases},$$

respectively. From the observations made in Subsection 2.3.1,

$$S\,[f(t)]|_\omega = j\frac{1}{2}\mathcal{F}\,[f_{\text{odd}}]|_\omega \qquad (2.4.4.3)$$

and

$$C\,[f(t)]|_\omega = \frac{1}{2}\mathcal{F}\,[f_{\text{even}}(t)]|_\omega. \qquad (2.4.4.4)$$

This shows that the (half-line) sine and cosine transforms can be treated as special cases of the standard Fourier transform. Indeed, by doing so it is possible to extend the class of functions that can be treated by sine and cosine transforms to include functions for which the integrals in (2.4.4.1) and (2.4.4.2) are not defined.

Example 2.4.4.1

Let $f(t) = t^2$ for $0 < t$. Formula (2.4.4.2) cannot be used to define $\mathcal{C}\left[f(t)\right]\big|_\omega$ because

$$\lim_{b \to \infty} \int_0^b t^2 \cos(\omega t)\, dt$$

does not converge. However, $f_{\text{even}}(t) = t^2$ for all values of t, and, using formula (2.4.4.4),

$$\mathcal{C}\left[f(t)\right]\big|_\omega = \frac{1}{2}\mathcal{F}\left[t^2\right]\big|_\omega = \frac{1}{2}\left[-2\pi\delta''(\omega)\right] = -\pi\delta''(\omega). \qquad \blacksquare$$

All the useful identities for the sine and cosine transforms can be derived through relations (2.4.4.3) and (2.4.4.4) from the corresponding identities for the standard Fourier transform.

Example 2.4.4.2   *Inversion formulas for the sine and cosine transforms*

Let $f(t)$, $F_S(\omega)$, and $f_{\text{odd}}(t)$ be as above, and let

$$F_{\text{odd}}(\omega) = \mathcal{F}\left[f_{\text{odd}}(t)\right]\big|_\omega .$$

According to equation (2.4.4.3)

$$F_{\text{odd}}(\omega) = -2j\,F_S(\omega).$$

Thus, for $0 < t$,

$$f(t) = \mathcal{F}^{-1}\left[F_{\text{odd}}(\omega)\right]\big|_t = -2j\mathcal{F}^{-1}\left[F_S(\omega)\right]\big|_t .$$

But, because $F_S(\omega)$ is an odd function of ω, the same arguments used in Subsection 2.3.1 yield

$$\mathcal{F}^{-1}\left[F_S(\omega)\right]\big|_t = \frac{1}{2\pi}\int_{-\infty}^{\infty} F_S(\omega)e^{j\omega t}\, d\omega = \frac{j}{\pi}\int_0^{\infty} F_S(\omega)\sin(\omega t)\, d\omega,$$

which, combined with the previous equation, gives the inversion formula for the sine transform,

$$f(t) = \frac{2}{\pi}\int_0^{\infty} F_S(\omega)\sin(\omega t)\, d\omega.$$

Precisely the same reasoning shows that the inversion formula for the cosine transform is

$$f(t) = \frac{2}{\pi}\int_0^{\infty} F_{\mathcal{C}}(\omega)\cos(\omega t)\, d\omega. \qquad \blacksquare$$

In using (2.4.4.3) and (2.4.4.4) to derive identities for the sine and cosine function it is important to keep in mind that if

$$f(0) = \lim_{t \to 0^+} f(t)$$

exists, then the even extension will be continuous at $t = 0$ with $f_{\text{even}}(0) = f(0)$, but the odd extension will have a jump discontinuity at $t = 0$ with a jump of $2f(0)$. This is why most of the sine and cosine transform analogs to the differentiation formulas of Subsection 2.2.11 include boundary values. Some of these identities are

$$\mathcal{S}\left[f'(t)\right]\big|_\omega = -\omega F_{\mathcal{C}}(\omega),$$

$$\mathcal{C}\left[f'(t)\right]\big|_\omega = \omega F_S(\omega) - f(0),$$

$$\mathcal{S}\left[f''(t)\right]\big|_\omega = \omega f(0) - \omega^2 F_S(\omega),$$

and

$$\mathcal{C}\left[f''(t)\right]\big|_\omega = -f'(0) - \omega^2 F_{\mathcal{C}}(\omega),$$

where $f'(t)$ and $f''(t)$ denote the generalized first and second derivatives of $f(t)$ for $0 < t$. (See also Subsection 2.3.16.)

2.4.5 *The discrete Fourier transform*

The discrete Fourier transform is a computational analog to the Fourier transform and is used when dealing with finite collections of sampled data rather than functions per se. Given an "Nth order sequence" of values, $\{f_0, f_1, f_2, \ldots, f_{N-1}\}$, the corresponding Nth order discrete transform is the sequence $\{F_0, F_1, F_2, \ldots, F_{N-1}\}$ given by the formula

$$F_n = \sum_{n=0}^{N-1} e^{-j\frac{2\pi}{N}nk} f_k. \tag{2.4.5.1}$$

This can also be written in matrix form, $\mathbf{F} = [\mathcal{F}_N]\mathbf{f}$, where

$$\mathbf{F} = \begin{pmatrix} F_0 \\ F_1 \\ F_2 \\ \vdots \\ F_{N-1} \end{pmatrix}, \qquad \mathbf{f} = \begin{pmatrix} f_0 \\ f_1 \\ f_2 \\ \vdots \\ f_{N-1} \end{pmatrix},$$

and

$$[\mathcal{F}_N] = \begin{bmatrix} 1 & 1 & 1 & \cdots & 1 \\ 1 & e^{-j\frac{2\pi}{N}} & e^{-j2\frac{2\pi}{N}} & \cdots & e^{-j(N-1)\frac{2\pi}{N}} \\ 1 & e^{-j2\frac{2\pi}{N}} & e^{-j2\cdot2\frac{2\pi}{N}} & \cdots & e^{-j2(N-1)\frac{2\pi}{N}} \\ \vdots & \vdots & \vdots & \ddots & \vdots \\ 1 & e^{-j(N-1)\frac{2\pi}{N}} & e^{-j2(N-1)\frac{2\pi}{N}} & \cdots & e^{-j(N-1)^2\frac{2\pi}{N}} \end{bmatrix}.$$

On occasion, the matrix, $[\mathcal{F}_N]$, is itself referred to as the Nth order discrete transform.

The inverse to formula (2.4.5.1) is given by

$$f_k = \frac{1}{N} \sum_{k=0}^{N-1} e^{j\frac{2\pi}{N}kn} F_n. \tag{2.4.5.2}$$

In matrix form this is $\mathbf{f} = [\mathcal{F}_N]^{-1}\mathbf{F}$, where $[\mathcal{F}_N]^{-1}$ is the matrix

$$\frac{1}{N} \begin{bmatrix} 1 & 1 & 1 & \cdots & 1 \\ 1 & e^{j\frac{2\pi}{N}} & e^{j2\frac{2\pi}{N}} & \cdots & e^{j(N-1)\frac{2\pi}{N}} \\ 1 & e^{j2\frac{2\pi}{N}} & e^{j2\cdot2\frac{2\pi}{N}} & \cdots & e^{j2(N-1)\frac{2\pi}{N}} \\ \vdots & \vdots & \vdots & \ddots & \vdots \\ 1 & e^{j(N-1)\frac{2\pi}{N}} & e^{j2(N-1)\frac{2\pi}{N}} & \cdots & e^{j(N-1)^2\frac{2\pi}{N}} \end{bmatrix}.$$

The similarity between the definitions for the discrete Fourier transforms, formulas (2.4.5.1) and (2.4.5.2), and formulas (2.3.11.4) and (2.3.11.5) should be noted. The discrete Fourier transforms can be treated as the regular Fourier transforms of corresponding regular periodic arrays generated from the sampled data.

Example 2.4.5.1

The matrices for the 4-th order discrete Fourier transforms are

$$[\mathcal{F}_N] = \begin{bmatrix} 1 & 1 & 1 & 1 \\ 1 & e^{-j\frac{2\pi}{4}} & e^{-j2\frac{2\pi}{4}} & e^{-j3\frac{2\pi}{4}} \\ 1 & e^{-j2\frac{2\pi}{4}} & e^{-j4\frac{2\pi}{4}} & e^{-j6\frac{2\pi}{4}} \\ 1 & e^{-j3\frac{2\pi}{4}} & e^{-j6\frac{2\pi}{4}} & e^{-j9\frac{2\pi}{4}} \end{bmatrix} = \begin{bmatrix} 1 & 1 & 1 & 1 \\ 1 & -j & -1 & j \\ 1 & -1 & 1 & -1 \\ 1 & j & -1 & -j \end{bmatrix}$$

and

$$[\mathcal{F}_N]^{-1} = \frac{1}{4} \begin{bmatrix} 1 & 1 & 1 & 1 \\ 1 & e^{j\frac{2\pi}{4}} & e^{j2\frac{2\pi}{4}} & e^{j3\frac{2\pi}{4}} \\ 1 & e^{j2\frac{2\pi}{4}} & e^{j4\frac{2\pi}{4}} & e^{j6\frac{2\pi}{4}} \\ 1 & e^{j3\frac{2\pi}{4}} & e^{j6\frac{2\pi}{4}} & e^{j9\frac{2\pi}{4}} \end{bmatrix} = \frac{1}{4} \begin{bmatrix} 1 & 1 & 1 & 1 \\ 1 & j & -1 & -j \\ 1 & -1 & 1 & -1 \\ 1 & -j & -1 & j \end{bmatrix}.$$

The discrete Fourier transform of $\{f_0, f_1, f_2, f_3\} = \{1, 2, 3, 4\}$ is given by

$$\begin{pmatrix} F_0 \\ F_1 \\ F_2 \\ F_3 \end{pmatrix} = \begin{bmatrix} 1 & 1 & 1 & 1 \\ 1 & -j & -1 & j \\ 1 & -1 & 1 & -1 \\ 1 & j & -1 & -j \end{bmatrix} \begin{pmatrix} 1 \\ 2 \\ 3 \\ 4 \end{pmatrix} = \begin{pmatrix} 10 \\ -2 + 2j \\ -2 \\ -2 - 2j \end{pmatrix},$$

and the discrete inverse Fourier transform of $\{F_0, F_1, F_2, F_3\} = \{10, -2 + 2j, -2, -2 - 2j\}$ is given by

$$\begin{pmatrix} f_0 \\ f_1 \\ f_2 \\ f_3 \end{pmatrix} = \frac{1}{4} \begin{bmatrix} 1 & 1 & 1 & 1 \\ 1 & j & -1 & -j \\ 1 & -1 & 1 & -1 \\ 1 & -j & -1 & j \end{bmatrix} \begin{pmatrix} 10 \\ -2 + 2j \\ -2 \\ -2 - 2j \end{pmatrix} = \begin{pmatrix} 1 \\ 2 \\ 3 \\ 4 \end{pmatrix}. \qquad \blacksquare$$

In practice the sample size, N, is often quite large and the computations of the discrete transforms directly from formulas (2.4.5.1) and (2.4.5.2) can be a time-consuming process even on fairly fast computers. For this reason it is standard practice to make heavy use of symmetries inherent in the computations of the discrete transforms for certain values of N (e.g., $N = 2^M$) to reduce the total number of calculations. Such implementations of the discrete Fourier transform are called "fast Fourier transforms" (FFTs).

2.4.6   Relations between the Laplace transform and the Fourier transform[5]

Attention in this subsection will be restricted to functions of t (and their transforms) that satisfy all of the following three conditions:

1. $f(t) = 0$ if $t < 0$.

2. $f(t)$ is piecewise continuous on $0 \leq t$.

3. For some real value of α, $f(t) = O(e^{\alpha t})$ as $\alpha \to \infty$.

It follows from the third condition that there is a minimum value of α_0, with $-\infty \leq \alpha_0 < \infty$, such that $f(t)e^{-\alpha t}$ is an exponentially decreasing function of t whenever $\alpha_0 < \alpha$. This minimal value of α_0 is called the "exponential order" of $f(t)$.

The Laplace transform of $f(t)$ is defined by

$$\mathcal{L}\left[f(t)\right]\big|_s = \int_{-\infty}^{\infty} f(t)e^{-st}\,dt. \tag{2.4.6.1}$$

The variable, s, in the transformed function may be any complex number whose real part is greater than the exponential order of $f(t)$.

There is a clear similarity between formula (2.4.6.1) defining the Laplace transform and the integral formula for the Fourier transform (formula [2.1.1.1]). Comparing the two immediately yields the formal relation

$$\mathcal{L}\left[f(t)\right]\big|_s = \int_{-\infty}^{\infty} f(t)e^{-j(-js)t}\,dt = \mathcal{F}\left[f(t)\right]\big|_{-js}.$$

Another, somewhat more useful relation is found by taking the Fourier transform of $f(t)e^{-xt}$ when x is greater than the order of $f(t)$:

$$\mathcal{F}\left[f(t)e^{-xt}\right]\big|_y = \int_{-\infty}^{\infty} f(t)e^{-(x+jy)t}\,dt = \mathcal{L}\left[f(t)\right]\big|_{x+jy}. \tag{2.4.6.2}$$

In particular,

$$\mathcal{F}\left[f(t)e^{-st}\right]\big|_0 = \mathcal{L}\left[f(t)\right]\big|_s.$$

The inversion formula for the Laplace transform can be quickly derived using relation (2.4.6.2). Let β be any real value greater than the exponential order of $f(t)$ and observe that, letting

$$F_{\mathcal{L}}(s) = \mathcal{L}\left[f(t)\right]\big|_s,$$

then, by relation (2.4.6.2)

$$F_{\mathcal{L}}(\beta + j\omega) = \mathcal{F}\left[f(t)e^{-\beta t}\right]\big|_\omega,$$

and so,

$$\begin{aligned}
f(t) &= e^{\beta t} f(t)e^{-\beta t} \\
&= e^{\beta t}\mathcal{F}^{-1}\left[\mathcal{F}\left[f(\tau)e^{-\beta \tau}\right]\big|_\omega\right]\big|_t \\
&= e^{\beta t}\frac{1}{2\pi}\int_{-\infty}^{\infty}\mathcal{F}\left[f(\tau)e^{-\beta \tau}\right]\big|_\omega e^{j\omega t}\,d\omega \\
&= \frac{1}{2\pi}\int_{-\infty}^{\infty}F_{\mathcal{L}}(\beta + j\omega)e^{(\beta + j\omega)t}\,d\omega.
\end{aligned}$$

---

[5]For a more complete discussion of the Laplace transform see Chapter 5 of this handbook.

This formula can be expressed in slightly more compact form as a contour integral in the complex plane,

$$f(t) = \mathcal{L}^{-1}\left[F_{\mathcal{L}}(s)\right]|_t = \frac{1}{2j\pi}\int_{z=\beta-j\infty}^{\beta+j\infty} F_{\mathcal{L}}(z)e^{zt}\,dz.$$

Alternatively, it can be left in terms of the Fourier inverse transform,

$$f(t) = \mathcal{L}^{-1}\left[F_{\mathcal{L}}(s)\right]|_t = e^{\beta t}\mathcal{F}^{-1}\left[F_{\mathcal{L}}(\beta+j\omega)\right]|_t.$$

2.5 Reconstruction of sampled signals

In practice a function is often known only by a sampling of its values at specific points. The following subsections describe when such a function can be completely reconstructed using its samples and how, using methods based on the Fourier transform, the values of the reconstructed function can be computed at arbitrary points.

2.5.1 Sampling theorem for band-limited functions

Assume $f(t)$ is a band-limited function with Fourier transform $F(\omega)$ (see Subsection 2.3.7). Let $2\Omega_0$ be the minimum bandwidth of $f(t)$, that is, Ω_0 is the smallest nonnegative value such that

$$F(\omega) = 0 \qquad \text{whenever } \Omega_0 < |\omega|.$$

The Nyquist interval, ΔT, and the Nyquist rate, ν, for $f(t)$ are defined by

$$\Delta T = \frac{\pi}{\Omega_0} \qquad \text{and} \qquad \nu = \frac{1}{\Delta T} = \frac{\Omega_0}{\pi}.$$

The sampling theorem for band-limited functions states that $f(t)$ (and hence, also $F(\omega)$ as well as the total energy in $f(t)$) can be completely reconstructed from a uniform sampling taken at the Nyquist rate or greater. More precisely, if $0 < \Delta t \le \Delta T$, then, letting $\Omega = \frac{\pi}{\Delta t}$,

$$f(t) = \sum_{n=-\infty}^{\infty} f(n\Delta t)\frac{\sin(\Omega(t-n\Delta t))}{\Omega(t-n\Delta t)} \tag{2.5.1.1}$$

and, taking the transform,

$$F(\omega) = \frac{\pi}{\Omega}\sum_{n=-\infty}^{\infty} f(n\Delta t)e^{-jn\Delta t\omega}p_\Omega(\omega), \tag{2.5.1.2}$$

where $p_\Omega(\omega)$ is the pulse function,

$$p_\Omega(\omega) = \begin{cases} 1, & \text{if } |\omega| < \Omega \\ 0, & \text{if } \Omega < |\omega| \end{cases}.$$

The energy in $f(t)$ is easily computed. Using equations (2.3.4.1), (2.5.1.2), and the fact that the exponentials in formula (2.5.1.2) are orthogonal on the interval $-\Omega < \omega < \Omega$,

$$E = \frac{1}{2\pi} \int_{-\infty}^{\infty} |F(\omega)|^2 \, d\omega$$

$$= \frac{1}{2\pi} \sum_{n=-\infty}^{\infty} \left(\frac{\pi}{\Omega}\right)^2 \int_{-\Omega}^{\Omega} f(n\Delta t) f^*(n\Delta t) e^{-jn\Delta t\omega} e^{jn\Delta t\omega} \, d\omega$$

$$= \frac{1}{2\pi} \sum_{n=-\infty}^{\infty} \left(\frac{\pi}{\Omega}\right)^2 |f(n\Delta t)|^2 2\Omega$$

$$= \Delta t \sum_{n=-\infty}^{\infty} |f(n\Delta t)|^2. \tag{2.5.1.3}$$

To see why formulas (2.5.1.1) and (2.5.1.2) are valid, let $\hat{F}(\omega)$ be the periodic extension of $F(\omega)$,

$$\hat{F}(\omega) = \begin{cases} F(\omega), & \text{if } -\Omega < \omega < \Omega \\ \hat{F}(\omega + 2\Omega), & \text{for all } \omega \end{cases}.$$

Observe that 2Ω is a bandwidth for $f(t)$, and so,

$$F(\omega) = \begin{cases} \hat{F}(\omega), & \text{if } |\omega| < \Omega \\ 0, & \text{if } \Omega < |\omega| \end{cases}. \tag{2.5.1.4}$$

This can be written more concisely using the pulse function as

$$F(\omega) = \hat{F}(\omega) p_\Omega(\omega).$$

From this it follows, using convolution, that

$$f(t) = \mathcal{F}^{-1}\left[\hat{F}(\omega) p_\Omega(\omega)\right]\Big|_t = \hat{f}(t) * \left(\frac{\sin(\Omega t)}{\pi t}\right), \tag{2.5.1.5}$$

where $\hat{f}(t)$ denotes the inverse transform of $\hat{F}(\omega)$. By formula (2.3.9.14),

$$\hat{f}(t) = \sum_{n=-\infty}^{\infty} C_n \delta(t - n\Delta\tau), \tag{2.5.1.6}$$

where

$$\Delta\tau = \frac{2\pi}{2\Omega} = \Delta t$$

and, using the above,

$$C_n = \frac{1}{2\Omega} \int_{-\Omega}^{\Omega} \hat{F}(\omega) e^{jn\Delta t\omega} \, d\omega$$

$$= \frac{\pi}{\Omega} \left(\frac{1}{2\pi} \int_{-\infty}^{\infty} F(\omega) e^{jn\Delta t\omega} \, d\omega\right)$$

$$= \Delta t f(n\Delta t).$$

Combining this last with equations (2.5.1.5) and (2.5.1.6), and using the shifting property of the delta function, yields

$$f(t) = \hat{f}(t) * \left(\frac{\sin(\Omega t)}{\pi t} \right)$$

$$= \sum_{n=-\infty}^{\infty} \Delta t f(n\Delta t) \delta(t - n\Delta t) * \left(\frac{\sin(\Omega t)}{\pi t} \right)$$

$$= \sum_{n=-\infty}^{\infty} \Delta t f(n\Delta t) \frac{\sin(\Omega(t - n\Delta t))}{\pi(t - n\Delta t)},$$

which is the same as formula (2.5.1.1).

2.5.2 Truncated sampling reconstructions of band-limited functions

Formula (2.5.1.1) employs an infinite number of samples of $f(t)$. Often this is impractical, and one must approximate $f(t)$ with the truncated version of formula (2.5.1.1),

$$f(t) \approx \sum_{n=-N}^{N} f(n\Delta t) \frac{\sin(\Omega(t - n\Delta t))}{\Omega(t - n\Delta t)}, \qquad (2.5.2.1)$$

where N is some positive integer. The pointwise error is

$$\varepsilon_N(t) = f(t) - \sum_{n=-N}^{N} f(n\Delta t) \frac{\sin(\Omega(t - n\Delta t))}{\Omega(t - n\Delta t)}.$$

If $f(t)$ is a band-limited function, then the sampling theorem implies that

$$\varepsilon_N(t) = \sum_{N < |n|} f(n\Delta t) \frac{\sin(\Omega(t - n\Delta t))}{\Omega(t - n\Delta t)},$$

and it can be shown that

$$|\varepsilon_N(t)|^2 \le \frac{E\Omega}{\pi} \sum_{N < |n|} \frac{\sin^2(\Omega(t - n\Delta t))}{\Omega^2(t - n\Delta t)^2}, \qquad (2.5.2.2)$$

where E is the energy in $f(t)$. In addition, if the samples are known to vanish sufficiently rapidly, then one can use

$$|\varepsilon_N(t)| \le \sqrt{\sum_{N < |n|} |f(n\Delta t)|^2}. \qquad (2.5.2.3)$$

This last bound is a uniform bound directly related to expression (2.5.1.3) for the energy of a band-limited function. It can be derived after observing that $\varepsilon_N(t)$ can be written as a proper integral,

$$\varepsilon_N(t) = \sum_{N<|n|} f(n\Delta t)\frac{\sin(\Omega(t-n\Delta t))}{\Omega(t-n\Delta t)}$$

$$= \sum_{N<|n|} f(n\Delta t)\frac{1}{2\Omega}\int_{-\Omega}^{\Omega} e^{-jn\Delta t\omega}e^{j\omega t}\,d\omega$$

$$= \frac{1}{2\Omega}\int_{-\Omega}^{\Omega}\left(\sum_{N<|n|} f(n\Delta t)e^{-jn\Delta t\omega}\right)e^{jt\omega}\,d\omega.$$

Using the Cauchy–Schwarz inequality,

$$|\varepsilon_N(t)|^2 = \frac{1}{4\Omega^2}\left|\int_{-\Omega}^{\Omega}\left(\sum_{N<|n|} f(n\Delta t)e^{-jn\Delta t\omega}\right)e^{jt\omega}\,d\omega\right|^2$$

$$\leq \frac{1}{4\Omega^2}\left(\int_{-\Omega}^{\Omega}\left|\sum_{N<|n|} f(n\Delta t)e^{-jn\Delta t\omega}\right|^2\,d\omega\right)\left(\int_{-\Omega}^{\Omega}|e^{jt\omega}|^2\,d\omega\right)$$

$$= \frac{1}{4\Omega^2}\left(\sum_{N<|n|}|f(n\Delta t)|^2 2\Omega\right)(2\Omega)$$

$$= \sum_{N<|n|}|f(n\Delta t)|^2,$$

as claimed by equation (2.5.2.3).

Example 2.5.2.1
Suppose $f(t)$ is a band-limited function (with bandwidth 2Ω) to be approximated on the interval $-L < t < L$. Suppose, further, that an upper bound, E_0, is known for the energy of $f(t)$. Let N be any integer such that $L < N\Delta t$. Then, for $-L < t < L$, using inequality (2.5.2.2) with well-known bounds,

$$|\varepsilon_N(t)|^2 \leq \frac{E\Omega}{\pi}\sum_{N<|n|}\frac{\sin^2(\Omega(t-n\Delta t))}{\Omega^2(t-n\Delta t)^2}$$

$$\leq \frac{2E_o}{\pi\Omega}\sum_{n=N+1}^{\infty}\frac{1}{(L-n\Delta t)^2}$$

$$\leq \frac{2E_o}{\pi\Omega}\int_{x=N}^{\infty}\frac{1}{(L-x\Delta t)^2}\,dx$$

$$= \frac{2E_0}{\pi\Omega}\frac{1}{\Delta t(N\Delta t-L)}$$

$$= \frac{2}{\pi^2(N\Delta t-L)}E_0.$$

Thus, to insure an error of less than $0.05\sqrt{E_0}$, it suffices to choose N satisfying

$$\left(\frac{2}{\pi^2(N\Delta t - L)}\right)^{\frac{1}{2}} < 0.05$$

or, equivalently,

$$\frac{800}{\pi^2} + L < N\Delta t. \qquad \blacksquare$$

Example 2.5.2.2

Suppose that $f(t)$ is a band-limited function (with bandwidth 2Ω) whose transform, $F(\omega)$, is known to be piecewise smooth and continuous. Assume further that, for some $A < \infty$, $|F'(\omega)| < A$ for all values of ω. Then, for each t,

$$|tf(t)| = \left|\mathcal{F}^{-1}\left[jF'(\omega)\right]\big|_t\right|$$

$$= \left|\frac{1}{2\pi}\int_{-\Omega}^{\Omega} jF'(\omega)e^{j\omega t}\,d\omega\right|$$

$$\leq \frac{1}{2\pi}\int_{-\Omega}^{\Omega} |F'(\omega)|\,d\omega$$

$$\leq A\frac{\Omega}{\pi}.$$

So, for each n,

$$|f(n\Delta t)|^2 \leq \left(\frac{A\Omega}{n\Delta t\pi}\right)^2 = \frac{1}{n^2}\frac{A^2}{\Delta t^4},$$

and inequality (2.5.2.3) becomes

$$|\varepsilon_N(t)|^2 \leq \sum_{N<|n|} \frac{1}{n^2}\frac{A^2}{\Delta t^4}$$

$$\leq \frac{2A^2}{\Delta t^4}\int_{x=N}^{\infty} \frac{1}{x^2}\,dx$$

$$= \frac{2A^2}{N\Delta t^4}. \qquad \blacksquare$$

2.5.3   *Reconstruction of sampled nearly band-limited functions*

Often, one must deal with a function, $f(t)$, which might not necessarily be band-limited, but is "nearly" band-limited, that is, letting

$$F_\Omega(\omega) = F(\omega)p_\Omega \qquad \text{and} \qquad f_\Omega(t) = \mathcal{F}^{-1}\left[F_\Omega(\omega)\right]\big|_t,$$

one can always choose $\Omega < \infty$ so that

$$\int_{-\infty}^{\infty} |F(\omega) - F_\Omega(\omega)|\,d\omega = \int_{\Omega<|\omega|} |F(\omega)|\,d\omega \qquad (2.5.3.1)$$

is as small as desired. Because

$$|f(t) - f_\Omega(t)| = \left|\mathcal{F}^{-1}\left[F(\omega) - F_\Omega(\omega)\right]\big|_t\right| \leq \frac{1}{2\pi}\int_{-\infty}^{\infty} |F(\omega) - F_\Omega(\omega)|\,d\omega,$$

it is clear that $f_\Omega(t)$ can also be made as close to $f(t)$ as desired by a suitable choice of Ω. Any value of 2Ω that makes (2.5.3.1) "sufficiently small" is called an effective bandwidth. For such a function it is reasonable to expect that if Ω is an effective bandwidth and $\Delta t = \frac{\pi}{\Omega}$, then the interpolation formula,

$$f_S(t) = \sum_{n=-\infty}^{\infty} f(n\Delta t)\frac{\sin(\Omega(t - n\Delta t))}{\Omega(t - n\Delta t)} \tag{2.5.3.2}$$

will be a good approximation to $f(t)$. Starting with the trivial observation that

$$f(t) = f_S(t) + f(t) - f_\Omega(t) + f_\Omega(t) - f_S(t),$$

one can derive

$$f(t) = f_S(t) + \mathcal{E}_0(t) - \mathcal{E}_\Sigma(t), \tag{2.5.3.3}$$

where

$$\mathcal{E}_0(t) = \frac{1}{2\pi}\int_{\Omega < |\omega|} F(\omega)e^{j\omega t}\,d\omega,$$

$$\mathcal{E}_\Sigma(t) = \sum_{n=-\infty}^{\infty} \varepsilon(n\Delta t)\frac{\sin(\Omega(t - n\Delta t))}{\Omega(t - n\Delta t)}$$

and

$$\varepsilon(n\Delta t) = \mathcal{F}^{-1}\left[(1 - p_\Omega(\omega))F(\omega)\right]|_{n\Delta t} = \frac{1}{2\pi}\int_{\Omega < |\omega|} F(\omega)e^{j\omega n\Delta t}\,d\omega.$$

Error estimates can be obtained from equation (2.5.3.3) provided it can be shown that $\varepsilon(n\,\Delta t)$ vanishes sufficiently rapidly as $n \to \infty$. As example 2.5.3.1 illustrates, finding such error estimates can be quite nontrivial.

Example 2.5.3.1
Suppose $f(t)$ is an infinitely differentiable, finite duration function with duration $2T$. Since $f(t)$ vanishes whenever $|t| \geq T$,

$$f_S(t) = \sum_{n=-N}^{N} f(n\Delta t)\frac{\sin(\Omega(t - n\Delta t))}{\Omega(t - n\Delta t)}, \tag{2.5.3.4}$$

where N is the integer satisfying

$$N\Delta t < T \leq (N + 1)\Delta t.$$

To avoid triviality, it may be assumed that $\Delta t < T$ and $N \geq 1$.

By continuity $f(t)$ and each of its derivatives must vanish whenever $|t| \geq T$. Also, for each positive integer, m, there is a finite A_m such that

$$\left|f^{(m)}(t)\right| \leq A_m$$

for all t. Thus, for each nonnegative integer, m,

$$\left|\omega^m F(\omega)\right| = \left|\mathcal{F}\left[f^{(m)}(t)\right]\big|_\omega\right|$$

$$= \left|\int_{-\infty}^{\infty} f^{(m)}(t)e^{-j\omega t}\,dt\right|$$

$$\leq \int_{-T}^{T} A_m \, dt$$

$$= 2A_m T. \tag{2.5.3.5}$$

Likewise, if $m \geq 2$,

$$|\omega^m F'(\omega)| = \left| \mathcal{F} \left[\frac{d^m}{dt^m} (t f(t)) \right] \right|_{\omega} = \left| \mathcal{F} \left[m f^{(m-1)}(t) + t f^{(m)}(t) \right] \right|_{\omega} \leq B_m \tag{2.5.3.6}$$

where

$$B_m = 2m A_{m-1} T + A_m T^2.$$

It follows from inequality (2.5.3.5) that $F(\omega)$ is absolutely integrable (and, hence, nearly band-limited) and that, for $m \geq 2$ and any t,

$$\frac{1}{2\pi} \left| \int_{\Omega < |\omega|} F(\omega) e^{j\omega t} \, d\omega \right| \leq \frac{1}{2\pi} \int_{\Omega < |\omega|} |F(\omega)| \, dt$$

$$\leq \frac{1}{\pi} \int_{\Omega}^{\infty} 2 A_m T \omega^{-m} \, d\omega$$

$$= \frac{2 A_m T}{(m-1)\pi} \Omega^{1-m}$$

$$= C_m \Delta t^{m-1} \tag{2.5.3.7}$$

where

$$C_m = \frac{2 A_m T}{(m-1)\pi^m}.$$

Thus, in particular, for any positive integer, k,

$$|\mathcal{E}_0(t)| = \frac{1}{2\pi} \left| \int_{\Omega < |\omega|} F(\omega) e^{j\omega t} \, d\omega \right| \leq C_{k+1} \Delta t^k. \tag{2.5.3.8}$$

Two bounds for $\varepsilon(n\Delta t)$ can be derived. First, using inequality (2.5.3.7),

$$|\varepsilon(n\Delta t)| = \frac{1}{2\pi} \left| \int_{\Omega < |\omega|} F(\omega) e^{j\omega n \Delta t} \, d\omega \right| \leq C_m \Delta t^{m-1} \tag{2.5.3.9}$$

provided $m \geq 2$. For $n \neq 0$, observe that

$$\varepsilon(n\Delta t) = \varepsilon_+(n\Delta t) + \varepsilon_-(n\Delta t),$$

where

$$\varepsilon_\pm(n\Delta t) = \pm \frac{1}{2\pi} \int_{\pm\Omega}^{\pm\infty} F(\omega) e^{j\omega n \Delta t} \, d\omega.$$

Using integration by parts and inequalities (2.5.3.5) and (2.5.3.6),

$$|\varepsilon_\pm(n\Delta t)| = \left| \frac{1}{2\pi} \int_{\pm\Omega}^{\pm\infty} F(\omega) e^{j\omega n \Delta t} \, d\omega \right|$$

$$= \frac{1}{2\pi} \left| \frac{j}{n\Delta t} F(\pm\Omega) e^{\pm j n \Delta t \Omega} - \frac{1}{jn\Delta t} \int_{\pm\Omega}^{\pm\infty} F'(\omega) e^{jn\Delta t\omega} \, d\omega \right|$$

$$\leq \frac{1}{2\pi} \left(\frac{2}{n\Delta t} A_m T \Omega^{-m} + \frac{1}{n\Delta t} \int_{\Omega}^{\infty} B_m \omega^{-m} \, d\omega \right)$$

$$= \frac{1}{2\pi n} \left(\frac{2}{\Delta t} A_m T \Omega^{-m} + \frac{1}{\Delta t (m-1)} B_m \Omega^{1-m} \right)$$

for any integer, $m \geq 2$. Because $\Delta t \Omega = \pi$ and $\Delta t \leq T$, this reduces to

$$|\varepsilon_{\pm}(n\Delta t)| \leq \frac{1}{2n} D_m \Delta t^{m-2},$$

where

$$D_m = \pi^{-m} \left(\frac{2}{\pi} A_m T^2 + \frac{1}{m-1} B_m \right).$$

Thus, for all $n \neq 0$ and $m \geq 2$,

$$|\varepsilon(n\Delta t)| \leq \frac{1}{n} D_m \Delta t^{m-2}. \tag{2.5.3.10}$$

Next, observe that

$$\mathcal{E}_{\Sigma}(t) = S_1(t) + S_2(t)$$

where

$$S_1(t) = \sum_{n=-2N-1}^{2N+1} \varepsilon(n\Delta t) \frac{\sin(\Omega(t - n\Delta t))}{\Omega(t - n\Delta t)}$$

and

$$S_2(t) = \sum_{2N+1 < |n|} \varepsilon(n\Delta t) \frac{\sin(\Omega(t - n\Delta t))}{\Omega(t - n\Delta t)}.$$

Using inequality (2.5.3.9),

$$|S_1(t)| \leq \sum_{n=-2N-1}^{2N+1} |\varepsilon(n\Delta t)| \left| \frac{\sin(\Omega(t - n\Delta t))}{\Omega(t - n\Delta t)} \right|$$

$$< |\varepsilon(0)| + |\varepsilon((2N+1)\Delta t)| + |\varepsilon(-(2N+1)\Delta t)|$$

$$+ \sum_{n=1}^{2N} (|\varepsilon(n\Delta t)| + |\varepsilon(-n\Delta t)|)$$

$$\leq 3C_{k+2} \Delta t^{k+1} + 2 \sum_{n=1}^{2N} C_{k+2} \Delta t^{k+1}$$

$$= C_{k+2}(3\Delta t + 4N\Delta t)\Delta t^k$$

$$\leq 7T C_{k+2} \Delta t^k$$

for any positive integer, k. Next, because of inequality (2.5.3.10) and the fact that $T <$

$(N + 1)\Delta t$, it follows that, for $|t| \leq T$ and $k \geq 1$,

$$|S_2(t)| \leq \sum_{2N+1 < |n|} |\varepsilon(n\Delta t)| \left| \frac{\sin(\Omega(t - n\Delta t))}{\Omega(t - n\Delta t)} \right|$$

$$\leq 2 \sum_{n=2N+2}^{\infty} \frac{1}{n} D_{k+1} \Delta t^{k-1} \frac{1}{\Omega(n\Delta t - |t|)}$$

$$\leq 2 \sum_{n=2N+2}^{\infty} \frac{1}{n} D_{k+1} \Delta t^{k-1} \frac{1}{\Omega \Delta t(n - (N+1))}$$

$$= \frac{2}{\pi} D_{k+1} \Delta t^{k-1} \sum_{n=2N+2}^{\infty} \frac{1}{n(n - N - 1)}.$$

But,

$$\sum_{n=2N+2}^{\infty} \frac{1}{n(n - N - 1)} < \int_{2N+1}^{\infty} \frac{1}{x(x - N - 1)} \, dx$$

$$= \frac{1}{N + 1} \ln \left| 2 + \frac{1}{N} \right|$$

$$< \frac{2}{N + 1}.$$

So,

$$|S_2(t)| < \frac{2}{\pi} D_{k+1} \Delta t^k \frac{1}{\Delta t} \frac{2}{N + 1} < \frac{4}{\pi T} D_{k+1} \Delta t^k.$$

Combining the bounds for $|S_1(t)|$ and $|S_2(t)|$ gives

$$|\mathcal{E}_\Sigma(t)| \leq |S_1(t)| + |S_2(t)| < E_k \Delta t^k \qquad (2.5.3.11)$$

for $|t| \leq T$ and $k \geq 1$, where

$$E_k = 7T C_{k+2} + \frac{4}{\pi T} D_{k+1}.$$

Combining (2.5.3.3), (2.5.3.8) and (2.5.3.11) gives an error estimate for using $f_S(t)$ as an approximation for $f(t)$ when $|t| \leq T$,

$$|f(t) - f_S(t)| \leq |\mathcal{E}_0(t)| + |\mathcal{E}_\Sigma(t)| < [C_{k+1} + E_k] \Delta t^k,$$

where k is any positive integer. In terms of the effective bandwidth, $\Omega = \frac{\pi}{\Delta t}$, this becomes

$$|f(t) - f_S(t)| = O(\Omega^{-k}),$$

confirming that $f_S(t)$ can be made to approximate $f(t)$ on $-T < t < T$ as accurately as desired by taking the effective bandwidth, Ω, sufficiently large. ∎

2.5.4 *Sampling theorem for finite duration functions*

Assume $f(t)$ is of finite duration with Fourier transform $F(\omega)$ (see Subsection 2.3.6). Let $2T_0$ be the minimum duration of $f(t)$, that is, T_0 is the smallest nonnegative value such that

$$f(t) = 0 \qquad \text{whenever } T_0 < |t|.$$

The sampling theorem for functions of finite duration states that $F(\omega)$ (and hence, also $f(t)$) can be reconstructed from a suitable uniform sampling in the frequency domain. More precisely, if $0 < \Delta\omega < \Delta\Omega$, where $\Delta\Omega$ denotes the "frequency Nyquist interval,"

$$\Delta\Omega = \frac{\pi}{T_0},$$

then, letting $T = \frac{\pi}{\Delta\omega}$,

$$F(\omega) = \sum_{n=-\infty}^{\infty} F(n\Delta\omega) \frac{\sin(T(\omega - n\Delta\omega))}{T(\omega - n\Delta\omega)},$$

and, taking the inverse transform,

$$f(t) = \sum_{n=-\infty}^{\infty} F(n\Delta\omega) \frac{\Delta\omega}{2\pi} e^{jn\Delta\omega t} p_T(t).$$

The energy in $f(t)$ is

$$E = \frac{\Delta\omega}{2\pi} \sum_{n=-\infty}^{\infty} |F(n\Delta\omega)|^2.$$

2.5.5 *Fundamental sampling formulas and Poisson's formula*

As long as either $f(t)$ or its Fourier transform, $F(\omega)$, is absolutely integrable and has a bounded derivative, then

$$\Delta t \sum_{n=-\infty}^{\infty} f(t - n\Delta t) = \sum_{n=-\infty}^{\infty} F(n\Delta\omega) e^{jn\Delta\omega t} \tag{2.5.5.1}$$

and

$$\frac{\Delta\omega}{2\pi} \sum_{n=-\infty}^{\infty} F(\omega - n\Delta\omega) = \sum_{n=-\infty}^{\infty} f(n\Delta t) e^{-jn\Delta t \omega} \tag{2.5.5.2}$$

where Δt and $\Delta\omega$ are positive constants with $\Delta t \Delta\omega = 2\pi$. Using these formulas it is possible to derive the sampling theorems for band-limited functions and for finite duration functions. While these formulas are not valid for periodic functions, they can be used to derive the classical Fourier series expansion for periodic functions and hence, can also be viewed as generalizations of the Fourier series expansion for periodic functions. Letting $t = 0$ in formula (2.5.5.1) yields Poisson's formula,

$$\Delta t \sum_{n=-\infty}^{\infty} f(n\Delta t) = \sum_{n=-\infty}^{\infty} F(n\Delta\omega). \tag{2.5.5.3}$$

These sampling formulas can be derived by a fairly straightforward use of properties of the delta and the comb functions along with the use of the convolution formulas of Subsection 2.2.9. Let

$$\phi(t) = f * \text{comb}_{\Delta t}(t).$$

Because of the properties of the delta functions making up the comb function,

$$\phi(t) = \sum_{n=-\infty}^{\infty} f * \delta(t - n\Delta t) = \sum_{n=-\infty}^{\infty} f(t - n\Delta t).$$

The Fourier transform of $\phi(t)$ is

$$\Psi(\omega) = \mathcal{F}\left[f * \mathrm{comb}_{\Delta t}(t)\right]\big|_{\omega}$$

$$= F(\omega)\Delta\omega\,\mathrm{comb}_{\Delta\omega}(\omega)$$

$$= \Delta\omega \sum_{n=-\infty}^{\infty} F(\omega)\delta(\omega - n\Delta\omega)$$

$$= \Delta\omega \sum_{n=-\infty}^{\infty} F(n\Delta\omega)\delta(\omega - n\Delta\omega).$$

Thus,

$$\Delta t \sum_{n=-\infty}^{\infty} f(t - n\Delta t) = \Delta t\phi(t)$$

$$= \mathcal{F}^{-1}\left[\Delta t\Psi(\omega)\right]\big|_{t}$$

$$= \mathcal{F}^{-1}\left[\Delta t\Delta\omega \sum_{n=-\infty}^{\infty} F(n\Delta\omega)\delta(\omega - n\Delta\omega)\right]\Bigg|_{t}$$

$$= 2\pi \sum_{n=-\infty}^{\infty} F(n\Delta\omega)\mathcal{F}^{-1}\left[\delta(\omega - n\Delta\omega)\right]\big|_{t}$$

$$= \sum_{n=-\infty}^{\infty} F(n\Delta\omega)e^{jn\Delta\omega t},$$

which is formula (2.5.5.1).

Similar computations yield formula (2.5.5.2).

Example 2.5.5.1   Evaluation of an infinite series

To evaluate

$$\sum_{n=-\infty}^{\infty} \frac{1}{1 + n^2},$$

observe that

$$\sum_{n=-\infty}^{\infty} \frac{1}{1 + n^2} = \sum_{n=-\infty}^{\infty} F(n\Delta\omega),$$

where

$$F(\omega) = \frac{1}{1 + \omega^2} \qquad \text{and} \qquad \Delta\omega = 1.$$

The Fourier inverse transform of $F(\omega)$ is $f(t) = \frac{1}{2}e^{-|t|}$, and so, by Poisson's formula,

$$\sum_{n=-\infty}^{\infty} \frac{1}{1+n^2} = \sum_{n=-\infty}^{\infty} F(n\Delta\omega)$$

$$= 2\pi \sum_{n=-\infty}^{\infty} f(n2\pi)$$

$$= 2\pi \sum_{n=-\infty}^{\infty} \frac{1}{2}e^{-|n2\pi|}$$

$$= 2\pi \left[\frac{1}{2} + \sum_{n=1}^{\infty} (e^{-2\pi})^n \right].$$

The last summation is simply a geometric series. Using the well-known formula for summing geometric series,

$$\sum_{n=-\infty}^{\infty} \frac{1}{1+n^2} = 2\pi \left[\frac{1}{2} + \frac{e^{-2\pi}}{1 - e^{-2\pi}} \right] = \pi \frac{1 + e^{-2\pi}}{1 - e^{-2\pi}}. \qquad \blacksquare$$

2.6   Linear systems

Much of signal processing can be described in terms of systems that can be readily analyzed using the Fourier transform. This section gives a brief introduction to such systems and how Fourier analysis is employed to study their behavior.

Mathematically, a system, S, is an operator that takes, as input, any function, $f_I(t)$, from the set of functions pertinent to the problem at hand (say, finite energy functions) and modifies the inputted function according to some fixed scheme to produce a corresponding function, $f_O(t)$, as output. This is denoted by either

$$S : f_I(t) \rightarrow f_O(t),$$

or

$$f_O(t) = S[f_I(t)].$$

As indicated, the input function and corresponding output function will, throughout this section, be denoted via the "I" and "O" subscripts. The output, $f_O(t)$, is also called the system's response to $f_I(t)$.

2.6.1   *Linear shift invariant systems*

A system, S, is said to be linear if every linear combination of inputs leads to the corresponding linear combination of outputs. More precisely, S is linear if, given any pair of inputs, $f_I(t)$ and $g_I(t)$, and any pair of constants, α and β, then

$$S[\alpha f_I(t) + \beta g_I(t)] = \alpha f_O(t) + \beta g_O(t).$$

A system, S, is said to be shift invariant if any shift in an input function leads to an identical shift in the output, that is, if

$$S[f_I(t - t_0)] = f_O(t - t_0)$$

for every real value of t_0 and every allowed input, $f_I(t)$. Other terms commonly used instead of "shift invariant" include "translation invariant," "time invariant," "stationary," and "fixed."

An LSI system is a system that is both linear and shift invariant. If S is an LSI system, then, using both linearity and shift invariance, the following string of equalities can be verified:

$$S[f_I * g_I(t)] = S\left[\int_{-\infty}^{\infty} f_I(s)g_I(t - s)\, ds\right]$$

$$= \int_{-\infty}^{\infty} f_I(s)S[g_I(t - s)]\, ds$$

$$= \int_{-\infty}^{\infty} f_I(s)g_O(t - s)\, ds$$

$$= f_I * g_O(t). \tag{2.6.1.1}$$

Given an LSI system, S, the system's impulse response function, usually denoted by $h(t)$, is the output corresponding to an inputted delta function,

$$h(t) = S[\delta(t)].$$

The transfer function of the system is the Fourier transform of the impulse response function,

$$H(\omega) = \mathcal{F}[h(t)]|_{\omega}.$$

Combining equation (2.6.1.1) with the fact that $f_I * \delta(t) = f_I(t)$ leads directly to the following important formula for computing the output of a system from any input:

$$f_O(t) = S[f_I(t)] = f_I * h(t). \tag{2.6.1.2}$$

Taking the Fourier transform gives the equally important formula

$$F_O(\omega) = F_I(\omega)H(\omega), \tag{2.6.1.3}$$

where $F_O(\omega)$ and $F_I(\omega)$ are the transforms

$$F_O(\omega) = \mathcal{F}[f_O(t)]|_{\omega} \qquad \text{and} \qquad F_I(\omega) = \mathcal{F}[f_I(t)]|_{\omega}.$$

Formulas (2.6.1.2) and (2.6.1.3) show that the effect of an LSI system on a signal is completely determined by either the system's impulse response function or the system's transfer function. One advantage of knowing the transfer function is that, in many cases, the transfer function provides better intuition on the effect the system has on inputted signals. Also, in many cases, the actual computations are easier using the transfer function instead of the impulse response function. Both advantages are illustrated in Example 2.6.1.1.

Example 2.6.1.1   Ideal low-pass filter
An ideal low-pass filter with cutoff frequency Ω (and zero delay) is an LSI system characterized by the transfer function

$$H(\omega) = p_\Omega(\omega) = \begin{cases} 1, & \text{if } |\omega| < \Omega \\ 0, & \text{if } \Omega < |\omega| \end{cases}.$$

The impulse response function is

$$h(t) = \mathcal{F}^{-1}\left[p_\Omega(\omega)\right]\big|_t = \frac{\sin(\Omega t)}{\pi t}.$$

Given an input signal, $f_I(t)$, with Fourier transform $F_I(\omega)$, the corresponding output, $f_O(t)$, is the inverse transform of

$$F_I(\omega)H(\omega) = \begin{cases} F(\omega), & \text{if } |\omega| < \Omega \\ 0, & \text{if } \Omega < |\omega| \end{cases}.$$

Clearly, this system passes, unaltered, the frequency components of $f_I(t)$ corresponding to frequencies below the cutoff while completely suppressing the frequency components of $f_I(t)$ corresponding to frequencies above the cutoff. For example, if

$$f_I(t) = \sin(\omega_0 t),$$

then

$$F_I(\omega) = \mathcal{F}[\sin(\omega_0 t)]\big|_\omega = -j\pi\left[\delta(\omega - \omega_0) - \delta(\omega + \omega_0)\right].$$

Thus,

$$F_O(\omega) = -j\pi\left[\delta(\omega - \omega_0) - \delta(\omega + \omega_0)\right]p_\Omega(\omega)$$

$$= -j\pi\left[p_\Omega(\omega_0)\delta(\omega - \omega_0) - p_\Omega(\omega_0)\delta(\omega + \omega_0)\right]$$

$$= \begin{cases} -j\pi\left[\delta(\omega - \omega_0) - \delta(\omega + \omega_0)\right], & \text{if } |\omega_0| < \Omega \\ 0, & \text{if } \Omega < |\omega_0| \end{cases}$$

and

$$f_O(t) = \mathcal{F}^{-1}\left[F_O(\omega)\right]\big|_t = \begin{cases} \sin(\omega_0 t), & \text{if } |\omega_0| < \Omega \\ 0, & \text{if } \Omega < |\omega_0| \end{cases}.$$

Alternatively, $f_O(t)$ could have been computed using the impulse response function,

$$f_O(t) = \sin(\omega_0 t) * \frac{\sin(\Omega t)}{\pi t} = \int_{-\infty}^{\infty} \frac{1}{\pi(t - s)}\sin(\omega_0 s)\sin(\Omega(t - s))\,ds. \qquad \blacksquare$$

In many applications it is convenient to write the transfer function in the form

$$H(\omega) = A(\omega)e^{-j\theta(\omega)}$$

where $A(\omega)$ and $\theta(\omega)$ are real-valued functions (called, respectively, the amplitude and phase of $H(\omega)$) with $A(\omega)$ often assumed to be nonnegative.

Example 2.6.1.2
In the simplest case, $A(\omega)$ is constant and $\phi(\omega)$ is linear,

$$A(\omega) = A_0 \qquad \text{and} \qquad \phi(\omega) = \tau_0\omega.$$

In this case,

$$f_O(t) = \mathcal{F}^{-1}\left[F_I(\omega)A_0 e^{-j\tau_0\omega}\right]\big|_t = A_0 f_I(t - \tau_0).$$

Thus, a system with transfer function

$$H(\omega) = A_0 e^{-j\tau_0 \omega}$$

amplifies each inputted signal by A_0 and delays it by τ_0.   ∎

2.6.2   *Reality and stability*

An LSI system is a "real" system if the output is a real-valued function whenever the input is a real-valued function. In practice, most physically defined systems can be assumed to be real. An equivalent condition for a system to be real is that the impulse response function, $h(t)$, of the system be real valued. By the discussion in Subsection 2.3.1, if the system is real and the transfer function is given by

$$H(\omega) = A(\omega) e^{-j\theta(\omega)},$$

where $A(\omega)$ and $\theta(\omega)$ are the amplitude and phase of $H(\omega)$, then

$$h(t) = \frac{1}{\pi} \int_0^\infty A(\omega) \cos(\omega t - \theta(\omega)) \, d\omega.$$

An LSI system is stable if there is a finite constant, B, such that

$$|f_O(t)| \le BM \qquad \text{for all } t$$

whenever

$$|f_I(t)| \le M \qquad \text{for all } t.$$

It can be shown that a system is stable if and only if its impulse response function, $h(t)$, is absolutely integrable and that, in this case,

$$B = \int_{-\infty}^\infty |h(t)| \, dt.$$

It follows from the discussion in Subsection 2.3.2 that if the transfer function, $H(\omega)$, is not bounded and continuous, then the system cannot be stable.

Example 2.6.2.1
Let S be the ideal low-pass filter of Example 2.6.1.1. Because the impulse response function,

$$h(t) = \frac{1}{\pi t} \sin(\Omega t),$$

is real, so is the system. This is obvious because, if a given input, $f_I(t)$, is real valued, so must be

$$f_O(t) = f_I * h(t) = \int_{-\infty}^\infty f_I(s) \frac{1}{\pi(t-s)} \sin(\Omega(t-s)) \, ds.$$

Because the transfer function,

$$H(\omega) = p_\Omega(\omega) = \begin{cases} 1, & \text{if } |\omega| < \Omega \\ 0, & \text{if } \Omega < |\omega| \end{cases},$$

is not continuous, the system cannot be stable. This is easily verified using the input function

$$f_I(t) = \begin{cases} +1, & \text{if } 0 \le \frac{1}{t} \sin(\Omega t) \\ -1, & \text{if } \frac{1}{t} \sin(\Omega t) < 0 \end{cases}.$$

Clearly,

$$|f_I(t)| \leq 1 \qquad \text{for all } t,$$

but

$$f_I(s)h(-s) = \left| \frac{1}{s} \sin(\Omega s) \right| \qquad \text{for all } s.$$

Thus,

$$f_O(0) = f_I * h(0)$$

$$= \int_{-\infty}^{\infty} f_I(s)h(0 - s) \, ds$$

$$= \int_{-\infty}^{\infty} \left| \frac{1}{s} \sin(\Omega s) \right| \, ds$$

$$= \infty.$$ ∎

2.6.3   *System response to complex exponentials and periodic functions*

Let S be an LSI system with impulse response function $h(t)$ and transfer function $H(\omega)$. Because $\mathcal{F}[e^{j\omega_0 t}]\big|_\omega = 2\pi \delta(\omega - \omega_0)$,

$$S\left[e^{j\omega_0 t}\right] = \mathcal{F}^{-1}\left[2\pi\delta(\omega - \omega_0)H(\omega)\right]\big|_t$$

$$= \mathcal{F}^{-1}\left[2\pi\delta(\omega - \omega_0)H(\omega_0)\right]\big|_t$$

$$= H(\omega_0)e^{j\omega_0 t}. \tag{2.6.3.1}$$

By this it is seen that the complex exponentials are eigenfunctions for S and that the transfer function gives the corresponding eigenvalues.

If $f_I(t)$ is a periodic function with period p and with Fourier series

$$\sum_{n=-\infty}^{\infty} c_n e^{jn\Delta\omega t},$$

where $\Delta\omega = \frac{2\pi}{p}$, then, from equation (2.6.3.1) and the linearity of the system,

$$S[f_I(t)] = \sum_{n=-\infty}^{\infty} c_n H(n\Delta\omega)e^{jn\Delta\omega t}. \tag{2.6.3.2}$$

In particular, for $\alpha > 0$,

$$S[\cos(\alpha t)] = \frac{1}{2}\left[H(\alpha)e^{j\alpha t} + H(-\alpha)e^{-j\alpha t}\right] \tag{2.6.3.3}$$

and

$$S[\sin(\alpha t)] = \frac{1}{2j}\left[H(\alpha)e^{j\alpha t} - H(-\alpha)e^{-j\alpha t}\right]. \tag{2.6.3.4}$$

If S is a real LSI system, then the imaginary parts of (2.6.3.3) and (2.6.3.4) must vanish. Using this fact it can be shown that

$$S[\cos(\alpha t)] = A(\alpha)\cos(\alpha t - \theta(\alpha))$$

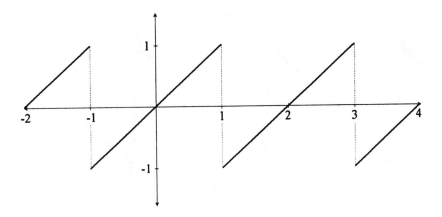

FIGURE 2.8
The saw function.

and

$$S[\sin(\alpha t)] = A(\alpha)\sin(\alpha t - \theta(\alpha)),$$

where $A(\omega)$ and $\theta(\omega)$ are the amplitude and phase of the transfer function,

$$H(\omega) = A(\omega)e^{-j\theta(\omega)}.$$

Example 2.6.3.1
Let S be the ideal low-pass filter from Example 2.6.1.1 with transfer function

$$H(\omega) = p_\Omega(\omega) = \begin{cases} 1, & \text{if } |\omega| < \Omega \\ 0, & \text{if } \Omega < |\omega| \end{cases}.$$

For this example assume $\Omega = (20 + \frac{1}{2})\pi$ and let $f_I(t)$ be the sawtooth function from Example 2.3.9.1. As seen in that example, $\Delta\omega = \pi$ and

$$f_I(t) = \sum_{\substack{n=-\infty \\ n\neq 0}}^{\infty} (-1)^n \frac{j}{n\pi} e^{jn\pi t}.$$

From this and equation (2.6.3.2) it follows that

$$f_O(t) = \sum_{\substack{n=-20 \\ n\neq 0}}^{20} (-1)^n \frac{j}{n\pi} e^{jn\pi t}.$$

The graphs of $f_I(t)$ and $f_O(t)$ are sketched in Figures 2.8 and 2.9, respectively. ∎

2.6.4 *Causal systems*

A function, $f(t)$, is said to be "causal" if

$$f(t) = 0, \qquad \text{whenever } t < 0.$$

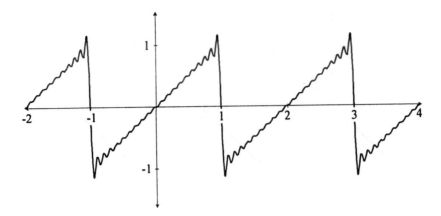

FIGURE 2.9
Low-pass filter output from a saw function input.

An LSI system, S, is said to be "causal" if the response of the system to every causal input is a causal output. By shift invariance, this is equivalent to defining S to be causal if

$$f_I(t) = 0, \qquad \text{whenever } t < t_0$$

implies that

$$f_O(t) = S[f_I(t)] = 0, \qquad \text{whenever } t < t_0$$

for any real value of t_0.

If S is a causal system, then its impulse response function, $h(t)$, must also be causal and formula (2.6.1.2) for computing the response of a system to an input $f_I(t)$ becomes

$$f_O(t) = \int_{-\infty}^{t} f_I(s)h(t-s)\,ds$$

or equivalently,

$$f_O(t) = \int_{0}^{\infty} f_I(t-s)h(s)\,ds.$$

If the input is also causal, then these further reduce to

$$f_O(t) = \int_{0}^{t} f_I(s)h(t-s)\,ds$$

and

$$f_O(t) = \int_{0}^{t} f_I(t-s)h(s)\,ds.$$

2.6.5 *Systems given by differential equations*

Often the output, $f_O(t)$, of a system, S, is a solution to a nonhomogeneous ordinary differential equation with the input being the nonhomogeneous part of the equation,

$$\sum_{n=0}^{N} A_n \frac{d^n}{dt^n}[f_O(t)] = f_I(t).$$

As long as the A_n's are constants, it is easily verified that S is an LSI system. The impulse response function, $h(t)$, must satisfy

$$\sum_{n=0}^{N} A_n \frac{d^n h}{dt^n} = \delta(t). \tag{2.6.5.1}$$

The general solution to (2.6.5.1) can be written

$$h_g(t) = h_p(t) + y_c(t),$$

where $y_c(t)$ is the general solution to the corresponding homogeneous equation,

$$\sum_{n=0}^{N} A_n \frac{d^n y}{dt^n} = 0, \tag{2.6.5.2}$$

and $h_p(t)$ is any particular solution to (2.6.5.1). After a particular solution is found the undetermined constants in $y_c(t)$ must be determined so that the resulting

$$h(t) = h_p(t) + y_c(t)$$

satisfies any additional constraints on the output (causality, finite energy, etc.).

The particular solution, $h_p(t)$, can be found by taking the Fourier transform of both sides of (2.6.5.1). Using identity (2.2.11.6) gives an equation for $H_p(\omega) = \mathcal{F}[h_p(t)]\big|_\omega$,

$$\sum_{n=0}^{N} A_n (j\omega)^n H_p(\omega) = 1. \tag{2.6.5.3}$$

Dividing through by

$$D(\omega) = \sum_{n=0}^{N} A_n (j\omega)^n$$

gives

$$H_p(\omega) = \frac{1}{D(\omega)}, \tag{2.6.5.4}$$

which is a rational function of ω. Taking the inverse transform of $H_p(\omega)$ (using, say, the approach described in Subsection 2.3.14) then yields $h_p(t)$.

Example 2.6.5.1 illustrates a common situation in which the obtained $h_p(t)$ already satisfies the additional conditions and thus, can be used directly as the impulse response function. In Example 2.6.5.2, $h_p(t)$ is not a valid output and, so, a nontrivial solution to the corresponding homogeneous equation must be added to obtain the impulse response function.

Example 2.6.5.1
Let the output, $f_O(t)$, corresponding to an input, $f_I(t)$, be given by the finite energy solution to

$$\frac{d^2 y}{dt^2} - y = f_I(t).$$

The solution to the corresponding homogeneous equation,

$$\frac{d^2 y}{dt^2} - y = 0,$$

is

$$y_c(t) = c_1 e^t + c_2 e^{-t},$$

while the impulse response function must satisfy

$$\frac{d^2h}{dt^2} - h = \delta(t). \qquad (2.6.5.5)$$

Letting $h_p(t)$ denote a particular solution, and taking the Fourier transform of both sides yields

$$-\omega^2 H_p(\omega) - H_p(\omega) = 1,$$

which, after some elementary algebra, reduces to

$$H_p(\omega) = \frac{-1}{1 + \omega^2}.$$

The inverse transform of this can be computed directly from tables:

$$h_p(t) = -\frac{1}{2} \mathcal{F}^{-1} \left[\frac{2}{1 + \omega^2} \right] \bigg|_t = -\frac{1}{2} e^{-|t|}.$$

The general solution to (2.6.5.5) is the sum of $h_p(t)$ and $y_c(t)$,

$$h_g(t) = -\frac{1}{2} e^{-|t|} + c_1 e^t + c_2 e^{-t},$$

but, clearly, the only way $h_g(t)$ can be a finite energy function is for $c_1 = c_2 = 0$. Thus, the impulse response function and the transfer function for this system are

$$h(t) = -\frac{1}{2} e^{-|t|}$$

and

$$H(\omega) = \frac{-1}{1 + \omega^2}. \qquad \blacksquare$$

Example 2.6.5.2
Assume $S : f_I(t) \to f_O(t)$ is a causal system for which the output satisfies

$$\frac{d^2 f_O}{dt^2} + f_O = f_I(t).$$

The solution to the corresponding homogeneous equation,

$$\frac{d^2 y}{dt^2} + y = 0,$$

is

$$y_c(t) = c_1 \cos(t) + c_2 \sin(t),$$

while the impulse response function must satisfy

$$\frac{d^2h}{dt^2} + h = \delta(t). \qquad (2.6.5.6)$$

Letting $h_p(t)$ denote a particular solution, and taking the Fourier transform of both sides yields

$$-\omega^2 H_p(\omega) + H_p(\omega) = 1$$

which, after some elementary algebra, reduces to

$$H_p(\omega) = \frac{1}{1 - \omega^2} = \frac{1}{2}\left[\frac{1}{\omega + 1} - \frac{1}{\omega - 1}\right].$$

Using either the tables or formula (2.3.14.1),

$$h_p(t) = \frac{1}{2}\left[\frac{j}{2}e^{-jt}\,\mathrm{sgn}(t) - \frac{j}{2}e^{jt}\,\mathrm{sgn}(t)\right] = \frac{1}{2}\sin(t)\,\mathrm{sgn}(t).$$

The general solution to (2.6.5.6) is then

$$h_g(t) = h_p(t) + y_c(t) = \frac{1}{2}\sin(t)\,\mathrm{sgn}(t) + c_1\cos(t) + c_2\sin(t).$$

Because S is a causal system and $\delta(t) = 0$ for $t < 0$, the impulse response function must vanish for negative values of t. Thus, c_1 and c_2 must be chosen so that for $t < 0$,

$$h_g(t) = \frac{1}{2}\sin(t)\,\mathrm{sgn}(t) + c_1\cos(t) + c_2\sin(t)$$

$$= \left(c_2 - \frac{1}{2}\right)\sin(t) + c_1\cos(t)$$

$$= 0.$$

Clearly, $c_1 = 0$, $c_2 = \frac{1}{2}$, and the impulse response function is

$$h(t) = \frac{1}{2}\sin(t)\,\mathrm{sgn}(t) + \frac{1}{2}\sin(t) = \sin(t)u(t).$$

The transfer function is

$$H(\omega) = \mathcal{F}\left[\sin(t)u(t)\right]\big|_\omega$$

$$= \frac{1}{2j}\left[\left(\pi\delta(\omega - 1) - j\frac{1}{\omega - 1}\right) - \left(\pi\delta(\omega + 1) - j\frac{1}{\omega + 1}\right)\right]$$

$$= \frac{1}{1 - \omega^2} + j\frac{\pi}{2}[\delta(\omega + 1) - \delta(\omega - 1)].$$

∎

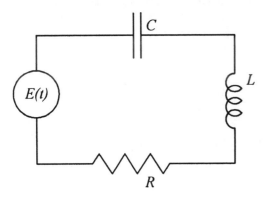

FIGURE 2.10
A simple RLC circuit.

2.6.6 *RLC circuits*

Consider the electric circuit sketched in Figure 2.10. This circuit consists of

> a resistor with a fixed resistance of R ohms,
>
> an inductor with a fixed inductance of L henries,
>
> a capacitor with a fixed capacitance of C farads, and
>
> a time-varying voltage supply providing a voltage of $E(t)$ volts.

The charge on the capacitor at time t will be denoted by $q(t)$ and the corresponding current in the circuit by $i(t)$. The charge and current are related by

$$i(t) = \frac{dq}{dt}.$$

By Kirchhoff's laws

$$L\frac{d^2q}{dt^2} + R\frac{dq}{dt} + \frac{1}{C}q = E(t). \tag{2.6.6.1}$$

For a physical circuit R must be positive and L and C cannot be negative. Also, it is reasonable to assume that no charge accumulates on the capacitor if no voltage has previously been provided. Thus, $q(t)$ can be viewed as the output corresponding to an input of $E(t)$ to a causal LSI system. The impulse response function, $h(t)$, to this system satisfies

$$L\frac{d^2h}{dt^2} + R\frac{dh}{dt} + \frac{1}{C}h = \delta(t). \tag{2.6.6.2}$$

   If the inductance and capacitance are nonzero, then straightforward computations, similar to those done in the examples of Subsections 2.6.5 and 2.3.14 lead to

$$h(t) = \frac{1}{2L\beta}[e^{\beta t} - e^{-\beta t}]e^{-\alpha t}u(t)$$

and

$$H(\omega) = \frac{j}{2L\beta}\left[\frac{1}{\omega - j(\alpha + \beta)} - \frac{1}{\omega - j(\alpha - \beta)}\right] = \frac{-C}{LC\omega^2 - jRC\omega - 1},$$

where

$$\alpha = \frac{R}{2L} \quad \text{and} \quad \beta = \sqrt{\left(\frac{R}{2L}\right)^2 - \frac{1}{LC}}.$$

It should be noted that, because the real part of $\alpha \pm \beta$ is positive, $h(t)$ is bounded by a decreasing exponential on $0 < t$. Hence, $h(t)$ is absolutely integrable and the system is stable.

In practice, the current, $i(t)$, is often of greater interest than the charge on the capacitor. Because

$$i(t) = \frac{dq}{dt} = \frac{d}{dt}(E * h(t)) = E * h'(t),$$

it follows that the current is given by a system with impulse response function

$$h_i(t) = h'(t) = \frac{1}{2L\beta}\left[(\beta - \alpha)e^{\beta t} + (\beta + \alpha)e^{-\beta t}\right]e^{-\alpha t}u(t)$$

and transfer function

$$H_i(\omega) = j\omega H(\omega) = \frac{-jC\omega}{LC\omega^2 - jRC\omega - 1}.$$

In either case the response of the system to the impulse function will depend on whether β has an imaginary component. If β does have a nonzero imaginary component the unit impulse response will be a sinusoidal function with an exponentially decreasing envelope. If the imaginary part of β is zero, then the response is simply a linear combination of decreasing exponentials.

2.6.7   Modulation and demodulation

Let $f(t)$ be any band-limited function with bandwidth 2Ω, and let ω_c and t_0 be real constants with $\Omega < \omega_c$. The product

$$g(t) = f(t)\cos(\omega_c t - t_0)$$

is the "modulation of the carrier signal, $\cos(\omega_t - t_0)$, by $f(t)$." The extraction of the modulating signal, $f(t)$, from the modulated signal provides an especially nice example of the application of Fourier analysis in signal processing.

To extract $f(t)$, first multiply $g(t)$ by the carrier signal. This gives

$$g(t)\cos(\omega_c t - t_0) = f(t)\cos^2(\omega_c t - t_0)$$

$$= f(t)\left[\frac{1}{2} + \frac{1}{2}\cos(2\omega_c t - 2t_0)\right]$$

$$= \frac{1}{2}f(t) + \frac{1}{2}f(t)\cos(2\omega_c t - 2t_0).$$

Using the basic identities, the Fourier transform of this is found to be

$$\frac{1}{2}F(\omega) + \frac{1}{2}F(\omega) * \left[\pi e^{-j4t_0\omega_c}\delta(\omega - 2\omega_c) + \pi e^{j4t_0\omega_c}\delta(\omega + 2\omega_c)\right]$$

$$= \frac{1}{2}F(\omega) + \frac{\pi}{2}\left[e^{-j4t_0\omega_c}F(\omega - 2\omega_c) + e^{j4t_0\omega_c}F(\omega + 2\omega_c)\right].$$

Sketches of $F(\omega)$, $F(\omega - 2\omega_c)$, and $F(\omega + 2\omega_c)$ are given in Figure 2.11. Observe that because

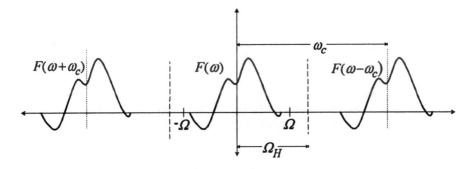

FIGURE 2.11
Translations of the transform of a band-limited function.

$f(t)$ is a band-limited function with bandwidth 2Ω and $\Omega < \omega_c$, if $H(\omega)$ is any band-limited function with bandwidth $2\Omega_H$ satisfying $\Omega_H < 2\omega_c - \Omega$, then

$$\left(\frac{1}{2}F(\omega) + \frac{\pi}{2}\left[e^{-j4t_0\omega_c}F(\omega - 2\omega_c) + e^{j4t_0\omega_c}F(\omega + 2\omega_c) \right] \right)H(\omega) = \frac{1}{2}F(\omega)H(\omega).$$

In particular if $H(\omega)$ is the perfect low-pass filter of Examples 2.6.1.1 and 2.6.3.1,

$$H(\omega) = \begin{cases} 1, & \text{if } |\omega| < \Omega_H \\ 0, & \text{if } |\omega| > \Omega_H \end{cases},$$

with

$$\Omega < \Omega_H < 2\omega_c - \Omega,$$

then

$$\left(\frac{1}{2}F(\omega) + \frac{\pi}{2}\left[e^{-j4t_0\omega_c}F(\omega - 2\omega_c) + e^{j4t_0\omega_c}F(\omega + 2\omega_c) \right] \right)H(\omega) = \frac{1}{2}F(\omega).$$

Thus, the signal

$$g(t) = f(t)\cos(\omega_c t - t_0)$$

can be perfectly demodulated (i.e., $f(t)$ can be completely extracted) by first multiplying the modulated signal by the carrier and then passing the result through an appropriate ideal low-pass filter.

2.7   Random variables

Because noise is an intrinsic factor in real-world systems, random variables play an important role in the mathematics of practical engineering problems. As illustrated in this section the Fourier transform is a useful tool in analyzing signals containing a significant random component and in extracting usable information from these signals.

2.7.1   Basic probability and statistics

A nonnegative function, $p(x)$, is a probability density function if it satisfies

$$\int_{-\infty}^{\infty} p(x)\,dx = 1.$$

Such a function is absolutely integrable and so, as noted in Subsection 2.3.2, its Fourier transform, $P(y) = \mathcal{F}[p(x)]|_y$, must be continuous and must satisfy $P(0) = 1$.

If x denotes the outcome of a random process governed by the probability density function $p(x)$ and if $-\infty \le a \le b \le \infty$, then

$$\int_{a}^{b} p(x)\,dx$$

is the probability that x is between a and b. The "mean" or "expected value" of x, is denoted by either μ or $E[x]$, and is given by the first moment of $p(x)$,

$$\mu = E[x] = \int_{-\infty}^{\infty} x p(x)\,dx.$$

The variance of x, denoted by either σ^2 or $\text{Var}[x]$, is the second moment of $p(x)$ about its mean,

$$\sigma^2 = \text{Var}[x] = \int_{-\infty}^{\infty} (x - \mu)^2 p(x)\,dx,$$

and the standard deviation, σ, is the square root of the variance. More generally, if $f(x)$ is any function of x then $f(x)$ is a random variable with expected value

$$\mu = E[f(x)] = \int_{-\infty}^{\infty} f(x) p(x)\,dx$$

and variance

$$\text{Var}[f(x)] = \int_{-\infty}^{\infty} |f(x) - \mu|^2 p(x)\,dx.$$

In particular, $\text{Var}[x] = E[|x - \mu|^2]$. It is easy to show that the variance of x, σ^2, is directly related to the first and second moments of $p(x)$,

$$\sigma^2 = E[x^2] - \mu^2 = \int_{-\infty}^{\infty} x^2 p(x)\,dx - \left(\int_{-\infty}^{\infty} x p(x)\,dx \right)^2.$$

It follows from the discussion in Subsections 2.2.11 and 2.2.12 that if $p(x)$ is a probability density function, then the corresponding mean, expected value of x^2, and variance can be computed from the density function's transform, $P(y) = \mathcal{F}[p(x)]|_y$, by

$$\mu = j P'(0), \tag{2.7.1.1}$$

$$E[x^2] = -P''(0), \tag{2.7.1.2}$$

and

$$\sigma^2 = [P'(0)]^2 - P''(0). \tag{2.7.1.3}$$

Example 2.7.1.1   The normal distribution
A normal (or Gaussian) probability distribution is given by the density function

$$p(x) = \sqrt{\frac{\alpha}{\pi}} e^{-\alpha(x - x_0)^2},$$

where $\alpha > 0$. Using the tables it is easily verified that

$$P(y) = \mathcal{F}[p(x)]|_y = \exp\left[-\frac{1}{4\alpha}y^2 - jx_0y\right].$$

Furthermore,

$$P(0) = 1,$$

$$P'(y) = -\left(\frac{1}{2\alpha}y + jx_0\right)P(y),$$

and

$$P''(y) = \frac{1}{4\alpha^2}\left[(y + j2\alpha x_0)^2 - 2\alpha\right]P(y).$$

Using formulas (2.7.1.1) through (2.7.1.3) to compute the mean and variance,

$$\mu = jP'(0) = -j(0 + jx_0)P(0) = x_0,$$

$$E[x^2] = -P''(0) = -\frac{1}{4\alpha^2}\left[(0 + j2\alpha x_0)^2 - 2\alpha\right]P(0) = x_0^2 + \frac{1}{2\alpha},$$

and

$$\sigma^2 = \left[P'(0)\right]^2 - P''(0) = \frac{1}{2\alpha}.$$

Replacing x_0 and α in the above formulas for $p(x)$ and $P(y)$ it follows that the normal probability distribution with mean μ and standard deviation σ is given by the density function

$$p(x) = \frac{1}{\sigma\sqrt{2\pi}}\exp\left[-\frac{1}{2}\left(\frac{x-\mu}{\sigma}\right)^2\right],$$

and that its Fourier transform is given by

$$P(y) = \exp\left[-\frac{1}{2}(\sigma y)^2 - j\mu y\right]. \qquad\blacksquare$$

Example 2.7.1.2   *The binomial distribution*

Consider a process consisting of n repetitions of an experiment with exactly two outcomes, "success" and "failure." Let p_0 be the probability of "success" and q_0 the probability of "failure" in one experiment (hence $p_0 + q_0 = 1$). Such a process is governed by the binomial probability density function

$$p(x) = \sum_{k=0}^{n}\binom{n}{k}p_0^k q_0^{n-k}\delta(x-k)$$

with

$$\int_a^b p(x)\,dx = \sum_{a<k<b}\binom{n}{k}p_0^k q_0^{n-k}$$

being the probability that the number of "successes," x, satisfies $a < x < b$.

In Example 2.3.10.3 the Fourier transform of this function was found to be

$$P(y) = \left(p_0 e^{-jy} + q_0\right)^n.$$

Assuming that $n > 1$,

$$P'(y) = -jnp_0 e^{-jy} \left(p_0 e^{-jy} + q_0\right)^{n-1}$$

and

$$P''(y) = -np_0 e^{-jy} \left(np_0 e^{-jy} + q_0\right) \left(p_0 e^{-jy} + q_0\right)^{n-2}.$$

Thus,

$$P'(0) = -jnp_0 \quad \text{and} \quad P''(0) = -np_0(np_0 + q_0).$$

So, using formulas (2.7.1.1) through (2.7.1.3) to compute the mean and variance,

$$\mu = jP'(0) = np_0,$$

$$E[x^2] = -P''(0) = -np_0(np_0 + q_0)$$

and

$$\sigma^2 = \left[P'(0)\right]^2 - P''(0) = np_0 q_0. \qquad \blacksquare$$

2.7.2   Multiple random processes and independence

Let x_1 and x_2 denote the outcomes of two random processes governed, respectively, by probability density functions $p_1(x_1)$ and $p_2(x_2)$, and with corresponding means μ_1 and μ_2, and corresponding standard deviations σ_1 and σ_2. Taken as a single pair, (x_1, x_2) can be viewed as the outcomes of a single two-dimensional random process. This process will be governed by a probability density function of two variables, $q(x_1, x_2)$. Given any $-\infty \leq a_1 \leq b_1 \leq \infty$ and $-\infty \leq a_2 \leq b_2 \leq \infty$, the probability that both

$$a_1 < x_1 < b_1 \quad \text{and} \quad a_2 < x_2 < b_2$$

is

$$\int_{a_1}^{b_1} \int_{a_2}^{b_2} q(x_1, x_2)\, dx_1\, dx_2.$$

In general, the relationship between the joint density function, $q(x_1, x_2)$, and the individual density functions, $p_1(x_1)$ and $p_2(x_2)$, depends strongly on the relationship that exists between the two random processes. If x_1 and x_2 are, in fact, the same, then

$$q(x_1, x_2) = p_1(x_1)\delta(x_2 - x_1). \qquad (2.7.2.1)$$

If the two random processes are completely independent of each other, then, for all values of x_1 and x_2,

$$q(x_1, x_2) = p_1(x_1)p_2(x_2). \qquad (2.7.2.2)$$

Example 2.7.2.1

Let x denote the number of heads resulting from a single toss of a fair coin, and let x_1 and x_2 be the number of heads reported by two perfectly accurate observers, each observing the single toss of a fair coin. The probability density function for x, $p(x)$, is well known to be

$$p(x) = \frac{1}{2}\delta(x) + \frac{1}{2}\delta(x - 1).$$

If both observers are observing the same coin toss then $x_1 = x_2$ and, according to formula (2.7.2.1), the joint probability density function is

$$q_{\text{same}}(x_1, x_2) = \left[\frac{1}{2}\delta(x_1) + \frac{1}{2}\delta(x_1 - 1) \right] \delta(x_2 - x_1). \qquad (2.7.2.3)$$

Note that if α is any real number and $\phi(x_1, x_2)$ is any two-dimensional test function, then

$$\int_{-\infty}^{\infty} \int_{-\infty}^{\infty} \delta(x_1 - \alpha)\delta(x_1 - x_2)\phi(x_1, x_2)\, dx_1\, dx_2$$

$$= \phi(\alpha, \alpha) = \int_{-\infty}^{\infty} \int_{-\infty}^{\infty} \delta(x_1 - \alpha, x_2 - \alpha)\phi(x_1, x_2)\, dx_1\, dx_2.$$

This shows that, in general,

$$\delta(x_1 - \alpha)\delta(x_1 - x_2) = \delta(x_1 - \alpha, x_2 - \alpha),$$

which, in turn, verifies that formula (2.7.2.3) is completely equivalent to the formula

$$q_{\text{same}}(x_1, x_2) = \frac{1}{2}\delta(x_1, x_2) + \frac{1}{2}\delta(x_1 - 1, x_2 - 1),$$

obtained by elementary probability theory.

On the other hand, if the two observers are observing two different tosses of the coin, then the value of x_1 and x_2 are independent of each other and the joint density function is

$$q_{\text{indep}}(x_1, x_2) = \left[\frac{1}{2}\delta(x_1) + \frac{1}{2}\delta(x_1 - 1) \right]\left[\frac{1}{2}\delta(x_2) + \frac{1}{2}\delta(x_2 - 1) \right]$$

$$= \frac{1}{4}\delta(x_1)\delta(x_2) + \frac{1}{4}\delta(x_1)\delta(x_2 - 1)$$

$$+ \frac{1}{4}\delta(x_1 - 1)\delta(x_2) + \frac{1}{4}\delta(x_1 - 1)\delta(x_2 - 1),$$

which agrees with the formula

$$q_{\text{indep}}(x_1, x_2) = \frac{1}{4}\left[\delta(x_1, x_2) + \delta(x_1, x_2 - 1) + \delta(x_1 - 1, x_2) + \delta(x_1 - 1, x_2 - 1) \right]$$

obtained by elementary probability theory. ∎

Formula (2.7.2.2) gives the mathematical definition for x_1 and x_2 being independent random variables. Assuming x_1 and x_2 are independent, the mean of the product is

$$E[x_1 x_2] = \int_{-\infty}^{\infty} \int_{-\infty}^{\infty} x_1 x_2 p_1(x_1) p_2(x_2)\, dx_1\, dx_2$$

$$= \int_{-\infty}^{\infty} x_1 p_1(x_1)\, dx_1 \int_{-\infty}^{\infty} x_2 p_2(x_2)\, dx_2$$

$$= \mu_1 \mu_2. \qquad (2.7.2.4)$$

Similar computations show that

$$\text{Var}[x_1 x_2] = \sigma_1^2 \sigma_2^2.$$

It should also be noted that the two-dimensional transform of the joint probability density is

$$Q(y_1, y_2) = \int_{-\infty}^{\infty} \int_{-\infty}^{\infty} p_1(x_1) p_2(x_2) e^{-j(x_1 y_1 + x_2 y_2)} \, dx_1 \, dx_2$$

$$= \int_{-\infty}^{\infty} p_1(x_1) e^{-j x_1 y_1} \, dx_1 \int_{-\infty}^{\infty} p_2(x_2) e^{-j x_2 y_2} \, dx_2$$

$$= P_1(y_1) P_2(y_2)$$

where $P_1(y_1)$ and $P_2(y_2)$ are the Fourier transforms of $p_1(x_1)$ and $p_2(x_2)$, respectively.

More generally, any number of random variables—$x_1, x_2, \ldots, x_n$—are considered to be independent if the probability density function for the vector $(x_1, x_2, \ldots, x_n)$ is the product of the density functions of the individual variables,

$$q(x_1, x_2, \ldots, x_n) = p_1(x_1) p_2(x_2) \cdots p_n(x_n).$$

If $x_1, x_2, \ldots, x_n$ are independent, then the n-dimensional Fourier transform of the joint density function is simply the product of the one-dimensional Fourier transforms of the individual density functions,

$$Q(x_1, x_2, \ldots, x_n) = P_1(x_1) P_2(x_2) \cdots P_n(x_n),$$

and the mean of the product of the variables and the corresponding variance are merely the products of the means and variances of the individual variables,

$$E[x_1 x_2 \cdots x_n] = \mu_1 \mu_2 \cdots \mu_n$$

and

$$\mathrm{Var}[x_1 x_2 \cdots x_n] = \sigma_1^2 \sigma_2^2 \cdots \sigma_n^2.$$

2.7.3   *Sums of random processes*

Let x_1 and x_2 denote the outcomes of two independent random processes governed, respectively, by probability density functions $p_1(x)$ and $p_2(x)$, and with corresponding means μ_1 and μ_2, and corresponding standard deviations σ_1 and σ_2. The sum of these two outcomes,

$$x_S = x_1 + x_2,$$

can be viewed as the outcome of another random process, which is governed by the probability density function

$$p_S(x) = \int_{-\infty}^{\infty} p_1(x - \xi) p_2(\xi) \, d\xi = p_1 * p_2(x).$$

If $P_1(y)$, $P_2(y)$, and $P_S(y)$ are the Fourier transforms of $p_1(x)$, $p_2(x)$, and $p_S(x)$, then, by identity (2.2.9.3),

$$P_S(y) = P_1(y) P_2(y).$$

Thus,

$$P_S(0) = P_1(0) P_2(0) = 1$$

and

$$P_S'(0) = P_1'(0) P_2(0) + P_1(0) P_2'(0) = P_1'(0) + P_2'(0).$$

From this last equation and equation (2.7.1.1) it immediately follows that the mean of x is the sum of the means of x_1 and x_2,

$$\mu_S = \mu_1 + \mu_2.$$

Likewise, computing $P_S''(0)$ and using equations (2.7.1.2) and (2.7.1.3) leads to

$$E[x_S^2] = E[x_1^2] + E[x_2^2] + 2\mu_1\mu_2$$

and

$$\sigma_S^2 = \sigma_1^2 + \sigma_2^2.$$

More generally, if

$$x_S = x_1 + x_2 + \cdots + x_N,$$

where each x_n denotes the outcome of an independent random process governed by a probability density function, $p_n(x)$, and with corresponding mean and standard deviation, μ_n and σ_n, then x_S is governed by the probability density function

$$p_S(x) = p_1 * p_2 * \cdots * p_N(x)$$

and has mean and variance

$$\mu_S = \mu_1 + \mu_2 + \cdots + \mu_N,$$

and

$$\sigma_S^2 = \sigma_1^2 + \sigma_2^2 + \cdots + \sigma_N^2.$$

If N is fairly large, the central limit theorem of probability theory states that under very general conditions,

$$p_S(x) \approx \frac{1}{\sigma_S\sqrt{2\pi}} \exp\left[-\frac{1}{2}\left(\frac{x - \mu_S}{\sigma_S}\right)^2\right]$$

or, equivalently, that

$$P_S(y) = \mathcal{F}[p_S(x)]|_y \approx \exp\left[-\frac{1}{2}(\sigma_S y)^2 - j\mu_S y\right].$$

In practice, the "noise" in a system is often the result of a large number of random processes each of which contributes a term to the total noise. According to the above discussion, it is not necessary to describe each source of noise accurately. Instead, the aggregate can be treated as a random process governed by a normal distribution.

2.7.4 *Random signals and stationary random signals*

A signal, $x(t)$, is "deterministic" if it can be treated, mathematically, as a well-defined function of t, that is, if for each value of t there is a single fixed value for $x(t)$. The signal is "random" if, instead, for each value of t, $x(t)$ must be treated as the outcome of a nontrivial random process.

Assume $x(t)$ is a random signal. For each value of t there is a corresponding probability density function, $p(x, t)$, with

$$\int_a^b p(x, t)\, dx$$

being the probability of $a < x(t) < b$. The corresponding mean and variance,

$$E[x(t)] = \mu(t) = \int_{-\infty}^{\infty} x p(x, t) \, dx$$

and

$$\text{Var}[x(t)] = \sigma^2(t) = \int_{-\infty}^{\infty} (x - \mu(t))^2 p(x, t) \, dx,$$

are deterministic functions of t. The Fourier transform of the density function is

$$P(y, t) = \int_{-\infty}^{\infty} p(x, t) e^{-jxy} \, dx.$$

If the statistical properties of the process generating $x(t)$ do not vary with t, then the process is said to be a "stationary" random process. The corresponding signal will also be called "stationary" though its value will certainly depend—in a random manner—on t. For a stationary random signal, $x(t)$, it is reasonable to expect that the long term time average of $x(t)$ will equal its mean, $E[x] = \mu$,

$$\mu = \lim_{T \to \infty} \frac{1}{2T} \int_{-T}^{T} x(t) \, dt. \tag{2.7.4.1}$$

Mathematically, it can be shown that, under fairly broad conditions, the probability that equation (2.7.4.1) is not correct for a given stationary random signal is vanishingly small. Likewise,

$$E[x^2(t)] = \lim_{T \to \infty} \frac{1}{2T} \int_{-T}^{T} |x(t)|^2 \, dt$$

and

$$\sigma^2 = \lim_{T \to \infty} \frac{1}{2T} \int_{-T}^{T} |x(t) - \mu|^2 \, dt.$$

Thus, stationary random signals (with finite means and variances) can be treated as finite power functions (see Subsection 2.3.8). Given a stationary random signal, $x(t)$, the corresponding average autocorrelation function is

$$\bar{\rho}_x(t) = \lim_{T \to \infty} \frac{1}{2T} \int_{-T}^{T} x^*(s) x(t + s) \, ds.$$

The average power is

$$\bar{\rho}_x(0) = \lim_{T \to \infty} \frac{1}{2T} \int_{-T}^{T} |x(s)|^2 \, ds$$

and the power spectrum is

$$P_x(\omega) = \mathcal{F}\left[\bar{\rho}_x(t)\right]\big|_{\omega} = \int_{-\infty}^{\infty} \bar{\rho}_x(t) e^{-j\omega t} \, dt$$

or, equivalently,

$$P_x(\omega) = \lim_{T \to \infty} \frac{1}{2T} \left| \int_{-T}^{T} x(t) e^{-j\omega t} \, dt \right|^2.$$

It should be recalled that one property of the average autocorrelation function is that

$$(\bar{\rho}_x(t))^* = \bar{\rho}_x(-t). \tag{2.7.4.2}$$

Thus, if $x(t)$ is a real random signal, then the average autocorrelation will be an even real-valued function. So, also, will the power spectrum.

2.7.5   Correlation of stationary random signals and independence

The average cross-correlation of two stationary random signals, $x(t)$ and $y(t)$, is

$$\bar{\rho}_{xy}(t) = \lim_{T \to \infty} \frac{1}{2T} \int_{-T}^{T} x^*(s) y(t+s) \, ds,$$

or, equivalently,

$$\bar{\rho}_{xy}(t) = E[x^* \Theta_{-t} y]$$

where, for any α, $\Theta_\alpha y$ denotes the translation of y by α,

$$\Theta_\alpha y(s) = y(s - \alpha).$$

The corresponding cross-power spectrum is

$$P_{xy}(\omega) = \mathcal{F}\left[\bar{\rho}_{xy}(t)\right]\big|_\omega = \int_{-\infty}^{\infty} \bar{\rho}_{xy}(t) e^{-j\omega t} \, dt.$$

It should be noted that

$$\left[\bar{\rho}_{xy}(t)\right]^* = \lim_{T \to \infty} \frac{1}{2T} \left[\int_{-T}^{T} x^*(s) y(s+t) \, ds \right]^*$$

$$= \lim_{T \to \infty} \frac{1}{2T} \int_{-T}^{T} x(s) y^*(s+t) \, ds$$

$$= \lim_{T \to \infty} \frac{1}{2T} \int_{-T}^{T} x(\sigma - t) y^*(\sigma) \, d\sigma$$

$$= \bar{\rho}_{yx}(-t). \tag{2.7.5.1}$$

From Schwarz's inequality it follows that

$$\left|\bar{\rho}_{xy}(t)\right|^2 \le \bar{\rho}_x(0) \bar{\rho}_y(0).$$

A somewhat more general statement, namely, that for any $-\infty \le a < b \le \infty$,

$$\left| \int_a^b P_{xy}(\omega) e^{j\omega t} \, d\omega \right|^2 \le \int_a^b P_x(\omega) \, d\omega \int_a^b P_y(\omega) \, d\omega$$

can also be proven. This shows that if the power spectrum of a signal vanishes on an interval, then so does the cross-power spectrum of that signal with any other signal.

The average cross-correlation indicates the extent to which the two processes are independent of each other. If, for example, $x(t)$ and $y(t)$ are generated by two completely independent stationary processes, then, following the discussion in Subsection 2.7.2,

$$\bar{\rho}_{xy}(t) = \mu_x^* \mu_y \qquad \text{for all } t$$

and

$$P_{xy}(\omega) = 2\pi \mu_x^* \mu_y \delta(\omega).$$

In particular, if one of the two independent processes has mean zero, then

$$\bar{\rho}_{xy}(t) = 0 \qquad \text{for all } t$$

and

$$P_{xy}(\omega) = 0.$$

On the other hand, if $y(t) = \gamma x(t - \tau)$ for some pair, γ and τ, of real values, then

$$\bar{\rho}_{xy}(t) = \lim_{T \to \infty} \frac{1}{2T} \int_{-T}^{T} x^*(s)\gamma x(s - \tau + t)\, ds.$$

Thus, even if the expected value of $x(t)$ is zero,

$$\bar{\rho}_{xy}(\tau) = \gamma \lim_{T \to \infty} \frac{1}{2T} \int_{-T}^{T} |x(s)|^2\, ds = \gamma \bar{\rho}_x(0),$$

and the cross-power spectrum, $P_{xy}(\omega)$, will not vanish.

Of particular interest is the average cross-correlation of a random signal, $x(t)$, with itself. This is the same as the average autocorrelation of $x(t)$ and indicates the extent to which the value of $x(t + \alpha)$ can be predicted from the value of $x(\alpha)$. If the expected value of $x(t)$ is zero and, for every α and nonzero value of t, $x(t)$ and $x(t + \alpha)$ are outcomes of completely independent random processes, then $x(t)$ is called "white noise." For such a signal there is a constant, P_0, such that

$$\bar{\rho}_x(t) = P_0 \delta(t)$$

and

$$P_x(\omega) = P_0.$$

2.7.6 *Systems and random signals*

Let S be a linear shift invariant system with impulse response function $h(t)$ and transfer function $H(\omega)$. Assume the input, $x(t)$, is a stationary random signal with mean μ_x, and let $y(t)$ be the corresponding output,

$$y(t) = S[x(t)].$$

The output is also a stationary random signal. It is related to the input by

$$y(t) = h * x(t)$$

or, equivalently, by

$$Y(\omega) = H(\omega)X(\omega),$$

where $X(\omega)$ and $Y(\omega)$ are, respectively, the Fourier transforms of $x(t)$ and $y(t)$. It should be easy to see that the expected value of the output is directly related to the expected value of the input,

$$\mu_y = S[\mu_x] = h * \mu_x = \mu_x \int_{-\infty}^{\infty} h(s)\, ds.$$

The auto- and cross-correlations of the input and output signals are related by

$$\bar{\rho}_y(t) = h * \bar{\rho}_{yx}(t), \tag{2.7.6.1}$$

$$\bar{\rho}_{xy}(t) = h * \bar{\rho}_x(t), \tag{2.7.6.2}$$

and

$$\bar{\rho}_y(t) = h * (h \star \bar{\rho}_x)(t). \tag{2.7.6.3}$$

Taking the Fourier transforms of these relations gives the corresponding relations for the power spectra,

$$P_y(\omega) = H(\omega)P_{yx}(\omega), \tag{2.7.6.4}$$

$$P_{xy}(\omega) = H(\omega)P_x(\omega),\tag{2.7.6.5}$$

and

$$P_y(\omega) = |H(\omega)|^2 P_x(\omega).\tag{2.7.6.6}$$

Derivations of identities (2.7.6.1) through (2.7.6.3) are relatively straightforward. For identity (2.7.6.1),

$$
\begin{aligned}
\bar\rho_y(t) &= \lim_{T\to\infty} \frac{1}{2T} \int_{-T}^{T} y^*(s)y(s+t)\,ds \\
&= \lim_{T\to\infty} \frac{1}{2T} \int_{-T}^{T} y^*(s)[h*x(s+t)]\,ds \\
&= \lim_{T\to\infty} \frac{1}{2T} \int_{-T}^{T} y^*(s) \int_{-\infty}^{\infty} h(\lambda)x(s+t-\lambda)\,d\lambda\,ds \\
&= \int_{-\infty}^{\infty} h(\lambda) \left(\lim_{T\to\infty} \frac{1}{2T} \int_{-T}^{T} y^*(s)x(s+t-\lambda)\,ds \right) d\lambda \\
&= \int_{-\infty}^{\infty} h(\lambda)\bar\rho_{yx}(t-\lambda)\,d\lambda \\
&= h*\bar\rho_{yx}(t).
\end{aligned}
$$

Derivations of (2.7.6.2) and (2.7.6.3) are similar with the derivation of equation (2.7.6.3) aided by identities (2.7.5.1) and (2.7.4.2).

Example 2.7.6.1
Let S be an LSI system with transfer function

$$H(\omega) = \frac{1}{j+\omega}.$$

Assume the input, $x(t)$, is white noise, that is, the power spectrum of $x(t)$ is a constant,

$$P_x(\omega) = P_0.$$

The corresponding output of the system, $y(t)$, will have power spectrum

$$P_y(\omega) = |H(\omega)|^2 P_x(\omega) = \frac{P_0}{1+\omega^2}$$

and autocorrelation

$$\bar\rho_y(t) = \mathcal{F}^{-1}\left[P_y(\omega)\right]\Big|_t = \frac{1}{2} P_0 e^{-|t|}.$$

The mean squared output is then

$$\lim_{T\to\infty} \frac{1}{2T} \int_{-T}^{T} |y(t)|^2\,dt = \bar\rho_y(0) = \frac{1}{2} P_0. \qquad \blacksquare$$

Example 2.7.6.2

Let S be the ideal low-pass filter from Example 2.6.1.1 with cutoff frequency Ω. The transfer function is

$$H(\omega) = p_\Omega(\omega) = \begin{cases} 1, & \text{if } |\omega| < \Omega \\ 0, & \text{if } \Omega < |\omega| \end{cases}.$$

The power spectrum of the output, $y(t)$, resulting from a white noise input, $x(t)$, with power spectrum $P_x(\omega) = P_0$ is

$$P_y(\omega) = |H(\omega)|^2 P_x(\omega) = P_0 p_\Omega(\omega),$$

and the autocorrelation of the output is

$$\bar{\rho}_y(t) = \mathcal{F}^{-1}[P_y(\omega)]\big|_t = P_0 \frac{1}{\pi t} \sin(\Omega t).$$

The mean squared output of the white noise is then

$$\lim_{T\to\infty} \frac{1}{2T} \int_{-T}^{T} |y(t)|^2 \, dt = \bar{\rho}_y(0) = P_0 \lim_{t\to 0} \frac{\sin(\Omega t)}{\pi t} = \frac{\Omega}{\pi} P_0.$$

Consider, now, an input of $f(t) + x(t)$ where $x(t)$ is the above white noise and $f(t)$ is a deterministic band-limited signal with bandwidth less that 2Ω. The corresponding output of this low-pass filter is $f(t) + y(t)$. The expected "intensity" of the output is

$$E\left[|f(t) + y(t)|^2\right] = E\left[|f(t)|^2 + f^*(t)y(t) + f(t)y^*(t) + |y(t)|^2\right]$$

$$= |f(t)|^2 + f^*(t)E[y(t)] + f(t)E[y^*(t)] + E\left[|y(t)|^2\right]$$

Because $x(t)$ comes from white noise,

$$E[y(t)] = E[S[x(t)]] = S[E[x(t)]] = S[0] = 0,$$

and, by the above,

$$E\left[|y(t)|^2\right] = \lim_{T\to\infty} \frac{1}{2T} \int_{-T}^{T} |y(t)|^2 \, dt = \frac{\Omega}{\pi} P_0.$$

Thus, the expected intensity of the output is

$$|f(t)|^2 + \frac{\Omega}{\pi} P_0,$$

and the ratio of the intensity of the deterministic signal to the intensity of the outputted noise (signal-to-noise ratio) is

$$\frac{\pi |f(t)|^2}{\Omega P_0}. \qquad\blacksquare$$

2.8   Partial differential equations

The Fourier transform is an especially useful tool for solving problems involving partial differential equations. To illustrate how the Fourier transform can be used in a variety of such problems, three different problems involving the partial differential equation describing heat flow are examined below.

2.8.1 *The one-dimensional heat equation*

The next few sections concern a uniform rod of some heat conducting material positioned on the X-axis between $x = \alpha$ and $x = \beta$. It is assumed that the sides of the rod are thermally insulated from the surroundings. The relevant material constants are

> $c =$ specific heat of the material,
>
> $\rho =$ the linear density of the material,

and

> $k =$ the thermal diffusivity.

Any heat sources (and sinks) are described by a density function, $f(x, t)$, where, for any $0 \leq t_1 < t_2$ and $\alpha \leq x_1 < x_2 \leq \beta$,

$$\int_{x_1}^{x_2} \int_{t_1}^{t_2} c\rho f(x, t) \, dt \, dx$$

is the total heat (in calories) generated in the rod between $x = x_1$ and $x = x_2$ during the period of time between $t = t_1$ and $t = t_2$.

The temperature distribution throughout the rod is described by

$$v(x, t) = \text{the temperature at time } t \text{ and position } x \text{ in the rod.}$$

Using basic thermodynamics it can be shown that $v(x, t)$ must satisfy the "one-dimensional heat equation,"

$$\frac{\partial v}{\partial t} - k \frac{\partial^2 v}{\partial x^2} = f(x, t) \tag{2.8.1.1}$$

for $\alpha < x < \beta$.

In Sections 2.8.2 through 2.8.4, the above equation is solved under various conditions. In each case the Fourier transform is taken with respect to the spatial variable, x, with (assuming an infinite rod, $\alpha = -\infty$ and $\beta = \infty$)

$$V = V(\xi, t) = \mathcal{F}_x \left[v(x, t) \right]|_\xi = \int_{-\infty}^{\infty} v(x, t) e^{-j\xi x} \, dx$$

and

$$F = F(\xi, t) = \mathcal{F}_x \left[f(x, t) \right]|_\xi = \int_{-\infty}^{\infty} f(x, t) e^{-j\xi x} \, dx.$$

Observe that

$$\mathcal{F}_x \left[\frac{\partial v}{\partial t} \right]\bigg|_\xi = \int_{-\infty}^{\infty} \frac{\partial v}{\partial t}\bigg|_{(x,t)} e^{-j\xi x} \, dx = \frac{\partial}{\partial t} \int_{-\infty}^{\infty} v(x, t) e^{-j\xi x} \, dx = \frac{\partial V}{\partial t}.$$

On the other hand, it is appropriate to use identity (2.2.11.6) to compute the transform (with respect to x) of any partial derivatives with respect to x. In particular,

$$\mathcal{F}_x \left[\frac{\partial^2 v}{\partial x^2} \right]\bigg|_\xi = (j\xi)^2 V(\xi, t) = -\xi^2 V(\xi, t).$$

Thus, taking the Fourier transform of equation (2.8.1.1) with respect to x yields

$$\frac{\partial V}{\partial t} + k\xi^2 V = F, \tag{2.8.1.2}$$

which can be treated as an ordinary first-order linear differential equation. From the elementary theory of ordinary differential equations the general solution to equation (2.8.1.2) is

$$V(\xi, t) = e^{-k\xi^2 t} \int_a^t e^{k\xi^2 \tau} F(x, \tau)\, d\tau + G(\xi)e^{-k\xi^2 t} \tag{2.8.1.3}$$

where a is any convenient value and $G(\xi)$ is an "arbitrary" function of ξ. The temperature distribution, $v(x, t)$, is then found by taking the inverse Fourier transform with respect to the "spatial frequency" variable, ξ.

2.8.2   *The initial value problem for heat flow on an infinite rod*

If the rod is infinite, there are no heat sources or sinks in the rod, and the initial temperature distribution is known to be given by $v_0(x)$, then $v(x, t)$ is the solution to the following system of equations:

$$\frac{\partial v}{\partial t} - k \frac{\partial^2 v}{\partial x^2} = 0$$

$$v(x, 0) = v_0(x).$$

Because $v_0(x)$ is the initial temperature distribution, it suffices to find $v(x, t)$ for $0 < t$.

Taking the Fourier transform with respect to x of each of the above equations yields the system of two equation,

$$\frac{\partial V}{\partial t} + k\xi^2 V = 0$$

$$V(\xi, 0) = V_0(\xi),$$

where

$$V_0(\xi) = \mathcal{F}\,[v_0(x)]|_\xi\,.$$

From formula (2.8.1.3) the general solution to the differential equation is

$$V(\xi, t) = G(\xi)e^{-k\xi^2 t}.$$

Plugging in the initial values,

$$G(\xi) = V(\xi, 0) = V_0(\xi),$$

shows that

$$V(\xi, t) = V_0(\xi)e^{-k\xi^2 t}.$$

The temperature distribution for all time is then found by taking the inverse transform (with respect to the spatial variable),

$$v(x, t) = \mathcal{F}_\xi^{-1}\left[V_0(\xi)e^{-kt\xi^2}\right]\Big|_x$$

$$= \mathcal{F}_\xi^{-1}[V_0(\xi)]\Big|_x * \mathcal{F}_\xi^{-1}\left[e^{-kt\xi^2}\right]\Big|_x$$

$$= v_0(x) * \left(\frac{1}{\sqrt{4\pi kt}} \exp\left[-\frac{1}{4kt}x^2\right]\right)$$

$$= \frac{1}{\sqrt{4\pi kt}} \int_{-\infty}^{\infty} v_0(s) \exp\left[-\frac{1}{4kt}(x - s)^2\right] ds.$$

2.8.3   *An infinite rod with heat sources and sinks*

For this problem it is assumed that the rod is infinite, and that the initial temperature distribution is $v(x, 0) = 0$. The source term, $f(x, t)$, is assumed to be nonzero for $0 < t$. Because the initial temperature distribution is a constant zero and it is only necessary to find $v(x, t)$ for $t > 0$, the following two assumptions may be made:

  1. for $t \leq 0$, $v(x, t) = 0$.
  2. for $t < 0$, $f(x, t) = 0$.

The heat flow problem is then one of solving the heat equation,

$$\frac{\partial v}{\partial t} - k \frac{\partial^2 v}{\partial x^2} = f(x, t) \qquad (2.8.3.1)$$

for all real values of x and t, subject to conditions (1) and (2).

   This problem is similar to that of finding the output to a causal LSI system. This suggests that it is convenient to find first the solution to

$$\frac{\partial h}{\partial t} - k \frac{\partial^2 h}{\partial x^2} = \delta(x)\delta(t) \qquad (2.8.3.2)$$

where $h(x, t)$ is assumed to vanish if $t < 0$. It is then relatively easy to verify that the solution to (2.8.3.2) is given by the two-dimensional convolution of $h(x, t)$ with $f(x, t)$. Because $f(s, \tau)$ vanishes for $\tau < 0$, this can be written

$$v(x, t) = f * h(x, t) = \int_{s=-\infty}^{\infty} \int_{\tau=0}^{\infty} f(s, \tau) h(x - s, t - \tau) \, ds \, d\tau. \qquad (2.8.3.3)$$

Taking the Fourier transform of equation (2.8.3.2) (with respect to the spatial variable) yields

$$\frac{\partial H}{\partial t} + k\xi^2 H = \delta(t) \qquad (2.8.3.4)$$

where, by the assumptions on $h(x, t)$,

$$H(\xi, t) = \mathcal{F}_\xi \left[h(x, t) \right]|_x$$

must vanish when $t < 0$. From (2.8.1.3) it follows that the solution to equation (2.8.3.4) is given by

$$H(\xi, t) = e^{-k\xi^2 t} \int_a^t e^{k\xi^2 \tau} \delta(\tau) \, d\tau + G(\xi) e^{-k\xi^2 t}.$$

It is convenient to take $a = -1$. Observe, then, that

$$\int_a^t e^{k\xi^2 \tau} \delta(\tau) \, d\tau = \left\{ \begin{array}{ll} 1, & \text{if } 0 < t \\ 0, & \text{if } t < 0 \end{array} \right\} = u(t).$$

Combining this with the fact that $H(\xi, t)$ vanishes for negative values of t gives

$$0 = H(\xi, -1) = e^{-k\xi^2(-1)}u(-1) + G(\xi)e^{-k\xi^2(-1)} = G(\xi)e^{k\xi^2},$$

implying that $G(\xi)$ vanishes and

$$H(\xi, t) = e^{-kt\xi^2}u(t).$$

Taking the inverse transform,

$$h(x, t) = \mathcal{F}_\xi^{-1}\left[e^{-kt\xi^2}u(t)\right]\Big|_x = \frac{1}{\sqrt{4\pi kt}}\exp\left[\frac{-x^2}{4kt}\right]u(t).$$

Formula (2.8.3.3) for the solution to the heat equation then becomes

$$v(x, t) = \int_{s=-\infty}^{\infty}\int_{\tau=0}^{\infty} f(s, \tau)\frac{1}{\sqrt{4\pi k(t-\tau)}}\exp\left[\frac{-(x-s)^2}{4k(t-\tau)}\right]u(t-\tau)\,d\tau\,ds$$

$$= \int_{s=-\infty}^{\infty}\int_{\tau=0}^{t} f(s, \tau)\frac{1}{\sqrt{4\pi k(t-\tau)}}\exp\left[\frac{-(x-s)^2}{4k(t-\tau)}\right]d\tau\,ds.$$

2.8.4   A boundary value problem for heat flow on a half-infinite rod

For this problem it is assumed that the rod occupies the positive X-axis, $0 < x$, that there are no sources or sinks of heat in the rod, and that the initial temperature throughout the rod is zero. At $x = 0$ the temperature is known to be given by some function, $\Theta(t)$, for $0 < t$. The temperature distribution function, $v(x, t)$, then must satisfy the following system of equations:

$$\frac{\partial v}{\partial t} - k\frac{\partial^2 v}{\partial x^2} = 0, \qquad 0 < x \text{ and } 0 < t$$
$$v(x, 0) = 0, \qquad 0 < x$$
$$v(0, t) = \Theta(t), \qquad 0 < t.$$

To apply the Fourier transform with respect to the spatial variable, $v(x, t)$ must be extended to a function on all of x. A review of relations (2.3.16.1) through (2.3.16.6) along with the observation that $v(0, t)$ is known, suggests that the odd extension,

$$\hat{v}(x, t) = \begin{cases} v(x, t), & \text{if } 0 < t \\ -v(-x, t), & \text{if } t < 0 \end{cases},$$

is appropriate. It is easily verified that $\hat{v}(x, t)$ satisfies

$$\frac{\partial \hat{v}}{\partial t} - k\frac{\partial^2 \hat{v}}{\partial x^2} = 0$$

for all $0 < t$ and $x \neq 0$. Combining this with relation (2.3.16.6),

$$D_{xx}\hat{v} = \frac{\partial^2 \hat{v}}{\partial x^2} + 2v(0, t)\delta'(x) = \frac{1}{k}\frac{\partial \hat{v}}{\partial t} + 2\Theta(t)\delta'(x),$$

gives the equation

$$\frac{\partial \hat{v}}{\partial t} - kD_{xx}\hat{v} = -2k\Theta(t)\delta'(x), \tag{2.8.4.1}$$

where $D_{xx}\hat{v}$ explicitly denotes the second generalized derivative of $\hat{v}(x, t)$ with respect to x. Equation (2.8.4.1) is valid for all x and is the same as equation (2.8.3.1) with

$$f(x, t) = -2k\Theta(t)\delta'(x).$$

Because $\hat{v}(x, t)$ also satisfies the same initial condition as assumed in Section 2.8.3, the solution derived in that section applies here,

$$\hat{v}(x, t) = \int_{s=-\infty}^{\infty}\int_{\tau=0}^{t}\left[-2k\Theta(\tau)\delta'(s)\right]\frac{1}{\sqrt{4\pi k(t-\tau)}}\exp\left[\frac{-(x-s)^2}{4k(t-\tau)}\right]d\tau\,ds.$$

Now,

$$\int_{-\infty}^{\infty}\delta'(s)e^{-\gamma(x-s)^2}\,ds = -\int_{-\infty}^{\infty}\delta(s)\frac{d}{ds}\left[e^{-\gamma(x-s)^2}\right]ds$$

$$= -\int_{-\infty}^{\infty}\delta(s)2\gamma(x-s)e^{-\gamma(x-s)^2}\,ds$$

$$= -2\gamma xe^{-\gamma x^2}.$$

Thus, for $0 < x$ and $0 < t$,

$$v(x, t) = \hat{v}(x, t)$$

$$= \int_{\tau=0}^{t}-2k\Theta(\tau)\frac{1}{\sqrt{4\pi k(t-\tau)}}\left(\int_{s=-\infty}^{\infty}\delta'(s)\exp\left[-\frac{(x-s)^2}{4k(t-\tau)}\right]ds\right)d\tau$$

$$= \int_{\tau=0}^{t}-2k\Theta(\tau)\frac{1}{\sqrt{4\pi k(t-\tau)}}\left(\frac{-2}{4k(t-\tau)}x\exp\left[\frac{-x^2}{4k(t-\tau)}\right]\right)d\tau$$

$$= \frac{x}{2\sqrt{\pi k}}\int_{\tau=0}^{t}\Theta(\tau)(t-\tau)^{-3/2}\exp\left[\frac{-x^2}{4k(t-\tau)}\right]d\tau.$$

Letting $\sigma = \frac{x}{2\sqrt{\pi k}}(t-\tau)^{-1/2}$, this simplifies somewhat to

$$v(x, t) = 2\int_{\sigma_0}^{\infty}\Theta\left(t - \frac{x^2}{4\pi k\sigma^2}\right)e^{-\pi\sigma^2}\,d\sigma,$$

where

$$\sigma_0 = \frac{x}{2\sqrt{\pi kt}}.$$

In particular, if the boundary temperature is constant, $\Theta(t) = \Theta_0$, then

$$v(x, t) = 2\Theta_0\int_{\sigma_0}^{\infty}e^{-\pi\sigma^2}\,d\sigma = \Theta_0\,\text{erf}\left(\frac{x}{2\sqrt{\pi kt}}\right).$$

2.9   Tables

TABLE 2.3
Fundamental Fourier identities.

If $f(t)$ and $G(\omega)$ are suitably integrable:

Integral Definitions:

$$F(\omega) = \mathcal{F}\left[f(t)\right]\big|_{\omega} = \int_{-\infty}^{\infty} f(t)e^{-j\omega t}\,dt$$

$$g(t) = \mathcal{F}^{-1}\left[G(\omega)\right]\big|_{t} = \frac{1}{2\pi}\int_{-\infty}^{\infty} G(\omega)e^{j\omega t}\,d\omega$$

Parseval's Equality:

$$\int_{-\infty}^{\infty} f(t)g^{*}(t)\,dt = \frac{1}{2\pi}\int_{-\infty}^{\infty} F(\omega)G^{*}(\omega)\,d\omega$$

Bessel's Equality:

$$\int_{-\infty}^{\infty} |f(t)|^{2}\,dt = \frac{1}{2\pi}\int_{-\infty}^{\infty} |F(\omega)|^{2}\,d\omega$$

For all transformable functions:

Linearity:

$$\mathcal{F}\left[\alpha f(t) + \beta g(t)\right]\big|_{\omega} = \alpha\mathcal{F}\left[f(t)\right]\big|_{\omega} + \beta\mathcal{F}\left[g(t)\right]\big|_{\omega}$$

$$\mathcal{F}^{-1}\left[\alpha F(\omega) + \beta G(\omega)\right]\big|_{t} = \alpha\,\mathcal{F}^{-1}[F(\omega)]\big|_{t} + \beta\mathcal{F}^{-1}\left[G(\omega)\right]\big|_{t}$$

Near Equivalence (Symmetry of Transforms):

$$\mathcal{F}^{-1}\left[\phi(x)\right]\big|_{y} = \frac{1}{2\pi}\mathcal{F}\left[\phi(-x)\right]\big|_{y} = \frac{1}{2\pi}\mathcal{F}\left[\phi(x)\right]\big|_{-y}$$

$$\mathcal{F}\left[\phi(x)\right]\big|_{y} = 2\pi\mathcal{F}^{-1}\left[\phi(-x)\right]\big|_{y} = 2\pi\mathcal{F}^{-1}\left[\phi(x)\right]\big|_{-y}$$

TABLE 2.4
Commonly used Fourier identities (α is any nonzero real number, $F(\omega) = \mathcal{F}\left[f(t)\right]\big|_{\omega}$ and $G(\omega) = \mathcal{F}\left[g(t)\right]\big|_{\omega}$).

$h(t)$	$H(\omega) = \mathcal{F}\left[h(t)\right]\big|_{\omega}$
$f(\alpha t)$	$\dfrac{1}{|\alpha|}F\left(\dfrac{\omega}{\alpha}\right)$
$f(t - \alpha)$	$e^{-j\alpha\omega}F(\omega)$
$e^{j\alpha t}f(t)$	$F(\omega - \alpha)$
$\cos(\alpha t)f(t)$	$\dfrac{1}{2}[F(\omega - \alpha) + F(\omega + \alpha)]$

TABLE 2.4
(continued)

$h(t)$	$H(\omega) = \mathcal{F}\left[h(t)\right]\big|_{\omega}$
$\sin(\alpha t) f(t)$	$\dfrac{1}{2j}\left[F(\omega - \alpha) - F(\omega + \alpha)\right]$
$\dfrac{df}{dt}$	$j\omega F(\omega)$
$\dfrac{d^n f}{dt^n}$	$(j\omega)^n F(\omega)$
$t f(t)$	$j\dfrac{dF}{d\omega}$
$t^n f(t)$	$j^n \dfrac{d^n F}{d\omega^n}$
$\dfrac{f(t)}{t}$	$-j\displaystyle\int_{\alpha}^{\omega} F(s)\,ds + c_\alpha$
$\displaystyle\int_{\alpha}^{t} f(s)\,ds$	$-j\dfrac{F(\omega)}{\omega} + c_\alpha \delta(\omega)$
$f(t)g(t)$	$\dfrac{1}{2\pi} F(\omega) * G(\omega)$
$f(t) * g(t)$	$F(\omega) G(\omega)$
$f(t) \star g(t)$	$F^*(\omega) G(\omega)$
$f^*(t) g(t)$	$\dfrac{1}{2\pi} F(\omega) \star G(\omega)$

TABLE 2.5
Fourier transforms of some common functions ($\alpha, \beta, \gamma, \lambda,$ and ν denote real numbers with $\alpha > 0$, $0 < \lambda < 1$, and $\nu > 0$).

$f(t)$	$F(\omega) = \mathcal{F}\left[f(t)\right]\big|_{\omega}$
$p_\alpha(t)$	$\dfrac{2}{\omega}\sin(\alpha\omega)$
$\mathrm{Rect}_{(\beta,\gamma)}(t)$	$\dfrac{j}{\omega}\left[e^{-j\gamma\omega} - e^{-j\beta\omega}\right]$
$e^{-(\alpha + j\beta)t} u(t)$	$\dfrac{1}{\alpha + j\beta + j\omega}$
$t^{\nu-1} e^{-(\alpha + j\beta)t} u(t)$	$\dfrac{\Gamma(\nu)}{(\alpha + j\beta + j\omega)^\nu}$
$e^{(\alpha + j\beta)t} u(-t)$	$\dfrac{1}{\alpha + j\beta - j\omega}$
$(-t)^{\nu-1} e^{(\alpha + j\beta)t} u(-t)$	$\dfrac{\Gamma(\nu)}{(\alpha + j\beta - j\omega)^\nu}$

TABLE 2.5
(continued)

$f(t)$	$F(\omega) = \mathcal{F}\left[f(t)\right]\big|_\omega$
$e^{-\alpha|t|}$	$\dfrac{2\alpha}{\alpha^2 + \omega^2}$
$e^{-\alpha t^2}$	$\sqrt{\dfrac{\pi}{\alpha}}\,\exp\left[-\dfrac{1}{4\alpha}\omega^2\right]$
$e^{-\alpha t^2 + \beta t}$	$\sqrt{\dfrac{\pi}{\alpha}}\,\exp\left[-\dfrac{1}{4\alpha}\omega^2 - j\dfrac{\beta}{2\alpha}\omega + \dfrac{\beta^2}{4\alpha}\right]$
$\dfrac{e^{-\lambda t}}{\alpha + j\beta + e^{-t}}$	$\pi(\alpha + j\beta)^{\lambda - 1 + j\omega}\,\csc(\pi\lambda + j\pi\omega)$
$\operatorname{sech}(\alpha t)$	$\dfrac{\pi}{\alpha}\operatorname{sech}\left(\dfrac{\pi}{2\alpha}\omega\right)$
$e^{\pm j\alpha t^2}$	$\sqrt{\dfrac{\pi}{\alpha}}\,\exp\left[\mp j\dfrac{1}{4\alpha}\left(\omega^2 - \alpha\pi\right)\right]$
$\dfrac{1}{t}\sin(\alpha t)$	$\pi p_\alpha(\omega)$
$\left(\dfrac{1}{t}\sin(\alpha t)\right)^2$	$\dfrac{\pi}{2}\left(2\alpha - |\omega|\right)p_{2\alpha}(\omega)$
$\dfrac{1}{|t|}\sin(\alpha t)$	$-j\operatorname{sgn}(\omega)\ln\left|\dfrac{|\omega| + \alpha}{|\omega| - \alpha}\right|$
1	$2\pi\delta(\omega)$
t^n	$j^n 2\pi\delta^{(n)}(\omega)$
$e^{j\beta t}$	$2\pi\delta(\omega - \beta)$
$\delta(t - \beta)$	$e^{-j\beta\omega}$
$\delta^{(n)}(t)$	$(j\omega)^n$
$\sin(\alpha t)$	$-j\pi\left[\delta(\omega - \alpha) - \delta(\omega + \alpha)\right]$
$\cos(\alpha t)$	$\pi\left[\delta(\omega - \alpha) + \delta(\omega + \alpha)\right]$
$\sin(\alpha t^2)$	$-\sqrt{\dfrac{\pi}{\alpha}}\,\sin\left[\dfrac{1}{4\alpha}\left(\omega^2 - \alpha\pi\right)\right]$
$\cos(\alpha t^2)$	$\sqrt{\dfrac{\pi}{\alpha}}\,\cos\left[\dfrac{1}{4\alpha}\left(\omega^2 - \alpha\pi\right)\right]$
$\operatorname{comb}_\alpha(t)$	$\dfrac{2\pi}{\alpha}\operatorname{comb}_{\frac{2\pi}{\alpha}}(\omega)$
$\operatorname{saw}(t)$	$j\displaystyle\sum_{\substack{n=-\infty \\ n\neq 0}}^{\infty}(-1)^n\dfrac{2}{n}\delta(\omega - n\pi)$
$\operatorname{sgn}(t)$	$-j\dfrac{2}{\omega}$

TABLE 2.5
(continued)

$f(t)$	$F(\omega) = \mathcal{F}[f(t)]\|_\omega$
$u(t)$	$\pi\delta(\omega) - j\dfrac{1}{\omega}$
$\dfrac{1}{t}$	$-j\pi\,\mathrm{sgn}(\omega)$
t^{-n}	$-j\pi\dfrac{(-j\omega)^{n-1}}{(n-1)!}\,\mathrm{sgn}(\omega)$
$\|t\|$	$-\dfrac{2}{\omega^2}$
$t^n\,\mathrm{sgn}(t)$	$(-j)^{n+1}\dfrac{2(n!)}{\omega^{n+1}}$
$\mathrm{ramp}(t)$	$j\pi\delta'(\omega) - \dfrac{1}{\omega^2}$
$t^n u(t)$	$j^n\pi\delta^{(n)}(\omega) + n!\left(\dfrac{-j}{\omega}\right)^{n+1}$
$\|t\|^{-1/2}$	$\sqrt{2\pi}\,\|\omega\|^{-1/2}$
$\|t\|^{\lambda-1}$	$2\Gamma(\lambda)\cos\left(\dfrac{\lambda\pi}{2}\right)\|\omega\|^{-\lambda}$
$t^{-1/2}J_{n+\frac{1}{2}}(t)$	$(-j)^n\sqrt{2\pi}\,P_n(\omega)p_1(\omega)$
$\mathrm{sgn}(t)J_0(t)$	$j\dfrac{2}{\sqrt{\omega^2-1}}\,\mathrm{sgn}(\omega)$

Notes: $\Gamma(x)$ = the Gamma function; $P_n(x)$ = the nth Legendre polynomial; $J_\nu(x)$ = the Bessel function of the first kin of order ν; $\mathrm{saw}(t)$ = the saw function of Example 2.3.9.1.

TABLE 2.6
Graphical representations of some Fourier transforms.

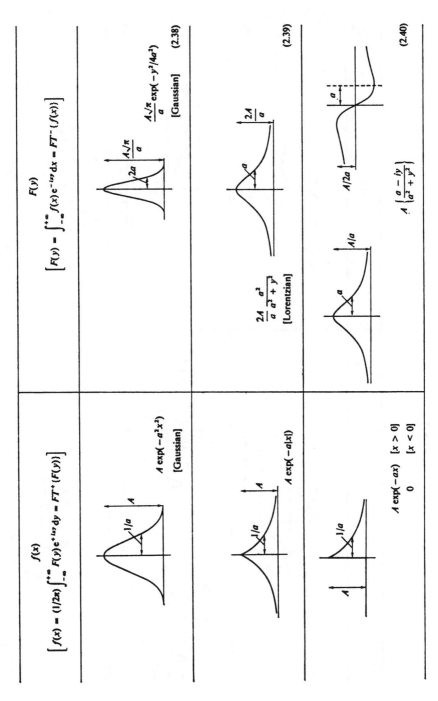

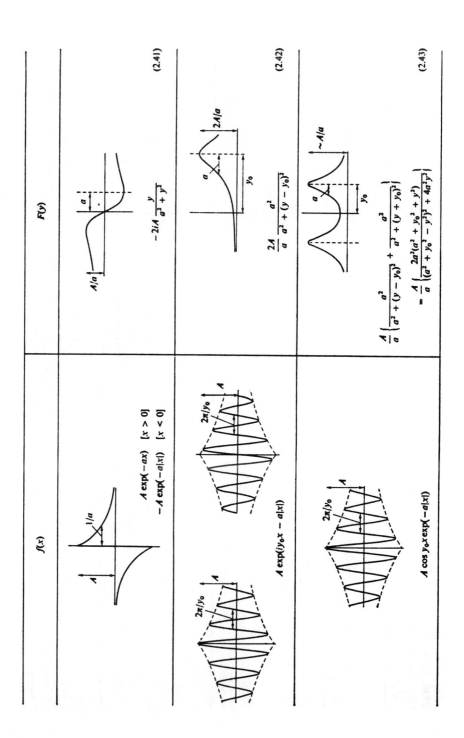

$f(x)$

$F(y)$

$A \exp(-ax)$ $[x > 0]$
$-A \exp(-a|x|)$ $[x < 0]$

$$-2iA\,\frac{y}{a^2 + y^2}$$

(2.41)

$A \exp(iy_0 x - a|x|)$

$$\frac{2A}{a}\,\frac{a^2}{a^2 + (y - y_0)^2}$$

(2.42)

$A \cos y_0 x \exp(-a|x|)$

$$\frac{A}{a}\left\{\frac{a^2}{a^2 + (y - y_0)^2} + \frac{a^2}{a^2 + (y + y_0)^2}\right\}$$

$$= \frac{A}{a}\left\{\frac{2a^2(a^2 + y_0^2 + y^2)}{(a^2 + y_0^2 - y^2)^2 + 4a^2 y^2}\right\}$$

(2.43)

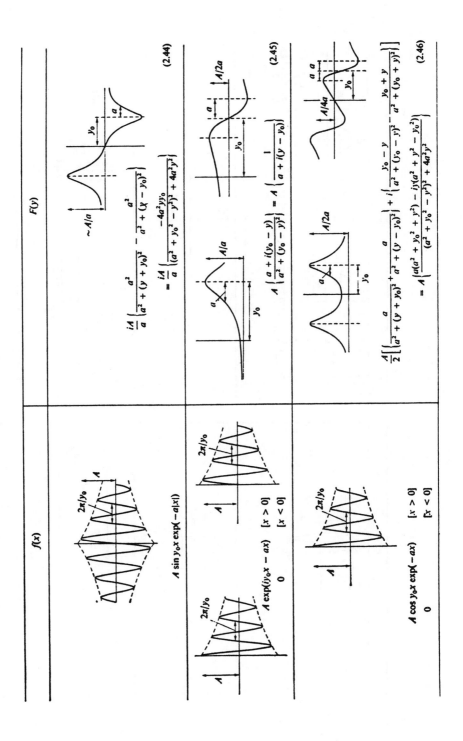

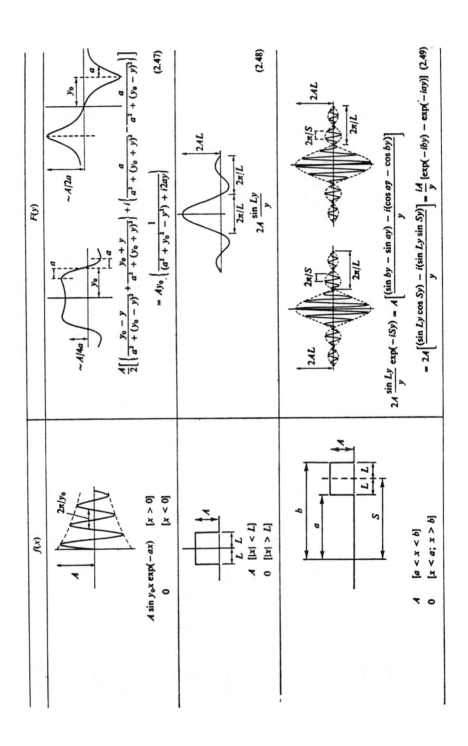

	$f(x)$	$F(y)$
(2.47)	$A \sin y_0 x \exp(-ax)$ $[x > 0]$ 0 $[x < 0]$	$\frac{A}{2}\left[\left(\frac{y_0 - y}{a^2 + (y_0 - y)^2} + \frac{y_0 + y}{a^2 + (y_0 + y)^2}\right) + i\left(\frac{a}{a^2 + (y_0 - y)^2} - \frac{a}{a^2 + (y_0 + y)^2}\right)\right]$ $= Ay_0\left(\frac{1}{(a^2 + y_0^2 - y^2) + i2ay}\right)$
(2.48)	A $[|x| < L]$ 0 $[|x| > L]$	$2A \dfrac{\sin Ly}{y}$
(2.49)	A $[a < x < b]$ 0 $[x < a;\; x > b]$	$2A \dfrac{\sin Ly}{y}\exp(-iSy) = A\left[\dfrac{\sin Ly \cos Sy}{y} - i\dfrac{\sin Ly \sin Sy}{y}\right]$ $= 2A\left[\dfrac{(\sin by - \sin ay)}{y} - i\dfrac{(\cos ay - \cos by)}{y}\right] = \dfrac{iA}{y}\left[\exp(-iby) - \exp(-iay)\right]$

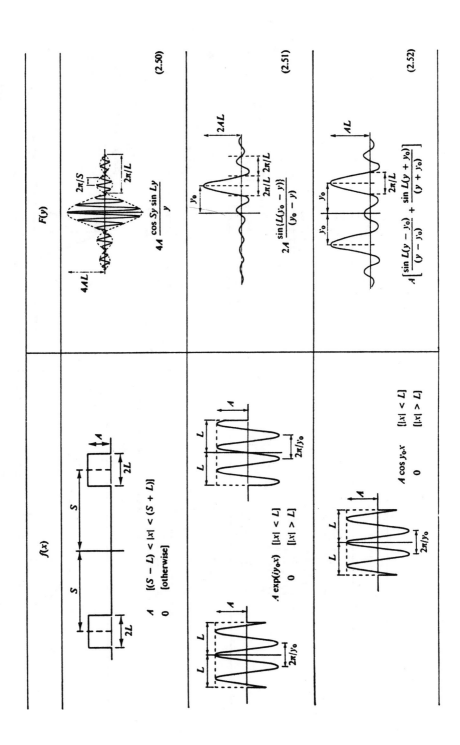

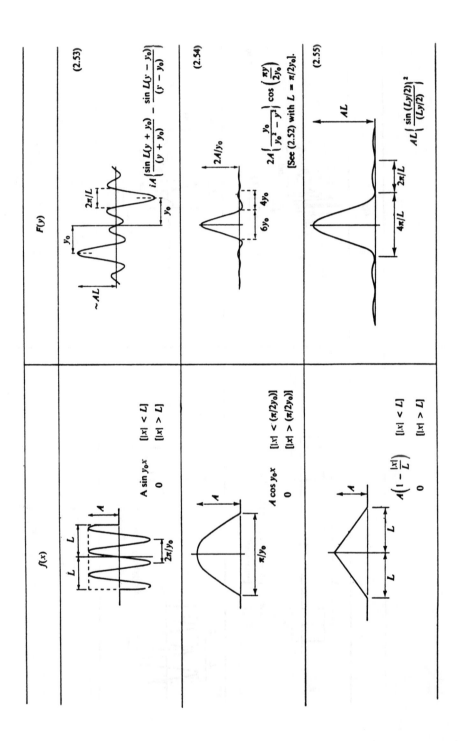

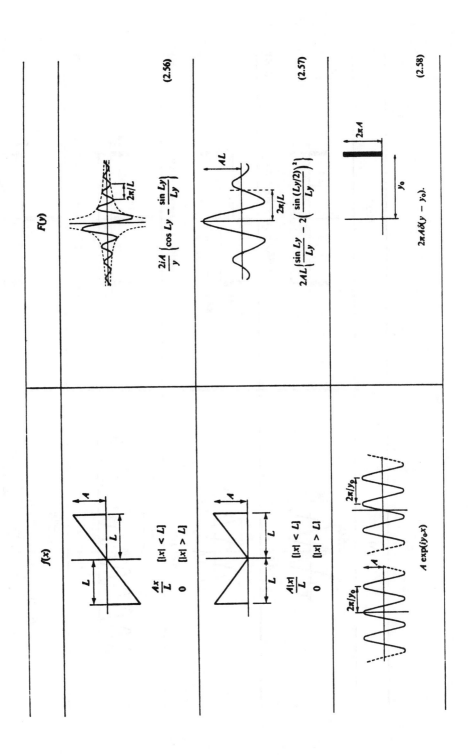

$f(x)$	$F(y)$
$A \cos y_0 x$	$\pi A\{\delta(y - y_0) + \delta(y + y_0)\}$ (2.59)
$A \sin y_0 x$	$\pi i A\{\delta(y + y_0) - \delta(y - y_0)\}$ (2.60)
$A \cos^2 y_0 x$	$\pi A\{\tfrac{1}{4}\delta(y + 2y_0) + \delta(y) + \tfrac{1}{4}\delta(y - 2y_0)\}$ (2.61)
$A \sin^2 y_0 x$	$\pi A\{-\tfrac{1}{4}\delta(y + 2y_0) + \delta(y) - \tfrac{1}{4}\delta(y - 2y_0)\}$ (2.62)

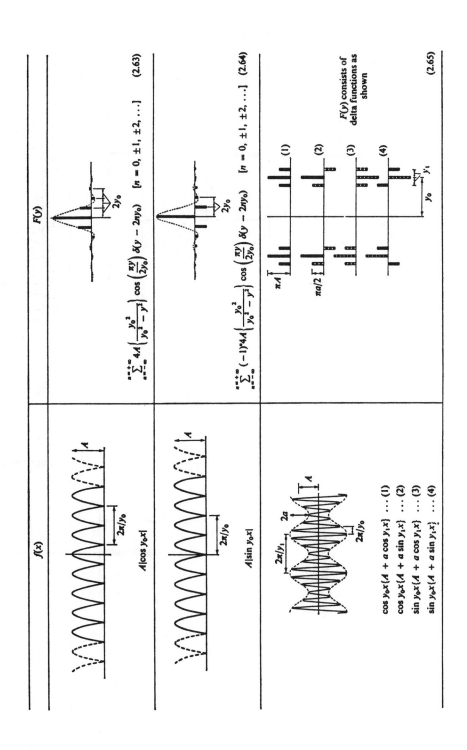

$f(x)$	$F(y)$

$A|\cos y_0 x|$

$$\sum_{n=-\infty}^{+\infty} 4A\left\{\frac{y_0^2}{y_0^2 - y^2}\right\} \cos\left(\frac{\pi y}{2y_0}\right) \delta(y - 2ny_0) \quad [n = 0, \pm 1, \pm 2, \ldots] \quad (2.63)$$

$A|\sin y_0 x|$

$$\sum_{n=-\infty}^{+\infty} (-1)^n 4A\left\{\frac{y_0^2}{y_0^2 - y^2}\right\} \cos\left(\frac{\pi y}{2y_0}\right) \delta(y - 2ny_0) \quad [n = 0, \pm 1, \pm 2, \ldots] \quad (2.64)$$

$\cos y_0 x\{A + a \cos y_1 x\} \quad \cdots (1)$
$\cos y_0 x\{A + a \sin y_1 x\} \quad \cdots (2)$
$\sin y_0 x\{A + a \cos y_1 x\} \quad \cdots (3)$
$\sin y_0 x\{A + a \sin y_1 x\} \quad \cdots (4)$

$F(y)$ consists of delta functions as shown

$$(2.65)$$

$f(x)$	$F(y)$	
$\exp(jy_0 x)(A + a \cos y_1 x)$	$2\pi\left\{A\delta(y - y_0) + \dfrac{a}{2}\delta(y - y_0 + y_1) + \dfrac{a}{2}\delta(y - y_0 - y_1)\right\}$	(2.66)
$\exp(jy_0 x)(A + a \sin y_1 x)$	$2\pi\left\{A\delta(y - y_0) + \dfrac{ia}{2}\delta(y - y_0 + y_1) - \dfrac{ia}{2}\delta(y - y_0 - y_1)\right\}$	(2.67)
$A\delta(x)$	A	(2.68)
$A\delta(x - x_0)$	$A \exp(-ix_0 y)$	(2.69)

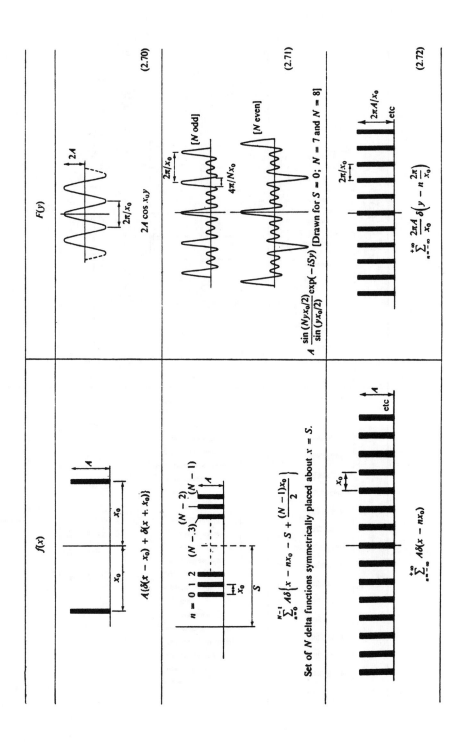

$f(x)$ $F(y)$

(2.70)

$$A\{\delta(x - x_0) + \delta(x + x_0)\}$$

$$2A\cos x_0 y$$

(2.71)

$$\sum_{n=0}^{N-1} A\delta\left\{x - nx_0 - S + \frac{(N-1)x_0}{2}\right\}$$

Set of N delta functions symmetrically placed about $x = S$.

$$A\frac{\sin(Nyx_0/2)}{\sin(yx_0/2)}\exp(-iSy) \quad \text{[Drawn for } S = 0; \ N = 7 \text{ and } N = 8\text{]}$$

[N odd]

[N even]

(2.72)

$$\sum_{n=-\infty}^{+\infty} A\delta(x - nx_0)$$

$$\sum_{n=-\infty}^{+\infty} \frac{2\pi A}{x_0}\delta\left(y - n\frac{2\pi}{x_0}\right)$$

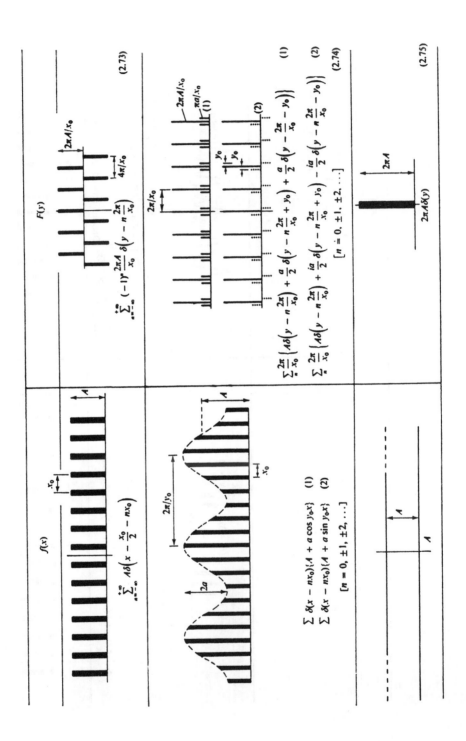

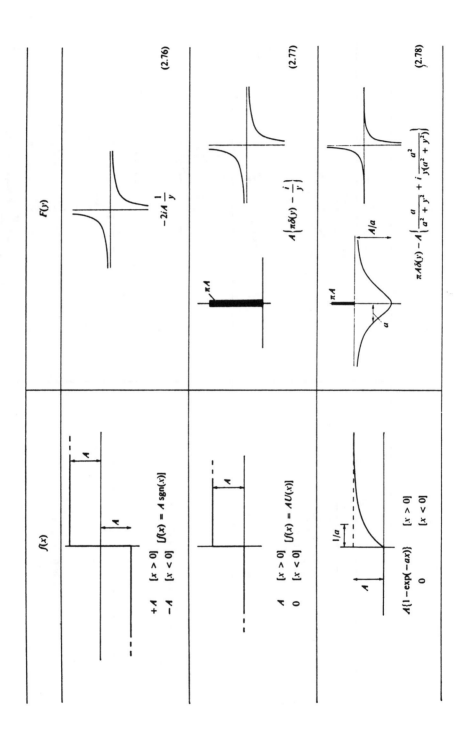

$f(x)$	$F(y)$
$+A$ $[x > 0]$ $[f(x) = A \operatorname{sgn}(x)]$ $-A$ $[x < 0]$	$-2iA\,\dfrac{1}{y}$ (2.76)
A $[x > 0]$ $[f(x) = AU(x)]$ 0 $[x < 0]$	$A\left\{\pi\delta(y) - \dfrac{i}{y}\right\}$ (2.77)
$A\{1-\exp(-ax)\}$ $[x > 0]$ 0 $[x < 0]$	$\pi A\delta(y) - A\left\{\dfrac{a}{a^2 + y^2} + i\,\dfrac{a^2}{y(a^2 + y^2)}\right\}$ (2.78)

$f(x)$	$F(y)$	
$\begin{cases} A & [\|x\| > L] \\ 0 & [\|x\| < L] \end{cases}$	$2\pi A \delta(y) - 2A \dfrac{\sin Ly}{y}$	(2.79)
$A \exp\{i(a \cos y_0 x + bx)\}$	$2\pi A \displaystyle\sum_{n=-\infty}^{+\infty} (i)^n J_n(a) \delta(y - b - ny_0)$	(2.80)
$A \exp\{i(a \sin y_0 x + bx)\}$	$2\pi A \displaystyle\sum_{n=-\infty}^{+\infty} J_n(a) \delta(y - b - ny_0)$	(2.81)

Note: $J_n(-a) = J_{-n}(a) = (-1)^n J_n(a)$. See Appendix H for some properties of Bessel functions.

$f(x)$	$F(y)$
$A \cos(a \sin y_0 x + bx)$	$\pi A \sum_{n=-\infty}^{+\infty} \{ J_n(a)\delta(y - b - ny_0) + J_n(a)\delta(y + b + ny_0) \}$ (2.82)
$A \cos(a \cos y_0 x + bx)$	$\pi A \sum_{n=-\infty}^{+\infty} \{ (+i)^n J_n(a)\delta(y - b - ny_0) + (-i)^n J_n(a)\delta(y + b + ny_0) \}$ (2.83)
$A \sin(a \sin y_0 x + bx)$	$i\pi A \sum_{n=-\infty}^{+\infty} \{ -J_n(a)\delta(y - b - ny_0) + J_n(a)\delta(y + b + ny_0) \}$ (2.84)

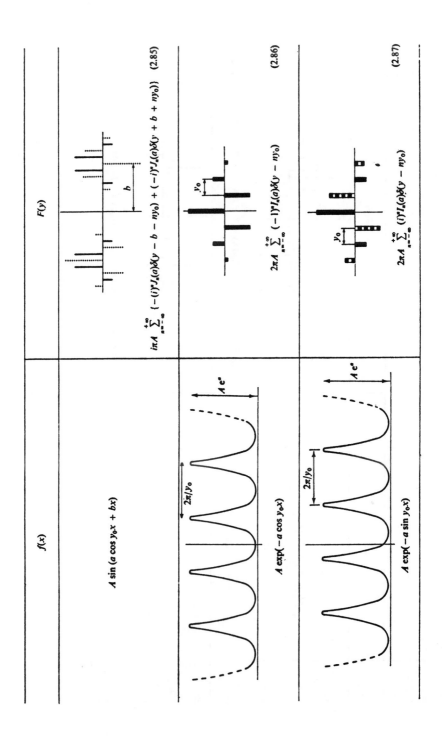

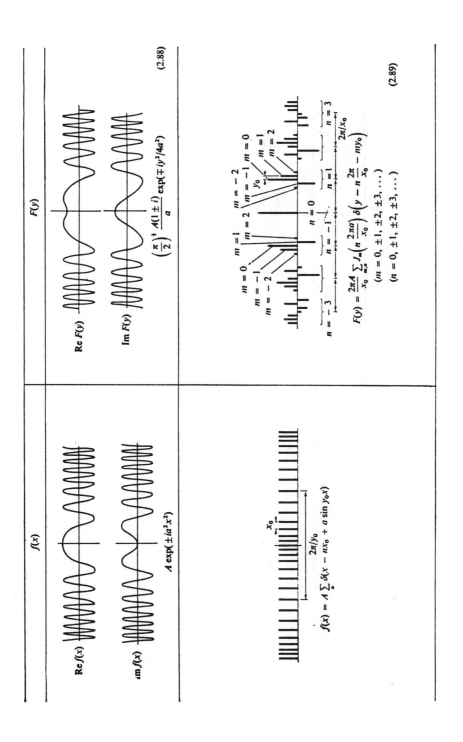

$$f(x)$$

Re $f(x)$

im $f(x)$

$$A \exp(\pm ia^2 x^2)$$

$$F(y)$$

Re $F(y)$

Im $F(y)$

$$\left(\frac{\pi}{2}\right)^{\frac{1}{2}} \frac{A(1 \pm i)}{a} \exp(\mp iy^2/4a^2) \tag{2.88}$$

$$f(x) = A \sum_n \delta(x - nx_0 + a \sin y_0 x)$$

$$F(y) = \frac{2\pi A}{x_0} \sum_{m,n} J_m\left(n\frac{2\pi a}{x_0}\right) \delta\left(y - n\frac{2\pi}{x_0} - my_0\right)$$

$$(m = 0, \pm 1, \pm 2, \pm 3, \ldots)$$

$$(n = 0, \pm 1, \pm 2, \pm 3, \ldots) \tag{2.89}$$

$f(x)$	$F(y)$

$$f(x) = h(x) \sum_{n=-\infty}^{+\infty} g(x - nx_0)$$

$$f(x) = \sum_{n=-\infty}^{+\infty} h(nx_0)g(x - nx_0)$$

$$F(y) = \frac{1}{x_0} \sum_{n=-\infty}^{+\infty} \left\{ G\!\left(\frac{n2\pi}{x_0}\right) H\!\left(y - \frac{n2\pi}{x_0}\right) \right\} \tag{2.90}$$

$$F(y) = \frac{1}{x_0} G(y) \sum_{n=-\infty}^{+\infty} H\!\left(y - \frac{n2\pi}{x_0}\right) \tag{2.91}$$

References

[1] Arsac, J. 1966. *Fourier Transforms and the Theory of Distributions*. Englewood Cliffs, N.J.: Prentice-Hall.

[2] Boyce, W., and DiPrima, R. 1977. *Elementary Differential Equations and Boundary Value Problems*. New York: John Wiley and Sons.

[3] Bracewell, R. 1965. *The Fourier Transform and Its Applications*. New York: McGraw-Hill.

[4] Champeney, D. C. 1973. *Fourier Transforms and Their Physical Applications*. New York: Academic Press.

[5] Champeney, D. C. 1987. *A Handbook of Fourier Theorems*. Cambridge, U.K.: Cambridge University Press.

[6] Erdélyi, A. (Ed.) 1954. *Tables of Integral Transforms (Bateman Manuscript Project)*. New York: McGraw-Hill.

[7] Holland, S. 1990. *Applied Analysis by the Hilbert Space Method*. New York: Marcel Dekker.

[8] Körner, T. W. 1988. *Fourier Analysis*. Cambridge, U.K.: Cambridge University Press.

[9] Papoulis, A. 1962. *The Fourier Integral and Its Application*. New York: McGraw-Hill.

[10] Papoulis, A. [1968]. 1986. *Systems and Transforms with Applications in Optics*. New York: McGraw-Hill. Reprinted, Malabar, FL: Robert E. Krieger Publishing Company.

[11] Pinsky, M. 1991. *Partial Differential Equations and Boundary-Value Problems with Applications*. New York: McGraw-Hill.

[12] Walker, J. S. 1988. *Fourier Analysis*. New York: Oxford University Press.

3

Sine and Cosine Transforms

Pat Yip

CONTENTS

3.1   Introduction

Transforms with cosine and sine functions as the transform kernels represent an important area of analysis. It is based on the so-called half-range expansion of a function over a set of cosine or sine basis functions. Because the cosine and the sine kernels lack the nice properties of an exponential kernel, many of the transform properties are less elegant and more involved than the corresponding ones for the Fourier transform kernel. In particular, the convolution property, which is so important in many applications, will be much more complex.

Despite these basic mathematical limitations, sine and cosine transforms have their own areas of applications. In spectral analysis of real sequences, in solutions of some boundary value problems, and in transform domain processing of digital signals, both cosine and sine transforms have shown their special applicability. In particular, the discrete versions of these transforms have found favor among the digital signal-processing community. Many data compression techniques now employ, in one way or another, the discrete cosine transform (DCT), which has been found to be asymptotically equivalent to the optimal Karhunen-Loeve transform (KLT) for signal decorrelation.

In this chapter, the basic properties of cosine and sine transforms are presented, together with some selected tranforms. To show the versatility of these transforms, several applications are discussed. Computational algorithms are also presented. The chapter ends with a table of sine and cosine transforms, which is not meant to be exhaustive. The reader is referred to the References for more details and for more exhaustive listings of the cosine and sine transforms.

0-8493-8342-0/96/$0.00 + $0.50
© 1996 by CRC Press, Inc.

3.2 The Fourier cosine transform (FCT)

3.2.1 *Definitions and relations to the exponential Fourier transforms*

Given a real- or complex-valued function $f(t)$, which is defined over the positive real line $t \geq 0$, for $\omega \geq 0$, the Fourier cosine transform of $f(t)$ is defined as

$$F_c(\omega) = \int_0^\infty f(t) \cos \omega t \, dt, \qquad \omega \geq 0, \tag{3.2.1}$$

subject to the existence of the integral. The definition is sometimes more compactly represented as an operator $\mathcal{F}_c$ applied to the function $f(t)$, so that

$$\mathcal{F}_c[f(t)] = F_c(\omega) = \int_0^\infty f(t) \cos \omega t \, dt. \tag{3.2.2}$$

The subscript c is used to denote the fact that the kernel of the transformation is a cosine function. The unit normalization constant used here provides for a definition for the inverse Fourier cosine transform, given by

$$\mathcal{F}_c^{-1}[F_c(\omega)] = \frac{2}{\pi} \int_0^\infty F_c(\omega) \cos \omega t \, d\omega, \qquad t \geq 0, \tag{3.2.3}$$

again subject to the existence of the integral used in the definition. The functions $f(t)$ and $F_c(\omega)$, if they exist, are said to form a Fourier cosine transform pair.

Because the cosine function is the real part of an exponential function of purely imaginary argument, that is,

$$\cos(\omega t) = \text{Re}[e^{j\omega t}] = \frac{1}{2}[e^{j\omega t} + e^{-j\omega t}], \tag{3.2.4}$$

it is easy to understand that there exists a very close relationship between the Fourier transform and the cosine transform. To see this relation, consider an even extension of the function $f(t)$ defined over the entire real line so that

$$f_e(t) = f(|t|), \qquad t \in R. \tag{3.2.5}$$

Its Fourier transform is defined as

$$\mathcal{F}[f_e(t)] = \int_{-\infty}^\infty f_e(t) e^{-j\omega t} \, dt, \qquad \omega \in R. \tag{3.2.6}$$

The integral in (3.2.6) can be evaluated in two parts over $(-\infty, 0]$ and $[0, \infty)$. Then using (3.2.5) and changing the integrating variable in the $(-\infty, 0]$ integral from t to $-t$, we have

$$\mathcal{F}[f_e(t)] = \left[\int_0^\infty f(t) e^{-j\omega t} \, dt + \int_0^\infty f(t) e^{j\omega t} \, dt \right] = 2 \int_0^\infty f(t) \cos \omega t \, dt,$$

by (3.2.4), and thus

$$\mathcal{F}[f_e(t)] = 2\mathcal{F}_c[f(t)], \qquad \text{if } f_e(t) = f(|t|). \tag{3.2.7}$$

Many of the properties of the Fourier cosine transforms can be derived from the properties of Fourier transforms of symmetric, or even, functions. Some of the basic properties and operational rules are discussed in Section 3.2.2.

3.2.2 Basic properties and operational rules

1. *Inverse Transformation:* As stated in (3.2.3), the inverse transformation is exactly the same as the forward transformation except for the normalization constant. This leads to the so-called Fourier cosine integral formula, which states that

$$f(t) = \frac{2}{\pi} \int_0^\infty F_c(\omega) \cos \omega t \, d\omega$$

$$= \frac{2}{\pi} \int_0^\infty \left[\int_0^\infty f(\tau) \cos \omega \tau \, d\tau \right] \cos \omega t \, d\omega. \qquad (3.2.8)$$

The sufficient conditions for the inversion formula (3.2.3) are that $f(t)$ be absolutely integrable in $[0, \infty)$ and that $f'(t)$ be piece-wise continuous in each bounded subinterval of $[0, \infty)$. In the range where the function $f(t)$ is continuous, (3.2.8) represents f. At the point t_0 where $f(t)$ has a jump discontinuity, (3.2.8) converges to the mean of $f(t_0 + 0)$ and $f(t_0 - 0)$, that is,

$$\frac{2}{\pi} \int_0^\infty \left[\int_0^\infty f(\tau) \cos(\omega \tau) \, d\tau \right] \cos(\omega t_0) \, d\omega = \frac{1}{2}[f(t_0 + 0) + f(t_0 - 0)]. \qquad (3.2.8')$$

2. *Transforms of Derivatives:* It is easy to show, because of the Fourier cosine kernel, that the transforms of even-order derivatives are reduced to multiplication by even powers of the conjugate variable ω, much as in the case of the Laplace transforms. For the second-order derivative, using integration by parts, we can show that,

$$\mathcal{F}_c[f''(t)] = \int_0^\infty f''(t) \cos(\omega t) \, dt$$

$$= -f'(0) - \omega^2 \int_0^\infty f(t) \cos \omega t \, dt$$

$$= -\omega^2 F_c(\omega) - f'(0) \qquad (3.2.9)$$

where we have assumed that $f(t)$ and $f'(t)$ vanish as $t \to \infty$. These form the sufficient conditions for (3.2.9) to be valid. As the transform is applied to higher order derivatives, corresponding conditions for higher derivatives of f are required for the operational rule to be valid. Here, we also assume that the function $f(t)$ and its derivative $f'(t)$ are continuous everywhere in $[0, \infty)$. If $f(t)$ and $f'(t)$ have a jump discontinuity at t_0 of magnitudes d and d' respectively, (3.2.9) is modified to

$$\mathcal{F}_c[f''(t)] = -\omega^2 F_c(\omega) - f'(0) - \omega d \sin \omega t_0 - d' \cos \omega t_0 \qquad (3.2.10)$$

Higher even-order derivatives of functions with jump discontinuities have similar operational rules that can be easily generalized from (3.2.10). For example, the Fourier cosine transfrom of the fourth-order derivative is

$$\mathcal{F}_c\left[f^{(iv)}(t)\right] = \omega^4 F_c(\omega) + \omega^2 f'(0) - f'''(0) \qquad (3.2.11)$$

if $f(t)$ is continuous to order three everywhere in $[0, \infty)$, and f, f', and f'' vanish as $t \to \infty$. If $f(t)$ has a jump discontinuity at t_0 to order three of magnitudes d, d', d'', and d''', then (3.2.11) is modified to

$$\mathcal{F}_c\left[f^{(iv)}(t)\right] = \omega^4 F_c(\omega) + \omega^2 f'(0) - f'''(0) + \omega^3 d \sin \omega t_0$$

$$+ \omega^2 d' \cos \omega t_0 - \omega d'' \sin \omega t_0 - d''' \cos \omega t_0 \qquad (3.2.12)$$

Here, and in (3.2.10), we have defined the magnitudes of the jump discontinuity at t_0 as

$$d = f(t_0 + 0) - f(t_0 - 0); \qquad d' = f'(t_0 + 0) - f'(t_0 - 0);$$

$$d'' = f''(t_0 + 0) - f''(t_0 - 0); \qquad d''' = f'''(t_0 + 0) - f'''(t_0 - 0). \quad (3.2.13)$$

For derivatives of odd order, the operational rules require the definition for the Fourier sine transform, given in Section 3.3. For example, the Fourier cosine transform of the first order derivative is given by

$$\mathcal{F}_c[f'(t)] = \int_0^\infty f'(t) \cos \omega t \, dt = -f(0) + \omega \int_0^\infty f(t) \sin \omega t \, dt$$

$$= \omega \mathcal{F}_s[f(t)] - f(0) = \omega F_s(\omega) - f(0), \qquad (3.2.14)$$

if f vanishes as $t \to \infty$, and where the operator $\mathcal{F}_s$ and the function $F_s(\omega)$ are defined in (3.3.1). When $f(t)$ has a jump discontinuity of magnitude d at $t = t_0$, (3.2.14) is modified to

$$\mathcal{F}_c[f'(t)] = \omega F_s(\omega) - f(0) - d \cos(\omega t_0). \qquad (3.2.15)$$

Generalization to higher odd-order derivatives with jump discontinuities is similar to that for even-order derivatives in (3.2.12).

3. *Scaling:* Scaling in the t domain translates directly to scaling in the ω domain. Expansion by a factor of a in t results in the contraction by the same factor in ω, together with a scaling down of the magnitude of the transform by the factor a. Thus, as we can show,

$$\mathcal{F}_c[f(at)] = \int_0^\infty f(at) \cos \omega t \, dt = \frac{1}{a} \int_0^\infty f(\tau) \cos \frac{\omega \tau}{a} \, d\tau, \qquad \text{by letting } \tau = at$$

$$= \frac{1}{a} F_c\left(\frac{\omega}{a}\right), \qquad a > 0. \qquad (3.2.16)$$

4. *Shifting:*

(a) Shifting in the t-domain: The shift-in-t property for the cosine transform is somewhat less direct compared with the exponential Fourier transform for two reasons. First, a shift to the left will require extending the definition of the function $f(t)$ onto the negative real line. Secondly, a shift-in-t in the transform kernel does not result in a constant phase factor as in the case of the exponential kernel.

If $f_e(t)$ is defined as the even extension of the function $f(t)$ such that $f_e(t) = f(|t|)$, and if $f(t)$ is piece-wise continous and absolutely integrable over $[0, \infty)$, then

$$\mathcal{F}_c[f_e(t+a) + f_e(t-a)] = \int_0^\infty [f_e(t+a) + f_e(t-a)] \cos \omega t \, dt$$

$$= \int_a^\infty f_e(\tau) \cos \omega(\tau + a) \, d\tau$$

$$+ \int_{-a}^\infty f_e(\tau) \cos \omega(\tau - a) \, d\tau.$$

By expanding the compound cosine functions and using the fact that the function $f_e(\tau)$ is even, these combine to give:

$$\mathcal{F}_c[f_e(t+a) + f_e(t-a)] = 2F_c(\omega) \cos a\omega, \qquad a > 0. \qquad (3.2.17)$$

This is sometimes called the kernel-product property of the cosine transform. In terms of the function $f(t)$, it can be written as:

$$\mathcal{F}_c[f(t+a) + f(|t-a|)] = 2F_c(\omega)\cos a\omega. \tag{3.2.18}$$

Similarly, the kernel-product $2F_c(\omega)\sin(a\omega)$ is related to the Fourier sine transform:

$$\mathcal{F}_s[f(|t-a|) - f(t+a)] = 2F_c(\omega)\sin a\omega, \qquad a > 0. \tag{3.2.19}$$

(b) Shifting in the ω-domain:

To consider the effect of shifting in ω by the amount of $\beta(>0)$, we examine the following,

$$F_c(\omega + \beta) = \int_0^\infty f(t)\cos(\omega + \beta)t\,dt$$

$$= \int_0^\infty f(t)\cos\beta t\cos\omega t\,dt - \int_0^\infty f(t)\sin\beta t\sin\omega t\,dt$$

$$= \mathcal{F}_c[f(t)\cos\beta t] - \mathcal{F}_s[f(t)\sin\beta t]. \tag{3.2.20}$$

Similarly,

$$F_c(\omega - \beta) = \mathcal{F}_c[f(t)\cos\beta t] + \mathcal{F}_s[f(t)\sin\beta t]. \tag{3.2.20'}$$

Combining (3.2.20) and (3.2.20') produces a shift-in-ω operational rule involving only the Fourier cosine transform as

$$\mathcal{F}_c[f(t)\cos\beta t] = \frac{1}{2}[F_c(\omega + \beta) + F_c(\omega - \beta)]. \tag{3.2.21}$$

More generally, for $a, \beta > 0$, we have,

$$\mathcal{F}_c[f(at)\cos\beta t] = \frac{1}{2a}\left[F_c\left(\frac{\omega + \beta}{a}\right) + F_c\left(\frac{\omega - \beta}{a}\right)\right]. \tag{3.2.22}$$

Similarly, we can easily derive:

$$\mathcal{F}_c[f(at)\sin\beta t] = \frac{1}{2a}\left[F_s\left(\frac{\omega + \beta}{a}\right) - F_s\left(\frac{\omega - \beta}{a}\right)\right]. \tag{3.2.22'}$$

5. *Differentiation in the ω domain:* Similar to differentiation in the t domain, the transform operation reduces a differentiation operation into multiplication by an appropriate power of the conjugate variable. In particular, even-order derivatives in the ω domain are transformed as:

$$F_c^{(2n)}(\omega) = \mathcal{F}_c[(-1)^n t^{2n} f(t)]. \tag{3.2.23}$$

We show here briefly, the derivation for $n = 1$:

$$F_c^{(2)}(\omega) = \frac{d^2}{d\omega^2}\int_0^\infty f(t)\cos\omega t\,dt$$

$$= \int_0^\infty f(t)\frac{d^2}{d\omega^2}\cos\omega t\,dt$$

$$= \int_0^\infty f(t)(-1)t^2\cos\omega t\,dt$$

$$= \mathcal{F}_c[(-1)t^2 f(t)].$$

For odd orders, these are related to Fourier sine transforms:

$$F_c^{(2n+1)}(\omega) = \mathcal{F}_s[(-1)^{n+1}t^{2n+1}f(t)]. \tag{3.2.24}$$

In both (3.2.23) and (3.2.24), the existence of the integrals in question is assumed. This means that $f(t)$ should be piece-wise continuous and that $t^{2n}f(t)$ and $t^{2n+1}f(t)$ should be absolutely integrable over $[0, \infty)$.

6. *Asymptotic behavior:* When the function $f(t)$ is piece-wise continuous and absolutely integrable over the region $[0, \infty)$, the Reimann-Lebesque theorem for Fourier series[1] can be invoked to provide the following asymptotic behavior of its cosine transform:

$$\lim_{\omega \to \infty} F_c(\omega) = 0. \tag{3.2.25}$$

7. *Integration:*

   (a) Integration in the t domain:

   Integration in the t domain is transformed to division by the conjugate variable, very similar to the cases of Laplace transforms and Fourier transforms, except the resulting transform is a Fourier sine transform. Thus,

   $$\mathcal{F}_c\left[\int_t^\infty f(\tau)\,d\tau\right] = \int_0^\infty \int_t^\infty f(\tau)\,d\tau \cos \omega t\,dt$$

   $$= \int_0^\infty \left[\int_0^\tau \cos \omega t\,dt\right] f(\tau)\,d\tau$$

   by reversing the order of integration. The inner integral results in a sine function and is the kernel for the Fourier sine transform. Therefore,

   $$\mathcal{F}_c\left[\int_t^\infty f(\tau)\,d\tau\right] = \frac{1}{\omega}\mathcal{F}_s[f(t)] = \frac{1}{\omega}F_s(\omega). \tag{3.2.26}$$

   Here again, $f(t)$ is subject to the usual sufficient conditions of being piece-wise continuous and absolutely integrable in $[0, \infty)$.

   (b) Integration in the ω domain:

   A similar and symmetric relation exists for integration in the ω-domain.

   $$\mathcal{F}_s^{-1}\left[\int_\omega^\infty F_c(\beta)\,d\beta\right] = -\frac{1}{t}f(t). \tag{3.2.27}$$

   Note that the integral transform inversion is of the Fourier sine type instead of the cosine type. Also the aysmptotic behavior of $F_c(\omega)$ has been invoked.

8. *The convolution property:* Let $f(t)$ and $g(t)$ be defined over $[0, \infty)$ and satisfy the sufficiency condition for the existence of F_c and G_c. If $f_e(t) = f(|t|)$ and $g_e(t) = g(|t|)$ are the even extensions of f and g respectively over the entire real line, then the convolution of f_e and g_e is given by:

$$f_e * g_e = \int_{-\infty}^\infty f_e(\tau)g_e(t - \tau)\,d\tau \tag{3.2.28}$$

---

[1]The Reimann-Lebesque theorem states that if a function $f(t)$ is piece-wise continuous over an interval $a < t < b$, then

$$\lim_{\gamma \to \infty} \int_a^b f(t) \cos \gamma t\,dt = \lim_{\gamma \to \infty} \int_a^b f(t) \sin \gamma t\,dt = 0.$$

where $*$ has been used to denote the convolution operation. It is easy to see that in terms of f and g, we have:

$$f_e * g_e = \int_0^\infty f(\tau)[g(t+\tau) + g(|t-\tau|)]\,d\tau \qquad (3.2.29)$$

which is an even function. Applying the exponential Fourier transform on both sides and using (3.2.7) and the convolution property of the exponential Fourier transform, we obtain the convolution property for the cosine transform:

$$2F_c(\omega)G_c(\omega) = \mathcal{F}_c\left\{\int_0^\infty f(\tau)[g(t+\tau) + g(|t-\tau|)]\,d\tau\right\}. \qquad (3.2.30)$$

In a similar way, the cosine transform of the convolution of odd extended functions is related to the sine transfroms. Thus,

$$2F_s(\omega)G_s(\omega) = \mathcal{F}_c\left\{\int_0^\infty f(\tau)[g(t+\tau) + g_o(t-\tau)]\,d\tau\right\}, \qquad (3.2.31)$$

where

$$\begin{aligned}
g_o(t) &= g(t) && \text{for } t > 0, \\
&= -g(-t) && \text{for } t < 0,
\end{aligned} \qquad (3.2.32)$$

is defined as the odd extension of the function $g(t)$.

3.2.3   *Selected Fourier cosine transforms*

In this section, the Fourier cosine transforms of some typical functions are given. Most are selected for their simplicity and application. For a more complete listing of cosine transforms, see Section 3.7 where a more extensive table is provided.

3.2.3.1   FCT of algebraic functions

1. *The unit rectangular function:*

$$\begin{aligned}
f(t) = U(t) - U(t-a), \qquad &\text{where } U(t) = 0 && \text{for } t < 0 \\
&\qquad\qquad\quad\ = 1 && \text{for } t > 0, \qquad (3.2.33)
\end{aligned}$$

is the Heaviside unit step function.

$$\mathcal{F}_c[f(t)] = \int_0^a \cos\omega t\,dt = \frac{1}{\omega}\sin\omega a. \qquad (3.2.34)$$

2. *The unit height tent function:*

$$\begin{aligned}
f(t) &= t/a && 0 < t < a, \\
&= (2a-t)/a && a < t < 2a, \\
&= 0 && t > 2a.
\end{aligned}$$

$$\begin{aligned}
\mathcal{F}_c[f(t)] &= \int_0^a \frac{t}{a}\cos\omega t\,dt + \int_a^{2a}\frac{2a-t}{a}\cos\omega t\,dt \\
&= \frac{1}{a\omega^2}[2\cos a\omega - \cos 2a\omega - 1]. \qquad (3.2.35)
\end{aligned}$$

3. *Delayed inverse:*

$$f(t) = U(t-a)/t.$$

$$\mathcal{F}_c[f(t)] = \int_a^\infty \frac{1}{t}\cos\omega t\,dt = \int_{a\omega}^\infty \frac{1}{\tau}\cos\tau\,d\tau = -\,\mathrm{Ci}(a\omega), \qquad (3.2.36)$$

where $\mathrm{Ci}(y) = -\int_y^\infty \frac{1}{\tau}\cos\tau\,d\tau$ is defined as the cosine integral function.

4. *The inverse square root:*

$$f(t) = 1/\sqrt{t},$$

$$\mathcal{F}_c[f(t)] = \int_0^\infty \frac{1}{\sqrt{t}}\cos\omega t\,dt = \sqrt{\frac{\pi}{2\omega}}. \qquad (3.2.37)$$

(3.2.37) is obtained by letting $t = z^2$, and considering the integral,

$$2\int \cos z^2\,dz$$

in the complex plane (see Appendix 1). Using contour integration around a pie-shape region with angle $\pi/4$, the result is obtained directly from the identity:

$$\int_0^\infty e^{jt^2}\,dt = \frac{1}{2}\sqrt{\frac{\pi}{2}}(1+j).$$

5. *Inverse linear function:*

$$f(t) = (\alpha + t)^{-1} \qquad |\arg(\alpha)| < \pi.$$

$$\mathcal{F}_c[f(t)] = \int_0^\infty (\alpha + t)^{-1}\cos\omega t\,dt$$

$$= -\cos\alpha\omega\,\mathrm{Ci}(\alpha\omega) - \sin\alpha\omega\,\mathrm{si}(\alpha\omega), \qquad (3.2.38)$$

(3.2.38) is obtained by shifting the integrating variable to $\alpha + t$, and then expanding the compound cosine function. Here, $\mathrm{si}(y)$ is related to the sine integral function $\mathrm{Si}(y)$, and is defined as:

$$\mathrm{si}(y) = -\int_y^\infty \frac{\sin x}{x}\,dx$$

$$= \int_0^y \frac{\sin x}{x}\,dx - \int_0^\infty \frac{\sin x}{x}\,dx = \mathrm{Si}(y) - (\pi/2). \qquad (3.2.39)$$

6. *Inverse quadratic functions:*

(a) $f(t) = (\alpha^2 + t^2)^{-1} \qquad \mathrm{Re}(\alpha) > 0.$

$$\mathcal{F}_c[f(t)] = \int_0^\infty (\alpha^2 + t^2)^{-1}\cos\omega t\,dt$$

$$= \frac{\pi}{2\alpha}e^{-\alpha\omega}, \qquad (3.2.40)$$

which is obtained also by a properly chosen contour integration over the upper half-plane.

(b) $f(t) = (a^2 - t^2)^{-1}$ $a > 0$,

$$\mathcal{F}_c[f(t)] = \text{P.V.} \int_0^\infty (a^2 - t^2)^{-1} \cos \omega t \, dt$$

$$= \frac{\pi}{2a} \sin a\omega \tag{3.2.41}$$

where "P.V." stands for "principal value" and the integral can be obtained by a proper contour integration in the complex plane.

(c) $f(t) = \dfrac{\beta}{\beta^2 + (\alpha - t)^2} + \dfrac{\beta}{\beta^2 + (\alpha + t)^2}$ $\text{Im} |\alpha| < \text{Re}(\beta)$,

$$\mathcal{F}_c[f(t)] = \int_0^\infty \left[\frac{\beta}{\beta^2 + (\alpha - t)^2} + \frac{\beta}{\beta^2 + (\alpha + t)^2} \right] \cos \omega t \, dt$$

$$= \pi \cos \alpha\omega \, e^{-\beta\omega} \tag{3.2.42}$$

where the integral can be obtained easily by considering a shift in t, applied to the result in (3.2.40).

(d) $f(t) = \dfrac{\alpha - t}{\beta^2 + (\alpha - t)^2} + \dfrac{\alpha + t}{\beta^2 + (\alpha + t)^2}$ $\text{Im} |\alpha| < \text{Re}(\beta)$,

$$\mathcal{F}_c[f(t)] = \int_0^\infty \left[\frac{\alpha - t}{\beta^2 + (\alpha - t)^2} + \frac{\alpha + t}{\beta^2 + (\alpha + t)^2} \right] \cos \omega t \, dt$$

$$= \pi \sin \alpha\omega \, e^{-\beta\omega} \tag{3.2.43}$$

which can be considered as the imaginary part of the contour integral needed in (3.2.42) when α and β are real and positive.

3.2.3.2 FCT of exponential and logarithmic functions

1. $f(t) = e^{-\alpha t}$ $\text{Re}(\alpha) > 0$.

$$\mathcal{F}_c[f(t)] = \int_0^\infty e^{-\alpha t} \cos \omega t \, dt = \frac{\alpha}{\alpha^2 + \omega^2} \tag{3.2.44}$$

which is identical to the Laplace transform of $\cos \omega t$.

2. $f(t) = \dfrac{1}{t}[e^{-\beta t} - e^{-\alpha t}]$ $\text{Re}(\alpha), \text{Re}(\beta) > 0$.

$$\mathcal{F}_c[f(t)] = \int_0^\infty \frac{1}{t}[e^{-\beta t} - e^{-\alpha t}] \cos \omega t \, dt$$

$$= \frac{1}{2} \ln \left(\frac{\alpha^2 + \omega^2}{\beta^2 + \omega^2} \right) \tag{3.2.45}$$

The result is easily obtained using the integration property of the Laplace transform in the phase plane.

3. $f(t) = e^{-\alpha t^2}$ $\text{Re}(\alpha) > 0$.

$$\mathcal{F}_c[f(t)] = \int_0^\infty e^{-\alpha t^2} \cos \omega t \, dt$$

$$= \frac{1}{2} \sqrt{\frac{\pi}{\alpha}} e^{-\omega^2/4\alpha} \tag{3.2.46}$$

This is easily seen as the result of the exponential Fourier transform of a Gaussian distribution.

4. $f(t) = \ln t[1 - U(t-1)]$

$$\mathcal{F}_c[f(t)] = \int_0^1 \ln t \cos \omega t \, dt$$

$$= -\frac{1}{\omega} \int_0^\omega \frac{\sin \tau}{\tau} \, d\tau = -\frac{1}{\omega} \text{Si}(\omega). \tag{3.2.47}$$

The result is obtained by integration by parts and a change of variables. The function $\text{Si}(\omega)$ is defined as the sine integral function given by:

$$\text{Si}(y) = \int_0^y \frac{\sin x}{x} \, dx \tag{3.2.48}$$

5. $f(t) = \dfrac{\ln \beta t}{(t^2 + \alpha^2)}$ $\text{Re}(\alpha) > 0$

$$\mathcal{F}_c[f(t)] = \int_0^\infty \frac{\ln \beta t}{(t^2 + \alpha^2)} \cos \omega t \, dt$$

$$= \frac{\pi}{4\alpha} \{ 2e^{-\alpha\omega} \ln(\alpha\beta) + e^{\alpha\omega} \text{Ei}(-\alpha\omega) - e^{-\alpha\omega} \overline{\text{Ei}}(\alpha\omega) \} \tag{3.2.49}$$

where $\text{Ei}(y)$ is the exponential integral function defined by,

$$\text{Ei}(y) = -\int_{-y}^\infty \frac{e^{-t}}{t} \, dt, \qquad |\arg(y)| < \pi,$$

and

$$\overline{\text{Ei}}(y) = (1/2)[\text{Ei}(y + j0) + \text{Ei}(y - j0)]. \tag{3.2.50}$$

The integral in (3.2.49) is evaluated using contour integration.

6. $f(t) = \ln \left| \dfrac{t+a}{t-a} \right|$, $a > 0$.

$$\mathcal{F}_c[f(t)] = \text{P.V.} \int_0^\infty \ln \left| \frac{t+a}{t-a} \right| \cos \omega t \, dt$$

$$= \frac{2}{\omega} [\text{si}(a\omega) \cos a\omega + \text{ci}(a\omega) \sin a\omega] \tag{3.2.51}$$

where $\text{si}(y)$ and $\text{ci}(y) = -\text{Ci}(y)$ are defined in (3.2.39) and (3.2.36), respectively. The result is obtained through integration by parts, and manifests the shift property of the cosine transform.

3.2.3.3   FCT of trigonometric functions

1. $f(t) = \dfrac{\sin at}{t}$ $a > 0$.

$$\mathcal{F}_c[f(t)] = \int_0^\infty \frac{\sin at}{t} \cos \omega t \, dt$$

$$= \pi/2 \quad \text{if } \omega < a,$$

$$= \pi/4 \quad \text{if } \omega = a,$$

$$= 0 \quad \text{if } \omega > a. \tag{3.2.52}$$

The result is obtained easily after some algebraic manipulations. It is, however, better understood as the result of the inverse Fourier transfrom of a sinc function, which is simply a rectangular window function, as is evident in (3.2.52).

2. $f(t) = e^{-\beta t} \sin at$, $\qquad a, \text{Re}(\beta) > 0$.

$$\mathcal{F}_c[f(t)] = \int_0^\infty e^{-\beta t} \sin at \cos \omega t \, dt$$

$$= \frac{1}{2}\left[\frac{a+\omega}{\beta^2 + (a+\omega)^2} + \frac{a-\omega}{\beta^2 + (a-\omega)^2} \right] \qquad (3.2.53)$$

The result can be easily understood as the Laplace transform of the function:

$$\frac{1}{2}[\sin(a+\omega)t + \sin(a-\omega)t].$$

3. $f(t) = e^{-\beta t} \cos \alpha t$, $\qquad \text{Re}(\beta) > |\text{Im}(\alpha)|$.

$$\mathcal{F}_c[f(t)] = \int_0^\infty e^{-\beta t} \cos \alpha t \cos \omega t \, dt$$

$$= \frac{\beta}{2}\left[\frac{1}{\beta^2 + (\alpha-\omega)^2} + \frac{1}{\beta^2 + (\alpha+\omega)^2} \right], \qquad (3.2.54)$$

which is the Laplace transform of the function $\frac{1}{2}[\cos(\alpha+\omega)t + \cos(\alpha-\omega)t]$.

4. $f(t) = \dfrac{t \sin at}{(t^2 + \beta^2)}$ $\qquad a, \text{Re}(\beta) > 0$.

$$\mathcal{F}_c[f(t)] = \int_0^\infty \frac{t \sin at}{(t^2 + \beta^2)} \cos \omega t \, dt$$

$$= \frac{\pi}{2} e^{-a\beta} \cosh \beta \omega \qquad \text{if } \omega < a$$

$$= -\frac{\pi}{2} e^{-\beta \omega} \sinh a\beta \qquad \text{if } \omega > a. \qquad (3.2.55)$$

The result is obtained by contour integration, as is the next cosine transform.

5. $f(t) = \dfrac{\cos at}{(t^2 + \beta^2)}$ $\qquad a, \text{Re}(\beta) > 0$.

$$\mathcal{F}_c[f(t)] = \int_0^\infty \frac{\cos at}{(t^2 + \beta^2)} \cos \omega t \, dt$$

$$= \frac{\pi}{2\beta} e^{-a\beta} \cosh \beta \omega \qquad \text{if } \omega < a,$$

$$= \frac{\pi}{2\beta} e^{-\beta \omega} \cosh a\beta \qquad \text{if } \omega > a. \qquad (3.2.56)$$

6. $f(t) = e^{-\beta t^2} \cos at$ $\qquad \text{Re}(\beta) > 0$.

$$\mathcal{F}_c[f(t)] = \int_0^\infty e^{-\beta t^2} \cos at \cos \omega t \, dt$$

$$= \frac{1}{2}\sqrt{\frac{\pi}{\beta}} e^{-(a^2+\omega^2)/4\beta} \cosh \frac{a\omega}{2\beta}. \qquad (3.2.57)$$

3.2.3.4 FCT of orthogonal polynomials

1. *Legendre polynomials:*

$$f(t) = P_n(1 - 2t^2) \qquad 0 < t < 1,$$
$$= 0 \qquad\qquad\qquad t > 1,$$

where the Legendre polynomial $P_n(x)$ is defined as

$$P_n(x) = \frac{1}{2^n n!} \frac{d^n}{dx^n}(x^2 - 1)^n, \qquad \text{for } |x| < 1 \text{ and } n = 0, 1, 2, \ldots$$

$$\mathcal{F}_c[f(t)] = \int_0^1 P_n(1 - 2t^2) \cos \omega t\, dt$$

$$= \frac{(-1)^n \pi}{2} J_{n+\frac{1}{2}}(\omega/2) J_{-n-\frac{1}{2}}(\omega/2), \qquad (3.2.58)$$

where $J_\nu(z)$ is the Bessel function of the first kind, and order ν, defined by

$$J_\nu(z) = \sum_{m=0}^{\infty} \frac{(-1)^m (z/2)^{\nu + 2m}}{\Gamma(m+1)\Gamma(\nu + m + 1)}, \qquad |z| < \infty, |\arg z| < \pi. \qquad (3.2.58')$$

2. *Chebyshev polynomials:*

$$f(t) = (a^2 - t^2)^{-1/2} T_{2n}(t/a) \qquad 0 < t < a,\ n = 0, 1, 2, \ldots$$
$$= 0, \qquad\qquad\qquad\qquad t > a,$$

where the Chebyshev polynomial is defined by,

$$T_n(x) = \cos(n \cos^{-1} x), \qquad n = 0, 1, 2, \ldots$$

$$\mathcal{F}_c[f(t)] = \int_0^a (a^2 - t^2)^{-1/2} T_{2n}(t/a) \cos \omega t\, dt$$

$$= (-1)^n (\pi/2) J_{2n}(a\omega), \qquad (3.2.59)$$

where $J_{2n}(x)$ is the Bessel function defined in $(3.2.58')$ with $\nu = 2n$.

3. *Laguerre polynomial:*

$$f(t) = e^{-t^2/2} L_n(t^2)$$

where $L_n(x)$ is the Laguerre polynomial defined by,

$$L_n(x) = \frac{e^x}{n!} \frac{d^n}{dx^n}(x^n e^{-x}), \qquad n = 0, 1, 2, \ldots$$

$$\mathcal{F}_c[f(t)] = \int_0^\infty e^{-t^2/2} L_n(t^2) \cos \omega t\, dt$$

$$= \sqrt{\frac{\pi}{2}} \frac{1}{n!} e^{-\omega^2/2} \{\text{He}_n(\omega)\}^2, \qquad (3.2.60)$$

where $\text{He}_n(x)$ is the Hermite polynomial given by,

$$\text{He}_n(x) = (-1)^n e^{x^2/2} \frac{d^n}{dx^n}\left(e^{-x^2/2}\right), \qquad n = 0, 1, 2, \ldots.$$

4. *Hermite polynomials:*

(a) $f(t) = e^{-t^2/2} \operatorname{He}_{2n}(t)$ $n = 0, 1, 2, \ldots$

$$\mathcal{F}_c[f(t)] = \int_0^\infty e^{-t^2/2} \operatorname{He}_{2n}(t) \cos \omega t \, dt$$

$$= (-1)^n \sqrt{\frac{\pi}{2}} e^{-\omega^2/2} \omega^{2n} \qquad (3.2.61)$$

which is obtained directly using the Rodriques formula for the Hermite polynomial given in (3) above.

(b) $f(t) = e^{-t^2/2} \{\operatorname{He}_n(t)\}^2$,

$$\mathcal{F}_c[f(t)] = \int_0^\infty e^{-t^2/2} \{\operatorname{He}_n(t)\}^2 \cos \omega t \, dt$$

$$= n! \sqrt{\frac{\pi}{2}} e^{-\omega^2/2} L_n(\omega^2), \qquad (3.2.62)$$

which shows a rare symmetry with (3.2.60).

3.2.3.5   FCT of some special functions

1. *The complementary error function:*

$$f(t) = t \operatorname{Erfc}(at) \qquad a > 0.$$

Here the complementary error function is defined as

$$\operatorname{Erfc}(x) = 1 - \operatorname{Erf}(x) = \frac{2}{\sqrt{\pi}} \int_x^\infty e^{-t^2} \, dt.$$

$$\mathcal{F}_c[f(t)] = \int_0^\infty t \operatorname{Erfc}(at) \cos \omega t \, dt$$

$$= \left[\frac{1}{2a^2} + \frac{1}{\omega^2}\right] e^{-\omega^2/4a^2} - \frac{1}{\omega^2}. \qquad (3.2.63)$$

2. *The sine integral function:*

$$f(t) = \operatorname{si}(at) \qquad a > 0,$$

where $\operatorname{si}(x)$ is defined in (3.2.39).

$$\mathcal{F}_c[f(t)] = \int_0^\infty \operatorname{si}(at) \cos \omega t \, dt$$

$$= -(1/2\omega) \ln \left|\frac{\omega + a}{\omega - a}\right|, \qquad \omega \neq a. \qquad (3.2.64)$$

Note certain amount of symmetry with (3.2.51).

3. *The cosine integral function:*

$$f(t) = \operatorname{Ci}(at) = -\operatorname{ci}(at) \qquad a > 0,$$

where ci(x) is defined in (3.2.36).

$$\mathcal{F}_c[f(t)] = \int_0^\infty \text{Ci}(at) \cos \omega t \, dt = 0 \qquad \text{for } 0 < \omega < a,$$

$$= -\pi/2\omega \qquad \text{for } \omega > a. \tag{3.2.65}$$

4. *The exponential integral function:*

$$f(t) = \text{Ei}(-at) \qquad a > 0,$$

where Ei$(-x)$ is defined by

$$\text{Ei}(-x) = -\int_x^\infty e^{-t}/t \, dt, \qquad |\arg(x)| < \pi.$$

$$\mathcal{F}_c[f(t)] = \int_0^\infty \text{Ei}(-at) \cos \omega t \, dt = -\frac{1}{\omega} \tan^{-1}(\omega/a). \tag{3.2.66}$$

5. *Bessel functions:* We list only a few here since a more comprehensive table is available in Chapter 9 on Henkel transforms.

(a) $f(t) = J_0(at) \qquad a > 0,$

where $J_n(x)$ is the Bessel function of the first kind defined in (3.2.58′).

$$\mathcal{F}_c[f(t)] = \int_0^\infty J_0(at) \cos \omega t \, dt$$

$$= (a^2 - \omega^2)^{-1/2} \qquad \text{for } 0 < \omega < a,$$

$$= \infty, \qquad \text{for } \omega = a,$$

$$= 0, \qquad \text{for } \omega > a. \tag{3.2.67}$$

(b) $f(t) = J_{2n}(at) \qquad a > 0.$

$$\mathcal{F}_c[f(t)] = \int_0^\infty J_{2n}(at) \cos \omega t \, dt$$

$$= (-1)^n (a^2 - \omega^2)^{-1/2} T_{2n}(\omega/a) \qquad \text{for } 0 < \omega < a,$$

$$= \infty, \qquad \text{for } \omega = a,$$

$$= 0, \qquad \text{for } \omega > a. \tag{3.2.68}$$

Here, $T_{2n}(x)$ is the Chebyshev polynomial defined in (3.2.59). Note the symmetry between this and (3.2.59).

(c) $f(t) = t^{-n} J_n(at) \qquad a > 0, \text{ and } n = 1, 2, \ldots$

$$\mathcal{F}_c[f(t)] = \int_0^\infty t^{-n} J_n(at) \cos \omega t \, dt$$

$$= \frac{\sqrt{\pi}}{\Gamma(n + 1/2)} (2a)^{-n} (a^2 - \omega^2)^{n-1/2}, \qquad 0 < \omega < a,$$

$$= 0, \qquad \omega > a. \tag{3.2.69}$$

Here, $\Gamma(x)$ is the gamma function defined by

$$\Gamma(x) = \int_0^\infty e^{-t} t^{x-1}\, dt. \tag{3.2.69'}$$

(d) $f(t) = Y_0(at) \qquad a > 0,$

where $Y_\nu(x)$ is the Bessel function of the second kind defined by:

$$Y_\nu(x) = \operatorname{cosec}(\nu\pi)[J_\nu(x)\cos(\nu\pi) - J_{-\nu}(x)] \tag{3.2.70}$$

$$\mathcal{F}_c[f(t)] = \int_0^\infty Y_0(at)\cos\omega t\, dt$$

$$= 0, \qquad\qquad \text{for } 0 < \omega < a,$$

$$= -(\omega^2 - a^2)^{-1/2} \qquad \text{for } \omega > a. \tag{3.2.70'}$$

(e) $f(t) = t^\nu Y_\nu(at) \qquad |\operatorname{Re}(\nu)| < 1/2, a > 0,$

$$\mathcal{F}_c[f(t)] = \int_0^\infty t^\nu Y_\nu(at)\cos\omega t\, dt$$

$$= -\sqrt{\pi}(2a)^\nu [\Gamma(1/2 - \nu)]^{-1}(\omega^2 - a^2)^{-\nu-1/2}, \qquad \omega > a,$$

$$= 0, \qquad \text{for } 0 < \omega < a. \tag{3.2.71}$$

3.2.4   *Examples on the use of some operational rules of FCT*

In this section, some simple examples on the use of operational rules of the FCT are presented. The examples are based on very simple functions and are intended to illustrate the procedure and the features in the FCT operational rules that have been discussed in Section 3.2.2.

3.2.4.1   Differentiation-in-*t*

Let $f(t)$ be defined as $f(t) = e^{-\alpha t}$, where $\operatorname{Re}(\alpha) > 0$. Then according to (3.2.44), its FCT is given by

$$F_c(\omega) = \frac{\alpha}{\alpha^2 + \omega^2}.$$

To obtain the FCT for $f''(t)$, we have according to the differentiation-in-*t* property, (3.2.9),

$$\mathcal{F}_c[f''(t)] = -\omega^2 F_c(\omega) - f'(0) = -\omega^2 \frac{\alpha}{\alpha^2 + \omega^2} + \alpha$$

$$= \frac{\alpha^3}{\alpha^2 + \omega^2} \tag{3.2.72}$$

This result is verified by noting that $f''(t) = \alpha^2 e^{-\alpha t}$, and that its FCT is given directly also by (3.2.72).

3.2.4.2   Differentiaion-in-*t* of functions with jump discontinuities

Consider the function $f(t) = tU(1 - t)$, which is sometimes called a ramp function. It has a jump discontinuity of $d = -1$ at $t = 1$. Its derivative is given by $f'(t) = U(1 - t)$, which also has a jump discontinuity at $t = 1$. Using the definition for FCT, we obtain

$$\mathcal{F}_c[f'(t)] = \mathcal{F}_c[U(1 - t)] = \frac{\sin\omega}{\omega}. \tag{3.2.73}$$

The FCT rule of differentiation with jump discontinuity (3.2.14) can also be applied to get

$$\mathcal{F}_c[f'(t)] = \omega F_s(\omega) - f(0) - d\cos(\omega t_0)$$

$$= \omega\left[-\frac{\cos\omega}{\omega} + \frac{\sin\omega}{\omega^2}\right] - (-1)\cos\omega, \quad \text{(because } d = -1, \text{ and } f(0) = 0.)$$

$$= \frac{\sin\omega}{\omega}, \qquad \text{as in (3.2.73).}$$

3.2.4.3   Shift-in-t, shift-in-ω, and the kernel product property

Let $f(t) = e^{-\alpha t}$, where $\text{Re}(\alpha) > 0$. The FCT of a positive shift in the t-domain is easy to obtain,

$$\mathcal{F}_c[f(t+a)] = e^{-\alpha a}\frac{\alpha}{\alpha^2 + \omega^2} \qquad a > 0. \tag{3.2.74}$$

To obtain the FCT of the function $f(|t - a|)$, one can apply the kernel product property in (3.2.18) to get:

$$\mathcal{F}_c[f(|t-a|)] = 2F_c(\omega)\cos a\omega - \mathcal{F}_c[f(t+a)].$$

Therefore,

$$\mathcal{F}_c[e^{-\alpha|t-a|}] = 2\frac{\alpha}{\alpha^2 + \omega^2}\cos a\omega - e^{-\alpha a}\frac{\alpha}{\alpha^2 + \omega^2}$$

$$= \frac{\alpha}{\alpha^2 + \omega^2}[2\cos a\omega - e^{-\alpha a}] \tag{3.2.75}$$

which is much easier than direct evaluation.

Equation (3.2.21) typifies the shift-in-ω property and when it is applied to the same function $f(t)$ above, we obtain,

$$\mathcal{F}_c(e^{-\alpha t}\cos\beta t) = \frac{1}{2}\left[\frac{\alpha}{\alpha^2 + (\omega + \beta)^2} + \frac{\alpha}{\alpha^2 + (\omega - \beta)^2}\right] \tag{3.2.76}$$

3.2.4.4   Differentiation-in-ω property

This property, (3.2.23), can often be used to generate FCTs for functions that are not listed in the tables. As an example, consider again the function $f(t) = e^{-\alpha t}$, where $\text{Re}(\alpha) > 0$. To obtain the FCT for the function $g(t) = t^2 e^{-\alpha t}$, we can use (3.2.23) on $F_c(\omega)$ for $f(t) = e^{-\alpha t}$. Thus,

$$F_c''(\omega) = -2\alpha\frac{\alpha^2 - 3\omega^2}{(\alpha^2 + \omega^2)^3}, \quad \text{because } F_c[e^{-\alpha t}] = \frac{\alpha}{\alpha^2 + \omega^2},$$

and

$$\mathcal{F}_c[t^2 e^{-\alpha t}] = 2\alpha\frac{\alpha^2 - 3\omega^2}{(\alpha^2 + \omega^2)^3} \text{ using (3.2.23) with } n = 1.$$

3.2.4.5   The convolution property

The convolution property for FCT is closely related to its kernel product property as illustrated by the following example.

Let $f(t) = e^{-\alpha t}$, $\text{Re}(\alpha) > 0$, and $g(t) = U(t) - U(t - a)$, $a > 0$. The FCTs of these functions are given respectively by,

$$F_c(\omega) = \frac{\alpha}{\alpha^2 + \omega^2}, \qquad \text{and} \qquad G_c(\omega) = \frac{\sin a\omega}{\omega}.$$

Thus $2F_c(\omega)G_c(\omega) = 2\left[\frac{\alpha}{\alpha^2+\omega^2}\right]\left[\frac{\sin a\omega}{\omega}\right]$. According to the convolution property (3.2.30), this is the FCT of the convolution defined as:

$$\int_0^\infty [U(\tau) - U(\tau - a)][e^{-\alpha(t+\tau)} + e^{-\alpha|t-\tau|}]\,d\tau. \tag{3.2.77}$$

Applying the operator $\mathcal{F}_c$ to (3.2.77) and integrating over t first, the kernel product property in the shift-in-t operation in (3.2.18) can be invoked to give,

$$\mathcal{F}_c\left\{\int_0^\infty [U(\tau) - U(\tau - a)][e^{-\alpha(t+\tau)} + e^{-\alpha|t-\tau|}]\,d\tau\right\}$$

$$= 2\int_0^\infty [U(\tau) - U(\tau - a)]\frac{\alpha}{\alpha^2 + \omega^2}\cos\omega\tau\,d\tau = 2\left[\frac{\alpha}{\alpha^2 + \omega^2}\right]\left[\frac{\sin a\omega}{\omega}\right],$$

as required.

3.3   The Fourier sine transform (FST)

3.3.1   *Definitions and relations to the exponential Fourier transforms*

Similar to the Fourier cosine transform, the Fourier sine transform of a function $f(t)$, which is piece-wise continuous and absolutely integrable over $[0, \infty)$, is defined by application of the operator $\mathcal{F}_s$ as:

$$F_s(\omega) = \mathcal{F}_s[f(t)] = \int_0^\infty f(t)\sin\omega t\,dt, \qquad \omega > 0. \tag{3.3.1}$$

The inverse operator $\mathcal{F}_s^{-1}$ is similarly defined:

$$f(t) = \mathcal{F}_s^{-1}[F_s(\omega)] = \frac{2}{\pi}\int_0^\infty F_s(\omega)\sin\omega t\,d\omega, \qquad t \geq 0, \tag{3.3.2}$$

subject to the existence of the integral. Functions $f(t)$ and $F_s(\omega)$ defined by (3.3.2) and (3.3.1) respectively are said to form a Fourier sine transform pair. It is noted in (3.2.3) and (3.3.2) for the inverse FCT and inverse FST that both transform operators have symmetric kernels and that they are involutary or unitary up to a factor of $\sqrt{(2/\pi)}$.

Fourier sine transforms are also very closely related to the exponential Fourier transform defined in (3.2.6). Using the property that

$$\sin\omega t = \text{Im}[e^{j\omega t}] = \frac{1}{2j}[e^{j\omega t} - e^{-j\omega t}], \tag{3.3.3}$$

one can consider the odd extension of the function $f(t)$ defined over $[0, \infty)$ as

$$f_o(t) = f(t) \qquad t \geq 0,$$
$$= -f(-t) \qquad t < 0.$$

Then the Fourier transform of $f_o(t)$ is

$$\mathcal{F}[f_o(t)] = \int_{-\infty}^\infty f_o(t)e^{-j\omega t}\,dt = -\int_0^\infty f(t)e^{j\omega t}\,dt + \int_0^\infty f(t)e^{-j\omega t}\,dt$$

$$= -2j\int_0^\infty f(t)\sin\omega t\,dt = -2j\mathcal{F}_s[f(t)],$$

and therefore,

$$\mathcal{F}_s[f(t)] = -\frac{1}{2j}\mathcal{F}[f_o(t)].\tag{3.3.4}$$

Equation (3.3.4) provides the relation between the FST and the exponential Fourier transform. As in the case for cosine transforms, many properties of the sine transform can be related to those for the Fourier transform through this equation. We shall present some properties and operational rules for FST in the next section.

3.3.2 Basic properties and operational rules

1. *Inverse Transformation:* The inverse transformation is exactly the same as the forward transformation except for the normalization constant. Combining the forward and inverse transformations leads to the Fourier sine integral formula, which states that,

$$f(t) = \frac{2}{\pi}\int_0^\infty F_s(\omega)\sin\omega t\, d\omega = \frac{2}{\pi}\int_0^\infty\left[\int_0^\infty f(\tau)\sin\omega\tau\, d\tau\right]\sin\omega t\, d\omega.\tag{3.3.5}$$

The sufficient conditions for the inversion formula (3.3.2) are the same as for the cosine transform. Where $f(t)$ has a jump discontinuity at $t = t_0$, (3.3.5) converges to the mean of $f(t_0 + 0)$ and $f(t_0 - 0)$.

2. *Transforms of Derivatives:* Derivatives transform in a fashion similar to FCT, even orders involving sine transforms only and odd orders involving cosine transforms only. Thus, for example,

$$\mathcal{F}_s[f''(t)] = -\omega^2 F_s(\omega) + \omega f(0)\tag{3.3.6}$$

and

$$\mathcal{F}_s[f'(t)] = -\omega F_c(\omega),\tag{3.3.7}$$

where $f(t)$ is assumed continuous to the first order.

For the fourth-order derivative, we apply (3.3.6) twice to obtain,

$$\mathcal{F}_s[f^{(iv)}(t)] = \omega^4 F_s(\omega) - \omega^3 f(0) + \omega f''(0),\tag{3.3.8}$$

if $f(t)$ is continuous at least to order three. When the function $f(t)$ and its derivatives have jump discontinuities at $t = t_0$, (3.3.8) is modified to become,

$$\mathcal{F}_s[f^{(iv)}(t)] = \omega^4 F_s(\omega) - \omega^3 f(0) + \omega f''(0) - \omega^3 d\cos\omega t_0$$

$$+\omega^2 d'\sin\omega t_0 + \omega d''\cos\omega t_0 - d'''\sin\omega t_0\tag{3.3.9}$$

where the jump discontinuities d, d', d'', and d''' are as defined in (3.2.13). Similarly, for odd-order derivatives, when the function $f(t)$ has jump discontinuities, the operational rule must be modified. For example, (3.3.7) will become:

$$\mathcal{F}_s[f'(t)] = -\omega F_c(\omega) + d\sin\omega t_0.\tag{3.3.7'}$$

Generalization to other orders and to more than one location for the jump discontinuities is straightforward.

3. *Scaling:* Scaling in the t-domain for the FST has exactly the same effect as in the case of FCT, giving,

$$\mathcal{F}_s[f(at)] = \frac{1}{a}F_s(\omega/a)\qquad a > 0.\tag{3.3.10}$$

4. *Shifting:*

(a) Shift in the t-domain:

As in the case of the Fourier cosine transform, we first define the even and odd extensions of the function $f(t)$ as,

$$f_e(t) = f(|t|), \qquad \text{and} \qquad f_o(t) = \frac{t}{|t|} f(|t|). \qquad (3.3.11)$$

Then it can be shown that:

$$\mathcal{F}_s[f_o(t+a) + f_o(t-a)] = 2F_s(\omega) \cos a\omega \qquad (3.3.12)$$

and

$$\mathcal{F}_c[f_o(t+a) - f_o(t-a)] = 2F_s(\omega) \sin a\omega; \qquad a > 0. \qquad (3.3.13)$$

These, together with (3.2.18) and (3.2.19), form a complete set of kernel-product relations for the cosine and the sine transforms.

(b) Shift in the ω-domain:

For a positive β shift in the ω-domain, it is easily shown that

$$F_s(\omega + \beta) = \mathcal{F}_s[f(t) \cos \beta t] + \mathcal{F}_c[f(t) \sin \beta t] \qquad (3.3.14)$$

and combining with the result for a negative shift, we get:

$$\mathcal{F}_s[f(t) \cos \beta t] = (1/2)[F_s(\omega + \beta) + F_s(\omega - \beta)]. \qquad (3.3.15)$$

More generally, for $a, \beta > 0$, we have,

$$\mathcal{F}_s[f(at) \cos \beta t] = (1/2a) \left[F_s \left(\frac{\omega + \beta}{a} \right) + F_s \left(\frac{\omega - \beta}{a} \right) \right]. \qquad (3.3.16)$$

Similarly, we can easily show that

$$\mathcal{F}_s[f(at) \sin \beta t] = -(1/2a) \left[F_c \left(\frac{\omega + \beta}{a} \right) - F_c \left(\frac{\omega - \beta}{a} \right) \right]. \qquad (3.3.17)$$

The shift-in-ω properties are useful in deriving some FCTs and FSTs. As well, because the quantities being transformed are modulated sinusoids, these are useful in applications to communication problems.

5. *Differentiation in the ω-domain:* The sine transform behaves in a fashion similar to the cosine transform when it comes to differentiation in the ω-domain. Even-order derivatives involve only sine transforms and odd-order derivatives involve only cosine transforms. Thus,

$$F_s^{(2n)}(\omega) = \mathcal{F}_s[(-1)^n t^{2n} f(t)],$$

and

$$F_s^{(2n+1)}(\omega) = \mathcal{F}_c[(-1)^n t^{2n+1} f(t)]. \qquad (3.3.18)$$

It is again assumed that the integrals in (3.3.18) exist.

6. *Asymptotic behavior:* The Reimann-Lebesque theorem gaurantees that any Fourier sine transform converges to zero as ω tends to infinity, that is,

$$\lim_{\omega \to \infty} F_s(\omega) = 0. \qquad (3.3.19)$$

7. *Integration:*

(a) Integration in the t-domain. In analogy to (3.2.26), we have

$$\mathcal{F}_s\left[\int_0^t f(\tau)\,d\tau\right] = (1/\omega)F_c(\omega) \tag{3.3.20}$$

provided $f(t)$ is piece-wise smooth and absolutely integrable over $[0, \infty)$.

(b) Integration in the ω-domain. As in the Fourier cosine transform, integration in the ω-domain results in division by t in the t-domain, giving,

$$\mathcal{F}_c^{-1}\left[\int_\omega^\infty F_s(\beta)\,d\beta\right] = (1/t)f(t) \tag{3.3.21}$$

in parallel with (3.2.27).

8. *The convolution property:* If functions $f(t)$ and $g(t)$ are piece-wise continuous and absolutely integrable over $[0, \infty)$, a convolution property involving $F_s(\omega)$ and $G_c(\omega)$ is

$$2F_s(\omega)G_c(\omega) = \mathcal{F}_s\left\{\int_0^\infty f(\tau)[g(|t - \tau|) - g(t + \tau)]\,d\tau\right\}. \tag{3.3.22}$$

Equivalently,

$$2F_s(\omega)G_c(\omega) = \mathcal{F}_s\left\{\int_0^\infty g(\tau)[f(t + \tau) + f_o(t - \tau)]\,d\tau\right\} \tag{3.3.23}$$

where $f_o(x)$ is the odd extension of the function $f(x)$ defined as in (3.3.11).

One can establish a convolution theorem involving only sine transforms. This is obtained by imposing an additional condition on one of the functions, say $g(t)$. We define the function $h(t)$ by,

$$h(t) = \int_t^\infty g(\tau)\,d\tau. \tag{3.3.24}$$

Then $g(t)$ must satisfy the condition that its integral $h(t)$ is absolutely integrable over $[0, \infty)$, so that the Fourier cosine transform of $h(t)$ exists. We note from (3.2.26) that

$$H_c(\omega) = (1/\omega)G_s(\omega). \tag{3.3.25}$$

Applying (3.3.22) to $f(t)$ and $h(t)$ yields immediately,

$$(2/\omega)F_s(\omega)G_s(\omega) = \mathcal{F}_s\left[\int_0^\infty f(\tau)\int_{|t-\tau|}^{t+\tau} g(\eta)\,d\eta\,d\tau\right] \tag{3.3.26}$$

noting that $g(t) = -h'(t)$.

Because the FSTs have properties and operation rules very similar to those for the FCTs, we refer the reader to Section 3.2.4 for simple examples on the use of these rules for FCTs.

3.3.3 *Selected Fourier sine transforms*

In this section, selected Fourier sine transforms are presented. These mostly correspond to those selected for the Fourier cosine transforms. It should be noted that because the sine and cosine transform kernels are related through differentiation, many of the Fourier sine transforms can be derived without direct computation by using differentiation properties listed in Sections 3.2.2 and 3.3.2. As before, we present first the FST of algebraic functions.

3.3.3.1   FST of algebraic functions

1. *The unit rectangular function:*

$$f(t) = U(t) - U(t - a), \qquad \text{where } U(t) \text{ is the Heaviside unit step function.}$$

$$\mathcal{F}_s[f(t)] = \int_0^a \sin \omega t \, dt = (1 - \cos \omega a)/\omega. \qquad (3.3.27)$$

2. *The unit height tent function:*

$$
\begin{aligned}
f(t) &= t/a, & 0 < t < a, \\
&= (2a - t)/a & a < t < 2a, \\
&= 0 & \text{otherwise.}
\end{aligned}
$$

$$\mathcal{F}_s[f(t)] = \int_0^a (t/a) \sin \omega t \, dt + \int_a^{2a} [(2a - t)/a] \sin \omega t \, dt$$

$$= \frac{1}{a\omega^2} [2 \sin a\omega - \sin 2a\omega]. \qquad (3.3.28)$$

3. *Delayed inverse:*

$$f(t) = (1/t) U(t - a).$$

$$\mathcal{F}_s[f(t)] = \int_a^\infty (1/t) \sin \omega t \, dt = \int_{a\omega}^\infty (1/\tau) \sin \tau \, d\tau = -\operatorname{si}(a\omega) \qquad (3.3.29)$$

where $\operatorname{si}(x)$ is the sine integral function defined in (3.2.39).

4. *The inverse square root:*

$$f(t) = 1/\sqrt{t}.$$

$$\mathcal{F}_s[f(t)] = \int_0^\infty \frac{1}{\sqrt{t}} \sin \omega t \, dt = \sqrt{\frac{\pi}{2\omega}} \qquad (3.3.30)$$

5. *The inverse linear function:*

$$f(t) = (\alpha + t)^{-1}, \qquad |\arg \alpha| < \pi.$$

$$\mathcal{F}_s[f(t)] = \int_0^\infty \frac{1}{\alpha + t} \sin \omega t \, dt$$

$$= \sin \omega\alpha \, \operatorname{Ci}(\omega\alpha) - \cos \omega\alpha \, \operatorname{si}(\omega\alpha). \qquad (3.3.31)$$

Here $\operatorname{Ci}(x)$ is the cosine integral function defined in (3.2.36).

6. *Inverse quadratic functions:*

(a) $f(t) = (t^2 + a^2)^{-1} \qquad a > 0.$

$$\mathcal{F}_s[f(t)] = \int_0^\infty \frac{1}{a^2 + t^2} \sin \omega t \, dt$$

$$= (1/2a)[e^{-a\omega} \overline{\operatorname{Ei}}(a\omega) - e^{a\omega} \operatorname{Ei}(-a\omega)] \qquad (3.3.32)$$

where $\operatorname{Ei}(x)$ and $\overline{\operatorname{Ei}}(x)$ are the exponential integral functions defined in (3.2.50).

Here, we note that (3.3.32) is related to the FCT of the function,

$$f(t) = -t(t^2 + a^2)^{-1}$$

by considering the derivative of (3.3.32) with respect to ω. Thus

$$\mathcal{F}_c[-t(t^2 + a^2)^{-1}] = (1/2)[e^{-a\omega}\overline{\text{Ei}}(a\omega) + e^{a\omega}\,\text{Ei}(-a\omega)] \tag{3.3.33}$$

(b) $f(t) = (a^2 - t^2)^{-1}$ $a > 0$.

$$\mathcal{F}_s[f(t)] = \text{P.V.} \int_0^\infty \frac{1}{a^2 - t^2} \sin \omega t \, dt$$

$$= [\sin a\omega\, \text{Ci}(a\omega) - \cos a\omega\, \text{Si}(a\omega)]/a, \tag{3.3.34}$$

where $\text{Ci}(x)$ and $\text{Si}(x)$ are the cosine and sine integral functions defined in (3.2.36) and (3.2.39) and "P.V." denotes the principal value of the integral. Again, we note that (3.3.34) is related to the FCT of the function,

$$f(t) = -t(a^2 - t^2)^{-1}.$$

Thus,

$$\mathcal{F}_c[-t(a^2 - t^2)^{-1}] = \cos a\omega\, \text{Ci}(a\omega) + \sin a\omega\, \text{Si}(a\omega). \tag{3.3.35}$$

(c) $f(t) = \dfrac{\beta}{\beta^2 + (a - t)^2} - \dfrac{\beta}{\beta^2 + (a + t)^2}$ $\text{Re}(\beta) > 0$.

$$\mathcal{F}_s[f(t)] = \int_0^\infty \left[\frac{\beta}{\beta^2 + (a - t)^2} - \frac{\beta}{\beta^2 + (a + t)^2} \right] \sin \omega t \, dt$$

$$= \pi \sin a\omega\, e^{-\beta\omega}. \tag{3.3.36}$$

(d) $f(t) = \dfrac{a + t}{\beta^2 + (a + t)^2} - \dfrac{a - t}{\beta^2 + (a - t)^2}$ $\text{Re}(\beta) > 0$.

$$\mathcal{F}_s[f(t)] = \int_0^\infty \left[\frac{a + t}{\beta^2 + (a + t)^2} - \frac{a - t}{\beta^2 + (a - t)^2} \right] \sin \omega t \, dt$$

$$= \pi \cos a\omega\, e^{-\beta\omega}. \tag{3.3.37}$$

We note here the symmetry among the transforms in (3.3.36), (3.3.37), and those in (3.2.43) and (3.2.42).

3.3.3.2   FST of exponential and logarithmic functions

1. $f(t) = e^{-\alpha t}$ $\text{Re}(\alpha) > 0$.

$$\mathcal{F}_s[f(t)] = \int_0^\infty e^{-\alpha t} \sin \omega t \, dt = \frac{\omega}{\alpha^2 + \omega^2} \tag{3.3.38}$$

which is also seen to be the Laplace transform of $\sin \omega t$.

2. $f(t) = \dfrac{e^{-\beta t} - e^{-\alpha t}}{t^2}$ $\text{Re}(\beta), \text{Re}(\alpha) > 0$.

$$\mathcal{F}_s[f(t)] = \int_0^\infty \frac{e^{-\beta t} - e^{-\alpha t}}{t^2} \sin \omega t \, dt$$

$$= \frac{\omega}{2} \ln \left(\frac{\alpha^2 + \omega^2}{\beta^2 + \omega^2} \right) + \alpha \tan^{-1} \left(\frac{\omega}{\alpha} \right) - \beta \tan^{-1} \left(\frac{\omega}{\beta} \right). \tag{3.3.39}$$

Equation (3.3.39) is seen to be related to the result (3.2.45) through the differentiation-in-ω property of the sine transform as defined in (3.3.18).

3. $f(t) = te^{-\alpha t^2}$ $\qquad |\arg(\alpha)| < \pi/2.$

$$\mathcal{F}_s[f(t)] = \int_0^\infty te^{-\alpha t^2} \sin \omega t \, dt$$

$$= \frac{1}{4}\sqrt{\frac{\pi}{\alpha^3}} \omega e^{-\omega^2/4\alpha}, \qquad (3.3.40)$$

which can also be related to the cosine transform in (3.2.46) using again the differentiation-in-ω property (3.3.18) of the sine transform.

4. $f(t) = \ln t[1 - U(t-1)]$

$$\mathcal{F}_s[f(t)] = \int_0^\infty \ln t[1 - U(t-1)] \sin \omega t \, dt$$

$$= -\frac{1}{\omega}[C + \ln \omega - \text{Ci}(\omega)], \qquad (3.3.41)$$

which is obtained easily through integration by parts. Here $C = 0.5772156649\ldots$ is the Euler constant and $\text{Ci}(x)$ is the cosine integral function.

5. $f(t) = \dfrac{t \ln bt}{(t^2 + a^2)}$ $\qquad a, b > 0.$

$$\mathcal{F}_s[f(t)] = \int_0^\infty \frac{t \ln bt}{(t^2 + a^2)} \sin \omega t \, dt$$

$$= \frac{\pi}{4}[2e^{-a\omega} \ln ab - e^{a\omega} \text{Ei}(-a\omega) - e^{-a\omega}\overline{\text{Ei}}(a\omega)] \qquad (3.3.42)$$

Note that (3.3.42) is related to (3.2.49) through the differentiation-in-ω property of the Fourier cosine transform as defined in (3.2.24).

6. $f(t) = \ln\left|\dfrac{t+a}{t-a}\right|$ $\qquad a > 0,$

$$\mathcal{F}_s[f(t)] = \int_0^\infty \ln\left|\frac{t+a}{t-a}\right| \sin \omega t \, dt$$

$$= \frac{\pi}{\omega} \sin a\omega. \qquad (3.3.43)$$

The result is obtained using integration by parts and the shift-in-t properties (3.3.11) to (3.3.13) of the sine transform.

3.3.3.3   FST of trigonometric functions

1. $f(t) = \dfrac{\sin at}{t}$ $\qquad a > 0,$

$$\mathcal{F}_s[f(t)] = \int_0^\infty \frac{\sin at}{t} \sin \omega t \, dt$$

$$= (1/2) \ln\left|\frac{\omega + a}{\omega - a}\right| \qquad (3.3.44)$$

This result is immediately understood when compared to (3.3.43), taking into account the normalization used in (3.3.1) and (3.3.2) for the definition of the Fourier sine transform.

2. $f(t) = \dfrac{e^{-\beta t}}{t} \sin \alpha t \qquad \text{Re}(\beta) > |\text{Im}(\alpha)|$

$$\mathcal{F}_s[f(t)] = \int_0^\infty \frac{e^{-\beta t}}{t} \sin \alpha t \sin \omega t \, dt$$

$$= (1/4) \ln \left(\frac{\beta^2 + (\omega + \alpha)^2}{\beta^2 + (\omega - \alpha)^2} \right) \qquad (3.3.45)$$

This result follows easily from the integration-in-ω property (3.2.27) as applied to the cosine transform in (3.2.53).

3. $f(t) = e^{-\beta t} \cos \alpha t \qquad \text{Re}(\beta) > |\text{Im}(\alpha)|$

$$\mathcal{F}_s[f(t)] = \int_0^\infty e^{-\beta t} \cos \alpha t \sin \omega t \, dt$$

$$= (1/2) \left[\frac{\omega - \alpha}{\beta^2 + (\omega - \alpha)^2} + \frac{\omega + \alpha}{\beta^2 + (\omega + \alpha)^2} \right], \qquad (3.3.46)$$

which is also recognized as the Laplace transform of the function $\cos \alpha t \sin \omega t$.

4. $f(t) = \dfrac{t \cos at}{(t^2 + \beta^2)} \qquad a, \text{Re}(\beta) > 0,$

$$\mathcal{F}_s[f(t)] = \int_0^\infty \frac{t \cos at}{(t^2 + \beta^2)} \sin \omega t \, dt$$

$$= -\frac{\pi}{2} e^{-a\beta} \sinh \beta \omega \qquad \omega < a,$$

$$= \frac{\pi}{2} e^{-\beta \omega} \cosh a\beta \qquad \omega > a. \qquad (3.3.47)$$

Note the symmetry of (3.3.47) with (3.2.55).

5. $f(t) = \dfrac{\sin at}{(t^2 + \beta^2)} \qquad a, \text{Re}(\beta) > 0,$

$$\mathcal{F}_s[f(t)] = \int_0^\infty \frac{\sin at}{(t^2 + \beta^2)} \sin \omega t \, dt$$

$$= \frac{\pi}{2\beta} e^{-a\beta} \sinh \beta \omega \qquad \omega < a,$$

$$= \frac{\pi}{2\beta} e^{-\beta \omega} \sinh a\beta \qquad \omega > a. \qquad (3.3.48)$$

The symmetry of (3.3.48) with (3.2.56) is apparent.

6. $f(t) = e^{-\beta t^2} \sin at \qquad \text{Re}(\beta) > 0.$

$$\mathcal{F}_s[f(t)] = \int_0^\infty e^{-\beta t^2} \sin at \sin \omega t \, dt$$

$$= \frac{1}{2} \sqrt{\frac{\pi}{\beta}} e^{-(\omega^2 + a^2)/4\beta} \sinh \frac{a\omega}{2\beta} \qquad (3.3.49)$$

similar to (3.2.57) for the cosine transform.

3.3.3.4   FST of orthogonal polynomials

1. *Legendre polynomial* (defined in [3.2.58]):

$$f(t) = P_n(1 - 2t^2)[1 - U(t - 1)] \qquad n = 0, 1, 2, \ldots$$

$$\mathcal{F}_s[f(t)] = \int_0^1 P_n(1 - 2t^2) \sin \omega t \, dt$$

$$= \frac{\pi}{2} \left[J_{n+1/2} \left(\frac{\omega}{2} \right) \right]^2 \qquad (3.3.50)$$

where $J_v(x)$ is the Bessel function of the first kind defined in (3.2.58′).

2. *Chebyshev polynomial* (defined in [3.2.59]):

$$f(t) = (a^2 - t^2)^{-1/2} T_{2n+1}(t/a)[1 - U(t - a)], \qquad n = 0, 1, 2, \ldots$$

$$\mathcal{F}_s[f(t)] = \int_0^a (a^2 - t^2)^{-1/2} T_{2n+1}(t/a) \sin \omega t \, dt$$

$$= (-1)^n \frac{\pi}{2} J_{2n+1}(a\omega). \qquad (3.3.51)$$

3. *Laguerre polynomials:*

$$f(t) = t^{2m} e^{-t^2/2} L_n^{2m+1}(t^2), \qquad m, n = 0, 1, 2, \ldots$$

$$\mathcal{F}_s[f(t)] = \int_0^\infty t^{2m} e^{-t^2/2} L_n^{2m+1}(t^2) \sin \omega t \, dt$$

$$= \sqrt{\frac{\pi}{2}} (n!)^{-1} (-1)^m e^{-\omega^2/2} \, \mathrm{He}_n(\omega) \, \mathrm{He}_{n+2m+1}(\omega) \qquad (3.3.52)$$

where $L_n^a(x) = \frac{e^x x^{-a}}{n!} \frac{d^n}{dx^n} (e^{-x} x^{n+a})$, is a Laguerre polynomial ($L_n^0(x) = L_n(x)$ as defined in [3.2.60]). Here, $\mathrm{He}_n(x)$ is the Hermite polynomial defined in (3.2.61).

4. *Hermite polynomials* (defined in [3.2.61]):

$$f(t) = e^{-t^2/2} \, \mathrm{He}_{2n+1}(\sqrt{2}t)$$

$$\mathcal{F}_s[f(t)] = \int_0^\infty e^{-t^2/2} \, \mathrm{He}_{2n+1}(\sqrt{2}t) \sin \omega t \, dt$$

$$= (-1)^n \sqrt{\frac{\pi}{2}} e^{-\omega^2/2} \, \mathrm{He}_{2n+1}(\sqrt{2}\omega). \qquad (3.3.53)$$

3.3.3.5   FST of some special functions

1. *The complementary error function* (defined in [3.2.63]):

$$f(t) = \mathrm{Erfc}(at) \qquad a > 0,$$

$$\mathcal{F}_s[f(t)] = \int_0^\infty \mathrm{Erfc}(at) \sin \omega t \, dt$$

$$= \frac{1}{\omega}[1 - e^{-\omega^2/4a^2}]. \qquad (3.3.54)$$

2. *The sine integral function* (defined in [3.2.39]):

$$f(t) = \text{si}(at) \qquad a > 0,$$

$$\mathcal{F}_s[f(t)] = \int_0^\infty \text{si}(at) \sin \omega t \, dt = 0 \qquad 0 \le \omega < a$$

$$= -\frac{\pi}{2\omega} \qquad \omega > a. \qquad (3.3.55)$$

Note the symmetry of (3.3.55) with (3.2.65).

3. *The cosine integral function* (defined in [3.2.36]):

$$f(t) = \text{Ci}(at) = -\text{ci}(at) \qquad a > 0$$

$$\mathcal{F}_s[f(t)] = \int_0^\infty \text{Ci}(at) \sin \omega t \, dt$$

$$= \frac{1}{2\omega} \ln \left| \frac{\omega^2}{a^2} - 1 \right|. \qquad (3.3.56)$$

4. *The exponential integral function* (defined in [3.2.66]):

$$f(t) = \text{Ei}(-at) \qquad a > 0$$

$$\mathcal{F}_s[f(t)] = \int_0^\infty \text{Ei}(-at) \sin \omega t \, dt = -\frac{1}{2\omega} \ln \left(\frac{\omega^2}{a^2} + 1 \right). \qquad (3.3.57)$$

5. *Bessel functions* (defined in [3.2.58′]):

   (a) $f(t) = J_0(at) \qquad a > 0$

$$\mathcal{F}_s[f(t)] = \int_0^\infty J_0(at) \sin \omega t \, dt = 0, \qquad\qquad 0 < \omega < a,$$

$$= (\omega^2 - a^2)^{-1/2} \qquad \omega > a. \qquad (3.3.58)$$

   (b) $f(t) = J_{2n+1}(at) \qquad a > 0$

$$\mathcal{F}_s[f(t)] = \int_0^\infty J_{2n+1}(at) \sin \omega t \, dt$$

$$= (-1)^n (a^2 - \omega^2)^{-1/2} T_{2n+1}(\omega/a) \qquad 0 < \omega < a,$$

$$= 0 \qquad\qquad \omega > a, \qquad (3.3.59)$$

where $T_n(x)$ is the Chebyshev polynomial defined in (3.2.59).

   (c) $f(t) = t^{-n} J_{n+1}(at) \qquad a > 0 \text{ and } n = 0, 1, 2, \ldots$

$$\mathcal{F}_s[f(t)] = \int_0^\infty t^{-n} J_{n+1}(at) \sin \omega t \, dt$$

$$= \frac{\sqrt{\pi}}{\Gamma(n+1/2)} \frac{1}{2^n a^{n+1}} \omega(a^2 - \omega^2)^{n-1/2}, \qquad 0 < \omega < a$$

$$= 0 \qquad\qquad \omega > a, \qquad (3.3.60)$$

where $\Gamma(x)$ is the gamma function defined in (3.2.69′).

(d) $f(t) = Y_0(at) \qquad a > 0.$

where $Y_\nu(x)$ is the Bessel function of the second kind (see [3.2.70]).

$$\mathcal{F}_s[f(t)] = \int_0^\infty Y_0(at) \sin \omega t \, dt$$

$$= \tfrac{2}{\pi}(a^2 - \omega^2)^{-1/2} \sin^{-1}\left(\tfrac{\omega}{a}\right), \qquad\qquad 0 < \omega < a$$

$$= \frac{2}{\pi}(\omega^2 - a^2)^{-1/2} \ln\left|\frac{\omega}{a} - \left(\frac{\omega^2}{a^2} - 1\right)^{1/2}\right|, \qquad \omega > a. \qquad (3.3.61)$$

(e) $f(t) = t^\nu Y_{\nu-1}(at) \qquad a > 0, |\operatorname{Re}(\nu)| < 1/2$

$$\mathcal{F}_s[f(t)] = \int_0^\infty t^\nu Y_{\nu-1}(at) \sin \omega t \, dt$$

$$= \frac{2^\nu a^{\nu-1}\sqrt{\pi}}{\Gamma(1/2 - \nu)}\omega(\omega^2 - a^2)^{-\nu-1/2} \qquad \omega > a,$$

$$= 0 \qquad\qquad\qquad\qquad 0 < \omega < a. \qquad (3.3.62)$$

As with the cosine transforms, more detailed results are found in the sections covering Henkel transforms.

3.4   The discrete sine and cosine transforms (DST and DCT)

In practical applications, the computations of the Fourier sine and cosine transforms are done with sampled data of finite duration. Because of the finite duration and the discrete nature of the data, much can be gained in theory and in ease of computation by formulating the corresponding discrete sine and cosine transforms (DST and DCT) directly. In what follows, we discuss the definitions and properties of the discrete sine and cosine transforms. It is possible to define four different types of each of the DCT and the DST (for details, see Rao and Yip 1990). We shall concentrate on Type I, which can be defined by simply discretizing the FST and FCT, within a finite rectangular window of unit height.

3.4.1   *Definitions of DCT and DST, and relations to FST and FCT*

Consider the transform kernel of the FCT given by

$$K_c(\omega, t) = \cos \omega t. \qquad (3.4.1)$$

Let $\omega_m = 2\pi m \Delta f$ and $t_n = n\Delta t$ be the sampled angular frequency and time, respectively. Here, Δf and Δt are the sample intervals for frequency and time, respectively. m and n are positive integers. The kernel in (3.4.1) can now be discretized as,

$$K_c(m, n) = K_c(\omega_m, t_n) = \cos(2\pi mn \Delta f \Delta t). \qquad (3.4.2)$$

If we further let $\Delta f \Delta t = 1/(2N)$, where N is a positive integer, we obtain the discrete cosine transform kernel:

$$K_c(m, n) = \cos(\pi mn/N) \qquad (3.4.3)$$

where $m, n = 0, 1, \ldots, N$. The transform kernel in (3.4.3) is the DCT kernel of Type I. It represents the mnth element in an $(N + 1) \times (N + 1)$ transformation matrix, which, with the proper normalization, provides the definition for the DCT transformation matrix $[C]$. These elements are,

$$[C]_{mn} = \sqrt{\frac{2}{N}} \left\{ k_m k_n \cos\left(\frac{mn\pi}{N}\right) \right\}, \qquad m, n = 0, 1, \ldots, N$$

where

$$k_i = 1 \qquad \text{for } i \neq 0 \text{ or } N$$

$$= 1/\sqrt{2} \qquad \text{for } i = 0 \text{ or } N. \tag{3.4.4}$$

The discretization can be viewed as taking a finite time duration and dividing it into N intervals of Δt each. Including the boundary points, there are $N + 1$ sample points to be considered. If the discrete $N + 1$ sample points are represented by a vector $\mathbf{x}$, the DCT of this vector is a vector $\mathbf{X}_c$ given by,

$$\mathbf{X}_c = [C]\mathbf{x} \tag{3.4.5}$$

which, in an element-by-element form, means

$$\mathbf{X}_c(m) = \sqrt{\frac{2}{N}} \sum_{n=0}^{N} k_m k_n \cos\left(\frac{mn\pi}{N}\right) \mathbf{x}(n). \tag{3.4.6}$$

It can be shown that $[C]$ is a unitary matrix. Thus, the inverse transformation is given by

$$\mathbf{x}(n) = \sqrt{\frac{2}{N}} \sum_{m=0}^{N} k_m k_n \cos\left(\frac{mn\pi}{N}\right) \mathbf{X}_c(m). \tag{3.4.7}$$

Vectors $\mathbf{X}_c$ and $\mathbf{x}$ are said to be a DCT pair.

Similar consideration in discretizing the FST kernel

$$K_s(\omega, t) = \sin \omega t \tag{3.4.8}$$

will lead to the definition of the $(N - 1) \times (N - 1)$ DST transform matrix, whose elements are given by

$$[S]_{mn} = \sqrt{\frac{2}{N}} \sin\left(\frac{mn\pi}{N}\right) \qquad m, n = 1, 2, \ldots, N - 1. \tag{3.4.9}$$

This matrix is also unitary and when it is applied to a data vector $\mathbf{x}$ of length $N - 1$, it produces a vector $\mathbf{X}_s$, whose elements are given by,

$$\mathbf{X}_s(m) = \sqrt{\frac{2}{N}} \sum_{n=1}^{N-1} \sin\left(\frac{mn\pi}{N}\right) \mathbf{x}(n). \tag{3.4.10}$$

The vectors $\mathbf{x}$ and $\mathbf{X}_s$ are said to form a DST pair. The inverse DST is given by

$$\mathbf{x}(n) = \sqrt{\frac{2}{N}} \sum_{m=1}^{N-1} \sin\left(\frac{mn\pi}{N}\right) \mathbf{X}_s(m). \tag{3.4.11}$$

It is evident in (3.4.7) and (3.4.11) that both DCT and DST are symmetric transforms. Both are obtained by discretizing a finite time duration into N equal intervals of Δt each, resulting in an $(N + 1) \times (N + 1)$ matrix for $[C]$ because the boundary elements are not zero, and resulting in an $(N - 1) \times (N - 1)$ matrix for $[S]$ because the boundary elements are zero.

3.4.2 Basic properties and operational rules

3.4.2.1 The unitarity property

Let $\mathbf{c}_m$ denote the mth column vector in the matrix $[C]$. Consider the inner product of two such vectors:

$$\langle \mathbf{c}_m, \mathbf{c}_n \rangle = \sum_{p=0}^{N} \left(\frac{2}{N} \right) k_m k_p \cos \left(\frac{mp\pi}{N} \right) k_p k_n \cos \left(\frac{pn\pi}{N} \right). \tag{3.4.12}$$

The summation can be carried out by defining the $2N$th primitive root of unity as

$$W_{2N} = e^{-j\pi/N} = \cos \left(\frac{\pi}{N} \right) - j \sin \left(\frac{\pi}{N} \right), \tag{3.4.13}$$

and applying it to the summation in (3.4.12). This gives

$$\langle \mathbf{c}_m, \mathbf{c}_n \rangle = \left(\frac{k_m k_n}{N} \right) \mathrm{Re} \left[\sum_{p=0}^{N-1} (W_{2N})^{-p(n-m)} + \sum_{p=1}^{N} (W_{2N})^{-p(n+m)} \right] \tag{3.4.14}$$

where $\mathrm{Re}[\cdot]$ denotes the real part of $[\cdot]$.

Considering the first summation in (3.4.14), and letting $\kappa = (n - m)$, the power series can be written as,

$$\sum_{p=0}^{N-1} \left(W_{2N}^{-\kappa} \right)^p = \frac{\left(1 - W_{2N}^{-N\kappa} \right)}{\left(1 - W_{2N}^{-\kappa} \right)}$$

$$= \{ 2[1 - \cos(\kappa\pi/N)] \}^{-1} \left\{ 1 - W_{2N}^{N\kappa} - W_{2N}^{\kappa} + W_{2N}^{-(N-1)\kappa} \right\}. \tag{3.4.15}$$

Similarly, the second series in (3.4.14) can be summed by letting $\lambda = (n + m)$,

$$\sum_{p=1}^{N} \left(W_{2N}^{-\lambda} \right)^p = \{ 2[1 - \cos(\lambda\pi/N)] \}^{-1} \left\{ W_{2N}^{-\lambda} - W_{2N}^{-(N+1)\lambda} - 1 + W_{2N}^{-N\lambda} \right\}. \tag{3.4.16}$$

Hence, for $m \neq n$, (i.e., $\kappa \neq 0$), the real part of (3.4.15) is

$$\mathrm{Re} \left[\sum_{p=0}^{N-1} \left(W_{2N}^{-\kappa} \right)^p \right] = \frac{[1 - (-1)^\kappa][1 - \cos(\kappa\pi/N)]}{\{ 2[1 - \cos(\kappa\pi/N)] \}} = [1 - (-1)^\kappa]/2,$$

and the real part of (3.4.16) is

$$\mathrm{Re} \left[\sum_{p=1}^{N} \left(W_{2N}^{-\lambda} \right)^p \right] = -\frac{[1 - (-1)^\lambda][1 - \cos(\lambda\pi/N)]}{\{ 2[1 - \cos(\lambda\pi/N)] \}} = -[1 - (-1)^\lambda]/2.$$

Combining these, and noting that κ and λ differ by $2m$, we obtain the orthogonality property for the inner product,

$$\langle \mathbf{c}_m, \mathbf{c}_n \rangle = 0 \qquad \text{for } m \neq n. \tag{3.4.17}$$

For $m = n \neq 0$ or N, the inner product is,

$$\langle \mathbf{c}_m, \mathbf{c}_n \rangle = (1/N)\,\mathrm{Re}\left[\sum_{p=0}^{N-1} 1 + \sum_{p=1}^{N} \left(W_{2N}^{-2m}\right)^p\right] = 1,$$

and for $m = n = 0$ or N, the inner product is,

$$\langle \mathbf{c}_m, \mathbf{c}_n \rangle = (1/2N)\,\mathrm{Re}\left(\sum_{p=0}^{N-1} 1 + \sum_{p=1}^{N} 1\right) = 1.$$

Therefore, the inner product satisfies the orthonormality condition,

$$\langle \mathbf{c}_m, \mathbf{c}_n \rangle = \delta_{mn}, \tag{3.4.18}$$

where δ_{mn} is the Kronecker delta and the DCT matrix $[C]$ is shown to be unitary.

Similar considerations can be applied to the DST matrix $[S]$ to show that it is also unitary.

3.4.2.2 Inverse transformation

As alluded to in Section 3.4.1, the unitary matrices $[C]$ and $[S]$ are symmetric and therefore the inverse transformations are exactly the same as the forward transformations, based on the above unitarity properties. Therefore,

$$[C]^{-1} = [C] \qquad \text{and} \qquad [S]^{-1} = [S]. \tag{3.4.19}$$

3.4.2.3 Scaling

Recall that in the discretization of the FCT, the time and frequency intervals are related by

$$\Delta f\,\Delta t = 1/2N \qquad \text{or } \Delta f = \frac{1}{2N\,\Delta t}. \tag{3.4.20}$$

Because the DCT and DST deal with discrete sample points, a scaling in time has no effect in the transform, except in changing the unit frequency interval in the transform domain. Thus, as Δt changes to $a\,\Delta t$, Δf changes to $\Delta f/a$, provided the number of divisions N remains the same. Hence, the properties (3.2.16) and (3.3.10) for the FCT and FST are retained, except for the $1/a$ factor, which is absent in the cases for DCT and DST.

Equation (3.4.20) may also be interpreted as giving the frequency resolution of a set of discrete data points, sampled at a time interval of Δt. Using $T = N\,\Delta t$ as the time duration of the sequence of data points, the frequency resolution for the transforms is,

$$\Delta f = \frac{1}{2T}. \tag{3.4.21}$$

3.4.2.4 Shift-in-t

Because the data are sampled, we obtain the shift-in-time properties of DCT and DST by examining the time shifts in units of Δt. Thus, if $\mathbf{x} = [x(0), x(1), \ldots, x(N)]^T$, we define the right-shifted sequence as $\mathbf{x}^+ = [x(1), x(2), \ldots, x(N+1)]^T$. Their corresponding DCTs are given by

$$\mathbf{X}_c = [C]\mathbf{x} \qquad \text{and} \qquad \mathbf{X}_c^+ = [C]\mathbf{x}^+. \tag{3.4.22}$$

The shift-in-time property seeks to relate $\mathbf{X}_c^+$ with $\mathbf{X}_c$. It turns out that it relates not only to $\mathbf{X}_c$ but also to $\mathbf{X}_s$, the DST of $\mathbf{x}$. This is to be expected because the shift-in-time properties of

FCT and FST are similarly related. It can be shown that the elements of $\mathbf{X}_c^+$ are given by

$$X_c^+(m) = \cos\left(\frac{m\pi}{N}\right) X_c(m) + k_m \sin\left(\frac{m\pi}{N}\right) X_s(m)$$

$$+ \sqrt{\frac{1}{N}} k_m \left[\left(-\frac{1}{\sqrt{2}}\right) \cos\left(\frac{m\pi}{N}\right) x(0) + \left(\frac{1}{\sqrt{2}} - 1\right) x(1)\right.$$

$$\left. + (-1)^m \left(\frac{1}{\sqrt{2}} - 1\right) \cos\left(\frac{m\pi}{N}\right) x(N) + (-1)^m \frac{1}{\sqrt{2}} x(N+1)\right]. \quad (3.4.23)$$

In (3.4.23), $X_c(m)$ and $X_s(m)$ are respectively the mth element of the DCT of the vector $[x(0), x(1), \ldots, x(N)]^T$ and the mth element of the DST of the vector $[x(1), x(2), \ldots, x(N-1)]^T$. While properties analogous to the so-called kernel-product properties for FCT in Section 3.2.2 may be developed, (3.4.23) is more practical in that it provides for a way of updating a DCT of a given dimension without having to recompute all the components. The corresponding result for DST is,

$$X_s^+(m) = \cos\left(\frac{m\pi}{N}\right) X_s(m) - \sin\left(\frac{m\pi}{N}\right) X_c(m)$$

$$+ \sqrt{\frac{2}{N}} \sin\left(\frac{m\pi}{N}\right) \left[\frac{1}{\sqrt{2}} x(0) - \left(1 - \frac{1}{\sqrt{2}}\right) (-1)^m x(N)\right]. \quad (3.4.24)$$

Here, it is noted that $X_c(m)$ are the elements of the DCT of the vector $[x(0), \ldots, x(N)]^T$.

3.4.2.5   The difference property

For discrete sequences, the difference operator replaces the differential operator for continuous sequences. The FCT and the FST of a derivative, therefore, are analogous to the DCT and the DST of the difference operator. We can define a difference vector $\mathbf{d}$ as:

$$\mathbf{d} = \mathbf{x}^+ - \mathbf{x} \quad (3.4.25)$$

where $\mathbf{x}^+$ is the right-shifted version of $\mathbf{x}$. It is clear that the DCT and the DST of $\mathbf{d}$ are simply given by

$$\mathbf{D}_c = \mathbf{X}_c^+ - \mathbf{X}_c \quad \text{and} \quad \mathbf{D}_s = \mathbf{X}_s^+ - \mathbf{X}_s. \quad (3.4.26)$$

As we can see from (3.4.26), the main operational advantage of the FCT and FST, namely that in the differentiation properties, have not carried over to the discrete cases. As well, properties with both integration-in-t and integration-in-ω are also lost in the discrete cases.

We conclude this section by mentioning that no simple convolution properties exist in the cases of DCT and DST. For finite sequences, it is possible to define a circular convolution for two periodic sequences or a linear convolution of two nonperiodic sequences. With these, certain convolution properties for some of the discrete cosine transforms may be developed. (For more details, the reader is referred to Rao and Yip 1990.) The results, however, are neither simple nor easy to apply.

3.4.3   *Relation to the Karhunen-Loeve transform (KLT)*

While the DCT and the DST discussed here are derived by discretizing the FCT and the FST, based on some unit time interval of Δt and some unit frequency interval of Δf, their forms are closely related to the Karhunen-Loeve transform (KLT) in digital signal processing. KLT is an optimal transform for digital signals in that it diagonalizes the auto-covariance matrix of a data vector. It completely decorrelates the signal in the transform domain, minimizes the mean squared errors (MSE) in data compression and packs the most energy (variance) in the fewest number of transform coefficients.

Consider a Markov-1 signal with correlation coefficient ρ. The $N \times N$ covariance matrix is a matrix $[A]$, which is real, symmetric, and Toeplitz. It is well known that a nonsingular symmetric Toeplitz matrix has an inverse of tri-diagonal form. In the case of the covariance matrix $[A]$ for a Markov-1 signal, we can write

$$[A]^{-1} = (1 - \rho^2)^{-1} \begin{pmatrix} 1 & -\rho & 0 & 0 & \cdots & \cdots & \cdots \\ -\rho & 1 + \rho^2 & -\rho & 0 & \cdots & \cdots & \cdots \\ \cdots & \cdots & \cdots & \cdots & \cdots & 1 + \rho^2 & -\rho \\ \cdots & \cdots & \cdots & \cdots & \cdots & -\rho & 1 \end{pmatrix}. \tag{3.4.27}$$

This matrix can be decomposed into a sum of two simpler matrices,

$$[A]^{-1} = [B] + [R]$$

where,

$$[B] = (1 - \rho^2)^{-1} \begin{pmatrix} 1 + \rho^2 & -\sqrt{2}\rho & 0 & \cdots & \cdots \\ -\sqrt{2}\rho & 1 + \rho^2 & -\rho & \cdots & \cdots \\ 0 & -\rho & 1 + \rho^2 & -\rho & \cdots \\ \cdots & \cdots & \cdots & \cdots & \cdots \\ \cdots & \cdots & \cdots & -\sqrt{2}\rho & 1 + \rho^2 \end{pmatrix}$$

and

$$[R] = (1 - \rho^2)^{-1} \begin{pmatrix} -\rho^2 & (\sqrt{2} - 1)\rho & 0 & \cdots & \cdots \\ (\sqrt{2} - 1)\rho & 0 & 0 & \cdots & \cdots \\ \cdots & \cdots & \cdots & 0 & (\sqrt{2} - 1)\rho \\ \cdots & \cdots & \cdots & (\sqrt{2} - 1)\rho & -\rho^2 \end{pmatrix}. \tag{3.4.28}$$

We note that $[R]$ is almost a null matrix, and can be considered so when N is very large. Thus the diagonalization of the matrix $[B]$ is asymptotically equivalent to the diagonalization of the matrix $[A]^{-1}$. Furthermore, it is well known that the similarity transformation that diagonalizes $[A]^{-1}$ will also diagonalize $[A]$. From these arguments, it is concluded that the transformation that diagonalizes $[B]$ will, asymptotically, diagonalize $[A]$. The transformation that diagonalizes $[B]$ depends on a three-terms recurrence relation that is exactly satisfied by the Chebyshev polynomials. With these, it can be shown that the matrix $[V]$ that will diagonalize $[B]$ and, in turn, also $[A]$ asymptotically, is defined by,

$$[V]_{mn} = k_n k_m \sqrt{\frac{2}{N - 1}} \cos\left(\frac{mn\pi}{(N - 1)}\right), \qquad m, n = 0, 1, \ldots, N - 1. \tag{3.4.29}$$

As can be seen in (3.4.29), these are the elements of the DCT matrix $[C]$, except that N has been replaced by $N - 1$. For large N, these are identical.

The foregoing has briefly demonstrated that for a Markov-1 signal, the diagonalization of the covariance matrix, which leads to the KLT, is provided by a transformation matrix $[V]$ which is almost identical to the DCT matrix $[C]$. This explains why the DCT performs so well in signal decorrelation, although it is signal independent. Similar arguments can be applied to the DST.

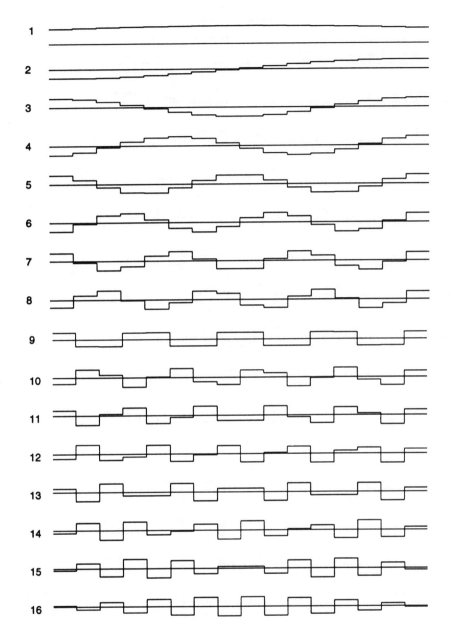

FIGURE 3.1
KLT Markov-1 signal $\rho = 0.95$, $N = 16$.

In Figure 3.1, the basis functions forming the KLT for $N = 16$ are shown. The signal is a Markov-1 signal with a correlation coefficient of $\rho = 0.95$. It is clear that the set of basis functions and hence the KLT is signal dependent, because they are the eigenvectors of the auto-covariance matrix of the signal vector.

In Figures 3.2 and 3.3, the basis functions for $N = 16$ of DCT and DST are shown. It is evident that they are very similar to the KLT basis functions. While it is true that the dimensions

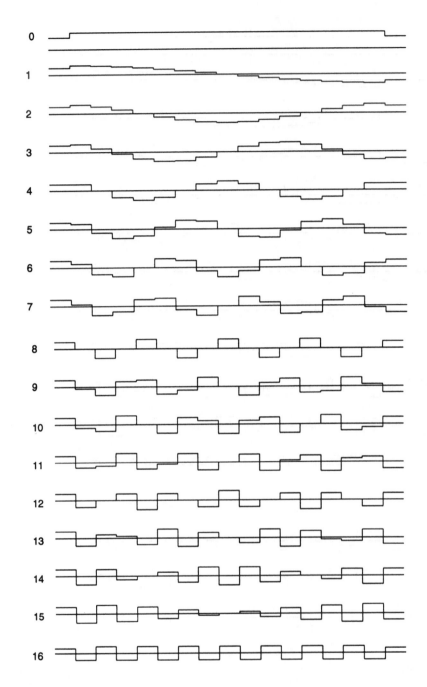

FIGURE 3.2
DCT $N = 16$.

of the spaces spanned by the KLT and the DCT and DST are different, it can be shown that as N increases, both discrete transforms will asymptotically approach KLT.

However, it is true that the similarity of the basis functions does not gaurantee the asymptotic behavior of the DCT and the DST, nor does it assure good performance. In applications, such as data compression and transform domain coding, the "variance distribution" of the transform

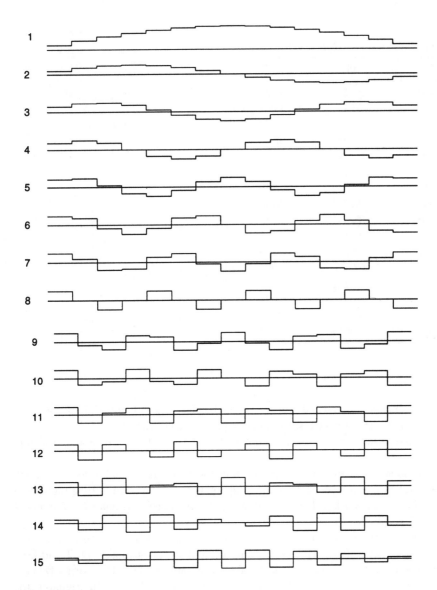

FIGURE 3.3
DST $N = 16$.

coefficients is an important criterion of performance. The variance of a transform coefficient is basically a measure of the information content of that coefficient. Therefore, the higher the variances are in a few transform coefficients, the more room there is for data compression in that transform domain.

Let $[A]$ be the data covariance matrix, and let $[T]$ be the transformation. Then, the covariance matrix in the transform domain, $[A]_T$, is given by,

$$[A]_T = [T][A][T]^{-1}. \qquad (3.4.30)$$

The diagonal elements of $[A]_T$ are the variances of the transform coefficients. In Table 3.1, comparisons are shown for the variance distributions of the DCT, the DST, and the DFT, based

TABLE 3.1
Variance Distributions for $N = 16$, $\rho = 0.9$.

i	DCT*	DST	DFT
0	9.835	9.218	9.835
1	2.933	2.640	1.834
2	1.211	1.468	1.834
3	0.581	0.709	0.519
4	0.348	0.531	0.519
5	0.231	0.314	0.250
6	0.166	0.263	0.250
7	0.129	0.174	0.155
8	0.105	0.153	0.155
9	0.088	0.110	0.113
10	0.076	0.099	0.113
11	0.068	0.078	0.091
12	0.062	0.071	0.091
13	0.057	0.061	0.081
14	0.055	0.057	0.081
15	0.053	0.054	0.078

*DCT is DCT-II here.

on a Markov-1 signal of $\rho = 0.9$ and $N = 16$. It can be clearly seen that both DCT and DST outperform DFT in using variance distribution as a performance criterion.

When the transformation $[T]$ in (3.4.30) is not the KLT, $[A]_T$ will not be diagonal. The nonzero off-diagonal elements in $[A]_T$ form a measure of the " residual correlation." The smaller the amount of residual correlation, the closer is the transform to being optimal. Figure 3.4 shows the residual correlation as a percentage of the total amount of correlation, for the transforms DCT, DST, and DFT, in a Markov-1 signal with $N = 16$. As can be seen, again DCT and DST outperform DFT generally.

There are other criteria of performance for a given transform, depending on what kind of signal processing is being done. However, using the KLT as a benchmark, DCT and DST are extremely good alternatives as signal independent, fast implementable transforms, because they are both asymptotic to the KLT. This asymptotic property of the discrete trigonometric transforms (particularly the DCT) has made them very important tools in digital signal processing. Although they are suboptimal, in the sense that they will not exactly diagonalize the data covariance matrix, they are signal independent and are computable using fast algorithms. KLT, though exactly optimal, is signal dependent and possesses no fast computational algorithm. Some typical applications are discussed in the next section.

3.5 Selected applications

This section contains some typical applications. We begin with fairly general applications to differential equations, and conclude with quite specific applications in the area of data compression. (See Churchill 1958 and Sneddon 1972 for more applications.)

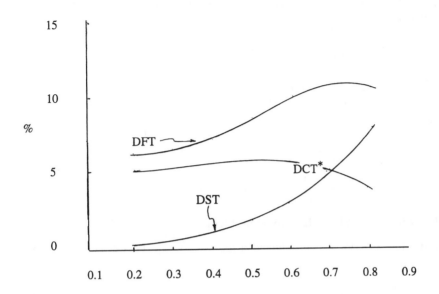

FIGURE 3.4
Percent Residual Correlation as a function of ρ, $N = 16$.

3.5.1 *Solution of differential equations*

3.5.1.1 One-dimensional boundary value problem

Consider the second-order differential equation,

$$y''(t) - h^2 y(t) = F(t) \qquad t \geq 0 \tag{3.5.1}$$

with boundary conditions: $y'(0) = 0$ and $y(\infty) = 0$, and

$$
\begin{aligned}
F(t) &= A &&\text{for } 0 < t < b \\
&= 0 &&\text{otherwise.}
\end{aligned}
$$

We note that $F(t)$ can be expressed in terms of a Heaviside step function. Thus,

$$F(t) = A[1 - U(t - b)]. \tag{3.5.2}$$

Here, we assume h, A and b to be constants. Applying the operator $\mathcal{F}_c$ to the differential equation and using the results in (3.2.9) and (3.2.34), we get

$$-\omega^2 Y_c - y'(0) - h^2 Y_c = \frac{A}{\omega} \sin \omega b. \tag{3.5.3}$$

Applying the boundary condition and solving for Y_c, we obtain

$$Y_c = -\frac{A}{\omega(\omega^2 + h^2)} \sin \omega b$$

$$= -\frac{A}{h^2} \left(\frac{\sin \omega b}{\omega} - \frac{\omega \sin \omega b}{\omega^2 + h^2} \right). \tag{3.5.4}$$

The inversion of Y_c can be accomplished with the use of (3.2.34), (3.2.55), and (3.2.3). Noting that the inverse FCT has a normalization factor of $2/\pi$, the solution for the original boundary value problem is given by,

$$y(t) = -\frac{A}{h^2}[1 - U(t - b) - e^{-hb} \cosh ht] \qquad t < b,$$

$$= -\frac{A}{h^2}[1 - U(t - b) + e^{-ht} \sinh hb] \qquad \text{for } t > b.$$

These can be re-written as

$$y(t) = \frac{A}{h^2}(e^{-hb} \cosh ht - 1) \qquad \text{for } t < b,$$

$$= -\frac{A}{h^2}e^{-ht} \sinh hb \qquad \text{for } t > b. \tag{3.5.5}$$

3.5.1.2   Two-dimensional boundary value problem

Consider a function $v(x, y)$, which is bounded for $x \geq 0, y \geq 0$. Let $v(x, y)$ satisfy the boundary value problem:

$$\frac{\partial^2 v}{\partial x^2} + \frac{\partial^2 v}{\partial y^2} = -h(x); \qquad \left.\frac{\partial v}{\partial x}\right|_{x=0} = 0, \qquad v(x, 0) = f(x). \tag{3.5.6}$$

We further assume that $\int_0^\infty h(x)\,dx = 0$, and that the function

$$p(x) = \int_x^\infty \left[\int_0^r h(t)\,dt\right] dr \tag{3.5.7}$$

exists and that the functions $p(x)$ and $f(x)$ have FCTs. We note from (3.5.7) that

$$p''(x) = -h(x) \qquad \text{and} \qquad p'(0) = 0,$$

leading to the following relation between their FCTs:

$$\omega^2 P_c(\omega) = H_c(\omega) \tag{3.5.8}$$

Applying $\mathcal{F}_c$ for the x variable in (3.5.6) reduces the partial differential equation to

$$-\omega^2 V_c(\omega, y) + \frac{\partial^2}{\partial y^2} V_c(\omega, y) = -\omega^2 P_c(\omega). \tag{3.5.9}$$

Because $V_c(\omega, y)$ is bounded for $y > 0$, (3.5.9) has the following solution,

$$V_c(\omega, y) = Ce^{-\omega y} + P_c(\omega) \tag{3.5.10}$$

where C is an arbitrary constant, to be determined by $v(x, 0) = f(x)$. In the ω-domain, this means,

$$V_c(\omega, 0) = F_c(\omega). \tag{3.5.11}$$

Thus,

$$V_c(\omega, y) = [F_c(\omega) - P_c(\omega)]e^{-\omega y} + P_c(\omega). \tag{3.5.12}$$

This can be inverted, and the solution in the (x, y) domain then is given by

$$v(x, y) = p(x) + \frac{1}{\pi} \int_0^\infty [f(t) - p(t)]\left[\frac{y}{(x + t)^2 + y^2} + \frac{y}{(x - t)^2 + y^2}\right] dt. \tag{3.5.13}$$

Here, we have made use of (3.2.44) and the convolution result of (3.2.30).

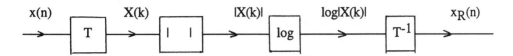

FIGURE 3.5
Block Diagram for Cepstral Analysis for $x(n)$.

3.5.1.3   Time-dependent one-dimensional boundary value problem

Consider the function $u(x, t)$, which is bounded for $x, t \geq 0$. Let this function satisfy the partial differential equation,

$$\frac{\partial u}{\partial t} - \frac{\partial^2 u}{\partial x^2} = h(x, t) \tag{3.5.14}$$

so that $u(x, 0) = f(x)$ and $u(0, t) = g(t)$ are the initial and boundary conditions.

Applying the FST for the variable x to (3.5.14) and assuming the existence of all the integrals involved, we obtain

$$\frac{\partial U_s}{\partial t} + \omega^2 U_s = \omega g(t) + H_s(\omega, t). \tag{3.5.15}$$

The solution for (3.5.15) is

$$U_s(\omega, t)e^{\omega^2 t} = \int_0^t [\omega g(\tau) + H_s(\omega, \tau)]e^{\omega^2 \tau} \, d\tau + C. \tag{3.5.16}$$

C is easily found to be $F_s(\omega)$ using the condition $U_s(\omega, 0) = F_s(\omega)$. With this, (3.5.16) can be inverse transformed by applying the operator $\mathcal{F}_s^{-1}$ to get

$$u(x, t) = \frac{2}{\pi} \int_0^\infty U_s(\omega, t) \sin \omega x \, d\omega. \tag{3.5.17}$$

We note here that, depending on the forms of the functions F_s and H_s, the inverse FST may be obtained by table look-up.

3.5.2   *Cepstral analysis in speech processing*

In cepstral analysis, a sequence is converted by a transform T, the logarithm of its absolute value is then taken and the cepstrum is then obtained by inverse transformation T^{-1}. Figure 3.5 shows the essential steps in cepstral analysis. Here, $\{x(n)\}$ is the input speech sequence, $\{X(k)\}$ is the transform sequence, and the output $\{x_R(n)\}$ is called the real cepstrum.

The transform may be any invertible transform. When T is an N-point DFT, the scheme can be implemented using the DCT. In the computation to obtain the real cepstrum using the DFT, the input sequence has to be padded with trailing zeros to double its length. However, a simple relation between the DFT and the DCT for real even sequences reduces the DFT to a DCT.

Let $x(n), n = 0, 1, 2, \ldots, M$ be the input speech sequence to be analysed. To obtain the real cepstrum $x_R(n)$ using DFT, the sequences is padded with zeros so that $x(n) = 0$, for

$n = M + 1, \ldots, 2M - 1$. If we consider a symmetric sequence $s(n)$ defined by,

$$s(n) = x(n) \qquad 0 < n < M,$$

$$= 2x(n) \qquad n = 0, M$$

$$= x(2M - n) \qquad M < n \leq 2M - 1, \qquad (3.5.18)$$

then the DFT of $s(n)$ can be obtained as

$$S_F(k) = 2 \left[x(0) + (-1)^k x(M) + \sum_{n=1}^{M-1} x(n) \cos\left(\frac{nk\pi}{M}\right) \right]. \qquad (3.5.19)$$

Equation (3.5.19) is clearly in the form of a DCT of the sequence $\{x(n)\}$ up to a constant factor of normalizaton. Now, because $\{s(n)\}$ is a symmetric real sequence, constructed out of $\{x(n)\}$, we have

$$S_F(k) = \text{Re}[X_F(k)]$$

where $\{X_F(k)\}$ is the 2M-point DFT of the zero-padded sequence. Combining this with (3.5.19) we see that

$$\text{Re}[X_F(k)] = 2[X_c(k)] \qquad (3.5.20)$$

where X_c is the $(M+1)$-point DCT of the speech sequence $\{x(n)\}$. Equation (3.5.20) is valid up to a normalization constant. Because direct sparse matrix factorization of the $(M+1) \times (M+1)$ DCT matrix is possible, fast algorithms exist for the computation of the DCT. This means that in order to obtain the real cepstrum of $\{x(n)\}$, there is no need to pad the sequence with trailing zeros, and the computation for $x_R(k)$ can be achieved through the use of the DCT of the sequence $\{x(n)\}$.

Rather than using DCT as a means of computing the DFT, the transform T in the cepstral analysis can directly be a DCT or a DST. It has been found that the performance of speech cepstral analysis using DCT and DST is comparable to the traditional DFT cepstral analysis.

3.5.3 *Data compression*

Data compression is an important application of transform coding when retrieval of a signal from a large database is required. Transform coefficients with large variances can be retained to represent significant features for pattern recognition, for example. Those with small variances, below a certain threshold, can be discarded. Such a scheme can be used in reducing the required bandwidth for purposes of transmission or storage.

The transforms used for these data compression purposes require maximal decorrelation of the data, with highest energy-packing efficiency possible (efficiency is defined as how much energy can be packed into the fewest number of transform coefficients). The ideal or optimal transform is the KLT, which will diagonalize the data covariance matrix and pack the most energy into the fewest transform coefficients. Unfortunately, KLT is data dependent, and has no known fast computational algorithm, and is therefore not practical. On the other hand, Markov models describe most of the data systems quite well, and suboptimal but asymptotically equivalent transforms such as the DCT and the DST are data independent, and implementable using fast algorithms. Therefore, in many applications, such as storage of electrocardiogram (ECG) or vector cardiogram (VCG) data, or video data transmission over telephone lines for video phones, suboptimal transforms such as the DCT are preferred over the optimal KLT. For such applications, depending upon the required fidelity of the reconstructed data, compression

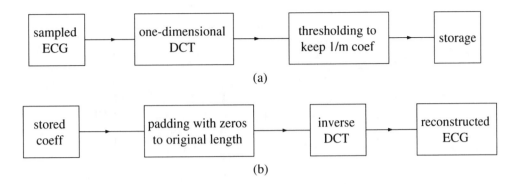

FIGURE 3.6
(a) Data compression for storage; (b) Reconstruction from compressed data.

ratios of up to 10:1 have been reported, and compression ratios of 3:1 to 5:1 using DCT for both ECG (one-dimensional) and VCG (two-dimensional) are commonplace.

Figures 3.6a and 3.6b show the block diagrams for processing, storage, and retrieval of a one-dimensional ECG, using m:1 compression ratio.

3.5.4   *Transform domain processing*

While discarding low variance coefficients in the DCT domain will provide data compression, certain details or desired features in the original data may be lost in the reconstruction. It is possible to remedy this partially by processing the transform coefficients before reconstruction. Adaptive processing can be applied based on some subjective criteria, such as in video phone applications. Coefficient quantization is another means of processing to minimize the effect of noise.

Other processing techniques such as subsampling (decimation) and up-sampling (interpolation) can also be performed in the DCT domain, effectively combining the operations of filtering and transform coding. Such processing techniques have been successfully employed to convert high definition TV signals to the standard NTSC TV signals.

One of the most popular digital signal processing tools is the adaptive least-mean-square (LMS) filtering. This can be done either in the time domain or in the transform domain. Figure 3.7 shows the block diagram for the adaptive DCT transform domain LMS filtering. Here $a_{n0}, a_{n1}, \ldots, a_{n,N-1}$ are the adaptive weights for the transform domain filter. The desired response is $\{r(n)\}$ and $\{y(n)\}$ is the filtered output. It has been found that such transform domain filtering speeds up the convergence of the LMS algorithm for speech-related applications such as spectral analysis and echo cancellation.

3.6   Computational algorithms

In actual computations of FCT and FST, the basic integrations are performed with quadratures. Because the data are sampled and the duration is finite, most of the quadratures can be implemented via matrix computations. The fact that the FST and the FCT are closely related to the Fourier transform translates directly to the close relations between the computation of the DCT and the DST with that of the DFT. Many algorithms have been developed for the DFT.

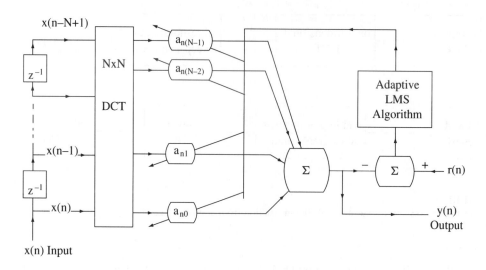

FIGURE 3.7
Adaptive transform domain LMS filtering.

The most well known among them is the Cooley-Tukey fast Fourier transform (FFT), which is often regarded as the single most important development in modern digital signal processing. More recently, there have been other algorithms such as the Winograd algorithm, which are based on prime-factor decomposition and polynomial factorization.

While DST and DCT can be computed using relations with DFT (thus using fast algorithms such as the Cooley-Tukey or the Winograd), the transform matrices have sufficient structure to be exploited directly, so that sparse factorizations can be applied to realize the transforms. The sparse factorization depends on the size of the transform, as well as the way permutations are applied to the data sequence. As a result, there are two distinct types of sparse factorizations, the decimation-in-time (DIT) algorithms and the decimation-in-frequency (DIF) algorithms. (DIT algorithms are of the Cooley-Tukey type while DIF algorithms are of the Sande-Tukey type).

In Section 3.6.1, the computations of FST and FCT using FFT are discussed. In Section 3.6.2, the direct fast computations of DCT and DST are presented. Both DIT and DIF algorithms are discussed. All algorithms discussed are radix-2 algorithms, where N, which is related to the sample size, is an integer power of two.

3.6.1   *FCT and FST algorithms based on FFT*

3.6.1.1   **FCT of real data sequence**

Let $\{x(n), n = 0, 1, \ldots, N\}$ be an $(N + 1)$-point sequence. Its DCT as defined in (3.4.6) is given by,

$$X_c(m) = \sqrt{\frac{2}{N}} \sum_{n=0}^{N} k_m k_n \cos\left(\frac{mn\pi}{N}\right) x(n),$$

where

$$k_n = 1 \qquad \text{for } n \neq 0 \text{ or } N$$
$$= 1/\sqrt{2} \qquad \text{for } n = 0 \text{ or } N.$$

Construct an even or symmetric sequence using $\{x(n)\}$ in the following way,

$$s(n) = x(n) \qquad 0 < n < N,$$
$$= 2x(n) \qquad n = 0, N,$$
$$= x(2N - n) \qquad N < n \leq 2N - 1. \tag{3.6.1}$$

Based on the fact that the Fourier transform of a real symmetric sequence is real and is related to the cosine transform of the half-sequence, it can be shown that the DFT of $\{s(n)\}$ is given by,

$$S_F(m) = 2\left[x(0) + (-1)^m x(N) + \sum_{n=1}^{N-1} \cos\left(\frac{mn\pi}{N}\right) x(n) \right]. \tag{3.6.2}$$

Thus the $(N+1)$-point DCT of $\{x(n)\}$ is the same as the $2N$-point DFT of the sequence $\{s(n)\}$, up to a normalization constant as indicated by (3.4.6). This means that the DCT of $\{x(n)\}$ can be computed using a $2N$-point FFT of $\{s(n)\}$. We note here that

$$S_F(m) = \sum_{n=0}^{2N-1} s(n) W_{2N}^{mn}, \tag{3.6.3}$$

where $W_{2N} = e^{-j2\pi/2N}$, the principal $2N$th root of unity, is used for defining the DFT.

It should be pointed out that the direct $2N$-point DFT of a real even sequence may be considered inefficient, because inherent complex arithmetics are used to produce real coefficents in the transform. However, it is well known that a real $2N$-point DFT can be implemented using an N-point DFT for a complex sequence. For details, the reader is referred to Chapter 2 on Fourier transforms.

3.6.1.2   FST of real data sequence

Let $\{x(n), n = 1, 2, \ldots, N - 1\}$ be an $(N - 1)$-point data sequence. Its DST as defined in (3.4.10) is given by,

$$X_s(m) = \sqrt{\frac{2}{N}} \sum_{n=1}^{N-1} \sin\left(\frac{mn\pi}{N}\right) x(n).$$

Construct a $(2N - 1)$-point odd or skew-symmetric sequence $\{s(n)\}$ using $\{x(n)\}$,

$$s(n) = x(n) \qquad 0 < n < N,$$
$$= 0 \qquad n = 0, N,$$
$$= -x(2N - n) \qquad N < n \leq 2N - 1. \tag{3.6.4}$$

The Fourier transform of a real skew-symmetric sequence is purely imaginary and is related to the sine transform of the half-sequence. Form this, it can be shown that the $2N$-point DFT of $\{s(n)\}$ in (3.6.4) is given by

$$S_F(m) = -2j \sum_{n=1}^{N-1} \sin\left(\frac{mn\pi}{N}\right) x(n). \tag{3.6.5}$$

Thus, the $2N$-point DFT of $\{s(n)\}$ is the same as the $(N-1)$-point DST of $\{x(n)\}$, up to a normalization constant. Again, $S_F(m)$ is as defined in (3.6.3) and the $2N$-point DFT for the real sequence can be implemented using an N-point DFT for a complex sequence.

3.6.2 Fast algorithms for DST and DCT, by direct matrix factorization

3.6.2.1 Decimation-in-time algorithms

These are Cooley-Tukey type algorithms, in which the time ordering of the input data sequence is permuted to allow for the sparse factorization of the transformation matrix. The essential idea is to reduce a size N transform matrix into a block diagonal form, in which each block is related to the same transform of size $N/2$. Recursively applying this procedure, one finally arrives at the basic 2×2 "butterfly." We present here the essential equations for this reduction and also the flow diagrams for the DIT computations of DCT and DST, in block form.

1. DIT algorithm for the DCT: Let

$$X_c(m) = \sum_{n=0}^{N} C_N^{mn} \tilde{x}(n), \qquad m = 0, 1, 2, \ldots, N, \tag{3.6.6}$$

be the DCT of the sequence $\{x(n)\}$ (i.e., $\tilde{x}(n)$ is $x(n)$ scaled by the normalization constant and the factor k_n, while $X_c(m)$ is scaled by k_m, as in [3.4.6]). Here we have simplified the notations using the definition

$$C_N^{mn} = \cos\left(\frac{mn\pi}{N}\right). \tag{3.6.7}$$

Equation (3.6.6) can be reduced to:

$$X_c(m) = g_c(m) + h_c(m),$$

$$X_c(N-m) = g_c(m) - h_c(m), \qquad \text{for } m = 0, 1, \ldots, N/2,$$

$$\text{and } X_c(N/2) = g_c(N/2). \tag{3.6.8}$$

Here, g_c and h_c are related to the DCT of size $N/2$, defined by the following equations,

$$g_c(m) = \sum_{n=0}^{N/2} C_{N/2}^{mn} \tilde{x}(2n), \quad \text{for } m = 0, 1, \ldots, N/2,$$

$$h_c(m) = \frac{1}{2C_N^m} \sum_{n=0}^{N/2} C_{N/2}^{mn} [\tilde{x}(2n+1) + \tilde{x}(2n-1)], \quad \text{for } m = 0, 1, \ldots, N/2 - 1,$$

$$\text{and } h_c(N/2) = 0, \quad \text{and where } \tilde{x}(N+1) \text{ is set to zero.} \tag{3.6.9}$$

We note that both $g_c(m)$ and $h_c(m)$ are DCTs of half the original size. This way, the size of the transform can be reduced by a factor of two at each stage. Some combinations of inputs to the lower order DCT are required as shown by the definition for $h_c(m)$, as well as some scaling of the output of the DCT transform. Figure 3.8 shows a signal flow graph for an $N = 16$ DCT. Note the reduction into two $N = 8$ DCTs in the flow diagram.

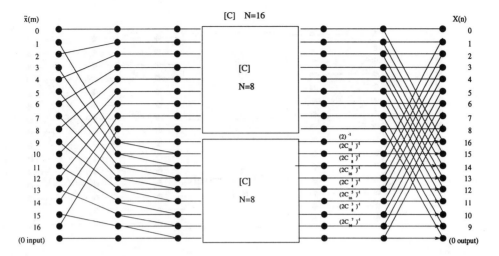

FIGURE 3.8
DIT DCT $N = 16$ flow graph $\longrightarrow (-1)$.

2. *DIT algorithm for DST:* Let

$$X_s(m) = \sum_{n=1}^{N-1} S_N^{mn} \tilde{x}(n), \qquad m = 1, 2, \ldots, N-1, \qquad (3.6.10)$$

be the DST of the sequence $\{x(n)\}$, (i.e., $\tilde{x}(n)$ is $x(n)$ that has been scaled with the proper normalization constant as required in [3.4.10]) and we have defined

$$S_N^{mn} = \sin\left(\frac{mn\pi}{N}\right). \qquad (3.6.11)$$

Following the same reasoning for the DIT algorithm for DCT, (3.6.10) can be reduced to

$$X_s(m) = g_s(m) + h_s(m),$$

$$X_s(N - m) = g_s(m) - h_s(m), \qquad \text{for } m = 1, 2, \ldots, N/2 - 1, \text{ and}$$

$$X_s(N/2) = \sum_{n=1}^{N/2-1} (-1)^n \tilde{x}(2n + 1). \qquad (3.6.12)$$

Here, $g_s(m)$ and $h_s(m)$ are defined as:

$$g_s(m) = \frac{1}{2C_N^m} \sum_{n=1}^{N/2-1} S_{N/2}^{mn}[\tilde{x}(2n + 1) + \tilde{x}(2n - 1)], \text{ and}$$

$$h_s(m) = \sum_{n=1}^{N/2-1} S_{N/2}^{mn} \tilde{x}(2n). \qquad (3.6.13)$$

As before, it can be seen that $g_s(m)$ and $h_s(m)$ are the DSTs of half the original size, one involving only the odd input samples, and the other involving only the even input samples. Figure 3.9 shows a DIT signal flow graph for the $N = 16$ DST. Note that it is reduced to two blocks of $N = 8$ DSTs.

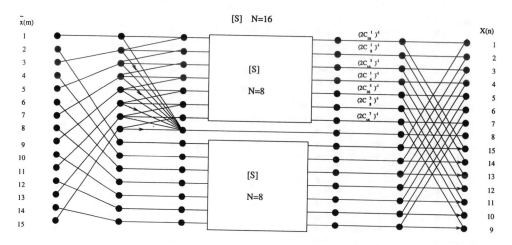

FIGURE 3.9
DIT DST $N = 16$ flow graph $\longrightarrow (-1)$.

3.6.2.2   Decimation-in-frequency algorithms

These are Sande-Tukey type algorithms in which the input sample sequence order is not permuted. Again, the basic principle is to reduce the size of the transform, at each stage of the computation, by a factor of two. It should be of no surprise that these algorithms are simply the conjugate versions of the DIT algorithms.

1. *The DIF algorithm for DCT:* In (3.6.6), consider the even ordered output points and the odd-ordered output points,

$$X_c(2m) = G_c(m), \qquad \text{for } m = 0, 1, \ldots, N/2, \text{ and}$$

$$X_c(2M + 1) = H_c(m) + H_c(m + 1), \qquad \text{for } m = 0, 1, \ldots, N/2 - 1. \text{ (3.6.14)}$$

Here,

$$G_c(m) = \sum_{n=0}^{N/2-1} [\tilde{x}(n) + \tilde{x}(N - n)]C_{N/2}^{mn} + (-1)^m \tilde{x}(N/2), \text{ and}$$

$$H_c(m) = \sum_{n=0}^{N/2-1} \frac{1}{2C_N^n}[\tilde{x}(n) - \tilde{x}(N - n)]C_{N/2}^{mn}. \tag{3.6.15}$$

As can be seen, both $G_c(m)$ and $H_c(m)$ are DCTs of size $N/2$. Therefore, at each stage of the computation, the size of the transform is reduced by a factor of two. The overall result is a sparse factorization of the original transform matrix. Figure 3.10 shows the signal flow graph for an $N = 16$ DIF type DCT.

2. *The DIF algorithm for DST:* The equation (3.6.11) can be split into even ordered and odd-ordered output points, where

$$X_s(m) = G_s(m),$$

$$\text{for } m = 1, 2, \ldots, N/2 - 1,$$

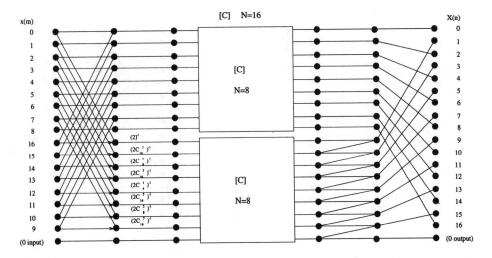

FIGURE 3.10
DIF DCT $N = 16$ flow graph $\longrightarrow (-1)$.

$$X_s(2m - 1) = H_s(m) + H_s(m - 1) + (-1)^{m+1}\tilde{x}(N/2),$$

$$\text{for } m = 1, 2, \ldots, N/2 - 1, \text{ and}$$

$$X_s(N - 1) = H_s(N/2 - 1) + (-1)^{N/2+1}\tilde{x}(N/2). \tag{3.6.16}$$

Here, the outputs $G_s(m)$ and $H_s(m)$ are defined by DSTs of half the original size as,

$$G_s(m) = \sum_{n=1}^{N/2-1} [\tilde{x}(n) - \tilde{x}(N - n)]S_{N/2}^{mn}, \text{ and}$$

$$H_s(m) = \sum_{n=1}^{N/2-1} \frac{1}{2C_N^n}[\tilde{x}(n) - \tilde{x}(N - n)]S_{N/2}^{mn}. \tag{3.6.17}$$

Figure 3.11 shows the signal flow graph for an $N = 16$ DIF type DST. Note that this flow graph is the conjugate of the flow graph shown in Figure 3.9.

3.7 Tables of transforms

This section contains tables of transforms for the FCT and the FST. They are not meant to be complete. For more details and a more complete listing of transforms, especially those of orthogonal and special functions, the reader is referred to the Bateman manuscripts (Erdelyi 1954). Section 3.7.3 contains a list of conventions and definitions of some special functions that have been referred to in the tables.

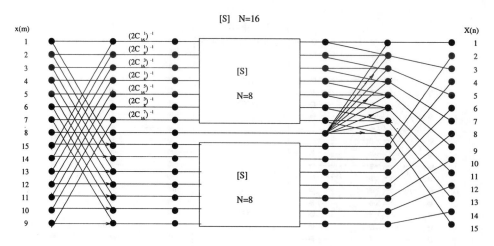

FIGURE 3.11
DIF DST $N = 16$ flow graph $\longrightarrow (-1)$.

3.7.1 *Fourier cosine transforms*

3.7.1.1 General properties

	$f(t)$	$F_c(\omega) = \int_0^\infty f(t) \cos \omega t \, dt \qquad \omega > 0$
1	$F_c(t)$	$(\pi/2) f(\omega)$
2	$f(at) \qquad a > 0$	$(1/a) F_c(\omega/a)$
3	$f(at) \cos bt \qquad a, b > 0$	$(1/2a)\left[F_c\left(\dfrac{\omega+b}{a}\right) + F_c\left(\dfrac{\omega-b}{a}\right)\right]$
4	$f(at) \sin bt \qquad a, b > 0$	$(1/2a)\left[F_s\left(\dfrac{\omega+b}{a}\right) - F_s\left(\dfrac{\omega-b}{a}\right)\right]$
5	$t^{2n} f(t)$	$(-1)^n \dfrac{d^{2n}}{d\omega^{2n}} F_c(\omega)$
6	$t^{2n+1} f(t)$	$(-1)^n \dfrac{d^{2n+1}}{d\omega^{2n+1}} F_s(\omega)$
7	$\int_0^\infty f(r)[g(t+r) + g(|t-r|)]\, dr$	$2 F_c(\omega) G_c(\omega)$
8	$\int_t^\infty f(r)\, dr$	$(1/\omega) F_s(\omega)$
9	$f(t+a) - f_o(t-a)$	$2 F_s(\omega) \sin a\omega \qquad a > 0$
10	$\int_0^\infty f(r)[g(t+r) - g_o(t-r)]\, dr$	$2 F_s(\omega) G_s(\omega)$

3.7.1.2   Algebraic functions

	$f(t)$	$F_c(\omega)$
1	$(1/\sqrt{t})$	$\sqrt{(\pi/2)}(1/\omega)^{1/2}$
2	$(1/\sqrt{t})[1 - U(t - 1)]$	$(2\pi/\omega)^{1/2}C(\omega)$
3	$(1/\sqrt{t})U(t - 1)$	$(2\pi/\omega)^{1/2}[1/2 - C(\omega)]$
4	$(t + a)^{-1/2} \qquad \lvert\arg a\rvert < \pi$	$(\pi/2\omega)^{1/2}\{\cos a\omega[1 - 2C(a\omega)]$ $+ \sin a\omega[1 - 2S(a\omega)]\}$
5	$(t - a)^{-1/2}U(t - a)$	$(\pi/2\omega)^{1/2}[\cos a\omega - \sin a\omega]$
6	$a(t^2 + a^2)^{-1} \qquad a > 0$	$(\pi/2)\exp(-a\omega)$
7	$t(t^2 + a^2)^{-1} \qquad a > 0$	$-1/2[e^{-a\omega}\overline{\mathrm{Ei}}(a\omega) + e^{a\omega}\,\mathrm{Ei}(a\omega)]$
8	$(1 - t^2)(1 + t^2)^{-2}$	$(\pi/2)\omega\exp(-\omega)$
9	$-t(t^2 - a^2)^{-1} \qquad a > 0$	$\cos a\omega\,\mathrm{Ci}(a\omega) + \sin a\omega\,\mathrm{Si}(a\omega)$

3.7.1.3   Exponential and logarithmic functions

	$f(t)$	$F_c(\omega)$
1	$e^{-at} \qquad \mathrm{Re}\,a > 0$	$a(a^2 + \omega^2)^{-1}$
2	$(1 + t)e^{-t}$	$2(1 + \omega^2)^{-2}$
3	$\sqrt{t}e^{-at} \qquad \mathrm{Re}\,a > 0$	$\dfrac{\sqrt{\pi}}{2}(a^2 + \omega^2)^{-3/4}\cos[3/2\tan^{-1}(\omega/a)]$
4	$e^{-at}/\sqrt{t} \qquad \mathrm{Re}\,a > 0$	$\sqrt{(\pi/2)}(a^2 + \omega^2)^{-1/2}$ $\bullet[(a^2 + \omega^2)^{1/2} + a]^{1/2}$
5	$t^n e^{-at} \qquad \mathrm{Re}\,a > 0$	$n![a/(a^2 + \omega^2)]^{n+1}$ $\bullet\sum_{2m=0}^{n+1}(-1)^m \begin{pmatrix} n+1 \\ 2m \end{pmatrix}\left(\dfrac{\omega}{a}\right)^{2m}$
6	$\exp(-at^2)/\sqrt{t} \qquad \mathrm{Re}\,a > 0$	$\pi(\omega/8a)^{1/2}\exp(-\omega^2/8a)$ $\bullet I_{-1/4}(-\omega^2/8a)$
7	$t^{2n}\exp(-a^2 t^2) \qquad \lvert\arg a\rvert < \pi/4$	$(-1)^n\sqrt{\pi}\,2^{-n-1}a^{-2n-1}$ $\bullet\exp[-(\omega/2a)^2]\,\mathrm{He}_{2n}(2^{-1/2}\omega/a)$
8	$t^{-3/2}\exp(-a/t) \qquad \mathrm{Re}\,a > 0$	$(\pi/a)^{1/2}\exp[-(2a\omega)^{1/2}]\cos(2a\omega)^{1/2}$
9	$t^{-1/2}\exp(-a/\sqrt{t}) \qquad \mathrm{Re}\,a > 0$	$(\pi/2\omega)^{1/2}[\cos(2a\sqrt{\omega}) - \sin(2a\sqrt{\omega})]$
10	$t^{-1/2}\ln t$	$-(\pi/2\omega)^{1/2}[\ln(4\omega) + C + \pi/2]$
11	$(t^2 - a^2)^{-1}\ln t \qquad a > 0$	$(\pi/2\omega)\{\sin(a\omega)[\mathrm{ci}(a\omega) - \ln a]$ $- \cos(a\omega)[\mathrm{si}(a\omega) - \pi/2]\}$

	$f(t)$	$F_c(\omega)$
12	$t^{-1}\ln(1+t)$	$(1/2)\{[\mathrm{ci}(\omega)]^2 + [\mathrm{si}(\omega)]^2\}$
13	$\exp(-t/\sqrt{2})\sin(\pi/4 + t/\sqrt{2})$	$(1+\omega^4)^{-1}$
14	$\exp(-t/\sqrt{2})\cos(\pi/4 + t/\sqrt{2})$	$\omega^2(1+\omega^4)^{-1}$
15	$\ln\dfrac{a^2+t^2}{1+t^2}\qquad a>0$	$(\pi/\omega)[\exp(-\omega)-\exp(-a\omega)]$
16	$\ln[1+(a/t)^2]\qquad a>0$	$(\pi/\omega)[1-\exp(-a\omega)]$

3.7.1.4 Trigonometric functions

	$f(t)$	$F_c(\omega)$
1	$t^{-1}e^{-t}\sin t$	$(1/2)\tan^{-1}(2\omega^{-2})$
2	$t^{-2}\sin^2(at)\qquad a>0$	$\begin{array}{ll}(\pi/2)(a-\omega/2) & \omega<2a\\ 0 & \omega>2a\end{array}$
3	$\left(\dfrac{\sin t}{t}\right)^n\qquad n=2,3,\dots$	$\begin{array}{ll}\dfrac{n\pi}{2^n}\displaystyle\sum_{r>0}^{r<(\omega+n)/2}\dfrac{(-1)^r(\omega+n-2r)^{n-1}}{r!(n-r)!}, & 0<\omega<n\\ 0 & n\le\omega\end{array}$
4	$\exp(-\beta t^2)\cos at\qquad \mathrm{Re}\,\beta>0$	$(1/2)(\pi/\beta)^{1/2}\exp\left(-\dfrac{a^2+\omega^2}{4\beta}\right)\cosh\left(\dfrac{a\omega}{2\beta}\right)$
5	$(a^2+t^2)^{-1}(1-2\beta\cos t+\beta^2)^{-1}$ $\mathrm{Re}\,a>0, |\beta|<1$	$(1/2)(\pi/a)(1-\beta^2)^{-1}(e^a-\beta)^{-1}$ $\bullet(e^{a-a\omega}+\beta e^{a\omega})\qquad 0\le\omega<1$
6	$\sin(at^2)\qquad a>0$	$(1/4)(2\pi/a)^{1/2}\left[\cos\left(\dfrac{\omega^2}{4a}\right)-\sin\left(\dfrac{\omega^2}{4a}\right)\right]$
7	$\sin[a(1-t^2)]\qquad a>0$	$-(1/2)(\pi/a)^{1/2}\cos[a+\pi/4+\omega^2/(4a)]$
8	$\cos(at^2)\qquad a>0$	$(1/4)(2\pi/a)^{1/2}\left[\cos\left(\dfrac{\omega^2}{4a}\right)+\sin\left(\dfrac{\omega^2}{4a}\right)\right]$
9	$\cos[a(1-t^2)]\qquad a>0$	$(1/2)(\pi/a)^{1/2}\sin[a+\pi/4+\omega^2/(4a)]$
10	$\tan^{-1}(a/t)\qquad a>0$	$(2\omega)^{-1}[e^{-a\omega}\,\mathrm{Ei}(a\omega)-e^{a\omega}\,\mathrm{Ei}(-a\omega)]$

3.7.2 *Fourier sine transforms*

3.7.2.1 General properties

	$f(t)$	$F_s(\omega)=\int_0^\infty f(t)\sin\omega t\,dt\qquad \omega>0$
1	$F_s(t)$	$(\pi/2)f(\omega)$
2	$f(at)\qquad a>0$	$(1/a)F_s(\omega/a)$
3	$f(at)\cos bt\qquad a,b>0$	$(1/2a)\left[F_s\left(\dfrac{\omega+b}{a}\right)+F_s\left(\dfrac{\omega-b}{a}\right)\right]$
4	$f(at)\sin bt\qquad a,b>0$	$-(1/2a)\left[F_c\left(\dfrac{\omega+b}{a}\right)-F_c\left(\dfrac{\omega-b}{a}\right)\right]$

	$f(t)$	$F_s(\omega) = \int_0^\infty f(t) \sin \omega t \, dt \qquad \omega > 0$
5	$t^{2n} f(t)$	$(-1)^n \dfrac{d^{2n}}{d\omega^{2n}} F_s(\omega)$
6	$t^{2n+1} f(t)$	$(-1)^{n+1} \dfrac{d^{2n+1}}{d\omega^{2n+1}} F_c(\omega)$
7	$\int_0^\infty f(r) \int_{|t-r|}^{t+r} g(s) \, ds \, dr$	$(2/\omega) F_s(\omega) G_s(\omega)$
8	$f_o(t+a) + f_o(t-a)$	$2 F_s(\omega) \cos a\omega$
9	$f_e(t-a) - f_e(t+a)$	$2 F_c(\omega) \sin a\omega$
10	$\int_0^\infty f(r)[g(|t-r|) - g(t+r)] \, dr$	$2 F_s(\omega) G_c(\omega)$

3.7.2.2   Algebraic functions

	$f(t)$	$F_s(\omega)$
1	$1/t$	$\pi/2$
2	$1/\sqrt{t}$	$(\pi/2\omega)^{1/2}$
3	$1/\sqrt{t}[1 - U(t-1)]$	$(2\pi/\omega)^{1/2} S(\omega)$
4	$(1/\sqrt{t}) U(t-1)$	$(2\pi/\omega)^{1/2}[1/2 - S(\omega)]$
5	$(t+a)^{-1/2} \qquad |\arg a| < \pi$	$(\pi/2\omega)^{1/2}\{\cos a\omega[1 - 2S(a\omega)] \\ \qquad - \sin a\omega[1 - 2C(a\omega)]\}$
6	$(t-a)^{-1/2} U(t-a)$	$(\pi/2\omega)^{1/2}(\sin a\omega + \cos a\omega)$
7	$t(t^2 + a^2)^{-1} \qquad a > 0$	$(\pi/2) \exp(-a\omega)$
8	$t(a^2 - t^2)^{-1} \qquad a > 0$	$-(\pi/2) \cos a\omega$
9	$t(a^2 + t^2)^{-2} \qquad a > 0$	$(\pi\omega/4a) \exp(-a\omega)$
10	$a^2[t(a^2 + t^2)]^{-1} \qquad a > 0$	$(\pi/2)[1 - \exp(-a\omega)]$
11	$t(4 + t^4)^{-1}$	$(\pi/4) \exp(-\omega) \sin \omega$

3.7.2.3   Exponential and logarithmic functions

	$f(t)$	$F_s(\omega)$
1	$e^{-at} \qquad \mathrm{Re}\, a > 0$	$\omega(a^2 + \omega^2)^{-1}$
2	$te^{-at} \qquad \mathrm{Re}\, a > 0$	$(2a\omega)(a^2 + \omega^2)^{-2}$
3	$t(1+at)e^{-at} \qquad \mathrm{Re}\, a > 0$	$(8a^3\omega)(a^2 + \omega^2)^{-3}$

	$f(t)$		$F_s(\omega)$
4	$e^{-at}\sqrt{t}$	$\mathrm{Re}\,a > 0$	$\sqrt{(\pi/2)}(a^2+\omega^2)^{-1/2}$ $\bullet[(a^2+\omega^2)^{1/2}-a]^{1/2}$
5	$t^{-3/2}e^{-at}$	$\mathrm{Re}\,a > 0$	$(2\pi)^{1/2}[(a^2+\omega^2)^{1/2}-a]^{1/2}$
6	$\exp(-at^2)$	$\mathrm{Re}\,a > 0$	$-j(1/2)(\pi/a)^{1/2}\exp(-\omega^2/4a)\,\mathrm{Erf}\left(\dfrac{j\omega}{2\sqrt{a}}\right)$
7	$t\exp(-t^2/4a)$	$\mathrm{Re}\,a > 0$	$2a\omega\sqrt{(\pi a)}\exp(-a\omega^2)$
8	$t^{-3/2}\exp(-a/t)$	$\lvert\arg a\rvert < \pi/2$	$(\pi/a)^{1/2}\exp[-(2a\omega)^{1/2}]\sin(2a\omega)^{1/2}$
9	$t^{-3/4}\exp(-a\sqrt{t})$	$\lvert\arg a\rvert < \pi/2$	$-(\pi/2)(a/\omega)^{1/2}[J_{1/4}(a^2/8\omega)$ $\bullet\cos(\pi/8+a^2/8\omega)+Y_{1/4}(a^2/8\omega)$ $\bullet\sin(\pi/8+a^2/8\omega)]$
10	$t^{-1}\ln t$		$-(\pi/2)[C+\ln\omega]$
11	$t(t^2-a^2)^{-1}\ln t$	$a > 0$	$-(\pi/2)\{\cos a\omega[\mathrm{Ci}(a\omega)-\ln a]$ $+\sin a\omega[\mathrm{Si}(a\omega)-\pi/2]\}$
12	$t^{-1}\ln(1+a^2t^2)$	$a > 0$	$-\pi\,\mathrm{Ei}(-\omega/a)$
13	$\ln\dfrac{t+a}{\lvert t-a\rvert}$	$a > 0$	$(\pi/\omega)\sin a\omega$

3.7.2.4 Trigonometric functions

	$f(t)$	$F_s(\omega)$
1	$t^{-1}\sin^2(at)\quad a>0$	$\begin{array}{ll}\pi/4 & 0<\omega<2a\\ \pi/8 & \omega=2a\\ 0 & \omega>2a\end{array}$
2	$t^{-2}\sin^2(at)\quad a>0$	$(1/4)(\omega+2a)\ln\lvert\omega+2a\rvert$ $+(1/4)(\omega-2a)\ln\lvert\omega-2a\rvert-(1/2)\omega\ln\omega$
3	$t^{-2}[1-\cos at]\quad a>0$	$(\omega/2)\ln\lvert(\omega^2-a^2)/\omega^2\rvert$ $+(a/2)\ln\lvert\omega+a)/(\omega-a)\rvert$
4	$\sin(at^2)\quad a>0$	$(\pi/2a)^{1/2}\{\cos(\omega^2/4a)C[\omega/(2\pi a)^{1/2}]$ $+\sin(\omega^2/4a)S[\omega/(2\pi a)^{1/2}]$
5	$\cos(at^2)\quad a>0$	$(\pi/2a)^{1/2}\{\sin(\omega^2/4a)C[\omega/(2\pi a)^{1/2}]$ $-\cos(\omega^2/4a)S[\omega/(2\pi a)^{1/2}]$
6	$\tan^{-1}(a/t)\quad a>0$	$(\pi/2\omega)[1-\exp(-a\omega)]$

3.7.3 *Notations and definitions*

1. $f(t)$: Piece-wise smooth and absolutely integrable function on the positive real line.

2. $F_c(\omega)$: The Fourier cosine transform of $f(t)$.

3. $F_s(\omega)$: The Fourier sine transform of $f(t)$.

4. $f_o(t)$: The odd extension of the function f over the entire real line.

5. $f_e(t)$: The even extension of the function f over the entire real line.

6. $C(\omega)$ is defined as the integral:

$$(2\pi)^{-1/2} \int_0^\omega t^{-1/2} \cos t \, dt.$$

7. $S(\omega)$ is defined as the integral:

$$(2\pi)^{-1/2} \int_0^\omega t^{-1/2} \sin t \, dt.$$

8. $\text{Ei}(x)$ is the exponential integral function defined as:

$$-\int_{-x}^\infty t^{-1} e^{-t} \, dt, \qquad |\arg(x)| < \pi.$$

9. $\overline{\text{Ei}}(x)$ is defined as $(1/2)[\text{Ei}(x + j0) + \text{Ei}(x - j0)]$.

10. $\text{Ci}(x)$ is the cosine integral function defined as

$$-\int_x^\infty t^{-1} \cos t \, dt.$$

11. $\text{Si}(x)$ is the sine integral function defined as

$$\int_0^x t^{-1} \sin t \, dt.$$

12. $I_\nu(z)$ is the modified Bessel function of the first kind defined as

$$\sum_{m=0}^\infty \frac{(z/2)^{\nu+2m}}{m! \Gamma(\nu + m + 1)}, \qquad |z| < \infty, |\arg(z)| < \pi.$$

13. $\text{He}_n(x)$ is the Hermite polynomial function defined as

$$(-1)^n \exp(x^2/2) \frac{d^n}{dx^n} [\exp(-x^2/2)].$$

14. C is the Euler constant defined as

$$\lim_{m \to \infty} \left[\sum_{n=1}^m (1/n) - \ln m \right] = 0.5772156649\ldots$$

15. $\text{ci}(x)$ and $\text{si}(x)$ are related to $\text{Ci}(x)$ and $\text{Si}(x)$ by the equations:

$$\text{ci}(x) = -\text{Ci}(x), \qquad \text{si}(x) = \text{Si}(x) - \pi/2.$$

16. $\text{Erf}(x)$ is the error function defined by

$$(2/\sqrt{\pi}) \int_0^x \exp(-t^2) \, dt.$$

17. $J_\nu(x)$ and $Y_\nu(x)$ are the Bessel functions for the first and second kind respectively,

$$J_\nu(x) = \sum_{m=0}^\infty (-1)^m \frac{(x/2)^{\nu+2m}}{m! \Gamma(\nu + m + 1)}$$

and

$$Y_\nu(x) = \text{cosec}\{\nu\pi [J_\nu(x) \cos \nu\pi - J_{-\nu}(x)]\}.$$

18. $U(t)$: is the Heaviside step function defined as

$$U(t) = 0 \qquad t < 0,$$
$$= 1 \qquad t > 0.$$

19. $\begin{pmatrix} m \\ n \end{pmatrix}$: is the binomial coefficient defined as $\frac{m!}{n!(m-n)!}$.

20. $\Gamma(x)$: is the Gamma function defined as,

$$\Gamma(x) = \int_0^\infty e^{-t} t^{x-1} dt.$$

References

[1] Churchill, R. V. 1958. *Operational Mathematics*, 3rd ed. New York: McGraw-Hill.

[2] Erdelyi, A. 1954. *Bateman Manuscript*, Vol. 1. New York: McGraw-Hill.

[3] Rao, K. R., and Yip, P. 1990. *Discrete Cosine Transform: Algorithms, Advantages, Applications*. Boston: Academic Press, Boston.

[4] Sneddon, I. N. 1972. *The Uses of Integral Transforms*. New York: McGraw-Hill.

4

The Hartley Transform

Kraig J. Olejniczak

CONTENTS

4.1   Introduction

The Hartley transform is an integral transformation that maps a **real-valued** temporal or spacial function into a **real-valued** frequency function via the kernel, $\operatorname{cas}(\nu x) \equiv \cos(\nu x) + \sin(\nu x)$. This novel symmetrical formulation of the traditional Fourier transform, attributed to Ralph Vinton Lyon Hartley in 1942 [1], leads to a parallelism that exists between the function of the original variable and that of its transform. Furthermore, the Hartley transform permits a function to be decomposed into two **independent** sets of sinusoidal components; these sets are represented in terms of positive and negative frequency components, respectively. This is in contrast to the complex exponential, $\exp(j\omega x)$, used in classical Fourier analysis. For periodic power signals, various mathematical forms of the familiar Fourier series come to mind. For aperiodic energy and power signals of either finite or infinite duration, the Fourier integral can be used. In either case, signal and systems analysis and design in the frequency domain using the Hartley transform may be deserving of increased awareness due necessarily to the existence of a fast algorithm that can substantially lessen the computational burden when compared to the classical complex-valued Fast Fourier Transform (FFT).

Throughout the remainder of this chapter, it is assumed that the function to be transformed is real valued. In most engineering applications of practical interest, this is indeed the case. However, in the case where complex-valued functions are of interest, they may be analyzed using the novel complex Hartley transform formulation presented in [10].

0-8493-8342-0/96/$0.00 + $0.50
©1996 by CRC Press, Inc.

4.2 Historical background

Ralph V. L. Hartley was born in Spruce Mountain, approximately 50 miles south of Wells, Nevada, in 1888. After graduating with the A.B. degree from the University of Utah in 1909, he studied at Oxford for three years as a Rhodes Scholar where he received the B.A. and B.Sc. degrees in 1912 and 1913, respectively. Upon completing his education, Hartley returned from England and began his professional career with the Western Electric Company engineering department (New York, NY) in September of the same year. It was here at AT&T's R&D unit that he became an expert on receiving sets and was in charge of the early development of radio receivers for the transatlantic radio telephone tests of 1915. His famous oscillating circuit, known as the Hartley oscillator, was invented during this work as well as a neutralizing circuit to offset the internal coupling of triodes that tended to cause singing.

During World War I, Hartley performed research on the problem of binaural location of a sound source. He formulated the accepted theory that direction was perceived by the phase difference of sound waves caused by the longer path to one ear than to the other. After the war, Hartley headed the research effort on repeaters and voice and carrier transmission. During this period, Hartley advanced Fourier analysis methods so that AC measurement techniques could be applied to telegraph transmission studies. In his effort to ensure some privacy for radio, he also developed the frequency-inversion system known to some as **greyqui hoy**.

In 1925, Hartley and his fellow research scientists and engineers became founding members of the Bell Telephone Laboratories when a corporate restructuring set R&D off as a separate entity. This change affected neither Hartley's position nor his work. R. V. L. Hartley was well known for his ability to clarify and arrange ideas into patterns that could be easily understood by others. In his paper entitled "Transmissions of Information" presented at the International Congress of Telegraphy and Telephony in Commemoration of Volta at Lake Como, Italy, in 1927, he stated the law that was implicitly understood by many transmission engineers at that time, namely, "the total amount of information which may be transmitted over such a system is proportional to the product of the frequency-range which it transmits by the time during which it is available for the transmission" [14, p. 554]. This contribution to information theory was later known by his name.

In 1929, Hartley gave up leadership of his research group due to illness. In 1939, he returned as a research consultant on transmission problems. During World War II he acted as a consultant on servomechanisms as applied to radar and fire control. R. V. L. Hartley, a fellow of the Institute of Radio Engineers (I.R.E.), the American Association for the Advancement of Science, the Physical and Acoustical Societies, and a member of the A.I.E.E., was awarded the I.R.E. Medal of Honor on January 24, 1946, "For his early work on oscillating circuits employing triode tubes and likewise for his early recognition and clear exposition of the fundamental relationship between the total amount of information which may be transmitted over a transmission system of limited band and the time required." Hartley was the holder of 72 patents that documented his contributions and developments. A transmission expert, he retired from Bell Laboratories in 1950 and died at the age of 81 on May 1, 1970.

4.3 Fundamentals of the Hartley transform

Perhaps one of Hartley's most long-lasting contributions was a more symmetrical Fourier integral originally developed for steady-state and transient analysis of telephone transmission system problems [1]. Although this transform remained in a quiescent state for over 40 years,

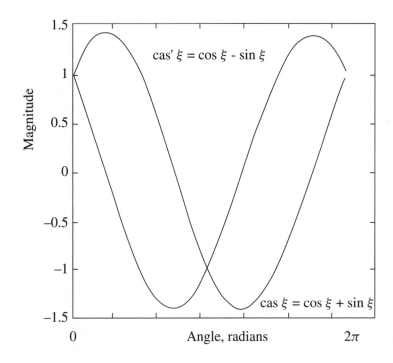

FIGURE 4.1
The cas function on the interval $[0, 2\pi]$.

the Hartley transform was rediscovered more than a decade ago by Wang [2, 3, 4, 6] and Bracewell [5, 7, 8], who authored definitive treatises on the subject.

The Hartley transform of a function $f(x)$ can be expressed as either

$$H(v) = \frac{1}{\sqrt{2\pi}} \int_{-\infty}^{\infty} f(x)\,\text{cas}(vx)\,dx \tag{4.3.1a}$$

or

$$H(f) = \int_{-\infty}^{\infty} f(x)\,\text{cas}(2\pi f x)\,dx \tag{4.3.1b}$$

where the angular or radian frequency variable v is related to the frequency variable f by $v = 2\pi f$ and

$$H(f) = \sqrt{2\pi}\,H(2\pi f) = \sqrt{2\pi}\,H(v). \tag{4.3.2}$$

Here the integral kernel, known as the cosine-and-sine or cas function, is defined as

$$\text{cas}(vx) \equiv \cos(vx) + \sin(vx)$$

$$\text{cas}(vx) = \sqrt{2}\sin\left(vx + \frac{\pi}{4}\right)$$

$$\text{cas}(vx) = \sqrt{2}\cos\left(vx - \frac{\pi}{4}\right). \tag{4.3.3}$$

Figure 4.1 depicts the cas function on the interval $[0, 2\pi]$. Additional properties of the cas function are shown in Tables 4.1 through 4.5 below.

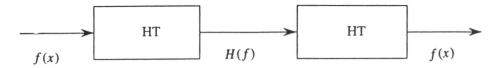

FIGURE 4.2
The self-inverse property associated with the Hartley transform.

The inverse Hartley transform may be defined as either

$$f(x) = \frac{1}{\sqrt{2\pi}} \int_{-\infty}^{\infty} H(\nu)\, \text{cas}(\nu x)\, d\nu \qquad (4.3.4a)$$

or

$$f(x) = \int_{-\infty}^{\infty} H(f)\, \text{cas}(2\pi f x)\, df. \qquad (4.3.4b)$$

The angular frequency variable ν, with units of radians per second, is equivalent to the frequency variable ω in the Fourier domain; however, it is used here to further distinguish $H(\nu)$, the Hartley transform of $f(x)$, from the Fourier transform of $f(x)$, $F(\omega)$. From Hartley's original formulation expressed in (4.3.1a) and (4.3.4a), it is clear that the inverse transformation (synthesis equation) calls for the identical integral operation as the direct transformation (analysis equation). The peculiar scaling coefficient $1/\sqrt{2\pi}$, chosen by Hartley for the direct and inverse transformations, is used to satisfy the self-inverse condition depicted in Figure 4.2. When the independent variable is angular frequency with units of radians per second, other coefficients may be used provided that the product of the direct and inverse transform coefficients is $1/2\pi$.

The existence of the Hartley transform of $f(x)$ given by (4.3.1) is equivalent to the existence of the Fourier transform of $f(x)$ given by

$$f(x) = \frac{1}{2\pi} \int_{-\infty}^{\infty} \int_{-\infty}^{\infty} f(\zeta) \cos \omega(x - \zeta)\, d\zeta\, d\omega. \qquad (4.3.5)$$

Equation (4.3.5) can also be equivalently expressed by the following three equations:

$$f(x) = \frac{1}{\sqrt{2\pi}} \int_{-\infty}^{\infty} [C(\omega) \cos \omega x + S(\omega) \sin \omega x]\, d\omega \qquad (4.3.6)$$

$$C(\omega) = \frac{1}{\sqrt{2\pi}} \int_{-\infty}^{\infty} f(x) \cos \omega x\, dx$$

$$= H^e(f) = \frac{H(f) + H(-f)}{2} \qquad (4.3.7)$$

$$S(\omega) = \frac{1}{\sqrt{2\pi}} \int_{-\infty}^{\infty} f(x) \sin \omega x\, dx$$

$$= H^o(f) = \frac{H(f) - H(-f)}{2} \qquad (4.3.8)$$

where $H^e(f)$ and $H^o(f)$ are the even and odd parts of the Hartley transform $H(f)$, respectively.

TABLE 4.1
Selected trigonometric properties of the cas function.

The cas function	$\operatorname{cas}\xi = \cos\xi + \sin\xi$
The cas function	$\operatorname{cas}\xi = \frac{1}{2}\left[(1+j)\exp(-j\xi) + (1-j)\exp(j\xi)\right]$
The complementary cas function	$\operatorname{cas}'\xi = \operatorname{cas}(-\xi) = \cos\xi - \sin\xi$
The complementary cas function	$\sqrt{2}\sin\left(\xi + \frac{3\pi}{4}\right) = \sqrt{2}\cos\left(\xi + \frac{\pi}{4}\right)$
Relation to cos	$\cos\xi = \frac{1}{2}\left[\operatorname{cas}\xi + \operatorname{cas}(-\xi)\right]$
Relation to sin	$\sin\xi = \frac{1}{2}\left[\operatorname{cas}\xi - \operatorname{cas}(-\xi)\right]$
Reciprocal relation	$\operatorname{cas}\xi = \frac{\csc\xi + \sec\xi}{\sec\xi\,\csc\xi}$
Product relation	$\operatorname{cas}\xi = \cot\xi\,\sin\xi + \tan\xi\,\cos\xi$
Function product relation	$\operatorname{cas}\tau\,\operatorname{cas}\upsilon = \cos(\tau - \upsilon) + \sin(\tau + \upsilon)$
Quotient relation	$\operatorname{cas}\xi = \frac{\cot\xi\,\sec\xi + \tan\xi\,\csc\xi}{\csc\xi\,\sec\xi}$
Double angle relation	$\operatorname{cas}2\xi = \operatorname{cas}^2\xi - \operatorname{cas}^2(-\xi)$
Indefinite integral relation	$\int \operatorname{cas}(\tau)\,d\tau = -\operatorname{cas}(-\tau) = -\operatorname{cas}'\tau$
Derivative relation	$\frac{d}{dt}\operatorname{cas}\tau = \operatorname{cas}(-\tau) = \operatorname{cas}'\tau$
Angle-sum relation	$\operatorname{cas}(\tau + \upsilon) = \cos\tau\,\operatorname{cas}\upsilon + \sin\tau\,\operatorname{cas}'\upsilon$
Angle-difference relation	$\operatorname{cas}(\tau - \upsilon) = \cos\tau\,\operatorname{cas}'\upsilon + \sin\tau\,\operatorname{cas}\upsilon$
Function-sum relation	$\operatorname{cas}\tau + \operatorname{cas}\upsilon = 2\,\operatorname{cas}\frac{1}{2}\left(\tau + \upsilon\right)\cos\frac{1}{2}\left(\tau - \upsilon\right)$
Function-difference relation	$\operatorname{cas}\tau - \operatorname{cas}\upsilon = 2\,\operatorname{cas}'\frac{1}{2}\left(\tau + \upsilon\right)\sin\frac{1}{2}\left(\tau - \upsilon\right)$

TABLE 4.2
Signs of the cas function.

Quadrant	cas
I	+
II	+ and −
III	−
IV	+ and −

TABLE 4.3
Variations of the cas function.

Quadrant	cas
I	$+1 \rightarrow +1$ with a maximum at $\dfrac{\pi}{4}$
II	$+1 \rightarrow -1$
III	$-1 \rightarrow -1$ with a minimum at $\dfrac{5\pi}{4}$
IV	$-1 \rightarrow +1$

TABLE 4.4
Trigonometric functions of some special angles.

Angle		cas
$0° = 0$		0
$30° =$	$\dfrac{\pi}{6}$	$\dfrac{1}{2}(\sqrt{3}+1)$
$45° =$	$\dfrac{\pi}{4}$	$\sqrt{2}$
$60° =$	$\dfrac{\pi}{3}$	$\dfrac{1}{2}(1+\sqrt{3})$
$90° =$	$\dfrac{\pi}{2}$	1
$120° =$	$\dfrac{2\pi}{3}$	$\dfrac{1}{2}(-1+\sqrt{3})$
$150° =$	$\dfrac{5\pi}{6}$	$\dfrac{1}{2}(1-\sqrt{3})$
$180° = \pi$		-1
$270° =$	$\dfrac{3\pi}{2}$	-1

Alternatively, (4.3.5) can be expressed as

$$f(x) = \frac{1}{\sqrt{2\pi}} \int_{-\infty}^{\infty} F(\omega)e^{j\omega x}\, d\omega \qquad (4.3.9a)$$

where

$$F(\omega) = \frac{1}{\sqrt{2\pi}} \int_{-\infty}^{\infty} f(x)e^{-j\omega x}\, dx. \qquad (4.3.9b)$$

Although the transform pair defined by (4.3.1a) and (4.3.4a) are equivalent to either (4.3.6)–(4.3.8) or (4.3.9), note that the variables x and v are symmetrically embedded in the former but in neither of the latter.

To derive (4.3.1a) and (4.3.4a), let

$$H(v) = C(\omega) + S(\omega)|_{\omega=v}, \qquad (4.3.10)$$

the linear combination of the cosine and sine transforms. Then, (4.3.4a) follows by linearity applied to (4.3.7) and (4.3.8). Because $C(\omega)$ and $S(\omega)$ are an even and odd function of ω, respectively, then

$$\frac{1}{\sqrt{2\pi}} \int_{-\infty}^{\infty} [C(\omega)\sin\omega x + S(\omega)\cos\omega x]\, d\omega = 0. \qquad (4.3.11)$$

When (4.3.11) is added to the right-hand side of (4.3.6), (4.3.1a) results. It is interesting to note that (4.3.5)–(4.3.8) are similar to (4.3.1a) and (4.3.4a) when $f(x)$ is real, in that $C(\omega)$ and $S(\omega)$ are real, as is $H(v)$ via (4.3.10). This is in stark contrast to the complex nature of (4.3.9b) when $f(x)$ is real.

The following expressions are used to further explain the physical nature of the Hartley transform. The functions $f(x) = f^e(x) + f^o(x),\ x > 0$ and $H(v) = H^e(v) + H^o(v),\ v > 0$

TABLE 4.5
The trigonometric function of an arbitrary angle.

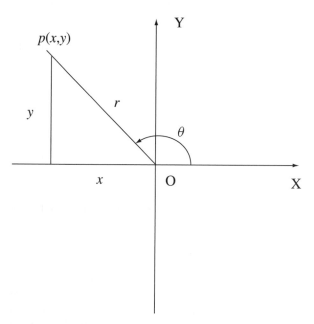

If one lets θ be any angle in the $x - y$ plane and $p(x, y)$ denotes any point on the terminal side of that angle, then denoting the positive distance from the origin to p as r,

$$\operatorname{cas} \theta = \cos \theta + \sin \theta = \frac{x}{r} + \frac{y}{r} = \frac{x + y}{\sqrt{x^2 + y^2}} \, .$$

can be resolved into their even and odd components as follows [1]:

$$f^e(x) = \frac{1}{2}[f(x) + f(-x)], \qquad x > 0 \tag{4.3.12}$$

$$= \frac{1}{\sqrt{2\pi}} \int_{-\infty}^{\infty} H^e(v) \cos vx \, dv$$

$$= \frac{1}{\sqrt{2\pi}} \int_{-\infty}^{\infty} H(v) \cos vx \, dv \tag{4.3.13}$$

$$f^o(x) = \frac{1}{2}[f(x) - f(-x)], \qquad x > 0 \tag{4.3.14}$$

$$= \frac{1}{\sqrt{2\pi}} \int_{-\infty}^{\infty} H^o(v) \sin vx \, dv$$

$$= \frac{1}{\sqrt{2\pi}} \int_{-\infty}^{\infty} H(v) \sin vx \, dv \tag{4.3.15}$$

$$H^e(v) = \frac{1}{2}[H(v) + H(-v)], \qquad v > 0 \tag{4.3.16}$$

$$= \frac{1}{\sqrt{2\pi}} \int_{-\infty}^{\infty} f^e(x) \cos vx \, dx$$

$$= \frac{1}{\sqrt{2\pi}} \int_{-\infty}^{\infty} f(x) \cos vx \, dx \tag{4.3.17}$$

$$H^o(v) = \frac{1}{2}[H(v) - H(-v)], \qquad v > 0 \tag{4.3.18}$$

$$= \frac{1}{\sqrt{2\pi}} \int_{-\infty}^{\infty} f^o(x) \sin vx \, dx$$

$$= \frac{1}{\sqrt{2\pi}} \int_{-\infty}^{\infty} f(x) \sin vx \, dx. \tag{4.3.19}$$

It is readily known that when the function to be transformed is real valued, then its Fourier transform exhibits Hermitian symmetry. That is,

$$F(-\omega) = F^*(\omega) \tag{4.3.20}$$

where the superscript $*$ denotes complex conjugation. This implies that the Fourier transform is overspecified because a dependency exists between transform values for positive and negative values of ω, respectively. This inherent redundancy is *not* present in the Hartley transform. Observe the effect of positive and negative values of v in (4.3.1a). Specifically, for negative values of v,

$$\cos(-vx) = \sqrt{2} \cos\left(-vx - \frac{\pi}{4}\right) = \sqrt{2} \cos\left(vx + \frac{\pi}{4}\right). \tag{4.3.21}$$

For positive values of v,

$$\cos(vx) = \sqrt{2} \sin\left(vx + \frac{\pi}{4}\right). \tag{4.3.22}$$

From (4.3.4a) it is clear that the function $f(x)$ is composed of an equal number of positive and negative frequency components. In light of the two equations above, it seems that any two components, one at v and the other at $-v$, vary as the cosine and sine of the same angle. Thus, whereas (4.3.7) and (4.3.8) represent a resolution into sine and cosine components, each of which is further decomposed into positive and negative frequencies, the Hartley transform of (4.3.1a) amalgamates these two resolutions into one. Equation (4.3.6) alludes to the fact that although $C(\omega)$ and $S(\omega)$ are each defined for positive and negative values of v, because of their respective symmetry properties, they are completely specified by their values over either half range alone. This is due to the Hermitian symmetry existing in the Fourier transform as shown by (4.3.20). Note in (4.3.1a) that $H(v)$ is a single function that contains no redundancy; the value of $H(v)$ for $v < 0$ is **independent** of that for $v > 0$. Therefore, $H(v)$ must be specified over the entire range of v.

Although *not* all time functions can be represented via the Fourier integral, for those functions where such a representation exists, there is a unique relationship between the function and its Fourier transform. This is possible if and only if the integral is convergent. Sufficient conditions (although not necessary) to guarantee convergence of the Fourier integral are the well known Dirichlet conditions, which are stated below for convenience.

1. $\int_{-\infty}^{\infty} |f(x)| \, dx < \infty$, that is, $f(x)$ is absolutely integrable.

2. $f(x)$ has a finite number of discontinuities over any finite interval.

3. $f(x)$ has a finite number of local maximum and local minimum points over any finite interval.

The above sufficient conditions include *most* finite-energy signals of engineering interest. Unfortunately, important signals such as periodic signals and the unit step function are not absolutely integrable. If we allow the Fourier transform, and thus the Hartley transform, to include the Dirac delta function, then even these signals can be handled using methods similar to those for finite-energy signals. This should not be surprising because the Hartley transform is simply a symmetrical representation of the Fourier transform.

4.3.1   The relationship between the Hartley and the sine and cosine transforms

The Hartley transform is trivially related to the cosine and sine transforms (see also Chapter 3) by the linear combination in (4.3.10) and to each transform individually using the fifth and sixth entries of Table 4.1, respectively.

4.3.2   The relationship between the Hartley and Fourier transforms

The Hartley transform is closely related to the familiar Fourier transform. It can be easily shown via (4.3.9b) that these transforms are related in a very simple way

$$H(\nu) = [\Re\{F(\omega)\} - \Im\{F(\omega)\}]_{\omega=\nu} \tag{4.3.23}$$

where

$$\Re\{F(\omega)\} = R(\omega) = H^e(\nu) = H^e(-\nu) \tag{4.3.24}$$

$$\Im\{F(\omega)\} = I(\omega) = -H^o(\nu) = H^o(-\nu) \tag{4.3.25}$$

and

$$H(f) = \frac{1+j1}{2}F(f) + \frac{1-j1}{2}F(-f)$$

$$H(f) = \frac{1}{2}e^{j\pi/4}F(f) + \frac{1}{2}e^{-j\pi/4}F^*(f). \tag{4.3.26}$$

The Fourier transform expressed in terms of the Hartley transform is

$$F(\omega) = \left[\frac{H(\nu) + H(-\nu)}{2} - j\frac{H(\nu) - H(-\nu)}{2} \right]_{\nu=\omega} \tag{4.3.27}$$

$$F(\omega) = [H^e(\nu) - jH^o(\nu)]_{\nu=\omega}, \tag{4.3.28}$$

or alternatively as

$$F(f) = \frac{1}{2}e^{-j\pi/4}H(f) + \frac{1}{2}e^{j\pi/4}H(-f). \tag{4.3.29}$$

To summarize, the Fourier transform is the even part of the Hartley transform plus negative j times the odd part; similarly, the Hartley transform is the real part plus the negative imaginary part of the Fourier transform. Equation (4.3.23) will be used most often by the engineer when computing the Hartley transform of an arbitrary time or spacial function when the Fourier transform is known or readily available via a table lookup; when this is not the case, direct evaluation of (4.3.1a) or (4.3.1b) is required.

4.3.3   The relationship between the Hartley and Hilbert transforms

The Hilbert transform (see also Chapter 7), $\hat{f}(x)$, of a function $f(x)$, is obtained by convolving $f(x)$ with the function $1/\pi x$. That is,

$$\hat{f}(x) = f(x) \uplus \frac{1}{\pi x} = \frac{1}{\pi} \int_{-\infty}^{\infty} \frac{f(\lambda)}{x - \lambda} d\lambda$$

where the integral is assumed to be taken of its principal value. Here, $\uplus$ denotes linear convolution (see Property 7 in Section 4.4 and Chapter 1, Section 3). The Fourier transform of $\hat{f}(x)$ is found by convolving $1/\pi x$ with the Fourier transform of $f(x)$, $F(f)$. Applying Property 7 yields $-j \operatorname{sgn} f F(f)$. The Hartley transform of $\hat{f}(x)$ is then found via (4.3.23).

Thus, a Hilbert transform simply shifts all positive-frequency components by -90 degrees and all negative-frequency components by $+90$ degrees. The amplitude always remains constant throughout this transformation.

4.3.4   The relationship between the Hartley and Laplace transforms

Because the Hartley transform is the symmetrical form of the classical Fourier transform defined in (4.3.9), it is most convenient to review how the Fourier transform relates to the one-sided or unilateral Laplace transform. Although the unilateral Laplace transform is concerned with time functions for $t > 0$, the Fourier transform includes both positive and negative time but falters with functions having finite average power because the concept of the Dirac delta function must be introduced.

For most functions of practical engineering significance, the conversion from the Laplace to the Fourier transform of $f(x)$ is quite straightforward. However, more difficult situations do exist but are rarely encountered in practical engineering problems; thus, these situations will not be discussed any further.

F(s) with poles in the left-half plane (LHP) only [9]

When the Laplace transform of a function $f(x)$ has no poles on the $j\omega$ axis and poles only in the LHP, the Fourier transform may be computed from the Laplace transform by simply substituting $s = j\omega$. These transforms include all finite-energy signals defined for positive time only. As an example, because

$$F(s) = L\{e^{-\alpha t} u(t)\} = \frac{1}{s + \alpha} \qquad \Re(s) > -\alpha$$

for all values of α, then if α is positive, the single pole of $F(s)$ resides in the LHP at $s = -\alpha$. Thus,

$$F(\omega) = F\{e^{-\alpha t} u(t)\} = F(s)|_{s=j\omega} = \frac{1}{s + \alpha}\bigg|_{s=j\omega} = \frac{1}{j\omega + \alpha}.$$

Lastly, to obtain the Hartley transform of $f(x)$, apply (4.3.23) to $F(\omega)$ above, or evaluate (4.3.4a) directly. Thus,

$$H(v) = \Re\left\{\frac{1}{\alpha + j\omega}\right\} - \Im\left\{\frac{1}{\alpha + j\omega}\right\} = \frac{\alpha + \omega}{\alpha^2 + \omega^2}.$$

F(s) with poles in the LHP and on the $j\omega$ axis [9]

When the function $f(x)$ has poles in the LHP and on the $j\omega$ axis, those terms with LHP poles are treated in the same manner as described above in the preceding paragraph. Each simple pole on the imaginary axis will result in two terms in the Fourier domain: one is obtained by substituting $s = j\omega$ and the other is found by the method of residues. The latter term results in a δ function having a strength of π times the residue at the pole. Mathematically, this is expressed as

$$F(\omega) = F(s)|_{s=j\omega} + \pi \sum_n k_n \delta(\omega - \omega_n). \qquad (4.3.30)$$

For example, consider the Laplace transform of the function $\cos \omega_0 t u(t)$. Via partial fraction expansion, $F(s)$ can be written as

$$F(s) = \frac{s}{s^2 + \omega_0^2} = \frac{\frac{1}{2}}{s + j\omega_0} + \frac{\frac{1}{2}}{s - j\omega_0}.$$

Invoking (4.3.30) leads to the following expression in the Fourier domain

$$F(\omega) = \frac{j\omega}{\omega_0^2 - \omega^2} + \frac{\pi}{2}[\delta(\omega + \omega_0) + \delta(\omega - \omega_0)].$$

Once again, to obtain the Hartley transform of $f(x)$, apply (4.3.23) to $F(\omega)$.

4.3.5 *The relationship between the Hartley and real Fourier transforms*

The real Fourier transform (RFT) of a real signal $f(x)$ of finite energy can be defined as

$$F(\Omega) = 2 \int_{-\infty}^{\infty} f(x) \cos[2\pi\Omega x + \Theta(\Omega)] \, dx \qquad (4.3.31)$$

where

$$\Theta(\Omega) = \begin{cases} 0, & \text{if } \Omega \geq 0 \\ \frac{\pi}{2}, & \text{if } \Omega < 0 \end{cases} \qquad (4.3.32)$$

and $\Omega = f$ is the frequency variable with units of Hertz.

The inverse RFT is given by

$$f(x) = \int_{-\infty}^{\infty} F(\Omega) \cos[2\pi\Omega x + \Theta(\Omega)] \, d\Omega. \qquad (4.3.33)$$

The transform pair (4.3.31) and (4.3.33) can also be written for $\Omega \geq 0$ as

$$F^e(\Omega) = 2 \int_{-\infty}^{\infty} f(x) \cos(2\pi\Omega x) \, dx \qquad (4.3.34)$$

$$F^o(\Omega) = 2 \int_{-\infty}^{\infty} f(x) \sin(2\pi\Omega x) \, dx \qquad (4.3.35)$$

and

$$f(x) = \int_0^{\infty} [F^e(\Omega) \cos(2\pi\Omega x) + F^o(\Omega) \sin(2\pi\Omega x)] \, d\Omega. \qquad (4.3.36)$$

Thus, $F(\Omega)$ equals $F^e(\Omega)$ for $\Omega \geq 0$, and $F^o(\Omega)$ for $\Omega < 0$. Note the similarity between (4.3.34) and (4.3.35) with (4.3.7) and (4.3.8).

The Hartley transform of $f(x)$ is related to the RFT by

$$\begin{bmatrix} H(f) \\ H(-f) \end{bmatrix} = \frac{1}{2} \begin{bmatrix} 1 & 1 \\ 1 & -1 \end{bmatrix} \begin{bmatrix} F^e(\Omega) \\ F^o(\Omega) \end{bmatrix}. \tag{4.3.37}$$

4.3.6 *The relationship between the Hartley and the complex and real Mellin transforms*

The Mellin transform is useful in scale-invariant image and speech recognition applications [10]. The complex Mellin transform is given by

$$F_M(s) = \int_0^\infty f(x) x^{s-1} \, dx \tag{4.3.38}$$

where the complex variable $s = \sigma + j\omega$. If one substitutes $\exp(-x)$ for the variable x, then (4.3.38) becomes

$$F_M(s) = \int_{-\infty}^\infty f'(x) e^{-xs} \, dx \tag{4.3.39}$$

where

$$f'(x) = f(e^{-x}).$$

Thus, from (4.3.39), the complex Mellin transform is the two-sided or bilateral Laplace transform of $f'(x)$.

Equation (4.3.39) can also be written as

$$F_M(\sigma + j\omega) = \int_{-\infty}^\infty f''(x) e^{-j\omega x} \, dx \tag{4.3.40}$$

where

$$f''(x) = f(e^{-x}) e^{-\sigma x}.$$

Thus, the complex Mellin transform is the Fourier transform of $f''(x)$. The Hartley transform of $f''(x) = f(e^{-x}) e^{-\sigma x}$ can then be found by direct application of (4.3.23).

The inverse complex Mellin transform can be written as

$$f(e^{-x}) = e^{\sigma x} \int_{-\infty}^\infty F_M(\sigma + j\omega) e^{j\omega x} \, df. \tag{4.3.41}$$

The real Mellin transform can be written as

$$F^e(\sigma, \omega) = 2 \int_{-\infty}^\infty f''(x) \cos \omega x \, dx \tag{4.3.42}$$

and

$$F^o(\sigma, \omega) = 2 \int_{-\infty}^\infty f''(x) \sin \omega x \, dx. \tag{4.3.43}$$

By analogy to (4.3.34) and (4.3.35), the Hartley transform of $f''(x)$ is related to the real Mellin transform by

$$\begin{bmatrix} H(f) \\ H(-f) \end{bmatrix} = \frac{1}{2} \begin{bmatrix} 1 & 1 \\ 1 & -1 \end{bmatrix} \begin{bmatrix} F^e(\sigma, \omega) \\ F^o(\sigma, \omega) \end{bmatrix}. \tag{4.3.44}$$

The inverse real Mellin transform is given by

$$f(e^{-x}) = e^{\sigma x} \int_0^\infty [F^e(\sigma, \omega) \cos \omega x + F^o(\sigma, \omega) \sin \omega x] \, df. \tag{4.3.45}$$

4.4   Elementary properties of the Hartley transform

In this section, several Hartley transform theorems are presented. These theorems are very useful for generating Hartley transform pairs as well as in signal and systems analysis. In most cases, proofs are presented; examples to illustrate their application are left to specific example problems contained later in this chapter.

Property 1   Linearity
If $f_1(x)$ and $f_2(x)$ have the Hartley transforms $H_1(f)$ and $H_2(f)$, respectively, then the sum $\alpha f_1(x) + \beta f_2(x)$ has the Hartley transform $\alpha H_1(f) + \beta H_2(f)$. This property is established as follows:

$$\int_{-\infty}^{\infty} [\alpha f_1(x) + \beta f_2(x)] \operatorname{cas}(2\pi f x)\, dx = \alpha \int_{-\infty}^{\infty} f_1(x) \operatorname{cas}(2\pi f x)\, dx$$

$$+ \beta \int_{-\infty}^{\infty} f_2(x) \operatorname{cas}(2\pi f x)\, dx$$

$$= \alpha H_1(f) + \beta H_2(f). \qquad (4.4.1)$$

Property 2   Power spectrum and phase
The power spectrum for a signal $f(x)$ can be expressed in the Fourier domain as

$$P(f) = |F(f)|^2 = \Re\{F(f)\}^2 + \Im\{F(f)\}^2.$$

The power spectrum can be obtained directly from the Hartley transform using (4.3.16)–(4.3.19) as follows:

$$P(f) = |F(f)|^2 = \Re\{F(f)\}^2 + \Im\{F(f)\}^2$$

$$= [H^e(f)]^2 + [-H^o(f)]^2$$

$$= \frac{1}{4}[H(f) + H(-f)]^2 + \frac{1}{4}[H(f) - H(-f)]^2$$

$$P(f) = \frac{[H(f)]^2 + [H(-f)]^2}{2}. \qquad (4.4.2)$$

The phase associated with the Fourier transform of $f(x)$ is well known; this is expressed as

$$\Phi(f) = \tan^{-1}\left[\frac{\Im\{F(f)\}}{\Re\{F(f)\}}\right] = \tan^{-1}\left[\frac{-H^o(f)}{H^e(f)}\right]$$

$$\Phi(f) = \tan^{-1}\left[\frac{H(-f) - H(f)}{H(f) + H(-f)}\right]. \qquad (4.4.3)$$

Note that the power spectrum $P(f)$ will always be even.

Property 3   Scaling/Similarity
If the Hartley transform of $f(x)$ is $H(f)$, then the Hartley transform of $f(kx)$ where k is a real constant greater than zero is determined by

$$\int_{-\infty}^{\infty} f(kx) \operatorname{cas}(2\pi f x)\, dx = \int_{-\infty}^{\infty} f(x') \operatorname{cas}\left(\frac{2\pi f x'}{k}\right) \frac{dx'}{k} = \frac{1}{k} H\left(\frac{f}{k}\right). \qquad (4.4.4)$$

For k negative, the limits of integration for the new variable $x' = kx$ are interchanged. Therefore, when k is negative, the last term in (4.4.4) becomes $(1/-k)H(f/k)$. The amalgamation of these two solutions can be expressed as follows: If $f(x)$ has the Hartley transform $H(f)$ then $f\left(\frac{k}{x}\right)$ has the Hartley transform $1/|k|H(f/k)$.

Property 4   Function reversal
If $f(x)$ and $H(f)$ are a Hartley transform pair, then the Hartley transform of $f(-x)$ is $H(-f)$. This is clearly seen when $k = -1$ is substituted into the last expression appearing in Property 3.

Property 5   Function shift/delay
If $f(x)$ is shifted in time by a constant T, then by substituting $x' = x - T$, the Hartley transform becomes

$$H(f) = \int_{-\infty}^{\infty} f(x') \, \text{cas}[2\pi f(x' + T)] \, dx'. \tag{4.4.5}$$

Notice that the basis function in (4.4.5) can be expanded using the appropriate entry of Table 4.1 in the following manner:

$$\text{cas}[2\pi f(x' + T)] = \cos(2\pi f x')\cos(2\pi f T) + \cos(2\pi f x')\sin(2\pi f T)$$

$$+ \sin(2\pi f x')\cos(2\pi f T) - \sin(2\pi f x')\sin(2\pi f T).$$

Expanding (4.4.5) into four integrals and grouping the first and third and second and fourth integrals, respectively, the final result is

$$H(f) = \cos(2\pi f T)H(f) + \sin(2\pi f T)H(-f). \tag{4.4.6}$$

Property 6   Modulation
If $f(x)$ is modulated by the sinusoid $\cos(2\pi f_0 x)$, then transforming to the Hartley space via (4.3.1b) yields

$$H(f) = \int_{-\infty}^{\infty} f(x)\cos(2\pi f_0 x)\,\text{cas}(2\pi f x)\,dx \tag{4.4.7}$$

$$H(f) = \int_{-\infty}^{\infty} f(x)\cos(2\pi f_0 x)\cos(2\pi f x)\,dx$$

$$+ \int_{-\infty}^{\infty} f(x)\cos(2\pi f_0 x)\sin(2\pi f x)\,dx.$$

Notice that if the function-product relations (i.e., $\cos \alpha \cos \beta$ and $\cos \alpha \sin \beta$) are expanded and grouped accordingly, the following relation results:

$$H(f) = \frac{1}{2}H(f - f_0) + \frac{1}{2}H(f + f_0). \tag{4.4.8}$$

Property 7   Convolution (⊎)
If $f_1(x)$ has the Hartley transform $H_1(f)$ and $f_2(x)$ has the Hartley transform $H_2(f)$, then $f_1(x) \uplus f_2(x)$ has the Hartley transform

$$\frac{1}{2}\left[H_1(f)H_2(f) + H_1(-f)H_2(f) + H_1(f)H_2(-f) - H_1(-f)H_2(-f)\right]. \tag{4.4.9}$$

To obtain this result directly, simply substitute the convolution integral

$$f_1(x) \uplus f_2(x) = \int_{-\infty}^{\infty} f_1(\lambda) f_2(x - \lambda) \, d\lambda \qquad (4.4.10)$$

into (4.3.1b) and utilize Property 5. The result is as follows:

$$H(f) = \int_{-\infty}^{\infty} [f_1(x) \uplus f_2(x)] \operatorname{cas}(2\pi f x) \, dx \qquad (4.4.11)$$

$$= \int_{-\infty}^{\infty} \left[\int_{-\infty}^{\infty} f_1(\lambda) f_2(x - \lambda) \, d\lambda \right] \operatorname{cas}(2\pi f x) \, dx$$

$$= \int_{-\infty}^{\infty} f_1(\lambda) \left[\int_{-\infty}^{\infty} f_2(x - \lambda) \operatorname{cas}(2\pi f x) \, dx \right] d\lambda.$$

Invoking the shift theorem,

$$= \int_{-\infty}^{\infty} f_1(t) \left[\cos(2\pi f \lambda) H_2(f) + \sin(2\pi f \lambda) H_2(-f) \right] d\lambda.$$

Factoring the $H_2(\cdot)$ term to the right and utilizing (4.3.12)–(4.3.19), the result follows. Note that (4.4.9) simplifies for the following symmetries:

- *if $f_1(x)$ and/or $f_2(x)$ is even, or if $f_1(x)$ is even and $f_2(x)$ is odd; or if $f_1(x)$ is odd and $f_2(x)$ is even, then $f_1(x) \uplus f_2(x) = H_1(f) H_2(f)$;*
- *if $f_1(x)$ is odd, then $f_1(x) \uplus f_2(x) = H_1(f) H_2(-f)$;*
- *if $f_2(x)$ is odd, then $f_1(x) \uplus f_2(x) = H_1(-f) H_2(f)$;*
- *if both functions are odd, then $f_1(x) \uplus f_2(x) = - H_1(f) H_2(f)$.*

*In most practical situations, it is possible to shift one of the functions entering into the convolution such that it exhibits even or odd symmetry. When this is possible, (4.4.9) simplifies to one **real** multiplication versus the single **complex** multiplication (= four real multiplications and three real additions) in the Fourier domain.*

Property 8   Autocorrelation ($\odot$)
If $f_1(x)$ has the Hartley transform $H_1(f)$ then the autocorrelation of $f_1(x)$ described by the equation below,

$$f_1(x) \odot f_1(x) = f_1(x) \uplus f_1(-x) = \int_{-\infty}^{\infty} f_1(\lambda) f_1(x + \lambda) \, d\lambda, \qquad (4.4.12)$$

has the Hartley transform

$$\frac{1}{2} \left[H_1(f)^2 + H_1(-f)^2 \right] = [H^e(f)]^2 + [H^o(f)]^2. \qquad (4.4.13)$$

Comparing equation (4.4.12) to (4.4.10), it is evident that the convolution and correlation integrals are closely related. Substituting the correlation integral of (4.4.12) into the direct Hartley transform and utilizing Property 5, the result is as follows:

$$H(f) = \int_{-\infty}^{\infty} \left[\int_{-\infty}^{\infty} f_1(\lambda) f_1(x + \lambda) \, d\lambda \right] \operatorname{cas}(2\pi f x) \, dx$$

$$= \int_{-\infty}^{\infty} f_1(x) \left[\int_{-\infty}^{\infty} f_1(x + \lambda) \operatorname{cas}(2\pi f x) \, dx \right] d\lambda \qquad (4.4.14)$$

Invoking the shift theorem with $T = -\lambda$,

$$= \int_{-\infty}^{\infty} f_1(x) \left[\cos(2\pi f \lambda) H_1(f) - \sin(2\pi f \lambda) H_1(-f)\right] d\lambda.$$

Factoring $H_1(\cdot)$ to the right and utilizing (4.3.12)–(4.3.19), the desired result follows.

Property 9   Product
If $f_1(x)$ is multiplied by a second function $f_2(x)$, then the product $f_1(x) \, f_2(x)$ is

$$\frac{1}{2} \left[H_1(f) \uplus H_2(f) + H_1(-f) \uplus H_2(f) + H_1(f) \uplus H_2(-f) - H_1(-f) \uplus H_2(-f)\right]$$

$$= H_1^e(f) \uplus H_2^e(f) - H_1^o(f) \uplus H_2^o(f) + H_1^e(f) \uplus H_2^o(f) + H_1^o(f) \uplus H_2^e(f)$$

Property 10   nth Derivative of a Function $\mathbf{f^{(n)}(x)}$
The nth derivative of a function $f(x)$ is

$$f^{(n)}(x) = cas' \frac{n\pi}{2} (2\pi f)^n H[(-1)^n f]. \tag{4.4.16}$$

This property is derived by recursive application of (4.3.23) to the Fourier transform of the function $df(x)/dx$ and its higher order derivatives.

A summary of the above properties appears in Table 4.6.

4.5   The Hartley transform in multiple dimensions

The Hartley transform also exists in multiple dimensions. For a function $f(x, y)$ the two-dimensional Hartley transform and its inverse is

$$H(\upsilon, \nu) = \int_{-\infty}^{\infty} \int_{-\infty}^{\infty} f(x, y) \operatorname{cas}[2\pi(\upsilon x + \nu y)] \, dx \, dy \tag{4.5.1}$$

$$f(x, y) = \int_{-\infty}^{\infty} \int_{-\infty}^{\infty} H(\upsilon, \nu) \operatorname{cas}[2\pi(\upsilon x + \nu y)] \, d\upsilon \, d\nu. \tag{4.5.2}$$

Although a three-dimensional Hartley transform exists, it is above and beyond the scope of this treatise. That is, the user will not typically utilize the higher dimension continuous-time integral. Therefore, the reader is referred to [8] for details.

4.6   Systems analysis using a Hartley series
representation of a temporal or spacial function

The Hartley series is an infinite series expansion of a periodic signal in which the orthogonal basis functions in the series are the cosine-and-sine function, $\operatorname{cas}(k\upsilon_0 t)$, where $\upsilon_0 = 2\pi f_0 = 2\pi/T_0$ is the fundamental radian frequency. This series formulation differs from the Fourier series in the selection of the basis functions; namely, the cas function versus the complex exponential, $\phi_k(t) = \exp(j2\pi kt/T_0)$, $k = 0, \pm1, \pm2, \ldots$ over the interval $t_0 \leq t \leq t_0 + T_0$

TABLE 4.6
A summary of Hartley transform theorems.

Theorem	$f(x)$	$H(f)$
Linearity	$f_1(x) + f_2(x)$	$H_1(f) + H_2(f)$
Power spectrum	$P(f) = \frac{1}{2}\{[H(f)]^2 + [H(-f)]^2\}$	$\Phi(f) = \tan^{-1}\left[\frac{H(-f) - H(f)}{H(f) + H(-f)}\right]$
Scaling/Similarity	$f(kx)$	$\lvert\frac{1}{k}\rvert H(\frac{f}{k})$
Reversal	$f(-x)$	$H(-f)$
Shift	$f(x - T)$	$H(f) = \cos(2\pi f T)H(f) + \sin(2\pi f T)H(-f)$
Modulation	$f(x)\cos(2\pi f_0 t)$	$H(f) = \frac{1}{2}H(f - f_0) + \frac{1}{2}H(f + f_0)$
Convolution	$f_1(x) \uplus f_2(x)$	$\frac{1}{2}[H_1(f)H_2(f) + H_1(-f)H_2(f)$ $+ H_1(f)H_2(-f) - H_1(-f)H_2(-f)]$
Autocorrelation	$f_1(x) \odot f_1(x)$	$\frac{1}{2}\left[H_1(f)^2 + H_1(-f)^2\right]$
Product	$f_1(x)f_2(x)$	$\frac{1}{2}[H_1(f) \uplus H_2(f) + H_1(-f) \uplus H_2(f)$ $+ H_1(f) \uplus H_2(-f) - H_1(-f) \uplus H_2(-f)]$
nth derivative	$f^{(n)}(x)$	$f^{(n)}(x) = \text{cas}'\,\frac{n\pi}{2}(2\pi f)^n H[(-1)^n f]$

where T_0 is the fundamental frequency of the periodic function. The Hartley series, so named as a result of the analogy drawn by Hartley to the Fourier series in [1], is capable of representing all functions in that interval providing they satisfy certain mathematical conditions developed by Dirichlet (see Section 4.3).

If a system is linear and its impulse response is available, then the response of this system to applied inputs can be found using the principles of linearity and superposition. If the forcing function or excitation is represented as a weighted sum of individual components, called basis functions, then it is only necessary to calculate the response of the system to each of these components and add them together. This method leads to the convolution integral that was presented in Chapter 1. Before proceeding with a mathematical description of a set of basis functions $\phi_k(t)$, consider a desired forcing function being represented as a sum of weighted (i.e., having different strengths) impulse functions. These impulse functions produce responses that are amplitude-scaled and time-shifted versions of the response to a unit impulse. Summing all responses to each impulse results in the total response of the system to the forcing function. It seems that the impulse function may be a type of basis function, and indeed it is.

There are a variety of basis functions that can be used for linear systems analysis. In addition to the impulse function, $\delta(t)$, one of the most familiar basis functions is the complex exponential $\phi(t) = \exp(j\omega_0 t)$ corresponding to the Fourier series. Another frequently used basis function is the complex exponential, $\phi(t) = \exp(st)$ where $s = \sigma + j\omega$ is a complex number. Clearly, the Fourier basis function is a specialization of $\exp(st)$ with $\sigma = 0$. When applications involve linear systems analysis, sinusoidal functions are a convenient choice for basis functions. The reason for this choice is that the sum or difference of two sinusoids of the same frequency is still a sinusoid, and the derivative or integral of a sinusoid is still a sinusoid. These characteristics lend themselves well to sinusoidal steady-state analysis using the phasor concept.

Before proceeding further, it is helpful to summarize briefly properties and characteristics of basis functions. A most desirable quality of a set of basis functions is known as finality of coefficients. Referring to the equation below,

$$x(t) \approx \sum_{n=-N}^{N} a_n \phi_n(t),$$
(4.6.1)

a function represented by a finite number of coefficients and basis functions in the form of a linear combination can always be more accurately described by adding additional terms (i.e., increasing N) to the linear combination without affecting any of the earlier coefficients. This desirable quality can be achieved if the basis functions are orthogonal over the time interval of interest (see also Chapter 1, Section 5).

DEFINITION 4.6.1   A set of functions $\{\phi_n\}, n = 0, \pm 1, \pm 2, \dots$ is an orthogonal set on the interval $a \le t \le b$ if for every $i \ne k$,

$$(\phi_i, \phi_k) = 0$$

where $(\cdot, \cdot)$ denotes the inner product. Here, the inner product of two functions f and g is defined as

$$(f, g) = \int_a^b f(t) g^*(t) \, dt.$$

Using the integral relationship for an inner product, the condition for orthogonality of basis functions is that for all k,

$$\int_\tau^{\tau+T_0} \phi_n(t) \phi_k^*(t) \, dt = \begin{cases} \lambda_k, & \text{if } k = n \\ 0, & \text{if } k \ne n \end{cases}$$
(4.6.2)

where $\phi_k^*(t)$ is the complex conjugate of $\phi_k(t)$ and the λ_k are real and $\lambda_k \ne 0$. If the basis functions are real, then $\phi_k^*(t)$ can be replaced by $\phi_k(t)$. Note that (4.6.2) can be expressed more compactly by the following notation:

$$(\phi_n(t), \phi_k^*(t)) = \lambda_k \delta_{nk} = \begin{cases} \lambda_k, & \text{if } k = n \\ 0, & \text{if } k \ne n \end{cases}$$
(4.6.3)

where δ_{nk} is the Kronecker delta function.

In order to calculate the coefficients a_n appearing in (4.6.1), the orthogonality property of the basis functions really demonstrates its desirable quality. If (4.6.1) is multiplied on both sides by $\phi_i^*(t)$, for any i, and then integrated over the specified interval t to $t + T$, the following results:

$$\int_\tau^{\tau+T_0} \phi_i^*(t) x(t) \, dt = \int_\tau^{\tau+T_0} \phi_i^*(t) \left[\sum_{n=-N}^{N} a_n \phi_n(t) \right] dt$$

$$= \sum_{n=-N}^{N} a_n \int_\tau^{\tau+T_0} \phi_i^*(t) \phi_n(t) \, dt.$$
(4.6.4)

From (4.6.2) above,

$$a_i = \frac{1}{\lambda_i} \int_\tau^{\tau+T_0} \phi_i^*(t) x(t) \, dt \tag{4.6.5}$$

when the basis functions are orthogonal. When the basis functions are complex, as in the case of the Fourier series, a complex-valued coefficient $a_i = \alpha_i$ will result. For real-valued signals of interest, the imaginary terms will always cancel.

Now that the coefficients to (4.6.1) have been calculated, is it possible to find a different set of coefficients that yield a better approximation to $x(t)$ for the same value of N? To investigate this question, it is necessary to measure the closeness of the approximation of (4.6.1) when N is finite and when N approaches infinity. One measure that is frequently used is the mean-squared error. This approach is generalized in detail for complex basis functions by minimizing the mean-squared error of the N-term truncation approximation to an infinite series.

The decomposition of a time function into a weighted linear combination of basis functions is an exact representation when the function is described by

$$f(t) = \sum_{k=-\infty}^{\infty} a_k \phi_k(t). \tag{4.6.6}$$

However, for practical numerical calculations it is computationally necessary to truncate the above sum to $2N$ terms. In this way, an approximation to the signal $f(t)$ may be calculated; this is guaranteed by the convergence properties of the Fourier series via the Riemann-Lebesgue Lemma. If we now denote the truncated linear combination of $2N$ basis functions by

$$\hat{f}(t) = \sum_{k=-N}^{N} \hat{a}_k \phi_k(t), \tag{4.6.7}$$

how can the possibly complex-valued weighting factors, $\hat{a}_k$, be selected in order to minimize the mean-squared error between $f(t)$ and $\hat{f}(t)$? Let the mean-squared error be represented by ϵ, then

$$0 \le \epsilon \le \|x - \hat{x}\|^2 = \left\| x - \sum_{k=1}^{N} \hat{a}_k \phi_k \right\|^2 = \left(x - \sum_k \hat{a}_k \phi_k, x - \sum_j \hat{a}_j \phi_j \right)$$

$$= (x, x) - \left(x, \sum_j \hat{a}_j \phi_j \right) - \left(\sum_k \hat{a}_k \phi_k, x \right) + \left(\sum_k \hat{a}_k \phi_k, \sum_j \hat{a}_j \phi_j \right)$$

$$= (x, x) - \sum_j \hat{a}_j^*(x, \phi_j) - \sum_k \hat{a}_k(\phi_k, x) + \sum_k \hat{a}_k \sum_j \hat{a}_j^*(\phi_k, \phi_j)$$

$$= (x, x) - \sum_j \hat{a}_j^* a_j - \sum_k \hat{a}_k a_k^* + \sum_k \hat{a}_k \sum_j \hat{a}_j^* \delta_{kj}. \tag{4.6.8}$$

Note that in the above step, the following results were used:

$$a_j = (x, \phi_j) \tag{4.6.9}$$

$$a_j^* = (x, \phi_j)^* = (\phi_j, x) \tag{4.6.10}$$

$$(\phi_k, \phi_j) = \delta_{kj}. \tag{4.6.11}$$

Utilizing only one set of subscripts and adding

$$\sum_j a_j a_j^* - \sum_j |a_j|^2$$

to the right-hand side of the previous equation,

$$= (x, x) - \sum_j |a_j|^2 - \sum_j \hat{a}_j a_j^* - \sum_j \hat{a}_j^* a_j + \sum_j \hat{a}_j \hat{a}_j^* + \sum_j a_j a_j^*$$

$$\epsilon \le (x, x) - \sum_j |a_j|^2 + \sum_j |\hat{a}_j - a_j|^2 \tag{4.6.12}$$

where

$$\sum_j |\hat{a}_j - a_j|^2 = \sum_j (\hat{a}_j - a_j)(\hat{a}_j^* - a_j^*) = \sum_j (\hat{a}_j - a_j)(\hat{a}_j - a_j)^*.$$

In (4.6.12), the first and second terms are independent of $\hat{a}_j$ and are strictly greater than or equal to zero. The "best choice" of $\hat{a}_j$, $j = 1, \ldots, N$ is chosen such that $\|x - \sum_{j=1}^N \hat{a}_j \phi_j\|$ is as small as possible. Therefore, choose $\hat{a}_j = a_j$. This results in the following:

$$0 \le \|x - \sum_{j=1}^N \hat{a}_j \phi_j\| = \|x\|^2 - \sum_j |a_j|^2.$$

From the above expression, the well-known Bessel's inequality is formed when N in the sum over j approaches ∞:

$$\sum_{j=1}^\infty |a_j|^2 \le \|x\|^2.$$

When Bessel's inequality is an exact equality, the familiar Parseval's equality results. From the results presented there, it can be concluded that the a_j of (4.6.9) are the best coefficients from the standpoint of minimizing the approximation error, ϵ, when only a finite number of terms are used. Thus, the use of orthogonal basis functions provide two desirable qualities: they guarantee the finality of coefficients and also the same coefficients minimize the mean-squared error of the function representation.

An additional property that is vitally important in the discussion of the Fourier series is the Riemann-Lebesgue Lemma. Briefly, this lemma states that supposing the function $f(t)$ is absolutely integrable on the interval (a, b), then

$$\int_a^b f(t) e^{j\omega t}\, dt \Longrightarrow 0 \qquad \text{as } |\omega| \Longrightarrow \infty.$$

Because this also implies that

$$\int_a^b f(t) \cos(\omega t)\, dt \Longrightarrow 0 \qquad \text{as } |\omega| \Longrightarrow \infty$$

and

$$\int_a^b f(t) \sin(\omega t)\, dt \Longrightarrow 0 \qquad \text{as } |\omega| \Longrightarrow \infty,$$

by linearity,

$$\int_a^b f(t) \operatorname{cas}(\omega t)\, dt \Longrightarrow 0 \qquad \text{as } |\omega| \Longrightarrow \infty.$$

Note that (a, b) may range from $-\infty$ to ∞. The importance of this result foreshadows the concept of a complete set of basis functions.

A set of basis functions is termed **complete** in the sense of mean convergence if the error in the approximation of $f(t)$ can be made arbitrarily small by making the value of N in (4.6.1) sufficiently large. That is,

$$\lim_{N \to \infty} ||f - S_N|| = 0$$

where $S_N(\cdot)$, $N = 1, 2, \ldots$ is the partial sum of piecewise continuous functions defined on the open interval (a, b). Also, it can be shown that a necessary and sufficient condition for an orthonormal set $\{\phi_n(t)\}$ to be complete is that for each function x considered, Parseval's equation

$$\sum_{n=1}^{\infty} (x, \phi_n)^2 = ||x||^2$$

must be satisfied. Note that $(x, \phi_n) = a_n$.

Now, attention turns to the analog of the complex Fourier series represented as follows:

$$x(t) = \sum_{n=-\infty}^{\infty} \alpha_n e^{jn\omega_0 t} \tag{4.6.13}$$

where

$$\alpha_n = \frac{1}{T_0} \int_{\tau}^{\tau+T_0} x(t) e^{-jn\omega_0 t} \, dt, \tag{4.6.14}$$

which can also be written as a single-sided series

$$x(t) = \frac{a_0}{2} + \sum_{n=1}^{\infty} [a_n \cos(n\omega_0 t) + b_n \sin(n\omega_0 t)] \tag{4.6.15}$$

by noting that

$$\alpha_{-n}^* = \alpha_n = \frac{1}{2}(a_n - jb_n)$$

from which

$$a_n = \alpha_n + \alpha_n^*$$

$$b_n = \alpha_n - \alpha_n^*.$$

The properties and use of the Fourier series (FS) are well known and well documented in the literature.

The set of basis functions used by the Hartley transform and in the Hartley series is the set $\{\phi_n(t)\}$, $n = 0, \pm1, \pm2, \ldots$ where $\{\phi_n(t)\} = \text{cas}(n\nu_0 t)$. This is an orthogonal set over the interval $t \leq \acute{t} \leq t + T_0$ and is capable of representing any time function that the Fourier series can in that interval. This set of time functions possesses a Fourier or Hartley series if the well-known Dirichlet conditions are met as presented in Section 4.3.

Let $\{\phi_n(t)\} = \text{cas}(n\nu_0 t)/\sqrt{2\pi}$, $n = 0, \pm1, \pm2, \ldots$ on the interval $[-\pi, \pi]$.

DEFINITION 4.6.2   *A set of functions $\{\phi_n\}$, $n = 0, \pm1, \pm2, \ldots$ is an "orthonormal set" on the interval $a \leq t \leq b$ if*

$$(\phi_i, \phi_k) = \delta_{ik} = \begin{cases} 1, & \text{if } i = k \\ 0, & \text{if } i \neq k. \end{cases}$$

CLAIM A set of functions $\{\phi_n(t)\}, n = 0, \pm1, \pm2, \ldots$ where $\{\phi_n(t)\} = \text{cas}(n\nu_0t)/\sqrt{2\pi}$ is an orthonormal set on the interval $-\pi \leq t \leq \pi$.

PROOF

$$(\phi_i, \phi_k) = \int_{-\pi}^{\pi} \phi_i(t)\phi_k^*(t)\,dt = \int_{-\pi}^{\pi} \phi_i(t)\phi_k(t)\,dt$$

$$= \int_{-\pi}^{\pi} \frac{1}{\sqrt{2\pi}}\,\text{cas}(i\nu_0t) \cdot \frac{1}{\sqrt{2\pi}}\,\text{cas}(k\nu_0t)\,dt$$

$$= \frac{1}{2\pi} \int_{-\pi}^{\pi} \text{cas}(i\nu_0t)\,\text{cas}(k\nu_0t)\,dt.$$

If each function of the integrand in the above equation is expanded to $\cos(\cdot) + \sin(\cdot)$ and then multiplied together, four terms result: $\cos(\cdot)\cos(\cdot)$, $\sin(\cdot)\sin(\cdot)$, and two cross products, $\cos(\cdot)\sin(\cdot)$ and $\sin(\cdot)\cos(\cdot)$. The integral of the two cross products are zero by the familiar orthogonality property for the cosine and sine functions, respectively. The other two integrands, when evaluated on the interval from $-\pi$ to π, equal 0 for $i \neq k$ and π when $i = k$. Therefore,

$$(\phi_i, \phi_k) = \begin{cases} 1, & \text{if } i = k \\ 0, & \text{if } i \neq k. \end{cases}$$

Thus, the basis functions $\{\phi_n\}$ are an orthonormal system on the interval $[-\pi, \pi]$. ∎

Let the periodic signal $x(t)$ with period T_0,

$$x(t + T_0) = x(t) \qquad \forall t$$

be written as an orthogonal series expansion (i.e., a linear combination possessing an orthogonal set of basis functions)

$$x(t) = \sum_{i=-\infty}^{\infty} \gamma_i\phi_i(t) \tag{4.6.16}$$

where $\phi_i(t)$ are orthogonal basis functions. It has been shown previously that

$$\phi_i(t) = \text{cas}(i\nu_0t)$$

is an orthogonal basis function over the interval $[t, t + T_0]$

$$\int_{\tau}^{\tau+T_0} \text{cas}(i\nu_0t)\,\text{cas}(k\nu_0t)\,dt = \begin{cases} T_0, & \text{if } i = k \\ 0, & \text{if } i \neq k \end{cases}$$

where $\nu_0 = 2\pi/T_0$. Therefore, the γ_i in (4.6.16) are readily obtained using the orthogonality property,

$$x(t)\,\text{cas}(k\nu_0t) = \sum_{i=-\infty}^{\infty} \gamma_i\,\text{cas}(i\nu_0t)\,\text{cas}(k\nu_0t)$$

$$\frac{1}{T_0} \int_{\tau}^{\tau+T_0} x(t)\,\text{cas}(k\nu_0t)\,dt = 0 + 0 + 0, \ldots, + \gamma_k + 0 + 0+, \ldots$$

This gives what will be termed the Hartley series (HS),

$$x(t) = \sum_{i=-\infty}^{\infty} \gamma_i\,\text{cas}(i\nu_0t)$$

$$v_0 = \frac{2\pi}{T_0}$$

$$\gamma_i = \frac{1}{T_0} \int_{\tau}^{\tau+T_0} x(t) \, \text{cas}(i \, v_0 t) \, dt. \tag{4.6.17}$$

It is a simple matter to show that

$$\gamma_k = \begin{cases} \Re\{\alpha_k\} - \Im\{\alpha_k\} & k \neq 0 \\ \alpha_k & k = 0. \end{cases} \tag{4.6.18}$$

Specifically, from (4.6.14) let

$$\Re\{\alpha_k\} = \frac{1}{T_0} \int_{\tau}^{\tau+T_0} x(t) \cos(k\omega_0 t) \, dt$$

$$\Im\{\alpha_k\} = \frac{-1}{T_0} \int_{\tau}^{\tau+T_0} x(t) \sin(k\omega_0 t) \, dt$$

then

$$\Re\{\alpha_k\} - \Im\{\alpha_k\} = \frac{1}{T_0} \int_{\tau}^{\tau+T_0} x(t) \, \text{cas}(k\omega_0 t) \, dt.$$

If $\omega_0 = v_0$, then the result follows. The FS coefficients are also related to the HS coefficients by

$$\alpha_i = E\{\gamma_i\} - j O\{\gamma_i\} \tag{4.6.19}$$

where $E\{\cdot\}$ and $O\{\cdot\}$ are the even and odd parts of a function,

$$E\{\theta_i\} = \frac{1}{2}(\theta_i + \theta_{-i}) \tag{4.6.20}$$

$$O\{\theta_i\} = \frac{1}{2}(\theta_i - \theta_{-i}). \tag{4.6.21}$$

As an example, the two-sided FS for the square wave

$$x(t) = \begin{cases} 1 & \frac{i}{2} - \frac{1}{4} < t < \frac{i}{2} + \frac{1}{4} & i = \text{even} \\ -1 & \frac{i}{2} - \frac{1}{4} < t < \frac{i}{2} + \frac{1}{4} & i = \text{odd} \end{cases}$$

is

$$x(t) = \cdots + \frac{2}{5\pi} e^{-j5\omega_0 t} - \frac{2}{3\pi} e^{-j3\omega_0 t} + \frac{2}{\pi} e^{-j\omega_0 t}$$

$$+ 0 + \frac{2}{\pi} e^{j\omega_0 t} - \frac{2}{3\pi} e^{j3\omega_0 t} + \frac{2}{5\pi} e^{j5\omega_0 t} - \cdots$$

and the HS is

$$x(t) = \cdots + \frac{2}{5\pi} \text{cas}(-5v_0 t) - \frac{2}{3\pi} \text{cas}(-3v_0 t) + \frac{2}{\pi} \text{cas}(-v_0 t)$$

$$+ 0 + \frac{2}{\pi} \text{cas}(v_0 t) - \frac{2}{3\pi} \text{cas}(3v_0 t) + \frac{2}{5\pi} \text{cas}(5v_0 t) - \cdots$$

where $v_0 = \omega_0 = 2\pi$ radians per second.

From (4.6.19)–(4.6.21), all of the familiar properties of the FS can be rewritten in terms of the HS. Table 4.7 summarizes some of these properties. A listing of the corresponding FS

TABLE 4.7
Selected properties of the Hartley series.

Function	Fourier series coefficients	Hartley series coefficients
Integral $\int f(t)\,dt$	$\frac{-j\alpha_i}{i\omega_0}$	$\frac{\gamma_{-i}}{i\nu_0}$
Derivative $\frac{df}{dt}$	$ji\omega_0\alpha_i$	$-i\nu_0\gamma_{-i}$
Convolution of f and h	$\alpha_i H(i\omega_0)$	$\frac{1}{2}\left[\gamma_i H(i\nu_0) + \gamma_i H(-i\nu_0)\right.$ $\left. + \gamma_{-i} H(i\nu_0) - \gamma_{-i} H(-i\nu_0)\right]$
Time reversal $f(-t)$	α_i^*	γ_{-i}

property is shown in the table for comparison. For purposes of this table, both the FS and HS of $h(t)$ and the function $h(t)$ itself are assumed to exist. For example, the table is read as follows: if a function $h(t)$ possesses an HS with coefficients γ_i, $-\infty < i < \infty$, then the HS of the indefinite integral

$$\int h(t)\,dt$$

has coefficients Γ_i where

$$\Gamma_i = \frac{\gamma_{-i}}{i\nu_0}.$$

The entry for convolution in Table 4.7 deserves special mention: If a periodic signal $f(t)$ is applied to a system with impulse response $h(t)$, the FS of the output (i.e., $f(t) \uplus h(t)$) has coefficients $\alpha_i H(i\omega_0)$ where $H(\omega)$ is the Fourier transform of $h(t)$; similarly, the HS coefficients are Γ_i,

$$\Gamma_i = \frac{1}{2}\left[\gamma_i H(i\nu_0) + \gamma_i H(-i\nu_0) + \gamma_{-i} H(i\nu_0) - \gamma_{-i} H(-i\nu_0)\right]. \tag{4.6.22}$$

Note that for the linear system response problem, the FS methodology requires one complex multiplication for each coefficient and the HS methodology requires four real multiplications. However, (4.6.22) can also be written as

$$\Gamma_i = \gamma_i H^e(i\nu_0) + \gamma_{-i} H^o(i\nu_0) \tag{4.6.23}$$

where $H^e(i\nu_0)$ and $H^o(i\nu_0)$ are the even and odd parts of $H(i\nu_0)$, respectively. Therefore, analogous use of the HS will require only two real multiplications in general, and for certain conditions of symmetry, the number of real multiplications reduces to one.

The HS methodology may be utilized wherever the FS methodology is applicable. The HS entails no complex quantities or calculations. Under certain types of symmetry, the HS simplifies (as does the FS); Table 4.8 summarizes these symmetries. The main properties of the HS are:

- HS coefficients γ_k are always real.
- for even functions $x(t) = x(-t)$, $\gamma_{-k} = \gamma_k$.
- for odd functions $x(t) = -x(-t)$, $\gamma_{-k} = -\gamma_k$.
- the HS exists when the FS exists.

Previously, it was mentioned that for some conditions of symmetry, the linear system response problem using the HS simplifies. The simplification refers to one real multiplication for

TABLE 4.8
Simplifications in the Hartley and Fourier series for symmetries in $f(t)$.

Symmetry	Fourier series coefficients	Hartley series coefficients
Even $f(t) = f(-t)$	$\Im\{\alpha_i\} = 0$ $\alpha_i = \alpha_{-i}$	$\gamma_i = \gamma_{-i}$
Odd $f(t) = -f(-t)$	$\Re\{\alpha_i\} = 0$ $\alpha_i = -\alpha_{-i}$	$\gamma_i = -\gamma_i$
Half-wave odd $f(t) = -f\left(t + \frac{T}{2}\right)$	$\alpha_i = 0$ for even i	$\gamma_i = 0$ for even i
None	$\alpha_{-i} = \alpha_i^*$	No simplification of γ_i

the calculation of each HS coefficient of the output. Such is the case when the input function $f(t)$ is an even function; in this case, $\gamma_i = \gamma_{-i}$ in (4.6.22) and (4.6.23) become

$$\Gamma_i = \gamma_i H(i\nu_0) \tag{4.6.24}$$

where $H(i\nu_0)$ is the sampled Hartley transform of the system impulse response, $H(\nu)$, and Γ_i are the HS coefficients of the system output. Similarly, for odd $f(t)$,

$$\Gamma_i = \gamma_i H(-i\nu_0). \tag{4.6.25}$$

The HS analysis technique closely parallels that of FS analysis; periodic excitation functions are resolved into a series where the system response to each term in the series is subsequently evaluated. The total response is the superposition of the individual responses. This method is applicable to a wide range of systems engineering problems, the only requirements being linearity of the system and existence of the HS (FS). Practical limitations often impose constraints in cases of high bandwidth signals: for such signals, the system models are often difficult to obtain accurately for frequencies far from the intended frequency band of operation. A particularly applicable area is that of industrial electric power distribution system analysis. This is the case because bandwidth limits of the power distribution system components naturally occur and thus limit the frequency range that must be accommodated in the system component models. Also, in power distribution applications, the periodicity of most currents and voltages and the behavior of these physical quantities virtually always insure the existence of the HS (FS). In industrial distribution circuits, although switched and pulsed load currents commonly occur, these currents are nonetheless periodic; the fundamental frequency of these phenomena is the "power frequency" (i.e., 50 or 60 Hz). In three-phase power quality assessment, often times unbalanced conditions make a full phase-by-phase analysis necessary; in such cases, the faster, all-real calculation offered by the HS is particularly attractive.

4.6.1 *Transfer function methodology and the Hartley series*

The familiar transfer function methodology used with periodic signals $x(t)$ that possess an FS is well known. If a signal $x(t)$ is applied as an input to a linear time-invariant (LTI) system whose frequency response matrix is $H(\omega)$, then the output $y(t)$,

$$y(t) = x(t) \uplus h(t)$$

$$h(t) = F^{-1}\{H(\omega)\}$$

may be calculated by superimposing the individual responses to the Fourier components $\alpha_i \exp(ji\omega_0 t)$,

$$x(t) = \sum_i \alpha_i e^{ji\omega_0 t}$$

where $\sum_i$ refers to the summation over $(-\infty, \infty)$. Thus the FS for $y(t)$,

$$y(t) = \sum_i A_i e^{ji\omega_0 t}$$

is readily found, noting

$$A_i = \alpha_i H(i\omega_0). \tag{4.6.26}$$

This is a simple consequence of LTI systems and the impulse-shifting property that is used with the FT of $\exp(ji\omega_0 t)$.

An analogous result occurs when periodic $x(t)$ is expressed as an HS. Consider $x(t)$ applied to an LTI system. Let $x(t)$ be written as an HS

$$x(t) = \sum_i \gamma_i \operatorname{cas}(i\nu_0 t)$$

and let the system possess an impulse response $h(t)$ whose HT is $H(\nu)$. Then, the output $y(t)$ has an HS

$$y(t) = \sum_i \Gamma_i \operatorname{cas}(i\nu_0 t)$$

where the HS coefficients are readily calculated from either (4.6.22) or (4.6.23).

4.6.2 The Hartley series applied to electric power quality assessment

Methodologies for electric power quality analysis and assessment have taken on renewed importance in recent years. This is due to two main factors: (1) the appearance of high-power switching devices and switched loads that can cause power quality problems at the distribution level and (2) the need for power quality at all power levels to avoid interference, excessive losses, and misoperation of critical loads. There are numerous fundamental issues to be resolved relating to quantifying power quality problems, instrumentation, and monitoring (especially in the environment of digital computer loads). The methodology used for the calculation of bus voltage and line current waveforms is also of salient importance.

Electric power quality assessment often involves the calculation of a bus voltage or line current. The HS methodology described above is applicable because:

1. Limited bandwidth of most electric distribution systems makes truncation of the HS practical at a reasonably low frequency (e.g., $\nu = \pm 19\nu_0$ or ± 7163 radians per second in a 60 Hz system, ± 5969 radians per second in a 50 Hz system).

2. Only real calculations are needed; thus microprocessor applications using elementary codes are possible.

3. Waveform symmetries inherent in the operation of electric power systems make further computational burden reduction possible.

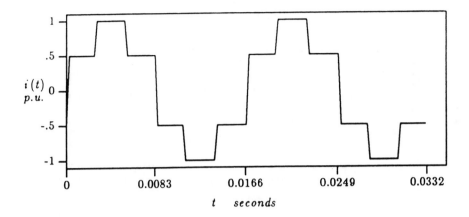

FIGURE 4.3
An ideal six-pulse rectifier load current, $i(t)$.

Of course, these advantages also apply to any electric circuit, but at least in (1) and (2), the electric power system application is a particularly appropriate application.

A brief example of the HS methodology is now presented. Consider a six-pulse rectifier that is injecting an ideal current waveform $i(t)$ into the system of Figure 4.9. This current is shown in Figure 4.3.

Working only in phase A (and using the time invariance property of the system to deduce the waveforms in the other two phases), and using $T_0 = 1$ as a normalized period, v_0 is 2π radians per second. Let the impedance characteristic of the network be that shown in Figure 4.4.

The one-sided FS of $i(t)$ is

$$i(t) = \sum_{\substack{k=6n\pm1 \\ n=\text{integer}}} b_k \sin(2\pi f k t)$$

$$b_{6n\pm1} = \frac{3}{(6n \pm 1)\pi}$$

where the sum in the equation above is carried over characteristic harmonics of order $6n \pm 1$, $n = 0, 1, 2, \ldots$, and the HS of $i(t)$ is

$$i(t) = \cdots - \frac{3}{7\pi} \operatorname{cas}(-14\pi t) - \frac{3}{5\pi} \operatorname{cas}(-10\pi t)$$

$$- \frac{3}{\pi} \operatorname{cas}(-2\pi t) + \frac{3}{\pi} \operatorname{cas}(2\pi t) + \frac{3}{5\pi} \operatorname{cas}(10\pi t) + \frac{3}{7\pi} \operatorname{cas}(14\pi t) + \cdots.$$

Note that $i(t)$ is odd (actually, odd half-wave symmetric) and has HS coefficients which behave like $\gamma_{-k} = -\gamma_k$. An important consequence of the fact that $i(t)$ is odd is that the convolution property in (4.6.22) collapses to one real multiplication,

$$\Gamma_i = \gamma_i Z_{18}(-iv_0).$$

In this example, the HS of $i(t)$ is calculated and used with samples of the transfer impedance $Z_{18}(v)$ to find the HS of $v_1(t)$. Here, $Z_{18}(v)$ is the HT of $z_{18}(t)$, the impulse response relating the current input at bus 8 to the voltage output at bus 1. The transfer impedance, $Z_{18}(v)$, is displayed in Figure 4.4 along with its Fourier analog $Z_{18}(\omega)$. The resulting nonsinusoidal

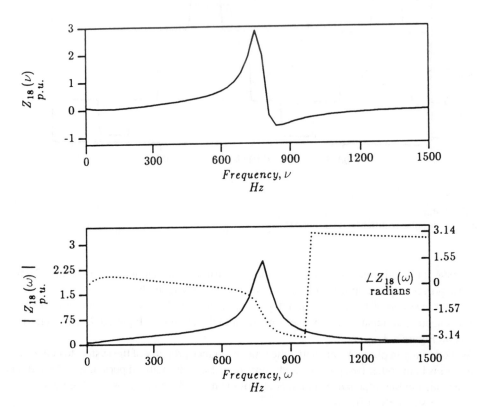

FIGURE 4.4
Transfer impedance between busses 1 and 8 in the Hartley and Fourier domains.

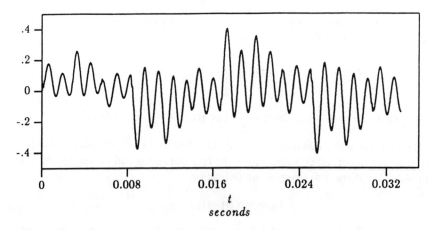

FIGURE 4.5
Substation nonsinusoidal bus voltage, $v_1(t)$, phase A to ground due to the current injection at bus 8.

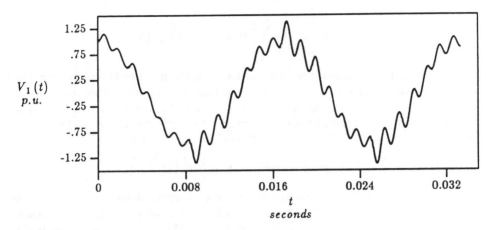

FIGURE 4.6
Superposition of the 60 Hz substation bus voltage and $v_1(t)$.

bus voltage, $v_1(t)$, produced by the rectifier load (i.e., only the component due to the rectifier; the 60 Hz component is excluded) is displayed in Figure 4.5. The superposition of this nonsinusoidal voltage on the 60 Hz substation bus voltage is depicted in Figure 4.6. Of course, only the phase A to neutral voltage (i.e., v_{an}) is calculated and the other phase voltages are deduced by shifting v_{an} by $\pm 1/3$ seconds.

4.7   Application of the Hartley transform via the fast Hartley transform

The discretized versions of the continuous Fourier and Hartley transform integrals may be put in an amenable form for digital computation. Consider the DFT and inverse DFT (IDFT) of a periodic function of period NT seconds,

$$F(k\Omega_\omega) = \sum_{n=0}^{N-1} f(nT)e^{-jk\Omega_\omega nT} \tag{4.7.1}$$

$$f(nT) = \frac{1}{N} \sum_{k=0}^{N-1} F(k\Omega_\omega)e^{jk\Omega_\omega nT} \tag{4.7.2}$$

where T is the sampling time resolution of the function $f(t)$, N is the number of points in the data sequence, and Ω_ω is the frequency resolution in radians per second,

$$\Omega_\omega = \frac{2\pi}{NT}.$$

Similarly, the discrete Hartley transform (DHT) is defined as

$$H(k\Omega_v) = \frac{1}{\sqrt{N}} \sum_{n=0}^{N-1} h(nT)\cos(k\Omega_v nT). \tag{4.7.3}$$

The inverse DHT (IDHT) is

$$h(nT) = \frac{1}{\sqrt{N}} \sum_{k=0}^{N-1} H(k\Omega_\nu) \cos(k\Omega_\nu nT) \qquad (4.7.4)$$

where $\Omega_\nu = \Omega_\omega$ is in radians per second. Note once again the ambiguity of the summation coefficients shown in these expressions. When working in radians per second, the product of the coefficients in the discrete case is $1/N$. The selection of $1/\sqrt{N}$ is made in the DHT and IDHT above so that the forward and inverse transforms satisfy the self-inverse property. As expected, the real-valued DHT is related to the DFT in a very simple way

$$H(k\Omega_\nu) = \Re\{F(k\Omega_\omega)\} - \Im\{F(k\Omega_\omega)\}. \qquad (4.7.5)$$

It should be noted that the the continuous-time Hartley integral, although necessary for theoretical development of the DHT, is typically not utilized in solving engineering problems. In fact, it is the existence of a fast Hartley transform (FHT) algorithm that has spurred the research and use of this transform for a large number of applications. In particular, the FHT is used in single- and multiple-dimension filtering applications where the computational burden is of great importance. To this end, the convolution property is of particular interest, and as such, will be discussed in some detail below.

The success of the DFT in the solution of engineering problems is largely due to the highly efficient algorithms that exist for evaluating both the DFT and IDFT. These are collectively known as the FFT algorithms. The FFT is an exact evaluation of the DFT. Numerous formulations exist when implementing the FFT; the most popular being either decimation in time or frequency and split-radix algorithms. At the crux of all FFT algorithms is the exploitation of the symmetry of $\exp\{-j[(2\pi kn)/N]\}$ in order to render the computational burden lower. The existence of these algorithms suggests that an analog exists for rapid calculation of the DHT. This is indeed the case.

At first glance, it seems strange that the N real values of the DHT can contain the same amount of information as the N complex values of the DFT, a total of $2N$ real numbers. Due to the Hermitian property of the DFT (i.e., $F(-k\Omega_\omega) = F^*(k\Omega_\omega)$), the DFT is redundant by a factor of two. The $N/2$ real numbers needed to specify the cosine transform and the $N/2$ needed for the sine transform combine to form a total of N DHT coefficients containing no degeneracy due to symmetry. This implies that the DFT, and therefore the FFT, is overspecified for performing linear filtering of real data. It should be noted that other efficient Fourier-based algorithms exist exclusively for real and real symmetric data. Other advantages of the DHT over the DFT are:

- the DHT avoids complex arithmetic;
- the DHT requires only half the memory storage for real data arrays versus complex data arrays;
- for a sequence of length N, the DHT performs $O(N \log_2 N)$ real operations versus the DFT $O(N \log_2 N)$ complex operations;
- the DHT performs fewer operations that may lead to fewer truncation and rounding errors from computer finite word length; and
- the DHT is its own inverse (i.e., it has a self-inverse).

For reasons of computational advantage either occurring through waveform symmetry or simply the use of only real quantities, the Hartley transform is recommended as a serious alternative to the Fourier transform for frequency-domain analysis. The salient disadvantage of the Hartley approach is that Fourier amplitude and phase information is not readily interpreted. This is

not a difficulty in many applications because this information is typically used as an intermediate stage toward a final goal. Where complex numbers are needed, they can be constructed
as a final step by (4.3.27) or (4.3.28).

Due to the cited advantages above, it is clear that the Hartley transform has much to offer
when engineering applications warrant digital filtering of real-valued signals. In particular,
the FHT should be used when either the computation time is to be minimized, for example, in
real-time signal processing. The minimization of computing time includes many other issues,
such as memory allocation, real versus complex variables, computing platforms, and so forth.
However, when one is interested in computing the Hartley transform or the convolution or correlation integral, the Hartley transform is the method of choice. In general, most engineering
applications based on the FFT can be reformulated in terms of the all-real FHT in order to
realize a computational advantage. This is due primarily to the vast amounts of research within
the past decade on FHT algorithm development as evidenced by the bibliography in [10]. A
voluminous number of applications exist for the Hartley transform [10], some of which are
listed below:

- fast convolution, correlation, interpolation, and extrapolation, finite-impulse response
  and multidimensional filter design;

- acoustics and speech processing, power spectrum and cepstrum analysis;

- numerical evaluation of the Laplace transform integral and Hilbert transformation;

- multidimensional optics and imaging compression applications; image reconstruction,
  decorrelation, coding, restoration, identification, lensless microwave imaging, pattern/
  image matching, feature extraction, and rough surface classification;

- power engineering applications, which include relaying, analysis of electrical transients,
  harmonic propagation in electric power systems, simulation of linear time-invariant electrical systems, and electromagnetic wave propagation in multiconductor transmission
  systems;

- biomedical applications, which include edge enhancement of digital radiographs, pattern recognition for real-time arrhythmias, autocorrelation to improve the 3-D plot of
  transcutaneous human electrogastrograms, functional angiographic images, electroen-
  cephalogram classification, and MRI applications;

- probability and number theory, which includes computing Wigner-Ville and pseudo-
  Wigner distributions, for example;

- artificial neural network applications, which include adaptive digital filtering, tracking
  problems, velocity estimation, and motion analysis;

- adaptive antenna arrays;

- geophysical applications;

- chemical applications; and

- astronomical applications.

Because many of the above applications involve the fast efficient evaluation of the convolution
integral, additional details are provided below.

4.7.1 *Convolution in the time and transform domains*

Transforms are a well-used mathematical tool in systems analysis. In engineering applications,
the one-sided Laplace transform (LT),

$$F(s) = \int_0^\infty f(t)e^{-st}\,dt, \tag{4.7.6}$$

$s = \alpha + j\omega$ and the two-sided Fourier transform (FT),

$$F(\omega) = \int_{-\infty}^{\infty} f(t)e^{-j\omega t}\,dt, \qquad (4.7.7)$$

are most widely used. Both the LT and FT have been widely used because of their special convolution property, which renders the convolution operation to a simple complex product in the transform domain.

Consider the two functions $h(t)$ and $x(t)$ and their transforms $H(\omega)$ and $X(\omega)$. Let $H(\omega)$ and $X(\omega)$ be multiplied to form the new function $Y(\omega)$ such that

$$Y(\omega) = H(\omega)X(\omega) \qquad (4.7.8)$$

and consider the inverse transform, $y(t)$. Here, $y(t)$ can be represented by the time-domain convolution integral

$$y(t) = \int_{-\infty}^{\infty} h(\xi)x(t - \xi)\,d\xi \qquad (4.7.9)$$

or

$$y(t) = h(t) \uplus x(t). \qquad (4.7.10)$$

The above convolution integral can be represented in a form convenient for digital implementation as follows:

$$y(n) = T \sum_{j=0}^{N-1} h(j)x(n - j)$$

$$y(n) = T \sum_{j=0}^{N-1} x(j)h(n - j) \qquad (4.7.11)$$

that is, as a sum of lagged products. The time required to compute $y(n)$ from either form in (4.7.11) is proportional to approximately N^2. If one computes the transforms of x and h, performs the complex multiplication of (4.7.8), and then computes the inverse transform of $Y(\omega)$, one requires a time proportional to $N \log_2 N$ if the FFT is utilized. The approximate ratio of computing the convolution in (4.7.8) to that of (4.7.11) is given by

$$\frac{N \log_2 N}{N^2} = \frac{\log_2 N}{N} = \frac{\rho}{N} \qquad (4.7.12)$$

where $N = 2^\rho$. For example, if $N = 2^{10}$, the FFT requires less than 1% of the normal computing time. Timing studies have shown that for N greater then about 28, the FFT method is at least an order of magnitude faster than the lagged products approach of (4.7.11).

The convolution property in the transform domain is of great interest. The familiar convolution property of the DFT has an analog in the Hartley space that suggests potential use of the DHT for the numerical solution of electric circuits problems. Many electric circuits problems involve the notion of convolution. A typical format of such problems is that an impedance is known, and a current associated with that impedance results in a voltage. In terms of the one-sided Laplace transform,

$$V(s) = Z(s)I(s). \qquad (4.7.13)$$

The impedance Z may be an open circuit driving point or a transfer impedance. Other problems are frequently encountered in electric circuit analysis, and many of these are of the same form as (4.7.13) with $Z(s)$ replaced by a transfer function, frequency response matrix or similar parameter. This is the case, for example, in the presence of power electronic loads and sources characterized by nonsinusoidal waveforms. "Quasi-periodic" transient inputs energizing a relaxed electric power system (i.e., zero initial conditions) at time 0^-, produce responses throughout the system that may be superimposed upon the sinusoidal steady-state solution. In the time domain, the impedance $Z(\omega)$ is, in fact, an impulse response, $z(t)$. In this context, $z(t)$ is the voltage response to an input that is a unit current impulse. The responses to these quasi-periodic inputs are found analytically via the convolution integral

$$v(t) = \int_{-\infty}^{\infty} z(t-\xi)i(\xi)\,d\xi = z(t) \uplus i(t) \tag{4.7.14}$$

and for causal systems

$$v(t) = \int_{-\infty}^{t} z(t-\xi)i(\xi)\,d\xi \tag{4.7.15}$$

where $z(\cdot)$ is the impulse response satisfying $z(t-\xi) = 0$ for $\xi > t$, that is, $z(t) = 0$ for $t < 0$. In (4.7.10) and (4.7.14), ($\uplus$) denotes conventional or linear convolution. The limits of integration may be changed to $[0, t]$ if $i(t)$ is a causal signal, that is, if $i(t) = 0$ for $t < 0$. Convolution in the time domain becomes a simple complex multiplication in the s-domain; this property makes the Laplace transform particularly attractive for systems analysis.

The familiar Fourier transform also possesses a similar convolution property

$$V(\omega) = \frac{1}{\sqrt{2\pi}} \int_{-\infty}^{\infty} v(t)e^{-j\omega t}\,dt \tag{4.7.16}$$

$$I(\omega) = \frac{1}{\sqrt{2\pi}} \int_{-\infty}^{\infty} i(t)e^{-j\omega t}\,dt \tag{4.7.17}$$

$$Z(\omega) = \frac{1}{\sqrt{2\pi}} \int_{-\infty}^{\infty} z(t)e^{-j\omega t}\,dt \tag{4.7.18}$$

$$F\{v(t)\} = F\{z(t) \uplus i(t)\} = V(\omega) = Z(\omega)I(\omega). \tag{4.7.19}$$

Note that in (4.7.16)–(4.7.19) above, the factor $1/\sqrt{2\pi}$ is often omitted in engineering work; when this factor is included in the transform, the inverse transform is

$$v(t) = \frac{1}{\sqrt{2\pi}} \int_{-\infty}^{\infty} V(\omega)e^{j\omega t}\,d\omega. \tag{4.7.20}$$

Analogously, a salient property of the Hartley transform for this application is that convolution is rendered to a simple sum of real products under the transform,

$$H\{v(t)\} = H\{z(t) \uplus i(t)\} = V(\nu). \tag{4.7.21}$$

Specifically,

$$V(\nu) = \frac{1}{2}[Z(\nu)I(\nu) + Z(-\nu)I(\nu) + Z(\nu)I(-\nu) - Z(-\nu)I(-\nu)] \tag{4.7.22}$$

$$= Z(\nu)\frac{[I(\nu)+I(-\nu)]}{2} + Z(-\nu)\frac{[I(\nu)-I(-\nu)]}{2} \tag{4.7.23}$$

$$= Z(\nu)I^e(\nu) + Z(-\nu)I^o(\nu) \tag{4.7.24}$$

$$= \frac{1}{2}[V_a(\nu) - V_a(-\nu) + V_b(\nu) + V_b(-\nu)] \tag{4.7.25}$$

where $V_a(v) = Z(v)I(v)$ and $V_b(v) = Z(v)I(-v)$. Thus, it is possible to solve a certain class of electric circuit problems using the Hartley transform.

As with the DFT/FFT, the DHT/FHT can be readily used for performing convolution. The DHT assumes periodicity of the function being transformed, that is, $H(k\Omega_v) = H[(N+k)\Omega_v]$. Therefore, $H(-k\Omega_v)$ for $-N \le k \le -1$, is equivalent to $H[(N-k)\Omega_v]$. When convolution is represented by a $\uplus$, linear or time-domain convolution is implied. In the frequency domain, as a result of the characteristic modulo N operations inherent in the DFT or DHT, a different form of convolution results in the time domain. Circular or cyclic convolution, denoted by ($\oplus$), in the time domain is the result of multiplication of two functions in the frequency domain. Let n represent the nth point of some finite duration sequence, then cyclic convolution in the time domain is expressed as

$$f_1(n) \oplus f_2(n) = \sum_{\tau=0}^{N-1} f_1(\underset{N}{\mod}(\tau)) f_2(\underset{N}{\mod}(n-\tau)) \qquad (4.7.26)$$

where τ and $n - \tau$ are depicted modulo N. The equivalent form of (4.7.22)–(4.7.25) in the Hartley domain is expressed as

$$V(k\Omega_v) = \frac{1}{2} [Z(k\Omega_v)I(k\Omega_v) + Z((N-k)\Omega_v)I(k\Omega_v)$$

$$+ Z(k\Omega_v)I((N-k)\Omega_v) - Z((N-k)\Omega_v)I((N-k)\Omega_v)] \qquad (4.8.27)$$

$$= Z(k\Omega_v)I^e(k\Omega_v) + Z((N-k)\Omega_v)I^o(k\Omega_v) \qquad (4.8.28)$$

$$= \frac{1}{2} [V_a(k\Omega_v) - V_a((N-k)\Omega_v) + V_b(k\Omega_v) + V_b((N-k)\Omega_v)] \qquad (4.8.29)$$

where $V_a(k\Omega_v) = Z(k\Omega_v)I(k\Omega_v)$ and $V_b(k\Omega_v) = Z(k\Omega_v)I((N-k)\Omega_v)$.

There are times when cyclic convolution is desired and other times when linear convolution is needed. Because both the DFT and DHT perform cyclic convolution, it would be unfortunate if methods for obtaining linear convolution by cyclic convolution were nonexistent. Fortunately, this is not the case. Linear convolution can be extracted from cyclic convolution, but at some expense. For finite duration sequences $f_1(n)$ and $f_2(n)$ of length M and L, respectively, their convolution is also finite in duration. In fact, the duration is $M + L - 1$. Therefore, a DFT or DHT of size $N \ge M + L - 1$ is required to represent the output sequence in the frequency domain without overlap. This implies that the N-point circular convolution of $f_1(n)$ and $f_2(n)$ must be equivalent to the linear convolution of $f_1(n)$ and $f_2(n)$. By increasing the length of both sequences to N points (i.e., by appending zeros), and then circularly convolving the resulting sequences, the end result is as if the two sequences were linearly convolved. Clearly with zero padding, the DHT can be used to perform linear filtering. It should be clear that aliasing results in the time domain if $N < M + L - 1$.

When N zero values are appended to a time sequence of N data samples, the $2N$-point DHT reduces to that of the N-point DHT at the even index values. The odd values of the $2N$ sequence represent the interpolated DHT values between the original N-point DHT values. The more zeros padded to the original N-point DHT, the more interpolation takes place on the sequence. In the limit, infinite zero padding may be viewed as taking the discrete-time Hartley transform of an N-point windowed data sequence. The prevalent misconception that zero padding improves the resolution of the sequence or additional information is obtained is well known. Zero padding does not increase the resolution of the transform made from a give finite sequence but simply provides an interpolated transform with a smoother appearance. The advantage of zero padding is that signal components with center frequencies that lie between

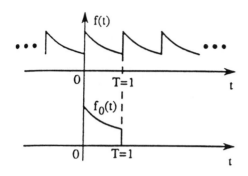
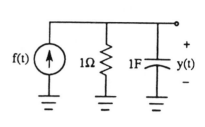

FIGURE 4.7
Injected load current into a simple RC network.

the N frequency bins of an unpadded DHT can now be discerned. Thus, the accuracy of estimating the frequency of spectral peaks is also enhanced with zero padding.

When comparing the number of real operations performed by (4.7.8) (with ω replaced by $k\Omega_\omega$) and (4.7.28) or (4.7.29), the DHT always offers a computational advantage of two as compared to the DFT method; in many (if not most) applications, currents in electrical engineering calculations exhibit symmetry, which results in a computational advantage of four favoring the Hartley method. In the case where $z(t)$ or $i(t)$ in (4.7.21) contains even symmetry, the four-term product of (4.7.27) reduces to $Z(k\Omega_v)I(k\Omega_v)$ or $V_a(k\Omega_v)$. If $z(t)$ or $i(t)$ is odd, then (4.7.27) degenerates to $Z(k\Omega_v)I((N-k)\Omega_v)$ or $V_b(k\Omega_v)$ and $Z((N-k)\Omega_v)I(k\Omega_v)$, respectively. That is, only one, versus the FFT's four real multiplications, is needed. The above symmetry conditions are more often the rule than the exception. Other symmetries exist as discussed by Bracewell for the Hartley transform [8].

As a brief example of the method, consider the periodic load current shown in Figure 4.7 having the description

$$f(t) = \sum_{m=-\infty}^{\infty} f_m(t) \tag{4.7.30}$$

where

$$f_0(t) = f(t)[u(t) - u(t-T)] = ee^{-t}[u(t) - u(t-1)] \tag{4.7.31}$$

$$f_m(t) = f_0(t - mT) \tag{4.7.32}$$

and $T = 1$. The transfer function, $H(s)$, for the RC network in Figure 4.7 is clearly $1/(s+1)$. (Note the system is initially relaxed—zero initial conditions.) If one denotes $y_m(t)$ as the zero-state response due to $f_m(t)$, then from the time shift property of a linear time-invariant system, $y_m(t) = y_0(t - mT)$. From the principle of superposition, $y(t) = \sum y_m(t)$. Thus, the crux of the problem is to find $y_0(t)$, the response for $0 \leq t < \infty$ due to the single pulse, $f_0(t)$, in Figure 4.7.

The convolution of $f_0(t)$ and the impulse response, $h(t)$, by the convolution integral is straightforward. In fact the response, $y_0(t)$, depicted in Figure 4.8, is readily calculated as

$$y_0(t) = \begin{cases} t\,e\,e^{-t}, & \text{if } 0 \leq t \leq 1 \\ e\,e^{-t}, & \text{if } t > 1 \end{cases} \tag{4.7.33}$$

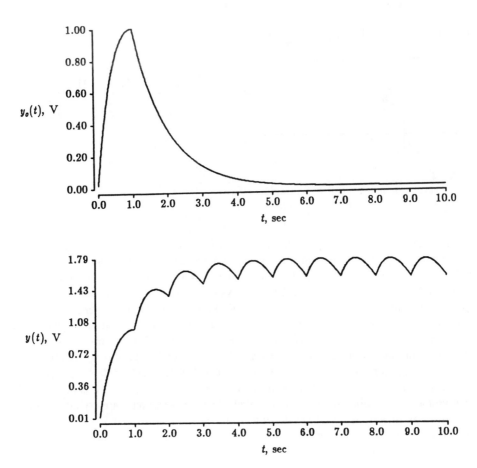

FIGURE 4.8
Output response (top) $y_0(t)$ to the input pulse $f_0(t)$ and (bottom) $y(t)$ to the input
$f(t) = \sum_{m=0}^{9} f_m(t) = f_0(t-m)$.

or alternatively by

$$y_0(t) = L^{-1}\{Y_0(s)\} = L^{-1}\left\{\frac{e-e^{-s}}{(s+1)^2}\right\} \tag{4.7.34}$$

where $Y_0(s) = F_0(s)H(s)$ and $F_0(s) = L\{f_0(t)\}$. The calculation of the steady-state system response to the input $f(t)$ in Figure 4.7 is more interesting. It is well known that for periodic $y(t)$,

$$y(t) = L^{-1}\left\{\frac{Y_0(s)}{1-e^{-sT}}\right\} = L^{-1}\{Y(s)\} \tag{4.8.35}$$

when $Y_0(s)$ represents the Laplace transform of any one period of $y(t)$ (i.e., $y_0(t)$). In general, the partial fraction expansion of $Y(s)$ is a nontrivial computation; how then does one solve for the response $y(t)$? Utilizing the assumed periodicity of the DHT (FHT), one can perform conventional convolution via circular convolution if provisions are made for aliasing (i.e., zero padding). This method of solution can be summarized by

$$y(t) = f(t) \oplus h(t) = \text{DHT}\{\text{DHT}\{f(t)\} \times \text{DHT}\{h(t)\}\} \qquad (4.8.36)$$

where $y(t)$ is shown in Figure 4.8. An FHT software program to compute the DHT efficiently can be found in the Appendix. Additional details concerning circular convolution of aperiodic inputs are discussed below.

4.7.2 An illustrative example

In this section, an illustrative example is presented from subtransmission and distribution engineering to illustrate the calculation of nonsinusoidal waveform propagation in an electric power system. Figure 4.9 displays the distribution network and the injected nonlinear load current into the network. The electrical load at bus 8 causes a nonsinusoidal current to propagate throughout the system and impact other loads in an unknown fashion.

The fast decay in this current results in high frequency signals in the network. An important consideration in this method is the selection of the sampling interval T and its effect on the maximum frequency component, $\Omega_{v,\max} = \pi/T$, represented in the simulation. Because power systems are essentially bandlimited due necessarily to system components designed to operate at or near the power frequency (e.g., distribution transformers), the nonlinear load current containing frequency components above $\Omega_{v,max}$ become negligible. That is, no matter how close the current approaches an impulse (e.g., a lightning strike), the significant energy components above $\Omega_{v,max}$ are multiplied in the transform domain by the system impedance frequency components that are asymptotically approaching zero. This can be seen by observing the Fourier magnitude, $|I_8(k\Omega_\omega)|$ and $|Z_{18}(k\Omega_\omega)|$, in Figures 4.10 and 4.11 for selected values of N (and thus T). The Hartley transform of $i(t)$, $I_8(k\Omega_v)$, is shown in Figure 4.12.

Load currents that decay rapidly are becoming less unusual with the advent of high-power semiconductor switches.

Referring to the system in Figure 4.9, the transformer at the load bus is modeled as a conventional T-equivalent. A lumped capacitance is used to model electrostatic coupling between the primary and secondary windings, and two lumped capacitances are used to model interwinding capacitance. Bus 1 is the substation bus and the negative-sequence impedance equivalent tie to the remainder of the network is shown as a shunt R–L series branch. The circuits shown between busses are all three-phase balanced, fixed series R–L branches, and frequency independent (i.e., $R = R(\omega)$). The latter assumption need not be made because frequency dependence may be included if required. The importance of frequency dependence should not be underestimated, particularly for cases in which significant energy components of the injection current spectrum lie above and beyond the 17th harmonic of 60 Hz, or approximately 1 kHz. Distributed parameter models can be readily represented as lumped parameters placed at the terminals of long lines. These refinements are quite important in actual applications, but they are omitted from this abbreviated example. If the injection current at bus 8 were "in phase" with the line to neutral voltage at that bus, the nonlinear device at bus 8 would be a source. Similarly, other phase values would result in different generation or load levels.

Each bus voltage was calculated using the Hartley transform simulation algorithm. These results were verified using an Euler predictor-trapezoidal corrector integration algorithm and time domain convolution implemented by (4.7.11). In order to choose an adequate time step, T, for calculating the "theoretical solution" by the predictor-corrector method, it was necessary to capture all system modes. The eigenvalues calculated by the International Mathematics and Statistical Library (IMSL) subroutine EVLRG are shown in Table 4.9. Routine EVLRG computes the eigenvalues of a real matrix by first balancing the matrix; second, orthogonal similarity transformations are used to reduce this balanced matrix to a real upper Hessenberg

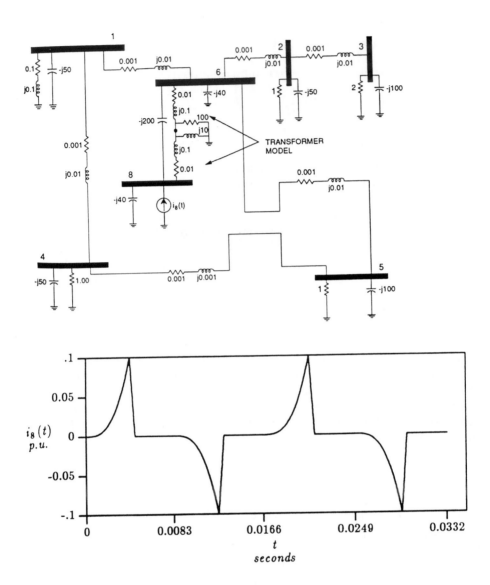

FIGURE 4.9
Load current injected at bus 8 of an example 8-bus distribution system.

matrix; third, the shifted QR algorithm is used to compute the eigenvalues of the Hessenberg matrix. This method is generally accepted as being most reliable.

In this example, the transfer impedance between the substation bus (bus 1) and bus 8 is of interest. Figures 4.13 and 4.14 display the DFT of $z_{18}(t)$, $Z_{18}(k\Omega_\omega)$. Of course, two graphs are required to illustrate this transfer impedance because the DFT is a complex transformation. Figure 4.15 shows the DHT of $z_{18}(t)$, $Z_{18}(k\Omega_v)$. One figure illustrates this real transform.

The resulting bus voltages due to the current injection at bus 8 are depicted in Figures 4.16 and 4.17.

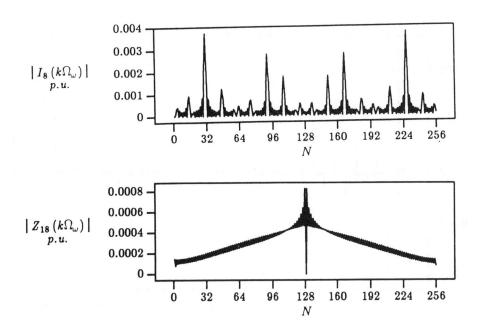

FIGURE 4.10
Fourier magnitude of the injected bus current, $i(t)$, and system impulse response, $z_{18}(t)$ for $N = 256$.

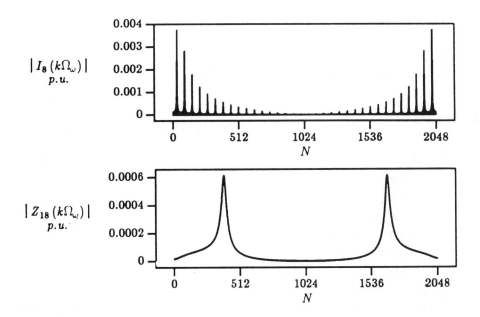

FIGURE 4.11
Fourier magnitude of the injected bus current, $i(t)$, and system impulse response, $z_{18}(t)$ for $N = 2048$.

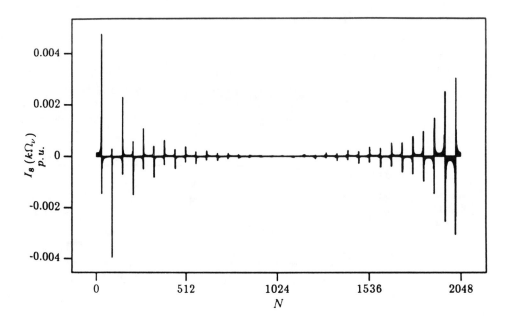

FIGURE 4.12
Hartley transform of the injected bus current, $i(t)$.

TABLE 4.9
Calculated eigenvalues for the example 8-bus power system.

λ_i	$\Re\{\lambda_i\}$	$\Im\{\lambda_i\}$
λ_1	−757 718	0
$\lambda_{2,3}$	− 13 548	±52 901
λ_4	− 11 988	0
$\lambda_{5,6}$	− 8 976	±19 575
$\lambda_{7,8}$	− 8 690	±50 131
$\lambda_{9,10}$	− 5 033	±38 050
$\lambda_{11,12}$	− 4 245	±43 600
λ_{13}	− 1 523	0
$\lambda_{14,15}$	−214	± 4 877
λ_{16}	−40	0
λ_{17}	−3	0

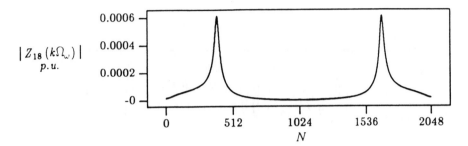

FIGURE 4.13
Fourier magnitude of the transfer impedance, $Z_{18}(k\Omega_\omega)$.

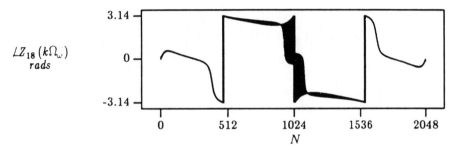

FIGURE 4.14
Fourier phase of the transfer impedance, $Z_{18}(k\Omega_\omega)$.

4.7.3 *Solution method for transient or aperiodic excitations*

Convolution of two finite duration waveforms is straightforward. One simply samples the two functions every T seconds and assumes that both sampled functions are periodic with period N. If the period is chosen according to that discussed earlier, there is no overlap in the resulting convolution. As long as N is chosen correctly, discrete convolution results in a periodic function where each period approximates the continuous convolution results of (4.7.14).

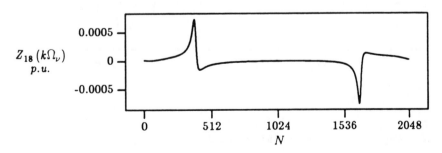

FIGURE 4.15
Hartley transform of the transfer impedance, $Z_{18}(k\Omega_\nu)$.

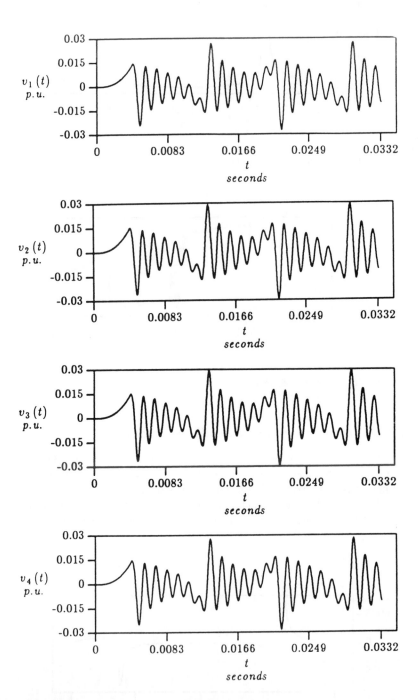

FIGURE 4.16

Resulting bus voltages due to the current injection at bus 8.

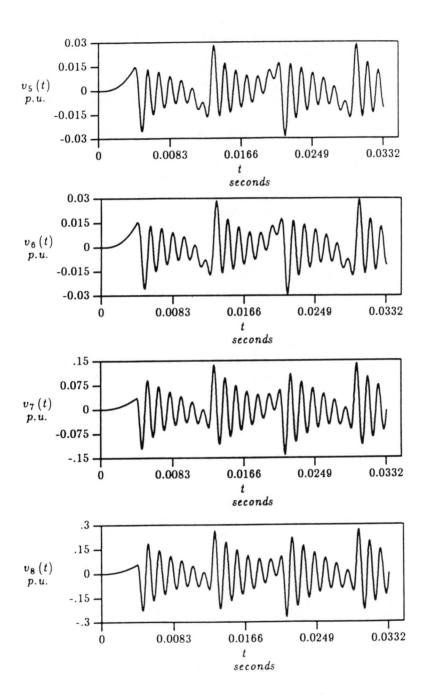

FIGURE 4.17
Resulting bus voltages due to the current injection at bus 8, continued.

The algorithm implemented by (4.7.36) assumed that the input is time limited and the system impulse response is band limited. That is, the periodic input is truncated to an integer multiple of its fundamental frequency and the system impulse response is of infinite duration. For stable systems, the system impulse response $z(t)$ must decrease to zero or to negligible values for large $|t|$. In reality, the system impulse response cannot be both time limited and band limited; therefore, one band limits in the frequency domain such that negligible signal energy exist for $t \geq T_0$.

The convolution of an aperiodic excitation with the system impulse response can be regarded as a periodic convolution of functions having an equal period. Through suitable modifications to the method presented in (4.7.36), one can use circular convolution to compute an aperiodic convolution when each function is zero everywhere outside some single time window of interest.

Let the functions $x(t)$ and $h(t)$ be convolved where both functions are finite in length. Let the larger sequence, $x(t)$, contain L discrete points and the smaller contain M discrete points. Then the resulting convolution of these functions can be obtained by circularly or cyclically convolving suitable zero-augmented functions. That is,

$$
x_{\text{pad}}(n) = \begin{cases} x(n + n_0), & \text{if } 1 \leq n \leq L \\ 0, & \text{if } L + 1 \leq n \leq N \\ x_{pad}(n + nN), & \text{otherwise} \end{cases} \qquad (4.7.37)
$$

where n_0 is the first point in the function window of interest and N is the smallest power of two greater than or equal to $M + L - 1$. Similarly for $h(t)$, simply replace x with h and L by M in (4.7.37). If one allows these zero-augmented functions to be periodic of period N, then the intervals of padded zeros disallow the two functions to be overlapped even though the convolution is a circular one. These periodic functions are formed by the superposition of the nonperiodic function shifted by all multiples of the fundamental period, T_0, where $T_0 = NT$. That is,

$$
f_p(t) = \sum_{k=-\infty}^{\infty} f(t + kT_0). \qquad (4.7.38)
$$

Thus, while the result is a periodic function (i.e., due to the assumed periodicity of the DHT/FHT), each period is an exact replica of the desired aperiodic convolution.

The relationship between the DHT and HT for finite duration waveforms is different when the input $i(t)$ is time limited. Because $i(t)$ is time limited, its Hartley transform cannot be band limited; therefore, sampling this function leads to aliasing in the frequency domain. It is necessary to choose the sampling interval T to be sufficiently small such that aliasing is reduced to an insignificant level.

If the number of samples of the time-limited waveform is chosen as N, then it is not necessary to window in the time domain. For this set of waveforms, the only error introduced is aliasing. Errors introduced by aliasing can be reduced by choosing T sufficiently small. This allows the DHT sample values to agree reasonably well with samples of the HT.

4.8 Table of Hartley transforms

Tables 4.10 and 4.11 contain the Hartley transforms of commonly encountered signals in engineering applications. When scanning the table entries, the Hartley transform entries seem to have more sophisticated expressions; this is usually the case. More exotic Hartley transforms may be generated in one of three ways. First, one can apply the elementary properties provided

TABLE 4.10
Hartley transforms of energy signals.

Description	$f(t)$	$F(f)$	$H(f)$
Rectangular pulse	$u(t + \frac{T}{2}) - u(t - \frac{T}{2})$	$T\frac{\sin \pi Tf}{\pi Tf} = T \operatorname{sinc} Tf$	Because $f(t)$ is even, $H(f) = F(f)$
Exponential	$\beta e^{-\alpha t} u(t)$	$\frac{\beta}{j\omega + \alpha}$	$\frac{\beta(\alpha + 2\pi f)}{\alpha^2 + (2\pi f)^2}$
Triangular	$1 - 2\frac{\|t\|}{T}, \|t\| < \frac{T}{2}$	$\frac{T}{2}\operatorname{sinc}^2(\frac{Tf}{2})$ $= \frac{1 - \cos \pi fT}{T\pi^2 f^2}$	Because $f(t)$ is even, $H(f) = F(f)$
Gaussian	$e^{-\alpha^2 t^2}$	$\frac{\sqrt{\pi}}{\alpha} e^{-(\pi^2 f^2/\alpha^2)}$	Because $f(t)$ is even, $H(f) = F(f)$
Double exp	$e^{-\alpha\|t\|}$	$\frac{2\alpha}{\alpha^2 + 4\pi^2 f^2}$	Because $f(t)$ is even, $H(f) = F(f)$
Damped sine	$e^{-\alpha t}\sin(\omega_0 t) u(t)$	$\frac{\omega_0}{(\alpha + j2\pi f)^2 + \omega_0^2}$	$\frac{\omega_0(\alpha^2 + \omega_0^2 - 4\pi^2 f^2 + 4\pi f\alpha)}{(\alpha^2 + \omega_0^2 - 4\pi^2 f^2)^2 + (4\pi f\alpha)^2}$
Damped cosine	$e^{-\alpha t}\cos(\omega_0 t) u(t)$	$\frac{\alpha + j2\pi f}{(\alpha + j2\pi f)^2 + \omega_0^2}$	$\frac{(\alpha - 2\pi f)(\alpha^2 + \omega_0^2 - 4\pi^2 f^2) + (4\pi f\alpha)(\alpha + 2\pi f)}{(\alpha^2 + \omega_0^2 - 4\pi^2 f^2)^2 + (4\pi f\alpha)^2}$
One-sided exp	$\frac{1}{\beta - \alpha}(e^{-\alpha t} - e^{-\beta t})u(t)$	$\frac{1}{(\alpha + j2\pi f)(\beta + j2\pi f)}$	$\frac{\alpha\beta - 2\pi f(\alpha + \beta + 2\pi f)}{[\alpha\beta - (2\pi f)^2]^2 + [2\pi f(\alpha + \beta)]^2}$
Cosine pulse	$\cos \omega_0 t \left[u\left(t + \frac{T}{2}\right) - u\left(t - \frac{T}{2}\right)\right]$	$\frac{T}{2}\left[\frac{\sin \pi T(f - f_0)}{\pi T(f - f_0)} + \frac{\sin \pi T(f + f_0)}{\pi T(f + f_0)}\right]$	Because $f(t)$ is even, $H(f) = F(f)$

TABLE 4.11
Hartley transforms of power signals.

Description	$f(t)$	$F(f)$	$H(f)$
Impulse	$K\delta(t)$	K	K
Constant	K	$K\delta(f)$	$K\delta(f)$
Unit step	$u(t)$	$\frac{1}{2}\delta(f) + \frac{1}{j2\pi f}$	$\frac{1}{2}\delta(f) + \frac{1}{2\pi f}$
Signum function	$\operatorname{sgn} t = \frac{t}{\|t\|}$	$\frac{1}{j\pi f}$	$\frac{1}{\pi f}$
Cosine wave	$\cos \omega_0 t$	$\frac{1}{2}[\delta(f - f_0) + \delta(f + f_0)]$	Because $f(t)$ is even, $H(f) = F(f)$
Sine wave	$\sin \omega_0 t$	$\frac{-j}{2}[\delta(f - f_0) - \delta(f + f_0)]$	$\frac{1}{2}[\delta(f - f_0) - \delta(f + f_0)]$
Impulse train	$\sum_{-\infty}^{\infty} \delta(t - nT)$	$\frac{1}{T}\sum_{-\infty}^{\infty} \delta(f - \frac{n}{T})$	Because $f(t)$ is even, $H(f) = F(f)$
Periodic wave	$\sum_{-\infty}^{\infty} \alpha_n e^{jn2\pi f_0 t}$	$\sum_{-\infty}^{\infty} \alpha_n \delta\left(f - \frac{n}{T}\right)$	$\sum_{-\infty}^{\infty} \gamma_n \delta\left(f - \frac{n}{T}\right)$
Complex sinusoid	$A e^{j\omega_0 t}$	$A\delta(f - f_0)$	$H(f) = F(f)$
Unit ramp	$tu(t)$	$\frac{j}{4\pi}\delta'(f) - \frac{1}{4\pi^2 f^2}$	$\frac{-1}{4\pi}\delta'(f) - \frac{1}{4\pi^2 f^2}$

in Section 4.4 to the entries of Tables 4.10 and 4.11; second, one can alternatively apply (4.3.23) to the Fourier transform entries of more comprehensive table listings such as those found in [11, 12, 13]; or third, use a DHT or FHT algorithm to evaluate numerically the Hartley transform of a discrete-time signal generated using a high-level computing language (e.g., FORTRAN, C, C++, etc.). A sample FHT algorithm, coded in the C programming language, is included in the Appendix.

Note that in the eighth entry of Table 4.11, α_n is a complex number representing the Fourier series expansion of the arbitrary periodic function. The value α_n is also equal to $1/T\, F_T(n/T)$ where $F_T(f)$ is the Fourier transform of $F(f)$ over a single period evaluated at n/T. Also, in that same entry, note that $\gamma_n = \Re\{\alpha_n\} - \Im\{\alpha_n\}$.

Appendix: A sample FHT program

```
/* Program FHT.C ******************************************************
/*
/* This FHT algorithm utilizes an efficient permutation algorithm
/* developed by David M. W. Evans.  Additional details may be found
/* in: IEEE Transaction on Acoustics, Speech, and Signal Processing,
/* vol. ASSP-35, n. 8, pp. 1120-1125, August 1987.
/*
/* This FHT algorithm, authored by Lakshmikantha S. Prabhu, is
/* optimized for the SPARC RISC platform.  Additional details may
/* be found in his M.S.E.E. thesis referenced below.
/*
/* L. S. Prabhu, "A Complexity-Based Timing Analysis of Fast
/* Real Transform Algorithms," Master's Thesis, University of
/* Arkansas, Fayetteville, AR, 72701-1201, 1993.
/*********************************************************************

/* This program assumes a maximum array length of 2^M = N where    */
/* M=9 and N=512.                                                   */
/* See Line 52 if the array length is increased.                   */

# include <stdio.h>
# include <math.h>
# define M   3
# define N   8
float* myFht();
main()
{
/* Read the integer values 1,...,N into the vector X[N].           */

        int i;
        float X[N];
        for( i = 0 ;i < N; i++ )
          X[i] = i+1;
        for( i = 0; i < N; i++ )
          printf("%f\n", X[i]);
          myFht(X,N,M);
          printf("\n");
        for ( i = 0; i < N; i++ )
```

```
            printf("%d: %f\n", i,X[i]/N);

/* It is assumed that the user divides by the integer N.              */
}
float*
myFht(x,n,m)
float* x;
int n,m;
{
int i,j,k,kk,l,l0,l1,l2,l3,l4,l5,m1,n1,n2,NN,s;
int diff = 0,diff2,gamma,gamma2=2,n2_2,n2_4,n_2,n_4,n_8,n_16;
int itemp,ntemp,phi,theta_by_2;
float ee,temp1,temp2,xtemp1,xtemp2;
float h_sec_b,x0,x1,x2,x3,x4,x5,xtemp;
double cc1,cc2,ss1,ss2;
double sine[257];
/*******************************************************************/
/* Digit reverse counter.                                          */
/*******************************************************************/
int powers_of_2[16],seed[256];
int firstj,log2_n, log2_seed_size;
int group_no,nn,offset;

log2_n = m >> 1;
nn = 2<< (log2_n - 1);
if( (m % 2) == 1 )
  log2_n = log2_n + 1;
  seed[0] = 0; seed[1] = 1;
  for(log2_seed_size = 2; log2_seed_size <= log2_n; log2_seed_size++)
    {
    for( i = 0; i < 2 <<(log2_seed_size - 2); i++)
      {
      seed[i] = 2 * seed[i];
      for(k = 1; k < 2; k++)
        seed[ i + k * (2 << (log2_seed_size - 1)>>1) ] = seed[i];
      }
    }
  for(offset = 1; offset < nn; offset++)
    {
    firstj = nn * seed[offset];
    i = offset; j = firstj;
    xtemp  = x[i];
    x[i] = x[j];
    x[j] = xtemp;
    for( group_no = 1; group_no < seed[offset]; group_no++)
      {
      i = i + nn; j = firstj + seed[group_no];
      xtemp = x[i];
      x[i] = x[j];
      x[j] = xtemp;
      }
    }
  j = 0;
  n1 = n - 1;
  n_16  = n>> 4;
```

```
  n`8 = n >> 3;
  n`4 = n >> 2;
  n`2 = n >> 1;
/**************************************************************************/
/* Start the transform computation with 2-point butterflies.        */
/**************************************************************************/
  for(i = 0;i < n; i += 2)
    {
    s = i+1;
    xtemp = x[i];
    x[i] += x[s];
    x[s] = xtemp - x[s];
    }
/**************************************************************************/
/* Now, the 4-point butterflies.                                    */
/**************************************************************************/
  for( i = 0; i < N; i += 4)
    {
    xtemp = x[i];
    x[i]  +=  x[i+2];
    x[i+2] = xtemp - x[i+2];
    xtemp = x[i+1];
    x[i+1]  += x[i+3];
    x[i+3] = xtemp - x[i+3];
    }
/**************************************************************************/
/* Sine table initialization.                                       */
/**************************************************************************/
  NN = n`4;
  sine[0] = 0;
  sine[n`16] = 0.382683432;
  sine[n`8] = 0.707106781;
  sine[3*n`16] = 0.923879533;
  sine[n`4] = 1.000000000;
  h`sec`b = 0.509795579;
  diff = n`16;
  theta`by`2 = n`4 >> 3;
  j = 0;
  while(theta`by`2 >= 1)
    {
    for( i = 0; i <= n`4; i += diff)
      {
      sine[j + theta`by`2] = h`sec`b * (sine[j] + sine[j + diff] );
      j = j + diff;
      }
    j = 0;
    diff = diff >> 1;
    theta`by`2 = theta`by`2 >>  1;
    h`sec`b = 1 / sqrt(2 + 1/h`sec`b);
    }
/**************************************************************************/
/* Other butterflies.                                               */
/**************************************************************************/
  for( i = 3; i <= m; i++ )
    {
```

```
      diff = 1; gamma = 0;
      ntemp = 0; phi = 2 << (m-i) >> 1;
      ss1 = sine[phi];
      cc1 = sine[n`4 - phi];
      n2 = 2 << (i-1);
      n2`2 = n2 >> 1;
      n2`4 = n2 >> 2;
      gamma2 = n2`4;
      diff2 = gamma2 + gamma2 - 1;
      itemp = n2`4;
      k=0;
/*********************************************************************/
/* Initial section of stages 3,4,...for which sines & cosines are   */
/* not required.                                                    */
/*********************************************************************/
      for(k = 0;  k < (2 << (m-i)>>1); k++)
        {
        l0 = gamma;
        l1 = l0 + n2`2;
        l3 = gamma2;
        l4 = gamma2 + n2`2;
        l5 = l1 + itemp;
        x0 = x[l0];
        x1 = x[l1];
        x3 = x[l3];
        x5 = x[l5];
        x[l0] = x0 + x1;
        x[l1] = x0 - x1;
        x[l3] = x3 + x5;
        x[l4] = x3 - x5;
        gamma = gamma + n2;
        gamma2 = gamma2 + n2;
        }
      gamma = diff;
      gamma2 = diff2;
/*********************************************************************/
/* Next sections of stages 3,4,...                                 */
/*********************************************************************/
      for( j = 1; j < 2 << (i-3); j++ )
        {
        for( k = 0; k < (2 << (m-i) >> 1); k++)
          {
          l0 = gamma;
          l1 = l0 + n2`2;
          l3 = gamma2;
          l4 = l3 + n2`2;
          x0 = x[l0];
          x1 = x[l1];
          x3 = x[l3];
          x4 = x[l4];
          x[l0] = x0 +  x1 * cc1 + x4 * ss1;
          x[l1] = x0 -  x1 * cc1 - x4 * ss1;
          x[l3] = x3 -  x4 * cc1 + x1 * ss1;
          x[l4] = x3 +  x4 * cc1 - x1 * ss1;
          gamma = gamma + n2;
```

```
        gamma2 = gamma2 + n2;
        }
    itemp = 0;
    phi = phi + ( 2 << (m-i) >> 1 );
    ntemp = (phi < n`4) ? 0 : n`4;
    ss1 = sine[phi - ntemp];
    cc1 = sine[n`4 - phi + ntemp];
    diff++;diff2-;
    gamma = diff;
    gamma2 = diff2;
    }
    }
}
```

Acknowledgments

The author would like to thank Mrs. Robert William Hartley and Dr. Sheldon Hochheiser, Senior Research Associate, AT&T Archives, for their assistance in accumulating the biographical information on R. V. L. Hartley. The assistance of R. N. Bracewell, G. T. Heydt, and Z. Wang is gratefully acknowledged.

References

[1] R. V. L. Hartley, A more symmetrical Fourier analysis applied to transmission problems, *Proc. of the I.R.E.*, 30, pp. 144–150, March 1942.

[2] Z. Wang, Harmonic analysis with a real frequency function—I. Aperiodic case, *Appl. Math. and Comput.*, 9, pp. 53–73, 1981.

[3] Z. Wang, Harmonic analysis with a real frequency function—II. Periodic and bounded case, *Appl. Math. and Comput.*, 9, pp. 153–163, 1981.

[4] Z. Wang, Harmonic analysis with a real frequency function—III. Data sequence, *Appl. Math. and Comput.*, 9, pp. 245–255, 1981.

[5] R. N. Bracewell, Discrete Hartley transform, *J. Opt. Soc. Amer.*, 73, pp. 1832–1835, December 1983.

[6] Z. Wang, Fast algorithms for the discrete W transform and for the discrete Fourier transform, *IEEE Trans. Acoust., Speech, Signal Process.*, ASSP-32, pp. 803–816, 1984.

[7] R. N. Bracewell, The fast Hartley transform, *Proc. IEEE*, 72, pp. 1010–1018, 1984.

[8] R. N. Bracewell, *The Hartley Transform*, Oxford University Press, New York, 1986.

[9] A. D. Poularikas and S. Seely, *Signals and Systems*, Second Edition, Krieger, Malabar, FL, 1994.

[10] K. J. Olejniczak and G. T. Heydt, eds., Special Section on the Hartley Transform, *Proc. IEEE*, 82, pp. 372–447, 1994.

[11] G. A. Campbell and R. M. Foster, *Fourier Integrals for Practical Applications*, Van Nostrand, Princeton, NJ, 1948.

[12] A. Erdélyi, *Tables of Integral Transforms*, Vol. 1, McGraw-Hill, New York, NY, 1954.

[13] W. Magnus and F. Oberhettinger, *Formulas and Theorems of the Special Functions of Mathematical Physics*, pp. 116–120, Chelsea, New York, NY, 1949.

[14] R. V. L. Hartley, Transmission of information, *The Bell System Technical Journal*, Vol. 7, pp. 535–563, July, 1928.

5

Laplace Transforms

Samuel Seely

CONTENTS

5.1 Introduction[1]

The Laplace transform has been introduced into the mathematical literature by a variety of procedures. Among these are: (a) in its relation to the Heaviside operational calculus; (b) as an extension of the Fourier integral; (c) by the selection of a particular form for the kernel in the general Integral transform; (d) by a direct definition of the Laplace transform; (e) as a mathematical procedure that involves multiplying the function $f(t)$ by $e^{-st}dt$ and integrating over the limits 0 to ∞. We will adopt this latter procedure.

Not all functions $f(t)$, where t is any variable, are Laplace transformable. For a function $f(t)$ to be Laplace transformable, it must satisfy the Dirichlet conditions—a set of sufficient but not necessary conditions. These are:

1. $f(t)$ must be piecewise continuous, that is, it must be single valued but can have a finite number of finite isolated discontinuities for $t > 0$.

2. $f(t)$ must be of exponential order, that is, $f(t)$ must remain less than $Me^{-a_o t}$ as t approaches ∞, where M is a positive constant and a_o is a real positive number.

For example, such functions as: $\tan \beta t$, $\cot \beta t$, e^{t^2} are not Laplace transformable. Given a

---

[1] All the contour integrations in the complex plane are counterclockwise.

0-8493-8342-0/96/$0.00 + $0.50
©1996 by CRC Press, Inc.

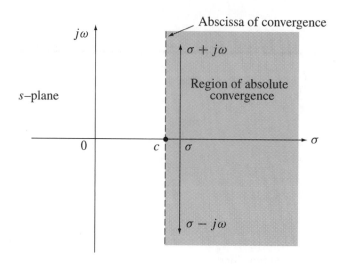

FIGURE 5.1
Path of integration for exponential order function.

function $f(t)$ that satisfies the Dirichlet conditions, then

$$F(s) = \int_0^\infty f(t)e^{-st}\,dt \qquad \text{written } \mathcal{L}\{f(t)\} \tag{1.1}$$

is called the Laplace transformation of $f(t)$. Here s can be either a real variable or a complex quantity. Observe the shorthand notation $\mathcal{L}\{f(t)\}$ to denote the Laplace transformation of $f(t)$. Observe also that only ordinary integration is involved in this integral.

To amplify the meaning of condition (2), we consider piecewise continuous functions, defined for all positive values of the variable t, for which

$$\lim_{t\to\infty} f(t)e^{-ct} = 0, \qquad c = \text{real constant.}$$

Functions of this type are known as functions of exponential order. Functions occuring in the solution for the time response of stable linear systems are of exponential order zero. Now we recall that the integral $\int_0^\infty f(t)e^{-st}\,dt$ converges if

$$\int_0^\infty |f(t)e^{-st}|\,dt < \infty, \qquad s = \sigma + j\omega.$$

If our function is of exponential order, we can write this integral as

$$\int_0^\infty |f(t)|e^{-ct}e^{-(\sigma-c)t}\,dt.$$

This shows that for σ in the range $\sigma > 0$ (σ is the abscissa of convergence) the integral converges, that is,

$$\int_0^\infty |f(t)e^{-st}|\,dt < \infty, \qquad \text{Re}(s) > c.$$

The restriction in this equation, namely, $\text{Re}(s) = c$, indicates that we must choose the path of integration in the complex plane as shown in Figure 5.1.

5.2 Laplace transform of some typical functions

We illustrate the procedure in finding the Laplace transform of a given function $f(t)$. In all cases it is assumed that the function $f(t)$ satisfies the conditions of Laplace transformability.

Example 5.2.1
Find the Laplace transform of the unit step function $f(t) = u(t)$, where $u(t) = 1$, $t > 0$, $u(t) = 0$, $t < 0$.

Solution By (1.1) we write

$$\mathcal{L}\{u(t)\} = \int_0^\infty u(t)e^{-st}\,dt = \int_0^\infty e^{-st}\,dt = -\frac{e^{-st}}{s}\bigg|_0^\infty = \frac{1}{s}. \tag{2.1}$$

The region of convergence is found from the expression $\int_0^\infty |e^{-st}|\,dt = \int_0^\infty e^{-\sigma t}\,dt < \infty$, which is the entire right half-plane, $\sigma > 0$. ∎

Example 5.2.2
Find the Laplace transform of the function $f(t) = 2\sqrt{\frac{t}{\pi}}$

$$F(s) = \frac{2}{\sqrt{\pi}} \int_0^\infty t^{\frac{1}{2}} e^{-st}\,dt. \tag{2.2}$$

To carry out the integration, define the quantity $x = t^{\frac{1}{2}}$, then $dx = \frac{1}{2}t^{-\frac{1}{2}}\,dt$, from which $dt = 2t^{\frac{1}{2}}\,dx = 2x\,dx$. Then

$$F(s) = \frac{4}{\sqrt{\pi}} \int_0^\infty x^2 e^{-sx^2}\,dx.$$

But the integral

$$\int_0^\infty x^2 e^{-sx^2}\,dx = \frac{\sqrt{\pi}}{4s^{3/2}}.$$

Thus, finally,

$$F(s) = \frac{1}{s^{3/2}}. \tag{2.3}$$

Example 5.2.3
Find the Laplace transform of $f(t) = \text{erfc}\,\frac{k}{2\sqrt{t}}$, where the error function, $\text{erf}\,t$, and the complementary error function, $\text{erfc}\,t$, are defined by

$$\text{erf}\,t = \frac{2}{\sqrt{\pi}} \int_0^t e^{-u^2}\,du, \qquad \text{erfc}\,t = \frac{2}{\sqrt{\pi}} \int_t^\infty e^{-u^2}\,du.$$

Solution Consider the integral

$$I = \frac{2}{\sqrt{\pi}} \int_0^\infty e^{-st} \left[\int_{\frac{\lambda}{\sqrt{t}}}^\infty e^{-u^2}\,du \right] dt \qquad \text{where } \lambda = \frac{k}{2}. \tag{2.4}$$

Change the order of integration, noting that $u = \frac{\lambda}{\sqrt{t}}, t = \frac{\lambda^2}{u^2}$

$$I = \frac{2}{\sqrt{\pi}} \int_0^\infty e^{-u^2} \left[\int_{\frac{\lambda^2}{u^2}}^\infty e^{-st} \right] dt\, du = \frac{2}{s\sqrt{\pi}} \int_0^\infty \exp\left(-u^2 - \frac{\lambda^2 s}{u^2}\right) du$$

The value of this intergral is known

$$= \frac{2}{s\sqrt{\pi}} \cdot \frac{\sqrt{\pi}}{2} e^{-2\lambda\sqrt{s}},$$

which leads to

$$\mathcal{L}\left\{\text{erfc}\, \frac{k}{2\sqrt{t}}\right\} = \frac{1}{s} \exp\left\{-k\sqrt{s}\right\}. \tag{2.5}$$

∎

Example 5.2.4
Find the Laplace transform of the function $f(t) = \sinh at$.

Solution   Express the function $\sinh at$ in its exponential form

$$\sinh at = \frac{e^{at} - e^{-at}}{2}.$$

The Laplace transform becomes

$$\mathcal{L}\{\sinh at\} = \frac{1}{2} \int_0^\infty \left[e^{-(s-a)t} - e^{-(s+a)t} \right] dt$$

$$= \frac{a}{s^2 - a^2}. \tag{2.6}$$

∎

A moderate listing of functions $f(t)$ and their Laplace transforms $F(s) = \mathcal{L}\{f(t)\}$ are given in Table 5.1, in the Appendix.

5.3   Properties of the Laplace transform

We now develop a number of useful properties of the Laplace transform; these follow directly from (1.1). Important in developing certain properties is the definition of $f(t)$ at $t = 0$, a quantity written $f(0+)$ to denote the limit of $f(t)$ as t approaches zero, assumed from the positive direction. This designation is consistent with the choice of function response for $t > 0$. This means that $f(0+)$ denotes the initial condition. Correspondingly, $f^{(n)}(0+)$ denotes the value of the nth derivative at time $t = 0+$, and $f^{(-n)}(0+)$ denotes the nth time integral at time $t = 0+$. This means that the direct Laplace transform can be written

$$F(s) = \lim_{\substack{R \to \infty \\ a \to 0+}} \int_a^R f(t)e^{-st}\, dt, \qquad R > 0,\ a > 0. \tag{3.1}$$

We proceed with a number of theorems.

THEOREM 5.3.1   *Linearity*

The Laplace transform of the linear sum of two Laplace transformable functions $f(t) + g(t)$ with respective abscissas of convergence σ_f and σ_g, with $\sigma_g > \sigma_f$, is

$$\mathcal{L}\{f(t) + g(t)\} = F(s) + G(s). \tag{3.2}$$

PROOF   From (3.1) we write

$$\mathcal{L}\{f(t) + g(t)\} = \int_0^\infty [f(t) + g(t)]e^{-st}\, dt = \int_0^\infty f(t)e^{-st}\, dt + \int_0^\infty g(t)e^{-st}\, dt,$$

$$\text{Re}(s) > \sigma_g.$$

Thus

$$\mathcal{L}\{f(t) + g(t)\} = F(s) + G(s).$$

As a direct extension of this result, for K_1 and K_2 constants,

$$\mathcal{L}\{K_1 f(t) + K_2 g(t)\} = K_1 F(s) + K_2 G(s). \tag{3.3}$$

∎

THEOREM 5.3.2   *Differentiation*

Let the function $f(t)$ be piecewise continuous with sectionally continuous derivatives $df(t)/dt$ in every interval $0 \le t \le T$. Also let $f(t)$ be of exponential order e^{ct} as $t \to \infty$. Then when $\text{Re}(s) > c$, the transform of $df(t)/dt$ exists and

$$\mathcal{L}\left\{\frac{df(t)}{dt}\right\} = s\mathcal{L}\{f(t)\} - f(0+) = sF(s) - f(0+). \tag{3.4}$$

PROOF   Begin with (3.1) and write

$$\mathcal{L}\left\{\frac{df(t)}{dt}\right\} = \lim_{T \to \infty} \int_0^T \frac{df(t)}{dt} e^{-st}\, dt.$$

Write the integral as the sum of integrals in each interval in which the integrand is continuous. Thus we write

$$\int_0^T e^{-st} f^{(1)}(t)\, dt = \int_0^{t_1} [\ ] + \int_{t_1}^{t_2} [\ ] + \cdots + \int_{t_{n-1}}^T [\ ].$$

Each of these integrals is integrated by parts by writing

$$u = e^{-st} \qquad du = -se^{-st}\, dt$$

$$dv = \frac{df}{dt}\, dt \qquad v = f$$

with the result

$$e^{-st} f(t)\big|_0^{t_1} + e^{-st} f(t)\big|_{t_1}^{t_2} + \cdots + e^{-st} f(t)\big|_{t_{n-1}}^T + s \int_0^T e^{-st} f(t)\, dt.$$

But $f(t)$ is continuous so that $f(t_1 - 0) = f(t_1 + 0)$, and so forth, hence

$$\int_0^T e^{-st} f^{(1)}(t)\, dt = -f(0+) + e^{-sT} f(T) + s \int_0^T e^{-st} f(t)\, dt.$$

However, with $\lim_{t\to\infty} f(t)e^{-st} = 0$ (otherwise the transform would not exist), then the theorem is established. ∎

THEOREM 5.3.3   *Differentiation*

Let the function $f(t)$ be piecewise continuous, have a continuous derivative $f^{(n-1)}(t)$ of order $n-1$ and a sectionally continuous derivative $f^{(n)}(t)$ in every finite interval $0 \le t \le T$. Also let $f(t)$ and all its derivatives through $f^{(n-1)}(t)$ be of exponential order e^{ct} as $t \to \infty$. Then the transform of $f^{(n)}(t)$ exists when $\mathrm{Re}(s) > c$ and it has the following form:

$$\mathcal{L}\{f^{(n)}(t)\} = s^n F(s) - s^{n-1} f(0+) - s^{n-2} f^{(1)}(0+) - \cdots - s^{n-1} f^{(n-1)}(0+). \qquad (3.5)$$

PROOF   The proof follows as a direct extension of the proof of Theorem 5.3.2. ∎

Example 5.3.1

Find $\mathcal{L}\{t^m\}$ where m is any positive integer.

Solution   The function $f(t) = t^m$ satisfies all the conditions of Theorem 5.3.3 for any positive c. Thus

$$f(0+) = f^{(1)}(0+) = \cdots = f^{(m-1)}(0+) = 0$$

$$f^{(m)}(t) = m!, \qquad f^{(m+1)}(t) = 0.$$

By (3.5) with $n = m + 1$ we have

$$\mathcal{L}\{f^{(m+1)}(t)\} = 0 = s^{m+1}\mathcal{L}\{t^m\} - m!.$$

It follows, therefore, that

$$\mathcal{L}\{t^m\} = \frac{m!}{s^{m+1}}. \qquad ∎$$

THEOREM 5.3.4   *Integration*

If $f(t)$ is sectionally continuous and has a Laplace transform, then the function $\int_0^t f(\xi)\, d\xi$ has the Laplace transform given by

$$\mathcal{L}\left\{\int_0^t f(\xi)\, d\xi\right\} = \frac{F(s)}{s} + \frac{1}{s} f^{(-1)}(0+). \qquad (3.6)$$

PROOF   Because $f(t)$ is Laplace transformable, its integral is written

$$\mathcal{L}\left\{\int_{-\infty}^t f(\xi)\, d\xi\right\} = \int_0^\infty \left[\int_{-\infty}^t f(\xi)\, d\xi\right] e^{-st}\, dt.$$

This is integrated by parts by writing

$$u = \int_{-\infty}^t f(\xi)\, d\xi \qquad du = f(\xi)\, d\xi = f(t)dt$$
$$dv = e^{-st} dt \qquad v = -\frac{1}{s} e^{-st}.$$

Then

$$\mathcal{L}\left\{\int_{-\infty}^{t} f(\xi)\, d\xi\right\} = \left[-\frac{e^{-st}}{s}\int_{-\infty}^{t} f(\xi)\, d\xi\right]\Bigg|_{0}^{\infty} + \frac{1}{s}\int_{0}^{\infty} f(t)e^{-st}\, dt$$

$$= \frac{1}{s}\int_{0}^{\infty} f(t)e^{-st}\, dt + \frac{1}{s}\int_{-\infty}^{0} f(\xi)\, d\xi$$

from which

$$\mathcal{L}\left\{\int_{0}^{t} f(\xi)\, d\xi\right\} = \frac{1}{s}F(s) + \frac{1}{s}f^{(-1)}(0+)$$

where $[f^{(-1)}(0+)/s]$ is the initial value of the integral of $f(t)$ at $t = 0+$. The negative number in the bracketed exponent indicates integration. ∎

Example 5.3.2
Deduce the value of $\mathcal{L}\{\sin at\}$ from $\mathcal{L}\{\cos at\}$ by employing Theorem 5.3.4.

Solution   By ordinary integration

$$\int_{0}^{t} \cos ax\, dx = \frac{\sin at}{a}.$$

From Theorem 5.3.4 we can write, knowing that $\mathcal{L}\{\cos at\} = \frac{s}{s^2+a^2}$,

$$\mathcal{L}\left\{\frac{\sin at}{a}\right\} = \frac{1}{s^2 + a^2}$$

so that

$$\mathcal{L}\{\sin at\} = \frac{a}{s^2 + a^2}. \qquad\qquad ∎$$

THEOREM 5.3.5
Division of the transform of a function by s corresponds to integration of the function between the limits 0 and t

$$\mathcal{L}^{-1}\left\{\frac{F(s)}{s}\right\} = \int_{0}^{t} f(\xi)\, d\xi$$

$$\mathcal{L}^{-1}\left\{\frac{F(s)}{s^2}\right\} = \int_{0}^{t}\int_{0}^{\xi} f(\lambda)\, d\lambda\, d\xi \qquad\qquad (3.7)$$

and so forth, for division by s^n, provided that $f(t)$ is Laplace transformable.

PROOF   The proof of this theorem follows from Theorem 5.3.4. ∎

THEOREM 5.3.6   *Multiplication by t*
If $f(t)$ is piecewise continuous and of exponential order, then each of the Laplace transforms:
$\mathcal{L}\{f(t)\}, \mathcal{L}\{tf(t)\}, \mathcal{L}\{t^2 f(t)\}, \ldots$ is uniformly convergent with respect to s when $s = c$, where
$\sigma > c$, and

$$\mathcal{L}\{t^n f(t)\} = (-1)^n \frac{d^n F(s)}{ds^n}. \qquad\qquad (3.8)$$

Further

$$\lim_{s \to \infty} \frac{d^n F(s)}{ds^n} = 0, \qquad \mathcal{L}_{s \to \infty}\{t^n f(t)\} = 0, \qquad n = 1, 2, 3, \ldots$$

PROOF It follows from (3.1) when this integral is uniformly convergent and the integral converges, that

$$\frac{\partial F(s)}{\partial s} = \int_0^\infty e^{-st}(-t) f(t)\, dt = \mathcal{L}\{-t f(t)\}.$$

Further, it follows that

$$\frac{\partial^2 F(s)}{\partial s^2} = \int_0^\infty e^{-st}(-t)^2 f(t)\, dt = \mathcal{L}\{t^2 f(t)\}.$$

Similar procedures follow for derivatives of higher order.   ∎

THEOREM 5.3.7 *Differentiation of a transform*
Differentiation of the transform of a function $f(t)$ corresponds to the multiplication of the function by $-t$; thus

$$\frac{d^n F(s)}{ds^n} = F^{(n)}(s) = \mathcal{L}\{(-t)^n f(t)\}, \qquad n = 1, 2, 3, \ldots. \tag{3.9}$$

PROOF This is a restatement of Theorem 5.3.6. This theorem is often useful for evaluating some types of integrals, and can be used to extend the table of transforms.   ∎

Example 5.3.3
Employ Theorem 5.3.7 to evaluate $\partial F(s)/\partial s$ for the function $f(t) = \sinh at$.

Solution   Initially we establish $\sinh at$

$$\mathcal{L}\{\sinh at\} = \int_0^\infty e^{-st}\left[\frac{e^{at} - e^{-at}}{2}\right] dt = \frac{a}{s^2 - a^2} = F(s).$$

By Theorem 5.3.7

$$\frac{\partial F(s)}{\partial s} = \int_0^\infty (-t) \sinh at\, e^{-st}\, dt = \frac{\partial}{\partial s}\left[\frac{a}{s^2 - a^2}\right] = -\frac{2as}{(s^2 - a^2)^2}$$

from which

$$\int_0^\infty e^{-st} \sinh at\, dt = \mathcal{L}\{t \sinh at\} = \frac{2as}{(s^2 - a^2)^2}.$$

We can, of course, differentiate $F(s)$ with respect to a. In this case Theorem 5.3.7 does not apply. However, the result is significant, and is

$$\frac{\partial F(s)}{\partial a} = \int_0^\infty e^{-st}(t \cosh at)\, dt = \mathcal{L}\{t \cosh at\} = \frac{\partial}{\partial a}\left[\frac{a}{s^2 - a^2}\right] = \frac{s^2 + a^2}{(s^2 - a^2)^2}.   ∎$$

THEOREM 5.3.8   *Complex integration*
If $f(t)$ is Laplace transformable and provided that $\lim_{t \to 0+} \frac{f(t)}{t}$ exists, the integral of the function $\int_s^\infty F(s)\,ds$ corresponds to the Laplace transform of the division of the function $f(t)$ by t,

$$\mathcal{L}\left\{\frac{f(t)}{t}\right\} = \int_s^\infty F(s)\,ds. \tag{3.10}$$

PROOF   Let $F(s)$ be piecewise continuous in each finite interval and of exponential order. Then

$$F(s) = \int_0^\infty e^{-st} f(t)\,dt$$

is uniformly convergent with respect to s. Consequently we can write for $\mathrm{Re}(s) > c$ and any $a > c$

$$\int_s^a F(s)\,ds = \int_s^a \int_0^\infty e^{-st} f(t)\,dt\,ds.$$

Express this in the form

$$= \int_0^\infty f(t) \int_s^a e^{-st}\,ds\,dt = \int_0^\infty \frac{f(t)}{t}(e^{-st} - e^{-at})\,dt.$$

Now if $f(t)/t$ has a limit as $t \to 0$, then the latter function is piecewise continuous and of exponential order. Therefore the last integral is uniformly convergent with respect to a. Thus as a tends to infinity

$$\int_s^\infty F(s)\,ds = \mathcal{L}\left\{\frac{f(t)}{t}\right\}. \qquad \blacksquare$$

THEOREM 5.3.9   *Time delay; real translation*
The substitution of $t - \lambda$ for the variable t in the transform $\mathcal{L}\{f(t)\}$ corresponds to the multiplication of the function $F(s)$ by $e^{-\lambda s}$, that is,

$$\mathcal{L}\{f(t - \lambda)\} = e^{-s\lambda} F(s). \tag{3.11}$$

PROOF   Refer to Figure 5.2, which shows a function $f(t)u(t)$ and the same function delayed by the time $t = \lambda$, where λ is a positive constant.

We write directly

$$\mathcal{L}\{f(t - \lambda)u(t - \lambda)\} = \int_0^\infty f(t - \lambda)u(t - \lambda)e^{-st}\,dt.$$

Now introduce a new variable $\tau = t - \lambda$. This converts this equation to the form

$$\mathcal{L}\{f(\tau)u(\tau)\} = e^{-s\lambda} \int_{-\lambda}^\infty f(\tau)u(\tau)e^{-s\tau}\,d\tau = e^{-s\lambda} \int_0^\infty f(\tau)e^{-s\tau}\,d\tau = e^{-s\lambda} F(s)$$

because $u(\tau) = 0$ for $-\lambda \le t \le 0$.

We would similarly find that

$$\mathcal{L}\{f(t + \lambda)u(t + \lambda)\} = e^{s\lambda} F(s). \tag{3.12}$$

$\blacksquare$

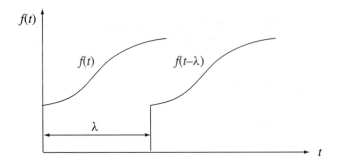

FIGURE 5.2
A function $f(t)$ at the time $t = 0$ and delayed time $t = \lambda$.

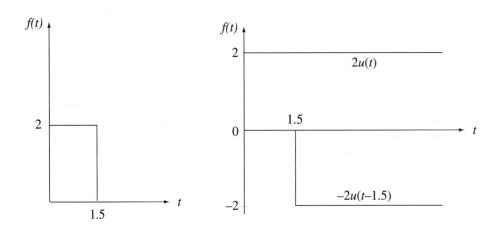

FIGURE 5.3
Pulse function and its equivalent representation.

Example 5.3.4
Find the Laplace transform of the pulse function shown in Figure 5.3.

Solution   Because the pulse function can be decomposed into step functions, as shown in Figure 5.3, its Laplace transform is given by

$$\mathcal{L}\{2[u(t) - u(t - 1.5)]\} = 2\left[\frac{1}{s} - \frac{1}{s}e^{-1.5s}\right] = \frac{2}{s}(1 - e^{1.5t})$$

where the translation property has been used.   ∎

THEOREM 5.3.10   Complex translation
The substitution of $s + a$ for s, where a is real or complex, in the function $F(s + a)$, corresponds to the Laplace transform of the product $e^{-at}f(t)$.

PROOF   We write

$$\int_0^\infty e^{-at} f(t) e^{-st} \, dt = \int_0^\infty f(t) e^{-(s+a)t} \, dt \qquad \text{for Re}(s) > c - \text{Re}(a),$$

which is

$$F(s+a) = \mathcal{L}\{e^{-at} f(t)\}. \tag{3.13}$$

In a similar way we find

$$F(s-a) = \mathcal{L}\{e^{at} f(t)\}. \tag{3.14}$$

∎

THEOREM 5.3.11   *Convolution*
The multiplication of the transforms of two sectionally continuous functions $f_1(t)$ ($= F_1(s)$) and $f_2(t)$ ($= F_2(s)$) corresponds to the Laplace transform of the convolution of $f_1(t)$ and $f_2(t)$

$$F_1(s) F_2(s) = \mathcal{L}\{f_1(t) * f_2(t)\} \tag{3.15}$$

where the asterisk $*$ is the shorthand designation for convolution.

PROOF   By definition, the convolution of two functions $f_1(t)$ and $f_2(t)$ is

$$f_1(t) * f_2(t) = \int_0^\infty f_1(t-\tau) f_2(\tau) \, d\tau = \int_0^\infty f_1(\tau) f_2(t-\tau) \, d\tau. \tag{3.16}$$

Thus

$$\mathcal{L}\{f_1(t) * f_2(t)\} = \mathcal{L}\left\{\int_0^\infty f_1(t-\tau) f_2(\tau) \, d\tau\right\}$$

$$= \int_0^\infty \left[\int_0^\infty f_1(t-\tau) f_2(\tau) \, d\tau\right] e^{-st} \, dt$$

$$= \int_0^\infty f_2(\tau) \, d\tau \int_0^\infty f_1(t-\tau) e^{-st} \, dt.$$

Now effect a change of variable, writing $t - \tau = \xi$ and therefore $dt = d\xi$, then

$$= \int_0^\infty f_2(\tau) d\tau \int_{-\tau}^\infty f_1(\xi) e^{-s(\xi+\tau)} \, d\xi.$$

But for positive time functions $f_1(\xi) = 0$ for $\xi < 0$, which permits changing the lower limit of the second integral to zero, and so

$$= \int_0^\infty f_2(\tau) e^{-s\tau} \, d\tau \int_0^\infty f_1(\xi) e^{-s\xi} \, d\xi,$$

which is

$$\mathcal{L}\{f_1(t) * f_2(t)\} = F_1(s) F_2(s). \qquad\qquad ∎$$

Example 5.3.5
Given $f_1(t) = t$ and $f_2(t) = e^{at}$, deduce the Laplace transform of the convolution $t * e^{at}$ by the use of Theorem 5.3.11.

Solution   Begin with the convolution

$$t * e^{at} = \int_0^t (t - \tau) e^{a\tau} \, d\tau = \frac{t e^{a\tau}}{a} \Big|_0^t - \left[\frac{\tau e^{a\tau}}{a} - \frac{e^{a\tau}}{a^2} \right]_0^t = \frac{1}{a^2}(e^{at} - at - 1).$$

Then

$$\mathcal{L}\{t * e^{at}\} = \frac{1}{a^2} \left(\frac{1}{s - a} - \frac{1}{s^2} - \frac{1}{s} \right) = \frac{1}{s^2} \frac{1}{(s - a)}.$$

By Theorem 5.3.11 we have

$$F_1(s) = \mathcal{L}\{f_1(t)\} = \mathcal{L}\{t\} = \frac{1}{s^2}, \qquad F_2(s) = \mathcal{L}\{f_2(t)\} = \mathcal{L}\{e^{at}\} = \frac{1}{s - a}.$$

and

$$\mathcal{L}\{t * e^{at}\} = \frac{1}{s^2} \frac{1}{(s - a)}. \qquad\qquad ▌$$

THEOREM 5.3.12
The multiplication of the transforms of three sectionally continuous functions $f_1(t)$, $f_2(t)$, and $f_3(t)$ corresponds to the Laplace transform of the convolution of the three functions

$$\mathcal{L}\{f_1(t) * f_2(t) * f_3(t)\} = F_1(s) F_2(s) F_3(s). \tag{3.17}$$

PROOF   This is an extension of Theorem 5.3.11. The result is obvious if we write

$$F_1(s) F_2(s) F_3(s) = \mathcal{L}\{f_1(t) * \mathcal{L}^{-1}\{F_2(s) F_3(s)\}\}. \qquad\qquad ▌$$

Example 5.3.6
Deduce the values of the convolution products: $1 * f(t)$; $1 * 1 * f(t)$.

Solution   By equations (3.14) and (3.16) we write directly

(a)   For $f_1(t) = 1$, $f_2(t) = f(t)$, $\mathcal{L}\{1 * f(t)\} = \frac{F(s)}{s} = \int_0^t f(\xi) \, d\xi$ by equation (3.7)

(b)   For $f_1(t) = 1$, $f_2(t) = 1$, $f_3(t) = f(t)$, $\mathcal{L}\{1 * 1 * f(t)\} = \frac{F(s)}{s^2} = \int_0^t \int_0^\xi f(\lambda) \, d\lambda \, d\xi$ ▌

THEOREM 5.3.13   *Frequency convolution—s-plane*
The Laplace transform of the product of two piecewise and sectionally continuous functions $f_1(t)$ and $f_2(t)$ corresponds to the convolution of their transforms, with

$$\mathcal{L}\{f_1(t) f_2(t)\} = \frac{1}{2\pi j} [F_1(s) * F_2(s)]. \tag{3.18}$$

PROOF   Begin by considering the following line integral in the z-plane:

$$f_2(t) = \frac{1}{2\pi j} \int_{C_2} F_2(z) e^{zt} \, dz, \qquad \sigma_2 = \text{axis of convergence.}$$

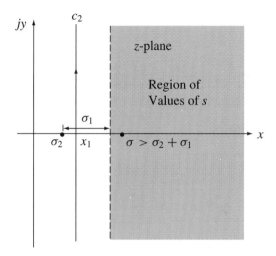

FIGURE 5.4
The contour C_2 and the allowed range of s.

This means that the contour intersects the x-axis at $x_1 > \sigma_2$ (see Figure 5.4) Then we have

$$\int_0^\infty f_1(t) f_2(t) e^{-st}\, dt = \frac{1}{2\pi j} \int_0^\infty f_1(t)\, dt \int_{C_2} F_2(z) e^{(z-s)t}\, dz.$$

Assume that the integral of $F_2(z)$ is convergent over the path of integration. This equation is now written in the form

$$\int_0^\infty f_1(t) f_2(t) e^{-st}\, dt = \frac{1}{2\pi j} \int_{\sigma_2 - j\infty}^{\sigma_2 + j\infty} F_2(z)\, dz \int_0^\infty f_1(t) e^{-(s-z)t}\, dt$$

$$= \frac{1}{2\pi j} \int_{\sigma_2 - j\infty}^{\sigma_2 + j\infty} F_2(z) F_1(s - z)\, dz \overset{\Delta}{=} \mathcal{L}\{f_1(t) f_2(t)\}. \qquad (3.19)$$

The Laplace transform of $f_1(t)$, the integral on the right, converges in the range $\mathrm{Re}(s - z) > \sigma_1$, where σ_1 is the abscissa of convergence of $f_1(t)$. In addition, $\mathrm{Re}(z) = \sigma_2$ for the z-plane integration involved in (3.18). Thus the abscissa of convergence of $f_1(t) f_2(t)$ is specified by

$$\mathrm{Re}(s) > \sigma_1 + \sigma_2. \qquad (3.20)$$

This situation is portrayed graphically in Figure 5.4 for the case when both σ_1 and σ_2 are positive. As far as the integration in the complex plane is concerned, the semicircle can be closed either to the left or to the right just so long as $F_1(s)$ and $F_2(s)$ go to zero as $s \to \infty$.

Based on the foregoing, we observe the following:

a. poles of $F_1(s - z)$ are contained in the region $\mathrm{Re}(s - z) < \sigma_1$;

b. poles of $F_2(z)$ are contained in the region $\mathrm{Re}(z) < \sigma_2$;

c. from (a) and (3.20) $\mathrm{Re}(z) > \mathrm{Re}(s - \sigma_1) > \sigma_2$;

d. poles of $F_1(s - z)$ lie to the right of the path of integration;

e. poles of $F_2(z)$ are to the left of the path of integration;

f. poles of $F_1(s - z)$ are functions of s whereas poles of $F_2(z)$ are fixed in relation to s. ∎

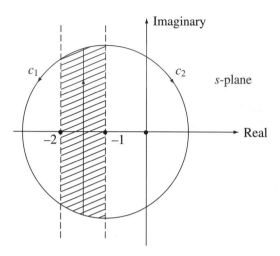

FIGURE 5.5
The contour for Example 5.3.7.

Example 5.3.7
Find the Laplace transform of the function $f(t) = f_1(t) f_2(t) = e^{-t} e^{-2t} u(t)$.

Solution From Theorem 5.3.13 and the absolute convergence region for each function, we have

$$F_1(s) = \frac{1}{s+1}, \qquad \sigma_1 > 1$$

$$F_2(s) = \frac{1}{s+2}, \qquad \sigma_2 > 2.$$

Further $f(t) = \exp[-(2+1)t]u(t)$ implies that $\sigma_f = \sigma_1 + \sigma_2 = 3$. We now write

$$F_2(z)F_1(s-z) = \frac{1}{z+2}\frac{1}{s-z+1} = \frac{1}{3+s}\frac{1}{z-(1+s)} - \frac{1}{3+s}\frac{1}{z+2}.$$

To carry out the integration dictated by equation (3.19) we use the contour shown in Figure 5.5. If we select contour C_1 and use the residue theorem, we obtain

$$F(s) = \frac{1}{2\pi j} \oint_{C_1} F_2(z)F_1(s-z)\,dz = -2\pi j\,\text{Res}[F_2(z)F_1(s-z)]|_{z=-2} = \frac{1}{s+3}.$$

The inverse of this transform is $\exp(-3t)$. If we had selected contour C_2, the residue theorem gives

$$F(s) = -\frac{1}{2\pi j} \oint_{C_2} F_2(z)F_1(s-z)\,dz = -2\pi j\,\text{Res}[F_2(z)F_1(s-z)]|_{z=1+s}$$

$$= -\left[-\frac{1}{s+3}\right] = \frac{1}{s+3}.$$

The inverse transform of this is also $\exp(-3t)$, as to be expected. ∎

THEOREM 5.3.14   Initial value theorem
Let $f(t)$ and $f^{(1)}(t)$ be Laplace transformable functions, then for case when $\lim s F(s)$ as $s \to \infty$ exists,

$$\lim_{s \to \infty} s F(s) = \lim_{t \to 0+} f(t). \qquad (3.21)$$

PROOF   Begin with equation (3.6) and consider

$$\lim_{s \to \infty} \int_0^\infty \frac{df}{dt} e^{-st}\, dt = \lim_{s \to \infty} [s F(s) - f(0+)].$$

Because $f(0+)$ is independent of s, and because the integral vanishes for $s \to \infty$, then

$$\lim_{s \to \infty} [s F(s) - f(0+)] = 0.$$

Furthermore, $f(0+) = \lim_{t \to 0+} f(t)$ so that

$$\lim_{s \to \infty} s F(s) = \lim_{t \to 0+} f(t).$$

If $f(t)$ has a discontinuity at the origin, this expression specifies the value of the impulse $f(0+)$. If $f(t)$ contains an impulse term, then the left-hand side does not exist, and the initial value property does not exist.   ∎

THEOREM 5.3.15   Final value theorem
Let $f(t)$ and $f^{(1)}(t)$ be Laplace transformable functions, then for $t \to \infty$

$$\lim_{t \to \infty} f(t) = \lim_{s \to 0} s F(s). \qquad (3.22)$$

PROOF   Begin with equation (3.6) and let $s \to 0$. Thus the expression

$$\lim_{s \to 0} \int_0^\infty \frac{df}{dt} e^{-st}\, dt = \lim_{s \to 0}[s F(s) - f(0+)].$$

Consider the quantity on the left. Because s and t are independent and because $e^{-st} \to 1$ as $s \to 0$, then the integral on the left becomes, in the limit

$$\int_0^\infty \frac{df}{dt}\, dt = \lim_{t \to \infty} f(t) - f(0+).$$

Combine the latter two equations to get

$$\lim_{t \to \infty} f(t) - f(0+) = \lim_{s \to 0} s F(s) - f(0+).$$

It follows from this that the final value of $f(t)$ is given by

$$\lim_{t \to \infty} f(t) = \lim_{s \to 0} s F(s).$$

This result applies if $F(s)$ possesses a simple pole at the origin, but it does not apply if $F(s)$ has imaginary axis poles, poles in the right half plane, or higher order poles at the origin.   ∎

Example 5.3.8
Apply the final value theorem to the following two functions:

$$F_1(s) = \frac{s + a}{(s + a)^2 + b^2}, \qquad F_2(s) = \frac{s}{s^2 + b^2}\ .$$

Solution   For the first function from $s F_1(s)$

$$\lim_{s \to 0} \frac{s(s+a)}{(s+a)^2 + b^2} = 0.$$

For the second function

$$s F(s) = \frac{s^2}{s^2 + b^2}.$$

However, this function has singularities on the imaginary axis at $s = \pm jb$, and the Final Value Theorem does not apply. ▌

The important properties of the Laplace transform are contained in Table 5.2 in the Appendix.

5.4   The inverse Laplace transform

We employ the symbol $\mathcal{L}^{-1}\{F(s)\}$, corresponding to the direct Laplace transform defined in (1.1), to denote a function $f(t)$ whose Laplace transform is $F(s)$. Thus we have the Laplace pair

$$F(s) = \mathcal{L}\{f(t)\}, \qquad f(t) = \mathcal{L}^{-1}\{F(s)\}. \tag{4.1}$$

This correspondence between $F(s)$ and $f(t)$ is called the inverse Laplace transformation of $f(t)$.

Reference to Table 5.1 shows that $F(s)$ is a rational function in s if $f(t)$ is a polynomial or a sum of exponentials. Further, it appears that the product of a polynomial and an exponential might also yield a rational $F(s)$. If the square root of t appears in $f(t)$ we do not get a rational function in s. Note also that a continuous function $f(t)$ may not have a continuous inverse transform.

Observe that the $F(s)$ functions have been uniquely determined for the given $f(t)$ function by (1.1). A logical question is whether a given time function in Table 5.1 is the only t-function that will give the corresponding $F(s)$. Clearly, Table 5.1 is more useful if there is a unique $f(t)$ for each $F(s)$. This is an important consideration because the solution of practical problems usually provides a known $F(s)$ from which $f(t)$ must be found. This uniqueness condition can be established using the inversion integral. This means that there is a one-to-one correspondence between the direct and the inverse transform. This means that if a given problem yields a function $F(s)$, the corresponding $f(t)$ from Table 5.1 is the unique result. In the event that the available tables do not include a given $F(s)$ we would seek to resolve the given $F(s)$ into forms that are listed in Table 5.1. This resolution of $F(s)$ is often accomplished in terms of a partial fraction expansion.

A few examples will show the use of the partial fraction form in deducing the $f(t)$ for a given $F(s)$.

Example 5.4.1
Find the inverse Laplace transform of the function

$$F(s) = \frac{s-3}{s^2 + 5s + 6}. \tag{4.2}$$

Solution   Observe that the denominator can be factored into the form $(s + 2)(s + 3)$. Thus $F(s)$ can be written in partial fraction form as

$$F(s) = \frac{s - 3}{(s + 2)(s + 3)} = \frac{A}{s + 2} + \frac{B}{s + 3} . \tag{4.3}$$

where A and B are constants that must be determined.

To evaluate A, multiply both sides of (4.3) by $(s + 2)$ and then set $s = -2$. This gives

$$A = F(s) \ (s + 2)|_{s=-2} = \left. \frac{s - 3}{s + 3} \right|_{s=-2} = -5$$

and $B(s + 2)/(s + 3)|_{s=-2}$ is identically zero. In the same manner, to find the value of B we multiply both sides of (4.3) by $(s + 3)$ and get

$$B = F(s)(s + 3)|_{s=-3} = \left. \frac{s - 3}{s + 2} \right|_{s=-3} = 6.$$

The partial fraction form of (4.3) is

$$F(s) = \frac{-5}{s + 2} + \frac{6}{s + 3} .$$

The inverse transform is given by

$$f(t) = \mathcal{L}^{-1}\{F(s)\} = -5\mathcal{L}^{-1}\left\{\frac{1}{s + 2}\right\} + 6\mathcal{L}^{-1}\left\{\frac{1}{s + 3}\right\} = -5e^{-2t} + 6e^{-3t}$$

where entry 8 in Table 5.1 is used.   ∎

Example 5.4.2
Find the inverse Laplace transform of the function

$$F(s) = \frac{s + 1}{[(s + 2)^2 + 1](s + 3)} .$$

Solution   This function is written in the form

$$F(s) = \frac{A}{s + 3} + \frac{Bs + C}{[(s + 2)^2 + 1]} = \frac{s + 1}{[(s + 2)^2 + 1](s + 3)} .$$

The value of A is deduced by multiplying both sides of this equation by $(s+3)$ and then setting $s = -3$. This gives

$$A = (s + 3)F(s)|_{s=-3} = \frac{-3 + 1}{(-3 + 2)^2 + 1} = -1.$$

To evaluate B and C, combine the two fractions and equate the coefficients of the like powers of s in the numerators. This yields

$$\frac{-1[(s + 2)^2 + 1] + (s + 3)(Bs + C)}{[(s + 2)^2 + 1](s + 3)} = \frac{s + 1}{[(s + 2)^2 + 1](s + 3)}$$

from which it follows that

$$-(s^2 + 4s + 5) + Bs^2 + (C + 3B)s + 2C = s + 1.$$

Combine like-powered terms to write

$$(-1 + B)s^2 + (-4 + C + 3B)s + (-5 + 3C) = s + 1.$$

Therefore

$$-1 + B = 0, \qquad -4 + C + 3B = 1, \qquad -5 + 3C = 1.$$

From these equations we obtain

$$B = 1, \qquad C = 2.$$

The function $F(s)$ is written in the equivalent form

$$F(s) = \frac{-1}{s+3} + \frac{s+2}{(s+2)^2 + 1}.$$

Now using Table 5.1, the result is

$$f(t) = -e^{-3t} + e^{-2t} \cos t, \qquad t > 0. \qquad \blacksquare$$

In many cases, $F(s)$ is the quotient of two polynomials with real coefficients. If the numerator polynomial is of the same or higher degree than the denominator polynomial, first divide the numerator polynomial by the denominator polynomial; the division is carried forward until the numerator polynomial of the remainder is one degree less than the denominator. This results in a polynomial in s plus a proper fraction. The proper fraction can be expanded into a partial fraction expansion. The result of such an expansion is an expression of the form

$$F'(s) = B_0 + B_1 s + \cdots + \frac{A_1}{s - s_1} + \frac{A_2}{s - s_2} + \cdots + \frac{A_{p1}}{s - s_p} + \frac{A_{p2}}{(s - s_p)^2} + \cdots$$

$$+ \frac{A_{pr}}{(s - s_p)^r}. \tag{4.4}$$

This expression has been written in a form to show three types of terms: polynomial, simple partial fraction including all terms with distinct roots, and partial fraction appropriate to multiple roots.

To find the constants $A_1, A_2, \ldots$ the polynomial terms are removed, leaving the proper fraction

$$F'(s) - (B_0 + B_1 s + \cdots) = F(s) \tag{4.5}$$

where

$$F(s) = \frac{A_1}{s - s_1} + \frac{A_{2+}}{s - s_2} + \cdots + \frac{A_k}{s - s_k} + \frac{A_{p1}}{s - s_p} + \frac{A_{p2}}{(s - s_p)^2} + \cdots + \frac{A_{pr}}{(s - s_p)^r}.$$

To find the constants A_k that are the residues of the function $F(s)$ at the simple poles s_k, it is only necessary to note that as $s \to s_k$ the term $A_k(s - s_k)$ will become large compared with all other terms. In the limit

$$A_k = \lim_{s \to s_k} (s - s_k) F(s). \tag{4.6}$$

Upon taking the inverse transform for each simple pole, the result will be a simple exponential of the form

$$\mathcal{L}^{-1}\left\{ \frac{A_k}{s - s_k} \right\} = A_k e^{s_k t}. \tag{4.7}$$

Note also that because $F(s)$ contains only real coefficients, if s_k is a complex pole with residue A_k, there will also be a conjugate pole s_k^* with residue A_k^*. For such complex poles

$$\mathcal{L}^{-1}\left\{\frac{A_k}{s - s_k} + \frac{A_k^*}{s - s_k^*}\right\} = A_k e^{s_k t} + A_k^* e^{s_k^* t}.$$

These can be combined in the following way:

$$\begin{aligned}
\text{response} &= (a_k + jb_k)e^{(\sigma_k + j\omega_k)t} + (a_k - jb_k)e^{(\sigma_k - j\omega_k)t} \\
&= e^{\sigma_k t}\left[(a_k + jb_k)(\cos\omega_k t + j\sin\omega_k t) + (a_k - jb_k)(\cos\omega_k t - j\sin\omega_k t)\right] \\
&= 2e^{\sigma_k t}(a_k \cos\omega_k t - b_k \sin\omega_k t) \\
&= 2A_k e^{\sigma_k t}\cos(\omega_k t + \theta_k)
\end{aligned} \tag{4.8}$$

where $\theta_k = \tan^{-1}(b_k/a_k)$ and $A_k = a_k/\cos\theta_k$.

When the proper fraction contains a multiple pole of order r, the coefficients in the partial-fraction expansion $A_{p1}, A_{p2}, \ldots, A_{pr}$ that are involved in the terms

$$\frac{A_{p1}}{(s - s_p)} + \frac{A_{p2}}{(s - s_p)^2} + \cdots + \frac{A_{pr}}{(s - s_p)^r}$$

must be evaluated. A simple application of (4.6) is not adequate. Now the procedure is to multiply both sides of (4.5) by $(s - s_p)^r$, which gives

$$\begin{aligned}
(s - s_p)^r F(s) = (s - s_p)^r &\left[\frac{A_1}{s - s_1} + \frac{A_2}{s - s_2} + \cdots + \frac{A_k}{s - s_k}\right] + A_{p1}(s - s_p)^{r-1} + \cdots \\
&+ A_{p(r-1)}(s - s_p) + A_{pr}
\end{aligned} \tag{4.9}$$

In the limit as $s = s_p$ all terms on the right vanish with the exception of A_{pr}. Suppose now that this equation is differentiated once with respect to s. The constant A_{pr} will vanish in the differentiation but $A_{p(r-1)}$ will be determined by setting $s = s_p$. This procedure will be continued to find each of the coefficients A_{pk}. Specifically, the procedure is specified by

$$A_{pk} = \frac{1}{(r-k)!}\left\{\frac{d^{r-k}}{ds^{r-k}}F(s)(s - s_p)^r\right\}_{s=s_p}, \qquad k = 1, 2, \ldots, r. \tag{4.10}$$

Example 5.4.3
Find the inverse transform of the following function:

$$F(s) = \frac{s^3 + 2s^2 + 3s + 1}{s^2(s + 1)}.$$

Solution   This is not a proper fraction. The numerator polynomial is divided by the denominator polynomial by simple long division. The result is

$$F(s) = 1 + \frac{s^2 + 3s + 1}{s^2(s + 1)}.$$

The proper fraction is expanded into partial fraction form

$$F(s) = \frac{s^2 + 3s + 1}{s^2(s + 1)} = \frac{A_{11}}{s} + \frac{A_{12}}{s^2} + \frac{A_2}{s + 1}.$$

The value of A_2 is deduced using (4.6)

$$A_2 = [(s+1)F(s)]_{s=-1} = \left.\frac{s^2 + 3s + 1}{s^2}\right|_{s=-1} = -1.$$

To find A_{11} and A_{12} we proceed as specified in (4.10)

$$A_{12} = [s^2 F(s)]_{s=0} = \left.\frac{s^2 + 3s + 1}{s+1}\right|_{s=0} = 1$$

$$A_{11} = \frac{1}{1!}\left\{\frac{d}{ds}s^2 F(s)\right\}_{s=0} = \frac{d}{ds}\left[\frac{s^2+3s+1}{s+1}\right]_{s=0} = \frac{s^2+3s+1}{(s+1)^2} + \left.\frac{2s+3}{s+1}\right|_{s=0} = 4.$$

Therefore

$$F(s) = 1 + \frac{4}{s} + \frac{1}{s^2} - \frac{1}{s+1}.$$

From Table 5.1 the inverse transform is

$$f(t) = \delta(t) + 4 + t - e^{-t}, \qquad \text{for } t \geq 0. \qquad \blacksquare$$

If the function $F(s)$ exists in proper fractional form as the quotient of two polynomials, we can employ the Heaviside expansion theorem in the determination of $f(t)$ from $F(s)$. This theorem is an efficient method for finding the residues of $F(s)$. Let

$$F(s) = \frac{P(s)}{Q(s)} = \frac{A_1}{s - s_1} + \frac{A_2}{s - s_2} + \cdots + \frac{A_k}{s - s_k}$$

where $P(s)$ and $Q(s)$ are polynomials with no common factors and with the degree of $P(s)$ less than the degree of $Q(s)$.

Suppose that the factors of $Q(s)$ are distinct constants. Then, as in (4.6) we find

$$A_k = \lim_{s \to s_k}\left[\frac{s - s_k}{Q(s)}P(s)\right].$$

Also, the limit $P(s)$ is $P(s_k)$. Now, because

$$\lim_{s \to s_k}\frac{s - s_k}{Q(s)} = \lim_{s \to s_k}\frac{1}{Q^{(1)}(s)} = \frac{1}{Q^{(1)}(s_k)},$$

then

$$A_k = \frac{P(s_k)}{Q^{(1)}(s_k)}.$$

Thus

$$F(s) = \frac{P(s)}{Q(s)} = \sum_{n=1}^{k}\frac{P(s_n)}{Q^{(1)}(s_n)} \cdot \frac{1}{(s - s_n)}. \qquad (4.11)$$

From this, the inverse transformation becomes

$$f(t) = \mathcal{L}^{-1}\left\{\frac{P(s)}{Q(s)}\right\} = \sum_{n=1}^{k}\frac{P(s_n)}{Q^{(1)}(s_n)}e^{s_n t}.$$

This is the Heaviside expansion theorem. It can be written in formal form:

THEOREM 5.4.1 *Heaviside expansion theorem*
If $F(s)$ is the quotient $P(s)/Q(s)$ of two polynomials in s such that $Q(s)$ has the higher degree and contains simple poles the factor $s - s_k$, which are not repeated, then the term in $f(t)$ corresponding to this factor can be written $\frac{P(s_k)}{Q^{(1)}(s_k)}e^{s_k t}$.

Example 5.4.4
Repeat Example 4.1 employing the Heaviside expansion theorem.

Solution   We write (4.2) in the form

$$F(s) = \frac{P(s)}{Q(s)} = \frac{s-3}{s^2+5s+6} = \frac{s-3}{(s+2)(s+3)}.$$

The derivative of the denominator is

$$Q^{(1)}(s) = 2s + 5$$

from which, for the roots of this equation,

$$Q^{(1)}(-2) = 1, \qquad Q^{(1)}(-3) = -1.$$

Hence

$$P(-2) = -5, \qquad P(-3) = -6.$$

The final value for $f(t)$ is

$$f(t) = -5e^{-2t} + 6e^{-3t}.$$

∎

Example 5.4.5
Find the inverse Laplace transform of the following function using the Heaviside expansion theorem:

$$\mathcal{L}^{-1}\left\{\frac{2s+3}{s^2+4s+7}\right\}.$$

Solution   The roots of the denominator are

$$s^2+4s+7 = \left(s+2+j\sqrt{3}\right)\left(s+2-j\sqrt{3}\right).$$

That is, the roots of the denominator are complex. The derivative of the denominator is

$$Q^{(1)}(s) = 2s + 4.$$

We deduce the values $P(s)/Q^{(1)}(s)$ for each root

For $s_1 = -2 - j\sqrt{3}$ $Q^{(1)}(s_1) = -j2\sqrt{3}$ $P(s_1) = -1 - j2\sqrt{3}$

For $s_2 = -2 + j/3$ $Q^{(1)}(s_2) = +j2\sqrt{3}$ $P(s_2) = -1 + j2\sqrt{3}.$

Then

$$f(t) = \frac{-1 - j2\sqrt{3}}{-j2\sqrt{3}} e^{(-2-j2\sqrt{3})t} + \frac{-1 + j2\sqrt{3}}{j2\sqrt{3}} e^{(-2+j2\sqrt{3})t}$$

$$= e^{-2t} \left[\frac{-1 - j2\sqrt{3}}{-j2\sqrt{3}} e^{-j2\sqrt{3}t} + \frac{-1 + j2\sqrt{3}}{j2\sqrt{3}} e^{j2\sqrt{3}t} \right]$$

$$= e^{-2t} \left[\frac{(e^{-j2\sqrt{3}t} - e^{j2\sqrt{3}t})}{j2\sqrt{3}} + (e^{-j2\sqrt{3}t} + e^{j2\sqrt{3}t}) \right]$$

$$= e^{-2t} \left(2\cos 2\sqrt{3}t - \frac{1}{\sqrt{3}} \sin 2\sqrt{3}t \right) \qquad \blacksquare$$

5.5 Solution of ordinary linear equations with constant coefficients

The Laplace transform is used to solve homogeneous and nonhomogeneous ordinary different equations or systems of such equations. To understand the procedure, we consider a number of examples.

Example 5.5.1
Find the solution to the following differential equation subject to prescribed initial conditions: $y(0+)$; $(dy/dt) + ay = x(t)$.

Solution   Laplace transform this differential equation. This is accomplished by multiplying each term by $e^{-st} dt$ and integrating from 0 to ∞. The result of this operation is

$$sY(s) - y(0+) + aY(s) = X(s),$$

from which

$$Y(s) = \frac{X(s)}{s+a} + \frac{y(0+)}{s+a}.$$

If the input $x(t)$ is the unit step function $u(t)$, then $X(s) = 1/s$ and the final expression for $Y(s)$ is

$$Y(s) = \frac{1}{s(s+a)} + \frac{y(0+)}{s+a}.$$

Upon taking the inverse transform of this expression

$$y(t) = \mathcal{L}^{-1}\{Y(s)\} = \mathcal{L}^{-1} \left(\frac{1}{a} \left[\frac{1}{s} - \frac{1}{s+a} \right] + \frac{y(0+)}{s+a} \right)$$

with the result

$$y(t) = \frac{1}{a}(1 - e^{-at}) + y(0+)e^{-at}. \qquad \blacksquare$$

Example 5.5.2

Find the general solution to the differential equation

$$\frac{d^2y}{dt^2} + 5\frac{dy}{dt} + 4y = 10$$

subject to zero initial conditions.

Solution   Laplace transform this differential equation. The result is

$$s^2Y(s) + 5sY(s) + 4Y(s) = \frac{10}{s}.$$

Solving for $Y(s)$, we get

$$Y(s) = \frac{10}{s(s^2 + 5s + 4)} = \frac{10}{s(s + 1)(s + 4)}.$$

Expand this into partial-fraction form, thus

$$Y(s) = \frac{A}{s + 1} + \frac{B}{s + 4} + \frac{C}{s}.$$

Then

$$A = Y(s)\,(s + 1)|_{s=-1} = \frac{10}{s(s + 4)}\bigg|_{s=-1} = -\frac{10}{3}$$

$$B = Y(s)\,(s + 4)|_{s=-4} = \frac{10}{s(s + 1)}\bigg|_{s=-4} = \frac{10}{12}$$

$$C = sY(s)|_{s=0} = \frac{10}{(s + 1)(s + 4)}\bigg|_{s=0} = \frac{10}{4}$$

and

$$Y(s) = 10\left[-\frac{1}{3(s + 1)} + \frac{1}{12(s + 4)} + \frac{1}{4s}\right].$$

The inverse transform is

$$x(t) = 10\left[-\frac{1}{3}e^{-t} + \frac{1}{12}e^{-4t} + \frac{1}{4}\right]. \qquad\blacksquare$$

Example 5.5.3

Find the velocity of the system shown in Figure 5.6a when the applied force is $f(t) = e^{-t}u(t)$. Assume zero initial conditions. Solve the same problem using convolution techniques. The input is the force and the output is the velocity.

Solution   The controlling equation is, from Figure 5.6b,

$$\frac{dv}{dt} + 5v + 4\int_0^t v\,dt = e^{-t}u(t).$$

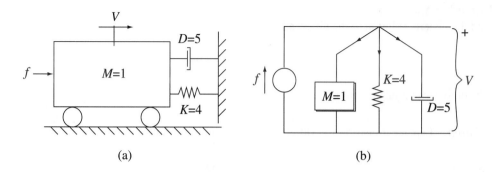

FIGURE 5.6
The mechanical system and its network equivalent.

Laplace transform this equation and then solve for $F(s)$. We obtain

$$V(s) = \frac{s}{(s+1)(s^2+5s+4)} = \frac{s}{(s+1)^2(s+4)}.$$

Write this expression in the form

$$V(s) = \frac{A}{s+4} + \frac{B}{s+1} + \frac{C}{(s+1)^2}$$

where

$$A = \left.\frac{s}{(s+1)^2}\right|_{s=-4} = -\frac{4}{9}$$

$$B = \frac{1}{1!}\frac{d}{ds}\left.\left(\frac{s}{s+4}\right)\right|_{s=-1} = \frac{4}{9}$$

$$C = \left.\frac{s}{s+4}\right|_{s=-1} = -\frac{1}{3}.$$

The inverse transform of $V(s)$ is given by

$$v(t) = -\frac{4}{9}e^{-4t} + \frac{4}{9}e^{-t} - \frac{1}{3}te^{-t}, \qquad t \geq 0.$$

To find $v(t)$ by the use of the convolution integral, we first find $h(t)$, the impulse response of the system. The quantity $h(t)$ is specified by

$$\frac{dh}{dt} + 5h + 4\int h\,dt = \delta(t)$$

where the system is assumed to be initially relaxed. The Laplace transform of this equation yields

$$H(s) = \frac{s}{s^2+5s+4} = \frac{s}{(s+4)(s+1)} = \frac{4}{3}\frac{1}{s+4} - \frac{1}{3}\frac{1}{s+1}.$$

The inverse transform of this expression is easily found to be

$$h(t) = \frac{4}{3}e^{-4t} - \frac{1}{3}e^{-t}, \qquad t \geq 0.$$

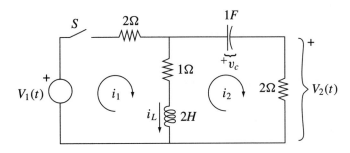

FIGURE 5.7
The circuit for Example 5.5.4.

The output of the system to the input $e^{-t}u(t)$ is written

$$v(t) = \int_{-\infty}^{\infty} h(\tau) f(t - \tau) \, d\tau = \int_{0}^{t} e^{-(t-\tau)} \left[\frac{4}{3} e^{-4\tau} - \frac{1}{3} e^{-\tau} \right] d\tau$$

$$= e^{-t} \left[\frac{4}{3} \int_{0}^{t} e^{-3\tau} \, d\tau - \frac{1}{3} \int_{0}^{t} d\tau \right] = e^{-t} \left[\frac{4}{3} \left(\frac{1}{-3} \right) e^{-3\tau} \Big|_{0}^{t} - \frac{1}{3} t \right]$$

$$= -\frac{4}{9} e^{-4t} + \frac{4}{9} e^{-t} - \frac{1}{3} t e^{-t}, \qquad t \geq 0.$$

This result is identical with that found using the Laplace transform technique. ∎

Example 5.5.4
Find an expression for the voltage $v_2(t)$ for $t > 0$ in the circuit of Figure 5.7. The source $v_1(t)$, the current $i_L(0-)$ through $L = 2H$, and the voltage $v_c(0-)$ across the capacitor $C = 1\ F$ at the switching instant are all assumed to be known.

Solution   After the switch is closed, the circuit is described by the loop equations

$$\left(3i_1 + \frac{2di_1}{dt} \right) - \left(1i_2 + \frac{2di_2}{dt} \right) = v_2(t)$$

$$-\left(1i_1 + \frac{2di_1}{dt} \right) + \left(3i_2 + \frac{2di_2}{dt} + \int i_2 dt \right) = 0$$

$$v_2(t) = 2i_2(t).$$

All terms in these equations are Laplace transformed. The result is the set of equations

$$(3 + 2s)I_1(s) - (1 + 2s)I_2(s) = V_1(s) + 2\left[i_1(0+) - i_2(0+) \right]$$

$$-(1 + 2s)I_1(s) + \left(3 + 2s + \frac{1}{s} \right) I_2(s) = 2[-i_1(0+) + i_2(0+)] - \frac{q_2(0+)}{s}$$

$$V_2(s) = 2I_2(s).$$

The current through the inductor is

$$i_L(t) = i_1(t) - i_2(t).$$

At the instant $t = 0+$

$$i_L(0+) = i_1(0+) - i_2(0+).$$

Also, because

$$\frac{1}{C}q_2(t) = \frac{1}{C}\int_{-\infty}^{t} i_2(t)\,dt$$

$$= \frac{1}{C}\lim_{t\to 0+}\int_{0}^{t} i_2(t)\,dt + \frac{1}{C}\int_{-\infty}^{0} i_2(t)\,dt = 0 + v_c(0-)$$

then

$$\frac{q_2(0+)}{C} \overset{\Delta}{=} v_c(0+) = v_c(0-) = i_2^{(-1)}(0) = \frac{q_2(0+)}{1}.$$

The equation set is solved for $I_2(s)$, which is written by Cramer's rule

$$I_2(s) = \frac{\begin{vmatrix} 3+2s & V_1(s)+2i_L(0+) \\ -(1+2s) & -2i_L(0+)-\frac{v_c(0+)}{s} \end{vmatrix}}{\begin{vmatrix} 3+2s & -(1+2s) \\ -(1+2s) & 3+2s+\frac{1}{s} \end{vmatrix}}$$

$$= \frac{(3+2s)\left[-2i_L(0+)-\frac{v_0(0+)}{s}\right] + (1+2s)[V_1(s)+2i_L(0+)]}{(3+2s)\left(\frac{2s^2+3s+1}{s}\right) - (1+2s)^2}$$

$$= \frac{-(2s^2+3s)v_c(0+) - 4si_L(0+) + (2s^2+s)V_1(s)}{8s^2+10s+3}.$$

Further

$$V_2(s) = 2I_2(s).$$

Then, upon taking the inverse transform

$$v_2(t) = 2\mathcal{L}^{-1}\{I_2(s)\}.$$

If the circuit contains no stored energy at $t = 0$, then $i_L(0+) = v_c(0+) = 0$ and now

$$v_2(t) = 2\mathcal{L}^{-1}\left\{\frac{(2s^2+s)V_1(s)}{8s^2+10s+3}\right\}.$$

For the particular case when $v_1 = u(t)$ so that $V_1(s) = 1/s$

$$v_2(t) = 2\mathcal{L}^{-1}\left\{\frac{2s+1}{8s^2+10s+3}\right\} = 2\mathcal{L}^{-1}\left\{\frac{2s+1}{8\left(s+\frac{1}{2}\right)(s+3/4)}\right\}$$

$$= \frac{1}{2}\mathcal{L}^{-1}\left\{\frac{1}{s+\frac{3}{4}}\right\} = \frac{1}{2}e^{-3t/4}, \qquad t \geq 0.$$

The validity of this result is readily confirmed because at the instant $t = 0+$ the inductor behaves as an open circuit and the capacitor behaves as a short circuit. Thus, at this instant,

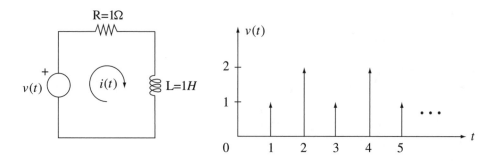

FIGURE 5.8
a) The circuit, b) the input pulse train.

the circuit appears as two equal resistors in a simple series circuit, and the voltage is shared equally. ∎

Example 5.5.5
The input to the RL circuit shown in Figure 5.8a is the recurrent series of impulse functions shown in Figure 5.8b. Find the output current.

Solution   The differential equation that characterizes the system is

$$\frac{di(t)}{dt} + i(t) = v(t).$$

For zero initial current through the inductor, the Laplace transform of the equation is

$$(s+1)I(s) = V(s).$$

Now, from the fact that $\mathcal{L}\{\delta(t)\} = 1$ and the shifting property of Laplace transforms, we can write the explicit form for $V(s)$, which is

$$V(s) = 2 + e^{-s} + 2e^{-2s} + e^{-3s} + 2e^{-4s} + \cdots$$

$$= (2 + e^{-s})(1 + e^{-2s} + e^{-4s} + \cdots)$$

$$= \frac{2 + e^{-s}}{1 - e^{-2s}}.$$

Thus, we must evaluate $i(t)$ from

$$I(s) = \frac{2 + e^{-s}}{1 - e^{-2s}} \frac{1}{s+1} = \frac{2}{(1 - e^{-2s})(s+1)} + \frac{e^{-s}}{(1 - e^{-2s})(s+1)}.$$

Expand these expressions into

$$I(s) = \frac{2}{s+1}(1 + e^{-2s} + e^{-4s} + e^{-6s} + \cdots) + \frac{1}{s+1}(e^{-s} + e^{-3s} + e^{-5s} + e^{-7s} + \cdots).$$

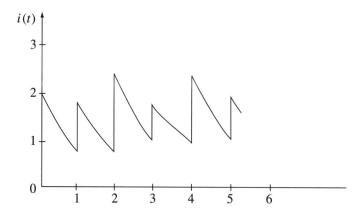

FIGURE 5.9
The response of the RL circuit to the pulse train.

The inverse transform of these expressions yields

$$i(t) = 2e^{-t}u(t) + 2e^{-(t-2)}u(t-2) + 2e^{-(t-4)}u(t-4) + \cdots$$

$$+ e^{-(t-1)}u(t-1) + e^{-(t-3)}u(t-3) + e^{-(t-5)}u(t-5) + \cdots$$

The result has been sketched in Figure 5.9.   ∎

5.6   The inversion integral

The discussion in Section 5.3 related the inverse Laplace transform to the direct Laplace transform by the expressions

$$F(s) = \mathcal{L}\{f(t)\} \tag{6.1a}$$

$$f(t) = \mathcal{L}^{-1}\{F(s)\}. \tag{6.1b}$$

The subsequent discussion indicated that the use of equation (6.1b) suggested that the $f(t)$ so deduced was unique; that there was no other $f(t)$ that yielded the specified $F(s)$. We found that although $f(t)$ represents a real function of the positive real variable t, the transform $F(s)$ can assume a complex variable form. What this means, of course, is that a mathematical form for the inverse Laplace transform was not essential for linear functions that satisfied the Dirichlet conditions. In some cases Table 5.1 is not adequate for many functions when s is a complex variable and an analytic form for the inversion process of (6.1b) is required.

To deduce the complex inversion integral, we begin with the Cauchy second integral theorem, which is written

$$\oint \frac{F(z)}{s - z} \, dz = j2\pi F(s)$$

where the contour encloses the singularity at s. The function $F(s)$ is analytic in the half-plane $\mathrm{Re}(s) \geq c$. If we apply the inverse Laplace transformation to the function s on both sides of

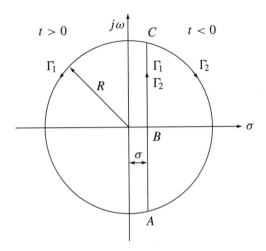

FIGURE 5.10
The path of integration in the s-plane.

this equation, we can write

$$j2\pi\mathcal{L}^{-1}\{F(s)\} = \lim_{\omega\to\infty} \int_{\sigma-j\omega}^{\sigma+j\omega} F(z)\mathcal{L}^{-1}\left\{\frac{1}{s-z}\right\} dz.$$

But $F(s)$ is the Laplace transform of $f(t)$; also, the inverse transform of $1/(s-z)$ is e^{zt}. Then it follows that

$$f(t) = \frac{1}{2\pi j} \lim_{\omega\to\infty} \int_{\sigma-j\omega}^{\sigma+j\omega} e^{zt} F(z)\,dz = \frac{1}{2\pi j} \int_{\sigma-j\infty}^{\sigma+j\infty} e^{zt} F(z)\,dz. \qquad (6.2)$$

This equation applies equally well to both the one-sided and the two-sided transforms.

It was pointed out in Section 5.1 that the path of integration (6.2) is restricted to value of σ for which the direct transform formula converges. In fact, for the two-sided Laplace transform, the region of convergence must be specified in order to determine uniquely the inverse transform. That is, for the two-sided transform, the regions of convergence for functions of time that are zero for $t > 0$, zero for $t < 0$, or in neither category, must be distinguished. For the one-sided transform, the region of convergence is given by σ, where σ is the abscissa of absolute convergence.

The path of integration in (6.2) is usually taken as shown in Figure 5.10 and consists of the straight line ABC displayed to the right of the origin by σ and extending in the limit from $-j\infty$ to $+j\infty$ with connecting semicircles. The evaluation of the integral usually proceeds by using the Cauchy integral theorem, which specifies that

$$f(t) = \frac{1}{2\pi j} \lim_{R\to\infty} \oint_{\Gamma_1} F(s)e^{st}\,ds$$

$$= \sum \text{residues of } F(s)e^{st} \text{ at the singularities to the left of } ABC; \quad t > 0. \qquad (6.3)$$

But the contribution to the integral around the circular path with $R \to \infty$ is zero, leaving the desired integral along the path ABC, and

$$f(t) = \frac{1}{2\pi j} \lim_{R \to \infty} \oint_{\Gamma_2} F(s) e^{st} \, ds$$

$$= \sum \text{residues of } F(s)e^{st} \text{ at the singularities to the right of } ABC; \quad t < 0. \quad (6.4)$$

We will present a number of examples involving these equations.

Example 5.6.1

Use the inversion integral to find $f(t)$ for the function

$$F(s) = \frac{1}{s^2 + w^2} .$$

Note that by entry 15 of Table 5.1, this is $\sin wt/w$.

Solution The inversion integral is written in a form that shows the poles of the integrand,

$$f(t) = \frac{1}{2\pi j} \oint \frac{e^{st}}{(\sigma + jw)(\sigma - jw)} \, ds.$$

The path chosen is Γ_1 in Figure 5.10. Evaluate the residues

$$\text{Res}\left[(s - jw)\frac{e^{st}}{s^2 + w^2} \right]_{s=jw} = \frac{e^{st}}{s + jw}\bigg|_{s=jw} = \frac{e^{jwt}}{2wj}$$

$$\text{Res}\left[(s + jw)\frac{e^{st}}{s^2 + w^2} \right]_{s=-jw} = \frac{e^{st}}{s - jw}\bigg|_{s=-jw} = \frac{e^{-jwt}}{-2wj} .$$

Therefore

$$f(t) = \sum \text{Res} = \frac{e^{jwt} - e^{-jwt}}{2jw} = \frac{\sin wt}{w} . \qquad \blacksquare$$

Example 5.6.2

Evaluate $\mathcal{L}^{-1}\{1/\sqrt{s}\}$.

Solution The function $F(s) = 1/\sqrt{s}$ is a double-valued function because of the square root operation. That is, if s is represented in polar form by $re^{j\theta}$, then $re^{j(\theta+2\pi)}$ is a second acceptable representation, and $\sqrt{s} = \sqrt{re^{j(\theta+2\pi)}} = -\sqrt{re^{j\theta}}$, thus showing two different values for $\sqrt{s}$. But a double-valued function is not analytic and requires a special procedure in its solution.

The procedure is to make the function analytic by restricting the angle of s to the range $-\pi < \theta < \pi$ and by excluding the point $s = 0$. This is done by constructing a branch cut along the negative real axis, as shown in Figure 5.11. The end of the branch cut, which is the origin in this case, is called a branch point. Because a branch cut can never be crossed, this essentially ensures that $F(s)$ is single valued. Now, however, the inversion integral (6.3)

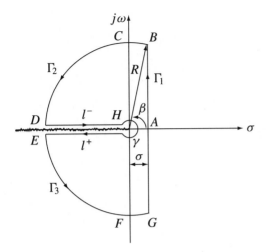

FIGURE 5.11
The integration contour for $\mathcal{L}^{-1}\{1/\sqrt{s}\}$.

becomes for $t > 0$

$$f(t) = \lim_{R \to \infty} \frac{1}{2\pi j} \int_{GAB} F(s)e^{st}\,ds = \frac{1}{2\pi j} \int_{\sigma - j\infty}^{\sigma + j\infty} F(s)e^{st}\,ds$$

$$= -\frac{1}{2\pi j}\left[\int_{BC} + \int_{\Gamma_2} + \int_{\ell-} + \int_{\gamma} + \int_{\ell+} + \int_{\Gamma_3} + \int_{FG} \right], \tag{6.5}$$

which does not include any singularity.

First we will show that for $t > 0$ the integrals over the contours BC and CD vanish as $R \to \infty$, from which $\int_{\Gamma_2} = \int_{\Gamma_3} = \Gamma_{BC} = \int_{FG} = 0$. Note from Figure 5.11 that $\beta = \cos^{-1}(\sigma/R)$ so that the integral over the arc BC is, because $|e^{j\theta}| = 1$,

$$|I| \leq \int_{BC}\left| \frac{e^{\sigma t}e^{jwt}}{R^{\frac{1}{2}}e^{j\theta/2}} j\,\mathrm{Re}^{j\theta}\,d\theta \right| = e^{\sigma t}\sqrt{R}\int_{\beta}^{\pi/2} d\theta = e^{\sigma t}\sqrt{R}\left(\frac{\pi}{2} - \cos^{-1}\frac{\sigma}{R} \right)$$

$$= e^{\sigma t}\sqrt{R}\sin^{-1}\frac{\sigma}{R}$$

But for small arguments $\sin^{-1}(\sigma/R) = \sigma/R$, and in the limit as $R \to \infty$, $I \to 0$. By a similar approach, we find that the integral over CD is zero. Thus the integrals over the contours Γ_2 and Γ_3, are also zero as $R \to \infty$.

For evaluating the integral over γ, let $s = re^{j\theta} = r(\cos\theta + j\sin\theta)$ and

$$\int_{\gamma} F(s)e^{st}\,ds = \int_{-\pi}^{\pi} \frac{e^{r(\cos\theta + j\sin\theta)}}{\sqrt{r}e^{j\theta/2}} jre^{j\theta}\,d\theta = 0 \qquad \text{as } r \to 0.$$

The remaining integrals in (6.5) are written

$$f(t) = -\frac{1}{2\pi j}\left[\int_{\ell-} F(s)e^{st}\,ds + \int_{\ell+} F(s)e^{st}\,ds \right]. \tag{6.6}$$

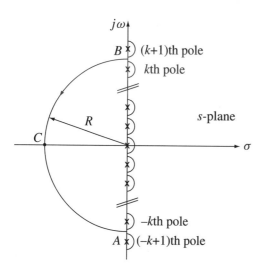

FIGURE 5.12
The pole distribution of the given function.

Along path $l-$, let $s = -u$; $\sqrt{s} = j\sqrt{u}$, and $ds = -du$, where u and $\sqrt{u}$ are real positive quantities. Then

$$\int_{\ell-} F(s)e^{st}\, ds = -\int_{\infty}^{0} \frac{e^{-ut}}{j\sqrt{u}}\, du = \frac{1}{j}\int_{0}^{\infty} \frac{e^{-ut}}{\sqrt{u}}\, du.$$

Along path $l+$, $s = -u$, $\sqrt{s} = -j/u$ (not $+j/u$), and $ds = -du$. Then

$$\int_{\ell+} F(s)e^{st}\, ds = -\int_{0}^{\infty} \frac{e^{-ut}}{-j\sqrt{u}}\, du = \frac{1}{j}\int_{0}^{\infty} \frac{e^{-ut}}{\sqrt{u}}\, du.$$

Combine these results to find

$$f(t) = -\frac{1}{2\pi j}\left[\frac{2}{j}\int_{0}^{\infty} u^{-\frac{1}{2}}e^{-ut}\, du\right] = \frac{1}{\pi}\int_{0}^{\infty} u^{-\frac{1}{2}}e^{-ut}\, du,$$

which is a standard form integral with the value

$$f(t) = \frac{1}{\pi}\sqrt{\frac{\pi}{t}} = \frac{1}{\sqrt{\pi t}}, \qquad t > 0. \qquad\qquad \blacksquare$$

Example 5.6.3
Find the inverse Laplace transform of the function

$$F(s) = \frac{1}{s(1 + e^{-s})}.$$

Solution   The integrand in the inversion integral $\frac{e^{st}}{s(1+e^{-s})}$ possesses simple poles at: $s = 0$ and $s = jn\pi$, $n = \pm1, \pm3, \pm\cdots$ (odd values). These are illustrated in Figure 5.12. We see that the function $e^{st}/s(1 + e^{-s})$ is analytic in the s-plane except at the simple poles at $s = 0$ and $s = jn\pi$. Hence the integral is specified in terms of the residues in the various poles.

We have, specifically,

$$\text{Res}\left\{\frac{se^{st}}{s(1+e^{-s})}\right\}\Big|_{s=0} = \frac{1}{2} \qquad \text{for } s = 0$$

$$\text{Res}\left\{\frac{(s-jn)e^{st}}{s(1+e^{-s})}\right\}\Big|_{s=jn} = \frac{0}{0} \qquad \text{for } s = jn\pi. \tag{6.7}$$

The problem we now face in this evaluation is that

$$\text{Res}\left\{(s-a)\frac{n(s)}{d(s)}\right\}\Big|_{s=a} = \frac{0}{0}$$

where the roots of $d(s)$ are such that $s = a$ cannot be factored. However, we know from complex function theory that

$$\frac{d[d(s)]}{ds}\Big|_{s=a} = \lim_{s\to a}\frac{d(s)-d(a)}{s-a} = \lim_{s\to a}\frac{d(s)}{s-a}$$

because $d(a) = 0$. Combine this result with the above equation to obtain

$$\text{Res}\left\{(s-a)\frac{n(s)}{d(s)}\right\}\Big|_{s=a} = \frac{n(s)}{\frac{d}{ds}[d(s)]}\Big|_{s=a}. \tag{6.8}$$

By combining (6.8) with (6.7) we obtain

$$\text{Res}\left\{\frac{e^{st}}{s\frac{d}{ds}(1+e^{-s})}\right\}\Big|_{s=jn\pi} = \frac{e^{jn\pi t}}{jn\pi} \qquad n \text{ odd.}$$

We obtain, by adding all of the residues,

$$f(t) = \frac{1}{2} + \sum_{n=-\infty}^{\infty}\frac{e^{jn\pi t}}{jn\pi}.$$

This can be rewritten as follows

$$f(t) = \frac{1}{2} + \left[\cdots + \frac{e^{-j3\pi t}}{-j3\pi} + \frac{e^{-j\pi t}}{-j\pi} + \frac{e^{j\pi t}}{j\pi} + \frac{e^{j3\pi t}}{3j\pi} + \cdots\right]$$

$$= \frac{1}{2} + \sum_{\substack{n=1 \\ n\text{ odd}}}^{\infty}\frac{2j\sin n\pi t}{jn\pi}.$$

This assumes the form

$$f(t) = \frac{1}{2} + \frac{2}{\pi}\sum_{k=1}^{\infty}\frac{\sin(2k-1)\pi t}{2k-1}. \tag{6.9}$$

As a second approach to a solution to this problem, we will show the details in carrying out the contour integration for this problem. We choose the path shown in Figure 5.12 that includes semicircular hooks around each pole, the vertical connecting line from hook to hook, and the semicircular path at $R \to \infty$. Thus we examine

$$f(t) = \frac{1}{2\pi j}\oint\frac{e^{st}}{s(1+e^{-s})}\,ds$$

$$= \frac{1}{2\pi j}\left[\underbrace{\int_{BCA}}_{I_1} + \underbrace{\int_{\text{vertical connecting lines}}}_{I_2} + \sum\underbrace{\int_{\text{Hooks}}}_{I_3} - \sum\text{Res}\right]. \tag{6.10}$$

We consider the several integrals in this equation.

Integral I_1. By setting $s = re^{j\theta}$ and taking into consideration that $\cos\theta = -\cos\theta$ for $\theta > \pi/2$, the integral $I_1 \to 0$ as $r \to \infty$.

Integral I_2. Along the Y-axis, $s = jy$ and

$$I_2 = j \int_{-\infty \atop r\to 0}^{\infty} \frac{e^{jyt}}{jy(1 + e^{-jy})} \, dy.$$

Note that the integrand is an odd function, whence $I_2 = 0$.

Integral I_3. Consider a typical hook at $s = jn\pi$. The result is

$$\lim_{r\to 0 \atop s\to jn\pi} \left[\frac{(s - jn)e^{st}}{s(1 + e^{-s})} \right] = \frac{0}{0}$$

This expression is evaluated (as for (6.7)) and yields $e^{jn\pi t/jn\pi}$. Thus for all poles,

$$I_3 = \frac{1}{2\pi j} \int_{-\frac{\pi}{2}}^{\frac{\pi}{2}} \frac{e^{st}}{s(1 + e^{-s})} \, ds = \frac{j\pi}{2\pi j} \left[\sum_{n=-\infty \atop n\,\text{odd}}^{\infty} \frac{e^{jn\pi t}}{jn\pi} + \frac{1}{2} \right] = \frac{1}{2} \left[\frac{1}{2} + \frac{2}{\pi} \sum_{n=1 \atop n\,\text{odd}}^{\infty} \frac{\sin n\pi t}{n} \right].$$

Finally, the residues enclosed within the contour are

$$\text{Res} \, \frac{e^{st}}{s(1 + e^{-s})} = \frac{1}{2} + \sum_{n=-\infty \atop n\,\text{odd}}^{\infty} \frac{e^{jn\pi t}}{jn\pi} = \frac{1}{2} + \frac{2}{\pi} \sum_{n=1 \atop n\,\text{odd}}^{\infty} \frac{\sin n\pi t}{n},$$

which is seen to be twice the value around the hooks. Then when all terms are included in (6.10), the final result is

$$f(t) = \frac{1}{2} + \frac{2}{\pi} \sum_{n=1 \atop n\,\text{odd}}^{\infty} \frac{\sin n\pi t}{n} = \frac{1}{2} + \frac{2}{\pi} \sum_{k=1}^{\infty} \frac{\sin(2k-1)\pi t}{2k-1}. \qquad \blacksquare$$

We now shall show that the direct and inverse transforms specified by (4.1) and listed in Table 5.1 constitute unique pairs. In this connection, we see that (6.2) can be considered as proof of the following theorem:

THEOREM 5.6.1

Let $F(s)$ be a function of a complex variable s that is analytic and of order $O(s^{-k})$ in the half-plane $\text{Re}(s) \geq c$, where c and k are real constants, with $k > 1$. The inversion integral (6.2) written $\mathcal{L}_t^{-1}\{F(s)\}$ along any line $x = \sigma$, with $\sigma \geq c$ converges to the function $f(t)$ that is independent of σ,

$$f(t) = \mathcal{L}_t^{-1}\{F(s)\}$$

whose Laplace transform is $F(s)$,

$$F(s) = \mathcal{L}\{f(t)\}, \qquad \text{Re}(s) \geq c.$$

In addition, the function $f(t)$ is continuous for $t > 0$ and $f(0) = 0$, and $f(t)$ is of the order $O(e^{ct})$ for all $t > 0$.

Suppose that there are two transformable functions $f_1(t)$ and $f_2(t)$ that have the same transforms

$$\mathcal{L}\{f_1(t)\} = \mathcal{L}\{f_2(t)\} = F(s).$$

The difference between the two functions is written $\phi(t)$

$$\phi(t) = f_1(t) - f_2(t)$$

where $\phi(t)$ is a transformable function. Thus

$$\mathcal{L}\{\phi(t)\} = F(s) - F(s) = 0.$$

Additionally

$$\phi(t) = \mathcal{L}_t^{-1}\{0\} = 0, \qquad t > 0.$$

Therefore this requires that $f_1(t) = f_2(t)$. The result shows that it is not possible to find two different functions by using two different values of σ in the inversion integral. This conclusion can be expressed as follows:

THEOREM 5.6.2
Only a single function $f(t)$ that is sectionally continuous, of exponential order, and with a mean value at each point of discontinuity, corresponds to a given transform $F(s)$.

5.7   Applications to partial differential equations

The Laplace transformations can be very useful in the solution of partial differential equations. A basic class of partial differential equations is applicable to a wide range of problems. However, the form of the solution in a given case is critically dependent on the boundary conditions that apply in any particular case. In consequence, the steps in the solution often will call on many different mathematical techniques. Generally, in such problems the resulting inverse transforms of more complicated functions of s occur than those for most linear systems problems. Often the inversion integral is useful in the solution of such problems. The following examples will demonstrate the approach to typical problems.

Example 5.7.1
Solve the typical heat conduction equation

$$\frac{\partial^2 \varphi}{\partial x^2} = \frac{\partial \varphi}{\partial t}, \qquad 0 < x < \infty, \quad t \geq 0 \tag{7.1}$$

subject to the conditions

C-1. $\varphi(x, 0) = f(x), \qquad t = 0$

C-2. $\frac{\partial \varphi}{\partial x} = 0, \qquad x = 0.$

Solution   Multiply both sides of (7.1) by $e^{-sx}\, dx$ and integrate from 0 to ∞.

$$\Phi(s, t) = \int_0^\infty e^{-sx} \varphi(x, t)\, dx.$$

Also

$$\int_0^\infty \frac{\partial^2 \varphi}{\partial x^2} e^{-sx}\, dx = s^2 \Phi(s, t) - s\varphi(0+) - \frac{\partial \varphi}{\partial x}(0+).$$

Equation (7.1) thus transforms, subject to C-2 and zero boundary conditions, to

$$\frac{d\Phi}{dt} - s^2\Phi = 0.$$

The solution to this equation is

$$\Phi = Ae^{s^2t}.$$

By an application of condition C-1, in transformed form, we have

$$\Phi = A = \int_0^\infty f(\lambda)e^{-s\lambda}\,d\lambda.$$

The solution, subject to C-1, is then

$$\Phi(s,t) = e^{+s^2t}\int_0^\infty f(\lambda)e^{-s\lambda}\,d\lambda.$$

Now apply the inversion integral to write the function in terms of x from s,

$$\varphi(x,t) = \frac{1}{2\pi j}\int_{-\infty}^\infty e^{+s^2t}\left[\int_0^\infty f(\lambda)e^{-s\lambda}\,d\lambda\right]e^{sx}\,ds$$

$$= \frac{1}{2\pi j}\int_{-\infty}^\infty f(\lambda)\,d\lambda\int_0^\infty e^{s^2t-s\lambda+sx}\,ds.$$

Note that we can write

$$s^2t - s(x-\lambda) = \left\{s\sqrt{t} - \frac{(x-\lambda)}{2\sqrt{t}}\right\}^2 - \frac{(x-\lambda)^2}{4t}.$$

Also write

$$s\sqrt{t} - \frac{(x-\lambda)}{2\sqrt{t}} = u.$$

Then

$$\varphi(x,t) = \frac{1}{2\pi j}\int_{-\infty}^\infty f(\lambda)\exp-\left[\frac{(x-\lambda)^2}{4t}\right]d\lambda\int_0^\infty e^{-u^2}\frac{du}{\sqrt{t}}.$$

But the integral

$$\int_0^\infty e^{-u^2}\,du = \sqrt{\pi}.$$

Thus the final solution is

$$\varphi(x,t) = \frac{1}{2\sqrt{\pi t}}\int_{-\infty}^\infty f(\lambda)e^{-\frac{(x-\lambda)^2}{4t}}\,d\lambda. \qquad\blacksquare$$

Example 5.7.2
A semi-infinite medium, initially at temperature $\varphi = 0$ throughout the medium, has the face $x = 0$ maintained at temperature φ_0. Determine the temperature at any point of the medium at any subsequent time.

Solution The controlling equation for this problem is

$$\frac{\partial^2 \varphi}{\partial x^2} = \frac{1}{K} \frac{\partial \varphi}{\partial t} \tag{7.2}$$

with the boundary conditions:

a. $\varphi = \varphi_0$ at $x = 0, t > 0$

b. $\varphi = 0$ at $t = 0, x > 0$.

To proceed, multiply both sides of equation (7.2) by $e^{-st} dt$ and integrate from 0 to ∞. The transformed form of equation (7.2) is

$$\frac{d^2 \Phi}{dx^2} - \left(\frac{s}{K} \Phi\right) = 0, \qquad K > 0.$$

The solution of this differential equation is

$$\Phi = A e^{-x\sqrt{\frac{s}{K}}} + B e^{x\sqrt{\frac{s}{K}}}.$$

But Φ must be finite or zero for infinite x; therefore $B = 0$, and

$$\Phi(s, x) = A e^{-\sqrt{\frac{s}{K}} x}.$$

Apply boundary condition (a) in transformed form, namely

$$\Phi(0, s) = \int_0^\infty e^{-st} \varphi_0 \, dt = \frac{\varphi_0}{s} \qquad \text{for } x = 0.$$

Therefore

$$A = \frac{\varphi_0}{s}$$

and the solution in Laplace transformed form is

$$\Phi(s, x) = \frac{\varphi_0}{s} e^{-\sqrt{\frac{s}{K}} x}. \tag{7.3}$$

To find $\varphi(x, t)$ requires that we find the inverse transform of this expression. This requires evaluating the inversion integral

$$\varphi(x, t) = \frac{\varphi_0}{2\pi j} \int_{\sigma-j\infty}^{\sigma+j\infty} \frac{e^{-x\sqrt{\frac{s}{K}}} e^{st}}{s} \, ds. \tag{7.4}$$

This integral has a branch point at the origin (see Figure 5.13). To carry out the integration we select a path such as that shown (see also Figure 5.11). The integral in (7.4) is written

$$\varphi(x, t) = \frac{\varphi_0}{2\pi j} \left[\int_{BC} + \int_{\Gamma_2} + \int_{l-} + \int_{\gamma} + \int_{l+} + \int_{\Gamma_3} + \int_{FG} \right].$$

As in Example 5.6.2

$$\int_{\Gamma_2} = \int_{\Gamma_3} = \int_{BC} = \int_{FG} = 0.$$

For the segments

$$\int_{l-} \quad \text{let } s = \rho e^{j\pi} \quad \text{and for} \quad \int_{l+} \quad \text{let } s = \rho e^{-j\pi}.$$

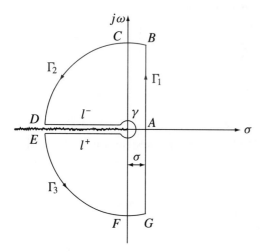

FIGURE 5.13
The path of integration.

Then for $\ell-$ and $\ell+$, writing this sum I_ℓ,

$$I_\ell = \frac{1}{2\pi j} \int_0^\infty e^{-st} \left[e^{jx\sqrt{\frac{s}{K}}} - e^{-jx\sqrt{\frac{s}{K}}} \right] \frac{ds}{s} = -\frac{1}{\pi} \int_0^\infty e^{-st} \sin x \sqrt{\frac{s}{K}} \frac{ds}{s}.$$

Write

$$u = \sqrt{\frac{s}{K}} \qquad s = ku^2, \qquad ds = 2ku\,du.$$

Then we have

$$I_\ell = -\frac{2}{\pi} \int_0^\infty e^{-Ku^2 t} \sin ux \frac{du}{u}.$$

This is a known integral that can be written

$$I_l = -\frac{2}{\sqrt{\pi}} \int_0^{\frac{x}{2\sqrt{Kt}}} e^{-u^2}\,du.$$

Finally, consider the integral over the hook,

$$I_\gamma = \frac{1}{2\pi j} \int_\gamma e^{st} \frac{e^{x\sqrt{s/K}}}{s}\,ds.$$

Let us write

$$s = re^{j\theta}, \qquad ds = jre^{j\theta}d\theta, \qquad \frac{ds}{s} = j\theta,$$

then

$$I_\gamma = \frac{j}{2\pi j} \int e^{tre^{j\theta}} e^{x\sqrt{\frac{r}{K}}e^{j\theta/2}}\,d\theta.$$

For $r \to 0$, $I_\gamma = \frac{j2\pi}{2\pi j} = \frac{2\pi j}{2\pi j}$, then $I_\gamma = 1$. Hence the sum of the integrals in (7.3) becomes

$$\varphi(t) = \varphi_0 \left[1 - \frac{2}{\sqrt{\pi}} \int_0^{\frac{x}{2\sqrt{Kt}}} e^{-u^2}\,du \right] = \varphi_0 \left[1 - \text{erf} \frac{x}{2\sqrt{Kt}} \right]. \tag{7.5}$$

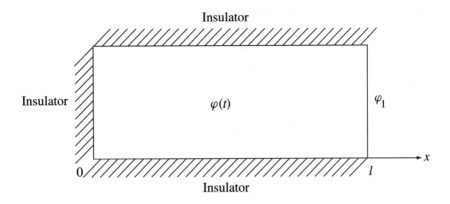

FIGURE 5.14
Details for Example 5.7.3.

Example 5.7.3
A finite medium of length l is at initial temperature φ_0. There is no heat flow across the boundary at $x = 0$, and the face at $x = l$ is then kept at φ_1 (see Figure 5.14). Determine the temperature $\varphi(t)$.

Solution   Here we have to solve

$$\frac{\partial^2 \varphi}{\partial x^2} = \frac{1}{k}\varphi$$

subject to the boundary conditions:

a. $\varphi = \varphi_0$ $t = 0$ $0 \le x \le l$
b. $\varphi = \varphi_1$ $t > 0$ $x = l$
c. $\frac{\partial \varphi}{\partial x} = 0$ $t > 0$ $x = 0$.

Upon Laplace transforming the controlling differential equation, we obtain

$$\frac{d^2 \Phi}{dx^2} - \frac{s}{k}\Phi = 0.$$

The solution is

$$\Phi = A'e^{-x\sqrt{\frac{s}{k}}} + B'e^{x\sqrt{\frac{s}{k}}} = A \cosh x\sqrt{\frac{s}{k}} + B \sinh x\sqrt{\frac{s}{k}}.$$

By condition c

$$\frac{d\Phi}{dx} = 0 \qquad x = 0 \qquad t > 0.$$

This imposes the requirement that $B = 0$ so that

$$\Phi = A \cosh x\sqrt{\frac{s}{k}}.$$

Now condition b is imposed. This requires that

$$\frac{\varphi_1}{s} = A \cosh l \sqrt{\frac{s}{k}} .$$

Thus by b and c

$$\Phi = \varphi_1 \frac{\cosh x \sqrt{\frac{s}{k}}}{s \cosh l \sqrt{\frac{s}{k}}} .$$

Now, to satisfy c we have

$$\Phi = \frac{\varphi_0}{s} - \frac{\varphi_0}{s} \frac{\cosh x \sqrt{\frac{s}{k}}}{\cosh l \sqrt{\frac{s}{k}}} .$$

Thus the final form of the Laplace transformed equation that satisfies all conditions of the problem is

$$\Phi = \frac{\varphi_0}{s} + \frac{\varphi_1 - \varphi_0}{s} \frac{\cosh x \sqrt{\frac{s}{k}}}{\cosh l \sqrt{\frac{s}{k}}} .$$

To find the expression for $\varphi(x, t)$, we must invert this expression. That is,

$$\varphi(x, t) = \varphi_0 + \frac{\varphi_1 - \varphi_0}{2\pi j} \int_{\sigma - j\infty}^{\sigma + j\infty} e^{st} \frac{\cosh x \sqrt{\frac{s}{k}}}{\cosh l \sqrt{\frac{s}{k}}} \frac{ds}{s} . \qquad (7.6)$$

The integrand is a single valued function of s with poles at $s = 0$ and $s = -k(\frac{2n-1}{2})^2 \frac{\pi^2}{l^2}$, $n = 1, 2, \ldots$. We select the path of integration that is shown in Figure 5.15. But the inversion integral over the path $BCA(= \Gamma) = 0$. Thus the inversion integral becomes

$$\frac{1}{2\pi j} \int_{\sigma - j\infty}^{\sigma + j\infty} e^{st} \frac{\cosh x \sqrt{\frac{s}{k}}}{\cosh l \sqrt{\frac{s}{k}}} \frac{ds}{s} .$$

By an application of the Cauchy integral theorem, we require the residues of the integrand at its poles. There results

$$\text{Res}|_{s=0} = 1$$

$$\text{Res}\Big|_{s=-k(n-\frac{1}{2})^2 \frac{\pi^2}{l^2}} = \frac{e^{-k(n-\frac{1}{2})^2 \frac{\pi^2}{l^2}} \cosh j \left(n - \frac{1}{2}\right) \frac{\pi x}{l}}{\left[s \frac{d}{ds} \{\cosh l \sqrt{\frac{s}{k}}\}_{s=-k(n-\frac{1}{2})^2 \frac{\pi^2}{l^2}} \right]} .$$

Combine these with (7.5) to write finally

$$\varphi(x, t) = \varphi_0 + \frac{4(\varphi_1 - \varphi_0)}{\pi} \sum_{n=1}^{\infty} \frac{(-1)^n}{2n - 1} l^{-k(n-\frac{1}{2})^2 \pi^2 / l^2} \cos \left[\left(n - \frac{1}{2}\right) \pi x / l\right]. \qquad (7.7)$$

∎

Example 5.7.4

A circular cylinder of radius a is initially at temperature zero. The surface is then maintained at temperature φ_0. Determine the temperature of the cylinder at any subsequent time t.

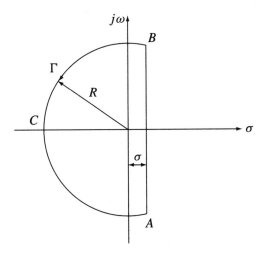

FIGURE 5.15
The path of integration for Example 5.7.3.

Solution The heat conduction equation in radial form is

$$\frac{\partial^2 \varphi}{\partial r^2} + \frac{1}{r}\frac{\partial \varphi}{\partial r} = \frac{1}{k}\frac{\partial \varphi}{\partial t}, \qquad 0 \le r < a, \qquad t > 0. \tag{7.8}$$

And for this problem the system is subject to the boundary conditions

C-1. $\varphi = 0$ $\quad t = 0$ $\quad 0 \le r < a$

C-2. $\varphi = \varphi_0$ $\quad t > 0$ $\quad r = a.$

To proceed, we multiply each term in the partial differential equation by $e^{-st}\,dt$ and integrate. We write

$$\int_0^\infty \varphi e^{-st}\,dt = \Phi(r, s)$$

Then (7.7) transforms to

$$k\left(\frac{d^2\Phi}{dr^2} + \frac{1}{r}\frac{d\Phi}{dr}\right) - s\Phi = 0,$$

which we write in the form

$$\frac{d^2\Phi}{dr^2} + \frac{1}{r}\frac{d\Phi}{dr} - \mu\Phi = 0, \qquad \mu = \sqrt{\frac{s}{k}}.$$

This is the Bessel equation of order 0 and has the solution

$$\Phi = AI_0(\mu r) + BN_0(\mu r).$$

However, the Laplace transformed form of C-1 when $z = 0$ imposes the condition $B = 0$ because $N_0(0)$ is not zero. Thus

$$\Phi = AI_0(\mu r).$$

The boundary condition C-2 requires $\Phi(r, a) = \frac{\varphi_0}{s}$ when $r = a$, hence

$$A = \frac{\varphi_0}{s}\frac{1}{I(\mu a)}$$

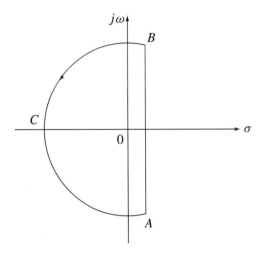

FIGURE 5.16
The path of integration for Example 5.7.4.

so that

$$\Phi = \frac{\varphi_0}{s} \frac{I_0(\mu r)}{I_0(\mu a)} .$$

To find the function $\varphi(r, t)$ requires that we invert this function. By an application of the inversion integral, we write

$$\varphi(r, t) = \frac{\varphi_0}{2\pi j} \int_{\sigma - j\infty}^{\sigma + j\infty} e^{\lambda t} \frac{I_0(\xi r)}{I_0(\xi a)} \frac{d\lambda}{\lambda}, \qquad \xi = \sqrt{\frac{\lambda}{k}} . \tag{7.9}$$

Note that $I_0(\xi r)/I_0(\xi a)$ is a single-valued function of λ. To evaluate this integral we choose as the path for this integration that shown in Figure 5.16. The poles of this function are at $\lambda = 0$ and at the roots of the Bessel function $J_0(\xi a)(= I_0(j\xi a))$; these occur when $J_0(\xi a) = 0$, with the roots for $J_0(\xi a) = 0$, namely $\lambda = -k\xi_1^2, -k\xi_2^2, \ldots$. The approximations for $I_0(\xi r)$ and $I_0(\xi a)$ show that when $n \to \infty$ the integral over the path BCA tends to zero. The resultant value of the integral is written in terms of the residues at zero and when $\lambda = k\xi_n^2$. These are

$$\text{Res} \big|_{=0} = 1$$

$$\text{Res} \bigg|_{k\xi_n^2} = \frac{\lambda d I_0(\xi a)}{d\lambda} \bigg|_{k\xi_n^2} .$$

Therefore

$$\varphi(r, t) = \varphi_0 \left[1 + \sum_n e^{-k\xi_n^2 t} \frac{J_0(\xi_n r)}{\lambda \frac{d}{d\lambda} J_0(\xi a) \big|_{\lambda = k\xi_n^2}} \right].$$

Further, $\lambda \frac{d}{d\lambda} I_0(\xi_n a) = \frac{1}{2} \xi a I_0^{(1)}(\xi a)$. Hence, finally,

$$\varphi(t) = \varphi_0 \left[1 + \frac{2}{a} \sum_{n=1}^{\infty} e^{-k\xi_n^2 t} \frac{J_0(\xi_n r)}{\xi_n J_0^{(1)}(\xi_n a)} \right]. \tag{7.10}$$

Example 5.7.5

A semi-infinite stretched string is fixed at each end. It is given an initial transverse displacement and then released. Determine the subsequent motion of the string.

Solution   This requires solving the wave equation

$$a^2 \frac{\partial^2 \varphi}{\partial x^2} = \frac{\partial^2 \varphi}{\partial t^2} \tag{7.11}$$

subject to the conditions

   C-1. $\varphi(x, 0) = f(x)$ $t = 0, \varphi(0, t) = 0$ $t > 0$

   C-2. $\lim_{x \to \infty} \varphi(x, t) = 0.$

To proceed, multiply both sides of (7.11) by $e^{-st} dt$ and integrate. The result is the Laplace-transformed equation

$$a^2 \frac{d^2 \Phi}{dx^2} = s^2 \Phi - s\varphi(0+), \qquad x > 0. \tag{7.12}$$

This transformed equation is subject to the transformed boundary conditions

   C-1. $\Phi(0, s) = 0$

   C-2. $\lim_{x \to \infty} \Phi(x, s) = 0.$

To solve (7.12) we will carry out a second Laplace transform, but this with respect to x, that is $\mathcal{L}\{\Phi(x, s)\} = N(z, s)$. Thus

$$N(z, s) = \int_0^\infty \Phi(x, s) e^{-zx} \, dx.$$

Apply this transformation to both members of (7.12) subject to $\Phi(0, s) = 0$. The result is

$$s^2 N(z, s) - s\Phi(z) = a^2 \left[z^2 N(z, s) - \frac{\partial \Phi}{\partial x}(0, s) \right], \qquad \Phi(z) = \mathcal{L}\{\varphi_0\}.$$

We denote $\frac{\partial \Phi}{\partial x}(0, s)$ by C. Then the solution of this equation is

$$N(z, s) = \frac{C}{z^2 - \frac{s^2}{a^2}} - \frac{s}{a^2} \Phi(z) \frac{1}{z^2 - \frac{s^2}{a^2}}.$$

The inverse transformation with respect to z is, employing convolution,

$$\Phi(x, s) = \frac{aC}{s} \sinh \frac{sx}{a} - \frac{1}{a} \int_0^x \varphi(\xi) \sinh \frac{s}{a}(x - \xi) \, d\xi.$$

To satisfy the condition $\lim_{x \to \infty} \Phi(x, s) = 0$ requires that the sinh terms be replaced by their exponential forms. Thus the factors

$$\sinh \frac{sx}{a} \to \frac{1}{2}, \qquad \sinh \frac{s}{a}(x - \xi) \to \frac{e^{-s\xi/a}}{2}, \qquad x \to \infty.$$

Then we have the expression

$$\Phi(x, s) = \frac{aC}{2s} - \frac{1}{2a} \int_0^x \varphi(\xi) e^{-s\xi/a} \, d\xi.$$

But for this function to be zero for $x \to \infty$ requires that

$$\frac{aC}{s} = \frac{1}{a} \int_0^\infty \varphi(\xi) e^{-s\xi/a} \, d\xi, \qquad x \to \infty.$$

Combine this result with $\Phi(x, s)$ to get

$$2a\Phi(x, s) = \int_0^\infty \varphi(\xi)e^{-s(\xi-x)/a}\, d\xi - \int_0^\infty \varphi(\xi)e^{-s(x+\xi)/a}\, d\xi + \int_0^x \varphi(\xi)e^{-s(x-\xi)/a}\, d\xi.$$

Each integral in this expression is integrated by parts. Here we write

$$u = \varphi(\xi), \qquad du = \varphi^{(1)}(\xi)d\xi; \qquad dv = e^{-\frac{s(\xi-x)}{a}}\, d\xi, \qquad v = \frac{a}{s}e^{-s(\xi-k)/a}.$$

The resulting integrations lead to

$$\Phi(x, s) = \frac{1}{s}\varphi(x) + \frac{1}{2s}\int_x^\infty \varphi^{(1)}(\xi)e^{-\frac{s(\xi-x)}{a}}\, d\xi - \frac{1}{2s}\int_0^\infty \varphi^{(1)}(\xi)e^{-\frac{s(x+\xi)}{a}}\, d\xi$$

$$-\frac{1}{2s}\int_0^\infty \varphi^{(1)}(\xi)e^{-\frac{s(x-\xi)}{a}}\, d\xi.$$

We note by entry 61, Table 5.1 that

$$\mathcal{L}^{-1}\left\{\frac{1}{s}e^{\frac{-s(\xi-x)}{a}}\right\} = 1 \qquad \text{when } at > \xi - x$$

$$= 0 \qquad \text{when } at < \xi - x.$$

This function of ξ vanishes except when $\xi \le x + at$. Thus

$$\mathcal{L}^{-1}\left\{\frac{1}{2}\int_x^\infty \varphi^{(1)}(\xi)e^{-s(\xi-x)/a}\, d\xi\right\} = \frac{1}{2}\int_x^{x+at} \varphi^{(1)}(\xi)\, d\xi = \frac{1}{2}\varphi(x + at) - \frac{1}{2}\varphi(x).$$

Proceed in the same way for the term

$$\mathcal{L}^{-1}\left\{\frac{1}{2}\int_0^\infty \varphi^{(1)}(\xi)e^{-s(x+\xi)/a}\, d\xi\right\} = 1 \qquad \text{when } at > x + \xi$$

$$= 0 \qquad \text{when } at < x + \xi.$$

Thus the second term becomes

$$\mathcal{L}^{-1}\left\{\frac{1}{2}\int_x^\infty \varphi^{(1)}(\xi)e^{-s(x+\xi)/a}\, d\xi\right\} = \frac{1}{2}\int_x^{x+at} \varphi^{(1)}(\xi)\, d\xi = \frac{1}{2}\varphi(x - at).$$

The final term becomes

$$-\frac{1}{2}\varphi(x) \qquad \text{when } at > x$$

$$-\frac{1}{2}\varphi(x) + \frac{1}{2}\varphi(x - at) \qquad \text{when } at < x.$$

The final result is

$$\varphi(x, t) = \frac{1}{2}[f(at + x) - f(at - x)] \qquad \text{when } t > x/a$$

$$= \frac{1}{2}[f(x + at) + f(x - at)] \qquad \text{when } t < x/a. \tag{7.13}$$

■

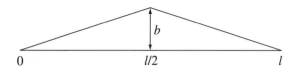

FIGURE 5.17
A stretched string plucked at its midpoint.

Example 5.7.6
A stretched string of length l is fixed at each end as shown in Figure 5.17. It is plucked at the midpoint and then released at $t = 0$. The displacement is b. Find the subsequent motion.

Solution   This problem requires the solution of

$$\frac{\partial^2 y}{\partial y^2} = c^2 \frac{\partial^2 y}{\partial t^2}, \qquad 0 < y < l, \qquad t > 0 \tag{7.14}$$

subject to

$$
\left.\begin{array}{lll}
1. & y = \frac{2bx}{l} & 0 < x < l/2 \\
2. & y = \frac{2b}{l}(l - x) & \frac{l}{2} < x < l
\end{array}\right\} t = 0
$$

$$
\begin{array}{llll}
3. & \frac{\partial y}{\partial t} = 0 & 0 < x < l & t = 0 \\
4. & y = 0 & x = 0; \quad x = l & t > 0.
\end{array}
$$

To proceed, multiply (7.13) by $e^{-st}\,dt$ and integrat in t. This yields

$$s^2 Y - sy(0) = c^2 \frac{d^2 Y}{dx^2}$$

or

$$c^2 \frac{d^2 Y}{dx^2} - s^2 Y = -sy(0) = -sf(x) \tag{7.15}$$

subject to $Y(0, s) = Y(l, s) = 0$. To solve this equation, we proceed as in Example 5.7.5; that is, we apply a transformation on x, namely $\mathcal{L}\{Y(x, s)\} = N(z, s)$. Thus

$$s^2 N(z, s) - sY(0) = c^2 \left[z^2 N(z, s) - \frac{y(0, s)}{x} \right].$$

This equation yields, writing $sY(0)$ as $\Phi(x, s)$,

$$N(z, s) = \frac{\Phi(z, s)}{z^2 - \frac{s^2}{c^2}}$$

The inverse transform is

$$Y(x, s) = \mathcal{L}^{-1}\{N(z, s)\} = \frac{\Phi(x, s)}{c \sinh \frac{s}{c}}$$

where

$$\Phi(x, s) = \sinh \frac{(l - x)s}{c} \int_0^x y(\xi) \sinh \frac{\xi s}{c}\, d\xi + \sinh \frac{xs}{c} \int_0^l y(\xi) \sinh \frac{(l - \xi)s}{c}\, d\xi.$$

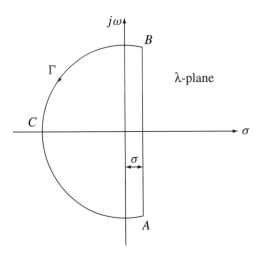

FIGURE 5.18
The path of integration for Example 5.7.6.

Combine these integrals with the known form of $f(x)$ in C-1 and C-2. Upon carrying out the integrations, the resulting forms become, with $k = \frac{s}{c}$,

$$\frac{lY}{2b} = \frac{x}{s} - \frac{c \sinh kx}{s^2 \cosh \frac{kl}{2}}, \qquad 0 \le x \le l/2$$

$$\frac{lY}{2b} = \frac{(l-x)}{s} - \frac{c \sinh k(l-x)}{s^2 \cosh \frac{kl}{2}}, \qquad \frac{l}{2} \le x \le l.$$

To find $y(t)$, we must invert these expressions. Note that symmetry exists and so we need consider only the first term. We use the inversion integral on the term $\frac{1}{s^2} \frac{\sinh kx}{\cosh \frac{kl}{2}}$. Thus we consider the integral

$$I = \frac{1}{2\pi j} \int_{\sigma - j\infty}^{\sigma + j\infty} e^{\lambda b} \frac{\sinh \frac{x\lambda}{c}}{\cosh \frac{x\lambda}{2c}} \frac{d\lambda}{\lambda^2}.$$

We choose the path in the λ-plane as shown in Figure 5.18. The value of the integral over path Γ is zero. Thus the value of the integral is given in terms of the residues. These occur at $\lambda = 0$ and at the values for which $\cosh \frac{\lambda}{2c} = 0$, which exist where

$$\frac{\lambda}{2c} = j \frac{2n-1}{l} \frac{\pi}{2} \qquad \text{or} \qquad \lambda = \pm j \frac{2n-1}{l} \pi c.$$

Thus we have, by the theory of residues

$$\text{Res}|_{\lambda=0} = \frac{x}{c}$$

$$\text{Res}|_{j \frac{2n-1}{l}\pi c} = e^{j(2n-1)\frac{\pi cx}{l}} \frac{\sinh j(2n-1)\frac{\pi x}{l}}{\frac{d}{d\lambda}\left[\lambda^2 \cosh \frac{\lambda l}{2c}\right]_{j\frac{(2n-1)}{l}\pi c}}.$$

These poles lead to

$$= (-1)^n \frac{2l}{\pi^2 c} \frac{\sin(2n-1)\frac{\pi x}{l}}{(2n-1)^2} e^{j(2n-1)\frac{\pi c}{l}t}.$$

Thus the poles at $\pm j(2n-1)\frac{\pi c}{l}$ lead to

$$= (-1)^n \frac{4l}{\pi^2 c} \frac{\sin(2n-1)\frac{\pi x}{l}\cos(2n-1)\frac{\pi c}{l}t}{(2n-1)^2} \; .$$

Then we have

$$\frac{\ell y}{2b} = x - \left\{ x + \frac{4l}{\pi^2} \sum_{n=1}^{\infty} \frac{(-1)^n}{(2n-1)^2} \sin(2n-1)\frac{\pi x}{l}\cos(2n-1)\frac{\pi ct}{l} \right\}$$

so that finally

$$y = \frac{8b}{\pi^2} \sum_{n=1}^{\infty} \frac{(-1)^{n-1}}{(2n-1)^2} \sin(2n-1)\frac{\pi x}{l}\cos(2n-1)\frac{\pi ct}{l}, \qquad 0 \le x \le \frac{l}{2} \; .$$

For the string for which $\frac{l}{2} \le x < l$, the corresponding expression is the same except that $(l-x)$ replaces x.

Note that this equation can be written, with $\eta = (2n-1)\frac{\pi}{l}$,

$$\sin \eta x \cos \eta ct = \frac{\sin \eta (x-ct) + \sin \eta (x+ct)}{2} \; ,$$

which shows the traveling wave nature of the solution. ∎

5.8 The bilateral or two-sided Laplace transform

In Section 5.1 we discussed the fact that the region of absolute convergence of the unilateral or one-sided Laplace transform is the region to the left of the abscissa of convergence. The situation for the two-sided Laplace transform is rather different; the region of convergence must be specified if we wish to invert a function $F(s)$ that was obtained using the bilateral Laplace transform. This requirement is necessary because different time signals might have the same Laplace transform but different regions of absolute convergence.

To establish the region of convergence write the bilateral Laplace transform in the form

$$F_2(s) = \int_{-\infty}^{\infty} e^{-st} f(t)\,dt = \int_{0}^{\infty} e^{-st} f(t)\,dt + \int_{-\infty}^{0} e^{-st} f(t)\,dt. \qquad (8.1)$$

If the function $f(t)$ is of exponential order $(e^{\sigma_1 t})$, then the region of convergence for $t > 0$ is $\mathrm{Re}(s) > \sigma_1$. If the function $f(t)$ for $t < 0$ is of exponential order $\exp(\sigma_2 t)$, then the region of convergence is $\mathrm{Re}(s) < \sigma_2$. Hence the function $F_2(s)$ exists and is analytic in the vertical strip defined by $\sigma_1 < \mathrm{Re}(s) < \sigma_2$, provided, of course, that $\sigma_1 < \sigma_2$. If $\sigma_2 > \sigma_1$, no region of convergence would exist and the inversion process could not be performed. This region of convergence is shown in Figure 5.19.

Example 5.8.1
Find the bilateral Laplace transform of the signals $f(t) = e^{-at}u(t)$ and $f(t) = -e^{-at}u(-t)$ and specify their regions of convergence.

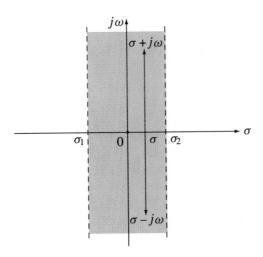

FIGURE 5.19
Region of convergence for the Bilateral Transform.

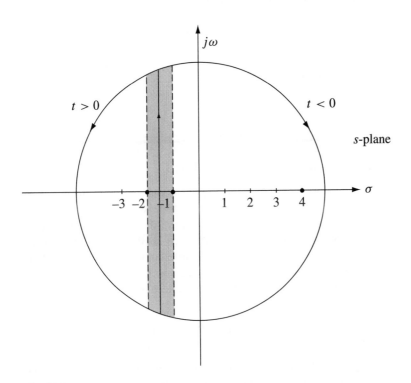

FIGURE 5.20
Illustrating Example 5.8.2.

Solution   Using the basic definition of the transform (8.1) we obtain

$$F_2(s) = \int_{-\infty}^{\infty} e^{-at} u(t) e^{-st} \, dt = \int_0^{\infty} e^{-(s+a)t} \, dt = \frac{1}{s+a}$$

and its region of convergence is $\text{Re}(s) > -a$.

For the second signal

$$F_2(s) = \int_{-\infty}^{\infty} -e^{-at}v(-t)e^{-st}\,dt = -\int_{-\infty}^{0} e^{-(s+a)t}\,dt = \frac{1}{s+a}$$

and its region of convergence is $\mathrm{Re}(s) < -a$.

Clearly, the knowledge of the region of convergence is necessary to find the time functions unambiguously.   ∎

Example 5.8.2

Find the function $f(t)$, if its Laplace transform is given by

$$F_2(s) = \frac{3}{(s-4)(s+1)(s+2)}, \qquad -2 < \mathrm{Re}(s) < -1.$$

Solution   The region of convergence and the paths of integration are shown in Figure 5.20. For $t > 0$ we close the contour to the left and we obtain

$$f(t) = \left.\frac{3e^{st}}{(s-4)(s+1)}\right|_{s=-2} = \frac{1}{2}e^{-2t}, \qquad t > 0.$$

For $t < 0$ the contour closes to the right, and now

$$f(t) = \left.\frac{3e^{st}}{(s-4)(s+2)}\right|_{s=-1} + \left.\frac{3e^{st}}{(s+1)(s+2)}\right|_{s=4} = -\frac{3}{5}e^{-t} + \frac{e^{4t}}{10}, \qquad t < 0. \qquad ∎$$

Appendix

TABLE 5.1
Laplace transform pairs.

	$F(s)$	$f(t)$
1	$\frac{1}{s}$	$u(t)$, unit step function
2	$\frac{1}{s^2}$	t
3	$\frac{1}{s^n}$ $(n = 1, 2, \ldots)$	$\frac{t^{n-1}}{(n-1)!}$
4	$\frac{1}{\sqrt{s}}$	$\frac{1}{\sqrt{\pi t}}$
5	$s^{-3/2}$	$2\sqrt{\frac{t}{\pi}}$
6	$s^{-[n+(1/2)]}$ $(n = 1, 2, \ldots)$	$\frac{2^n t^{n-(1/2)}}{1 \cdot 3 \cdot 5 \cdots (2n-1)\sqrt{\pi}}$
7	$\frac{\Gamma(k)}{s^k}$ $(k \geq 0)$	t^{k-1}
8	$\frac{1}{s-a}$	e^{at}
9	$\frac{1}{(s-a)^2}$	te^{at}
10	$\frac{1}{(s-a)^n}$ $(n = 1, 2, \ldots)$	$\frac{1}{(n-1)!}t^{n-1}e^{at}$
11	$\frac{\Gamma(k)}{(s-a)^k}$ $(k \geq 0)$	$t^{k-1}e^{at}$
12	$\frac{1}{(s-a)(s-b)}$	$\frac{1}{a-b}(e^{at} - e^{bt})$
13	$\frac{s}{(s-a)(s-b)}$	$\frac{1}{a-b}(ae^{at} - be^{bt})$
14	$\frac{1}{(s-a)(s-b)(s-c)}$	$-\frac{(b-c)e^{at}+(c-a)e^{bt}+(a-b)e^{ct}}{(a-b)(b-c)(c-a)}$
15	$\frac{1}{s^2+a^2}$	$\frac{1}{a}\sin at$
16	$\frac{s}{s^2+a^2}$	$\cos at$
17	$\frac{1}{s^2-a^2}$	$\frac{1}{a}\sinh at$
18	$\frac{s}{s^2-a^2}$	$\cosh at$
19	$\frac{1}{s(s^2+a^2)}$	$\frac{1}{a^2}(1 - \cos at)$
20	$\frac{1}{s^2(s^2+a^2)}$	$\frac{1}{a^3}(at - \sin at)$
21	$\frac{1}{(s^2+a^2)^2}$	$\frac{1}{2a^3}(\sin at - at\cos at)$
22	$\frac{s}{(s^2+a^2)^2}$	$\frac{t}{2a}\sin at$

	$F(s)$	$f(t)$
23	$\frac{s^2}{(s^2+a^2)^2}$	$\frac{1}{2a}(\sin at + at\cos at)$
24	$\frac{s^2-a^2}{(s^2+a^2)^2}$	$t\cos at$
25	$\frac{s}{(s^2+a^2)(s^2+b^2)}\,(a^2 \neq b^2)$	$\frac{\cos at - \cos bt}{b^2-a^2}$
26	$\frac{1}{(s-a)^2+b^2}$	$\frac{1}{b}e^{at}\sin bt$
27	$\frac{s-a}{(s-a)^2+b^2}$	$e^{at}\cos bt$
27.1	$\frac{1}{[(s+a)^2+b^2]^n}$	$\frac{-e^{-at}}{4^{n-1}b^{2n}}\sum_{r=1}^{n}\begin{pmatrix}2n-r-1\\n-1\end{pmatrix}(-2t)^{r-1}\frac{d^r}{dt^r}[\cos(bt)]$
27.2	$\frac{s}{[(s+a)^2+b^2]^n}$	$\frac{e^{-at}}{4^{n-1}b^{2n}}\left\{\sum_{r=1}^{n}\begin{pmatrix}2n-r-1\\n-1\end{pmatrix}\right.$ $(-2t)^{r-1}\frac{d^r}{dt^r}[a\cos(bt)+b\sin(bt)]$ $-2b\sum_{r=1}^{n-1}r\begin{pmatrix}2n-r-2\\n-1\end{pmatrix}$ $\left.(-2t)^{r-1}\frac{d^r}{dt^r}[\sin bt]\right\}$
28	$\frac{3a^2}{s^3+a^3}$	$e^{-at}-e^{(at)/2}\left(\cos\frac{at\sqrt{3}}{2}-\sqrt{3}\sin\frac{at\sqrt{3}}{2}\right)$
29	$\frac{4a^3}{s^4+4a^4}$	$\sin at\cosh at - \cos at\sinh at$
30	$\frac{s}{s^4+4a^4}$	$\frac{1}{2a^2}\sin at\sinh at$
31	$\frac{1}{s^4-a^4}$	$\frac{1}{2a^3}(\sinh at - \sin at)$
32	$\frac{s}{s^4-a^4}$	$\frac{1}{2a^2}(\cosh at - \cos at)$
33	$\frac{8a^3s^2}{(s^2+a^2)^3}$	$(1+a^2t^2)\sin at - \cos at$
34*	$\frac{1}{s}\left(\frac{s-1}{s}\right)^n$	$L_n(t)=\frac{e^t}{n!}\frac{d^n}{dt^n}(t^ne^{-t})$
35	$\frac{s}{(s-a)^{3/2}}$	$\frac{1}{\sqrt{\pi t}}e^{at}(1+2at)$
36	$\sqrt{s-a}-\sqrt{s-b}$	$\frac{1}{2\sqrt{\pi t^3}}(e^{bt}-e^{at})$
37	$\frac{1}{\sqrt{s}+a}$	$\frac{1}{\sqrt{\pi t}}-ae^{a^2t}\,\mathrm{erfc}(a\sqrt{t})$
38	$\frac{\sqrt{s}}{s-a^2}$	$\frac{1}{\sqrt{\pi t}}+ae^{a^2t}\,\mathrm{erfc}(a\sqrt{t})$
39	$\frac{\sqrt{s}}{s+a^2}$	$\frac{1}{\sqrt{\pi t}}-\frac{2a}{\sqrt{\pi}}e^{-a^2t}\int_0^{a\sqrt{t}}e^{\lambda^2}\,d\lambda$
40	$\frac{1}{\sqrt{s}(s-a^2)}$	$\frac{1}{a}e^{a^2t}\,\mathrm{erfc}(a\sqrt{t})$
41	$\frac{1}{\sqrt{s}(s+a^2)}$	$\frac{2}{a\sqrt{\pi}}e^{-a^2t}\int_0^{a\sqrt{t}}e^{\lambda^2}\,d\lambda$
42	$\frac{b^2-a^2}{(s-a^2)(b+\sqrt{s})}$	$e^{a^2t}[b-a\,\mathrm{erf}(a\sqrt{t})]-be^{b^2t}\,\mathrm{erfc}(b\sqrt{t})$
43	$\frac{1}{\sqrt{s}(\sqrt{s}+a)}$	$e^{a^2t}\,\mathrm{erfc}(a\sqrt{t})$
44	$\frac{1}{(s+a)\sqrt{s+b}}$	$\frac{1}{\sqrt{b-a}}e^{-at}\,\mathrm{erf}(\sqrt{b-a}\sqrt{t})$

*$L_n(t)$ is the Laguerre polynomial of degree n.

	$F(s)$	$f(t)$
45	$\frac{b^2-a^2}{\sqrt{s}(s-a^2)(\sqrt{s}+b)}$	$e^{a^2 t}\left[\frac{b}{a}\operatorname{erf}(a\sqrt{t})-1\right]+e^{b^2 t}\operatorname{erfc}(b\sqrt{t})$
46*	$\frac{(1-s)^n}{s^{n+(1/2)}}$	$\frac{n!}{(2n)!\sqrt{\pi t}}H_{2n}(\sqrt{t})$
47	$\frac{(1-s)}{s^{n+(3/2)}}$	$-\frac{n!}{\sqrt{\pi}(2n+1)!}H_{2n+1}(\sqrt{t})$
48†	$\frac{\sqrt{s+2a}}{\sqrt{s}}-1$	$ae^{-at}[I_1(at)+I_0(at)]$
49	$\frac{1}{\sqrt{s+a}\sqrt{s+b}}$	$e^{-(1/2)(a+b)t}I_0\left(\frac{a-b}{2}t\right)$
50	$\frac{\Gamma(k)}{(s+a)^k(s+b)^k}(k\geq 0)$	$\sqrt{\pi}\left(\frac{t}{a-b}\right)^{k-(1/2)}e^{-(1/2)(a+b)t}I_{k-(1/2)}\left(\frac{a-b}{2}t\right)$
51	$\frac{1}{(s+a)^{1/2}(s+b)^{3/2}}$	$te^{-(1/2)(a+b)t}\left[I_0\left(\frac{a-b}{2}t\right)+I_1\left(\frac{a-b}{2}t\right)\right]$
52	$\frac{\sqrt{s+2a}-\sqrt{s}}{\sqrt{s+2a}+\sqrt{s}}$	$\frac{1}{t}e^{-at}I_1(at)$
53	$\frac{(a-b)^k}{(\sqrt{s+a}+\sqrt{s+b})^{2k}}(k>0)$	$\frac{k}{t}e^{-(1/2)(a+b)t}I_k\left(\frac{a-b}{2}t\right)$
54	$\frac{(\sqrt{s+a}+\sqrt{s})^{-2\nu}}{\sqrt{s}\sqrt{s+a}}(\nu>-1)$	$\frac{1}{a^\nu}e^{-(1/2)(at)}I_\nu\left(\frac{1}{2}at\right)$
55	$\frac{1}{\sqrt{s^2+a^2}}$	$J_0(at)$
56	$\frac{(\sqrt{s^2+a^2}-s)^\nu}{\sqrt{s^2+a^2}}(\nu>-1)$	$a^\nu J_\nu(at)$
57	$\frac{1}{(s^2+a^2)^k}(k>0)$	$\frac{\sqrt{\pi}}{\Gamma(k)}\left(\frac{t}{2a}\right)^{k-(1/2)}J_{k-(1/2)}(at)$
58	$(\sqrt{s^2+a^2}-s)^k(k>0)$	$\frac{ka^k}{t}J_k(at)$
59	$\frac{(s-\sqrt{s^2-a^2})^\nu}{\sqrt{s^2-a^2}}(\nu>-1)$	$a^\nu I_\nu(at)$
60	$\frac{1}{(s^2-a^2)^k}(k>0)$	$\frac{\sqrt{\pi}}{\Gamma(k)}\left(\frac{t}{2a}\right)^{k-(1/2)}I_{k-(1/2)}(at)$
61	$\frac{e^{-ks}}{s}$	$S_k(t)=\begin{cases}0 & \text{when } 0<t<k \\ 1 & \text{when } t>k\end{cases}$
62	$\frac{e^{-ks}}{s^2}$	$\begin{cases}0 & \text{when } 0<t<k \\ t-k & \text{when } t>k\end{cases}$
63	$\frac{e^{-ks}}{s^\mu}(\mu>0)$	$\begin{cases}0 & \text{when } 0<t<k \\ \frac{(t-k)^{\mu-1}}{\Gamma(\mu)} & \text{when } t>k\end{cases}$
64	$\frac{1-e^{-ks}}{s}$	$\begin{cases}1 & \text{when } 0<t<k \\ 0 & \text{when } t>k\end{cases}$
65	$\frac{1}{s(1-e^{-ks})}=\frac{1+\coth\frac{1}{2}ks}{2s}$	$S(k,t)=n$ when $(n-1)k<t<nk(n=1,2,\ldots)$
66	$\frac{1}{s(e^{ks}-a)}$	$\begin{cases}0 \text{ when } 0<t<k \\ 1+a+a^2+\cdots+a^{n-1} \\ \quad \text{when } nk<t<(n+1)k(n=1,2,\ldots)\end{cases}$
67	$\frac{1}{s}\tanh ks$	$M(2k,t)=(-1)^{n-1}$ when $2k(n-1)<t<2kn$ $(n=1,2,\ldots)$
68	$\frac{1}{s(1+e^{-ks})}$	$\frac{1}{2}M(k,t)+\frac{1}{2}=\frac{1-(-1)^n}{2}$ when $(n-1)k<t<nk$

* $H_n(x)$ is the hermite polynomial, $H_n(x)=e^{x^2}\frac{d^n}{dx^n}(e^{-x^2})$.

† $I_n(x)=i^{-n}J_n(ix)$, where J_n is Bessel's function of the first kind.

	$F(s)$	$f(t)$
69*	$\frac{1}{s^2}\tanh ks$	$H(2k, t)$
70	$\frac{1}{s\sinh ks}$	$2S(2k, t+k) - 2 = 2(n-1)$ $\quad$ when $(2n-3)k < t < (2n-1)k \ (t>0)$
71	$\frac{1}{s\cosh ks}$	$M(2k, t+3k) + 1 = 1 + (-1)^n$ $\quad$ when$(2n-3)k < t < (2n-1)k \ (t>0)$
72	$\frac{1}{s}\coth ks$	$2S(2k, t) - 1 = 2n - 1$ $\quad$ when $2k(n-1) < t < 2kn$
73	$\frac{k}{s^2+k^2}\coth\frac{\pi s}{2k}$	$\lvert \sin kt \rvert$
74	$\frac{1}{(s^2+1)(1-e^{-\pi s})}$	$\begin{cases} \sin t & \text{when } (2n-2)\pi < t < (2n-1)\pi \\ 0 & \text{when } (2n-1)\pi < t < 2n\pi \end{cases}$
75	$\frac{1}{s}e^{-k/s}$	$J_0(2\sqrt{kt})$
76	$\frac{1}{\sqrt{s}}e^{-k/s}$	$\frac{1}{\sqrt{\pi t}}\cos 2\sqrt{kt}$
77	$\frac{1}{\sqrt{s}}e^{k/s}$	$\frac{1}{\sqrt{\pi t}}\cosh 2\sqrt{kt}$
78	$\frac{1}{s^{3/2}}e^{-k/s}$	$\frac{1}{\sqrt{\pi k}}\sin 2\sqrt{kt}$
79	$\frac{1}{s^{3/2}}e^{k/s}$	$\frac{1}{\sqrt{\pi k}}\sinh 2\sqrt{kt}$
80	$\frac{1}{s^\mu}e^{-k/s}\,(\mu > 0)$	$\left(\frac{t}{k}\right)^{(\mu-1)/2} J_{\mu-1}(2\sqrt{kt})$
81	$\frac{1}{s^\mu}e^{k/s}\,(\mu > 0)$	$\left(\frac{t}{k}\right)^{(\mu-1)/2} I_{\mu-1}(2\sqrt{kt})$
82	$e^{-k\sqrt{s}}\,(k > 0)$	$\frac{k}{2\sqrt{\pi t^3}}\exp\left(\frac{k^2}{4t}\right)$
83	$\frac{1}{s}e^{-k\sqrt{s}}\,(k \geq 0)$	$\operatorname{erfc}\left(\frac{k}{2\sqrt{t}}\right)$
84	$\frac{1}{\sqrt{s}}e^{-k\sqrt{s}}\,(k \geq 0)$	$\frac{1}{\sqrt{\pi t}}\exp\left(-\frac{k^2}{4t}\right)$
85	$s^{-3/2}e^{-k\sqrt{s}}\,(k \geq 0)$	$2\sqrt{\frac{t}{\pi}}\exp\left(-\frac{k^2}{4t}\right) - k\operatorname{erfc}\left(\frac{k}{2\sqrt{t}}\right)$
86	$\frac{ae^{-k\sqrt{s}}}{s(a+\sqrt{s})}\,(k \geq 0)$	$-e^{ak}e^{a^2 t}\operatorname{erfc}\left(a\sqrt{t}+\frac{k}{2\sqrt{t}}\right) + \operatorname{erfc}\left(\frac{k}{2\sqrt{t}}\right)$
87	$\frac{e^{-k\sqrt{s}}}{\sqrt{s}(a+\sqrt{s})}\,(k \geq 0)$	$e^{ak}e^{a^2 t}\operatorname{erfc}\left(a\sqrt{t}+\frac{k}{2\sqrt{t}}\right)$
88	$\frac{e^{-k\sqrt{s(s+a)}}}{\sqrt{s(s+a)}}$	$\begin{cases} 0 & \text{when } 0 < t < k \\ e^{-(1/2)(at)}I_0\left(\frac{1}{2}a\sqrt{t^2-k^2}\right) & \text{when } t > k \end{cases}$
89	$\frac{e^{-k\sqrt{s^2+a^2}}}{\sqrt{s^2+a^2}}$	$\begin{cases} 0 & \text{when } 0 < t < k \\ J_0(a\sqrt{t^2-k^2}) & \text{when } t > k \end{cases}$
90	$\frac{e^{-k\sqrt{s^2-a^2}}}{\sqrt{s^2-a^2}}$	$\begin{cases} 0 & \text{when } 0 < t < k \\ I_0(a\sqrt{t^2-k^2}) & \text{when } t > k \end{cases}$
91	$\frac{e^{-k(\sqrt{s^2+a^2}-s)}}{\sqrt{s^2+a^2}}\,(k \geq 0)$	$J_0(a\sqrt{t^2+2kt})$
92	$e^{-ks} - e^{-k\sqrt{s^2+a^2}}$	$\begin{cases} 0 & \text{when } 0 < t < k \\ \frac{ak}{\sqrt{t^2-k^2}}J_1(a\sqrt{t^2-k^2}) & \text{when } t > k \end{cases}$

$^*H(2k, t) = k + (r - k)(-1)^n$ where $t = 2kn + r$; $0 \leq r < 2k$; $n = 0, 1, 2, \ldots$.

	$F(s)$	$f(t)$
93	$e^{-k\sqrt{s^2+a^2}} - e^{-ks}$	$\begin{cases} 0 & \text{when } 0 < t < k \\ \dfrac{ak}{\sqrt{t^2-k^2}} I_1(a\sqrt{t^2-k^2}) & \text{when } t > k \end{cases}$
94	$\dfrac{a^\nu e^{-k\sqrt{s^2-a^2}}}{\sqrt{s^2+a^2}(\sqrt{s^2+a^2}+s)^\nu}\,(\nu > -1)$	$\begin{cases} 0 & \text{when } 0 < t < k \\ \left(\dfrac{t-k}{t+k}\right)^{(1/2)\nu} J_\nu(a\sqrt{t^2-k^2}) & \text{when } t > k \end{cases}$
95	$\dfrac{1}{s}\log s$	$\Gamma'(1) - \log t\,[\Gamma'(1) = -0.5772]$
96	$\dfrac{1}{s^k}\log s\,(k > 0)$	$t^{k-1}\left\{\dfrac{\Gamma'(k)}{[\Gamma(k)]^2} - \dfrac{\log t}{\Gamma(k)}\right\}$
97	$\dfrac{\log s}{s-a}\,(a > 0)$	$e^{at}[\log a - \text{Ei}(-at)]$
98	$\dfrac{\log s}{s^2+1}$	$\cos t\,\text{Si}(t) - \sin t\,\text{Ci}(t)$
99	$\dfrac{s\log s}{s^2+1}$	$-\sin t\,\text{Si}(t) - \cos t\,\text{Ci}(t)$
100	$\dfrac{1}{s}\log(1 + ks)\,(k > 0)$	$-\text{Ei}\left(-\dfrac{t}{k}\right)$
101	$\log\dfrac{s-a}{s-b}$	$\dfrac{1}{t}(e^{bt} - e^{at})$
102	$\dfrac{1}{s}\log(1 + k^2s^2)$	$-2\,\text{Ci}\left(\dfrac{t}{k}\right)$
103	$\dfrac{1}{s}\log(s^2 + a^2)\,(a > 0)$	$2\log a - 2\,\text{Ci}(at)$
104	$\dfrac{1}{s^2}\log(s^2 + a^2)\,(a > 0)$	$\dfrac{2}{a}[at\log a + \sin at - at\,\text{Ci}(at)]$
105	$\log\dfrac{s^2+a^2}{s^2}$	$\dfrac{2}{t}(1 - \cos at)$
106	$\log\dfrac{s^2-a^2}{s^2}$	$\dfrac{2}{t}(1 - \cosh at)$
107	$\arctan\dfrac{k}{s}$	$\dfrac{1}{t}\sin kt$
108	$\dfrac{1}{s}\arctan\dfrac{k}{s}$	$\text{Si}(kt)$
109	$e^{k^2s^2}\text{erfc}(ks)\,(k > 0)$	$\dfrac{1}{k\sqrt{\pi}}\exp\left(-\dfrac{t^2}{4k^2}\right)$
110	$\dfrac{1}{s}e^{k^2s^2}\text{erfc}(ks)\,(k > 0)$	$\text{erf}\left(\dfrac{t}{2k}\right)$
111	$e^{ks}\text{erfc}(\sqrt{ks})\,(k > 0)$	$\dfrac{\sqrt{k}}{\pi\sqrt{t}(t+k)}$
112	$\dfrac{1}{\sqrt{s}}\text{erfc}(\sqrt{ks})$	$\begin{cases} 0 & \text{when } 0 < t < k \\ (\pi t)^{-1/2} & \text{when } t > k \end{cases}$
113	$\dfrac{1}{\sqrt{s}}e^{ks}\text{erfc}(\sqrt{ks})\,(k > 0)$	$\dfrac{1}{\sqrt{\pi(t+k)}}$
114	$\text{erf}\left(\dfrac{k}{\sqrt{s}}\right)$	$\dfrac{1}{\pi t}\sin(2k\sqrt{t})$
115	$\dfrac{1}{\sqrt{s}}e^{k^2/s}\text{erfc}\left(\dfrac{k}{\sqrt{s}}\right)$	$\dfrac{1}{\sqrt{\pi t}}e^{-2k\sqrt{t}}$
115.1	$-e^{as}\text{Ei}(-as)$	$\dfrac{1}{t+a};\,(a > 0)$
115.2	$\dfrac{1}{a} + se^{as}\text{Ei}(-as)$	$\dfrac{1}{(t+a)^2};\,(a > 0)$
115.3	$\left[\dfrac{\pi}{2} - \text{Si}(s)\right]\cos s + \text{Ci}(s)\sin s$	$\dfrac{1}{t^2+1}$

Si = sine integral; Ci = cosine integral; Ei = error integral.

	$F(s)$	$f(t)$
116*	$K_0(ks)$	$\begin{cases} 0 & \text{when } 0 < t < k \\ (t^2 - k^2)^{-1/2} & \text{when } t > k \end{cases}$
117	$K_0(k\sqrt{s})$	$\frac{1}{2t}\exp\left(-\frac{k^2}{4t}\right)$
118	$\frac{1}{s}e^{ks}K_1(ks)$	$\frac{1}{k}\sqrt{t(t+2k)}$
119	$\frac{1}{\sqrt{s}}K_1(k\sqrt{s})$	$\frac{1}{k}\exp\left(-\frac{k^2}{4t}\right)$
120	$\frac{1}{\sqrt{s}}e^{k/s}K_0\left(\frac{k}{s}\right)$	$\frac{2}{\sqrt{\pi t}}K_0(2\sqrt{2kt})$
121	$\pi e^{-ks}I_0(ks)$	$\begin{cases} [t(2k-t)]^{-1/2} & \text{when } 0 < t < 2k \\ 0 & \text{when } t > 2k \end{cases}$
122	$e^{-ks}I_1(ks)$	$\begin{cases} \frac{k-t}{\pi k\sqrt{t(2k-t)}} & \text{when } 0 < t < 2k \\ 0 & \text{when } t > 2k \end{cases}$

*$K_n(x)$ is Bessel's function of the second kind for the imaginary argument.

Several additional transforms, especially those involving other Bessel functions, can be found in the following sources:

"Fourier Integrals for Practical Applications," G. A. Campbell and R. M. Foster, Van Nostrand, 1948. In these tables, only those entries containing the condition $0 < g$ or $k < g$, where g is our t, are Laplace transforms.

"Formulaire pour le calcul symbolique," N. W. McLachlan and P. Humbert, Gauthier–Villars, Paris, 1947.

"Tables of Integral Transforms," Bateman Manuscript Project, California Institute of Technology, A. Erdélyi and W. Magnus, Eds., McGraw-Hill, 1954; based on notes left by Harry Bateman.

TABLE 5.2
Properties of Laplace transforms.

Number	Theorem	$f(t)$	$F(s)$
1	Linearity	$K_1 f_1(t) + K_2 f_2(t)$	$K_1 F_1(s) + K_2 F_2(s)$
2	Time differentiation	$\dfrac{df(t)}{dt}$	$sF(s) - f(0+)$
3	Time differentiation	$\dfrac{d^n f(t)}{dt^n}$	$s^n F(s) - s^{n-1}f(0+) - s^{n-2}f^{(1)}(0+)$ $- \cdots - f^{(n-1)}(0+)$
4	Integration	$\displaystyle\int_\infty^t f(\xi)\,d\xi$	$\dfrac{F(s)}{s} + \dfrac{1}{s}f^{(-1)}(0+),$ $f^{(-1)}(0+) = \displaystyle\lim_{t=0^+}\int_{-\infty}^t f(\xi)\,d\xi$
5	Real definite integration	$\displaystyle\int_0^t f(\xi)\,d\xi$	$\dfrac{F(s)}{s}$
		$\displaystyle\int_0^t \int_0^\xi f(\lambda)\,d\lambda\,d\xi$	$\dfrac{F(s)}{s^2}$
6	Multiplication by t	$tf(t)$	$-\dfrac{dF(s)}{ds}$

TABLE 5.2
Properties of Laplace transforms.

Number	Theorem	$f(t)$	$F(s)$
7	Differentiation of transform	$(-t)^n f(t)$	$\dfrac{d^n F(s)}{ds^n} = F^{(n)}(s)$
8	Division by t	$\dfrac{f(t)}{t}$	$\displaystyle\int_0^\infty F(s)\,ds$
9	Time shifting	$f(t \pm \lambda)u(t \pm \lambda)$	$e^{\mp s\lambda} f(s)$
	Real translation		
10	Complex translation	$e^{\pm at} f(t)$	$F(s \mp a)$
11	Convolution $-t$ plane	$f_1(t) * f_2(t)$	$F_1(s) F_2(s)$
12	Convolution $-t$ plane	$f_1(t) * f_2(t) * f_3(t)$	$F_1(s) F_2(s) F_3(s)$
13	Convolution $-s$ plane	$f_1(t) f_2(t)$	$\dfrac{1}{2\pi j}[F_1(s) * F_2(s)]$
14	Initial value theorem	$\lim\limits_{t \to 0+} f(t)$	$\lim\limits_{s \to \infty} s F(s)$
15	Final value theorem	$\lim\limits_{t \to \infty} f(t)$	$\lim\limits_{s \to 0} s F(s)$

References

[1] R. V. Churchill, *Modern Operational Mathematics in Engineering*, McGraw-Hill, 1944.

[2] J. Irving and N. Mullineux, *Mathematics in Physics and Engineering*, Academic Press, 1959.

[3] H. S. Carslaw and J. C. Jaeger, *Operational Methods in Applied Mathematics*, Dover Publications, 1963.

[4] W. R. LePage, *Complex Variables and the Laplace Transform for Engineers*, McGraw-Hill, 1961.

[5] R. E. Bolz and G. L. Turve, eds., *CRC Handbook of Tables for Applied Engineering Science*, 2d ed., CRC Press, 1973.

[6] A. D. Poularikas and S. Seely, *Signals and Systems*, corrected 2d ed. Krieger Publishing Co., 1994.

6

The Z-Transform

Alexander D. Poularikas

CONTENTS

6.1   Introduction

The Z-transform is a powerful method for solving difference equations and, in general, to represent discrete systems. Although applications of Z-transforms are relatively new, the essential features of this mathematical technique date back to the early 1730s when DeMoivre introduced the concept of a generating function that is identical with that for the Z-transform. Recently, the development and extensive applications of the Z-transform are much enhanced as a result of the use of digital computers.

A. One-sided Z-transform

6.2   The Z-transform and discrete functions

Let $f(t)$ be defined for $t \geq 0$. The Z-transform of the sequence $\{f(nT)\}$ is given by

$$Z\{f(nT)\} \doteq F(z) = \sum_{n=0}^{\infty} f(nT)z^{-n} \tag{2.1}$$

where T, the sampling time, is a positive number.[1]

To find the values of z for which the series converges we use the ratio test or the root test. The ratio test states that a series of complex numbers

$$\sum_{n=0}^{\infty} a_n$$

with limit

$$\lim_{n \to \infty} \left| \frac{a_{n+1}}{a_n} \right| = A \tag{2.2}$$

converges absolutely if $A < 1$ and diverges if $A > 1$. For $A = 1$ the series may or may not converge.

The root test states that if

$$\lim_{n \to \infty} \sqrt[n]{|a_n|} = A \tag{2.3}$$

then the series converges absolutely if $A < 1$, and diverges if $A > 1$, and may converge or diverge if $A = 1$.

More generally, the series converges absolutely if

$$\overline{\lim_{n \to \infty}} \sqrt[n]{|a_n|} < 1 \tag{2.4}$$

where $\overline{\lim}$ denotes the **greatest** limit points of $\overline{\lim_{n \to \infty}} |f(nT)|^{1/n}$, and diverges if

$$\overline{\lim_{n \to \infty}} \sqrt[n]{|a_n|} > 1 \tag{2.5}$$

If we apply the root test in (2.1) we obtain the convergence condition

$$\overline{\lim_{n \to \infty}} \sqrt[n]{|f(nT)z^{-n}|} = \overline{\lim_{n \to \infty}} \sqrt[n]{|f(nT)||z^{-1}|^n} < 1$$

---

[1] The symbol $\doteq$ means equal by definition.

or

$$|z| > \varlimsup_{n \to \infty} \sqrt[n]{|f(nT)|} = R \qquad (2.6)$$

where R is known as the **radius of convergence** for the series. Therefore, the series will converge absolutely for all points in the z-plane that lie **outside** the circle of radius R, and is centered at the origin (with the possible exception of the point at infinity). This region is called the **region of convergence** (ROC).

Example
The radius of convergence of $f(nT) = e^{-anT}u(nT)$, a positive number, is

$$|z^{-1}e^{-aT}| < 1 \qquad \text{or} \qquad |z| > e^{-aT}$$

The Z-transform of $f(nT) = e^{-anT}u(nT)$ is

$$F(z) = \sum_{n=0}^{\infty} f(nT)z^{-n} = \sum_{n=0}^{\infty}(e^{-aT}z^{-1})^n = \frac{1}{1 - e^{-aT}z^{-1}}$$

If $a = 0$

$$F(z) = \sum_{n=0}^{\infty} u(nT)z^{-n} = \frac{1}{1 - z^{-1}} = \frac{z}{z - 1} \qquad \blacksquare$$

Example
The function $f(nT) = a^{nT}\cos nT\omega\, u(nT)$ has the Z-transform

$$F(z) = \sum_{n=0}^{\infty} a^{nT}\frac{e^{jnT\omega} + e^{-jnT\omega}}{2}z^{-n} = \frac{1}{2}\sum_{n=0}^{\infty}(a^T e^{jT\omega}z^{-1})^n + \frac{1}{2}\sum_{n=0}^{\infty}(a^T e^{-jT\omega}z^{-1})^n$$

$$= \frac{1}{2}\frac{1}{1 - a^T e^{jT\omega}z^{-1}} + \frac{1}{2}\frac{1}{1 - a^T e^{-jT\omega}z^{-1}} = \frac{1 - a^T z^{-1}\cos T\omega}{1 - 2a^T z^{-1}\cos T\omega + a^{2T}z^{-2}}.$$

The ROC is given by the relations

$$|a^T e^{jT\omega}z^{-1}| < 1 \qquad \text{or} \qquad |z| > |a^T|$$

$$|a^T e^{-jT\omega}z^{-1}| < 1 \qquad \text{or} \qquad |z| > |a^T|$$

Therefore the ROC is $|z| > |a^T|$. $\blacksquare$

6.3 Properties of the Z-transform

Linearity

If there exists transforms of sequences $\mathcal{Z}\{c_i f_i(nT)\} = c_i F_i(z)$, c_i are complex constants, with radii of convergence $R_i > 0$ for $i = 0, 1, 2, \ldots, \ell$ (ℓ finite), then

$$\mathcal{Z}\left\{\sum_{i=0}^{\ell} c_i f_i(nT)\right\} = \sum_{i=0}^{\ell} c_i F_i(z) \qquad |z| > \max R_i \qquad (3.1)$$

Shifting property

$$\mathcal{Z}\{f(nT - kT)\} = z^{-k}F(z), \qquad f(-nT) = 0 \qquad n = 1, 2, \ldots \tag{3.2}$$

$$\mathcal{Z}\{f(nT - kT)\} = z^{-k}F(z) + \sum_{n=1}^{k} f(-nT)z^{-(k-n)} \tag{3.3}$$

$$\mathcal{Z}\{f(nT + kT)\} = z^{k}F(z) - \sum_{n=0}^{k-1} f(nT)z^{k-n} \tag{3.4}$$

$$\mathcal{Z}\{f(nT + T)\} = z[F(z) - f(0)] \tag{3.4a}$$

Example
To find the Z-transform of $y(nT)$ we proceed as follows:

$$\frac{d^2y(t)}{dt^2} = x(t), \qquad \frac{y(nT) - 2y(nT - T) + y(nT - 2T)}{T^2} = x(nT),$$

$$Y(z) - 2[z^{-1}Y(z) + y(-T)z^{-0}] + z^{-2}Y(z) + y(-T)z^{-1} + y(-2T)z^{-0} = X(z)T^2$$

or

$$Y(z) = \frac{2y(-T) - y(-T)z^{-1} - y(-2T) + X(z)T^2}{1 - 2z^{-1} + z^{-2}} \qquad \blacksquare$$

Time scaling

$$\mathcal{Z}\{a^{nT}f(nT)\} = F(a^{-T}z) = \sum_{n=0}^{\infty} f(nT)(a^{-T}z)^{-n} \tag{3.5}$$

Example

$$\mathcal{Z}\{\sin \omega nT u(nT)\} = \frac{z \sin \omega T}{z^2 - 2z \cos \omega T + 1} \qquad |z| > 1,$$

$$\mathcal{Z}\{e^{-n} \sin \omega nT u(nT)\} = \frac{e^{+1}z \sin \omega T}{e^{+2}z^2 - 2e^{+1}z \cos \omega T + 1} \qquad |z| > e^{-1} \qquad \blacksquare$$

Periodic sequence

$$\mathcal{Z}\{f(nT)\} = \frac{z^N}{z^N - 1}\mathcal{Z}\{f_1(nT)\} = \frac{z^N}{z^N - 1}F_1(z), \qquad f_1(nT) = \text{first period} \tag{3.6}$$

$$N \text{ is the number of time units in a period}, \qquad |z| > R$$

where R is the radius of convergence of $F_1(z)$.

PROOF

$$\mathcal{Z}\{f(nT)\} = \mathcal{Z}\{f_1(nT)\} + \mathcal{Z}\{f_1(nT - NT)\} + \mathcal{Z}\{f_1(nT - 2NT)\} + \cdots$$

$$= F_1(z) + z^{-N}F_1(z) + z^{-2N}F_1(z) + \cdots$$

$$= F_1(z)\frac{1}{1 - z^{-N}} = \frac{z^N}{z^N - 1}F_1(z)$$

For finite sequence of K terms

$$F(z) = F_1(z)\frac{1 - z^{-N(K+1)}}{1 - z^{-N}} \tag{3.6a}$$

∎

Multiplication by n and nT

R is the radius of convergence of $F(z)$.

$$\mathcal{Z}\{nf(nT)\} = -z\frac{dF(z)}{dz} \tag{3.7}$$

$$\mathcal{Z}\{nTf(nT)\} = -Tz\frac{dF(z)}{dz} \qquad |z| > R$$

PROOF

$$\sum_{n=0}^{\infty} nTf(nT)z^{-n} = Tz\sum_{n=0}^{\infty} f(nT)\left[-\frac{d}{dz}z^{-n}\right] = -Tz\frac{d}{dz}\left[\sum_{n=0}^{\infty} f(nT)z^{-n}\right]$$

$$= -Tz\frac{dF(z)}{dz}$$

∎

Example

$$\mathcal{Z}\{u(n)\} = \frac{z}{z - 1}, \qquad \mathcal{Z}\{nu(n)\} = -z\frac{d}{dz}\left(\frac{z}{z - 1}\right) = \frac{z}{(z - 1)^2},$$

$$\mathcal{Z}\{n^2 u(n)\} = -z\frac{d}{dz}\left(\frac{z}{(z - 1)^2}\right) = \frac{z(z^2 - 1)}{(z - 1)^4}$$

∎

Convolution

If $\mathcal{Z}\{f(nT)\} = F(z)$ $|z| > R_1$ and $\mathcal{Z}\{h(nT)\} = H(z)$ $|z| > R_2$ then

$$\mathcal{Z}\{f(nT) * h(nT)\} = \mathcal{Z}\left\{\sum_{m=0}^{\infty} f(mT)h(nT - mT)\right\} = F(z)H(z) \qquad |z| > \max(R_1, R_2)$$

$$\tag{3.8}$$

PROOF

$$\mathcal{Z}\{f(nT) * h(nT)\} = \sum_{n=0}^{\infty} \left[\sum_{m=0}^{\infty} f(mT)h(nT - mT) \right] z^{-n}$$

$$= \sum_{m=0}^{\infty} f(mT) \sum_{n=0}^{\infty} h(nT - mT)z^{-n}$$

$$= \sum_{m=0}^{\infty} f(mT) \sum_{r=-m}^{\infty} h(rT)z^{-r}z^{-m}$$

$$= \sum_{m=0}^{\infty} f(mT)z^{-m} \sum_{r=0}^{\infty} h(rT)z^{-r} = F(z)H(z).$$

The value of $h(nT)$ for $n < 0$ is zero. ∎

Additional relations of convolution are

$$\mathcal{Z}\{f(nT) * h(nT)\} = F(z)H(z) = \mathcal{Z}\{h(nT) * f(nT)\} = F(z)H(z) \qquad (3.8a)$$

$$\mathcal{Z}\{\{f(nT) + h(nT)\} * \{g(nT)\}\} = \mathcal{Z}\{f(nT) * g(nT)\} + Z\{h(nT) * g(nT)\}$$

$$= F(z)G(z) + H(z)G(z) \qquad (3.8b)$$

$$\mathcal{Z}\{\{f(nT) * h(nT)\} * g(nT)\} = \mathcal{Z}\{f(nT) * \{h(nT) * g(nT)\}\} = F(z)H(z)G(z) \qquad (3.8c)$$

Example
The Z-transform of the output of the discrete system $y(n) = \frac{1}{2}y(n-1) + \frac{1}{2}x(n)$, when the input is the unit step function $u(n)$ given by $Y(z) = H(z)U(z)$. The Z-transform of the difference equation with a delta function input $\delta(n)$ is

$$H(z) - \frac{1}{2}z^{-1}H(z) = \frac{1}{2} \quad \text{or} \quad H(z) = \frac{1}{2}\frac{1}{1 - \frac{1}{2}z^{-1}} = \frac{1}{2}\frac{z}{z - \frac{1}{2}}$$

Therefore the output is given by

$$Y(z) = \frac{1}{2}\frac{z}{z - \frac{1}{2}}\frac{z}{z - 1} \qquad \blacksquare$$

Example
Find the $f(n)$ if

$$F(z) = \frac{z^2}{(z - e^{-a})(z - e^{-b})} \qquad a, b \text{ are constants.}$$

From this equation we obtain

$$f_1(n) = \mathcal{Z}^{-1}\left\{\frac{z}{(z - e^{-a})}\right\} = e^{-an}, \qquad f_2(n) = \mathcal{Z}^{-1}\left\{\frac{z}{(z - e^{-b})}\right\} = e^{-bn}$$

Therefore

$$f(n) = f_1(n) * f_2(n) = \sum_{m=0}^{n} e^{-am} e^{-b(n-m)} = e^{-bn} \sum_{m=0}^{n} e^{-(a-b)m}$$

$$= e^{-bn} \frac{1 - e^{-(a-b)(n+1)}}{1 - e^{-(a-b)}} \qquad \blacksquare$$

Initial value

$$f(0) = \lim_{z \to \infty} F(z) \tag{3.9}$$

The above value is obtained from the definition of the Z-transform. If $f(0) = 0$, we obtain $f(1)$ as the limit

$$\lim_{z \to \infty} z F(z) \tag{3.9a}$$

Final value

$$\lim_{n \to \infty} f(n) = \lim_{z \to 1} (z - 1) F(z) \qquad \text{if } f(\infty) \text{ exists} \tag{3.10}$$

PROOF

$$\mathcal{Z}\{f(k+1) - f(k)\} = \lim_{n \to \infty} \sum_{k=0}^{n} [f[(k+1)] - f(k)] z^{-k}$$

$$z F(z) - z f(0) - F(z) = (z - 1) F(z) - z f(0) = \lim_{n \to \infty} \sum_{k=0}^{n} [f[(k+1)] - f(k)] z^{-k}$$

By taking the limit as $z \to 1$ the above equation becomes

$$\lim_{z \to 1} (z - 1) F(z) - f(0) = \lim_{n \to \infty} \sum_{k=0}^{n} [f[(k+1)] - f(k)]$$

$$= \lim_{n \to \infty} \{f(1) - f(0) + f(2) - f(1) + \cdots$$

$$+ f(n) - f(n-1) + f(n+1) - f(n)\}$$

$$= \lim_{n \to \infty} \{-f(0) + f(n+1)\}$$

$$= -f(0) + f(\infty)$$

which is the required result. $\blacksquare$

Example

If $F(z) = 1/[(1 - z^{-1})(1 - e^{-1}z^{-1})]$ with $|z| > 1$ then

$$f(0) = \lim_{z \to \infty} F(z) = \frac{1}{\left(1 - \frac{1}{\infty}\right)\left(1 - e^{-1}\frac{1}{\infty}\right)} = 1$$

$$\lim_{n \to \infty} f(n) = \lim_{z \to 1}(z - 1)\frac{1}{(1 - z^{-1})(1 - e^{-1}z^{-1})} = \lim_{z \to 1}\frac{z^2}{(z - e^{-1})} = \frac{1}{(1 - e^{-1})} \qquad ∎$$

Multiplication by $(nT)^k$

$$\mathcal{Z}\{n^k T^k f(nT)\} = -Tz\frac{d}{dz}\mathcal{Z}\{(nT)^{k-1} f(nT)\} \qquad k > 0 \text{ and is an integer} \qquad (3.11)$$

As a corollary to this theorem we can deduce

$$\mathcal{Z}\{n^{(k)} f(n)\} = z^{-k}\frac{d^k F(z)}{d(z^{-1})^k}, \qquad n^{(k)} = n(n - 1)(n - 2)\cdots(n - k + 1) \qquad (3.11a)$$

The following relations are also true:

$$\mathcal{Z}\{(-1)^k n^{(k)} f(n - k + 1)\} = z\frac{d^k F(z)}{dz^k} \qquad (3.11b)$$

$$\mathcal{Z}\{n(n + 1)(n + 2)\cdots(n + k - 1) f(n)\} = (-1)^k z^k\frac{d^k F(z)}{dz^k} \qquad (3.11c)$$

Example

$$\mathcal{Z}\{n\} = -z\frac{d}{dz}\left(\frac{z}{z - 1}\right) = \frac{z}{(z - 1)^2},$$

$$\mathcal{Z}\{n^2\} = -z\frac{d}{dz}\mathcal{Z}\{n\} = -z\frac{d}{dz}\frac{z}{(z - 1)^2} = \frac{z(z + 1)}{(z - 1)^3},$$

$$\mathcal{Z}\{n^3\} = -z\frac{d}{dz}\frac{z(z + 1)}{(z - 1)^3} = \frac{z(z^2 + 4z + 1)}{(z - 1)^4} \qquad ∎$$

Initial value of $f(nT)$

$$\mathcal{Z}\{f(nT)\} = f(0T) + f(T)z^{-1} + f(2T)z^{-2} + \cdots = F(z)$$

$$f(0T) = \lim_{z \to \infty} F(z) \qquad |z| > R \qquad (3.12)$$

Final value for $f(nT)$

$$\lim_{n\to\infty} f(nT) = \lim_{z\to 1}(z-1)F(z) \qquad f(\infty T) \text{ exists} \tag{3.13}$$

Example

For the function

$$F(z) = \frac{1}{(1-z^{-1})(1-e^{-T}z^{-1})} \qquad |z| > 1$$

we obtain

$$f(0T) = \lim_{z\to\infty} F(z) = \frac{1}{\left(1-\frac{1}{\infty}\right)\left(1-\frac{e^{-T}}{\infty}\right)} = 1$$

$$\lim_{n\to\infty} f(nT) = \lim_{z\to 1}(z-1)\frac{z}{z-1}\frac{z}{1-e^{-T}} = \frac{1}{1-e^{-T}} \qquad\blacksquare$$

Complex conjugate signal

$$F(z) = \sum_{n=0}^{\infty} f(nT)z^{-n} \qquad |z| > R \quad \text{or} \quad F(z^*) = \sum_{n=0}^{\infty} f(nT)(z^*)^{-n} \quad \text{or}$$

$$F^*(z^*) = \sum_{n=0}^{\infty} f^*(nT)z^{-n} = \mathcal{Z}\{f^*(nT)\}$$

Hence

$$\mathcal{Z}\{f^*(nT)\} = F^*(z^*) \qquad |z| > R \tag{3.14}$$

Transform of product

If

$$\mathcal{Z}\{f(nT)\} = F(z) \qquad |z| > R_f$$
$$\mathcal{Z}\{h(nT)\} = H(z) \qquad |z| > R_h$$

then

$$\mathcal{Z}\{g(nT)\} \doteq \mathcal{Z}\{f(nT)h(nT)\}$$

$$= \sum_{n=0}^{\infty} f(nT)h(nT)z^{-n}$$

$$= \frac{1}{2\pi j}\oint_C F(\tau)H\left(\frac{z}{\tau}\right)\frac{d\tau}{\tau} \qquad |z| > R_f R_h \tag{3.15}$$

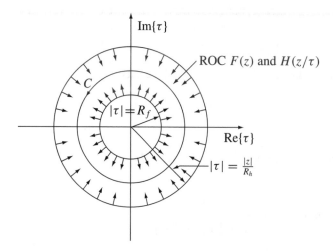

FIGURE 3.1

where C is a simple contour encircling counterclockwise the origin with (see Figure 3.1)

$$R_f < |\tau| < \frac{|z|}{R_h} \tag{3.15a}$$

PROOF   The integration is performed in the positive sense along the circle, inside which lie all the singular points of the function $F(\tau)$ and outside which lie all the singular points of the function $H(z/\tau)$. From (3.15) we write

$$G(z) = \frac{1}{2\pi j} \oint_C F(\tau) \sum_{n=0}^{\infty} h(nT) \left(\frac{z}{\tau}\right)^{-n} \frac{d\tau}{\tau} \tag{3.16}$$

which converges uniformly for some choice of contour C and values of z. From (3.16) we must have

$$\left| R_h \left(\frac{z}{\tau}\right)^{-1} \right| < 1 \qquad \text{or} \qquad \left| \frac{z}{\tau} \right| > R_h \qquad \text{or} \qquad |\tau| < \frac{|z|}{R_h} \tag{3.17}$$

so that the sum in (3.16) converges. Because $|z| > R_f$ and τ takes the place of z, then (3.16) implies that

$$|\tau| > R_f \tag{3.18}$$

$$R_f < |\tau| < \frac{|z|}{R_h} \tag{3.19}$$

and also

$$R_f R_h < |z|.$$

Figure 3.1 shows the region of convergence.

The integral is solved with the aid of the residue theorem, which yields in this case

$$G(z) = \sum_{i=1}^{K} \operatorname{res}_{\tau=\tau_i} \left\{ \frac{F(\tau) H(z/\tau)}{\tau} \right\} \tag{3.20}$$

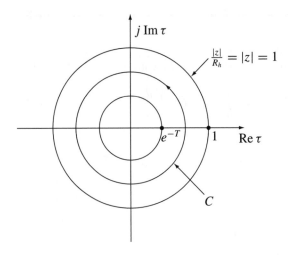

FIGURE 3.2

where K is the number of different poles τ_i $(i = 1, 2, \ldots, K)$ of the function $F(\tau)/\tau$. For the residue at the pole τ_i of multiplicity m of the function $F(\tau)/\tau$ we have

$$\operatorname{res}_{\tau=\tau_i} \left\{ \frac{F(\tau)H(z/\tau)}{\tau} \right\} = \frac{1}{(m-1)!} \lim_{\tau \to \tau_i} \frac{d^{m-1}}{d\tau^{m-1}} \left[(\tau - \tau_i)^m \frac{F(\tau)H\left(\frac{z}{\tau}\right)}{\tau} \right] \qquad (3.21)$$

Hence for a simple pole, $m = 1$, we obtain

$$\operatorname{res}_{\tau=\tau_i} \left\{ \frac{F(\tau)H(z/\tau)}{\tau} \right\} = \lim_{\tau \to \tau_i} (\tau - \tau_i) \left\{ \frac{F(\tau)H\left(\frac{z}{\tau}\right)}{\tau} \right\} \qquad (3.22)$$

∎

Example
See Figure 3.2 for graphical representation of the complex integration.

$$\mathcal{Z}\{nT\} \doteq H(z) = \frac{z}{(z-1)^2}T \qquad |z| > 1, \qquad \mathcal{Z}\{e^{-nT}\} \doteq F(z) = \frac{z}{z - e^{-T}} \qquad |z| > e^{-T}$$

Hence

$$\mathcal{Z}\{nTe^{-nT}\} = \frac{1}{2\pi j} \oint_C T \frac{z}{\tau(\tau - e^{-T})\left(\frac{z}{\tau} - 1\right)^2}\, d\tau.$$

The contour must have a radius $|\tau|$ of the value $e^{-T} < |\tau| < |z| = 1$ and we have from (3.22)

$$\mathcal{Z}\{nTe^{-nT}\} = \operatorname{res}_{\tau=e^{-T}} \left\{ (\tau - e^{-T})T \frac{z\tau}{(\tau - e^{-T})(z - \tau)^2} \right\} = T \frac{ze^{-T}}{(z - e^{-T})^2}$$

From (3.11)

$$\mathcal{Z}\{nTe^{-nT}\} = -Tz\frac{d}{dz}\left(\frac{1}{1 - e^{-T}z^{-1}} \right) = T \frac{ze^{-T}}{(z - e^{-T})^2}$$

and verifies the complex integration approach. ∎

Parseval's theorem

If $\mathcal{Z}\{f(nT)\} = F(z)$, $|z| > R_f$ and $\mathcal{Z}\{h(nT)\} = H(z)$, $|z| > R_h$ with $|z| = 1 > R_f R_h$, then

$$\sum_{n=0}^{\infty} f(nT)h(nT) = \frac{1}{2\pi j} \oint_C F(z)H(z^{-1})\frac{dz}{z} \qquad (3.23)$$

where the contour is taken counterclockwise.

PROOF   From (3.15) set $z = 1$ and change the dummy variable τ to z.   ∎

Example
$f(nT) = e^{-nT}u(nT)$ has the following Z-transform:

$$F(z) = \frac{1}{1 - e^{-T}z^{-1}} \qquad |z| > e^{-T}$$

From (3.23) and with C a unit circle ($R_f = e^{-T} < 1$)

$$\sum_{n=0}^{\infty} f(nT)f(nT) = \frac{1}{2\pi j}\oint_C \frac{1}{1 - e^{-T}z^{-1}}\frac{1}{1 - e^{-T}z}\frac{dz}{z} = \frac{1}{2\pi j}\oint_C \frac{1}{z - e^{-T}}\frac{e^T}{e^T - z}dz$$

$$= \frac{2\pi j}{2\pi j}\sum_i \text{residues} = \frac{e^T}{e^T - e^{-T}} = \frac{1}{1 - e^{-2T}} \qquad ∎$$

Correlation

Let the Z-transform of the two sequences $\mathcal{Z}\{f(nT)\} = F(z)$ and $\mathcal{Z}\{h(nT)\} = H(z)$ exist for $|z| > 1$. Then the **cross correlation** is given by

$$g(nT) \doteq f(nT) \otimes h(nT) = \sum_{m=0}^{\infty} f(mT)h(mT - nT) = \lim_{z \to 1+} \sum_{m=0}^{\infty} f(mT)h(mT - nT)z^{-m}$$

$$= \lim_{z \to 1+} \mathcal{Z}\{f(mT)h(mT - nT)\}$$

But $\mathcal{Z}\{h(mT - nT)\} = z^{-n}H(z)$ and therefore (see [3.15])

$$g(nT) = \lim_{z \to 1+} \frac{1}{2\pi j}\oint_C F(\tau)\left(\frac{z}{\tau}\right)^{-n} H\left(\frac{z}{\tau}\right)\frac{d\tau}{\tau}$$

$$= \frac{1}{2\pi j}\oint_C F(\tau)H\left(\frac{1}{\tau}\right)\tau^{n-1}\,d\tau \qquad n \geq 1 \qquad (3.24)$$

This relation is the inverse Z-transform of $g(nT)$ and hence

$$\mathcal{Z}\{g(nT)\} \doteq \mathcal{Z}\{f(nT) \otimes h(nT)\} = F(z)H\left(\frac{1}{z}\right) \qquad \text{for } |z| = 1 \qquad (3.25)$$

If $f(nT) = h(nT)$ for $n \geq 0$ the **autocorrelation sequence** is

$$g(nT) \doteq f(nT) \otimes h(nT)$$

$$= \sum_{m=0}^{\infty} f(mT) f(mT - nT)$$

$$= \frac{1}{2\pi j} \oint_C F(\tau) F\left(\frac{1}{\tau}\right) \tau^{n-1} d\tau \tag{3.26}$$

and hence

$$G(z) = \mathcal{Z}\{g(nT)\} = \mathcal{Z}\{f(nT) \otimes h(nT)\} = F(z) F\left(\frac{1}{z}\right) \tag{3.27}$$

If we set $n = 0$ we obtain the Parseval's theorem in the same form it was developed above.

Example
The sequence $f(nT) = e^{-nT}$, $n \geq 0$, has the Z-transform

$$\mathcal{Z}\{e^{-nT}\} = \frac{z}{z - e^{-T}} \qquad |z| > e^{-T}$$

The autocorrelation is given by (3.26) in the form

$$G(z) \doteq \mathcal{Z}\{f(nT) \otimes f(nT)\} = \frac{z}{z - e^{-T}} \frac{\frac{1}{z}}{\frac{1}{z} - e^{-T}} = -\frac{z}{z - e^{-T}} \frac{e^T}{z - e^T}$$

The function is regular in the region $e^{-T} < |z| < e^T$. Using the residue theorem from (3.24) we obtain

$$g(nT) = \sum_{i=1}^{K} \text{res}_{\tau=\tau_i} \left\{ F(\tau) H\left(\frac{1}{\tau}\right) \right\} \tau^{n-1} \tag{3.28}$$

where τ_i are all poles of the integrand inside the circle $|\tau| = 1$. Similarly from (3.27)

$$g(nT) = \sum_{i=1}^{K} \text{res}_{\tau=\tau_i} \left\{ F(\tau) F\left(\frac{1}{\tau}\right) \tau^{n-1} \right\} \tag{3.29}$$

where τ_i are the poles included inside the unit circle.   ∎

Example
From the previous example we obtain (only the root inside the unit circle)

$$-\frac{1}{2\pi j} \oint_C \frac{z}{z - e^{-T}} \frac{e^T}{z - e^T} z^{n-1} dz = -\text{res}_{z=e^{-T}} \left\{ \frac{z e^T}{z - e^T} z^{n-1} \right\} = \frac{e^{2T}}{e^{2T} - 1} e^{-Tn}$$

which is equal to the autocorrelation of $f(nT) = e^{-nT} u(nT)$. Using the summation definitions we obtain

$$\sum_{m=0}^{\infty} e^{-mT} u(mT) e^{-T(m-n)} u(mT - nT) = e^{Tn} \sum_{m=n}^{\infty} e^{-2mT}$$

$$= e^{Tn} (e^{-2nT} + e^{-2nT} e^{-2T} + e^{-2nT} e^{-4T} + \cdots)$$

$$= e^{-nT} (1 + e^{-2T} + (e^{-2T})^2 + \cdots)$$

$$= e^{-nT} \frac{1}{1 - e^{-2T}} = e^{-nT} \frac{e^{2T}}{e^{2T} - 1} \qquad ∎$$

Z-transforms with parameters

$$\mathcal{Z}\left\{\frac{\partial}{\partial a}f(nT,a)\right\} = \frac{\partial}{\partial a}F(z,a) \tag{3.30}$$

$$\mathcal{Z}\left\{\lim_{a\to a_0}f(nT,a)\right\} = \lim_{a\to a_0}F(z,a) \tag{3.31}$$

$$\mathcal{Z}\left\{\int_{a_0}^{a_1}f(nT,a)\,da\right\} = \int_{a_0}^{a_1}F(z,a)\,da \qquad \text{finite integral} \tag{3.32}$$

Table 1 in the Appendix contains the Z-transform properties for positive-time sequences.

6.4 Inverse Z-transform

The **inverse Z-transform** provides the object function from its given transform. We use the symbolic solution

$$f(nT) = \mathcal{Z}^{-1}\{F(z)\} \tag{4.1}$$

To find the inverse transform we may proceed as follows:

1. Use tables.
2. Decompose the expression into simpler partial forms, which are included in the tables.
3. If the transform is decomposed into a product of partial sums, the resulting object function is obtained as the convolution of the partial object function.
4. Use the inversion integral.

Power series method

When $F(z)$ is analytic for $|z| > R$ (and at $z = \infty$), the value of $f(nT)$ is obtained as the coefficient of z^{-n} in the power series expansion (Taylor's series of $F(z)$ as a function of z^{-1}). For example, if $F(z)$ is the ratio of two polynomials in z^{-1}, the coefficients $f(0T), \ldots, f(nT)$ are obtained as follows:

$$F(z) = \frac{p_0 + p_1 z^{-1} + p_2 z^{-2} + \cdots + p_n z^{-n}}{q_0 + q_1 z^{-1} + q_2 z^{-2} + \cdots + q_n z^{-n}} = f(0T) + f(T)z^{-1} + f(2T)z^{-2} + \cdots \tag{4.2}$$

where

$$p_0 = f(0T)q_0$$

$$p_1 = f(1T)q_0 + f(0T)q_1$$

$$\vdots$$

$$p_n = f(nT)q_0 + f[(n-1)T]q_1 + f[(n-2)T]q_2 + \cdots + f(0T)q_n \tag{4.3}$$

The same can be accomplished by synthetic division.

Example

$$F(z) = \frac{1 + z^{-1}}{1 + 2z^{-1} + 3z^{-2}} = \frac{z^2 + z}{z^2 + 2z + 3} = 1 - z^{-1} - z^{-2} + 5z^{-3} + \cdots \qquad |z| > \sqrt{6}$$

From (4.3): $1 = f(0T) \cdot 1$ or $f(0T) = 1$, $1 = f(1T) \cdot 1 + 1 \cdot 2$ or $f(1T) = -1$, $0 = f(2T) \cdot 1 + f(1T) \cdot 2 + f(0T) \cdot 3$ or $f(2T) = +2 - 3 = -1$, $0 = f(3T) \cdot 1 + f(2T)2 + f(1T)3 + f(0T) \cdot 0$ or $f(3T) = 2 + 3 = 5$, and so forth. ▮

Partial fraction expansion

If $F(z)$ is a rational function of z and analytic at infinity it can be expressed as follows:

$$F(z) = F_1(z) + F_2(z) + F_3(z) + \cdots \tag{4.4}$$

and, therefore,

$$f(nT) = \mathcal{Z}^{-1}\{F_1(z)\} + \mathcal{Z}^{-1}\{F_2(z)\} + \mathcal{Z}^{-1}\{F_3(z)\} + \cdots \tag{4.5}$$

For an expansion of the form

$$F(z) = \frac{F_1(z)}{(z - p)^n} = \frac{A_1}{z - p} + \frac{A_2}{(z - p)^2} + \cdots + \frac{A_n}{(z - p)^n} \tag{4.6}$$

the constants A_i are given by

$$A_n = (z - p)^n F(z)\big|_{z=p}$$

$$A_{n-1} = \frac{d}{dz}[(z - p)^n F(z)]\Big|_{z=p}$$

$$\vdots$$

$$A_{n-k} = \frac{1}{k!}\frac{d^k}{dz^k}[(z - p)^n F(z)]\big|_{z=p}$$

$$\vdots$$

$$A_1 = \frac{1}{(n-1)!}\frac{d^{n-1}}{dz^{n-1}}[(z - p)^n F(z)]\big|_{z=p} \tag{4.7}$$

Example
Let

$$F(z) = \frac{1 + 2z^{-1} + z^{-2}}{1 - \frac{3}{2}z^{-1} + \frac{1}{2}z^{-2}} = \frac{z^2 + 2z + 1}{z^2 - \frac{3}{2}z + \frac{1}{2}} = 1 + \frac{7}{2}z^{-1} + \frac{23}{4}z^{-2} + \cdots \qquad |z| > 1$$

Also

$$F(z) = 1 + \frac{\frac{7}{2}z + \frac{1}{2}}{(z - 1)\left(z - \frac{1}{2}\right)} = 1 + \frac{A}{z - 1} + \frac{B}{z - \frac{1}{2}}$$

from which we find that

$$A = \left. \frac{(z-1)\left(\frac{7}{2}z+\frac{1}{2}\right)}{(z-1)\left(z-\frac{1}{2}\right)} \right|_{z=1} = 8$$

and

$$B = \left. \frac{\left(z-\frac{1}{2}\right)\left(\frac{7}{2}z+\frac{1}{2}\right)}{(z-1)\left(z-\frac{1}{2}\right)} \right|_{z=1/2} = -\frac{9}{2}$$

Hence

$$F(z) = 1 + \frac{8}{z-1} - \frac{9}{2}\frac{1}{z-\frac{1}{2}}$$

and therefore its inverse transform is $f(nT) = \delta(nT) + 8u(nT - T) - \frac{9}{2}(\frac{1}{2})^{n-1}u(nT - T)$ with ROC $|z| > 1$. ∎

Example
a) If

$$F(z) = \frac{z^2+1}{(z-1)(z-2)} = A + \frac{Bz}{z-1} + \frac{Cz}{z-2} \qquad |z| > 2$$

then we obtain

$$A = \frac{0+1}{(0-1)(0-2)} = \frac{1}{2},$$

$$B = \left. \frac{1}{z}\frac{z^2+1}{(z-2)} \right|_{z=1} = -2,$$

and

$$C = \left. \frac{1}{z}\frac{z^2+1}{(z-1)} \right|_{z=2} = \frac{5}{2}$$

Hence

$$F(z) = \frac{1}{2} - 2\frac{z}{z-1} + \frac{5}{2}\frac{z}{z-2}$$

and its inverse is $f(nT) = \frac{1}{2}\delta(nT) - 2u(nT) + \frac{5}{2}(2)^n u(nT)$.

b) If

$$F(z) = \frac{z+1}{(z-1)(z-2)} = \frac{A}{z-1} + \frac{B}{z-2}$$

then we obtain

$$A = \left. \frac{z+1}{(z-2)} \right|_{z=1} = -2$$

and

$$B = \left. \frac{z+1}{(z-1)} \right|_{z=2} = 3$$

Hence

$$F(z) = -2\frac{1}{(z-1)} + 3\frac{1}{(z-2)}$$

and

$$f(nT) = -2u(nT - T) + 3(2)^{n-1}u(nT - T)$$

with ROC $|z| > 2$. ∎

Example

If $F(z) = \frac{z^2+1}{(z+1)(z-1)^2} = \frac{A}{z+1} + \frac{B}{z-1} + \frac{C}{(z-1)^2}$ with $|z| > 1$, then we find

$$A = \frac{z^2 + 1}{(z - 1)^2}\bigg|_{z=-1} = \frac{1}{2},$$

$$C = \frac{z^2 + 1}{z + 1}\bigg|_{z=1} = 1.$$

To find B we set any value of z (small for convenience) in the equality. Hence with say $z = 2$ we obtain

$$\frac{z^2 + 1}{(z + 1)(z - 1)^2}\bigg|_{z=2} = \frac{1}{2}\frac{1}{z+1}\bigg|_{z=2} + B\frac{1}{z - 1}\bigg|_{z=2} + \frac{1}{(z - 1)^2}\bigg|_{z=2}$$

or $B = 1/2$. Therefore $F(z) = \frac{1}{2}\frac{1}{z+1} + \frac{1}{2}\frac{1}{z-1} + \frac{1}{(z-1)^2}$ and its inverse transform is $f(nT) = \frac{1}{2}(-1)^{n-1}u(nT - T) + \frac{1}{2}u(nT - T) + (nT - T)u(nT - T)$ with ROC $|z| > 1$. ∎

Example

The function $F(z) = z^3/(z-1)^2$ with $|z| > 1$ can be expanded as follows: $F(z) = z+2+\frac{3z-2}{(z-1)^2}$ or $F(z) = z+2+\frac{3z-2}{(z-1)^2} = z+2+\frac{A}{z-1}+\frac{B}{(z-1)^2}$. Therefore we obtain $B = \frac{(3z-2)(z-1)^2}{(z-1)^2}\bigg|_{z=1} = 1$. Set any value of z (e.g., $z = 2$) in the above equality we obtain

$$2 + 2 + \frac{3 \cdot 2 - 2}{(2 - 1)^2} = 2 + 2 + A\frac{1}{2 - 1} + \frac{1}{(2 - 1)^2} \text{ or } A = 3$$

Hence

$$F(z) = z + 2 + \frac{3}{z - 1} + \frac{1}{(z - 1)^2}$$

and its inverse transform is

$$f(nT) = \delta(nT + T) + 2\delta(nT) + 3u(nT - T) + (nT - T)u(nT - T)$$

with ROC $|z| > 1$. ∎

Tables 3 and 4 are useful for finding the inverse transforms.

Inverse transform by integration

If $F(z)$ is a regular function in the region $|z| > R$, then there exists a single sequence $\{f(nT)\}$ for which $\mathcal{Z}\{f(nT)\} = F(z)$, namely,

$$f(nT) = \frac{1}{2\pi j} \oint_C F(z)z^{n-1} \, dz = \sum_{i=1}^{K} \text{res}_{z=z_i} \left\{ F(z)z^{n-1} \right\} \qquad n = 0, 1, 2, \dots \qquad (4.8)$$

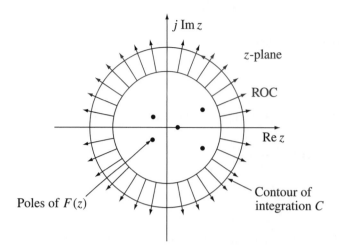

FIGURE 4.1

The contour C encloses all the singularities of $F(z)$ as shown in Figure 4.1 and it is taken in a counterclockwise direction.

Simple poles

If $F(z) = H(z)/G(z)$, then the residue at the singularity $z = a$ is given by

$$\lim_{z \to a}(z-a)F(z)z^{n-1} = \lim_{z \to a}\left[(z-a)\frac{H(z)}{G(z)}z^{n-1}\right] \tag{4.9}$$

Multiple poles

The residue at the pole z_i with multiplicity m of the function $F(z)z^{n-1}$ is given by

$$\text{res}_{z=z_i} F(z)z^{n-1} = \frac{1}{(m-1)!}\lim_{z \to z_i}\frac{d^{m-1}}{dz^{m-1}}[(z-z_i)^m F(z)z^{n-1}] \tag{4.10}$$

Simple poles not factorable

The residue at the singularity a_m is

$$F(z)z^{n-1}\Big|_{z=a_m} = \frac{H(z)}{\frac{dG(z)}{dz}}z^{n-1}\Bigg|_{z=a_m} \tag{4.11}$$

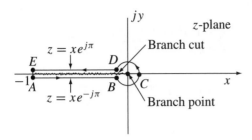

FIGURE 4.2

$F(z)$ is irrational function of z

Let $F(z) = [(z + 1)/z]^\alpha$, where α is a real noninteger. By (4.8) we write

$$f(nT) = \frac{1}{2\pi j} \oint_C \left(\frac{z + 1}{z}\right)^\alpha z^{n-1} \, dz$$

where the closed contour C is that shown in Figure 4.2.

It can easily be shown that at the limit as $z \to 0$ the integral around the small circle BCD is zero (set $z = re^{j\theta}$ and take the limit $r \to 0$). Also, the integral along EA is also zero. Because along AB $z = xe^{-j\pi}$ and along DE $z = xe^{j\pi}$, which implies that x is positive, we obtain

$$f(nT) = \frac{1}{2\pi j} \left[\int_1^0 \left(\frac{xe^{-j\pi} + 1}{xe^{-j\pi}}\right)^\alpha x^{n-1} e^{-j\pi n} \, dx + \int_0^1 \left(\frac{xe^{j\pi} + 1}{xe^{j\pi}}\right)^\alpha x^{n-1} e^{j\pi n} \, dx \right]$$

$$= \frac{1}{2\pi j} \left[-\int_0^1 (1 - x)^\alpha x^{n-1-\alpha} e^{-j\pi(n-\alpha)} \, dx + \int_0^1 (1 - x)^\alpha x^{n-1-\alpha} e^{j\pi(n-\alpha)} \, dx \right]$$

$$= \frac{\sin[(n - \alpha)\pi]}{\pi} \int_0^1 x^{n-1-\alpha} (1 - x)^\alpha \, dx \tag{4.12}$$

But the beta function is given by

$$B(m, k) = \frac{\Gamma(m)\Gamma(k)}{\Gamma(m + k)} = \int_0^1 x^{m-1}(1 - x)^{k-1} \, dx \tag{4.13}$$

and hence

$$f(nT) = \frac{\sin[(n - \alpha)\pi]}{\pi} \frac{\Gamma(n - \alpha)\Gamma(\alpha + 1)}{\Gamma(n + 1)} \tag{4.14}$$

But

$$\Gamma(m)\Gamma(1 - m) = \frac{\pi}{\sin \pi m} \tag{4.15}$$

and therefore

$$f(nT) = \frac{\Gamma(n - \alpha)\Gamma(\alpha + 1)}{\Gamma(n + 1)} \frac{1}{\Gamma(n - \alpha)\Gamma(\alpha - n + 1)} = \frac{\Gamma(\alpha + 1)}{\Gamma(n + 1)\Gamma(\alpha - n + 1)} \tag{4.16}$$

The Taylor's expansion of $F(z)$ is given as follows:

$$F(z) = \left(\frac{z+1}{z}\right)^\alpha = (1+z^{-1})^\alpha = \sum_{n=0}^\infty \frac{1}{n!} \frac{d^n(1+z^{-1})^\alpha}{(dz^{-1})^n}\bigg|_{z^{-1}=0} z^{-n}$$

$$= \sum_{n=0}^\infty \frac{1}{n!} \alpha(\alpha-1)(\alpha-2)\cdots(\alpha-n+1)z^{-n} \tag{4.17}$$

But

$$\Gamma(\alpha+1) = \alpha(\alpha-1)(\alpha-2)\cdots(\alpha-n+1)\Gamma(\alpha-n+1), \qquad \Gamma(n+1) = n! \tag{4.18}$$

and, therefore, (4.17) becomes

$$F(z) = \sum_{n=0}^\infty \frac{\Gamma(\alpha+1)}{\Gamma(n+1)\Gamma(\alpha-n+1)} z^{-n} \tag{4.19}$$

The above equation is a Z-transform expansion and hence the function $f(nT)$ is that given in (4.16).

Example
To find the inverse of the transform

$$F(z) = \frac{(z-1)}{(z+2)\left(z-\frac{1}{2}\right)} \qquad |z| > 2$$

we proceed with the following approaches:

1. **By Fraction Expansion**

$$\frac{(z-1)}{(z+2)\left(z-\frac{1}{2}\right)} = \frac{A}{z+2} + \frac{B}{z-\frac{1}{2}},$$

$$A = \frac{(z-1)}{\left(z-\frac{1}{2}\right)}\bigg|_{z=-2} = \frac{6}{5}, \qquad B = \frac{(z-1)}{(z+2)}\bigg|_{z=\frac{1}{2}} = -\frac{1}{5}$$

$$f(nT) = \mathcal{Z}^{-1}\left\{\frac{6}{5}\frac{1}{z+2} - \frac{1}{5}\frac{1}{z-\frac{1}{2}}\right\} = \frac{6}{5}(-2)^{n-1} - \frac{1}{5}\left(\frac{1}{2}\right)^{n-1} \qquad n \geq 1$$

2. **By Integration**

$$f(nT) = \operatorname{res}_{z=-2}\left\{(z+2)\frac{z-1}{(z+2)\left(z-\frac{1}{2}\right)}z^{n-1}\right\} + \operatorname{res}_{z=\frac{1}{2}}\left\{\left(z-\frac{1}{2}\right)\frac{z-1}{(z+2)\left(z-\frac{1}{2}\right)}z^{n-1}\right\}$$

$$= \frac{6}{5}(-2)^{n-1} - \frac{1}{5}\left(\frac{1}{2}\right)^{n-1} \qquad n \geq 1$$

3. **By Power Expansion**

$$\frac{z-1}{z^2+\frac{3}{2}z-1} = z^{-1} - \frac{5}{2}z^{-2} + \frac{19}{4}z^{-3} + \cdots = z^{-1}\left(1 - \frac{5}{2}z^{-1} + \frac{19}{4}z^{-2} + \cdots\right)$$

The multiplier z^{-1} indicates one time-unit shift and, hence, $\{f(nT)\} = \{1, -\frac{5}{2}, \frac{19}{4}, \ldots\}$
$n = 1, 2, \ldots$. $\blacksquare$

Example

1. **By Expansion**

   If $F(z)$ has the region of convergence $|z| > 5$, then

   $$F(z) = \frac{5z}{(z-5)^2} = \frac{5z}{z^2 - 10z + 25} = 5z^{-1} + 50z^{-2} + 375z^{-3} + \cdots$$

   $$= 0 \cdot 5^0 z^{-0} + 1 \cdot 5z^{-1}$$

   $$+ 2 \cdot 5^2 z^{-2} + 3 \cdot 5^3 z^{-3} + \cdots$$

   Hence $f(nT) = n5^n \; n = 0, 1, 2, \ldots$, which sometimes is difficult to recognize using the expansion method.

2. **By Fraction Expansion**

   $$F(z) = \frac{5z}{(z-5)^2} = \frac{Az}{z-5} + \frac{Bz^2}{(z-5)^2},$$

   $$B = \left. \frac{5}{z} \right|_{z=5} = 1,$$

   $$\frac{5 \times 6}{(6-1)^2} = \frac{A \times 6}{6-5} + \frac{6^2}{(6-5)^2} \qquad \text{or} \qquad A = -1.$$

   Hence

   $$F(z) = -\frac{z}{z-5} + \frac{z^2}{(z-5)^2}$$

   and $f(nT) = -(5)^n + (n+1)5^n = n5^n, n \geq 0$.

3. **By Integration**

   $$\frac{1}{(2-1)!} \frac{d^{2-1}}{dz^{2-1}} \left[(z-5)^2 \frac{5z}{(z-5)^2} z^{n-1} \right]\Bigg|_{z=5} = 5nz^{n-1}\big|_{z=5} = n5^n, \qquad n \geq 0. \quad \blacksquare$$

Figure 4.3 shows the relation between pole location and type of poles and the behavior of causal signals. m stands for pole multiplicity. Table 5 gives the Z-transform of a number of sequences.

B. Two-sided Z-transform

6.5 The Z-transform

If a function $f(z)$ is defined by $-\infty < t < \infty$, then the Z-transform of its discrete representation $f(nT)$ is given by

$$Z_{II}\{f(nT)\} \doteq F(z) = \sum_{n=-\infty}^{\infty} f(nT) z^{-n} \qquad R_+ < |z| < R_- \qquad (5.1)$$

Single Real Poles—Causal Signals

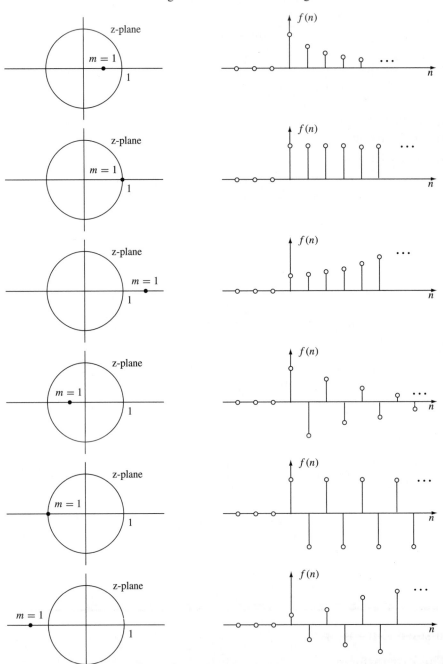

FIGURE 4.3

Double Real Poles—Causal Signals

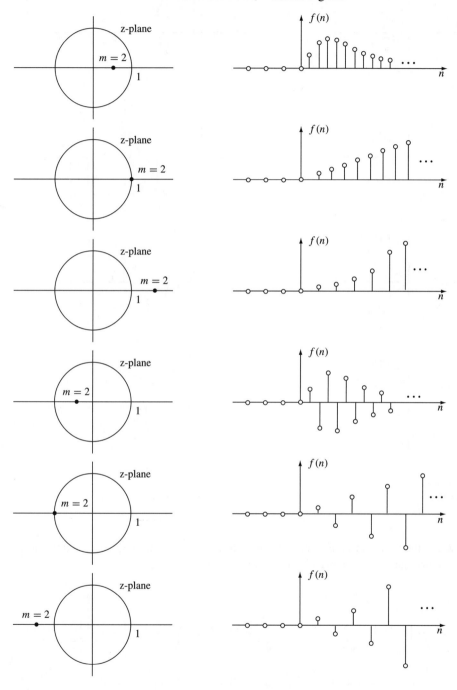

FIGURE 4.3
(continued)

Complex-Conjugate Poles—Causal Signals

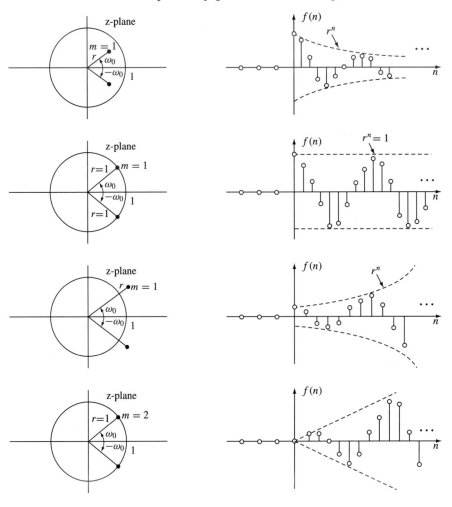

FIGURE 4.3
(continued)

where R_+ is the radius of convergence for the positive time of the sequence, and R_- is the radius of convergence for the negative time of the sequence.

Example

$$F(z) = \mathcal{Z}_{II}\left\{e^{-|nT|}\right\} = \sum_{n=-\infty}^{-1} e^{nT}z^{-n} + \sum_{n=0}^{\infty} e^{-nT}z^{-n} = \sum_{n=-\infty}^{0} e^{nT}z^{-n} - 1 + \sum_{n=0}^{\infty} e^{-nT}z^{-n}$$

$$= \sum_{n=0}^{\infty} e^{-nT}z^{n} - 1 + \sum_{n=0}^{\infty} e^{-nT}z^{-n} = \frac{1}{1 - e^{-nT}z} - 1 + \frac{1}{1 - e^{-nT}z^{-1}}$$

The first sum (negative time) converges if $|e^{-T}z| < 1$ or $|z| < e^{T}$. The second sum (positive time) converges if $|e^{-T}z^{-1}| < 1$ or $e^{-T} < |z|$. Hence the region of convergence is $R_+ = e^{-T} < |z| < R_- = e^{T}$. The two poles of $F(z)$ are $z = e^{T}$ and $z = e^{-T}$. ∎

Example

The Z-transform of the functions of $u(nT)$ and $-u(-nT - T)$ are

$$\mathcal{Z}_{II}\{u(nT)\} = \sum_{n=0}^{\infty} u(nT)z^{-n} = \frac{1}{1 - z^{-1}} = \frac{z}{z - 1} \qquad |z| > 1$$

$$\mathcal{Z}_{II}\{-u(-nT - T)\} = -\sum_{n=-\infty}^{-1} u(-nT - T)z^{-n}$$

$$= -\left[\sum_{n=-\infty}^{0} z^{-n} - 1\right]$$

$$= 1 - \sum_{n=0}^{\infty} z^{n} = 1 - \frac{1}{1 - z} = \frac{z}{z - 1} \qquad |z| < 1$$

Although their Z-transform is identical their ROC is different. Therefore, to find the inverse Z-transform the region of convergence must also be given. ∎

Figure 5.1 shows signal characteristics and their corresponding region of convergence.

Assuming that the algebraic expression for the Z-transform $F(z)$ is a rational function and that $f(nT)$ has finite amplitude, except possibly at infinities, the properties of the region of convergence are

1. The ROC is a ring or disc in the z-plane and centered at the origin, and $0 \leq R_+ < |z| < R_- \leq \infty$.

2. The Fourier transform converges also absolutely if and only if the ROC of the Z-transform of $f(nT)$ includes the unit circle.

3. No poles exist in the ROC.

4. The ROC of a finite sequence $\{f(nT)\}$ is the entire z-plane except possibly for $z = 0$ or $z = \infty$.

5. If $f(nT)$ is right handed, $0 \leq n < \infty$, the ROC extends outward from the outermost pole of $F(z)$ to infinity.

6. If $f(nT)$ is left handed, $-\infty < n < 0$, the ROC extends inward from the innermost pole of $F(z)$ to zero.

7. An infinite-duration two-sided sequence $\{f(nT)\}$ has a ring as its ROC, bounded on the interior and exterior by a pole. The ring contains no poles.

8. The ROC must be a connected region.

6.6 Properties

Linearity

The proof is similar to the one-sided Z-transform.

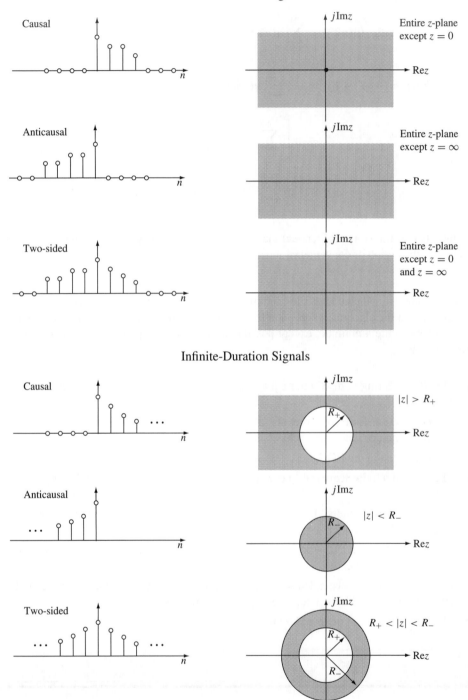

FIGURE 5.1

Shifting

$$\mathcal{Z}_{II}\{f(nT \pm kT)\} = z^{\pm k} F(z) \tag{6.1}$$

PROOF

$$\mathcal{Z}_{II}\{f(nT - kT)\} = \sum_{n=-\infty}^{\infty} f(nT - kT)z^{-n} = z^{-k} \sum_{m=-\infty}^{\infty} f(mT)z^{-m}$$

The last step results from setting $m = n - k$. Proceed similarly for the positive sign. The ROC of the shifted functions is the same as that of the unfinished function except at $z = 0$ for $k > 0$ and $z = \infty$ for $k < 0$.   ∎

Example
To find the transfer function of the system $y(nT) - y(nT - T) + 2y(nT - 2T) = x(nT) + 4x(nT - T)$ we take the Z-transform of both sides of the equation. Hence we find

$$Y(z) - z^{-1}Y(z) + 2z^{-2}Y(z) = X(z) + 4z^{-1}X(z)$$

or

$$H(z) = \frac{Y(z)}{X(z)} = \frac{1 + 4z^{-1}}{1 - z^{-1} + 2z^{-2}} \qquad\qquad ∎$$

Example
Consider the Z-transform

$$F(z) = \frac{1}{z - \frac{1}{2}} \qquad |z| > \frac{1}{2}$$

Because the pole is inside the ROC, it implies that the function is causal. We next write the function in the form

$$F(z) = z^{-1}\frac{z}{z - \frac{1}{2}} = z^{-1}\frac{1}{1 - \frac{1}{2}z^{-1}} \qquad |z| > \frac{1}{2}$$

which indicates that it is a shifted function (because of the multiplier z^{-1}). Hence the inverse transform is $f(n) = (\frac{1}{2})^{n-1}u(n - 1)$ because the inverse transform of $1/(1 - \frac{1}{2}z^{-1})$ is equal to $(\frac{1}{2})^n$.   ∎

Scaling

If

$$\mathcal{Z}_{II}\{f(nT)\} = F(z) \qquad R_+ < |z| < R_-$$

then

$$\mathcal{Z}_{II}\{a^{nT} f(nT)\} = F(a^{-T}z) \qquad |a^T|R_+ < |z| < |a^T|R_- \tag{6.2}$$

PROOF

$$\mathcal{Z}_{II}\{a^{nT} f(nT)\} = \sum_{n=-\infty}^{\infty} a^{nT} f(nT)z^{-n} = \sum_{n=-\infty}^{\infty} f(nT)(a^{-T}z)^{-n} = F(a^{-T}z)$$

Because the ROC of $F(z)$ is $R_+ < |z| < R_-$, the ROC of $F(a^{-T}z)$ is

$$R_+ < |a^{-T}z| < R_- \qquad \text{or} \qquad R_+|a^T| < |z| < |a^T|R_- \qquad \blacksquare$$

Example
If the Z-transform of $f(nT) = \exp(-|nT|)$ is

$$F(z) = \frac{1}{1 - e^{-nT}z} + \frac{1}{1 - e^{-nT}z^{-1}} - 1 \qquad e^{-T} < |z| < e^T$$

then the Z-transform of $g(nT) = a^{nT}f(nT)$ is

$$G(z) = \frac{1}{1 - e^{-nT}a^{-T}z} + \frac{1}{1 - e^{-nT}a^Tz^{-1}} - 1 \qquad a^Te^{-T} < |z| < e^Ta^T \qquad \blacksquare$$

Time reversal

If

$$\mathcal{Z}_{II}\{f(nT)\} = F(z) \qquad R_+ < |z| < R_-$$

then

$$\mathcal{Z}_{II}\{f(-nT)\} = F(z^{-1}) \qquad \frac{1}{R_-} < |z| < \frac{1}{R_+} \qquad (6.3)$$

PROOF

$$\mathcal{Z}_{II}\{f(-nT)\} = \sum_{n=-\infty}^{\infty} f(-nT)z^{-n} = \sum_{n=-\infty}^{\infty} f(nT)(z^{-1})^{-n} = F(z^{-1})$$

and

$$R_+ < |z^{-1}| < R_- \qquad \text{or} \qquad |z| > \frac{1}{R_-} \qquad \text{and} \qquad |z| < \frac{1}{R_+}$$

The above means that if z_0 belongs to the ROC of $F(z)$ then $1/z_0$ is in the ROC of $F(z^{-1})$. The reflection in the time domain corresponds to inversion in the z-domain. $\blacksquare$

Example
The Z-transform of $f(n) = u(n)$ is $z/(z - 1)$ for $|z| > 1$. Therefore, the Z-transform of $f(-n) = u(-n)$ is

$$\frac{\frac{1}{z}}{\frac{1}{z} - 1} = \frac{1}{1 - z}$$

Also from the definition of the Z-transform we write

$$\mathcal{Z}\{u(-n)\} = \sum_{n=-\infty}^{0} z^{-n} - 1 = \sum_{n=0}^{\infty} z^n - 1 = \frac{1}{1 - z} - 1 = \frac{z}{z - 1} \qquad \blacksquare$$

Multiplication by nT

If

$$\mathcal{Z}_{II}\{f(nT)\} = F(z) \qquad R_+ < |z| < R_-$$

then

$$\mathcal{Z}_{II}\{nTf(nT)\} = -zT\frac{dF(z)}{dz} \qquad R_+ < |z| < R_- \tag{6.4}$$

PROOF   A Laurent series can be differentiated term-by-term in its ROC and the resulting series has the same ROC. Therefore we have

$$\frac{dF(z)}{dz} = \frac{d}{dz}\sum_{n=-\infty}^{\infty} f(nT)z^{-n} = \sum_{n=-\infty}^{\infty} -nf(nT)z^{-n-1} \qquad \text{for } R_+ < |z| < R_-$$

Multiply both sides by $-zT$

$$-zT\frac{dF(z)}{dz} = \sum_{n=-\infty}^{\infty} nTf(nT)z^{-n} = \mathcal{Z}\{nTf(nT)\} \qquad \text{for } R_+ < |z| < R_- \qquad \blacksquare$$

Example
If $F(z) = \log(1 + az^{-1})$ $|z| > |a|$, then

$$\frac{dF(z)}{dz} = \frac{-az^{-2}}{1 + az^{-1}} \qquad \text{or} \qquad -z\frac{dF(z)}{dz} = az^{-1}\frac{1}{1 - (-a)z^{-1}} \qquad |z| > |a|$$

The z^{-1} implies a time shift, and the inverse transform of the fraction is $(-a)^n$. Hence the inverse transform is $a(-a)^{n-1}u(n-1)$. From the differentiation property (with $T = 1$) we obtain

$$nf(n) = a(-a)^{n-1}u(n-1) \qquad \text{or} \qquad f(n) = (-1)^{n-1}\frac{a^n}{n}u(n-1) \qquad \blacksquare$$

Example
If $f(nT) = au(nT)$ then its Z-transform is $F(z) = a/(1 - z^{-1})$ for $|z| > 1$. Therefore

$$\mathcal{Z}\{nTau(nT)\} = -zTa\frac{dF(z)}{dz} = aT\frac{z}{(z-1)^2} \qquad |z| > 1 \qquad \blacksquare$$

Convolution

If

$$\mathcal{Z}_{II}\{f_1(nT)\} = F_1(z) \qquad \text{and} \qquad \mathcal{Z}_{II}\{f_2(nT)\} = F_2(z)$$

then

$$F(z) = \mathcal{Z}_{II}\{f_1(nT) * f_2(nT)\} = F_1(z)F_2(z) \tag{6.5}$$

The ROC of $F(z)$ is, at least, the intersection of that for $F_1(z)$ and $F_2(z)$.

PROOF

$$F(z) = \sum_{n=-\infty}^{\infty} f(nT)z^{-n} = \sum_{n=-\infty}^{\infty} \left[\sum_{m=-\infty}^{\infty} f_1(mT) f_2(nT - mT) \right] z^{-n}$$

$$= \sum_{m=-\infty}^{\infty} f_1(mT) \left[\sum_{n=-\infty}^{\infty} f_2(nT - mT)z^{-n} \right]$$

$$= \sum_{m=-\infty}^{\infty} f_1(mT)z^{-m} F_2(z) = F_1(z)F_2(z)$$

where the shifting property was invoked. ∎

Example
The Z-transform of the convolution of $e^{-n}u(n)$ and $u(n)$ is

$$\mathcal{Z}_{II}\{(e^{-n}u(n)) * u(n)\} = \mathcal{Z}\left\{ \sum_{m=0}^{n} e^{-m}u(n-m) \right\} = \mathcal{Z}\{e^{-n}\}\mathcal{Z}\{u(n)\} = \frac{z}{z-e^{-1}}\frac{z}{z-1}$$

Also from the convolution definition we find

$$\mathcal{Z}\left\{ \sum_{m=0}^{n} e^{-m}u(n-m) \right\} = \mathcal{Z}\left\{ \frac{1 - e^{-n-1}}{1 - e^{-1}} \right\}$$

$$= \mathcal{Z}\left\{ \frac{1}{1-e^{-1}} - \frac{e^{-1}}{1-e^{-1}}e^{-n} \right\}$$

$$= \frac{1}{1-e^{-1}}\left(\frac{z}{z-1} - e^{-1}\frac{z}{z-e^{-1}} \right)$$

$$= \frac{z^2}{(z-1)(z-e^{-1})}$$

which verifies the convolution property. The ROC for $e^{-n}u(n)$ is $|z| > e^{-1}$ and the ROC of $u(n)$ is $|z| > 1$. The ROC of $e^{-n}u(n) * u(n)$ is the intersection of these two ROCs, and hence the ROC is $|z| > 1$. ∎

Example
The convolution of $f_1(n) = \{2, 1, -3\}$ for $n = 0, 1,$ and 2, and $f_2(n) = \{1, 1, 1, 1\}$ for $n = 0,$ 1, 2, and 3 is

$$G(z) = F_1(z)F_2(z) = (2 + z^{-1} - 3z^{-2})(1 + z^{-1} + z^{-2} + z^{-3}) = 2 + 3z^{-1} - 2z^{-4} - 3z^{-5}$$

which indicates that the output is $g(n) = \{2, 3, 0, 0, -2, -3\}$ which can easily be found by simply convoluting $f_1(n)$ and $f_2(n)$. ∎

Correlation

If

$$\mathcal{Z}_{II}\{f_1(nT)\} = F_1(z) \qquad \text{and} \qquad \mathcal{Z}_{II}\{f_2(nT)\} = F_2(z)$$

then

$$\mathcal{Z}_{II}\{r_{f_1 f_2}(\ell T)\} \doteq \mathcal{Z}_{II}\{f_1(nT) \otimes f_2(nT)\} = \mathcal{Z}_{II}\left\{ \sum_{n=-\infty}^{\infty} f_1(nT) f_2(nT - \ell T) \right\}$$

$$= R_{f_1 f_2}(z) = F_1(z) F_2(z^{-1}) \tag{6.6}$$

The ROC of $R_{f_1 f_2}(z)$ is at least the intersection of that for $F_1(z)$ and $F_2(z^{-1})$.

PROOF   But $r_{f_1 f_2}(\ell T) = f_1(T\ell) * f_2(-T\ell)$ and, hence, from the convolution property and the time-reversal property $R_{f_1 f_2}(z) = F_1(z) F_2(z^{-1})$.   ∎

Example

The transform of the autocorrelation sequencing $f(nT) = a^{nT} u(n)$, $-1 < a < 1$ is

$$R_{ff}(z) \doteq \mathcal{Z}_{II}\{r_{ff}(\ell T)\} = F(z) F(z^{-1})$$

But

$$F(z) = \frac{1}{1 - a^T z^{-1}} \qquad |z| > |a|^T \qquad \text{causal signal}$$

and

$$F(z^{-1}) = \frac{1}{1 - a^T z} \qquad |z| < \frac{1}{|a|^T} \quad \text{anticausal signal}$$

Hence

$$R_{ff}(z) = \frac{1}{1 - a^T(z + z^{-1}) + a^{2T}} \qquad \text{ROC } |a|^T < |z| < \frac{1}{|a|^T}$$

Because the ROC of $R_{ff}(z)$ is a ring, it implies that $r_{ff}(\ell T)$ is a two-sided signal.
   We can proceed to find the autocorrelation first

$$r_{ff}(nT) = \sum_{m=n}^{\infty} a^{mT} a^{(m-n)T} = a^{-nT} \sum_{m=0}^{\infty} a^{2Tm} - a^{-nT} \sum_{m=0}^{n-1} a^{2Tm}$$

$$= a^{-nT} \frac{1}{1 - a^{2T}} - a^{-nT} \frac{1 - a^{2Tn}}{1 - a^{2T}} = \frac{a^{nT}}{1 - a^{2T}} \qquad n \geq 0$$

$$r_{ff}(nT) = \sum_{m=0}^{\infty} a^{mT} a^{(m-n)T} = a^{-nT} \frac{1}{1 - a^{2T}} \qquad n \leq 0$$

and then compare by inverting the function $F(z) F(z^{-1})$.   ∎

Multiplication by e^{-anT}

If

$$\mathcal{Z}_{II}\{f(nT)\} = F(z) \qquad R_+ < |z| < R_-$$

then

$$\mathcal{Z}_{II}\{e^{-anT}f(nT)\} = F(e^{aT}z) \qquad |e^{-aT}|R_+ < |z| < |e^{-aT}|R_- \tag{6.7}$$

PROOF

$$\mathcal{Z}_{II}\{e^{-anT}f(nT)\} = \sum_{n=-\infty}^{\infty} f(nT)(e^{aT}z)^{-n} = F(e^{aT}z) \qquad R_+ < |e^{aT}z| < R_- \qquad \blacksquare$$

Frequency translation

If the region of convergence of $F(z)$ includes the unit circle and $g(nT) = e^{j\omega_0 nT}f(nT)$, then

$$G(\omega) = F(\omega - \omega_0) \tag{6.8}$$

PROOF   From (6.7) $G(z) = F(e^{-j\omega_0 T}z)$ and has the same region of convergence as $F(z)$ because $|\exp(j\omega_0 T)| = 1$. Therefore

$$G(\omega) = G(z)|_{z=e^{j\omega T}} = F(e^{j(\omega-\omega_0)T}) = F(\omega - \omega_0) \qquad \blacksquare$$

Product

If

$$\mathcal{Z}_{II}\{f(nT)\} = F(z) \qquad R_{+f} < |z| < R_{-f} \tag{6.9}$$

$$\mathcal{Z}_{II}\{h(nT)\} = H(z) \qquad R_{+h} < |z| < R_{-h} \tag{6.10}$$

$$g(nT) = f(nT)h(nT)$$

then

$$\mathcal{Z}_{II}\{f(nT)h(nT)\} \doteq G(z) = \sum_{n=-\infty}^{\infty} f(nT)h(nT)z^{-n}$$

$$= \frac{1}{2\pi j}\oint_C F(\tau)H\left(\frac{z}{\tau}\right)\frac{d\tau}{\tau} \qquad R_{+f}R_{+h} < |z| < R_{-f}R_{-h} \tag{6.11}$$

where C is any simple closed curve encircling the origin counterclockwise with

$$\max\left(R_{+f}, \frac{|z|}{R_{-h}}\right) < |\tau| < \min\left(R_{-f}, \frac{|z|}{R_{+h}}\right) \tag{6.12}$$

PROOF   The series in (6.11) will converge to an analytic function $G(z)$ for $R_{+g} < |z| < R_{-g}$. Using the root test (see Section 6.2) we obtain

$$R_{+g} = \varlimsup_{n\to\infty}(|f(nT)h(nT)|)^{1/n}$$

$$\leq \varlimsup_{n\to\infty}(|f(nT)|)^{1/n}\varlimsup_{n\to\infty}(|h(nT)|)^{1/n} = R_{+f}R_{+h} \tag{6.13}$$

for positive n. However,

$$F(z) = \sum_{n=-\infty}^{0} f(nT)z^{-n} = \sum_{n=0}^{\infty} f(-nT)z^n \tag{6.14}$$

and this series converges if

$$|z| < \frac{1}{\overline{\lim_{n \to \infty}} (|f(-nT)|)^{1/n}} = R_{-f} \tag{6.15}$$

Hence

$$R_{-g} = \frac{1}{\overline{\lim_{n \to \infty}} (|f(-nT)h(-nT)|)^{1/n}}$$

$$\geq \frac{1}{\overline{\lim_{n \to \infty}} (|f(-nT)|)^{1/n} \, \overline{\lim_{n \to \infty}} (|h(-nT)|)^{1/n}}$$

$$\geq R_{-f}R_{-h} \tag{6.16}$$

Replacing $f(nT)$ in the summation of (6.11) by its inversion formula (4.8), we find

$$G(z) = \sum_{n=-\infty}^{\infty} \frac{1}{2\pi j} \oint_C F(\tau)\tau^n \frac{d\tau}{\tau} h(nT)z^{-n} = \frac{1}{2\pi j} \oint_C F(\tau) \sum_{n=-\infty}^{\infty} h(nT) \left(\frac{z}{\tau}\right)^{-n} \frac{d\tau}{\tau} \tag{6.17}$$

The interchange of the sum and integral is justified if the integrand converges uniformly for some choice of C and z. The contour must be chosen so that

$$R_{+f} < |\tau| < R_{-f} \tag{6.18}$$

If

$$R_{+h} < \left|\frac{z}{\tau}\right| < R_{-h} \qquad \text{or} \qquad \frac{|z|}{R_{-h}} < |\tau| < \frac{|z|}{R_{+h}} \tag{6.19}$$

the series in the intergand of (6.17) will converge uniformly to $H(z/\tau)$, and otherwise will diverge. Figure 6.1 shows the region of convergence for $F(\tau)$ and $H(z/\tau)$. From (6.18) and (6.19) we obtain

$$\frac{|z|}{R_{-h}} < R_{-f} \qquad \text{or} \qquad |z| < R_{-f}R_{-h}$$

$$\frac{|z|}{R_{+h}} > R_{+f} \qquad \text{or} \qquad |z| > R_{+f}R_{+h}$$

or equivalently

$$R_{+f}R_{+h} < |z| < R_{-f}R_{-h} \tag{6.20}$$

When z satisfies the above equation, the intersection of the domain identified by (6.18) and (6.19) is

$$(R_{+f} < |\tau| < R_{-f}) \cap \left(\frac{|z|}{R_{-h}} < |\tau| < \frac{|z|}{R_{+h}}\right) = \max\left(R_{+f}, \frac{|z|}{R_{-h}}\right)$$

$$< |\tau| < \min\left(R_{-f}, \frac{|z|}{R_{+h}}\right) \tag{6.21}$$

The contour must be located inside the intersection.

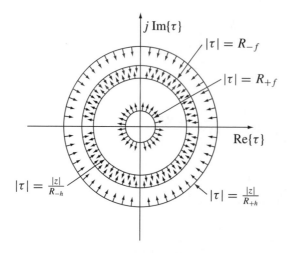

FIGURE 6.1

When signals are causal, $R_{-f} = R_{-h} = \infty$ and the conditions (6.20) and (6.21) reduce to

$$R_{+f} R_{+h} < |z| \tag{6.22}$$

$$R_{+f} < |\tau| < \frac{\tau}{R_{+h}} \tag{6.23}$$

Hence, all of the poles of $F(\tau)$ lie inside the contour and all the poles of $H(z/\tau)$ lie outside the contour. ▌

Example
The Z-transform of $u(nT)$ is

$$F(z) = \frac{1}{1 - z^{-1}} \qquad |z| > 1 = R_{+f}, \ R_{-f} = \infty$$

and the Z-transform of $h(nT) = \exp(-|nT|)$ is

$$H(z) = \frac{1 - e^{-2T}}{(1 - e^{-T}z^{-1})(1 - e^{-T}z)} \qquad R_{+h} = e^{-T} < |z| < e^T = R_{-h}$$

But $R_{-f} = \infty$ and hence from (6.11) $1 \cdot \exp(-T) < |z| < \infty$. The contour must lie in the region $\max(1, |z|e^{-T}) < |\tau| < \min(-\infty, |z|e^T)$ as given by (6.21). The pole-zero configuration and the contour are shown in Figure 6.2. If we choose $|z| > e^T$ then the contour is that shown in the figure. Therefore, (6.11) becomes

$$\mathcal{Z}_{II}\{u(nT)h(nT)\} \doteq G(z) = \frac{1}{2\pi j} \oint_C \frac{1}{1 - \tau^{-1}} \frac{1 - e^{-2T}}{\left(1 - e^{-T}\frac{\tau}{z}\right)\left(1 - e^{-T}\frac{z}{\tau}\right)} \frac{d\tau}{\tau}$$

The poles of $H(z/\tau)$ are at $\tau = z\exp(-T)$ and $\tau = z\exp(T)$. Hence the contour encloses the poles $\tau = 1$ and $\tau = z\exp(-T)$. Applying the residue theorem next we obtain

$$G(z) = \frac{1}{1 - e^{-T}z^{-1}} \qquad |z| > e^{-T}$$

which has the inverse function $g(nT) = e^{-nT}u(nT)$, as expected. ▌

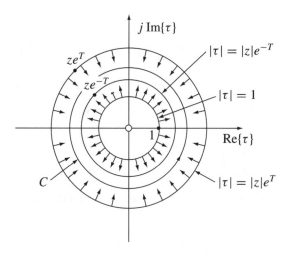

FIGURE 6.2

Parseval's theorem

If

$$\mathcal{Z}_{II}\{f(nT)\} = F(z) \qquad R_{+f} < |z| < R_{-f}$$

$$\mathcal{Z}_{II}\{h(nT)\} = H(z) \qquad R_{+h} < |z| < R_{-h} \tag{6.24}$$

with

$$R_{+f}R_{+h} < |z| = 1 < R_{-f}R_{-h} \tag{6.25}$$

then we have

$$\sum_{n=-\infty}^{\infty} f(nT)h(nT) = \frac{1}{2\pi j} \oint_C F(z)H(z^{-1})\frac{dz}{z} \tag{6.26}$$

where the contour encircles the origin with

$$\max\left(R_{+f}, \frac{1}{R_{-h}}\right) < |z| < \min\left(R_{-f}, \frac{1}{R_{+h}}\right) \tag{6.27}$$

PROOF   In (6.11) and (6.12) set $z = 1$ and replace the dummy variable τ with z to obtain (6.26) and (6.27).   ∎

For complex signals Parseval's relation (6.26) is modified as follows:

$$\sum_{n=-\infty}^{\infty} f(nT)h^*(nT) = \frac{1}{2\pi j} \oint_C F(z)H^*\left(\frac{1}{z^*}\right)\frac{dz}{z} \tag{6.28}$$

If $f(nT)$ and $h(nT)$ converge on the unit circle, we can use the unit circle as the contour. We then obtain

$$\sum_{n=-\infty}^{\infty} f(nT)h^*(nT) = \frac{1}{\omega_s} \int_{-\omega_s/2}^{\omega_s/2} F(e^{j\omega T})H^*(e^{j\omega T})\, d\omega \qquad \omega_s = \frac{2\pi}{T} \tag{6.29}$$

where we set $z = e^{j\omega T}$. If $f(nT) = h(nT)$ then

$$\sum_{n=-\infty}^{\infty} |f(nT)|^2 = \frac{1}{\omega_s} \int_{-\omega_s/2}^{\omega_s/2} |F(e^{j\omega T})|^2 \, d\omega \qquad (6.30)$$

Example

The Z-transform of $f(nT) = \exp(-nT)u(nT)$ is $F(z) = 1/(1 - e^{-T}z^{-1})$ for $|z| > e^{-T}$. From (6.26) we obtain

$$\sum_{n=-\infty}^{\infty} f^2(nT) = \sum_{n=0}^{\infty} f^2(nT) = \frac{1}{2\pi j} \oint_C \frac{1}{1 - e^{-T}z^{-1}} \frac{1}{1 - e^{-T}z} \frac{dz}{z}$$

From (6.27) we see that $\max(e^{-T}, 0) < |z| < \min(\infty, e^T)$. The contour encircles the pole at $z = e^{-T}$ so that

$$\sum_{n=0}^{\infty} f^2(nT) = \text{res} \left\{ \left[\frac{z - e^{-T}}{(z - e^{-T})(1 - e^{-T}z)} \right] \right\}_{z=e^{-T}} = \frac{1}{1 - e^{-2T}}$$

Also we find directly

$$\sum_{n=0}^{\infty} e^{-nT}e^{-nT} = \sum_{n=0}^{\infty} e^{-2nT} = (1 + e^{-2T} + (e^{-2T})^2 + \cdots) = \frac{1}{1 - e^{-2T}} \qquad \blacksquare$$

Complex conjugate signal

If

$$\mathcal{Z}_{II}\{f(nT)\} = F(z) \qquad R_{+f} < |z| < R_{-f}$$

then

$$\mathcal{Z}_{II}\{f^*(nT)\} = F^*(z^*) \qquad R_{+f} < |z| < R_{-f} \qquad (6.31)$$

PROOF   By definition we have

$$F(z) = \sum_{n=-\infty}^{\infty} f(nT)z^{-n}$$

Replacing z with z^* and taking the conjugate of both sides of the above equation we obtain (6.31). $\blacksquare$

6.7   Inverse Z-transform

Power series expansion

The inverse Z-transform in operational form is given by

$$f(nT) = \mathcal{Z}_{II}^{-1}\{F(z)\}$$

If $F(z)$ corresponds to a causal signal, then the signal can be found by dividing the denominator into the numerator to generate a power series in z^{-1} and recognizing that $f(nT)$ is the coefficient of z^{-n}. Similarly, if it is known that $f(nT)$ is zero for positive time (n positive), the value of $f(nT)$ can be found by dividing the denominator into the numerator to generate a power series in z.

Example

If $F(z) = [z(z + 1)]/(z^2 - 2z + 1) = (1 + z^{-1})/(1 - 2z^{-1} + z^{-2})$ and the ROC is $|z| > 1$, then

$$
1 - 2z^{-1} + z^{-2} \overline{)\smash{\sqrt{1 + z^{-1}}}} \quad \frac{1 + 3z^{-1} + 5z^{-2} + 7z^{-3} + \cdots}{}
$$

$$
\frac{1 - 2z^{-1} + z^{-2}}{3z^{-1} - z^{-2}}
$$

$$
\frac{3z^{-1} - 6z^{-2} + 3z^{-3}}{5z^{-2} - 3z^{-3}}
$$

$$\cdots$$

and by continuing the division we recognize that

$$
f(nT) = \begin{cases} 0 & n < 0 \\ (2n + 1) & n \geq 0 \end{cases}
$$

If $f(nT)$ is known to be zero for positive n, that the ROC is $|z| < 1$, then

$$
z^{-2} - 2z^{-1} + 1 \overline{)\smash{\sqrt{z^{-1} + 1}}} \quad \frac{z + 3z^2 + 5z^3 + \cdots}{}
$$

$$
\frac{z^{-1} - 2 + z}{3 - z}
$$

$$
\frac{3 - 6z + 3z^2}{5z - 3z^2}
$$

$$\cdots$$

This series is recognized as

$$
f(nT) = \begin{cases} -(2n + 1) & n < 0 \\ 0 & n \geq 0 \end{cases}
$$

Example

If $F(z) = \log(1 + 2z^{-1})$, $|z| > 2$, then using power series expansion for $\log(1 + x)$, with $|x| < 1$, we obtain

$$
F(z) = \sum_{n=1}^{\infty} \frac{(-1)^{n+1} 2^n z^{-n}}{n}
$$

which indicates that

$$
f(nT) = \begin{cases} (-1)^{n+1} \frac{2^n}{n} & n \geq 0 \\ 0 & n \leq 0 \end{cases}
$$

∎

In general, any **improper** rational function ($M \geq N$) can be expressed as

$$F(z) = \frac{N(z)}{D(z)} = \frac{b_0 + b_1 z^{-1} + \cdots + b_M z^{-M}}{1 + a_1 z^{-1} + \cdots + a_N z^{-N}}$$

$$= c_0 + c_1 z^{-1} + \cdots + c_{M-N} z^{-(M-N)} + \frac{N_1(z)}{D(z)} \tag{7.1}$$

where the inverse Z-transform of the polynomial can easily be found by inspection.
    A **proper** function ($M < N$) is of the form

$$F(z) = \frac{N(z)}{D(z)} = \frac{b_0 + b_1 z^{-1} + \cdots + b_M z^{-M}}{1 + a_1 z^{-1} + \cdots + a_N z^{-N}} \qquad a_N \neq 0, \ M < N$$

or

$$F(z) = \frac{N(z)}{D(z)} = \frac{b_0 z^N + b_1 z^{N-1} + \cdots + b_M z^{N-M}}{z^N + a_1 z^{N-1} + \cdots + a_N} \tag{7.2}$$

Because $N > M$, the function

$$\frac{F(z)}{z} = \frac{b_0 z^{N-1} + b_1 z^{N-2} + \cdots + b_M z^{N-M-1}}{z^N + a_1 z^{N-1} + \cdots + a_N} \tag{7.3}$$

is always a proper function.

Partial fraction expansion

Distinct poles If the poles $p_1, p_2, \ldots, p_N$ of a proper function $F(z)$ are all different, then
we expand it in the form

$$\frac{F(z)}{z} = \frac{A_1}{z - p_1} + \frac{A_2}{z - p_2} + \cdots + \frac{A_N}{z - p_N} \tag{7.4}$$

where all A_i are unknown constants to be determined.
    The inverse Z-transform of the kth term of (7.4) is given by

$$Z^{-1}\left\{ \frac{1}{1 - p_k z^{-1}} \right\} = \begin{cases} (p_k)^n u(nT) & \text{if ROC} : |z| > |p_k| \text{ (causal signal)} \\ -(p_k)^n u(-nT - T) & \text{if ROC} : |z| < |p_k| \text{ (anticausal signal)} \end{cases} \tag{7.5}$$

If the signal is causal, the ROC is $|z| > p_{\max}$, where $p_{\max} = \max\{|p_1|, |p_2|, \ldots, |p_N|\}$. In this
case all terms in (7.4) result in causal signal components.

Example
    a)   If $F(z) = z(z + 3)/(z^2 - 3z + 2)$ with $|z| > 2$ then

$$\frac{F(z)}{z} = \frac{z + 3}{(z - 2)(z - 1)} = \frac{A_1}{z - 2} + \frac{A_2}{z - 1}$$

$$A_1 = \frac{(z + 3)(z - 2)}{(z - 2)(z - 1)}\bigg|_{z=2} = 5, \qquad A_2 = \frac{(z + 3)(z - 1)}{(z - 2)(z - 1)}\bigg|_{z=1} = -4$$

Therefore

$$F(z) = 5\frac{z}{z - 2} - 4\frac{z}{z - 1} \qquad \text{or} \qquad f(nT) = 5(2)^n - 4(1)^n \qquad n \geq 0$$

b) If $F(z) = z(z + 3)/(z^2 - 3z + 2)$ with $1 < |z| < 2$, then following exactly the same procedure

$$F(z) = 5\frac{z}{z - 2} - 4\frac{z}{z - 1}$$

However, the pole at $z = 2$ belongs to the negative-time sequence and the pole at $z = 1$ belongs to the positive-time sequence. Hence

$$f(nT) = \begin{cases} -4(1)^n & n \geq 0 \\ -5(2)^n & n \leq -1 \end{cases}$$

∎

Example

To determine the inverse Z-transform of $F(z) = 1/(1 - 1.5z^{-1} + 0.5z^{-2})$ if a) ROC : $|z| > 1$, b) ROC : $|z| < 0.5$, and c) ROC : $0.5 < |z| < 1$, we proceed as follows:

$$F(z) = \frac{z^2}{z^2 - 1.5z + 0.5} = \frac{z^2}{(z - 1)\left(z - \frac{1}{2}\right)} = A + \frac{Bz}{z - 1} + \frac{Cz}{z - \frac{1}{2}}$$

or

$$F(z) = 2\frac{z}{z - 1} - \frac{z}{z - \frac{1}{2}}$$

a) $f(nT) = 2(1)^n - (1/2)^n, n \geq 0$ because both poles are outside the region of convergence $|z| > 1$ (inside the unit circle).

b) $f(nT) = -2(1)^n u(-nT - T) + (1/2)^n u(-nT - T), n \leq -1$ because both poles are outside the region of convergence (outside the circle $|z| = 0.5$).

c) Pole at $1/2$ provides the causal part and the pole at 1 provides the anticausal. Hence

$$f(nT) = -2(1)^n u(-nT - T) - \left(\frac{1}{2}\right)^n u(nT) \qquad -\infty < n < \infty$$

∎

Multiple poles If $F(z)$ has repeated poles, we must modify the form of the expansion. Suppose $F(z)$ has a pole of multiplicity m at $z = p_i$. Then one form of expansion is of the form

$$A_1\frac{z}{z - p_i} + A_2\frac{z^2}{(z - p_i)^2} + \cdots + A_m\frac{z^m}{(z - p_i)^m} \tag{7.6}$$

The following example shows how to find A_i's.

Example

Let the transfer function of each of two cascade systems be $1/(1 - 1/2z^{-1})$. If the input to this system is the unit step function $1/(1 - z^{-1})$, then its output is

$$F(z) = \frac{1}{1 - z^{-1}} \frac{1}{\left(1 - \frac{1}{2}z^{-1}\right)^2} = \frac{z^3}{(z - 1)\left(z - \frac{1}{2}\right)^2}$$

$$= A_0 + \frac{A_1 z}{z - 1} + \frac{A_2 z}{z - \frac{1}{2}} + \frac{A_3 z^2}{\left(z - \frac{1}{2}\right)^2} \qquad |z| > 1$$

If we set $z = 0$ in both sides we find that $A_0 = 0$. Next we find A_3 by multiplying both sides by $(z - 1/2)^2$ and setting $z = 1/2$. Hence

$$A_3 = \left. \frac{z^3 \left(z - \frac{1}{2}\right)^2}{z^2(z - 1)\left(z - \frac{1}{2}\right)^2} \right|_{z=\frac{1}{2}} = \frac{\frac{1}{2}}{\frac{1}{2} - 1} = -1$$

and then we write

$$\frac{z^3}{(z - 1)\left(z - \frac{1}{2}\right)^2} = \frac{A_1 z}{z - 1} + \frac{A_2 z}{z - \frac{1}{2}} - \frac{z^2}{\left(z - \frac{1}{2}\right)^2}$$

$$= \frac{A_1 z \left(z - \frac{1}{2}\right)^2 + A_2 z(z - 1)\left(z - \frac{1}{2}\right) - z^2(z - 1)}{(z - 1)\left(z - \frac{1}{2}\right)^2}$$

$$= \frac{(A_1 + A_2 - 1)z^3 + \left(1 - \frac{3}{2}A_2 - A_1\right)z^2 + \left(\frac{1}{4}A_1 + \frac{1}{2}A_2\right)z}{(z - 1)\left(z - \frac{1}{2}\right)^2}$$

Equating coefficients of equal powers we obtain the system

$$A_1 + A_2 - 1 = 1, \qquad 1 - A_1 - \frac{3}{2}A_2 = 0, \qquad A_1 = 4, \qquad \text{and} \qquad A_2 = -2$$

Hence

$$\frac{z^3}{(z - 1)\left(z - \frac{1}{2}\right)^2} = 4\frac{z}{z - 1} - 2\frac{z}{z - \frac{1}{2}} - \frac{z^2}{\left(z - \frac{1}{2}\right)^2}$$

and the output is

$$f(nT) = 4(1)^n - 2\left(\frac{1}{2}\right)^n - (n + 1)\left(\frac{1}{2}\right)^n \qquad n \geq 0 \qquad \blacksquare$$

Another form of expansion of a proper function (the degree of the denominator is one less than the numerator) is of the form

$$\frac{A_1}{z - p_i} + \frac{A_2 z}{(z - p_i)^2} + \frac{A_3 z(z + p_i)}{(z - p_i)^3} \tag{7.7}$$

and the following example explains its use (see Table 4 in the Appendix).

Example
Using the previous example for $F(z)$ with $|z| > 1$ we obtain

$$F(z) = \frac{z^3}{(z - 1)\left(z - \frac{1}{2}\right)^2} = 1 + \frac{2z^2 - \frac{5}{4}z + \frac{1}{4}}{(z - 1)\left(z - \frac{1}{2}\right)^2}$$

$$= 1 + \frac{A_1}{z - 1} + \frac{A_2}{\left(z - \frac{1}{2}\right)} + \frac{A_3 z}{\left(z - \frac{1}{2}\right)^2}$$

Hence

$$A_1 = \left. \frac{\left(2z^2 - \frac{5}{4}z + \frac{1}{4}\right)(z - 1)}{(z - 1)\left(z - \frac{1}{2}\right)^2} \right|_{z=1} = 4,$$

$$A_3 = \frac{1}{z} \frac{\left(2z^2 - \frac{5}{4}z + \frac{1}{4}\right)\left(z - \frac{1}{2}\right)^2}{(z-1)\left(z - \frac{1}{2}\right)^2}\Bigg|_{z=\frac{1}{2}} = -\frac{1}{2},$$

$$A_2 = -\frac{3}{2}$$

where A_2 was found by setting an arbitrary value of z, that is, $z = -1$, in both sides of the equation. Therefore, the inverse Z-transform is given by

$$f(nT) = \begin{cases} \delta(n) & n = 0 \\ 4(1)^{n-1} - \frac{3}{2}\left(\frac{1}{2}\right)^{n-1} - \frac{1}{2}n\left(\frac{1}{2}\right)^{n-1} & n \geq 1 \end{cases}$$ ∎

Example

Now let us assume the same example but with $|z| < 1/2$. This indicates that the output signal is anticausal. Hence from

$$F(z) = 4\frac{z}{z-1} - 2\frac{z}{z - \frac{1}{2}} - \frac{z^2}{\left(z - \frac{1}{2}\right)^2}$$

and Table 3, we obtain

$$f(nT) = -4(1)^n + 2\left(\frac{1}{2}\right)^n + (n+1)\left(\frac{1}{2}\right)^n \qquad n \leq -1$$

Similarly from

$$F(z) = 1 + 4\frac{1}{z-1} - \frac{3}{2}\frac{1}{z - \frac{1}{2}} - \frac{1}{2}\frac{z}{\left(z - \frac{1}{2}\right)^2}$$

and Table 4, we obtain

$$f(nT) = \begin{cases} \delta(n) & n = 0 \\ -4(1)^{n-1} + \frac{3}{2}\left(\frac{1}{2}\right)^{n-1} + \frac{1}{2}n\left(\frac{1}{2}\right)^{n-1} & n \leq -1 \end{cases}$$ ∎

Integral inversion formula

THEOREM 7.1

If

$$F(z) = \sum_{m=-\infty}^{\infty} f(mT)z^{-m} \tag{7.8}$$

converges to an analytic function in the annular domain $R_+ < |z| < R_-$, then

$$f(nT) = \frac{1}{2\pi j}\oint_C F(z)z^n\frac{dz}{z} \tag{7.9}$$

where C is any simple closed curve separating $|z| = R_+$ from $|z| = R_-$ and it is traced in the counterclockwise direction.

PROOF   Multiply (7.8) by z^{n-1} and integrate around C. Then

$$\frac{1}{2\pi j}\oint_C F(z)z^n\frac{dz}{z} = \sum_{m=-\infty}^{\infty} f(mT)\frac{1}{2\pi j}\oint_C z^{n-m}\frac{dz}{z} \tag{7.10}$$

Set $z = \mathrm{Re}^{j\theta}$ with $R_+ < R < R_-$ to obtain

$$\frac{1}{2\pi j} \oint_C z^{n-m} \frac{dz}{z} = \frac{1}{2\pi j} \int_0^{2\pi} R^{n-m-1} e^{j\theta(n-m-1)} R j e^{j\theta} \, d\theta$$

$$= \frac{1}{2\pi} R^k \int_0^{2\pi} e^{j\theta k} \, d\theta$$

$$= \begin{cases} 1 & k = 0 \\ 0 & \text{elsewhere} \end{cases} \tag{7.11}$$

Hence the summation on the right hand side of (7.10) reduces to $f(nT)$. ▌

Let $\{a_k\}$ be the set of poles of $F(z)z^{n-1}$ inside the contour C and $\{b_k\}$ be the set of poles of $F(z)z^{n-1}$ outside C in a finite region of the z-plane. By Cauchy's residue theorem

$$f(nT) = \sum_k \mathrm{Res}\{F(z)z^{n-1}, a_k\} \qquad n \geq 0 \tag{7.12}$$

$$f(nT) = -\sum_k \mathrm{Res}\{F(z)z^{n-1}, b_k\} \qquad n < 0 \tag{7.13}$$

Example
Let

$$F(z) = \frac{1}{(1 - z^{-1})(1 - a^T z^{-1})} \qquad a < 1, \ |z| > 1$$

The function $F(z)z^{n-1} = z^{n+1}/(z-1)(z-a^T)$ has two poles enclosed by C for $n \geq 0$. Hence

$$f(nT) = \mathrm{Res}\{F(z)z^{n-1}, 1\} + \mathrm{Res}\{F(z)z^{n-1}, a\}$$

$$= \frac{1}{1 - a^T} + \frac{a^{(n+1)T}}{a^T - 1} \qquad n \geq 0 \qquad\qquad ▌$$

Example
Let

$$F(z) = \frac{1 - 0.8^2}{(1 - 0.8z)(1 - 0.8z^{-1})} \qquad 0.8 < |z| < 0.8^{-1}$$

For $n \geq 0$ the contour C encloses only the pole $z = 0.8$ of the function $F(z)z^{n-1}$. Therefore

$$f(nT) = \mathrm{Res}\{F(z)z^{n-1}\}\big|_{z=0.8} = \frac{(1 - 0.8^2)z^n(z - 0.8)}{(1 - 0.8z)(z - 0.8)}\bigg|_{z=0.8} = 0.8^n \qquad n \geq 0$$

For $n < 0$ only the pole $z = 1/0.8$ is outside C. Hence

$$f(nT) = -\mathrm{Res}\left\{F(z)z^{n-1}\right\}\big|_{z=1/0.8}$$

$$= -\frac{(1 - 0.8^2)0.8^{-1}z^n(z - 0.8^{-1})}{-(1 - 0.8^{-1})(z - 0.8)}\bigg|_{z=0.8^{-1}} = 0.8^{-n} \qquad n \leq -1 \qquad ▌$$

The residue for a multiple pole of order k at z_0 is given by

$$\mathrm{Res}\{F(z)z^{n-1}\}\big|_{z=z_0} = \lim_{z \to z_0} \frac{1}{(k-1)!} \frac{d^{k-1}}{dz^{k-1}} \left[(z - z_0)^k F(z)z^{n-1}\right] \tag{7.14}$$

C. Applications

6.8   Solutions of difference equations with constant coefficients

Based on the relation

$$\mathcal{Z}\{f(n-m)\} = \sum_{\ell=-m}^{-1} f(\ell)z^{-(\ell+m)} = z^{-m} F(z) \tag{8.1}$$

where $\mathcal{Z}\{f(n)\} = F(z)$, we can solve a difference equation of the form

$$\sum_{k=0}^{N} a_k y(n-k) = \sum_{k=0}^{L} b_k f(n-k) \tag{8.2}$$

using the Z-transform approach.

Example

To find the solution to $y(n) = y(n-1) + 2y(n-2)$ with initial conditions $y(0) = 1$ and $y(1) = 2$ we proceed as follows:

From the difference equation

$$y(0) = y(-1) + 2y(-2) = 1$$

$$y(1) = y(0) + 2y(-1) = 2$$

Hence $y(-1) = \frac{1}{2}$ and $y(-2) = \frac{1}{4}$. The Z-transform of the difference equation is given by

$$Y(z) = \sum_{\ell=-1}^{-1} y(\ell)z^{-(\ell+1)} + z^{-1} Y(z) + 2\left(\sum_{\ell=-2}^{-1} y(\ell)z^{-(\ell+2)} + z^{-2} Y(z)\right)$$

$$= y(-1) + z^{-1} Y(z) + 2(y(-2) + y(-1)z^{-1} + z^{-2} Y(z))$$

$$= \frac{1}{2} + z^{-1} Y(z) + \frac{1}{2} + z^{-1} + 2z^{-2} Y(z) = 1 + z^{-1} + z^{-1} Y(z) + 2z^{-2} Y(z)$$

Hence

$$Y(z) = \frac{1}{1 - z^{-1} - 2z^{-2}} + \frac{z^{-1}}{1 - z^{-1} - 2z^{-2}} = \frac{z^2}{z^2 - z - 2} + \frac{z}{z^2 - z - 2}$$

and

$$\mathcal{Z}^{-1}\{Y(z)\} \doteq y(n) = \mathcal{Z}^{-1}\left\{\frac{z^2}{z^2 - z - 2}\right\} + \mathcal{Z}^{-1}\left\{\frac{z}{z^2 - z - 2}\right\} \qquad \blacksquare$$

Example

The solution of the difference equation $y(n) - ay(n-1) = u(n)$ with initial condition $y(-1) = 2$ and $|a| < 1$ proceeds as follows:

$$Y(z) - ay(-1) - az^{-1} Y(z) = \frac{z}{z-1}$$

$$Y(z) = \frac{2a}{1 - az^{-1}} + \frac{z}{z-1}\frac{1}{1 - az^{-1}} = \frac{2a}{1 - az^{-1}} + \frac{z^2}{(z-1)(z-a)}$$

$$= \frac{2a}{1 - az^{-1}} + \frac{1}{1-a}\frac{1}{1 - z^{-1}} + \frac{a}{a-1}\frac{1}{1 - az^{-1}}$$

Hence, the inverse Z-transform gives

$$y(n) = \underbrace{2a \cdot a^n}_{\text{zero input}} + \underbrace{\frac{1}{1-a}u(n) + \frac{a}{a-1}a^n}_{\text{zero state}} = \underbrace{\frac{1}{1-a}u(n)}_{\text{steady state}} + \underbrace{\frac{2a-1}{a-1}a^{n+1}}_{\text{transient}} \qquad n \geq 0 \qquad \blacksquare$$

6.9   Analysis of linear discrete systems

Transfer function

From (8.2) we obtain the **transfer function** by ignoring initial conditions. The result is

$$H(z) = \frac{Y(z)}{F(z)} = \frac{\sum_{k=0}^{L} b_k z^{-k}}{\sum_{k=0}^{N} a_k z^{-k}} = \text{transfer function} \qquad (9.1)$$

where $H(z)$ is the **transform of the impulse response of a discrete system**.

Stability

Using the convolution relation between input and output of a discrete system we obtain

$$|y(n)| = \left| \sum_{k=0}^{n} h(k) f(n-k) \right| \leq M \sum_{k=0}^{\infty} |h(k)| < \infty \qquad (9.2)$$

where M is the maximum value of $f(n)$. The above inequality specifies that a discrete system is stable if to a finite input the absolute sum of its impulse response is finite. From the properties of the Z-transform, the ROC of the impulse response satisfying (9.2) is $|z| > 1$. Hence all the poles of $H(z)$ of a stable system lie inside the unit circle.

The modified Schur-Cohn criterion establishes if the zeros of the denominator of the rational transfer function $H(z) = N(z)/D(z)$ are inside or outside the unit circle.

The first step is to form the polynomial

$$D_{rp}(z) = z^N D(z^{-1}) = d_0 z^N + \cdots + d_{N-1} z + d_N$$

where $D(z^{-1}) = d_0 + \cdots + d_{N-1} z^{N-1} + d_N z^N$. This $D_{rp}(z)$ is called the **reciprocal polynomial** associated with $D(z)$. The roots of $D_{rp}(z)$ are the reciprocals of the roots of $D(z)$ and $|D_{rp}(z)| = |D(z)|$ on the unit circle. Next, we must divide $D_{rp}(z)$ by $D(z)$ starting at the high power and obtain the quotient $\alpha_0 = d_0/d_N$ and remainder $D_{1rp}(z)$ of degree $N-1$ or less, so that

$$\frac{D_{rp}(z)}{D(z)} = \alpha_0 + \frac{D_{1rp}(z)}{D(z)}$$

The division is repeated with $D_{1rp}(z)$ and its reciprocal polynomial $D_1(z)$ and the sequence $\alpha_0, \alpha_1, \ldots, \alpha_{N-2}$ is generated according to the rule

$$\frac{D_{krp}(z)}{D_k(z)} = \alpha_k + \frac{D_{(k+1)rp}(z)}{D_k(z)} \qquad \text{for } k = 0, 1, 2, \ldots, N-2$$

The zeros of $D(z)$ are all inside the unit circle (stable system) if and only if the following three conditions are satisfied:

1. $D(1) > 0$

2. $D(-1) \begin{cases} < 0 & N \text{ odd} \\ > 0 & N \text{ even} \end{cases}$

3. $|\alpha_k| < 1$ for $k = 0, 1, \ldots, N - 2$

Check conditions (1) and (2) before proceeding to (3). If they are not satisfied, the system is unstable.

Example

$$D(z) = z^3 - 0.2z^2 + z - 0.2, \qquad D_{rp}(z) = -0.2z^3 + z^2 - 0.2z + 1$$

$$\alpha_0 = \frac{-0.2z^3 + z^2 - 0.2z + 1}{z^3 - 0.2z^2 + z - 0.2} = -0.2 + \frac{0.8z^2 + 0.96}{D(z)}, \qquad \alpha_1 = \frac{0.96z^2 + 0.96}{0.96z^2 + 0.96} = 1$$

Because $|\alpha_1| = 1$, condition (3) is not satisfied and the system is unstable. ∎

The transfer function of a feedback system with forward (open-loop) gain $D(z)G(z)$ and unit feedback gain is given by

$$H(z) = \frac{D(z)G(z)}{1 + D(z)G(z)}$$

Assuming that all the individual systems are causal and have rational transfer function, the open-loop gain $D(z)G(z)$ can be written as

$$D(z)G(z) = \frac{A(z)}{B(z)}$$

where

$$A(z) = a_L z^L + \cdots + a_0, \qquad B(z) = z^M + b_{M-1}z^{M-1} + \cdots + b_0, \qquad L \leq M$$

Hence the total transfer function becomes

$$H(z) = \frac{A(z)}{B(z) + A(z)}$$

which indicates that the system will be stable if $B(z) + A(z)$ or $1 + D(z)G(z)$ has zeros inside the unit circle.

Causality

A system is causal if $h(n) = 0$ for $n < 0$. From the properties of the Z-transform, $H(z)$ is regular in the ROC and at the infinity point. For rational functions the numerator polynomial has to be at most of the same degree as the polynomial in the denominator.

The Paley-Wiener theorem provides the necessary and sufficient conditions that a frequency response characteristic $H(\omega)$ must satisfy in order for the resulting filter to be causal.

PALEY-WIENER THEOREM

If $h(n)$ has finite energy and $h(n) = 0$ for $n < 0$, then

$$\int_{-\pi}^{\pi} |\ell n|H(\omega)|| \, d\omega < \infty$$

Conversely, if $|H(\omega)|$ is square integrable and if the above integral is finite, then we can associate with $|H(\omega)|$ a phase response with $\varphi(\omega)$ so that the resulting filter with frequency response

$$H(\omega) = |H(\omega)|e^{j\varphi(\omega)}$$

is causal.

The relationship between the real and imaginary parts of an absolutely summable, causal, and real sequence is given by the relation

$$H_i(\omega) = -\frac{1}{2\pi} \int_{-\pi}^{\pi} H_r(\lambda) \cot \frac{\omega - \lambda}{2} \, d\lambda$$

which is known as the **discrete Hilbert transform**.

Summary of Causality

1. $H(\omega)$ cannot be zero except at a finite set of points.

2. $|H(\omega)|$ cannot be constant in any finite range of frequencies.

3. The transition from pass band to stop band cannot be infinitely sharp.

4. The real and imaginary parts of $H(\omega)$ are independent and are related by the discrete Hilbert transform.

5. $|H(\omega)|$ and $\varphi(\omega)$ cannot be chosen arbitrarily.

Frequency characteristics

With input $f(n) = e^{j\omega n}$, the output is

$$y(n) = \sum_{k=0}^{\infty} h(k)e^{j\omega(n-k)} = e^{j\omega n} \sum_{k=0}^{\infty} h(k)e^{-j\omega k} = e^{j\omega n} H(e^{j\omega}) \tag{9.3}$$

where

$$H(e^{j\omega}) = H(z)|_{z=e^{j\omega}} = H_r(e^{j\omega}) + jH_i(e^{j\omega}) = A(\omega)e^{j\varphi(\omega)} \tag{9.4}$$

$$A(\omega) = \left[H_r^2(e^{j\omega}) + H_i^2(e^{j\omega})\right]^{1/2} = \text{amplitude response} \tag{9.5}$$

$$\varphi(\omega) = \tan^{-1}\left[H_i(e^{j\omega})/H_r(e^{j\omega})\right] = \text{phase response} \tag{9.6}$$

$$\tau(\omega) = -\frac{d\varphi(\omega)}{d\omega} = -\text{Re}\left\{z\frac{d}{dz}\ell n H(z)\right\}\bigg|_{z=e^{j\omega}} = \begin{array}{l} \text{group delay} \\ \text{characteristic} \end{array} \tag{9.7}$$

Because $H(e^{j\omega}) = H(e^{j(\omega+2\pi k)})$ it implies that the **frequency characteristics of discrete systems are periodic with period 2π.**

Z-transform and discrete Fourier transform (DFT)

If $x(n)$ has a finite duration of length N or less, the sequence can be recovered from its N-point DFT. Hence its Z-transform is uniquely determined by its N-point DFT. Hence, we find

$$X(z) = \sum_{n=0}^{N-1} x(n)z^{-n} = \sum_{n=0}^{N-1} \left[\frac{1}{N} \sum_{k=0}^{N-1} X(k)e^{j2\pi kn/N} \right] z^{-n}$$

$$= \frac{1}{N} \sum_{k=0}^{N-1} X(k) \sum_{n=0}^{N-1} \left(e^{j2\pi k/N} z^{-1} \right)^n = \frac{1-z^{-N}}{N} \sum_{k=0}^{N-1} \frac{X(k)}{1-e^{j2\pi k/N} z^{-1}} \tag{9.8}$$

Set $z = e^{j\omega}$ (evaluated on the unit circle) to find

$$X(\omega) = \frac{1-e^{-j\omega N}}{N} \sum_{k=0}^{N-1} \frac{X(k)}{1-e^{-j(\omega-2\pi k/N)}} \tag{9.9}$$

$X(\omega)$ is the Fourier transform of the finite-duration sequence in terms of its DFT.

6.10   Digital filters

Infinite impulse response (IIR) filters

A discrete, linear, and time invariant system can be described by a higher-order difference equation of the form

$$y(n) - \sum_{k=1}^{N} a_k y(n-k) = \sum_{k=0}^{M} b_k x(n-k) \tag{10.1}$$

Taking the Z-transform of the above equation and solving for the ratio $Y(z)/X(z)$ we obtain

$$H(z) = \frac{Y(z)}{X(z)} = \frac{\sum_{k=0}^{M} b_k z^{-k}}{1 - \sum_{k=1}^{N} a_k z^{-k}} \tag{10.2}$$

The block diagram representation of (10.1), in the form of the following pair of equations:

$$v(n) = \sum_{k=0}^{M} b_k x(n-k) \tag{10.3}$$

$$y(n) = \sum_{k=1}^{N} a_k y(n-k) + v(n) \tag{10.4}$$

is shown in Figure 10.1. Each appropriate rearrangement of the block diagram represents a **different** computational algorithm for implementing the **same** system.

Figure 10.1 can be viewed as an implementation of $H(z)$ through the decomposition

$$H(z) = H_2(z)H_1(z) = \left(\frac{1}{1 - \sum_{k=1}^{N} a_k z^{-k}} \right) \left(\sum_{k=0}^{M} b_k z^{-k} \right) \tag{10.5}$$

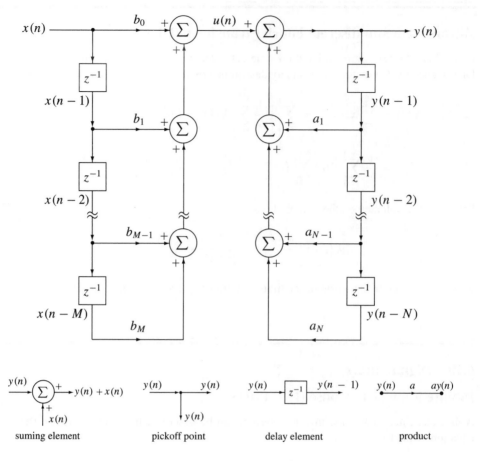

FIGURE 10.1

or through the pair of equations

$$V(z) = H_1(z)X(z) = \left(\sum_{k=0}^{M} b_k z^{-k}\right) X(z) \tag{10.6}$$

$$Y(z) = H_2(z)V(z) = \left(\frac{1}{1 - \sum_{k=1}^{N} a_k z^{-k}}\right) V(z) \tag{10.7}$$

If we rearrange (10.5), we can create the following two equations:

$$W(z) = H_2(z)X(z) = \left(\frac{1}{1 - \sum_{k=1}^{N} a_k z^{-k}}\right) X(z) \tag{10.8}$$

$$Y(z) = H_1(z)W(z) = \left(\sum_{k=1}^{M} b_k z^{-k}\right) W(z) \tag{10.9}$$

The last two equations are presented graphically in Figure 10.2 ($M = N$).

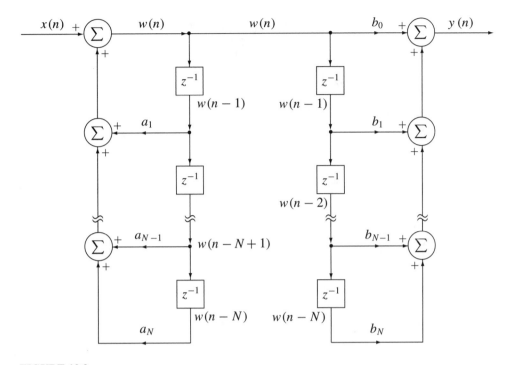

FIGURE 10.2

The time domain of Figure 10.2 is the pair of equations

$$w(n) = \sum_{k=1}^{N} a_k w(n-k) + x(n) \tag{10.10}$$

$$y(n) = \sum_{k=0}^{M} b_k w(n-k) \tag{10.11}$$

Because the two internal branches of Figure 10.2 are identical, they can be combined in one branch so that Figure 10.3 results. Figure 10.1 represents the **direct form I** of the general Nth-order system and Figure 10.3 is often referred to as the **direct form II** or **canonical direct form** implementation.

Finite impulse responses (FIR) filters

For causal FIR systems, the difference equation describing such a system is given by

$$y(n) = \sum_{k=0}^{M} b_k x(n-k) \tag{10.12}$$

which is recognized as the discrete convolution of $x(n)$ with the impulse response

$$h(n) = \begin{cases} b_n & n = 0, 1, \ldots, M \\ 0 & \text{otherwise} \end{cases} \tag{10.13}$$

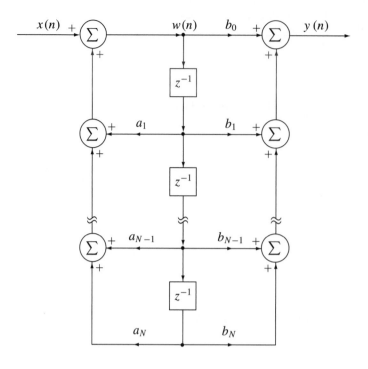

FIGURE 10.3

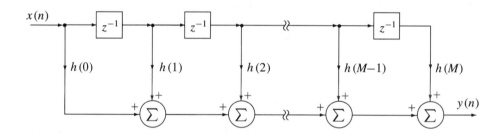

FIGURE 10.4

The direct form I and direct form II structures are shown in Figures 10.4 and 10.5. Because of the chain of delay elements across the top of the diagram, this structure is also referred to as a **tapped delay line** structure or a **transversal filter** structure.

6.11   Linear, time-invariant, discrete-time, dynamical systems

The mathematical models describing dynamical systems are almost always of finite-order difference equations. If we know the initial conditions at $t = t_0$ their behavior can be uniquely determined for $t \geq t_0$. To see how to develop a dynamic, let us consider the example below.

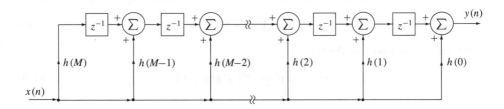

FIGURE 10.5

Example

Let a discrete system with input $v(n)$ and output $y(n)$ be described by the difference equation

$$y(n) + 2y(n-1) + y(n-2) = v(n) \tag{11.1}$$

If $y(n_0 - 1)$ and $y(n_0 - 2)$ are the initial conditions for $n > n_0$, then $y(n)$ can be found recursively from (11.1). Let us take the pair $y(n-1)$ and $y(n-2)$ as the state of the system at time n. Let us call the vector

$$\underline{x}(n) = \begin{bmatrix} x_1(n) \\ x_2(n) \end{bmatrix} = \begin{bmatrix} y(n-2) \\ y(n-1) \end{bmatrix} \tag{11.2}$$

the **state vector** for the system. From the definition above we obtain

$$x_1(n+1) = y(n+1-2) = y(n-1) = x_2(n) \tag{11.3}$$

and

$$x_2(n+1) = y(n) = v(n) - y(n-2) - 2y(n-1) \tag{11.4}$$

or

$$x_2(n+1) = v(n) - x_1(n) - 2x_2(n) \tag{11.5}$$

Equations (11.3) and (11.5) can be written in the form

$$\begin{bmatrix} x_1(n+1) \\ x_2(n+1) \end{bmatrix} = \begin{bmatrix} 0 & 1 \\ -1 & -2 \end{bmatrix} \begin{bmatrix} x_1(n) \\ x_2(n) \end{bmatrix} + \begin{bmatrix} 0 \\ 1 \end{bmatrix} v(n) \tag{11.6}$$

or

$$\underline{x}(n+1) = \underline{A}\,\underline{x}(n) + \underline{B}v(n) \tag{11.7}$$

But (11.4) can be written in the form

$$y(n) = v(n) - x_1(n) - 2x_2(n) = \begin{bmatrix} -1 & -2 \end{bmatrix} \begin{bmatrix} x_1(n) \\ x_2(n) \end{bmatrix} + v(n)$$

or

$$y(n) = \underline{C}\,\underline{x} + v(n) \tag{11.8}$$

Hence the system can be described by vector-matrix difference equation (11.7) and an output equation (11.8) rather than by the second-order difference equation (11.1). ∎

A time-invariant, linear, and discrete dynamic system is described by the state equation

$$\underline{x}(nT + T) = \underline{A}\,\underline{x}(nT) + \underline{B}\,\underline{v}(nT) \tag{11.9}$$

and the output equation is of the form

$$\underline{y}(nT) = \underline{C}\,\underline{x}(nT) + \underline{D}\,\underline{v}(nT) \tag{11.10}$$

where

$$\underline{x}(nT) = N\text{-dimensional column vector}$$

$$\underline{v}(nT) = M\text{-dimensional column vector}$$

$$\underline{y}(nT) = R\text{-dimensional column vector}$$

$$\underline{A} = N \times N \text{ nonsingular matrix}$$

$$\underline{B} = N \times M \text{ matrix}$$

$$\underline{C} = R \times N \text{ matrix}$$

$$\underline{D} = R \times M \text{ matrix}$$

When the input is identically zero (11.9) reduces to

$$\underline{x}(nT + T) = \underline{A}\,\underline{x}(nT) \tag{11.11}$$

so that

$$\underline{x}(nT + 2T) = \underline{A}\,\underline{x}(nT + T) = \underline{A}\,\underline{A}\,\underline{x}(nT) = \underline{A}^2\underline{x}(nT)$$

and so on. In general we have

$$\underline{x}(nT + kT) = \underline{A}^k\underline{x}(nT) \qquad k > 0 \tag{11.12}$$

The **state transition matrix** from n_1T to n_2T $(n_2 > n_1)$ is given by

$$\underline{\varphi}(n_2T, n_1T) = \underline{A}^{n_2 - n_1} \tag{11.13}$$

This is a function only of the time difference $n_2T - n_1T$. Therefore, it is customary to name the matrix

$$\underline{\varphi}(nT) = \underline{A}^n \tag{11.14}$$

the state transition matrix with the understanding that $n = n_2 - n_1$. It follows that the system states at two times, n_2T and n_1T, are related by the relation

$$\underline{x}(n_2T) = \underline{\varphi}(n_2T, n_1T)\underline{x}(n_1T) \tag{11.15}$$

when the input is zero. From (11.13) we obtain the following relationships:

a)

$$\underline{\varphi}(nT, nT) = \underline{I} = \text{identity matrix} \tag{11.16}$$

b)

$$\underline{\varphi}(n_2T, n_1T) = \underline{\varphi}^{-1}(n_1T, n_2T) \tag{11.17}$$

c)

$$\underline{\varphi}(n_3T, n_2T)\underline{\varphi}(n_2T, n_1T) = \underline{\varphi}(n_3T, n_1T) \tag{11.18}$$

If the input is not identically zero and $\underline{x}(nT)$ is known, then the progress (later states) of the system can be found recursively from (11.9). Proceeding with the recursion we obtain

$$\underline{x}(nT + 2T) = \underline{A}\,\underline{x}(nT + T) + \underline{B}\,\underline{v}(nT + T)$$

$$= \underline{A}\,\underline{A}\,\underline{x}(nT) + \underline{A}\,\underline{B}\,\underline{v}(nT) + \underline{B}\,\underline{v}(nT + T)$$

$$= \underline{\varphi}(nT + 2T, nT)\underline{x}(nT) + \underline{\varphi}(nT + 2T, nT + T)\underline{B}\,\underline{v}(nT) + \underline{B}\,\underline{v}(nT + T)$$

In general for $k > 0$ we have the solution

$$\underline{x}(nT + kT) = \underline{\varphi}(nT + kT, nT)\underline{x}(nT) + \sum_{i=n}^{n+k-1} \underline{\varphi}(nT + kT, iT + T)\underline{B}\,\underline{v}(iT) \tag{11.19}$$

From (11.15), when the input is zero, we obtain the relation

$$\underline{x}(n_2T) = \underline{\varphi}(n_2T - n_1T)\underline{x}(n_1T) = \underline{A}^{n_2-n_1}\underline{x}(n_1T) \tag{11.20}$$

According to (11.19), the solution to the dynamic system when the input is not zero is given by

$$\underline{x}(nT + kT) = \underline{\varphi}(nT + kT - nT)\underline{x}(nT) + \sum_{i=n}^{n+k-1} \underline{\varphi}[(n + k - i - 1)T)]\underline{B}\,\underline{v}(iT) \tag{11.21}$$

or

$$\underline{x}(nT + kT) = \underline{\varphi}(kT)\underline{x}(nT) + \sum_{i=n}^{n+k-1} \underline{\varphi}[(n + k - i - 1)T)]\underline{B}\,\underline{v}(iT) \qquad k > 0 \tag{11.22}$$

To find the solution using the Z-transform method, we define the one-sided Z-transform of an $R \times S$ matrix function $\underline{f}(nT)$ as the $R \times S$ matrix

$$\underline{F}(z) = \sum_{n=0}^{\infty} \underline{f}(nT)z^{-n} \tag{11.23}$$

The elements of $\underline{F}(z)$ are the transforms of the corresponding elements of $\underline{f}(nT)$. Taking the Z-transform of both sides of the state equation (11.9) we find

$$z\underline{X}(z) - z\underline{x}(0) = \underline{A}\,\underline{X}(z) + \underline{B}\,\underline{V}(z)$$

or

$$\underline{X}(z) = (z\underline{I} - \underline{A})^{-1}z\underline{x}(0) + (z\underline{I} - \underline{A})^{-1}\underline{B}\,\underline{V}(z) \tag{11.24}$$

From the output equation (11.10) we see that

$$\underline{Y}(z) = \underline{C}\,\underline{X}(z) + \underline{D}\,\underline{V}(z) \tag{11.25}$$

The state of the system $\underline{x}(nT)$ and its output $\underline{y}(nT)$ can be found for $n \geq 0$ by taking the inverse Z-transform of (11.24) and (11.25).

For a zero input, (11.24) becomes

$$\underline{X}(z) = (z\underline{I} - \underline{A})^{-1}z\underline{x}(0) \tag{11.26}$$

so that

$$\underline{x}(nT) = \mathcal{Z}^{-1}\{(z\underline{I} - \underline{A})^{-1}z\}\underline{x}(0) \tag{11.27}$$

If we let $n_1 = 0$ and $n_2 = n$ then (11.20) becomes

$$\underline{x}(nT) = \underline{\varphi}(nT)\underline{x}(0) = \underline{A}^n x(0) \tag{11.28}$$

Comparing (11.27) and (11.28) we observe that

$$\underline{\varphi}(nT) = \underline{A}^n = \mathcal{Z}^{-1}\{(z\underline{I} - \underline{A})^{-1}z\} \qquad n \geq 0 \tag{11.29}$$

or equivalently,

$$\underline{\Phi}(z) = \mathcal{Z}\{\underline{A}^n\} = (z\underline{I} - \underline{A})^{-1}z \tag{11.30}$$

The Z-transform provides a straightforward method for calculating the state transition matrix. Next combine (11.30) and (11.24) to find

$$\underline{X}(z) = \underline{\Phi}(z)\underline{x}(0) + \underline{\Phi}(z)z^{-1}\underline{B}\,\underline{V}(z) \tag{11.31}$$

By applying the convolution theorem and the fact that

$$\mathcal{Z}^{-1}\{\underline{\Phi}(z)z^{-1}\} = \underline{\varphi}(nT - T)u(nT - T) \tag{11.32}$$

the inverse Z-transform of (11.31) is given by

$$\underline{x}(kT) = \underline{\varphi}(kT)\underline{x}(0) + \sum_{i=0}^{k-1}\underline{\varphi}[(k - i - 1)T)]\underline{B}\,\underline{v}(iT) \tag{11.33}$$

The above equation is identical to (11.22) with $n = 0$.

The behavior of the system with zero input depends on the location of the poles of

$$\underline{\Phi}(z) = (z\underline{I} - \underline{A})^{-1}z \tag{11.34}$$

Because

$$(z\underline{I} - \underline{A})^{-1} = \frac{\text{adj}(z\underline{I} - \underline{A})}{\det(z\underline{I} - \underline{A})} \tag{11.35}$$

where $\text{adj}(\cdot)$ denotes the regular adjoint in matrix theory, these poles can only occur at the roots of the polynomial

$$D(z) = \det(z\underline{I} - \underline{A}) \tag{11.36}$$

$D(z)$ is known as the **characteristic polynomial** for $\underline{A}$ (for the system) and its roots are known as the **characteristic values** or **eigenvalues** of $\underline{A}$. If all roots are inside the unit circle, the system is stable. If even one root is outside the unit circle, the system is unstable.

Example
Consider the system

$$\begin{bmatrix} x_1(nT + T) \\ x_2(nT + T) \end{bmatrix} = \begin{bmatrix} 0 & 2 \\ 0.22 & 2 \end{bmatrix}\begin{bmatrix} x_1(nT) \\ x_2(nT) \end{bmatrix} + \begin{bmatrix} 0 \\ 1 \end{bmatrix}v(nT)$$

$$y(nT) = \begin{bmatrix} 0.22 & 2 \end{bmatrix}\begin{bmatrix} x_1(nT) \\ x_2(nT) \end{bmatrix} + v(nT)$$

For this system we have

$$\underline{A} = \begin{bmatrix} 0 & 2 \\ 0.22 & 2 \end{bmatrix}, \qquad \underline{B} = \begin{bmatrix} 0 \\ 1 \end{bmatrix}, \qquad \underline{C} = [\ 0.22 \quad 2\ ], \qquad \underline{D} = [1]$$

The characteristic polynomial is

$$D(z) = \det(z\underline{I} - \underline{A}) = \det\left[\begin{bmatrix} z & 0 \\ 0 & z \end{bmatrix} - \begin{bmatrix} 0 & 2 \\ 0.22 & 2 \end{bmatrix}\right] = \det\begin{bmatrix} z & -2 \\ -0.22 & z-2 \end{bmatrix}$$

$$= z(z-2) - 0.44 = z^2 - 2z - 0.44 = (z-2.2)(z+0.2)$$

Hence we obtain (see [11.34])

$$\underline{\Phi}(z) = \frac{z}{(z-2.2)(z+0.2)}\begin{bmatrix} z-2 & 2 \\ 0.22 & z \end{bmatrix} = \begin{bmatrix} \frac{z(z-2)}{(z-2.2)(z+0.2)} & \frac{2z}{(z-2.2)(z+0.2)} \\ \frac{0.22z}{(z-2.2)(z+0.2)} & \frac{z^2}{(z-2.2)(z+0.2)} \end{bmatrix}$$

Because $D(z)$ has a root outside the unit circle at 2.2, the system is unstable. Taking the inverse transform we find that

$$\underline{\varphi}(nT) = \begin{bmatrix} \frac{1}{12}(2.2)^n + \frac{11}{12}(-0.2)^n & \frac{5}{6}(2.2)^n - \frac{5}{6}(-0.2)^n \\ \frac{11}{120}(2.2)^n - \frac{11}{120}(-0.2)^n & \frac{11}{12}(2.2)^n + \frac{1}{12}(-0.2)^n \end{bmatrix} \qquad n \geq 0$$

To check, set $n = 0$ to find $\underline{\varphi}(0) = \underline{I}$ and $\underline{\varphi}(T) = \underline{A}$.

Let $\underline{x}(0) = \underline{0}$ and the input be , the unit impulse $v(nT) = \delta(nT)$ so that $V(z) = 1$. Hence, according to (11.31)

$$\underline{X}(z) = \underline{\Phi}(z)z^{-1}\underline{B}\ V(z) = \frac{1}{(z-2.2)(z+0.2)}\begin{bmatrix} z-2 & 2 \\ 0.22 & z \end{bmatrix}\begin{bmatrix} 0 \\ 1 \end{bmatrix}$$

$$= \frac{1}{(z-2.2)(z+0.2)}\begin{bmatrix} 2 \\ z \end{bmatrix}$$

The inverse Z-transform gives

$$\underline{x}(nT) = \frac{5}{6}\begin{bmatrix} (2.2)^{n-1} - (-0.2)^{n-1} \\ \frac{1}{2}(2.2)^n - \frac{1}{2}(-0.2)^n \end{bmatrix} \qquad n > 0$$

and the output is given by

$$y(nT) = \underline{C}\ \underline{x}(nT) + \underline{D}v(nT)$$

$$= \begin{cases} 1 & n = 0 \\ \frac{5}{12}(2.2)^{n+1} - \frac{5}{12}(-0.2)^{n+1} & n > 0 \end{cases}$$

6.12   Z-transform and random processes

Power spectral densities

The Z-transform of the autocorrelation function $R_{xx}(\tau) = E\{x(t+\tau)x(t)\}$ sampled uniformly at nT times is given by

$$S_{xx}(z) = \sum_{n=-\infty}^{\infty} R_{xx}(nT)z^{-n} \tag{12.1}$$

where the Fourier transform of $R_{xx}(\tau)$ is designated by $S_{xx}(\omega)$. The **sampled power spectral density** for $x(nT)$ is defined to be

$$S_{xx}(e^{j\omega T}) = S_{xx}(z)|_{z=e^{j\omega T}} = \sum_{n=-\infty}^{\infty} R_{xx}(nT)e^{-j\omega nT} \tag{12.2}$$

However, from the sampling theorem we have

$$S_{xx}(e^{j\omega T}) = \frac{1}{T}\sum_{n=-\infty}^{\infty} S_{xx}(\omega - n\omega_s), \qquad \omega_s = 2\pi/T \tag{12.3}$$

Because $S_{xx}(\omega)$ is real, nonnegative, and even, it follows from (12.3) that $S_{xx}(e^{j\omega T})$ is also real, nonnegative, and even. If the envelope of $R_{xx}(\tau)$ decays exponentially for $|\tau| > 0$, then the region of convergence for $S_{xx}(z)$ includes the unit circle. If $R_{xx}(\tau)$ has undamped periodic components the series in (12.2) converges in the distribution sense that contains impulse function.

The average power in $x(nT)$ is

$$E\{x^2(nT)\} = R_{xx}(0) = \frac{1}{2\pi j}\oint_C S_{xx}(z)\frac{dz}{z} \tag{12.4}$$

where C is a simple, closed contour lying in the region of convergence and the integration is taken in counterclockwise sense. If C is the unit circle, then

$$R_{xx}(0) = \frac{1}{\omega_s}\int_{-\omega_s/2}^{\omega_s/2} S_{xx}(e^{j\omega T})\,d\omega \qquad \omega_s = \frac{2\pi}{T} \tag{12.5}$$

$$S_{xx}(e^{j\omega T})\frac{d\omega}{\omega_s} = \text{average power in } d\omega \tag{12.6}$$

$S_{xy}(z)$ is called the **cross power spectral density** for two jointly wide-sense stationary processes $x(t)$ and $y(t)$. It is defined by the relation

$$S_{xy}(z) = \sum_{n=-\infty}^{\infty} R_{xy}(nT)z^{-n} \tag{12.7}$$

Because $R_{xy}(nT) = R_{yx}(-nT)$ it follows that

$$S_{xy}(z) = S_{xy}(z^{-1}), \qquad S_{xx}(z) = S_{xx}(z^{-1}) \tag{12.8}$$

Equivalently, we have

$$S_{xx}(e^{j\omega T}) = S_{xx}(e^{-j\omega T}) \tag{12.9}$$

If $S_{xx}(z)$ is a rational polynomial it can be factored in the form

$$S_{xx}(z) = \frac{N(z)}{D(z)} = \gamma^2 G(z)G(z^{-1}) \tag{12.10}$$

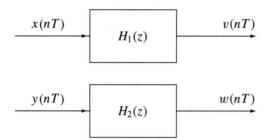

FIGURE 12.1

where

$$G(z) = \frac{\Pi_{k=1}^{L}(1 - \alpha_k z^{-1})}{\Pi_{k=1}^{M}(1 - \beta_k z^{-1})} = \frac{\sum_{k=0}^{L} a_k z^{-k}}{\sum_{k=0}^{M} b_k z^{-k}}$$

$$\gamma^2 > 0, \qquad |a_k| < 1, \qquad |b_k| < 1, \qquad a_k \text{ and } b_k \text{ are real}$$

Linear discrete-time filters

Let $R_{xx}(nT)$, $R_{yy}(nT)$, and $R_{xy}(nT)$ be known. Let two systems have transfer functions $H_1(z)$ and $H_2(z)$, respectively. The output of these filters, when the inputs are $x(nT)$ and $y(nT)$ (see Figure 12.1), are

$$v(nT) = \sum_{k=-\infty}^{\infty} h_1(kT)x(nT - kT) \tag{12.11}$$

$$w(nT) = \sum_{k=-\infty}^{\infty} h_2(kT)y(nT - kT) \tag{12.12}$$

Let $n = n + m$ in (12.11), multiply by $y(nT)$, and take the ensemble average to find

$$R_{vy}(mT) = \sum_{k=-\infty}^{\infty} h_1(kT)E\{x(mT + nT - kT)y(nT)\}$$

$$= \sum_{k=-\infty}^{\infty} h_1(kT)R_{xy}(mT - kT) \tag{12.13}$$

Hence, by taking the Z-transform we obtain

$$S_{vy}(z) = H_1(z)S_{xy}(z) \tag{12.14}$$

Similarly from (12.12) we obtain

$$R_{vw}(mT) = \sum_{k=-\infty}^{\infty} h_2(kT)R_{vy}(mT + kT) \tag{12.15}$$

and

$$S_{vw}(z) = H_2(z^{-1})S_{vy}(z) \tag{12.16}$$

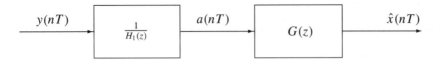

FIGURE 12.2

From (12.14) and (12.16) we obtain

$$S_{vw}(z) = H_1(z)H_2(z^{-1})S_{xy}(z) \tag{12.17}$$

Also, for $x(nT) = y(nT)$ and $h_1(nT) = h_2(nT) = h(nT)$, (12.17) becomes

$$S_{vv}(z) = H(z)H(z^{-1})S_{xx}(z) \tag{12.18}$$

and

$$S_{vv}\left(e^{j\omega T}\right) = H\left(e^{j\omega T}\right) H\left(e^{-j\omega T}\right) S_{xx}\left(e^{j\omega T}\right)$$

$$= \left|H\left(e^{j\omega T}\right)\right|^2 S_{xx}\left(e^{j\omega T}\right) \tag{12.19}$$

Optimum linear filtering

Let $y(nT)$ be an observed wide-sense stationary process and $x(nT)$ be a desired wide-sense stationary process. The process $y(nT)$ could be the result of the desired signal $x(nT)$ and a noise signal $v(nT)$. It is desired to find a system with transfer function $H(z)$ such that the error $e(nT) = x(nT) - \hat{x}(nT) = x(nT) - \mathcal{Z}^{-1}\{Y(z)H(z)\}$ is minimized. Referring to Figure 12.2 and to (12.18), we can write

$$S_{aa}(z) = \frac{1}{H_1(z)H_1(z^{-1})}S_{yy}(z) = \gamma^2 \tag{12.20}$$

where $a(nT)$ is taken as white noise (uncorrelated process). We therefore can write

$$R_{aa}(mT) = \gamma^2\delta(mT) \tag{12.21}$$

The signal $a(nT)$ is known as the **innovation** process associated with $y(nT)$. From Figure 12.2 we obtain

$$\hat{x}(nT) = \sum_{k=-\infty}^{\infty} g(kT)a(nT - kT) \tag{12.22}$$

The mean square error is given by

$$E\{e^2(nT)\} = E\left\{\left[x(nT) - \sum_{k=-\infty}^{\infty} g(kT)a(nT - kT)\right]^2\right\}$$

$$= E\{x^2(nT)\} - 2E\left\{\sum_{k=-\infty}^{\infty} g(kT)x(nT)a(nT - kT)\right\}$$

$$+ E\left\{\left[\sum_{k=-\infty}^{\infty} g(kT)a(nT - kT)\right]^2\right\}$$

$$= R_{xx}(0) - 2\sum_{k=-\infty}^{\infty} g(kT)R_{xa}(kT) + \gamma^2 \sum_{k=-\infty}^{\infty} g^2(kT)$$

$$= R_{xx}(0) + \sum_{k=-\infty}^{\infty}\left[\gamma g(kT) - \frac{R_{xa}(kT)}{\gamma}\right]^2 - \frac{1}{\gamma^2}\sum_{k=-\infty}^{\infty} R_{xa}^2(kT)$$

To minimize the error we must set the quantity in the brackets equal to zero. Hence

$$g(nT) = \frac{1}{\gamma^2}R_{xa}(nT) \qquad -\infty < n < \infty$$

and its Z-transform is

$$G(z) = \frac{1}{\gamma^2}S_{xa}(z)$$

but from (12.17) (because $v(nT) = x(nT)$ implies that $H_1(z) = 1$) we have

$$S_{xy}(z) = H_1(z^{-1})S_{xa}(z) \qquad \text{or} \qquad S_{xa}(z) = \frac{S_{xy}(z)}{H_1(z^{-1})} \tag{12.23}$$

$$G(z) = \frac{1}{\gamma^2}\frac{S_{xy}(z)}{H_1(z^{-1})} \tag{12.24}$$

From Figure 12.2 the optimum filter is given by (see also [12.20])

$$H(z) = \frac{1}{H_1(z)}G(z) = \frac{S_{xy}(z)}{\gamma^2 H_1(z)H_1(z^{-1})} = \frac{S_{xy}(z)}{S_{yy}(z)} \tag{12.25}$$

The mean square error for an optimum filter is

$$E\{e^2(nT)\} = R_{xx}(0) - \frac{1}{\gamma^2}\sum_{k=-\infty}^{\infty} R_{xa}^2(kT) \tag{12.26}$$

Applying Parseval's theorem in the above equation, we obtain

$$E\{e^2(nT)\} = \frac{1}{2\pi j}\oint_C \left[S_{xx}(z) - \frac{1}{\gamma^2}S_{xa}(z)S_{xa}(z^{-1})\right]\frac{dz}{z}$$

$$= \frac{1}{2\pi j}\oint_C \left[S_{xx}(z) - \frac{S_{xy}(z)S_{xy}(z^{-1})}{S_{yy}(z)}\right]\frac{dz}{z}$$

$$= \frac{1}{2\pi j}\oint_C [S_{xx}(z) - H(z)S_{xy}(z^{-1})]\frac{dz}{z} \tag{12.27}$$

where C can be the unit circle.

6.13 Relationship between the Laplace and Z-transform

The one-sided Laplace transform and its inverse are given by the following two equations:

$$F(s) \doteq \mathcal{L}\{f(t)\} = \int_0^\infty f(t)e^{-st}\,dt \qquad \mathrm{Re}\{s\} > \sigma_c \tag{13.1}$$

$$f(t) = \mathcal{L}^{-1}\{F(s)\} = \frac{1}{2\pi j}\int_{c-j\infty}^{c+j\infty} F(s)e^{st}\,ds \qquad c > \sigma_c \tag{13.2}$$

where σ_c is the abscissa of convergence.

The Laplace transform of a sampled function

$$f_s(t) = f(t)\sum_{k=-\infty}^\infty \delta(t - nT) \doteq f(t)\,\mathrm{comb}_T(t) = \sum_{k=-\infty}^\infty f(nT)\delta(t - nT) \tag{13.3}$$

is given by

$$F_s(s) \doteq \mathcal{L}\{f_s(t)\} = \sum_{k=-\infty}^\infty f(nT)e^{-nTs} \tag{13.4}$$

because

$$\mathcal{L}\{\delta(t - nT)\} = \int_{-\infty}^\infty \delta(t - nT)e^{-st}\,dt = e^{-snT} \tag{13.5}$$

From (13.4) we obtain

$$F(z) = F_s(s)|_{s=T^{-1}\ell n z} \tag{13.6}$$

and hence

$$F(z)|_{z=e^{Ts}} = F_s(s) \doteq \mathcal{L}\{f_s(t)\} = \mathcal{L}\{f(t)\,\mathrm{comb}_T(t)\} \tag{13.7}$$

If the region of convergence for $F(z)$ includes the unit circle, $|z| = 1$, then

$$F_s(\omega) = F(z)|_{z=e^{j\omega T}} = \sum_{n=-\infty}^\infty f(nT)e^{-j\omega nT} \tag{13.8}$$

$$F_s(s + j\omega_s) = F_s(s) = \text{periodic} \qquad \omega_s = \frac{2\pi}{T} \tag{13.9}$$

The knowledge of $F_s(s)$ in the strip $-\omega_s/2 < \omega \le \omega_s/2$ determines $F_s(s)$ for all s. The transformation $z = e^{sT}$ maps this strip uniquely onto the complex z-plane. Therefore, $F(z)$ contains all the information in $F_s(s)$ without redundancy. Letting $s = \sigma + j\omega$, then

$$z = e^{\sigma T}e^{j\omega T} \tag{13.10}$$

Because $|z| = e^{\sigma T}$, we obtain

$$|z| = \begin{cases} < 1 & \sigma < 0 \\ = 1 & \sigma = 0 \\ > 1 & \sigma > 0 \end{cases} \tag{13.11}$$

Therefore we have the following correspondence between the s- and z-planes:

1. Points in the left half of the s-plane are mapped inside the unit circle in the z-plane.

2. Points on the $j\omega$-axis are mapped onto the unit circle.

3. Points in the right half of the s-plane are mapped outside the unit circle.

4. Lines parallel to the $j\omega$-axis are mapped into circles with radius $|z| = e^{\sigma T}$.

5. Lines parallel to the σ-axis are mapped into rays of the form $\arg z = \omega T$ radians from $z = 0$.

6. The origin of the s-plane corresponds to $z = 1$.

7. The σ-axis corresponds to the positive $u = \operatorname{Re} z$-axis.

8. As ω varies between $-\omega_s/2$ and $\omega_s/2$, $\arg z = \omega T$ varies between $-\pi$ and π radians.

Let $f(t)$ and $g(t)$ be causal functions with Laplace transforms $F(s)$ and $G(s)$ that converge absolutely for $\operatorname{Re} s > \sigma_f$ and $\operatorname{Re} s > \sigma_g$, respectively; then

$$\mathcal{L}\{f(t)g(t)\} = \frac{1}{2\pi j} \int_{c-j\infty}^{c+j\infty} F(p)G(s-p)\,dp \tag{13.12}$$

The contour is parallel to the imaginary axis in the complex p-plane with

$$\sigma = \operatorname{Re} s > \sigma_f + \sigma_g \qquad \text{and} \qquad \sigma_f < c < \sigma - \sigma_g \tag{13.13}$$

With this choice the poles $G(s-p)$ lie to the right of the integration path.

For causal $f(t)$, its sampling form is given by

$$f_s(t) = f(t) \sum_{n=0}^{\infty} \delta(t-nT) \doteq f(t)\,\mathrm{comb}_T(t) = \sum_{n=0}^{\infty} f(nT)\delta(t-nT) \tag{13.14}$$

If

$$g(t) = \mathrm{comb}_T(t) \doteq \sum_{n=0}^{\infty} \delta(t-nT) \tag{13.15}$$

then its Laplace transform is

$$G(s) = \mathcal{L}\{g(t)\} = \sum_{n=0}^{\infty} e^{-nTs} = \frac{1}{1-e^{-Ts}} \qquad \operatorname{Re} s > 0 \tag{13.16}$$

Because $\sigma_g = 0$, then (13.12) becomes

$$F_s(s) = \frac{1}{2\pi j} \int_{c-j\infty}^{c+j\infty} \frac{F(p)}{1-e^{-(s-p)T}}\,dp \qquad \sigma > \sigma_f,\ \sigma_f < c < \sigma \tag{13.17}$$

The distance p in Figure 13.1 is given by

$$p = c + Re^{j\theta} \qquad \pi/2 \le \theta \le 3\pi/2 \tag{13.18}$$

If the function $F(p)$ is analytic for some $|p|$ greater than a finite number R_0 and has a zero at infinity, then in the limit as $R \to \infty$ the integral along the path BDA is identically zero and the integral along the path AEB averages to $F_s(s)$. The contour $C_1 + C_2$ encloses all the poles of $F(p)$. Because of these assumptions $F(p)$ must have a Laurent series expansion of the form

$$F(p) = \frac{a_{-1}}{p} + \frac{a_{-2}}{p2} + \cdots = \frac{a_{-1}}{p} + \frac{Q(p)}{p2} \qquad |p| > R_0 \tag{13.19}$$

$Q(p)$ is analytic in this domain and

$$|Q(p)| < M < \infty \qquad |p| > R_0 \tag{13.20}$$

Therefore from (13.19)

$$a_{-1} = \lim_{p \to \infty} pF(p) \tag{13.21}$$

From the initial value theorem

$$a_{-1} = f(0+) \tag{13.22}$$

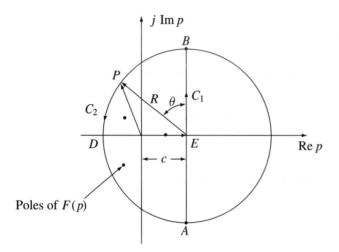

FIGURE 13.1

Applying Cauchy's residue theorem to (13.17) we obtain

$$F_s(s) = \sum_k \mathrm{Res}\left\{\frac{F(p)}{1 - e^{pT}e^{-sT}}\right\}\Bigg|_{p=p_k} - \lim_{R\to\infty}\frac{1}{2\pi j}\int_{C_2}\frac{F(p)}{1 - e^{pT}e^{-sT}}\,dp \qquad (13.23)$$

where $\{p_k\}$ are the poles of $F(p)$ and $\sigma = \mathrm{Re}\{s\} > \sigma_f$.

Introducing (13.22) and (13.19) into the above equation it can be shown (see Jury, 1973)

$$F_s(s) = \sum_k \mathrm{Res}\left\{\frac{F(p)}{1 - e^{pT}e^{-sT}}\right\}\Bigg|_{p=p_k} - \frac{f(0+)}{2} \qquad (13.24)$$

By letting $z = e^{sT}$ the above equation becomes

$$F(z) = F_s(s)|_{s=\frac{1}{T}\ell nz} = \sum_k \mathrm{Res}\left\{\frac{F(p)}{1 - e^{pT}z^{-1}}\right\}\Bigg|_{p=p_k} - \frac{f(0+)}{2}, \qquad |z| > e^{\sigma_f T} \qquad (13.25)$$

Example
The Laplace transform of $f(t) = tu(t)$ is $1/s^2$. The integrand $|te^{-\sigma t}e^{-j\omega t}| < \infty$ for $\sigma > 0$ implies that the region of convergence is $\mathrm{Re}\{s\} > 0$. Because $f(t)$ has a double pole at $s = 0$, (13.25) becomes

$$F(z) = \mathrm{Res}\left\{\frac{1}{p^2(1 - e^{pT}z^{-1})}\right\}\Bigg|_{p=0} - \frac{0}{2}$$

$$= \frac{d}{dp}\frac{p^2}{p^2(1 - e^{pT}z^{-1})}\Bigg|_{p=0} = \frac{Tz^{-1}}{(1 - z^{-1})^2} \qquad ∎$$

Example
The Laplace transform of $f(t) = e^{-aT}u(t)$ is $1/(s + a)$. The ROC is $\mathrm{Re}s > -a$ and from (13.25) we obtain

$$F(z) = \mathrm{Res}\left\{\frac{1}{(p + a)(1 - e^{pT}z^{-1})}\right\}\Bigg|_{p=-a} - \frac{1}{2} = \frac{1}{1 - e^{-aT}z^{-1}} - \frac{1}{2}$$

The inverse transform is

$$f(nT) = -\frac{1}{2}\delta(n) + e^{-anT}u(nT)$$

If we had proceeded to find the Z-transform from $f(nT) = \exp(-anT)u(nT)$ we would have found $F(z) = 1/(1 - e^{-aT}z^{-1})$. Hence to make a causal signal $f(t)$ consistent with $F(s)$ and the inversion formula, $f(0)$ should be assigned the value $f(0+)/2$.

It is conventional in calculating with the Z-transform of causal signals to assign the value of $f(0+)$ to $f(0)$. With this convention the formula for calculating $F(z)$ from $F(s)$ reduces to

$$F(z) = \sum_k \text{Res}\left\{\frac{F(p)}{1 - e^{pT}z^{-1}}\right\}\Bigg|_{p=p_k}, \qquad |z| > e^{\sigma_f T} \qquad (13.26)$$

∎

6.14   Relationship to the Fourier transform

The sampled signal can be represented by

$$f_s(t) = \sum_{n=-\infty}^{\infty} f(nT)\delta(t - nT) \qquad (14.1)$$

with corresponding Laplace and Fourier transforms

$$F_s(s) = \sum_{n=-\infty}^{\infty} f(nT)e^{-snT} \qquad (14.2)$$

$$F_s(\omega) = \sum_{n=-\infty}^{\infty} f(nT)e^{-j\omega nT} \qquad (14.3)$$

If we set $z = e^{sT}$ in the definition of the Z-transform we see that

$$F_s(s) = F(z)|_{z=e^{sT}} \qquad (14.4)$$

If the region of convergence for $F(z)$ includes the unit circle, $|z| = 1$, then

$$F_s(\omega) = F(z)|_{z=e^{j\omega T}} \qquad (14.5)$$

Because $F_s(s)$ is periodic with period $\omega_s = 2\pi/T$, we need only consider the strip $-\omega_s/2 < \omega \le \omega_s/2$, which uniquely determines $F_s(s)$ for all s. The transformation $z = \exp(sT)$ maps this strip uniquely onto the complex z-plane so that $F(z)$ contains all the information in $F_s(s)$ without the redundancy.

References

[1] R. A. Gabel and R. A. Roberts, *Signals and Linear Systems*, John Wiley and Sons, New York, 1980.

[2] H. Freeman, *Discrete-Time Systems*, John Wiley and Sons, New York, 1965.

[3] E. I. Jury, *Theory and Application of the Z-Transform Method*, Krieger Publishing Co., Melbourne, FL, 1973.

[4] A. D. Poularikas and S. Seely, *Signals and Systems*, reprinted second edition, Krieger Publishing Co., Melbourne, FL, 1994.

[5] S. A. Tretter, *Introduction to Discrete-Time Signal Processing*, John Wiley and Sons, New York, 1976.

[6] R. Vich, *Z-Transform Theory and Applications*, D. Reidel Publishing Co., Boston, 1987.

Appendix: Tables

TABLE 1
Z-transform properties for positive-time sequences.

1. Linearity

$$\mathcal{Z}\{c_i f_i(nT)\} = c_i F_i(z) \qquad |z| > R_i, \ c_i \text{ are constants}$$

$$\mathcal{Z}\left\{\sum_{i=0}^{\ell} c_i f_i(nT)\right\} = \sum_{i=0}^{\ell} c_i F_i(z) \qquad |z| > \max R_i$$

2. Shifting Property

$$\mathcal{Z}\{f(nT - kT)\} = z^{-k} F(z), \qquad f(-nT) = 0 \qquad \text{for } n = 1, 2, \ldots$$

$$\mathcal{Z}\{f(nT - kT)\} = z^{-k} F(z) + \sum_{n=1}^{k} f(-nT) z^{-(k-n)}$$

$$\mathcal{Z}\{f(nT + kT)\} = z^{k} F(z) - \sum_{n=0}^{k-1} f(nT) z^{k-n}$$

$$\mathcal{Z}\{f(nT + T)\} = z[F(z) - f(0)]$$

3. Time Scaling

$$\mathcal{Z}\{a^{nT} f(nT)\} = F(a^{-T} z) = \sum_{n=0}^{\infty} f(nT)(a^{-T} z)^{-n} \qquad |z| > a^T$$

4. Periodic Sequence

$$\mathcal{Z}\{f(nT)\} = \frac{z^N}{z^N - 1} F_{(1)}(z) \qquad |z| > R$$

$$N = \text{number of time units in a period}$$
$$R = \text{radius of convergence of } F_{(1)}(z)$$
$$F_{(1)}(z) = \text{Z-transform of the first period}$$

TABLE 1
(continued)

---

5. Multiplication by n and nT

$$\mathcal{Z}\{nf(nT)\} = -z\frac{dF(z)}{dz} \qquad |z| > R$$

$$\mathcal{Z}\{nTf(nT)\} = -zT\frac{dF(z)}{dz} \qquad |z| > R$$

$$R = \text{radius of convergence of } F(z)$$

6. Convolution

$$\mathcal{Z}\{f(nT)\} = F(z) \qquad |z| > R_1$$

$$\mathcal{Z}\{h(nT)\} = H(z) \qquad |z| > R_2$$

$$\mathcal{Z}\{f(nT) * h(nT)\} = F(z)H(z) \qquad |z| > \max(R_1, R_2)$$

7. Initial Value

$$f(0T) = \lim_{z \to \infty} F(z) \qquad |z| > R$$

8. Final Value

$$\lim_{n \to \infty} f(nT) = \lim_{z \to 1}(z-1)F(z) \qquad \text{if } f(\infty T) \text{ exists}$$

9. Multiplication by $(nT)^k$

$$\mathcal{Z}\{n^k T^k f(nT)\} = -Tz\frac{d}{dz}\mathcal{Z}\{(nT)^{k-1}f(nT)\} \qquad k > 0 \text{ and is an integer}$$

10. Complex Conjugate Signals

$$\mathcal{Z}\{f(nT)\} = F(z) \qquad |z| > R$$

$$\mathcal{Z}\{f^*(nT)\} = F^*(z^*) \qquad |z| > R$$

11. Transform of Product

$$\mathcal{Z}\{f(nT)\} = F(z) \qquad |z| > R_f$$

$$\mathcal{Z}\{h(nT)\} = H(z) \qquad |z| > R_h$$

$$\mathcal{Z}\{f(nT)h(nT)\} = \frac{1}{2\pi j}\oint_C F(\tau)H\left(\frac{z}{\tau}\right)\frac{d\tau}{\tau}, \qquad |z| > R_f R_h, \; R_f < |\tau| < \frac{|z|}{R_h}$$

$$\text{counterclockwise integration}$$

12. Parseval's Theorem

$$\mathcal{Z}\{f(nT)\} = F(z) \qquad |z| > R_f$$

$$\mathcal{Z}\{h(nT)\} = H(z) \qquad |z| > R_h$$

$$\sum_{n=0}^{\infty} f(nT)h(nT) = \frac{1}{2\pi j}\oint_C F(z)H(z^{-1})\frac{dz}{z} \qquad |z| = 1 > R_f R_h$$

$$\text{counterclockwise integration}$$

13. Correlation

$$f(nT) \otimes h(nT) = \sum_{m=0}^{\infty} f(mT)h(mT - nT) = \frac{1}{2\pi j}\oint_C F(\tau)H\left(\frac{1}{\tau}\right)\tau^{n-1}d\tau \qquad n \geq 1$$

Both $f(nT)$ and $h(nT)$ must exist for $|z| > 1$. The integration is taken in counterclockwise direction.

TABLE 1
(continued)

---

14. Transforms with Parameters

$$\mathcal{Z}\left\{\frac{\partial}{\partial a}f(nT,a)\right\} = \frac{\partial}{\partial a}F(z,a)$$

$$\mathcal{Z}\left\{\lim_{a\to a_0}f(nT,a)\right\} = \lim_{a\to a_0}F(z,a)$$

$$\mathcal{Z}\left\{\int_{a_0}^{a_1}f(nT,a)\,da\right\} = \int_{a_0}^{a_1}F(z,a)\,da \quad \text{finite interval}$$

TABLE 2
Z-transform properties for positive- and negative-time sequences.

---

1. Linearity

$$\mathcal{Z}_{II}\left\{\sum_{i=0}^{\ell}c_i f_i(nT)\right\} = \sum_{i=0}^{\ell}c_i F_i(z) \qquad \max R_{i+} < |z| < \min R_{i-}$$

2. Shifting Property

$$\mathcal{Z}_{II}\{f(nT\pm kT)\} = z^{\pm k}F(z) \qquad R_+ < |z| < R_-$$

3. Scaling

$$\mathcal{Z}_{II}\{f(nT)\} = F(z) \qquad R_+ < |z| < R_-$$
$$\mathcal{Z}_{II}\{a^{nT}f(nT)\} = F(a^{-T}z) \qquad |a^T|R_+ < |z| < |a^T|R_-$$

4. Time Reversal

$$\mathcal{Z}_{II}\{f(nT)\} = F(z) \qquad R_+ < |z| < R_-$$
$$\mathcal{Z}_{II}\{f(-nT)\} = F(z^{-1}) \qquad \frac{1}{R_-} < |z| < \frac{1}{R_+}$$

5. Multiplication by nT

$$\mathcal{Z}_{II}\{f(nT)\} = F(z) \qquad R_+ < |z| < R_-$$
$$\mathcal{Z}_{II}\{nTf(nT)\} = -zT\frac{dF(z)}{dz} \qquad R_+ < |z| < R_-$$

6. Convolution

$$\mathcal{Z}_{II}\{f_1(nT)*f_2(nT)\} = F_1(z)F_2(z)$$
$$\text{ROC }F_1(z)\cup\text{ROC }F_2(z) \qquad \max(R_{+f_1},R_{+f_2}) < |z| < \min(R_{-f_1},R_{-f_2})$$

TABLE 2
(continued)

---

7. Correlation

$$R_{f_1 f_2}(z) = \mathcal{Z}_{II}\{f_1(nT) \otimes f_2(nT)\} = F_1(z)F_2(z^{-1})$$

$$\text{ROC } F_1(z) \cup \text{ROC } F_2(z^{-1}) \qquad \max(R_{+f_1}, R_{+f_2}) < |z| < \min(R_{-f_1}, R_{-f_2})$$

8. Multiplication by e^{-anT}

$$\mathcal{Z}_{II}\{f(nT)\} = F(z) \qquad R_+ < |z| < R_-$$

$$\mathcal{Z}_{II}\{e^{-anT} f(nT)\} = F(e^{aT}z) \qquad \left|e^{-aT}\right| R_+ < |z| < \left|e^{-aT}\right| R_-$$

9. Frequency Translation

$$G(\omega) = \mathcal{Z}_{II}\{e^{j\omega_0 nT} f(nT)\} = G(z)|_{z=e^{j\omega T}} = F\left(e^{j(\omega-\omega_0)T}\right) = F(\omega - \omega_0)$$

$$\text{ROC of } F(z) \text{ must include the unit circle}$$

10. Product

$$\mathcal{Z}_{II}\{f(nT)\} = F(z) \qquad R_{+f} < |z| < R_{-f}$$

$$\mathcal{Z}_{II}\{h(nT)\} = H(z) \qquad R_{+h} < |z| < R_{-h}$$

$$\mathcal{Z}_{II}\{f(nT)h(nT)\} = G(z) = \frac{1}{2\pi j} \oint_C F(\tau)H\left(\frac{z}{\tau}\right)\frac{d\tau}{\tau}, \qquad R_{+f}R_{+h} < |z| < R_{-f}R_{-h}$$

$$\max\left(R_{+f}, \frac{|z|}{R_{-h}}\right) < |\tau| < \min\left(R_{-f}, \frac{|z|}{R_{+h}}\right)$$

$$\text{counterclockwise integration}$$

11. Parseval's Theorem

$$\mathcal{Z}_{II}\{f(nT)\} = F(z) \qquad R_{+f} < |z| < R_{-f}$$

$$\mathcal{Z}_{II}\{h(nT)\} = H(z) \qquad R_{+h} < |z| < R_{-h}$$

$$\sum_{n=-\infty}^{\infty} f(nT)h(nT) = \frac{1}{2\pi j} \oint_C F(z)H(z^{-1})\frac{dz}{z} \qquad R_{+f}R_{+h} < |z| = 1 < R_{-f}R_{-h}$$

$$\max\left(R_{+f}, \frac{1}{R_{-h}}\right) < |z| < \min\left(R_{-f}, \frac{1}{R_{+h}}\right)$$

$$\text{counterclockwise integration}$$

12. Complex Conjugate Signals

$$\mathcal{Z}_{II}\{f(nT)\} = F(z) \qquad R_{+f} < |z| < R_{-f}$$

$$\mathcal{Z}_{II}\{f^*(nT)\} = F^*(z^*) \qquad R_{+f} < |z| < R_{-f}$$

TABLE 3
Inverse transforms of the partial fractions of $F(z)$.

Partial Fraction Term	Inverse Transform Term If $F(z)$ Converges Absolutely for Some $|z| > |a|$
$\dfrac{z}{z-a}$	$a^k, \quad k \geq 0$
$\dfrac{z^2}{(z-a)^2}$	$(k+1)a^k, \quad k \geq 0$
$\dfrac{z^3}{(z-a)^3}$	$\dfrac{1}{2}(k+1)(k+2)a^k, \quad k \geq 0$
$\vdots$	$\vdots$
$\dfrac{z^n}{(z-a)^n}$	$\dfrac{1}{(n-1)!}(k+1)(k+2)\cdots(k+n-1)a^k, \quad k \geq 0$

Partial Fraction Term	Inverse Transform Term If $F(z)$ Converges Absolutely for Some $|z| < |a|$
$\dfrac{z}{z-a}$	$-a^k, \quad k \leq -1$
$\dfrac{z^2}{(z-a)^2}$	$-(k+1)a^k, \quad k \leq -1$
$\dfrac{z^3}{(z-a)^3}$	$-\dfrac{1}{2}(k+1)(k+2)a^k, \quad k \leq -1$
$\vdots$	$\vdots$
$\dfrac{z^n}{(z-a)^n}$	$-\dfrac{1}{(n-1)!}(k+1)(k+2)\cdots(k+n-1)a^k, \quad k \leq -1$

TABLE 4
Inverse transforms of the partial fractions of $F_i(z)$.[a]

Elementary Transform Term $F_i(z)$	Corresponding Time Sequence	
	(I) $F_i(z)$ converges for $|z| > R_c$	(II) $F_i(z)$ converges for $|z| < R_c$
1. $\dfrac{1}{z-a}$	$a^{k-1}\|_{k \geq 1}$	$-a^{k-1}\|_{k \leq 0}$
2. $\dfrac{z}{(z-a)^2}$	$ka^{k-1}\|_{k \geq 1}$	$-ka^{k-1}\|_{k \leq 0}$
3. $\dfrac{z(z+a)}{(z-a)^3}$	$k^2 a^{k-1}\|_{k \geq 1}$	$-k^2 a^{k-1}\|_{k \leq 0}$
4. $\dfrac{z(z^2 + 4az + a^2)}{(z-a)^4}$	$k^3 a^{k-1}\|_{k \geq 1}$	$-k^3 a^{k-1}\|_{k \leq 0}$

[a]The function must be a proper function

TABLE 5

Z-transform pairs.[a]

Number	Discrete Time-Function $f(n), n \geq 0$	z-Transform $\mathcal{F}(z) = z[f(n)], \lvert z \rvert > R$ $= \sum_{n=0}^{\infty} f(n)z^{-n}$
1	$u(n) = \begin{cases} 1, & \text{for } n \geq 0 \\ 0, & \text{otherwise} \end{cases}$	$\dfrac{z}{z-1}$
2	$e^{-\alpha n}$	$\dfrac{z}{z - e^{-\alpha}}$
3	n	$\dfrac{z}{(z-1)^2}$
4	n^2	$\dfrac{z(z+1)}{(z-1)^3}$
5	n^3	$\dfrac{z(z^2 + 4z + 1)}{(z-1)^4}$
6	n^4	$\dfrac{z(z^3 + 11z^2 + 11z + 1)}{(z-1)^5}$
7	n^5	$\dfrac{z(z^4 + 26z^3 + 66z^2 + 26z + 1)}{(z-1)^6}$
8	n^k [†]	$(-1)^k D^k \left(\dfrac{z}{z-1} \right); D = z\dfrac{d}{dz}$
9	$u(n-k)$	$\dfrac{z^{-k+1}}{z-1}$
10	$e^{-\alpha n} f(n)$	$\mathcal{F}(e^{\alpha} z)$
11	$n^{(2)} = n(n-1)$	$2\dfrac{z}{(z-1)^3}$
12	$n^{(3)} = n(n-1)(n-2)$	$3!\dfrac{z}{(z-1)^4}$
13	$n^{(k)} = n(n-1)(n-2)\ldots(n-k+1)$	$k!\dfrac{z}{(z-1)^{k+1}}$
14	$n^{[k]} f(n), n^{[k]} = n(n+1)(n+2)\ldots(n+k-1)$	$(-1)^k z^k \dfrac{d^k}{dz^k}[\mathcal{F}(z)]$
15	$(-1)^k n(n-1)(n-2)\ldots(n-k+1)f_{n-k+1}$ [‡]	$z\mathcal{F}^{(k)}(z), \mathcal{F}^{(k)}(z) = \dfrac{d^k}{dz^k}\mathcal{F}(z)$
16	$-(n-1)f_{n-1}$	$\mathcal{F}^{(1)}(z)$
17	$(-1)^k(n-1)(n-2)\ldots(n-k)f_{n-k}$	$\mathcal{F}^{(k)}(z)$
18	$nf(n)$	$-z\mathcal{F}^{(1)}(z)$
19	$n^2 f(n)$	$z^2\mathcal{F}^{(2)}(z) + z\mathcal{F}^{(1)}(z)$

[a] Source: E. I. Jury, *Theory and Application of the Z-Transform Method*, New York, John Wiley & Sons, Inc., 1964. With permission.

[†] Table IV represents entries for k up to 10.

[‡] It may be noted that f_n is the same as $f(n)$

TABLE 5
(continued)

Number	Discrete Time-Function $f(n), n \geq 0$	z-Transform $\mathcal{F}(z) = z[f(n)], |z| > R$ $= \sum_{n=0}^{\infty} f(n)z^{-n}$
20	$n^3 f(n)$	$-z^3 \mathcal{F}^{(3)}(z) - 3z^2 \mathcal{F}^{(2)}(z) - z\mathcal{F}^{(1)}(z)$
21	$\dfrac{c^n}{n!}$	$e^{c/z}$
22	$\dfrac{(\ln c)^n}{n!}$	$c^{1/z}$
23	$\dbinom{k}{n} c^n a^{k-n}, \quad \dbinom{k}{n} = \dfrac{k!}{(k-n)!n!}, n \leq k$	$\dfrac{(az+c)^k}{z^k}$
24	$\dbinom{n+k}{k} c^n$	$\dfrac{z^{k+1}}{(z-c)^{k+1}}$
25	$\dfrac{c^n}{n!}, \quad (n = 1, 3, 5, 7, \ldots)$	$\sinh\left(\dfrac{c}{z}\right)$
26	$\dfrac{c^n}{n!}, \quad (n = 0, 2, 4, 6, \ldots)$	$\cosh\left(\dfrac{c}{z}\right)$
27	$\sin(\alpha n)$	$\dfrac{z \sin \alpha}{z^2 - 2z \cos \alpha + 1}$
28	$\cos(\alpha n)$	$\dfrac{z(z - \cos \alpha)}{z^2 - 2z \cos \alpha + 1}$
29	$\sin(\alpha n + \psi)$	$\dfrac{z^2 \sin \psi + z \sin(\alpha - \psi)}{z^2 - 2z \cos \alpha + 1}$
30	$\cosh(\alpha n)$	$\dfrac{z(z - \cosh \alpha)}{z^2 - 2z \cosh \alpha + 1}$
31	$\sinh(\alpha n)$	$\dfrac{z \sinh \alpha}{z^2 - 2z \cosh \alpha + 1}$
32	$\dfrac{1}{n}, \quad n > 0$	$\ln \dfrac{z}{z - 1}$
33	$\dfrac{1 - e^{-\alpha n}}{n}$	$\alpha + \ln \dfrac{z - e^{-\alpha}}{z - 1}, \quad \alpha > 0$
34	$\dfrac{\sin \alpha n}{n}$	$\alpha + \tan^{-1} \dfrac{\sin \alpha}{z - \cos \alpha}, \quad \alpha > 0$
35	$\dfrac{\cos \alpha n}{n}, \quad n > 0$	$\ln \dfrac{z}{\sqrt{z^2 - 2z \cos \alpha + 1}}$
36	$\dfrac{(n+1)(n+2)\ldots(n+k-1)}{(k-1)!}$	$\left(1 - \dfrac{1}{z}\right)^k, \quad k = 2, 3, \ldots$
37	$\displaystyle\sum_{m=1}^{n} \dfrac{1}{m}$	$\dfrac{z}{z-1} \ln \dfrac{z}{z-1}$
38	$\displaystyle\sum_{m=0}^{n-1} \dfrac{1}{m!}$	$\dfrac{e^{1/z}}{z-1}$

TABLE 5
(continued)

Number	Discrete Time-Function $f(n), n \geq 0$	z-Transform $\mathcal{F}(z) = z[f(n)], |z| > R$ $= \sum_{n=0}^{\infty} f(n)z^{-n}$
39	$\dfrac{(-1)^{(n-p)/2}}{2^n \left(\frac{n-p}{2}\right)! \left(\frac{n+p}{2}\right)!}$, for $n \geq p$ and $n - p = $ even $= 0$, for $n < p$ or $n - p = $ odd	$J_p(z^{-1})$
40	$\left\{ \begin{array}{ll} \binom{\alpha}{n/k} b^{n/k}, & n = mk, \quad (m = 0, 1, 2, \ldots) \\ = 0 & n \neq mk \end{array} \right\}$	$\left(\dfrac{z^k + b}{z^k}\right)^\alpha$
41	$a^n P_n(x) = \dfrac{a^n}{2^n n!} \left(\dfrac{d}{dx}\right)^n (x^2 - 1)^n$	$\dfrac{z}{\sqrt{z^2 - 2xaz + a^2}}$
42	$a^n T_n(x) = a^n \cos(n \cos^{-1} x)$	$\dfrac{z(z - ax)}{z^2 - 2xaz + a^2}$
43	$\dfrac{L_n(x)}{n!} = \sum_{r=0}^{\infty} \binom{n}{r} \dfrac{(-x)^r}{r!}$	$\dfrac{z}{z - 1} e^{-x/(z-1)}$
44	$\dfrac{H_n(x)}{n!} = \sum_{k=0}^{[n/2]} \dfrac{(-1)^{n-k} x^{n-2k}}{(n - 2k)! 2^k}$	$e^{-x/z - 1/2z^2}$
45	$a^n P_n^m(x) = a^n (1 - x^2)^{m/2} \left(\dfrac{d}{dx}\right)^m P_n(x), m = $ integer	$\dfrac{(2m)!}{2^m m!} \dfrac{z^{m+1}(1 - x^2)^{m/2} a^m}{(z^2 - 2xaz + a^2)^{m+1/2}}$
46	$\dfrac{L_n^m(x)}{n!} = \left(\dfrac{d}{dx}\right)^m \dfrac{L_n(x)}{n!}$, $m = $ integer	$\dfrac{(-1)^m z}{(z - 1)^{m+1}} e^{-x/(z-1)}$
47	$-\dfrac{1}{n} z^{-1} \left[\dfrac{\mathcal{F}'(z)}{\mathcal{F}(z)} - \dfrac{\mathcal{G}'(z)}{\mathcal{G}(z)} \right]$, where $\mathcal{F}(z)$ and $\mathcal{G}(z)$ are rational polynomials in z of the same order	$\ln \dfrac{\mathcal{F}(z)}{\mathcal{G}(z)}$
48	$\dfrac{1}{m(m + 1)(m + 2) \ldots (m + n)}$	$(m - 1)! z^m \left[e^{1/z} - \sum_{k=0}^{m-1} \dfrac{1}{k! z^k} \right]$
49	$\dfrac{\sin(\alpha n)}{n!}$	$e^{\cos \alpha / z} \cdot \sin\left(\dfrac{\sin \alpha}{z}\right)$
50	$\dfrac{\cos(\alpha n)}{n!}$	$e^{\cos \alpha / z} \cdot \cos\left(\dfrac{\sin \alpha}{z}\right)$
51	$\sum_{k=0}^{n} f_k g_{n-k}$	$\mathcal{F}(z)\mathcal{G}(z)$
52	$\sum_{k=0}^{n} k f_k g_{n-k}$	$-\mathcal{F}^{(1)}(z)\mathcal{G}(z), \mathcal{F}^{(1)}(z) = \dfrac{d\mathcal{F}(z)}{dz}$
53	$\sum_{k=0}^{n} k^2 f_k g_{n-k}$	$\mathcal{F}^{(2)}(z)\mathcal{G}(z)$
54	$\dfrac{\alpha^n + (-\alpha)^n}{2\alpha^2}$	$\dfrac{1}{z^2 - \alpha^2}$
55	$\dfrac{\alpha^n - \beta^n}{\alpha - \beta}$	$\dfrac{z}{(z - \alpha)(z - \beta)}$

TABLE 5
(continued)

Number	Discrete Time-Function $f(n), n \geq 0$	z-Transform $\mathcal{F}(z) = z[f(n)], |z| > R$ $= \sum_{n=0}^{\infty} f(n)z^{-n}$
56	$(n+k)^{(k)}$	$k!z^k \dfrac{z}{(z-1)^{k+1}}$
57	$(n-k)^{(k)}$	$k!z^{-k} \dfrac{z}{(z-1)^{k+1}}$
58	$\dfrac{(n \mp k)^{(m)}}{m!} e^{\alpha(n-k)}$	$\dfrac{z^{1 \mp k} e^{m\alpha}}{(z-e^\alpha)^{m+1}}$
59	$\dfrac{1}{n} \sin \dfrac{\pi}{2} n$	$\dfrac{\pi}{2} + \tan^{-1} \dfrac{1}{z}$
60	$\dfrac{\cos \alpha(2n-1)}{2n-1}, \quad n > 0$	$\dfrac{1}{4\sqrt{z}} \ln \dfrac{z + 2\sqrt{z}\cos\alpha + 1}{z - 2\sqrt{z}\cos\alpha + 1}$
61	$\dfrac{\gamma^n}{(\gamma-1)^2} + \dfrac{n}{1-\gamma} - \dfrac{1}{(1-\gamma)^2}$	$\dfrac{z}{(z-\gamma)(z-1)^2}$
62	$\dfrac{\gamma + a_0}{(\gamma-1)^2}\gamma^n + \dfrac{1+a_0}{1-\gamma}n + \left(\dfrac{1}{1-\gamma} - \dfrac{a_0+1}{(1-\gamma)^2}\right)$	$\dfrac{z(z+a_0)}{(z-\gamma)(z-1)^2}$
63	$a^n \cos \pi n$	$\dfrac{z}{z+a}$
64	$e^{-\alpha n} \cos an$	$\dfrac{z(z - e^{-\alpha}\cos a)}{z^2 - 2ze^{-\alpha}\cos a + e^{-2\alpha}}$
65	$e^{-\alpha n} \sinh(an + \psi)$	$\dfrac{z^2 \sinh\psi + ze^{-\alpha}\sinh(a-\psi)}{z^2 - 2ze^{-\alpha}\cosh a + e^{-2\alpha}}$
66	$\dfrac{\gamma^n}{(\gamma-\alpha)^2 + \beta^2} + \dfrac{(\alpha^2+\beta^2)^{n/2}\sin(n\theta+\psi)}{\beta[(\alpha-\gamma)^2 + \beta^2]^{1/2}}$ $\theta = \tan^{-1}\dfrac{\beta}{\alpha}$ $\psi = \tan^{-1}\dfrac{\beta}{\alpha-\gamma}$	$\dfrac{z}{(z-\gamma)[(z-\alpha)^2 + \beta^2]}$
67	$\dfrac{n\gamma^{n-1}}{(\gamma-1)^3} - \dfrac{3\gamma^n}{(\gamma-1)^4}$ $+ \dfrac{1}{2}\left[\dfrac{n(n-1)}{(1-\gamma)^2} - \dfrac{4n}{(1-\gamma)^3} + \dfrac{6}{(1-\gamma)^4}\right]$	$\dfrac{z}{(z-\gamma)^2(z-1)^3}$
68	$\displaystyle\sum_{v=0}^{k} (-1)^v \binom{k}{v} \dfrac{(n+k-v)^{(k)}}{k!} e^{\alpha(n-v)}$	$\dfrac{z(z-1)^k}{(z-e^\alpha)^{k+1}}$
69	$\dfrac{f(n)}{n}$	$\displaystyle\int_z^\infty p^{-1}\mathcal{F}(p)\,dp + \lim_{n \to 0} \dfrac{f(n)}{n}$
70	$\dfrac{f_{n+2}}{n+1}, \quad \begin{array}{l} f_0 = 0 \\ f_1 = 0 \end{array}$	$z \displaystyle\int_z^\infty \mathcal{F}(p)\,dp$

TABLE 5
(continued)

Number	Discrete Time-Function $f(n), n \geq 0$	z-Transform $\mathcal{F}(z) = z[f(n)], |z| > R$ $= \sum_{n=0}^{\infty} f(n)z^{-n}$
71	$\dfrac{1 + a_0}{(1 - \gamma)[(1 - \alpha)^2 + \beta^2]}$ $+ \dfrac{(\gamma + a_0)\gamma^n}{(\gamma - 1)[(\gamma - \alpha)^2 + \beta^2]}$ $+ \dfrac{[\alpha^2 + \beta^2]^{n/2}[(a_0 + \alpha)^2 + \beta^2]^{1/2}}{\beta[(\alpha - 1)^2 + \beta^2]^{1/2}[(\alpha - \gamma)^2 + \beta^2]^{1/2}},$ $\times \sin(n\theta + \psi + \lambda)$ $\psi = \psi_1 + \psi_2, \psi_1 = -\tan^{-1}\dfrac{\beta}{\alpha - 1}, \theta = \tan^{-1}\dfrac{\beta}{\alpha}$ $\lambda = \tan^{-1}\dfrac{\beta}{a_0 + \alpha}, \psi_2 = -\tan^{-1}\dfrac{\beta}{\alpha - \gamma}$	$\dfrac{z(z + a_0)}{(z - 1)(z - \gamma)[(z - \alpha)^2 + \beta^2]}$
72	$(n+1)e^{\alpha n} - 2ne^{\alpha(n+1)} + e^{\alpha(n-2)}(n-1)$	$\left(\dfrac{z-1}{z-e^\alpha}\right)^2$
73	$(-1)^n \dfrac{\cos \alpha n}{n}, \quad n > 0$	$\ln \dfrac{z}{\sqrt{z^2 + 2z\cos\alpha + 1}}$
74	$\dfrac{(n+k)!}{n!} f_{n+k}, \quad f_n = 0, \text{ for } 0 \leq n < k$	$(-1)^k z^{2k} \dfrac{d^k}{dz^k}[\mathcal{F}(z)]$
75	$\dfrac{f(n)}{n+h}, \quad h > 0$	$z^h \displaystyle\int_z^\infty p^{-(1+h)} \mathcal{F}(p)\, dp$
76	$-na^n \cos\dfrac{\pi}{2}n$	$\dfrac{2a^2 z^2}{(z^2 + a^2)^2}$
77	$na^n \dfrac{1 + \cos\pi n}{2}$	$\dfrac{2a^2 z^2}{(z^2 - a^2)^2}$
78	$a^n \sin\dfrac{\pi}{4}n \cdot \dfrac{1 + \cos\pi n}{2}$	$\dfrac{a^2 z^2}{z^4 + a^4}$
79	$a^n \left(\dfrac{1 + \cos\pi n}{2} - \cos\dfrac{\pi}{2}n\right)$	$\dfrac{2a^2 z^2}{z^4 - a^4}$
80	$\dfrac{P_n(x)}{n!}$	$e^{xz^{-1}} J_0(\sqrt{1 - x^2}z^{-1})$
81	$\dfrac{P_n^{(m)}(x)}{(n+m)!}, m > 0, P_n^m = 0, \text{ for } n < m$	$(-1)^m e^{xz^{-1}} J_m(\sqrt{1 - x^2}z^{-1})$
82	$\dfrac{1}{(n+\alpha)^\beta}, \quad \alpha > 0, \operatorname{Re}\beta > 0$	$\Phi(z^{-1}, \alpha, \beta)$, where $\Phi(1, \beta, \alpha) = \zeta(\beta, \alpha)$ = generalized Rieman-Zeta function
83	$a^n \left(\dfrac{1 + \cos\pi n}{2} + \cos\dfrac{\pi}{2}n\right)$	$\dfrac{2z^4}{z^4 - a^4}$

TABLE 5
(continued)

Number	Discrete Time-Function $f(n), n \geq 0$	z-Transform $\mathcal{F}(z) = z[f(n)], |z| > R$ $= \sum_{n=0}^{\infty} f(n)z^{-n}$
84	$\dfrac{c^n}{n}, \quad (n = 1, 2, 3, 4, \ldots)$	$\ln z - \ln(z - c)$
85	$\dfrac{c^n}{n}, \quad n = 2, 4, 6, 8, \ldots$	$\ln z - \dfrac{1}{2}\ln(z^2 - c^2)$
86	$n^2 c^n$	$\dfrac{cz(z + c)}{(z - c)^3}$
87	$n^3 c^n$	$\dfrac{cz(z^2 + 4cz + c^2)}{(z - c)^4}$
88	$n^k c^n$	$-\dfrac{d\mathcal{F}(z/c)}{dz}, \qquad \mathcal{F}(z) = z[n^{k-1}]$
89	$-\cos\dfrac{\pi}{2}n \displaystyle\sum_{i=0}^{(n-2)/4} \binom{n/2}{2i+1} a^{n-2-4i}(a^4 - b^4)^i$	$\dfrac{z^2}{z^4 + 2a^2 z^2 + b^4}$
90	$n^k f(n), \quad k > 0$ and integer	$-z\dfrac{d}{dz}\mathcal{F}_1(z), \ \mathcal{F}_1(z) = z[n^{k-1}f(n)]$
91	$\dfrac{(n-1)(n-2)(n-3)\ldots(n-k+1)}{(k-1)!}a^{n-k}$	$\dfrac{1}{(z-a)^k}$
92	$\dfrac{k(k-1)(k-2)\ldots(k-n+1)}{n!}$	$\left(1 + \dfrac{1}{z}\right)^k$
93	$na^n \cos bn$	$\dfrac{[(z/a)^3 + z/a]\cos b - 2(z/a)^2}{[(z/a)^2 - 2(z/a)\cos b + 1]^2}$
94	$na^n \sin bn$	$\dfrac{(z/a)^3 \sin b - (z/a)\sin b}{[(z/a)^2 - 2(z/a)\cos b + 1]^2}$
95	$\dfrac{na^n}{(n+1)(n+2)}$	$\dfrac{z(a - 2z)}{a^2}\ln\left(1 - \dfrac{a}{z}\right) - \dfrac{2}{a}z$
96	$\dfrac{(-a)^n}{(n+1)(2n+1)}$	$2\sqrt{z/a}\,\tan^{-1}\sqrt{a/z} - \dfrac{z}{a}\ln\left(1 + \dfrac{a}{z}\right)$
97	$\dfrac{a^n \sin\alpha n}{n+1}$	$\dfrac{z\cos\alpha}{a}\tan^{-1}\dfrac{a\sin\alpha}{z - a\cos\alpha}$ $+\dfrac{z\sin\alpha}{2a}\ln\dfrac{z^2 - 2az\cos\alpha + a^2}{z^2}$
98	$\dfrac{a^n \cos(\pi/2)n \sin\alpha(n+1)}{n+1}$	$\dfrac{z}{4a}\ln\dfrac{z^2 + 2az\sin\alpha + a^2}{z^2 - 2az\sin\alpha + a^2}$
99	$\dfrac{1}{(2n)!}$	$\cosh(z^{-1/2})$
100	$\binom{-\frac{1}{2}}{n}(-a)^n$	$\sqrt{z/(z-a)}$
101	$\binom{-\frac{1}{2}}{\frac{n}{2}}a^n \cos\dfrac{\pi}{2}n$	$\dfrac{z}{\sqrt{z^2 - a^2}}$

TABLE 5
(continued)

Number	Discrete Time-Function $f(n), n \geq 0$	z-Transform $\mathcal{F}(z) = z[f(n)], |z| > R$ $= \sum_{n=0}^{\infty} f(n)z^{-n}$
102	$\dfrac{B_n(x)}{n!}$ $B_n(x)$ are Bernoulli polynomials	$\dfrac{e^{x/z}}{z(e^{1/z} - 1)}$
103	$W_n(x) \doteq$ Tchebycheff polynomials of the second kind	$\dfrac{z^2}{z^2 - 2xz + 1}$
104	$\left| \sin \dfrac{n\pi}{m} \right|, \quad m = 1, 2, \ldots$	$\dfrac{z \sin \pi/m}{z^2 - 2z \cos \pi/m + 1} \dfrac{1 + z^{-m}}{1 - z^{-m}}$
105	$Q_n(x) = \sin(n \cos^{-1} x)$	$\dfrac{z}{z^2 - 2xz + 1}$

7

Hilbert Transforms

Stefan L. Hahn

CONTENTS

0-8493-8342-0/96/$0.00 + $0.50
© 1996 by CRC Press, Inc.

7.0   Foreword

The Hilbert transformations are of widespread interest, because they are applied in the theoretical description of many devices and systems and directly implemented in the form of Hilbert analog or digital filters (transformers). Let us quote some important applications of Hilbert transformations:

1. The complex notation of harmonic signals in the form of Euler's equation $\exp(j\omega t) = \cos(\omega t) + \mathrm{j}\sin(\omega t)$ has been used in electrical engineering since the 1890s and nowadays is commonly applied in the theoretical description of various, not only electrical, systems. This complex notation had been introduced before Hilbert derived his transformations. However, $\sin(\omega t)$ is the Hilbert transform of $\cos(\omega t)$, and the complex signal $\exp(j\omega t)$ is a precursor of a wide class of complex signals called analytic signals.

2. The concept of the analytic signal (Gabor 1946 [11]) of the form $\psi(t) = u(t) + jv(t)$, where $v(t)$ is the Hilbert transform of $u(t)$, extends the complex notation to a wide class of signals for which the Fourier transform exists. The notion of the analytic signal is widely used in the theory of signals, circuits, and systems. A device called the Hilbert transformer (or filter), which produces at the output the Hilbert transform of the input signal, finds many applications, especially in modern digital signal processing.

3. The real and imaginary parts of the transmittance of a linear and causal two-port system form a pair of Hilbert transforms. This property finds many applications.

4. Recently 2-D and multidimensional Hilbert transformations have been applied to define 2-D and multidimensional complex signals, opening the door for applications in multidimensional signal processing [13].

7.1   Basic definitions

The Hilbert transformation of a one-dimensional real signal (function) $u(t)$ is defined by the integral

$$v(t) = \frac{-1}{\pi} P \int_{-\infty}^{\infty} \frac{u(\eta)}{\eta - t} \, d\eta = \frac{1}{\pi} P \int_{-\infty}^{\infty} \frac{u(\eta)}{t - \eta} \, d\eta \qquad (7.1.1)$$

and the inverse Hilbert transformation is

$$u(t) = \frac{1}{\pi} P \int_{-\infty}^{\infty} \frac{v(\eta)}{\eta - t} \, d\eta = \frac{-1}{\pi} P \int_{-\infty}^{\infty} \frac{v(\eta)}{t - \eta} \, d\eta \qquad (7.1.2)$$

where P stands for principal value of the integral. For convenience, two conventions of the sequence of variables in the denominator are given; both have been used in studies. The left-hand formulae will be used in this chapter. The following terminology is applied: The algorithm, that is, the right-hand side of Eq. (7.1.1) or Eq. (7.1.2), is called "transformation," and the specific result for a given function, that is, the left-hand side of Eq. (7.1.1) or Eq. (7.1.2), is called the "transform." The above definitions of Hilbert transformations are conveniently written in the convolution notations

$$v(t) = u(t) * \frac{1}{\pi t} \tag{7.1.3}$$

$$u(t) = -v(t) * \frac{1}{\pi t} \tag{7.1.4}$$

The integrals in definition (7.1.1) are improper because the integrand goes to infinity for $\eta = t$. Therefore, the integral is defined as the Cauchy Principal Value (sign P) of the form

$$v(t) = \lim_{\substack{\varepsilon \Rightarrow 0 \\ A \Rightarrow \infty}} \frac{-1}{\pi} \left(\int_{-A}^{-\varepsilon} + \int_{\varepsilon}^{A} \frac{u(\eta)}{\eta - t} \, d\eta \right) \tag{7.1.5}$$

Using numerical integration in the sense of the Cauchy Principal Value with uniform sampling of the integrand, the origin $\eta = 0$ should be positioned exactly at the center of the sampling interval. The limit $\varepsilon \Rightarrow 0$ is substituted by a given value of the sampling interval and the limit $A \Rightarrow \infty$ by a given value of A. The accuracy of the numerical integration increases with smaller sampling intervals and larger values of A.

The Hilbert transformation was originally derived by Hilbert in the frame of the theory of analytic functions. The theory of Hilbert transformations is closely related to Fourier transformation of signals of the form

$$U(\omega) = \int_{-\infty}^{\infty} u(t)e^{-j\omega t} \, dt; \qquad \omega = 2\pi f \tag{7.1.6}$$

The complex function $U(\omega)$ is called the Fourier spectrum or Fourier image of the signal $u(t)$ and the variable $f = \omega/2\pi$, the Fourier frequency. The inverse Fourier transformation is

$$u(t) = \int_{-\infty}^{\infty} U(\omega)e^{j\omega t} \, df \tag{7.1.7}$$

The pair of transforms (7.1.6) and (7.1.7) may be denoted

$$u(t) \overset{F}{\Longleftrightarrow} U(\omega) \tag{7.1.8}$$

called a Fourier pair. Similarly the Hilbert transformations (7.1.1) and (7.1.2) may be denoted

$$u(t) \overset{H}{\Longleftrightarrow} v(t) \tag{7.1.9}$$

forming a Hilbert pair of functions. Contrary to other transformations, the Hilbert transformation does not change the domain. For example, the function of a time variable t (or of any other variable x) is transformed to a function of the same variable, while the Fourier transformation changes a function of time into a function of frequency.

The Fourier transform (see also Chapter 2) of the kernel of the Hilbert transformation, that is, $\Theta(t) = 1/(\pi t)$ (see Eqs. [7.1.3] and [7.1.4]) is

$$\Theta(t) = \frac{1}{\pi t} \overset{F}{\Longleftrightarrow} -j \, \mathrm{sgn}(\omega) \tag{7.1.10}$$

with the signum function (distribution) defined as follows:

$$\mathrm{sgn}(\omega) = \begin{cases} +1 & \omega > 0 \\ 0 & \omega = 0 \\ -1 & \omega < 0 \end{cases} \tag{7.1.11}$$

The multiplication to convolution theorem of the Fourier analysis yields the following spectrum of the Hilbert transform

$$v(t) \stackrel{F}{\Longleftrightarrow} V(\omega) = -j \operatorname{sgn}(\omega) U(\omega) \tag{7.1.12}$$

that is, the spectrum of the signal $u(t)$ should be multiplied by the operator $-j \operatorname{sgn}(\omega)$. This relation enables the calculation of the Hilbert transform using the inverse Fourier transform of the spectrum defined by Eq. (7.1.12), that is, using the following algorithm:

$$u(t) \stackrel{F}{\Rightarrow} U(\omega) \Rightarrow V(\omega) = -j\operatorname{sgn}(\omega) U(\omega) \stackrel{F^{-1}}{\Rightarrow} v(t) \tag{7.1.13}$$

where the symbols F and F^{-1} denote the Fourier and inverse Fourier transformations respectively. In practice, the algorithms of DFT (Discrete Fourier Transform) or FFT (Fast Fourier Transform) can be applied (see Section 7.20).

7.2   The analytic functions aspect of Hilbert transformations

The complex signal whose imaginary part is the Hilbert transform of the real signal is called the **analytic signal**. The simplest example is the harmonic complex signal given by Euler's formula $\psi(t) = \exp(j\omega t) = \cos(\omega t) + j \sin(\omega t)$. A more general form of the analytic signal was defined in 1946 by Gabor [11]. The term "analytic" is used in the meaning of a complex function $\Psi(z)$ of a complex variable $z = t + j\tau$, which is defined as follows [39]:

Consider a plane with rectangular coordinates (t, τ) (called **C** plane or **C** "space") and take a domain **D** in this plane. If we define a rule connecting to each point in **D** a complex number ψ, we defined a complex function $\psi(z), z \in \mathbf{D}$. This function may be regarded as a complex function of two real variables:

$$\psi(z) = \psi(t, \tau) = u(t, \tau) + jv(t, \tau) \tag{7.2.1}$$

in the domain $\mathbf{D} \in \mathbf{R}^2$ ($\mathbf{R}^2$ is Euclidean plane or "space"). The complete derivative of the function $\psi(z)$ has the form

$$d\psi = \frac{\partial \psi}{\partial z} dz + \frac{\partial \psi}{\partial z^*} dz^* \tag{7.2.2}$$

where $z^* = t - j\tau$ is the complex conjugate and the partial derivatives are:

$$\frac{\partial \psi}{\partial z} = \frac{1}{2} \left(\frac{\partial \psi}{\partial t} - j \frac{\partial \psi}{\partial \tau} \right); \qquad \frac{\partial \psi}{\partial z^*} = \frac{1}{2} \left(\frac{\partial \psi}{\partial t} + j \frac{\partial \psi}{\partial \tau} \right) \tag{7.2.3}$$

The function $\psi(z) = u(t, \tau) + jv(t, \tau)$ is called the **analytic function** in the domain **D** if and only if $u(t, \tau)$ and $v(t, \tau)$ are **continuously differentiable**. It can be shown that this requirement is satisfied, if $\partial \psi / \partial z^* = 0$. This complex equation may be substituted by two real equations:

$$\frac{\partial u}{\partial t} = \frac{\partial v}{\partial \tau}; \qquad \frac{\partial u}{\partial \tau} = -\frac{\partial v}{\partial t} \tag{7.2.4}$$

called the **Cauchy-Riemann equations**. These equations should be satisfied if the function $\psi(z)$ is analytic in the domain $z \in \mathbf{D}$. For example, the complex function

$$\psi(z) = \frac{1}{\alpha - jz} = u(t, \tau) + jv(t, \tau) \tag{7.2.5}$$

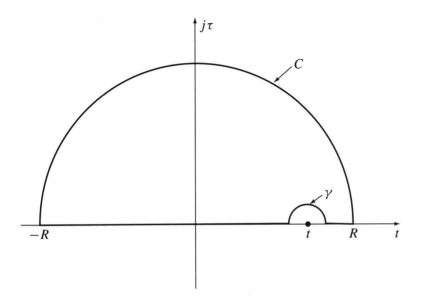

FIGURE 7.2.1
A half-circle in the complex plane $(t, j\tau)$.

is analytic because

$$u(t, \tau) = \frac{(\alpha + \tau)}{(\alpha + \tau)^2 + t^2}; \qquad v(t, \tau) = \frac{t}{(\alpha + \tau)^2 + t^2} \qquad (7.2.6)$$

and the differentiation

$$\frac{\partial u(t, \tau)}{\partial t} = \frac{\partial v(t, \tau)}{\partial \tau} = \frac{-2t(\alpha + \tau)}{[(\alpha + \tau)^2 + t^2]^2} \qquad (7.2.7)$$

verifies the Cauchy-Riemann equations.

It was shown by Cauchy that if z_0 is a point inside a closed contour $C \in \mathbf{D}$ such that $\psi(z_0)$ is analytic inside and on C, then (see also Appendix 1)

$$\psi(z_0) = \frac{1}{2\pi j} \int_C \frac{\psi(z)}{z - z_0} \, dz \qquad (7.2.8)$$

This is a contour integral in the $(t, j\tau)$ plane. Let us take the contour in the form of a half-circle as shown in Figure 7.2.1. The change of variable $y = z - z_0$ produces an alternative form of the Cauchy integral:

$$\psi(z_0) = \frac{1}{2\pi j} \int_C \frac{\psi(y + z_0)}{y} \, dy \qquad (7.2.9)$$

The **analytic signal** is defined as a complex function of the real variable t in the form:

$$\psi(t) = u(t, 0) + jv(t, 0) \qquad (7.2.10)$$

obtained by substituting in Eq. (7.2.1) $\tau = 0$. Therefore, the analytic signal (complex signal) presents the values of the analytic function $\psi(z_0)$ taken along the real axis. However, the choice of this axis is a matter of convention. Alternatively the analytic signal could also be defined along the imaginary axis. It can be shown that the contour integral vanishes along the half-circle R, as $R \Rightarrow \infty$, and the analytic signal is defined by the integral

$$\psi(t, 0) = \lim_{R \Rightarrow \infty} \frac{1}{2\pi j} P \int_{-R}^{R} \frac{\psi(\eta)}{\eta - t} d\eta \qquad (7.2.11)$$

The symbol P denotes the Cauchy Principal Value and the integral has the form

$$\lim \int_{-R}^{R} = \int_{-R}^{t-\varepsilon} + \int_{t+\varepsilon}^{R} + \int_{\gamma} \qquad (7.2.12)$$

$$R \Rightarrow \infty, \ \varepsilon \Rightarrow 0, \ \gamma \Rightarrow 0$$

It has been shown that the real and imaginary parts of the analytic signal are given by the integrals (a Hilbert pair)

$$v(t, 0) = \frac{-1}{\pi} P \int_{-\infty}^{\infty} \frac{u(\eta, 0)}{\eta - t} d\eta \qquad (7.2.13)$$

$$u(t, 0) = \frac{1}{\pi} P \int_{-\infty}^{\infty} \frac{v(\eta, 0)}{\eta - t} d\eta \qquad (7.2.14)$$

The only difference between the above integrals and those defined by Eqs. (7.1.1) and (7.1.2) consists in notation (deleting zeros in paranthesis). Therefore, the real and imaginary parts of the analytic signal

$$\psi(t) = u(t) + jv(t) \qquad (7.2.15)$$

form a Hilbert pair of functions. For example, inserting $\tau = 0$ in Eq. (7.2.6) yields the Hilbert pair

$$u(t) = \frac{\alpha}{\alpha^2 + t^2} \overset{H}{\Longleftrightarrow} v(t) = \frac{t}{\alpha^2 + t^2} \qquad (7.2.16)$$

The signal $u(t)$ is called the Cauchy signal and $v(t)$ is its Hilbert transform.

A real signal $u(t)$ may be written in terms of analytic signals

$$u(t) = \frac{\psi(t) + \psi^*(t)}{2} \qquad (7.2.17)$$

and its Hilbert transform is

$$v(t) = \frac{\psi(t) - \psi^*(t)}{2j} \qquad (7.2.18)$$

where $\psi^*(t) = u(t) - jv(t)$ is the conjugate analytic signal. Notice, that the above formulae present a generalization of Euler's formulae

$$\cos(\omega t) = \frac{e^{j\omega t} + e^{-j\omega t}}{2} \qquad (7.2.19)$$

$$\sin(\omega t) = \frac{e^{j\omega t} - e^{-j\omega t}}{2j} \qquad (7.2.20)$$

7.3 Spectral description of the Hilbert transformation: one-sided spectrum of the analytic signal

Any real signal $u(t)$ may be decomposed into a sum

$$u(t) = u_e(t) + u_o(t) \tag{7.3.1}$$

where the **even term** is defined as

$$u_e(t) = \frac{u(t) + u(-t)}{2} \tag{7.3.2}$$

and the **odd term**

$$u_o(t) = \frac{u(t) - u(-t)}{2} \tag{7.3.3}$$

The decomposition is **relative**, i.e., changes with the shift of the origin of the coordinate $t' = t - t_o$. In general, the Fourier image of $u(t)$ defined by Eq. (7.1.6) is a complex function

$$U(\omega) = U_{\text{Re}}(\omega) + jU_{\text{Im}}(\omega) \tag{7.3.4}$$

where the real part is given by the cosine transform

$$U_{\text{Re}}(\omega) = \int_{-\infty}^{\infty} u_e(t) \cos(\omega t)\, dt \tag{7.3.5}$$

and the imaginary part by the sine transform

$$U_{\text{Im}}(\omega) = -\int_{-\infty}^{\infty} u_o(t) \sin(\omega t)\, dt \tag{7.3.6}$$

The multiplication of the Fourier image by the operator $-j\,\text{sgn}(\omega)$ changes the real part of the spectrum to the imaginary one and vice versa (see Eq. [7.1.12]). The spectrum of the Hilbert transform is

$$V(\omega) = V_{\text{Re}}(\omega) + jV_{\text{Im}}(\omega) \tag{7.3.7}$$

where

$$V_{\text{Re}}(\omega) = -j\,\text{sgn}(\omega)[jU_{\text{Im}}(\omega)] = \text{sgn}(\omega)U_{\text{Im}}(\omega) \tag{7.3.8}$$

and

$$V_{\text{Im}}(\omega) = -\,\text{sgn}(\omega)U_{\text{Re}}(\omega) \tag{7.3.9}$$

Therefore, the Hilbert transformation changes any even term to an odd term and any odd term to an even term. The Hilbert transforms of harmonic functions are

$$H[\cos(\omega t)] = \sin(\omega t) \tag{7.3.10}$$

$$H[\sin(\omega t)] = -\cos(\omega t) \tag{7.3.11}$$

$$H[e^{j\omega t}] = -j\,\text{sgn}(\omega)e^{j\omega t} = \text{sgn}(\omega)e^{j(\omega t - 0.5\pi)} \tag{7.3.12}$$

Therefore, the Hilbert transformation changes any cosine term to a sine term and any sine term to a reversed signed cosine term. Because $\sin(\omega t) = \cos(\omega t - 0.5\pi)$ and $-\cos(\omega t) = \sin(\omega t - 0.5\pi)$, the Hilbert transformation in the time domain corresponds to a phase lag by -0.5π (or $-90°$) of all harmonic terms of the Fourier image (spectrum). Using the complex notation of the Fourier transform, the multiplication of the spectral function $U(\omega)$ by the operator $-j\,\text{sgn}(\omega)$ provides a 90° phase lag at all positive frequencies and a 90° phase lead at all negative frequencies. A linear two-port network with a transfer function $H(\omega) = -j\,\text{sgn}(\omega)$ is called an ideal **Hilbert transformer** or filter. Such a filter cannot be exactly realized because of constraints imposed by causality (details in Section 7.21).

The Fourier image of the analytic signal

$$\psi(t) = u(t) + jv(t) \tag{7.3.13}$$

is one-sided. We have:

$$u(t) \overset{H}{\Longleftrightarrow} v(t); \qquad u(t) \overset{F}{\Longleftrightarrow} U(\omega); \qquad v(t) \overset{F}{\Longleftrightarrow} -j\,\text{sgn}(\omega)U(\omega). \tag{7.3.14}$$

Therefore,

$$\psi(t) \overset{F}{\Longleftrightarrow} U(\omega) + j[-j\,\text{sgn}(\omega)U(\omega)] = [1 + \text{sgn}(\omega)]U(\omega) \tag{7.3.15}$$

where

$$1 + \text{sgn}(\omega) = \begin{cases} 2 & \text{for } \omega > 0 \\ 1 & \text{for } \omega = 0 \\ 0 & \text{for } \omega < 0 \end{cases} \tag{7.3.16}$$

The Fourier image of the analytic signal is doubled at positive frequencies and cancelled at negative frequencies with respect to $U(\omega)$. For the conjugate signal $\psi^*(t) = u(t) - jv(t)$ the Fourier image is doubled at negative frequencies and cancelled at positive frequencies.

Examples

1. Consider the analytic signal $e^{j\omega_0 t} = \cos(\omega_0 t) + j\sin(\omega_0 t)$. We have:

$$\cos(\omega_0 t) \overset{H}{\Longleftrightarrow} \sin(\omega_0 t); \qquad \omega_0 = 2\pi f_0$$

$$\cos(\omega_0 t) \overset{F}{\Longleftrightarrow} 0.5[\delta(f + f_0) + \delta(f - f_0)]$$

$$\sin(\omega_0 t) \overset{F}{\Longleftrightarrow} 0.5j[\delta(f + f_0) - \delta(f - f_0)]$$

$$e^{j\omega_0 t} \overset{F}{\Longleftrightarrow} \delta(f - f_0)$$

The spectra are shown in Figure 7.3.1.

2. Consider the analytic signal $\psi(t) = \frac{1}{1+t^2} + j\frac{t}{1+t^2}$. We have:

$$\frac{1}{1+t^2} \overset{H}{\Longleftrightarrow} \frac{t}{1+t^2}$$

$$\frac{1}{1+t^2} \overset{F}{\Longleftrightarrow} \pi e^{-|\omega|}; \qquad \frac{t}{1+t^2} \overset{F}{\Longleftrightarrow} -j\,\text{sgn}(\omega)\pi e^{|\omega|}$$

$$\psi(t) \overset{F}{\Longleftrightarrow} [1 + \text{sgn}(\omega)]\pi e^{-|\omega|}$$

The signals and spectra are shown in Figure 7.3.2. ∎

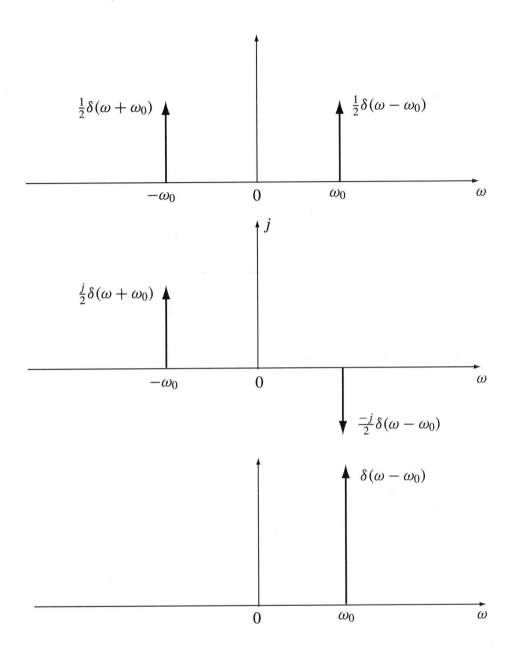

FIGURE 7.3.1
The spectra of $\cos(\omega_0 t)$, $\sin(\omega t_0)$ and of the analytic signal $e^{j\omega_0 t}$.

Derivation of Hilbert transforms using Hartley transforms

(See also Chapter 4.) Alternatively, the Hilbert transform may be derived using a special Fourier transformation known as Hartley transformation; it is given by the integral

$$U_{\text{Ha}}(\omega) = \int_{-\infty}^{\infty} u(t)\,\text{cas}(\omega t)\,dt; \qquad \omega = 2\pi f \qquad (7.3.17)$$

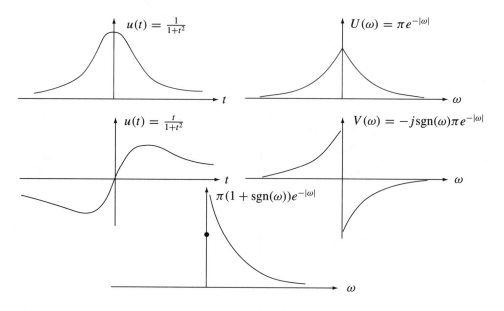

FIGURE 7.3.2
The Cauchy pulse, its Hilbert transform, and the corresponding spectra and the spectrum of the analytic signal $\psi(t) = 1/(1 - jt)$.

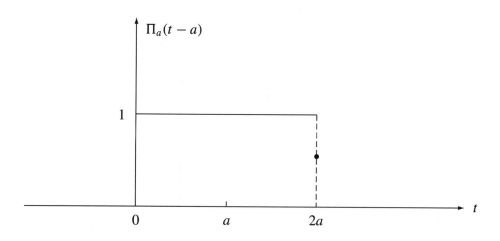

FIGURE 7.3.3
One-sided square pulse.

and the inverse Hartley transformation is

$$u(t) = \int_{-\infty}^{\infty} U_{\text{Ha}}(\omega) \cos(\omega t)\, df \qquad (7.3.18)$$

where $\text{cas}(\omega t) = \cos(\omega t) + \sin(\omega t)$. The Hartley spectral function was denoted by the index Ha because in this chapter the index H denotes the Hilbert transform. Consider the Hartley pair

$$u(t) \overset{\text{Ha}}{\Longleftrightarrow} U_{\text{Ha}}(\omega) \qquad (7.3.19)$$

The Hartley spectral function of the Hilbert transform is

$$V_{\text{Ha}}(\omega) = \text{sgn}(\omega)U_{\text{Ha}}(-\omega) \tag{7.3.20}$$

Therefore, the Hilbert transform is given by the inverse Hartley transformation

$$v(t) = \int_{-\infty}^{\infty} \text{sgn}(\omega)U_{\text{Ha}}(-\omega)\,\text{cas}(\omega t)\,df \tag{7.3.21}$$

Example
Consider the one-sided square pulse $\Pi_a(t-a)$ (see Fig. 7.3.3). The Hartley transform of this pulse is

$$U_{\text{Ha}}(\omega) = \int_0^{2a} [\cos(\omega t) + \sin(\omega t)]\,dt = 2a\left[\frac{\sin(2\omega a)}{2\omega a} + \frac{\sin^2(\omega a)}{\omega a}\right]$$

The spectrum of the Hilbert transform given by Eq. (7.3.20) is

$$V_{\text{Ha}}(\omega) = 2a\,\text{sgn}(\omega)\left[\frac{\sin(2\omega a)}{2\omega a} - \frac{\sin^2(\omega a)}{\omega a}\right]$$

The inverse Hartley transformation of this spectrum is

$$\int_{-\infty}^{\infty} 2a\,\text{sgn}(\omega)\left[\frac{\sin(2\omega a)}{2\omega a} - \frac{\sin^2(\omega a)}{\omega a}\right][\cos(\omega t) + \sin(\omega t)]\,df$$

Notice that the integrals of products of opposite symmetry equal zero and the integration yields

$$v(t) = \frac{1}{\pi}\ln\left|\frac{t}{t-2a}\right|$$

(see Eq. [7.4.7]). ∎

7.4   Examples of derivation of Hilbert transforms

1. The harmonic signal $u(t) = \cos(\omega t)$; $\omega = 2\pi f$, where f is a constant. The Hilbert transform of the periodic cosine signal using the defining integral (7.1.1) is:

$$H[\cos(\omega t)] = v(t) = \frac{-1}{\pi}P\int_{-\infty}^{\infty}\frac{\cos(\omega\eta)}{\eta - t}\,d\eta \tag{7.4.1}$$

The change of variable $y = \eta - t$, $dy = d\eta$ yields

$$v(t) = \frac{-1}{\pi}P\int_{-\infty}^{\infty}\frac{\cos[\omega(y+t)]}{y}\,dy \tag{7.4.2}$$

$$= \frac{-1}{\pi}\left\{\cos(\omega t)P\int_{-\infty}^{\infty}\frac{\cos(\omega y)}{y}\,dy - \sin(\omega t)P\int_{-\infty}^{\infty}\frac{\sin(\omega y)}{y}\,dy\right\}$$

The integrals inside the brackets are:

$$P\int_{-\infty}^{\infty}\frac{\cos(\omega y)}{y}\,dy = 0; \qquad P\int_{-\infty}^{\infty}\frac{\sin(\omega y)}{y}\,dy = \pi \tag{7.4.3}$$

Therefore, $v(t) = \sin(\omega t)$. The same derivation for the function $u(t) = \sin(\omega t)$ yields $v(t) = -\cos(\omega t)$.

2. The two-sided symmetric unipolar square pulse:

$$u(t) = \Pi_a(t) = \begin{cases} 1 & \text{for } |t| < a \\ 0.5 & \text{for } |t| = a \\ 0 & \text{for } |t| > a \end{cases} \qquad (7.4.4)$$

The Hilbert transform of this pulse is

$$v(t) = H[\Pi_a(t)] = \frac{-1}{\pi} P \int_{-\infty}^{\infty} \frac{\Pi_a(\eta)}{\eta - t} d\eta$$

$$= \lim_{\varepsilon \Rightarrow 0} \left\{ \frac{-1}{\pi} \int_{-a}^{t-\varepsilon} \frac{d\eta}{\eta - t} - \frac{1}{\pi} \int_{t+\varepsilon}^{a} \frac{d\eta}{\eta - t} \right\}$$

$$= \lim_{\varepsilon \Rightarrow 0} \left\{ -\frac{1}{\pi} \ln(\eta - t) \Big|_{-a}^{t-\varepsilon} - \frac{1}{\pi} \ln(\eta - t) \Big|_{t+\varepsilon}^{a} \right\} \qquad (7.4.5)$$

The insertion of the limits of integration yields

$$v(t) = \frac{1}{\pi} \ln \left| \frac{t+a}{t-a} \right| \qquad (7.4.6)$$

The square pulse and its Hilbert transform are shown in Figure 7.4.1. Notice that the support of the square pulse is limited within the interval $|t| \leq a$, while the support of the Hilbert transform is infinite. This statement applies to all Hilbert transforms of functions of limited support. Of course, the inverse Hilbert transformation of the logarithmic function (7.4.6) restores the square pulse of limited support. The change of variable $t' = t - a$ (time shift of the pulse) yields the Hilbert transform of a one-sided square pulse

$$H[\Pi_a(t-a)] = \frac{1}{\pi} \ln \left| \frac{t}{t-2a} \right| \qquad (7.4.7)$$

3. The Hilbert transform of a constant function $u(t) = u_0$ equals zero. This is easily seen from Eq. (7.4.6) at the limit $a \Rightarrow \infty$. The mean value of a function is given by the integral

$$u_0 = \lim_{T \Rightarrow \infty} \frac{1}{T} \int_{-T/2}^{T/2} u(t) \, dt \qquad (7.4.8)$$

Therefore, the Hilbert transform of a function $u(t) = u_0 + u_1(t)$ is

$$H[u_0 + u_1(t)] = H[u_1(t)] \qquad (7.4.9)$$

that is, in electrical terminology the Hilbert transformation cancels the DC term u_0.

4. Consider the Gaussian pulse and its Fourier image

$$e^{-\pi t^2} \overset{F}{\Longleftrightarrow} e^{-\pi f^2}; \qquad \omega = 2\pi f \qquad (7.4.10)$$

Because for this signal the Hilbert transform defined by the integral (7.1.1) has no closed form, it is convenient to derive the Hilbert transform using the inverse Fourier transformation of the Fourier image (Eq. [7.4.10]). This inverse transform has the form

$$v(t) = \int_{-\infty}^{\infty} -j \, \text{sgn}(\omega) e^{-\pi f^2} e^{j\omega t} \, df \qquad (7.4.11)$$

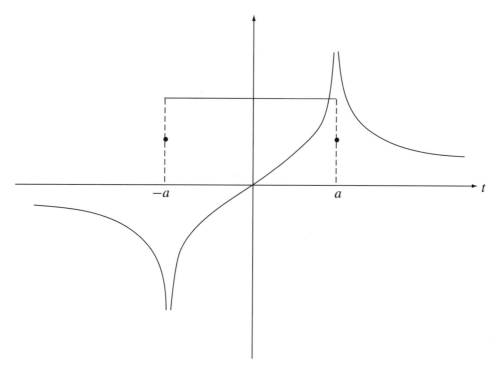

FIGURE 7.4.1
The square pulse $\Pi_a(t)$ and its Hilbert transform.

Because the integrand is an odd function, this integral has the simplified form

$$v(t) = 2 \int_0^\infty e^{-\pi f^2} \sin(\omega t)\, df \qquad (7.4.12)$$

This integral has no closed solution and may be represented by a power series defining a function called $Ei(t)$. However, in the days of proliferation of computers it is much simpler to find a numerical solution of this integral. The Gaussian pulse and its Hilbert transform computed using Eq. (7.4.12) are shown in Figure 7.4.2.

7.5 Definition of the Hilbert transformation by using a distribution

It is well known that the concept of the delta function, the unit step, and similar functions extend the class of functions for which the Fourier transform exists. The formal Fourier integral theory restricts the functions to those satisfying Dirichlet's conditions, including the requirement of finite energy (finite value of the integral of the square). The mathematicians eliminated this restriction by introducing the concept of a distribution, giving a rigorous foundation to the notions of the delta pulse, the signum function, and so forth. The notion of a distribution is not unique, because there exist several definitions. The most accepted theory of distributions was formulated by Schwartz [35], who used the concept of a functional. Another useful approach was formulated by Mikusinski [23] who used sequences of approximating functions.

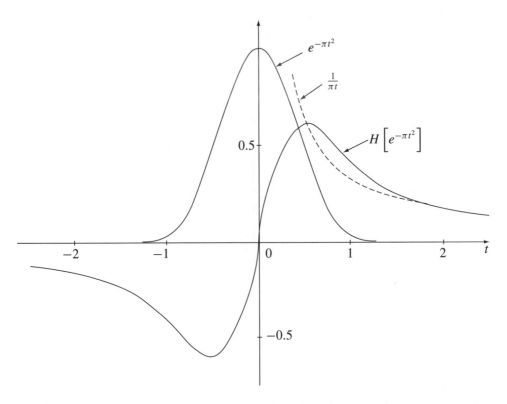

FIGURE 7.4.2
The Gaussian pulse and its Hilbert transform.

Equation (7.3.15) shows that the Fourier image of analytic signals is one-sided. A good example is the one-sided spectrum given by the doubled unit step $2\,\mathbf{1}(f)$ defined as a distribution in the Fourier frequency domain. This distribution may be decomposed into the even and odd parts:

$$2\,\mathbf{1}(f) = \mathbf{1} + \mathrm{sgn}(f) \tag{7.5.1}$$

where $\mathbf{1}$ is a constant distribution and $\mathrm{sgn}(f)$ is a signum distribution. The inverse Fourier transformation of this unit step is given by the integral

$$\psi_\delta(t) = \int_{-\infty}^{\infty} 2\,\mathbf{1}(f)e^{j\omega t}\,df; \qquad \omega = 2\pi f \tag{7.5.2}$$

or by the integral

$$\psi_\delta(t) = 2\int_0^\infty e^{j\omega t}\,df \tag{7.5.3}$$

which defines the complex delta distribution of the form

$$\psi_\delta(t) = \delta(t) + jP\frac{1}{\pi t} = F^{-1}[2\,\mathbf{1}(f)] \tag{7.5.4}$$

with P the Cauchy Principal Value. We observe, that the delta distribution and the kernel of the Hilbert transformation are forming a Hilbert pair

$$\delta(t) \overset{\mathrm{H}}{\Longleftrightarrow} P\frac{1}{\pi t} \tag{7.5.5}$$

where the Fourier images are:

$$\delta(f) \overset{F}{\Longleftrightarrow} 1; \qquad \Theta(t) = P\frac{1}{\pi t} \overset{F}{\Longleftrightarrow} -j \operatorname{sgn}(\omega) \tag{7.5.6}$$

Therefore, the kernel of the Hilbert transformations (7.1.3) and (7.1.4) denoted by $\Theta(t)$ has been redefined as a distribution in the form of the Hilbert transform of the delta pulse (distribution).

The **analytic signal** (7.2.15) may be defined in the form of a **convolution** of a given function (or distribution) $u(t)$ with the **complex delta distribution**, that is,

$$\psi(t) = \psi_\delta(t) * u(t) = \left[\delta(t) + jP\frac{1}{\pi t}\right] * u(t) \tag{7.5.7}$$

Indeed, the well-known alternative definition of the delta distribution is

$$u(t) = u(t) * \delta(t) \tag{7.5.8}$$

This is a convolution equation. The application of the theorem about the Hilbert transform of a convolution (see Table 7.7.3) yields the following two alternative forms of $H[u(t)]$:

$$v(t) = u(t) * \frac{1}{\pi t}; \qquad v(t) = v(t) * \delta(t) \tag{7.5.9}$$

The complex delta distribution may be defined alternatively using approximating functions. A convenient choice is the Cauchy signal (see Eq. [7.2.16]):

$$\psi_\delta(t) = \lim_{\alpha \to 0}\left[\frac{\alpha}{\pi(\alpha^2 + t^2)} + j\frac{t}{\pi(\alpha^2 + t^2)}\right] \tag{7.5.10}$$

(see Fig. 7.5.1). The division by π is needed to get the integral of the real part equal to 1. In terms of this representation, the distribution $\Theta(t) = 1/\pi t$ equals zero for $t = 0$. The real and imaginary parts of the complex delta distribution, as for any analytic signal, are orthogonal, that is, the integral of their product equals zero, that is,

$$P\int_{-\infty}^{\infty} \frac{\delta(t)}{\pi t}\, dt = 0 \tag{7.5.11}$$

7.6 Hilbert transforms of periodic signals

A real function (signal) $u_p(t)$ is periodic if there is some interval T (the period) for which

$$u_p(t) = u_p(t + kT) \tag{7.6.1}$$

for all t in $(-\infty, \infty)$, where k is an integer $(-\infty, \infty)$. The fundamental frequency is $f = 1/T$ and the fundamental angular frequency is $\omega = 2\pi f = 2\pi/T$. The periodic function may be alternatively defined using a periodic repetition of a so-called **generating function** $u_T(t)$. This repetition is represented by the infinite series

$$u_p(t) = \sum_{k=-\infty}^{\infty} u_T(t - kT) \tag{7.6.2}$$

where the generating function is

$$u_T(t) = \begin{cases} u_p(t) & \text{in the interval } t_0, t_0 + T \\ 0 & \text{otherwise} \end{cases} \tag{7.6.3}$$

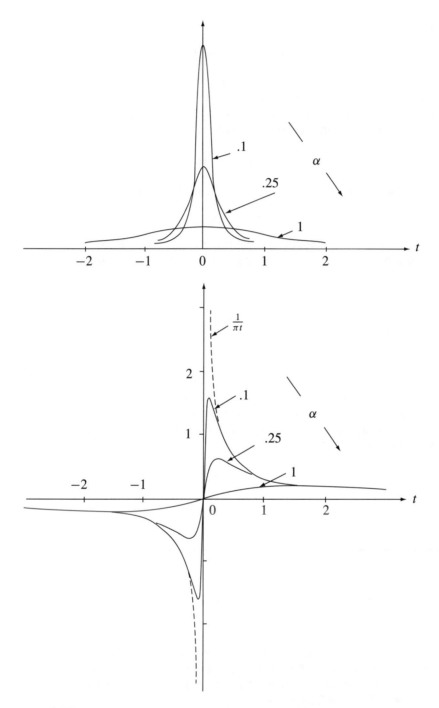

FIGURE 7.5.1
The approximation of the delta pulse $\delta(t)$ and its Hilbert transform $\Theta(t) = 1/(\pi t)$ by Cauchy pulses and its Hilbert transforms (see Fig. 7.3.2).

Using the well-known shifting property of the convolution of a given function with the delta pulse, the periodic function (7.6.2) may be written in the form

$$u(t) = u_T(t) * \sum_{k=-\infty}^{\infty} \delta(t - kT) \qquad (7.6.4)$$

that is, the generating function is convolved with the periodic sequence of delta pulses well known from the sampling theory.

Three different methods of derivation of the Hilbert transform of periodic functions are presented here:

1. A method using Fourier series.

2. Direct derivation in the form of infinite products.

3. The convolution with a cotangent periodic function.

First method

The periodic function may be expanded into a Fourier series

$$u(t) = U_0 + \sum_{n=1}^{\infty} U_n \cos(n\omega_0 t + \Phi_n); \qquad \omega_0 = \frac{2\pi}{T} \tag{7.6.5}$$

The number of terms of this series may be finite or infinite. Because $H[\cos(n\omega t + \Phi)] = \sin(n\omega t + \Phi)$, the Hilbert transform of the periodic function $u_p(t)$ is given by the Fourier series

$$u_p(t) = \sum_{n=1}^{\infty} U_n \sin(n\omega_0 t + \Phi_n) \tag{7.6.6}$$

Notice the cancellation of the constant term U_0 (in electrical terminology the DC term). If the Fourier series is given using the complex notation

$$u_p(t) = \sum_{n=-\infty}^{\infty} C_n e^{jn\omega_0 t} \tag{7.6.7}$$

where the complex coefficient C_n is given by the integral

$$C_n = \frac{1}{T} \int_{-T/2}^{T/2} u_p(t) e^{-jn\omega_0 t} \, dt \tag{7.6.8}$$

then the Hilbert transform has the form (see Eq. [7.3.12])

$$v_p(t) = \sum_{n=-\infty}^{\infty} -j \, \text{sgn}(n) C_n e^{jn\omega_0 t}; \qquad \omega_0 > 0 \tag{7.6.9}$$

Again, the constant term is eliminated ($\text{sgn}(0) = 0$).

Example

Consider the Fourier series of the periodic square wave given by the formula $u_p(t) = \text{sgn}[\cos(\omega t)]$ ($\omega = 2\pi f$ – a constant):

$$u_p(t) = \frac{4}{\pi} \left[\cos(\omega t) - \frac{1}{3} \cos(3\omega t) + \frac{1}{5} \cos(5\omega t) - \frac{1}{7} \cos(7\omega t) + \cdots \right] \tag{7.6.10}$$

The Hilbert transform has the form

$$v_p(t) = \frac{4}{\pi} \left[\sin(\omega t) - \frac{1}{3} \sin(3\omega t) + \frac{1}{5} \sin(5\omega t) - \frac{1}{7} \sin(7\omega t) + \cdots \right] \tag{7.6.11}$$

Figure 7.6.1 shows the the signals represented by the Fourier series (7.6.10) and (7.6.11) truncated at the 5th harmonic and at a much higher harmonic term. We observe the Gibb's peaks for the cosine series. Because in the limit, the energy of the Gibbs peaks equals zero (a zero function), the Gibbs peaks disappear for the sine series. ∎

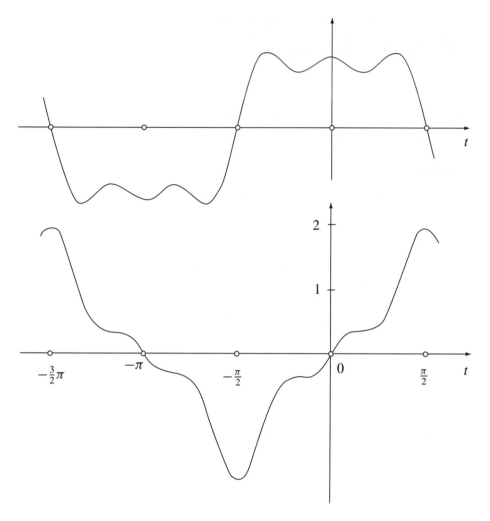

FIGURE 7.6.1

(a) The waveforms given by the truncation of the Fourier series of a square wave at the fifth harmonic number and of the corresponding Hilbert transform. ***Continued***.

Second method

The derivation of the Hilbert transform of a periodic signal directly in the time domain (or any other domain) using the basic integral definition of the Hilbert transformation given by Eq. (7.1.1) has the form of the infinite sum of integrals over successive periods. Only one of these integrals includes the pole of the kernel $1/(\pi t)$. For example, the Hilbert transform of the periodic square wave (see Fig. 7.6.2a) has the form

$$
v_p(t) = -\frac{1}{\pi} \left\{ \cdots \int_{-5b}^{-3b} \frac{d\eta}{\eta - t} - \int_{-3b}^{-b} \frac{d\eta}{\eta - t} \right.
$$

$$
\left. + \lim_{\varepsilon \Rightarrow 0} \left[\int_{-b}^{t-\varepsilon} \frac{d\eta}{\eta - t} + \int_{t+\varepsilon}^{b} \frac{d\eta}{n - t} \right] - \int_{b}^{3b} \frac{d\eta}{\eta - t} + \int_{3b}^{5b} \frac{d\eta}{\eta - t} - \cdots \right\} \quad (7.6.12)
$$

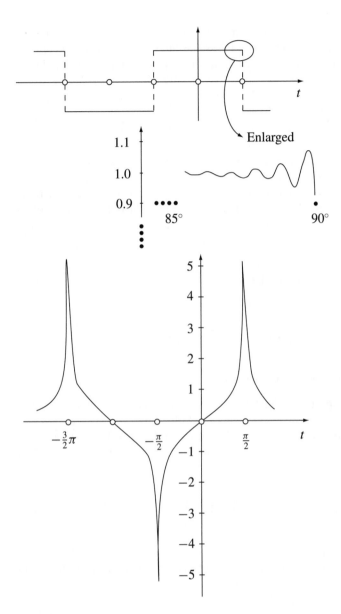

FIGURE 7.6.1
(b) Analogous waveforms by the truncation at a high harmonic number.

where $b = T/4$. The result of this integration has the form

$$v_p(t) = \frac{2}{\pi} \left\{ \ln \left| \frac{\Pi_{m=1} [2m - 1 - (1)^m x]}{\Pi_{m=1} [2m - 1 + (1)^m x]} \right| \right\} \qquad (7.6.13)$$

where $x = 4t/T$ and $m = 1, 2, 3, \ldots$. The first terms of the infinite products are

$$v_p(t) = \frac{2}{\pi} \ln \left| \frac{(1 + x)(3 - x)(5 + x)(7 - x) \ldots}{(1 - x)(3 + x)(5 - x)(7 + x) \ldots} \right| \qquad (7.6.14)$$

The infinite products in the above formulas are convergent. Using the numerical evaluation of Eq. (7.6.14) we have to truncate the products having the same number of terms in the nominator and denominator. For the odd square wave $u_p(t) = \text{sgn}[\sin(\omega t)]$ (see Fig. 7.6.2b), Eq. [7.6.14] changes to

$$v_p(t) = \frac{2}{\pi} \ln \left| \frac{y(4 - y^2)(16 - y^2)(36 - y^2)\cdots}{(1 - y^2)(9 - y^2)(25 - y^2)(7 - y)\cdots} \right|; \qquad y = 2x = 2t/T \qquad (7.6.15)$$

Notice that the denominator has been truncated so that a half-term of $(49 - y^2) = (7 - y)(7 + y)$ is deleted. This is needed to obtain a symmetrical truncation. Using a computer, the quotients in Eqs. (7.6.14) or (7.6.15) should be calculated using one term of the nominator divided by one term of the denominator. Otherwise there is a danger of entering in the overflow range of the computer ("number too big"). Let us recall that the harmonic functions have a representation in the form of infinite series

$$\sin(z) = z \prod_{k=1}^{\infty} \left(1 - \frac{z^2}{k^2 \pi^2} \right) \qquad (7.6.16)$$

$$\cos(z) = \prod_{k=1}^{\infty} \left(1 - \frac{4z^2}{(2k - 1)^2 \pi^2} \right) \qquad (7.6.17)$$

Third method: Cotangent Hilbert transformations

The cotangent form of the Hilbert transformation of periodic functions may be conveniently derived starting with the convolution equation (7.6.4). The Hilbert transform of a convolution of two functions equals the convolution of the Hilbert transform of one function (arbitrary choice) with the original of the other function (see Table 7.7.3). The Hilbert transform of the delta sampling sequence is

$$\delta_p(t) = \sum_{k=-\infty}^{\infty} \delta(t - kT) \overset{\text{H}}{\Longleftrightarrow} \Theta_p(t) = \frac{1}{T} \sum_{k=-\infty}^{\infty} \cot\left[\frac{\pi}{T}(t - kT) \right] \qquad (7.6.18)$$

This Hilbert pair is shown in Figure 7.6.3. The derivation is given at the end of this section. The insertion of this Hilbert transform in the convolution equation (7.6.4) yields the following form of the Hilbert transform of periodic functions:

$$v_p(t) = u_T(t) * \frac{1}{T} \sum_{k=-\infty}^{\infty} \cot\left[\frac{\pi}{T}(t - kT) \right] \qquad (7.6.19)$$

where $u_T(t)$ is the generating function defined by Eq. (7.6.3). Contrary to Fourier series, Eq. (7.6.19) has a closed integral form and for many generating functions a closed analytic solution. If the analytic solution does not exist, a numerical evaluation of the convolution yields the desired Hilbert transform.

Example

Consider again the square wave of Figure 7.6.1. The generating function is

$$u_T(t) = \begin{cases} \text{sgn}[\cos(\omega t)] & \text{for } |t| \leq 0.5T; \qquad \omega = \frac{2\pi}{T} \\ 0 & \text{otherwise} \end{cases} \qquad (7.6.20)$$

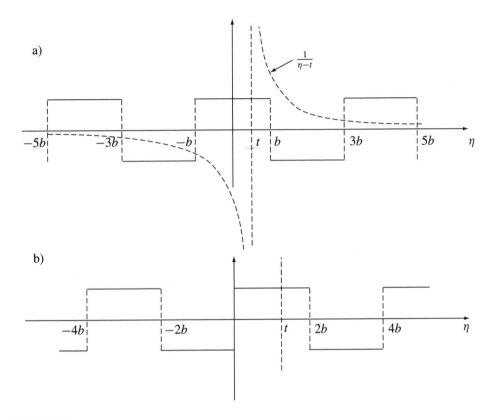

FIGURE 7.6.2
Illustration to the derivation of the Hilbert transform of a square wave.

This generating function equals -1 in the intervals $-T/2$ to $-T/4$ and $T/4$ to $T/2$ and equals 1 in the interval $-T/4$ to $T/4$. The insertion of the integration intervals (Cauchy Principal Value)

$$- \Big|_{-T/2}^{-T/4} + \Big|_{-T/4}^{t-\varepsilon} + \Big|_{t+\varepsilon}^{T/4} - \Big|_{T/4}^{T/2} \tag{7.6.21}$$

into the integral

$$\frac{1}{T} \int \cot\left[\frac{\pi}{T}(\tau - t)\right] d\tau = \frac{1}{\pi} \ln\left|\sin\left[\frac{\pi}{T}(\tau - t)\right]\right| \tag{7.6.22}$$

yields the following form of the Hilbert transform of the square wave

$$v_p(t) = \frac{2}{\pi} \ln\left|\frac{\sin\left[\frac{\pi}{T}\left(\frac{T}{4} - t\right)\right]}{\sin\left[\frac{\pi}{T}\left(\frac{T}{4} + t\right)\right]}\right| \tag{7.6.23}$$

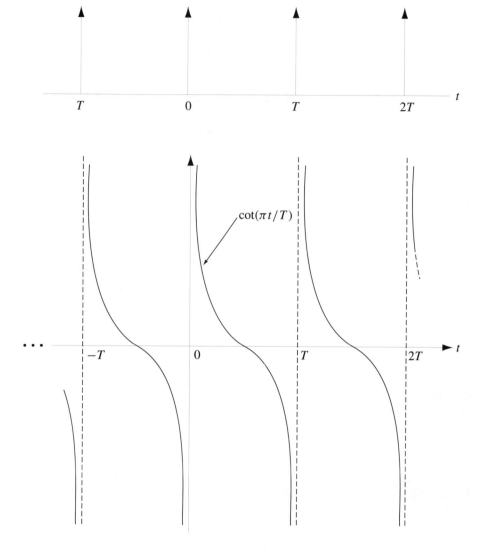

FIGURE 7.6.3
The periodic sequence of delta pulses and its Hilbert transform.

Using trigonometric relations we get the Hilbert pair

$$\text{sgn}[\cos(\omega t)] \overset{\text{H}}{\Longleftrightarrow} \frac{2}{\pi} \ln |\tan(\omega t/2 + \pi/4)| \tag{7.6.24}$$

Similarly it may be shown that

$$\text{sgn}[\sin(\omega t)] \overset{\text{H}}{\Longleftrightarrow} \frac{2}{\pi} \ln |\tan(\omega t/2)| \tag{7.6.25}$$

The Hilbert transform of the periodic delta sequence given by Eq. (7.6.18) may be derived as follows: We start with the Hilbert pair

$$\delta(t) \xLeftrightarrow{\text{H}} \frac{1}{\pi t} \tag{7.6.26}$$

The support of the Hilbert transform $1/\pi t$ is infinite. Therefore, in the interval of one period, for example the interval from 0 to T, there is a summation of succesive tails of functions $\Theta_n(t) = 1/[\pi(t - nT)]$, that is, the generating function of the Hilbert transform of the delta sampling sequence is

$$\Theta_T(t) = \sum_{n=-\infty}^{\infty} \frac{1}{\pi(t - nT)} = \frac{1}{T} \cot(\pi t/T) \tag{7.6.27}$$

that is, the infinite sum converges to the cotangent function. The repetition of this generating function yields the periodic Hilbert transform of the delta sampling sequence of the form

$$\Theta_p(t) = \sum_{k=-\infty}^{\infty} \Theta_T(t - kT) = \frac{1}{T} \sum_{k=-\infty}^{\infty} \cot\left[\frac{\pi}{T}(t - kT)\right] \tag{7.6.28}$$

This sequence may be also written in the convolution form

$$\Theta_p(t) = \frac{1}{T} \cot(\pi t/T) * \sum_{k=-\infty}^{\infty} \delta(t - kT) \tag{7.6.29}$$

The generating function $\Theta_T(t)$ (Eq. [7.6.27]) may be alternatively derived using Fourier transforms. The well-known Fourier pair is

$$\sum_{k=-\infty}^{\infty} \delta(t - kT) \xLeftrightarrow{F} \frac{1}{T} \sum_{k=-\infty}^{\infty} \delta(f - k/T) \tag{7.6.30}$$

The multiplication of this Fourier image by the operator $-j\,\text{sgn}(f)$ yields the Fourier image of the generating function $\Theta_T(t)$:

$$\Theta_T(t) \xLeftrightarrow{F} \frac{1}{T} \sum_{n=-\infty}^{\infty} -j\,\text{sgn}(f)\delta(f - n/T) \tag{7.6.31}$$

The inverse Fourier transform of this spectrum yields:

$$\Theta_T(t) = \frac{j}{T} \sum_{n=-\infty}^{-1} e^{-j2\pi nt/T} - \frac{j}{T} \sum_{n=1}^{\infty} e^{j2\pi nt/T} = \frac{2}{T} \sum_{n=1}^{\infty} \sin(2\pi nt/T) \tag{7.6.32}$$

The insertion of the relation (in the distribution sense)

$$\sum_{n=1}^{\infty} \sin(nx) = \frac{1}{2} \cot(x/2) \tag{7.6.33}$$

yields $\Theta(t)$ given by the formula (7.6.27). Notice that the derivation of the periodic Hilbert transform $\Theta_p(t)$ involves two summations. The first yields the generating function $\Theta_T(t)$ and the second gives the periodic repetition of this function.

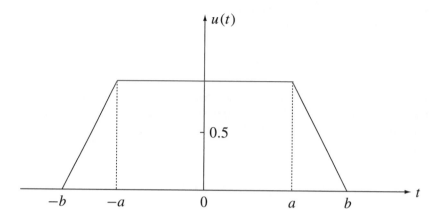

FIGURE 7.7.1
A trapezoidal pulse.

7.7 Tables listing selected Hilbert pairs and properties of Hilbert transformations

Table 7.7.1 presents the Hilbert transforms of some selected aperiodic signals and the two basic periodic harmonic signals $\cos(\omega t)$ and $\sin(\omega t)$. The Hilbert transforms of selected other periodic signals are listed in Table 7.7.2. The knowledge of the Hilbert transforms listed in these tables and the application of various properties of the Hilbert transformation listed in Table 7.7.3 enables an easy derivation of a large variety of Hilbert transforms. Applications of the properties listed in these tables are given in Sections 7.8 to 7.14. Sections 7.8 to 7.14 include selected derivations and applications of the properties of Hilbert transformations.

7.8 Linearity, iteration, autoconvolution, and energy equality

The Hilbert transformation is linear, and if a complicated waveform can be decomposed into a sum of simpler waveforms, then the summation of the Hilbert transforms of each term yields the desired transform. For example, the waveform of Figure 7.8.1a may be decomposed into a sum of two rectangular pulses. Therefore, the Hilbert transform of this waveform is (see Table 7.7.1)

$$v(t) = H\left[\Pi_a(t) + \Pi_b(t)\right] = \widehat{\Pi}_a(t) + \widehat{\Pi}_b(t)$$
$$= \frac{1}{\pi}\left\{\ln\left|\frac{t+a}{t-a}\right| + \ln\left|\frac{t+b}{t-b}\right|\right\} = \frac{1}{\pi}\ln\left|\frac{(t+b)(t+a)}{(t-b)(t-a)}\right| \tag{7.8.1}$$

Let us derive in a similar way the Hilbert transform of the "ramp" pulse shown in Figure 7.8.1b. We decompose this pulse into a sum of one-sided square pulse and one-sided

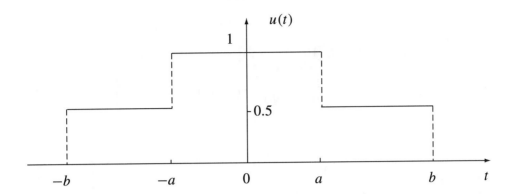

FIGURE 7.8.1
(a) A pulse given by the summation of two square pulses $\Pi_a(t) + \Pi_b(t)$. ***Continued***.

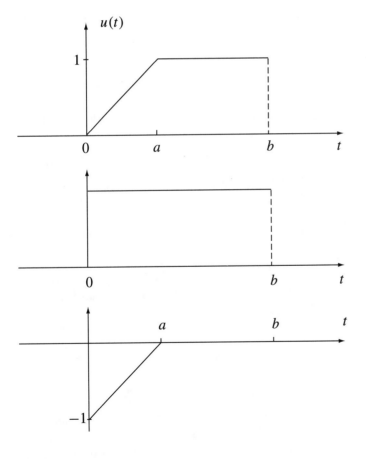

FIGURE 7.8.1
(b) The "ramp" pulse and its decomposition in two pulses.

inverse triangle. The summation of Eq. (7.4.7) and of Table 7.7.1 yields

$$H[\text{"ramp"}] = H\left[\Pi_{b/2}(t - b/2)\right] - H[1(t)\,\text{tri}(t)]$$

$$= \frac{1}{\pi}\left\{\ln\left|\frac{t}{t-b}\right| - (1 - t/a)\ln\left|\frac{t}{t+a}\right| - 1\right\} \tag{7.8.2}$$

TABLE 7.7.1
Selected useful Hilbert pairs. *Continued following page*.

Number	Name	$u(t)$	$v(t)$
1	sine	$\sin(\omega t)$	$-\cos(\omega t)$
2	cosine	$\cos(\omega t)$	$\sin(\omega t)$
3	exponential harmonic	$e^{j\omega t}$	$-j\,\text{sgn}(\omega)e^{j\omega t}$
4	square pulse	$\Pi_a(t)^\dagger$	$\frac{1}{\pi}\ln\left|\frac{t+a}{t-a}\right|$
5	bipolar pulse	$\Pi_a(t)\,\text{sgn}(t)$	$\frac{1}{\pi}\ln\left|1 - (a/t)^2\right|$
6	double triangle	$t\,\Pi_a(t)\,\text{sgn}(t)$	$\frac{t}{\pi}\ln\left|1 - (a/t)^2\right|$
7	triangle tri(t)	$\begin{array}{ll}1 - |t/a|; & |t| \leq a \\ 0; & |t| > a\end{array}$	$\frac{-1}{\pi}\left\{\ln\left|\frac{t-a}{t+a}\right| + \frac{t}{a}\ln\left|\frac{t^2}{t^2-a^2}\right|\right\}$
8	one-sided triangle	$1(t)\,\text{tri}(t)$	$\frac{1}{\pi}\left\{(1 - t/a)\ln\left|\frac{t}{t+a}\right| + 1\right\}$
9	trapezoid pulse	waveform‡	$\frac{-1}{\pi}\left\{\frac{b}{b-a}\ln\left|\frac{(a+t)(b-t)}{(b+t)(a-t)}\right|\right.$ $\left.+ \frac{t}{b-a}\ln\left|\frac{a^2-t^2}{b^2-t^2}\right| + \ln\left|\frac{a-t}{a+t}\right|\right\}$
10	Cauchy pulse	$\frac{\alpha}{\alpha^2+t^2};\quad \alpha > 0$	$\frac{t}{\alpha^2+t^2}$
11	Gaussian pulse	$e^{-\pi t^2}$	$2\int_0^\infty e^{-\pi f^2}\,\sin(\omega t)\,df$ $\omega = 2\pi f$
12	parabolic pulse	$\begin{array}{ll}1 - |t/a|^2; & |t| \leq a \\ 0; & |t| > a\end{array}$	$\frac{-1}{\pi}\left\{\left[1 - (t/a)^2\right]\ln\left|\frac{t-a}{t+a}\right| - \frac{2t}{a}\right\}$
13	symmetric exponential	$e^{-a|t|}$	$2\int_0^\infty \frac{2a}{a^2+\omega^2}\sin(\omega t)\,df$ or $\frac{-1}{\pi}\{\exp(-a|t|)E(-a|t|)$ $- \exp(a|t|)E(a|t|)\}$ where $E(x) = \int_X^\infty \frac{\exp(-\tau)}{\tau}\,d\tau$
14	antisymmetric exponential	$\text{sgn}(t)e^{-a|t|}$	$-2\int_0^\infty \frac{2\omega}{a^2+\omega^2}\cos(\omega t)\,df$
15	one-sided exponential	$1(t)e^{-a|t|}$	$2\int_0^\infty \frac{a\sin(\omega t)-\omega\cos(\omega t)}{a^2+\omega^2}\,df$
16	sinc pulse	$\frac{\sin(at)}{at}$	$\frac{\sin^2(at/2)}{(at/2)} = \frac{1-\cos(at)}{at}$
17	video test pulse	$\begin{array}{ll}\cos^2(\pi t/2a); & |t| \leq a \\ 0; & |t| > a\end{array}$	$2\int_0^\infty \frac{2a^2}{4a^2-\omega^2}\frac{\sin(\pi\omega/2a)}{\omega}\sin(\omega t)\,df$
18	constant	a	zero

†See Figure 7.4.1

‡See Figure 7.7.1

TABLE 7.7.1
Continued.

Hyperbolic Functions: Approximation by summation of Cauchy
signals (see Hilbert pairs 10 and 43)

Number	$u(t)$	$v(t)$
19	$\tanh(t) = 2 \sum_{\eta=0}^{\infty} \frac{t}{(\eta+0.5)^2\pi^2+t^2}$	$-2\pi \sum_{\eta=0}^{\infty} \frac{(\eta+0.5)}{(\eta+0.5)^2\pi^2+t^2}$

Notice the infinite energy of the functions $\tanh(t)$, $\coth(t)$, and $\text{cosech}(t)$.
The part of finite energy of $\tanh(t)$ is:

20	$\text{sgn}(t) - \tanh(t);$	$\pi\delta(t) + 2\pi \sum_{\eta=0}^{\infty} \frac{(\eta+0.5)}{(\eta+0.5)^2\pi^2+t^2}$
21	$\coth(t) = \frac{1}{t} + 2\sum_{\eta=1}^{\infty} \frac{t}{(\eta\pi)^2+t^2};$	$-\pi\delta(t) + 2\pi \sum_{\eta=1}^{\infty} \frac{t}{(\eta\pi)^2+t^2}$
22	$\text{sech}(t) = -2\pi \sum_{\eta=0}^{\infty}(-1)^{(\eta-1)} \frac{(\eta+0.5)}{(\eta+0.5)^2\pi^2+t^2};$	$-2\sum_{\eta=0}^{\infty}(-1)^{(\eta-1)} \frac{t}{(\eta+0.5)^2\pi^2+t^2}$
23	$\text{cosech}(t) = \frac{1}{t} - 2\sum_{\eta=1}^{\infty}(-1)^{(\eta-1)} \frac{t}{(\eta\pi)^2+t^2};$	$-\pi\delta(t) + 2\pi \sum_{\eta=1}^{\infty}(-1)^{(\eta-1)} \frac{\eta}{(\eta\pi)^2+t^2}$

Hyperbolic Functions by inverse Fourier transformation ($\omega = 2\pi f$)

24	$\text{sgn}(t) - \tanh(at/2);$ $\text{Re}\, a > 0;$	$2\int_0^{\infty} \left[\frac{2\pi}{a\sinh(\pi\omega/a)} - \frac{2}{\omega}\right]\cos(\omega t)\, df$
25	$\coth(at/2) - \text{sgn}(t);$	$2\int_0^{\infty} \left[\frac{2\pi}{a}\coth(\pi\omega/a) - \frac{2}{\omega}\right]\cos(\omega t)\, df$
26	$\text{sech}(at/2);$	$2\int_0^{\infty} \frac{2\pi}{a\cosh(\pi\omega/2a)}\sin(\omega t)\, df$
27	$\text{cosech}(at/2);$	$-2\int_0^{\infty} \frac{2\pi}{a}\tanh(\pi\omega/2a)\cos(\omega t)\, df$
28	$\text{sech}^2(at/2);$	$2\int_0^{\infty} \frac{2\pi\omega}{a\sinh(\pi\omega/2a)}\sin(\omega t)\, df$

Delta distribution, $1/\pi t$ distribution and its derivatives

---

29	$\delta(t)$	$1/\pi t$
30	$1/\pi t$	$-\delta(t)$
31	$\delta^{(1)}(t)$	$-1/\pi t^2$
32	$1/\pi t^2$	$\delta^{(1)}(t)$
33	$\delta^{(2)}(t)$	$2/\pi t^3$
34	$1/\pi t^3$	$-0.5\delta^{(2)}(t)$
35	$\delta^{(3)}(t)$	$-6/\pi t^4$
36	$1/\pi t^4$	$(1/6)\delta^{(3)}(t)$
37	$u(t)\delta(t)$	$v(t) = (1/\pi t)u(0)$

Equality of convolutions:

38	$\delta(t) = \delta(t) * \delta(t)$	$\delta(t) = -(1/\pi t) * (1/\pi t)$
39	$\delta^{(1)}(t) = \delta^{(1)}(t) * \delta(t)$	$\delta^{(1)}(t) = (1/\pi t^2) * (1/\pi t)$
40	$\delta^{(2)}(t) = \delta^{(1)}(t)*\delta^{(1)}(t)$	$\delta^{(2)}(t) = -(1/\pi t^2) * (1/\pi t^2)$
41	$\delta^{(3)}(t) = \delta^{(3)}(t) * \delta(t) =$ $\delta^{(2)}(t) * \delta^{(1)}(t)$	$\delta^{(3)}(t) = (6/\pi t^4) * (1/\pi t) =$ $(2/\pi t^3) * (1/\pi t^2)$

TABLE 7.7.1
Continued.

Approximating functions to the above distributions

---

42    $\int \delta(a,t)\,dt = \frac{1}{\pi} \tan^{-1}(t/a)$; $\int \Theta(a,t)\,dt = \frac{\ln(a^2+t^2)}{2\pi}$

43    $\delta(a,t) = \frac{1}{\pi} \frac{a}{a^2+t^2}$; $\Theta(a,t) = \frac{1}{\pi} \frac{t}{a^2+t^2}$

44    $\delta^{(1)}(a,t) = \frac{1}{\pi} \frac{-2at}{(a^2+t^2)^2}$; $\Theta^{(1)}(a,t) = \frac{1}{\pi} \frac{a^2-t^2}{(a^2+t^2)^2}$

45    $\delta^{(2)}(a,t) = \frac{1}{\pi} \frac{6at^2-2a^2}{(a^2+t^2)^3}$; $\Theta^{(2)}(a,t) = \frac{1}{\pi} \frac{2t^3-6at^2}{(a^2+t^2)^3}$

46    $\delta^{(3)}(a,t) = \frac{1}{\pi} \frac{24a^3t-24at^2}{(a^2+t^2)^4}$; $\Theta^{(3)}(a,t) = \frac{1}{\pi} \frac{-6t^2+36a^2t^2-6a^4}{(a^2+t^2)^4}$

Trigonometric functions

---

47    $\frac{\sin(at)}{t}$ $\frac{1-\cos(at)}{t}$

48    $\frac{\cos(at)}{t}$ $-\pi\delta(t) + \frac{\sin(at)}{t}$

49    $\frac{\sin(at)}{t^2}$ $-\pi a\delta(t) + \frac{1-\cos(at)}{t^2}$

50    $\frac{\cos(at)}{t^2}$ $\pi\delta^{(1)}(t) - \frac{a}{t} + \frac{\sin(at)}{t^2}$

51    $\frac{\sin(at)}{t^3}$ $\pi a\delta^{(1)}(t) - \frac{a^2}{2t} + \frac{1-\cos(at)}{t^3}$

52    $\frac{\cos(at)}{t^3}$ $-\frac{\pi}{2}\delta^{(2)}(t) + \frac{a^2\pi}{2}\delta(t) - \frac{a}{t^2} + \frac{\sin(at)}{t^3}$

Iteration

Iteration of the Hilbert transformation two times yields the original signal with the reverse sign, and the iteration four times restores the original signal $u(t)$. In the Fourier frequency domain the n-time iteration is translated to the n-time multiplication by the operator $-j\,\mathrm{sgn}(\omega)$. We have $(-j\,\mathrm{sgn}(\omega))^2 = -1$, $(-j\,\mathrm{sgn}(\omega))^3 = j\,\mathrm{sgn}(\omega)$ and $(-j\,\mathrm{sgn}(\omega))^4 = 1$. In analog or digital signal processing the Hilbert transform is produced approximately and with a delay. The n-time iteration is implemented using a series connection of Hilbert filters (see Section 7.21) and the time delay increases n-times.

Autoconvolution and energy equality

The energy of a real signal $u(t) \overset{F}{\Longleftrightarrow} U(\omega)$ is given by the integrals

$$E_u = \int_{-\infty}^{\infty} u^2(t)\,dt = \int_{-\infty}^{\infty} |U(\omega)|^2\,df; \qquad \omega = 2\pi f \qquad (7.8.3)$$

The above equality of the energy defined in the time domain and Fourier frequency domain is called **Parseval's theorem**. The squared magnitude of the Fourier image of the Hilbert transform $v(t) = H[u(t)] \overset{F}{\Longleftrightarrow} V(\omega) = -j\,\mathrm{sgn}(\omega)U(\omega)$ is

$$|V(\omega)|^2 = |-j\,\mathrm{sgn}(\omega)U(\omega)|^2 = |U(\omega)|^2 \qquad (7.8.4)$$

that is, the energy of the Hilbert transform is given by the integrals

$$E_v = \int_{-\infty}^{\infty} v^2(t)\,dt = \int_{-\infty}^{\infty} |U(\omega)|^2\,df \qquad (7.8.5)$$

Therefore, the energies E_u and E_v are equal. This property of a pair of Hilbert transforms may be used to check the algorithms of numerical evaluation of Hilbert transforms. A large discrepancy $\Delta E = E_v - E_u$ indicates a fault in the program. A small discrepancy may be

TABLE 7.7.2
Selected useful Hilbert pairs of periodic signals.

Number	Name	$u_p(t)$	$v_p(t)$
1	sampling sequence	$\sum_{k=-\infty}^{\infty} \delta(t - kT)$	$\frac{1}{T} \sum_{k=-\infty}^{\infty} \cot[(\pi/T)(t - kT)]$
2	even square wave	$\text{sgn}[\cos(\omega t)]$ $\omega = 2\pi/T$	$(2/\pi) \ln|\tan(\omega t/2 + \pi/4)|$
3	odd square wave	$\text{sgn}[\sin(\omega t)]$ $\omega = 2\pi/T$	$(2/\pi) \ln|\tan(\omega t/2)|$
4	squared cosine	$\cos^2(\omega t)$	$0.5 \sin(2\omega t)$
5	squared sine	$\sin^2(\omega t)$	$-0.5 \sin(2\omega t)$
6	cube cosine	$\cos^3(\omega t)$	$\frac{3}{4} \sin(\omega t) + \frac{1}{4} \sin(3\omega t)$
7	cube sine	$\sin^3(\omega t)$	$-\frac{3}{4} \cos(\omega t) + \frac{1}{4} \cos(3\omega t)$
8		$\cos^4(\omega t)$	$\frac{1}{2} \sin(2\omega t) + \frac{1}{8} \sin(4\omega t)$
9		$\sin^4(\omega t)$	$\frac{-1}{2} \sin(2\omega t) + \frac{1}{8} \sin(4\omega t)$
10		$e^{\pm j\omega t}$	$\mp j\, \text{sgn}(\omega)e^{\pm j\omega t}$
11	product	$\cos(at + \varphi)\cos(bt + \psi)$ $0 < a < b;$ φ, ψ are constants	$\cos(at + \varphi)\sin(bt + \psi)$
12	Fourier series	$U_0 + \sum_{k=1}^{n} U_k \cos(k\omega t + \phi_k)$	$\sum_{k=1}^{n} U_k \sin(k\omega t + \phi_k)$
13	any periodic function	$u_T(t) * \sum_{k=-\infty}^{\infty} \delta(t - kT)^a$	$u_T(t) * \frac{1}{T} \sum_{k=-\infty}^{\infty} \cot[(\pi/T)(t - kT)]$

$^a u_T(t)$ is the generating function (see Eq. [7.7.3]).

used as a measure of the accuracy. Notice that the Hilbert transformation cancels the mean value of the signal. Therefore, the energy (or the power) of this term is rejected.

The signals forming a Hilbert pair are **orthogonal**, that is, the mutual energy defined by the integral

$$\int_{-\infty}^{\infty} u(t)v(t)\, dt = 0 \tag{7.8.6}$$

equals zero. The **autoconvolution** of the signal $u(t)$ is defined by the integral

$$\rho_{u-u}(t) = u(t) * u(t) = \int_{-\infty}^{\infty} u(\tau)u(t - \tau)\, d\tau \tag{7.8.7}$$

The autoconvolution equality theorem for a Hilbert pair of signals has the form:

$$\rho_{u-u}(t) = -\rho_{v-v}(t) \tag{7.8.8}$$

that is, the autoconvolutions of $u(t)$ and $v(t)$ have the same waveform and differ only by sign.

TABLE 7.7.3
Properties of the Hilbert transformation.

Number	Name	Original or Inverse Hilbert Transform	Hilbert Transform
1	notations	$u(t)$ or $H^{-1}[v]$	$v(t)$ or $\hat{u}(t)$ or $H[u]$
2	time domain definitions	$\begin{cases} u(t) = \frac{1}{\pi} \int_{-\infty}^{\infty} \frac{v(\eta)}{\eta-t}\, d\eta \\ u(t) = \frac{-1}{\pi t} * v(t) \end{cases}$ or	$\begin{aligned} v(t) &= \frac{-1}{\pi} \int_{-\infty}^{\infty} \frac{u(\eta)}{\eta-t}\, d\eta \\ v(t) &= \frac{1}{\pi t} * u(t) \end{aligned}$
3	change of symmetry	$u(t) = u_{1e}(t) + u_{2o}(t)^*$;	$v(t) = v_{1o}(t) + v_{2e}(t)$
4	Fourier spectra	$u(t) \overset{F}{\Longleftrightarrow} U(\omega) = U_e(\omega) + j\, U_o(\omega)$; $U(\omega) = j\,\mathrm{sgn}(\omega) V(\omega)$;	$v(t) \overset{F}{\Longleftrightarrow} V(\omega) = V_e(\omega) + j V_o(\omega)$ $V(\omega) = -j\,\mathrm{sgn}(\omega) U(\omega)$

For even functions the Hilbert transform is odd:

$$U_e(\omega) = 2 \int_0^{\infty} u_{1e}(t) \cos(\omega t)\, dt \qquad\qquad v_o(t) = 2 \int_0^{\infty} U_e(\omega) \sin(\omega t)\, df$$

For odd functions the Hilbert transform is even:

$$U_o(\omega) = -2 \int_0^{\infty} u_{2o}(t) \sin(\omega t)\, dt \qquad\qquad v_e(t) = 2 \int_0^{\infty} U_o(\omega) \cos(\omega t)\, df$$

Number	Name	Original or Inverse Hilbert Transform	Hilbert Transform
5	linearity	$a u_1(t) + b u_2(t)$	$a v_1(t) + b v_2(t)$
6	scaling and time reversal	$u(at)$; $a > 0$ $u(-at)$	$v(at)$ $-v(-at)$
7	time shift	$u(t-a)$	$v(t-a)$
8	scaling and time shift	$u(bt-a)$	$v(bt-a)$

Number	Name	Original or Inverse Hilbert Transform	Fourier image
9	iteration	$H[u(t)] = v(t)$ $H[H[u]] = -u(t)$ $H[H[H[u]]] = -v(t)$ $H[H[H[Hu]]]] = u(t)$	$-j\,\mathrm{sgn}(\omega) U(\omega)$ $[-j\,\mathrm{sgn}(\omega)]^2\ U(\omega)$ $[-j\,\mathrm{sgn}(\omega)]^3\ U(\omega)$ $[-j\,\mathrm{sgn}(\omega)]^4\ U(\omega)$

10	time derivatives	First option
		$\dot{u}(t) = \frac{-1}{\pi t} * \dot{v}(t)$ $\qquad\qquad \dot{v}(t) = \frac{1}{\pi t} * \dot{u}(t)$
		Second option
		$\dot{u}(t) = \left[\frac{d}{dt}(-1/\pi t)\right] * v(t) \qquad\qquad \dot{v}(t) = \left[\frac{d}{dt}(1/\pi t)\right] * u(t)$

Number	Name	Original or Inverse Hilbert Transform	Hilbert Transform
11	convolution	$u_1(t) * u_2(t) =$ $-v_1(t) * v_2(t)$	$u_1(t) * v_2(t) =$ $v_1(t) * u_2(t)$
12	autoconvolution equality	$\int u(\tau)u(t-\tau)\, d\tau = -\int v(\tau)v(t-\tau)\, d\tau$ for $\tau = 0$ energy equality	
13	multiplication by t	$t u(t)$	$t v(t) - \int_{-\infty}^{\infty} u(\tau)\, d\tau$
14	multiplication of signals with non-overlapping spectra	$u_1(t)$ (low pass signal) $u_1(t)u_2(t)$	$u_2(t)$ (high pass signals) $u_1(t)v_2(t)$
15	analytic signal	$\psi(t) = u(t) + j H[u(t)]$	$H[\psi(t)] = -j\psi(t)$

$^*e$ = even; o = odd

TABLE 7.7.3
Continued.

16	product of analytic signals	$\psi(t) = \psi_1(t)\psi_2(t)$	$H[\psi(t)] = \psi_1(t)H[\psi_2(t)] = H[\psi_1(t)]\psi_2(t)$
17	nonlinear transformations	$u(x)$	$v(x)$
17a	$x = \frac{c}{bt+a}$	$u_1(t) = u\left[\frac{c}{bt+a}\right]$	$v_1(t) = v\left[\frac{c}{bt+a}\right] - \frac{1}{\pi}P\int_{-\infty}^{\infty}\frac{u(t)}{t}\,dt$
17b	$x = a + \frac{b}{t}$	$u_1(t) = u\left[a + \frac{b}{t}\right]$	$v_1(t) = \frac{b}{a}\left\{v\left[a + \frac{b}{t}\right] - v(a)\right\}$

Notice that the nonlinear transformation may change the signal $u(t)$ of finite energy to a signal $u_1(t)$ of infinite energy. P is the Cauchy Principal Value.

18	Asymptotic value as $t \Rightarrow \infty$ for even functions of finite support:		
	$u_e(t) = u_e(-t)$		$\lim_{t\Rightarrow\infty}|v_o(t)| = \frac{1}{\pi t}\int_S u_e(t)\,dt^a$

[a] S is support of $u_e(t)$

PROOF   Let us apply the convolution to multiplication theorem of Fourier analysis to both sides of the equality (7.8.8). We get the Fourier pairs

$$\rho_{u-u}(t) = u(t) * u(t) \overset{F}{\Longleftrightarrow} U^2(\omega) \tag{7.8.9}$$

$$\rho_{v-v}(t) = v(t) * v(t) \overset{F}{\Longleftrightarrow} [-j\,\text{sgn}(\omega)U(\omega)]^2 = -U^2(\omega) \tag{7.8.10}$$

We have shown that the functions $\rho_{u-u}(t)$ and $-\rho_{v-v}(t)$ have the same waveforms, because they have equal Fourier transforms.   ∎

Examples

1. It is really amazing to observe the result of calculation of the autoconvolutions of some Hilbert pairs. Consider the Hilbert pair $\delta(t) \overset{H}{\Longleftrightarrow} \frac{1}{\pi t}$. Because the autoconvolution of the delta pulse is $\delta(t) = \delta(t) * \delta(t)$ (see Section 7.5), the autoconvolution equality yields the surprising result

$$\delta(t) = -\frac{1}{\pi t} * \frac{1}{\pi t} \tag{7.8.11}$$

that is, the autoconvolution of the function (distribution) $\frac{1}{\pi t}$ of infinite support yields the delta pulse of a point support. Figure 7.8.2 shows the result of a numerical approximate calculation of the autoconvolution (7.8.11).

2. Consider the square pulse and its Hilbert transform

$$\Pi_a(t) \overset{H}{\Longleftrightarrow} \frac{1}{\pi}\ln\left|\frac{t+a}{t-a}\right| \tag{7.8.12}$$

The waveforms are shown in Figure 7.4.1 The autoconvolution of the square pulse is a $\text{tri}(t)$ (triangle) pulse of doubled support (see Fig. 7.8.3a). Again, the autoconvolution

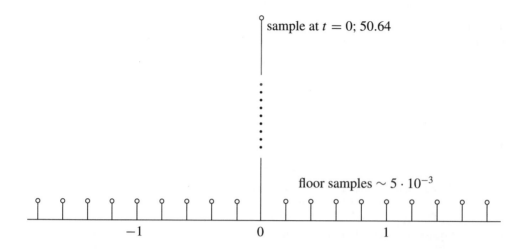

FIGURE 7.8.2
The discrete delta pulse obtained by numerical computing of the autoconvolution $-1/(\pi t) * 1/(\pi t)$.

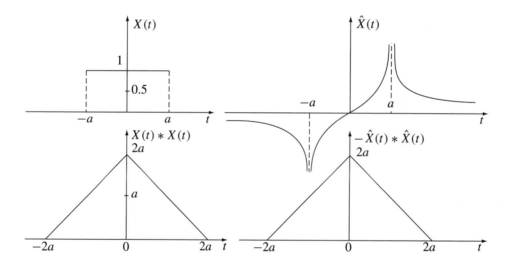

FIGURE 7.8.3
(a) An example of the autoconvolution equality. Left: The square pulse and its autoconvolution. Right: The Hilbert transform of the square pulse and its autoconvolution. ***Continued.***

of the logarithmic function of infinite support defined by Eq. (7.8.12), which has infinite peaks at points $|t| = a$, yields the triangle pulse of finite support. Indeed, we have

$$\text{tri}(t) = -\frac{1}{\pi^2} \left\{ \ln\left|\frac{t+a}{t-a}\right| * \ln\left|\frac{t+a}{t-a}\right| \right\} \tag{7.8.13}$$

Figure 7.8.3b shows the result of a numerical evaluation of the above autoconvolution. ∎

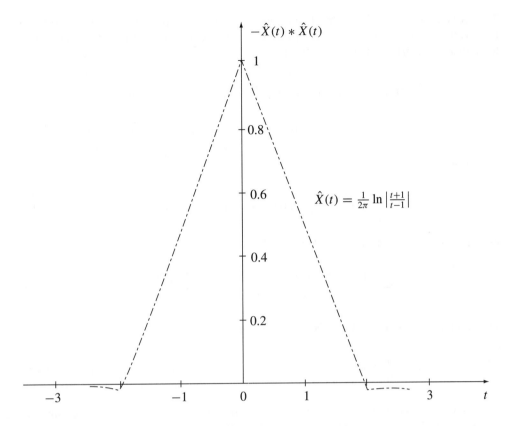

FIGURE 7.8.3
(b) The result of numerical computing of the autoconvolution of the Hilbert transform.

7.9   Differentiation of Hilbert pairs

Consider a Hilbert pair $u(t) \stackrel{H}{\Longleftrightarrow} v(t)$. Differentiation of both sides gives a new Hilbert pair:

$$\dot{u}(t) \stackrel{H}{\Longleftrightarrow} \dot{v}(t) \tag{7.9.1}$$

Therefore, differentiation is a useful tool for creating new Hilbert pairs. Obviously, the operation can be repeated to get the next Hilbert pairs:

$$\frac{d^n u}{dt^n} \stackrel{H}{\Longleftrightarrow} \frac{d^n v}{dt^n} \tag{7.9.2}$$

Because the signal $\psi(t) = u(t) + jv(t)$ is an analytic function, in principle all of its derivatives exist [39].

Consider the convolution notation of the Hilbert transformations:

$$u(t) = \frac{-1}{\pi t} * v(t) \stackrel{H}{\Longleftrightarrow} v(t) = \frac{1}{\pi t} * u(t) \tag{7.9.3}$$

The derivative of a convolution has **two options**: the convolution of the derivative of the first term with the second term, or the convolution of the first term with the derivative of the second term; that is, the **first option** has the form

$$\dot{u}(t) = -\frac{d}{dt}(1/\pi t) * v(t) \xoverset{H}{\Longleftrightarrow} \dot{v}(t) = \frac{d}{dt}(1/\pi t) * u(t)$$

$$= [1/(\pi t^2)] * v(t) = [-1/(\pi t^2)] * u(t) \tag{7.9.4}$$

and the **second option** is

$$\dot{u}(t) = -\frac{1}{\pi t} * \dot{v}(t) \xoverset{H}{\Longleftrightarrow} \dot{v}(t) = \frac{1}{\pi t} * \dot{u}(t) \tag{7.9.5}$$

PROOF   The Hilbert integrals (7.1.1) and (7.1.2) are

$$v(t) = \frac{-1}{\pi}P\int_{-\infty}^{\infty}\frac{u(\eta)}{\eta - t}\,d\eta; \qquad u(t) = \frac{1}{\pi}P\int_{-\infty}^{\infty}\frac{v(\eta)}{\eta - t}\,d\eta \tag{7.9.6}$$

The differentiation of these integrals with respect to t yields

$$\dot{v}(t) = \frac{1}{\pi}P\int_{-\infty}^{\infty}\frac{u(\eta)}{(\eta - t)^2}\,d\eta; \qquad \dot{u}(t) = \frac{-1}{\pi}P\int_{-\infty}^{\infty}\frac{v(\eta)}{(\eta - t)^2}\,d\eta \tag{7.9.7}$$

These integrals have in the convolution notation the form (7.9.4). The change of variable $y = \eta - t$ yields the following form of the Hilbert integrals:

$$v(t) = \frac{-1}{\pi}P\int_{-\infty}^{\infty}\frac{u(y + t)}{y}\,dy; \qquad u(t) = \frac{1}{\pi}P\int_{-\infty}^{\infty}\frac{v(y + t)}{y}\,dy \tag{7.9.8}$$

and the differentiation yields

$$\dot{v}(t) = \frac{-1}{\pi}P\int_{-\infty}^{\infty}\frac{\dot{u}(y + t)}{y}\,dy; \qquad \dot{u}(t) = \frac{1}{\pi}P\int_{-\infty}^{\infty}\frac{\dot{v}(y + t)}{y}\,dy \tag{7.9.9}$$

These integrals have in the convolution notation the form (7.9.5).

Very illustrative is the same proof in terms of the frequency domain representation:

$$v(t) = \frac{1}{\pi t} * u(t) \xoverset{F}{\Longleftrightarrow} -j\,\mathrm{sgn}(\omega)U(\omega) \tag{7.9.10}$$

Time domain differentiation corresponds to the multiplication of the Fourier image by the differentiation operator $j\omega$. Therefore,

$$\dot{v}(t) \xoverset{F}{\Longleftrightarrow} j\omega[-j\,\mathrm{sgn}(\omega)U(\omega)] \tag{7.9.11}$$

However, the operator $j\omega$ may be arbitrarily assigned to the first or second factor of the product in parentheses. In the time domain this arbitrary choice corresponds to the two options of the convolution.   ∎

TABLE 7.9.1

Hilbert transforms of the derivatives of the Cauchy signal $u(t) = 1/(1+t^2)$.

Notations: $u^{(n)}$; $v^{(n)}$; $\psi^{(n)} = d^n/dt^n$

n	Signal	Hilbert Transform	Analytic Signal	Energy
	$u^{(n)}$	$v^{(n)}$	$\psi^{(n)}$	
0	$\frac{1}{1+t^2}$	$\frac{t}{1+t^2}$	$\frac{1}{1-jt}$	$\frac{\pi}{2}$
1	$\frac{-2t}{(1+t^2)^2}$	$\frac{1-t^2}{(1+t^2)^2}$	$\frac{j}{(1-jt)^2}$	$\frac{\pi}{4}$
2	$2\frac{3t^2-1}{(1+t^2)^3}$	$2\frac{t^3-3t}{(1+t^2)^3}$	$\frac{-2}{(1-jt)^3}$	$\frac{3\pi}{4}$
3	$-6\frac{4t^3-4t}{(1+t^2)^4}$	$-6\frac{t^4-6t^2+1}{(1+t^2)^4}$	$\frac{-6j}{(1-jt)^4}$	$\frac{45}{8}\pi$
4	$24\frac{5t^4-10t^2+1}{(1+t^2)^5}$	$24\frac{t^5-10t^3+5t}{(1+t^2)^5}$	$\frac{24}{(1-jt)^5}$	$\frac{315}{4}\pi$

. . . .

| n | | $\frac{(-j)^n n!}{(1-jt)^{n+1}}$ | *) | |

*) $\text{Energy} = \int_0^\infty \frac{n!\,dt}{(1+t^2)^{n+1}} = \frac{(n!)^2}{2}\frac{1\,3\,5}{4\,6} \frac{(2n-1)}{2n}\frac{\pi}{2}$

Example 1

Consider the Hibert pair

$$\delta(t) \overset{H}{\Longleftrightarrow} \frac{1}{\pi t} \tag{7.9.12}$$

The derivatives are:

$$\dot\delta(t) \overset{H}{\Longleftrightarrow} \frac{d}{dt}(1/\pi t) = -\frac{1}{\pi t^2} \tag{7.9.13}$$

The derivative $\dot\delta(t)$ and hence the function $d/dt(1/\pi t)$ are defined in the distribution sense (notation FP $1/(\pi t^2)$, where FP denotes "finite part of" [35]). The energy of these signals is infinite. ∎

Example 2

Consider the Hilbert pair

$$u(t) = \frac{1}{1+t^2} \overset{H}{\Longleftrightarrow} \frac{t}{1+t^2} = v(t) \tag{7.9.14}$$

Let us differentiate n-times both sides of this equation. In this way we find an infinite series of Hilbert transform pairs as shown in Table 7.9.1. The derivations are simpler by using the differentiation of the analytic signal

$$\psi(t) = u(t) + jv(t) = \frac{1}{1-jt} \tag{7.9.15}$$

and determining the real and imaginary parts of the derivatives in the form of Hilbert pairs.

The waveforms of the first four terms of the Hilbert pairs of Table 7.9.1 are shown in Figure 7.9.1a,b. The energy was normalized to unity by division of the amplitudes by the SQR of energy. The Cauchy pulse may serve as the function approximating the delta pulse (see Eq. [7.5.10]). Therefore, the derivatives of the Cauchy-Hilbert pair may serve as the

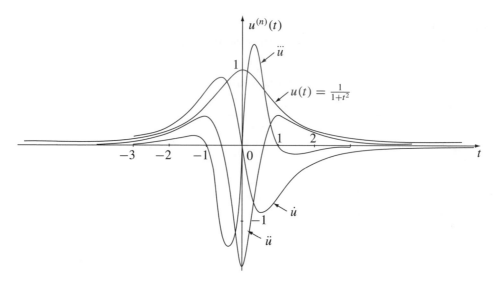

FIGURE 7.9.1
(a) The waveforms of the Cauchy pulse and of its derivatives. *Continued*.

approximating functions defining the derivatives of the complex delta distribution. For example

$$\dot{\delta}(t) = \lim_{\alpha \Rightarrow 0}\left[\frac{1}{\pi}\frac{-2\alpha t}{(\alpha^2 + t^2)^2}\right] \overset{H}{\Longleftrightarrow} \dot{\Theta}(t) = \lim_{\alpha \Rightarrow 0}\left[\frac{1}{\pi}\frac{\alpha^2 - t^2}{(\alpha^2 + t^2)^2}\right] \tag{7.9.16}$$

(see Table 7.7.1, No. 42 to 46).   ∎

7.10   Differentiation and multiplication by t: Hilbert transforms of Hermite polynomials and functions

Consider the Gaussian Fourier pair:

$$e^{-t^2} \overset{F}{\Longleftrightarrow} \pi^{0.5}e^{-\pi^2 f^2} \tag{7.10.1}$$

The succesive differentiation of the Gaussian pulse $\exp(-t^2)$ generates the n-th order **Hermite polynomial** (see Table 7.10.1). The Hermite polynomials are defined by the formula (see also Chapter 1)

$$H_n(t) = (-1)^n e^{t^2}\frac{d^n}{dt^n}e^{-t^2} \tag{7.10.2}$$

$$n = 0, 1, 2, \ldots; \qquad t \in \pm\infty$$

(roman H is used to denote the Hermite polynomial in distinction from the italic H for the

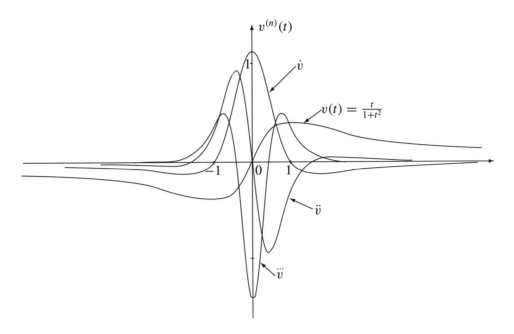

FIGURE 7.9.1
(b) The waveforms of the corresponding Hilbert transforms.

Hilbert transform). The Hermite polynomials are also defined by the recursion formula

$$H_n(t) = 2tH_{n-1}(t) - 2(n-1)H_{n-2}(t); \qquad n = 1, 2, \ldots \qquad (7.10.2)$$

The first terms of the Hermite polynomials weighted by the generating function $\exp(-t^2)$ and their Hilbert transforms are listed in Table 7.10.1. The Hilbert transform of the first term was calculated using the frequency domain method represented by the Hilbert pair (see Table 7.7.1, the Hilbert transform of the Gaussian pulse)

$$e^{-t^2} \overset{\mathrm{H}}{\Longleftrightarrow} 2\pi^{0.5} \int_{-\infty}^{\infty} e^{-\pi^2 f^2} \sin(\omega t)\, df; \qquad \omega = 2\pi f \qquad (7.10.3)$$

The next terms are obtained by calculating the successive time derivatives of both sides of this Hilbert pair. For example, the second term is

$$2te^{-t^2} \overset{\mathrm{H}}{\Longleftrightarrow} -2\pi^{0.5} \int_{-\infty}^{\infty} \omega e^{-\pi^2 f^2} \cos(\omega t)\, df \qquad (7.10.4)$$

The value of the energy of successive terms is listed in the last column of Table 7.10.1. The waveforms are shown in Figure 7.10.1. Each Hilbert pair in Table 7.10.1 is a pair of orthogonal functions. However, the weighted Hermite polynomials do not form a set of orthogonal functions, that is, the integral of the product

$$\int_{-\infty}^{\infty} e^{-2t^2} H_n(t) H_m(t)\, dt \neq 0 \qquad \text{for } n \neq m \qquad (7.10.5)$$

TABLE 7.10.1
Weighted Hermite Polynomials and Their Hilbert Transforms.

Notation: $u = \exp(-t^2)$

	Hermite Polynomial	Hilbert Transform	Energy
n	$(-1)^n d^n u/dt^n = H_n u$	$H(H_n u)$	E
0	$(1)u$	$2\sqrt{\pi} \int_0^\infty \exp(-\pi^2 f^2) \sin(\omega t)\, df$	$\sqrt{\pi/2}$
1	$(2t)u$	$-2\sqrt{\pi} \int_0^\infty \omega \exp(-\pi^2 f^2) \cos(\omega t)\, df$	$\sqrt{\pi/2}$
2	$(4t^2 - 2)u$	$-2\sqrt{\pi} \int_0^\infty \omega^2 \exp(-\pi^2 f^2) \sin(\omega t)\, df$	$3\sqrt{\pi/2}$
3	$(8t^3 - 12t)u$	$2\sqrt{\pi} \int_0^\infty \omega^3 \exp(-\pi^2 f^2) \cos(\omega t)\, df$	$15\sqrt{\pi/2}$
4	$(16t^4 - 48t^2 + 12)u$	$2\sqrt{\pi} \int_0^\infty \omega^4 \exp(-\pi^2 f^2) \sin(\omega t)\, df$	$105\sqrt{\pi/2}$
5	$(32t^5 - 160t^3 + 120t)u$	$-2\sqrt{\pi} \int_0^\infty \omega^5 \exp(-\pi^2 f^2) \cos(\omega t)\, df$	$945\sqrt{\pi/2}$
n	$H_n u =$	$(-1)^n 2\sqrt{\pi} \int_0^\infty \omega^n \exp(-\pi^2 f^2) \sin(\omega t + n\pi/2)\, df$	
	$(-1)^n [2t H_{n-1}(t) - 2(n-1) H_{n-2}(t)$		

Energy $= \int_{-\infty}^\infty u^2 H_n^2\, dt = \int_{-\infty}^\infty [H(u H_n)]^2\, dt = 1 \times 3 \times 5 \times \cdots \times |2n - 1| \times \sqrt{\pi/2}$

differs from zero for $n \neq m$. The Hermite polynomials can be orthogonalized by replacing the weighting function $\exp(-t^2)$ by $\exp(-2t^2)$, because

$$\int_{-\infty}^\infty e^{-t^2} H_n(t) H_m(t) = \begin{cases} 0 & \text{for } n = m \\ 2^n n! \pi^{0.5} & \text{for } n \neq m \end{cases} \tag{7.10.6}$$

Therefore, the functions denoted by small italic $h(t)$

$$h_n(t) = (2^n n!)^{-0.5} \pi^{-0.25} e^{-t^2/2} H_n(t); \qquad n = 0, 1 \ldots \tag{7.10.7}$$

are forming an orthonormal (energy is equal unity) set of functions called **Hermite functions**. Let us derive the Hilbert transforms of the Hermite functions. Combining the Eqs. (7.10.2) and (7.10.6) we get the following recurrency:

$$h_n(t) = \left\{ \frac{2(n-1)!}{n} \right\}^{0.5} t h_{n-1}(t) - (n-1) \left\{ \frac{(n-2)!}{n!} \right\}^{0.5} h_{n-2}(t) \tag{7.10.8}$$

The Hilbert transforms $H[h_n(t)]$ may be derived using the multiplication by t theorem (see Table 7.7.3):

$$tu(t) \overset{H}{\Longleftrightarrow} tv(t) - \frac{1}{\pi} \int_{-\infty}^\infty u(\tau)\, d\tau \tag{7.10.9}$$

PROOF The formula (7.1.1) yields

$$H[tu(t)] = -\frac{1}{\pi} \int_{-\infty}^\infty \frac{\eta u(\eta)}{\eta - t}\, d\eta \tag{7.10.10}$$

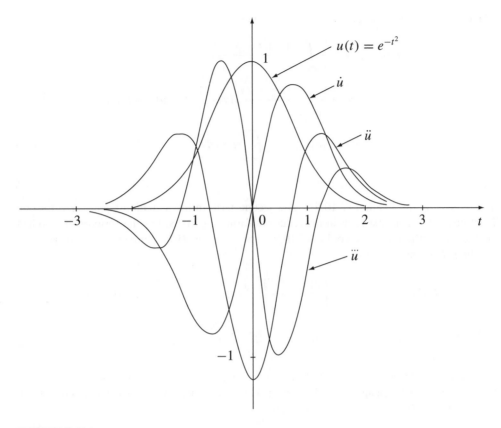

FIGURE 7.10.1
(a) The waveforms of Hermite polynomials. *Continued*.

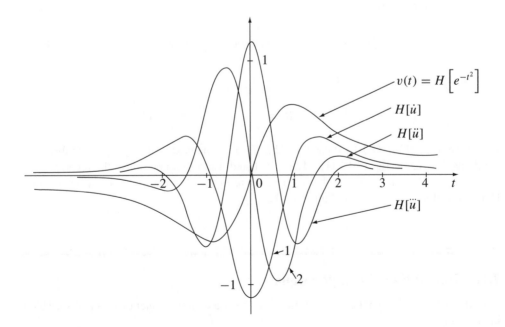

FIGURE 7.10.1
(b) The waveforms of the corresponding Hilbert transforms.

The insertion of the new variable $y = \eta - t$ gives

$$H[tu(t)] = -\frac{1}{\pi} \int_{-\infty}^{\infty} \frac{(y+t)u(y+t)}{y} \, dy$$

$$= -\frac{1}{\pi} \int_{-\infty}^{\infty} \frac{tu(y+t)}{y} \, dy - \frac{1}{\pi} \int_{-\infty}^{\infty} u(y+t) \, dy$$

$$= t H[u(t)] - \frac{1}{\pi} \int_{-\infty}^{\infty} u(\tau) \, d\tau \qquad (7.10.11)$$

$\blacksquare$

This is exactly the relation (7.10.9). The second term in this equation equals zero for odd functions $u(t)$. The first term in the recurrent formula (7.10.8) has the form of the product $th(t)$ enabling the application of Eq. (7.10.9). Therefore, the Hilbert transforms of the Hermite functions $h_n(t)$ have the form:

$$H\big[h_n(t)\big] = v_n(t) = \left[\frac{2(n-1)!}{n!}\right]^{0.5} \left[tv_{n-1}(t) - \frac{1}{\pi} \int_{-\infty}^{\infty} u_{n-1}(\tau) \, d\tau\right]$$

$$-(n-1)\left[\frac{(n-2)!}{n!}\right]^{0.5} v_{n-2}(t) \qquad (7.10.12)$$

To derive the Hilbert transforms of Hermiten functions, we have to derive by any method the first term $v_0(t)$ and then apply the above recurrency. Let us use the frequency domain method. The function $h_0(t)$ and its Fourier image are:

$$h_0(t) = \pi^{-0.25} \exp\left(-t^2/2\right) \overset{F}{\Longleftrightarrow} (4\pi)^{-0.25} \exp\left[-2(\pi f)^2\right] \qquad (7.10.13)$$

By using Eq. (7.4.12) we obtain:

$$H\left[h_0(t)\right] = v_0(t) = 2(4\pi)^{0.25} \int_0^{\infty} e^{-2\pi^2 f^2} \sin(\omega t) \, df \qquad (7.10.14)$$

Introducing the abbreviated notation ($\omega = 2\pi f$)

$$b = \pi^{0.25}, \qquad g(t) = \int_0^{\infty} e^{-2\pi^2 f^2} \sin(\omega t) \, df \qquad (7.10.15)$$

we get the form of Eq. (7.10.13) used in Table 7.10.2. The next terms $v_1, v_2, \ldots$ in this table are derived by using Eq. (7.10.11). They are listed using two notations: the recurrent and nonrecurrent. The waveforms of the first four terms of the Hermite functions $h_n(t)$ and their Hilbert transforms are shown in Figure 7.10.2.

7.11 Integration of analytic signals

Consider the analytic signal defined by Eq. (7.2.15) as a complex function of a real variable t in the form

$$\psi(t) = u(t) + j \, v(t) \qquad (7.11.1)$$

TABLE 7.10.2
Hilbert Transforms of Orthonormal Hermite Functions (Energy $= 1$).

Notations: $h_0(t), h_1(t), \ldots \Rightarrow h_0, h_1, \ldots;$ $v_0(t), v_1(t), \ldots \Rightarrow v_0, v_1, \ldots$
$g(t) = \int_0^\infty e^{-2\pi^2 f^2} \sin(2\pi f t)\, df;$ $a = \pi^{-0.25} e^{-t^2/2};$ $b = \pi^{0.25}$

Hermite Functions $h_n(t)$	Hilbert Transforms $v_n(t)$
Recurrent Notation	
$h_0 = a$	$v_0 = 2\sqrt{2}\,bg(t)$
$h_1 = \sqrt{2}\,th_0$	$v_1 = \sqrt{2}\left[tv_0\,\frac{\sqrt{2b}}{\pi}\right]$
$h_2 = th_1 - \sqrt{1/2}\,h_0$	$v_2 = tv_1 - \sqrt{1/2}\,v_0$
$h_3 = \sqrt{2/3}\,[th_2 - h_1]$	$v_3 = \sqrt{2/3}\left[tv_2 - \frac{b}{\pi} - v_1\right]$
$h_4 = \sqrt{1/2}\,th_3 - \sqrt{3/4}\,h_2$	$v_4 = \sqrt{1/2}\,tv_3 - \sqrt{3/4}\,v_2$
$h_5 = \sqrt{2/5}\,th_4 - \sqrt{4/5}\,h_3$	$v_5 = \sqrt{2/5}\left[tv_4 - \frac{\sqrt{3}b}{2\pi}\right] - \sqrt{4/5}\,v_3$
. . . .	
$h_n = \sqrt{\frac{2(n-1)!}{n!}}\,th_{n-1}+$ $(n-1)\sqrt{\frac{(n-2)!}{n!}}\,h_{n-2}$	$v_n = \sqrt{\frac{2(n-1)}{n!}}\,[\,tv_{n-1}+$ $-\frac{1}{\pi}\int h_{n-1}(\tau)\,d\tau\,] - (n-1)\sqrt{\frac{(n-2)!}{n!}}\,v_{n-2}$
Nonrecurrent Notation	
$h_0 = a1$	$2\sqrt{2}\,bg(t)$
$h_1 = \sqrt{2}\,at$	$2b\left[2\,tg(t) - \pi^{-1}\right]$
$h_2 = \frac{a}{\sqrt{8}}\left(4t^2 - 2\right)$	$2b\left[\left(2t^2 - 1\right)g(t) - t\pi^{-1}\right]$
$h_3 = \frac{a}{\sqrt{48}}(8t^3 - 12t)$	$\sqrt{8/3}\,b\left[\left(2t^3 - 3t\right)g(t) - \frac{t^2}{\pi} + \frac{1}{2\pi}\right]$
$h_4 = \frac{a}{\sqrt{384}}\left(16t^4 - 48t^2 + 12\right)$	$\sqrt{4/3}\,b\left[\left(2t^4 - 6t^2 + 1.5\right)g(t) - \frac{t^3}{\pi} + \frac{2t}{\pi}\right]$
$h_5 = \frac{a}{\sqrt{3840}}\left(32t^5 - 160t^3 + 120t\right)$	$\sqrt{8/15}\,b\left[\left(2t^5 - 10t^3 + 7.5\right)g(t) + -\frac{(t^4 - 4t^2) + 1.75}{\pi}\right]$
. . . .	
$h_n(t) = \frac{a}{\sqrt{2^n n!}}\,H_n(t),$	

$$H_n(t) = 2t H_{n-1}(t) - 2(n-1) H_{n-2}(t)$$

n	0	1	2	3	4	5	$\ldots$
$\int_{-\infty}^{\infty} h_n(\tau)\,d\tau$	$\sqrt{2}\,b$	0	b	0	$\sqrt{3/4}\,b$	0	$\ldots$

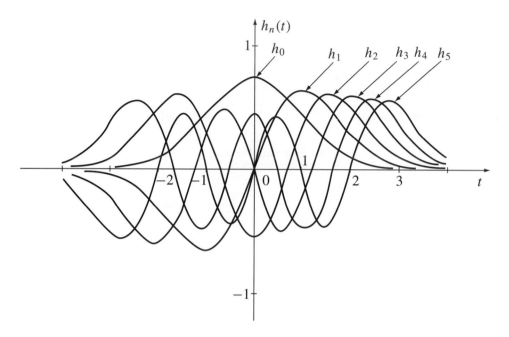

FIGURE 7.10.2
(a) Waveforms of Hermite functions. *Continued*.

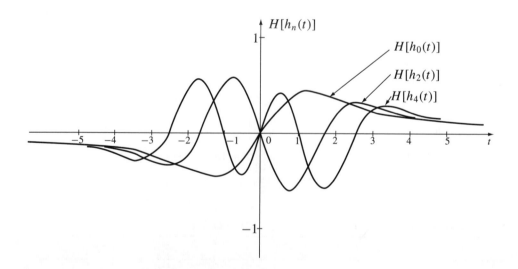

FIGURE 7.10.2
(b) Waveforms of the corresponding Hilbert transforms.

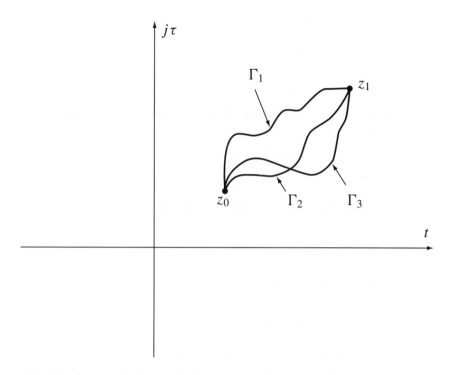

FIGURE 7.11.1
Passes of integration in the complex plane $(t, j\tau)$.

This function is integrable in the Riemann sense in the interval $[\alpha, \beta]$ if and only if the functions $u(t)$ and $v(t)$ are integrable, that is,

$$\Phi(t) = \int_\alpha^\beta \psi(t)\, dt = \int_{\substack{\alpha \\ \alpha \le t \le \beta}}^\beta u(t)\, dt + j \int_\alpha^\beta v(t)\, dt \qquad (7.11.2)$$

Let us define

$$\Phi(t) = U(t) + jV(t) \qquad (7.11.3)$$

The functions $U(t)$ and $V(t)$ are forming a Hilbert pair only if $\Phi(z)$ is an analytic function of a complex variable $z = t + j\tau$. Therefore, let us give without a proof the following theorem:

If the function $\psi(z) = u(t, \tau) + j\, v(t, \tau)$ is analytic in a simply connected domain **D**, then the function

$$\Phi(z) = \int_{z_0}^z \psi(z)\, dz \qquad (7.11.4)$$

is also analytic, and the derivative $\Phi'(z) = \psi(z)$. The integral (7.11.4) is defined as a path integral in the plane (t, τ), and in the domain **D** the integral depends on z and z_0 but not on the particular path Γ connecting them (Figure 7.11.1) [39].

If function (7.11.1) is continuous in the interval $[\alpha, \beta]$, then the function defined by the integral

$$\Phi(t) = \int_\alpha^t \psi(t)\, dt; \qquad \alpha \le t \le \beta \qquad (7.11.5)$$

is called the primary function, or antiderivative of $\Psi(t)$, and has in the interval $[\alpha, \beta]$ a continuous derivative $\Phi'(t) = \psi(t)$ and the relation holds

$$\int_{\alpha}^{\beta} \psi(t)\, dt = \Phi(t) \Big|_{\alpha}^{\beta} = \Phi(\beta) - \Phi(\alpha) \tag{7.11.6}$$

Example
The function e^{jt} has in the interval $(-\infty, \infty)$ the primary function $e^{jt}/j + c$, where c is any complex constant. We have

$$\int_{0}^{\pi/2} e^{jt}\, dt = \frac{e^{jt}}{j} \Big|_{0}^{\pi/2} = \frac{e^{\pi j/2} - 1}{j} = 1 + j \qquad \blacksquare$$

If the analytic function has a representation in the form of a power series:

$$\psi(z) = \sum_{n=0}^{\infty} d_n \, (z - z_0)^n \tag{7.11.7}$$

its integral must have a power series in the form:

$$\Phi(z) = a + \sum_{n=0}^{\infty} \frac{d_n}{n+1} \, (z - z_0)^{n+1} \tag{7.11.8}$$

This means that the power series representation can be integrated term-by-term.

Integration in the time domain can be converted by using the Fourier transforms into integration in the frequency domain. For instance, the function $u(t)$ can be integrated using the Fourier pairs

$$u(t) \overset{F}{\Longleftrightarrow} U(\omega) \tag{7.11.9}$$

$$\int_{-\infty}^{t} u(t)\, dt \overset{F}{\Longleftrightarrow} U(\omega) \left[\frac{\delta(f)}{2} + \frac{1}{j\omega} \right] \tag{7.11.10}$$

$$\omega = 2\pi f$$

The term $[\delta(f)/2]U(\omega)$ is equal to $(1/2)U(0)$ and the term $1/j\omega$ is the well-known integration operator. The same algorithm may be used to integrate the Hilbert transform $v(t)$.

Example
Consider the analytic function of the complex variable $z = t + j\tau$

$$\psi(z) = \frac{1}{\pi} \frac{1}{\alpha - jz} = \frac{1}{\pi} \frac{1}{\alpha + \tau - jt} \tag{7.11.11}$$

where α is a real constant ($\alpha > 0$). We get

$$\psi(z) = \psi(t, \tau) = u(t, \tau) + j\, v(t, \tau) \tag{7.11.12}$$

where

$$u(t, \tau) = \frac{1}{\pi} \frac{\alpha + \tau}{(\alpha + \tau)^2 + t^2} \tag{7.11.13}$$

and

$$v(t, \tau) = \frac{1}{\pi} \frac{t}{(\alpha + \tau)^2 + t^2} \tag{7.11.14}$$

Let us integrate the function (7.11.11) in the interval $[-a, t]$ where $a > 0$ is a real constant. Hence, we find

$$\Phi(z) = \int_{-a}^{t} \frac{1}{\pi} \frac{dz}{\alpha - jz} = \frac{j}{\pi} \text{Ln}(\alpha - jz) \Big|_{-a}^{t} = \frac{j}{\pi} \text{Ln}(\alpha + \tau - jt) \Big|_{-a}^{t} \qquad (7.11.15)$$

The insertion of the limits of integration and change of coordinates from rectangular to polar yields

$$\Phi(t, \tau) = \frac{1}{\pi} \left[\tan^{-1} \left(\frac{t}{\alpha + \tau} \right) + \tan^{-1} \left(\frac{a}{\alpha + \tau} \right) \right] + \frac{j}{2\pi} \text{Ln} \left[\frac{(\alpha + \tau)^2 + t^2}{(\alpha + \tau)^2 + a^2} \right] \qquad (7.11.16)$$

Because $\arg(\alpha - jz)$ is only determined to within a constant multiple of 2π, the function $(1/\pi) \text{Ln}(\alpha - jz)$ is not single valued (Notation Ln instead of ln). To prevent any winding of the integration path around $z = -j\alpha$, let us make a cut extending from the point $z = -j\alpha$ to infinity. Then $\Phi(z)$ is analytic in the remaining part of the z-plane and satisfies the Cauchy-Riemann equations (see also Appendix 1). ∎

Example

Consider a signal represented by the product:

$$u(t) = \text{sgn}(t) \Pi_a(t) \qquad (7.11.17)$$

where $\Pi_a(t)$ is defined by Eq. (7.4.4) and $\text{sgn}(t)$ by Eq. (7.1.11). We have the Fourier pair

$$0.5 \, \text{sgn}(t) \Pi_a(t) \overset{F}{\Longleftrightarrow} \frac{1 - \cos(\omega a)}{j\omega} \qquad (7.11.18)$$

The above Fourier spectrum is easy to derive by decomposing $u(t)$ into right-sided and reverse sign left-sided square pulses and adding the spectra of these pulses. In a similar way we can derive the Hilbert transform by adding the two Hilbert transforms defined by Eq. (7.4.7). The resulting Hilbert pair is

$$0.5 \, \text{sgn}(t) \Pi_a(t) \overset{H}{\Longleftrightarrow} \frac{1}{2\pi} \ln \left| \frac{t^2}{t^2 - a^2} \right| \qquad (7.11.19)$$

Let us integrate the signal $u(t)$ by frequency domain integration. We get the spectrum of the primary function using the operator $1/j\omega$:

$$U_p(\omega) = \frac{1}{j\omega} \frac{1 - \cos(\omega a)}{j\omega} = \frac{-1 + \cos(\omega a)}{\omega^2} \qquad (7.11.20)$$

The primary function of $u(t)$ is the inverse Fourier transform of Eq. (7.11.20) and has the form of a reverse signed triangle pulse.

$$-\frac{a}{2} \text{tri}(t) \overset{F}{\Longleftrightarrow} \frac{-1 + \cos(\omega a)}{\omega^2} \qquad (7.11.21)$$

The signal $\text{tri}(t)$ is defined in Table 7.7.1 and its Hilbert transform is

$$-\frac{a}{2} \text{tri}(t) \overset{H}{\Longleftrightarrow} \frac{1}{2\pi} \left\{ a \ln \left| \frac{t - a}{t + a} \right| + t \ln \left| \frac{t^2}{t^2 - a^2} \right| \right\} \qquad (7.11.22)$$

∎

7.12   Multiplication of signals with nonoverlapping spectra

Consider a signal of the form of the product

$$u(t) = f(t)g(t) \tag{7.12.1}$$

where $f(t)$ is a low-pass and $g(t)$ a high-pass signal. The Fourier spectra of these signals do not overlap, that is, if

$$f(t) \overset{F}{\Longleftrightarrow} F(\omega) \tag{7.12.2}$$

$$g(t) \overset{F}{\Longleftrightarrow} G(\omega) \tag{7.12.3}$$

then ($\omega = 2\pi f$)

$$|F(f)| = 0 \qquad \text{for } |f| > W \tag{7.12.4}$$

$$|G(f)| = 0 \qquad \text{for } |f| < W \tag{7.12.5}$$

as shown in Figure 7.12.1. In terms of Fourier methods, the Hilbert transform of the product $u(t) = f(t)g(t)$ may be derived using **the multiplication-convolution theorem** of the form (see also Chapter 2)

$$f(t)g(t) \overset{F}{\Longleftrightarrow} \int_{-\infty}^{\infty} F(f - u)G(u)\, du \tag{7.12.6}$$

The multiplication of the spectrum by $-j\,\text{sgn}(f)$ (see Eq. (7.1.12)) yields the spectrum of the Hilbert transform

$$H[f(t)g(t)] \overset{F}{\Longleftrightarrow} -j\,\text{sgn}(f) \int_{-\infty}^{\infty} F(f - u)G(u)\, du \tag{7.12.7}$$

However the product $f(t)H[g(t)]$ and its Fourier transform are

$$f(t)H[g(t)] \overset{F}{\Longleftrightarrow} \int_{-\infty}^{\infty} F(f - u)[-j\,\text{sgn}(u)G(u)]\, du \tag{7.12.8}$$

One can show [4], that the right-hand sides of (7.12.7) and (7.12.8) are identical. Therefore, the left-hand sides are identical too, and

$$H[f(t)g(t)] = f(t)H[g(t)] \tag{7.12.9}$$

This equation presents Bedrosian's theorem: Only the high-pass signal in the product of low-pass and high-pass signals gets Hilbert transformed [4].

Example

Consider a signal in the form of the amplitude-modulated harmonic function:

$$u(t) = A(t)\cos(\Omega t + \Phi); \qquad \Omega = 2\pi F \tag{7.12.10}$$

$$A(t) \overset{F}{\Longleftrightarrow} C_A(f) \tag{7.12.11}$$

and the magnitude of $C_A(f)$ is low-pass limited:

$$|C_A(f)| = 0 \text{ for } f \geq F \tag{7.12.12}$$

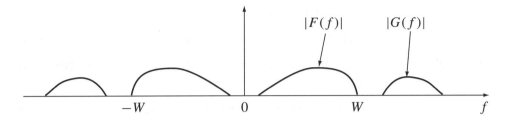

FIGURE 7.12.1
Non-overlapping Fourier spectra of two signals.

By using Bedrosian's theorem we get

$$v(t) = H[u(t)] = A(t)\sin(\Omega t + \Phi) \tag{7.12.13}$$

Therefore, the amplitude-modulated signal (7.12.10) is a real part of the the analytic signal:

$$\Psi(t) = A(t)e^{j(\Omega t + \Phi)} \tag{7.12.14}$$

and has a geometrical representation in the form of a phasor of instantaneous amplitude $A(t)$ and rotating with a constant angular velocity Ω. Bedrosian's theorem was extended by Nuttal and Bedrosian [25] to include "frequency-translated" analytic signals. The condition, which applies to vanishing spectra at negative frequencies, can be applied more generally to signals whose Fourier spectra satisfy the condition

$$F(\omega) = F\big[\Psi_1(t)\big] = 0, \qquad \omega < -a$$

$$G(\omega) = f\big[\Psi_2(t)\big] = 0, \qquad \omega > a \tag{7.12.15}$$

where a is an arbitrary positive constant. The extension of Bedrosian's theorem for multidimensional signals is given in Section 7.22. ∎

7.13   Multiplication of analytic signals

The Hilbert transform of the analytic signal is given by the formula

$$H[\psi(t)] = H\big[u(t) + j\,H[u(t)]\big] = H[u(t)] - j\,u(t) = -j\psi(t) \tag{7.13.1}$$

where the formula $H[H[u(t)]] = -u(t)$ (iteration) (see Table 7.7.3) has been applied. The Hilbert transform of the product of two analytic signals is given by the formula

$$H\left[\psi_1(t)\psi_2(t)\right] = \psi_1(t)H\left[\psi_2(t)\right] = \psi_2(t)H\left[\psi_1(t)\right] \tag{7.13.2}$$

that is, the Hilbert transformation should be applied to **one term of the product only** (to the first or the second).

PROOF   The product of two analytic functions is an analytic function [39]. Therefore, if

$$\psi(t) = \psi_1(t)\psi_2(t) \tag{7.13.3}$$

where $\psi_1(t)$ and $\psi_2(t)$ are analytic signals, then using Eq. (7.13.1) we get:

$$H[\psi(t)] = -j\psi(t) = -j\psi_1(t)\psi_2(t) \qquad (7.13.4)$$

However, the operator $-j$ may be assigned either to $\psi_1(t)$ or $\psi_2(t)$. The application of Eq. (7.13.2) yields two options:

$$H[\psi(t)] = H[\psi_1(t)]\psi_2(t); \qquad H[\psi] = \psi_1(t)H[\psi_2(t)] \qquad (7.13.5)$$

Let us apply Eqs. (7.13.1) and (7.13.5) to find the Hilbert transforms of the n-th power of the analytic signal. We get

$$H\left[\psi^2(t)\right] = \psi(t)H[\psi(t)] = -j\psi^2(t) \qquad (7.13.6)$$

$$H\left[\psi^n(t)\right] = \psi^{n-1}(t)H[\psi(t)] = -j\psi^n(t) \qquad (7.13.7)$$

∎

Example

Let us find the Hilbert transform of

$$\psi^2(t) = (1 - jt)^{-2} \qquad (7.13.8)$$

The application of Eq. (7.13.1) gives

$$H[\psi(t)] = -j(1 - jt)^{-1} \qquad (7.13.9)$$

and Eq. (7.13.6) yields

$$H\left[\psi^2(t)\right] = (1 - jt)^{-1}\left[-j(1 - jt)^{-1}\right] = -j(1 - jt)^{-2} \qquad ∎$$

Equation (7.13.1) has a generalized form given by the formula

$$H[\psi(at)] = -j\,\mathrm{sgn}(a)\psi(at) \qquad (7.13.10)$$

where a is a real positive or negative constant. The negative sign of a may be interpreted as time reversal. For example, the Hilbert transform of $\exp(j\omega t)$ is

$$H\left(e^{j\omega t}\right) = -j\,\mathrm{sgn}(\omega)e^{j\omega t}$$

where ω may be positive or negative.

7.14 Hilbert transforms of Bessel functions of the first kind

The Bessel functions (see also Chapter 1) are the solution of the second order Bessel differential equation:

$$z^2\psi''(z) + z\psi'(z) + \left(z^2 - \lambda^2\right)\psi(z) = 0 \qquad (7.14.1)$$

where $\psi(z)$ is a complex function of a complex variable $z = t + j\tau$ and λ is a complex constant. If $\lambda = n$, where n is an integer $(0, 1, 2, \ldots)$, and $z = t$, we get the solution in the form of Bessel functions of the first kind of the order n denoted $J_n(t)$. They find numerous applications in signal and system theory. For example, they are used to calculate the Fourier spectra of frequency modulated signals.

The substitution in Eq. (7.14.1) of a solution in the form of a series $J_n(t) = \sum_{m=0}^{\infty} a_m x^m$ gives the power series representation

$$J_n(t) = \sum_{k=0}^{\infty} \frac{(-1)^k}{k!(n-k)!}(t/2)^{n+2k}; \qquad -\infty < t < \infty \tag{7.14.2}$$

The computation of the Bessel functions by means of this power series is inconvenient. Due to the truncation of the series at some value of k, we get divergence for large values of t. It is possible to apply Eq. (7.14.2) up to $t < t_1$ and calculate the values for $t > t_1$ using the asymptotic formula

$$J_n(t) = \frac{2}{\pi t} \sin\left(t - \frac{\pi n}{2} + \frac{\pi}{4}\right) + \frac{r(t)}{t\sqrt{t}} \tag{7.14.3}$$

The term $r(t)$ is a limited function for $t \Rightarrow \infty$. However, it is much **easier** to compute the Bessel functions and its Hilbert transforms using **integral forms**, as decribed below.

Let us start with the periodic complex function $\exp(jt \sin(\varphi))$ and its Hilbert transform. We have a Hilbert pair

$$e^{jt \sin(\varphi)} \overset{\text{H}}{\Longleftrightarrow} H\left[e^{jt \sin(\varphi)}\right] = -j \, \text{sgn}[\sin(\varphi)] e^{jt \sin(\varphi)} \tag{7.14.4}$$

The Fourier series expansion of the left-hand side is

$$e^{jt \sin \varphi} = \sum_{n=-\infty}^{\infty} J_n(t) e^{jn\varphi} \tag{7.14.5}$$

The Bessel functions, that is, the coefficients of this series are given by the integral:

$$J_n(t) = \frac{1}{2\pi} \int_{-\pi}^{\pi} e^{j(t \sin \varphi - n\varphi)} \, d\varphi \tag{7.14.6}$$

The odd-ordered Bessel functions are odd functions of the argument t, while the even-ordered are even functions and

$$J_{-n}(t) = (-1)^n J_n(t) \tag{7.14.7}$$

In fact, the integral of the imaginary part of Eq. (7.14.6) equals zero, and due to the evenness of the real part of the integrand we have

$$J_n(t) = \frac{1}{\pi} \int_{0}^{\pi} \cos[t \sin \varphi - n\varphi] \, d\varphi \tag{7.14.8}$$

This formula enables very efficient calculation of Bessel functions $J_n(t)$ using numerical integration. The number of integration steps may be halved using two separate integrals:

$$J_{2n}(t) = \frac{2}{\pi} \int_{0}^{\pi/2} \cos(t \sin \varphi) \cos(2n\varphi) \, d\varphi \tag{7.14.9}$$

$$J_{2n+1}(t) = \frac{2}{\pi} \int_{0}^{\pi 2} \sin(t \sin \varphi) \sin[(2n+1)\varphi] \, d\varphi \tag{7.14.10}$$

The real part of the Fourier series (7.14.5) is

$$\cos(t \sin \varphi) = J_0(t) + 2 \sum_{n=1}^{\infty} J_{2n}(t) \cos(2n\varphi) \tag{7.14.11}$$

and the imaginary part is

$$\sin(t \sin \varphi) = 2 \sum_{n=1}^{\infty} J_{2n-1}(t) \sin[(2n-1)\varphi] \tag{7.14.12}$$

Inserting $\varphi = \pi/2$ gives the well-known formulae:

$$\cos t = J_0(t) - 2J_2(t) + 2J_4(t) - \cdots \tag{7.14.13}$$

$$\sin t = 2J_1(t) - 2J_3(t) + \cdots \tag{7.14.14}$$

The following recursion formula is very useful

$$(2n/t)J_n(t) = J_{n-1}(t) + J_{n+1}(t) \tag{7.14.15}$$

The derivative of a Bessel function is also given by the recursion formula

$$2\dot{J}_n(t) = J_{n-1}(t) - J_{n+1}(t) \tag{7.14.16}$$

For example

$$\dot{J}_0(t) = 0.5[J_{-1}(t) - J_1(t)] = -J_1(t) \tag{7.14.17}$$

(we used Eq. (7.14.7)).

The left-hand side of Eq. (7.14.4) was expanded in the Fourier series (7.14.5). Similarly, due to the linearity of the Hilbert transformation, the right-hand side may be expanded in the Fourier series

$$H\left[e^{jt \sin \varphi}\right] = -j \, \text{sgn}(\varphi) e^{jt \sin \varphi} = \sum_{n=-\infty}^{\infty} \hat{J}_n(t) e^{jn\varphi} \tag{7.14.18}$$

where $\hat{J}_n(t) = H[J_n(t)]$ are the Hilbert transforms of the Bessel functions. For these functions we have the relation

$$\hat{J}_{-n}(t) = (-1)^{n+1} \hat{J}_n(t) \tag{7.14.19}$$

because the Hilbert transforms of odd functions are even, and vice versa (compare with Eq. [7.14.7]). The functions $\hat{J}_n(t)$, that is, the coefficients of the Fourier series (7.14.18), are given by the integral

$$\hat{J}_n(t) = \frac{1}{2\pi} \int_{-\pi}^{\pi} H\left[e^{j[t \sin \varphi - n\varphi]}\right] d\varphi \tag{7.14.20}$$

As in Eq. (7.14.6), the integral of the imaginary part equals zero and due to the evenness of the real part we have

$$\hat{J}_n(t) = \frac{1}{\pi} \int_{0}^{\pi} \sin[t \sin \varphi - n\varphi] \, d\varphi \tag{7.14.21}$$

Notice that the integrand is even, because it is multiplied by sgn(φ) (see Eq. [7.14.18]). As before, using numerical integration, the Hilbert transforms of the Bessel functions can be easily computed. The first five Bessel functions and their Hilbert transforms computed using Eqs. (7.14.8) and (7.14.21) are shown in Figure 7.14.1.

Let us derive the Hilbert transforms of the Bessel functions $J_n(t)$ using Fourier transforms. The Fourier transform of the function $J_0(t)$ is

$$J_0(t) \xLeftrightarrow{F} C_0(f) = \begin{cases} \frac{2}{(1-\omega^2)^{0.5}} & \text{for } |\omega| \leq 1 \\ 0 & \text{for } |\omega| > 1 \end{cases} \tag{7.14.22}$$

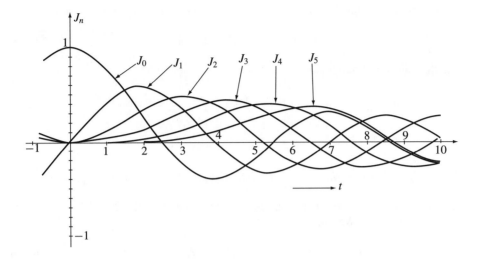

FIGURE 7.14.1
(a) Waveforms of the first five Bessel functions $J_n(t)$. *Continued*.

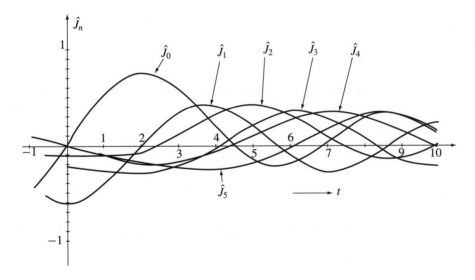

FIGURE 7.14.1
(b) Waveforms of the corresponding Hilbert transforms.

PROOF   Let us find the inverse transform of this spectrum:

$$J_0(t) = \frac{1}{2\pi} \int_{-1}^{1} \frac{2}{(1-\omega^2)^{0.5}} \cos(\omega t)\, d\omega = \left\{ \begin{array}{l} \omega = \sin(\varphi) \\ d\omega = \cos(\varphi)d\varphi \end{array} \right\}$$

$$= \frac{1}{\pi} \int_{0}^{\pi} \cos[t \sin \varphi]\, d\varphi \qquad\qquad (7.14.23)$$

(See Eq. [7.14.8].)   The Fourier transforms of higher-order Bessel functions can be calculated using the recursion formula (7.14.16) and frequency domain differentiation. We have

$$J_{n+1}(t) = J_{n-1}(t) - 2\dot{J}_n(t) \tag{7.14.24}$$

obtaining the following Fourier pairs

$$J_0(t) \xLeftrightarrow{F} C_0(f) = \frac{2}{(1-\omega^2)^{0.5}} \tag{7.14.25}$$

$$J_1(t) = -\dot{J}_0(t) \xLeftrightarrow{F} -j\omega C_0(f) \tag{7.14.26}$$

$$J_2(t) = J_0(t) - 2\dot{J}_1(t) \xLeftrightarrow{F} C_0(f) - 2j\omega C_1(f) \tag{7.14.27}$$

Successive application of the reccurency gives the Fourier spectra of the Bessel functions $J_n(t)$ tabulated in Table 7.14.1. We find that

$$J_n(t) \xLeftrightarrow{F} C_n(f) = (-j)^n 2^{n-1} T_n(t) C_0(f) \tag{7.14.28}$$

where $C_0(f)$ is defined by Eq. (7.14.25) and $T_n(t)$ is a Tchebycheff polynomial defined by the formula

$$T_n(t) = \cos[n \cos^{-1}(t)]; \quad n = 0, 1, 2, \ldots \tag{7.14.29}$$

A recursion formula can be applied

$$T_{n+1}(t) - 2t T_n(t) + T_{n-1}(t) = 0; \qquad n = 1, 2, \ldots \tag{7.14.30}$$

Because we derived the analytical expressions for the Fourier images of the Bessel functions, the use of inverse Fourier transformations enables the evaluation of either the Bessel function $J_n(t)$ or its Hilbert transform $\hat{J}_n(t)$. For example

$$J_0(t) = \frac{1}{\pi} \int_0^1 \frac{2}{(1-\omega^2)^{0.5}} \cos(\omega t) \, d\omega \tag{7.14.31}$$

and the Hilbert transform is

$$\hat{J}_0(t) = H\left[J_0(t)\right] = \frac{1}{\pi} \int_0^1 \frac{2}{(1-\omega^2)^{0.5}} \sin(\omega t) \, d\omega \tag{7.14.32}$$

Hence we have an analytic signal

$$\psi_0(t) = J_0(t) + j\hat{J}_0(t) \tag{7.14.33}$$

Equations (7.14.31) and (7.14.32) may be regarded as alternative definitions of the Bessel functions $J_0(t)$ and $\hat{J}_0(t)$. However, the computation by means of the integrals (7.14.8) and (7.14.9) ($n = 0$) gives much better accuracy with a given number of integration steps.

The expressions for the Fourier images of Bessel functions and their Hilbert transforms derived using these images are listed in Table 7.14.1. If needed, the Fourier spectra enable the the derivation of the coefficients of the power series representation of $J_n(t)$ and $\hat{J}_n(t)$. Starting with the power series for $J_n(t)$ given by Eq. (7.14.2) let us derive the power series for $\hat{J}_n(t)$. We start with the expression defining the Taylor series

$$\hat{J}_n(t) = \sum_{n=0}^{\infty} \frac{\hat{J}_n^{(n)}(t=0)}{n!} t^n \tag{7.14.34}$$

TABLE 7.14.1
Fourier and Hilbert Transforms of Bessel Functions of the First Kind.

Bessel Function	Fourier Transform			Hilbert Transform
$J_n(t)$	$C_n(f)$			$\hat{J}_n(t) = H[J_n(t)]$
$J_0(t)$	C_0	$= \frac{2}{(1-\omega^2)^{0.5}};$	$\lvert\omega\rvert < 1$	$\frac{1}{\pi}\int_0^1 C_0(f)\sin(\omega t)\,d\omega$
		$= 0;$	$\lvert\omega\rvert > 0$	
$J_1(t)$	$C_1 = -j\omega C_0$			$-\frac{1}{\pi}\int_0^1 \lvert C_1(f)\rvert \cos(\omega t)\,d\omega$
$J_2(t)$	$C_2 = -(2\omega^2 - 1)C_0$			$-\frac{1}{\pi}\int_0^1 \lvert C_2(f)\rvert \sin(\omega t)\,d\omega$
$J_3(t)$	$C_3 = j(4\omega^3 - 3\omega)C_0$			$\frac{1}{\pi}\int_0^1 \lvert C_3(f)\rvert \cos(\omega t)\,d\omega$
$J_4(t)$	$C_4 = (8\omega^4 - 8\omega^2 + 1)C_0$			$\frac{1}{\pi}\int_0^1 \lvert C_4(f)\rvert \sin(\omega t)\,d\omega$
$J_5(t)$	$C_5 = -j(16\omega^5 - 20\omega^3 + 5\omega)C_0$			$-\frac{1}{\pi}\int_0^1 \lvert C_5(f)\rvert \cos(\omega t)\,d\omega$
$J_6(t)$	$C_6 = -(32\omega^6 - 48\omega^4 + 18\omega^2 - 1)C_0$			$-\frac{1}{\pi}\int_0^1 \lvert C_6(f)\rvert \sin(\omega t)\,d\omega$

. . . .

$J_n(t)$	$C_n = (-j)^n 2^{n-1} T_n(\omega)C_0$	$\frac{(-1)^{n/2}}{\pi}\int_0^1 \lvert C_n(f)\rvert \sin(\omega t)\,d\omega$
		for $n = 0, 2, 4, \ldots$
		$\frac{(-1)^{(n+1)/2}}{\pi}\int_0^1 \lvert C_n(f)\rvert \cos(\omega t)\,d\omega$
		for $n = 1, 3, 5, \ldots$

$T_n(\omega) = \cos[n\cos^{-1}(\omega)]$ is the Tchebycheff polynomial

The derivatives $\hat{J}_n^{(n)}(t)$ ($t = 0$) can be obtained by differentiation of the integrand of the integrals listed in Table 7.14.1. By inserting $t = 0$, we obtain

$$\hat{J}_0(0) = \frac{1}{\pi}\int_0^1 \frac{2d\omega}{(1-\omega^2)^{0.5}}\sin(0) = 0$$

$$\hat{J}_0^{(1)}(0) = \frac{1}{\pi}\int_0^1 \frac{2\omega d\omega}{(1-\omega^2)^{0.5}}\cos(0) = \frac{2}{\pi}$$

$$\hat{J}_0^{(2)}(0) = -\frac{1}{\pi}\int_0^1 \frac{2\omega^2 d\omega}{(1-\omega^2)^{0.5}}\sin(0) = 0$$

$$\hat{J}_0^{(3)}(0) = -\frac{1}{\pi}\int_0^1 \frac{2\omega^3 d\omega}{(1-\omega^2)^{0.5}}\cos(0) = \frac{-4}{3\pi} \tag{7.14.35}$$

where (1), (2), ... denote the order of the derivative. Continuing the differentiation using Eq. (7.14.34), we get the following power series:

$$\hat{J}_0(t) = \frac{2}{\pi}\left[t - \frac{1}{9}t^3 + \frac{1}{225}t^5 - \frac{2}{33075}t^7 \right.$$

$$\left. + \cdots + \frac{(-1)^{(3+n)/2}2^{n-2}}{n!(1.3.5..n)}t^n + \cdots\right] \tag{7.14.36}$$

In the same way one can derive the power series of higher order Hilbert transforms of the Bessel functions. ∎

7.15 The instantaneous amplitude, complex phase, and complex frequency of analytic signals

Signal theory needs precise definitions of various quantities such as the **instantaneous amplitude**, instantaneous **phase**, and instantaneous **frequency** of a given signal and many other quantities. Let us recall that **neither definition is true or false**. If we define something, we simply propose to make an **agreement** to use a specific name in the sense of the definition. When using this name, for instance, "instantaneous frequency," we should never forget what we have defined. The history of signal theory contains examples of misunderstanding, when various authors applied the same name—instantaneous frequency—to different definitions and then tried to discuss which is true or false. Such a discussion is **meaningless**. Of course one may discuss which definition has advantages or disadvantages from a specific point of view or whether it is compatible with other definitions or existing knowledge. The notions of the instantaneous amplitude, instantaneous phase, and instantaneous frequency of the analytic signal $\psi(t) = u(t) + jv(t)$ may be uniquely and conveniently defined introducing the notion of a phasor rotating in the Cartesian (u, v) plane, as shown in Figure 7.15.1. The change of coordinates from rectangular (u, v) to polar (A, φ) gives

$$u(t) = A(t)\cos[\varphi(t)] \tag{7.15.1}$$

$$v(t) = A(t)\sin[\varphi(t)] \tag{7.15.2}$$

$$\psi(t) = A(t)e^{j\varphi(t)} \tag{7.15.3}$$

We define the instantaneous amplitude of the analytic signal equal to the length of the phasor (radius vector) A:

$$A(t) = \sqrt{u^2(t) + v^2(t)} \tag{7.15.4}$$

and define the instantaneous phase of the analytic signal equal to the instantaneous angle

$$\varphi(t) = \mathrm{Tan}^{-1}\frac{v(t)}{u(t)} \tag{7.15.5}$$

The notation with capital T indicates the multibranch character of the Tan^{-1} function, as shown in Figure 7.15.2. As time elapses, the phasor rotates in the (u, v) plane and its instantaneous angular speed defines the **instantaneous angular frequency** of the analytic signal given by the time derivative

$$\dot{\varphi}(t) = \Omega(t) = 2\pi F(t) \tag{7.15.6}$$

or

$$\Omega(t) = \frac{d}{dt}\tan^{-1}\frac{v(t)}{u(t)} = \frac{u(t)\dot{v}(t) - v(t)\dot{u}(t)}{u^2(t) + v^2(t)} \tag{7.15.7}$$

Notice the anticlock direction of rotation for positive angular frequencies. The instantaneous frequency is defined by the formula

$$F(t) = \frac{\Omega(t)}{2\pi} = \frac{1}{2\pi}\dot{\varphi}(t) \tag{7.15.8}$$

Summarizing, using the notion of the analytic signal, we defined the instantaneous amplitude, phase, and frequency. A number of different definitions of the notion of instantaneous amplitude, phase, and frequency have developed over the years. There are many pairs of functions $A(t)$ and $\varphi(t)$, which inserted into Eq. (7.15.1) reconstruct a given signal $u(t)$, for example,

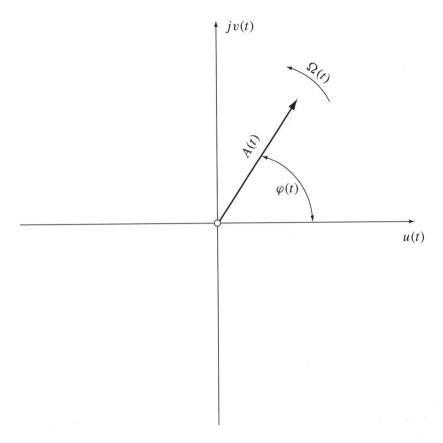

FIGURE 7.15.1
A phasor in the Cartesian (u, v) plane representing the analytic signal $\psi(t) = u(t) + jv(t) = A(t)e^{j\varphi(t)}$.

functions defining a phasor in the phase plane $[u(t), \dot{u}(t)]$. But only the analytic signal has the unique feature of having a one-sided Fourier spectrum. Let us recall that a real signal and its Hilbert transform are given in terms of analytic signals by Eqs. (7.2.17) and (7.2.18) (see Section 7.2). Figure 7.15.3 shows the geometrical representation of these formulae in the form of two phasors of a length $0.5A(t)$ and opposite direction of rotation, positive for $\psi(t)$ and negative for $\psi^*(t)$. Equation (7.15.5) defines the instantaneous frequency of a signal regardless of the bandwidth. It is sometimes believed that the notion of instantaneous frequency has a physical meaning only for narrow-band signals (high-frequency (HF) modulated signals). However, using adders, multipliers, dividers, Hilbert filters, and differentiators, it is possible to implement a frequency demodulator used for wide-band signals, for example, speech signals, the algorithm defined by Eq. (7.15.7). Modern VLSI enables efficient implementation of such frequency demodulators at reasonable cost.

Instantaneous complex phase and complex frequency

Signal and systems theory widely uses the Laplace transformation of a real signal $u(t)$ of the form

$$U(s) = \int_0^\infty u(t)e^{-st}\, dt \tag{7.15.9}$$

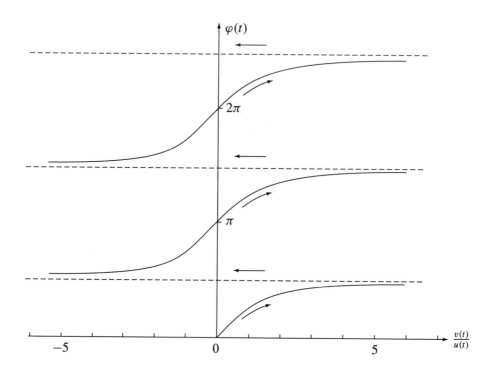

FIGURE 7.15.2
The multi-branch function $\varphi(t) = \tan^{-1}[u(t)/v(t)]$. As time elapses (arrows) they are jumps from one branch to a next branch.

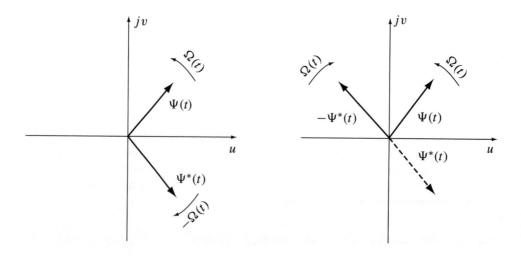

FIGURE 7.15.3
A pair of conjugate phasors representing the Eqs. (7.2.17) and (7.2.18).

where

$$s = \alpha + j\omega; \qquad \omega = 2\pi f$$

is a time-independent complex frequency (α and ω are real). The exponential kernel e^{-st} has the form of a harmonic wave with an exponentially decaying amplitude, that is, its instantaneous amplitude is

$$A(t) = e^{-\alpha t} \tag{7.15.10}$$

The notion of the complex frequency has been generalized by this author in 1964 by defining a complex instantaneous variable frequency using the notion of the analytic signal [12]. It is convenient to define the **instantaneous complex frequency** as the time derivative of a complex phase. The instantaneous complex phase of the analytic signal $\psi(t)$ is defined by the formula

$$\Phi_c(t) = \text{Ln}[\psi(t)] \tag{7.15.11}$$

Capital L denotes the multibranch character of the logarithmic function of the complex function $\psi(t)$. The insertion of the polar form of the analytic signal (see Eq. [7.15.3]) yields:

$$\Phi_c(t) = \text{Ln}[A(t)] + j\varphi(t) \tag{7.15.12}$$

The instantaneous complex frequency is defined by the derivative

$$s(t) = \dot{\Phi}_c(t) = \frac{\dot{A}(t)}{A(t)} + j\omega(t) \tag{7.15.13}$$

or

$$s(t) = \alpha(t) + j\omega(t) \tag{7.15.14}$$

where

$$\alpha(t) = \frac{\dot{A}(t)}{A(t)} \tag{7.15.15}$$

is the instantaneous radial frequency (a measure of the radial velocity representing the speed of changes of the radius or amplitude of the phasor), and

$$\omega(t) = \dot{\varphi}(t) \tag{7.15.16}$$

is the instantaneous angular frequency. Equation (7.15.15) has the form of a first-order differential equation. The solution of this equation yields the following form of the instantaneous amplitude

$$A(t) = A_0 e^{\int_0^t \alpha(t)\,dt} \tag{7.15.17}$$

A_0 is the value of the amplitude at the momement $t = 0$. Let us introduce the notation

$$\beta(t) = \int_0^t \alpha(t)\,dt \tag{7.15.18}$$

Using this notation the complex phase can be written as

$$\Phi_c(t) = \ln A_0 + \beta(t) + j\varphi(t) \tag{7.15.19}$$

or

$$\Phi_c(t) = \ln A_0 + \int_0^t s(t)\,dt + j\Phi_0 \tag{7.15.20}$$

Φ_0 is the integration constant or the angular position of the phasor at $t = 0$. The introduction of the concept of a complex constant $\psi_0 = A_0 e^{j\Phi_0}$ gives the following form of the analytic signal

$$\Psi(t) = \psi_0 e^{\int_0^t s(t)\, dt} \tag{7.15.21}$$

Examples

1. Consider the analytic signal given by Eq. (7.5.10):

$$\psi_\delta(t) = \underbrace{\frac{\alpha}{\pi(\alpha^2 + t^2)}}_{u(t)} + j \underbrace{\frac{t}{\pi(\alpha^2 + t^2)}}_{v(t)} \tag{7.15.22}$$

The polar form of this signal is

$$\psi_\delta(t) = \underbrace{\frac{1}{\pi\sqrt{\alpha^2 + t^2}}}_{A(t)} \exp\underbrace{\left[j\tan^{-1}(t/\alpha)\right]}_{\varphi(t)} \tag{7.15.23}$$

Therefore, the instantaneous complex phase is

$$\Phi_c(t) = \text{Ln}\underbrace{\left\{\frac{1}{\pi\sqrt{\alpha^2 + t^2}}\right\}}_{\beta(t)} + j\underbrace{\tan^{-1}(t/\alpha)}_{\varphi(t)} \tag{7.15.24}$$

and the instantaneous complex frequency is

$$s(t) = \dot{\Phi}_c(t) = \underbrace{\frac{-t}{\alpha^2 + t^2}}_{\alpha(t)} + j\underbrace{\frac{\alpha}{\alpha^2 + t^2}}_{\omega(t)} \tag{7.15.25}$$

Because in the limit $\alpha \Rightarrow 0$ the signal (7.15.22) approximates the complex delta distribution (see Eq. [7.5.4]), the instantaneous complex phase of this distribution is

$$\Phi_{c\delta}(t) = \text{Ln}\underbrace{\frac{1}{\pi|t|}}_{A(t)} + j\underbrace{0.5\pi\,\text{sgn}(t)}_{\varphi(t)} \tag{7.15.26}$$

and the complex frequency is

$$s_\delta(t) = \underbrace{\frac{-1}{t}}_{\alpha(t)} + j\underbrace{\pi\delta(t)}_{\omega(t)} \tag{7.15.27}$$

2. Consider the analytic signal

$$\psi(t) = \underbrace{\frac{\sin(at)}{at}}_{u(t)} + j\underbrace{\frac{\sin^2(0.5at)}{0.5at}}_{v(t)} \tag{7.15.28}$$

where $u(t)$ is the well-known interpolating function of the sampling theory. Equations (7.15.4) and (7.15.5) yield, using trigonometric relations, the polar form of this signal:

$$\psi(t) = \underbrace{\left|\frac{\sin(0.5at)}{0.5at}\right|}_{A(t)} \exp\underbrace{(jat/2)}_{\varphi(t)} \tag{7.15.29}$$

Therefore, the instantaneous complex phase is

$$\psi(t) = \mathrm{Ln} \underbrace{\left| \frac{\sin(0.5at)}{0.5at} \right|}_{A(t)} + jat/2 \tag{7.15.30}$$

and the instantaneous complex frequency

$$s(t) = \frac{a}{2} \cot(0.5at) - \frac{1}{t} + j\frac{a}{2} \tag{7.15.31}$$

In conclusion, the interpolating function may be regarded as a signal of a variable amplitude and a constant angular frequency $\omega = a/2$.

3. The classic complex notation of a frequency or phase modulated signal (Carson and Fry, 1937) has the form [41]

$$\psi(t) = A_0 e^{j[\Omega_0 t + \Phi_0 + \varphi(t)]}; \qquad \Omega_0 = 2\pi F_0 \tag{7.15.32}$$

where $\varphi(t)$ represents the angle modulation. The whole argument of the exponential function $\Phi(t) = \Omega_0 t + \Phi_0 + \varphi(t)$ defines the instantaneous phase and its derivative, the instantaneous frequency

$$F(t) = \frac{1}{2\pi} \frac{d\Phi}{dt} = F_0 + \frac{1}{2\pi} \frac{d\varphi}{dt} \tag{7.15.33}$$

The signal (7.15.32) is represented by a phasor in the plane $(\cos[(\Phi(t)], \sin[\Phi(t)])$, as shown in Figure 7.15.4. These definitions of the instantaneous phase and frequency differ from the definition using the analytic signal that is represented by a phasor in the $(\cos[\Phi(t)], H(\cos[\Phi(t)])$ plane, because $\sin[\Phi(t)]$ is not the Hilbert transform of $\cos[\Phi(t)]$ and the signal (7.15.32) is not an analytic function. However, it may be nearly analytic if the carrier frequency is large. If the spectra of the functions $\cos[\varphi(t)]$ and $\sin[\varphi(t)]$ have a limited low-pass support of a highest frequency $|W| < |F_0|$, then Bedrosian's theorem (see Section 7.12) may be applied and

$$H \{\cos[\Omega_0 t + \Phi_0 + \varphi(t)]\} = \cos[\varphi(t)]H[\cos(\Omega_0 t + \Phi_0)]$$

$$- \sin[\varphi(t)]H[\sin(\Omega_0 t + \Phi_0)]$$

$$= \sin[\Omega_0 t + \Phi_0 + \varphi(t)] \tag{7.15.34}$$

In the case of harmonic modulation with $\varphi(t) = \beta \sin(\omega t)$, where β is the modulation index, the spectra of the functions $\cos[\varphi(t)]$ and $\sin[\varphi(t)]$ are given by the Fourier series

$$\cos[\beta \sin(\omega t)] = J_0(\beta) + 2 \sum_{n=1}^{\infty} J_{2n}(\beta) \cos(2n\omega t) \tag{7.15.35}$$

$$\sin[\beta \sin(\omega t)] = 2 \sum_{n=1}^{\infty} J_{2n-1}(\beta) \sin[(2n-1)\omega t] \tag{7.15.36}$$

and this is not a pair of Hilbert transforms (see Section 7.6). Although the number of terms of the series is infinite, the number of significant terms is limited and for a good approximation Bedrosian's theorem may be applied for large values of F_0. Further comments are given in [25]. ∎

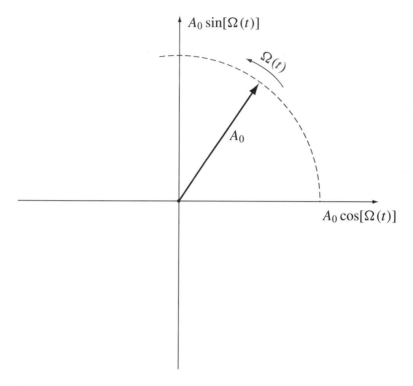

FIGURE 7.15.4
A phasor representing a frequency (or phase) modulated signal.

7.16   Hilbert transforms in modulation theory

This section is devoted to the theory of analog modulation of a harmonic carrier $u_c(t) = A_0 \cos(2\pi F_0 t + \Phi_0)$ with emphasis on the role of Hilbert transformation, analytic signals, and complex frequencies. The theory of amplitude and angle modulation is mentioned briefly in favor of a more detailed description of the theory of single side-band modulations. The last are conveniently defined using Hilbert transforms. Many modulators are implemented using Hilbert filters, mostly digital filters, because nowadays modulated signals can be conveniently generated digitally and converted into analog signals.

The concept of the modulation function of a harmonic carrier

The complex notation of signals is widely used in modern modulation theory. The harmonic carrier is written in the form of the analytic signal

$$\Psi_c(t) = A_0 e^{j(\Omega_0 t + \Phi_0)} \tag{7.16.1}$$

Analog modulation is the operation of continuous change of one or more of the three parameters of the carrier: the amplitude A_0, the frequency F_0, or the phase Φ_0, resulting in amplitude, frequency, or phase modulation. The complex modulated signal has a convenient representation in the form of a product [3]

$$\Psi(t) = \gamma(t)\Psi_c(t) = A_0\gamma(t)e^{j(\Omega_0 t + \Phi_0)} \tag{7.16.2}$$

The function $\gamma(t)$ is called the **modulation function**. It is a function of the modulating signal (the message) $x(t)$, that is, $\gamma(t) = \gamma[x(t)]$. Any kind of modulation, for example, amplitude, frequency, or phase modulation, is represented by a specific real or complex modulation function. We shall investigate models of modulating signals for which the Fourier transform exists and is given by the Fourier pair

$$x(t) \overset{F}{\Longleftrightarrow} X(\omega); \qquad \omega = 2\pi f \tag{7.16.3}$$

The frequency band containing the terms of the spectrum $X(\omega)$ is called the **baseband**. In general, the modulation function is a nonlinear function of the variable x, and the spectrum of the modulation function differs from $X(\omega)$ and is represented by the Fourier pair:

$$\gamma(t) \overset{F}{\Longleftrightarrow} \Gamma(\omega) \tag{7.16.4}$$

The nonlinear transformations of the spectrum may have a complicated analytic representation. Usually only approximate determination of the spectrum is possible. The approximations are easier to perform if the energy of the modulating signal is nonuniformly distributed and concentrated in the low-frequency part of the baseband, for example, the energy of voice, music, or TV signals. Usually it is possible to find the terms of $\Gamma(\omega)$ for harmonic modulating signals. In special cases, if the modulation function is proportional to the message, that is, $\gamma(t) = mx(t)$ (m is a constant) we have

$$\Gamma(f) = mX(f) \tag{7.16.5}$$

The initial phase of the carrier Φ_0 is of importance only if we deal with two or more modulated carriers of the same frequency, for example, by summation or multiplication of modulated signals. It is convenient to write the modulated signal in the form

$$\psi(t) = A_0 \gamma(t) e^{j\Phi_0} e^{j\Omega_0 t} \tag{7.16.6}$$

and define a **modified modulation function** in the form of the product

$$\gamma_1(t) = \gamma(t) e^{j\Phi_0} \tag{7.16.7}$$

The new Fourier spectrum is

$$\gamma_1(t) \overset{F}{\Longleftrightarrow} \Gamma_1(\omega) = \Gamma(\omega) e^{j\Phi_0} \tag{7.16.8}$$

We observe that the spectra (7.16.4) and (7.16.8) have the same magnitude and differ only by the phase relations. Notice that the spectrum $\Gamma_1(\omega)$ is defined at zero carrier frequency and the spectrum of the modulated signal is obtained by shifting this spectrum from zero to carrier frequency by the Fourier shift operator $e^{j\Omega_0 t}$. This approach enables us to study the spectra of modulated signals at zero carrier frequency.

Examples of modulation functions:

The modulation function for a linear full-carrier AM has the form

$$\gamma(t) = 1 + mx(t); \qquad |mx(t)| < 1 \tag{7.16.9}$$

The number 1 represents the carrier term. Therefore, the modulation function for balanced modulation (suppressed carrier) has the simple form

$$\gamma(t) = mx(t) \tag{7.16.10}$$

Therefore, the spectra of the message and of the modulation function are to within the scale

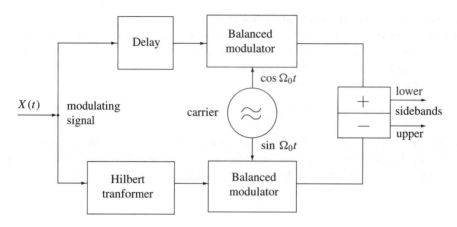

FIGURE 7.16.1
Block diagram of a SSB modulator (phase method) implementing Eq. (7.16.16).

factor m, the same. The message may be written in the form (see Eq. [7.2.17])

$$x(t) = \frac{\psi_x(t) + \psi_x^*(t)}{2} \qquad (7.16.11)$$

This formula shows that the upper sideband of the AM signal is represented by the analytic signal $\psi_x(t)$ of a one-sided spectrum at positive frequencies and the lower sideband by the conjugate analytic signal $\psi_x^*(t)$ of a one-sided spectrum at negative frequencies. The sidebands have the geometric form of two conjugate phasors (see Fig. 7.15.3). The instantaneous amplitude of the phasors is

$$A(t) = \frac{m}{2} |\psi_x(t)| = \frac{m}{2} \sqrt{x^2(t) + (\hat{x}(t))^2} \qquad (7.16.12)$$

$(\hat{x}(t) = H[x(t)])$ and the instantaneous angular frequency is

$$\omega_x(t) = \pm \frac{d}{dt} \tan^{-1}\left[\frac{\hat{x}(t)}{x(t)}\right] \qquad (7.16.13)$$

Therefore, a single sideband represents a signal with simultaneous amplitude and phase modulation. The multiplication of $\psi_x(t)$ or $\psi_x^*(t)$ with the complex carrier (Fourier shift operator) $e^{j\Omega_0 t}$ yields the high-frequency analytic signals. The upper sideband ($\Phi_0 = 0$) is (with $mA_0 = 2$)

$$\psi_{\text{upper}}(t) = \psi_x(t)e^{j\Omega_0 t} \qquad (7.16.14)$$

with the modulation function $\psi_x(t)$, and the lower sideband is

$$\psi(t) = \psi_x^*(t)e^{j\Omega_0 t} \qquad (7.16.15)$$

with the conjugate modulation function $\psi_x^*(t)$. The above signals represent the complex form of single side-band (SSB) AM. The real notation of these signals is

$$u_{\text{SSB}}(t) = x(t)\cos(\Omega t) \mp \hat{x}(t)\sin(\Omega t) \qquad (7.16.16)$$

with the minus sign for the upper sideband and plus sign for the lower one. The products $x(t)\cos(\Omega_0 t)$ and $\hat{x}(t)\sin(\Omega_0 t)$ represent double side-band (DSB) compressed carrier AM signals. Therefore, an SSB modulator may be implemented, as shown in Figure 7.16.1.

The **angle modulation** is represented by the exponential modulation function of the form

$$\gamma(t) = e^{j\varphi[x(t)]} \tag{7.16.17}$$

Therefore, the complex signal representation of the angle modulation has the form

$$\Psi(t) = A_0 e^{j[\Omega_0 t + \varphi(t)]} \tag{7.16.18}$$

where φ is a function of the modulating signal $x(t)$. In general, this complex signal may be only approximately analytic (see Section 7.15, example 3). In the case of a linear phase modulation, the modulation function has the form

$$\gamma(t) = e^{jmx(t)} \tag{7.16.19}$$

and for the linear frequency modulation

$$\gamma(t) = e^{jm \int_{-\infty}^{t} x(t)\, dt} \tag{7.16.20}$$

The Fourier spectrum of the modulation function is given by the integral

$$\Gamma(\omega) = \int_{-\infty}^{\infty} e^{j\varphi[x(t)]} e^{-j\omega t}\, dt \tag{7.16.21}$$

If for a specific function $\varphi[x(t)]$ the closed form of this integral does not exist, a numerical integration may be applied. In the simplest case of linear phase modulation with a harmonic modulating signal the modulation function (7.16.19) has the form

$$\gamma(t) = e^{j\beta \sin(\omega_0 t)} \tag{7.16.22}$$

where β is the modulation index (in radians). The Fourier series expansion of this complex periodic function has the form:

$$\gamma(t) = J_0(\beta) + \sum_{n=1}^{\infty} J_{2n}(\beta) \cos(2n\omega_0 t) + j \sum_{n=1}^{\infty} J_{2n-1}(\beta) \sin\left[(2n-1)\omega_0 t\right] \tag{7.16.23}$$

Using Euler's formulae (see Eqs. [7.2.19], [7.2.20]), this modulation function becomes

$$\gamma(t) = J_0(\beta) + \sum_{n=1}^{\infty} J_{2n}(\beta) \left[\frac{e^{j2n\omega_0 t} + e^{-j2n\omega_0 t}}{2} \right]$$

$$+ \sum_{n=1}^{\infty} J_{2n-1}(\beta) \left[\frac{e^{j(2n-1)\omega_0 t} - e^{-j(2n-1)\omega_0 t}}{2} \right] \tag{7.16.24}$$

Because the exponentials in the time domain are represented by delta functions in the frequency domain ($e^{\pm jn\omega_0 t} \overset{F}{\Longleftrightarrow} \delta(f \mp nf_0)$), the spectrum of the modulation function (zero carrier frequency) has the form shown in Figure 7.16.2 ($\beta = 4$).

Generalized single side-band modulations

The SSB AM signal defined by Eqs. (7.16.14) and (7.16.15) is an example of many other possible single side-band modulations. Any kind of modulation of a harmonic carrier is called

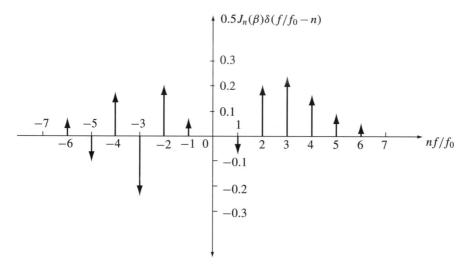

FIGURE 7.16.2
The spectrum of a phase modulated signal translated to zero carrier frequency, i.e., of the modulation function. Phase deviation $\beta = 4$ radians.

single side-band modulation if the modulation function is an analytic signal of a one-sided spectrum at positive frequencies for the upper sideband and at negative frequencies for the lower sideband. Therefore, the modulation function should have the form

$$\gamma(t) = \gamma_x(t) + j\hat{\gamma}_x(t) = A(t)e^{j\phi(t)} \tag{7.16.25}$$

where $\gamma_x(t) \overset{H}{\Longleftrightarrow} \hat{\gamma}_x(t)$. Let us use here the notion of the **instantaneous complex phase** defined by Eq. (7.15.11) of the form

$$\phi_c(t) = \ln A(t) + j\phi(t) \tag{7.16.26}$$

The modulation function (7.16.25) can be written in the form

$$\gamma(t) = e^{\phi_c(t)} = e^{\ln[A(t)] + j\phi(t)} \tag{7.16.27}$$

that is, the instantaneous amplitude is written in the exponential form

$$A(t) = e^{\ln[A(t)]} \tag{7.16.28}$$

We now put the question: under what conditions are $\gamma(t)$ and simultaneously $\phi_c(t)$ analytic? That is, when is not only the relation (7.16.25), but also the relation

$$\ln A(t) \overset{H}{\Longleftrightarrow} \phi(t) = \widehat{\text{Ln}}[A(t)] \tag{7.16.29}$$

satisfied? The answer comes from the dual (time domain) version of the **Paley-Wiener criterion** [28]

$$\int_{-\infty}^{\infty} \frac{|\operatorname{Ln}[A(t)]|}{1 + t^2} \, dt < \infty \qquad (7.16.30)$$

which should be satisfied. Let us remember that $A(t)$ is defined as a nonnegative function of time. The Paley-Wiener criterion is equivalent to a requirement that $A(t)$ should not approach zero faster than any exponential function. This is a property of each signal with finite bandwidth that is of any practical signal.

CSSB: compatible single side-band modulation

The CSSB signal has the same instantaneous amplitude as the conventional DSB full-carrier AM signal, that is, of the form

$$A(t) = A_o \left(1 + mx(t)\right); \qquad mx(t) < 1 \qquad (7.16.31)$$

and can be demodulated by a conventional linear diode demodulator (but not by a synchronous detector). The one-sided spectrum of the CSSB signal is achieved by a simultaneous specific phase modulation. The analytic modulation function should satisfy the requirement (7.16.29) and has the form

$$\gamma(t) = [1 + mx(t)]e^{j\hat{\ln}[1 + mx(t)]} \qquad (7.16.32)$$

Figure 7.16.3 shows a block diagram of a modulator producing a high-frequency CSSB signal implemented by the use of Eq. (7.16.32). This modulation function guarantees the exact cancellation of the undesired sideband. Using digital implementation, the level of the undesired sideband depends only on design. The bandwidth of the nonlinear logarithmic device, the Hilbert filter and phase modulator, should be several times wider than the bandwidth of the input signal. In practice it should be three to four times larger than the baseband. The instantaneous amplitude $A(t)$ should never fall to zero because the logarithm of zero equals minus infinity. Tradeoff is needed between the smallest value of A and the phase deviation.

The spectrum of the CSSB signal

It may be a surprise that the **bandwidth** of the one-sided spectrum of the CSSB signal **is limited**. If the spectrum of the modulating signal exists in the interval $-W < f < W$, then the spectrum of the modulation function exists in the interval $0 < f < 2W$. Seemingly, the bandwidths of the CSSB and DSB AM signals are equal. However, the spectra of many messages such as speech or video signals are nonuniform, with significant terms concentrated at the lower part of the baseband. This enables us to transmit the CSSB signal in a smaller band, for example, from F_0 to $F_0 + W$ instead to $F_0 + 2W$, at the cost of some distortions enforced by the truncation of insignificant terms of the spectrum. Let us investigate the spectra and distortions using the model of a wide-band modulating signal given in the form of the Fourier series

$$x(t) = \sum_{k=0}^{N} (-1)^k C_{2k+1} \cos[(2k + 1)\omega_0 t]; \qquad \omega_0 = 2\pi f_0 \qquad (7.16.33)$$

For $C_{2k+1} = 1/(2k + 1)$ this modulating signal is a truncated Fourier series of a square wave. Its bandwidth equals $W = (2N + 1)f_0$. The insertion of this signal in Eq. (7.16.31) yields a

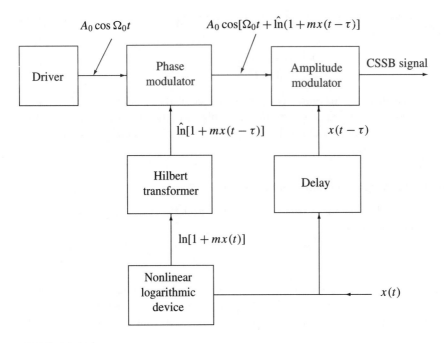

FIGURE 7.16.3
Diagram of the modulator producing the Compatible Single Side-band AM signal.

periodic modulation function given by the Fourier series

$$\gamma(t) = \sum_{k=0}^{4N+2} A_k e^{jk\omega_0 t} \tag{7.16.34}$$

The truncation of this series at the term $4N + 2$ is not arbitrary, because it will be shown that terms for $k > 4N + 2$ vanish. Therefore, the **bandwidth of** $\gamma(t)$ equals exactly **2W**. To give the evidence, let us insert $x(t)$ given by Eq. (7.16.33) in Eq. (7.16.32). The square of the instantaneous amplitude of so-defined modulation function is

$$A^2(t) = \left[1 + m \sum_{k=0}^{N} (-1)^k C_{2k+1} \cos\left[(2k + 1)\omega_0 t\right] \right]^2 \tag{7.16.35}$$

The highest term of this Fourier series has the harmonic number $4N + 2$. Analogously, the square of the instantaneous amplitude of the modulation function (7.16.34) is

$$A^2(t) = \left[\sum_{k=0}^{4N+2} A_k \cos(k\omega_0 t) \right]^2 + \left[\sum_{k=1}^{4N+2} A_k \sin(k\omega_0 t) \right]^2 \tag{7.16.36}$$

However, the functions (7.16.34) and (7.16.35) should be equal. Therefore, they should have the same coefficients of the Fourier series. The comparison of these coefficients yields a set of $4N+3$ equations. The solution of these equations yields the coefficients $A_0, A_1, A_2, \ldots, A_{2N+2}$ as functions of the modulation index m and the amplitudes C_{2k+1} of the modulating signal (7.16.31).

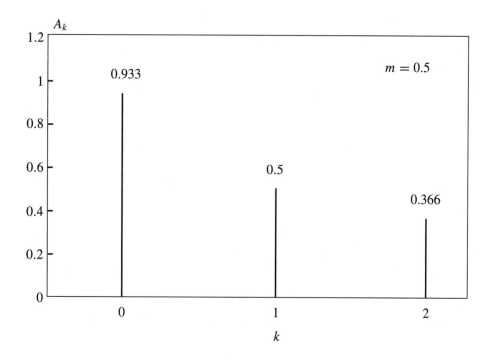

FIGURE 7.16.4
Example of a spectrum of the CSSB AM signal with a cosine envelope.

Example
1. For the harmonic modulating signal $x(t) = \cos(\omega_0 t)$, $N = 0$, $C_1 = 1$ and $C_{2k+1} = 0$ for $k > 0$. The comparison of the squares of the instantaneous amplitudes yields three equations

$$A_0^2 + A_1^2 + A_2^2 = \left(1 + m^2/2\right) C_1 \tag{7.16.37}$$

$$A_0 A_1 + A_1 A_2 = m C_1 \tag{7.16.38}$$

$$A_0 A_2 = (m C_1)^2/4 \tag{7.16.39}$$

The solution of these equations yields ($C_1 = 1$): The amplitude of the zero frequency carrier

$$A_0 = 0.5 + 0.5\sqrt{1 - m^2} \tag{7.16.40}$$

The amplitude of the first sideband

$$A_1 = m \tag{7.16.41}$$

and the amplitude of the second sideband

$$A_2 = 0.5 - 0.5\sqrt{1 - m^2} \tag{7.16.42}$$

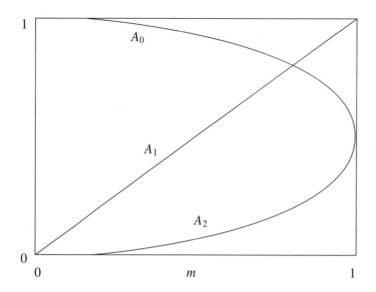

FIGURE 7.16.5
The dependence of the three terms of the spectrum depicted in Figure 7.16.5 on the modulation index m.

Figure 7.16.4 shows an example of the spectrum of the CSSB signal and Figure 7.16.5 the dependence of the amplitudes on m.

2. For the modulating signal $x(t) = C_1 \cos(\omega_0 t) - C_3 \cos(3\omega_0 t)$, $N = 1$, and $C_{2k+1} = 0$ for $k > 1$. We get seven equations of the form

$$\sum_{k=0}^{6} A_k^2 = 1 + \frac{m^2}{2}\left(C_1^2 + C_3^2\right) \tag{7.16.43}$$

$$\sum_{k=0}^{5} A_k A_{k+1} = mC_1; \qquad \sum_{k=0}^{4} A_k A_{k+2} = \frac{m^2}{2}\left(0.5C_1^2 - C_1 C_3\right) \tag{7.16.44}$$

$$\sum_{k=0}^{3} A_k A_{k+3} = -mC_3; \qquad \sum_{k=0}^{2} A_k A_{k+4} = \frac{-m^2}{2}C_1 C_3 \tag{7.16.45}$$

$$\sum_{k=0}^{1} A_k A_{k+5} = 0; \qquad \sum_{k=0}^{0} A_k A_{k+6} = \frac{m^2}{4}C_3^2 \tag{7.16.46}$$

The solutions of these equations yield the seven terms of the spectrum of the CSSB signal. In practice it is simpler to find these terms applying any numerical method of determination of the coefficients of the Fourier series expansion of the modulation function (7.16.32). However, the above set of equations gives the evidence that the spectrum has a finite number of terms. The above equations may be used to control the accuracy of numerical calculations. Notice that Eqs. (7.16.37) and (7.16.43) have the form of power equality equations.

Let us quote three other modulation functions generating CSSB AM signals. The analytic modulation function of the form

$$\gamma(t) = \sqrt{1 + mx(t)}\, e^{j\frac{1}{2}\hat{\ln}[1+mx(t)]} \tag{7.16.47}$$

uses the square root of the instantaneous amplitude of an AM signal. Its spectrum is exactly one-sided. A squaring demodulator should be applied at the receiver. The phase deviation equals one-half of the phase deviation of the function (7.16.32). Some years ago Kahn implemented a CSSB modulator using the modulation function [17]

$$\gamma(t) = [1 + mx(t)]\, e^{j\,\tan^{-1}\frac{m\hat{x}(t)}{1+mx(t)}} \tag{7.16.48}$$

Similarly Villard (1948) implemented a modulator using another modulation function [40]

$$\gamma(t) = (1 + mx(t))\, e^{jm\hat{x}(t)} \tag{7.16.49}$$

The last two modulation functions are not exactly analytic and their spectra are only approximately one-sided. ∎

Compatible single side-band modulation for angle detectors

The modulation function of a single side-band modulation compatible with a linear phase detector has the form

$$\gamma(t) = e^{-\beta\hat{x}(t)+j\beta x(t)} \tag{7.16.50}$$

and the modulation function of a single side-band modulation compatible with a linear frequency demodulator has the form

$$\gamma(t) = e^{-m_f H[\int x(t)\,dt]+jm_f\int x(t)\,dt} \tag{7.16.51}$$

where β and m_f are modulation indexes of phase or frequency modulation (in radians). The above modulation functions are analytic. Therefore, their spectra are exactly one-sided due to the simultaneous amplitude and angle modulation. Notice the exponential amplitude modulation function. For large modulation indexes the required dynamic range of the amplitude modulator is extremely large. An example is the modulating signal $x(t) = \sin(\omega_0 t)$. Here, the instantaneous amplitude has the form $A(t) = \exp[\beta\cos(\omega_0 t)]$ and is shown in Figure 7.16.7 for $\beta = 2$ radians. Figure 7.16.8 shows the amplitudes of the one-sided spectrum in dependence of β.

7.17 Hilbert transforms in the theory of linear systems: Kramers-Kronig relations

The notions of **impedance**, **admittance**, and **transfer function** are commonly used to describe the properties of **linear**, **time-invariant** (LTI) systems. If the signal at the input port of the LTI system varies in time as $\exp(j\omega t)$, the signal at the output is a sine wave of the same frequency with a different amplitude and phase. In other words, the LTI conserves the waveform of sine signals. A pure sine waveform is a mathematical entity. However, it is easy to generate physical quantities that vary in time practically as $\exp(j\omega t)$. Signal generators producing nearly ideal sine waves are widely used in many applications, including precise measurements

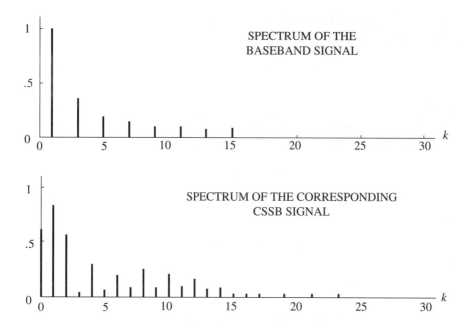

FIGURE 7.16.6
The spectrum of the CSSB AM signal with an envelope given by the Fourier series of a square wave
truncated at the fifteenth harmonic number.

of the behavior of circuits and systems. The transfer function of the LTI system is defined as
a quotient of the output and input analytic signals

$$H(j\omega) = \frac{\psi_2(t)}{\psi_1(t)} = \frac{A_2 e^{j(\omega t + \varphi_2)}}{A_1 e^{j(\omega t + \varphi_1)}} \qquad (7.17.1)$$

This transfer function describes the steady-state, input-output relations. Theoretically, the
input sine wave should be applied at the time at minus infinity. In practice, the steady state
arrives, if the transients die out. The transfer function is time independent, because the term
$\exp(j\omega t)$ may be deleted from the nominator and denominator of Eq. (7.17.1).

The frequency domain description by means of the transfer function can be converted into
the time-domain description using the Fourier transformation. A response of the LTI system
to the delta pulse, that is, the **impulse response**, is defined by the Fourier pair:

$$h(t) = \delta(t) * h(t) \overset{F}{\Longleftrightarrow} 1 H(j\omega) = H(j\omega) \qquad (7.17.2)$$

where $\delta(t) \overset{F}{\Longleftrightarrow} 1$.

Causality

All physical systems are causal. Causality implies that any response of a system at the time
t, depends only on excitations at earlier times. For this reason the impulse response of a
causal system is one-sided, that is, $h(t) = 0$ for $t < 0$. But one-sided time signals have
analytic spectra (see Section 7.3). Therefore, the spectrum of the impulse response given by
Eq. (7.17.2), and thus the transfer function of a causal system, is an analytic function of the
complex frequency $s = \alpha + j\omega$. The analytic transfer function

$$H(s) = A(\alpha, \omega) + j B(\alpha, \omega) \qquad (7.17.3)$$

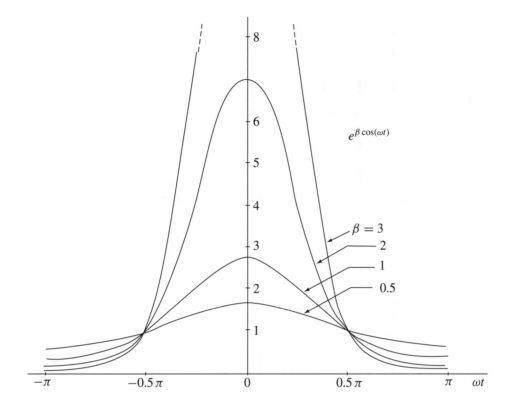

FIGURE 7.16.7
Envelope of the compatible with a linear FM detector single side-band FM signal. β-modulation index in radians.

satisfies the Cauchy-Riemann equations (see Eq. [7.2.4])

$$\frac{\partial A}{\partial \alpha} = \frac{\partial B}{\partial \omega}; \qquad \frac{\partial A}{\partial \omega} = -\frac{\partial B}{\partial \alpha} \tag{7.17.4}$$

and the real and imaginary parts ($\alpha = 0$) of the transfer function form a Hilbert pair:

$$A(\omega) = \frac{-1}{\pi} P \int_{-\infty}^{\infty} \frac{B(\lambda)}{\lambda - \omega} \, d\lambda \tag{7.17.5}$$

$$B(\omega) = \frac{1}{\pi} P \int_{-\infty}^{\infty} \frac{A(\lambda)}{\lambda - \omega} \, d\lambda \tag{7.17.6}$$

A one-sided impulse response can be regarded as a sum of **noncausal** even and odd parts (see Eqs. [7.3.2] and [7.3.3])

$$h(t) = h_e(t) + h_o(t) \tag{7.17.7}$$

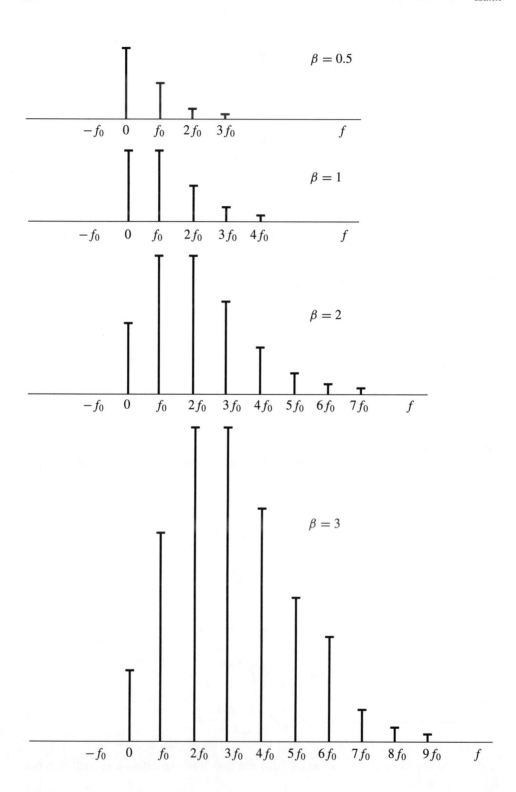

FIGURE 7.16.8
One-sided spectrum of the modulation function of the compatible with a linear detector FM signal.

Because $h(t)$ is real, we have the following Fourier pairs:

$$h_e(t) = \frac{1}{2}[h(t) + h(-t)] \overset{F}{\Longleftrightarrow} A(\omega) \tag{7.17.8}$$

$$h_o(t) = \frac{1}{2}[h(t) - h(-t)] \overset{F}{\Longleftrightarrow} jB(\omega) \tag{7.17.9}$$

The causality of $h(t)$ yields the relations

$$h_e(t) = \text{sgn}(t)h_o(t) \tag{7.17.10}$$

$$h_o(t) = \text{sgn}(t)h_e(t) \tag{7.17.11}$$

These products are the time-domain representation of the convolution integrals (7.17.5) and (7.17.6) (convolution to multiplication theorem).

Physical realizability of transfer functions

The Hilbert relations between real and imaginary parts of transfer functions are valid for physically realizable transfer functions. The terminology "physically realizable" may be misleading, because a transfer function given by a closed algebraic form is a mathematical representation of a model of a circuit built using ideal inductances, capacitances, and resistors or amplifiers. Such models are a theoretical, approximate description of physical systems. The physical realizability of a particular transfer function in the sense of circuit (or systems) theory is defined by means of causality. A general question of whether a particular amplitude characteristic can be realized by a causal system (filter) is answered by the Paley-Wiener criterion. Consider a specific magnitude of a transfer function $|H(j\omega)|$ (an even function of ω). It can be realized by means of a causal filter if and only if the integral

$$\int_{-\infty}^{\infty} \frac{\ln|H(j\omega)|}{1 + \omega^2} \, d\omega < \infty \tag{7.17.12}$$

is bounded [28]. Then a phase function exists such that the impulse response $h(t)$ is causal. The Paley-Wiener criterion is satisfied only if the support of $|H(j\omega)|$ is unbounded, otherwise $|H(j\omega)|$ would be equal to zero over finite intervals of frequency resulting in infinite values of the logarithm ($\ln|H(j\omega)| = -\infty$).

Minimum phase property

Transfer functions satisfying the Paley-Wiener criterion have a general form:

$$H(j\omega) = H_\varphi(j\omega)H_{ap}(j\omega) \tag{7.17.13}$$

where $H_\varphi(j\omega)$ is called a **minimum phase transfer function** and $H_{ap}(j\omega)$ is an all-pass transfer function. The minimum phase transfer function

$$H_\varphi(j\omega) = |H(j\omega)|e^{j\varphi(\omega)} = A_\varphi(\omega) + jB_\varphi(\omega) \tag{7.17.14}$$

has a minimum phase lag $\varphi(\omega)$ for a given magnitude characteristic. The minimum phase transfer function $H_\varphi(s)$ has all the zeros lying in the left half-plane (i.e., $\alpha < 0$) of the s-plane. The minimum phase transfer function is analytic and its real and imaginary parts form a Hilbert pair

$$A_\varphi(\omega) \overset{H}{\Longleftrightarrow} -B_\varphi(\omega) \tag{7.17.15}$$

An important feature of the minimum phase transfer function is that the **propagation function**

$$\gamma(s) = \ln[H(s)] = \beta(\alpha, \omega) + j\varphi(\alpha, \omega) \tag{7.17.16}$$

is analytic in the right half-plane. It is so because all zeros are in the left half-plane, and because we postulate stability, all poles are in the left half-plane, too. Then the real and imaginary part of the propagation function form a Hilbert pair:

$$\varphi(\omega) = \frac{-1}{\pi} P \int_{-\infty}^{\infty} \frac{\beta(\lambda)}{\lambda - \omega} d\lambda = \frac{-1}{\pi} P \int_{-\infty}^{\infty} \frac{\ln|H(j\lambda)|}{\lambda - \omega} d\lambda \tag{7.17.17}$$

$$\beta(\omega) = \frac{1}{\pi} P \int_{-\infty}^{\infty} \frac{\varphi(\lambda)}{\lambda - \omega} \tag{7.17.18}$$

These relations can be converted to take the form of the well-known **Bode phase-integral** theorem:

$$\varphi(\omega_0) = \frac{\pi}{2} \left|\frac{d\beta}{du}\right|_o + \frac{1}{\pi} P \int_{-\infty}^{\infty} \left[\left|\frac{d\beta}{du}\right| - \left|\frac{d\beta}{du}\right|_o\right] \ln[\coth|u/2|] \, du \tag{7.17.19}$$

where $u = \ln(\omega/\omega_0)$ is the normalized logarithmic frequency scale, and $d\beta/du$ is the slope of the β-curve in a ln-ln scale. The Bode formula shows that for the minimum-phase transfer functions the phase depends on the slope of the β-curve (β is the damping coefficient). The factor $\ln[\coth|u/2|]$ is peaked at $u = 0$ (or $\omega = \omega_0$) and hence the phase at a given ω_0 is mostly influenced by the slope $d\beta/du$ in the vicinity of ω_0.

The all-pass part of the **nonminimum phase transfer function** defined by Eq. (7.17.13) may be written in the form:

$$H_{ap}(j\omega) = e^{j\psi(\omega)} \tag{7.17.20}$$

Therefore, the total phase function has two terms:

$$\arg[H(j\omega)] = \varphi(\omega) + \psi(\omega) \tag{7.17.21}$$

where $\varphi(\omega)$ is the minimum phase and $\psi(\omega)$ the nonminimum phase part of the total phase.

Amplitude-phase relations in DLTI systems

A discrete, linear, and time-invariant system (DLTI) is characterized by the Z-pair (see also Chapter 6)

$$h(i) \stackrel{Z}{\Longleftrightarrow} H(z); \qquad z = e^{j\psi} \tag{7.17.22}$$

The sequence $h(i)$ $(i = 0, 1, 2, \ldots)$ is the impulse response of the system to the excitation by the **Kronecker delta** and $H(z)$ is the one-sided Z transform of the impulse response called the transfer function (or frequency characteristic) of the DLTI system, a function of the **dimensionless** normalized frequency $\psi = 2\pi f/f_s$, where f is the actual frequency and f_s the sampling frequency. For causal systems the impulse response is one-sided ($h(i) = 0$ for $i < 0$). The transfer function $H(e^{j\psi})$ is periodic with the period equal to 2π. This periodic function may be expanded into a Fourier series

$$H(e^{j\psi}) = \sum_{i=-\infty}^{\infty} h(i)e^{-j\psi i} = \sum_{i=0}^{\infty} h(i)e^{-j\psi i} \tag{7.17.23}$$

The Fourier coefficients $h(i)$ are equal to the terms of the impulse response and are given by the Fourier integral:

$$h(i) = \frac{1}{2\pi} \int_{-\pi}^{\pi} H\left(e^{j\psi}\right) e^{j\psi i} \, d\psi \tag{7.17.24}$$

In general, the transfer function is a complex quantity

$$H(e^{j\psi}) = A(\psi) + jB(\psi) \tag{7.17.25}$$

Analogously to Eq. (7.17.7) the causal impulse response $h(i)$ can be regarded as a sum of two noncausal even and odd parts of the form

$$h(i) = h(0) + h_e(i) + h_0(i) \tag{7.17.26}$$

The even part is defined by the equation

$$h_e(i) = 0.5[h(i) + h(-i)]; \qquad |i| > 0 \tag{7.17.27}$$

and the odd part by the equation:

$$h_0(i) = 0.5[h(i) - h(-i)] \tag{7.17.28}$$

Let us write the Fourier series (7.17.23) term-by-term. We get

$$H(e^{j\psi}) = h(0) + \sum_{i=1}^{\infty} h(i) \cos(\psi i) - j \sum_{i=1}^{\infty} h(i) \sin(\psi i) \tag{7.17.29}$$

The comparison of Eqs. (7.17.25) and (7.17.29) shows that

$$A(\psi) = h(0) + \sum_{i=1}^{\infty} h(i) \cos(\psi i) = h(0) + F^{-1}[h_e(i)] \tag{7.17.30}$$

and

$$B(\psi) = -\sum_{i=1}^{\infty} h(i) \sin(\psi i) = F^{-1}[h_0(i)] \tag{7.17.31}$$

and we have a Hilbert pair:

$$A(\psi) \overset{\text{H}}{\Longleftrightarrow} -B(\psi) \tag{7.17.32}$$

We used the relations $H[h(0)] = 0$ and $H[\cos(\psi i)] = \sin(\psi i)$. Because $A(\psi)$ and $B(\psi)$ are periodic functions of ψ, we may apply the cotangent form of the Hilbert transform (see Section 7.6).

$$B(\psi) = \frac{1}{2\pi} P \int_{-\pi}^{\pi} A(\Theta) \cot[(\Theta - \psi)/2] \, d\Theta \tag{7.17.33}$$

and

$$A(\psi) = h(0) - \frac{1}{2\pi} P \int_{-\pi}^{\pi} B(\Theta) \cot[(\Theta - \psi)/2] \, d\Theta \tag{7.17.34}$$

Minimum phase property in DLTI systems

Analogous to Eqs. (7.17.13) and (7.17.14) the transfer function of the DLTI system may be written in the form:

$$H(z) = H_\varphi(z) H_{ap}(z) \tag{7.17.35}$$

where $H_\varphi(z)$ satisfies the constraints of a **minimum phase transfer function**, that is, has all the zeros inside the unit circle of the z-plane and $H_{ap}(z)$ is an **all-pass function** consisting of a cascade of factors of the form:

$$H_{ap}(z) = \left[z^{-1} - z_i\right] / \left[1 - z_i^* z^{-1}\right] \tag{7.17.36}$$

The all-pass function has a magnitude of one, hence, $H(z)$ and $H_\varphi(z)$ have the same magnitude. $H_\varphi(z)$ differs from $H(z)$ in that the zeros of $H(z)$, lying outside the unit circle at points $z = 1/z_i$, are reflected inside the unit circle at $z = z_i^*$. Let us take the complex logarithm of $H_\varphi(e^{j\psi})$:

$$\ln\left[H_\varphi\left(e^{j\psi}\right)\right] = \ln\left|H_\varphi\left(e^{j\psi}\right)\right| + j\arg\left[H_\varphi\left(e^{j\psi}\right)\right] \tag{7.17.37}$$

and analogous to Eqs. (7.17.17) and (7.17.18), we have a Hilbert pair

$$\ln\left|H_\varphi\left(e^{j\psi}\right)\right| = \ln[h(0)] - \frac{1}{2\pi}P\int_{-\pi}^{\pi}\arg\left[H_\varphi\left(e^{j\Theta}\right)\right]\cot[(\Theta - \psi)/2] \tag{7.17.38}$$

$$\arg\left[H_\varphi\left(e^{j\Theta}\right)\right] = \frac{1}{2\pi}P\int_{-\pi}^{\pi}\log\left|H\left(e^{j\Theta}\right)\right|\cot[(\Theta - \psi)/2]\,d\Theta \tag{7.17.39}$$

It can be proved that the relations (7.17.38) and (7.17.39) are valid for transfer functions with zeros on the unit circle. In general, a stable and causal system has all its poles inside, while its zeros may lie outside the unit circle. However, starting from a nonminimum-phase transfer function, a minimum-phase function can be constructed by reflecting those zeros lying outside the unit circle, inside it.

The Kramers-Kronig relations in linear macroscopic continuous media

The amplitude-phase relations of the circuit theory are known in the macroscopic theory of continuous lossy media as the Kramers-Kronig relations [18], [19]. Almost all media display some degree of frequency dependence of some parameters, called **dispersion**. Let us take the example of a linear and isotropic electromagnetic medium. The simplest constitutive macroscopic relations describing this medium are [32]

$$D = \varepsilon\varepsilon_o E = (1 + \chi_e)\varepsilon_o E \tag{7.17.40}$$

$$B = \mu\mu_o H = (1 + \chi_m)\mu_o H \tag{7.17.41}$$

and

$$P = \chi_e\varepsilon_o E \tag{7.17.42}$$

$$M = \chi_m H \tag{7.17.43}$$

where $E[V/m]$ is the electric field vector, $H[A/m]$ is the magnetic field vector, $D[C/m^2]$ is the electric displacement, $B[Wb/m^2]$ is the magnetic induction, $\mu_o = 4\pi 10^{-7}[Hy/m]$ the permeability, $\varepsilon_0 = 1/36\pi 10^{-9}[F/m]$ the permittivity of free space, and ε, μ, χ_m and χ_e are dimensionless constants. The vectors P and M are called polarization and magnetization of the medium. If we substitute the electrostatic field vector E with a field varying in time as $\exp(j\omega t)$, then the properties of the medium are described by the frequency-dependent complex susceptibility

$$\chi(j\omega) = \chi'(\omega) - j\chi''(\omega) \tag{7.17.44}$$

where χ' is an even and χ'' an odd function of ω. The imaginary term χ'' represents the conversion of electric energy into heat, that is, losses of the medium. In fact, $\chi(j\omega)$ plays the same role as the transfer function in circuit theory and is defined by the equation:

$$\chi(j\omega) = \frac{P_m e^{j(\omega t + \varphi)}}{\varepsilon_o E_m e^{j\omega t}} = \frac{P_m}{\varepsilon_o E_m} e^{j\varphi} \tag{7.17.45}$$

Let us apply Fourier spectral methods to examine Eqs. (7.17.42) and (7.17.45). We consider a disturbance $E(t)$ given by the Fourier pair

$$E(t) \overset{F}{\Longleftrightarrow} \mathbf{X}_E(j\omega) \tag{7.17.46}$$

The response $P(t)$ is represented by the Fourier pair:

$$P(t) \overset{F}{\Longleftrightarrow} \mathbf{X}_p(j\omega) \tag{7.17.47}$$

where

$$\mathbf{X}_p(j\omega) = \varepsilon_o \chi(j\omega)\mathbf{X}_E(j\omega) \tag{7.17.48}$$

The multiplication-convolution theorem yields the time-domain solution:

$$P(t) = \varepsilon_o \int_{-\infty}^{\infty} h(\tau)E(t-\tau)\,d\tau \tag{7.17.49}$$

where $h(t)$ is given by the Fourier pair

$$h(t) \overset{F}{\Longleftrightarrow} \chi(j\omega) \tag{7.17.50}$$

is the "impulse response" of the medium, that is, the response to the excitation $\delta(t)$. For any physical medium the impulse response is causal. This is possible if $\chi(j\omega)$ is analytic. Therefore, its real and imaginary parts form a Hilbert pair

$$\chi''(\omega) = -\frac{1}{\pi} P \int_{-\infty}^{\infty} \frac{\chi'(\eta)}{\eta - \omega}\,d\eta \tag{7.17.51}$$

$$\chi'(\omega) = \frac{1}{\pi} P \int_{-\infty}^{\infty} \frac{\chi''(\eta)}{\eta - \omega}\,d\eta \tag{7.17.52}$$

These relations are known as the Kramers-Kronig relations and are a direct consequence of causability. They apply for many media; for example, in optics, the real and imaginary parts of the complex reflection coefficient form a Hilbert pair.

The concept of signal delay in Hilbertian sense

Consider a signal and its Fourier transform:

$$x(t) \overset{F}{\Longleftrightarrow} X(j\omega) \tag{7.17.53}$$

Let us assume that the Fourier spectrum $X(j\omega)$ may be written in the form of a product defined by Eq. (7.17.13)

$$X(j\omega) = X_1(j\omega)X_2(j\omega) \tag{7.17.54}$$

where $X_1(j\omega)$ fulfills the constraints of a minimum-phase function and $X_2(j\omega)$ is an "all-pass" function of the magnitude equal to one and the phase function $\psi(\omega)$, that is, $X_2(j\omega) = e^{j\psi(\omega)}$. The application of the convolution-multiplication theory yields the convolution

$$x(t) = x_1(t) * x_2(t) \tag{7.17.55}$$

where $x_1(t) \overset{F}{\Longleftrightarrow} X_1(j\omega)$ is defined as a minimum-phase signal satisfying relations (7.17.17) and (7.17.18), that is,

$$\arg[X_1(j\omega)] \overset{H}{\Longleftrightarrow} \ln|X_1(j\omega)| \tag{7.17.56}$$

and the signal

$$x_2(t) \overset{F}{\Longleftrightarrow} X_2(j\omega) = e^{j\psi(\omega)} \tag{7.17.57}$$

is defined as the nonmimimum-phase part of the signal $x(t)$. Let us formulate the following definitions:

Definition 1
The minimum phase signal $x_1(t)$ has a zero delay in the Hilbert sense.

Definition 2
The delay of the signal relative to the moment $t = 0$ is defined by a specific property of the signal $x_2(t)$. Krylov and Ponomariev [20] used the name "ambiguity function" for $x_2(t)$, and proposed to define the delay by the position of its maximum. Another possibility is to define the delay using the position of the center of gravity of $x_2(t)$.

Examples
  1. If the function $x_2(t) = \delta(t)$ the delay equals zero because

$$x(t) = x_1(t) * \delta(t) = x_1(t) \tag{7.17.58}$$

  2. If the function $x_2(t) = \delta(t - t_0)$ the delay equals t_0 because

$$x(t) = x_1(t) * \delta(t - t_0) = x_1(t - t_0) \tag{7.17.59}$$

  3. Consider a phase-delayed harmonic signal and its Fourier image:

$$\cos(\omega_0 t - \varphi_0) \overset{F}{\Longleftrightarrow} \pi\delta(\omega + \omega_0)e^{j\varphi_0} + \pi\delta(\omega - \omega_0)e^{-j\varphi_0} \tag{7.17.60}$$

or

$$\cos\omega_0 t * \delta\left(t - \frac{\varphi_0}{\omega_0}\right) \overset{F}{\Longleftrightarrow} \pi[\delta(\omega + \omega_0) + \delta(\omega - \omega_0)]e^{-j\varphi_0\,\mathrm{sgn}\,\omega} \tag{7.17.61}$$

Evidently the "ambiguity function" $x_2(t)$ is

$$x_2(t) = \delta\left(t - \frac{\varphi_0}{\omega_0}\right) \overset{F}{\Longleftrightarrow} e^{-j\varphi_0\,\mathrm{sgn}\,\omega} \tag{7.17.62}$$

and the time delay is of course $t_0 = \varphi_0/\omega_0$, as we could expect.

  4. Consider the series connection of the first-order low-pass with the transfer function

$$X_1(j\omega) = \frac{1}{1 + j\omega\tau} \tag{7.17.63}$$

and the first-order all-pass with the phase function of the form

$$\arg[X_2(j\omega)] = \tan^{-1}\frac{2\omega\tau}{(\omega\tau)^2 - 1} \tag{7.17.64}$$

The impulse response of the low-pass is:

$$x_1(t) = \mathbf{F}^{-1}\left[\frac{1}{1 + j\omega\tau}\right] = \mathbf{1}(t)e^{-t/\tau} \tag{7.17.65}$$

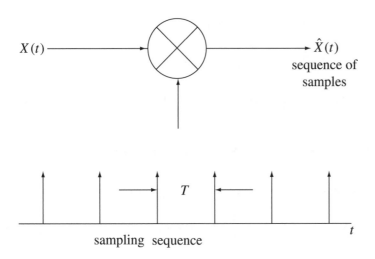

FIGURE 7.18.1
A method of generation of a sequence $\hat{x}(t)$ of samples of the analog signal $x(t)$.

and satisfies the definition of the minimum-phase signal. The impulse response of the all-pass plays here the role of the "ambiguity function" and has the form:

$$x_2(t) = \mathbf{F}^{-1}\left[\exp\frac{-2\omega\tau}{\omega^2\tau^2 - 1}\right] = \mathbf{1}(t)\frac{2}{\tau}e^{-t/\tau} - \delta(t)$$

We observe that the maximum of $x_2(t)$ is at $t = 0$. However, we expect that the all-pass introduces some delay. In this case it would be advisable to define the delay using the center of gravity of the signal $x_2(t)$.   ∎

7.18   Hilbert transforms in the theory of sampling

The generation of a sequence of samples of a continuous signal (sampling) and the recovery of this signal from its samples (interpolation) is a widely used procedure in modern signal processing and communications techniques. Basic and advanced theory of sampling and interpolation is presented in many textbooks. This section presents the role of Hilbert transforms in the theory of sampling and interpolation. Figure 7.18.1, for reference, is the usual means by which the sequence of samples is produced. In general, the sampling pulses may be nonequidistant. However, this section presents the role of Hilbert transforms in the basic **WKS** (Wittaker, Kotielnikow, Shannon) theory of periodic sampling and interpolation.

The periodic sequence of sampling pulses may be written in the form (see Eq. [7.6.4])

$$p(t) = p_T(t) * \sum_{k=-\infty}^{\infty} \delta(t - kT) \tag{7.18.1}$$

where $p_T(t)$ defines the waveform of the sampling pulse (the generating function of the periodic sequence of pulses) and $f = 1/T$ is the sampling frequency. From the point of view of the presentation of the role of Hilbert transforms in sampling and interpolation, it is sufficient to use the delta sampling sequence inserting $p_T(t) = \delta(t)$. The **delta sampling sequence** is given by the formula (remember that $\delta(t) * \delta(t) = \delta(t)$)

$$p(t) = \sum_{k=-\infty}^{\infty} \delta(t - kT) \tag{7.18.2}$$

For convenience, let us write here the Hilbert transform of this sampling sequence (see Section 7.6, Eq. [7.6.18])

$$\sum_{k=-\infty}^{\infty} \delta(t - nT) \overset{\text{H}}{\Longleftrightarrow} \frac{1}{T} \sum_{k=-\infty}^{\infty} \cot[(\pi/T)(t - kT)] \tag{7.18.3}$$

The Fourier image of the delta sampling sequence is given by another periodic delta sequence

$$\sum_{k=-\infty}^{\infty} \delta(t - kT) \overset{F}{\Longleftrightarrow} \frac{1}{T} \sum_{k=-\infty}^{\infty} \delta(f - k/T) \tag{7.18.4}$$

The sampler produces as an output a sequence of samples given by the formula

$$x_s(t) = \sum_{k=-\infty}^{\infty} x(kT)\delta(t - kT) \tag{7.18.5}$$

that is, a sequence of delta functions weighted by the samples of the signal $x(t)$. Let us recall the basic WKS sampling theorem. Consider a signal $x(t)$ and its Fourier image $X(f)$, $\omega = 2\pi f$. If the Fourier image is low-pass band limited, that is, $|X(jf)| = 0$ for $|f| > W$, then $x(t)$ is completely determined by the sequence of its samples taken at the moments t_k spaced $T = 1/2W$ apart. The sampling frequency $f_s = 2W$ is called the Nyquist rate. The multiplication to convolution theorem yields the spectrum of the sequence of samples

$$X_s(jf) = X(jf) * \frac{1}{T} \sum_{k=-\infty}^{\infty} \delta(f - k/T) \tag{7.18.6}$$

Figure 7.18.2 shows an example of a low-pass band-limited spectrum of a signal $x(t)$ and the well-known three spectra of the sequence of samples: The spectrum of oversampled signal with no aliasing with the sampling frequency $f_s = 1/T > W$, the limit case with $f_s = 2W$ and the spectrum of undersampled signal with $f_s < 2W$. Notice that the sequence of samples given by Eq. (7.18.5) may be regarded as a model of a signal with pulse amplitude modulation (PAM). The original signal $x(t)$ may be recovered by filtering this PAM signal using the ideal noncausal and physically unrealizable low-pass filter defined by the transfer function

$$Y(jf) = \begin{cases} 1 & \text{for } |f| < W \\ 0.5 & \text{for } |f| = W \\ 0 & \text{for } |f| > W \end{cases} \tag{7.18.7}$$

The noncausal impulse response of this filter is

$$h(t) = F^{-1}[Y(jf)] = 2W\frac{\sin(2\pi Wt)}{2\pi Wt} \tag{7.18.8}$$

and is called the **interpolatory function**. The total response is a sum of responses to succeeding samples giving the well-known **interpolatory expansion** ($f_s = 2W$):

$$x(t) = \sum_{k=-\infty}^{\infty} x\left[\frac{k}{2W}\right] \frac{\sin\left[2\pi W\left(t - \frac{k}{2W}\right)\right]}{2\pi W\left(t - \frac{k}{2W}\right)} \tag{7.18.9}$$

The summation exactly restores the original signal $x(t)$. In the following text the argument of the interpolatory function will be written using the notation

$$2\alpha(t, k) = 2\pi W \left(t - \frac{k}{2W} \right) \tag{7.18.10}$$

giving the following form of the interpolation expansion

$$x(t) = \sum_{k=-\infty}^{\infty} x \left(\frac{k}{2W} \right) \frac{\sin[2\alpha(t, k)]}{2\alpha(t, k)} \tag{7.18.11}$$

Notice that the sampling of the function $x(t) = a$ (a constant) yields the formula

$$\sum_{k=-\infty}^{\infty} \frac{\sin[2\alpha(t, k)]}{2\alpha(t, k)} = 1 \tag{7.18.12}$$

This equation may be used to calculate the accuracy of the interpolation due to any truncation of the summation.

The Whittaker's interpolatory function and its Hilbert transform are forming the Hilbert pair

$$\frac{\sin[2\alpha(t, k)]}{2\alpha(t, k)} \overset{H}{\Longleftrightarrow} \frac{\sin^2[\alpha(t, k)]}{\alpha(t, k)} \tag{7.18.13}$$

Therefore, the interpolatory expansion of the Hilbert transform $H[x(t)] = \hat{x}(t)$, due to the linearity property, is given by the formula

$$\hat{x}(t) = \sum_{k=-\infty}^{\infty} x \left[\frac{k}{2W} \right] \frac{\sin^2[\alpha(t, k)]}{\alpha(t, k)} \tag{7.18.14}$$

This formula may be applied to calculate the Hilbert transforms of low-pass signals using their samples. The transfer function of the low-pass Hilbert filter (transformer) is given by the Fourier transform of the impulse response given by the right-hand side of Eq. (7.18.13):

$$\begin{aligned} Y_H(jf) &= F \left[\frac{\sin^2(\alpha, k)}{\alpha(t, k)} \right] \\ &= -j \, \text{sgn}(f) Y(jf) \\ &= \begin{cases} j & \text{for } |f| < W \\ 0 & \text{for } f = 0; \qquad 0.5 \text{ for } |f| = W \\ -j & \text{for } |f| > W \end{cases} \end{aligned} \tag{7.18.15}$$

The sampling of the function $x(t) = a$ yields

$$\sum_{k=-\infty}^{\infty} \frac{\sin^2[\alpha(t, k)]}{\alpha(t, k)} = 0 \tag{7.18.16}$$

The expansion of the analytic signal $\Psi(t) = x(t) + j\hat{x}(t)$ using interpolatory functions has the form

$$\psi(t) = \sum_{k=-\infty}^{\infty} x \left[\frac{k}{2W} \right] \left[\frac{\sin[2\alpha(t, k)]}{2\alpha(t, k)} + j \frac{\sin^2[\alpha(t, k)]}{\alpha(t, k)} \right] \tag{7.18.17}$$

and using trigonometric identities we get the following form of the interpolatory expansion of the analytic signal:

$$\psi(t) = -j \sum_{k=-\infty}^{\infty} x \left[\frac{k}{2W} \right] \frac{e^{j2\alpha(t,k)} - 1}{2\alpha(t, k)} \tag{7.18.18}$$

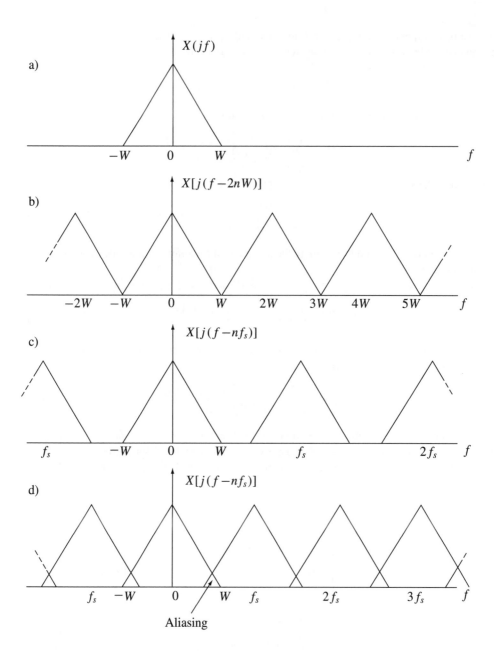

FIGURE 7.18.2
(a) A band-limited low-pass spectrum of a signal, (b) the corresponding spectrum of the sequence of samples with Nyquist rate of sampling $f_s = 2W$, (c) spectrum by oversampling $f_s > 2W$, and (d) spectrum by undersampling $f_s < 2W$ showing the aliasing of the sidebands.

Band-pass filtering of the low-pass sampled signal

Consider the ideal band-pass with a physically unrealizable transfer function in the form of a "spectral window" as shown in Figure 7.18.3. The impulse response of this filter is:

$$h(t) = 2\,(f_2 - f_1)\,\frac{\sin\,[\pi\,(f_2 - f_1)]}{\pi(f_2 - f_1)t}\,\cos\,[\pi(f_1 + f_2)t] \qquad (7.18.19)$$

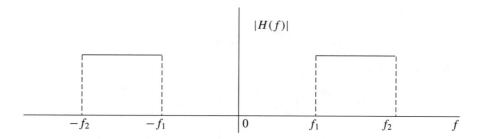

FIGURE 7.18.3
The magnitude of the transfer function of an ideal band-pass.

The insertion $f_1 = W$ and $f_2 = 3W$ yields

$$h(t) = 4W \frac{\sin(2\pi W t)}{2\pi W t} \cos(4\pi W t) \tag{7.18.20}$$

If the sequence of samples of the signal $x(t)$ is applied to the input of this band-pass, the output signal $z(t)$ is given by the interpolatory expansion of the form

$$z(t) = \sum_{k=-\infty}^{\infty} \left\{ x\left(k/(2W)\right) \frac{\sin[2\alpha(t,k)]}{2\alpha(t,k)} \cos[4\alpha(t,k)] \right\} \tag{7.18.21}$$

where $\alpha(t,k)$ is given by Eq. (7.18.10). We obtained the compressed-carrier amplitude-modulated signal of the form

$$z(t) = x(t) \cos(4\pi W t) \tag{7.18.22}$$

with a carrier frequency $2W$. Therefore, the AM-balanced modulator may be implemented using a sampler and a band-pass. Multiplication of the carrier frequency is possible using band-pass filters with $f_1 = 3W$ and $f_2 = 5W$ or $f_1 = 5W$ and $f_2 = 7W, \ldots$. The conclusion is that in principle one may multiply the carrier frequency of AM signals getting undistorted sidebands (envelope). The comparison of Eqs. (7.18.11) and (7.18.22) enables us to write the signal $z(t)$ in the form:

$$z(t) = \left\{ \sum_{k=-\infty}^{\infty} x\left[\frac{k}{2W}\right] \frac{\sin[2\alpha(t,k)]}{2\alpha(t,k)} \right\} \cos(4\pi W t) \tag{7.18.23}$$

and because $\cos(4\pi W t - k2\pi) = \cos(4\pi W t)$, in the form

$$z(t) = \sum_{k=-\infty}^{\infty} x\left[\frac{k}{2W}\right] \frac{\sin[2\alpha(t,k)]}{2\alpha(t,k)} \cos(4\pi W t) \tag{7.18.24}$$

Analogously, a single side-band AM signal may be produced by band-pass filtering of the sequence of samples using a filter with $f_1 = 2W$ and $f_2 = 3W$ (upper sideband). The impulse response of this filter is:

$$h(t) = 2W \frac{\sin(\pi W t)}{\pi W t} \cos(5\pi W t) \tag{7.18.25}$$

and the interpolatory expansion is:

$$z_{SSB}(t) = \sum_{k=-\infty}^{\infty} x\left[\frac{k}{2W}\right] \frac{\sin[\alpha(t,k)]}{\alpha(t,k)} \cos[5\alpha(t,k)] \tag{7.18.26}$$

This SSB signal may be written in the standard form given by Eq. (7.16.16) (see Section 7.16).

$$z_{SSB}(t) = x(t)\cos(4\pi W t) - \hat{x}(t)\sin(4\pi W t) \tag{7.18.27}$$

Let us derive the above form starting with Eq. (7.18.25). Using the trigonometric identity $\cos(5\alpha) = \cos\alpha\cos(4\alpha) - \sin\alpha\sin(4\alpha)$, Eq. (7.18.26) becomes:

$$z_{SBB}(t) = \sum_{k=-\infty}^{\infty} x\left[\frac{k}{2W}\right] \left\{ \frac{\sin[2\alpha(t,k)]}{2\alpha(t,k)} \cos[4\alpha(t,k)] \right.$$

$$\left. - \frac{\sin^2[\alpha(t,k)]}{\alpha(t,k)} \sin[4\alpha(t,k)] \right\} \tag{7.18.28}$$

It may be shown in the same manner as before that Eqs. (7.18.27) and (7.18.28) have identical left-hand sides.

Sampling of band-pass signals

Consider a band-pass signal $f(t)$ with the spectrum limited in band $f_1 < |f| < f_2 = f_1 + W$ (see Fig. 7.18.4). In general, a so-called second-order sampling should be applied to recover, using interpolation, the signal $f(t)$. However, it may be shown that alternatively, first-order sampling at the rate W may be applied with simultaneous sampling of the signal $f(t)$ and of its Hilbert transform $H[f(t)] = \hat{f}(t)$. The following interpolation formula has to be applied to recover the signal using the sequences of samples $f(k/W)$ and $\hat{f}(k/W)$.

$$f(t) = \sum_{n=-\infty}^{\infty} f\left(\frac{n}{W}\right) s\left(t - \frac{n}{W}\right) + \hat{f}\left(\frac{n}{W}\right) \hat{s}\left(t - \frac{n}{W}\right) \tag{7.18.29}$$

where the interpolating functions are given by the impulse response of the band-pass

$$s(t) = \frac{\sin(\pi W t)}{\pi W t} \cos\left[2\pi\left(f_1 + \frac{W}{2}\right)t\right] \tag{7.18.30}$$

and of a band-pass Hilbert filter (see Section 7.21).

$$\hat{s}(t) = \frac{\sin(\pi W t)}{\pi W t} \sin\left[2\pi\left(f_1 + \frac{W}{2}\right)t\right] \tag{7.18.31}$$

7.19   The definition of electrical power in terms of Hilbert transforms and analytic signals

The problem of efficient energy transmission from the source to the load is of importance in electrical systems. Usually the voltage and current waveforms may be regarded as sinusoidal.

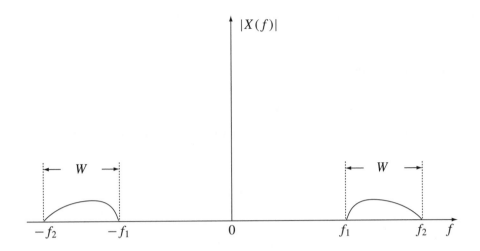

FIGURE 7.18.4
The magnitude of the spectrum of a band-pass signal.

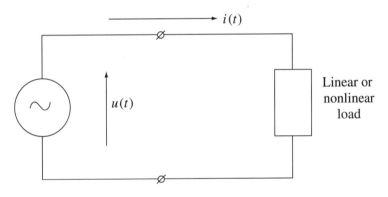

FIGURE 7.19.1
An electrical one-port where $u(t)$ is the instantaneous voltage and $i(t)$ the instantaneous current.

However, many loads are nonlinear and therefore nonsinusoidal cases should be investigated. In many applications the voltages and currents are nearly periodic, unperiodic, or even random. Therefore, some generalizations of theories developed for periodic cases are needed.

Consider an electrical one-port (linear or nonlinear) as shown in Figure 7.19.1. The **instantaneous power** is defined by the equation:

$$P(t) = u(t)i(t) \qquad (7.19.1)$$

where $u(t)$ is the instantaneous voltage across the load and $i(t)$ the instantaneous current in the load. We arbitrarily assign a positive sign to P if the energy $P(t)dt$ is delivered from the source to the load and a negative sign for the opposite direction. The above formal definition of power involves **all limitations** associated with the definition of voltage, current, and the electrical one-port.

Let us introduce the notion of **quadrature instantaneous power** defined by the equation

$$Q(t) = u(t)\hat{i}(t) = -\hat{u}(t)i(t) \tag{7.19.2}$$

where $\hat{u}$ and $\hat{i}$ are Hilbert transforms of the voltage and current waveforms.

Harmonic waveforms of voltage and current

Consider the classical case of a linear load with sine waveforms of $u(t)$ and $i(t)$. We have

$$u(t) = U \cos(\omega t + \varphi_u) \tag{7.19.3}$$

$$i(t) = J \cos(\omega t + \varphi_i) \tag{7.19.4}$$

The instantaneous power is:

$$P(t) = UJ \cos(\omega t + \varphi_u) \cos(\omega t + \varphi_i) \tag{7.19.5}$$

The Fourier series expansion of $P(t)$ is:

$$P(t) = 0.5UJ \cos(\varphi_i - \varphi_u) + 0.5UJ \left[\cos\left[2(\omega t + \varphi_i)\right] \cos(\varphi_i - \varphi_u) \right.$$

$$\left. - \sin\left[2(\omega t + \varphi_i)\right] \sin(\varphi_i - \varphi_u) \right] \tag{7.19.6}$$

The instantaneous quadrature power is:

$$Q(t) = UJ \cos(\omega t + \varphi_u) \sin(\omega t + \varphi_i) \tag{7.19.7}$$

The Fourier series expansion of $Q(t)$ is:

$$Q(t) = 0.5UJ \sin(\varphi_i - \varphi_u) + 0.5UJ \left[\sin\left[2(\omega t + \varphi_i)\right] \cos(\varphi_i - \varphi_u) \right.$$

$$\left. + \cos\left[2(\omega t + \varphi_i)\right] \sin(\varphi_i - \varphi_u) \right] \tag{7.19.8}$$

The mean value of $P(t)$ defined by the equation

$$\bar{P} = \frac{1}{T} \int_0^T P(t)\, dt = 0.5UJ \cos(\varphi_i - \varphi_u); \qquad \omega = \frac{2\pi}{T} \tag{7.19.9}$$

is called the **active power** and it is a measure of the unilateral energy transfer from the source to the load. The mean value of the quadrature power $Q(t)$ defined by the equation

$$\bar{Q} = \frac{1}{T} \int_0^T Q(t)\, dt = 0.5UJ \sin(\varphi_i - \varphi_u) \tag{7.19.10}$$

is called the **reactive power**. The value of the reactive power depends on energy that is delivered periodically back and forth between the source and the load, with no net transfer. The waveform of the instantaneous power given by Eq. (7.19.6) is shown in Figure 7.19.2 (for convenience $\varphi_u = 0$). The energy transfer from the source to the load is given by the integral

$$E_+ = \frac{1}{\omega} \int_{-\pi/2}^{\pi/2 - \varphi_i} UI \cos(\omega t) \cos(\omega t + \varphi_i)\, d\omega t = \frac{UI}{2\omega} \left[(\pi - \varphi) \cos\varphi + \sin\varphi\right] \tag{7.19.11}$$

and the energy transfer from the load to the source during the remaining part of the half-period is

$$E_- = \frac{1}{\omega} \int_{\pi/2 - \varphi_i}^{\pi/2} UI \cos(\omega t) \cos(\omega t + \varphi_i)\, d\omega t = \frac{UI}{2\omega} \left[\varphi \cos\varphi - \sin\varphi\right] \tag{7.19.12}$$

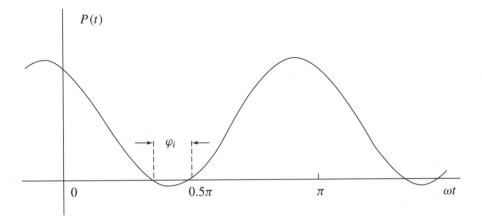

FIGURE 7.19.2
The waveform of the instantaneous power given by Eq. (7.19.5).

Therefore, the net energy transfer toward the load is

$$E = E_+ - E_- = \frac{UIT}{4} \cos(\varphi_i) \tag{7.19.13}$$

The division of this energy by $0.5T$ gives the mean value of the power equal to the active power. However, the division of E_- by $0.5T$ yields

$$\bar{P}_- = \frac{2E}{T} [\varphi \cos \varphi_i - \sin \varphi_i] \tag{7.19.14}$$

and this mean power differs from the reactive power defined by Eq. (7.19.10). Therefore, the notions of active and reactive power differ considerably. The active power equals the time-independent or constant term of the instantaneous power given by the Fourier series (7.19.6) while the reactive power equals the amplitude of the quadrature (or sine) term of (7.19.6). Notice that in the Fourier series (7.19.8) the role of both quantities is reversed. Let us recall that the quantity

$$S = 0.5UJ = U_{\text{RMS}} J_{\text{RMS}} \tag{7.19.15}$$

is called the **apparent power** and the quantity

$$\rho = \cos(\varphi_i - \varphi_u) = \frac{\bar{P}}{S} \tag{7.19.16}$$

is called the **power factor**. The power factor may be regarded as a normalized correlation coefficient of the voltage and current signals while $\sin(\varphi_i - \varphi_u) = SQR(1-\rho^2)$ may be called the anticorrelation coefficient. The quantities S, $\bar{P}$, and $\bar{Q}$ satisfy the relation

$$S^2 = \bar{P}^2 + \bar{Q}^2 \tag{7.19.17}$$

The notion of complex power

Consider the analytic (complex) form of the voltage and current harmonic signals defined by Eqs. (7.19.3) and (7.19.4). We have $\psi_u(t) = U \exp(\omega t + \varphi_u)$ and $\psi_i(t) = J \exp(\omega t + \varphi_i)$.

The complex power is defined by the equation:

$$S = \frac{1}{2}\psi_u(t)\psi_i^*(t) = 0.5UJ\exp\left[j\left(\varphi_i - \varphi_u\right)\right] \tag{7.19.18}$$

In the following text the symbol S will be used to denote the complex power. We have

$$S = P + jQ = |S|\exp[j(\varphi_i - \varphi_u)] \tag{7.19.19}$$

The real part of S equals the active power and the imaginary part equals the reactive power. The module of the complex power equals the apparent power and the argument equals the phase angle $\varphi_i - \varphi_u$.

Generalization of the notion of power

The above-described well known notions of apparent, active, and reactive power were in the past generalized by several authors for nonsinusoidal cases and later for signals with finite average power. The nonsinusoidal periodic waveforms of $u(t)$ and $i(t)$ may be described in the frequency domain by the Fourier series:

$$u(t) = U_0 + \sum_{n=1}^{N} U_n \cos\left(n\omega t + \varphi_{un}\right) \tag{7.19.20}$$

$$i(t) = I_0 + \sum_{n=1}^{N} J_n \cos\left(n\omega t + \varphi_{in}\right) \tag{7.19.21}$$

where ω is a constant equal to the fundamental angular frequency, $\omega = 2\pi/T$, and T is the period. Some or even all harmonics of the voltage waveform may not be included in the current waveform and vice versa. The active power may be defined using the same equation (7.19.9) as for sinusoidal waveforms. Inserting Eqs. (7.19.20) and (7.19.21) into (7.19.9) yields

$$\bar{P} = U_0 J_0 + \sum 0.5 U_n J_n \cos\left(\varphi_{in} - \varphi_{un}\right) \tag{7.19.22}$$

The summation involves terms included in both waveforms. Analogously, the reactive power is defined using Eq. (7.19.10):

$$\bar{Q} = \sum 0.5 U_n J_n \sin\left(\varphi_{in} - \varphi_{un}\right) \tag{7.19.23}$$

This definition of the reactive power was proposed in 1927 by Budeanu [6] and is nowadays commonly accepted. It has been sometimes criticized as "lacking of physical meaning." Another definition of reactive power was introduced by Fryze [10] who proposed to resolve the current waveform in two components:

$$i(t) = i_p(t) + i_q(t) \tag{7.19.24}$$

The "in-phase" component is given by the relation

$$i_p(t) = \frac{\frac{1}{T}\int_0^T iu\,dt}{\frac{1}{T}\int_0^T u^2\,dt}u(t) = \frac{\bar{P}}{U_{\text{RMS}}^2}u(t) \tag{7.19.25}$$

U_{RMS} is the root mean square (RMS) value of the voltage. The "quadrature" component is

$$i_q = i - i_p \tag{7.19.26}$$

and satisfies the orthogonality property

$$\int_0^T i_q i_p \, dt = 0 \tag{7.19.27}$$

This orthogonality yields for the RMS values:

$$I_{\text{RMS}}^2 = I_{p,\text{RMS}}^2 + I_{q,\text{RMS}}^2 \tag{7.19.28}$$

The reactive power is defined by the product

$$Q = U_{\text{RMS}} I_{q,\text{RMS}} \tag{7.19.29}$$

The comparison of Budeanu's and Fryze's definitions of the reactive power shows how misleading it is to apply the same name, "reactive power," for notions having **different** definitions. Let us illustrate this statement with an example. A source of a cosine voltage is loaded with the ideal diode with a nonlinear characteristic (see Fig. 7.19.3)

$$\begin{aligned} i = Gu \qquad & \text{if } u > 0 \\ i = 0 \qquad & \text{if } u < 0 \end{aligned} \tag{7.19.30}$$

The current has the waveform of a half-wave rectified cosine (see Fig. 7.19.3a) and may be resolved into the "in-phase" and "quadrature" components. The Fourier series expansion of the current has the form

$$i(t) = \frac{U}{\pi}\left[1 + \frac{\pi}{2}\cos(\omega t) + \frac{2}{3}\cos(2\omega t) - \frac{2}{15}\cos(4\omega)t + \frac{2}{35}\cos(6\omega t) - \cdots \right] \tag{7.19.31}$$

The "in-phase" component is:

$$i_p(t) = \frac{U}{2}\cos(\omega t) \tag{7.19.32}$$

and the Fourier series of the "quadrature" component (full-wave rectified cosine) is:

$$i_q(t) = \frac{U}{\pi}\left[1 + \frac{2}{3}\cos(2\omega t) - \frac{2}{15}\cos(4\omega t) + \frac{2}{35}\cos(6\omega t) - \cdots \right] \tag{7.19.33}$$

The reactive power defined by Eq. (7.19.23) equals zero while the reactive power defined by Eq. (7.19.29) equals

$$Q = \frac{U^2}{8} \tag{7.19.34}$$

However the instantaneous power (Figure 7.19.3) is always positive so there is no energy oscillating back and forth between the source and load. Therefore, we should expect that the reactive power equals zero. This requirement is satisfied using Budeanu's definition but not Fryze's definition.

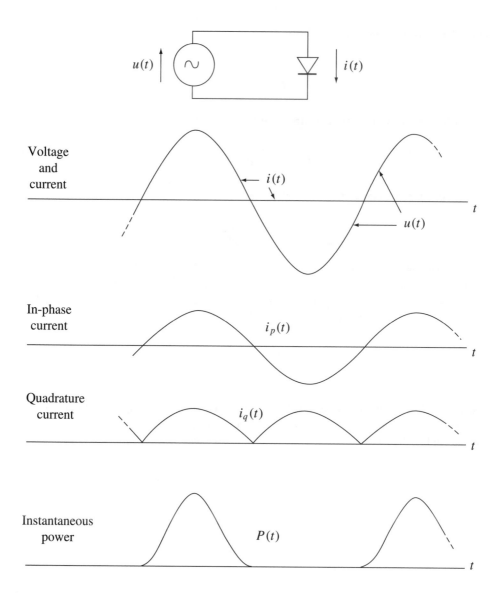

FIGURE 7.19.3
(a) A source of sine voltage loaded with a diode, (b) the voltage and current waveforms, (c) the in-phase component of the current, (d) the quadrature component of the current, and (e) the waveform of the instantaneous power.

Generalization of the notion of power for signals with finite average power

A generalized theory of electric power by use of Hilbert transforms was presented by Nowomiejski [24]. He considered voltages and currents with **finite average power**, that is, finite RMS defined by the equations

$$U_{\text{RMS}} = \sqrt{\lim_{T \Rightarrow \infty} \frac{1}{2T} \int_{-T}^{T} u^2(t)\,dt} \tag{7.19.35}$$

$$I_{\text{RMS}} = \sqrt{\lim_{T \Rightarrow \infty} \frac{1}{2T} \int_{-T}^{T} i^2(t)\,dt} \tag{7.19.36}$$

The apparent power is defined as

$$S = U_{\text{RMS}} I_{\text{RMS}} \tag{7.19.37}$$

and the active and reactive powers are defined by means of the relations:

$$\bar{P} = \lim_{T \Rightarrow \infty} \frac{1}{2T} \int_{-T}^{T} u(t) i(t)\,dt \tag{7.19.38}$$

and

$$\bar{Q} = \lim_{T \Rightarrow \infty} \frac{1}{2T} \int_{-T}^{T} u(t) i(t)\,dt \tag{7.19.39}$$

or

$$\bar{Q} = -\lim_{T \Rightarrow \infty} \frac{1}{2T} \int_{-T}^{T} \hat{u}(t) i(t)\,dt \tag{7.19.40}$$

where ˆ indicates the Hilbert transform. Nowomiejski has not explicitly defined the notion of the quadrature power (see Eq. [7.19.2]) but in fact the integrand in Eqs. (7.19.39) and (7.19.40) equals $Q(t)$. However, a new quantity called **distortion power** was defined. Generally, for each value of T the identity

$$\int_{-T}^{T} u^2(t)\,dt \int_{-T}^{T} i^2(t)\,dt = \left| \int_{-T}^{T} u(t) i(t)\,dt \right|^2$$
$$+ \int_{-T}^{T} \int_{-T}^{T} \frac{1}{2} [u(t) i(\tau) - u(\tau) i(t)]^2\,dt\,d\tau \tag{7.19.41}$$

holds true, and because the limit exists

$$S^2 = \left\{ \lim_{T \Rightarrow \infty} \frac{1}{2T} \int_{-T}^{T} u^2(t)\,dt \right\} \left\{ \lim_{T \Rightarrow \infty} \frac{1}{2T} \int_{-T}^{T} i^2(t)\,dt \right\} \tag{7.19.42}$$

the quantity D, called distortion power, may be defined by means of the equation

$$\bar{D} = \sqrt{\lim_{T \Rightarrow \infty} \left[\frac{1}{2T} \right]^2 \int_{-T}^{T} \int_{-T}^{T} \frac{1}{2} [u(t) i(\tau) - u(\tau) i(t)]^2\,dt\,d\tau} \tag{7.19.43}$$

Based on Eq. (7.19.41) we arrive at:

$$S^2 = \bar{P}^2 + \bar{D}^2 \tag{7.19.44}$$

In the case

$$i(t) = \text{const}\, u(t) \tag{7.19.45}$$

the "quadrature" component defined by Eq. (7.19.19) equals zero and the distortion power equals zero, too. Otherwise, the distortion power is given by

$$\bar{D} = U \sqrt{\lim_{T \Rightarrow \infty} \frac{1}{2T} \int_{-T}^{T} i_q^2(t)\, dt} \tag{7.19.46}$$

Let us define a power factor ρ_D using the relation:

$$\rho_D = \frac{\bar{P}}{\bar{P}^2 + \bar{D}^2} \tag{7.19.47}$$

The power factor is a measure of the efficiency of the utilization of the power supplied to the load being equal to unity only, if the distorsion power $D = 0$. The cross-correlation of the instantaneous voltage and current waveforms is defined by the integral

$$\rho_{ui}(\tau) = \lim_{T \Rightarrow \infty} \frac{1}{2T} \int_{-T}^{T} u(t) i(t - \tau)\, dt \tag{7.19.48}$$

This function enables us to introduce the frequency domain interpretations of the above-defined powers. The **cross-power spectrum** $\Theta(\omega)$ is defined by the Fourier pair

$$\rho_{u-i}(\tau) \overset{F}{\Longleftrightarrow} \Theta(\omega) \tag{7.19.49}$$

It may be shown that the active power is given by the integral of the power spectrum

$$\bar{P} = \frac{1}{2\pi} \int_{-\infty}^{\infty} \Theta(\omega)\, d\omega \tag{7.19.50}$$

In general, $\Theta(\omega)$ is a complex function, but the integral of the odd imaginary part equals zero. The reactive power is given by

$$\bar{Q} = \frac{1}{2\pi j} \int_{-\infty}^{\infty} \operatorname{sgn}(\omega) \Theta(\omega)\, d\omega \tag{7.19.51}$$

Hence, the **complex power** is

$$S = \bar{P} + j\bar{Q} = \frac{1}{2\pi} \int_{-\infty}^{\infty} [1 + \operatorname{sgn}(\omega)]\Theta(\omega)\, d\omega \tag{7.19.52}$$

Because the integrand presents a one-sided complex power spectrum, the complex power is an analytic function and $\bar{P}$ and $\bar{Q}$ form a pair of Hilbert transforms. If at least one of the signals $u(t)$ or $i(t)$ does not contain a constant component, Eq. (7.19.52) reduces to the form

$$S = \bar{P} + j\bar{Q} = \frac{1}{\pi} \int_{0}^{\infty} \Theta(\omega)\, d\omega \tag{7.19.53}$$

Notice that the Wiener-Khinchin relation (7.19.49) holds for stationary and ergodic processes. If the load presents a linear, time-invariant, and strictly stable system defined by the Fourier pair

$$h(t) \overset{F}{\Longleftrightarrow} H(j\omega) \tag{7.19.54}$$

(where $h(t)$ is the impulse response and $H(j\omega)$ is the transfer function) then the autocorrelation function of the voltage and its power spectrum are given by the Fourier pair

$$\rho_{u-u}(\tau) \overset{F}{\Longleftrightarrow} \Phi(\omega) \tag{7.19.55}$$

and the RMS values of the voltage and current have the form

$$U_{\text{RMS}}^2 = \rho_{u-u}(0); \qquad I_{\text{RMS}}^2 = \frac{1}{2\pi} \int_{-\infty}^{\infty} \Phi(\omega) |H(j\omega)|^2 \, d\omega \qquad (7.19.56)$$

and the complex power is given by

$$S = \bar{P} + j\bar{Q} = \frac{1}{\pi} \int_0^{\infty} \Phi(\omega) H^*(\omega) \, d\omega \qquad (7.19.57)$$

7.20   The discrete Hilbert transformation

The theory and applications of the DHT (Discrete Hilbert Transformation) are closely tied with the principles of digital signal processing [26], [30]. Because discrete transforms will be included in another handbook in this series, this section presents only basic concepts. The formulas for the DFT and for the Z-transformation are given to fix the notations, because there are various notations (definitions) of the DFT.

For reference, let us recall the Fourier transformations given by Eqs. (7.1.6) or (7.1.7) and defined using the exponential kernels $\exp(-j2\pi ft)$ and $\exp(j2\pi ft)$, respectively ($\omega = 2\pi f$). In digital signal processing a time signal $u(t)$ is substituted by a sequence of samples $u(i)$. Therefore, in the DFT the time variable t is replaced by the discrete integer variable i, $0 \leq i \leq N-1$, where N is the length of the sequence. The discrete signal has the form of a sequence of samples $u(0), u(1), u(2), \ldots, u(N-1)$. The DFT of this sequence is defined by the formula

$$U(k) = \sum_{i=0}^{N-1} u(i) e^{-jw}; \qquad w = 2\pi i k / N \qquad (7.20.1)$$

where k is a discrete integer frequency variable, $0 \leq k \leq N-1$. The discrete spectrum is periodic, that is, $U(k) = U(k+N) = U(k+2N) \ldots$. The inverse transformation denoted DFT^{-1} has the form

$$u(i) = \frac{1}{N} \sum_{k=0}^{N-1} U(k) e^{jw} \qquad (7.20.2)$$

The sequence generated by this inverse transformation is periodic, that is, $u(i) = u(i+N) = u(i+2N) = \ldots$. Usually of interest is the basic period. The comparison of Fourier integrals with the DFT shows that integration is replaced by summation and the exponential kernel $\exp(\pm j\omega t)$ is replaced by $\exp(\pm jw)$. The discrete Fourier pair may be shortened to

$$u(i) \xleftrightarrow{\text{DFT}} U(k) \qquad (7.20.3)$$

In general, for real sequences $u(i)$ the spectral function $U(k)$ is complex, that is,

$$U(k) = U_{re}(k) + j U_{im}(k) \qquad (7.20.4)$$

The real part is defined by the cosine DFT of the form

$$U_{re}(k) = \sum_{i=0}^{N-1} u(i) \cos(w) \qquad (7.20.5)$$

and the imaginary part by the sine DFT of the form

$$U_{im}(k) = -\sum_{i=0}^{N-1} u(i) \sin(w) \tag{7.20.6}$$

A given sequence may be resolved in two parts

$$u(i) = u_e(i) + u_0(i) \tag{7.20.7}$$

where for even values of N the even and odd parts are given by

$$u_e(i) = \frac{u(N/2 + i) + u(N/2 - i)}{2}; \qquad u_0(i) = \frac{u(N/2 + i) - u(N/2 - i)}{2} \tag{7.20.8}$$

with $N/2 \leq i \leq N - 1$. The cosine transform depends only on $u_e(i)$ and the sine transform on $u_0(i)$. Using the complex form (7.20.4) the inverse DFT may be written in the form

$$u(i) = \frac{1}{N} \sum_{k=0}^{N-1} [U_{re}(k) \cos(w) - U_{im}(k) \sin(w)] \tag{7.20.9}$$

The one-sided Z-transformation of the sequence $u(i)$ is defined by the formula (see also Chapter 6)

$$U(z) = \sum_{i=0}^{N-1} u(i) z^{-i} \tag{7.20.10}$$

where the complex frequency variable $z = x + jy$ is continuous differently than the discrete frequencies used in the DFTs. We shall denote the Z-pair by

$$u(i) \overset{Z}{\Longleftrightarrow} U(z) \tag{7.20.11}$$

The discrete one-sided convolution is defined by the equation

$$y(i) = \sum_{m=0}^{N-1} h(i - m) u(m) \tag{7.20.12}$$

and if $h(i) \overset{Z}{\Longleftrightarrow} H(z)$ and $u(i) \overset{Z}{\Longleftrightarrow} U(z)$ then the well-known convolution to multiplication property yields the Z-pair

$$y(i) \overset{Z}{\Longleftrightarrow} H(z) U(z) \tag{7.20.13}$$

Because the DFT is periodic, it is a periodic function of the normalized frequency

$$\psi = 2\pi k/N \tag{7.20.14}$$

The basic period equals the interval $0 \leq \psi < 2\pi$, the next period is $2\pi \leq \psi < 4\pi$, and so forth. The DFT equals the Z-transform for values of z given by

$$z = e^{j\psi}; \qquad \psi = 2\pi k/N \tag{7.20.15}$$

that is, equally spaced on the unit circle of the z-plane (see Figure 7.20.1). The half-period $0 \leq \psi < \pi$ (upper half-circle) is classified as positive frequencies and the other half-period, $\pi \leq \psi < 2\pi$, as negative frequencies. The insertion of Eq. (7.20.15) in (7.20.13) yields the k-domain form of the multiplication to convolution theorem

$$y(i) = u(i) * h(i) \overset{DFT}{\Longleftrightarrow} U(k) H(k) \tag{7.20.16}$$

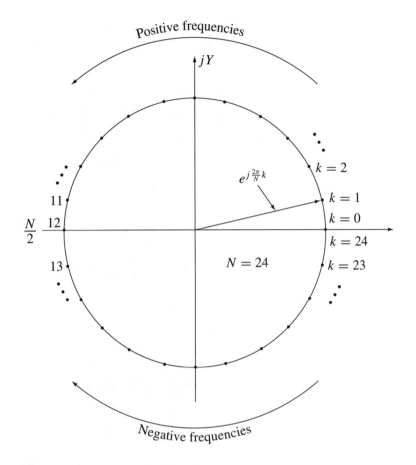

FIGURE 7.20.1
The unit circle in the $z = x + jy$ plane (see Eq. (7.20.15)), $N = 24$.

The discrete equivalent of the delta pulse is the Kronecker delta sample

$$\delta_K(i) = \begin{cases} 1 & \text{for } i = 0 \\ 0 & \text{for } i \neq 0 \end{cases} \tag{7.20.17}$$

The impulse response of the DLTI system defined as the response to the δ_K sample and the transfer function $H(z)$ of the system are forming a Z-pair

$$h(i) \overset{Z}{\Longleftrightarrow} H(z) \tag{7.20.18}$$

The insertion $z = e^{j\psi}$ ($\psi = 2\pi k/N$) yields the relation

$$h(i) \overset{\text{DFT}}{\Longleftrightarrow} H(k) \tag{7.20.19}$$

The transfer function of an ideal discrete Hilbert filter is defined by the equation (N even) [7]

$$H(k) = \begin{cases} -j & \text{for } k = 1, 2, \ldots, N/2 - 1 \\ 0 & \text{for } k = 0 \text{ and } N/2 \\ j & \text{for } k = N/2 + 1, N/2 + 2, \ldots, N - 1 \end{cases} \tag{7.20.20}$$

This transfer function may be written in the closed form

$$H(k) = -j \operatorname{sgn}(N/2 - k) \operatorname{sgn}(k) \tag{7.20.21}$$

where

$$\operatorname{sgn}(x) = \begin{cases} 1 & \text{for } x > 0 \\ 0 & \text{for } x = 0 \\ -1 & \text{for } x < 0 \end{cases} \tag{7.20.22}$$

The output sequence $v(i)$ of the Hilbert filter by a given input sequence $u(i)$ defines the discrete Hilbert pair

$$u(i) \overset{\text{DHT}}{\Longleftrightarrow} v(i) \tag{7.20.23}$$

The impulse response of the Hilbert filter is given by the inverse DFT of $H(k)$:

$$h(i) = \frac{1}{N} \sum_{k=0}^{N-1} H(k)e^{jw} = \frac{1}{N} \sum_{k=0}^{N-1} -j \operatorname{sgn}(N/2 - k) \operatorname{sgn}(k)e^{jw} = \frac{1}{N} \sum_{k=0}^{N-1} \sin(w) \tag{7.20.24}$$

($w = 2\pi i k/N$). The closed form of this sum is (see Fig. 7.20.2)

$$h(i) = \frac{2}{N} \sin^2(\pi i/2) \cot(\pi i/N) \tag{7.20.25}$$

Therefore, the impulse response is given by the samples of the cotangent function (compare with Eq. [7.6.27]) with the even samples ($i = 0, 2, 4, \ldots, N$) cancelled by the term $\sin^2(\pi i/2)$. The convolution to multiplication theorem (7.20.16) yields the DHT in the form of the convolution:

$$v(i) = -u(i) \otimes h(i) = -u(i) \otimes \frac{2}{N} \sin^2(\pi i/2) \cot(\pi i/N) \tag{7.20.26}$$

where the sign $\otimes$ denotes a so-called circular convolution. This convolution may be written in the form

$$v(i) = \sum_{r=0}^{N-1} h(i - r)u(r) \tag{7.20.27}$$

Concluding, the DHT of a given sequence $u(i)$ may be calculated using the above circular convolution or alternatively via the DFT using the algorithm:

$$u(i) \overset{\text{DFT}}{\Rightarrow} U(k) \Rightarrow V(k) = -j \operatorname{sgn}(N/2 - k) \operatorname{sgn}(k)U(k) \overset{\text{DFT}^{-1}}{\Rightarrow} v(i) \tag{7.20.28}$$

Both algorithms give exactly the same result. Of course, the convolution algorithm is faster, because it involves only a single summation. However, the DFT may be replaced by the FFT.

The above formulas apply for even values of N. If N is odd, the transfer function of the Hilbert filter has the form

$$H(k) = \begin{cases} -j & \text{for } k = 1, 2, \ldots, (N-1)/2 \\ 0 & \text{for } k = 0 \\ j & \text{for } k = N/2 + 1, (N+1)/2, \ldots, N - 1 \end{cases} \tag{7.20.29}$$

and the impulse response is

$$h(i) = \frac{2}{N} \sum_{k=1}^{(N-1)/2} \sin(2\pi i k/N); \qquad i = 0, 1, 2, \ldots, N - 1 \tag{7.20.30}$$

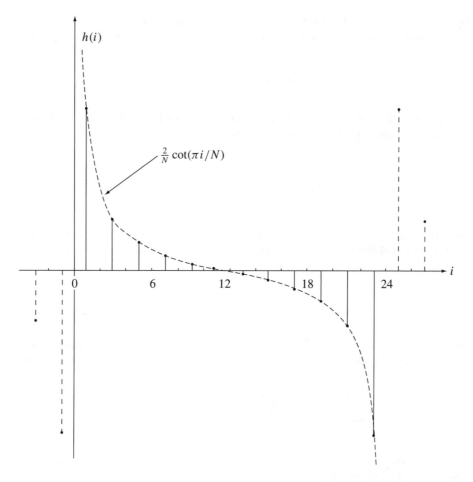

FIGURE 7.20.2
The noncausal impulse response of a Hilbert filter (see Eq. (7.20.5)), $N = 24$.

or

$$h(i) = \frac{1}{N}\left[1 - \frac{\cos(\pi i)}{\cos(\pi i/N)}\cot(\pi i/N)\right]$$

(7.20.31)

Properties of the DFT and DHT illustrated with examples

Parseval's theorem

Consider the discrete Fourier pair $u(i) \overset{\text{DFT}}{\Longleftrightarrow} U(k)$. The discrete form of the Parseval's energy (or power) equality has the form

$$E[u(i)] = \sum_{i=0}^{N-1} |u(i)|^2 = \frac{1}{N}\sum_{k=0}^{N-1} |U(k)|^2$$

(7.20.32)

This equation may be used to check the correctness of calculations of DFTs and DHTs. However, the energies of the sequences $u(i)$ and its DHT, $v(i)$, may differ, that is, in general,

$$E[u(i)] \neq E[v(i)] \tag{7.20.33}$$

The explanation is given by Eq. (7.20.27). The operator $-j\,\mathrm{sgn}(N/2 - k)\,\mathrm{sgn}(k)$ cancels the spectral terms $U(0)$ and $U(N/2)$. The term $U(0)$ has the form

$$U(0) = \sum_{i=0}^{N-1} u(i) = N u_{\mathrm{DC}} \tag{7.20.34}$$

where u_{DC} is the mean value of the signal sequence $u(i)$, or in electrical terminology, the DC term. The algorithm of DHT cancels this term. Therefore, the sequence $v(i)$ is defined by the DHT pair

$$u_{\mathrm{AC}}(i) \overset{\mathrm{DHT}}{\Longleftrightarrow} v(i) \tag{7.20.35}$$

where $u_{\mathrm{AC}}(i) = u(i) - u_{\mathrm{DC}}$ is the alternate current component of the signal sequence (with DC term removed). The energies of the sequences $u_{\mathrm{AC}}(i)$ and $v(i)$ are given by the equation

$$\sum_{i=1}^{N-1} |u_{\mathrm{AC}}(i)|^2 = \sum_{i=1}^{N-1} |v(i)|^2 + \frac{|U(N/2)|^2}{N} \tag{7.20.36}$$

that is, the energies differ by the energy of the spectral term $U(N/2)$ and only if this term equals zero are both energies equal.

Example

Consider the signal given by a Kronecker delta $u(0) = \delta_K(i)$ and $u(i) = 0$ for $i \geq 1$, $N = 8$. This sequence and its DFT are shown in Figure 7.20.3a,b. The circular convolution (7.20.26) yields in this case

$$v(i) = -\delta_K(i) * \sin^2(\pi i/2)\cot(\pi i/N) \tag{7.20.37}$$

that is, the following sequence

i	0	1	2	3	4	5	6	7
$v(i)$	0	$\cot(\pi/8)$	0	$\cot(\pi/4)$	0	$-\cot(\pi/4)$	0	$-\cot(\pi/8)$

where $\cot(\pi/8) = (\sqrt{2}+1)/4 = 0.6035\ldots$ and $\cot(\pi/4) = (\sqrt{2}-1)/4 = 0.1035\ldots$. The sequence $v(i)$ and its DFT are shown in Figure 7.20.3c,d. The DC term defined by Eq. (7.20.34) is $u_{\mathrm{DC}} = 1/N = 0.125$. For convenience, Figure 7.20.3e,f shows the sequence $u_{\mathrm{AC}}(i)$ and its DFT. The energies are: $E[u(i)] = 1$, $E[u_{\mathrm{AC}}(i)] = 1 - 1^2/N = 0.875$, $E[v(i)] = 1 - 1^2/N - 1^2/N = 1 - 2/N = 0.75$. ∎

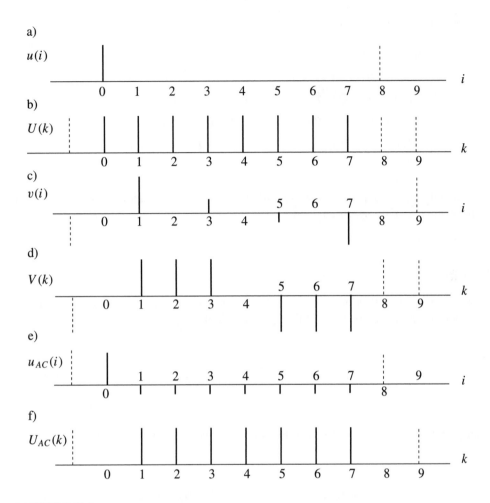

FIGURE 7.20.3
(a) The sequence $u(i)$ consisting of a single sample $\delta_K(i)$, (b) its spectrum $U(k)$ given by the DFT, (c) the samples of the discrete Hilbert transform, (d) the corresponding spectrum $V(k)$, (e) the samples of the AC component of $u(i)$, and (f) the corresponding spectrum $U_{AC}(k)$.

Shifting property

Consider the discrete Fourier pair $u(i) \overset{\text{DFT}}{\Longleftrightarrow} U(k)$. It can be shown that

$$u(i + m) \overset{\text{DFT}}{\Longleftrightarrow} e^{j2\pi mk/N} U(k) \tag{7.20.38}$$

where m is an integer.

Example
The spectrum of Figure 7.20.3b is real with all samples equal to 1. The shifted-by-one interval $(m = 1)$ delta pulse and its spectrum are

$$\delta_K(i - m) \overset{\text{DFT}}{\Longleftrightarrow} e^{-j2\pi k/N} \tag{7.20.39}$$

This spectrum is complex and of the form

k	0	1	2	3	4	5	6	7
$U_{re}(k)$	1	$\sqrt{2}/2$	0	$-\sqrt{2}/2$	-1	$-\sqrt{2}/2$	0	$\sqrt{2}/2$
$U_{im}(k)$	0	$-\sqrt{2}/2$	-1	$-\sqrt{2}/2$	0	$\sqrt{2}/2$	1	$\sqrt{2}/2$
$|U(k)|$	1	1	1	1	1	1	1	1

This example shows the general rule that shift changes in phase relations will have no effect on the magnitude of the spectrum. ∎

Linearity

Consider the discrete Fourier pairs $u_1(i) \overset{\text{DFT}}{\Longleftrightarrow} U_1(k)$ and $u_2(i) \overset{\text{DFT}}{\Longleftrightarrow} U_2(k)$. Due to the linearity property the summation of the sequencies yields

$$au_1(i) + bu_2(i) \overset{\text{DFT}}{\Longleftrightarrow} aU_1(k) + bU_2(k) \qquad (7.20.40)$$

where a and b are constants. The linearity property applies also for the DHTs:

$$au_1(i) + bu_2(i) \overset{\text{DHT}}{\Longleftrightarrow} av_1(i) + bv_2(i) \qquad (7.20.41)$$

Example
Consider the sequence of two deltas $u(i) = \delta_K(i) + \delta_K(i-1)$ for $i = 0$ and 1 and $u(i) = 0$ for $1 < i \leq N - 1$, $N = 8$. The DFT of this sequence may be obtained by adding to each term of the real part of the spectrum given by Eq. (7.20.39) the number 1, that is, the terms of the spectrum of $\delta_K(i)$ (see Fig. 7.20.3b). This yields the complex spectrum

k	0	1	2	3	4	5	6	7
$U_{re}(k)$	2	$1+\sqrt{2}/2$	1	$1-\sqrt{2}/2$	0	$1-\sqrt{2}/2$	1	$1+\sqrt{2}/2$
$U_{im}(k)$	0	$-\sqrt{2}/2$	-1	$-\sqrt{2}/2$	0	$\sqrt{2}/2$	1	$\sqrt{2}/2$
$|U(k)|$	2	$1.847\ldots$	$\sqrt{2}$	$0.765\ldots$	0	$0.765\ldots$	$\sqrt{2}$	$1.847\ldots$

Notice that the term $U(N/2) = U(4)$ equals zero. Therefore, the energies $E[u_{AC}(i)] = E[v(i)] = 2 - 2^2/N = 1.5$ are equal. The DC term is $u_{DC} = 2/N = 0.25$. ∎

Example

Consider the sequence

$$u(i) = e^{-0.05\pi[(N-1)/2-i]^2}; \qquad N = 16 \tag{7.20.42}$$

representing a sampled Gaussian pulse as shown in Figure 7.20.4a. Figure 7.20.4b,c shows the DFT of this pulse and the DHT calculated via the DFT. The DC term equals $u_{DC} = 0.2795\ldots$. The energies are: $E[u(i)] = 3.1622\ldots, E[u_{AC}(i)] = E[v(i)] = 1.9122\ldots$, that is, the energy difference is negligible due to the negligible value of the term $U(N/2)$.

The complex analytic discrete sequence

A sequence of complex samples of a signal and its discrete Hilbert transform does not represent an analytic signal in the sense of the definition of the analytic function. However, it is possible to define the analytic sequence of the form of a sequence of samples

$$\psi(i) = u(i) + jv(i) \tag{7.20.43}$$

where $v(i)$ is the DHT of $u(i)$. Let us derive the spectrum of the sequence $\psi(i)$. If $u(i) \overset{\text{DFT}}{\Longleftrightarrow} U(k)$, then the spectrum of $v(i)$ is given by Eq. (7.20.28), and due to the linearity property, the spectrum of the complex sequence $\psi(i)$ is

$$\psi(i) \overset{\text{DFT}}{\Longleftrightarrow} U(k) + j[-j\,\mathrm{sgn}(N/2 - k)\,\mathrm{sgn}(k)]U(k)$$

that is,

$$\psi(i) \overset{\text{DFT}}{\Longleftrightarrow} [1 + \mathrm{sgn}(N/2 - k)\,\mathrm{sgn}(k)]U(k) \tag{7.20.44}$$

The spectrum is doubled at positive frequencies and canceled at negative frequencies. ∎

Example

Consider the signals and spectra of Figure 7.20.3. Figure 7.20.5 shows the real spectra of the delta pulse and its DHT and the resulting spectrum of the complex sequence. The terms of the spectrum of $u(i)$ are canceled at negative frequencies and doubled at positive frequencies. The DC term, that is, $U(0)$ is unaltered. The property that analytic sequences have a one-sided spectrum makes it possible to implement anti-aliasing schemes of sampling. ∎

The bilinear transformation and the cotangent form of Hilbert transformations

The transfer function of an analog LTI system is defined as the quotient of the output-to-input analytic signals (see Eq. [7.17.1]), and if analytical, is an analytic function of the complex frequency $s = \alpha + j\omega$. Similarly, the transfer function of the DLTI system defined by Eq. (7.20.18), if analytical, is an analytic function of the complex variable $z = x + jy$. Let us study the problem of a **conformal mapping** of the s-plane into the z-plane by means of the **bilinear transformations** defined by the formulae

$$z = \frac{1+s}{1-s} \tag{7.20.45}$$

and

$$s = \frac{z-1}{z+1} \tag{7.20.46}$$

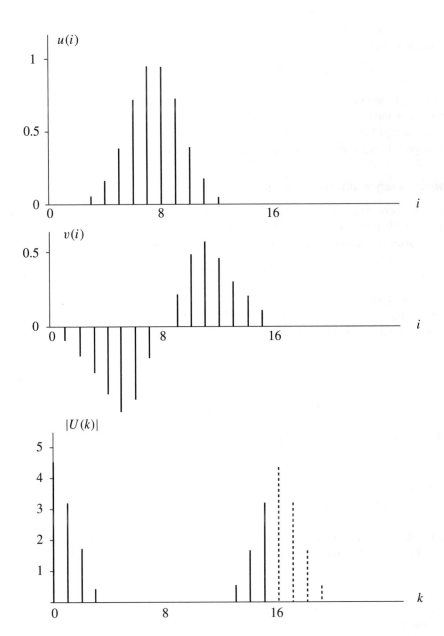

FIGURE 7.20.4
(a) A sequence of samples of a Gaussian pulse, (b) the samples of the DHT, and (c) the samples of the magnitude of the DFT of the Gaussian pulse.

where s is a normalized complex frequency (normalised $s = s/f_s = s\Delta t$, where f_s is the sampling frequency and Δt the sampling period). Inserting $s = \alpha + j\omega$ into Eq. (7.20.45) and equating the real and imaginary parts yields:

$$x = \frac{1 - \alpha^2 - \omega^2}{(1 - \alpha)^2 + \omega^2}; \qquad y = \frac{2\omega}{(1 - \alpha)^2 + \omega^2} \qquad (7.20.47)$$

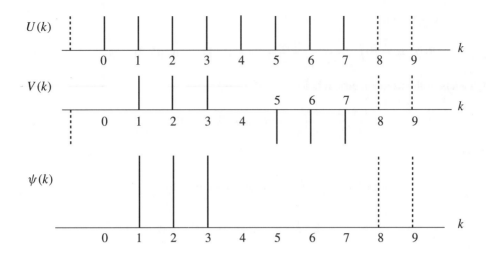

FIGURE 7.20.5
(a,b) The spectra $U(k)$ and $V(k)$ of Figure 7.20.3. (c) The corresponding spectrum of the analytic sequence.

These equations are mapping a family of orthogonal lines $\alpha = \text{const.}$ and $\omega = \text{const.}$ of the s-plane into a family of orthogonal circles of the z-plane, as shown in Figure 7.20.6. The magnitude of the variable is $|z| = \text{SQR}(x^2 + y^2)$ giving

$$|z| = \sqrt{\frac{(1+\alpha)^2 + \omega^2}{(1-\alpha)^2 + \omega^2}} \tag{7.20.48}$$

and the argument

$$\psi = \arg(z) = \tan^{-1}\left[\frac{2\omega}{1 - \alpha^2 - \omega^2}\right] \tag{7.20.49}$$

This equation defines the nonlinear dependence between the angular frequency ω and the normalized frequency ψ defined by the representation $z = e^{j\psi}$ (see Eq. [7.20.15]). For $s = j\omega$, that is, $\alpha = 0$, Eq. (7.20.49) takes the form of a quadratic equation

$$\tan(\psi)\omega^2 + 2\omega - \tan(\psi) = 0 \tag{7.20.50}$$

The roots of this equation are

$$\omega = \tan(\psi/2) \tag{7.20.51}$$

and

$$\omega = -\cot(\psi/2) \tag{7.20.52}$$

Let us use these nonlinear relations to derive a new form of Hilbert transformations. We start with the Hilbert transformation

$$B(\omega) = -\frac{1}{\pi} P \int_{-\infty}^{\infty} \frac{A(\eta)}{\eta - \omega} \, d\eta \tag{7.20.53}$$

Let us introduce the notations

$$\eta = \tan(\phi/2); \qquad \omega = \tan(\psi/2) \tag{7.20.54}$$

and $d\eta = 0.5[1 + \tan^2(\phi/2)]d\phi$. We get

$$B(\psi) = \frac{1}{\pi} P \int_{-\pi}^{\pi} \frac{A[\tan(\phi/2)]}{\tan(\phi/2) - \tan(\psi/2)} 0.5 \left[1 + \tan^2(\phi/2)\right] d\phi \qquad (7.20.55)$$

By means of the trigonometric relation

$$\frac{1 + \tan^2(\phi/2)}{\tan(\phi/2) - \tan(\psi/2)} = \tan(\phi/2) + \cot[(\phi - \psi)/2] \qquad (7.20.56)$$

we get

$$B(\psi) = -\frac{1}{2\pi} \int_{-\pi}^{\pi} A[\tan(\phi/2)] \tan(\phi/2) d\phi$$

$$-\frac{1}{2\pi} \int_{-\pi}^{\pi} A[\tan(\phi/2)] \cot[(\phi - \psi)/2] \, d\phi \qquad (7.20.57)$$

If we start with the inverse Hilbert transformation

$$A(\omega) = \frac{1}{\pi} P \int_{-\infty}^{\infty} \frac{B(\eta)}{\eta - \omega} \, d\eta \qquad (7.20.58)$$

the same derivation gives

$$A(\psi) = \frac{1}{2\pi} \int_{-\pi}^{\pi} B[\tan(\phi/2] \tan(\phi/2) \, d\phi$$

$$+\frac{1}{2\pi} \int_{-\pi}^{\pi} B[\tan(\phi/2)] \cot[(\phi - \psi)/2] \, d\phi \qquad (7.20.59)$$

The first term of Eq. (7.20.57) is a constant depending only on the even part of $A[\tan(\phi/2)]$, while the first term of Eq. (7.20.59) depends only on the odd part of $B[\tan(\psi/2)]$.

If we use instead of Eq. (7.20.51) the next root defined by Eq. (7.20.52), then Hilbert transformations (7.20.57) and (7.20.59) have the alternative form:

$$B(\psi) = \frac{-1}{2\pi} \int_0^{2\pi} A[-\cot(\phi/2)] \cot(\phi/2) \, d\phi$$

$$-\frac{1}{2\pi} \int_0^{2\pi} A[-\cot(\phi/2)] \cot[(\phi - \psi)/2] \, d\phi \qquad (7.20.60)$$

$$A(\psi) = \frac{1}{2\pi} \int_0^{2\pi} B[-\cot(\phi/2)] \cot(\phi/2) \, d\phi$$

$$+\frac{1}{2\pi} \int_0^{2\pi} B[-\cot(\phi/2)] \cot[(\phi - \psi)/2] \, d\phi \qquad (7.20.61)$$

The Hilbert transforms in the cotangent form are periodic functions of the variable ψ.

Example
Consider the square function

$$A(\omega) = \begin{cases} 1 & \text{for } |\omega| < a \\ 0.5 & \text{for } |\omega| = a \\ 0 & \text{for } |\omega| > a \end{cases} \qquad (7.20.62)$$

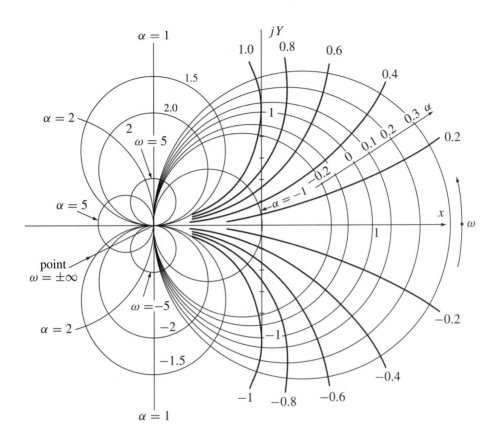

FIGURE 7.20.6
The mapping of the s-plane, $s = \alpha + j\omega$, into the z-plane, $z = x + jy$, defined by Eq. (7.20.47).

Introducing $\omega = \tan(\Psi/2)$ gives

$$A[\tan(\psi/2)] = \begin{cases} 1 & \text{for } |\psi| < \psi_p = 2\tan^{-1}(a) \\ 0.5 & \text{for } |\psi| = \psi_p \\ 0 & \text{for } |\psi| > \psi_p \end{cases} \tag{7.20.63}$$

The Hilbert transform defined by Eq. (7.20.61) is here

$$B(\psi) = -\frac{1}{2\pi} \int_{-\psi_p}^{\psi_p} \tan\left[\frac{\phi}{2}\right] d\phi - \frac{1}{2\pi} \int_{-\psi_p}^{\psi_p} \cot\left[\frac{\phi - \psi}{2}\right] d\phi \tag{7.20.64}$$

The first integral equals zero and the result of the second integration (Cauchy Principal Value (CPV) value) is

$$B(\psi) = \frac{1}{\pi} \ln \left| \frac{\sin\frac{\psi_p + \psi}{2}}{\sin\frac{\psi_p - \psi}{2}} \right| \tag{7.20.65}$$

Figure 7.20.7 shows $B(\psi)$ for two values of ψ_p: 0.4π and 0.1π corresponding to the normalized frequencies $\omega \cong .726$ and 0.155. The functions $A(\psi)$ and $B(\psi)$ are periodic with the period of 2π. ∎

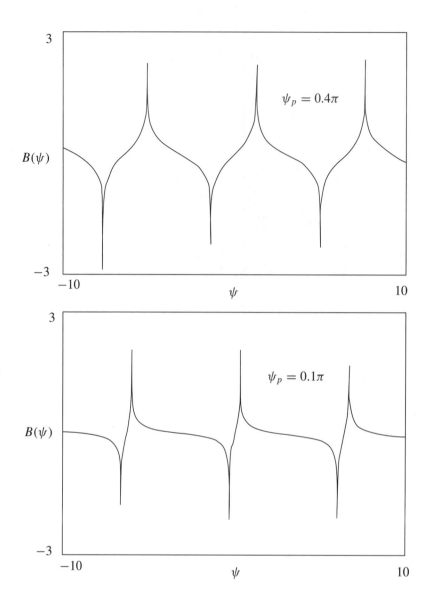

FIGURE 7.20.7
The function $B(\psi)$ given by Eq. (7.20.65).

7.21   Hilbert transformers (filters)

The **Hilbert transformer**, also called a **quadrature filter** or wide-band 90° phase shifter, is a device in the form of a linear two-port whose output signal is a Hilbert transform of the input signal. Hilbert transformers find numerous applications, for example, in radar systems, single side-band modulators, speech processing, measurement systems, schemes of sampling band-pass signals and many other systems. They are implemented as analog or digital filters.

The transfer function of the ideal analog Hilbert filter is (see Eq. [7.1.10])

$$H(jf) = F[1/(\pi t)] = |H(jf)|e^{j\varphi(f)} = -j\,\mathrm{sgn}(f) \qquad (7.21.1)$$

Hence, the transfer function is given by

$$H(jf) = \begin{cases} -j & \text{for } f > 0 \\ 0 & \text{for } f = 0 \\ j & \text{for } f < 0 \end{cases} \qquad (7.21.2)$$

The magnitude is $|H(jf)| = 1$ and the phase function is:

$$\varphi(f) = \arg[H(jf)] = -(\pi/2)\,\mathrm{sgn}(f) \qquad (7.21.3)$$

Notice that the convention with a $+$ sign by $\varphi(f)$ results in a negative slope of the phase function. The last equation explains the terminology "quadrature filter" or "wide-band 90° phase shifter." The ideal Hilbert filter is noncausal and physically unrealizable. Causality implies the introduction of an infinite delay. In any practical implementation of the Hilbert filter the output signal is a delayed and more or less distorted Hilbert transform of the input signal. The spectrum of the input signal should be band-limited between the low-frequency edge f_1 and high-frequency edge f_2 of the pass-band. The necessary delay depends only on f_1. Inside the pass-band $W = f_2 - f_1$, it is possible to get an approximate version of the transfer function defined by Eq. (7.21.1). Good approximations require sophisticated methods of design and implementations.

Hilbert transformers can be implemented in the form of analog or digital convolvers using the time definition of the Hilbert transforms given by Eqs. (7.1.3) and (7.1.4) (analog convolutions) or by Eq. (7.20.26) (discrete circular convolution). Another implementation uses so-called quadrature filters.

The performance of analog Hilbert transformers depends on design and alignment. Having in mind that ideal alignment is impossible and that even by good initial alignment it is detoriated by aging and various physical changes, for example, temperature, humidity, pressure, vibrations, and others, the use of extremely sophisticated design methods and implementations may be unreasonable. Differently, the performance of digital Hilbert transformers may depend only on design.

Because the magnitude of the transfer function defined by Eq. (7.21.2) equals 1, all-pass filters are frequently used in analog and digital implementations of Hilbert transformers.

Phase-splitter Hilbert transformers

Analog Hilbert transformers are mostly implemented in the form of a phase splitter consisting of two parallel all-pass filters with a common input port and separated output ports, as shown in Figure 7.21.1. The transfer functions of the all-pass filters are:

$$Y_1(jf) = e^{j\varphi_1(f)}; \qquad Y_2(jf) = e^{j\varphi_2(f)} \qquad (7.21.4)$$

The magnitude of both functions equals 1. The antisymmetry of the phase functions allows us to consider only the positive frequency part. The phase differrence of the harmonic signals at the output ports of the phase splitter should be:

$$\delta(f) = \varphi_1(f) - \varphi_2(f) = -\pi/2; \qquad \text{all } f > 0 \qquad (7.21.5)$$

The realization of this requirement is possible in a limited frequency band between the low-frequency edge f_1 and the high-frequency edge f_2, as shown in Figures 7.21.6 to 7.21.9. Therefore, the spectrum of the input signal should be band limited between f_1 and f_2. Due to

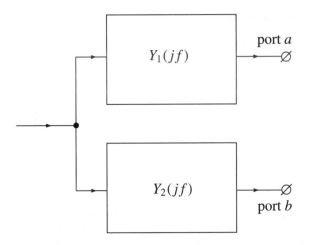

FIGURE 7.21.1
A phase splitter Hilbert transformer, where $Y_1(jf)$ and $Y_2(jf)$ are all-pass transfer functions.

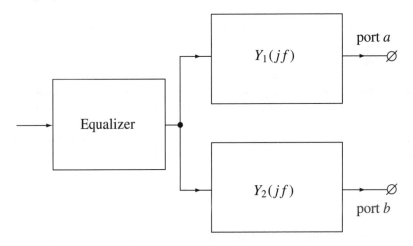

FIGURE 7.21.2
The series connection of a phase equalizer and the Hilbert transformer of Figure 7.21.1.

unavoidable amplitude and phase errors, the output signals of the phase splitter approximately are forming a Hilbert pair. The phase functions of the all-pass filters defined by Eq. (7.21.4) should be inside the band $W = f_2 - f_1$, approximately linear in the logarithmic frequency scale, but are nonlinear in a linear scale. This nonlinearity introduces phase distortions. Therefore, the output signals are forming a distorted in relation to the input signal Hilbert pair. The distortions can be removed using a suitable phase equalizer connected in series to the input port, as shown in Figure 7.21.2. By proper phase equalization the output signals are forming an undistorted pair of Hilbert transforms.

Analog all-pass filters

Hilbert transformers in the form of phase splitters are implemented using all-pass filters. A convenient choice is the all-pass consisting of two complementary filters, a low-pass and a high-pass, as shown in Figure 7.21.3a. The impedance $Z(j\omega) = X(j\omega)$ is a loss-less one-port (pure reactance). The transfer function of this all-pass has the form:

$$H(j\omega) = \frac{R - jX(\omega)}{R + jX(\omega)}; \qquad \omega = 2\pi f \tag{7.21.6}$$

The magnitude of this function equals one for all f and the phase function is

$$\varphi(\omega) = \arg\left[(R - jX)^2\right] = \tan^{-1}\left[\frac{-2RX}{R^2 - X^2}\right] \tag{7.21.7}$$

The insertion $X = 1/\omega C$ (see Fig. 7.21.3b) yields the phase function of a first-order all-pass

$$\varphi(y) = \tan^{-1}\left[\frac{-2y}{1 - y^2}\right]; \qquad y = \omega RC = \omega\tau \tag{7.21.8}$$

The insertion $X = \omega L - 1/\omega C$ (see Fig. 7.21.3c) yields a phase function of a second-order all-pass

$$\varphi(y) = \tan^{-1}\left[\frac{2(1 - y^2)qy}{(1 - y^2)^2 - q^2 y^2}\right] \tag{7.21.9}$$

where $y = \omega/\omega_r$, $\omega_r = 1/\sqrt{LC}$ and $q = \omega_r RC = R\sqrt{C/L}$. The phase functions defined by Eqs. (7.21.8) and (7.21.9) are shown in Figure 7.21.4 in linear and logarithmic frequency scales. The second-order function best shows linearity in the logarithmic scale for $q = 4$. Notice that the phase functions are continuous if we remove the phase jumps by π by changing the branch of a multiple-valued $\tan^{-1}$ function, similar to that in Figure 7.15.2. To get a wider frequency range of Hilbert transformers, higher order all-passes have to be applied. But more practical is the use of a series connection of first-order all-passes with appropriate staggering of the individual phase functions. For a given frequency band $W = f_2 - f_1$, optimum staggering yields the smallest value of the RMS phase error. The local value of the phase error is defined as a difference between $\delta(f)$ given by Eq. (7.21.5) and $-\pi/2$. Therefore the local error is

$$\varepsilon(f) = \delta(f) + \pi/2 \tag{7.21.10}$$

The design methods of $90°$ phase splitters were described by Dome [9] in 1946. Later Darlington [8], Orchard [27], Weaver [38] and Saraga [33] described design methods based on a Chebyshev approximation of a desired phase error. Tables and diagrams of these approximations can be found in Bedrosian [2].

A simple method of design of Hilbert phase splitters

Analog Hilbert transformers are designed using models of a given filter consisting of loss-less capacitors, low-loss inductors, ideal resistors, and ideal operational amplifiers. More accurate models that take into account spurious capacitances, inductances, and other spurious effects are sophisticated and rarely applied at the design stage. The alignment of circuits with an accuracy better than 0.5–1% is difficult to achieve. Having in mind the above arguments, the required accuracy of design of the parameters of the phase splitter is limited. Therefore, the simple method of design using a personal computer may be effective in many applications and is presented here.

The method consists of two steps. In the first step, the phase function $\varphi_1(f)$, given by Eq. (7.21.4), is linearized in the logarithmic frequency scale. In the second step, the phase function $\varphi_2(f)$ is obtained by shifting the function $\varphi_1(f)$ in order to get a minimum value of the RMS phase error defined by Eq. (7.21.5). The lower and upper frequency edges f_1 and f_2 are chosen as abcissae at which the error function diverges. The method is illustrated by four examples of design of Hilbert transformers given by the circuit models in Figure 7.21.5.

a)

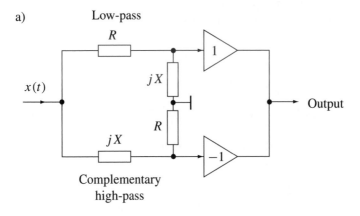

b)

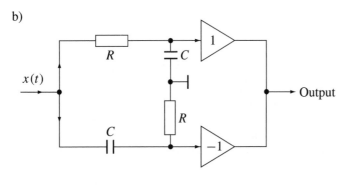

c)

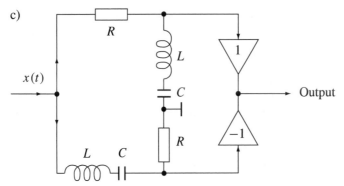

FIGURE 7.21.3
An all-pass consisting of (a) a low-pass and a complementary high-pass, (b) a first-order RC low-pass and complementary CR high-pass, and (c) a second-order RLC low-pass and complementary RLC high-pass.

Example

First example: The Hilbert transformer of this example is implemented using two first-order all-pass filters (see Fig. 7.21.5a). The phase function of the first filter is (see Eq. [7.21.8])

$$\varphi_1(f) = \tan^{-1}\left[\frac{-2y}{y^2 - 1}\right]; \qquad y = 2\pi f RC = 2\pi f \tau \qquad (7.21.11)$$

The first step is abandoned because $\varphi_1(f)$ has no degree of freedom for linearization. In the

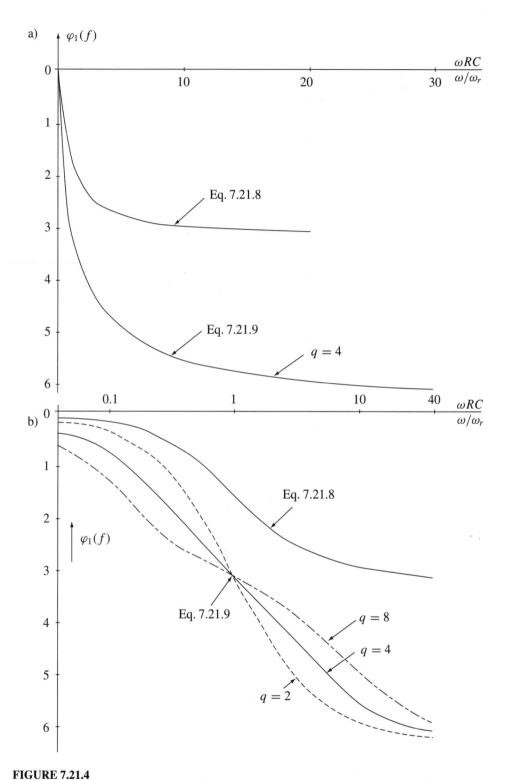

FIGURE 7.21.4
(a) Nonlinear phase functions of the first-order all-pass given by Eq. (7.21.8) and the second-order all-pass given by Eq. (7.21.9). (b) The same functions in a logarithmic frequency scale. The second-order function shows best linearity for $q = 4$.

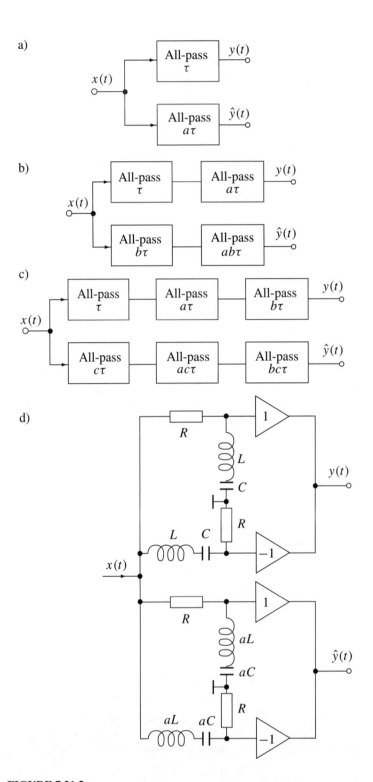

FIGURE 7.21.5
The phase splitter Hilbert transformer using (a) first-order all-pass filters, (b) a series connection of two first-order all-passes, (c) three first-order all-passes, and (d) second-order all-passes.

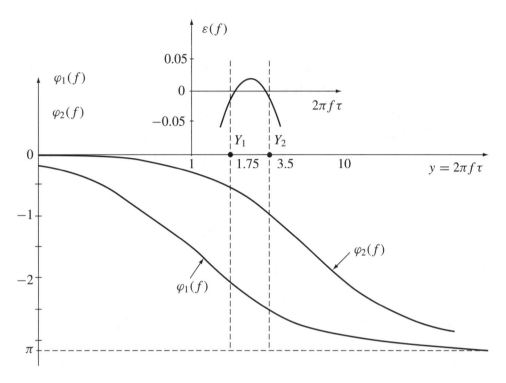

FIGURE 7.21.6
The phase functions and the phase error of the Hilbert transformer of Figure 7.21.5a.

second step we have to find the shift parameter denoted a in the phase function

$$\varphi_2(f) = \tan^{-1}\left[\frac{-2ay}{a^2y^2 - 1}\right] \tag{7.21.12}$$

giving the minimum RMS phase error. The functions $\varphi_1(f)$, $\varphi_2(f)$, and the error function $\varepsilon(f)$ are shown in Figure 7.21.6. Simple computer calculations yield the value of $a = 0.167$ giving the normalized frequency edges $y_1 = 1,75$ and $y_2 = 3,5$ and the RMS phase error $\varepsilon_{\text{RMS}} = 0.012$. The pass-band equals one octave.

Second example: The phase splitter of this example is implemented using two first-order all-pass filters in each chain (see Figure 7.21.5b). The phase function of the first filter is

$$\varphi_1(f) = \tan^{-1}\left[\frac{-2y}{y^2 - 1}\right] + \tan^{-1}\left[\frac{-2ay}{a^2y^2 - 1}\right] \tag{7.21.13}$$

In the first step we have to find the shift parameter a to get the best linearity of $\varphi_1(f)$ in the logarithmic scale. Small changes of a introduce a tradeoff between the RMS phase error and the pass-band of the Hilbert transformer. In the second step we have to find the value of the shift parameter b in the phase function

$$\varphi_2(f) = \tan^{-1}\left[\frac{2by}{b^2y^2 - 1}\right] + \tan^{-1}\left[\frac{2aby}{a^2b^2y^2 - 1}\right] \tag{7.21.14}$$

yielding the minimum of the RMS phase error. Figure 7.21.7 shows an example with $a = 0.08$ and $b = 0.24$ giving the normalized edge frequencies $y_1 = 1.6$ and $y_2 = 30$ ($f_2/f_1 = 18.75$ or more than 4 octaves) with $\varepsilon_{\text{RMS}} = 0.016$.

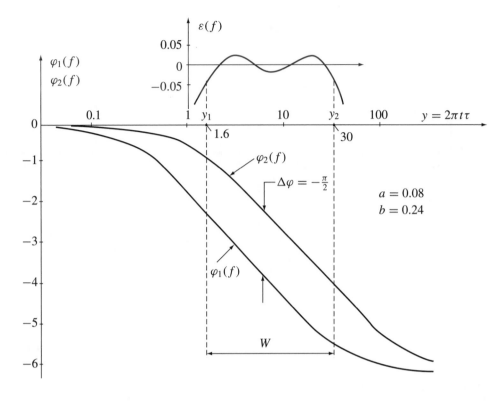

FIGURE 7.21.7
The phase functions and the phase error of the Hilbert transformer of Figure 7.21.5b.

Third example: The phase splitter consists of three first-order all-passes in each chain (see Figure 7.21.5c). The phase functions are

$$\varphi_1(f) = \tan^{-1}\left[\frac{-2y}{y^2 - 1}\right] + \tan^{-1}\left[\frac{-2ay}{a^2y^2 - 1}\right] + \tan^{-1}\left[\frac{-2by}{b^2y^2 - 1}\right] \tag{7.21.15}$$

and

$$\varphi_2(f) = \tan^{-1}\left[\frac{-2cy}{c^2y^2 - 1}\right] + \tan^{-1}\left[\frac{-2cay}{c^2a^2y^2 - 1}\right] + \tan^{-1}\left[\frac{2cby}{c^2b^2y^2 - 1}\right] \tag{7.21.16}$$

Good linearity of the phase function $\varphi_1(f)$ depend on the shift parameters a and b. The first step yields $a = 0.08$ and $b = 0.008$. In the second step the parameter $c = 0.24$ yields the minimum value of the RMS phase error. Figure 7.21.8 shows the phase functions and the error distribution $\varepsilon(f)$. The RMS phase error is $\varepsilon_{\text{RMS}} = 0.025$. The edge frequencies are $y_1 = 1.8$, $y_2 = 300$ giving $f_2/f_1 = 166$ (more than 7 octaves). A smaller phase error may be achieved at the cost of the frequency range.

Fourth example: The phase splitter consists of one second-order all-pass in each chain (see Fig. 7.21.5d). The phase functions are:

$$\varphi_1(f) = \tan^{-1}\left[\frac{2(1 - y^2)qy}{(1 - y^2)^2 - q^2y^2}\right] \tag{7.21.17}$$

$$\varphi_2(f) = \tan^{-1}\left[\frac{2(1 - a^2y^2)qay}{(1 - a^2y^2)^2 - q^2a^2y^2}\right] \tag{7.21.18}$$

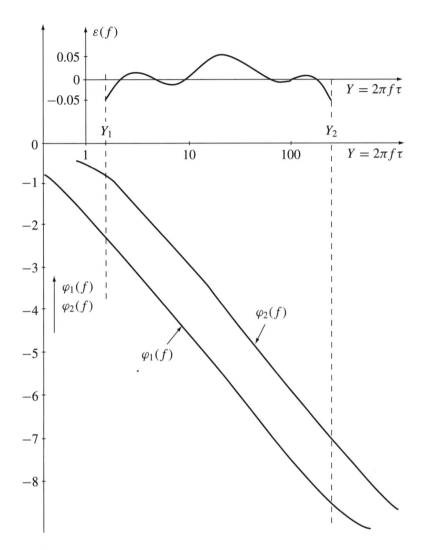

FIGURE 7.21.8
The phase functions and the phase error of the Hilbert transformer of Figure 7.21.5c.

Good linearity of $\varphi_1(f)$ yields the value $q = 4$ (see Figure 7.21.9). The minimum value of the RMS phase error yields the shift parameter $a = 0.232$. The phase functions and the error distribution are shown in Figure 7.21.4. The edge frequencies are $y_1 = 0.5$ and $y_2 = 9$ giving $f_2/f_1 = 18$ with $\varepsilon_{RMS} = 0.0186$. The bandwidth is about the same as in the second example with two first order all-passes in each chain.   ∎

Delay, phase distortions, and equalization

The phase functions of the all-pass filters used to implement the Hilbert transformer are, disregarding the small phase errors, linear in the logarithmic frequency scale, but nonlinear in a linear frequency scale. Let us investigate the phase distortions due to that nonlinearity for the Hilbert filter of the second example. Consider a wide-band test signal given by the Fourier

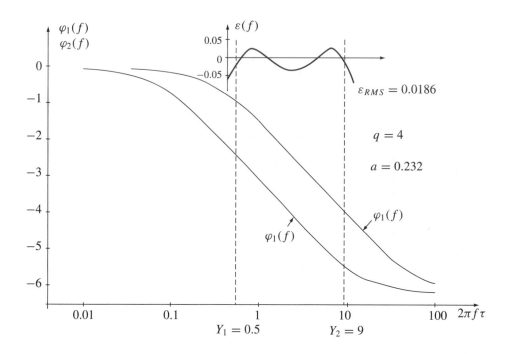

FIGURE 7.21.9
The phase functions and the phase error of the Hilbert transformer of Figure 7.21.5d.

series of a square wave truncated at the seventh harmonic term:

$$x(t) = \frac{4}{\pi}\left[\sin(\omega_1 t) + \frac{1}{3}\sin(3\omega_1 t) + \frac{1}{5}\sin(5\omega_1 t) + \frac{1}{7}\sin(7\omega_1 t)\right] \tag{7.21.19}$$

where $\omega_1 = 2\pi f_1 = 1.75/\tau$ was chosen near the low-frequency edge of the pass-band W. The spectrum of this signal is enclosed inside W. The waveforms of this signal and its Hilbert transform are shown in Figure 7.21.10a. The phase-distorted Hilbert pair at the output ports of the phase splitter is shown in Figure 7.21.10b. The phase distortions can be removed by connecting a phase equalizer in series to the input port, predistorting the input signal (see the waveform of Figure 7.21.10d). The required phase function of the equalizer may have the form

$$\varphi_{\text{equalizer}}(f) = \varphi_L(f) - \varphi_2(f) \tag{7.21.20}$$

where $\varphi_2(f)$ is given by Eq. (7.21.14) and

$$\varphi_L(f) = \varphi_2(f_0) + \left.\frac{d\varphi_2(f)}{df}\right|_{f=f_0}(f - f_0) \tag{7.21.21}$$

is a linear phase function tangential to $\varphi_2(f)$ at $f = f_0$. Figure 7.21.11 shows the phase function of the equalizer for three different values of the abcissae f_0. Figure 7.21.10c shows the delayed and practically undistorted output waveforms of the equalized Hilbert transformer with $f_0 = 0$. **The delay** is given by the slope of the phase function

$$t_0 = \left.\frac{d\varphi_2(f)}{df}\right|_{f_0=0} = -2\tau b(1+a) \tag{7.21.22}$$

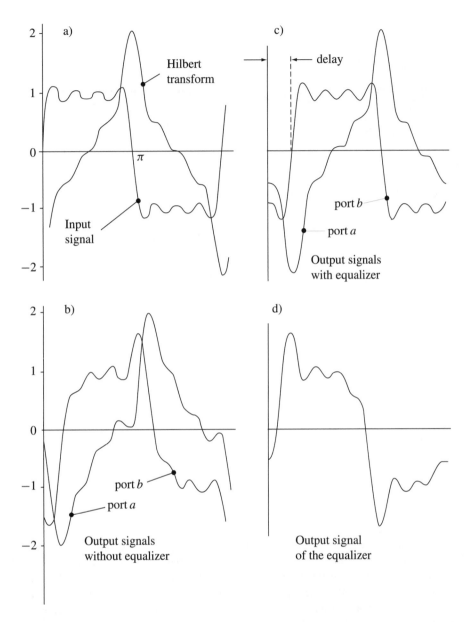

FIGURE 7.21.10
The waveform given by (a) the truncated Fourier series (7.21.15) and of its Hilbert transform, (b) the distorted Hilbert pair at the output with no equalization, (c) the equalized undistorted and delayed Hilbert pair, and (d) the input signal predistorted by the equalizer.

giving the delay $t_0 = 0.5065$ sec ($\tau = 1$). Another method of linearization of the phase function is given in [21].

Hilbert transformers with tapped delay-line filters

Tapped delay-line filters often referred to as transversal filters may be used as phase equalizers. Such a filter enables the approximation of a given transmittance $Y(jf)$ with a desired accuracy.

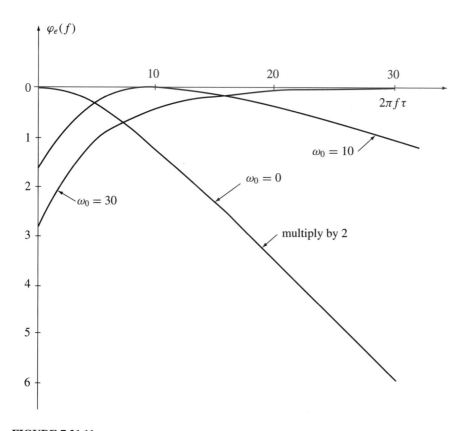

FIGURE 7.21.11
The phase functions of the equalizer given by Eq. (7.21.10) for the phase function $\varphi_2(f)$ given by Eq. (7.21.14).

Therefore, a Hilbert filter may be implemented using a tapped delay line [15], [34] (see Fig. 7.21.12). If the spectrum of the input signal is band-pass limited such that $X(f) = 0$ for $|f| > W$, then the transfer function of the ideal Hilbert transformer given by Eq. 7.21.2 may be truncated at $|f| = W$. The tapped delay-line Hilbert filter may be designed using a periodic repetition of this truncated function, as shown in Figure 7.21.13. The expansion of this function in a Fourier series yields, using truncation, the following approximate form of the transfer function

$$Y_N(jf) = 2j \sum_{i=1}^{(N-1)/2} b(i)\sin[i2\pi f t_0]; \qquad t_0 = \frac{1}{2W} \tag{7.21.23}$$

with

$$b(i) = -\frac{2}{\pi i}\sin^2\left[\frac{\pi i}{2}\right] \tag{7.21.24}$$

Different from the implementations of Hilbert transformers with all-pass filters, where the design amplitude error equals zero and the phase error is distributed over the pass-band, here the roles are interchanged. The amplitude error is distributed over the pass-band and there is no phase error (linear phase). The RMS amplitude ripple decreases with the increasing number of

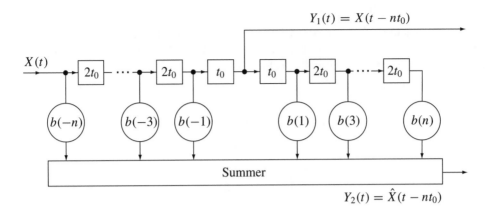

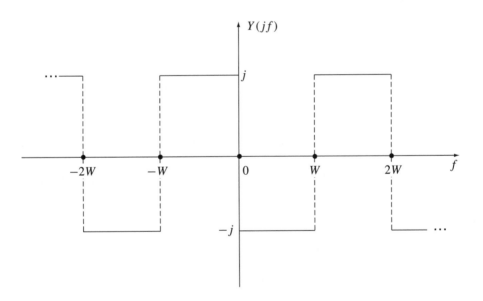

FIGURE 7.21.12
A tapped delay line Hilbert transformer.

FIGURE 7.21.13
A truncated at $\pm W$ and periodically repeated transfer function of an ideal Hilbert transformer (see Eq. [7.21.2]).

tapes of the delay line (increasing number of coefficients $b(n)$). The transversal Hilbert transformer, disregarding the small distortions due to the amplitude ripple, produces at the output a delayed undistorted signal and its Hilbert transform. However, analog implementations are rarely used in favor of digital implementations in the form of FIR (Finite Impulse Response) Hilbert transformers.

Band-pass Hilbert transformers

The transfer function of a band-pass Hilbert transformer may be defined as the frequency-translated transfer function of a low-pass Hilbert transformer. The transfer function of an ideal low-pass with linear phase is given by the formula

$$H_{LP}(jf) = \Pi[f/(2W)]e^{-j2\pi f\tau} \tag{7.21.25}$$

where τ is the time delay and $\Pi(x)$ has the form

$$\Pi(x) = \begin{cases} 1 & \text{for } |x| < 0.5 \\ 0.5 & \text{for } |x| = 0.5 \\ 0 & \text{for } |x| > 0.5 \end{cases} \tag{7.21.26}$$

This is illustrated in Figure 7.21.14. The impulse response of this filter is:

$$h_{LP}(t) = F^{-1}[H_{LP}(jf)] = 2W\frac{\sin X}{X} \tag{7.21.27}$$

where $X = 2\pi W(t - \tau)$. The response, as shown in Figure 7.21.15 is noncausal, but for large delays τ is nearly causal. The transfer function of the Hilbert transformer derived from Eq. (7.21.25) is given by

$$H_H(jf) = H_{LP}(jf)e^{-j[0.5\pi \, \text{sgn}(f) + 2\pi f\tau]} \tag{7.21.28}$$

as illustrated in Figure 7.21.14a,c. The impulse response of such a Hilbert transformer is

$$h_H(t) = F^{-1}[H_H(jf)] = \frac{1}{\pi(t - \tau)}[1 - \cos 2\pi W(t - \tau)] \tag{7.21.29}$$

or

$$h_H(t) = \frac{2\sin^2[\pi W(t - \tau)]}{\pi(t - \tau)} \tag{7.21.30}$$

This is illustrated in Figure 7.21.15b. If W goes to infinity the mean value of $h_H(t)$ taken over the period $T = 1/W$ approximates the distribution $1/(\pi(t - \tau))$. The transfer function of an ideal band-pass filter is given by

$$H_{BP}(jf) = \left\{\Pi\left[\frac{f + f_0}{2W}\right] + \Pi\left[\frac{f - f_0}{2W}\right]\right\}e^{-j2\pi f\tau} \tag{7.21.31}$$

This is illustrated in Figure 7.21.16a,b. The impulse response of this filter is

$$h_{BP}(t) = 2W\frac{\sin X}{X}\cos[2\pi f_0(t - \tau)] \tag{7.21.32}$$

and is shown in Figure 7.21.17a. The transfer function of an ideal band-pass Hilbert transformer derived from the transfer function (7.21.31) is

$$H_{HBP}(jf) = H_{BP}(jf)\exp\{-j0.5\pi[\text{sgn}(f + f_0) + \text{sgn}(f - f_0)]\} \tag{7.21.33}$$

This is illustrated in Figure 7.21.16a,c. The impulse response of this Hilbert transformer is

$$h_{HBP}(t) = \frac{2\sin^2[\pi W(t - \tau)]}{\pi(t - \tau)}\cos[2\pi f_0(t - \tau)] \tag{7.21.34}$$

and is shown in Figure 7.21.17b.

Consider the response of the band-pass Hilbert transformer to a band-pass signal $u_1(t) = x(t)\cos(2\pi f_0 t)$ where $x(t)$ has no spectral terms for $|f| > W$ and $f_0 > W$. This response has the form

$$u_2(t) = \hat{x}(t - \tau)\cos[2\pi f_0(t - \tau)] \tag{7.21.35}$$

that is, the modulating signal $x(t)$ is replaced by the delayed version of its Hilbert transform. Notice that due to Bedrosian's theorem the Hilbert transform of the input signal (see Section 7.12) has the form

$$u_2(t) = x(t - \tau)\sin[2\pi f_0(t - \tau)] \tag{7.21.36}$$

that is, only the carrier is Hilbert transformed, compared to signal (7.21.35), for which the envelope is transformed. The transfer function of a band-pass producing at the output the Hilbert transform in agreement with Bedrosian's theorem is given by the equation

$$H_{\text{BBP}}(jf) = -j\,\text{sgn}(f)H_{\text{BP}}(jf) \tag{7.21.37}$$

where $H_{\text{BP}}(jf)$ is given by Eq. (7.21.31) and is shown in Figure 7.21.18.

A possible implementation of a band-pass Hilbert transformer defined by Eq. (7.21.31) is shown in Figure 7.21.19. It consists of a linear phase lower side-band band-pass, analogous upper side-band band-pass, and a substractor. Figure 7.21.20 shows the implementation of such a Hilbert transformer by use of a SAW (Surface Acoustic Wave) filter.

Generation of Hilbert transforms using SSB filtering

The Hilbert transform of a given signal may be obtained by band-pass filtering of a double side-band AM signal. The SSB signal has the form (see Section 7.16)

$$u_{\text{SSB}}(t) = x(t)\cos(2\pi F_0 t) \mp \hat{x}(t)\sin(2\pi F_0 t) \tag{7.21.38}$$

where F_0 is the carrier frequency. Such a signal can be obtained by band-pass filtering of a double side-band AM signal. A synchronous demodulator using the quadrature carrier $\sin(2\pi F_0 t)$ generates at his output the Hilbert transform $\hat{x}(t)$.

Digital Hilbert transformers

The ideal discrete-time Hilbert transformer is defined as an all-pass with a pure imaginary transfer function, that is, if

$$\begin{aligned} H\left(e^{j\psi}\right) &= H_r(\psi) + jH_i(\psi), \qquad \text{then} \\ H_r(\psi) &= 0, \qquad \text{all } f \end{aligned} \tag{7.21.39}$$

and

$$H\left(e^{j\psi}\right) = jH_i(\psi) = \begin{cases} -j & 0 < \psi < \pi \\ 0 & \psi = 0,\ |\psi| = \pi \\ j & -\pi < \psi < 0 \end{cases} \tag{7.21.40}$$

or in another equivalent notation

$$H\left(e^{j\psi}\right) = -j\,\text{sgn}(\sin\psi) = -\text{sgn}(\sin\psi)e^{j\pi/2} = |H(\psi)|\,e^{j\,\arg H(\psi)} \tag{7.21.41}$$

The magnitude (see Fig. 7.21.21) has the form

$$|H(\psi)| = |\text{sgn}(\sin\psi)| = \begin{cases} 1, & 0 < |\psi| < \pi \\ 0, & \psi = 0,\ |\psi| = \pi \end{cases} \tag{7.21.42}$$

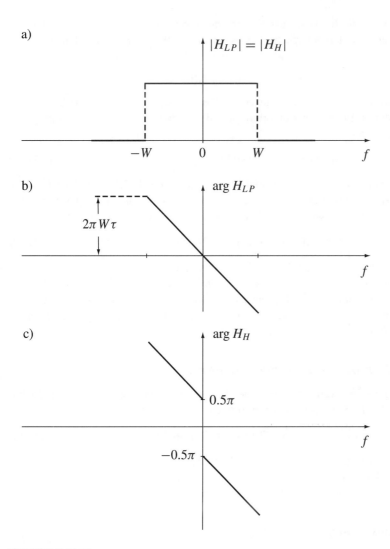

FIGURE 7.21.14
The transfer function of the ideal low-pass: (a) magnitude, (b) linear phase function, and (c) phase
function of a Hilbert transformer derived from the low-pass function.

and the phase function is

$$\arg[H(\psi)] = -(\pi/2)\,\mathrm{sgn}(\sin\psi) \tag{7.21.43}$$

Notice that $\psi = 2\pi f_n$, where $f_n = f/f_s$ is a frequency normalized against the sampling
frequency f_s. The basic period has the interval from $-\pi$ to π corresponding to the values
of f_n from -0.5 to 0.5. The noncausal infinite range impulse response of the ideal Hilbert
transformer has the form of the antisymmetric sequence

$$h(i) = (2/\pi i)\sin^2(i\pi/2) \tag{7.21.44}$$

If one allows the addition of a linear phase term in the ideal frequency response introducing
a frequency-independent group delay (in samples), then the transfer function (7.21.40) takes

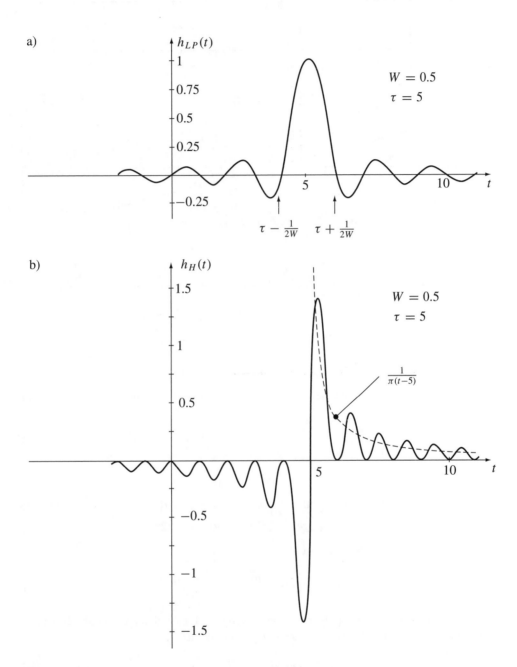

FIGURE 7.21.15
Impulse responses of (a) the low-pass and (b) the corresponding Hilbert transformer. Transfer functions are shown in Figure 7.21.14.

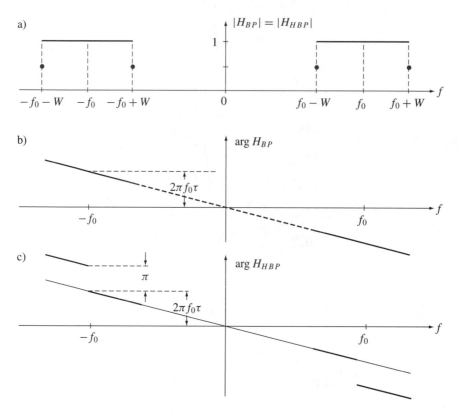

FIGURE 7.21.16
The transfer functions of an ideal band-pass filter and of the corresponding Hilbert transformer: (a) the magnitude, (b) the phase function of the bandpass, and (c) the Hilbert transformer.

the form

$$H\left(e^{j\psi}\right) = \begin{cases} -je^{-j\psi\tau} & 0 < \psi < \pi \\ 0 & \psi = 0, \ |\psi| = \pi \\ je^{-j(\psi - 2\pi)\tau} & \pi < \psi < 2\pi \end{cases} \tag{7.21.45}$$

and the impulse response takes the form

$$h(i) = (2/\pi)\frac{\sin^2[(\pi/2)(i - \tau)]}{i - \tau} \tag{7.21.46}$$

The important feature of the impulse response given by (7.21.44) is that even-numbered samples are exactly zero, and that the samples are antisymmetric, that is,

$$h(i) = -h(-i); \qquad i = 0, 1, \ldots \tag{7.21.47}$$

Methods of design

There are several methods of realizing digital Hilbert transformers. The three basic implementations are:

1. The **FIR** (Finite Impulse Response) Hilbert transformer

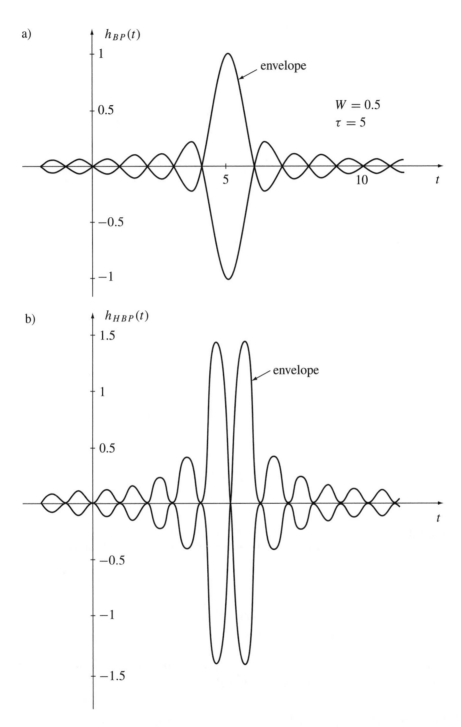

FIGURE 7.21.17
The envelopes of the impulse responses of (a) a bandpass and (b) the Hilbert transformer. Transfer functions are shown in Figure 7.21.16.

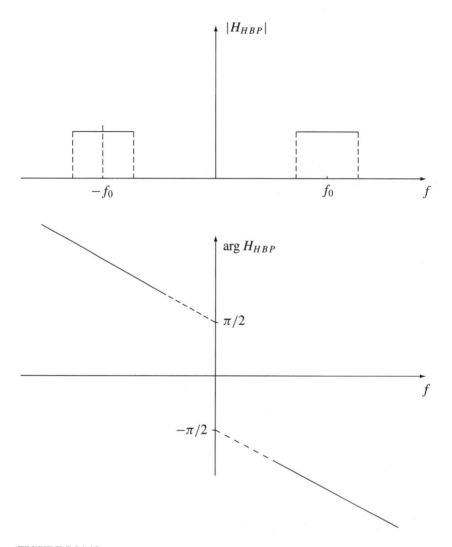

FIGURE 7.21.18
The transfer function of a Hilbert transformer that transforms the carrier signal with no change of the waveform of the envelope.

2. The **IIR** (Infinite Impulse Response) Hilbert transformer
3. Digital phase splitter Hilbert transformer

It is possible to realize a differentiating Hilbert transformer that produces at the output the derivative of the Hilbert transform of the input signal.

FIR Hilbert transformers [25], [31]

The FIR Hilbert transformer is a digital version of the taped delay line Hilbert transformer (see Section 7.21). Its structure is shown in Figure 7.21.22. The string of z^{-1} delays acts as a discrete taped delay line. Such a filter is inherently stable and its impulse response is given by the coefficients (gains) $h(0), h(1), h(2), \ldots, h(i), \ldots, h(N-1)$, that is, has the length of

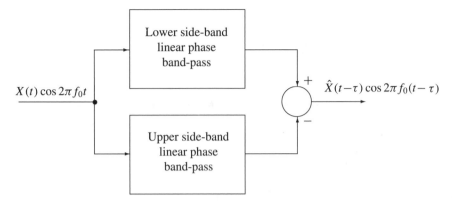

FIGURE 7.21.19
The implementation of the Hilbert transformer of the transfer function defined in Figure 7.21.16 using two bandpass filters.

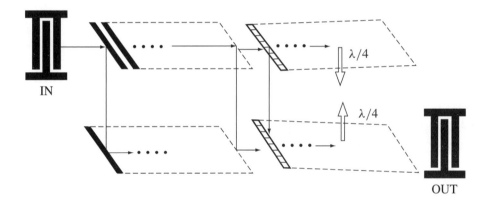

FIGURE 7.21.20
A SAW filter implementing the band-pass Hilbert transformer of Figure 7.21.19.

N samples. An example of the impulse response of the FIR Hilbert transformer is shown in Figure 7.21.23b, where for convenience, N is an odd number. This causal impulse response is obtained by a truncation and shifting by $(N-1)/2$ samples of the infinite impulse response of the ideal Hilbert transformer shown in Figure 7.21.23a. The transfer function of the Hilbert filter defined by the causal impulse response (see Eqs. [7.20.10] and [7.20.11]) is given by the Z-transform

$$H_H(i_1) = \sum_{i_1=0}^{N-1} h_1(i_1) z^{-i_1} \qquad (7.21.48)$$

where i_1 is the discrete coordinate given in the Eq. (7.21.49). The shifted causal impulse response $h_1(i_1)$ and the noncausal impulse $h(i)$ of Figure 7.21.23a satisfy the relation

$$h_1[i + (N-1)/2] = h(i); \qquad i_1 = i + (N-1)/2 \qquad (7.21.49)$$

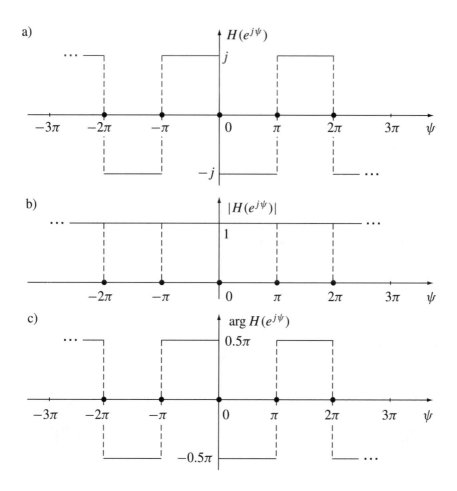

FIGURE 7.21.21
An ideal discrete-time Hilbert transformer's (a) transfer function, (b) magnitude, and (c) phase function.

The insertion of $z = e^{j\psi}$ in Eq. (7.21.48) (see Eq. [7.20.15]) and using (7.21.49) yields

$$H_H\left(e^{j\psi}\right) = e^{-j\psi(N-1)/2} \sum_{i=-(N-1)/2}^{(N-1)} h(i)e^{-j\psi i} \qquad (7.21.50)$$

Using Euler's formula for the sine function and the relation $h(i) = -h(-i)$, this transfer function takes the form

$$H_H\left(e^{j\psi}\right) = e^{-j\psi(N-1)/2} \sum_{i=1}^{(N-1)/2} -j2h(i)\sin(\psi i) \qquad (7.21.51)$$

Because every second sample of the impulse response equals zero, the summation should be written: from $i = 0$ to $(N-1)/2$ step 2. Let us denote this by

$$G\left(e^{j\psi}\right) = - \sum_{i=1}^{(N-1)/2} 2h(i)\sin(\psi i) \qquad (7.21.52)$$

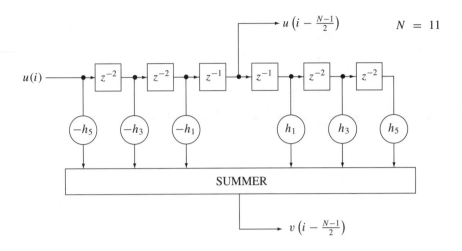

FIGURE 7.21.22
The structure of the FIR Hilbert transformer.

This function has the form of a Fourier series and defines the amplitude of the transfer function of the Hilbert transformer (this is not the magnitude because $G(j^{j\psi})$ has positive and negative values). An example is shown in Figure 7.21.24. The normalized dimensionless pass-band of this Hilbert transformer is given by the edge frequencies

$$W_\psi = \psi_2 - \psi_1 = \pi - 2\Delta \tag{7.21.53}$$

Because $\psi = 2\pi f/f_s$, where f is the frequency in [Hz] and f_s is the sampling frequency, the pass-band in [Hz] is

$$W_f[\text{Hz}] = \frac{\pi - 2\Delta}{2\pi} f_s \tag{7.21.54}$$

The pass-band increases and the amplitude δ of the ripple decreases with increasing length N of the impulse response, that is, at the cost of the delay, which equals $(N-1)/2$ samples. The amplitude ripple in the pass-band depends on the coefficients $h(i)$ in the Fourier series (7.21.52). Let us consider three cases:

1. The coefficients $h(i)$ are given by the Fourier series of an odd square periodic function of the form

$$G\left(e^{j\psi}\right) = -\frac{4}{\pi}\left[\sin(\psi i) + \frac{1}{3}\sin(3\psi i) + \frac{1}{5}\sin(5\psi i) + \right.$$

$$\left. \cdots + \frac{1}{2i+1}\sin[\psi i(N-1)/2]\right] \tag{7.21.55}$$

   corresponding to the truncation of the Fourier series by a rectangular window. This yields a nonequiripple amplitude distribution with "Gibbs peaks" at the edges of the pass-band, as shown in Figure 7.21.24.

2. The coefficients $h(i)$ in the above Fourier series are changed using an appropriate spectral window function, for example, Blackman, Hamming, or Kaiser windows. This yields a more uniform amplitude ripple.

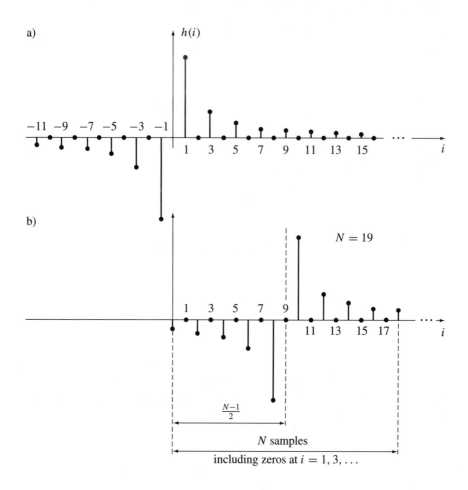

FIGURE 7.21.23
Impulse responses of (a) the ideal discrete time Hilbert transformer (see Eq. [7.21.44]) and (b) a FIR Hilbert transformer given by the truncation and shifting of the impulse response shown in (a).

3. The coefficients $h(i)$ are calculated to obtain an equiripple amplitude distribution in the pass-band in the mini-max or Tchebycheff sense, for example using the Parks-McClellan algorithm [22]. Figure 7.21.25 shows an example for $N = 19$. The product of N and Δ is given by the asymptotic relation derived by Kaiser [29]

$$N\Delta \cong 0.61 \log_{10} \delta \qquad (7.21.56)$$

Concluding, the FIR Hilbert transformer has a linear-phase characteristic and an amplitude ripple in the pass-band depending on design. Odd values of N are preferred. Design with even N is possible but inconvenient because all the impulse response coefficients are non-zero and the frequency response cannot have the required symmetry. A symmetric FIR Hilbert transformer (odd N) may be eventually derived from corresponding designs of symmetric half-band FIR filters [16].

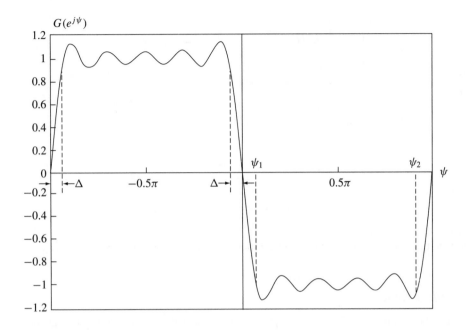

FIGURE 7.21.24
The $G(e^{j\psi})$ function of a FIR Hilbert transformer defined by the Fourier series (7.21.55).

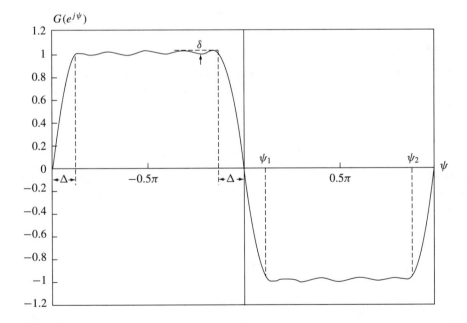

FIGURE 7.21.25
The equiripple $G(e^{j\psi})$ function of a FIR Hilbert transformer designed in the mini-max or Tchebycheff sense.

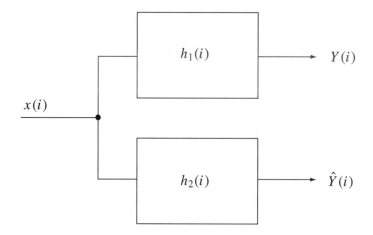

FIGURE 7.21.26
The discrete-time (or digital) version of the phase splitter Hilbert transformer of Figure 7.21.1.

Digital phase splitters

A digital Hilbert transformer may be implemented in the form of a digital phase splitter as shown in Figure 7.21.26. The transfer functions of the all-pass filters may be derived directly from the analog transfer functions by use of the bilinear frequency transformation (see Section 7.20). Details of the procedure can be found in any textbook on digital filters. All basic properties of the analog implementation are conserved. Without a phase equalizer the output Hilbert pair is distorted in reference to the input signal. Nonlinearity of the bilinear transformation introduces some tradeoffs not present in the analog case.

IIR Hilbert transformers

IIR Hilbert transformers may be derived using noncausal generalized half-band filters. Generalized half-band filters are derived by modifying the conventional elliptic filter design so that all poles of the half-band filter lie on the imaginary axis.The IIR ideal half-band transfer function proposed by Ansari [1] has the form

$$H_{\text{HB}}(z) = 1 + z^{-1} G\left(z^2\right) \tag{7.21.57}$$

where $G(z^2)$ is an all-pass filter with unit magnitude. The ideal example of this transfer function is shown in Figure 7.21.27a. Let us show that the transfer function of an ideal IIR Hilbert transformer is given by

$$H_H(z) = z^{-1} G\left(-z^2\right) \tag{7.21.58}$$

This is illustrated step-by-step in Figure 7.21.27. The term

$$F(z) = z^{-1} G\left(z^2\right) \tag{7.21.59}$$

is an all-pass with $F(e^{j\psi})$ shown in Figure 7.21.27b. It has a unit magnitude and a phase function equal to zero in the pass-band and $\pm\pi$ in the stop-band as shown in Figure 7.21.27d. This phase function can be written in the form

$$\Phi(\psi) = 0.5\pi \left[\text{sgn}\left(\sin(2\psi)\right) - \text{sgn}(\psi)\right] \tag{7.21.60}$$

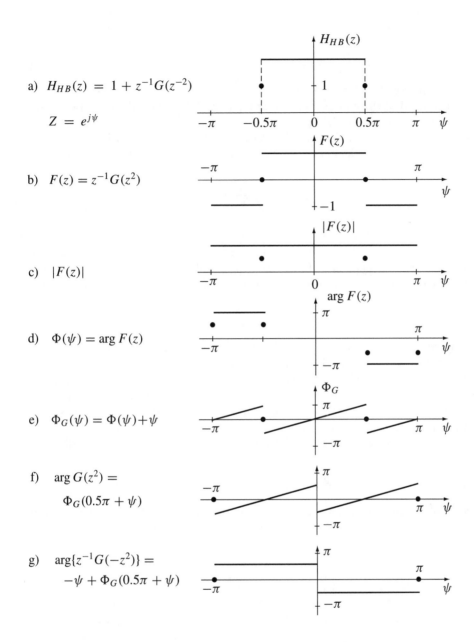

a) $H_{HB}(z) = 1 + z^{-1}G(z^{-2})$

$Z = e^{j\psi}$

b) $F(z) = z^{-1}G(z^2)$

c) $|F(z)|$

d) $\Phi(\psi) = \arg F(z)$

e) $\Phi_G(\psi) = \Phi(\psi) + \psi$

f) $\arg G(z^2) =$
$\Phi_G(0.5\pi + \psi)$

g) $\arg\{z^{-1}G(-z^2)\} =$
$-\psi + \Phi_G(0.5\pi + \psi)$

FIGURE 7.21.27
Step-by-step derivation of the IIR transfer function of a Hilbert transformer defined by Eq. (7.21.58),
starting from the transfer function of the ideal half-band filter given by (7.21.57).

The term $F(z)$ can be written in the form ($z = e^{j\psi}$)

$$F(e^{j\psi}) = e^{-j\psi}e^{j\Phi_G(\psi)} = e^{j\Phi(\psi)} \tag{7.21.61}$$

where

$$\Phi_G(\psi) = \Phi(\psi) + \Psi \tag{7.21.62}$$

This phase function is shown in Figure 7.21.27e. Because $z^2 = e^{j2\psi}$ and $-z^2 = e^{j2(0.5\pi+\psi)}$ we have

$$H_H\left(e^{j\psi}\right) = e^{-j\psi}e^{j\Phi_G(0.5\pi+\psi)} \qquad (7.21.63)$$

The phase function $\Phi_G(0.5\pi + \psi)$ is shown in Figure 7.21.27f and is the same as the phase function of the ideal Hilbert transformer (see Eq. [7.21.43]) and finally the phase function of $H_H(z)$ is shown in Figure 7.21.27g.

Differently than FIR transformers, the above IIR Hilbert transformer is designed with an equiripple phase function and exact amplitude. The explicit form of the noncausal transfer function may have the form

$$H(z) = z^{-1}\sum_{i=1}^{N}\frac{1 - a_iz^2}{z^2 - a_i} \qquad (7.21.64)$$

where N is an integer. Let us present an example: Consider the IIR Hilbert transformer with the low-frequency edge $\Psi_1 = 0.02\pi$, the high-frequency edge $\psi_2 = 0.98\pi (\Delta = 0.02\pi)$, and with the required amplitude of the phase ripple $|\Delta\Phi| \leq 0.01\pi$. The following relation between the phase ripple and the stop-band amplitude ripple of the half-band filter was derived [1]

$$\delta = \sin(0.5\Delta\phi) \qquad (7.21.65)$$

Inserting $\Delta\Phi = 0.01\pi$ gives $\delta = 0.0157$. The design procedure described in Ref. [1a] was applied to find the filter coefficients $a(i)$ giving $a(1) = 5.36078$, $a(2) = 1.2655$, $a(3) = 0.94167$, and $a(4) = 0.53239$. The insertion of the coefficients in Eq. (7.21.64) enabled the calculation of the phase function shown in Figure 7.21.28. The phase error has a symmetric distribution around $\Psi = 0.5\pi$, that is, half of the sampling frequency. The pass-band of this Hilbert transformer covers about 4.5 octaves. The phase ripple in the pass-band may be eliminated using half-band Butterworth IIR filters. For this kind of filter the coefficients $a(i)$ in Eq. (7.21.64) are given by a simple formula

$$a_i = \tan^2[\pi i/(2N + 1)]; \qquad i = 1, 2, \ldots, N \qquad (7.21.66)$$

Figure 7.21.29 shows a family of maximum flat phase functions for $N = 2, 4$, and 6. The pass-band depends considerably on the permissible phase error at the edges. The edge frequencies for edge errors 0.1% and 1% are given in the table:

N	2		4		6	
edge error	ψ_1	ψ_2	ψ_1	ψ_2	ψ_1	ψ_2
0.1%	0.36	0.64	0.24	0.76	0.165	0.835
1%	0.265	0.735	0.165	0.835	0.115	0.885

The widest pass-band for $N = 6$ and 1% error covers about 3 octaves and the smallest for $N = 2$ and 0.1% only 1 octave. The frequency around which the phase function is maximally flat equals $\psi = 0.5\pi$. It can be shifted by use of a suitable digital-to-digital frequency transformation. The disadvantage of the Butterworth filter is that the ratio of the first to the last coefficient is very large and increases with N.

Differentiating Hilbert transformers

The differentiating Hilbert transformer is defined as a linear system the output of which is the derivative of the Hilbert transform of the input signal. In principle, a differentiating Hilbert transformer may be implemented as a cascade connection of a differentiator and a Hilbert transformer as shown in Figure 7.21.30. However, it may be designed as a specialized FIR

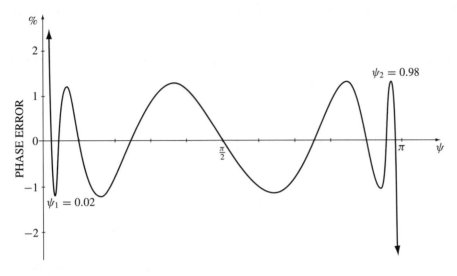

FIGURE 7.21.28
An example of the equi-ripple phase function of the IIR Hilbert transformer.

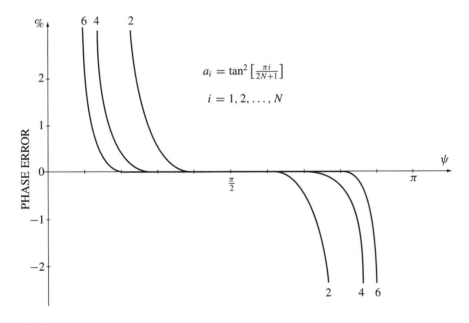

FIGURE 7.21.29
Phase errors of Butterworth IIR Hilbert transformers.

filter. Due to the cascade connection, the transfer function of the differentiating discrete Hilbert transformer is given by the product of the transfer function of the discrete Hilbert transformer given by Eq. 7.21.45 and the transfer function of the ideal discrete differentiator of the form [29]

$$H_D\left(e^{j\psi}\right) = \begin{cases} j\psi e^{-j\psi t} & 0 < \psi < \pi \\ 1 & |\psi| = k\pi; \qquad k = 0, 1, 2 \\ j(\psi - 2\pi)e^{-j(\psi - 2\pi)\tau} & \pi < \psi < 2\pi \end{cases} \qquad (7.21.67)$$

or using equivalent notation

$$H_D\left(e^{j\psi}\right) = \left[\psi + \text{sgn}\left(\sin(\psi)\right) - 1\right] e^{j[0.5\pi - \psi\tau - \pi\tau(\text{sgn}(\sin(\psi)) - 1)]}$$

$$0 < \psi < 2\pi \qquad (7.21.68)$$

The product of both transfer functions may be written in the form

$$H_{\text{HD}}\left(e^{j\psi}\right) = \left[\psi + \pi(\text{sgn}(\sin(\psi)) - 1\right] \text{sgn}(\sin(\psi))e^{-j[2\psi\tau - \pi\tau(\text{sgn}(\sin(\psi)) - 1)]} \qquad (7.21.69)$$

The magnitude and the phase function of this transfer function of the differentiating Hilbert transformer are shown in Figure 7.21.31. The inverse Fourier transform with ($\tau = 0$) yields the noncausal, even, and of infinite duration impulse response:

$$h_{\text{HD}}(i) = \begin{cases} a(i) = -[2\sin^2(0.5\pi i)]/(\pi i^2); & i \neq 0 \\ 0.5\pi; & i = 0 \end{cases} \qquad (7.21.70)$$

This is illustrated in Figure 7.21.32a. The design method is the same as for the discrete Hilbert filter. The impulse response should be truncated to include N samples and shifted by $\tau = (N - 1)/2$ samples as shown in Figure 7.21.32b. The $G(e^{j\psi})$ function defined by Eq. (7.21.52) here takes the form

$$G_{\text{HD}}\left(e^{j\psi}\right) = [\psi + \pi(\text{sgn}(\sin(\psi)) - 1)]\,\text{sgn}(\sin(\psi)) : \qquad 0 < \psi < 2\pi \qquad (7.21.71)$$

and in this case is equal to the magnitude of the transfer function (see Fig. 7.21.31). The truncated Fourier series is:

$$G_{\text{HDT}}\left(e^{j\psi}\right) = 0.5\pi - \sum_{i=1}^{(N-1)/2} 2a(i)\cos(\psi i) \qquad (7.21.72)$$

Compare this function with the analogous function of the FIR Hilbert transformer (see Eq. [7.21.52]). The design methods to get the desired amplitude ripple are the same as described in the three points following Eq. (7.21.52). However, the Fourier series given by Eq. (7.21.53) takes for the differentiating Hilbert transformer the form

$$G_{\text{HDT}}\left(e^{j\psi}\right) = \frac{\pi}{2} - \left(\frac{4}{\pi}\right)\left[\cos\psi + \left(\frac{1}{9}\right)\cos 3\psi + \left(\frac{1}{25}\right)\cos 5\psi + \cdots\right] \qquad (7.21.73)$$

This Fourier series differs by three important features from the series given by Eq. (7.21.53). First, it converges faster (coefficients $1/i^2$ instead $1/i$) and second, there are no Gibbs peaks at the edges of the pass-band. Third, the function is unipolar with the mean value equal to $\pi/2$. An example of the magnitude designed in the minimax sense is shown in Figure 7.21.33. The coefficients are given in Table 7.21.1.

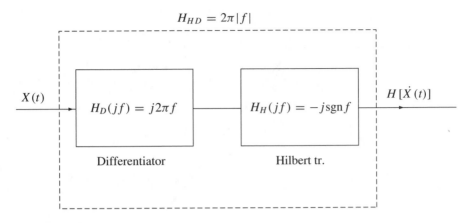

FIGURE 7.21.30
A cascade connection of a Hilbert transformer and a differentiating filter.

TABLE 7.21.1

	Pass-band edges				
	$\psi_1 = 0;$		$\psi_2 = \pi$	$\psi_1 = 0.2\pi;$	$\psi_2 = 0.8\pi^*$
	$N = 7$	$N = 11$	$N = 19$	$N = 11$	$N = 19$
$a(1)$	0.6426919	0.6388893	0.6373537	0.6068935	0.6184231
$a(3)$	0.0997952	0.07348499	0.0715001	0.0450341	0.05423771
$a(5)$		0.459263	0.0263422	0.00615878	0.0118350*
$a(7)$			0.0140561		0.00260689
$a(9)$			0.0020769		0.0003889

(Data from a paper of Čižek, 1989, URSI, ISSSE, *corrected by this author.)

7.22 Multidimensional Hilbert transformations

Multidimensional transformations are applied in modern multidimensional digital signal processing. The theory of complex notation of multidimensional signals uses multidimensional Hilbert transforms. These are the reasons basic definitions and properties of multidimensional Hilbert transformations are presented here. As in the one-dimensional case, the theory of Hilbert transformations is closely tied with multidimensional Fourier transformations.

Let us define the n-dimensional signal $u(\mathbf{x})$ as a function of the n-dimensional variable $\mathbf{x} = \{x_1, x_2, \ldots, x_n\}$, an n-dimensional real column vector. For example, a single frame of a video black-and-white signal may be described by the 2-D signal $u(x_1, x_2)$.

Evenness and oddness of N-dimensional signals

Let us remember that the 1-D signal may be resolved in a sum of the even and odd parts (see Eqs. [7.3.2] and [7.3.3]). Therefore, it has two degrees of freedom concerning evenness or oddness. In general, the n-dimensional real signal $u(\mathbf{x})$ has 2^n degrees of freedom in this respect. For example, a 2-D function may be resolved into a sum of four terms:

$$u(x_2, x_1) = u_{ee}(x_2, x_1) + u_{eo}(x_2, x_1) + u_{oe}(x_2, x_1) + u_{oo}(x_2, x_1) \tag{7.22.1}$$

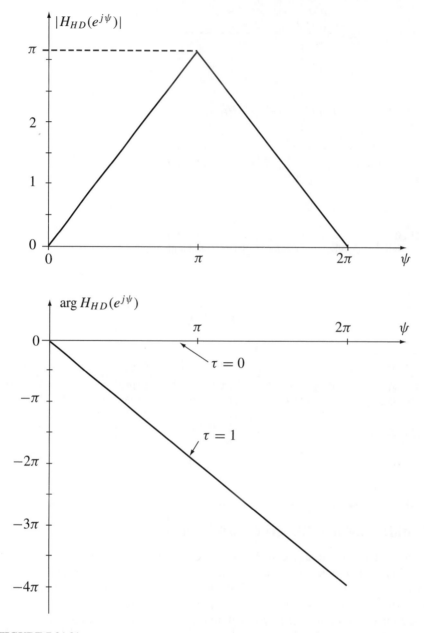

FIGURE 7.21.31
The transfer function of a differentiating Hilbert transformer: (a) magnitude and (b) phase function.

where the indices "e" and "o" indicate evenness or oddness in respect to the variables x_1 and x_2. Notice that the indices "ee," "eo," "oe," and "oo" are written in the natural order of binary numbers using "e" $= 0$ (zero) and "o" $= 1$, that is, 00, 01, 10, 11. The even-even part is given by:

$$u_{ee}(x_2, x_1) = \frac{u(x_2, x_1) + u(x_2, -x_1) + u(-x_2, x_1) + u(-x_2, -x_1)}{4} \qquad (7.22.2)$$

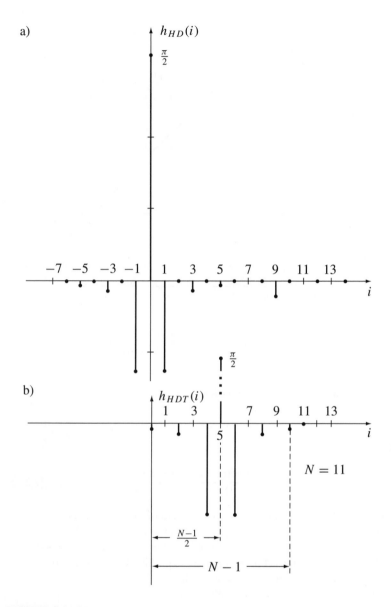

FIGURE 7.21.32
The impulse responses of the differentiating Hilbert transformer: (a) noncausal ideal and (b) truncated
and causal.

the even-odd part by:

$$u_{eo}(x_2, x_1) = \frac{u(x_2, x_1) - u(x_2, -x_1) + u(-x_2, x_1) - u(-x_2, -x_1)}{4} \qquad (7.22.3)$$

the odd-even part by:

$$u_{oe}(x_2, x_1) = \frac{u(x_2, x_1) + u(x_2, -x_1) - u(-x_2, x_1) - u(-x_2, -x_1)}{4} \qquad (7.22.4)$$

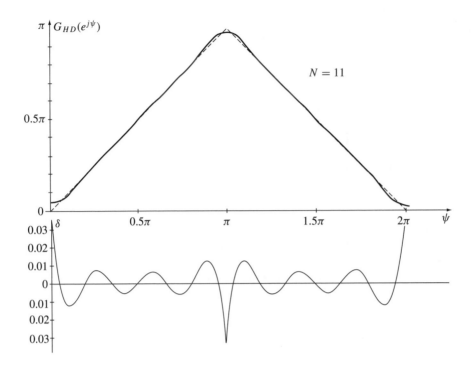

FIGURE 7.21.33
The $G(e^{j\psi})$ function of a FIR differentiating Hilbert transformer.

and the odd-odd part by:

$$u_{oo}(x_2, x_1) = \frac{u(x_2, x_1) - u(x_2, -x_1) - u(-x_2, x_1) + u(-x_2, -x_1)}{4} \tag{7.22.5}$$

We used a reversed order of the indices, that is, (x_2, x_1) instead of (x_1, x_2) and as before used the order 00, 01, 10, 11. The sign of a given term in the nominators of Eqs. (7.22.2) to (7.22.5) is equal to the product of the signs of odd indexed variables. If only one variable is odd, as in Eqs. (7.22.3) or (7.22.4), its sign decides. For example, in Eq. (7.22.4) we have $-u(-x_2, x_1)$ and $-u(-x_2, -x_1)$ because only the variable x_2 is odd indexed.

A 3-D function may be resolved into a sum of eight terms

$$u(x_3, x_2, x_1) = u_{eee} + u_{eeo} + u_{eoe} + u_{eoo} + u_{oee} + u_{oeo} + u_{ooe} + u_{ooo} \tag{7.22.6}$$

Using the same rules as above we get:

$$u_{eee}(x_3, x_2, x_1)$$
$$= \frac{u(x_3, x_2, x_1) + u(x_3, x_2, -x_1) + u(x_3, -x_2, x_1) + u(x_3, -x_2, -x_1)}{16}$$
$$+ \frac{u(-x_3, x_2, x_1) + u(-x_3, x_2, -x_1) + u(-x_3, -x_2, x_1) + u(-x_3, -x_2, -x_1)}{16} \tag{7.22.7}$$
$$u_{eeo}(x_3, x_2, x_1)$$
$$= \frac{u(x_3, x_2, x_1) - u(x_3, x_2, -x_1) + u(x_3, -x_2, x_1) - u(x_3, -x_2, -x_1)}{16}$$

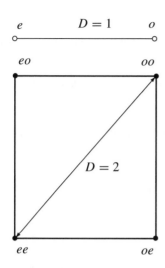 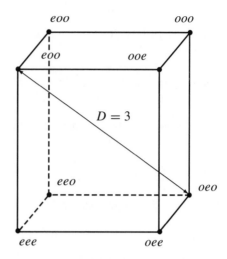

FIGURE 7.22.1
The geometrical interpretation of the "distance" concerning the evenness and oddness of 1-D, 2-D, and 3-D functions.

$$+ \frac{u\left(-x_3, x_2, x_1\right) - u\left(-x_3, x_2, -x_1\right) + u\left(-x_3, -x_2, x_1\right) - u\left(-x_3, -x_2, -x_1\right)}{16} \quad (7.22.8)$$

....

$$u_{ooe}\left(x_3, x_2, x_1\right)$$

$$= \frac{u\left(x_3, x_2, x_1\right) + u\left(x_3, x_2, -x_1\right) - u\left(x_3, -x_2, x_1\right) - u\left(x_3, -x_2, -x_1\right)}{16}$$

$$+ \frac{-u\left(-x_3, x_2, x_1\right) - u\left(-x_3, x_2, -x_1\right) + u\left(-x_3, -x_2, x_1\right) + u\left(-x_3, -x_2, -x_1\right)}{16} \quad (7.22.9)$$

$$u_{ooo}\left(x_3, x_2, x_1\right)$$

$$= \frac{u\left(x_3, x_2, x_1\right) - u\left(x_3, x_2, -x_1\right) - u\left(x_3, -x_2, x_1\right) + u\left(x_3, -x_2, -x_1\right)}{16}$$

$$+ \frac{-u\left(-x_3, x_2, x_1\right) + u\left(-x_3, x_2, -x_1\right) + u\left(-x_3, -x_2, x_1\right) - u\left(-x_3, -x_2, -x_1\right)}{16} \quad (7.22.10)$$

It is possible to introduce a geometric interpretation of the decomposition of a function into even and odd terms, as shown in Figure 7.22.1, and to define the "distance" between the terms. The distance D is:

$$1 > D > n \quad (7.22.11)$$

For example, the distance between f_e and f_o or between f_{eo} and f_{oo} equals 1, between f_{ee} and f_{oo} equals 2.

n-D Hilbert transformations

The n-dimensional (n-D) Hilbert transformation of the n-dimensional function $u(\mathbf{x})$ is defined by the n-fold integral [37]

$$v(\mathbf{x}) = H_n[u(\mathbf{x})] = \frac{1}{\pi^n} P \int_{-\infty}^{\infty} \cdots \int_{-\infty}^{\infty} \frac{u(\mathbf{H})}{\prod_{k=1}^{n}(x_k - \eta_k)} d\mathbf{H}; \quad \begin{aligned} \mathbf{H} &= \{\eta_1, \eta_2, \ldots, \eta_n\} \\ d\mathbf{H} &= d\eta_1, d\eta_2, \ldots d\eta_n \end{aligned}$$

(7.22.12)

where P denotes the Cauchy principal value and H_n the operator of the n-D Hilbert transformation. The inverse transformation is

$$u(\mathbf{x}) = H_n^{-1}[v(\mathbf{x})] = \frac{(-1)^n}{\pi^n} P \int_{-\infty}^{\infty} \cdots \int_{-\infty}^{\infty} \frac{v(\mathbf{H})}{\prod_{k=1}^{n}(x_k - \eta_k)} d\mathbf{H}$$

(7.22.13)

The n-dimensional Hilbert pair will be denoted by

$$u(\mathbf{x}) \overset{n-H}{\Longleftrightarrow} v(\mathbf{x})$$

(7.22.14)

Analogous to the 1-D case, the n-dimensional Hilbert transformation changes the indices of the terms in equations, such as Eqs. (7.22.1) or (7.22.6), from even to odd and from odd to even. Similar to the 1-D case, the n-D Hilbert transformation may be derived from the n-dimensional Cauchy integral

$$f(z) = \frac{1}{(2\pi i)^n} \int_{\Gamma} \frac{f(\zeta) d\zeta}{(\zeta - \mathbf{z})} \quad \begin{aligned} \mathbf{z} &= \{z_1, \ldots z_n\} \\ \zeta &= \{\zeta_1, \ldots \zeta_n\} \end{aligned}$$

(7.22.15)

where $\Gamma = \partial D_1 \times \cdots \times \partial D_n$ is an n-D surface, being the bound of ∂D, where the region D has the form of the Cartesian product $D = D_1 \times \cdots \times D_n$.

2-D Hilbert transformations

The 2-D Hilbert transformation is given by Eq. (7.22.12) with $n = 2$ and has the form [36]

$$v(\mathbf{x}) = H[u(\mathbf{x})] = \frac{1}{\pi^2} P \int_{-\infty}^{\infty} \int_{-\infty}^{\infty} \frac{u(\eta_1, \eta_2)}{(x_1 - \eta_1)(x_2 - \eta_2)} d\eta_1 \, d\eta_2$$

(7.22.16)

and because n is even, the inverse Hilbert transformation has the same form:

$$u(\mathbf{x}) = H[v(\mathbf{x})] = \frac{1}{\pi^2} P \int_{-\infty}^{\infty} \int_{-\infty}^{\infty} \frac{v(\eta_1, \eta_2)}{(x_1 - \eta_1)(x_2 - \eta_2)} d\eta_1 \, d\eta_2$$

(7.22.17)

The 2-D Hilbert transformation may be written using the convolution notation:

$$v(x_1, x_2) = u(x_1, x_2) * * \frac{1}{\pi^2 x_1 x_2}$$

(7.22.18)

$$u(x_1, x_2) = v(x_1, x_2) * * \frac{1}{\pi^2 x_1 x_2}$$

(7.22.19)

Partial Hilbert transformations

The partial Hilbert transformation of the n-D function $u(\mathbf{x})$, $\mathbf{x} = \{x_1, x_2, \ldots, x_n\}$ is defined as the Hilbert transformation in respect to a part of the variables. For example, the partial transformation of a 2-D function in respect to x_1 has the form

$$v_1(x_1, x_2) = \frac{1}{\pi} P \int_{-\infty}^{\infty} \frac{u(\eta_1, x_2)}{(x_1 - \eta_1)} d\eta_1$$

(7.22.20)

and in respect to the variable x_2

$$v_2(x_1, x_2) = \frac{1}{\pi} P \int_{-\infty}^{\infty} \frac{u(x_1, \eta_2)}{(x_2 - \eta_2)} \, d\eta_2 \tag{7.22.21}$$

For 3-D functions it is possible to derive three first-order partial Hilbert transforms denoted v_1, v_2, v_3 and three second-order Hilbert transforms denoted v_{12}, v_{13}, v_{23}. For example,

$$v_1(x_1, x_2, x_3) = \frac{1}{\pi} P \int_{-\infty}^{\infty} \frac{u(\eta_1, x_2, x_3)}{(x_1 - \eta_1)} \, d\eta_1 \tag{7.22.22}$$

and

$$v_{12}(x_1, x_2, x_3) = \frac{1}{\pi} P \int_{-\infty}^{\infty} \int_{-\infty}^{\infty} \frac{u(\eta_1, \eta_2, x_3)}{(x_1 - \eta_1)(x_2 - \eta_2)} \, d\eta_1 \, d\eta_2 \tag{7.22.23}$$

Spectral description of n-D Hilbert transformations

The n-dimensional Fourier transformation of $u(\mathbf{x})$ is defined by the n-fold integral

$$U(\mathbf{\Omega}) = F_n[u(\mathbf{x})] = \int_{-\infty}^{\infty} \cdots \int_{-\infty}^{\infty} u(\mathbf{x}) \exp(-j\mathbf{\Omega}^T \mathbf{x}) \, d\mathbf{x} \tag{7.22.24}$$

where $\mathbf{\Omega} = \{\omega_1, \omega_2, \ldots, \omega_n\}$ is the n-dimensional column vector of Fourier frequencies. The index "T" denotes transpose. Therefore, the **exponential kernel** of the n-D Fourier transformation has the form

$$\exp(-j\mathbf{\Omega}^T \mathbf{x}) = e^{-j(\omega_1 x_1 + \omega_2 x_2 + \cdots + \omega_n x_n)} \tag{7.22.25}$$

The inverse Fourier transformation is defined by the n-fold integral

$$u(\mathbf{x}) = F_n^{-1}[U(\mathbf{\Omega})] = \int_{-\infty}^{\infty} \cdots \int_{-\infty}^{\infty} U(\mathbf{\Omega}) \exp(j\mathbf{\Omega}^T \mathbf{x}) \, df_1 \, df_2 \ldots df_n \tag{7.22.26}$$

where $df_i = d\omega_i / 2\pi$ $(i = 1, 2, \ldots, n)$. The n-D Fourier pair may be denoted

$$u(\mathbf{x}) \stackrel{n-F}{\Longleftrightarrow} U(\mathbf{\Omega}) \tag{7.22.27}$$

The n-D Fourier image of the Hilbert transform is given by the formula

$$V(\mathbf{\Omega}) = F_n\{H_n[u(\mathbf{x})]\} = (-j)^n \left[\prod_{k=1}^{n} \text{sgn}(\omega_k)\right] U(\mathbf{\Omega}) \tag{7.22.28}$$

PROOF   By the definition given by Eq. (7.22.24)

$$F_n\{H_n[u(\mathbf{x})]\} = \frac{1}{\pi^n} \int_{-\infty}^{\infty} \left\{\int_{-\infty}^{\infty} \frac{U(\mathbf{H})}{\prod_{k=1}^{n}(x_k - \eta_k)} \, d\mathbf{H}\right\} \exp(-j\mathbf{\Omega}^T \mathbf{x}) \, d\mathbf{x} \tag{7.22.29}$$

where $\mathbf{H} = \{\eta_1, \eta_2, \ldots, \eta_n\}, d\mathbf{H} = d\eta_1, d\eta_2, \ldots, d\eta_n$ and for convenience the n-fold integrals are denoted by a single integral sign. Formally

$$F_n\{H_n[u(\mathbf{x})]\} = \frac{1}{\pi^n} \int_{-\infty}^{\infty} u(\mathbf{H}) \, d\mathbf{H} \int_{-\infty}^{\infty} \frac{\exp(-j\mathbf{\Omega}^T \mathbf{x})}{\prod_{k=1}^{n}(x_k - \eta_k)} \, d\mathbf{x}$$

$$= \int_{-\infty}^{\infty} u(\mathbf{H}) \, d\mathbf{H} \prod_{k=1}^{n} \left(\frac{1}{\pi} \int_{-\infty}^{\infty} \frac{\exp(-j\omega_k x_k)}{x_k - \eta_k} \, dx_k\right) \tag{7.22.30}$$

In the one-dimensional case we have

$$H[\exp(-j\omega x)] = j\,\mathrm{sgn}(\omega)\exp(-j\omega x) \tag{7.22.31}$$

Hence

$$F_n\{H_n[u(\mathbf{x})]\} = (-j)^n \left[\prod_{k=1}^{n}\mathrm{sgn}(\omega_k)\right]\int_{-\infty}^{\infty} u(\mathbf{x})\exp(-j\Omega^{\mathrm{T}}\mathbf{x})\,d\mathbf{x} \tag{7.22.32}$$

and this is equal to Eq. (7.22.28). ■

This equation enables the calculation of the n-D Hilbert transform using the inverse Fourier transform of the spectrum given by the above equation, that is, using the algorithm

$$u(\mathbf{x}) \overset{n-F}{\Rightarrow} U(\Omega) \Rightarrow V(\Omega) = (-j)^n \left[\prod_{k=1}^{n}\mathrm{sgn}(\omega_k)\right] U(\Omega) \overset{2-F^{-1}}{\Rightarrow} v(\mathbf{x}) \tag{7.22.33}$$

For example, in the 2-D case the Hilbert transform is given by

$$v(x_1, x_2) = \int_{-\infty}^{\infty}\int_{-\infty}^{\infty} -\,\mathrm{sgn}(\omega_1)\,\mathrm{sgn}(\omega_2)\,U(\omega_1, \omega_2)\,e^{j(\omega_1 x_1 + \omega_2 x_2)}\,df_1\,df_2 \tag{7.22.34}$$

($\omega_1 = 2\pi f_1, \omega_2 = 2\pi f_2$). This general formula may be simplified, if the signal given by Eq. (7.22.1) has only a part of the four terms (see Appendix 7.22).

Example   [14]

Consider the 2-D Gaussian signal and its Fourier image given by the Fourier pair

$$u(x_1, x_2) = e^{-\pi(x_1^2 + x_2^2)} \overset{2-F}{\Longleftrightarrow} U(\omega_1, \omega_2) = e^{-\pi(f_1^2 + f_2^2)} \tag{7.22.35}$$

where $\omega_1 = 2\pi f_1$ and $\omega_2 = 2\pi f_2$. The Fourier image of the Hilbert transform is

$$V(\omega_1, \omega_2) = -\,\mathrm{sgn}(\omega_1)\,\mathrm{sgn}(\omega_2)e^{-\pi(f_1^2 + f_2^2)} \tag{7.22.36}$$

Because this spectral function is real and odd-odd, the 2-D inverse Fourier transformation takes the simplified form (see Appendix 7.22)

$$v(x_1, x_2) = 4\int_{0}^{\infty}\int_{0}^{\infty} e^{-\pi(f_1^2 + f_2^2)}\sin(\omega_1 x_1)\sin(\omega_2 x_2)\,df_1\,df_2 \tag{7.22.37}$$

■

n-D Hilbert transforms of separable functions

The n-dimensional function $u(\mathbf{x})$ is said to be **separable** in the coordinates $\mathbf{x} = \{x_1, x_2, \ldots, x_n\}$ if it is given by the product of 1-D functions

$$u(\mathbf{x}) = f_1(x_1)f_2(x_2)\ldots f_n(x_n) \tag{7.22.38}$$

Let us denote by $g_1(x_1), g_2(x_2), \ldots, g_n(x_n)$ the Hilbert transforms of the terms of this product. Because the Fourier image of a separable function is a separable function of the Fourier coordinates $\Omega = \{\omega_1, \omega_2, \ldots, \omega_n\}$ the inverse Fourier transform is a product of 1-D integrals. Therefore, the Hilbert transform of a separable function has the form

$$v(\mathbf{x}) = g_1(x_1)g_2(x_2) \ldots g_n(x_n) \tag{7.22.39}$$

Analogously, the partial Hilbert transforms of $u(\mathbf{x})$ are separable functions, for example, a first-order partial transform is

$$v_1(\mathbf{x}) = g_1(x_1) f_2(x_2) \ldots f_n(x_n) \tag{7.22.40}$$

and a second order partial transform is

$$v_{12}(\mathbf{x}) = g_1(x_1)g_2(x_2) f_3(x_3) \ldots f_n(x_n) \tag{7.22.41}$$

Examples

1. The 2-D delta pulse is a separable distribution of the form

$$\delta(x_1, x_2) = \delta(x_1)\,\delta(x_2) \tag{7.22.42}$$

Because $\delta(x_1) \overset{H}{\Longleftrightarrow} 1/(\pi x_1)$ and $\delta(x_2) \overset{H}{\Longleftrightarrow} 1/(\pi x_2)$, the Hilbert transform of the 2-D delta pulse has the form

$$\delta(x_1, x_2) \overset{2\text{-}H}{\Longleftrightarrow} 1/\left(\pi^2 x_1 x_2\right) \tag{7.22.43}$$

and the partial transforms are

$$v_1(x_1, x_2) = \delta(x_2)/(\pi x_1);$$

$$v_2(x_1, x_2) = \delta(x_1)/(\pi x_2) \tag{7.22.44}$$

2. The 2-D signal has the form

$$u(x_1, x_2) = \Pi_a(x_1)\,\Pi_b(x_2) \tag{7.22.45}$$

The Hilbert transform takes the form (see Table 7.7.1, No. 4)

$$v(x_1, x_2) = \frac{1}{\pi^2} \ln\left|\frac{x_1 + a}{x_1 - a}\right| \ln\left|\frac{x_2 + b}{x_2 - b}\right| \tag{7.22.46}$$

and the partial transforms are

$$v_1(x_1, x_2) = \frac{1}{\pi} \ln\left|\frac{x_1 + a}{x_1 - a}\right| \Pi_b(x_2);$$

$$v_2(x_1, x_2) = \frac{1}{\pi}\Pi_a(x_1) \ln\left|\frac{x_2 + b}{x_2 - b}\right| \tag{7.22.47}$$

3. Derivation of the Hilbert Transform of a nonseparable function defined by the equation

$$\begin{aligned}
u(x_1, x_2) &= \tfrac{h}{a}(a - |x_1| - |x_2|); & (x_1, x_2) \in S \\
u(x_1, x_2) &= 0; & (x_1, x_2) \notin S
\end{aligned} \tag{7.22.48}$$

The support S is shown in Figure 7.22.2b. This function has the geometric form of a pyramid (Figure 7.22.2a). The Hilbert transform of this function is

$$
\begin{aligned}
v(x_1, x_2) = \frac{h}{\pi^2 a} \Bigg\{ & P \int_0^a \int_0^{a-x_2} \frac{a - \eta - \gamma}{(x_1 - \eta)(x_2 - \gamma)} \, d\gamma \, d\eta \\
& + P \int_0^a \int_{-a+x_2}^0 \frac{a - \eta - \gamma}{(x_1 - \eta)(x_2 - \gamma)} \, d\gamma \, d\eta \\
& + P \int_{-a}^0 \int_0^{a+x_2} \frac{a - \eta - \gamma}{(x_1 - \eta)(x_2 - \gamma)} \, d\gamma \, d\eta \\
& + P \int_{-a}^0 \int_{-a-x_2}^0 \frac{a - \eta - \gamma}{(x_1 - \eta)(x_2 - \gamma)} \, d\gamma \, d\eta \Bigg\}
\end{aligned} \qquad (7.22.49)
$$

The integration yields:

$$
\begin{aligned}
v(x_1, x_2) = \frac{h}{\pi^2 a} \Bigg\{ & 2x_1 \ln |x_1| \ln \left| \frac{x_2 - a}{x_2 + a} \right| \\
& + P \int_0^a \frac{(x_1 + \gamma - a) \ln |x_1 + \gamma - a| + (x_1 + a - \gamma) \ln |x_1 + a - \gamma|}{x_2 - \gamma} \, d\gamma \\
& + P \int_{-a}^0 \frac{(x_1 - a - \gamma) \ln |x_1 - a - \gamma| + (a + x_1 + \gamma) \ln |a + x_1 + \gamma|}{x_2 - \gamma} \, d\gamma \Bigg\}
\end{aligned} \qquad (7.22.50)
$$

The one-dimensional integrals do not have a closed solution and a numerical integration should be applied. Notice that the support of the Hilbert transform (7.22.50) is infinite, which is different than the finite support of $u(x_1, x_2)$. The partial Hilbert transforms, defined by Eqs. (7.22.20) and (7.22.21) are:

$$
\begin{aligned}
v_1(x_1, x_2) = \frac{h}{2a} [& -2x_1 \ln |x_1| + (x_1 + x_2 - a) \ln |x_1 + x_2 - a| \\
& + (x_1 + a - x_2) \ln |x_1 + a - x_2|]
\end{aligned} \qquad (7.22.51)
$$

$$
\begin{aligned}
v_2(x_1, x_2) = \frac{h}{2a} [& -2y \ln |x_2| + (y + x_1 - a) \ln |x_2 + x_1 - a| \\
& + (x_2 + a - x_1) \ln |x_2 + a - x_1|]
\end{aligned} \qquad (7.22.52)
$$

The supports of these functions are shown in Figure 7.22.2c. They are infinite in one dimension and finite in a band $(-a, a)$ in the second dimension. ∎

Properties of 2-D Hilbert transformations

Selected properties of 2-D Hilbert transformations are summarized in Table 7.22.1.

Orthogonality

The terms of the 1-D Hilbert pair form a pair of orthogonal functions satisfying the condition

$$
E = \int u(t) v(t) \, dt = 0 \qquad (7.22.53)
$$

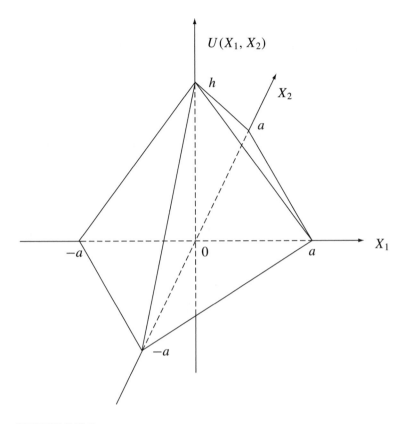

FIGURE 7.22.2
(a) The pyramid pulse.

that is, the **mutual energy** of both signals equals zero. In general, the terms of the 2-D Hilbert
pair are not orthogonal, that is, the mutual energy

$$E = \int \int u(x_1, x_2) v(x_1, x_2) \, dx \, dx \neq 0 \tag{7.22.54}$$

does not equal zero. However, this integral equals zero for 2-D separable signals and for nonseparable signals with certain symmetry, for example, the pyramid signal defined by Eq. (7.22.48).
For separable signals the above double integral takes the form of a product of single integrals,
each of which equals zero.

Example
Consider the 2-D Gaussian signal of the form [14]

$$u(x_1, x_2) = \left[2\pi \sigma_1 \sigma_2 \, \text{SQR} \left(1 - \rho^2 \right) \right]^{-1}$$

$$\times \, \exp \left\{ - \left(1 - \rho^2 \right)^{-1} \left[(x_1/\sigma_1)^2 + (x_2/\sigma_2)^2 - 2\rho x_1 x_2 / (\sigma_1 \sigma_1) \right] \right\} \tag{7.22.55}$$

This function is well known in probability theory. It is a separable function if the parameter
$\rho = 0$. Otherwise, for $0 < \rho < 1$ it is a nonseparable function. Its Hilbert transform may be
calculated using the inverse Fourier transform of the Fourier image, which has the form

$$U(\omega_1, \omega_2) = \exp \left\{ -0.5 \left[\sigma_1^2 \omega_1^2 + \sigma_2^2 \omega_2^2 + 2\rho \sigma_1 \sigma_2 \omega_1 \omega_2 \right] \right\} \tag{7.22.56}$$

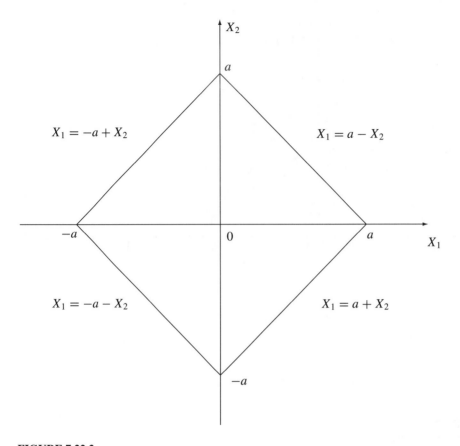

FIGURE 7.22.2
(b) The support of the pyramid pulse.

Because U is a real function, the inverse Fourier transformation has the simplified form

$$v(x_1, x_2) = \int_{-\infty}^{\infty} \int_{-\infty}^{\infty} - \text{sgn}(\omega_1) \, \text{sgn}(\omega_2) \, U(\omega_1, \omega_2) \cos(\omega_1 x_1 + \omega_2 x_2) \, df_1 \, df_2 \quad (7.22.57)$$

and the partial Hilbert transforms are given by the integrals

$$v_1(x_1, x_2) = \int_{-\infty}^{\infty} \int_{-\infty}^{\infty} - \text{sgn}(\omega_1) \, U(\omega_1, \omega_2) \sin(\omega_1 x_1 + \omega_2 x_2) \, df_1 \, df_2 \quad (7.22.58)$$

$$v_2(x_1, x_2) = \int_{-\infty}^{\infty} \int_{-\infty}^{\infty} - \text{sgn}(\omega_2) \, U(\omega_1, \omega_2) \sin(\omega_1 x_1 + \omega_2 x_2) \, df_1 df_2 \quad (7.22.59)$$

Because these integrals cannot be expressed in the closed form, a numerical integration scheme must be used for their approximation. Figure 7.22.3 shows the equal-value contour lines of the Gaussian function $u(x_1, x_2)$ and the total and partial Hilbert transforms ($\rho = 0.5$, $\sigma_1 = 1$, $\sigma_2 = 2$). The numerical integration yields the value of the mutual energy $E \cong 0.25$ (relative to the signal energy). E equals zero only if $\rho = 0$, that is, for separable Gaussian signals. ∎

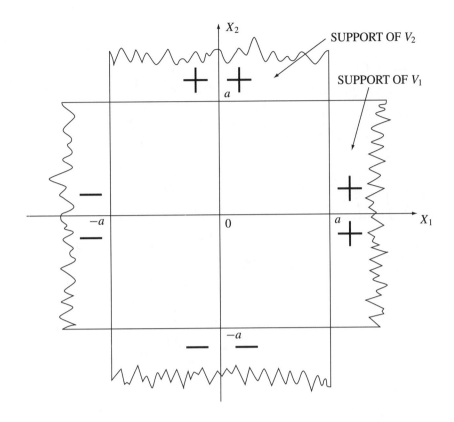

FIGURE 7.22.2
(c) The supports of the partial Hilbert transforms of this pulse.

Stark's extension of Bedrosian's theorem [37]

Bedrosian's theorem defines the Hilbert transform of a product of low-pass and high-pass signals. Stark formulated an extension of this theorem for 2-D signals. A 2-D function $u_{LP}(x_1, x_2) \overset{2-F}{\Longleftrightarrow} U_{LP}(\omega_1, \omega_2)$ is said to be low-pass with cutoff vector $\Omega_0 = \{\omega_{10}, \omega_{20}\}$ if

$$\max |\omega_1| = \omega_{10} \qquad \text{and} \qquad \max |\omega_2| = \omega_{20}$$

$$\text{all } \omega_1 \in \text{supp } U_{LP}(\omega_1, \omega_2) \qquad \text{all } \omega_2 \in \text{supp } U_{LP}(\omega_1, \omega_2) \qquad (7.22.60)$$

where supp U_{LP} denotes the support of the Fourier image , that is, the set of points for which $U_{LP}(\omega_1, \omega_2)$ is not zero. Analogously, the function $u_{HP}(x_1, x_2) \overset{2-F}{\Longleftrightarrow} U_{HP}(\omega_1, \omega_2)$ is said to be high-pass with cutoff vector $\Omega_0 = \{\omega_{10}, \omega_{20}\}$ if

$$\min |\omega_1| > \omega_{10} \qquad \text{and} \qquad \min |\omega_2| > \omega_{20}$$

$$\text{all } \omega_1 \in \text{supp } U_{HP}(\omega_1, \omega_2) \qquad \text{all } \omega_2 \in \text{supp } U_{HP}(\omega_1, \omega_2) \qquad (7.22.61)$$

The signals u_{LP} and u_{HP} are said to be strongly spectrally separable if the conditions (7.22.60) and (7.22.61) are satisfied. We say that the functions u_{LP} and u_{HP} are spectrally disjointed if they have nonoverlapping supports. However, spectral disjointedness may not coincide with strong separability, as shown in Figure 7.22.4. Stark's extension of Bedrosian's theorem has

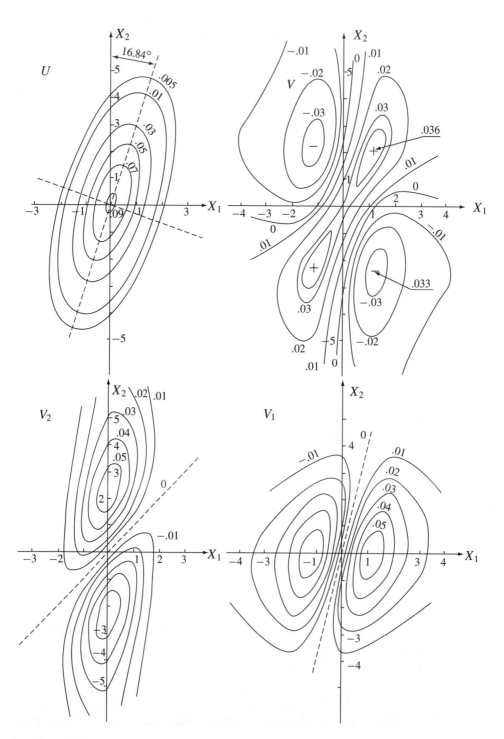

FIGURE 7.22.3
The elliptical equal-value contours of a nonseparable Gaussian function (see Eq. [7.22.55]) and of the Hilbert transforms v, v_1, and v_2, where $\rho = 0.5$, $\sigma_1 = 1$, $\sigma_2 = 2$.

TABLE 7.22.1

Number	Name	Original Signal or the Inverse Transform	Hilbert Transform
1	notations	$u(x_1, x_2) = H_2^{-1}[v(x_1, x_2)];$	$v(x_1, x_2) = H_2[u(x_1, x_2)]$
2	signal domain definitions	$u(x_1, x_2) =$ $1/(\pi^2 x_1 x_2) ** v(x_1, x_2);^a$	$v(x_1, x_2) =$ $1/(\pi^2 x_1 x_2) ** u(x_1, x_2)^b$
3	Fourier spectra	$u(x_1, x_2) \overset{2-F}{\Longleftrightarrow} U(\omega_1, \omega_2) =;$ $- \operatorname{sgn}(\omega_1) \operatorname{sgn}(\omega_2) V(\omega_1, \omega_2)$	$v(x_1, x_2) \overset{2-F}{\Longleftrightarrow} V(\omega_1, \omega_2) =$ $- \operatorname{sgn}(\omega_1) \operatorname{sgn}(\omega_2) U(\omega_1, \omega_2)$
4	linearity	$a u_a(x_1, x_2) + b u_b(x_1, x_2);$	$a v_a(x_1, x_2) + b v_b(x_1, x_2)$
5	change of symmetry	$u_{ee} + u_{oo} + u_{eo} + u_{oe};^c$	$v_{oo} + v_{ee} - v_{oe} - v_{eo}{}^c$
6	iteration	$v(x_1, x_2) \overset{2-H}{\Rightarrow} u(x_1, x_2)$	
7	energy equality	$\int \int u^2(x_1, x_2) \, dx_1 \, dx_2 = \int \int v^2(x_1, x_2) \, dx_1 \, dx_2$	
8	product of low-pass and high-pass signals with strongly separated spectra	$u_{LP}(x_1, x_2)$ $u_{HP}(x_1, x_2)$ $H_2(u_{LP} u_{HP}) \qquad =$	low-pass signal high-pass signal $u_{LP}[H_2(u_{HP})]$
9	separable functions	$f_1(x_1) \overset{H}{\Longleftrightarrow} g_1(x_1); \; f_2(x_2) \overset{H}{\Longleftrightarrow} g_2(x_2)$	
	Total Hilbert tr.	$u(x_1, x_2) = f_1 f_2$	$v(x_1, x_2) = g_1 g_2$
	Partial Hilbert tr.	$v_1(x_1, x_2) = g_1 f_2$	$v_2(x_1, x_2) = f_1 g_2$

^{a} See Eq. (7.22.17)

^{b} See Eq. (7.22.16)

^{c} indices: e-even, o-odd

the form

$$H_2 [u_{LP}(x_1, x_2) u_{HP}(x_1, x_2)] = u_{LP}(x_1, x_2) H_2 [u_{HP}(x_1, x_2)] \qquad (7.22.62)$$

that is, only the high-pass term of the product is transformed.

Appendix 7.22.1

Consider the 2-D signal given by Eq. (7.22.1). Its 2-D Fourier transform is

$$U(\omega_1, \omega_2) = \int_{-\infty}^{\infty} \int_{-\infty}^{\infty} u(x_1, x_2) e^{-j(\omega_1 x_1 + \omega_2 x_2)} \, dx_1 \, dx_2$$

$$= U_{Re} + j U_{Im}$$

$$= U_{ee} - U_{oo} - j(U_{eo} + U_{oe}) \qquad (7.22.63)$$

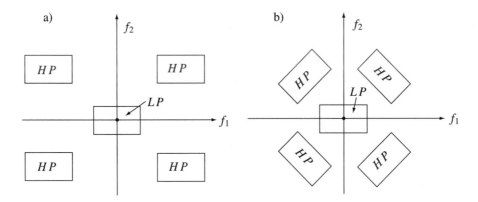

FIGURE 7.22.4
(a) LP is the support of the spectrum of a low-pass signal; HP is the support of the spectrum of a
high-pass signal strongly separable from the low-pass. (b) Analogous spectra with spectral
disjointedness.

TABLE 7.22.2
2-D total and partial Hilbert transforms.

Notations: $u(x_1, x_2)$—original signal, $v(x_1, x_2)$—total Hilbert transform, $v_1(x_1, x_2)$ or $v_2(x_1, x_2)$—partial
Hilbert transforms

Number	Name	Original signal, Total, and Partial Hilbert transforms	
1	delta	$u = \delta(x_1, x_2)$ $v_1 = \delta(x_2)/(\pi x_1)$	$v = 1/(\pi^2 x_1 x_2)$ $v_2 = \delta(x_1)/(\pi x_2)$
2	Gaussian pulse	$u = e^{-\pi(x_1^2 + x_2^2)}$ $v = 4 \int_{-\infty}^{\infty} \int_{-\infty}^{\infty} e^{-\pi(f_1^2 + f_2^2)} \sin(\omega_1 x_1) \sin(\omega_2 x_2)\, df_1\, df_2$ $v_1 = 4 \int_{-\infty}^{\infty} \int_{-\infty}^{\infty} e^{-\pi(f_1^2 + f_2^2)} \sin(\omega_1 x_1) \cos(\omega_2 x_2)\, df_1\, df_2$ $v_2 = 4 \int_{-\infty}^{\infty} \int_{-\infty}^{\infty} e^{-\pi(f_1^2 + f_2^2)} \cos(\omega_1 x_1) \sin(\omega_2 x_2)\, df_1\, df_2$	$\omega_1 = 2\pi f_1;\ \omega_2 = 2\pi f_2$
3	Cauchy pulse	$u = \frac{ab}{(a^2 + x_1^2)(b^2 + x_2^2)}$ $v_1 = \frac{x_1 b}{(a^2 + x_1^2)(b^2 + x_2^2)}$	$v = \frac{x_1 x_2}{(a^2 + x_1^2)(b^2 + x_2^2)}$ $v_2 = \frac{a x_2}{(a^2 + x_1^2)(b^2 + x_2^2)}$
4	Cube pulse	$u = \Pi_a(x)\Pi_b(x_2)$ $v_1 = \frac{1}{\pi} \ln\left|\frac{x_1 + a}{x_1 - a}\right| \Pi_b(x_2)$	$v = \frac{1}{\pi^2} \ln\left|\frac{x_1 + a}{x_1 - a}\right| \ln\left|\frac{x_2 + b}{x_2 - b}\right|$ $v_2 = \frac{1}{\pi} \ln\left|\frac{x_2 + b}{x_2 - b}\right| \Pi_a(x_1)$
5	Sinc pulse	$u = \frac{\sin(ax_1)\sin(bx_2)}{abx_1x_2}$ $v_1 = 2\frac{\sin^2(ax_1/2)\sin(bx_2)}{abx_1x_2}$	$v = 4\frac{\sin^2(ax_1/2)\sin^2(bx_2/2)}{abx_1x_2}$ $v_2 = 2\frac{\sin(ax_1)\sin^2(bx_2/2)}{abx_1x_2}$
6	nonseparable Gaussian pulse	see Eqs. (7.22.55) to (7.22.59)	
7	pyramid pulse	see Eqs. (7.22.48) to (7.22.52)	
8	2-D periodic signals	see Table 7.22.3	

TABLE 7.22.3
2-D total and partial Hilbert transforms of periodic functions.

Notations: $u(x_1, x_2)$—original signal, $v(x_1, x_2)$—total Hilbert transform, $v_1(x_1, x_2)$ or $v_2(x_1, x_2)$—partial Hilbert transforms

Number	Original signal, total, and partial Hilbert transforms	
1	$u = \cos(\omega_1 x_1)\cos(\omega_2 x_2)$ $v_1 = \sin(\omega_1 x_1)\cos(\omega_2 x_2)$	$v = \sin(\omega_1 x_1)\sin(\omega_2 x_2)$ $v_2 = \cos(\omega_1 x_1)\sin(\omega_2 x_2)$
2	$u = \sin(\omega_1 x_1)\sin(\omega_2 x_2)$ $v_1 = -\cos(\omega_1 x_1)\sin(\omega_2 x_2)$	$v = \cos(\omega_1 x_1)\cos(\omega_2 x_2)$ $v_2 = -\sin(\omega_1 x_1)\cos(\omega_2 x_2)$
3	$u = \cos(\omega_1 x_1 + \omega_2 x_2)$ $v_1 = \sin(\omega_1 x_1 + \omega_2 x_2)$	$v = -\cos(\omega_1 x_1 + \omega_2 x_2)$ $v_2 = \sin(\omega_1 x_1 + \omega_2 x_2)$
4	$u = \sin(\omega_1 x_1 + \omega_2 x_2)$ $v_1 = -\cos(\omega_1 x_1 + \omega_2 x_2)$	$v = -\sin(\omega_1 x_1 + \omega_2 x_2)$ $v_2 = -\cos(\omega_1 x_1 + \omega_2 x_2)$
5	$u = e^{j(\omega_1 x_1 + \omega_2 x_2)}$ $v_1 = -j\,\mathrm{sgn}(\omega_1)e^{j(\omega_1 x_1 + \omega_2 x_2)};$	$v = -\,\mathrm{sgn}(\omega_1)\,\mathrm{sgn}(\omega_2)e^{j(\omega_1 x_1 + \omega_2 x_2)}$ $v_2 = -j\,\mathrm{sgn}(\omega_2)e^{j(\omega_1 x_1 + \omega_2 x_2)}$
6	$u = \mathrm{sgn}[\cos(\omega_1 x_1)]\,\mathrm{sgn}[\cos(\omega_2 x_2)]$ (2-D square wave) $v = (4/\pi^2)\ln|\tan(\omega_2 x_1/2 + \pi/4)|\ln|\tan(\omega_1 x_2/2 + \pi/4)|$ $v_1 = (2/\pi)\ln|\tan(\omega_1 x_1/2 + \pi/4)|\,\mathrm{sgn}[\cos(\omega_2 x_2)]$ $v_2 = (2/\pi)\ln|\tan(\omega_2 x_2/2 + \pi/4)|\,\mathrm{sgn}[\cos(\omega_1 x_1)]$	
7	$u = \sum_{n=-\infty}^{\infty}\sum_{m=-\infty}^{\infty}\delta(x_1 - na, x_2 - mb)$ (2-D delta sampling sequence) $v = \frac{1}{ab}\sum_{n=-\infty}^{\infty}\sum_{m=-\infty}^{\infty}\cot[(\pi/a)(x_1 - na)]\cot[(\pi/b)(x_2 - mb)]$ $v_1 = \frac{1}{a}\sum_{n=-\infty}^{\infty}\sum_{m=-\infty}^{\infty}\cot[(\pi/a)(x_1 - na)]\delta(x_2 - mb)$ $v_2 = \frac{1}{b}\sum_{n=-\infty}^{\infty}\sum_{m=-\infty}^{\infty}\cot[(\pi/b)(x_2 - mb)]\delta(x_1 - na)$	

TABLE 7.22.4
n-D Hilbert transforms of harmonic functions.

n	Function	Hilbert Transform
1	$\cos(\omega t)$ $\sin(\omega t)$	$\sin(\omega t)$ $-\cos(\omega t)$
2	$\cos(\omega_1 x_1 + \omega_2 x_2)$ $\sin(\omega_1 x_1 + \omega_2 x_2)$	$-\cos(\omega_1 x_1 + \omega_2 x_2)$ $-\sin(\omega_1 x_1 + \omega_2 x_2)$
3	$\cos(\omega_1 x_1 + \omega_2 x_2 + \omega_3 x_3)$ $\sin(\omega_1 x_1 + \omega_2 x_2 + \omega_3 x_3)$	$-\sin(\omega_1 x_1 + \omega_2 x_2 + \omega_3 x_3)$ $\cos(\omega_1 x_1 + \omega_2 x_2 + \omega_3 x_3)$
4	$\cos(\omega_1 x_1 + \omega_2 x_2 + \omega_3 x_3 + \omega_4 x_4)$ $\sin(\omega_1 x_1 + \omega_2 x_2 + \omega_3 x_3 + \omega_4 x_4)$	$\cos(\omega_1 x_1 + \omega_2 x_2 + \omega_3 x_3 + \omega_4 x_4)$ $\sin(\omega_1 x_1 + \omega_2 x_2 + \omega_3 x_3 + \omega_4 x_4)$
1	$e^{j\omega t}$	$-j\,\mathrm{sgn}(\omega)e^{j\omega t}$
2	$e^{j(\omega_1 x_1 + \omega_2 x_2)}$	$-\,\mathrm{sgn}(\omega_1)\,\mathrm{sgn}(\omega_2)e^{j(\omega_1 x_1 + \omega_2 x_2)}$
3	$e^{j(\omega_1 x_1 + \omega_2 x_2 + \omega_3 x_3)}$	$-j\,\mathrm{sgn}(\omega_1)\,\mathrm{sgn}(\omega_2)\,\mathrm{sgn}(\omega_3)e^{j(\omega_1 x_1 + \omega_2 x_2 + \omega_3 x_3)}$
4	$e^{j(\omega_1 x_1 + \omega_2 x_2 + \omega_3 x_3 + \omega_4 x_4)}$	$\mathrm{sgn}(\omega_1)\,\mathrm{sgn}(\omega_2)\,\mathrm{sgn}(\omega_3)\,\mathrm{sgn}(\omega_4)e^{j(\omega_1 x_1 + \omega_2 x_2 + \omega_3 x_3 + \omega_4 x_4)}$

where

$$U_{ee}(\omega_1, \omega_2) = \int_{-\infty}^{\infty} \int_{-\infty}^{\infty} u_{ee}(x_1, x_2) \cos(\omega_1 x_1) \cos(\omega_2 x_2) \, dx_1 \, dx_2 \qquad (7.22.64)$$

$$U_{oo}(\omega_1, \omega_2) = \int_{-\infty}^{\infty} \int_{-\infty}^{\infty} u_{oo}(x_1, x_2) \sin(\omega_1 x_1) \sin(\omega_2 x_2) \, dx_1 \, dx_2 \qquad (7.22.65)$$

$$U_{oe}(\omega_1, \omega_2) = \int_{-\infty}^{\infty} \int_{-\infty}^{\infty} u_{oe}(x_1, x_2) \sin(\omega_1 x_1) \cos(\omega_2 x_2) \, dx_1 \, dx_2 \qquad (7.22.66)$$

$$U_{eo}(\omega_1, \omega_2) = \int_{-\infty}^{\infty} \int_{-\infty}^{\infty} u_{eo}(x_1, x_2) \cos(\omega_1 x_1) \sin(\omega_2 x_2) \, dx_1 \, dx_2 \qquad (7.22.67)$$

Using computer programs for numerical integration, the insertion in these integrals of $u(x_1, x_2)$ instead of $u_{ee}, u_{oo}, \ldots$ does not change the result because the trigonometric kernels are selecting the right terms themselves. Because the Fourier image of the Hilbert transform is

$$V(\omega_1, \omega_2) = -\operatorname{sgn}(\omega_1) \operatorname{sgn}(\omega_2)[U_{ee} - U_{oo} - j(U_{oe} + U_{eo})] \qquad (7.22.68)$$

the inverse Fourier transform yields the Hilbert transform

$$v(x_1, x_2) = v_{ee} + v_{oo} - v_{oe} - v_{eo} \qquad (7.22.69)$$

where due to the symmetry conditions, the terms of v may be given by one-sided integrals

$$V_{ee}(x_1, x_2) = 4 \int_{0}^{\infty} \int_{0}^{\infty} U_{ee}(\omega_1, \omega_2) \cos(\omega_1 x_1) \cos(\omega_2 x_2) \, df_1 \, df_2 \qquad (7.22.70)$$

$$V_{oo}(x_1, x_2) = 4 \int_{0}^{\infty} \int_{0}^{\infty} U_{oo}(\omega_1, \omega_2) \sin(\omega_1 x_1) \sin(\omega_2 x_2) \, df_1 \, df_2 \qquad (7.22.71)$$

$$V_{oe}(x_1, x_2) = 4 \int_{0}^{\infty} \int_{0}^{\infty} U_{oe}(\omega_1, \omega_2) \sin(\omega_1 x_1) \cos(\omega_2 x_2) \, df_1 \, df_2 \qquad (7.22.72)$$

$$V_{eo}(x_1, x_2) = 4 \int_{0}^{\infty} \int_{0}^{\infty} U_{eo}(\omega_1, \omega_2) \cos(\omega_1 x_1) \sin(\omega_2 x_2) \, df_1 \, df_2 \qquad (7.22.73)$$

Two-Dimensional Hilbert Transformers

The transfer function of the ideal "noncausal" 2-D Hilbert transformer is given by a product of 1-D transfer functions (see Eqs. [7.21.11] to [7.21.13]). Therefore,

$$H_{2-H}(f_1, f_2) = [-j \operatorname{sgn}(f_1)][-j \operatorname{sgn}(f_2)] = |H_{2-H}| e^{j\Phi(f_1, f_2)} \qquad (7.22.74)$$

The magnitude equals 1 and the phase function is

$$\Phi(f_1, f_2) = -\frac{\pi}{2} \operatorname{sgn}(f_1) - \frac{\pi}{2} \operatorname{sgn}(f_2) \qquad (7.22.75)$$

7.23   Multidimensional complex signals

Short historical review

The complex notation of harmonic signals in the form of Euler's equation $e^{j\omega t} = \cos(\omega t) + j\sin(\omega t)$ was introduced to electrical engineering at the end of the nineteenth century (E. Kennedy and C. Steinmetz) and soon proliferated to many science and enginneering disciplines. Restating the equation in the form $\cos(\omega t) = 0.5(e^{j\omega t} + e^{-j\omega t})$ introduces the concept of **negative frequencies**, commonly used in modern Fourier spectral analysis. In 1946, Gabor [11] introduced the extension of the complex notation of time signals in the form of the **analytic signal** $\psi(t) = u(t) + jv(t)$, where $v(t)$ is the Hilbert transform of $u(t)$ (see Section 7.3). It has the unique feature that its Fourier transform is one-sided. In 1964 this author [12] used the analytic signal to define the notion of the instantaneous complex frequency. This section presents how the complex notation of signals and the notion of the analytic signal can be **generalized** for multidimensional signals. This generalization has been recently developed by this author [13].

The definition of the multidimensional complex signal

Let us remember that no definition is "true" or "false." However, it is very desirable that a definition of the n-dimensional complex signal satisfy certain requirements. The basic requirement is the **compatibility** with the 1-D case, that is, with the definition of the analytic signal. Many other requirements may be formulated, such as usefulness in applications. The definition of the multidimensional complex signal introduced in [13] is based on the frequency domain description of the multidimensional signals given by the Fourier pair

$$u(\mathbf{x}) \stackrel{n-F}{\Longleftrightarrow} U(\boldsymbol{\Omega}) \tag{7.23.1}$$

where $\mathbf{x} = \{x_1, x_2, \ldots, x_n\}$ and $\boldsymbol{\Omega} = \{\omega_1, \omega_2, \ldots, \omega_n\}$ are n-dimensional real column vectors (see Section 7.22). Let us remember that the **kernels** of the n-D Fourier transformations are: In 1-D $e^{\pm j\omega t}$, in 2-D $e^{\pm j(\omega_1 x_1 + \omega_2 x_2)}$, in 3-D $e^{\pm j(\omega_1 x_1 + \omega_2 x_2 + \omega_3 x_3)}$, and in n-D $e^{\pm j(\omega_1 x_1 + \cdots + \omega_n x_n)}$. These kernels have the form of complex signals of a constant amplitude $A = 1$ and **linear phase** in respect to the variables $x_1, x_2, \ldots, x_n$. Therefore, a compatible definition of a multidimensional complex signal should define n-dimensional complex harmonic signals in the form of the above kernels of the n-D Fourier transformation. Because the 1-D complex analytic signal has a one-sided spectrum at positive frequencies, let us define the n-dimensional complex signal using the inverse Fourier transform of its spectrum cancelled at all **orthants** of the Fourier frequencies space except in the first orthant. In 1-D this space has two **half-axes**, in 2-D four **quadrants**, in 3-D eight **octants**, and in general 2^n **orthants**. Mathematicians denote the orthant with all the axis of positive sign by R^+. Therefore, the n-D complex signal is defined by the Fourier pair

$$\psi(\mathbf{x}) \stackrel{n-F}{\Longleftrightarrow} \Gamma(\boldsymbol{\Omega}) = 2^n \mathbf{1}(\boldsymbol{\Omega}) U(\boldsymbol{\Omega}) \tag{7.23.2}$$

The cancellation of the spectrum in all but the first orthant is achieved by multiplication of the n-D Fourier image by the n-**D unit step function** (or distribution) $\mathbf{1}(\boldsymbol{\Omega})$ defined by the formula ($\boldsymbol{\Omega} = \{\omega_1, \omega_2, \ldots, \omega_n\}$)

$$\mathbf{1}(\boldsymbol{\Omega}) = \begin{cases} 1 & \text{all } \omega > 0 \\ 0.5 & \text{all } \omega = 0 \\ 0 & \text{all } \omega < 0 \end{cases} \tag{7.23.3}$$

TABLE 7.23.1

The n-D complex signals and its Fourier spectra.

n-D	Complex Signal	Fourier Image	
1-D	$\psi_1 = u + jv$	$\Gamma(\omega) = [1 + \text{sgn}(\omega)]U(\omega)$	(7.23.9)
2-D	$\psi_1 = u - v + j(v_1 + v_2)$	$\Gamma(\omega_1, \omega_2) = U(\omega_1, \omega_2)$ $\times [1 + \text{sgn}(\omega_1) + \text{sgn}(\omega_2) + \text{sgn}(\omega_1)\,\text{sgn}(\omega_2)]$	(7.23.10)
3-D	$\psi_1 = u - v_{12} - v_{13} - v_{23} + j(v_1 + v_2 + v_3 - v)$	$\Gamma(\omega_1, \omega_2, \omega_1) = U(\omega_1, \omega_2, \omega_3)$ $\times [1 + \text{sgn}(\omega_1) + \text{sgn}(\omega_2) + \text{sgn}(\omega_3)$ $+ \text{sgn}(\omega_1)\,\text{sgn}(\omega_2) + \text{sgn}(\omega_1)\,\text{sgn}(\omega_3)$ $+ \text{sgn}(\omega_2)\,\text{sgn}(\omega_3) + \text{sgn}(\omega_1)\,\text{sgn}(\omega_2)\,\text{sgn}(\omega_3)]$	(7.23.11)
	$\vdots$	$\vdots$	
n-D	$\psi_1 = F^{-1}[\Gamma(\mathbf{\Omega})]$	$\Gamma(\mathbf{\Omega}) = 2^n \mathbf{1}(\mathbf{\Omega})U(\mathbf{\Omega})$	(7.23.12)

u is the original signal, v is its total Hilbert transform, v_1, v_2, v_3 are the first-order partial Hilbert transforms, v_{12}, v_{13}, v_{23} are the second-order partial Hilbert transforms. For notational ease, the dependence of $u, v, v_1, \ldots$ on $x_1, x_2, x_3, \ldots$ is omitted. The index 1 in ψ_1 indicates a complex signal with a single orthant spectrum in the first orthant (in R^+).

The numerical factor 2^n is used to normalize the energy of the complex signal. The n-D unit step may be written in the form of a product of 1-D unit steps, that is, given by the formula

$$\mathbf{1}(\mathbf{\Omega}) = \mathbf{1}(\omega_1) \otimes \mathbf{1}(\omega_2) \otimes \cdots \otimes \mathbf{1}(\omega_n) \tag{7.23.4}$$

where $\otimes$ denotes a **tensor product** of distributions. In the following text we will suppress the symbol $\otimes$ because here it has a pure formal meaning. The 1-D unit step may be written in the form $\mathbf{1}(\omega) = 0.5[1 + \text{sgn}(\omega)]$ (see Eq. [7.3.16]). The insertion of this form in Eq. (7.23.4) yields

$$\mathbf{1}(\mathbf{\Omega}) = [0.5 + 0.5\,\text{sgn}(\omega_1)][0.5 + 0.5\,\text{sgn}(\omega_2)] \ldots [0.5 + 0.5\,\text{sgn}(\omega_n)] \tag{7.23.5}$$

The application of the convolution to multiplication theorem of Fourier analysis to the spectrum $\Gamma(\mathbf{\Omega})$ defined by Eq. (7.23.2) yields the signal domain definition of the n-D complex signal in the form of the n-fold convolution

$$\psi(\mathbf{x}) = \psi_\delta(\mathbf{x}) * \cdots * u(\mathbf{x}) \tag{7.23.6}$$

where the signal $\psi_\delta(\mathbf{x})$ is given by the inverse Fourier transform of the unit step, that is,

$$\psi_\delta(\mathbf{x}) \stackrel{n-F}{\Longleftrightarrow} 2^n \mathbf{1}(\mathbf{\Omega}) \tag{7.23.7}$$

The n-D delta pulse (distribution) $\delta(\mathbf{x}) = \delta(x_1)\delta(x_2) \ldots \delta(x_n)$ may be defined by the inverse Fourier transform of the spectrum $U(\mathbf{\Omega}) = \mathbf{1}$, that is, by the Fourier pair

$$\delta(\mathbf{x}) \stackrel{n-F}{\Longleftrightarrow} \mathbf{1} \tag{7.23.8}$$

Therefore, the signal $\psi_\delta(\mathbf{x})$ defines the **n-D complex delta distribution** (see the 1-D case, Section 7.5, Eq. [7.5.7] and a more detailed description in the next part of this section). Notice that Eqs. (7.23.2) and (7.23.6) uniquely define the n-D complex signal due to the uniqueness theorem of the Fourier analysis.

To get the structure of the n-D complex signal $\psi(\mathbf{x})$ let us insert in the spectrum $\Gamma(\mathbf{\Omega})$ defined by Eq. (7.23.2) the developed form of the multiple product given by Eq. (7.23.5), as shown in Table 7.23.1. The real part of the complex signals ψ_1 in Table 7.23.1 corresponds to the spectral terms obtained by multiplication of $U(\Omega)$ by $1, \mathrm{sgn}(\omega_1)\,\mathrm{sgn}(\omega_2), \ldots, \mathrm{sgn}(\omega_1)\,\mathrm{sgn}(\omega_2)\,\mathrm{sgn}(\omega_3)$ $\mathrm{sgn}(\omega_4), \ldots$, that is, by a product of an even number of signum functions, and the imaginary part by $\mathrm{sgn}(\omega_1), \ldots, \mathrm{sgn}(\omega_1)\,\mathrm{sgn}(\omega_2)\,\mathrm{sgn}(\omega_3), \ldots$, that is, by a product of an odd number of signum functions.

Example

Consider the 2-D harmonic signal $u = \cos(\omega_1 x_1)\cos(\omega_2 x_2)$. The Hilbert transforms are (see Table 7.22.3) $v = \sin(\omega_1 x_1)\sin(\omega_2 x_2)$, $v_1 = \sin(\omega_1 x_1)\cos(\omega_2 x_2)$, $v_2 = \cos(\omega_1 x_1)\sin(\omega_2 x_2)$. The insertion of u, v, v_1, and v_2 into Eq. (7.23.10) yields the complex signal

$$\psi_1(x_1, x_2) = \cos(\omega_1 x_1)\cos(\omega_2 x_2) - \sin(\omega_1 x_1)\cos(\omega_2 x_2)$$
$$+ j[\sin(\omega_1 x_1)\cos(\omega_2 x_2) + \cos(\omega_1 x_1)\sin(\omega_2 x_2)] \qquad (7.23.13)$$

The application of standard trigonometric relations yields

$$\psi_1(x_1, x_2) = e^{j(\omega_1 x_1 + \omega_2 x_2)} \qquad (7.23.14)$$

Notice that the real part of this signal equals $u - v = \cos(\omega_1 x_1 + \omega_2 x_2)$ and is not equal to u. The n-D generalization of the above signal is $u(\mathbf{x}) = \cos(\omega_1 x_1)\cos(\omega_2 x_2)\ldots\cos(\omega_2 x_2)$, yielding the complex signal

$$\psi(\mathbf{x}) = e^{j(\omega_1 x_1 + \omega_2 x_2 + \cdots + \omega_n x_n)} \qquad (7.23.15)$$

This formula gives evidence that the important requirement of compatibility of the definition of a multidimensional complex signal with the 1-D case is satisfied by complex signals with single orthant spectra. ∎

Example

Consider the 2-D delta pulse distribution $\delta(x_1, x_2) = \delta(x_1)\delta(x_2)$. The Hilbert transforms are given in Table 7.23.2 and Eq. (7.23.10) yields the following form of the 2-D complex delta distribution

$$\psi_\delta(x_1, x_2) = \delta(x_1, x_2) - 1/(\pi^2 x_1 x_2) + j[\delta(x_2)/(\pi x_1) + \delta(x_1)/(\pi x_2)] \qquad (7.23.16)$$

The insertion in Eq. (7.23.11) of the appropriate Hilbert transforms of the 3-D delta pulse $\delta(x_1, x_2, x_3)$ yields the following form of the 3-D complex delta distribution

$$\psi_\delta(x_1, x_2, x_3) = \delta(x_1, x_2, x_3) - \frac{\delta(x_3)}{\pi^2 x_1 x_2} - \frac{\delta(x_2)}{\pi^2 x_1 x_3} - \frac{\delta(x_1)}{\pi^2 x_2 x_3}$$
$$+ j\left(\frac{\delta(x_2, x_3)}{\pi x_1} + \frac{\delta(x_1, x_3)}{\pi x_2} + \frac{\delta(x_1, x_2)}{\pi x_3} - \frac{1}{\pi^3 x_1 x_2 x_3}\right) \qquad (7.23.17)$$

∎

Conjugate 2-D complex signals

The 2-D complex signal defined by Eq. (7.23.10) has the single quadrant spectrum in the first quadrant. Let us define 2-D complex signals with single quadrant spectra in succesive quadrants. The accepted numeration of the quadrants is shown in Figure 7.23.1. The so-defined complex signals and their spectra are shown in Table 7.23.2.

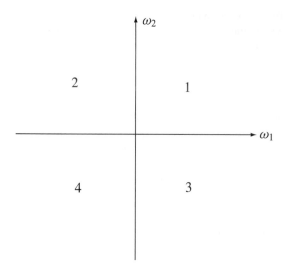

FIGURE 7.23.1
The numeration of quadrants (see Appendix).

TABLE 7.23.2
2-D complex signals with single-quadrant spectra in successive quadrants of the Fourier frequency
plane (ω_1, ω_2).

Quadrant	Complex Signal	Fourier Image
1	$\psi_1 = u - v + j(v_1 + v_2)$	$\Gamma_1(\omega_1, \omega_2) = 4\,\mathbf{1}(\omega_1, \omega_2)U(\omega_1, \omega_2) = $ $U[1 + \mathrm{sgn}(\omega_1) + \mathrm{sgn}(\omega_2) + \mathrm{sgn}(\omega_1)\,\mathrm{sgn}(\omega_2)]$
2	$\psi_2 = u + v - j(v_1 - v_2)$	$\Gamma_2(\omega_1, \omega_2) = 4\,\mathbf{1}(-\omega_1, \omega_2)U(\omega_1, \omega_2) = $ $U[1 - \mathrm{sgn}(\omega_1) + \mathrm{sgn}(\omega_2) - \mathrm{sgn}(\omega_1)\,\mathrm{sgn}(\omega_2)]$
3	$\psi_3 = u + v + j(v_1 - v_2)$	$\Gamma_3(\omega_1, \omega_2) = 4\,\mathbf{1}(\omega_1, -\omega_2)U(\omega_1, \omega_2) = $ $U[1 + \mathrm{sgn}(\omega_1) - \mathrm{sgn}(\omega_2) - \mathrm{sgn}(\omega_1)\,\mathrm{sgn}(\omega_2)]$
4	$\psi_4 = u - v - j(v_1 + v_2)$	$\Gamma_3(\omega_1, \omega_2) = 4\,\mathbf{1}(\omega_1, -\omega_2)U(\omega_1, \omega_2) = $ $U[1 - \mathrm{sgn}(\omega_1) - \mathrm{sgn}(\omega_2) + \mathrm{sgn}(\omega_1)\,\mathrm{sgn}(\omega_2)]$
	Two pairs of conjugate signals: $\psi_1 = \psi_4^*$ and $\psi_3 = \psi_2^*$	

Notations: See Table 7.23.1.

Local (or "instantaneous") amplitudes, phases, and complex frequencies

Let us write the complex signals of Table 7.23.2 in polar coordinates:

$$\psi_1(x_1, x_2) = \psi_4^*(x_1, x_2) = A_1(x_1, x_2)e^{j\Phi_1(x_1, x_2)} \qquad (7.23.18)$$

$$\psi_3(x_1, x_2) = \psi_2^*(x_1, x_2) = A_2(x_1, x_2)e^{j\Phi_2(x_1, x_2)} \qquad (7.23.19)$$

This representation defines the **local** (or "instantaneous") **amplitudes**

$$A_1(x_1, x_2) = \mathrm{SQR}\left\{[u(x_1, x_2) - v(x_1, x_2)]^2 + [v_1(x_1, x_2) + v_2(x_1, x_2)]^2\right\} \quad (7.23.20)$$

$$A_2(x_1, x_2) = \mathrm{SQR}\left\{[u(x_1, x_2) + v(x_1, x_2)]^2 + [v_1(x_1, x_2) - v_2(x_1, x_2)]^2\right\} \quad (7.23.21)$$

and the **local** (or "instantaneous") **phases**

$$\Phi_1(x_1, x_2) = \tan^{-1} \frac{v_1(x_1, x_2) + v_2(x_1, x_2)}{u(x_1, x_2) - v(x_1, x_2)} \tag{7.23.22}$$

$$\Phi_2(x_1, x_2) = \tan^{-1} \frac{v_1(x_1, x_2) - v_2(x_1, x_2)}{u(x_1, x_2) + v(x_1, x_2)} \tag{7.23.23}$$

of the real signal $u(x_1, x_2)$. Analogous to the 1-D case (see Section 7.15, Eq. [7.15.12]) let us define the **complex phases**

$$\Phi_{1c}(x_1, x_2) = \text{Ln } \psi_1(x_1, x_2) \tag{7.23.24}$$

$$\Phi_{2c}(x_1, x_2) = \text{Ln } \psi_3(x_1, x_2) \tag{7.23.25}$$

and the **partial** instantaneous complex frequencies

$$s_{1x_1}(x_1, x_2) = \frac{\partial \Phi_{1c}(x_1, x_2)}{\partial x_1} = \alpha_{1x_1}(x_1, x_2) + j\omega_{1x_1}(x_1, x_2) \tag{7.23.26}$$

$$s_{1x_2}(x_1, x_2) = \frac{\partial \Phi_{1c}(x_1, x_2)}{\partial x_2} = \alpha_{1x_2}(x_1, x_2) + j\omega_{1x_2}(x_1, x_2) \tag{7.23.27}$$

$$s_{2x_1}(x_1, x_2) = \frac{\partial \Phi_{2c}(x_1, x_2)}{\partial x_1} = \alpha_{2x_1}(x_1, x_2) + j\omega_{2x_1}(x_1, x_2) \tag{7.23.28}$$

$$s_{2x_2}(x_1, x_2) = \frac{\partial \Phi_{2c}(x_1, x_2)}{\partial x_2} = \alpha_{2x_2}(x_1, x_2) + j\omega_{2x_2}(x_1, x_2) \tag{7.23.29}$$

In Eqs. (7.23.26) and (7.23.28) x_2 is a parameter and the complex frequencies are defined along the lines parallel to the x_1 axis. Similarly, Eq. (7.23.27) and (7.23.29) define complex frequencies parallel to the x_2 axis.

For separable 2-D signals (see Eq. [7.22.38]), the amplitudes A_1 and A_2 defined by Eqs. (7.23.20) and (7.23.21) are equal and given by the formula

$$A(x_1, x_2) = \text{SQR}[u^2 + v^2 + v_1^2 + v_2^2] \tag{7.23.30}$$

and the phases (7.23.22) and (7.23.23) are

$$\Phi_1(x_1, x_2) = \varphi_1(x_1) + \varphi_2(x_2) \tag{7.23.31}$$

$$\Phi_2(x_1, x_2) = \varphi_1(x_1) - \varphi_2(x_2) \tag{7.23.32}$$

where $\varphi_1 = \tan^{-1}(g_1/f_1)$ and $\varphi_2 = \tan^{-1}(g_2/f_2)$ (see Eq. [7.22.38]). The complex frequencies have, for separable signals, the simplified form

$$s_{1x_1} = s_{2x_1} = \alpha_1 + j\omega_1 \tag{7.23.33}$$

$$s_{1x_2} = s_{2x_2}^* = \alpha_2 + j\omega_2 \tag{7.23.34}$$

Example
Consider the 2-D signal of the form

$$u(x_1, x_2) = \frac{\sin(ax_1) \sin(bx_2)}{abx_1x_2} \tag{7.23.35}$$

The insertion of this signal and its Hilbert transforms v, v_1, and v_2 (see Table 7.22.2) in Eq. (7.23.30) using certain trigonometric relations yields

$$A(x_1, x_2) = \left| \frac{1 - \cos(ax_1)}{\sin(ax_1)} \right| \left| \frac{1 - \cos(bx_2)}{\sin(bx_2)} \right| \qquad (7.23.36)$$

The phase functions (7.23.31) and (7.23.32) take the form

$$\Phi_1(x_1, x_2) = \tan^{-1} \frac{1 - \cos(ax_1)}{\sin(ax_1)} + \tan^{-1} \frac{1 - \cos(bx_2)}{\sin(bx_2)} = \frac{a}{2}x_1 + \frac{b}{2}x_2 \quad (7.23.37)$$

$$\Phi_2(x_1, x_2) = \tan^{-1} \frac{1 - \cos(ax_1)}{\sin(ax_1)} - \tan^{-1} \frac{1 - \cos(bx_2)}{\sin(bx_2)} = \frac{a}{2}x_1 - \frac{b}{2}x_2 \quad (7.23.38)$$

The local partial angular frequencies defined by the imaginary parts of (7.23.33) and (7.23.34) are $\omega_1 = a/2$ and $\omega_2 = b/2$. The local amplitude (7.23.36) is a product of local amplitudes of the separable terms of the signal (7.23.35), and the phase is a sum (or difference) of phases of these terms. The phase functions in this example are linear (a constant slope if we remove the jumps of the multibranch $\tan^{-1}$ function), giving constant values of the angular frequencies. ∎

The relations between real and complex notation

In one dimension we have the following well-known relations:

$$u(t) = \frac{\psi(t) + \psi^*(t)}{2}; \qquad v(t) = \frac{\psi(t) - \psi^*(t)}{2j} \qquad (7.23.39)$$

In two dimensions the corresponding relations become (see Table 7.23.2)

$$u(x_1, x_2) = \frac{\psi_1 + \psi_2 + \psi_3 + \psi_4}{4} \qquad (7.23.40)$$

$$v(x_1, x_2) = \frac{\psi_1 - \psi_2 + \psi_3 - \psi_4}{4} \qquad (7.23.41)$$

$$v_1(x_1, x_2) = \frac{\psi_1 - \psi_2 + \psi_3 - \psi_4}{4j} \qquad (7.23.42)$$

$$v_2(x_1, x_2) = \frac{\psi_1 + \psi_2 - \psi_3 - \psi_4}{4j} \qquad (7.23.43)$$

Using the relations $\psi_1 = \psi_4^*$ and $\psi_3 = \psi_2^*$, the real part of the complex signal ψ_1 takes the form

$$u - v = 0.5(\psi_1 + \psi_4) = 0.5(\psi_1 + \psi_1^*) \qquad (7.23.44)$$

and the real part of ψ_3 is

$$u + v = 0.5(\psi_2 + \psi_3) = 0.5(\psi_2 + \psi_2^*) \qquad (7.23.45)$$

Notice that the spectra of these two signals exist in two quadrants of the Fourier frequency plane. The insertion of the polar representations (7.23.18) and (7.23.19) into (7.23.44) and (7.23.45) yield

$$u - v = A_1 \cos(\Phi_1) \qquad (7.23.46)$$

$$u + v = A_2 \cos(\Phi_2) \qquad (7.23.47)$$

The summation (or substraction) yields the following relations

$$u(x_1, x_2) = \frac{A_1 \cos(\Phi_1) + A_2 \cos(\Phi_2)}{2} \qquad (7.23.48)$$

$$v(x_1, x_2) = \frac{A_2 \cos(\Phi_2) - A_1 \cos(\Phi_1)}{2} \qquad (7.23.49)$$

and analogous derivation yields

$$v_1(x_1, x_2) = \frac{A_1 \sin(\Phi_1) + A_2 \sin(\Phi_2)}{2} \qquad (7.23.50)$$

$$v_2(x_1, x_2) = \frac{A_2 \sin(\Phi_2) - A_1 \sin(\Phi_1)}{2} \qquad (7.23.51)$$

For separable signals these relations have the simplified form

$$u(x_1, x_2) = A \cos[\varphi_1(x_1)] \cos[\varphi_2(x_2)] \qquad (7.23.52)$$

$$v(x_1, x_2) = A \sin[\varphi_1(x_1)] \sin[\varphi_2(x_2)] \qquad (7.23.53)$$

$$v_1(x_1, x_2) = A \sin[\varphi_1(x_1)] \cos[\varphi_2(x_2)] \qquad (7.23.54)$$

$$v_2(x_1, x_2) = A \cos[\varphi_1(x_1)] \sin[\varphi_2(x_2)] \qquad (7.23.55)$$

In three dimensions the number of octants equals eight and the relation between the real and complex notation becomes (we applied the method of numeration of octants given in Appendix 7.23.1).

$$u(x_1, x_2, x_3) = \frac{\psi_1 + \psi_2 + \psi_3 + \psi_4 + \psi_5 + \psi_6 + \psi_7 + \psi_8}{8} \qquad (7.23.56)$$

Using the relations $\psi_1 = \psi_8^* = A_1 e^{j\Phi_1}$, $\psi_2 = \psi_7^* = A_2 e^{j\Phi_2}$, $\psi_3 = \psi_6^* = A_3 e^{j\Phi_3}$, and $\psi_4 = \psi_5^* = A_4 e^{j\Phi_4}$, the above formula takes the form

$$(x_1, x_2, x_3) = \frac{A_1 \cos(\Phi_1) + A_2 \cos(\Phi_2) + A_3 \cos(\Phi_3) + A_4 \cos(\Phi_4)}{8} \qquad (7.23.57)$$

For separable signals all amplitudes are equal and the phase functions are

$$\Phi_1(x_1, x_2, x_3) = \varphi_1(x_1) + \varphi_2(x_2) + \varphi_3(x_3) \qquad (7.23.58)$$

$$\Phi_2(x_1, x_2, x_3) = -\varphi_1(x_1) + \varphi_2(x_2) + \varphi_3(x_3) \qquad (7.23.59)$$

$$\Phi_3(x_1, x_2, x_3) = \varphi_1(x_1) - \varphi_2(x_2) + \varphi_3(x_3) \qquad (7.23.60)$$

$$\Phi_4(x_1, x_2, x_3) = -\varphi_1(x_1) - \varphi_2(x_2) + \varphi_3(x_3) \qquad (7.23.61)$$

The insertion of these phase functions in Eq. (7.23.57) yields

$$u(x_1, x_2, x_3) = A \cos[\varphi_1(x_1)] \cos[\varphi_2(x_2)] \cos[\varphi_3(x_3)] \qquad (7.23.62)$$

Similar formulae for v, v_1, and v_2 may be easily derived or written directly by comparison with the 2-D case. In general, for the n-D separable signal of the form $u(\mathbf{x}) = \Pi_{k=1}^n a_k f(x_k)$ this formula takes the form $u(\mathbf{x}) = A \Pi_{k=1}^n \cos[\varphi(x_k)]$, where $A = \Pi_{k=1}^n a_k$. If all a_k's are equal to a, then $A = a^n$ or $a = \sqrt[n]{A}$.

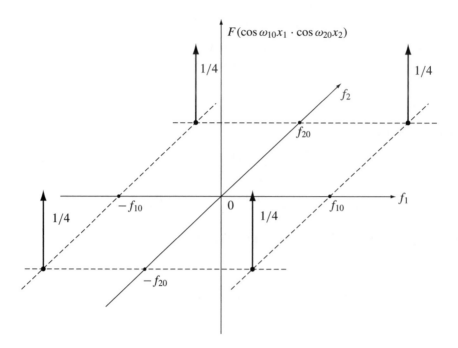

FIGURE 7.23.2
The Fourier spectrum of the 2-D harmonic signal $u(x_1, x_2) = \cos(\omega_{10}x_1)\cos(\omega_{20}x_2)$.

Example

Consider again the signal $u(x_1, x_2) = \cos(\omega_{10}x_1)\cos(\omega_{20}x_2)$ of the previous example. The four complex signals of Table 7.23.2 and their spectra are:

Quadrant	Complex signal	Fourier image
1	$\psi_1 = e^{j(\omega_{10}x_1 + \omega_{20}x_2)}$	$\delta(\omega_1 - \omega_{10}, \omega_2 - \omega_{20})$
2	$\psi_2 = e^{j(-\omega_{10}x_1 + \omega_{20}x_2)}$	$\delta(\omega_1 + \omega_{10}, \omega_2 - \omega_{20})$
3	$\psi_3 = e^{j(\omega_{10}x_1 - \omega_{20}x_2)}$	$\delta(\omega_1 - \omega_{10}, \omega_2 + \omega_{20})$
4	$\psi_1 = e^{j(-\omega_{10}x_1 - \omega_{20}x_2)}$	$\delta(\omega_1 + \omega_{10}, \omega_2 + \omega_{20})$

The spectrum of the signal u is shown in Figure 7.23.2. The insertion of these functions in Eqs. (7.23.40) to (7.23.51) gives the verifications of these relations.   ▐

2-D modulation theory

The 1-D modulated signal has the complex representation in the form of a product of the **modulation function** and the complex harmonic carrier (see Section 7.16, Eq. [7.16.6]). From the formal point of view, the concept of the modulation function can be extended to multidimensional modulating signals with multidimensional harmonic carriers. The n-D complex harmonic carrier has the form

$$\Psi_c(\mathbf{x}) = A_0 e^{j(\omega_{10}x_1 + \varphi_1 + \omega_{20}x_2 + \varphi_2 + \cdots + \omega_{n0}x_n + \varphi_n)} \tag{7.23.63}$$

We define the n-D modulated signal in the form of a product

$$\psi(\mathbf{x}) = \gamma(\mathbf{x})\psi_c(\mathbf{x}) \tag{7.23.64}$$

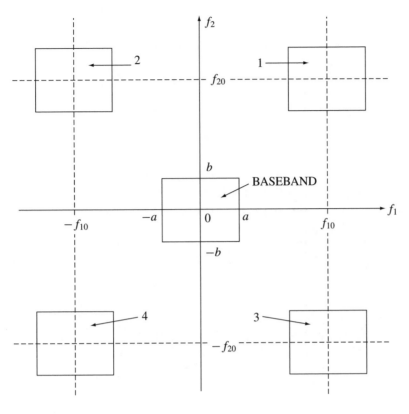

FIGURE 7.23.3
The supports of the spectrum of a 2-D carrier with 2-D amplitude modulation.

where $\gamma(\mathbf{x}) = f[u(\mathbf{x})]$ is called the *n*-**D modulation function** and $f[u(\mathbf{x})]$ is a function of the *n*-D message $u(\mathbf{x})$. As in the 1-D case, the function f defines a specific type of modulation. The 2-D modulated signal has the form

$$\Psi(x_1, x_2) = \gamma(x_1, x_2) A_0 e^{j(\omega_{10}x_1 + \omega_{20}x_2)} \tag{7.23.65}$$

where for convenience, the phases $\varphi_1 = \varphi_2 = 0$.

Example
Consider a 2-D low-pass message and its Fourier image

$$u(x_1, x_2) \stackrel{2-F}{\Longleftrightarrow} U(j\omega_1, j\omega_2) \tag{7.23.66}$$

with the base-band spectrum band limited such that $U(j\omega_1, j\omega_2) = 0$ for $|\omega_1| > a$ and

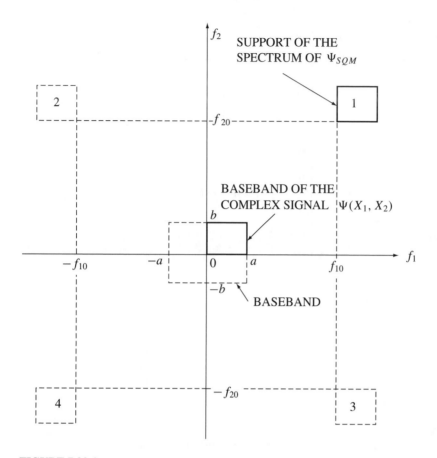

FIGURE 7.23.4
The supports of the spectrum of a 2-D signal with Single Quadrant Modulation (SQM).

$|\omega_2| > b$. The modulation function of the 2-D suppressed carrier amplitude modulation is

$$\gamma_{AM}(x_1, x_2) = mu(x_1, x_2) \tag{7.23.67}$$

Figure 7.23.3 shows the spectra of the base-band signal u and of the modulated signal.

The 2-D equivalent of the 1-D SSB modulation is the **single quadrant modulation (SQM)**. The modulation function is given by the inverse Fourier transform of the base-band single quadrant spectrum

$$\gamma_{SQM}(x_1, x_2) = F^{-1}[41(\omega_1, \omega_2)U(\omega_1, \omega_2)]$$

$$= u(x_1, x_2) - v(x_1, x_2) + j[v_1(x_1, x_2) + v_2(x_1, x_2)] \tag{7.23.68}$$

that is, in the form of the complex signal (7.23.10) in Table 7.23.1 The insertion of this modulation function in Eq. (7.23.55) yields the complex SQM signal

$$\Psi_{SQM}(x_1, x_2) = \{u(x_1, x_2) - v(x_1, x_2) + j[v_1(x_1, x_2) + v_2(x_1, x_2)]\} A_0 e^{j(\omega_{10}x_1 + \omega_{20}x_2)}$$

$$\tag{7.23.69}$$

and its real notation is

$$u_{\text{SQM}}(x_1, x_2) = u \cos(\omega_{10}x_1) \cos(\omega_{20}x_2) + v \sin(\omega_{10}x_1) \sin(\omega_{20}x_2) +$$

$$-v_1 \sin(\omega_{10}x_1) \cos(\omega_{20}x_2) - v_2 \cos(\omega_{10}x_1) \sin(\omega_{20}x_2) \quad (7.23.70)$$

Figure 7.23.4 shows the supports of the spectra of γ_{SQM}, Ψ_{SQM}, and u_{SQM}. ∎

7.23.1   *Appendix: A method of labeling orthants*

The applied numbering method of successive orthants is the following: We assign the binary number zero to a plus sign, and the binary number 1 to a minus sign of the variable ω. For example, the unit step function $\mathbf{1}(\omega_4, -\omega_3, -\omega_2, \omega_1)$ corresponds to the binary number 0110. If the decimal-coded binary number is a, we assign to the given orthant the decimal number $l = a + 1$. So we have in four dimensions:

$l = a + 1$	binary	$\omega_4,$	$\omega_3,$	$\omega_2,$	ω_1
		sign of the Ω axis			
1	0000	+	+	+	+
2	0001	+	+	+	−
3	0010	+	+	−	+
4	0011	+	+	−	−
. . . .					
16	1111	−	−	−	−

References

[1]   Ansari, R. IIR discrete-time Hilbert transformers. *IEEE Trans. ASSP.* vol. ASSP-35, No. 8, Aug. 1987, pp. 1116–1119.

[1a] Ansari, R. Elliptic filter design for generalized half-band filters. vol. ASSP-33, No. , Oct. 1985, pp. 1146–1150.

[2]   Bedrosian, S. D. Normalised design of 90° phase-difference networks. *Trans. of the Inst. Radio Engrs* CT-7, 1960, pp. 128–136.

[3]   Bedrosian, E. The analytic signal representation of modulated waveforms. *Proc. IEEE* vol. 50, No. 10, Oct. 1962, pp. 2071–2076.

[4]   Bedrosian, E. A product theorem for Hilbert transforms. *Proc. IEEE (Lett.)* vol. 51, No. 5, May 1963, pp. 868–869.

[5]   Bedrosian, E., and Stark, H. Comments on "An extension of the Hilbert transform product theorem," *Proc. IEEE* vol. 60, Feb. 1972, pp. 228–229.

[6]   Budeanu, C. Reactive and fictitous powers. *Romanian National Inst. publ.* No. 2, 1927, Bucharest, Romania.

[7]   Čižek, V. Discrete Hilbert transform. *IEEE Trans.* vol. AU-18, No. 4, Dec. 1970, pp. 340–343.

[8]   Darlington, S. Realization of a constant phase differrence. *Bell Syst. Tech. J.* 29, 1950, pp. 94–104.

[9] Dome, R. B. Wide-band phase shift networks. *Electronics* 19, Dec. 1946, pp. 112–115.

[10] Fryze, S., Wirk—und Blind—und Scheinleistung in Elektrischen Stromkreisen mit nichtsinusoidalenformigen Verlauf von Strom und Spannung. *Elektrotech. Z.* 53 (25), 1932, pp. 596–599 (Active, reactive, and apparent power in electrical circuits with non-sinusoidals currents and voltages).

[11] Gabor, D. Theory of communications. *J. Inst. EE*, Pt. III, vol. 93, Nov. 1946, pp. 429–457.

[12] Hahn, S. L. Complex variable frequency electric circuit theory. *Proc. IEEE (Lett.)* vol. 52, No. 6, June 1964, pp. 735–736.

[13] Hahn, S. L. Multidimensional complex signals with single-orthant spectra. *Proc. IEEE* vol. 80, No. 8, Aug. 1992, pp. 1287–1300.

[14] Hahn, S. L. Amplitudes, phases and complex frequencies of 2-D Gaussian signals. *Bulletin of the Polish Acad. Sci.* vol. 40, No. 3, 1992, pp. 289–311.

[15] Herrman, O. Transversalfilter zur Hilbert-Transformation. *Arch. Elektr. Übertr.* 23, Heft 12, 1969, pp. 581–587 (Transversal filters as Hilbert transformers).

[16] Jackson, L. B. On the relationship between digital Hilbert transformers and certain low-pass filters. *IEEE Trans. ASSP*, vol. 23, No. 8, Aug. 1975, pp. 381–383.

[17] Kahn, L.R. Compatible single sideband. *Proc. IRE* vol.49, Oct. 1961, pp. 1503–1527.

[18] Kramers, H. A. *Phys. Z.* 30, 1929, p. 522.

[19] Kronig, R. de L. *Journ. Opt. Soc. Amer.* vol. 12, 1926, p. 547.

[20] Krylov, W. W., and Ponomariev, D. M. Definition of the signal delay in linear two-ports using the Hilbert transform (in russian). *Radiotechnika i Elektronika* No. 5, 1980, pp. 204–206.

[21] Klyagin, L. Ye. Linearization of the phase characteristics of wide-band phase shifters (translated from russian). *Telecomm. and Radio Eng.* Part 2 (USA) 1967, pp. 82–87.

[22] McClellan, J. H. A computer program for designing optimum FIR linear phase digital filters. *IEEE Trans.* vol. AU-21, No. 6, Dec. 1973, pp. 506–526.

[23] Mikusinski, J. The elementary theory of distributions. *Rozprawy Matematyczne XII*, Warsaw: PWN, 1957.

[24] Nowomiejski, Z. Generalized theory of electric power, *Archiv. f. Elektrotechnik* vol. 63, 1981, pp. 177–182. 1972, pp. 1361–1362.

[25] Nuttal, A. and Bedrosian, E. On the quadrature approximation of the Hilbert transform of modulated signals. *Proc. IEEE (Lett.)* vol. 54, Oct. 1966, pp. 1458–1459.

[26] Oppenheim, A. V. and Schafer, R. W. *Digital Signal Processing.* New Jersey: Prentice-Hall, Inc., 1975.

[27] Orchard, H. J. The synthesis of RC networks to have prescribed transfer functions. *Proc. IRE* vol. 39, April 1951, pp. 428–432.

[28] Paley, E. A. C. and Wiener, N. Fourier transforms in the complex domain. *Amer. Math. Soc. Colloqium Publicaton* 19, New York, 1934.

[29] Rabiner, L. R. and Schafer, R. W. On the behaviour of minimax FIR digital Hilbert transformers. *Bell Syst. Tech. J.* vol. 53, No. 2, Feb. 1974, pp. 363–390, and On the behaviour of minimax relative error FIR differentiators, pp. 333–362.

[30] Rabiner, L. R. and Gold, B. *Theory and Applications of Digital Signal Processing.* New Jersey: Prentice-Hall, Inc., 1975.

[31] Rabiner, L. R. et al. FIR digital filter design techniques using weighted Chebyshev approximation. *Proc. IEEE* vol. 63, No. 4, Apr. 1975, pp. 595–610.

[32] Robinson, F. N. H. *Macroscopic Electromagnetism*. Pergamon Press, 1973.

[33] Saraga, W. The design of wide-band two-phase networks. *Proc. Inst. Electron. Engrs* 38, 1950, pp. 754–770.

[34] Schneider, W. Quadraturfilter nach der methode der angezapften Laufzeitketten. *Telefunken-Ztg.* 40, 1967 (Quadrature filter using the method of a tapped-delay line).

[35] Schwartz, L. *Methodes mathematique pour les science physique*. Paris: Hermann, 1965.

[36] Stark, H. and Tuteur, F. B. Modern electrical communications. *Theory and systems*. New Jersey: Prentice-Hall, Inc., 1979.

[37] Stark, H. An extension of the Hilbert transform product theorem. *Proc. IEEE* vol. 59, No. 9, Sept. 1971, pp. 1359–1360.

[38] Weaver, D. K. Design of RC wide-band 90-degree phase difference networks. *Proc. IRE* vol. 42, April 1954, pp. 671–676.

[39] Weinberger, H. F. *Partial Differential Equations with Complex Variables and Transform Methods*. New York: Blaisdell Publishing Co., 1965.

[40] Villard, O. G., Jr. Composite amplitude and phase modulation. *Electronics* 21, Nov. 1948, pp. 86–89.

[41] Carson, J. R., and Fry, C. F., Variable frequency electric circuit theory with application to the theory of frequency modulation, *Bell Syst. Techn. J.*, vol. 6, No. 4, 1954, pp. 513–530.

8

Radon and Abel Transforms

Stanley R. Deans

CONTENTS

8.1 Introduction

The Austrian mathematician Johann Radon (1887–1956) wrote a classic paper in 1917, "Über die Bestimmung von Funktionen durch ihre Integralwerte längs gewisser Mannigfaltigkeiten" (On the determination of functions from their integrals along certain manifolds) [Radon, 1917]. This work forms the foundation for what we now call the Radon transform. English translations are available in the monograph by Deans [1983, 1993] and the translation by Parks [1986]. The problem of determining a function $f(x, y)$ from knowledge of its line integrals (the two-dimensional case), or a function $f(x, y, z)$ from integrals over planes (the three-dimensional case) arises in widely diverse fields. These include medical imaging, astronomy, crystallography, electron microscopy, geophysics, optics, and material science. In these applications

0-8493-8342-0/96/$0.00 + $0.50

©1996 by CRC Press, Inc.

the central aim is to obtain certain information about the internal structure of an object either by passing some probe (such as x-rays) through the object or by using information from the source itself when it is self-emitting, such as an organ in the body that contains a radioactive isotope, or perhaps the interior of the Earth when motions occur. Comprehensive reviews of these and other applications are contained in Brooks and Di Chiro [1976], Scudder [1978], Barrett [1984], Chapman [1987], and Deans [1983, 1993].

The general problem of unfolding internal structure of an object by observations of projections is known as the problem of **reconstruction from projections**. Many situations arise when it is possible to determine (reconstruct) various structural properties of an object or substance by methods that utilize projected information and leave the object in an essentially undamaged state. The Radon transform and its inversion forms the mathematical framework common to a large class of these problems. This problem of reconstructing a function from knowledge of its projections emerges naturally in fields so diverse that those working in one area seldom communicate with their counterparts in the other areas. This was especially true prior to the advent of computerized tomography in the 1970s. As a consequence, there is an interesting history of the independent development of applications of the Radon transform by individuals who were not aware of the original work by Radon in 1917, or of contemporary work in other fields. Those interested in pursuing these historical matters can consult Cormack [1973, 1982, 1984], Barrett, Hawkins, and Joy [1983], and Deans [1985, 1993].

Also, the Radon transform has varying degrees of relevance in three Nobel prizes: (Medicine 1979, Allan M. Cormack and Godfrey N. Hounsfield) [Di Chiro and Brooks, 1979, 1980], [Cormack, 1980], and [Hounsfield, 1980]: (Chemistry 1982, Aaron Klug) [Caspar and DeRosier, 1982]: (Chemistry 1991, Richard R. Ernst) [Amato, 1991].

As short a time as a decade ago the Radon transform was known by very few engineers and scientists. Only those working directly on reconstruction from projections in one of the major areas of application had knowledge of this transform. Today, the Radon transform is widely known by working scientists in medicine, engineering, physical science, and mathematics. It has made its way into the image processing texts [Kak, 1984, 1985], [Kak and Slaney, 1988], [Jain, 1989], [Jähne, 1993], and is widely appreciated in many diverse areas; among the best known include: medical imaging [Herman, 1980], [Macovski, 1983], [Natterer, 1986], [Swindell and Webb, 1988], [Parker, 1990], [Russ, 1992], [Cho, Jones, and Singh, 1993]; optics and holographic interferometry [Vest, 1979]; geophysics [Claerbout, 1985], [Chapman, 1987], [Ruff, 1987], [Bregman, Bailey and Chapman, 1989]; radio astronomy [Bracewell, 1979]; and pure mathematics [Grinberg and Quinto, 1990], [Gindikin and Michor, 1994].

The purpose of this chapter is to review (and illustrate with examples) important properties of Radon and Abel transforms and indicate some of the applications, along with important sources for applications. Because the Abel transform is a special case of the Radon transform, most of the discussion is for the more general transform. This is especially important to keep in mind for applications where the Abel transform can be used. Section 8.10 is devoted to Abel integral equations and Abel transforms. The formal connection between Abel and Radon transforms is made in Section 8.11; the reader primarily interested in Abel transforms may want to look at those two sections first.

The overall goal is to provide the reader with basic material that can be used as a foundation for understanding current research that makes use of the transforms. A conscientious attempt is made to present essential mathematical material in a way that is easily understood by anyone having a basic knowledge of Fourier transforms. In keeping with this goal, the emphasis will be on the two-dimensional and three-dimensional cases. The extension to higher dimensions will be mentioned at various times, especially when the extension is rather obvious. For the most part, derivations are kept as simple and intuitive as possible. Reference is made to more rigorous discussions and abstract applications. The same policy is followed for highly technical problems related to sampling and numerical implementation of inversion algorithms. These are ongoing research problems that lie a level above the basic treatment presented here. Section 8.1.1 contains a brief summary of how the chapter is organized. An attempt is made to cross reference the various sections, so the reader interested in a given topic can go directly to that topic without having to read everything that precedes. Finally, it is to be noted that liberal use is made of material contained in books by the author on the same subject [Deans, 1983, 1993].

8.1.1   *Organization of the chapter*

Section 8.2 is devoted mainly to fundamental definitions, concepts, and spaces. The definitions are given several ways and for various dimensions to make it easier for the reader to make connection with usage in the current literature. The section on probes, structure, and transforms outlines the connection of the Radon transform to physical applications. A very important theorem known as the central-slice theorem serves to relate three spaces of special importance: feature space, Radon space, and Fourier space. A proof is provided for the two-dimensional case, and an example is given to illustrate how a function transforms among the three spaces.

Some of the most basic properties of the Radon transform are presented in Section 8.3 and compared with the corresponding properties for the Fourier transform. These properties are used many times throughout the sections that follow.

A brief, but important, discussion of the Radon transform of a linear transformation is in Section 8.4. This provides the foundation for powerful methods to calculate transforms of various functions. In Section 8.5 this idea is combined with the basic properties to illustrate, by several examples, just how the Radon transform works when applied to certain special functions. These examples are selected to bring out subtle points that emerge when actually computing a transform.

More advanced topics on derivatives and the transform are in Section 8.6. This work serves as background for transforms involving Hermite polynomials in Section 8.7 and Laguerre polynomials in Section 8.8.

The important problem of inversion is initiated in Section 8.9. Details are given for two and three dimensions, and the foundation is provided for some of the currently utilized inversion methods outlined in sections that follow.

Abel transforms and Abel-type integral equations are discussed in Section 8.10. Four different types of Abel transforms are defined along with the corresponding inverses. Inter-relationships among the transforms are illustrated along with several useful examples. A rule is given to establish a method for finding Abel transforms from extensive tables of Riemann-Liouville and Weyl (fractional) integrals. The way the Radon and Fourier transforms relate to the Abel and Hankel transforms is developed in Section 8.11. An important observation is that the Abel transform is a special case of the Radon transform. Examples are given to demonstrate the connection for specific cases.

The earlier work on inversion is supplemented in Section 8.12 by some methods that form the basis for modern algorithms for numerical inversion of discrete data using backprojection and convolution methods. Diagrams that clearly illustrate the various options are included in this section.

Series methods for inversion are discussed in Section 8.13, with emphasis on two and three dimensions. Special attention is given to functions defined on the unit disk in feature space. Several examples are provided to illustrate both techniques and the connection with earlier sections.

The Parseval relation for the Radon transform is given in Section 8.14 for the general n-dimensional case. A useful example in two dimensions serves to highlight the difference between the Fourier and Radon cases.

Extensions and emerging concepts are mentioned briefly in Section 8.15. An especially exciting area involves the use of the wavelet transform to facilitate inversion of the Radon transform.

Finally, Appendix A contains a compilation of formulas and special functions used throughout the chapter, and a list of selected Radon and Abel transforms appears in Appendix B.

8.1.2 Remarks about notation

The Radon transform is defined on real Euclidean space for two and higher dimensions. Many results are just as easy to obtain for the n-dimensional transform as for the two-dimensional transform. However, most illustrations (and applications) of the transform are easier in two or three dimensions. Consequently, several equivalent notations are appropriate for vectors. Various notations are given here and the policy throughout the entire discussion is to change freely from one notation to the other with absolutely no apology.

Both component and matrix notations will be used. In component notation, all of the following expressions are used,

$$\mathbf{x} = \mathbf{r} = (x, y) \qquad \mathbf{x} = (x_1, x_2) \qquad \mathbf{y} = (y_1, y_2).$$

In matrix notation these would be

$$\mathbf{x} = \mathbf{r} = \begin{pmatrix} x \\ y \end{pmatrix} \qquad \mathbf{x} = \begin{pmatrix} x_1 \\ x_2 \end{pmatrix} \qquad \mathbf{y} = \begin{pmatrix} y_1 \\ y_2 \end{pmatrix}.$$

Similar notations are used for three dimensions by appending z or x_3 or y_3. For the n-dimensional case we use

$$\mathbf{x} = (x_1, \dots, x_n) \qquad \mathbf{y} = (y_1, \dots, y_n),$$

or the equivalent matrix form. When there is no confusion about which variables are being integrated, the abbreviated notation

$$\int f(\mathbf{x}) \, d\mathbf{x} \equiv \int_{-\infty}^{\infty} \cdots \int_{-\infty}^{\infty} f(x_1, \dots, x_n) \, dx_1 \dots dx_n$$

will be used for integration over all space.

8.2 Definitions

In a discussion of the Radon transform it is convenient to identify three spaces. These spaces are designated by **feature space**, **Radon space**, and **Fourier space**.

Feature space is just Euclidean space in two, three, or n dimensions, designated by 2D, 3D, or nD. This is where the spatial distribution f of some physical property is defined. Radon space and Fourier space designate the spaces for the corresponding transforms of this distribution. Functions in feature space that represent the distribution are designated by $f(x, y)$, $f(x, y, z)$, and $f(x_1, \dots, x_n)$, depending on the dimension of the transform. For the purposes of this presentation, these functions f are selected from some nice class of functions, such as the class of infinitely differentiable (C^∞) functions with compact support or rapidly decreasing C^∞ functions [Schwartz, 1966]. This assumption serves well for the current discussion; however, it can be relaxed in more general treatments [Gel'fand, Graev, and Vilenkin, 1966], [Lax and Phillips, 1970, 1979], [Helgason, 1980], [Grinberg and Quinto, 1990], [Mikusiński and Zayed, 1993], [Gindikin and Michor, 1994].

The transformation from one space to another can be represented symbolically as a mapping operation. Let $\mathcal{R}$ be the operator that transforms f to Radon space. If the corresponding function in Radon space is designated by $\check{f}$, the mapping operation is expressed by

$$\check{f} = \mathcal{R}f. \tag{8.2.1}$$

In a similar way, the transformation to Fourier space is written

$$\tilde{f} = \mathcal{F}f. \tag{8.2.2}$$

These operations will be made more precise in the next sections where explicit definitions are given for various dimensions.

8.2.1 *Two dimensions*

The Radon transform of the function $f(x, y)$ is defined as the line integral of f for all lines ℓ defined by the parameters ϕ and p, illustrated in Figure 8.1. There are several ways this can be expressed. In terms of integrals along ℓ,

$$\check{f}(p, \phi) = \int_{-\infty}^{\infty} f(\mathbf{r}) \, d\ell, \tag{8.2.3}$$

where $\mathbf{r} = (x, y)$ is a general position vector. Another way to write this is to define the unit vector $\boldsymbol{\xi} = (\cos\phi, \sin\phi)$ and the perpendicular vector $\boldsymbol{\xi}' = (-\sin\phi, \cos\phi)$, then the position vector is given by $\mathbf{r} = p\,\boldsymbol{\xi} + t\,\boldsymbol{\xi}'$ and (Note that $r^2 = p^2 + t^2$)

$$\check{f}(p, \boldsymbol{\xi}) = \int_{-\infty}^{\infty} f(p\,\boldsymbol{\xi} + t\,\boldsymbol{\xi}') \, dt. \tag{8.2.4}$$

An equivalent definition making use of the delta function (see Chapter 1) is most convenient for the current discussion,

$$\check{f}(p, \phi) = \int_{-\infty}^{\infty} \int_{-\infty}^{\infty} f(x, y)\, \delta(p - x\cos\phi - y\sin\phi)\, dx\, dy. \tag{8.2.5}$$

Note that due to the property of the delta function and the fact that the normal form for the equation of the line ℓ is given by $p = x\cos\phi + y\sin\phi$, the integral over the plane reduces to a line integral in agreement with the previous definitions. A slightly different form proves especially useful for generalization to higher dimensions. In terms of the vectors $\mathbf{r}$ and $\boldsymbol{\xi}$,

$$\check{f}(p, \boldsymbol{\xi}) = \int_{-\infty}^{\infty} \int_{-\infty}^{\infty} f(\mathbf{r})\, \delta(p - \boldsymbol{\xi} \cdot \mathbf{r})\, dx\, dy, \tag{8.2.6}$$

where $\boldsymbol{\xi} \cdot \mathbf{r} = \xi_1 x + \xi_2 y = x\cos\phi + y\sin\phi$.

It is important to understand that $\check{f}$ is **not** defined on a circular polar coordinate system. The appropriate space is on the surface of a half-cylinder. Consider an infinite cylinder of radius unity. Let the parameter p measure length along the cylinder from $-\infty$ to $+\infty$, and let the angle ϕ measure the angle of rotation with respect to an arbitrary reference position. A point on an arbitrary cross-section of the cylinder is represented by (p, ϕ) as illustrated in Figure 8.2.

Observe that from the definition of the transform, if $\check{f}$ is known for $-\infty < p < \infty$, then only values of ϕ in the range $0 \le \phi < \pi$ are needed. To verify this, recall that the delta function is even $\delta(x) = \delta(-x)$, and the change $\phi \to \phi + \pi$ corresponds to $\boldsymbol{\xi} \to -\boldsymbol{\xi}$. Hence, the coordinates $(-p, \phi)$ and $(p, \phi + \pi)$ denote the same point in Radon space. Likewise, the

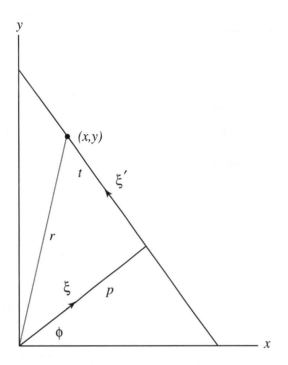

FIGURE 8.1
Coordinates in feature space used to define the Radon transform. The equation of the line is given by
$p = x \cos \phi + y \sin \phi$.

function $\check{f}$ is completely defined for $0 \le p < \infty$ and $0 \le \phi < 2\pi$. More will be said about properties of $\check{f}$ in Section 8.3.

Now, suppose we unroll the half-cylinder in Figure 8.2. The resulting surface is a plane with points represented by (p, ϕ) on a rectangular grid. It is convenient to let p vary along the vertical axis and ϕ along the horizontal axis, restricted to the range 0 to π. This construction is especially useful for illustrations because the values of $\check{f}$ can be represented as a surface in the third dimension perpendicular to this plane. Also, note that for most practical applications the object of interest in feature space does not extend to infinity. Suppose $f(\mathbf{r}) = 0$ for $|\mathbf{r}| > R$, where R is finite. It follows that $\check{f} = 0$ for $|p| > R$, and p varies on a finite interval.

To help interpret (8.2.6), let $f(x, y)$ represent the density (in 2D) for some finite mass distributed throughout the plane. (Here we are considering a special case of the more general result in nD discussed by Gel'fand, Graev, and Vilenkin [1966].) If $\mathcal{M}(p, \boldsymbol{\xi})$ denotes the total mass in the region $\boldsymbol{\xi} \cdot \mathbf{r} < p$, then

$$\mathcal{M}(p, \boldsymbol{\xi}) = \iint_{\boldsymbol{\xi} \cdot \mathbf{r} < p} f(x, y) \, dx \, dy = \iint f(x, y) \, \mathcal{U}(p - \boldsymbol{\xi} \cdot \mathbf{r}) \, dx \, dy,$$

where $\mathcal{U}(\cdot)$ denotes the unit step function. Now from the relation $\frac{\partial \mathcal{U}(p)}{\partial p} = \delta(p)$ for generalized functions, the above equation becomes

$$\frac{\partial \mathcal{M}(p, \boldsymbol{\xi})}{\partial p} = \int_{-\infty}^{\infty} \int_{-\infty}^{\infty} f(\mathbf{r}) \, \delta(p - \boldsymbol{\xi} \cdot \mathbf{r}) \, dx \, dy = \mathcal{R}\{f(x, y)\}. \tag{8.2.7}$$

This result shows that if $f(x, y)$ denotes a density with which a finite mass is distributed

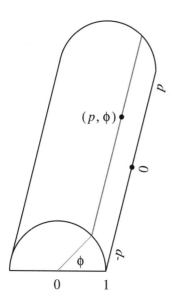

FIGURE 8.2
Coordinates in Radon space on the surface of a cylinder.

throughout space, its Radon transform is

$$\check{f}(p, \boldsymbol{\xi}) = \frac{\partial \mathcal{M}(p, \boldsymbol{\xi})}{\partial p}.$$

where $\mathcal{M}(p, \boldsymbol{\xi})$ is the mass in the half-space $\boldsymbol{\xi} \cdot \mathbf{r} < p$, and the derivative with respect to p is assumed to exist. It is important to observe that to have complete knowledge of the Radon transform one must know the mass distribution for all values of the variables p and $\boldsymbol{\xi}$. If the transform is found for only selected values of these variables, we may call the result a **sample** of the Radon transform. The next example illustrates this idea.

Example 1

Find a sample of the Radon transform for the case shown in Figure 8.3, for the case where the mass is proportional to the area. For simplicity, let the proportionality constant be unity. The equation of the line specified in the figure is $x = p$ and the angle is $\phi = 0$. The required sample is found from

$$\check{f} = \frac{\partial A}{\partial p},$$

where A is the area in the neighborhood of the line $x = p$. This example is simple enough to yield, by simple calculus for finding areas, an explicit expression for A as a function of p,

$$A(p) = 2 \int_0^p \sqrt{1 - x^2}\, dx.$$

It follows that $\check{f} = 2\sqrt{1 - p^2}.$ ∎

In this example, it is worth noting that although a sample of the Radon transform is found, the result has relevance to the entire Radon transform for circular symmetry. More will be

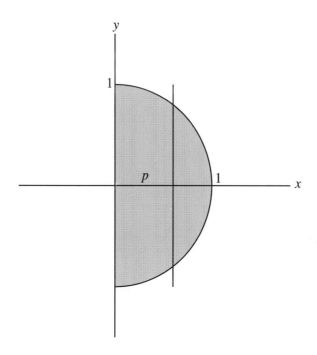

FIGURE 8.3
A semicircle of unit radius. The equation of the line is $x = p$.

said about this in several of the sections that follow. Also, observe that $\check{f}$ depends on how A changes with p where the derivative is taken, and not on how much area lies to the left or right of the line $x = p$.

From (8.2.3) the Radon transform can also be defined by

$$\check{f}(p, \phi) = \int_{\boldsymbol{\xi} \cdot \mathbf{r} = p} f(x, y) \, ds, \tag{8.2.8}$$

where the integration is taken along the line $\boldsymbol{\xi} \cdot \mathbf{r} = p$ and ds is an infinitesimal element on the line. Observe specifically that each line can be uniquely specified by the two coordinates ϕ and p.

In terms of rotated coordinates of Figure 8.4, equations (8.2.5) and (8.2.8) can be expressed in the form (with $x = p \cos\phi - t \sin\phi$, $y = p \sin\phi + t \cos\phi$)

$$\check{f}(p, \phi) = \int_{-\infty}^{\infty} f(p \cos\phi - t \sin\phi, \, p \sin\phi + t \cos\phi) \, dt. \tag{8.2.9}$$

This reflects a rotation of the coordinate axes by ϕ such that the p axis is perpendicular to the original line $\boldsymbol{\xi} \cdot \mathbf{r} = p$. The above equation can also be interpreted as follows: If $f_\phi(p, t)$ is the representation of $f(x, y)$ with respect to the rotated coordinate system, then $\check{f}_\phi(p)$ is the integral of $f_\phi(p, t)$ with respect to t for fixed ϕ. That is

$$\check{f}_\phi(p) = \int_{-\infty}^{\infty} f_\phi(p, t) \, dt, \tag{8.2.10}$$

where $f_\phi(p, t) = f(p \cos\phi - t \sin\phi, \, p \sin\phi + t \cos\phi)$. The interpretation given here covers those cases where the Radon transform is treated as a function of a single variable p with the

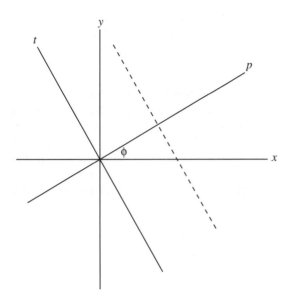

FIGURE 8.4
Rotated coordinates so the line of integration (dashed) is perpendicular to the p axis.

angle $\phi = \Phi$ viewed as a parameter. In this case the functions of p for various values of Φ are called the **projections** of $f(x, y)$ at angle Φ.

8.2.2 Three dimensions

The definition given by (8.2.6) is easy to extend to three dimensions. Let the line ℓ be replaced by a plane, and let the vector $\boldsymbol{\xi}$ be a unit vector from the origin such that the vector $p\boldsymbol{\xi}$ is perpendicular to the plane. That is, the perpendicular distance from the origin to the plane is p and the vector $\boldsymbol{\xi}$ defines the direction. Now, the equation of the plane is given by $p = \boldsymbol{\xi} \cdot \mathbf{r}$, where the position vector is extended to three dimensions, $\mathbf{r} = (x, y, z)$. The Radon transform of this function is given by

$$\check{f}(p, \boldsymbol{\xi}) = \int_{-\infty}^{\infty} \int_{-\infty}^{\infty} \int_{-\infty}^{\infty} f(\mathbf{r})\, \delta(p - \boldsymbol{\xi} \cdot \mathbf{r})\, dx\, dy\, dz. \tag{8.2.11}$$

Here, it is understood that the integral is over all planes defined by the equation $p = \boldsymbol{\xi} \cdot \mathbf{r}$.

8.2.3 Higher dimensions

The extension to higher dimensions is accomplished by defining the position vector $\mathbf{r} = (x_1, \ldots, x_n)$, extending the unit vector $\boldsymbol{\xi}$ to n dimensions, and integrating over all hyperplanes with equation given by $p = \boldsymbol{\xi} \cdot \mathbf{r}$,

$$\check{f}(p, \boldsymbol{\xi}) = \int_{-\infty}^{\infty} \cdots \int_{-\infty}^{\infty} f(\mathbf{r})\, \delta(p - \boldsymbol{\xi} \cdot \mathbf{r})\, dx_1 \ldots dx_n. \tag{8.2.12}$$

Although we do not emphasize use of the transform in higher dimensions in this discussion, it should be noted that the nD version is just a natural extension of the 3D transform. And, as might be expected, most of the major properties and theorems are just logical extensions of the corresponding results for two and three dimensions [Ludwig, 1966], [Helgason, 1980].

8.2.4 *Probes, structure, and transforms*

The Radon transform encompasses the appropriate mathematical formalism for solving a large class of practical problems related to reconstruction from projections. This is easy to see by the following considerations. Suppose there exists some physical probe that is capable of producing a projection (profile) that approximates a cumulative measurement of some property of the internal structure of an object. For a fixed angle ϕ this corresponds to knowledge of $\check{f}$ at each point along a line on the cylinder of Figure 8.2. We say that the distribution (represented by f) of some physical property of the object is measured by the probe to produce the indicated profile. The correspondence is that:

$$[\text{physical probe}] \text{ acting on (Distribution)} \rightarrow \text{Profile}$$

corresponds to

$$[\text{Radon transform}] \text{ acting on } (f) \rightarrow \check{f}_\Phi$$

for a fixed value of the angle $\phi = \Phi$. Here, the notation $\check{f}_\Phi$ is used to emphasize that a single profile serves only to determine a sample of the function $\check{f}$. A complete determination of $\check{f}$ requires the measurement of the profiles for all angles $0 \le \phi < \pi$.

In applications, typical probes include x-rays, gamma rays, visible light, microwaves, electrons, protons, heavy ions, sound waves, and magnetic resonance signals. These probes are used to obtain information about a wide variety of internal distributions: various types of attenuation coefficients, various densities, isotope distributions, index of refraction distributions, solar microwave distributions, radar brightness distributions, synthetic seismograms, and electron momentum in solids. References for applications and reviews of applications are given in Section 8.1.

8.2.5 *Transforms between spaces, central-slice theorem*

The general result is that the nD Fourier transform $\mathcal{F}_n$ of $f(\mathbf{r})$ is equivalent to the Radon transform of $f(\mathbf{r})$ followed by a 1D Fourier transform $\mathcal{F}_1$ on the variable p. This can be represented by the diagram

$$\boxed{\text{Feature space}} \xrightarrow{\mathcal{R}} \boxed{\text{Radon space}}$$
$$\mathcal{F}_n \searrow \qquad \swarrow \mathcal{F}_1$$
$$\boxed{\text{Fourier space}}$$

Or, in operator equation form

$$\mathcal{F}_1 \mathcal{R} f = \mathcal{F}_1 \check{f} = \mathcal{F}_n f = \tilde{f}. \tag{8.2.13}$$

This result is important enough to have a special name. It is known as the **central-slice theorem**, very nicely illustrated and discussed by Swindell and Barrett [1977]. This designation follows from the observation that the 1D Fourier transform of a projection of f for a fixed angle is a slice of the nD Fourier transform of f for the same fixed angle. A proof is given for $n = 2$. The extension to higher dimensions is not difficult.

Start with the 2D Fourier transform,

$$\tilde{f}(u, v) = \int_{-\infty}^{\infty} \int_{-\infty}^{\infty} f(x, y) e^{-i2\pi(ux+vy)} dx \, dy. \tag{8.2.14}$$

By using the delta function this can be rewritten as

$$\tilde{f}(u, v) = \int_{-\infty}^{\infty} dx \int_{-\infty}^{\infty} dy \int_{-\infty}^{\infty} ds \, f(x, y) e^{-i2\pi s} \delta(s - ux - vy).$$

Next, interchange the order of integration and let $s = qp$ with $q > 0$. This gives

$$\tilde{f}(u, v) = q \int_{-\infty}^{\infty} dp \int_{-\infty}^{\infty} dx \int_{-\infty}^{\infty} dy \, f(x, y) e^{-i2\pi qp} \delta(qp - ux - vy).$$

In Fourier space, let $u = q \cos\phi$ and $v = q \sin\phi$. Then the variable q can be factored from the delta function by use of the general property (see Chapter 1, Section 2) $\delta(ax) = \delta(x)/|a|$, and

$$\tilde{f}(u, v) = \int_{-\infty}^{\infty} dp \, e^{-i2\pi qp} \int_{-\infty}^{\infty} dx \int_{-\infty}^{\infty} dy \, f(x, y) \delta(p - x\cos\phi - y\sin\phi).$$

The integral over the (x, y) plane is just the Radon transform of f from (8.2.5), and the desired result follows easily

$$\tilde{f}(q\cos\phi, q\sin\phi) = \int_{-\infty}^{\infty} \check{f}(p, \phi) e^{-i2\pi qp} dp. \tag{8.2.15}$$

It is interesting to observe the simple result obtained if the coordinates are selected such that the angle ϕ is fixed and equal to zero, $\Phi = 0$, then

$$\tilde{f}(q, 0) = \int_{-\infty}^{\infty} \check{f}(p, 0) e^{-i2\pi qp} dp. \tag{8.2.16}$$

By thinking about what the last two equations mean it should be clear that the 1D Fourier transform of a projection of f for fixed angle $\phi = \Phi$ is a slice of the 2D Fourier transform of f, and this slice in Fourier space is defined by the angle Φ. One further remark is in order here. This result is sometimes referred to as the **projection-slice theorem**; however, for higher dimensions this designation may have a slightly different meaning as used by Mersereau and Oppenheim [1974]. To avoid confusion, in the current presentation (8.2.13) is called the nD form of the central-slice theorem. For historical purposes, it is to be noted that Bracewell [1956] derived and used this theorem without prior knowledge of the theory of the Radon transform.

Example 2

A simple example is useful to illustrate the use of transforms between spaces. Suppose the feature space function is the two-dimensional Gaussian

$$f(x, y) = e^{-x^2-y^2}.$$

First, compute the Fourier transform. Let $x = r\cos\theta$ and $y = r\sin\theta$, then the polar form of (8.2.14) is given by

$$\tilde{f}(u, v) = \int_0^{\infty} dr \, r \, e^{-r^2} \int_0^{2\pi} d\theta \, e^{[-i2\pi qr\cos(\theta-\phi)]},$$

with $u = q\cos\phi$ and $v = q\sin\phi$. The integral over θ is given by $2\pi J_0(2\pi qr)$, where J_0 is a Bessel function of order zero (see Chapter 1, Section 5.6). The remaining integral is a Hankel transform of order zero. It follows that

$$\tilde{f} = 2\pi \int_0^\infty r\, e^{-r^2}\, J_0(2\pi qr)\, dr = \pi e^{-\pi^2 q^2}.$$

Now use (8.2.12) in the form

$$\check{f} = \mathcal{F}^{-1}\tilde{f},$$

to obtain (see Appendix A)

$$\check{f}(p,\xi) = \pi \int_{-\infty}^\infty e^{-\pi^2 q^2}\, e^{i2\pi qp}\, dq = \sqrt{\pi}\, e^{-p^2}. \tag{8.2.17}$$

∎

In this example the path from feature space to Radon space was taken through Fourier space for purposes of illustration. Actually, in this case, it is easier to compute the Radon transform directly; see Example 1 in Section 8.5.

8.3 Basic properties

Important properties of the Radon transform follow directly from the definition. These properties can be compared with the corresponding properties of the Fourier transform discussed in detail by Bracewell [1986]. In this section these basic properties (theorems) are given for the 2D case, along with the corresponding results for the Fourier transform. The slight loss in generality suffered by using 2D illustrations is compensated for by being able to show details that are familiar from a knowledge of elementary calculus. It proves useful to keep the notation for the components of the unit vector ξ as simple as possible. Rather than always using $(\cos\phi, \sin\phi)$ for these components, the notation (ξ_1, ξ_2) is often convenient, where it is understood that

$$\xi_1 = \cos\phi, \qquad \xi_2 = \sin\phi, \qquad \text{and} \qquad \xi_1^2 + \xi_2^2 = 1. \tag{8.3.1}$$

This means that for the discussion in this section (8.2.5) may be modified to read

$$\check{f}(p, \xi_1, \xi_2) = \int_{-\infty}^\infty \int_{-\infty}^\infty f(x, y)\, \delta(p - x\xi_1 - y\xi_2)\, dx\, dy. \tag{8.3.2}$$

The 2D Fourier transform is still given by (8.2.14). Also, in the following discussion it is always assumed that the transforms actually exist. The reader interested in examples can look ahead to Section 8.5 where several of these basic properties are used to illustrate ways to find transforms. The reader should also consult Chapter 2 for detail exposition of the Fourier transform properties.

8.3.1 *Linearity*

The Radon and Fourier transforms are both linear. If $f(x, y)$ and $g(x, y)$ are functions in feature space, then for any constants a and b,

$$\mathcal{R}[af + bg] = a\check{f} + b\check{g} \tag{8.3.3}$$

and

$$\mathcal{F}[af + bg] = a\tilde{f} + b\tilde{g}. \tag{8.3.4}$$

8.3.2 *Similarity*

If $\mathcal{R}f(x, y) = \check{f}(p, \xi_1, \xi_2)$, then for arbitrary constants a and b the Radon transform of $f(ax, by)$ is given by

$$\mathcal{R}f(ax, by) = \frac{1}{|ab|} \check{f}\left(p, \frac{\xi_1}{a}, \frac{\xi_2}{b}\right). \tag{8.3.5}$$

This follows immediately by making the change of variables $x' = ax$ and $y' = by$ in the expression

$$\int_{-\infty}^{\infty} \int_{-\infty}^{\infty} f(ax, by)\, \delta(p - x\xi_1 - y\xi_2)\, dx\, dy.$$

The corresponding scaling equation for Fourier transforms: If $\mathcal{F}f(x, y) = \tilde{f}(u, v)$ then

$$\mathcal{F}f(ax, by) = \frac{1}{|ab|} \tilde{f}\left(\frac{u}{a}, \frac{v}{b}\right). \tag{8.3.6}$$

8.3.3 *Symmetry*

A similar technique can be applied to give an important **symmetry** property. Examine the expression

$$\check{f}(ap, a\boldsymbol{\xi}) = \int_{-\infty}^{\infty} \int_{-\infty}^{\infty} f(x, y)\, \delta(ap - ax\xi_1 - ay\xi_2)\, dx\, dy.$$

The constant a can be factored from the delta function to yield

$$\check{f}(ap, a\boldsymbol{\xi}) = |a|^{-1}\, \check{f}(p, \boldsymbol{\xi}). \tag{8.3.7}$$

If $a = -1$ this demonstrates that the Radon transform is an even homogeneous function of degree -1,

$$\check{f}(-p, -\boldsymbol{\xi}) = \check{f}(p, \boldsymbol{\xi}). \tag{8.3.8}$$

Another useful form for the symmetry property is

$$\check{f}(p, s\boldsymbol{\xi}) = |s|^{-1} \check{f}\left(\frac{p}{s}, \boldsymbol{\xi}\right). \tag{8.3.9}$$

8.3.4 *Shifting*

Given that $\mathcal{R}f(x, y) = \check{f}(p, \boldsymbol{\xi})$, then for arbitrary constants a and b the Radon transform of $f(x - a, y - b)$ is found by

$$\mathcal{R}f(x - a, y - b) = \check{f}(p - a\xi_1 - b\xi_2,\, \boldsymbol{\xi}). \tag{8.3.10}$$

As in the previous case the proof follows immediately by introducing a change of variables. Let $x' = x - a$ and $y' = y - b$ in the expression

$$\int_{-\infty}^{\infty} \int_{-\infty}^{\infty} f(x - a, y - b)\, \delta(p - x\xi_1 - y\xi_2)\, dx\, dy.$$

The corresponding theorem for the Fourier transform is a little different, involving a phase change

$$\mathcal{F} f(x - a, y - b) = e^{-i2\pi(au+bv)} \, \tilde{f}(u, v). \tag{8.3.11}$$

8.3.5 Differentiation

Details of the derivation are given for the Radon transform of $\partial f / \partial x$. Other results follow directly by using the same method. First note that

$$\frac{\partial f}{\partial x} = \lim_{\epsilon \to 0} \frac{f[x + (\epsilon/\xi_1), y] - f(x, y)}{\epsilon/\xi_1}.$$

Now take the Radon transform of both sides and apply (8.3.10) with $a = -\epsilon/\xi_1$ and $b = 0$ to get

$$\mathcal{R} \frac{\partial f}{\partial x} = \xi_1 \lim_{\epsilon \to 0} \frac{\check{f}(p + \epsilon, \boldsymbol{\xi}) - \check{f}(p, \boldsymbol{\xi})}{\epsilon}.$$

By definition of a partial derivative it follows that

$$\mathcal{R} \frac{\partial f}{\partial x} = \xi_1 \frac{\partial \check{f}(p, \boldsymbol{\xi})}{\partial p}. \tag{8.3.12a}$$

Likewise, differentiation with respect to y yields

$$\mathcal{R} \frac{\partial f}{\partial y} = \xi_2 \frac{\partial \check{f}(p, \boldsymbol{\xi})}{\partial p}. \tag{8.3.12b}$$

Using the same approach, the second derivatives are given by

$$\mathcal{R} \frac{\partial^2 f}{\partial x^2} = \xi_1^2 \frac{\partial^2 \check{f}(p, \boldsymbol{\xi})}{\partial p^2}$$

$$\mathcal{R} \frac{\partial^2 f}{\partial x \partial y} = \xi_1 \xi_2 \frac{\partial^2 \check{f}(p, \boldsymbol{\xi})}{\partial p^2} \tag{8.3.13}$$

$$\mathcal{R} \frac{\partial^2 f}{\partial y^2} = \xi_2^2 \frac{\partial^2 \check{f}(p, \boldsymbol{\xi})}{\partial p^2}.$$

The derivative theorems for the 2D Fourier transform are

$$\mathcal{F} \frac{\partial f}{\partial x} = 2\pi i u \, \tilde{f}(u, v), \qquad \mathcal{F} \frac{\partial f}{\partial y} = 2\pi i v \, \tilde{f}(u, v), \tag{8.3.14}$$

and

$$\mathcal{F} \frac{\partial^2 f}{\partial x^2} = -4\pi^2 u^2 \, \tilde{f}(u, v)$$

$$\mathcal{F} \frac{\partial^2 f}{\partial x \partial y} = -4\pi^2 uv \, \tilde{f}(u, v) \tag{8.3.15}$$

$$\mathcal{F} \frac{\partial^2 f}{\partial y^2} = -4\pi^2 v^2 \, \tilde{f}(u, v).$$

8.3.6   *Convolution*

The convolution of two functions f and g is commonly designated by $f * g$, regardless of the dimension. Here, this convention is modified slightly to emphasize the distinction between convolution in one and two dimensions. We write two-dimensional convolution as

$$f ** g = \int_{-\infty}^{\infty} \int_{-\infty}^{\infty} f(x', y') \, g(x - x', y - y') \, dx' \, dy'. \tag{8.3.16}$$

The Fourier convolution theorem is very simple, yielding a simple product in Fourier space,

$$\mathcal{F}(f ** g) = \tilde{f}(u, v) \, \tilde{g}(u, v). \tag{8.3.17}$$

The corresponding theorem for the Radon transform is considerably more complicated. If $f = g ** h$, then the Radon transform of f is given by a one-dimensional convolution in Radon space, rather than a simple product as in the Fourier case,

$$\check{f}(p, \boldsymbol{\xi}) = \mathcal{R}(g ** h) = \check{g} * \check{h} = \int_{-\infty}^{\infty} \check{g}(\tau, \boldsymbol{\xi}) \, \check{h}(p - \tau, \boldsymbol{\xi}) \, d\tau. \tag{8.3.18}$$

The proof follows by applying the definition followed by some tricky manipulations with double integrals and delta functions. The details are given by Deans [1983, 1993].

8.4   Linear transformations

A practical method for finding Radon transforms involves making a change of variables. This approach can be related to the Radon transform of a function of a linear transformation of coordinates. Here, inner products are designated by

$$\boldsymbol{\xi} \cdot \mathbf{x} = \xi_1 x_1 + \xi_2 x_2 + \cdots + \xi_n x_n. \tag{8.4.1}$$

Or, in matrix notation

$$\boldsymbol{\xi}^{\top} \mathbf{x} = \begin{pmatrix} \xi_1 & \xi_2 & \cdots & \xi_n \end{pmatrix} \begin{pmatrix} x_1 \\ x_2 \\ \vdots \\ x_n \end{pmatrix} = \xi_1 x_1 + \xi_2 x_2 + \cdots + \xi_n x_n, \tag{8.4.2}$$

where $\top$ means transpose.

Let A be a nonsingular $n \times n$ matrix with real elements, then a change of coordinates follows by matrix multiplication

$$\mathbf{y} = \mathrm{A}\mathbf{x}. \tag{8.4.3}$$

An important identity, in matrix notation, is

$$\boldsymbol{\xi}^{\top} \mathbf{y} = \boldsymbol{\xi}^{\top} \mathrm{A} \mathbf{x} = (\mathrm{A}^{\top} \boldsymbol{\xi})^{\top} \mathbf{x}, \tag{8.4.4}$$

and the same identity in the "dot" notation is

$$\boldsymbol{\xi} \cdot \mathbf{y} = \boldsymbol{\xi} \cdot \mathrm{A}\mathbf{x} = \mathrm{A}^{\top} \boldsymbol{\xi} \cdot \mathbf{x}. \tag{8.4.5}$$

Because A is nonsingular, the inverse exists. For convenience, let $\mathrm{B} = \mathrm{A}^{-1}$, then $\mathbf{x} = \mathrm{B}\mathbf{y}$.

The Radon transform of $f(\text{A}\mathbf{x})$ follows:

$$\mathcal{R}\, f(\text{A}\mathbf{x}) = \int f(\text{A}\mathbf{x})\, \delta(p - \boldsymbol{\xi} \cdot \mathbf{x})\, d\mathbf{x}$$

$$= |\det \text{B}| \int f(\mathbf{y})\, \delta(p - \boldsymbol{\xi} \cdot \text{B}\mathbf{y})\, d\mathbf{y}$$

$$= |\det \text{B}| \int f(\mathbf{y})\, \delta(p - \text{B}^{\top}\boldsymbol{\xi} \cdot \mathbf{y})\, d\mathbf{y}$$

$$= |\det \text{B}|\ \check{f}(p, \text{B}^{\top}\boldsymbol{\xi}). \tag{8.4.6}$$

The term $|\det \text{B}|$ appears because the Jacobian of the transformation is just the magnitude of the determinant of the matrix B. Because $\text{A} = \text{B}^{-1}$, an equivalent result is

$$\mathcal{R}\, f(\text{B}^{-1}\mathbf{x}) = |\det \text{B}|\ \check{f}(p, \text{B}^{\top}\boldsymbol{\xi}). \tag{8.4.7}$$

A word of caution is in order here. It may be that $\text{B}^{\top}$ is not a unit vector. In such case, it is a good idea to define s equal to the magnitude of the vector $\text{B}^{\top}\boldsymbol{\xi}$ and observe that

$$\boldsymbol{\mu} = \frac{\text{B}^{\top}\boldsymbol{\xi}}{s} \tag{8.4.8}$$

is a unit vector. Now from the results of Section 8.3.3 the right side of (8.4.7) becomes

$$|\det \text{B}|\ \check{f}(p, \text{B}^{\top}\boldsymbol{\xi}) = |\det \text{B}|\ \check{f}(p, s\boldsymbol{\mu}) = \frac{|\det \text{B}|}{s}\ \check{f}\left(\frac{p}{s}, \boldsymbol{\mu}\right). \tag{8.4.9}$$

Finally, we have the useful result that

$$\mathcal{R}\, f(\text{B}^{-1}\mathbf{x}) = \frac{|\det \text{B}|}{s}\ \check{f}\left(\frac{p}{s}, \boldsymbol{\mu}\right), \qquad \text{with } s = |\text{B}^{\top}\boldsymbol{\xi}|. \tag{8.4.10}$$

There are two important special cases that deserve attention. First, suppose B is orthogonal. Then $\text{B}^{-1} = \text{B}^{\top} = \text{A}$, with $|\det \text{B}| = 1$, and

$$\mathcal{R}\, f(\text{A}\mathbf{x}) = \check{f}(p, \text{A}\boldsymbol{\xi}) \tag{8.4.11}$$

where $\text{A}\boldsymbol{\xi}$ is a unit vector. The other special case is for A equal to a multiple of the identity. If $\text{A} = c\text{I}$ with c real, then $\text{B} = \text{A}^{-1} = c^{-1}\text{I}$, and

$$\mathcal{R}\, f(c\mathbf{x}) = \frac{1}{|c|^{n}}\ \check{f}\left(p, \frac{\boldsymbol{\xi}}{c}\right) = \frac{1}{|c|^{n-1}}\ \check{f}(cp, \boldsymbol{\xi}). \tag{8.4.12}$$

8.5 Finding transforms

In this section some simple examples are worked out in detail to illustrate the use of the various formulas developed in the previous sections. These examples demonstrate how to find transforms and point out pitfalls that sometimes occur during a calculation. The definite integrals that occur in the calculations are tabulated in Appendix A.

Example 1
Recall from Example 2 in Section 8.2 that the Radon transform of

$$f(x, y) = e^{-x^2 - y^2}$$

was found by going through Fourier space to yield

$$\check{f}(p, \boldsymbol{\xi}) = \sqrt{\pi}\, e^{-p^2}.$$

In this example the Radon transform is calculated directly. Suppose the matrix A from Section 8.4 is given in terms of the components of the unit vector $\boldsymbol{\xi} = (\cos\phi, \sin\phi)$,

$$A = \begin{pmatrix} \xi_1 & \xi_2 \\ -\xi_2 & \xi_1 \end{pmatrix}.$$

Now define the components of the transformed vector by

$$\begin{pmatrix} u \\ v \end{pmatrix} = A \begin{pmatrix} x \\ y \end{pmatrix} = \begin{pmatrix} \xi_1 x + \xi_2 y \\ -\xi_2 x + \xi_1 y \end{pmatrix}.$$

Observe that A is orthogonal and (8.4.11) applies. Also, note that $u^2 + v^2 = x^2 + y^2$ and $u = \xi_1 x + \xi_2 y$. It follows that

$$\mathcal{R}f(A\mathbf{x}) = \mathcal{R}f(u, v) = \int_{-\infty}^{\infty} \int_{-\infty}^{\infty} e^{-u^2 - v^2} \delta(p - u)\, du\, dv = e^{-p^2} \int_{-\infty}^{\infty} e^{-v^2}\, dv = \sqrt{\pi}\, e^{-p^2}.$$

Because this result is not dependent on $\boldsymbol{\xi}$, or equivalently ϕ, it follows that

$$\mathcal{R}\{e^{-x^2 - y^2}\} = \sqrt{\pi}\, e^{-p^2}. \tag{8.5.1}$$

The lack of dependence on ϕ is certainly expected because the Gaussian is symmetric and centered at the origin.   ∎

Example 2
Extend the result in the previous example to three dimensions. Let the orthogonal transformation matrix be selected as

$$A = \begin{pmatrix} \xi_1 & \xi_2 & \xi_3 \\ \dfrac{-\xi_1 \xi_2}{s} & s & \dfrac{-\xi_2 \xi_3}{s} \\ \dfrac{-\xi_3}{s} & 0 & \dfrac{\xi_1}{s} \end{pmatrix}$$

where $s = (\xi_1^2 + \xi_3^2)^{1/2}$ and $|\boldsymbol{\xi}| = 1$. If the components of the transformed vector are given by (u, v, w), then after the substitutions are made in (8.4.11) the transform is given by the integral

$$\int_{-\infty}^{\infty} \int_{-\infty}^{\infty} \int_{-\infty}^{\infty} e^{-u^2 - v^2 - w^2} \delta(p - u)\, du\, dv\, dw = \pi\, e^{-p^2}.$$

The final result above is obtained by use of the delta function and the evaluation of the two remaining Gaussian integrals over v and w. Once again by the invariance argument it follows that

$$\mathcal{R}\{e^{-x^2 - y^2 - z^2}\} = \pi\, e^{-p^2}. \tag{8.5.2}$$

∎

Example 3
If the results of the previous example are extended to n dimensions, then

$$\mathcal{R}\{\exp(-x_1^2 - \cdots - x_n^2)\} = (\sqrt{\pi})^{n-1} e^{-p^2}. \tag{8.5.3}$$

∎

Example 4
Start with $f(x, y) = \exp(-x^2 - y^2)$ and apply (8.4.12) with $n = 2$ and $c = 1/\sigma\sqrt{2}$. This yields the Radon transform of the symmetric Gaussian probability density function. Note that

$$f(A\mathbf{x}) = \exp\left(-\frac{x^2}{2\sigma^2} - \frac{y^2}{2\sigma^2}\right)$$

and

$$\frac{1}{c}\check{f}(cp, \boldsymbol{\xi}) = \sigma\sqrt{2\pi}\, e^{-p^2/2\sigma^2}.$$

An overall division by $2\pi\sigma^2$ yields the standard form,

$$\mathcal{R}\left\{\frac{1}{2\pi\sigma^2}\exp\left(-\frac{x^2}{2\sigma^2} - \frac{y^2}{2\sigma^2}\right)\right\} = \frac{1}{\sigma\sqrt{2\pi}}\exp\left(-\frac{p^2}{2\sigma^2}\right). \tag{8.5.4}$$

∎

Example 5
The problem here is to find the Radon transform of

$$\exp\left[-\left(\frac{x}{a}\right)^2 - \left(\frac{y}{b}\right)^2\right]$$

with both a and b real. Again, the starting function is selected to be

$$f(x, y) = e^{-x^2 - y^2}.$$

Now we use (8.4.10) with

$$B = \begin{pmatrix} a & 0 \\ 0 & b \end{pmatrix}, \qquad B^{-1} = \begin{pmatrix} \frac{1}{a} & 0 \\ 0 & \frac{1}{b} \end{pmatrix}, \qquad |\det B| = |ab|.$$

In this example

$$B^{\mathsf{T}}\boldsymbol{\xi} = \begin{pmatrix} a\cos\phi \\ b\sin\phi \end{pmatrix}$$

is not a unit vector, having magnitude

$$s = (a^2\cos^2\phi + b^2\sin^2\phi)^{1/2}.$$

With these observations, (8.4.10) yields

$$\mathcal{R}\left\{\exp\left[-\left(\frac{x}{a}\right)^2 - \left(\frac{y}{b}\right)^2\right]\right\} = \frac{|ab|\sqrt{\pi}}{s}\exp\left(-\frac{p^2}{s^2}\right). \tag{8.5.5}$$

Note that once the symmetry is lost in feature space the angle ϕ appears in the transform. ∎

Example 6

Use the similarity theorem to obtain (8.5.5). Application of (8.3.5) with

$$f(x, y) = e^{-x^2-y^2} \quad \text{and} \quad \check{f}(p, \boldsymbol{\xi}) = \sqrt{\pi}\, e^{-p^2}$$

yields

$$\mathcal{R} f\left(\frac{x}{a}, \frac{y}{b}\right) = |ab| \check{f}\left(p, \frac{\xi_1}{a}, \frac{\xi_2}{b}\right).$$

This is not in the desired form, so we let $\boldsymbol{\mu} = (a\xi_1/s, b\xi_2/s)$ with s defined as in the previous example so $\boldsymbol{\mu}$ is a unit vector. Now the right side of the above equation becomes

$$|ab| \check{f}(p, s\boldsymbol{\mu}) = \frac{|ab|}{s} \check{f}\left(\frac{p}{s}, \boldsymbol{\mu}\right) = \frac{|ab|\sqrt{\pi}}{s} \exp\left(-\frac{p^2}{s^2}\right), \tag{8.5.6}$$

as in the previous example.   ∎

Example 7

Find the Radon transform of the characteristic function of a unit disk, sometimes called the cylinder function, cyl(r). This function is given by

$$f(x, y) = \begin{cases} 1, & \text{for } x^2 + y^2 \le 1 \\ 0, & \text{for } x^2 + y^2 > 1. \end{cases} \tag{8.5.7a}$$

By inspection, the transform is given by the length of a chord at a distance p from the center and is independent of the angle ϕ,

$$\check{f}(p, \phi) = \begin{cases} 2(1 - p^2)^{1/2}, & \text{for } p \le 1 \\ 0, & \text{for } p > 1. \end{cases} \tag{8.5.7b}$$

∎

Example 8

Find the Radon transform of the characteristic function of an ellipse where f is given by

$$f(x, y) = \begin{cases} 1, & \text{for } (x/a)^2 + (y/b)^2 \le 1 \\ 0, & \text{for } (x/a)^2 + (y/b)^2 > 1. \end{cases} \tag{8.5.8a}$$

If the matrix B is selected as in Example 5 above, then from the result in Example 7 it follows immediately that

$$\check{f}(p, \phi) = \begin{cases} \frac{2|ab|}{s}\left[1 - \left(\frac{p}{s}\right)^2\right]^{1/2}, & \text{for } \frac{|p|}{s} \le 1 \\ \\ 0, & \text{for } \frac{|p|}{s} > 1. \end{cases} \tag{8.5.8b}$$

where $s = (a^2 \cos^2 \phi + b^2 \sin \phi)^{1/2}$.   ∎

Example 9

Use the method of Example 1 to find a general expression for the Radon transform of a function defined on the unit disk, and zero outside the unit disk. From the matrix A in Example 1, it follows that (see Figure 8.4 with $(p, t) \rightarrow (u, v)$)

$$x = u \cos \phi - v \sin \phi \qquad \text{and} \qquad y = u \sin \phi + v \cos \phi.$$

Therefore

$$\check{f}(p, \phi) = \int_{\text{disk}} f(x, y)\, \delta(p - x \cos \phi - y \sin \phi)\, dx\, dy$$

$$= \int_{\text{disk}} f(u \cos \phi - v \sin \phi, u \sin \phi + v \cos \phi)\, \delta(p - u)\, du\, dv.$$

After the integration over u,

$$\check{f}(p, \phi) = \int_{-\sqrt{1-p^2}}^{\sqrt{1-p^2}} f(p \cos \phi - v \sin \phi, p \sin \phi + v \cos \phi)\, dv. \qquad (8.5.9)$$

∎

Example 10

Find the Radon transform over the unit square, situated as indicated in Figure 8.5. It is adequate to consider the transform for $0 < \phi \le \pi/4$.

$$\check{f}(p, \phi) = \begin{cases} \dfrac{p}{\sin \phi \cos \phi} & \text{for region 1, } 0 < p < \sin \phi \\[2mm] \sec \phi & \text{for region 2, } \sin \phi < p < \cos \phi \\[2mm] \dfrac{\sin \phi + \cos \phi - p}{\sin \phi \cos \phi} & \text{for region 3, } \cos \phi < p < \sin \phi + \cos \phi. \end{cases} \qquad (8.5.10a)$$

By symmetry, for $\pi/4 \le \phi < \pi/2$,

$$\check{f}(p, \phi) = \begin{cases} \dfrac{p}{\sin \phi \cos \phi} & \text{for region 1, } 0 < p < \cos \phi \\[2mm] \csc \phi & \text{for region 2, } \cos \phi < p < \sin \phi \\[2mm] \dfrac{\sin \phi + \cos \phi - p}{\sin \phi \cos \phi} & \text{for region 3, } \sin \phi < p < \sin \phi + \cos \phi. \end{cases} \qquad (8.5.10b)$$

∎

Example 11

The shift theorem from Section 8.3.4 can be written as

$$\mathcal{R} f(\mathbf{x} - \mathbf{a}) = \check{f}(p - p_0, \boldsymbol{\xi}), \qquad \text{with } p_0 = \boldsymbol{\xi} \cdot \mathbf{a}.$$

Apply this equation with

$$\mathbf{a} = (a, b) \qquad \text{and} \qquad p_0 = a \cos \phi + b \sin \phi$$

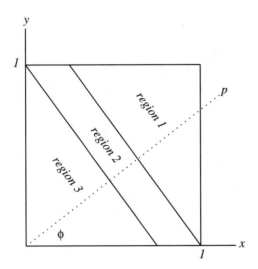

FIGURE 8.5
Coordinates for unit square with regions defined as p varies along dotted line.

to the result of Example 4 above. This gives the transform of a 2D Gaussian density function

$$\mathcal{R}\left\{\frac{1}{2\pi\sigma^2}\exp\left(-\frac{(x-a)^2}{2\sigma^2}-\frac{(y-b)^2}{2\sigma^2}\right)\right\} = \frac{1}{\sigma\sqrt{2\pi}}\exp\left(-\frac{(p-p_0)^2}{2\sigma^2}\right). \qquad (8.5.11)$$

Again, note that the loss of rotational symmetry about the origin in feature space causes the function in Radon space to have explicit dependence on the angle ϕ.   ∎

Example 12
In the previous example, if the limit $\sigma \to +0$ is taken, both sides are convergent δ sequences. This observation yields

$$\mathcal{R}\left\{\delta(x-a)\,\delta(y-b)\right\} = \delta(p-p_0), \qquad (8.5.12)$$

with $p_0 = a\cos\phi + b\sin\phi$. This result also follows easily by substitution of

$$f(x,y) = \delta(x-a)\,\delta(y-b) \equiv \delta(x-a,\,y-b)$$

into the definition of the Radon transform, Section (8.2.5). This example has some special significance, because it demonstrates how an impulse function centered at (a, b) in feature space transforms to Radon space. In Radon space (p, ϕ) there is an impulse function everywhere along a sinusoidal curve with the equation of the curve given by

$$p = a\cos\phi + b\sin\phi. \qquad (8.5.13)$$

An illustration is given in Figure 8.6 for $p = 2\cos\phi + \sin\phi$.   ∎

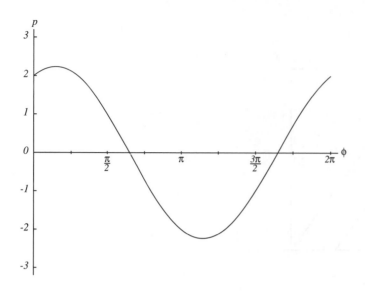

FIGURE 8.6
The impulse function maps to a sinusoidal curve.

Example 13

Another way to approach the transform of the delta function is to observe that for the delta function centered at the origin $\delta(x, y) = \delta(p, t)$. Then, in the rotated system (see Figure 8.4) it follows that

$$\int_{-\infty}^{\infty} \delta(p, t)\, dt = \delta(p).$$

This result can be used to obtain the transform of the shifted delta function. By use of Section 8.3.4 it follows that (8.5.12) holds. If $\phi_0 = \tan^{-1}\frac{b}{a}$ and $r_0 = \sqrt{a^2 + b^2}$, then $p_0 = r_0 \cos(\phi_0 - \phi)$ and

$$\mathcal{R}\{\delta(x - a, y - b)\} = \delta\left[p - r_0 \cos(\phi_0 - \phi)\right].$$

As with the example in Figure 8.6, the region of support for the delta function is a sinusoidal curve in Radon space. ∎

Example 14

Find the Radon transform of a finite-extended delta function (see Figure 8.7a),

$$f(x, y) = \begin{cases} \delta(x - p_0), & \text{for } |y| < L/2 \\ 0, & \text{for } |y| \geq L/2. \end{cases}$$

We write

$$\check{f}(p, \phi) = \int_{-\frac{L}{2}}^{\frac{L}{2}} \delta(p \cos \phi + t \sin \phi - p_0)\, dt = \begin{cases} L\,\delta(p - p_0), & \text{for } \phi = 2n\pi \\ L\,\delta(p + p_0), & \text{for } \phi = (2n + 1)\pi. \end{cases}$$

However, if the angle ϕ is different from a multiple of π, then we obtain (see Chapter 1, Section 2)

$$\check{f}(p, \phi) = \int \delta(p \cos \phi + t \sin \phi - p_0)\, dt = \begin{cases} |\sin \phi|^{-1}, & \text{for } |p - p_0 \cos \phi| \le |\frac{L}{2} \sin \phi| \\ 0, & \text{otherwise.} \end{cases}$$

The inequality can be deduced from the geometry of Figure 8.7b. The region of support is shown in Figure 8.7c and the transform is illustrated in Figure 8.7d. ∎

This example illustrates a useful property of the Radon transform, namely, its ability to serve as an instrument for the detection of line segments in images. A slightly more general version of this example is given by Deans [1985].

Example 15

Find the Radon transform of the cylinder function defined in Example 7 displaced at the point (x_0, y_0) as shown in Figure 8.8a. The solution follows immediately from the solution of Example 7 combined with the shifting property in Section 8.3.4. Also, the solution can be deduced from the geometry in Figure 8.8a. When $d = 1$, the length $t = 0$; also, for p such that the line of integration passes through the cylinder,

$$t = 2\sqrt{1 - [p - r_0 \cos(\phi_0 - \phi)]^2}.$$

Further, when ϕ varies, the values p can assume follow from the geometry. The transform is

$$\check{f}(p, \phi) = \begin{cases} 2\sqrt{1 - [p - r_0 \cos(\phi_0 - \phi)]^2}, & -1 + r_0 \cos(\phi_0 - \phi) \le p \\ & \le 1 + r_0 \cos(\phi_0 - \phi) \\ 0, & \text{otherwise.} \end{cases}$$

Figure 8.8b shows the (sinusoidal) region of support of the transform. ∎

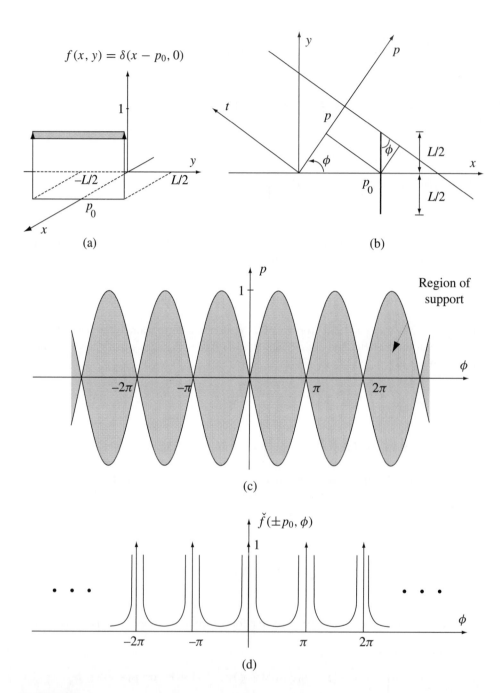

FIGURE 8.7
Radon transform of finite-extended delta function.

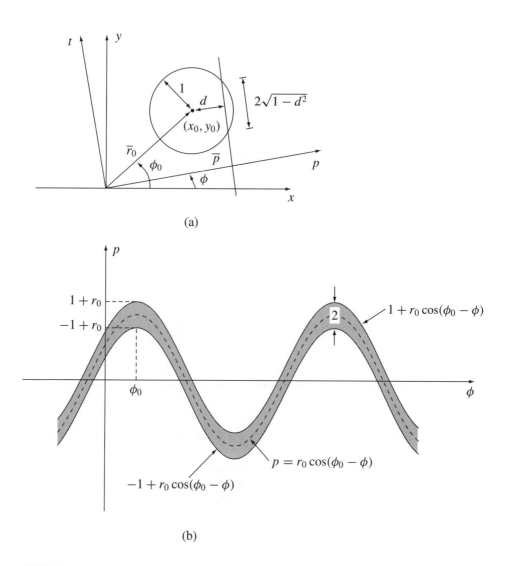

FIGURE 8.8
Displaced cylinder function and region of support of the transform.

Example 16

Suppose the points in feature space lie along a line defined by parameters p_0 and ϕ_0 as indicated in Figure 8.9. All of these collinear points map to sinusoidal curves in Radon space; moreover, these curves all intersect at the same point (p_0, ϕ_0) in Radon space. By selecting an appropriate threshold and only plotting values of $\check{f}$ above the threshold it follows that a single point in Radon space serves to identify a line of collinear points in feature space. It is in this sense that the Radon transform is sometimes regarded as a line-to-point transformation. This idea has been used by various authors interested in detecting lines in digital images: [Duda and Hart, 1972], [Shapiro and Iannino, 1979]. When the Radon transform is used in this fashion it is often referred to as the Hough transform after the work of Hough [1962].   ∎

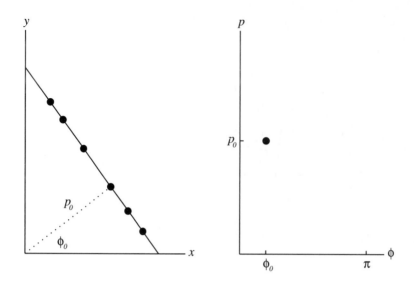

FIGURE 8.9
After thresholding, a single point in Radon space corresponds to a line in feature space.

8.6 More on derivatives

In Section 8.3.5 basic equations were given for the Radon transform of derivatives in two dimensions. Clearly, these results can be generalized and it is useful to do that, especially in connection with using the Radon transform in connection with partial differential equations and series expansions. Another use of derivatives is related to the derivatives of the Radon transform. Both of these cases are covered in this section.

8.6.1 *Transform of derivatives*

Let $f(\mathbf{x}) = f(x_1, \ldots, x_n)$. The generalization of (8.3.12) is

$$\mathcal{R}\left\{\frac{\partial f}{\partial x_k}\right\} = \xi_k \frac{\partial \check{f}(p, \boldsymbol{\xi})}{\partial p}, \tag{8.6.1}$$

where ξ_k is the kth component of the unit vector $\boldsymbol{\xi}$. The linearity property (8.3.3) can be used to find the transform of the sum

$$\sum_{k=1}^{n} a_k \frac{\partial f}{\partial x_k}$$

for arbitrary constants a_k. If the constants are components of the vector $\mathbf{a}$, then

$$\mathcal{R}\left\{\sum_{k=1}^{n} a_k \frac{\partial f}{\partial x_k}\right\} = (\mathbf{a} \cdot \boldsymbol{\xi}) \frac{\partial \check{f}(p, \boldsymbol{\xi})}{\partial p}. \tag{8.6.2}$$

Example 1
Let $n = 3$, and let ∇ be the gradient operator $(\partial/\partial x_1, \partial/\partial x_2, \partial/\partial x_3)$. Now (8.6.2) is interpreted as the Radon transform of a directional derivative.

$$\mathcal{R}\{\mathbf{a} \cdot \nabla f\} = (\mathbf{a} \cdot \boldsymbol{\xi}) \frac{\partial \check{f}(p, \boldsymbol{\xi})}{\partial p}. \tag{8.6.3}$$

∎

Another obvious generalization from Section 8.3.5 is

$$\mathcal{R}\left\{\frac{\partial^2 f}{\partial x_l \, \partial x_k}\right\} = \xi_l \, \xi_k \frac{\partial^2 \check{f}(p, \boldsymbol{\xi})}{\partial p^2}. \tag{8.6.4}$$

Consequently, for arbitrary constant vectors $\mathbf{a}$ and $\mathbf{b}$,

$$\mathcal{R}\left\{\sum_{l=1}^{n} \sum_{k=1}^{n} a_l b_k \frac{\partial^2 f}{\partial x_l \partial x_k}\right\} = (\mathbf{a} \cdot \boldsymbol{\xi})(\mathbf{b} \cdot \boldsymbol{\xi}) \frac{\partial^2 \check{f}(p, \boldsymbol{\xi})}{\partial p^2}. \tag{8.6.5}$$

Example 2
There is a very important special case of the last equation. Suppose the product $a_l b_k$ reduces to the Kronecker delta,

$$a_l b_k = \delta_{lk} = \begin{cases} 1, & \text{for } l = k \\ 0, & \text{for } l \neq k. \end{cases}$$

Now the operator is just the Laplacian operator

$$\nabla^2 = \frac{\partial^2}{\partial x_1^2} + \cdots + \frac{\partial^2}{\partial x_n^2}$$

and

$$\mathcal{R}\{\nabla^2 f(\mathbf{x})\} = |\boldsymbol{\xi}|^2 \frac{\partial^2 \check{f}(p, \boldsymbol{\xi})}{\partial p^2} = \frac{\partial^2 \check{f}(p, \boldsymbol{\xi})}{\partial p^2}. \tag{8.6.6}$$

Note that $|\boldsymbol{\xi}| = 1$ has been used. ∎

Results of this type have been used by John [1955] in applications of the Radon transform to partial differential equations.

Example 3
Suppose f is a function of both time and space variables. For example, if $n = 3$, then $f = f(x, y, x; t)$. The wave equation in three dimensions is given by

$$\frac{\partial^2 f}{\partial x^2} + \frac{\partial^2 f}{\partial y^2} + \frac{\partial^2 f}{\partial z^2} = \frac{\partial^2 f}{\partial t^2}. \tag{8.6.7}$$

Because the operator $\mathcal{R}$ does not involve time, it must commute with the time derivative operator $\partial/\partial t$. Thus the Radon transform of the wave equation yields

$$\frac{\partial^2 \check{f}}{\partial p^2} = \frac{\partial^2 \check{f}}{\partial t^2}, \tag{8.6.8}$$

where it is understood that $\check{f}$ now depends on time, $\check{f} = \check{f}(p, \boldsymbol{\xi}; t) = \mathcal{R} f(x, y, z; t)$. The important significance is that the wave equation in three spatial dimensions has been reduced to a wave equation in one spatial dimension. ∎

8.6.2   *Derivatives of the transform*

Here we investigate what happens when $\check{f}$ is differentiated with respect to one of the components of the unit vector $\boldsymbol{\xi}$. To facilitate this an identity related to derivatives of the delta function is needed. First, note that

$$\frac{\partial}{\partial y}\delta(x - y) = -\frac{\partial}{\partial x}\delta(x - y)$$

and if y is replaced by ay,

$$\frac{\partial}{\partial(ay)}\delta(x - ay) = \frac{1}{a}\frac{\partial}{\partial y}\delta(x - ay) = -\frac{\partial}{\partial x}\delta(x - ay).$$

In n dimensions

$$\frac{\partial}{\partial y_j}\delta(\mathbf{x} - \mathbf{y}) = -\frac{\partial}{\partial x_j}\delta(\mathbf{x} - \mathbf{y}).$$

From these equations it is easy to see that

$$\frac{\partial}{\partial \eta_j}\delta(p - \boldsymbol{\eta}\cdot\mathbf{x}) = -x_j\frac{\partial}{\partial p}\delta(p - \boldsymbol{\eta}\cdot\mathbf{x}). \tag{8.6.9}$$

This identity is in terms of $\boldsymbol{\eta}\cdot\mathbf{x}$ where $\boldsymbol{\eta}$ must not be restricted to being a unit vector; however, the desired derivatives are in terms of components of the unit vector $\boldsymbol{\xi}$. The way to deal with this is to take derivatives with respect to components of $\boldsymbol{\eta}$ and then evaluate the results at $\boldsymbol{\eta} = \boldsymbol{\xi}$. This prescription is followed starting with

$$\check{f}(p, \boldsymbol{\eta}) = \int f(\mathbf{x})\,\delta(p - \boldsymbol{\eta}\cdot\mathbf{x})\,d\mathbf{x},$$

$$\frac{\partial \check{f}}{\partial \xi_k} = \left[\frac{\partial \check{f}(p, \boldsymbol{\eta})}{\partial \eta_k}\right]_{\boldsymbol{\eta}=\boldsymbol{\xi}} = \left[\int f(\mathbf{x})\frac{\partial}{\partial \eta_k}\delta(p - \boldsymbol{\eta}\cdot\mathbf{x})\,d\mathbf{x}\right]_{\boldsymbol{\eta}=\boldsymbol{\xi}}$$

$$= -\frac{\partial}{\partial p}\int x_k f(\mathbf{x})\,\delta(p - \boldsymbol{\xi}\cdot\mathbf{x})\,d\mathbf{x}.$$

This gives the desired formula,

$$\frac{\partial \check{f}}{\partial \xi_k} = \left[\frac{\partial}{\partial \eta_k}\mathcal{R}\{f(\mathbf{x})\}\right]_{\boldsymbol{\eta}=\boldsymbol{\xi}} = -\frac{\partial}{\partial p}\mathcal{R}\{x_k f(\mathbf{x})\}. \tag{8.6.10}$$

CONVENTION   *Whenever the transformed function $\check{f}$ is differentiated with respect to a component of the unit vector $\boldsymbol{\xi}$ it is understood that*

$$\frac{\partial \check{f}(p, \boldsymbol{\xi})}{\partial \xi_k} \equiv \left[\frac{\partial \check{f}(p, \boldsymbol{\eta})}{\partial \eta_k}\right]_{\boldsymbol{\eta}=\boldsymbol{\xi}}. \tag{8.6.11}$$

The following example clearly illustrates the need for caution when taking derivatives of $\check{f}$.

Example 4

Start with

$$f(x, y) = e^{-x^2-y^2} \qquad \text{and} \qquad \check{f}(p, \xi) = \sqrt{\pi}\, e^{-p^2}.$$

Apply the scaling relation (8.3.9) with

$$\eta = s\,\xi \qquad \text{and} \qquad s = (\eta_1^2 + \eta_2^2)^{1/2},$$

to obtain

$$\check{f}(p, \eta) = \check{f}(p, s\,\xi) = \frac{\sqrt{\pi}}{s}\, e^{-p^2/s^2}.$$

Now use

$$\frac{\partial}{\partial \eta_k} = \frac{\partial s}{\partial \eta_k} \frac{\partial}{\partial s}, \qquad (k = 1, 2),$$

to get

$$\frac{\partial \check{f}}{\partial \eta_k} = \sqrt{\pi}\, \frac{\eta_k}{s} \frac{\partial}{\partial s} \left(s^{-1} e^{-p^2/s^2} \right)$$

$$= \sqrt{\pi}\, \frac{\eta_k}{s^5} (2p^2 - s^2)\, e^{-p^2/s^2}.$$

The desired derivative is found when this expression is evaluated at $\eta = \xi$, or equivalently for $s = 1$,

$$\frac{\partial \check{f}}{\partial \xi_k} = \sqrt{\pi}\, \xi_k (2p^2 - 1)\, e^{-p^2}. \tag{8.6.12}$$

The significance of this result becomes more apparent when compared with Example 3 in Section 8.7. ∎

Example 5

In Example 5 of Section 8.13 it is shown that the Radon transform of $x^2 + y^2$ confined to the unit disk and zero outside the disk is given by

$$\mathcal{R}\{x^2 + y^2\} = \frac{2}{3}\sqrt{1 - p^2}\, (1 + 2p^2),$$

and in Example 7 of the same section

$$\mathcal{R}\{x(x^2 + y^2)\} = \frac{2}{3} p\sqrt{1 - p^2}\, (1 + 2p^2) \cos\phi.$$

It is left as an exercise for the reader to demonstrate that (8.6.10) is satisfied by this pair of transforms. That is, verify that

$$\left[\frac{\eta_1}{s} \frac{\partial \check{f}(p, \eta)}{\partial s} \right]_{\eta=\xi} = -\frac{\partial}{\partial p} \mathcal{R}\{x(x^2 + y^2)\},$$

where

$$\check{f}(p, \eta) = \frac{2}{3} s^{-4} \sqrt{s^2 - p^2}\, (s^2 + 2p^2).$$

This can be done by showing that both sides reduce to

$$\frac{2}{3} \cos\phi \, (1 - p^2)^{-1/2} (8p^4 - 4p^2 - 1). \qquad ∎$$

There are some rather obvious generalizations for derivatives of higher order. These results follow immediately by differentiating (8.6.10); it is understood that the convention (8.6.11) always applies. For second derivatives

$$\frac{\partial^2 \check{f}(p, \boldsymbol{\xi})}{\partial \xi_l \, \partial \xi_k} = \frac{\partial^2}{\partial p^2} \mathcal{R} \{x_l \, x_k \, f(\mathbf{x})\}. \tag{8.6.13}$$

For higher derivatives the procedure is to differentiate this expression. For example, one of the third derivatives is given by

$$\frac{\partial^3 \check{f}(p, \boldsymbol{\xi})}{\partial \xi_l \, \partial \xi_k^2} = -\frac{\partial^3}{\partial p^3} \mathcal{R} \{x_l \, x_k^2 \, f(\mathbf{x})\}. \tag{8.6.14}$$

Note that there is an alternating sign, $+$ for even derivatives and $-$ for odd derivatives. One final example is given here. Additional examples involving derivatives are given in Section 8.7.

Example 6

If $f = f(x, y)$ is a 2D function, a generalization of (8.6.14) to arbitrarily high derivatives provides a method for finding many additional transforms of functions in two dimensions,

$$\frac{\partial^{l+k} \check{f}(p, \boldsymbol{\xi})}{\partial \xi_1^l \, \partial \xi_2^k} = \left(-\frac{\partial}{\partial p}\right)^{l+k} \mathcal{R} \{x^l \, y^k \, f(x, y)\}. \tag{8.6.15}$$

∎

8.7 Hermite polynomials

In this section the discussion is confined to two dimensions, and the components of $\boldsymbol{\xi}$ are written as $(\xi_1, \xi_2) = (\cos \phi, \sin \phi)$ to emphasize the dependence of the transform on ϕ. The extension to higher dimensions does not involve complications except that the formulas contain more variables. The previous section on derivatives can be used to find transforms of functions of the form

$$H_l(x) \, H_k(y) \, e^{-x^2 - y^2}$$

where H_l and H_k are Hermite polynomials of order l and k, respectively. More information on these polynomials is contained in Appendix A of this chapter and in Chapter 1, Section 5.

We start with the Rodrigues formula for Hermite polynomials [Rainville (1960)],

$$e^{-x^2} H_l(x) = (-1)^l \left(\frac{\partial}{\partial x}\right)^l e^{-x^2}. \tag{8.7.1}$$

A similar formula holds for the variable y. When these are combined the joint formula is

$$H_l(x) \, H_k(y) \, e^{-x^2 - y^2} = (-1)^{l+k} \left(\frac{\partial}{\partial x}\right)^l \left(\frac{\partial}{\partial y}\right)^k e^{-x^2 - y^2}. \tag{8.7.2}$$

From the methods developed in Section 8.6 we deduce that

$$\mathcal{R} \left\{ \left(\frac{\partial}{\partial x}\right)^l \left(\frac{\partial}{\partial y}\right)^k f(x, y) \right\} = (\cos \phi)^l \, (\sin \phi)^k \left(\frac{\partial}{\partial p}\right)^{l+k} \check{f}(p, \boldsymbol{\xi}).$$

By using this derivative relation, it follows that the Radon transform of the Rodrigues formula gives

$$\mathcal{R}\left\{H_l(x)\,H_k(y)\,e^{-x^2-y^2}\right\} = (\cos\phi)^l\,(\sin\phi)^k\left(\frac{\partial}{\partial p}\right)^{l+k}\sqrt{\pi}\,e^{-p^2}.$$

By application of the Rodrigues formula in one variable to the right side of this equation, the basic formula for transforms of Hermite polynomials is

$$\mathcal{R}\left\{H_l(x)\,H_k(y)\,e^{-x^2-y^2}\right\} = \sqrt{\pi}\,(\cos\phi)^l\,(\sin\phi)^k\,e^{-p^2}\,H_{l+k}(p). \tag{8.7.3}$$

The importance of the last equation becomes more apparent after observing that members of the sequence

$$1, x, y, x^2, xy, y^2, \ldots, x^l y^k, \ldots$$

can be expressed in terms of Hermite polynomials. Some examples are given to illustrate the way transforms of members of this sequence are found.

Example 1
Find the Radon transform of $xy^2 e^{-x^2-y^2}$. From Appendix A,

$$xy^2 = \frac{1}{8}H_1(x)H_2(y) + \frac{1}{4}H_1(x)H_0(y).$$

It follows immediately from the fundamental relation (8.7.3) that

$$\mathcal{R}\left\{xy^2 e^{-x^2-y^2}\right\} = \frac{\sqrt{\pi}}{8}[\cos\phi\sin^2\phi\,H_3(p) + 2\cos\phi\,H_1(p)]e^{-p^2}. \tag{8.7.4}$$

This result can be modified by using explicit expressions for the Hermite polynomials, from Appendix A,

$$\mathcal{R}\left\{xy^2\,e^{-x^2-y^2}\right\} = \frac{\sqrt{\pi}}{2}\,e^{-p^2}[2p^3\,\cos\phi\sin^2\phi + p\cos\phi(1 - 3\sin^2\phi)]. \tag{8.7.5}$$

∎

Example 2
The method used for the previous example can be applied to obtain some basic results; then other theorems can be applied to get easy extensions. The linear property is especially useful. Given that

$$\mathcal{R}\left\{x\,e^{-x^2-y^2}\right\} = \sqrt{\pi}\,p\,e^{-p^2}\,\cos\phi. \tag{8.7.6}$$

By just changing x to y and $\cos\phi$ to $\sin\phi$ it follows that

$$\mathcal{R}\left\{y\,e^{-x^2-y^2}\right\} = \sqrt{\pi}\,p\,e^{-p^2}\,\sin\phi. \tag{8.7.7}$$

Now, by linearity

$$\mathcal{R}\left\{(x + y)\,e^{-x^2-y^2}\right\} = \sqrt{\pi}\,p\,e^{-p^2}\,(\cos\phi + \sin\phi). \tag{8.7.8}$$

The same technique can be applied to obtain:

$$\mathcal{R}\left\{x^2\,e^{-x^2-y^2}\right\} = \frac{\sqrt{\pi}}{2}(2p^2\cos^2\phi + \sin^2\phi)\,e^{-p^2}; \tag{8.7.9}$$

$$\mathcal{R}\left\{y^2\,e^{-x^2-y^2}\right\} = \frac{\sqrt{\pi}}{2}(2p^2\sin^2\phi + \cos^2\phi)\,e^{-p^2}; \tag{8.7.10}$$

$$\mathcal{R}\left\{(x^2+y^2)\,e^{-x^2-y^2}\right\} = \frac{\sqrt{\pi}}{2}(2p^2+1)\,e^{-p^2}. \tag{8.7.11}$$

∎

Example 3

It is instructive to relate the transforms in the last example to earlier results. We focus attention on formula (8.7.6),

$$\mathcal{R}\left\{x\,e^{-x^2-y^2}\right\} = \sqrt{\pi}\,p\,e^{-p^2}\,\cos\phi.$$

From Example 4 of Section 8.6.2 with $k = 1$,

$$\frac{\partial \check{f}}{\partial \xi_1} = \sqrt{\pi}\,\cos\phi\,(2p^2-1)\,e^{-p^2}.$$

Now, from formula (8.6.10) it should be true that this is the same as

$$-\frac{\partial}{\partial p}\left\{\sqrt{\pi}\,p\,e^{-p^2}\right\},$$

and, of course, the consistency is verified by doing the differentiation. This explicitly demonstrates that

$$\frac{\partial \check{f}}{\partial \xi_1} = -\frac{\partial}{\partial p}\,\mathcal{R}\left\{x\,e^{-x^2-y^2}\right\}. \qquad\qquad ∎$$

Example 4

It is easy to find the extension of (8.7.3) for scaled variables. By use of (8.3.5) with $a = b = c$,

$$\mathcal{R}\left\{H_l(cx)\,H_k(cy)\,e^{-c^2(x^2+y^2)}\right\} = \frac{\sqrt{\pi}}{c}\,(\cos\phi)^l\,(\sin\phi)^k\,e^{-c^2p^2}\,H_{l+k}(cp). \tag{8.7.12}$$

∎

8.8   Laguerre polynomials

Here, a very brief introduction to transforms of Laguerre polynomials is given. A much more extensive treatment is given by Deans [1983, 1993], where several examples and applications are provided. Additional applications are contained in the work by Maldonado and Olsen [1966] and Louis [1985]. As in the previous section, the discussion is confined to two dimensions and the angle ϕ appears explicitly in the transform. The approach is the same as with the Hermite polynomials. We start with the Rodrigues formula for the Laguerre polynomials [Szegö, 1939], [Rainville, 1960],

$$e^{-t}t^l\,L_k^l(t) = \frac{1}{k!}\left(\frac{\partial}{\partial t}\right)^k e^{-t}t^{l+k}, \tag{8.8.1}$$

and derive a generalized expression that accommodates the Radon transform,

$$\left(\frac{\partial}{\partial x} \pm i \frac{\partial}{\partial y}\right)^l \left(\frac{\partial^2}{\partial x^2} + \frac{\partial^2}{\partial y^2}\right)^k e^{-x^2-y^2} = (-1)^{l+k} 2^{2k+l} k! (x \pm iy)^l\, e^{-x^2-y^2}\, L_k^l(x^2 + y^2).$$

$$(8.8.2)$$

From the two previous sections, the Radon transform of the left side is

$$\mathcal{R}\left\{\left(\frac{\partial}{\partial x} \pm i \frac{\partial}{\partial y}\right)^l \left(\frac{\partial^2}{\partial x^2} + \frac{\partial^2}{\partial y^2}\right)^k e^{-x^2-y^2}\right\} = (-1)^{2k+l} \sqrt{\pi}\, e^{\pm il\phi}\, e^{-p^2} H_{l+2k}(p),$$

leading to the expression

$$\mathcal{R}\left\{(-1)^k 2^{2k+l} k! (x \pm iy)^l\, e^{-x^2-y^2} L_k^l(x^2 + y^2)\right\} = \sqrt{\pi}\, e^{\pm il\phi}\, e^{-p^2} H_{l+2k}(p). \qquad (8.8.3)$$

A more standard form is obtained by making substitutions,

$$x^2 + y^2 = r^2, \qquad (x \pm iy)^l = (x^2 + y^2)^{l/2}\, e^{\pm il\theta} \qquad \text{with } \theta = \tan^{-1}\left(\frac{y}{x}\right),$$

and defining a normalization constant by

$$N_k^l = \frac{1}{2^{2k+l}} \left[\frac{1}{k!(l+k)!}\right]^{1/2}. \qquad (8.8.4)$$

These changes lead to the standard form for the transform of expressions that involve Laguerre polynomials,

$$\mathcal{R}\left\{(-1)^k \left[\frac{k!}{(l+k)!}\right]^{1/2} r^l\, e^{\pm il\theta} L_k^l(r^2)\right\} = N_k^l\, e^{\pm il\phi} e^{-p^2} H_{l+2k}(p). \qquad (8.8.5)$$

8.9   Inversion

Inversion of the Radon transform is especially important because it yields information about an object in feature space when some probe has been used to produce projection data. This inversion is the solution of the problem of "reconstruction from projections" when the projections can be interpreted as the Radon transform of some function in feature space.

There are several routes that can be followed to go from Radon space to feature space. The direct route illustrated by the diagram

is probably the most difficult to derive and certainly the most difficult to implement in practical situations; however, see the alternative method used by Nievergelt [1986]. The direct method is discussed in some detail by John [1955] and Deans [1983, 1993].

For those already familiar with Fourier transforms, the route through Fourier space pioneered by Bracewell [1956] may be easier. Other important early references include Helgason [1965] and Ludwig [1966]. The route from feature space to Fourier space and the route from Radon space to Fourier space is discussed in Section 8.2.5. The basic ideas presented there can be used to derive formulas for the inverse Radon transform.

It turns out that there is a fundamental difference between inversion in even dimension and inversion in odd dimension. Although this may seem a bit strange at first, it is something that is quite common in the study of partial differential equations and Green's function for the wave equation: [Morse and Feshbach, 1953] and [Wolf, 1979]. This difference is discussed in connection with the Radon transform by Shepp [1980], Barrett [1984], Berenstein and Walnut [1994], and Olson and DeStefano [1994]. The important observation is that the operations required for the inverse in two dimensions are global; the transform must be known over all of Radon space. By contrast, in three dimensions, because derivatives are required, the inversion operations are local. Hence, the procedure here is to give separate derivations for two and three dimensions. It is not very much more difficult to do the derivation for general even and odd dimensions; however, it is a bit easier to follow the specific cases. And, after all, these are the most important for applications anyway. The method used is patterned after that used by Barrett [1984] and Deans [1985].

8.9.1   *Two dimensions*

The notation is the same as used previously for vectors, $\mathbf{x} = (x, y)$ and $\boldsymbol{\xi} = (\cos\phi, \sin\phi)$. The coordinates in Fourier space are designated by $(u, v) = (q\cos\phi, q\sin\phi) = q\,\boldsymbol{\xi}$. The starting point is (8.2.15),

$$\tilde{f}(q\,\boldsymbol{\xi}) = \mathcal{F}_1 \mathcal{R} f(x, y) \tag{8.9.1}$$

along with the observation that f is given by the inverse two-dimensional Fourier transform,

$$f(x, y) = \mathcal{F}_2^{-1} \tilde{f}(u, v).$$

In polar form,

$$f(x, y) = \int_{-\infty}^{\infty} dq\,|q| \int_0^{\pi} d\phi\, \tilde{f}(q\,\boldsymbol{\xi})\, e^{i2\pi q\,\boldsymbol{\xi}\cdot\mathbf{x}}$$

$$= \int_0^{\pi} d\phi \left[\int_{-\infty}^{\infty} dq\,|q|\tilde{f}(q\,\boldsymbol{\xi})\, e^{i2\pi qp} \right]_{p=\boldsymbol{\xi}\cdot\mathbf{x}}. \tag{8.9.2}$$

Now the term in square brackets is the inverse one-dimensional Fourier transform of the product $|q|\tilde{f}$ and this is to be evaluated at $p = \boldsymbol{\xi} \cdot \mathbf{x}$. The convolution theorem for Fourier transforms can be used to obtain

$$\mathcal{F}^{-1}\left\{ |q|\tilde{f}(q\,\boldsymbol{\xi}) \right\} = \mathcal{F}^{-1}\left\{ |q| \right\} * \mathcal{F}^{-1}\left\{ \tilde{f}(q\,\boldsymbol{\xi}) \right\}.$$

From Section 8.2.5 the last term on the right is just the Radon transform $\check{f}(p, \boldsymbol{\xi})$. This observation leads to

$$f(x, y) = \int_0^{\pi} d\phi \left[\check{f}(p, \boldsymbol{\xi}) * \mathcal{F}^{-1}\left\{ |q| \right\} \right]_{p=\boldsymbol{\xi}\cdot\mathbf{x}}. \tag{8.9.3}$$

The inverse Fourier transform in this equation is interpreted in terms of generalized functions to give [Lighthill, 1962], [Bracewell, 1986]

$$\mathcal{F}^{-1}\left\{ |q| \right\} = \mathcal{F}^{-1}\left\{ 2\pi i q \right\} * \mathcal{F}^{-1}\left\{ \frac{\operatorname{sgn} q}{2\pi i} \right\}.$$

Here, we have written

$$|q| = q\,\operatorname{sgn} q = 2\pi i q\,\frac{\operatorname{sgn} q}{2\pi i},$$

where

$$
\operatorname{sgn} q = \begin{cases} +1, & \text{for } q > 0 \\ 0, & \text{for } q = 0 \\ -1, & \text{for } q < 0. \end{cases} \tag{8.9.4}
$$

The methods needed to work with these inverse Fourier transforms is given by Lighthill [1962] and Bracewell [1986]. By use of the derivative theorem

$$
\mathcal{F}^{-1}\{2\pi i q\} = \delta'(p),
$$

where the prime denotes first order derivative with respect to variable p. The other transform is given in terms of a Cauchy principal value,

$$
\mathcal{F}^{-1}\left\{\frac{\operatorname{sgn} q}{2\pi i}\right\} = \frac{1}{2\pi^2}\, \mathcal{P}\left(\frac{1}{p}\right).
$$

It follows that

$$
\mathcal{F}^{-1}\{|q|\} = \delta'(p) * \frac{1}{2\pi^2}\, \mathcal{P}\left(\frac{1}{p}\right).
$$

Now, (8.9.3) becomes

$$
f(x, y) = \frac{1}{2\pi^2}\int_0^\pi d\phi \left[\check{f}(p, \boldsymbol{\xi}) * \delta'(p) * \mathcal{P}\left(\frac{1}{p}\right)\right]_{p=\boldsymbol{\xi}\cdot\mathbf{x}}. \tag{8.9.5}
$$

By using the derivative theorem for convolution and the properties of the delta function,

$$
\check{f}(p, \boldsymbol{\xi}) * \delta'(p) = \frac{\check{f}(p, \boldsymbol{\xi})}{\partial p} * \delta(p) = \frac{\check{f}(p, \boldsymbol{\xi})}{\partial p}.
$$

It is convenient to use the subscript notation for partial derivatives and write

$$
\check{f}_p(p, \boldsymbol{\xi}) \equiv \frac{\check{f}(p, \boldsymbol{\xi})}{\partial p}.
$$

Now the term in square brackets in (8.9.5) can be written as

$$
\left[\check{f}_p(p, \boldsymbol{\xi}) * \mathcal{P}\left(\frac{1}{p}\right)\right]_{p=\boldsymbol{\xi}\cdot\mathbf{x}} = \left[\mathcal{P}\int_{-\infty}^\infty \frac{\check{f}_t(t, \boldsymbol{\xi})}{p-t}\, dt\right]_{p=\boldsymbol{\xi}\cdot\mathbf{x}} = -\mathcal{P}\int_{-\infty}^\infty \frac{\check{f}_t(t, \boldsymbol{\xi})}{t-\boldsymbol{\xi}\cdot\mathbf{x}}\, dt.
$$

Note that t is a dummy variable in the last integral, and can be replaced by p to agree with earlier notation. The final formula follows by substituting this result in (8.9.5) to get

$$
f(x, y) = \frac{-1}{2\pi^2}\, \mathcal{P}\int_0^\pi d\phi \int_{-\infty}^\infty \frac{\check{f}_p(p, \boldsymbol{\xi})}{p-\boldsymbol{\xi}\cdot\mathbf{x}}\, dp. \tag{8.9.6}
$$

Here, the Cauchy principal value is related to the integral over p. It has been placed outside for convenience. Sometimes the $\mathcal{P}$ is dropped altogether; in this case it is "understood" that the singular integral is interpreted in terms of the Cauchy principal value.

The inversion formula (8.9.6) can be expressed in terms of a Hilbert transform (see also Chapter 7). The Hilbert transform of $f(t)$ is defined by Sneddon [1972] and Bracewell [1986],

$$\mathcal{H}_i\left[f(t); t \to x\right] = \frac{1}{\pi} \int_{-\infty}^{\infty} \frac{f(t)\,dt}{t-x}, \tag{8.9.7}$$

where the Cauchy principal value is understood. Thus the inversion formula can be written as

$$f(x, y) = \frac{-1}{2\pi} \int_0^\pi \mathcal{H}_i\left[\check{f}_p(p, \boldsymbol{\xi}); p \to \boldsymbol{\xi} \cdot \mathbf{x}\right] d\phi. \tag{8.9.8}$$

For reasons that will become apparent in the subsequent discussion it is extremely desirable to make the following definition for the Hilbert transform of the derivative of some function, say g,

$$\bar{g}(t) = \frac{-1}{4\pi} \mathcal{H}_i\left[g_p(p); p \to t\right] \qquad \text{for } n = 2. \tag{8.9.9}$$

If this is done, the inversion formula for $n = 2$, is given by

$$f(x, y) = 2 \int_0^\pi d\phi \,[\bar{\check{f}}(t, \boldsymbol{\xi})]_{t=\boldsymbol{\xi}\cdot\mathbf{x}}. \tag{8.9.10}$$

8.9.2 *Three dimensions*

The inversion formula in three dimensions is actually easier to derive, because no Hilbert transforms emerge. The path through Fourier space is used again with the unit vector $\boldsymbol{\xi}$ given in terms of the polar angle θ and azimuthal angle ϕ,

$$\boldsymbol{\xi} = (\sin\theta\cos\phi, \sin\theta\sin\phi, \cos\theta).$$

The feature space function $f(\mathbf{x}) = f(x, y, z)$ is found from the inverse 3D Fourier transform,

$$f(\mathbf{x}) = \mathcal{F}_3^{-1}\tilde{f}(q\boldsymbol{\xi}) = \int_0^\infty dq\, q^2 \int_{|\boldsymbol{\xi}|=1} d\boldsymbol{\xi}\, \tilde{f}(q\boldsymbol{\xi})\, e^{i2\pi q\,\boldsymbol{\xi}\cdot\mathbf{x}}. \tag{8.9.11}$$

Here, the integral over the unit sphere is indicated by

$$\int_{|\boldsymbol{\xi}|=1} d\boldsymbol{\xi} = \int_0^{2\pi} d\phi \int_0^\pi \sin\theta\, d\theta.$$

Now recall that $\tilde{f}$ is given by the 1D Fourier transform of $\check{f}$, and from the symmetry properties of $\check{f}$ the integral over q from 0 to ∞ can be replaced by one-half the integral from $-\infty$ to ∞. Hence

$$f(\mathbf{x}) = \frac{1}{2} \int_{|\boldsymbol{\xi}|=1} d\boldsymbol{\xi} \left[\int_{-\infty}^\infty dq\, q^2 \tilde{f}(q\boldsymbol{\xi})\, e^{i2\pi qp}\right]_{p=\boldsymbol{\xi}\cdot\mathbf{x}}$$

$$= \frac{1}{2} \int_{|\boldsymbol{\xi}|=1} d\boldsymbol{\xi}\, \mathcal{F}^{-1}\left[q^2 \tilde{f}(q\boldsymbol{\xi})\right]_{p=\boldsymbol{\xi}\cdot\mathbf{x}}.$$

Now from the inverse of the 1D derivative theorem

$$\mathcal{F}^{-1}\left[q^2 \tilde{f}\right] = \frac{-1}{4\pi} \frac{\partial^2 \check{f}}{\partial p^2} = \frac{-1}{4\pi} \check{f}_{pp},$$

one form of the inversion formula is

$$f(\mathbf{x}) = \frac{-1}{8\pi^2} \int_{|\boldsymbol{\xi}|=1} d\boldsymbol{\xi} \left[\check{f}_{pp}(p, \boldsymbol{\xi})\right]_{p=\boldsymbol{\xi}\cdot\mathbf{x}}. \tag{8.9.12}$$

Another form for (8.9.12) comes from the observation that for any function of $\boldsymbol{\xi} \cdot \mathbf{x}$

$$\nabla^2 \psi(\boldsymbol{\xi} \cdot \mathbf{x}) = |\boldsymbol{\xi}|^2 \left[\psi_{pp}(p)\right]_{p=\boldsymbol{\xi} \cdot \mathbf{x}} = \left[\psi_{pp}(p)\right]_{p=\boldsymbol{\xi} \cdot \mathbf{x}}.$$

The last equality follows because $\boldsymbol{\xi}$ is a unit vector. These observations lead to the inversion formula

$$f(\mathbf{x}) = \frac{-1}{8\pi^2} \nabla^2 \int_{|\boldsymbol{\xi}|=1} \check{f}(\boldsymbol{\xi} \cdot \mathbf{x}, \boldsymbol{\xi}) \, d\boldsymbol{\xi}. \tag{8.9.13}$$

8.10 Abel transforms

In this section we focus attention on a particular class of singular integral equations and how transforms known as Abel transforms emerge. Actually, it is convenient to define four different Abel transforms. Although all of these transforms are called Abel transforms at various places in the literature, there is no agreement regarding the numbering. Consequently, an arbitrary decision is made here in that respect. There is an intimate connection with the Radon transform; however, that discussion is delayed until Section 8.11. There are some very good recent references devoted primarily to Abel integral equations, Abel transforms, and applications. The monograph by Gorenflo and Vessella [1991] is especially recommended for both theory and applications. Also, the chapter by Anderssen and de Hoog [1990] contains many applications along with an excellent list of references. A recent book by Srivastava and Bushman [1992] is valuable for convolution integral equations in general. Other general references include Kanwal [1971], Widder [1971], Churchill [1972], Doetsch [1974], and Knill [1994]. Another valuable resource is the review by Lonseth [1977]. His remarks on page 247 regarding Abel's contributions "back in the springtime of analysis" are required reading for those who appreciate the history of mathematics. Other references to Abel transforms and relevant resource material are contained in Section 8.11 and in the following discussion.

8.10.1 *Singular integral equations, Abel type*

An integral equation is called singular if either the range of integration is infinite or the kernel has singularities within the range of integration. Singular integral equations of Volterra type of the first kind are of the form [Tricomi, 1985]

$$g(x) = \int_0^x k(x, y) f(y) \, dy \qquad x > 0, \tag{8.10.1}$$

where the kernel satisfies the condition $k(x, y) \equiv 0$ if $y > x$. If $k(x, y) = k(x - y)$, then the equation is of convolution type. The type of kernel of interest here is

$$k(x - y) = \frac{1}{(x - y)^\alpha} \qquad 0 < \alpha < 1.$$

This leads to an integral equation of Abel type,

$$g(x) = \int_0^x \frac{f(y)}{(x - y)^\alpha} \, dy = f(x) * \frac{1}{x^\alpha}, \qquad x > 0, \qquad 0 < \alpha < 1. \tag{8.10.2}$$

Integral equations of the type in (8.10.2) were studied by the Norwegian mathematician Niels H. Abel (1802–1829) with particular attention to the connection with the tautochrone problem. This work by Abel [1823, 1826a,b] served to introduce the subject of integral equations. The connection with the tautochrone problem emerges when $\alpha = 1/2$ in the integral equation. This is the problem of determining a curve through the origin in a vertical plane such that the time required for a massive particle to slide without friction down the curve to the origin is independent of the starting position. It is assumed that the particle slides freely from rest under the action of its weight and the reaction of the curve (smooth wire) that constrains its movement. Details of this problem are discussed by Churchill [1972] and Widder [1971].

One way to solve (8.10.1) when $k(x, y) = k(x - y)$ is by use of the Laplace transform (see Chapter 5); this yields

$$G(s) = F(s) K(s). \tag{8.10.3}$$

The solution for $F(s)$ can be written in two forms,

$$F(s) = \frac{G(s)}{K(s)} = [sG(s)] \left[\frac{1}{s K(s)} \right] \tag{8.10.4}$$

The second form is used when the inverse Laplace transform of $1/K(s)$ does not exist.

Example 1
Solve equation (8.10.2) for $f(x)$. From (8.10.3) and Laplace transform tables (Chapter 5),

$$G(s) = \mathcal{L}\{f(x)\} \mathcal{L} \left\{ \frac{1}{x^\alpha} \right\} = F(s) s^{\alpha-1} \Gamma(1 - \alpha).$$

To find $F(s)$ we must invert the equation

$$F(s) = \frac{s}{\Gamma(\alpha)\Gamma(1 - \alpha)} \left[\Gamma(\alpha) s^{-\alpha} G(s) \right].$$

The inversion yields

$$f(x) = \mathcal{L}^{-1} \left\{ \frac{s}{\Gamma(\alpha)\Gamma(1 - \alpha)} \left[\Gamma(\alpha) s^{-\alpha} G(s) \right] \right\}$$

$$= \mathcal{L}^{-1} \left\{ \frac{s}{\Gamma(\alpha)\Gamma(1 - \alpha)} \mathcal{L} \left\{ \int_0^x (x - y)^{\alpha-1} g(y) \, dy \right\} \right\}.$$

By invoking the property $df(x)/dx = \mathcal{L}^{-1}\{s\mathcal{L}\{f(x)\}\}$ the above equation becomes

$$f(x) = \frac{\sin \alpha \pi}{\pi} \frac{d}{dx} \int_0^x (x - y)^{\alpha-1} g(y) \, dy. \tag{8.10.5}$$

Here, use is made of the gamma function identity

$$\Gamma(\alpha)\Gamma(1 - \alpha) = \frac{\pi}{\sin \alpha \pi}. \qquad\blacksquare$$

Another form of (8.10.4) can be found if $g(y)$ is differentiable. One way to find this other solution is to use integration by parts, $\int u \, dv = uv - \int v \, du$, with $u = g(y)$ and $dv = (x - y)^{\alpha-1} dy$,

$$\int_0^x (x - y)^{\alpha-1} g(y) \, dy = \frac{g(+0) x^\alpha}{\alpha} + \frac{1}{\alpha} \int_0^x (x - y)^\alpha g'(y) \, dy.$$

When this expression is multiplied by $\sin \alpha \pi / \pi$ and differentiated with respect to x the alternative expression for (8.10.5) follows,

$$f(x) = \frac{\sin \alpha \pi}{\pi} \left[\frac{g(+0)}{x^{1-\alpha}} + \int_0^x \frac{g'(y)}{(x - y)^{1-\alpha}} \, dy \right]. \tag{8.10.6}$$

REMARK   It is tempting to take a quick look at (8.10.2) and assume that $g(0) = 0$. This is wrong! The proper interpretation is to do the integral first and then take the limit as $x \to 0$ through positive values. This is why we have written $g(+0)$ in (8.10.6).   ∎

The above equation (8.10.6) also follows by taking into consideration the convolution properties and derivatives for the Laplace transform. We observe that (8.10.4) can be written in two alternative forms,

$$F(s) = s[G(s)\,H(s)] = [sG(s)]\,[H(s)],$$

where $H(s)$ is defined by

$$H(s) = \frac{1}{sK(s)}.$$

The inversion gives

$$f(x) = \frac{d}{dx} \int_0^x g(y)\,h(x - y)\,dy, \tag{8.10.7a}$$

or

$$f(x) = g(0)h(x) + \int_0^x g'(y)\,h(x - y)\,dy. \tag{8.10.7b}$$

The previous equations can be used to solve an integral equation of the form

$$g(x) = \int_0^x f(y)\,k(x^2 - y^2)\,dy. \tag{8.10.8}$$

After making the substitutions,

$$x = u^{\frac{1}{2}}, \qquad y = v^{\frac{1}{2}}, \qquad f_1(v) = \frac{1}{2}f(v^{\frac{1}{2}})v^{-\frac{1}{2}}, \qquad g_1(u) = g(u^{\frac{1}{2}}),$$

equation (8.10.8) becomes

$$g_1(u) = \int_0^u f_1(v)\,k(u - v)\,dv. \tag{8.10.9}$$

This equation is identical to (8.10.1) with $k(x, y) = k(x - y)$ and the solution is given by (8.10.7) with k replaced by h,

$$f_1(u) = \frac{d}{du} \int_0^u g_1(v)\,h(u - v)\,dv$$

$$= g_1(0)h(u) + \int_0^u g_1'(v)\,h(u - v)\,dv,$$

where $h(x) = \mathcal{L}^{-1}\{1/sK(s)\}$. Using the substitutions in reverse gives

$$\frac{f(x)}{2x} = \frac{1}{2x}\frac{d}{dx} \int_0^x g(y)h(x^2 - y^2)\,2y\,dy,$$

or

$$f(x) = 2\frac{d}{dx} \int_0^x y\,g(y)\,h(x^2 - y^2)\,dy, \tag{8.10.10a}$$

and if the derivative of g exists,

$$f(x) = 2xg(0)h(x^2) + 2x \int_0^x g'(y)\,h(x^2 - y^2)\,dy. \tag{8.10.10b}$$

Example 2

Find the solution of (8.10.8) if the kernel is $k(x) = x^{-\alpha}$ and $0 < \alpha < 1$. With this kernel the equation to be solved is

$$g(x) = \int_0^x \frac{f(y)\,dy}{(x^2 - y^2)^\alpha}. \tag{8.10.11a}$$

We need the inverse Laplace transform of $H(s) = 1/sK(s)$. From

$$K(s) = \int_0^\infty e^{-sx} x^{-\alpha}\,dx = \Gamma(1 - \alpha)s^{\alpha - 1},$$

it follows that

$$h(x) = \mathcal{L}^{-1}\left\{\frac{1}{\Gamma(1 - \alpha)}\frac{1}{s^\alpha}\right\} = \frac{1}{\Gamma(\alpha)\Gamma(1 - \alpha)}x^{\alpha - 1} = \frac{\sin\alpha\pi}{\pi}x^{\alpha - 1}.$$

Now the solution follows directly from (8.10.10a),

$$f(x) = \frac{2\,\sin\alpha\pi}{\pi}\frac{d}{dx}\int_0^x \frac{y\,g(y)\,dy}{(x^2 - y^2)^{1 - \alpha}}. \tag{8.10.11b}$$

∎

Example 3

Apply (8.10.10b) to find an alternative expression for the inverse (8.10.11b). Note that

$$h(x^2) = \frac{\sin\alpha\pi}{\pi}x^{2\alpha - 2}$$

follows from Example 2. Consequently, the desired equation is

$$f(x) = \frac{2\,\sin\alpha\pi}{\pi}\left[g(0)x^{2\alpha - 1} + x\int_0^x \frac{g'(y)\,dy}{(x^2 - y^2)^{1 - \alpha}}\right]. \tag{8.10.11c}$$

∎

There are other integral equations, similar to the one in Example 2, that are of particular interest here. The relevant results are given without proof. The derivations are very similar to the procedures used above. A transform pair related to the pair of (8.10.11a,b) is

$$g(x) = \int_x^\infty \frac{f(y)\,dy}{(y^2 - x^2)^\alpha}, \qquad 0 < \alpha < 1, f(\infty) = 0, \tag{8.10.12a}$$

and

$$f(x) = -\frac{2\,\sin\alpha\pi}{\pi}\frac{d}{dx}\int_x^\infty \frac{y\,g(y)\,dy}{(y^2 - x^2)^{1 - \alpha}}. \tag{8.10.12b}$$

Another pair of interest is

$$g(x) = 2\int_x^\infty \frac{y\,f(y)\,dy}{(y^2 - x^2)^\alpha} \qquad 0 < \alpha < 1, f(\infty) = 0, \tag{8.10.13a}$$

and

$$f(x) = -\frac{\sin\alpha\pi}{\pi}\int_x^\infty \frac{g'(y)\,dy}{(y^2 - x^2)^{1 - \alpha}}. \tag{8.10.13b}$$

8.10.2 *Some Abel transform pairs*

If the choice $\alpha = \frac{1}{2}$ is made in equations (8.10.11), (8.10.12), and (8.10.13) the resulting transforms are known as Abel transforms. In order, these are designated by $\mathcal{A}_1\{f\}$, $\mathcal{A}_2\{f\}$, and $\mathcal{A}_3\{f\}$. The numerical designation is not standard, and some authors leave α in the equations. With the exception of a constant factor, Sneddon [1972] uses the same notation for $\mathcal{A}_1\{f\}$ and $\mathcal{A}_2\{f\}$. Bracewell [1986] introduces only $\mathcal{A}_3\{f\}$, and uses the notation $\mathcal{A}\{f\}$. This is the transform most directly related to the Radon transform. It is discussed in much more detail in Section 8.11. Also, for completeness we add a fourth transform. It is related to Riemann-Liouville (fractional) integrals of order $1/2$, discussed in Section 8.10.3.

Explicitly, the transforms are designated by

$$\hat{f}_1(x) \equiv \mathcal{A}_1\{f_1(r); x\} = \int_0^x \frac{f_1(r)\,dr}{(x^2 - r^2)^{\frac{1}{2}}}, \qquad x > 0 \tag{8.10.14a}$$

$$\hat{f}_2(x) \equiv \mathcal{A}_2\{f_2(r); x\} = \int_x^\infty \frac{f_2(r)\,dr}{(r^2 - x^2)^{\frac{1}{2}}}, \qquad x > 0 \tag{8.10.14b}$$

$$\hat{f}_3(x) \equiv \mathcal{A}_3\{f_3(r); x\} = 2\int_x^\infty \frac{r\,f_3(r)\,dr}{(r^2 - x^2)^{\frac{1}{2}}}, \qquad x > 0 \tag{8.10.14c}$$

$$\hat{f}_4(x) \equiv \mathcal{A}_4\{f_4(r); x\} = 2\int_0^x \frac{r\,f_4(r)\,dr}{(x^2 - r^2)^{\frac{1}{2}}}, \qquad x > 0. \tag{8.10.14d}$$

Note the change from $y \to r$ to agree with the short tables of transforms given in Appendix B. Also note the change $g \to \hat{f}$, and the use of subscripts to keep track of which transform is being applied.

The corresponding inversion expressions are:

$$f_1(r) = \frac{2}{\pi}\frac{d}{dr}\int_0^r \frac{x\,\hat{f}_1(x)\,dx}{(r^2 - x^2)^{\frac{1}{2}}} \tag{8.10.15a}$$

$$f_2(r) = -\frac{2}{\pi}\frac{d}{dr}\int_r^\infty \frac{x\,\hat{f}_2(x)\,dx}{(x^2 - r^2)^{\frac{1}{2}}} \tag{8.10.15b}$$

$$f_3(r) = -\frac{1}{\pi r}\frac{d}{dr}\int_r^\infty \frac{x\,\hat{f}_3(x)\,dx}{(x^2 - r^2)^{\frac{1}{2}}} \tag{8.10.15c}$$

$$f_4(r) = \frac{1}{\pi r}\frac{d}{dr}\int_0^r \frac{x\,\hat{f}_4(x)\,dx}{(r^2 - x^2)^{\frac{1}{2}}}. \tag{8.10.15d}$$

There are alternative ways to write the above inverses. The results can be verified by integrating by parts before taking the derivative with respect to r:

$$f_1(r) = \frac{2\,\hat{f}_1(0)}{\pi} + \frac{2r}{\pi}\int_0^r \frac{\hat{f}_1'(x)\,dx}{(r^2 - x^2)^{\frac{1}{2}}} \tag{8.10.16a}$$

$$f_2(r) = -\frac{2r}{\pi}\int_r^\infty \frac{\hat{f}_2'(x)\,dx}{(x^2 - r^2)^{\frac{1}{2}}} \tag{8.10.16b}$$

$$f_3(r) = -\frac{1}{\pi} \int_r^\infty \frac{\hat{f}_3'(x)\,dx}{(x^2 - r^2)^{\frac{1}{2}}} \tag{8.10.16c}$$

$$f_4(r) = \frac{\hat{f}_4(0)}{\pi r} + \frac{1}{\pi} \int_0^r \frac{\hat{f}_4'(x)\,dx}{(r^2 - x^2)^{\frac{1}{2}}}. \tag{8.10.16d}$$

In these equations it is assumed that the transform vanishes at infinity, $\hat{f}(\infty) \equiv 0$, and the prime means derivative with respect to x.

There is yet another form that is useful for f_3. The result comes from a study of the Radon transform [Deans, 1983, 1993]:

$$f_3(r) = -\frac{1}{\pi}\frac{d}{dr}\int_r^\infty \frac{r\,\hat{f}_3(x)\,dx}{x(x^2 - r^2)^{\frac{1}{2}}}. \tag{8.10.17}$$

To verify that this indeed reduces to (8.10.16c), let the integration by parts be done in (8.10.17) with

$$u = r\hat{f}_3(x), \qquad du = r\hat{f}'(x)dx, \qquad v = \frac{1}{r}\cos^{-1}\frac{r}{x}, \qquad dv = \frac{dx}{x(x^2 - r^2)^{\frac{1}{2}}}.$$

After doing the integration by parts, take the derivative with respect to r to get (8.10.16c).

Some important observations

From the definitions of the transforms $\mathcal{A}_i$ it follows that

$$\mathcal{A}_3\{f(r)\} = 2\mathcal{A}_2\{rf(r)\} \tag{8.10.18a}$$

$$\mathcal{A}_4\{f(r)\} = 2\mathcal{A}_1\{rf(r)\} \tag{8.10.18b}$$

$$\mathcal{A}_4\{r^{-1}f_1(r)\} = 2\hat{f}_1(x) \tag{8.10.18c}$$

$$\mathcal{A}_3\{r^{-1}f_2(r)\} = 2\hat{f}_2(x) \tag{8.10.18d}$$

$$f_1(r) \equiv \mathcal{A}_1^{-1}\{\hat{f}_1(x)\} = \frac{2}{\pi}\frac{d}{dr}\mathcal{A}_1\{x\hat{f}_1(x)\} \tag{8.10.18e}$$

$$f_2(r) \equiv \mathcal{A}_2^{-1}\{\hat{f}_2(x)\} = -\frac{2}{\pi}\frac{d}{dr}\mathcal{A}_2\{x\hat{f}_2(x)\}. \tag{8.10.18f}$$

These equations (along with obvious variations) can be used to find transforms and inverse transforms. A few samples are provided in the examples of Section 8.10.4.

8.10.3 *Fractional integrals*

The Abel transforms are related to the Riemann-Liouville and Weyl (fractional) integrals of order $1/2$; these are discussed along with an extensive tabulation in Chapter 13 of Erdélyi et al. [1954]. In the notation of this reference, the Riemann-Liouville integral is given by

$$g(y; \mu) = \frac{1}{\Gamma(\mu)} \int_0^y f(x)\,(y - x)^{\mu-1}\,dx, \tag{8.10.19}$$

and the Weyl integral is given by

$$h(y; \mu) = \frac{1}{\Gamma(\mu)} \int_y^\infty f(x)\,(x - y)^{\mu-1}\,dx. \tag{8.10.20}$$

Now in (8.10.19) let $\mu = 1/2$, make the replacement $y \to x^2$, and change the variable of integration $x = r^2$ to obtain

$$\sqrt{\pi}\, g\left(x^2, \frac{1}{2}\right) = 2 \int_0^x \frac{r\, f(r^2)\, dr}{(x^2 - r^2)^{\frac{1}{2}}}.$$

Clearly, this form of the Riemann-Liouville integral can be converted to (8.10.14d) by the appropriate replacements. By a similar argument, the Weyl integral (8.10.20) can be converted to (8.10.14c). This leads to the following useful rule for finding Abel transforms $\mathcal{A}_3$ and $\mathcal{A}_4$ from the tables in Chapter 13 of Erdélyi et al. [1954].

RULE

1. *Replace:* $\mu \to \frac{1}{2}$.
2. *Replace:* $x \to r^2$ *(column on left).*
3. *Replace:* $y \to x^2$ *and multiply the transform by* $\sqrt{\pi}$ *(column on right).*

It is easy to verify that this rule works by its application to cases that yield results quoted in Appendix B for $\mathcal{A}_3$. Verification of the rule for $\mathcal{A}_4$ follows immediately from the use of standard integral tables. Although the rule works most directly for the $\mathcal{A}_3$ and $\mathcal{A}_4$ transforms, it can be extended to apply to finding $\mathcal{A}_1$ and $\mathcal{A}_2$ transforms by use of the formulas in equation (8.10.18). Finally, it is interesting to note that these integrals lead to an interpretation for fractional differentiation and fractional integration. A good resource for details on this concept is the monograph by Gorenflo and Vessella [1991].

8.10.4   *Some useful examples*

We close this section with a few useful examples. These are especially valuable for those concerned with the analytic computation of Abel transforms or inverse Abel transforms.

Example 4

Consider the Abel transform

$$\hat{f}_1(x) = \mathcal{A}_1\{a - r\} = \frac{\pi a}{2} - x.$$

This is a simple case where $\hat{f}_1(x)$ is not zero at $x = 0$; here, $\hat{f}_1(0) = \pi a/2$ and $\hat{f}_1'(x) = -1$. If (8.10.16a) is used to verify the transform, the calculation is

$$f_1(r) = \frac{2}{\pi} \frac{\pi a}{2} + \frac{2r}{\pi} \int_0^r \frac{-dx}{(r^2 - x^2)^{\frac{1}{2}}} = a - r.$$

Verification of this inverse for (8.10.15a) follows by using the appropriate integral formulas from Appendix A, and application of the derivative with respect to r:

$$\frac{2}{\pi} \frac{d}{dr} \int_0^r \frac{\frac{1}{2}\pi ax - x^2}{(r^2 - x^2)^{\frac{1}{2}}}\, dx = a - r.$$

From (8.10.18c) we know the transform

$$\mathcal{A}_4\{r^{-1}(a - r)\} = \pi a - 2x.$$

Inversion formulas (8.10.15d) and (8.10.16d) apply for this case.   ∎

Example 5

It is instructive to apply inversion formulas (8.10.15c), (8.10.16c), and (8.10.17) to the same problem. From Appendix B, we use

$$\mathcal{A}_3\{\chi(r/a)\} = 2(a^2 - x^2)^{\frac{1}{2}}\chi(x/a).$$

Application of (8.10.15c) gives

$$-\frac{1}{\pi r}\frac{d}{dr}\int_r^a \frac{2x(a^2 - x^2)^{\frac{1}{2}}\,dx}{(x^2 - r^2)^{\frac{1}{2}}} = -\frac{2}{\pi r}\frac{d}{dr}\int_r^a \frac{x(a^2 - x^2)\,dx}{(a^2 - x^2)^{\frac{1}{2}}(x^2 - r^2)^{\frac{1}{2}}}$$

$$= -\frac{2}{\pi r}\frac{d}{dr}\int_r^a \frac{a^2 x\,dx}{(a^2 - x^2)^{\frac{1}{2}}(x^2 - r^2)^{\frac{1}{2}}}$$

$$+ \frac{2}{\pi r}\frac{d}{dr}\int_r^a \frac{x^3\,dx}{(a^2 - x^2)^{\frac{1}{2}}(x^2 - r^2)^{\frac{1}{2}}}$$

$$= -\frac{2}{\pi r}\frac{d}{dr}\left(\frac{a^2\pi}{2}\right) + \frac{2}{\pi r}\frac{d}{dr}\left(a^2 + r^2\right)\frac{\pi}{4} = 0 + 1 = 1.$$

Application of (8.10.16c) gives

$$-\frac{1}{\pi}\int_r^a \frac{-2x\,dx}{(a^2 - x^2)^{\frac{1}{2}}(x^2 - r^2)^{\frac{1}{2}}} = \frac{2}{\pi}\int_r^a \frac{x\,dx}{(a^2 - x^2)^{\frac{1}{2}}(x^2 - r^2)^{\frac{1}{2}}} = 1.$$

Application of (8.10.17) gives

$$-\frac{1}{\pi}\frac{d}{dr}\int_r^a \frac{2r(a^2 - x^2)^{\frac{1}{2}}\,dx}{x(x^2 - r^2)^{\frac{1}{2}}} = -\frac{2}{\pi}\frac{d}{dr}\int_r^a \frac{r(a^2 - x^2)\,dx}{x(a^2 - x^2)^{\frac{1}{2}}(x^2 - r^2)^{\frac{1}{2}}}$$

$$= -\frac{2}{\pi}\frac{d}{dr}\int_r^a \frac{ra^2\,dx}{x(a^2 - x^2)^{\frac{1}{2}}(x^2 - r^2)^{\frac{1}{2}}}$$

$$+ \frac{2}{\pi}\frac{d}{dr}\int_r^a \frac{rx\,dx}{(a^2 - x^2)^{\frac{1}{2}}(x^2 - r^2)^{\frac{1}{2}}}$$

$$= -\frac{2}{\pi}\frac{d}{dr}(a^2 r)\left(\frac{\pi}{2ar}\right) + \frac{2}{\pi}\frac{d}{dr}\left(\frac{r\pi}{2}\right) = 0 + 1 = 1.$$

Evaluation of the various integrals above follows from material in Appendix A. ∎

Example 6

The following Bessel function identities are used in this example.

$$\frac{\partial}{\partial x}\{x^\nu J_\nu(bx)\} = bx^\nu J_{\nu-1}(bx). \tag{8.10.21a}$$

$$\frac{\partial}{\partial x}\{x^{-\nu}J_\nu(bx)\} = -bx^{-\nu}J_{\nu+1}(bx). \tag{8.10.21b}$$

It follows from the formulas

$$\frac{\pi}{2}J_0(bx) = \int_0^x \frac{\cos br\,dr}{(x^2 - r^2)^{\frac{1}{2}}}, \qquad \frac{\pi}{2}J_0(bx) = \int_x^\infty \frac{\sin br\,dr}{(r^2 - x^2)^{\frac{1}{2}}},$$

for the Bessel function J_0 that

$$\mathcal{A}_1\{\cos br\} = \frac{\pi}{2}\, J_0(bx),$$

and

$$\mathcal{A}_2\{\sin br\} = \frac{\pi}{2}\, J_0(bx).$$

Differentiation of the previous two expressions with respect to the parameter b yields the formulas

$$\mathcal{A}_1\{r \sin br\} = \frac{\pi x}{2}\, J_1(bx),$$

and

$$\mathcal{A}_2\{r \cos br\} = -\frac{\pi x}{2}\, J_1(bx).$$

From formula (8.10.18e) with $\hat{f}_1(x) = \sin bx$,

$$\frac{2}{\pi}\frac{d}{dt}\mathcal{A}_1\{x \sin bx\} = \frac{2}{\pi}\frac{d}{dt}\left\{\frac{1}{2}\pi t\, J_1(bt)\right\} = bt\, J_0(bt).$$

This means that

$$\mathcal{A}_1^{-1}\{\sin bx\} = bt\, J_0(bt),$$

or equivalently

$$\mathcal{A}_1\{r\, J_0(br)\} = \frac{\sin bx}{b}.$$

And by the same technique, from (8.10.18f)

$$\mathcal{A}_2\{r\, J_0(br)\} = \frac{\cos bx}{b}.$$

From (8.10.18f) with $\hat{f}_2(x) = x^{-1}\sin bx$,

$$\mathcal{A}_2^{-1}\{x^{-1}\sin bx\} = -\frac{2}{\pi}\frac{d}{dt}\mathcal{A}_2\{\sin bx\} = -\frac{2}{\pi}\frac{d}{dt}\left\{\frac{1}{2}\pi\, J_0(bt)\right\} = b\, J_1(bt),$$

or

$$\mathcal{A}_2\{J_1(br)\} = \frac{\sin bx}{bx}.$$

From the formulas developed above for the $\mathcal{A}_2$ transforms and (8.10.18a),

$$\mathcal{A}_3\{\cos br\} = -\pi x\, J_1(bx),$$

$$\mathcal{A}_3\{r^{-1}\sin br\} = \pi\, J_0(bx),$$

$$\mathcal{A}_3\{J_0(br)\} = \frac{2\cos bx}{b},$$

and

$$\mathcal{A}_3\{r^{-1}J_1(br)\} = \frac{2\sin bx}{bx}.$$

∎

Additional formulas similar to those in the previous example are contained in Sneddon [1972] and Gorenflo and Vessella [1991]. These authors also make use of the formulas of this example to make the connection between the Abel transform and the Hankel transform. This connection is also discussed in Section 8.11 in the more general context of the Radon transform.

Example 7

Use the rule in Section 8.10.3 to compute

$$\mathcal{A}_4\{r^{2\nu-2}\}.$$

From item (7) of Table 13.1, Riemann-Liouville fractional integrals, of Erdélyi et al. [1954],

$$\mathcal{A}_4\{r^{2\nu-2}\} = \frac{\sqrt{\pi}\,\Gamma(\nu)}{\Gamma(\nu+\frac{1}{2})}\,x^{2\nu-1}.$$

A special case is provided by $\nu = 2$. This leads to the expression

$$2\int_0^x \frac{r^3\,dr}{(x^2-r^2)^{\frac{1}{2}}} = \frac{4x^3}{3}. \qquad\blacksquare$$

8.11 Related transforms and symmetry, Abel and Hankel

The direct connection of the Radon transform and the Fourier transform is used extensively throughout earlier sections of this chapter. Several other transforms are also related to the Radon transform. Some of these are related by circumstances that involve some type of symmetry. The Abel and Hankel transforms emerge naturally in this context. Other related transforms follow more naturally from considerations of orthogonal function series expansions. In this section some of these relations are explored and examples provided to help illustrate the connections.

8.11.1 *Abel transform*

The Abel transform is closely connected with a generalization of the tautochrone problem. This is the problem of determining a curve through the origin in a vertical plane such that the time required for a particle to slide without friction down the curve to the origin is independent of the starting position. It was the generalization of this problem that led Abel to introduce the subject of integral equations (see Section 8.10). More recent applications of Abel transforms in the area of holography and interferometry with phase objects (of practical importance in aerodynamics, heat and mass transfer, and plasma diagnostics) are discussed by Vest [1979], Schumann, Zürcher, and Cuche [1985], and Ladouceur and Adiga [1987]. A very good description of the relation of the Abel and Radon transform to the problem of determining the refractive index from knowledge of a holographic interferogram is provided by Vest [1979]; in particular, see Chapter 6, where many references to original work are cited. Minerbo and Levy [1969], Sneddon [1972], and Bracewell [1986] also contain useful material on the Abel transform. Many other references are contained in Section 8.10.

Suppose the feature space function $f(x, y)$ is rotationally symmetric and depends only on $(x^2 + y^2)^{1/2}$. Now, knowledge of one set of projections, for any angle ϕ, serves to define the Radon transform for all angles. For simplicity, let $\phi = 0$ in the definition (8.2.5). Then

$\check{f}(p, \phi) = \check{f}(p, 0)$; because there is no dependence on angle there is no loss of generality by writing this as $\check{f}(p)$. With these modifications taken into account the definition becomes

$$\check{f}(p) = \int_{-\infty}^{\infty} \int_{-\infty}^{\infty} f\left(\sqrt{x^2 + y^2}\right) \delta(p - x)\, dx\, dy$$

$$= \int_{-\infty}^{\infty} f\left(\sqrt{p^2 + y^2}\right) dy$$

$$= 2 \int_{0}^{\infty} f\left(\sqrt{p^2 + y^2}\right) dy.$$

Clearly, because p appears only as p^2, the function $\check{f}(p)$ is even and it is sufficient to always choose $p > 0$. A change of variables $r^2 = (p^2 + y^2)$ yields

$$\check{f}(p) = 2 \int_{|p|}^{\infty} \frac{r\, f(r)}{\left(r^2 - p^2\right)^{1/2}}\, dr.$$

This equation is just the defining equation for the Abel transform [Bracewell, 1986], designated by

$$f_A(p) = \mathcal{A}\{f(r)\} = 2 \int_{|p|}^{\infty} \frac{r\, f(r)}{\left(r^2 - p^2\right)^{1/2}}\, dr. \tag{8.11.1}$$

The absolute value can be removed if p is restricted to $p > 0$ and $f_A(p) = f_A(-p)$.

REMARK ABOUT NOTATION   The Abel transform used here is $\mathcal{A}_3$ of Section 8.10; that is, $\mathcal{A} \equiv \mathcal{A}_3$.   ∎

The Abel transform can be inverted by using the Laplace transform, Section 8.10, or by using the Fourier transform [Bracewell, 1986]. For purposes of illustration, the method employed by Barrett [1984] is used here. Equation (8.2.13), with $n = 2$, coupled with the observation that the Radon transform operator $\mathcal{R} = \mathcal{A}$ when $f(x, y)$ has rotational symmetry, becomes

$$\mathcal{F}_1 \mathcal{A} f = \mathcal{F}_2 f. \tag{8.11.2}$$

Moreover, for rotationally symmetric functions, the $\mathcal{F}_2$ operator is just the Hankel transform operator of order zero, $\mathcal{H}_0$. (More on the Hankel transform appears in Section 8.12 and in Chapter 9.) This means that

$$\mathcal{F}_2 f = f_H(q) = 2\pi \int_{0}^{\infty} f(r)\, J_0(2\pi q r)\, r\, dr.$$

From the observation that $\mathcal{F}_2 = \mathcal{H}_0$, and from the reciprocal property of the Hankel transform, $\mathcal{H}_0 = \mathcal{H}_0^{-1}$, we have

$$\mathcal{H}_0 f = \mathcal{F}_1 f_A,$$

or

$$f = \mathcal{H}_0^{-1} \mathcal{F}_1 f_A = \mathcal{H}_0 \mathcal{F}_1 f_A.$$

It follows that the inverse Abel transform operator is given by

$$\mathcal{A}^{-1} = \mathcal{H}_0 \mathcal{F}_1. \tag{8.11.3}$$

From (8.11.3) the first step in finding the inverse Abel transform is to determine the Fourier transform of f_A,

$$\mathcal{F}f_A = \int_{-\infty}^{\infty} f_A(p) \, e^{-i2\pi qp} \, dp = 2 \int_0^{\infty} f_A(p) \cos(2\pi qp) \, dp.$$

The last step follows because $f_A(p)$ is an even function. Integration by parts gives

$$\mathcal{F}f_A = \frac{-1}{\pi q} \int_0^{\infty} f_A'(p) \, \sin(2\pi qp) \, dp,$$

where it is assumed that $f_A(p) \to 0$ as $p \to 0$. The prime means differentiation with respect to p. Now the inverse of (8.11.1) is given by

$$f(r) = 2\pi \int_0^{\infty} dq \, q \, J_0(2\pi qr) \left(\frac{-1}{\pi q} \right) \int_0^{\infty} f_A'(p) \, \sin(2\pi qp) \, dp,$$

or, after simplification and interchanging the order of integration,

$$f(r) = -2 \int_0^{\infty} dp \, f_A'(p) \int_0^{\infty} dq \, \sin(2\pi qp) \, J_0(2\pi qr).$$

The integral over q is tabulated [Gradshteyn, Ryzhik, and Jeffrey, 1994]; it vanishes for $0 < p < r$ and gives

$$\frac{1}{2\pi} \left(p^2 - r^2 \right)^{-1/2} \qquad \text{for } 0 < r < p.$$

Hence the inverse is found from

$$f(r) = -\frac{1}{\pi} \int_r^{\infty} f_A'(p) \left(p^2 - r^2 \right)^{-1/2} dp. \tag{8.11.4}$$

This equation and (8.11.1) are an Abel transform pair. Other forms for the inversion are given in Section 8.10. It may be useful to observe that, for rotationally symmetric functions, if the angle ϕ in the Radon transform is chosen $\phi = 0$, then the p that appears in these formulas is just the same as x, the projection of the radius r on the horizontal axis. For this reason, in many discussions of the Abel transform the variable p used here is replaced by the variable x. This notation is used in Section 8.10 and in Appendix B.

Because the Abel transform is a special case of the Radon transform all of the various basic theorems for the Radon transform apply to the Abel transform. One way to make use of this is to apply the theory of the Radon transform to obtain general results. Then observe that for all rotationally symmetric functions the same results apply to the Abel transform. Some examples of Radon transforms already worked out illustrate the idea.

Example 1
Consider Example 1 in Section 8.5. The feature space function has the required rotational symmetry, so it follows immediately that the corresponding Abel transform is

$$\mathcal{A}\{e^{-r^2}\} = \sqrt{\pi} \, e^{-p^2}. \tag{8.11.5}$$

From Example 7 of that same section, if $\chi(r)$ represents the characteristic function of a unit disk, then

$$\mathcal{A}\{\chi(r)\} = \begin{cases} 2(1 - p^2)^{1/2}, & \text{for } p < 1 \\ 0, & \text{for } p > 1. \end{cases} \tag{8.11.6}$$

Another rotationally symmetric case worked out for the Radon transform is from the last part of Example 2 in Section 8.7. The corresponding Abel transform is

$$\mathcal{A}\{r^2 e^{-r^2}\} = \frac{\sqrt{\pi}}{2}(2p^2 + 1)e^{-p^2}. \tag{8.11.7}$$

∎

Example 2

In some cases it is just as easy to apply the definition of the Abel directly; for example, the transform of $(a^2 + r^2)^{-1}$ is given by

$$\mathcal{A}\left\{(a^2 + r^2)^{-1}\right\} = 2\int_p^\infty \frac{r\,dr}{(r^2 - p^2)^{1/2}(r^2 + a^2)}.$$

The change of variables $z^2 = r^2 + a^2$ leads to a form that is easy to evaluate see (Appendix A),

$$\mathcal{A}\left\{(a^2 + r^2)^{-1}\right\} = \frac{\pi}{\left(p^2 + a^2\right)^{1/2}}. \tag{8.11.8}$$

∎

Example 3

Suppose the desired transform is of $(1 - r^2)^{1/2}$ restricted to the unit disk, or

$$f(r) = (1 - r^2)^{1/2}\chi(r).$$

One way to do this is to find the Radon transform of this function, and identify the result with the Abel transform. From the definition of the Radon transform, taking $\phi = 0$, and restricting the integral to the unit disk D,

$$\check{f}(r, \phi) = \int_D \left(1 - x^2 - y^2\right)^{1/2}\delta(p - x)\,dx\,dy.$$

The integral over x is easy using the delta function, and the remaining integral over y is accomplished by observing that over the unit disk $y^2 + p^2 = 1$, thus

$$\check{f} = \int_{-\sqrt{1-p^2}}^{\sqrt{1-p^2}} \left[(1 - p^2) - y^2\right]^{1/2}\,dy.$$

This integral can be evaluated by use of trigonometric substitution or from integral tables (Appendix A). The result is the Abel transform

$$\check{f} = \mathcal{A}\left\{(1 - r^2)^{1/2}\chi(r)\right\} = \frac{\pi}{2}\left(1 - p^2\right)\chi(p). \tag{8.11.9}$$

Now suppose it is desired to scale this result to a disk of radius a. The scaling can be accomplished by application of Section 8.3.2 in the form

$$\mathcal{R}f\left(\frac{x}{a}, \frac{y}{a}\right) = a^2\check{f}(p, a\,\xi) = a\check{f}\left(\frac{p}{a}, \xi\right).$$

The scaled Abel transform follows, with $r \to r/a$,

$$\mathcal{A}\left\{\left(1 - \frac{r^2}{a^2}\right)^{1/2}\chi\left(\frac{r}{a}\right)\right\} = \frac{\pi a}{2}\left(1 - \frac{p^2}{a^2}\right)\chi\left(\frac{p}{a}\right),$$

or

$$A\left\{(a^2 - r^2)^{1/2}\,\chi\left(\frac{r}{a}\right)\right\} = \frac{\pi}{2}\,(a^2 - p^2)\,\chi\left(\frac{p}{a}\right).\qquad(8.11.10)$$

▌

By following the approach used in the last example, it is possible to find a whole class of Abel transforms. These are listed in Appendix B. More results for Abel transforms appear in sections that follow, especially in the section on transforms restricted to the unit disk.

8.11.2 *Hankel transform*

See Chapter 9 for details about Hankel transforms. By using an approach similar to that in Section 8.11.1 it is possible to find the connection between the Hankel transform of order ν and the Radon transform. Note that throughout this discussion, if $\nu = 0$ the results here correspond to results for the Abel transform. Let the feature space function be given by a rotationally symmetric function multiplied by $e^{i\nu\theta}$,

$$f(x, y) = f(r)\,e^{i\nu\theta}.$$

The polar form of the two-dimensional Fourier transform is given by

$$\tilde{f}(q, \phi) = \int_0^{2\pi} \int_0^\infty e^{i\nu\theta}\, e^{-i2\pi qr\cos(\theta - \phi)}\, r\, f(r)\, dr\, d\theta.$$

Now, after the change of variables $\beta = \theta - \phi$, followed by an interchange of the order of integration,

$$\tilde{f}(q, \phi) = e^{i\nu\phi} \int_0^\infty dr\, rf(r) \int_0^{2\pi} d\beta\, e^{i(\nu\beta - 2\pi qr\cos\beta)}.$$

The integral over β can be related to a Bessel function identity from Appendix A to yield

$$\tilde{f}(q, \phi) = 2\pi\, e^{i\nu\phi}\, e^{-i\nu\pi/2} \int_0^\infty f(r)\, J_\nu(2\pi qr)\, r\, dr.$$

This is where the Hankel transform of order ν comes in, by definition,

$$\mathcal{H}_\nu\{f(r)\} = 2\pi \int_0^\infty f(r)\, J_\nu(2\pi qr)\, r\, dr.\qquad(8.11.11)$$

Thus,

$$\tilde{f}(q, \phi) = (-i)^\nu\, e^{i\nu\phi}\, \mathcal{H}_\nu\{f(r)\}.\qquad(8.11.12)$$

This equation can be related to the Radon transform by first finding the Radon transform of f, and then applying the Fourier transform as indicated in (8.2.13). In polar form,

$$\check{f}(p, \phi) = \int_0^{2\pi} \int_0^\infty e^{i\nu\theta} f(r)\, \delta[p - r\cos(\theta - \phi)]\, r\, dr\, d\theta.$$

Once again, the change of variables $\beta = \theta - \phi$ is employed to obtain

$$\check{f}(p, \phi) = e^{i\nu\phi} \int_0^\infty dr\, rf(r) \int_0^{2\pi} d\beta\, e^{i\nu\beta}\, \delta(p - r\cos\beta).$$

The integration over β in this expression has been discussed by many authors, including Cormack [1963, 1964] and Barrett [1984], where details can be found leading to

$$\check{f}(p, \phi) = 2 e^{i\nu\phi} \int_{|p|}^{\infty} f(r) \, T_\nu \left(\frac{p}{r}\right) \left(1 - \frac{p^2}{r^2}\right)^{-1/2} dr. \qquad (8.11.13)$$

Some of the more useful properties of the Tchebycheff polynomials of the first kind T_ν are given in Appendix A. For more details see the summary by Arfken [1985] and the interesting discussion by Van der Pol and Weijers [1934].

It is useful to identify a Tchebycheff transform by

$$\mathcal{T}_\nu\{f(r)\} = 2 \int_{|p|}^{\infty} f(r) \, T_\nu \left(\frac{p}{r}\right) \left(1 - \frac{p^2}{r^2}\right)^{-1/2} dr. \qquad (8.11.14)$$

Then,

$$\check{f}(p, \phi) = e^{i\nu\phi} \mathcal{T}_\nu\{f(r)\}.$$

The Fourier transform of (8.11.13) must be equal to (8.11.12). It follows that the Hankel transform is given in terms of the Radon transform by

$$(-i)^\nu \, e^{i\nu\phi} \, \mathcal{H}_\nu\{f(r)\} = \mathcal{F} \mathcal{R} f = e^{i\nu\phi} \mathcal{T}_\nu\{f(r)\}. \qquad (8.11.15)$$

Or, in terms of the Tchebycheff transform, because the $e^{i\nu\phi}$ term cancels,

$$\mathcal{H}_\nu\{f(r)\} = i^\nu \mathcal{F} \mathcal{T}_\nu\{f(r)\}. \qquad (8.11.16)$$

Note that an operator identity follows immediately,

$$\mathcal{H}_\nu = i^\nu \mathcal{F} \mathcal{T}_\nu. \qquad (8.11.17)$$

This relation between the Hankel transform and the Fourier transform of the Radon transform is a useful expression because it serves as the starting point for finding Hankel transforms without having to do integrals over Bessel functions. Several authors have made contributions in this area. For applications and references to the literature see Hansen [1985], Higgins and Munson [1987, 1988] and Suter [1991].

In this section we have concentrated on how the Hankel transform relates to the Radon transform. A logical extension of some of the ideas presented in this discussion appear in Section 8.13 on circular harmonic decomposition.

8.11.3 *Spherical symmetry, three dimensions*

An interesting generalization of the above cases arises when the function $f(x, y, z)$ has spherical symmetry. In this case the Radon transform of f can be found by letting both the polar angle θ and the azimuthal angle ϕ be zero. Now the unit vector $\boldsymbol{\xi} = (0, 0, 1)$, and formula (8.2.7) is given by

$$\check{f}(p) = \int_{-\infty}^{\infty} \int_{-\infty}^{\infty} \int_{-\infty}^{\infty} f\left(\sqrt{x^2 + y^2 + z^2}\right) \delta(p - z) \, dx \, dy \, dz$$

$$= \int_{-\infty}^{\infty} \int_{-\infty}^{\infty} f\left(\sqrt{x^2 + y^2 + p^2}\right) dx \, dy$$

$$= \int_{0}^{2\pi} \int_{0}^{\infty} f\left(\sqrt{\rho^2 + p^2}\right) \rho \, d\rho \, d\phi$$

$$= 2\pi \int_0^\infty f\left(\sqrt{\rho^2 + p^2}\right) \rho \, d\rho.$$

In these equations the transformation $x = \rho \cos \phi$, $y = \rho \cos \phi$ is used. One more transformation, $\rho^2 + p^2 = r^2$, leads to

$$\check{f}(p) = 2\pi \int_p^\infty f(r) r \, dr, \qquad p > 0. \tag{8.11.18}$$

Note that the lower limit follows from $r = (p^2)^{1/2}$ when $\rho = 0$. The interesting point is that for this highly symmetric case, the original function f can be found by differentiation,

$$\frac{d\check{f}(p)}{dp} = -2\pi p f(p).$$

In this equation the variable p is actually a dummy variable, and it can be replaced by r,

$$f(r) = \frac{-1}{2\pi r} \check{f}'(r). \tag{8.11.19}$$

This same result can be found directly from the inversion methods of Section 8.9.2. Also, Barrett [1984] does the same derivation and he makes the interesting observation that (8.11.19) was given in the optics literature by Vest and Steel [1978], but was actually known much earlier by Du Mond [1929] in connection with Compton scattering, and by Stewart [1957] and Mijnarends [1967] in connection with positron annihilation.

8.12   Methods of inversion

The inversion formulas given by Radon [1917] and the formulas given in Section 8.9 serve only as a beginning for an applied problem. This point is emphasized by Shepp and Kruskal [1978]. The main problem is that these formulas are rigorously valid for an infinite number of projections, and in practical situations the projections are a discrete set. This discrete nature of the projections gives rise to subtle and difficult questions. Most of these are related in some way to the "Indeterminacy Theorem" by Smith, Solmon, and Wagner [1977]. After a little rephrasing the theorem establishes that: **A function $f(x, y)$ with compact support is uniquely determined by an infinite set of projections, but not by any finite set of projections.** This clearly means that uniqueness must be sacrificed in applications. Experience with known images shows that this is not so serious if one can come close to the actual $\check{f}$ and then apply an approximate reconstruction algorithm. Moreover, some encouragement comes from another theorem by Hamaker, Smith, Solmon, and Wagner [1980]. The main thrust of this theorem is that arbitrarily good approximations to f can be found by utilization of an arbitrarily large number of projections. Perhaps the way to express all of this is to say: Even though you can't win you must never give up!

There are several other considerations about inversion. The inverse Radon transform is technically an **ill-posed problem**. Small errors in knowledge of the function $\check{f}$ can lead to very large errors in the reconstructed function f. Hence problems of stability, ill-posedness, accuracy, resolution, and optimal methods of sampling must be addressed when working with experimental data. These are obviously very important problems, and the subject of ongoing research. A thorough discussion would have to be highly technical and inappropriate for inclusion here. For those concerned with these matters, the papers by Lindgren and Rattey [1981], Rattey and Lindgren [1981], Louis [1984], Madych and Nelson [1986], Hawkins, and Barrett [1986], Hawkins, Leichner, and Yang [1988], Kruse [1989], Madych [1990], Faridani [1991], Faridani, Ritman, and Smith [1992], Maass [1992], Desbat [1993], Natterer [1993], Olson and DeStefano [1994], and the books by Herman [1980] and Natterer [1986] are good starting points for methods and references to other important work. Good examples illustrating many of the difficulties encountered when dealing with real data along with defects in the reconstructed image associated with the performance of various algorithms are given in Chapter 7 of the book by Russ [1992].

There are several methods that serve as the basis for the development of algorithms that can be viewed as discrete implementations of the inversion formula. Our purpose here is to present several of these along with reference to their implementation. Those interested in more detail and other flow charts may want to see Barrett and Swindell [1977] and Deans [1983, 1993]. The first topic below, the operation of **backprojection**, is an essential step in some of the reconstruction algorithms. Also, this operation is closely related to the adjoint of the Radon transform, discussed in Section 8.14. More on inversion methods is contained in Section 8.13 on series.

8.12.1   Backprojection

Let $G(p, \phi)$ be an arbitrary function of a radial variable p and angle ϕ. The backprojection operation is defined by replacing p by $x \cos \phi + y \sin \phi$ and integrating over the angle ϕ, to obtain a function of x and y,

$$g(x, y) = \mathcal{B}\, G(p, \phi) = \int_0^\pi G(x \cos \phi + y \sin \phi, \phi)\, d\phi. \tag{8.12.1}$$

Note: From the definition of the backprojection operator it follows that the inversion formula (8.9.10) can be written as

$$f(x, y) = 2\, \mathcal{B}\, \bar{\bar{f}}(t, \phi). \tag{8.12.2}$$

8.12.2   Backprojection of the filtered projections

The algorithm known as the **filtered backprojection algorithm** is presently the optimum computational method for reconstructing a function from knowledge of its projections. This algorithm can be considered as an approximate method for computer implementation of the inversion formula for the Radon transform. Unfortunately, there is some confusion associated with the name, because the filtering of the projections is done **before** the backprojection operation. Hence a better name is the one chosen for the title of this section. There are several ways to derive the basic formula for this algorithm. Because we want to emphasize its relation to the inversion formula, the starting point is (8.12.2). First, rewrite that equation as

$$f = 2\, \mathcal{B}\mathcal{F}^{-1}\mathcal{F}\, \bar{\bar{f}}. \tag{8.12.3}$$

Here, the identity operator for the 1D Fourier transform is used. Now, making use of various operations from Section 8.9.1, we obtain

$$f = \frac{2\mathcal{B}}{4\pi^2}\mathcal{F}^{-1}\left\{ \mathcal{F}\frac{\partial}{\partial p}\left[\frac{1}{p} * \check{f}(p, \phi) \right] \right\}$$

$$= \frac{\mathcal{B}}{2\pi^2}\mathcal{F}^{-1}\left\{ (i2\pi k)\left[\mathcal{F}\left(\frac{1}{p} \right) \right]\left[\mathcal{F}\check{f}(p, \phi) \right] \right\}$$

$$= \frac{\mathcal{B}}{2\pi^2}\mathcal{F}^{-1}\left\{ (i2\pi k)(i\pi \,\mathrm{sgn}\, k)\mathcal{F}\check{f}(p, \phi) \right\}$$

$$= \mathcal{B}\mathcal{F}^{-1}\left\{|k|\mathcal{F}\check{f}(p,\phi)\right\}. \tag{8.12.4}$$

The inverse Fourier transform operation converts a function of k to a function of some other radial variable, say s. This observation leads to a natural definition; for convenience of notation, define

$$F(s,\phi) = \mathcal{F}^{-1}\left\{|k|\mathcal{F}\check{f}(p,\phi)\right\} = \mathcal{F}^{-1}\left\{|k|\tilde{\check{f}}(k,\phi)\right\}. \tag{8.12.5}$$

Now the feature space function is recovered by backprojection of F,

$$f(x,y) = \mathcal{B}\, F(s,\phi) = \int_0^\pi F(x\cos\phi + y\sin\phi, \phi)\, d\phi. \tag{8.12.6}$$

The beautiful part of this formula is that the need to use the Hilbert transform has been eliminated. From a computational viewpoint this is a real plus. For additional information on computationally efficient algorithms based on these equations, see Rowland [1979] and Lewitt [1983].

Convolution methods

Due to the presence of the $|k|$ in (8.12.5) the story is not over. This causes a problem with numerical implementation due to the behavior for large values of k. It would be desirable to have a well-behaved function, say g, such that $\mathcal{F}g = |k|$. Then (8.12.5) could be modified to read

$$F(s,\phi) = \mathcal{F}^{-1}\left[(\mathcal{F}g)(\mathcal{F}\check{f})\right].$$

And, by the convolution theorem,

$$F(s,\phi) = \check{f} * g = \int_{-\infty}^{\infty} \check{f}(p,\phi)g(s-p)\, dp. \tag{8.12.7}$$

A function g such that $\mathcal{F}g = |k|$ can be found, but is not well behaved. In fact, it is a singular distribution [Lighthill, 1962]. In view of these difficulties a slight compromise is in order. Rather than looking for a function whose Fourier transform equals $|k|$, try to find a well-behaved function with a Fourier transform that approximates $|k|$. The usual approach is to define a filter function in terms of a window function; that is, let

$$\mathcal{F}g = |k|w(k). \tag{8.12.8}$$

Then (8.12.7) can be used to find the function F used in the backprojection equation. One advantage of this approach is that there is no need to find the Fourier transform of the projection data $\check{f}$; however, it is necessary to compute the convolver function

$$g(s) = \mathcal{F}^{-1}\left\{|k|w(k)\right\} \tag{8.12.9}$$

before implementing (8.12.7). This signal space convolution approach is discussed in some detail by Rowland [1979]. An approach directly aimed toward computer implementation is in Rosenfeld and Kak [1982]. Excellent practical discussions of windows and filters are given by Harris [1978] and by Embree and Kimble [1991].

Frequency space implementation

It should be noted that there are times when it is desirable to implement the filter in Fourier space and use (8.12.5) in the form

$$F(s,\phi) = \mathcal{F}^{-1}\left\{|k|w(k)\mathcal{F}\check{f}(p,\phi)\right\} = \mathcal{F}^{-1}\left\{|k|w(k)\tilde{\check{f}}(k,\phi)\right\}, \tag{8.12.10}$$

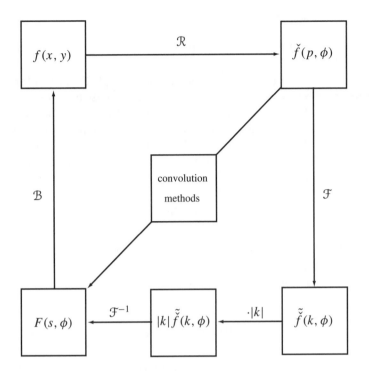

FIGURE 8.10
Filtered backprojection, convolution.

to approximate F before backprojecting. This has been emphasized by Budinger, Gullberg, and Huesman [1979] for data where noise is an important consideration.

A diagram of the options associated with the algorithm for backprojection of the filtered projections is given in Figure 8.0.

8.12.3 *Filter of the backprojections*

In this approach to reconstruction the backprojection operation is applied first and the filtering or convolution comes last. When the backprojection operator is applied to the projections the result is a blurred image that is related to the true image by a two-dimensional convolution with $1/r$. Let this blurred image of the backprojected projections be designated by

$$b(x, y) = \mathcal{B}\check{f}(p, \phi)$$

$$= \int_0^\pi \check{f}(x \cos \phi + y \sin \phi, \phi) \, d\phi. \tag{8.12.11}$$

The true image is related to b by

$$b(x, y) = f(x, y) ** \frac{1}{r} = \int_{-\infty}^\infty \int_{-\infty}^\infty \frac{f(x', y') \, dx' \, dy'}{\left[(x - x')^2 + (y - y')^2\right]^{1/2}}. \tag{8.12.12}$$

This is not an obvious result; it can be deduced by considering (8.2.13) in the form

$$\check{f} = \mathcal{F}_1^{-1} \mathcal{F}_2 f.$$

Apply the backprojection operator to obtain

$$b = \mathcal{B}\check{f} = \mathcal{B}\check{f} = \mathcal{B}\mathcal{F}_1^{-1}\mathcal{F}_2 f. \tag{8.12.13}$$

(In this section subscripts on the Fourier transform operator are shown explicitly to avoid any possible confusion.) There is a subtle point lurking in this equation. Suppose the 2D Fourier transform of f produces $\tilde{f}(u, v)$. The inverse 1D operator $\mathcal{F}_1^{-1}$ is understood to operate on a radial variable in Fourier space. This means $\tilde{f}(u, v)$ must be converted to polar form, say $\tilde{f}(q, \phi)$ before doing the inverse 1D Fourier transform. The variable q is the radial variable in Fourier space, $q^2 = u^2 + v^2$. If we designate the inverse 1D Fourier transform of $\tilde{f}(q, \phi)$ by $f(s, \phi)$, then

$$b(x, y) = \mathcal{B}f(s, \phi) = \mathcal{B}\int_{-\infty}^{\infty} \tilde{f}(q, \phi)e^{i2\pi sq}\, dq.$$

Explicitly, the backprojection operation with $s \rightarrow x\cos\phi + y\sin\phi$ gives

$$b(x, y) = \int_0^{\pi} \int_{-\infty}^{\infty} dq\, \tilde{f}(q, \phi)\, e^{i2\pi q(x\cos\phi + y\sin\phi)}$$

$$= \int_0^{2\pi} \int_0^{\infty} q^{-1}\tilde{f}(q, \phi)\, e^{i2\pi qr\cos(\theta - \phi)}\, q\, dq\, d\phi,$$

where the replacements $x = r\cos\theta$ and $y = r\sin\theta$ have been made, and the radial integral is over positive values of q. We observe that the expression on the right is just the inverse 2D Fourier transform,

$$b(x, y) = \mathcal{F}_2^{-1}\left\{|q|^{-1}\,\tilde{f}\right\}, \tag{8.12.14}$$

and from the convolution theorem

$$b(x, y) = \left[\mathcal{F}_2^{-1}\left\{\tilde{f}\right\}\right] * * \left[\mathcal{F}_2^{-1}\left\{|q|^{-1}\right\}\right]. \tag{8.12.15}$$

The last term on the right is just the Hankel transform of $|q|^{-1}$ that gives $|r|^{-1}$, and the other term yields $f(x, y)$. These substitutions immediately verify (8.12.12).

The desired algorithm follows by taking the 2D Fourier transform of (8.12.14),

$$\mathcal{F}_2\, b(x, y) = |q|^{-1}\tilde{f}(u, v),$$

or

$$\tilde{f}(u, v) = |q|\,\mathcal{F}_2\, b.$$

Application of $\mathcal{F}_2^{-1}$ to both sides of this equation, along with the replacement $b = \mathcal{B}\check{f}$, yields the basic reconstruction formula for filter of the backprojected projections.

$$f(x, y) = \mathcal{F}_2^{-1}\left\{|q|\,\mathcal{F}_2\,\mathcal{B}\check{f}\right\}. \tag{8.12.16}$$

Just as in the previous section a window function can be introduced, but this time it must be a 2D function. Let

$$\tilde{g}(u, v) = |q|w(u, v).$$

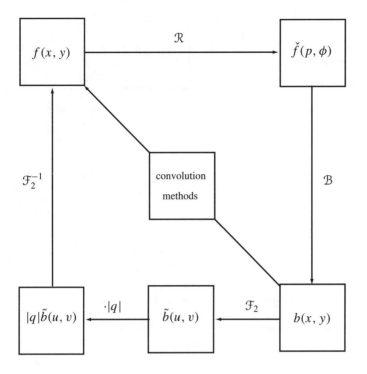

FIGURE 8.11
Filter of backprojections and convolution, $q = \sqrt{u^2 + v^2}$.

Now (8.12.16) becomes

$$f(x, y) = \mathcal{F}_2^{-1} \left\{ \tilde{g}\, \mathcal{F}_2\, \mathcal{B}\, \check{f} \right\}$$

$$= \left[\mathcal{F}_2^{-1}\{\tilde{g}\}\right] ** \left[\mathcal{B}\check{f}\right]$$

$$= g(x, y) ** b(x, y). \tag{8.12.17}$$

Once the window function is selected, g can be found in advance by calculating the inverse 2D Fourier transform, and the reconstruction is accomplished by a 2D convolution with the backprojection of the projections.

Options for implementation of these results are illustrated in Figure 8.11. Important references for applications and numerical implementation of this algorithm are Bates and Peters [1971], Smith, Peters, and Bates [1973], Gullberg [1979], and Budinger, Gullberg, and Huesman [1979].

8.12.4 *Direct Fourier method*

The direct Fourier method follows immediately from the central-slice theorem, Section 8.2.5, in the form

$$f = \mathcal{F}_2^{-1} \mathcal{F}_1 \check{f}. \tag{8.12.18}$$

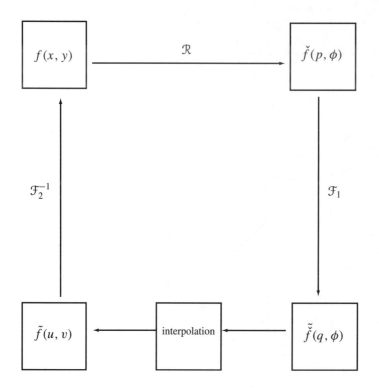

FIGURE 8.12
Direct Fourier method.

The important point is that the 1D Fourier transform of the projections produces $\tilde{f}(q, \phi)$ defined on a polar grid in Fourier space. An interpolation is needed to get $\tilde{f}(u, v)$ and then apply $\mathcal{F}_2^{-1}$ to recover $f(x, y)$. The procedure is illustrated in Figure 8.12. Although this appears to be the simplest inversion algorithm, it turns out that there are computational problems associated with the interpolation and there is a need to do a 2D inverse Fourier transform. For a detailed discussion see: Mersereau [1976], Stark, Woods, Paul, and Hingorani [1981], and Sezan and Stark [1984].

8.12.5 *Iterative and algebraic reconstruction techniques*

The so-called algebraic reconstruction techniques (ART) form a large family of reconstruction algorithms. They are iterative procedures that vary depending on how the discretization is performed. There is a high computational cost associated with ART, but there are some advantages too: Standard numerical analysis techniques can be applied to a wide range of problems and ray configurations, and a priori information can be incorporated in the solution. Details about various methods, the history, and extensive reference to original work is provided by Herman [1980], Rosenfeld and Kak [1982], and Natterer [1986]. Also, the discrete Radon transform and its inversion is described by Beylkin [1987] and Kelly and Madisetti [1993], where both the forward and inverse transforms are implemented using standard methods of linear algebra.

8.13 Series

There are many series approaches to finding an approximation to the original feature space function f when given sufficient information about the corresponding function $\check{f}$ in Radon space. The particular method selected usually depends on the physical situation and the quality of the data. The purpose of this section is to present some of the more useful approaches, and observe that the basic ideas developed here carry over to other series techniques not discussed.

The approach is to give details for some of the 2D cases and quote results and references for higher dimensional cases. The first method discussed, the circular harmonic expansion, is the method used by Cormack [1963, 1964] in his now famous work that many regard as the beginning of modern computed tomography.

8.13.1 *Circular harmonic decomposition*

The basic ideas developed in Section 8.11.2 can be extended to obtain the major results. First, note that in polar coordinates in feature space, functions that represent physical situations are periodic with period 2π. This immediately leads to a consideration of expanding the function in a Fourier series. If $f(x, y)$ is written as $f(r, \theta)$, then the decomposition is

$$f(r, \theta) = \sum_l h_l(r) \, e^{il\theta}. \tag{8.13.1}$$

The sum is understood to be from $-\infty$ to ∞, and the Fourier coefficient h_l is given by

$$h_l(r) = \frac{1}{2\pi} \int_0^{2\pi} f(r, \theta) e^{-il\theta} \, d\theta. \tag{8.13.2}$$

The Radon transform of f can also be expanded in a Fourier series of the same form,

$$\check{f}(p, \phi) = \sum_l \check{h}_l(p) \, e^{il\phi}, \tag{8.13.3}$$

where

$$\check{h}_l(p) = \frac{1}{2\pi} \int_0^{2\pi} \check{f}(p, \phi) e^{-il\phi} \, d\phi, \qquad p \geq 0, \tag{8.13.4a}$$

and

$$\check{h}_l(-p) = (-1)^l \, \check{h}_l(p). \tag{8.13.4b}$$

The connection between the Fourier coefficients in the two spaces can be determined by taking the Radon transform of f, as given by (8.13.1). The polar form of the transform gives

$$\check{f}(p, \phi) = \sum_l \int_0^{2\pi} \int_0^\infty e^{il\theta} h_l(r) \, \delta[p - r\cos(\theta - \phi)] r \, dr \, d\theta.$$

Now, the change of variables $\beta = \theta - \phi$ leads to an expression similar to one obtained in Section 8.11.2,

$$\check{f}(p, \phi) = \sum_l e^{il\phi} \int_0^\infty dr \, r h_l(r) \int_0^{2\pi} d\beta \, e^{il\beta} \, \delta(p - r\cos\beta). \tag{8.13.5}$$

From the linear independence of the functions $e^{il\phi}$, it follows by comparison of (8.13.3) and (8.13.5) that

$$\check{h}_l(p) = \int_0^\infty dr \, r h_l(r) \int_0^{2\pi} d\beta \, e^{il\beta} \, \delta(p - r\cos\beta).$$

From (8.11.13) this gives the connection between the Fourier coefficients in terms of a Tchebycheff transform,

$$\check{h}_l(p) = 2 \int_p^\infty h_l(r) \, T_l\left(\frac{p}{r}\right) \left(1 - \frac{p^2}{r^2}\right)^{-1/2} dr, \qquad p \geq 0. \tag{8.13.6a}$$

One form of the inverse is

$$h_l(r) = -\frac{1}{\pi r} \int_r^\infty \check{h}_l'(p) T_l\left(\frac{p}{r}\right) \left(\frac{p^2}{r^2} - 1\right)^{-1/2} dp, \qquad r > 0. \tag{8.13.6b}$$

Here the prime means derivative with respect to p.

The inverse (8.13.6b) can be found by various techniques. These include use of the Mellin transform, contour integration, and orthogonality properties of the Tchebycheff polynomials of the first and second kinds. The method used by Barrett [1984] is easy to follow, and he provides extensive reference to other derivations and some of the subtleties related to the stability and uniqueness of the inverse. The problem with this expression for the inverse is that T_l increases exponentially as $l \to \infty$ and $\check{h}_l$ is a rapidly oscillating function. The integration of the product of these two functions leads to severe cancellations and numerical instability. For a further discussion of stability, uniqueness, and other forms for the inverse, see Hansen [1981], Hawkins and Barrett [1986], and Natterer [1986]. Additional details on the circular harmonic Radon transform are given by Chapman and Cary [1986].

Extension to higher dimensions

The extension to higher dimensions is presented in detail by Ludwig [1966]. Other relevant references include Deans [1978, 1979] and Barrett [1984]. The nD counterpart of the transform pair given by (8.13.6) is a Gegenbauer transform pair for the radial functions,

$$\check{h}_l(p) = \frac{(4\pi)^\nu \Gamma(l+1)\Gamma(\nu)}{\Gamma(l+2\nu)} \int_p^\infty r^{2\nu} h_l(r) C_l^\nu\left(\frac{p}{r}\right) \left(1 - \frac{p^2}{r^2}\right)^{\nu - \frac{1}{2}} dr, \tag{8.13.7a}$$

and

$$h_l(r) = \frac{(-1)^{2\nu+1} \Gamma(l+1)\Gamma(\nu)}{2\pi^{\nu+1} \Gamma(l+2\nu) r} \int_r^\infty \check{h}_l^{(2\nu+1)}(p) C_l^\nu\left(\frac{p}{r}\right) \left(\frac{p^2}{r^2} - 1\right)^{\nu - \frac{1}{2}} dp. \tag{8.13.7b}$$

In these equations, $r \geq 0$, $p \geq 0$, $\check{h}_l^{(2\nu+1)} = (d/dp)^{2\nu+1} \check{h}_l(p)$, $\check{h}_l(-p) = (-1)^l \check{h}_l(p)$, and ν is related to the dimension n by $\nu = (n-2)/2$.

The Gegenbauer polynomials C_l^ν are orthogonal over the interval $[-1, +1]$ [Rainville, 1960] and [Szegö, 1939]. This leads to questions about the integration in (8.13.7b). And, just as mentioned in connection with (8.13.6b), this formula is not practical for numerical implementation. However, the integral can be understood because it is possible to define Gegenbauer functions $G_l^\nu(z)$ analytic in the complex z plane cut from -1 to $-\infty$. For a discussion and proofs, see Durand, Fishbane, and Simmons [1976].

Three dimensions

The 3D version of the expansion (8.13.1) is in terms of the real orthonormal spherical harmonics $S_{lm}(\omega)$, discussed by Hochstadt [1971],

$$f(r, \omega) = \sum_{l,m} A_{lm} h_l(r) S_{lm}(\omega). \tag{8.13.8}$$

The A_{lm} are real constants, and ω is a 3D unit vector,

$$\omega = (\sin\theta\cos\phi, \sin\theta\sin\phi, \cos\theta).$$

The corresponding expansion in Radon space is

$$\check{f}(p, \xi) = \sum_{l,m} A_{lm}\check{h}_l(r)S_{lm}(\xi). \tag{8.13.9}$$

It follows from the orthogonality of the spherical harmonics that

$$A_{lm}\check{h}_l(p) = \int_{|\xi|=1} \check{f}(p, \xi)S_{lm}(\xi)\,d\xi, \tag{8.13.10}$$

where $d\xi$ is the surface element on a unit sphere. The Gegenbauer transform equation (8.13.7) reduces to a Legendre transform for $n = 3$, $\nu = \frac{1}{2}$, and the radial functions satisfy

$$\check{h}_l(p) = 2\pi \int_p^\infty r h_l(r) P_l\left(\frac{p}{r}\right) dr, \tag{8.13.11a}$$

$$h_l(r) = \frac{1}{2\pi r} \int_r^\infty \check{h}_l''(p) P_l\left(\frac{p}{r}\right) dp. \tag{8.13.11b}$$

The spherical harmonics $Y_{lm}(\theta, \phi)$, discussed by Arfken [1985], are probably more familiar to engineers and physicists. These can be used in place of the S_{lm} suggested here. However, various properties (real, orthonormal, symmetry) of the S_{lm} make them more suitable for use in connection with problems involving the general nD Radon transform [Ludwig, 1966].

For the 3D case one possible connection is given by

$$S_{lm} = \begin{cases} \dfrac{Y_{lm} + Y_{lm}^*}{\sqrt{2}}, & \text{for } m = 1, 2, \ldots, l \\[2mm] Y_{l0}, & \text{for } m = 0 \\[2mm] \dfrac{Y_{lm} - Y_{lm}^*}{i\sqrt{2}}, & \text{for } m = -1, -2, \ldots, -l, \end{cases}$$

where $Y_{l,-m} = (-1)^m Y_{lm}^*$. Note that under the parity operation

$$(x \to -x, \ y \to -y \ z \to -z),$$

the well known result $Y_{lm} \to (-1)^l Y_{lm}$ carries over to the S_{lm}, giving

$$S_{lm}(-\omega) = (-1)^l S_{lm}(\omega).$$

8.13.2   *Orthogonal functions on the unit disk*

In most practical reconstruction problems the function in feature space is confined to a finite region. This region can always be scaled to fit inside a unit disk. Hence, the development of an orthogonal function expansion on the unit disk holds promise as a useful approach for inversion using series methods. (In this connection, note that when the problem is confined to the unit disk the infinite upper limit on all integrals in the previous section can be replaced by unity.) Orthogonal polynomials that have been used for many years in optics are especially good candidates. These are the Zernike polynomials; a standard reference is Born and Wolf [1975]; also see Chapter 1. A more recent reference, Kim and Shannon [1987], contains a graphic library of 37 selected Zernike expansion terms. One reason why these functions are desirable is that their transforms ($\mathcal{R}$ and $\mathcal{F}$) lead to orthogonal function expansions in **both** Radon and Fourier space. This choice for basis functions in reconstruction has been discussed by Cormack [1964], Marr [1974], Zeitler [1974], and Hawkins and Barrett [1986], and examples similar to those given here are given by Deans [1983, 1993].

The approach is to assume that $f(x, y)$ can be approximated by a sum of monomials of the form $x^k y^j$. Then $x^k y^j$ can be written as r^{k+j} multiplied by some function of $\sin \theta$ and $\cos \theta$. This leads to the consideration of an expansion of the form

$$f(r, \theta) = \sum_{l=-\infty}^{\infty} h_l(r)\, e^{il\theta} = \sum_{s=0}^{\infty} \sum_{l=-\infty}^{\infty} A_{ls} Z^{|l|}_{|l|+2s}(r) e^{il\theta}, \qquad (8.13.12)$$

in terms of complex constants A_{ls} and Zernike polynomials $Z_m^l(r)$, with $m = |l| + 2s$. The Radon transform of this expression can be found exactly, and it contains the same constants. These constants are evaluated in Radon space, and the feature space function is found by the expansion (8.13.12). There are several subtle points associated with this process, and it is useful to break the problem into separate parts. First, we discuss relevant properties of the Zernike polynomials, and give some simple examples. This is followed with the transform to Radon space, and more examples. Next, the expression for the constants A_{ls} is found in terms of $\check{f}$, which is assumed known from experiment. Finally, to emphasize that this application also extends to Fourier space, the transform to Fourier space is illustrated, along with some observations regarding three different orthonormal basis sets.

Zernike polynomials

The Zernike polynomials (see Chapter 1, Section 5) can be found by orthogonalizing the powers

$$r^l, \ r^{l+2}, \ r^{l+4}, \ \cdots$$

with weight function r over the interval $[0, 1]$. The exponent l is a nonnegative integer. The resulting polynomial $Z_m^l(r)$ is of degree $m = l + 2s$ and it contains no powers of r less than l. The polynomials are even if l is even and odd if l is odd. This leads to an important symmetry relation,

$$Z_m^l(-r) = (-1)^l\, Z_m^l(r). \qquad (8.13.13)$$

The orthogonality condition is given by

$$\int_0^1 Z_{l+2s}^l(r)\, Z_{l+2t}^l(r)\, r\, dr = \frac{1}{2(l + 2s + 1)} \delta_{st}. \qquad (8.13.14)$$

It follows that the expansion coefficients are given by

$$A_{ls} = \frac{2(l + 2s + 1)}{2\pi} \int_0^{2\pi} \int_0^1 f(r \cos \theta, r \sin \phi) Z_{l+2s}^l(r)\, e^{-il\theta} r\, dr\, d\theta. \qquad (8.13.15a)$$

In this equation $l \geq 0$. To find the expansion coefficient for negative values of l, use the complex conjugate,

$$A_{-l,s} = A_{ls}^*. \qquad (8.13.15b)$$

Some simple examples are useful to gain an understanding of just how the expansion works. A short table of Zernike polynomials is given in Appendix A. Methods for extending the table and many other properties are given by Born and Wolf [1975].

Example 1

Let the feature space function be given by $f(x, y) = y$ in the unit circle and zero outside the circle. Thus, in terms of r,

$$f(x, y) = r \sin \theta.$$

Here the degree is 1 and $|l| + 2s \leq 1$. The series expansion (8.13.12) reduces to

$$f(x, y) = A_{00} Z_0^0 + A_{10} Z_1^1 e^{i\theta} + A_{-10} Z_1^1 e^{-i\theta}.$$

This case is easy enough to do by inspection of the table of Zernike polynomials in Appendix A. The coefficients are $A_{00} = 0$, $A_{10} = \frac{1}{2i}$, $A_{-10} = -\frac{1}{2i}$. This choice gives

$$f(x, y) = r \frac{e^{i\theta} - e^{-i\theta}}{2i} = Z_1^1(r) \sin \theta. \qquad \blacksquare$$

Example 2

This time let $f(x, y) = xy$, so $f(r, \theta) = r^2 \cos \theta \sin \theta$. It follows immediately from the angular part of the integral in (8.13.15a) that the only nonzero coefficients are given by $A_{20} = \frac{1}{4i}$ and $A_{-20} = -\frac{1}{4i}$. This leads to the expansion

$$f(x, y) = \left(A_{20} e^{2i\theta} + A_{-20} e^{-2i\theta} \right) Z_2^2(r),$$

or

$$f(x, y) = r^2 \frac{e^{2i\theta} - e^{-2i\theta}}{4i} = r^2 \cos \theta \sin \theta. \qquad \blacksquare$$

Example 3

Let $f(x, y) = x(x^2 + y^2)$. Now, changing to r and θ gives $f(r, \theta) = r^3 \cos \theta$. It is tempting to take a quick look at the table and say the expansion must contain A_{30} and Z_3^3 because this polynomial is equal to r^3. This is **not** the correct thing to do! A quick inspection of the angular part of (8.13.15a) reveals that A_{30} vanishes. The nonzero constants are $A_{11} = A_{-11} = \frac{1}{6}$, and $A_{10} = A_{-10} = \frac{1}{3}$. This gives the correct expansion

$$f(x, y) = \frac{1}{3} Z_3^1(r) \cos \theta + \frac{2}{3} Z_1^1(r) \cos \theta = r^3 \cos \theta. \qquad \blacksquare$$

Transform of the Zernike polynomials

We need to find the Radon transform of a function of the form

$$f(x, y) = Z_m^l(r) e^{il\theta}.$$

It is adequate to consider $l \geq 0$, because the negative case follows by complex conjugation. The angular part transforms to $e^{il\phi}$ and the radial part must satisfy (8.13.6a) with upper limit 1,

$$\check{h}_l(p) = 2 \int_p^1 Z_m^l(r) \, T_l\left(\frac{p}{r}\right) \left(1 - \frac{p^2}{r^2}\right)^{-1/2} dr, \qquad p \geq 0. \qquad (8.13.16)$$

There are various ways to evaluate this integral, and the details are not shown here. The method used by Zeitler [1974] and Deans [1983, 1993] makes use of the path through Fourier space to find the transformed function in Radon space. The important result is that the orthogonal set of Zernike polynomials transforms to the orthogonal set of Tchebycheff polynomials of the second kind,

$$\mathcal{R}\left\{ Z_m^l(r) e^{il\theta} \right\} = \frac{2}{m+1} \sqrt{1 - p^2} \, U_m(p) e^{il\phi}, \qquad (8.13.17)$$

with $m = l + 2s$. Basic properties of the U_m are given in Appendix A, and summaries are given by Arfken [1985] and Erdélyi et al. [1953].

The Radon transform of (8.13.12) follows immediately by use of (8.13.17),

$$\check{f}(p, \phi) = \sum_{s=0}^{\infty} \sum_{l=-\infty}^{\infty} A_{ls} \frac{2}{|l| + 2s + 1} \sqrt{1 - p^2}\, U_{|l|+2s}(p) e^{il\phi}. \tag{8.13.18}$$

Some more examples serve to illustrate how the method developed here relates to transforms found in earlier sections when the function is confined to the unit disk. Also, these examples are designed to point out ways certain pitfalls can be avoided.

Example 4

If $f(x, y) = 1$ on the unit disk and zero elsewhere, the expansion in terms of Zernike polynomials is just $f = Z_0^0$, with $A_{00} = 1$. From (8.13.18), $\check{f} = 2\sqrt{1 - p^2}$, because $U_0 = 1$. Note that this is just another way of doing Example 7 in Section 8.5. ∎

Example 5

If $f(x, y) = x^2 = r^2 \cos^2 \theta = \frac{1}{2} r^2 (1 + \cos 2\theta)$ on the unit disk, then

$$f(x, y) = \frac{1}{4}(Z_0^0 + Z_2^0) + \frac{1}{2} Z_2^2 \cos 2\theta.$$

This serves to identify the coefficients A_{ls} and by use of (8.13.18)

$$\check{f} = \sqrt{1 - p^2}\left[\frac{1}{4}(2U_0 + \frac{2}{3}U_2) + \frac{1}{3}U_2 \cos 2\phi\right].$$

After simplification,

$$\mathcal{R}\{x^2\} = \sqrt{1 - p^2}\left[2p^2 \cos^2 \phi + \frac{2}{3}(1 - p^2) \sin^2 \phi\right].$$

Now note that if $f(x, y) = y^2 = \frac{1}{2} r^2 (1 - \cos 2\theta)$, the change is $(\cos \phi \leftrightarrow \sin \phi)$ in the equation for $\mathcal{R}\{x^2\}$, and

$$\mathcal{R}\{y^2\} = \sqrt{1 - p^2}\left[2p^2 \sin^2 \phi + \frac{2}{3}(1 - p^2) \cos^2 \phi\right].$$

Finally, by linearity, the transform of $f(x, y) = x^2 + y^2$ is given by the sum of the above transforms

$$\mathcal{R}\{x^2 + y^2\} = \frac{2}{3}\sqrt{1 - p^2}\,(2p^2 + 1). \qquad ∎$$

Example 6

Let $f(x, y) = 1 - r^2$ on the unit disk. By using the methods of the earlier examples in this section

$$f = Z_0^0 - \frac{1}{2}(Z_0^0 + Z_2^0) = \frac{1}{2}Z_0^0 - \frac{1}{2}Z_2^0.$$

From (8.13.18)

$$\check{f} = \frac{1}{2} 2\sqrt{1 - p^2}\, U_0 - \frac{1}{2}\frac{2}{3}\sqrt{1 - p^2}\, U_2.$$

Or, after substitution for U_0 and U_2 from Appendix A,

$$\check{f} = \frac{4}{3}(1 - p^2)\sqrt{1 - p^2}.$$

Another way to obtain this is to use Examples 4 and 5 and linearity. ∎

Example 7

For $f(x, y) = x(x^2 + y^2)$ as in Example 3, it follows from knowing the A_{ls} that

$$\check{f} = \frac{1}{3}\sqrt{1 - p^2}\left[\frac{1}{2}U_3 + 2U_1\right]\cos\phi$$

$$= \frac{2}{3}p\,(2p^2 + 1)\sqrt{1 - p^2}\cos\phi. \qquad ∎$$

Example 8

It may be worthwhile to emphasize that there are certain transforms that cannot be found by a naive application of the Zernike polynomials. To illustrate, suppose $f(x, y) = x\sqrt{x^2 + y^2}$. Although this has the form $f = xr = Z_2^2\cos\theta$, it is not a simple sum over monomials $x^k y^j$, and the method of this section does not apply. The transform can be found by use of the technique in Example 9 of Section 8.5. The solution is

$$\check{f}(p, \phi) = 2p\cos\phi\left[\frac{1}{2}\sqrt{1 - p^2} + \frac{p^2}{2}\log\left(\frac{1 + \sqrt{1 - p^2}}{p}\right)\right].$$

Clearly, this does not follow by Zernike decomposition of xr. ∎

Evaluation in Radon space

In the previous section, (8.13.18) was used to find Radon transforms when the constants A_{ls} can be determined by knowing the feature space function. Here the idea is to determine the same constants by knowledge of the Radon space function $\check{f}$. It is easy to solve for the constants directly from (8.13.18). Multiply both sides by $e^{-il'\phi}U_{l'+2t}$ and integrate over p and ϕ. Then use the orthogonality equation for the U_m in Appendix A to find the constants,

$$A_{ls} = \frac{|l| + 2s + 1}{2\pi^2}\int_0^{2\pi}\int_{-1}^1\check{f}(p, \phi)e^{-il\phi}U_{|l|+2s}\,dp\,d\phi. \qquad (8.13.19)$$

Example 9

The simplest test of (8.13.19) is for the inverse of the problem of Example 4. We assume that $\check{f} = 2\sqrt{1 - p^2}$ with $l = s = 0$, then $f = 1$ on the unit disk,

$$A_{00} = \frac{1}{2\pi^2}\int_0^{2\pi}\int_{-1}^1 2\sqrt{1 - p^2}\,dp\,d\phi$$

$$= \frac{2}{\pi}\int_{-1}^1\sqrt{1 - p^2}\,dp = 1. \qquad ∎$$

Transform to Fourier space

The Radon transform of the basis set given in (8.13.17) transformed one orthogonal set to another orthogonal set. It is interesting to examine the Fourier transform of the basis set. It turns out that this also leads to another orthogonal set. Details are given by Zeitler [1974] and Deans [1983, 1993]. The important result is that

$$\mathcal{F}_2\left\{Z_{l+2s}^l(r)e^{il\theta}\right\} = (-i)^l(-1)^s e^{il\phi}\,\frac{J_{l+2s+1}(2\pi q)}{q}. \tag{8.13.20}$$

This equation is obtained using the symmetric form of the Fourier transform (see [8.2.14]). These Bessel functions are orthogonal with respect to weight function q^{-1}, and have been studied by Wilkins [1948],

$$\int_0^\infty J_{|l|+2s+1}(q)\, J_{|l|+2t+1}(q)\, q^{-1}\, dq = \frac{\delta_{st}}{2(|l|+2s+1)}.$$

The Fourier space version of (8.13.12) is

$$\tilde{f}(q,\phi) = \sum_{s=0}^\infty \sum_{l=-\infty}^\infty (-i)^l(-1)^s A_{ls}\, e^{il\phi}\,\frac{J_{|l|+2s+1}(2\pi q)}{q}. \tag{8.13.21}$$

Example 10
The Fourier transform of the characteristic function of the unit disk, Example 4, with $A_{00}=1$ and $l=s=0$, is given by $J_1(2\pi q)/q$. ∎

Example 11
For the function in Example 6, the expansion (8.13.21), with $A_{00}=\frac{1}{2}$ and $A_{01}=-\frac{1}{2}$, yields

$$\mathcal{F}_2\left\{1-r^2\right\} = \frac{J_1(2\pi q)}{2q} + \frac{J_3(2\pi q)}{2q} = \frac{J_2(2\pi q)}{\pi q^2}.$$

The last equality follows from the Bessel function identity

$$J_{n-1}(z) + J_{n+1}(z) = \frac{2n}{z} J_n(z)$$

with $n=2$ and $z=2\pi q$. ∎

Example 12
Repeat Example 5 with transforms to Fourier space using (8.13.21).

$$\mathcal{F}_2\left\{x^2\right\} = \frac{J_1(2\pi q)}{4q} - \frac{J_3(2\pi q)}{4q} - \frac{J_3(2\pi q)}{2q}\cos 2\phi.$$

$$\mathcal{F}_2\left\{y^2\right\} = \frac{J_1(2\pi q)}{4q} - \frac{J_3(2\pi q)}{4q} + \frac{J_3(2\pi q)}{2q}\cos 2\phi.$$

$$\mathcal{F}_2\left\{x^2+y^2\right\} = \frac{J_1(2\pi q)}{2q} - \frac{J_3(2\pi q)}{2q} = \frac{J_1(2\pi q)}{q} - \frac{J_2(2\pi q)}{\pi q^2}.$$

The last part follows from the identity in Example 11. Also, note that the result for x^2+y^2 follows directly from Examples 10 and 11 and linearity. ∎

Some final observations

It is possible to find orthogonal function expansions that transform to each other in all three spaces. In feature space the Zernike polynomials, defined on the unit disk, are orthogonal with weight function r over the interval $0 \leq r \leq 1$. In Radon space the Tchebycheff polynomials of the second kind emerge, orthogonal on the interval $-1 \leq p \leq 1$ with weight function $\sqrt{1 - p^2}$. These are both defined on finite intervals and consequently, as is to be expected, in Fourier space the interval is infinite, $0 \leq q \leq \infty$. The orthogonal functions are no longer polynomials; they are orthogonal Bessel functions with weight function q^{-1}. The orthogonality integrals over the three spaces, including the angles, are given by:

$$\int_0^{2\pi} \int_0^1 \left[Z_{|l|+2s}^{|l|}(r)e^{il\theta} \right]^* Z_{|l'|+2s'}^{|l'|}(r)e^{il'\theta} \, r \, dr \, d\theta = \frac{\pi}{|l| + 2s + 1} \delta_{ll'} \, \delta_{ss'}, \qquad (8.13.22a)$$

$$\int_0^{2\pi} \int_{-1}^1 \left[U_{|l|+2s}^{|l|}(p)e^{il\phi} \right]^* U_{|l'|+2s'}^{|l'|}(p)e^{il'\phi} \, \sqrt{1 - p^2} \, dp \, d\phi = \pi^2 \, \delta_{ll'} \, \delta_{ss'}, \qquad (8.13.22b)$$

$$\int_0^{2\pi} \int_0^\infty \left[J_{|l|+2s+1}(q)e^{il\phi} \right]^* J_{|l'|+2s'+1}(q)e^{il'\theta} \, q^{-1} \, dq \, d\phi = \frac{\pi}{|l| + 2s + 1} \delta_{ll'} \, \delta_{ss'}. \quad (8.13.22c)$$

8.14   Parseval relation

In the notation of Section 8.1.2, let inner products in nD be designated by

$$\langle f, g \rangle = \int f^*(\mathbf{x}) g(\mathbf{x}) \, d\mathbf{x}.$$

If the nD Fourier transforms of f and g are designated by $\tilde{f}$ and $\tilde{g}$, the Parseval relation for the Fourier transform is given by

$$\langle f, g \rangle = \langle \tilde{f}, \tilde{g} \rangle. \tag{8.14.1}$$

The integral on the right is over all Fourier space. If $g = f$, then the integrals are normalization integrals. This guarantees that if f is normalized to unity, then its Fourier transform is also normalized to unity.

The corresponding expression for the Radon transform is considerably more complicated, and we need to extend some of the previous work in order to give a general result. First, define the **adjoint** for the Radon transform. If inner products in Radon space are designated by square brackets, then $\mathcal{R}^\dagger$ is the adjoint in the sense that

$$\langle f, \mathcal{R}^\dagger G \rangle = [\mathcal{R}f, G]. \tag{8.14.2}$$

Here, G is a function of the variables in Radon space, $G = G(p, \boldsymbol{\xi})$, and the adjoint operator $\mathcal{R}^\dagger$ converts G to a function of $\mathbf{x}$, designated by $g(\mathbf{x}) = \mathcal{R}^\dagger G(p, \boldsymbol{\xi})$. For example, in 2D the adjoint is just two times the backprojection operator, $\mathcal{R}^\dagger = 2\mathcal{B}$.

Example 1
It is instructive to see how (8.14.2) comes from the definitions,

$$\langle f, \mathcal{R}^\dagger G \rangle = \int d\mathbf{x} \, f(\mathbf{x}) g(\mathbf{x})$$

$$= \int d\mathbf{x} \, f(\mathbf{x}) \int_{|\boldsymbol{\xi}|=1} d\boldsymbol{\xi} \, G(\boldsymbol{\xi} \cdot \mathbf{x}, \boldsymbol{\xi})$$

$$= \int d\mathbf{x}\, f(\mathbf{x}) \int_{|\xi|=1} d\xi \int_{-\infty}^{\infty} dp\, G(p,\xi)\, \delta(p - \xi \cdot \mathbf{x})$$

$$= \int_{|\xi|=1} d\xi \int_{-\infty}^{\infty} dp \int d\mathbf{x}\, f(\mathbf{x})\delta(p - \xi \cdot \mathbf{x})G(p,\xi)$$

$$= \int_{|\xi|=1} d\xi \int_{-\infty}^{\infty} dp\, \check{f}(p,\xi)\, G(p,\xi)$$

$$= [\mathcal{R}f, G]. \qquad\blacksquare$$

The significance of this result is more apparent after making a generalization of Section 8.9 to include the nD inversion formula. Define the operator Υ to cover both the even and odd dimension cases [Ludwig, 1966], [Deans, 1983, 1993]:

$$\bar{g}(t) = \Upsilon g = \begin{cases} N_n \left[\left(\frac{\partial}{\partial p}\right)^{n-1} g(p) \right]_{p=t} & n \text{ odd} \\[2ex] \dfrac{N_n}{i} \left[\mathcal{H}_i \left\{ \left(\frac{\partial}{\partial p}\right)^{n-1} g(p) \right\} \right]_{p=t} & n \text{ even,} \end{cases} \qquad (8.14.3)$$

where $N_n = \frac{1}{2}(2\pi i)^{1-n}$. This reduces to (8.9.9) for $n = 2$ and to (8.9.12) for $n = 3$.

With this definition the inversion formula for the Radon transform is given by

$$f = \mathcal{R}^\dagger \bar{\check{f}} = \mathcal{R}^\dagger \Upsilon \check{f} = \mathcal{R}^\dagger \Upsilon \mathcal{R} f. \qquad (8.14.4)$$

This leads to the operator identity **operating in feature space**,

$$I = \mathcal{R}^\dagger \Upsilon \mathcal{R}. \qquad (8.14.5)$$

By starting with

$$\mathcal{R}^\dagger G = g = Ig = \mathcal{R}^\dagger \Upsilon \mathcal{R}\mathcal{R}^\dagger G,$$

it follows that the identity **operating on functions in Radon space** is given by

$$I = \Upsilon \mathcal{R}\mathcal{R}^\dagger. \qquad (8.14.6)$$

When equations (8.14.2) and (8.14.6) are combined we obtain the desired form for the Parseval relation for the Radon transform,

$$\langle f, g \rangle = \langle f, \mathcal{R}^\dagger G \rangle$$

$$= [\mathcal{R}f, IG]$$

$$= [\mathcal{R}f, \Upsilon \mathcal{R}\mathcal{R}^\dagger G]$$

$$= [\mathcal{R}f, \Upsilon \mathcal{R}g]$$

$$= [\check{f}, \Upsilon \check{g}] \qquad (8.14.7)$$

An important special case is for $g = f$, then

$$\langle f, f \rangle = [\check{f}, \Upsilon \check{f}]. \qquad (8.14.8)$$

Example 2
Verify the Parseval relation (8.14.8) explicitly in all three spaces for $f(x, y) = e^{-x^2 - y^2}$. This looks simple, but it demonstrates the difficulty of dealing with Radon space compared with feature space and Fourier space.

Feature space:

$$\langle f, f \rangle = \int_{-\infty}^{\infty} \int_{-\infty}^{\infty} e^{-x^2 - y^2} e^{-x^2 - y^2} \, dx \, dy$$

$$= \int_{-\infty}^{\infty} \int_{-\infty}^{\infty} e^{-2x^2 - 2y^2} \, dx \, dy$$

$$= \frac{\sqrt{\pi}}{\sqrt{2}} \frac{\sqrt{\pi}}{\sqrt{2}} = \frac{\pi}{2}.$$

Fourier space: Note that $q^2 = u^2 + v^2$. Then

$$\langle \tilde{f}, \tilde{f} \rangle = \int_{-\infty}^{\infty} \int_{-\infty}^{\infty} \pi e^{-\pi^2(u^2 - v^2)} \pi e^{-\pi^2(u^2 - v^2)} \, du \, dv$$

$$= \pi^2 \int_{-\infty}^{\infty} \int_{-\infty}^{\infty} e^{-2\pi^2 u^2} e^{-2\pi^2 u^2} \, du \, dv$$

$$= \pi^2 \frac{\sqrt{\pi}}{\sqrt{2\pi^2}} \frac{\sqrt{\pi}}{\sqrt{2\pi^2}} = \frac{\pi}{2}.$$

Radon space: Verification in Radon space is not as easy as the other two cases due to the presence of the Hilbert transform. The entire calculation is shown in detail, because there are some tricky parts. First note that $\partial \check{f} / \partial p = -2\sqrt{\pi} \, p \, e^{-p^2}$. Then (8.14.8) is

$$[\check{f}, \Upsilon \check{f}] = \left[\sqrt{\pi} \, e^{-p^2}, \frac{-1}{4\pi} \mathcal{H}_i \left(\frac{\partial \check{f}}{\partial p} \right) \right]$$

$$= \left[\sqrt{\pi} \, e^{-p^2}, \left(\frac{-1}{4\pi} \right) (-2\sqrt{\pi}) \mathcal{H}_i \left(p \, e^{-p^2} \right) \right]$$

$$= \frac{1}{2} \left[e^{-p^2}, \mathcal{H}_i \left(p \, e^{-p^2} \right) \right].$$

Because there is no angle dependence, the integral $\int_{|\xi|} d\xi = 2\pi$. Hence the last inner product becomes

$$[\check{f}, \Upsilon \check{f}] = \frac{2\pi}{2} \int_{-\infty}^{\infty} dp \, e^{-p^2} \frac{1}{\pi} \int_{-\infty}^{\infty} ds \frac{s e^{-s^2}}{s - p}$$

$$= \int_{-\infty}^{\infty} dp \, e^{-p^2} \int_{-\infty}^{\infty} ds \frac{s e^{-s^2}}{s - p}.$$

Now the problem is to demonstrate that this double integral yields $\pi/2$. Change the order of integration to get

$$[\check{f}, \Upsilon \check{f}] = \int_{-\infty}^{\infty} ds \, s e^{-s^2} \int_{-\infty}^{\infty} dp \frac{e^{-p^2}}{s - p}.$$

From page 227 of Davis and Rabinowitz [1984] this becomes

$$[\check{f}, \Upsilon \check{f}] = \int_{-\infty}^{\infty} ds \, se^{-s^2} \sqrt{\pi} \, se^{-s^2} \int_{-1}^{1} dp \, e^{s^2 p^2}.$$

Another change in the order of integration followed by evaluation of the definite integrals (Appendix A) yields the desired result,

$$[\check{f}, \Upsilon \check{f}] = \sqrt{\pi} \int_{-1}^{1} dp \int_{-\infty}^{\infty} ds \, s^2 \, e^{-(2-p^2)s^2}$$

$$= \frac{\sqrt{\pi}\sqrt{\pi}}{2} \int_{-1}^{1} \frac{dp}{\left(2 - p^2\right)^{3/2}}$$

$$= \pi \int_{0}^{1} \frac{dp}{\left(2 - p^2\right)^{3/2}}$$

$$= \frac{\pi}{2}.$$

∎

8.15   Generalizations and wavelets

Mathematical generalizations of the Radon transform and some of the more technical applications are discussed in the recent publications edited by Grinberg and Quinto [1990] and Gindikin and Michor [1994]. There are many other references, and the reader interested in some of the more abstract treatments will find these two books good entry points to the literature.

A generalization that has important applicability in the area of image reconstruction in nuclear medicine is known as the **attenuated** Radon transform. One way to define this transform is to modify (8.2.4) to read

$$\check{f}_{\mu}(p, \phi) = \int_{-\infty}^{\infty} f(p\,\boldsymbol{\xi} + t\,\boldsymbol{\xi}') \exp\left[-\int_{t}^{\infty} \mu(p\,\boldsymbol{\xi} + s\,\boldsymbol{\xi}')\,ds\right] dt. \tag{8.15.1}$$

If the attenuation term μ is a constant, say μ_0, that vanishes outside a finite region, then this equation reduces to what is often referred to as the **exponential** Radon transform,

$$\check{f}_{\mu_0}(p, \phi) = \int_{-\infty}^{\infty} e^{\mu_0 t} f(p\,\boldsymbol{\xi} + t\,\boldsymbol{\xi}')\,dt. \tag{8.15.2}$$

These transforms are fundamental in single photon emission tomography (SPECT), and to a lesser degree in positron emission tomography (PET) where corrections can be introduced to compensate for attenuation [Budinger, Gullberg, and Huesman, 1979]. For details see Natterer [1979, 1986], Tretiak and Metz [1980], Clough and Barrett [1983], Hawkins, Leichner, and Yang [1988], Hazou and Solmon [1989], and Nievergelt [1991].

One of the most recent, and certainly one of the most exciting new developments is the use of the wavelet transform in connection with the Radon transform. The application of wavelets to inversion of the Radon transform has been investigated by Kaiser and Streater [1992]. They make use of a change of variables to connect a generalized version of the Radon transform to a continuous wavelet transform. Work along related lines was done by Holschneider [1991] where the inverse wavelet transform is used to obtain a pointwise and uniformly convergent inversion formula for the Radon transform.

Berenstein and Walnut [1994] use the theory of the continuous wavelet transform to derive inversion formulas for the Radon transform. The inversion formula they obtain is "local" in even dimensions in the following sense (stated for 2D): To recover f to a given accuracy in a circle of radius r about a point (x_0, y_0) it is sufficient to know only those projections through a circle of radius $r + \alpha$ about (x_0, y_0) for some $\alpha > 0$. The accuracy increases as α increases. In a related paper, Walnut [1992] demonstrates how the Gabor and wavelet transforms relate to the Radon transform. He finds inversion formulas for $\check{f}$ based on Gabor and wavelet expansions by a direct method and by the filtered backprojection method.

More work on wavelet localization of the Radon transform is in the paper by Olson and DeStefano [1994]. As mentioned in Section 8.9, they point out that one problem with the Radon transform in two dimensions (most relevant in medical imaging) is that the inversion formula is globally dependent upon the line integral of the object function f. A fundamental important aspect of their work is that they are able to develop a stable algorithm that uses properties of wavelets to "essentially localize" the Radon transform. This means collect line integrals that pass through the region of interest, plus a small number of integrals not through the region.

The work by Donoho [1992] on nonlinear solution of linear inverse problems by wavelet-vaguelette decomposition is relevant to the inversion of Abel-type transforms and Radon transforms. This method serves as a substitute for the singular value decomposition of an inverse problem, and applies to a large class of ill-posed inverse problems.

Another important applied generalization is related to fan beam and cone beam tomography. Recent work in these areas can be found in papers by Natterer [1993], Kudo and Saito [1990, 1991], Rizo et al. [1991], Gullberg et al. [1991], and in the book by Natterer [1986].

In recent work by Wood and Barry [1994], the Wigner distribution is combined with the Radon transform to facilitate the analysis of multicomponent linear FM signals. These authors provide several references to other applications of this combined transform, now known as the Radon-Wigner transform.

Appendix A: Functions and formulas

Various functions and formulas are recorded here for the convenience of the reader. (Also, see Chapter 1 and Appendices.) The information here can be found in standard sources. In particular, those used here include Abramowitz and Stegun [1972], Arfken [1985], Born and Wolf [1975], Erdélyi et al. [1953], Gradshteyn et al. [1994], Lide [1993], Rainville [1960], and Szegö [1939].

A.1   *Tchebycheff polynomials: first kind:* $T_l(x)$

Definitions

$$T_l(x) = \cos(l \arccos x), \qquad 0 < x < 1$$

$$T_l(x) = \cosh(l \cosh^{-1} x), \qquad 1 < x < \infty$$

$$T_l(x) = \frac{1}{2}\left(x + \sqrt{x^2 - 1}\right)^l + \frac{1}{2}\left(x - \sqrt{x^2 - 1}\right)^l, \qquad 0 < x < \infty$$

$$T_l(1) = 1, \qquad T_l(0) = \cos\frac{l\pi}{2}, \qquad T_l(-x) = (-1)^l T_l(x)$$

Orthogonality

$$\int_{-1}^{1} T_l(x)T_m(x)(1-x^2)^{-1/2}dx = \begin{cases} 0, & \text{for } l \neq m \\ \frac{\pi}{2}, & \text{for } l = m \neq 0 \\ \pi, & \text{for } l = m = 0 \end{cases}$$

Recurrence and Derivatives

$$T_{l+1} = 2xT_l - T_{l-1}$$

$$(1-x^2)T_l' = lT_{l-1} - lxT_l$$

$$(1-x^2)T_l'' - xT_l' + l^2 T_l = 0$$

First Few

$$T_0 = 1$$

$$T_1 = x$$

$$T_2 = 2x^2 - 1$$

$$T_3 = 4x^3 - 3x$$

Useful Integrals

$$\int_a^b \frac{T_l(x/a)T_l(x/b)\,dx}{x\,(b^2-x^2)^{\frac{1}{2}}(x^2-a^2)^{\frac{1}{2}}} = \frac{\pi}{2ab}$$

$$\int_a^b \frac{dx}{x\,(b^2-x^2)^{\frac{1}{2}}(x^2-a^2)^{\frac{1}{2}}} = \frac{\pi}{2ab}$$

$$\int_a^b \frac{x\,dx}{(b^2-x^2)^{\frac{1}{2}}(x^2-a^2)^{\frac{1}{2}}} = \frac{\pi}{2}$$

$$\int_a^b \frac{x^3\,dx}{(b^2-x^2)^{\frac{1}{2}}(x^2-a^2)^{\frac{1}{2}}} = \frac{\pi}{4}(a^2+b^2)$$

A.2   *Tchebycheff polynomials: second kind:* $U_l(x)$

Definitions

$$U_{l-1}(x) = \frac{\cos(l \arccos x)}{\sqrt{1 - x^2}}, \qquad 0 < x < 1$$

$$U_{l-1}(x) = \frac{\sinh(l \cosh^{-1} x)}{\sqrt{x^2 - 1}}, \qquad 1 < x < \infty$$

$$U_{l-1}(x) = \frac{\left(x + \sqrt{x^2 - 1}\right)^l - \left(x - \sqrt{x^2 - 1}\right)^l}{2\sqrt{x^2 - 1}}, \qquad 0 < x < \infty, \qquad x \neq 1$$

$$U_l(-x) = (-1)^l U_l(x), \qquad U_l(1) = l + 1, \qquad U_l(0) = \cos \frac{l\pi}{2}$$

Orthogonality

$$\int_{-1}^{1} U_l(x) U_m(x)(1 - x^2)^{1/2} dx = \frac{\pi}{2} \delta_{lm}$$

Recurrence and Derivatives

$$U_{l+1} = 2x U_l - U_{l-1}$$

$$(1 - x^2) U_l' = (l + 1) U_{l-1} - l x U_l$$

$$(1 - x^2) U_l'' - 3x U_l' + l(l + 2) U_l = 0$$

First Few

$$U_0 = 1$$

$$U_1 = 2x$$

$$U_2 = 4x^2 - 1$$

$$U_3 = 8x^3 - 4x$$

$$U_4 = 16x^4 - 12x^2 + 1$$

Miscellaneous Connections

$$U_{l-1} = \frac{1}{l} T_l', \qquad l \geq 1$$

$$T_l = U_l - x U_{l-1}, \qquad l \geq 1$$

$$(1 - x^2) U_l = x T_{l+1} - T_{l+2}$$

$$\frac{T_l}{\sqrt{x^2-1}} - U_{l-1} = \frac{\left(x - \sqrt{x^2-1}\right)^l}{\sqrt{x^2-1}} = \frac{\left(x + \sqrt{x^2-1}\right)^{-l}}{\sqrt{x^2-1}}, \qquad x \neq 1$$

A.3　*Hermite polynomials:* $H_l(x)$

Generating Function

$$e^{2xt-t^2} = \sum_{l=0}^{\infty} \frac{H_l(x)t^l}{l!}$$

Orthogonality

$$\int_{-\infty}^{\infty} H_l(x)H_m(x)e^{-x^2}\,dx = \sqrt{\pi}\, 2^l\, l\, \delta_{lm}$$

Recurrence and Derivatives

$$H_{l+1} = 2x\,H_l - 2l\,H_{l-1}$$

$$H_l' = 2l\,H_{l-1}$$

$$H_l'' - 2x\,H_l' + 2l\,H_l = 0$$

Special Values

$$H_l(x) = (-1)^l H_l(-x)$$

$$H_{2l}(0) = (-1)^l \frac{(2l)!}{l!}$$

$$H_{2l+1}(0) = 0$$

First Few

$$H_0 = 1$$

$$H_1 = 2x$$

$$H_2 = 4x^2 - 2$$

$$H_3 = 8x^3 - 12x$$

$$H_4 = 16x^4 - 48x^2 + 12$$

Reverse Expansions

$$x^0 = H_0$$

$$x^1 = \frac{1}{2}H_1$$

$$x^2 = \frac{1}{4}(H_2 + 2H_0)$$

$$x^3 = \frac{1}{8}(H_3 + 6H_1)$$

$$x^4 = \frac{1}{16}(H_4 + 12H_2 + 12H_0)$$

A.4   *Zernike polynomials:* $Z_m^l(r)$

Definition

The Zernike polynomials can be defined in terms of the more general Jacobi polynomials $P_n^{(\alpha,\beta)}(z)$ by

$$Z_{l+2s}^l(r) = r^{|l|} P_s^{(0,|l|)}(2r^2 - 1).$$

An extensive discussion of the Zernike polynomials is given by Born and Wolf [1975]. Jacobi polynomials are discussed by the other references cited at the beginning of this appendix.

First Few

$$Z_0^0 = 1$$

$$Z_1^1 = r$$

$$Z_2^0 = 2r^2 - 1$$

$$Z_2^2 = r^2$$

$$Z_3^1 = 3r^3 - 2r$$

$$Z_3^3 = r^3$$

$$Z_4^0 = 6r^4 - 6r^2 + 1$$

$$Z_4^2 = 4r^4 - 3r^2$$

$$Z_4^4 = r^4$$

A.5 *Selected integral formulas*

$$\int_{-\infty}^{\infty} e^{-\alpha x^2} \, dx = \sqrt{\frac{\pi}{\alpha}}$$

$$\int_{-\infty}^{\infty} x^2 e^{-\alpha x^2} \, dx = \frac{1}{2\alpha} \sqrt{\frac{\pi}{\alpha}}$$

$$\int_{-a}^{a} \sqrt{a^2 - x^2} \, dx = \frac{a\pi}{2}$$

$$\int \frac{x \, dx}{\sqrt{a^2 - x^2}} = -\sqrt{a^2 - x^2}$$

$$\int \frac{dx}{\sqrt{x^2 \pm a^2}} = \log(x + \sqrt{x^2 \pm a^2})$$

$$\int \frac{dx}{x\sqrt{x^2 - a^2}} = \frac{1}{a} \cos^{-1}\left(\frac{a}{x}\right)$$

$$\int \frac{dx}{x^2\sqrt{x^2 - a^2}} = \frac{\sqrt{x^2 - a^2}}{a^2 x}$$

$$\int \frac{dx}{\left(a^2 - x^2\right)^{3/2}} = \frac{x}{a^2\sqrt{a^2 - x^2}}$$

$$\int \frac{dx}{\sqrt{a^2 - x^2}} = \sin^{-1}\left(\frac{x}{a}\right)$$

$$\int \frac{x^2 \, dx}{\sqrt{a^2 - x^2}} = -\frac{x}{2}\sqrt{a^2 - x^2} + \frac{a^2}{2} \sin^{-1}\left(\frac{x}{a}\right)$$

$$\int \frac{x^3 \, dx}{\sqrt{a^2 - x^2}} = \frac{1}{3}\sqrt{(a^2 - x^2)^3} - a^2\sqrt{a^2 - x^2}$$

$$\int \frac{dx}{x^2\sqrt{a^2 - x^2}} = -\frac{\sqrt{a^2 - x^2}}{a^2 x}$$

$$\int \frac{\sqrt{a^2 - x^2} \, dx}{x^2} = -\frac{\sqrt{a^2 - x^2}}{x} - \sin^{-1}\left(\frac{x}{a}\right)$$

$$\int_{0}^{\pi/2} \cos^n x \, dx = \frac{1 \cdot 3 \cdot 5 \cdots (n-1)}{2 \cdot 4 \cdot 6 \cdot 8 \cdots n} \frac{\pi}{2}, \qquad \text{for } n \text{ even integer}$$

$$\int_{0}^{\pi/2} \cos^n x \, dx = \frac{2 \cdot 4 \cdot 6 \cdots (n-1)}{1 \cdot 3 \cdot 5 \cdot 7 \cdots n}, \qquad \text{for } n \text{ odd integer}$$

Appendix B: A short list of Abel and Radon transforms

The list of transforms recorded here is by no means complete. It contains some of the more common and useful transforms that can be found in closed form. Other Radon and Abel transforms are scattered throughout this chapter. The notation for the Abel transforms is the same as in Section 8.10.2, and the notation for the 2D Radon transform is the same as in other parts of the chapter.

Also, just to remind the user of these tables, the sinc function is defined by

$$\operatorname{sinc} x = \frac{\sin \pi x}{\pi x},$$

and the characteristic function for the unit disk, designated by $\chi(r)$ is defined by

$$\chi(r) = \begin{cases} 1, & \text{for } 0 \le r \le 1 \\ 0, & \text{for } r > 1. \end{cases}$$

The complete elliptic integral of the first kind is designated by $F(\frac{1}{2}\pi, t)$ and the complete elliptic integral of the second kind is designated by $E(\frac{1}{2}\pi, t)$. A good source for these is the tabulation by Gradshteyn et al. [1994]. The constant $C(n)$ in the table for $\mathcal{A}_3$ is $C(n) = 2\int_0^{\pi/2} \cos^n x\, dx$, with $n \ge 1$; it can be calculated from Appendix A. Bessel functions of the first kind J_ν, and second kind N_ν (Neumann functions) conform to the standard definitions in Arfken [1985] and Gradshteyn et al. [1994]. In these tables, $a > 0$ and $b > 0$.

Abel Transforms $\mathcal{A}_1$

$f(r)$	$\mathcal{A}_1\{f(r); x\}$
$\chi(r/a)$	$\sin^{-1}\left(\frac{a}{x}\right), \quad x > a$
$\delta(r - a)$	$(x^2 - a^2)^{-\frac{1}{2}}, \quad x > a$
$(a^2 - r^2)^{-\frac{1}{2}}$	$a^{-1} F\left(\frac{\pi}{2}, \frac{x}{a}\right), \quad x < a$
$(a^2 - r^2)^{\frac{1}{2}}$	$a E\left(\frac{\pi}{2}, \frac{x}{a}\right), \quad x < a$
$r^2 (a^2 - r^2)^{-\frac{1}{2}}$	$a\left[F\left(\frac{\pi}{2}, \frac{x}{a}\right) - E\left(\frac{\pi}{2}, \frac{x}{a}\right)\right], \quad x < a$
$a - r$	$\frac{1}{2}\pi a - x, \quad x < a$
$\cos br$	$\frac{1}{2}\pi J_0(bx)$
$r \sin br$	$\frac{1}{2}\pi x J_1(bx)$
$r J_0(br)$	$b^{-1} \sin bx$
$J_\nu(br)$	$\frac{1}{2}\pi \left[J_{\frac{\nu}{2}}\left(\frac{bx}{2}\right)\right]^2$
$r^{\nu+1} J_\nu(br)$	$\pi^{\frac{1}{2}} (2b)^{-\frac{1}{2}} x^{\nu+\frac{1}{2}} J_{\nu+\frac{1}{2}}(bx)$

Abel Transforms $\mathcal{A}_2$

$f(r)$	$\mathcal{A}_2\{f(r); x\}$
$\chi(r/a)$	$\log\left(\frac{a+\sqrt{a^2-x^2}}{x}\right), \quad x < a$
$\delta(r-a)$	$(a^2-x^2)^{-\frac{1}{2}}, \quad x < a$
$(a^2-r^2)^{-\frac{1}{2}}\chi(r/a)$	$a^{-1}F\left(\frac{1}{2}\pi, t\right), \quad x < a$
$(a^2-r^2)^{\frac{1}{2}}\chi(r/a)$	$a\left[F\left(\frac{1}{2}\pi, t\right) - E\left(\frac{1}{2}\pi, t\right)\right], \quad x < a$
$r^2(a^2-r^2)^{-\frac{1}{2}}\chi(r/a)$	$a E\left(\frac{1}{2}\pi, t\right), \quad x < a$
$(a-r)\chi(r/a)$	$\log\left(\frac{a+\sqrt{a^2-x^2}}{x}\right) - \sqrt{a^2-x^2}, \quad x < a$
$\sin br$	$\frac{1}{2}\pi J_0(bx)$
$r\cos br$	$-\frac{1}{2}\pi x J_1(bx)$
$r J_0(br)$	$b^{-1}\cos bx$

Note: $t = a^{-1}\sqrt{a^2-x^2}$.

Abel Transforms $\mathcal{A}_3$

$f(r)$	$\mathcal{A}_3\{f(r); x\}$
$(a^2-r^2)^{-\frac{1}{2}}\chi(r/a)$	$\pi\chi(x/a)$
$\chi(r/a)$	$2(a^2-x^2)^{\frac{1}{2}}\chi(x/a)$
$(a^2-r^2)^{\frac{1}{2}}\chi(r/a)$	$\frac{1}{2}\pi(a^2-x^2)\chi(x/a)$
$(a^2-r^2)\chi(r/a)$	$\frac{4}{3}(a^2-x^2)^{\frac{3}{2}}\chi(x/a)$
$(a^2-r^2)^{\frac{3}{2}}\chi(r/a)$	$\frac{3\pi}{8}(a^2-x^2)^2\chi(x/a)$
$(a^2-r^2)^2\chi(r/a)$	$\frac{15}{16}(a^2-x^2)^{\frac{5}{2}}\chi(x/a)$
$(a^2-r^2)^{\frac{n-1}{2}}\chi(r/a)$	$C(n)(a^2-x^2)^{\frac{n}{2}}\chi(x/a)$
$(a^2+r^2)^{-1}$	$\pi(a^2+x^2)^{-\frac{1}{2}}$
$(a^2+r^2)^{-\frac{3}{2}}$	$2(a^2+x^2)^{-1}$
e^{-r^2}	$\sqrt{\pi}\,e^{-x^2}$
$r^2 e^{-r^2}$	$\frac{1}{2}\sqrt{\pi}\,(2x^2+1)\,e^{-x^2}$
$\text{sinc } 2ar$	$\frac{1}{2a}J_0(2\pi ax)$
$\cos br$	$-\pi x J_1(bx)$
$J_0(br)$	$2b^{-1}\cos bx$
$r^{-1}J_1(br)$	$2(bx)^{-1}\sin bx$
$r^{-1}J_\nu(br)$	$-\pi J_{\frac{\nu}{2}}\left(\frac{bx}{2}\right) N_{\frac{\nu}{2}}\left(\frac{bx}{2}\right)$
$r^{-1}N_\nu(br)$	$\frac{1}{2}\pi\left[J_{\frac{\nu}{2}}\left(\frac{bx}{2}\right)\right]^2 - \frac{1}{2}\pi\left[N_{\frac{\nu}{2}}\left(\frac{bx}{2}\right)\right]^2$

Radon Transforms

$f(x, y)$	$\check{f}(p, \phi)$
$e^{-x^2-y^2}$	$\sqrt{\pi}\, e^{-p^2}$
$(x^2 + y^2)\, e^{-x^2-y^2}$	$\frac{1}{2}\sqrt{\pi}\,(2p^2 + 1)\, e^{-p^2}$
$x\, e^{-x^2-y^2}$	$\sqrt{\pi}\, e^{-p^2} \cos\phi$
$y\, e^{-x^2-y^2}$	$\sqrt{\pi}\, e^{-p^2} \sin\phi$
$x^2\, e^{-x^2-y^2}$	$\frac{1}{2}\sqrt{\pi}\,(2p^2\cos^2\phi + \sin^2\phi)e^{-p^2}$
$y^2\, e^{-x^2-y^2}$	$\frac{1}{2}\sqrt{\pi}\,(2p^2\sin^2\phi + \cos^2\phi)e^{-p^2}$
$\exp\left[-(\frac{x}{a})^2 - (\frac{y}{a})^2\right]$	$\frac{|ab|\sqrt{\pi}}{s}\exp[-(\frac{p}{s})^2]$
$\delta(x - a)\,\delta(y - b)$	$\delta(p - p_0)$
$\chi(r)$	$2(1 - p^2)^{\frac{1}{2}}\chi(p)$
$\chi(\text{ellipse})$	(See Example 8, Section 8.5)
$\chi(\text{square})$	(See Example 10, Section 8.5)
$x^2\chi(r)$	$(1 - p^2)^{\frac{1}{2}}[2p^2\cos^2\phi + \frac{2}{3}(1 - p^2)\sin^2\phi]$
$y^2\chi(r)$	$(1 - p^2)^{\frac{1}{2}}[2p^2\sin^2\phi + \frac{2}{3}(1 - p^2)\cos^2\phi]$
$(x^2 + y^2)\chi(r)$	$\frac{2}{3}(1 - p^2)^{\frac{1}{2}}(2p^2 + 1)$

The following notation is used in the above table,

$$s = (a^2\cos^2\phi + b^2\sin^2\phi)^{\frac{1}{2}}, \qquad r = \sqrt{x^2 + y^2}, \qquad p_0 = a\cos\phi + b\sin\phi.$$

Formulas for Radon transforms involving Hermite polynomials, Laguerre polynomials, and Zernike polynomials appear in Sections 8.7, 8.8, and 8.13, respectively.

References

[1] Abel, N. H. Solution de quelques problèmes à l'aide d'intégrales définies, 1823. In *Oeuvres Complètes*, Vol. 1, pp. 11–27, Christiania, Oslo, 1881.

[2] Abel, N. H. Résolution d'un problème de mecanique, 1826a. In *Oeuvres Complètes*, Vol. 1, pp. 97–101, Christiania, Oslo, 1881.

[3] Abel, N. H. Auflösung einer mechanischen Aufgabe. *Journal für die reine und angewandte Mathematik*, 1, pp. 153–157, 1826b.

[4] Abramowitz, M., and Stegun, A. eds. *Handbook of Mathematical Functions with Formulas, Graphs, and Mathematical Tables*, National Bureau of Standards Applied Mathematics Series 55, 10th printing. U. S. Government Printing Office, Washington, DC, 1972.

[5] Amato, I. Nobel Prizes '91. *Science*, 254, pp. 518–519, 1991.

[6] Anderssen, R. S., and de Hoog, F. R. Abel integral equations. In *Numerical Solution of Integral Equations*, ed. M. A. Golberg, pp. 373–410. Plenum Press, New York and London, 1990.

[7] Arfken, G. *Mathematical Methods for Physicists*, 3rd ed. Academic Press, San Diego, CA, 1985.

[8] Barrett, H. H. The Radon transform and its applications. In *Progrsss in Optics*, ed. E. Wolfe, vol. 21, pp. 219–286, 1984. Elsevier, Amsterdam.

[9] Barrett, H. H., Hawkins, W. G., and Joy, M. L. Historical note on computed tomography. *Radiology*, 147, p. 72, 1983.

[10] Barrett, H. H., and Swindell, W. Analog reconstruction methods for transaxial tomography. *Trans. IEEE*, 65, pp. 89–107, 1977.

[11] Bates, R. H. T., and Peters, T. M. Towards improvements in tomography. *N. Z. J. Sci.* 14, pp. 883–896, 1971.

[12] Berenstein, C., and Walnut D. Local inversion of the Radon transform in even dimensions using wavelets. In *75 Years of Radon Transform*, eds. S. Gindikin and P. Michor, International Press, Cambridge, MA, 1994.

[13] Beylkin, G. Discrete Radon transform. *IEEE Trans. Acoust., Speech, Signal Processing* 35(2), pp. 162–172, 1987.

[14] Born, M., and Wolf, E. *Principles of Optics*, 5th ed. Pergamon Press, New York, 1975.

[15] Bracewell, R. N. Strip integration in radio astronomy. *Aust. J. Phys.* 9, pp. 198–217, 1956.

[16] Bracewell, R. N. Image reconstruction in radio astronomy. In *Image Reconstruction from Projections*, ed. G. T. Herman, pp. 81–104. Topics in Applied Physics, Vol. 32. Springer-Verlag, New York, 1979.

[17] Bracewell, R. N. *The Fourier Transform and Its Applications*, 2nd ed. Revised. McGraw-Hill, New York, 1986.

[18] Bregman, N. D., Bailey, R. C., and Chapman, C. H. Crosshole seismic tomography. *Geophysics*, 54, pp. 200–215, 1989.

[19] Brooks, R. A., and Di Chiro, G. Principles of computer assisted tomography (CAT) in radiographic and radioisotopic imaging. *Phys. Med. Biol.*, 21, pp. 689–732, 1976.

[20] Budinger, T. F., Gullberg, G. T., and Huesman, R. H. Emission computed tomography. In *Image Reconstruction from Projections*, ed. G. T. Herman, pp. 147–246. Topics in Applied Physics, Vol. 32. Springer-Verlag, New York, 1979.

[21] Caspar, D. L. D., and De Rosier, D. J. The 1982 Nobel Prize in Chemistry. *Science*, 218, pp. 653–655, 1982.

[22] Chapman, C. H. The Radon transform and seismic tomography. In *Seismic Tomography*, ed. G. Nolet, pp. 25–47. D. Reidel, Dordrecht, Holland, 1987.

[23] Chapman, C. H., and Cary, P. W. The circular harmonic Radon transform. *Inverse Problems*, 2, pp. 23–49, 1986.

[24] Cho, Z. H., Jones, J. P., and Singh, M. *Foundations of Medical Imaging*, Wiley, New York, 1993.

[25] Churchill, R. V. *Operational Mathematics*, 3rd ed. McGraw-Hill, New York, 1972.

[26] Claerbout, J. F. *Imaging the Earth's Interior*, Blackwell Scientific, Palo Alto, CA, 1985.

[27] Clough, A. V., and Barrett, H. H. Attenuated Radon and Abel transforms. *J. Opt. Soc. Am.*, 73(11), pp. 1590–1595, 1983.

[28] Cormack, A. M. Representation of a function by its line integrals, with some radiological applications. *J. Appl. Phys.*, 34, pp. 2722–2727, 1963.

[29] Cormack, A. M. Representation of a function by its line integrals, with some radiological

applications. II. *J. Appl. Phys.*, 35, pp. 2908–2913, 1964.

[30] Cormack, A. M. Reconstruction of densities from their projections, with applications in radiological physics. *Phys. Med. Biol.*, 18, pp. 195–207, 1973.

[31] Cormack, A. M. Nobel Prize address, Dec. 8, 1979. Early two-dimensional reconstruction and recent topics stemming from it. *Med. Phys.*, 7, pp. 277–282, 1980.

[32] Cormack, A. M. Computed tomography: some history and recent developments. *Proc. Symposia Appl. Math.*, 27, pp. 35–42, 1982.

[33] Cormack, A. M. Radon's problem—old and new. *SIAM–AMS Proc., Inverse Problems*, 14, pp. 33–39, American Mathematical Society, Providence, RI, 1984.

[34] Davis, P. J., and Rabinowitz, P. *Methods of Numerical Integration*, 2nd ed. Academic Press, Orlando, FL, 1984.

[35] Deans, S. R. A Unified Radon inversion formula. *J. Math. Phys.*, 19(11), pp. 2346–2349, 1978.

[36] Deans, S. R. Gegenbauer transforms via the Radon transform. *SIAM J. Math. Anal.* 10(3), pp. 577–585, 1979.

[37] Deans, S. R. *The Radon Transform and Some of Its Applications*, Wiley-Interscience, New York, 1983.

[38] Deans, S. R. The Radon transform. In *Mathematical Analysis of Physical Systems*, ed. R. E. Mickens, pp. 81–133. Van Nostrand Reinhold, New York, 1985.

[39] Deans, S. R. *The Radon Transform and Some of Its Applications*, Revised edition. Krieger, Malbar, FL, 1993.

[40] Desbat, L. Efficient sampling on coarse grids in tomography. *Inverse Problems*, 9, pp. 251–269, 1993.

[41] Di Chiro, G., and Brooks, R. A. The 1979 Nobel Prize in physiology or medicine. *Science*, 205, pp. 1060–1062, 1979.

[42] Di Chiro, G., and Brooks, R. A. The 1979 Nobel Prize in physiology or medicine. *J. Comput. Assisted Tomog.*, 4, pp. 241–245, 1980.

[43] Doetsch, G. *Introduction to the Theory and Application of the Laplace Transformation*, Springer-Verlag, New York, 1974.

[44] Donoho, D. L. Nonlinear solution of linear inverse problems by wavelet-vaguelette decomposition, Tech. Rept., Statistics, Stanford, 1992, 1992.

[45] Duda, R. O., and Hart, P. E. Use of the Hough transform to detect lines and curves in pictures. *Commun. Assoc. Comput. Mach.*, 15, pp. 11–15, 1972.

[46] Du Mond, J. W. M. Compton modified line structure and its relation to the electron theory of solid bodies. *Phys. Rev.*, 33(5), pp. 643–658, 1929.

[47] Durand, L., Fishbane, P. M., and Simmons, L. M. Jr. Expansion formulas and addition theorems for Gegenbauer functions. *J. Math. Phys.* 17(11), pp. 1933–1948, 1976.

[48] Embree, P. M., and Kimball, B. *C Language Algorithms for Digital Signal Processing*, Prentice-Hall, Englewood Cliffs, NJ, 1991.

[49] Erdélyi, A., Magnus, W., Oberhettinger, F., and Tricomi, F. G. *Higher Transcendental Functions*, Vol. II. McGraw-Hill, New York, 1953.

[50] Erdélyi, A., Magnus, W., Oberhettinger, F., and Tricomi, F. G. *Tables of Integral Transforms*, Vol. II. McGraw-Hill, New York, 1954.

[51] Faridani, A. Reconstruction from efficiently sampled data in parallel-beam computed tomography. In *Inverse Problems and Imaging*, ed. G. F. Roach, 68–102. Pitman Research

Notes in Mathematics, Vol. 245. Longman Press, London, 1991.

[52] Faridani, A., Ritman, E. L., and Smith, K. T. Local tomography. *SIAM J. Appl. Math.*, 52(2), pp. 459–484, and Examples of local tomography. *SIAM J. Appl. Math.*, 52(4), pp. 1193–1198, 1992.

[53] Gel'fand, I. M., Graev, M. I., and Vilenkin, N. Ya. *Generalized Functions*, Vol. 5. Academic, New York, 1966.

[54] Gindikin, S., and Michor, P. *75 Years of Radon Transform*, ed. S. Gindikin and P. Michor. International Press, Cambridge, MA, 1994.

[55] Gorenflo, R., and Vessella, S. *Abel Integral Equations*, Lecture Notes in Mathematics 1461, ed. A. Dold, B. Eckmann, and F. Takens, Springer-Verlag, Berlin, 1991.

[56] Gradshteyn, I. S., Ryzhik, I. M., and Jeffrey, A. *Table of Integrals, Series, and Products*, 5th ed. Academic Press, San Diego, CA, 1994.

[57] Grinberg, E., and Quinto, E. T. Preface. In *Contemporary Mathematics*, vol. 113: *Integral Geometry and Tomography*, ed. E. Grinberg and E. T. Quinto. American Mathematical Society, Providence, RI, 1990.

[58] Gullberg, G. T. The reconstruction of fan-beam data by filtering the back-projection. *Comput. Graph. Image Proc.*, 10, pp. 30–47, 1979.

[59] Gullberg, G. T., Christian, P. E., Zeng, G. L., Datz, F. L., and Morgan, H. T. Cone beam tomography of the heart using single-photon emission-computed tomography. *Invest. Radiol.*, 26, pp. 681–688, 1991.

[60] Hamaker C., Smith, K. T., Solmon, D. C., and Wagner, S. L. The divergent beam x-ray transform. *Rocky Mountain J. Math.*, 10, pp. 253–283, 1980.

[61] Hansen, E. W. Circular harmonic image reconstruction: experiments. *Applied Optics*, 20, pp. 2266–2274, 1981.

[62] Hansen, E. W. Fast Hankel transform algorithm. *IEEE Trans. Acoust., Speech, Signal Processing*, 33, pp. 666–671, 1985. [Erratum, 34, pp. 623–624 (1986).]

[63] Harris, F. J. On the use of windows for harmonic analysis with the discrete Fourier transform. *Proc. IEEE*, 66(1), pp. 51–83, 1978.

[64] Hawkins, W. G., and Barrett, H. H. A numerically stable circular harmonic reconstruction algorithm. *SIAM J. Numer. Anal.*, 23: 873–890, 1986.

[65] Hawkins, W. G., Leichner, P. K., Yang, N-C. The circular harmonic transform for SPECT reconstruction and boundary conditions on the Fourier transform of the sinogram. *IEEE Trans. on Medical Imaging*, 7, pp. 135–148, 1988.

[66] Hazou, I. A., and Solmon, D. C. Filtered-backprojection and the exponential Radon transform. *J. Math. Anal. Appl.*, 141, 109–119, 1989.

[67] Helgason, S. The Radon transform on Euclidean spaces, compact two-point homogeneous spaces and Grassmann manifolds. *Acta Math.*, 113, pp. 153–180, 1965.

[68] Helgason, S. *The Radon Transform*, Birkhäuser, Boston, 1980.

[69] Herman, G. T. *Image Reconstruction from Projections*, Academic Press, New York, 1980.

[70] Higgins, W. E., and Munson, D. C. Jr. An algorithm for computing general integer-order Hankel transforms. *IEEE Trans. Acoust., Speech, Signal Processing*, 35, pp. 86–97, 1987.

[71] Higgins, W. E., and Munson, D. C. Jr. A Hankel transform approach to tomographic image reconstruction. *IEEE Trans. Medical Imaging*, 7, pp. 59–72, 1988.

[72] Hochstadt, H. *The Functions of Mathematical Physics*, Wiley, New York, 1971.

[73] Holschneider, M. Inverse Radon transforms through inverse wavelet transforms. *Inverse Problems*, 7, pp. 853–861, 1991.

[74] Hough, P. V. C. Method for recognizing complex patterns. U. S. Patent 3 096 654, 1962.

[75] Hounsfield, G. N. Nobel Prize address, Dec. 8, 1979. Computed medical imaging. *Med. Phys.*, 7, pp. 283–290, 1980. (also in *J. Comput. Assisted Tomog.*, 4, pp. 665–674 and *Science*, 210, pp. 22–28.)

[76] Jähne, B. *Digital Image Processing*, 2nd ed. Springer-Verlag, Berlin, 1993.

[77] Jain, A. K. *Fundamentals of Digital Image Processing*, Prentice-Hall, Englewood Cliffs, NJ, 1989.

[78] John, F. *Plane Waves and Spherical Means Applied to Partial Differential Equations*, Interscience, New York, 1955.

[79] Kaiser, G., and Streater, R. F. Windowed Radon transforms, analytic signals, and the wave function. In *Wavelets: A Tutorial in Theory and Applications*, ed. C. K. Chui, pp. 399–441. Academic Press, San Diego, CA, 1992.

[80] Kak, A. C. Image reconstructions from projections. In *Digital Image Processing Techniques*, ed. M. P. Ekstrom, pp. 111–167. Academic Press, Orlando, FL, 1984.

[81] Kak, A. C. Tomographic imaging with diffracting and nondiffracting sources. In *Array Signal Processing*, ed. S. Haykin, pp. 351–428. Prentice-Hall, Englewood Cliffs, NJ, 1985.

[82] Kak, A. C., and Slaney, M. *Principles of Computerized Tomographic Imaging*, IEEE Press, New York, 1988.

[83] Kanwal, R. P. *Linear Integral Equations*, Academic Press, New York, 1971.

[84] Kelley, B. T., and Madisetti, V. K. The fast discrete Radon transform—I: Theory. *IEEE Trans. Image Proc.*, 2(3), pp. 382–400, 1993.

[85] Kim, C-J., and Shannon, R. R. Catalog of Zernike polynomials. In *Applied Optics and Optical Engineering*, Vol. 10, ed. R. R. Shannon and J. C. Wyant, pp. 193–221. Academic Press, San Diego, CA, 1987.

[86] Knill, O. Diagonalization of Abel's integral operator. *SIAM J. Appl. Math.*, 54(5), pp. 1250–1253, 1994.

[87] Kruse, H. Resolution of reconstruction methods in computerized tomography. *SIAM J. Sci. Statist. Comput.*, 10, pp. 447–474, 1989.

[88] Kudo, H., and Saito, T. Feasible cone beam scanning methods for exact reconstruction in three-dimensional tomography. *J. Opt. Soc. Am. A.*, 7, pp. 2169–2183, 1990.

[89] Kudo, H., and Saito, T. Sinogram recovery with the method of convex projections for limited-data reconstruction in computed tomography. *J. Opt. Soc. Am. A.*, 8, pp. 1148–1160, 1991.

[90] Ladouceur, H. D., and Adiga, K. C. The application of Abel inversion to combustion diagnostics. In *Some Topics on Inverse Problems*, ed. P. C. Sabatier, pp. 369–390. World Scientific, Singapore, 1987.

[91] Lax, P. D., and Phillips, R. S. The Paley-Wiener theorem for the Radon transform. *Comm. Pure Appl. Math.*, 23, pp. 409–424, 1970.

[92] Lax, P. D., and Phillips, R. S. Translation representation for the solution of the non-Euclidean wave equation. *Comm. Pure Appl. Math.*, 32, pp. 617–667, 1979. [Correction. *Comm. Pure Appl. Math.*, 33, p. 685, 1980.]

[93] Lewitt, R. M. Reconstruction algorithms: transform methods. *Proc. IEEE*, 71, pp. 390–408, 1983

[94] Lide, D. R., ed. *Handbook of Chemistry and Physics*, 74th ed. CRC Press, Inc., Boca Raton, FL, 1993.

[95] Lindgren, A. G., and Rattey, P. A. The inverse discrete Radon transform with applications to tomographic imaging using projection data. In *Advances in Electronics and Electron Physics*, Vol. 56, ed. C. Marton, pp. 359–410. Academic Press, New York, 1981.

[96] Lighthill, M. J. *Fourier Analysis and Generalized Functions*, Cambridge University Press, Cambridge, 1962.

[97] Lonseth, A. T. Sources and applications of integral equations. *SIAM Review*, 19, pp. 241–278, 1977.

[98] Louis, A. K. Orthogonal function series expansions and the null space of the Radon transform. *SIAM J. Math. Anal.*, 15, pp. 621–633, 1984.

[99] Louis, A. K. Laguerre and computerized tomography: consistency conditions and stability of the Radon transform. In *Polynômes Orthogonaux et Applications*, ed. C. Brezinski et al., pp. 524–531. Lecture Notes in Mathematics 1171. Springer-Verlag, Berlin, 1985.

[100] Ludwig, D. The Radon transform on Euclidean space. *Comm. Pure Appl. Math.*, 19, pp. 49–81, 1966.

[101] Maass, P. The interior Radon transform. *SIAM J. Appl. Math.* 52(3), pp. 710–724, 1992.

[102] Macovski, A. *Medical Imaging Systems.* Prentice-Hall, Englewood Cliffs, NJ, 1983.

[103] Madych, W. R., and Nelson, S. A. Reconstruction from restricted Radon transform data: resolution and ill-conditionedness. *SIAM J. Math. Anal.*, 17, pp. 1447–1453, 1986.

[104] Madych, W. R. Summability and approximate reconstruction from Radon transform data. In *Contemporary Mathematics*, vol. 113: *Integral Geometry and Tomography*, ed. E. Grinberg and E. T. Quinto, 189–219. American Mathematical Society, Providence, RI, 1990.

[105] Maldonado, C. D., and Olsen, H. N. New method for obtaining emission coefficients from emitted spectral intensities. Part II: Asymmetrical sources. *J. Opt. Soc. Am.*, 56, pp. 1305–1313, 1966.

[106] Marr, R. B. On the reconstruction of a function on a circular domain from a sampling of its line integrals. *J. Math. Anal. Appl.*, 45, pp. 357–374, 1974.

[107] Mersereau, R. M. Direct Fourier transform techniques in 3-D image reconstruction. *Comput. Biol. Med.*, 6, pp. 247–258, 1976.

[108] Mersereau, R. M., and Oppenheim, A. V. Digital reconstruction of multidimensional signals from their projections. *Proc. IEEE*, 62, pp. 1319–1338, 1974.

[109] Mijnarends, P. E. Determination of anisotropic momentum distribution in positron annihilation. *Phys. Rev.* 160, pp. 512–519, 1967.

[110] Mikusiński, P., and Zayed, A. An extension of the Radon transform. In *Generalized Functions and their Applications*, ed. R. S. Pathak, pp. 141–147. Plenum Press, New York, 1993.

[111] Minerbo, G. N., and Levy, M. E. Inversion of Abel's integral equation by means of orthogonal polynomials. *SIAM J. Num. Anal.*, 6(4), pp. 598–616, 1969.

[112] Morse, P. M., and Feshbach, H. *Methods of Theoretical Physics*, McGraw-Hill, New York, 1953.

[113] Natterer, F. On the inversion of the attenuated Radon transform. *Numer. Math.*, 32, pp. 431–438, 1979.

[114] Natterer, F. *The Mathematics of Computerized Tomography*, Wiley, New York, 1986.

[115] Natterer, F. Sampling in fan beam tomography. *SIAM J. Appl. Math.*, 53(2), pp. 358–380, 1993.

[116] Nievergelt, Y. Elementary inversion of Radon's transform. *SIAM Review*, 28(1), pp. 79–84, 1986.

[117] Nievergelt, Y. Elementary inversion of the exponential x-ray transform. *IEEE Trans. Nucl. Sci.*, 38(2), pp. 873–876, 1991.

[118] Olson, T., and DeStefano, J. Wavelet localization of the Radon transform. *IEEE Trans. Sig. Proc.*, 42(8), pp. 2055–2067, 1994.

[119] Parker, J. A. *Image Reconstruction in Radiology*, CRC Press, Inc., Boca Raton, FL, 1990.

[120] Parks, P. C. On the determination of functions from their integral values along certain manifolds. (Translated by P. C. Parks) *IEEE Trans. on Medical Imaging*, 5, pp. 170–176, 1986.

[121] Radon, J. Über die Bestimmung von Funktionen durch ihre Integralwerte längs gewisser Mannigfaltigkeiten. *Berichte Sachsische Akademie der Wissenchaften, Leipzig, Mathematische-Physikalische Klasse*, vol. 69, pp. 262–267, 1917.

[122] Rainville, E. D. *Special Functions*, Chelsea, New York, 1960.

[123] Rattey, P. A., and Lindgren, A. G. Sampling the 2D Radon transform. *IEEE Trans. Acoust., Speech, Signal Processing*, 29(5), pp. 994–1002, 1981.

[124] Rizo, P., Grangeat, P., Sire, P., Lemasson, P., and Melennec, P. Comparison of two three-dimensional x-ray cone-beam-reconstruction algorithms with circular source trajectories. *J. Opt. Soc. Am. A*, 8, pp. 1639–1648, 1991.

[125] Rosenfeld, A., and Kak, A. C. *Digital Picture Processing*, 2nd ed., Vols. I and II. Academic Press, New York, 1982.

[126] Rowland, S. W. Computer implementation of image reconstruction formulas. In *Image Reconstruction from Projections*, ed. G. T. Herman, Topics in Applied Physics Vol. 32. Springer-Verlag, New York, 1979.

[127] Ruff, J. L. Tomographic imaging of seismic sources. In *Seismic Tomography*, ed. G. Nolet, pp. 339–366. D. Reidel, Dordrecht, Holland, 1987.

[128] Russ, J. C. *The Image Processing Handbook*, CRC Press, Inc., Boca Raton, FL, 1992.

[129] Schumann, W., Zürcher, J.-P., and Cuche, D. *Holography and Deformation Analysis*, Springer-Verlag, Berlin, 1985.

[130] Schwartz, L. *Mathematics for the Physical Sciences*, Addison-Wesley, Reading, MA, 1966.

[131] Scudder, H. J. Introduction to computer aided tomography. *Proc. IEEE*, 66, pp. 628–637, 1978.

[132] Sezan, M. I., and Stark, H. Tomographic image reconstruction from incomplete view data by convex projections and direct Fourier inversion. *IEEE Trans. Med. Imag.*, 3, pp. 91–98, 1984.

[133] Shapiro, S. D., and Iannino, A. Geometric constructions for predicting Hough transform performance. *IEEE Trans. Pattern Anal. Machine Intell.*, 1, pp. 310–317, 1979.

[134] Shepp, L. A. Computerized tomography and nuclear magnetic resonance. *J. Comput. Assisted Tomog.*, 4(1), pp. 94–107, 1980.

[135] Shepp, L. A., and Kruskal, J. B. Computerized tomography: The new medical x-ray technology. *Am. Math. Monthly*, 85: 420–439, 1978.

[136] Smith, K. T., Solmon, D. C., and Wagner, S. L. Practical and mathematical aspects of the problem of reconstructing objects from radiographs. *Bull. Am. Math. Soc.*, 83, pp. 1227–1270, 1977.

[137] Smith, P. R., Peters, T. M., and Bates, R. H. T. Image reconstruction from finite numbers of projections. *J. Phys. A: Math. Nucl. Gen.*, 6, pp. 361–382, 1973.

[138] Sneddon, I. N. *The Use of Integral Transforms*, McGraw-Hill, New York, 1972.

[139] Srivastava, H. M., and Bushman, R. G. *Theory and Applications of Convolution Integral Equations*, Kluwer Academic Publishers, Dordrecht, The Netherlands, 1992.

[140] Stark, H., Woods, J. W., Paul, I., and Hingorani, R. An investigation of computerized tomography by direct Fourier inversion and optimum interpolation. *IEEE Trans. Biomed. Eng.*, 28, pp. 496–505, 1981.

[141] Stewart, A. T. Momentum distribution of metallic electrons by positron annihilation. *Canadian J. of Phys.*, 35, pp. 168–183, 1957.

[142] Suter, B. W. Fast Nth-order Hankel transform algorithm. *IEEE Trans. Signal Processing*, 39, pp. 532–536, 1991.

[143] Swindell, W., and Barrett, H. H. Computerized tomography. *Physics Today*, 30(12), pp. 32–41, 1977.

[144] Swindell, W., and Webb, S. X-ray transmission computed tomography. In *The Physics of Medical Imaging*, ed. S. Webb, pp. 98–127. Institute of Physics Publishing, Bristol and Philadelphia, 1988.

[145] Szegö, G. *Orthogonal Polynomials*, American Mathematical Society Colloquium Publications, Vol. 23. American Mathematical Society, Providence, RI, 1939.

[146] Tretiak, O. J., and Metz, C. The exponential Radon transform. *SIAM J. Appl. Math.*, 39, pp. 341–354, 1980.

[147] Tricomi, F. G. *Integral Equations*, Dover Publications, New York, 1985.

[148] Van der Pol, B., and Weijers, Th. Tchebycheff polynomials and their relation to circular functions, Bessel functions and Lissajous-figures. *Physica*, 1, pp. 78–96, 1934.

[149] Vest, C. M. *Holographic Interferometry*, Wiley, New York, 1979.

[150] Vest, C. M., and Steel, D. G. Reconstruction of spherically symmetric objects from slit-imaged emission: Application to spatially resolved spectroscopy. *Optics Letters*, 3, pp. 54–56, 1978.

[151] Walnut, D. Applications of Gabor and wavelet expansions to the Radon transform. In *Probabilistic and Stochastic Methods in Analysis, with Applications*, ed. J. S. Byrnes et al., 187–205. Kluwer Academic Publishers, The Netherlands, 1992.

[152] Widder, D. V. *An Introduction to Transform Theory*, Academic Press, New York, 1971.

[153] Wilkins, J. E. Jr. Neumann series of Bessel functions. *Transactions Am. Math. Soc.*, 64, pp. 359–385, 1948.

[154] Wolf, K. B. *Integral Transforms in Science and Engineering*, Plenum Press, New York, 1979.

[155] Wood, J. C., and Barry, D. T. Tomographic time-frequency analysis and its application

toward time-varying filtering and adaptive kernel design for multicomponent linear-FM signals. *IEEE Trans. Sig. Proc.*, 42(8), pp. 2094–2104, and Linear signal synthesis using the Radon-Wigner transform. *IEEE Trans. Sig. Proc.*, 42(8), pp. 2105–2111, 1994.

[156] Zeitler, E. The reconstruction of objects from their projections. *Optik.*, 39, pp. 396–415, 1974.

9

The Hankel Transform

Robert Piessens

ABSTRACT Hankel transforms are integral transformations whose kernels are Bessel functions. They are sometimes referred to as Bessel transforms. When we are dealing with problems that show circular symmetry, Hankel transforms may be very useful. Laplace's partial differential equation in cylindrical coordinates can be transformed into an ordinary differential equation by using the Hankel transform. Because the Hankel transform is the two-dimensional Fourier transform of a circularly symmetric function, it plays an important role in optical data processing.

CONTENTS

9.1 Introductory definitions and properties

Bessel functions are solutions of the differential equation

$$x^2 y'' + xy' + (x^2 - p^2)y = 0 \tag{1}$$

where p is a parameter.

0-8493-8342-0/96/$0.00 + $0.50

Equation (1) can be solved using series expansions. The Bessel function $J_p(x)$ of the first kind and of order p is defined by

$$J_p(x) = \left(\frac{1}{2}x\right)^p \sum_{k=0}^{\infty} \frac{\left(-\frac{1}{4}x^2\right)^k}{k!\Gamma(p+k+1)}. \tag{2}$$

The Bessel function $Y_p(x)$ of the second kind and of order p is another solution that satisfies

$$W(x) = \det \begin{bmatrix} J_p(x) & Y_p(x) \\ J'_p(x) & Y'_p(x) \end{bmatrix} = \frac{2}{\pi x}.$$

Properties of Bessel function have been studied extensively (see [7, 22, 26]).

Elementary properties of the Bessel functions are:

1. *Asymptotic forms.*

$$J_p(x) \sim \sqrt{\frac{2}{\pi x}} \cos\left(x - \frac{1}{2}p\pi - \frac{1}{4}\pi\right), \qquad x \to \infty. \tag{3}$$

2. *Zeros.* $J_p(x)$ and $Y_p(x)$ have an infinite number of real zeros, all of which are simple, with the possible exception of $x = 0$. For nonnegative p the sth positive zero of $J_p(x)$ is denoted by $j_{p,s}$. The distance between two consecutive zeros tends to π: $\lim_{s \to \infty} (j_{p,s+1} - j_{p,s}) = \pi$.

3. *Integral representations.*

$$J_p(x) = \frac{\left(\frac{1}{2}x\right)^p}{\pi^{1/2}\Gamma(p+1/2)} \int_0^{\pi} \cos(x\cos\theta)\sin^{2p}\theta\, d\theta. \tag{4}$$

If p is a positive integer or zero, then

$$J_p(x) = \frac{1}{\pi} \int_0^{\pi} \cos(x\sin\theta - p\theta)\, d\theta$$

$$= \frac{j^{-n}}{\pi} \int_0^{\pi} e^{jx\cos\theta}\cos(p\theta)\, d\theta. \tag{5}$$

4. *Recurrence relations.*

$$J_{p-1}(x) - \frac{2p}{x}J_p(x) + J_{p+1}(x) = 0 \tag{6}$$

$$J_{p-1}(x) - J_{p+1}(x) = 2J'_p(x) \tag{7}$$

$$J'_p(x) = J_{p-1}(x) - \frac{p}{x}J_p(x) \tag{8}$$

$$J'_p(x) = -J_{p+1}(x) + \frac{p}{x}J_p(x). \tag{9}$$

5. *Hankel's repeated integral.* Let $f(r)$ be an arbitrary function of the real variable r, subject to the condition that

$$\int_0^{\infty} f(r)\sqrt{r}\, dr$$

is absolutely convergent. Then for $p \geq -1/2$

$$\int_0^\infty s\,ds \int_0^\infty f(r) J_p(sr) J_p(su) r\,dr = \frac{1}{2}[f(u+) + f(u-)] \qquad (10)$$

provided that $f(r)$ satisfies certain Dirichlet conditions.

For a proof, see [26]. The reader should also refer to Chapter 1, Section 5.6 for more information regarding Bessel functions.

9.2 Definition of the Hankel transform

Let $f(r)$ be a function defined for $r \geq 0$. The νth order Hankel transform of $f(r)$ is defined as

$$F_\nu(s) \equiv \mathcal{H}_\nu\{f(r)\} \equiv \int_0^\infty r f(r) J_\nu(sr)\,dr. \qquad (11)$$

If $\nu > -1/2$, Hankel's repeated integral immediately gives the inversion formula

$$f(r) = \mathcal{H}_\nu^{-1}\{F_\nu(s)\} \equiv \int_0^\infty s F_\nu(s) J_\nu(sr)\,ds. \qquad (12)$$

The most important special cases of the Hankel transform correspond to $\nu = 0$ and $\nu = 1$. Sufficient but not necessary conditions for the validity of (11) and (12) are

1. $f(r) = O(r^{-k})$, $r \to \infty$ where $k > 3/2$.
2. $f'(r)$ is piecewise continuous over each bounded subinterval of $[0, \infty)$.
3. $f(r)$ is defined as $[f(r+) + f(r-)]/2$.

These conditions can be relaxed.

9.3 Connection with the Fourier transform

We consider the two-dimensional Fourier transform of a function $\varphi(x, y)$, which shows a circular symmetry. This means that $\varphi(r \cos\theta, r \sin\theta) \equiv f(r, \theta)$ is independent of θ.

The Fourier transform of φ is

$$\Phi(\zeta, \eta) = \frac{1}{2\pi} \int_{-\infty}^\infty \int_{-\infty}^\infty f(x, y) e^{-j(x\zeta, y\eta)}\,dx\,dy. \qquad (13)$$

We introduce the polar coordinates

$$x = r\cos\theta, \qquad y = r\sin\theta$$

and

$$\zeta = s\cos\varphi, \qquad \eta = s\sin\varphi.$$

We have then

$$\phi(s\cos\varphi, s\sin\varphi) \equiv F(s,\varphi) = \frac{1}{2\pi}\int_0^\infty r\,dr \int_0^{2\pi} e^{-jrs\cos(\theta-\varphi)} f(r)\,d\theta$$

$$= \frac{1}{2\pi}\int_0^\infty rf(r)\,dr \int_0^{2\pi} e^{-jrs\cos\alpha}\,d\alpha$$

$$= \int_0^\infty rf(r)J_0(rs)\,dr.$$

This result shows that $F(s,\varphi)$ is independent of φ, so that we can write $F(s)$ instead of $F(s,\varphi)$. Thus, the two-dimensional Fourier transform of a circularly symmetric function is, in fact, a Hankel transform of order zero.

This result can be generalized: the N-dimensional Fourier transform of a circularly symmetric function of N variables is related to the Hankel transform of order $N/2 - 1$. If $f(r,\theta)$ depends on θ, we can expand it into a Fourier series

$$f(r,\theta) = \sum_{n=-\infty}^{\infty} f_n(r)e^{jn\theta} \tag{14}$$

and, similarly

$$F(s,\varphi) = \frac{1}{2\pi}\int_0^\infty r\,dr \int_0^\infty e^{-jrs\cos(\theta-\varphi)} f(r,\theta)\,d\theta = \sum_{n=-\infty}^{\infty} F_n(s)e^{jn\varphi} \tag{15}$$

where

$$f_n(r) = \frac{1}{2\pi}\int_0^{2\pi} f(r,\theta)e^{-jn\theta}\,d\theta \tag{16}$$

and

$$F_n(s) = \frac{1}{2\pi}\int_0^{2\pi} F(s,\varphi)e^{-jn\varphi}\,d\varphi. \tag{17}$$

Substituting (15) into (17) and using (14) we obtain

$$F_n(s) = \frac{1}{(2\pi)^2}\int_0^{2\pi} e^{-jn\varphi}\,d\varphi \int_0^{2\pi} d\theta \int_0^\infty f(r,\theta)e^{jsr\cos(\theta-\varphi)} r\,dr$$

$$= \frac{1}{(2\pi)^2}\int_0^{2\pi} e^{-jn\varphi}\,d\varphi \int_0^\infty r\,dr \int_0^{2\pi} e^{jsr\cos(\theta-\varphi)}\,d\theta \times \sum_{m=-\infty}^{\infty} f_m(r)e^{jm\theta}$$

$$= \frac{1}{(2\pi)}\int_0^\infty r\,dr \int_0^{2\pi} e^{-jn\alpha}e^{jsr\cos\alpha} f_n(r)\,d\alpha$$

$$= \int_0^\infty rf_n(r)J_n(sr)\,dr$$

$$= \mathcal{H}_n\{f_n(r)\}.$$

In a similar way, we can derive

$$f_n(r) = \mathcal{H}_n\{F_n(s)\}. \tag{18}$$

9.4   Properties and examples

Hankel transforms do not have as many elementary properties as do the Laplace or the Fourier transforms.

For example, because there is no simple addition formula for Bessel functions, the Hankel transform does not satisfy any simple convolution relation.

1. Derivatives. Let

$$F_\nu(s) = \mathcal{H}_\nu\{f(x)\}.$$

Then

$$G_\nu(s) = \mathcal{H}_\nu\{f'(x)\} = s\left[\frac{\nu+1}{2\nu}F_{\nu-1}(s) - \frac{\nu-1}{2\nu}F_{\nu+1}(s)\right]. \tag{19}$$

PROOF

$$G_\nu(s) = \int_0^\infty x f'(x) J_\nu(sx)\,dx$$

$$= [x f(x) J_\nu(sx)]_0^\infty - \int_0^\infty f(x)\frac{d}{dx}[x J_\nu(sx)]\,dx.$$

In general, the expression between the brackets is zero, and

$$\frac{d}{dx}[x J_\nu(sx)] = \frac{sx}{2\nu}[(\nu+1)J_{\nu-1}(sx) - (\nu-1)J_{\nu+1}(sx)].$$

Hence we have (19).   ∎

2. The Hankel transform of the Bessel differential operator. The Bessel differential operator

$$\Delta_\nu \equiv \frac{d^2}{dr^2} + \frac{1}{r}\frac{d}{dr} - \left(\frac{\nu}{r}\right)^2 = \frac{1}{r}\frac{d}{dr}r\frac{d}{dr} - \left(\frac{\nu}{r}\right)^2$$

is derived from the Laplacian operator

$$\nabla^2 = \frac{\partial^2}{\partial r^2} + \frac{1}{r}\frac{\partial}{\partial r} + \frac{1}{r^2}\frac{\partial^2}{\partial\theta^2} + \frac{\partial^2}{\partial z^2}$$

after separation of variables in cylindrical coordinates (r, θ, z).

Let $f(r)$ be an arbitrary function with the property that $\lim_{r\to\infty} f(r) = 0$. Then

$$\mathcal{H}_\nu\{\Delta_\nu f(r)\} = -s^2\mathcal{H}_\nu\{f(r)\}. \tag{20}$$

This result shows that the Hankel transform may be a useful tool in solving problems with cylindrical symmetry and involving the Laplacian operator.

PROOF   Integrating by parts we have

$$\mathcal{H}_\nu\{\Delta_\nu f(r)\} = \int_0^\infty \left[\frac{d}{dr} r \frac{df}{dr} - \frac{\nu^2}{r} f(r) \right] J_\nu(sr)\, dr$$

$$= \int_0^\infty \left[s^2 J_\nu''(sr) + \frac{s}{x} J_\nu'(sr) - \frac{\nu^2}{r^2} J_\nu(sr) \right] f(r) r\, dr$$

$$= -s^2 \int_0^\infty r f(r) J_\nu(rs)\, dr$$

$$= -s^2 \mathcal{H}_\nu\{f(r)\}.$$

This property is the principal one for applications of the Hankel transforms to solving differential equations [2, 3, 24, 25, 28].   ∎

3. Similarity.

$$\mathcal{H}_\nu\{f(ar)\} = \frac{1}{a^2} F_\nu\left(\frac{s}{a}\right). \tag{21}$$

4. Division by r.

$$\mathcal{H}_\nu\{r^{-1} f(r)\} = \frac{s}{2\nu} [F_{\nu-1}(s) + F_{\nu+1}(s)]. \tag{22}$$

5.

$$\mathcal{H}_\nu\left\{ r^{\nu-1} \frac{d}{dr}[x^{1-\nu} f(r)] \right\} = -s F_{\nu-1}(s). \tag{23}$$

6.

$$\mathcal{H}_\nu\left\{ r^{-\nu-1} \frac{d}{dr}[r^{\nu+1} f(r)] \right\} = s F_{\nu+1}(s). \tag{24}$$

7. Parseval's theorem. Let

$$F_\nu(s) = \mathcal{H}_\nu\{f(r)\}$$

and

$$G_\nu(s) = \mathcal{H}_\nu\{g(r)\}.$$

Then

$$\int_0^\infty F_\nu(s) G_\nu(s) s\, ds = \int_0^\infty F_\nu(s) s\, ds \int_0^\infty r\, g(r) J_\nu(sr)\, dr$$

$$= \int_0^\infty r\, g(r)\, dr \int_0^\infty s F_\nu(s) J_\nu(sx)\, ds$$

$$= \int_0^\infty r\, g(r) f(r)\, dr. \tag{25}$$

Example 9.1
From the Fourier pair (see Chapter 2) $\mathcal{F}\{e^{-a(x^2+y^2)}\} = (\pi/a) e^{-(\zeta^2+\eta^2)/4a}$ and the Fourier transform relationship $\mathcal{F}\{f(\sqrt{x^2+y^2})\} = 2\pi F_0(\sqrt{\zeta^2+\eta^2}) \equiv 2\pi F_0(s)$ we obtain the Hankel transform

$$\mathcal{H}\{e^{-ar^2}\} = \frac{1}{2a} e^{-s^2/4a}, \qquad a > 0. \qquad \blacksquare$$

Example 9.2
From the relationship $\int_0^a r J_0(sr)\,dr = \int_0^a (1/s)(d/dr)[r J_1(sr)] = [a J_1(as)]/s$ (see Chapter 1, Section 5.6) we conclude that

$$\mathcal{H}_0\{p_a(r)\} = \frac{a J_1(as)}{s}$$

where $p_a(r) = 1$ for $|r| < a$ and zero otherwise. ∎

Example 9.3
From the identity $\int_0^\infty J_0(sr)\,dr = 1/s$, $s > 0$ (see Chapter 1, Section 5.6), we obtain

$$\mathcal{H}_0\left\{\frac{1}{r}\right\} = \frac{1}{s}.$$

∎

Example 9.4
Since $\int_0^\infty r\delta(r - a) J_0(sr)\,dr = a J_0(as)$ (see Chapter 1, Section 2.4), we obtain

$$\mathcal{H}\{\delta(r - a)\} = a J_0(as), \qquad a > 0$$

and because of symmetry

$$\mathcal{H}\{a J_0(ar)\} = \delta(s - a), \qquad a > 0.$$

∎

CONVOLUTION IDENTITY
*Let $f_1(r)$ and $f_2(r)$ have Hankel transforms $F_1(s)$ and $F_2(s)$ respectively. From Example 4.1
above we have*

$$\mathcal{F}\left\{\iint_{-\infty}^{\infty} f_1\left(\sqrt{x_1^2 + y_1^2}\right) f_2\left(\sqrt{(x - x_1)^2 + (y - y_1)^2}\right) dx_1\,dy_1\right\} = 4\pi^2 F_1(s) F_2(s).$$

Hence, we have

$$H_0\{f_1(r) \star\star f_2(r)\} = \frac{1}{2\pi}\mathcal{F}_{(2)}\{f_1(r) \star\star f_2(r)\} = 2\pi F_1(s) F_2(s).$$

Therefore, to find the inverse Hankel transform of $2\pi F_1(s) F_2(s)$, we convolve $f_1(\sqrt{x^2 + y^2})$
with $f_2(\sqrt{x^2 + y^2})$, and in the answer we replace $\sqrt{x^2 + y^2}$ by r. We can also write the above
relationship in the form

$$\mathcal{H}_0\{2\pi f_1(r) f_2(r)\} = F_1(s) \star\star F_2(s).$$

Example 9.5
If $f_1(r) = f_2(r) = [J_1(ar)]/r$ then from the convolution identity above we obtain

$$\mathcal{H}_0\left\{2\pi \frac{J_1^2(ar)}{r^2}\right\} = \frac{1}{a} p_a(s) \star\star p_a(s)$$

where

$$p_a(s) \star\star p_a(s) = \left(2\cos^{-1}\frac{s}{2a} - \frac{s}{a}\sqrt{1 - \frac{s^2}{4a^2}}\right) a^2.$$

Hence

$$\mathcal{H}_0 \left\{ 2\pi \frac{J_1^2(ar)}{r^2} \right\} = \left(2\cos^{-1} \frac{s}{2a} - \frac{s}{a}\sqrt{1 - \frac{s^2}{4a^2}} \right) p_{2a}(s)$$

$$p_{2a}(s) = \begin{cases} 1 & |s| \leq 2a \\ 0 & \text{otherwise.} \end{cases}$$

∎

Example 9.6

From the definition, the Hankel transform of $r^\nu h(a - r)$, $a > 0$ is given by

$$\mathcal{H}_\nu\{r^\nu h(a - r)\} = \int_0^a r^{\nu+1} J_\nu(sr)\, dr = \frac{1}{s^{\nu+2}} \int_0^{as} x^{\nu+1} J_\nu(x)\, dx$$

since $h(a - r)$ is the unit step function with value equal to 1 for $r \leq a$ and 0 for $r > a$. But $\int t^\nu J_{\nu-1}(t)\, dt = t^\nu J_\nu(t) + C$ (see Chapter 1, Section 5.6, Table 5-6.1) and hence

$$\mathcal{H}_\nu\{r^\nu h(a - r)\} = \frac{(as)^{\nu+1}}{s^{\nu+2}} J_{\nu+1}(as) = \frac{a^{\nu+1}}{s} J_{\nu+1}(ap), \qquad a > 0,\ \nu > -\frac{1}{2}.$$

∎

Example 9.7

The Hankel transform of $r^{\nu-1}e^{-ar}$, $a > 0$ is given by

$$\mathcal{H}_\nu\{r^{\nu-1}e^{-ar}\} = \int_0^\infty r^\nu e^{-ar} J_\nu(sr)\, dr = \frac{1}{s^{\nu+1}} \int_0^\infty t^\nu J_\nu(t) e^{-\frac{a}{s}t}\, dt$$

$$= \frac{1}{s^{\nu+1}} \mathcal{L}\left\{ t^\nu J_\nu(t); p = \frac{a}{s} \right\}$$

where we set $t = rs$ and $\mathcal{L}$ is the Laplace transform operator (see also Chapter 5). But

$$t^\nu J_\nu(t) = \sum_{n=0}^\infty \frac{(-1)^n t^{2n+2\nu}}{n!\,\Gamma(n + \nu + 1)2^{2n+\nu}}$$

and hence

$$\mathcal{L}\{t^\nu J_\nu(t); p\} = \sum_{n=0}^\infty \frac{(-1)^n}{n!\,\Gamma(n + \nu + 1)2^{2n+\nu}} \mathcal{L}\{t^{2n+2\nu}; p\}$$

$$= \sum_{n=0}^\infty \frac{(-1)^n \Gamma(2n + 2\nu + 1)}{n!\,\Gamma(n + \nu + 1)2^{2n+\nu} p^{2n+2\nu+1}}.$$

From Chapter 1, Section 2.5, the duplication formula of the gamma function gives the relationship

$$\frac{\Gamma(2n + 2\nu + 1)}{\Gamma(n + \nu + 1)} = \frac{1}{\sqrt{\pi}} 2^{2n+2\nu} \Gamma\left(n + \nu + \frac{1}{2} \right)$$

and therefore the Laplace transform relation becomes

$$\mathcal{L}\{t^\nu J_\nu(t); p\} = \frac{2^\nu}{\sqrt{\pi}\, p^{2\nu+1}} \sum_{n=0}^\infty \frac{(-1)^n \Gamma(n + \nu + \frac{1}{2})}{n!} \left(\frac{1}{p^2} \right)^n.$$

The last series can be summed by using properties of the binomial series

$$(1 + x)^{-b} = \sum_{n=0}^\infty \binom{-b}{n} x^n = \sum_{n=0}^\infty \frac{(-1)^n \Gamma(n + b)}{n!\,\Gamma(b)} x^n, \qquad |x| < 1$$

where the relation

$$\binom{-b}{n} = \frac{(-1)^n b(b+1)\cdots(b+n-1)}{n!} = \frac{(-1)^n \Gamma(n+b)}{n!\Gamma(b)}$$

was used. The Laplace transform now becomes

$$\mathcal{L}\{t^\nu J_\nu(t); p\} = \frac{2^\nu \Gamma(\nu + \frac{1}{2})}{\sqrt{\pi}(p^2+1)^{\nu+\frac{1}{2}}} = \frac{2^\nu \Gamma(\nu + \frac{1}{2})}{\sqrt{\pi}\left[\left(\frac{a}{s}\right)^2 + 1\right]^{\nu+\frac{1}{2}}}, \qquad \mathrm{Re}(p) > 1$$

and hence

$$\mathcal{H}_\nu\{r^{\nu-1}e^{-ar}\} = \frac{1}{s^{\nu+1}} \frac{2^\nu \Gamma(\nu + \frac{1}{2})}{\sqrt{\pi}\left[\left(\frac{a}{s}\right)^2 + 1\right]^{\nu+\frac{1}{2}}} = \frac{s^\nu 2^\nu \Gamma(\nu + \frac{1}{2})}{\sqrt{\pi}(a^2 + s^2)^{\nu+\frac{1}{2}}}, \qquad \nu > -\frac{1}{2}.$$

If we set $\nu = 0$ and $a = 0$ in the above equation, we obtain the results of Example 4.3. If we set $\nu = 0$, we obtain

$$\mathcal{H}_0\{r^{-1}e^{-ar}\} = \frac{1}{\sqrt{a^2 + s^2}}, \qquad a > 0. \qquad \blacksquare$$

Example 9.8
The Hankel transform $\mathcal{H}_0\{e^{-ar}\}$ is given by

$$\mathcal{H}_0\{e^{-ar}\} = \mathcal{L}\{r J_0(sr); r \to a\} = -\frac{d}{da}\left[(s^2 + a^2)^{-\frac{1}{2}}\right]$$

$$= \frac{a}{[s^2 + a^2]^{3/2}}, \qquad a > 0$$

since multiplication by r corresponds to differentiation in the Laplace transform domain. $\blacksquare$

Example 9.9
From Chapter 1, Section 5.6, we have the following identity:

$$\frac{d^2 r_n(x)}{dx^2} + \frac{1}{x}\frac{dr_n(x)}{dx} + r_n(x) = 2n r_{n+1}(x),$$

where $r_n(x) = J_n(x)/x^n$.

Using the Hankel transform property of the Bessel operator we obtain the relationship

$$(1 - s^2) R_n(s) = 2n R_{n+1}(s)$$

or

$$R_{n+1}(s) = \frac{1 - s^2}{2n} R_n(s) = \cdots = \frac{(1 - s^2)^n}{2^n n!} R_1(s).$$

But from Example 4.2, $\mathcal{H}_0\left\{\frac{J_1(r)}{r}\right\} = p_1(s)$ and hence

$$\mathcal{H}_0\left\{\frac{J_n(r)}{r^n}\right\} = \frac{(1 - s^2)^{n-1}}{2^{n-1}(n-1)!} p_1(s)$$

where $p_1(s)$ is a pulse of width 2 centered at $s = 0$. $\blacksquare$

Example 9.10

If the impulse response of a linear space invariant system is $h(r)$ and the input to the system is $f(r)$, then its output is $g(r) = f(r) \star \star h(r)$ and, hence,

$$G(s) = 2\pi F(s) H(s).$$

Since $\mathcal{H}_0\{J_0(ar)\} = [\delta(s - a)]/a$ (see Example 9.4) and $\varphi(s)\delta(s - a) = \varphi(a)\delta(s - a)$, we conclude that if the input is $f(r) = J_0(ar)$ then

$$G(s) = \frac{2\pi}{a}\delta(s - a) H(s) = \frac{2\pi H(a)}{a}\delta(s - a).$$

Therefore, the output is

$$g(r) = 2\pi H(a) J_0(ar). \qquad \blacksquare$$

9.5   Applications

9.5.1   *The electrified disc*

Let v be the electric potential due to a flat circular electrified disc, with radius $R = 1$, the center of the disc being at the origin of the three-dimensional space and its axis along the z-axis.

In polar coordinates, the potential satisfies Laplace's equation

$$\nabla^2 v \equiv \frac{\partial^2 v}{\partial r^2} + \frac{1}{r}\frac{\partial v}{\partial r} + \frac{\partial^2 v}{\partial z^2} = 0. \tag{26}$$

The boundary conditions are

$$v(r, 0) = v_0, \qquad 0 \leq r < 1 \tag{27}$$

$$\frac{\partial v}{\partial z}(r, 0) = 0, \qquad r > 1. \tag{28}$$

In (27), v_0 is the potential of the disc. Condition (28) arises from the symmetry about the plane $z = 0$.

Let

$$V(s, z) = \mathcal{H}_0\{v(r, z)\}$$

so that

$$\mathcal{H}_0\{\nabla^2 v\} = -s^2 V(s, z) + \frac{\partial^2 V}{\partial z^2}(s, z) = 0.$$

The solution of this differential equation is

$$V(s, z) = A(s)e^{-sz} + B(s)e^{sz}$$

where A and B are functions that we have to determine using the boundary conditions.

Because the potential vanishes as z tends to infinity, we have $B(s) \equiv 0$. By inverting the Hankel transform we have

$$v(r, z) = \int_0^\infty s A(s)e^{-sz} J_0(sr)\,ds. \tag{29}$$

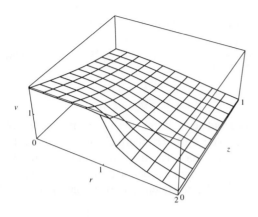

FIGURE 9.1
Electrical potential due to an electrified disc.

The boundary conditions are now

$$v(r, 0) = \int_0^\infty s A(s) J_0(rs)\, ds = v_0, \qquad 0 \le r < 1 \tag{30}$$

$$\frac{\partial v}{\partial z}(r, 0) = \int_0^\infty s^2 A(s) J_0(rs)\, ds = 0, \qquad r > 1. \tag{31}$$

Using entries (8) and (9) of Table 9.1 (see Section 9.11) we see that $A(s) = \sin s / s^2$ so that

$$v(r, z) = \frac{2v_0}{\pi} \int_0^\infty \frac{\sin s}{s} e^{-sz} J_0(sr)\, ds. \tag{32}$$

In Figure 9.1, the graphical representation of $v(r, z)$ for $v_0 = 1$ is depicted on the domain $0 \le r \le 2, 0 \le z \le 1$. The evaluation of $v(r, z)$ requires numerical integration.

Equations (30) and (31) are special cases of the more general pair of equations

$$\int_0^\infty f(t) t^{2\alpha} J_\nu(xt)\, dt = a(x), \qquad 0 \le x < 1 \tag{33}$$

$$\int_0^\infty f(t) J_\nu(xt)\, dt = 0, \qquad x > 1 \tag{34}$$

where $a(x)$ is given and $f(x)$ is to be determined.

The solution of (29) can be expressed as a repeated integral (see Luke [12]):

$$f(x) = \frac{2^{-\alpha} x^{1-\alpha}}{\Gamma(\alpha+1)} \int_0^1 s^{-\nu-\alpha} J_{\nu+\alpha}(xs) \frac{d}{ds} \int_0^s a(t) t^{\nu+1} (s^2 - t^2)^\alpha\, dt\, ds, \quad -1 < \alpha < 0 \tag{35}$$

$$f(x) = \frac{(2x)^{1-\alpha}}{\Gamma(\alpha)} \int_0^1 s^{-\nu-\alpha+1} J_{\nu+\alpha}(xs) \int_0^s a(t) t^{\nu+1} (s^2 - t^2)^{\alpha-1}\, dt\, ds, \quad 0 < \alpha < 1. \tag{36}$$

If $a(x) = x^\beta$, and $\alpha < 1, 2\alpha + \beta > -3/2, \alpha + \nu > -1, \nu > -1$ then

$$f(x) = \frac{\Gamma\left(1 + \frac{\beta+\nu}{2}\right) x^{-(2\alpha+\beta+1)}}{2^\alpha \Gamma\left(1 + \alpha + \frac{\beta+\nu}{2}\right)} \int_0^x t^{\alpha+\beta+1} J_{\nu+\alpha}(t)\, dt. \tag{37}$$

With $\beta = \nu$ and $\alpha < 1, \alpha + \nu > -1, \nu > -1$ further simplification is possible:

$$f(x) = \frac{\Gamma(\nu + 1)}{(2x)^\alpha \Gamma(\nu + \alpha + 1)} J_{\nu+\alpha+1}(x). \tag{38}$$

9.5.2   Heat conduction

Heat is supplied at a constant rate Q per unit area and per unit time through a circular disc of radius a in the plane $z = 0$, to the semi-infinite space $z > 0$. The thermal conductivity of the space is K. The plane $z = 0$ outside the disc is insulated. The mathematical model of this problem is very similar to that of Section 5.1. The temperature is denoted by $v(r, z)$. We have again the Laplace equation (26) in polar coordinates, but the boundary conditions are now

$$-K\frac{\partial v(r, z)}{\partial z} = Q, \qquad r < a, z = 0$$
$$= 0, \qquad r > a, z = 0. \tag{39}$$

The Hankel transform of the differential equation is again

$$\frac{\partial^2 V}{\partial z^2}(s, z) - s^2 V(s, z) = 0. \tag{40}$$

We can now transform also the boundary condition, using formula (3) in Table 9.1:

$$-K\frac{\partial V}{\partial z}(s, 0) = Qa\, J_1(as)/s. \tag{41}$$

The solution of (39) must remain finite as z tends to infinity. We have

$$V(s, z) = A(s)e^{-sz}.$$

Using condition (41) we can determine

$$A(s) = Qa\, J_1(as)/(Ks^2).$$

Consequently, the temperature is given by

$$v(r, z) = \frac{Qa}{K} \int_0^\infty e^{-sz} J_1(as) J_0(rs)s^{-1}\, ds. \tag{42}$$

9.5.3   The Laplace equation in the halfspace $z > 0$, with a circularly symmetric Dirichlet condition at $z = 0$

We try to find the solution $v(r, z)$ of the boundary value problem

$$\begin{cases} \dfrac{\partial^2 v}{\partial r^2} + \dfrac{1}{r}\dfrac{\partial v}{\partial r} + \dfrac{\partial^2 v}{\partial z^2} = 0, & z > 0, 0 < r < \infty \\ v(r, 0) = f(r). \end{cases} \tag{43}$$

Taking the Hankel transform of order 0 yields

$$\frac{\partial^2 V}{\partial z^2}(s, z) - s^2 V(s, z) = 0$$

and

$$V(s, 0) = \int_0^\infty rf(r) J_0(sr)\, dr.$$

The solution is

$$V(s, z) = e^{-sz} \int_0^\infty r f(r) J_0(sr) \, dr$$

so that

$$v(r, z) = \int_0^\infty s \, e^{-sz} J_0(sr) \, ds \int_0^\infty p \, f(p) J_0(sp) \, dp. \qquad (44)$$

For the special case

$$f(r) = h(a - r)$$

where $h(r)$ is the unit step function, we have the solution

$$v(r, z) = a \int_0^\infty e^{-sz} J_0(sr) J_1(as) \, ds. \qquad (45)$$

9.5.4 An electrostatic problem

The electrostatic potential $Q(r, z)$ generated in the space between two grounded horizontal plates at $z = \pm \ell$ by a point charge q at $r = 0, z = 0$ shows a singular behavior at the origin. It is given by

$$v(r, z) = \varphi(r, z) + q(r^2 + z^2)^{-1/2} \qquad (46)$$

where $\varphi(r, z)$ satisfies Laplace's equation (26). The boundary conditions are

$$\varphi(r, \pm \ell) + q(r^2 + \ell^2)^{-1/2} = 0. \qquad (47)$$

Taking the Hankel transform of order 0, we obtain

$$\frac{\partial^2 \Phi}{\partial z^2}(s, z) - s^2 \Phi(s, z) = 0 \qquad (48)$$

$$\Phi(s, \pm \ell) = -\frac{q \, e^{-s\ell}}{s} \qquad (49)$$

(see formula (18) in Table 9.1).

The solution is

$$A(s)e^{-sz} + B(s)e^{sz}$$

where $A(s)$ and $B(s)$ must satisfy

$$A(s)e^{+s\ell} + B(s)e^{-s\ell} = -\frac{q \, e^{-s\ell}}{s}$$

$$A(s)e^{-s\ell} + B(s)e^{s\ell} = -\frac{q \, e^{-s\ell}}{s}.$$

Hence

$$A(s) = B(s) = -\frac{q \, e^{-s\ell}}{2s \cosh(s\ell)}$$

and

$$\Phi(s, z) = -\frac{q \, e^{-s\ell}}{s} \frac{\cosh(sz)}{\cosh(s\ell)}.$$

Hence

$$\varphi(r, z) = \frac{q}{\sqrt{r^2 + z^2}} - q \int_0^\infty e^{-s\ell} \frac{\cosh(sz)}{\cosh(s\ell)} J_0(sr) \, ds. \qquad (50)$$

9.6   The finite Hankel transform

We consider the integral transformation

$$F_v(\alpha) = H_v\{f, \alpha\} = \int_0^1 r f(r) J_v(\alpha r)\, dr. \tag{51}$$

A property of this transformation is that

$$H_v(\Delta_v f, \alpha) = -\alpha^2 F_v(\alpha) + \left[J_v(\alpha) f'(1) - \alpha J_v'(\alpha) f(1) \right]$$

where Δ_v is the Bessel differential operator.

If α is equal to the sth positive zero $j_{v,s}$ of $J_v(x)$ we have

$$H_v(\Delta_v f, j_{v,s}) = -j_{v,s}^2 H_v(f, j_{v,s}) + j_{v,s} J_{v+1}(j_{v,s}) f(1).$$

If α is equal to the sth positive root $\beta_{v,s}$ of

$$h J_v(x) + x J_v'(x) = 0$$

where h is a nonnegative constant, we have

$$H_v(\Delta_v f, \beta_{v,s}) = -\beta_{v,s}^2 H_v(f, \beta_{v,s}) + J_v(\beta_{v,s}) \left[h f(1) + f'(1) \right].$$

The transformation (51) with $\alpha = j_{v,s}$, $s = 1, 2, \ldots$ is the finite Hankel transform. It maps the function $f(r)$ into the vector $(F_v(j_{v,1}), F_v(j_{v,2}), F_v(j_{v,3}) \ldots)$. The inversion formula can be obtained from the well-known theory of Fourier-Bessel series

$$f(r) = 2 \sum_{s=1}^{\infty} \frac{F_v(j_{v,s})}{J_{v+1}^2(j_{v,s})} J_v(j_{v,s} r). \tag{52}$$

The transformation (51) with $\alpha = \beta_{v,s}$, $s = 1, 2, \ldots$ is the modified finite Hankel transform. The inversion formula is

$$f(r) = 2 \sum_{s=1}^{\infty} \frac{\beta_{v,s}^2 F_v(\beta_{v,s})}{h^2 + \beta_{v,s}^2 - v^2} \frac{J_v(\beta_{v,s} r)}{J_v^2(\beta_{v,s})}. \tag{53}$$

If $h = 0$, $\beta_{v,s}$ is the sth positive zero of $J_v'(x)$, denoted by $j_{v,s}'$.

Formulas for the computation of $j_{v,s}$ and $j_{v,s}'$ are given by Olver [15]. Values of $j_{v,s}$ and $j_{v,s}'$ are tabulated in [1]. A Fortran program for the computation of $j_{v,s}$ and $j_{v,s}'$ is given in [18].

Application
We calculate the temperature $v(r, t)$ at time t of a long solid cylinder of unit radius. The initial temperature is unity, and radiation takes place at the surface into the surrounding medium maintained at zero temperature.

The mathematical model of this problem is the diffusion equation in polar coordinates

$$\frac{\partial^2 v}{\partial r^2} + \frac{1}{r} \frac{\partial v}{\partial r} = \frac{\partial v}{\partial t}, \qquad 0 \le r < 1, \ \ t > 0 \tag{54}$$

The initial condition is

$$v(r, 0) = 1, \qquad 0 \le r \le 1. \tag{55}$$

The radiation at the surface of the cylinder is described by the mixed boundary condition

$$\frac{\partial v}{\partial r}(1,t) = -hv(1,t) \tag{56}$$

where h is a positive constant.

Transformation of (54) by the modified finite Hankel transform yields

$$\frac{dV}{dt}(\beta_{0,s},t) = -\beta_{0,s}^2 V(\beta_{0,s},t) \tag{57}$$

where

$$V(\alpha,t) = \int_0^1 rv(r,t)J_0(\alpha r)\,dr$$

so that

$$V(\alpha,0) = \int_0^1 rJ_0(\alpha r)\,dr = \frac{J_1(\alpha)}{\alpha}. \tag{58}$$

The solution of (57), with the initial condition (58), is

$$V(\beta_{0,s},t) = \frac{J_1(\beta_{0,s})}{\beta_{0,s}}e^{-\beta_{0,s}^2 t}.$$

Using the inversion formula, we obtain

$$v(r,t) = 2\sum_{j=1}^{\infty} e^{-\beta_{0,s}^2 t}\frac{\beta_{0,s}J_1(\beta_{0,s})}{h^2+\beta_{0,s}^2}\frac{J_0(\beta_{0,s}r)}{J_0^2(\beta_{0,s})}. \tag{59}$$

9.7   Related transforms

For some applications, Hankel transforms with a more general kernel may be useful. We give one example.

We consider the cylinder function

$$Z_\nu(s,r) = J_\nu(sr)Y_\nu(s) - Y_\nu(sr)J_\nu(s). \tag{60}$$

Using this function as a kernel, we can construct the following transform pair:

$$F_\nu(s) = \int_1^\infty rf(r)Z_\nu(s,r)\,dr \tag{61}$$

$$f(r) = \int_0^\infty sF_\nu(s)\frac{Z_\nu(s,r)}{J_\nu^2(s)+Y_\nu^2(s)}\,ds. \tag{62}$$

The inversion formula follows immediately from Weber's integral theorem (see Watson [26]):

$$\int_1^\infty u\,du \int_0^\infty f(s)Z_\nu(r,u)Z_\nu(s,u)s\,ds = \frac{1}{2}\left[J_\nu^2(r)+Y_\nu^2(r)\right]\left[f(r+)+f(r-)\right]. \tag{63}$$

For this reason, we will refer to (61) and (62) as the Weber transform. This transform has the following important property:

If

$$f(x) = g''(x) + \frac{1}{x}g'(x) - \frac{v^2}{x^2}g(x) \tag{64}$$

then

$$F_v(s) = -s^2 G_v(s) - \frac{2}{\pi}g(1). \tag{65}$$

We may expect that this transform is useful for solving Laplace's equation in cylindrical coordinates, with a boundary condition at $r = 1$.

Example

We want to compute the steady-state temperature $u(r, z)$ in a horizontal infinite homogeneous slab of thickness 2ℓ, through which there is a vertical circular hole of radius 1. The horizontal faces are held at temperature zero and the circular surface in the hole is at temperature T_0.

The mathematical model is

$$\frac{\partial^2 u}{\partial r^2} + \frac{1}{r}\frac{\partial u}{\partial r} + \frac{\partial^2 u}{\partial z^2} = 0$$

$$u(r, \ell) = u(r, -\ell) = 0$$

$$u(1, z) = T_0. \tag{66}$$

Taking the Weber transform of order zero, we have

$$\frac{\partial^2 U_0}{\partial z^2}(s, z) - s^2 U_0(s, z) = \frac{2}{\pi}T_0.$$

The solution of this ordinary differential equation, satisfying the boundary condition, is

$$U_0(s, z) = \frac{2T_0}{\pi s^2}\left[\frac{\cosh sz}{\cosh s\ell} - 1\right].$$

Consequently, we have

$$u(r, z) = \frac{2T_0}{\pi}\int_0^\infty \frac{1}{s}\left[\frac{\cosh sz}{\cosh s\ell} - 1\right]\frac{Z_0(s, r)}{J_0^2(s) + Y_0^2(s)}\,ds \qquad (r > 1). \tag{67}$$

9.8   Need of numerical integration methods

When using the Hankel transform for solving partial differential equations, the solution is found as an integral of the form

$$I(a, p, v) = \int_0^a J_v(px)f(x)\,dx \tag{68}$$

where a is a positive real number or infinite. In most cases, analytical integration of (68) is impossible, and numerical integration is necessary. But integrals of type (68) are difficult to evaluate numerically if

1. the product ap is large

2. a is infinite

3. $f(x)$ shows a singular or oscillatory behavior.

In cases 1 and 2 the difficulties arise from the oscillatory behavior of $J_\nu(x)$ and they grow when the oscillations become stronger.

We give here a survey of numerical methods that are especially suited for the evaluation of $I(a, p, \nu)$ when ap is large or a is infinite. We restrict ourselves to cases where $f(x)$ is smooth, or where $f(x) = x^\alpha g(x)$ where $g(x)$ is smooth and α is a real number.

9.9 Computation of Bessel function integrals over a finite interval

Integral (68) can be written as

$$I(a, p, \nu) = a^{\alpha+1} \int_0^1 x^\alpha J_\nu(apx)g(ax)\,dx. \tag{69}$$

We assume that $\alpha + \nu > -1$. If $(\alpha + \nu)$ is not an integer, then there is an algebraic singularity of the integrand at $x = 0$. If ap is large, then the integrand is strongly oscillatory.

If $(\alpha + \nu)$ is an integer and ap is small, classical numerical integration methods, such as Romberg integration, Clenshaw-Curtis integration or Gauss-Legendre integration (see Davis and Rabinowitz [4]) are applicable. If $(\alpha + \nu)$ is not an integer, and ap is small, the only difficulty is the algebraic singularity at $x = 0$, and Gauss-Jacobi quadrature or Iri-Moriguti-Takesawa (IMT) integration [4] can be used. If ap is large, special methods should be applied that take into account the oscillatory behavior of the integrand. We describe two methods here.

9.9.1 *Integration between the zeros of $J_\nu(x)$*

We denote the sth positive zero of $J_\nu(x)$ by $j_{\nu,s}$ and we set $j_{\nu,0} = 0$. Then

$$I(a, p, \nu) = \sum_{k=1}^N (-1)^{k+1} I_k + \int_{j_{\nu,N}/p}^a J_\nu(px)f(x)\,dx \tag{70}$$

where

$$I_k = \int_{j_{\nu,k-1}/p}^{j_{\nu,k}/p} |J_\nu(px)|f(x)\,dx \tag{71}$$

and where N is the largest natural number for which $j_{\nu,N} \le ap$. This means that N is large when ap is large.

Using a transformation attributed to Longman [10], the summation in equation (70) can be written as

$$S = \sum_{k=1}^N (-1)^{k+1} I_k = \frac{1}{2}I_1 - \frac{1}{4}\Delta I_1 + \frac{1}{8}\Delta^2 I_1 + \cdots + (-1)^{p-1}2^{-p}\Delta^{p-1}I_1$$

$$+ \ (-1)^{N-1}\left[\frac{1}{2}I_N + \frac{1}{4}\Delta I_{N-1} + \frac{1}{8}\Delta^2 I_{N-2} + \cdots + 2^{-p}\Delta^{p-1}I_{n-p+1}\right]$$

$$+ \ 2^{-p}(-1)^p\left[\Delta^p I_1 - \Delta^p I_2 + \Delta^p I_3 - \cdots + (-1)^{N-1-p}\Delta^p I_{N-p}\right].$$

Assuming now that N and p are large and that high-order differences are small, the last bracket may be neglected and

$$S \simeq \frac{1}{2}I_1 - \frac{1}{4}\Delta I_1 + \frac{1}{8}\Delta^2 I_1 - \cdots$$

$$+ (-1)^{N-1}\left[\frac{1}{2}I_N + \frac{1}{4}\Delta I_{N-1} + \frac{1}{8}\Delta^2 I_{N-2} + \cdots\right]. \tag{72}$$

The summations in (72) may be truncated as soon as the terms are small enough. For the evaluation of $I_k, k = 1, 2, \ldots$, classical integration methods (e.g., Lobatto's rule) can be used, but special Gauss quadrature formulas (see Piessens [16]) are more efficient. If the integral I_1 has an algebraic singularity at $x = 0$, then the Gauss-Jacobi rules or the IMT rule are recommended.

9.9.2 *Modified Clenshaw-Curtis quadrature*

The Clenshaw-Curtis quadrature method is a well-known and efficient method for the numerical evaluation of an integral I with a smooth integrand. This method is based on a truncated Chebyshev series approximation of the integrand. However, when the integrand shows a singular or strongly oscillatory behavior, the classical Clenshaw-Curtis method is not efficient or even applicable, unless it is modified in an appropriate way, taking into account the type of difficulty of the integrand. We call this method then a modified Clenshaw-Curtis method (MCC method). The principle of the MCC method is the following: The integration interval is mapped onto $[-1, +1]$ and the integrand is written as the product of a smooth function $g(x)$ and a weight function $w(x)$ containing the singularities or the oscillating factors of the integrand, that is,

$$I = \int_{-1}^{+1} w(x)g(x)\,dx. \tag{73}$$

The smooth function is then approximated by a truncated series of Chebyshev polynomials

$$g(x) \simeq \sum_{k=0}^{N}{}' c_k T_k(x), \qquad -1 \le x \le 1. \tag{74}$$

Here the symbol $\sum'$ indicates that the first term in the sum must be halved. For the computation of the coefficients c_k in equation (74) several good algorithms, based on the Fast Fourier Transform, are available.

The integral in (73) can now be approximated by

$$I \simeq \sum_{k=0}^{N}{}' c_k M_k \tag{75}$$

where

$$M_k = \int_{-1}^{+1} w(x)T_k(x)\,dx$$

are called modified moments.

The integration interval may also be mapped onto $[0, 1]$ instead of $[-1, 1]$, but then the shifted Chebyshev polynomials $T_k^\star(x)$ are to be used.

We now consider the computation of the integral (69),

$$I = \int_0^1 x^\alpha J_\nu(\omega x)g(x)\,dx. \tag{76}$$

If

$$g(x) \simeq \sum_{k=0}^{N}{}' c_k T_k'(x) \qquad (77)$$

then

$$I \simeq \sum_{k=0}^{N}{}' c_k M_k(\omega, \nu, \alpha) \qquad (78)$$

where

$$M_k(\omega, \nu, \alpha) = \int_0^1 x^\alpha J_\nu(\omega x) T_k^\star(x)\, dx. \qquad (79)$$

These modified moments satisfy the following homogeneous, linear, nine-term recurrence relation:

$$\frac{\omega^2}{16} M_{k+4} + \left[(k+3)(k+3+2\alpha) + \alpha^2 - \nu^2 - \frac{\omega^2}{4} \right] M_{k+2}$$

$$+ \left[4(\nu^2 - \alpha^2) - 2(k+2)(2\alpha - 1) \right] M_{k+1}$$

$$- \left[2(k^2 - 4) + 6(\nu^2 - \alpha^2) - 2(2\alpha - 1) - \frac{3\omega^2}{8} \right] M_k$$

$$+ \left[4(\nu^2 - \alpha^2) + 2(k-2)(2\alpha - 1) \right] M_{k-1}$$

$$+ \left[(k-3)(k-3-2\alpha) + \left(\alpha^2 - \nu^2 - \frac{\omega^2}{4} \right) \right] M_{k-2} + \frac{\omega^2}{16} M_{k-4} = 0 . \qquad (80)$$

Because of the symmetry of the recurrence relation of the shifted Chebyshev polynomials, it is convenient to define

$$T_{-k}^\star(x) = T_k^\star(x), \qquad k = 1, 2, 3, \ldots$$

and consequently

$$M_{-k}(\omega, \nu, \alpha) = M_k(\omega, \nu, \alpha).$$

To start the recurrence relation with $k = 0, 1, 2, 3, \ldots$ we need only M_0, M_1, M_2, and M_3. Using the explicit expressions of the shifted Chebyshev polynomials we obtain

$$M_0 = G(\omega, \nu, \alpha)$$

$$M_1 = 2G(\omega, \nu, \alpha + 1) - G(\omega, \nu, \alpha)$$

$$M_2 = 8G(\omega, \nu, \alpha + 2) - 8G(\omega, \nu, \alpha + 1) + G(\omega, \nu, \alpha)$$

$$M_3 = 32G(\omega, \nu, \alpha + 3) - 48G(\omega, \nu, \alpha + 2) + 18G(\omega, \nu, \alpha + 1) - G(\omega, \nu, \alpha) \qquad (81)$$

where

$$G(\omega, \nu, \alpha) = \int_0^1 x^\alpha J_\nu(\omega x)\, dx. \qquad (82)$$

Because

$$\omega^2 G(\omega, \nu, \alpha + 2) = \left[\nu^2 - (\alpha + 1)^2 \right] G(\omega, \nu, \alpha)$$

$$+ (\alpha + \nu + 1) J_\nu(\omega) - \omega J_{\nu-1}(\omega) \qquad (83)$$

we need only $G(\omega, \nu, \alpha)$ and $G(\omega, \nu, \alpha + 1)$.

Luke [12] has given the following formulas:

1. a Neumann series expansion that is suitable for small ω

$$G(\omega, \nu, \alpha) = \frac{2}{\omega(\alpha + \nu + 1)} \sum_{k=0}^{\infty} \frac{(\nu + 2k + 1)\left(\frac{\nu - \alpha + 1}{2}\right)}{\left(\frac{\nu + \alpha + 3}{2}\right)_k} J_{\nu + 2k + 1}(\omega) \qquad (84)$$

2. an asymptotic expansion that is suitable for large ω

$$G(\omega, \nu, \alpha) = \frac{2^\alpha}{\omega^{\alpha + 1}} \frac{\Gamma\left(\frac{\nu + \alpha + 1}{2}\right)}{\Gamma\left(\frac{\nu - \alpha + 1}{2}\right)} - \sqrt{\frac{2}{\pi \omega^3}} (g_1 \cos\theta + g_2 \sin\theta) \qquad (85)$$

where

$$\theta = \omega - \nu\pi/2 + \pi/4$$

and

$$g_1 \sim \sum_{k=0}^{\infty} (-1)^k a_{2k} \omega^{-2k}, \qquad \omega \to \infty$$

$$g_2 \sim \sum_{k=0}^{\infty} (-1)^k a_{2k+1} \omega^{-2k-1}, \qquad \omega \to \infty$$

$$a_k = \frac{(1/2 - \nu)_k (1/2 + \nu)_k}{2^k k!} b_k$$

$$b_0 = 1$$

$$b_{k+1} = 1 + \frac{2(k+1)(\alpha - k - 1/2)}{(\nu - k - 1/2)(\nu + k + 1/2)} b_k.$$

If α and ν are integers, the following formulas are useful [1]:

$$\int_0^1 J_{2\nu}(\omega x)\, dx = \int_0^1 J_0(\omega x)\, dx - \frac{2}{\omega} \sum_{k=0}^{\nu-1} J_{2k+1}(\omega)$$

$$\int_0^1 J_{2\nu+1}(\omega x)\, dx = \frac{1 - J_0(\omega)}{\omega} - \frac{2}{\omega} \sum_{k=1}^{\nu} J_{2k}(\omega).$$

For the evaluation of

$$\int_0^1 J_0(\omega x)\, dx$$

Chebyshev series approximations are given by Luke [11]. We now discuss the numerical aspect of the recurrence formula (80). The numerical stability of forward recursion depends on the asymptotic behavior of $M_k(\omega, \nu, \alpha)$ and of eight linearly independent solutions $y_{i,k}$, $i = 1, 2, \ldots, 8$, $k \to \infty$.

Using the asymptotic theory of Fourier integrals, we find

$$|y_{1,k}| \sim k^{-2}$$

$$|y_{2,k}| \sim k^{-4}$$

$$|y_{3,k}| \sim k^{-2(\alpha+1)-2\nu}$$

$$|y_{4,k}| \sim k^{-2(\alpha+1)+2\nu}, \qquad \text{if } \nu \neq 0$$

$$\sim k^{-2(\alpha+1)} \ell nk, \qquad \text{if } \nu = 0$$

$$|y_{5,k}| \sim |y_{6,k}| \sim \left(\frac{\omega}{4k}\right)^k e^k k^\alpha$$

$$|y_{7,k}| \sim |y_{8,k}| \sim \left(\frac{4k}{\omega}\right)^k e^{-k} k^\alpha \tag{86}$$

and

$$M_k(\omega, \nu, \alpha) \sim -\frac{1}{2} J_\nu(\omega) k^{-2}$$

$$+ (-1)^k 2^{-3\nu-2\alpha-1} \frac{\omega^\nu}{\Gamma(\nu+1)} \cos\left[\pi(\alpha+1)\right] \Gamma(2\alpha+2) k^{-2\alpha-2\nu-2}. \tag{87}$$

The asymptotically dominant solutions are $y_{7,k}$ and $y_{8,k}$. The asymptotically minimal solutions are $y_{5,k}$ and $y_{6,k}$. We may conclude that forward and backward recursion are asymptotically unstable. However, the instability of forward recursion is less pronounced if $k \leq \omega/2$. Indeed, practical experiments demonstrate that $M_k(\omega, \nu, \alpha)$ can be computed accurately using forward recursion for $k \leq \omega/2$. For $k > \omega/2$ the loss of significant figures increases and forward recursion is no longer applicable. In that case, Oliver's algorithm [14] has to be used. This means that (80) has to be solved as a boundary value problem with six initial values and two end values. The solution of this boundary value problem requires the solution of a linear system of equations having a band structure.

An important advantage of the MCC method is that the function evaluations of g, needed for the computation of the coefficients c_k of the Chebyshev series expansion, are independent of the value of ω. Consequently, the same function evaluations may be used for different values of ω, and have to be computed only once.

Numerical examples can be found in [19] and [20].

9.10   Computation of Bessel function integrals over an infinite interval

In this section we consider methods for the computation of

$$I(p, \nu) = \int_0^\infty J_\nu(px) f(x) \, dx. \tag{88}$$

9.10.1 *Integration between the zeros of $J_\nu(x)$ and convergence acceleration*

We have

$$I(p, \nu) = \sum_{k=1}^{\infty} (-1)^{k+1} I_k \tag{89}$$

where

$$I_k = \int_{j_{\nu,k-1}/p}^{j_{\nu,k}/p} |J_\nu(px)| f(x)\, dx. \tag{90}$$

Using Euler's transformation [4], the convergence of series (89) can be accelerated

$$I(p, \nu) = \frac{1}{2} I_1 - \frac{1}{4} \Delta I_1 + \frac{1}{8} \Delta^2 I_1 - \frac{1}{16} \Delta^3 I_1 + \cdots. \tag{91}$$

It is not always desirable to start the convergence acceleration with I_1, but with some later term, say I_m, so that

$$I(p, \nu) = \int_0^{j_{\nu,m-1}/p} J_\nu(px) f(x)\, dx + (-1)^{m-1} \left[\frac{1}{2} I_m - \frac{1}{4} \Delta I_m + \frac{1}{8} \Delta^2 I_m - \cdots \right].$$

Other convergence accelerating methods, for example the ϵ-algorithm [23], are also applicable (for an example, see [21]).

9.10.2 *Transformation into a double integral*

Substituting the integral expression

$$J_\nu(x) = 2 \frac{(x/2)^\nu}{\Gamma(\nu + 1/2)\sqrt{\pi}} \int_0^1 (1 - t^2)^{\nu - 1/2} \cos(xt)\, dt \tag{92}$$

into (88) and changing the order of integration, we obtain

$$I(p, \nu) = \frac{2(p/2)^\nu}{\Gamma(\nu + 1/2)\sqrt{\pi}} \int_0^1 (1 - t^2)^{\nu - 1/2} F(t)\, dt \tag{93}$$

where

$$F(t) = \int_0^\infty x^\nu f(x) \cos(pxt)\, dx. \tag{94}$$

We assume that the integral in (94) is convergent. If we want to evaluate (93) using an N-point Gauss-Jacobi rule, then we have to compute the Fourier integral (94) for N values of t. Because $F(t)$ shows a peaked or even a singular behavior especially when $f(x)$ is slowly decaying, a large enough N has to be chosen.

This method is closely related to Linz's method [8], which is based on the Abel transformation of $I(p, \nu)$.

9.10.3 *Truncation of the infinite interval*

If a is an arbitrary positive real number, we can write

$$I(p, \nu) = \int_0^a J_\nu(px) f(x)\, dx + R(p, a) \tag{95}$$

where

$$R(p, a) = \int_a^\infty J_\nu(px) f(x)\, dx. \tag{96}$$

The first integral in the right side of (95) can be computed using the methods of Section 9.9.
   If a is sufficiently large and f is strongly decaying, then we may neglect $R(p, a)$.
   If

$$f(x) \sim \frac{c_1}{x} + \frac{c_2}{x^2} + \cdots \tag{97}$$

is an asymptotic series approximation which is sufficiently accurate in the interval $[a, \infty)$,
then

$$R(p, a) \simeq \sum_k c_k \int_a^\infty \frac{J_\nu(px)}{x^k} \, dx. \tag{98}$$

Longman [9] has tabulated the values of the integrals in (98) for some values of ν and ap.
   Using Hankel's asymptotic expansion [1], for $x \to \infty$

$$J_\nu(px) \sim \sqrt{\frac{2}{\pi px}} \, [P_\nu(px) \cos \chi - Q_\nu(px) \sin \chi] \tag{99}$$

where $\chi = px - (\nu/2 + 1/4)\pi$, and where $P_\nu(x)$ and $Q_\nu(x)$ can be expressed as a well-known
asymptotic series, $R(p, a)$ can be written as the sum of two Fourier integrals.
   Especially, if $a = (8 + \nu/2 + 1/4)\pi/p$, we have

$$R(p, a) = \int_0^\infty \sqrt{\frac{2}{\pi p(u + a)}} P_\nu(p(u + a)) f(u + a) \cos pu \, du$$

$$- \int_0^\infty \sqrt{\frac{2}{\pi p(u + a)}} Q_\nu(p(u + a)) f(u + a) \sin pu \, du. \tag{100}$$

For the computation of the Fourier integrals in (100), tailored methods are available [4].

9.11   Tables of Hankel transforms

Table 9.1 lists the Hankel transform of some particular functions for the important special case
$\nu = 0$. Table 9.2 lists Hankel transforms of general order ν. In these tables, $h(x)$ is the unit
step function, I_ν and K_ν are modified Bessel functions, L_0 and H_0 are Struve functions, and
Ker and Kei are Kelvin functions as defined in Abramowitz and Stegun [1]. Extensive tables
are given by Erdélyi et al. [6], Ditkin and Prudnikov [5], Luke [11], Wheelon [27], Sneddon
[24], and Oberhettinger [13].

TABLE 9.1
Hankel transforms of order 0.

	$f(r)$	$F_0(s) = \mathcal{H}_0\{f(r)\}$
(1)	$\dfrac{1}{r}$	$\dfrac{1}{s}$
(2)	$r^{-\mu}, \qquad 1/2 < \mu < 2$	$2^{1-\mu} \dfrac{\Gamma(1 - \frac{\mu}{2})}{\Gamma(\frac{\mu}{2})} \dfrac{1}{s^{2-\mu}}$
(3)	$h(a - r)$	$\dfrac{a}{s} J_1(as)$
(4)	e^{-ar}	$\dfrac{a}{(s^2 + a^2)^{3/2}}$
(5)	$\dfrac{e^{-ar}}{r}$	$\dfrac{1}{\sqrt{s^2 + a^2}}$
(6)	$\dfrac{e^{-ar}}{r^2}$	$\dfrac{1}{2} \log \dfrac{\sqrt{s^2 + a^2} - a}{\sqrt{s^2 + a^2} + a}$
(7)	$\dfrac{1 - e^{ar}}{r^2}$	$\log\left(\dfrac{a + \sqrt{a^2 + s^2}}{s} \right)$
(8)	$\dfrac{\sin r}{r}$	$\dfrac{1}{\sqrt{1 - s^2}}, \quad s < 1$ $0, \qquad\qquad s > 1$
(9)	$\dfrac{\sin r}{r^2}$	$\dfrac{\pi}{2}, \qquad s \leq 1$ $\arcsin \dfrac{1}{s}, \quad s > 1$
(10)	$\dfrac{\sin(ar)}{r^2 + b^2}$	$\dfrac{\pi}{2} e^{-ab} I_0(bs)$
(11)	$\dfrac{\cos(ar)}{r^2 + b^2}$	$\cosh(ab) K_0(bs)$
(12)	$e^{-a^2 r^2}$	$\dfrac{e^{-s^2/4a^2}}{2a^2}$
(13)	$\dfrac{1}{r(r + a)}$	$\dfrac{\pi}{2} [\mathbf{H}_0(as) - Y_0(as)]$
(14)	$\dfrac{1}{r^2 + a^2}$	$K_0(as)$
(15)	$\dfrac{1}{r(r^2 + a^2)}$	$\dfrac{\pi}{2a} [I_0(as) - \mathbf{L}_0(as)]$
(16)	$\dfrac{1}{1 + r^4}$	$- \operatorname{Kei}(s)$
(17)	$\dfrac{r^3}{1 + r^4}$	$\operatorname{Ker}(s)$
(18)	$\dfrac{1}{\sqrt{r^2 + a^2}}$	$\dfrac{e^{-sa}}{s}$
(19)	$\dfrac{1}{\sqrt{r^4 + a^4}}$	$K_0(as/\sqrt{2}) J_0(as/\sqrt{2})$
(20)	$\dfrac{1 - J_0(ar)}{r^2}$	$\log \dfrac{a}{s}, \quad s \leq a$ $0, \qquad\quad s \geq a$
(21)	$\dfrac{a}{r} J_1(ar)$	$1, \quad \text{if } 0 < s < a$ $0, \quad \text{if } s > a$
(22)	$\dfrac{1}{r} J_0(2\sqrt{ar})$	$\dfrac{1}{s} J_0\left(\dfrac{a}{s} \right)$

TABLE 9.2
Hankel transforms of general order ν.

	$f(r)$	$F_\nu(s) = \mathcal{H}_\nu\{f(r)\}$
(1)	$\dfrac{1}{r}$	$\dfrac{1}{s}$
(2)	$r^{-\mu}, \dfrac{1}{2} < \mu < \nu + 2$	$\dfrac{2^{1-\mu}}{s^{2-\mu}} \dfrac{\Gamma(\frac{\nu+2-\mu}{2})}{\Gamma(\frac{\nu+\mu}{2})}$
(3)	$x^\nu(a^2 - r^2)^\mu h(a-r),$ $\mu > -1$	$2^\mu a^{\mu+\nu+1} s^{-\mu-1}\Gamma(\mu+1)J_{\nu+\mu+1}(as)$
(4)	$\dfrac{\sin ar}{r}$	$\dfrac{1}{(s^2-a^2)^{1/2}}\sin\left(\nu\arcsin\left(\dfrac{a}{s}\right)\right) \quad s > a$ $\dfrac{1}{(a^2-s^2)^{1/2}}\dfrac{s^\nu}{(a+(a^2-s^2)^{1/2})^\nu} \quad s < a$
(5)	$\dfrac{\sin ar}{r^2}$	$\dfrac{\nu^{-1}s^\nu}{\left(a+\sqrt{a^2-s^2}\right)^\nu}\sin\dfrac{\nu\pi}{2} \quad s \le a$ $\nu^{-1}\sin\left(\nu\arcsin\left(\dfrac{a}{s}\right)\right) \quad s > a$
(6)	$\dfrac{e^{-ar}}{r}$	$\dfrac{(\sqrt{s^2+a^2}-a)^\nu}{s^\nu\sqrt{s^2+a^2}}$
(7)	$\dfrac{e^{-ar}}{r^2}$	$\dfrac{(\sqrt{s^2+a^2}-a)^\nu}{\nu s^\nu}$
(8)	$r^{\nu-1}e^{-ar}$	$\dfrac{(2s)^\nu\Gamma(\nu+1/2)}{(s^2+a^2)^{\nu+1/2}\sqrt{\pi}}$
(9)	$r^\nu e^{-ar}$	$\dfrac{2a(2s)^\nu\Gamma(\nu+3/2)}{(s^2+a^2)^{\nu+3/2}\sqrt{\pi}}$
(10)	$e^{-a^2r^2}r^\nu$	$\dfrac{s^\nu}{(2a^2)^{\nu+1}}\exp\left(-\dfrac{s^2}{4a^2}\right)$
(11)	$e^{-a^2r^2}r^\mu$	$\dfrac{\Gamma((\nu+\mu+2)/2)\left(\frac{1}{2}\frac{s}{a}\right)^\nu}{2a^{\mu+2}\Gamma(\nu+1)} \times {}_1F_1\left(\dfrac{\nu+\mu+2}{2}; \nu+1; -\dfrac{s^2}{4a^2}\right)$
(12)	$\dfrac{r^\nu}{(r^2+a^2)^{\mu+1}}$	$\dfrac{s^\mu a^{\nu-\mu}}{2^\mu\Gamma(\mu+1)}K_{\nu-\mu}(as)$
(13)	$\dfrac{r^\nu}{(r^4+4a^4)^{\nu+\frac{1}{2}}}$	$\dfrac{\left(\frac{1}{2}s\right)^\nu\sqrt{\pi}}{(2a)^{2\nu}\Gamma\left(\nu+\frac{1}{2}\right)}J_\nu(as)K_\nu(as)$
(14)	$\dfrac{r^{\nu+2}}{(r^4+4a^4)^{\nu+\frac{1}{2}}}$	$\dfrac{\left(\frac{1}{2}s\right)^\nu\sqrt{\pi}}{2(2a)^{2\nu-2}\Gamma\left(\nu+\frac{1}{2}\right)}J_{\nu-1}(as)K_{\nu-1}(as)$
(15)	$r^{\mu-\nu}J_\mu(ar)$	$0 \qquad\qquad\qquad 0 < s < a$ $\dfrac{2^{\mu-\nu+1}a^\mu(s^2-a^2)^{\nu-\mu-1}}{s^\nu\Gamma(\nu-\mu)} \quad a < s$

References

[1] M. Abramowitz and I. N. Stegun, *Handbook of Mathematical Functions*, Dover Publications, New York, 1965.

[2] R. N. Bracewell, *The Fourier Transform and Its Applications*, McGraw-Hill, New York, 1978.

[3] B. Davies, *Integral Transforms and Their Applications*, Springer-Verlag, New York, 1978.

[4] P. J. Davis and P. Rabinowitz, *Methods of Numerical Integration*, Academic Press, New York, 1975.

[5] V. A. Ditkin and A. P. Prudnikov, *Integral Transforms and Operational Calculus*, Pergamon Press, Oxford, 1965.

[6] A. Erdélyi, W. Magnus, F. Oberhettinger, and F. G. Tricomi, *Tables of Integral Transforms*, Vol. 2, McGraw-Hill, New York, 1954.

[7] A. Gray and G. B. Mathews, *A Treatise on Bessel Functions and Their Applications to Physics*, Dover Publications, New York, 1966.

[8] P. Linz, A method for computing Bessel function integrals, *Math. Comp.*, 20, pp. 504–513, 1972.

[9] I. M. Longman, A short table of $\int_x^\infty J_0(t)t^{-n}\,dt$ and $\int_x^\infty J_1(t)t^{-n}\,dt$, *Math. Tables Aids Comp.*, 13, pp. 306–311, 1959.

[10] I. M. Longman, A method for the numerical evaluation of finite integrals of oscillatory functions, *Math. Comp.*, 14, pp. 53–59, 1960.

[11] Y. L. Luke, *The Special Functions and their Approximations*, Vol. 2, Academic Press, New York, 1969.

[12] Y. L. Luke, *Integrals of Bessel Functions*, McGraw-Hill, New York, 1962.

[13] F. Oberhettinger, *Table of Bessel Transforms*, Springer-Verlag, Berlin, 1971.

[14] J. Oliver, The numerical solution of linear recurrence relations, *Num. Math.*, 11, pp. 349–360, 1968.

[15] F. W. J. Olver, *Bessel Functions, Part III: Zeros and Associated Values*, Cambridge University Press, Cambridge, UK, 1960.

[16] R. Piessens, Gaussian quadrature formulas for integrals involving Bessel functions, Microfiche section of *Math. Comp.*, 26, 1972.

[17] R. Piessens, Automatic computation of Bessel function integrals, *Comp. Phys. Comm.*, 25, pp. 289–295, 1982.

[18] R. Piessens, On the computation of zeros and turning points of Bessel functions, in E. A. Lipitakis (ed.), *Advances on Computer Mathematics and Its Applications*, World Scientific, Singapore, pp. 53–57, 1993.

[19] R. Piessens and M. Branders, Modified Clenshaw-Curtis method for the computation of Bessel function integrals, *Bit 23*, pp. 370–381, 1983.

[20] R. Piessens and M. Branders, A survey of numerical methods for the computation of Bessel function integrals, *Rend. Sem. Mat. Univers. Politecn. Torino*, Fascicolo speciale, pp. 250–265, 1985.

[21] R. Piessens, E. de Doncker-Kapenga, C. W. Ueberhuber, and D. K. Kahaner, *Quadpack, A Subroutine Package for Automatic Integration*, Springer-Verlag, Berlin, 1983.

[22] F. E. Relton, *Applied Bessel Functions*, Dover Publications, New York, 1965.

[23] D. Shanks, Non-linear transformations of divergent and slowly convergent sequences, *J. Math. Phys.*, 34, pp. 1–42, 1955.

[24] I. N. Sneddon, *The Use of Integral Transforms*, McGraw-Hill, New York, 1972.

[25] C. J. Tranter, *Integral Transforms in Mathematical Physics*, Methuens & Co, London, 1951.

[26] G. N. Watson, *A Treatise on the Theory of Bessel Functions*, Cambridge University Press, Cambridge, UK, 1966.

[27] A. D. Wheelon, *Table of Summable Series and Integrals Involving Bessel Functions*, Holden-Day, San Francisco, 1968.

[28] K. B. Wolf, *Integral Transforms in Science and Engineering*, Plenum Press, New York, 1979.

10

Wavelet Transform

Yunlong Sheng

CONTENTS

ABSTRACT The wavelet transform is a new mathematical tool developed mainly since the middle of the 1980s. It is efficient for local analysis of nonstationary and fast transient signals. Similar to the short-time Fourier transform, the Wigner distribution and the ambiguity function, the wavelet transform is a mapping of the signal to the time-scale joint representation. The temporal aspect of the signals is preserved. The wavelet transform provides multiresolution analysis with dilated windows. The high frequency analysis is done using narrow windows and the low frequency analysis is done using wide windows. The wavelet transform is a constant-Q analysis.

The basis functions of the wavelet transform, the wavelets, are generated from a basic wavelet function by dilations and translations. They satisfy an admissible condition so that the original signal can be reconstructed by the inverse wavelet transform. The wavelets also satisfy the regularity condition so that the wavelet coefficients decrease quickly with decrease of the scale. The wavelet transform is not only local in time but also in frequency.

To reduce the time-bandwidth product of the wavelet transform output, the discrete wavelet transform with wavelets of discrete dilations and translations can be used. The orthonormal wavelet transform is implemented in the multiresolution analysis framework, based on the scaling functions. The discrete translates of the scaling functions form an orthonormal basis at each resolution level. The wavelet basis is generated from the scaling function basis. The two bases are mutually orthogonal at each resolution level. The scaling function is an averaging function so that the orthogonal projection of a function onto the scaling function basis is an averaged approximation. The orthogonal projection onto the wavelet basis is the difference between two approximations at two adjacent resolution levels. Both the scaling functions and the wavelets satisfy the orthonormality conditions and the regularity conditions.

The discrete orthonormal wavelet series decomposition and reconstruction are computed in the multiresolution analysis framework with two recurring discrete conjugate quadrature mirror filters that are, in fact, the two-band, paraunitary, perfect reconstruction quadrature mirror filters, developed in the subband coding theory, with the additional regularity. The tree algorithm operating the discrete wavelet transform requires only $O(L)$ operations where L is the length of the data vector. The time-bandwidth product of the wavelet transform output is only slightly increased with respect to that of the signal.

The wavelet transform is powerful for multiresolution local spectrum analysis of nonstationary signals, such as sound, radar, sonar, seismic, and electrocardiographic signals, and for image compression, image processing, and pattern recognition.

In this chapter, all integrations extend from $-\infty$ to ∞, if not stated otherwise. The formulation of the wavelet transform in this chapter is one-dimensional. The wavelet transform can be easily generalized to any dimensions.

10.1   Introduction

10.1.1   *Continuous wavelet transform*

Let $\mathbf{L}$ denote the vector space of measurable, square-integrable functions. The continuous wavelet transform of a function $f(t) \in \mathbf{L}$ is a decomposition of $f(t)$ into a set of basis functions $h_{s,\tau}(t)$, called the wavelets:

$$W_f(s, \tau) = \int f(t) h_{s,\tau}^*(t)\, dt \qquad (10.1.1)$$

where $*$ denotes the complex conjugate. However, most wavelets are real valued. The wavelets are generated from a single basic wavelet $h(t)$ by scaling and translation:

$$h_{s,\tau}(t) = \frac{1}{\sqrt{s}} h\left(\frac{t - \tau}{s}\right) \qquad (10.1.2)$$

where s is the scale factor and τ is the translation factor. We usually consider only positive scale factor $s > 0$. The wavelets are dilated when the scale $s > 1$ and are contracted when $s < 1$. The wavelets $h_{s,\tau}(t)$ generated from the same basic wavelet have different scales s and locations τ, but all have the identical shape.

The constant $s^{-1/2}$ in the expression (10.1.2) of the wavelets is for energy normalization. The wavelets are normalized as

$$\int |h_{s,\tau}(t)|^2\, dt = \int |h(t)|^2\, dt = 1 \qquad (10.1.3)$$

so that all the wavelets scaled by the factor s have the same energy. The wavelets can also be normalized in terms of the amplitude

$$\int |h_{s,\tau}(t)|\, dt = 1 \qquad (10.1.4)$$

In this case, the normalization constant is $1/s$ instead of $s^{-1/2}$, and the wavelets are generated as

$$h_{s,\tau}(t) = \frac{1}{s} h\left(\frac{t - \tau}{s}\right) \qquad (10.1.5)$$

In this chapter we use only the wavelets normalized by Eq. (10.1.3).

Substituting Eq. (10.1.2) into Eq. (10.1.1) we write the wavelet transform of $f(t)$ as a correlation between the signal and the scaled wavelets $h(t/s)$ as follows:

$$w_f(s, \tau) = \frac{1}{\sqrt{s}} \int f(t) h^* \left(\frac{t - \tau}{s} \right) dt \qquad (10.1.6)$$

The Fourier transforms of the wavelets are

$$H_{s,\tau}(\omega) = \int \frac{1}{\sqrt{s}} h \left(\frac{t - \tau}{s} \right) \exp(-j\omega t) \, dt$$

$$= \sqrt{s} H(s\omega) \exp(-j\omega \tau) \qquad (10.1.7)$$

where $H(\omega)$ is the Fourier transform of the basic wavelet $h(t)$. In the frequency domain the wavelets are scaled by $1/s$ and are multiplied by a phase factor $\exp(-j\omega\tau)$ and by a normalization factor $s^{1/2}$. The amplitude of the scaled wavelets is proportional to $s^{-1/2}$ in the time domain and is proportional to $s^{1/2}$ in the frequency domain. When the wavelets are normalized in terms of amplitude, as shown in Eq. (10.1.5), the Fourier transforms of the wavelets with different scales have the same amplitude.

Equation (10.1.7) shows a well-known concept that a dilatation t/s ($s > 1$) of a function in the time domain produces a contraction $s\omega$ of its Fourier transform. The term $1/s$ is equivalent to the frequency ω. However, we prefer the term scale to the term frequency for the wavelet transform. The term frequency is reserved for the Fourier transform.

Because the wavelet transform is the correlation between a function $f(t)$ and the scaled wavelets, it can be written as the inverse Fourier transform of the product of the conjugate Fourier transforms $H(\omega)$ of the wavelets and the Fourier transform $F(\omega)$ of $f(t)$:

$$W_f(s, \tau) = \frac{\sqrt{s}}{2\pi} \int F(\omega) H^*(s\omega) \exp(j\omega\tau) \, d\omega \qquad (10.1.8)$$

The Fourier transforms of the wavelets are referred to as the wavelet transform filters. The impulse responses of the wavelet transform filters, $\sqrt{s} H(s\omega)$, are the scaled wavelets $h(t/s)$. Therefore, the wavelet transform is a bank of wavelet transform filters with different scales s.

In the definition of the wavelet transform, the wavelet basis functions are not specified. This is a difference between the wavelet transform and the Fourier transform, or other transforms. The theory of wavelet transform deals with the general properties of the wavelets and the wavelet transform, such as the admissible condition and the regularity. The wavelet bases are built to satisfy those basic properties. The wavelets can be analytical or numerical, orthonormal or nonorthonormal, continuous or discrete. One can choose or even build himself a proper wavelet basis for a specific application. Therefore, it is not enough to talk only about the wavelet transform of a function. One must always discuss the wavelet transform with respect to a basic wavelet function.

The most important properties of the wavelets are the admissibility and the regularity conditions. As we shall see below, according to the admissible condition, the wavelets must oscillate to have their mean value equal to zero. According to the regularity condition, the wavelets have exponential decay so that their first low-order moments are equal to zero. Hence, in the time domain, the wavelets are just like waves that oscillate and vanish, as described by the name wavelets. The wavelet transform is a local operator in the time domain.

The wavelet transform of a one-dimensional signal is a two-dimensional function of scale s and time τ. At first glance, the time-bandwidth product of the wavelet transform output is the square of that of the input signal. However, many applications, such as signal compression and pattern classification, require describing a signal with a minimum number of parameters. Therefore, one wants the wavelet transform coefficients to decay quickly with increasing of

$1/s$. The regularity condition of the wavelets can insure that the wavelet transform is local in the frequency domain.

10.1.2 *Time-frequency space analysis*

The wavelet transform of a one-dimensional function is a two-dimensional function of the scale s and the time shift τ, which is a representation of the function in the time-scale space and is referred to as the time-scale joint representation. The time-scale wavelet representation is equivalent to the time-frequency joint representation, familiar in the short-time Fourier transform, and so forth, for analysis of nonstationary and fast transient signals.

The wavelet transform is of particular interest for analysis of nonstationary and fast transient signals. The wavelet transform provides an alternative to the classical short-time Fourier transform or the Gabor transform and is more efficient than the short-time Fourier transform. As the wavelet transform is local in both time and frequency domains, one can use fewer wavelet transform coefficients to describe short-duration and nonstationary signals.

Nonstationary signals

Signals are stationary if their properties do not change during the course of the signals. The concept of stationarity is well defined in the theory of stochastic processes. A stochastic process is called strict-sense stationary if its statistical properties are invariant to a shift of the origin of the time axis. A stochastic process is called wide-sense (or weak) stationary if its second-order statistics depends only on time difference.

The properties of nonstationary signals change with time. Most signals in nature are nonstationary. Examples of nonstationary signals are speech, radar, sonar, seismic, and electrocardiographic signals, and music. Two-dimensional images are also nonstationary with edges, textures, and deterministic objects at different locations. The nonstationary signals are, in general, characterized by their local features rather than by their global ones. For example, a human face is recognized by the local features of hairs, eyes, mouth, and so on.

An example of the time-frequency joint representation is the notation of music score. The frequency spectra of music notes change with time. A piece of music can be described accurately by air pressure as a function of time. It can be equally accurately described by the Fourier transform of the pressure function. Neither of the two signal representations would be useful for a musician who wants to perform a certain piece. Musicians prefer a two-dimensional plot, with time and logarithmic frequency as axes. The music scores tell them when and what notes should be played. The music scores are, in fact, a time-frequency joint representation.

The time-frequency joint representation has an intrinsic limitation: The product of the resolutions in time and in frequency is limited by the uncertainty principle:

$$\Delta t \, \Delta \omega \geq \frac{1}{2} \tag{10.1.9}$$

This is also referred to as Heisenberg inequality, familiar in quantum mechanics and important for time-frequency joint representation. A signal cannot be represented as a point in the time-frequency space. We can only determine its position within a rectangle of $\Delta t \, \Delta \omega$ in the time-frequency space.

Limitation of the Fourier analysis

The Fourier transform is widely used in signal analysis and processing. When the signals are periodic and sufficiently regular, the Fourier coefficients converge quickly. For nonperiodic signals, the Fourier integral gives a continuous spectrum. The Fast Fourier Transform (FFT) permits efficient Fourier analysis.

The Fourier transform is not satisfactory for analyzing signals whose spectra vary with time. The Fourier transform is a decomposition of a signal into two series of orthogonal functions $\cos \omega t$ and $j \sin \omega t$ with $j = (-1)^{1/2}$. Those Fourier basis functions are of infinite duration along the time axis. They are perfectly local in frequency, but are global in time. A signal may be reconstructed from its Fourier components, which are the Fourier basis functions of infinite duration multiplied by the Fourier coefficients of the signal. Any signal that we are interested in is, however, of finite extent. Outside that finite duration, the Fourier components of the signal are nonzero, but their sum is equal to zero. A short pulse that is local in time is not local in frequency. Its Fourier coefficients decay slowly with frequency. The reconstruction of the pulse from its Fourier coefficients depends heavily on the cancellation of high-frequency Fourier components and, therefore, is sensitive to high-frequency noise.

The Fourier spectrum analysis is global in time and is basically not suitable to analyze nonstationary and fast varying signals. Many temporal aspects of the signal, such as the start and end of a finite signal and the instant of appearance of a singularity in a transient signal, are not presented by the Fourier spectrum. The Fourier transform does not provide any information regarding the time evolution of spectral characteristics of the signal. The short-time Fourier transform, or Gabor transform, the bilinear Wigner distribution, and the ambiguity function are usually used to overcome the drawback of the Fourier analysis for nonstationary and fast transient signals.

10.1.3   *Short-time Fourier transform*

The most intuitive way to analyze a nonstationary signal is to perform a time-dependent spectral analysis. A nonstationary signal is divided into a sequence of time segments in which the signal may be considered as quasi-stationary. Then, the Fourier transform is applied to each of the local segments of the signal.

DEFINITION
The short-time Fourier transform is associated with a window of fixed width. Gabor, in 1940, was the first to introduce the short-time Fourier transform [1], which is known as the sliding-window Fourier transform. The transform is defined as

$$S_f(\omega', \tau) = \int f(t) g^*(t - \tau) \exp(-j\omega' t) \, dt \qquad (10.1.10)$$

where $g(t)$ is a square integrable short-time window. The window has a fixed width and is shifted along the time axis by a factor τ.

The Gabor transform may also be regarded as an inner product between the signal and a set of kernel functions, called the Gabor functions: $g(t - \tau) \exp(j\omega' t)$. The Gabor function basis is generated from a basic window function $g(t)$ by translations along the time axis by τ. The phase modulations $\exp(j\omega' t)$ correspond to translations of the Gabor function spectrum along the frequency axis by ω'. The Fourier transform of the basic Gabor function $g(t) \exp(j\omega' t)$ is expressed as

$$\int g(t) \exp(j\omega' t) \exp(-j\omega t) \, dt = G(\omega - \omega') \qquad (10.1.11)$$

The Fourier transform $G(\omega)$ of the basic window function $g(t)$ is shifted along the frequency axis by ω'. The short-time Fourier transform of a one-dimensional signal is a complex valued function of two real parameters: time τ and frequency ω' in the two-dimensional time-frequency space.

When the time and frequency are continuous variables, the signal $f(t)$ may be reconstructed completely by integrating the Gabor functions multiplied with the short-time Fourier transform coefficients:

$$f(t) = \frac{1}{2\pi} \int \int S_f(\omega', \tau) g(t - \tau) \exp(j\omega't) \, d\omega' \, d\tau \qquad (10.1.12)$$

and this holds for any chosen window $g(t)$. The inverse short-time Fourier transform may be proved by the following calculation:

$$\int \int S_f(\omega', \tau) g(t - \tau) \exp(j\omega't) \, d\omega' \, d\tau$$

$$= \int \int \int f(t') g^*(t' - \tau) \exp(-j\omega't') g(t - \tau) \exp(j\omega't) \, d\omega' \, d\tau \, dt'$$

$$= \int \int 2\pi \delta(t' - t) g^*(t' - \tau) g(t - \tau) f(t') \, d\tau \, dt'$$

$$= 2\pi f(t) \int \int |g(t - \tau)|^2 \, d\tau$$

$$= 2\pi f(t) \qquad (10.1.13)$$

provided that the window function is normalized as:

$$\int |g(t)|^2 \, dt = 1 \qquad (10.1.14)$$

Spectrogram and sonagram

The short-time Fourier transform can be considered as a Fourier analysis of the short-time windowed signal. In the two-dimensional time-frequency joint representation, the vertical stripes of the complex valued short-time Fourier transform coefficients, $S_f(\omega', \tau)$, correspond to the Fourier spectra of the windowed signal with the window shifted to given times $\tau = \tau_1, \tau_2, \ldots$. If one takes the square modulus of the short-time Fourier transform, the vertical stripes of the time-frequency joint representation are the spectral powers of the quasi-stationary segments of the signal and are referred to as the spectrogram:

$$|S_f(\omega', \tau)|^2 = \left| \int f(t) g^*(t - \tau) \exp(-j\omega'\tau) \, dt \right|^2 \qquad (10.1.15)$$

The short-time Fourier transform can also be considered as a bank of band-pass filters. The horizontal stripes of the short-time Fourier transform coefficients $S_f(\omega', \tau)$ correspond to the inner products between the signal at all times and the Gabor functions $g^*(t-\tau) \exp(j\omega't)$ at the given frequencies $\omega' = \omega_1, \omega_2, \ldots$. Multiplying the two sides on Eq. (10.1.10) by $\exp(j\omega'\tau)$, we can rewrite the $S_f(\omega', \tau) \exp(j\omega'\tau)$ as the correlation between the signal $f(t)$ and the basic Gabor function $g(t) \exp(j\omega't)$:

$$S_f(\omega', \tau) \exp(j\omega'\tau) = f(t) * [g(t) \exp(j\omega't)] \qquad (10.1.16)$$

where $*$ denotes the correlation. Therefore the signal at all times is filtered with a bank of filters $G(\omega - \omega')$ in the frequency domain. The Fourier transform $G(\omega)$ of the window $g(t)$ is usually still a window and is therefore a band-pass filter. The $G(\omega - \omega')$ is a bank of band-pass filters, $G(\omega)$, shifted along the frequency axis by ω'. We write the correlation on the right-hand side of Eq. (10.1.16) in the frequency domain and take square modulus of the equation, which yields

$$|S_f(\omega', \tau)|^2 = \left| \int_{-\infty}^{\infty} F(\omega) G^*(\omega - \omega') \exp(j\omega\tau) \, d\omega \right|^2 \tag{10.1.17}$$

For a given ω', the horizontal stripes of the time-frequency joint representation are the correlation intensities of the band-pass filter bank, $G(\omega - \omega')$, and are referred to as the sonagram.

Time and frequency resolution

In the short-time Fourier transform the signal is multiplied by a sliding window to become a sequence of segments. Multiplying the signal by the window localizes the signal in the time domain, but results in a convolution of the signal spectrum with the spectrum of the window, that is, a blurring of the signal in the frequency domain. The narrower the window, the better we localize the signal and the poorer we localize its spectrum. The width Δt of the window $g(t)$ in the time domain and the bandwidth $\Delta\omega$ of the window $G(\omega)$ in the frequency domain are defined respectively as

$$\Delta t^2 = \frac{\int t^2 |g(t)|^2 \, dt}{\int |g(t)|^2 \, dt} \qquad \Delta\omega^2 = \frac{\int \omega^2 |G(\omega)|^2 \, d\omega}{\int |G(\omega)|^2 \, d\omega} \tag{10.1.18}$$

where the denominator is the energy of $g(t)$.

Consider now the ability of the short-time Fourier transform to discriminate between two pure sinusoids. The two sinusoids can be discriminated only if they are more than $\Delta\omega$ apart in the frequency domain. Thus, the resolution in the frequency domain of the short-time Fourier transform is determined by $\Delta\omega$. Similarly, two pulses in time can be discriminated only if they are more than Δt apart. More important is that once a window has been chosen for the short time Fourier transform, the time and frequency resolutions given by Eq. (10.1.18) are fixed over the entire time-frequency plane. The short-time Fourier transform is a fixed-window Fourier transform.

Gaussian window

The time-bandwidth product $\Delta t \Delta \omega$ must obey the uncertainty principle. We can only trade time resolution for frequency resolution or vice versa. Gabor proposed the Gaussian function as the window function. The Fourier transform of the Gaussian window is still Gaussian, which has a minimum spread. We have

$$g(t) = \frac{1}{s} \exp\left(-\frac{\pi t^2}{s^2}\right) \qquad \text{and} \qquad G(\omega) = \frac{1}{\sqrt{\pi}} \exp(-s^2 \omega^2 / 4\pi)$$

A simple calculation shows that

$$\Delta t^2 = \frac{\int t^2 |g(t)|^2 \, dt}{\int |g(t)|^2 \, dt} = \frac{s^2}{4\pi} \qquad \Delta\omega^2 = \frac{\int \omega^2 |G(\omega)|^2 \, d\omega}{\int |G(\omega)|^2 \, d\omega} = \frac{\pi}{s^2} \tag{10.1.19}$$

Hence, the Gaussian windows have the minimum time-bandwidth product determined by the uncertainty principle because according to Eq. (10.1.19)

$$\Delta t \Delta \omega = \frac{1}{2} \tag{10.1.20}$$

The short-time Fourier analysis depends critically on the choice of the window. Its application requires a priori information concerning the time evolution of the signal properties in order to make a priori choice of the window function. Once a window is chosen, the width of the window along both time and frequency axes are fixed in the entire time-frequency plane.

Discrete short-time Fourier transform

When the translation factors of the Gabor functions along the time and the frequency axes τ and ω' take discrete values, $\tau = n\tau_0$ and $\omega' = m\omega_0$ with m and $n \in \mathbf{Z}$, the discrete Gabor functions are written as

$$g_{m,n}(t) = g(t - n\tau_0) \exp(-jm\omega_0 t) \qquad (10.1.21)$$

and their Fourier transforms are

$$G_{m,n}(\omega) = G(\omega - m\omega_0) \exp[-j(\omega - m\omega_0)n\tau_0] \qquad (10.1.22)$$

The discrete Gabor transform is

$$S_f(m, n) = \int_{-\infty}^{\infty} f(t) g^*(t - n\tau_0) \exp(-jm\omega_0 t) \, dt \qquad (10.1.23)$$

The signal $f(t)$ can still be recovered from the coefficients $S_f(m, n)$, provided that ω_0 and τ_0 are suitably chosen. Gabor's original choice was $\omega_0 \tau_0 = 2\pi$.

If the window function is normalized as shown in Eq. (10.1.14) and is also centered to the origin in the time-frequency space:

$$\int t|g(t)|^2 \, dt = 0 \qquad \int \omega |G(\omega)|^2 \, d\omega = 0 \qquad (10.1.24)$$

the locations of the Gabor functions in the time-frequency space are determined by

$$\int t|g_{m,n}(t)|^2 \, dt = \int t|g(t - n\tau_0)|^2 \, dt = n\tau_0 \qquad (10.1.25)$$

and

$$\int \omega |G_{m,n}(\omega)|^2 \, d\omega = \int \omega |G(\omega - m\omega_0)|^2 \, d\omega = m\omega_0 \qquad (10.1.26)$$

The discrete Gabor function set will be represented by a regular lattice with the equal intervals τ_0 and ω_0 in the time-frequency space.

10.1.4   *Wigner distribution and ambiguity functions*

The Wigner distribution function [2] and the ambiguity function are second-order transforms or bilinear transforms that perform the mapping of signals into the time-frequency space.

Wigner distribution function

The Wigner distribution function [2] is an alternative to the short-time Fourier transform for nonstationary and transient signal analysis. The Wigner distribution of a function $f(t)$ is defined in the time domain as

$$W_f(\tau, \omega) = \int f\left(\tau + \frac{t}{2}\right) f^*\left(\tau - \frac{t}{2}\right) \exp(-j\omega t) \, dt \qquad (10.1.27)$$

that is, the Fourier transform of the product $f(\tau + t/2) f^*(\tau - t/2)$ between the dilated function $f(t/2)$ and the dilated and inverted function $f^*(-t/2)$. The product is shifted along the time axis by τ. The Wigner distribution is a complex valued function in the time-frequency space and is a time-frequency joint representation of the signal. In the frequency domain the Wigner distribution function is expressed as

$$W_f(\tau, \omega) = \frac{1}{2\pi} \int F\left(\omega + \frac{\xi}{2}\right) F^*\left(\omega - \frac{\xi}{2}\right) \exp(j\tau\xi) \, d\xi \tag{10.1.28}$$

where $F(\omega)$ is the Fourier transform of $f(t)$.

The inverse relations of the Wigner distribution function can be obtained from the inverse Fourier transforms of Eqs. (10.1.27) and (10.1.28). With changing of variables $t_1 = \tau + t/2$ and $t_2 = \tau - t/2$, the inverse Fourier transform of the Wigner distribution of Eq. (10.1.27) gives

$$f(t_1)f^*(t_2) = \frac{1}{2\pi} \int W_f\left(\frac{t_1 + t_2}{2}, \omega\right) \exp[j(t_1 - t_2)\omega] \, d\omega \tag{10.1.29}$$

Similarly, the inverse Fourier transform of Eq. (10.1.28) gives

$$F(\omega_1)F^*(\omega_2) = \int W_f\left(\tau, \frac{\omega_1 + \omega_2}{2}\right) \exp[-j(\omega_1 - \omega_2)\tau] \, d\tau \tag{10.1.30}$$

The signal $f(t)$ can be recovered from the inverse Wigner distribution function. Let $t_1 = t$ and $t_2 = 0$; Eq. (10.1.29) becomes

$$f(t)f^*(0) = \frac{1}{2\pi} \int W_f\left(\frac{t}{2}, \omega\right) \exp(j\omega t) \, d\omega \tag{10.1.31}$$

where $f^*(0)$ is a constant. Hence, the function $f(t)$ is reconstructed from the inverse Fourier transform of the Wigner distribution function $W_f(t/2, \omega)$, dilated in the time domain.

For the basic properties of the Wigner distribution function we mention that the projections of $W_f(\tau, \omega)$ along the τ-axis in the time-frequency space gives the square modulus of $F(\omega)$, because according to Eq. (10.1.28) the projection along the τ-axis is

$$\int W_f(\tau, \omega) \, d\tau = \frac{1}{2\pi} \int \int F\left(\omega + \frac{\xi}{2}\right) F^*\left(\omega - \frac{\xi}{2}\right) \exp(j\tau\xi) \, d\tau \, d\xi$$

$$= |F(\omega)|^2 \tag{10.1.32}$$

The projection of $W_f(\tau, \omega)$ along the ω-axis gives the square modulus of $f(t)$, because according to Eq. (10.1.28) the projection along the ω-axis is

$$\int W_f(\tau, \omega) \, d\omega = \int \int f\left(\tau + \frac{t}{2}\right) f^*\left(\tau - \frac{t}{2}\right) \exp(-j\omega t) \, dt \, d\omega$$

$$= 2\pi |f(t)|^2 \tag{10.1.33}$$

Also, there is the conservation of energy of the Wigner distribution in the time-frequency joint representation, because

$$\frac{1}{2\pi} \int W_f(\tau, \omega) \, d\tau \, d\omega = \frac{1}{2\pi} \int |F(\omega)|^2 \, d\omega = \int |f(t)|^2 \, dt \tag{10.1.34}$$

Ambiguity function

The ambiguity function is also a mapping of a transient time function $f(t)$ into the time-frequency space. The ambiguity function is defined in the time domain as [3]

$$A_f(t, \omega) = \int f\left(\tau + \frac{t}{2}\right) f^*\left(\tau - \frac{t}{2}\right) \exp(-j\omega\tau) \, d\tau \tag{10.1.35}$$

In the frequency domain, the ambiguity function is expressed as

$$A_f(\tau, \omega) = \frac{1}{2\pi} \int F\left(\xi + \frac{\omega}{2}\right) F^*\left(\xi - \frac{\omega}{2}\right) \exp(j\tau\xi)\, d\xi \qquad (10.1.36)$$

The ambiguity function can be viewed as a time-frequency autocorrelation function of the signal with the time delay t and the Doppler frequency shift ω. The ambiguity function has found wide applications for radar signal processing.

According to definitions (10.1.27) and (10.1.35) the double Fourier transform of the product $f(\tau + t/2)f^*(\tau - t/2)$ with respect to both variables t and τ gives the relation between the Wigner distribution function and the ambiguity function

$$\int A_f(t, \omega) \exp(-j\omega t)\, dt = \int W_f(\tau, \omega) \exp(-j\omega\tau)\, d\tau \qquad (10.1.37)$$

The cross ambiguity function is defined as the Fourier transform of the product $f(\tau)g^*(\tau)$ of two functions $f(\tau)$ and $g(\tau)$

$$A(t, \omega) = \int f\left(\tau + \frac{t}{2}\right) g^*\left(\tau - \frac{t}{2}\right) \exp(-j\omega\tau)\, d\tau \qquad (10.1.38)$$

High value of $A(t, \omega)$ means that the two functions are ambiguous. The function $g(\tau)$ can also be considered as a window function of fixed width that is shifted along the time axis by t. Hence, the cross-ambiguity function is the fixed-window short-time Fourier transform.

The cross-Wigner-distribution function is defined as

$$W(\tau, \omega) = \int f\left(\tau + \frac{t}{2}\right) g^*\left(\tau - \frac{t}{2}\right) \exp(-j\omega t)\, dt \qquad (10.1.39)$$

which can be seen as the Fourier transform of the signal $f(t)$ dilated by a factor of two and multiplied with an inverted window $g(-t)$ that is also dilated by a factor of two and shifted by τ.

Both the ambiguity function and the Wigner distribution function are useful for active and passive transient signal analysis. Both transforms are bilinear transforms. However, the mapping of a summation of signals $f_1(t) + f_2(t)$ into the time-frequency space with the ambiguity function or with the Wigner distribution function produces cross-product interference terms that might be a nuisance in the projections in the time-frequency space and in the reconstruction of the signal.

10.1.5 *Multiresolution wavelet analysis*

Bank of multiresolution filters

The wavelet transform is the correlation between a function and the dilated wavelets. For a given scale, the wavelet transform is performed with a wavelet transform filter $\sqrt{s}\,H(s\omega)$ in the Fourier domain, whose impulse response is the scaled wavelet $h(t/s)$. In the time-scale joint representation the horizontal stripes of the wavelet transform coefficients are the correlations between the signal and the wavelets $h(t/s)$ at given scales $s = s_1, s_2, \ldots$. When the scale is small the wavelet is concentrated in time, and the wavelet analysis has a detailed view of the signal. When the scale increases the wavelet becomes spread out in time, and the wavelet analysis takes into account the long-time behavior of the signal. Hence, the wavelet transform is a bank of multiresolution filters.

Figure 10.1 shows a typical wavelet multiresolution analysis for an electrical power-system transient signal. The signal is decomposed with different resolutions corresponding to different scale factors of the wavelets. The signal components in multiple frequency bands and the times

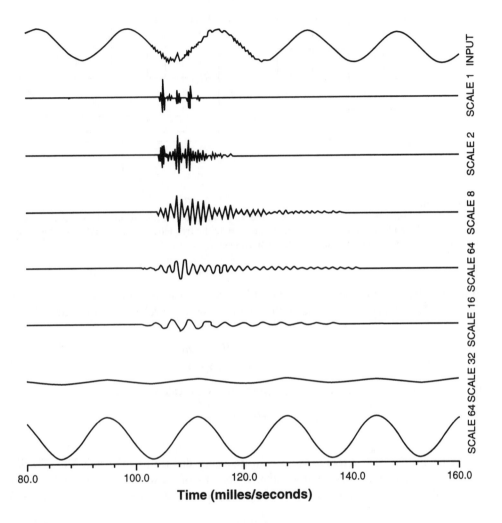

FIGURE 10.1
Multiresolution wavelet analysis of a transient signal in the electrical power system (see explanation in Section 10.9.1) [18].

of occurrence of those components are well presented in the figure. This figure is a time-scale joint representation, with the vertical axis in each discrete scale representing the amplitude of wavelet components. More detailed discussion will be given in Section 10.9.1.

Constant fidelity analysis

The dilations of the wavelets permit the wavelet analysis to zoom in on discontinuities, singularities, and edges and to zoom out for a global view. This is a unique property of the wavelet transform, important for nonstationary and fast transient signal analysis. In the time-scale joint representation the wavelet transform zooms in and gives, for example, the arrival time of an edge signal. The fixed-window short-time Fourier transform does not have this ability.

With the bank of multiresolution wavelet transform filters, the signal is divided into different frequency bands. In each band the signal is analyzed with a resolution matched to the scales

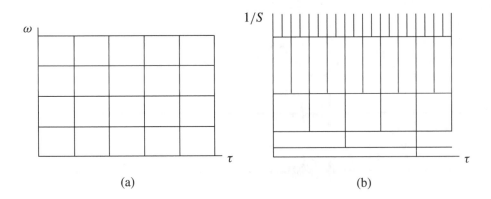

FIGURE 10.2
Coverage of the time-frequency space with (a) the short-time Fourier transform, where $\Delta\omega$ and Δt are fixed in the whole plane; (b) the wavelet transform, where the frequency bandwidth $\Delta\omega$ increases and the time resolution Δt improves with increase of $\Delta(1/s)$.

of the wavelets. As the wavelet is localized in time, the wavelet transform filter has a finite extension in frequency. The bandwidth $\Delta\omega$ of the basic wavelet transform filter $H(\omega)$ is defined as

$$(\Delta\omega)^2 = \frac{\int \omega^2 |H(\omega)|^2 \, d\omega}{\int |H(\omega)|^2 \, d\omega} \tag{10.1.40}$$

Hence, the bandwidth of the wavelet transform filter of a scale s is equal to

$$(\Delta\omega)_s^2 = \frac{\int \omega^2 |H(s\omega)|^2 \, d\omega}{\int |H(s\omega)|^2 \, d\omega} = \frac{1}{s^2}(\Delta\omega)^2 \tag{10.1.41}$$

The fidelity factor Q refers to the central frequency divided by the bandwidth of a filter that is the inverse of the relative bandwidth. In the case of the wavelet transform the relative bandwidths of the wavelet transform filters are constant because according to Eqs. (10.1.41)

$$\frac{1}{Q} = \frac{(\Delta\omega)_s}{1/s} = (\Delta\omega) \tag{10.1.42}$$

that is independent of the scale s. Hence, the wavelet transform is a constant-Q analysis. At low frequency, corresponding to a large scale factor s, the wavelet transform filter has a small bandwidth and a broad time window with a low time resolution. At high frequency, corresponding to a small scale factor s, the wavelet transform filter has a wide bandwidth, which implies a narrow time window with high time resolution. The time resolution of the wavelet analysis increases with the frequency of the signal. This adaptive window property is desirable for time-frequency analysis.

When the constant-Q relation (10.1.42) is satisfied, the bandwidth $\Delta\omega$ and, therefore, Δt change with the center frequency $1/s$ of the wavelet transform filter. The product $\Delta\omega\Delta t$ still satisfies the uncertainty principle (10.1.9). The time resolution Δt can be arbitrarily good at small scale, and the frequency resolution $\Delta\omega$ can be arbitrarily good at large scale. Figure 10.2 shows the coverage of the time-scale space for the wavelet transform and, as a comparison, that for the short-time Fourier transform.

Figure 10.3 shows the real part of the cos-Gaussian wavelets in comparison with the Gabor transform basis functions. Both functions consist of a cos function with a Gaussian window.

The basic wavelet is

$$h(t) = \frac{1}{\sqrt{2\pi}} \cos(\omega_0 t) \exp\left(\frac{-t^2}{2}\right) \tag{10.1.43}$$

where $\omega_0 = 5$. The wavelets $h_{m,n}(t)$ are generated from $h(t)$ by Eq. (10.1.2):

$$h_{s,\tau} = \frac{1}{\sqrt{s}} h\left(\frac{t - \tau}{s}\right)$$

with the discrete scale factor $s = 2^m$ and the discrete translation factor $(\tau/s) = n$. The basic window function for the Gabor transform is the Gaussian function

$$g(t) = \frac{1}{\sqrt{2\pi}} \exp\left(-\frac{t^2}{2}\right) \tag{10.1.44}$$

The discrete Gabor function $g_{m,n}(t)$ are generated from $g(t)$ by Eq. (10.1.21)

$$g_{m,n}(t) = g(t - n\tau_0) \exp(-jm\omega_0 t)$$

where $\omega_0 = \pi$.

From Figure 10.3a we see that the wavelets are with the dilated window. All the dilated wavelets contain the same number of oscillations. The wavelet transform performs multiresolution analysis by doing high-frequency analysis for narrow-windowed signals and low-frequency analysis for wide-windowed signals. This constant-Q analysis property makes the wavelet transform surpass the fixed-window short-time Fourier transform for analysis of local property of signals.

Figure 10.3b show a comparison between the wavelet transform and the Gabor transform for a step-function input. The wavelets are with the dilated windows. The Gabor functions are with windows of fixed width. The time-scale joint representation $\log s - t$ of the wavelet transform and the time-frequency joint representation $\log \omega - t$ of the Gabor transform are also shown. The wavelet transform with very small scale s and very narrow window is able to "zoom in" on the discontinuity and to indicate the arrival time of the step signal.

10.2   Properties of the wavelets

In this section we discuss some general properties of the wavelets. One basic property is related to the reconstruction of the signal from its wavelet transform. This property involves the resolution of identity, the energy conservation in the time-scale space, and the wavelet admissible condition. We shall find that any function that has finite energy and is square integrable and satisfies the wavelets admissible condition can be a wavelet. The second basic property is that the wavelet transform should be a local operator in both time and frequency domains. The regularity condition is usually imposed on wavelets.

10.2.1   *Admissible condition*

Resolution of identity

The wavelet transform of a one-dimensional signal is a two-dimensional time-scale joint representation. No information should be lost during the wavelet transform. Hence, the resolution

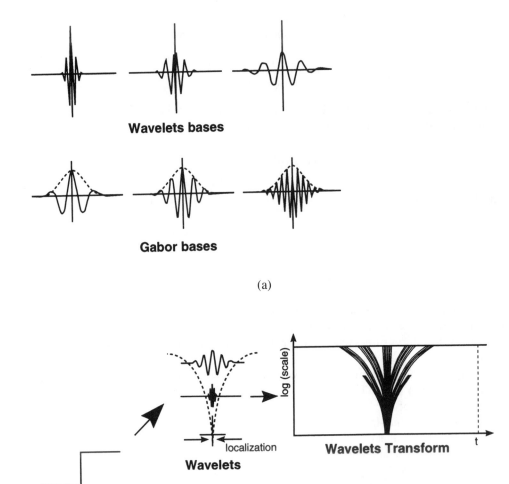

(a)

(b)

FIGURE 10.3

(a) The cos-Gaussian wavelets $h_{m,n}(t)$ and the real-part of the Gabor functions $g_{m,n}(t)$, with the scale factor $s = 2^m$ and the translation factor $\tau = ns$ for different values of m. The wavelets have a dilated window. The Gabor functions have a window with fixed width. (b) Time-scale joint representation of the wavelet transform and time-frequency joint representation of the Gabor transform for a step function. [22, 23]

of identity must be satisfied, which is expressed as

$$\int \frac{ds}{s^2} \int d\tau \langle f_1, h_{s,\tau} \rangle \langle h_{s,\tau}, f_2 \rangle = c_h \langle f_1, f_2 \rangle \qquad (10.2.1)$$

where $\langle , \rangle$ denotes the inner product and the constant c_h will be defined by Eq. (10.2.4). The left-hand side is basically an energy measure in the time-scale space and the right-hand side is the same energy measure in the time domain. Let us first examine the left-hand side of Eq. (10.2.1). The extra factor $1/s^2$ in the integral is the Haar invariant measure, owing to the time-scale space differential elements, $d\tau d(1/s) = d\tau ds/s^2$. We have assumed that only positive dilation $s > 0$ is accepted in our analysis. Using the expression (10.1.8) in the Fourier domain of the wavelet transform, we write the left-hand side of the resolution of identity as

$$\int \frac{ds}{s^2} \int d\tau \langle f_1, h_{s,t} \rangle \langle h_{s,\tau}, f_2 \rangle$$

$$= \frac{1}{4\pi^2} \int \frac{ds}{s^2} \int d\tau \int \int s F_1(\omega_1) H^*(s\omega_1) F_2^*(\omega_2) H(s\omega_2) e^{j\tau(\omega_1 - \omega_2)} \, d\omega_1 \, d\omega_2$$

$$= \frac{1}{2\pi} \int \int F_1(\omega_1) F_2^*(\omega_1) |H(s\omega_1)|^2 \frac{ds}{s} \, d\omega_1$$

$$= \frac{c_h}{2\pi} \int F_1(\omega_1) F_2^*(\omega_1) \, d\omega_1 \qquad (10.2.2)$$

If we let $\omega = s\omega_1$ then $ds = d\omega/|\omega_1|$, so that ds and $d\omega$ are of the same sign. Because $s > 0$, $ds/s = d\omega/|\omega|$. Then we define a constant

$$c_h = \int |H(\omega)|^2 \frac{d\omega}{|\omega|}$$

According to the well-known property of the Fourier transform, we have

$$\frac{1}{2\pi} \int F_1(\omega_1) F_2^*(\omega_1) \, d\omega_1 = \int f_1(t) f_2^*(t) \, dt = \langle f_1, f_2 \rangle \qquad (10.2.3)$$

Hence, the resolution of identity is satisfied on the condition that

$$c_h = \int \frac{|H(\omega)|^2}{|\omega|} \, d\omega < +\infty \qquad (10.2.4)$$

Admissible condition

Condition (10.2.4) is called the admissible condition of the wavelet. The admissible condition implies that the Fourier transform of the wavelet must have a zero component at the zero frequency:

$$|H(\omega)|^2 \big|_{\omega=0} = 0 \qquad (10.2.5)$$

Hence, the wavelets are inherently band-pass filters in the Fourier domain. In the time domain the wavelet must be oscillatory, like a wave, to have a zero-integrated area, or a zero-mean value:

$$\int h(t) \, dt = 0 \qquad (10.2.6)$$

Energy conservation

When $f_1 = f_2$, the resolution of identity (10.2.1) becomes

$$\int\int |W_f(s, \tau)|^2 \, d\tau \frac{ds}{s^2} = c_h \int |f(t)|^2 \, dt \tag{10.2.7}$$

This is the energy-conservation relation of the wavelet transform, equivalent to the Parseval energy relation of the Fourier transform.

Inverse wavelet transform

From the resolution of identity we have directly

$$f(t) = \frac{1}{c_h} \int\int W_f(s, \tau) \frac{1}{\sqrt{s}} h\left(\frac{t - \tau}{s}\right) d\tau \frac{ds}{s^2} \tag{10.2.8}$$

This is the inverse wavelet transform. The right-hand side of Eq. (10.2.8) is an integral in time-scale space of the wavelets $h_{s,\tau}(t)$ weighted by the wavelet transform coefficients $W_f(s, \tau)$. The function $f(t)$ is then recovered from the inverse wavelet transform.

The wavelet transform analysis is a decomposition of a function into a linear combination of the wavelets. The coefficients $W_f(s, \tau)$ in this combination are the inner products between the function and the wavelets. The wavelet transform coefficients indicate how close the function $f(t)$ is to a particular basis function $h_{s,\tau}(t)$. The inverse wavelet transform, shown in Eq. (10.2.8), tells us that the original signal may be synthesized by summing up all the projections of the signal onto the wavelets. In this sense, the continuous wavelet transform behaves like an orthogonal transform. We call this property of the continuous wavelet transform the quasi-orthogonality. Obviously, the set of the wavelet basis functions $h_{s,\tau}(t)$ with continuously varying scaling and shift is not orthogonal, but is very redundant.

From the reconstruction equation we have

$$W_f(s_0, \tau_0) = \int\int W_f(s, \tau) K(s_0, s; \tau_0, \tau) \, d\tau \frac{ds}{s^2} \tag{10.2.9}$$

where the reproducing kernel

$$K(s_0, s; \tau_0, \tau) = \frac{1}{c_h} \frac{1}{\sqrt{ss_0}} \int h^*\left(\frac{t - \tau_0}{s_0}\right) h\left(\frac{t - \tau}{s}\right) dt \tag{10.2.10}$$

is not zero with continuously varying factors s_0, s, τ_0, and τ and describes the intrinsic redundancy between the values of the wavelets at (s, τ) and at (s_0, τ_0).

Any square integrable function satisfying the admissible condition may be a wavelet function. When the wavelets satisfy the admissible condition, the signal can be recovered by the inverse wavelet transform. No information is lost.

10.2.2 *Regularity*

In practice, we require the wavelet function to have additional properties than the admissible condition. The wavelet transform of a one-dimensional function is two-dimensional; the wavelet transform of a two-dimensional function is four-dimensional. As a consequence, we would have an explosion of the time-bandwidth product with the wavelet transform, which is in contradiction with the restrictions of many applications, such as data compression and pattern classification, where the signals need to be described efficiently by few coefficients.

We usually impose some regularity conditions on the wavelet functions, so that the wavelet transform coefficients decrease quickly with decreasing of the scale s and increasing of $1/s$. For this purpose, the basic wavelet $h(t)$ and its Fourier transform $H(\omega)$ should have some smoothness and concentration in both time and frequency domains. The wavelet transform should be a local operator in both time and frequency domains.

According to the Fourier domain expression (10.1.8) of the wavelet transform, a fast decrease of the wavelet transform coefficients corresponds to a localization of $H(\omega)$ in the frequency domain. In the frequency domain, we already have $H(0) = 0$ from the admissible condition and we want $H(\omega)$ to vanish above a certain frequency by imposing regularity on the wavelets. Thus, the wavelet transform filter $H(\omega)$ is intrinsically a bandpass filter.

For the sake of simplicity, let $\tau = 0$ and consider the convergence to zero of the wavelet transform coefficients with increasing of $1/s$. We expand the signal $f(t)$ into the Taylor series at $t = 0$ until order n. The wavelet transform coefficients become [4]

$$
\begin{aligned}
W_f(s, 0) &= \frac{1}{\sqrt{s}} \int f(t) h^* \left(\frac{t}{s} \right) dt \\
&= \frac{1}{\sqrt{s}} \left[\sum_{p=0}^{n} f^{(p)}(0) \int \frac{t^p}{p!} h \left(\frac{t}{s} \right) dt + \int R(t) h \left(\frac{t}{s} \right) dt \right]
\end{aligned}
\tag{10.2.11}
$$

where the remainder

$$
R(t) = \int_0^t \frac{(t - t')^n}{n!} f^{(n+1)}(t') \, dt'
$$

and $f^{(p)}$ denotes the pth derivative. Denoting the moments of the wavelets by M_p

$$
M_p = \int t^p h(t) \, dt
\tag{10.2.12}
$$

it is easy to show that the last term in the right-hand side of Eq. (10.2.11), which is the wavelet transform of the remainder, decreases as s^{n+2}. We have then a finite development as

$$
\begin{aligned}
W_f(s, 0) &= \frac{1}{\sqrt{s}} \left[f(0) M_0 s + \frac{f'(0)}{1!} M_1 s^2 + \frac{f''(0)}{2!} M_2 s^3 \right. \\
&\quad \left. + \cdots + \frac{f^{(n)}(0)}{n!} M_n s^{n+1} + O(s^{n+2}) \right]
\end{aligned}
\tag{10.2.13}
$$

According to the admissible condition, $M_0 = 0$. The first term in the right-hand side of Eq. (10.2.13) is then zero. The speed of convergence to zero of the wavelet transform coefficients $W_f(s, \tau)$ with decreasing of the scale s or increasing of $1/s$ is then determined by the first nonzero moment of the basic wavelet $h(t)$. If the first $n + 1$ moments of order up to n of $h(t)$ are zero:

$$
M_p = \int t^p h(t) \, dt = 0 \qquad \text{for} \qquad p = 0, 1, 2, \ldots, n
\tag{10.2.14}
$$

According to Eq. (10.2.13) the wavelet transform coefficient $W_f(s, \tau)$ decays as fast as s^{n+2} for a smooth signal $f(t)$. Thus, the regularity of the wavelet $h(t)$ leads to localization of the wavelet transform in the frequency domain.

The wavelet satisfying condition (10.2.14) is called the wavelet of order n. In frequency domain, this condition is equivalent to the derivatives of the Fourier transform of the wavelet $h(t)$ up to order n to be equal to zero at the zero frequency $\omega = 0$:

$$
H^{(p)}(0) = 0 \qquad \text{for} \qquad p = 0, 1, 2, \ldots, n
\tag{10.2.15}
$$

The Fourier transform of the wavelet has zero of order $n + 1$ in the frequency domain about $\omega = 0$.

10.2.3 *Linear transform property*

By definition the wavelet transform is a linear operation. Given a function $f(t)$, the wavelet transform $W_f(s, \tau)$ satisfies the following relations

   i. linear superposition without the cross-terms

$$W_{f_1+f_2}(s, \tau) = W_{f_1}(s, \tau) + W_{f_2}(s, \tau) \tag{10.2.16}$$

  ii. translation

$$W_{f(t-t_0)}(s, \tau) = W_{f(t)}(s, \tau - t_0) \tag{10.2.17}$$

  iii. rescale

$$W_{\alpha^{1/2}f(\alpha t)}(s, \tau) = W_{f(t)}(\alpha s, \alpha \tau) \tag{10.2.18}$$

Different from the standard Fourier transform and other transforms, the wavelet transform is not ready for closed-form solution apart from some very simple functions such as:

   i. for $f(t) = 1$, from definition (10.1.6) and the admissible condition of the wavelets, Eq. (10.2.6), we have

$$W_f(s, \tau) = 0 \tag{10.2.19}$$

    The wavelet transform of a constant is equal to zero.

  ii. for a sinusoidal function $f(t) = \exp(j\omega_0 t)$, we have directly from the Fourier transform of the wavelets, Eq. (10.1.7)

$$W_f(s, \tau) = \sqrt{s} H^*(s\omega_0) \exp(j\omega_0 \tau) \tag{10.2.20}$$

    The wavelet transform of a sinusoidal function is a sinusoidal function of the time shift τ. Its modulus $|W_f(s, \tau)|$ depends only on the scale s.

  iii. for $f(t) = t$, we have

$$W_f(s, \tau) = \frac{1}{\sqrt{s}} \int th^*\left(\frac{t-\tau}{s}\right) dt$$

$$= s^{3/2} \int th^*(t-\tau') dt$$

$$= \frac{s^{3/2}}{j} \left. \frac{dH^*(\omega)}{d\omega}\right|_{\omega=0} \tag{10.2.21}$$

Hence, if the wavelet $h(t)$ is regular and of order $n \geq 1$ so that the derivatives of first order of its Fourier transform is equal to zero at $\omega = 0$, the wavelet transform of $f(t) = t$ is equal to zero.

For most functions the wavelet transforms have no analytical solutions and can be calculated only by numerical computer or calculated by optical analog computer. The optical continuous wavelet transform is based on the explicit definition of the wavelet transform, Eqs. (10.1.6) and (10.1.8), and is implemented using a bank of optical wavelet transform filters in the Fourier plane of an optical correlator [21].

The discrete and orthonormal wavelet transform can be computed with the fast wavelet transform tree algorithm, just as the discrete Fourier transform is computed with the FFT algorithm. The fast computation is an important advantage of the discrete orthonormal wavelet transform.

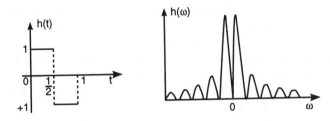

FIGURE 10.4
Haar basic wavelet $h(t)$ and its Fourier transform $H(\omega)$ [21].

10.2.4   *Examples of the wavelets*

In this section we give some examples of wavelets. Most of them are continuous wavelets. Examples of discrete orthonormal wavelets will be given in Section 10.7.

Haar wavelet

The Haar wavelet was historically introduced by Haar [5] in 1910 and it is a bipolar step function:

$$h(t) = \begin{cases} 1 & \text{when } 0 < t < 1/2 \\ -1 & \text{when } 1/2 < t < 1 \\ 0 & \text{otherwise} \end{cases} \tag{10.2.22}$$

We define the rectangular function as

$$\text{rect}(t) = \begin{cases} 1 & \text{when } -1/2 < t < 1/2 \\ 0 & \text{otherwise} \end{cases} \tag{10.2.23}$$

The Haar wavelet can be written as a correlation between a rectangle function rect($2t$) and two delta functions

$$h(t) = \text{rect}\left[2\left(t - \frac{1}{4}\right)\right] - \text{rect}\left[2\left(t - \frac{3}{4}\right)\right]$$

$$= \text{rect}(2t) * \left[\delta\left(t - \frac{1}{4}\right) - \delta\left(t - \frac{3}{4}\right)\right] \tag{10.2.24}$$

The Haar wavelet is a real function, antisymmetric with respect to $t = 1/2$, as shown in Figure 10.4. The wavelet admissible condition (10.2.6) is satisfied. The Fourier transform of the Haar wavelet is complex valued and is equal to a sinc function modulated by a sine function

$$H(\omega) = 2j \exp\left(-j\frac{\omega}{2}\right) \text{sinc}\left(\frac{\omega}{4}\right) \sin\left(\frac{\omega}{4}\right)$$

$$= 4j \exp\left(-j\frac{\omega}{2}\right) \frac{1 - \cos\frac{\omega}{2}}{\omega} \tag{10.2.25}$$

whose amplitude $|H(\omega)|$ is even and symmetric to $\omega = 0$. That is a band-pass filter, as shown in Figure 10.4. The phase factor $\exp(-j\omega/2)$ is related only to the fact that the $h(t)$ is antisymmetric with respect to $t = 1/2$.

The Haar wavelet transform involves a bank of multiresolution filters that yield the correlations between the signal and the Haar wavelets scaled by factor s. The Haar wavelet transform is a local operation in the time domain. The time resolution depends on the scale s. When the signal is constant, the Haar wavelet transform is equal to zero. The amplitude of the Haar wavelet transform has high peak values when there are discontinuities of the signal.

The Haar wavelet is irregular and discontinuous in time. The first-order moment of the Haar wavelet is not zero. According to Eq. (10.2.25), the amplitude of the Fourier spectrum of the Haar wavelet converges to zero very slowly as $1/\omega$. According to Eq. (10.2.13), the Haar wavelet transform decays with increasing of $1/s$ as $(1/s)^{-3/2}$.

The set of discrete dilations and translations of the Haar wavelets constitute the simplest orthonormal wavelet basis. We shall use the Haar wavelets as an example of the orthonormal wavelet basis in Sections 10.5 and 10.6. However, the Haar wavelet transform has not found many practical applications because of its poor localization property in the frequency domain.

Morlet wavelet

This wavelet was used by Martinet, Morlet, and Grossmann [6] for analysis of sound patterns. The Morlet's basic wavelet function is a multiplication of the Fourier basis with a Gaussian window:

$$h(t) = \exp(j\omega_0 t) \exp\left(-\frac{t^2}{2}\right) \qquad (10.2.26)$$

Its real part is a Cos-Gaussian and the imaginary part is a Sin-Gaussian function. The Morlet wavelet is similar to the Gabor function in the short-time Fourier transform with Gaussian window. The difference is that in the Morlet wavelet basis the Gaussian window can be dilated together with the Fourier basis function, while in the Gabor transform the window has a fixed width.

The Cos-Gaussian wavelets are real functions. Their Fourier transform consists of two Gaussian functions shifted to ω_0 and $-\omega_0$, respectively:

$$H(\omega) = \sqrt{\frac{\pi}{2}} \left\{ \exp\left[-\frac{(\omega - \omega_0)^2}{2}\right] + \exp\left[-\frac{(\omega + \omega_0)^2}{2}\right] \right\} \qquad (10.2.27)$$

that are real positive, even, and symmetric to the origin $\omega = 0$. The Gaussian window is perfectly local in both time and frequency domains and achieves the minimum time-bandwidth product determined by the uncertainty principle, as shown by Eq. (10.1.20). The Cos-Gaussian wavelets are band-pass filters in frequency domain. They converge to zero like the Gaussian function as the frequency increases. Figure 10.5 shows the Cos-Gaussian wavelet and its Fourier spectrum.

The Morlet wavelets do not satisfy the wavelet admissible condition, because

$$H(0) \neq 0 \qquad (10.2.28)$$

which leads to $c_h = +\infty$. But the value of $H(0)$ is very close to zero provided that the ω_0 is sufficiently large. When $\omega_0 = 5$, for example,

$$H(0) = \sqrt{2\pi} \exp\left(-\frac{25}{2}\right)$$

which is of 10^{-5} order of magnitude and can be practically considered as zero in numerical computations.

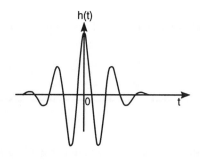

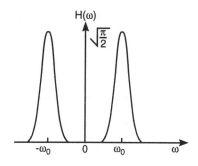

FIGURE 10.5
Cos-Gaussian wavelet $h(t)$ and its Fourier transform $H(\omega)$ [21].

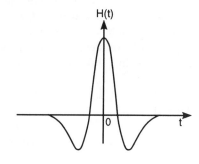

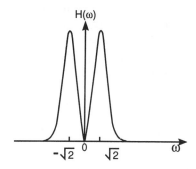

FIGURE 10.6
Mexican-hat wavelet $h(t)$ and its Fourier transform $H(\omega)$ [21].

Mexican-hat wavelet

The Mexican-hat-like wavelet was first introduced by Gabor and it is the second-order derivative of the Gaussian function [7]:

$$h(t) = (1 - t^2) \exp\left(-\frac{t^2}{2}\right) \tag{10.2.29}$$

The Mexican-hat wavelet is even and real valued. The wavelet admissible condition is satisfied. The Fourier transform of the Mexican-hat wavelet is

$$H(\omega) = -\omega^2 \exp\left(-\frac{\omega^2}{2}\right) \tag{10.2.30}$$

which is also even and real valued, as shown in Figure 10.6.

The Gaussian function is perfectly local in both time and frequency domains and is indefinitely derivable. In fact, a derivative of any order n of the Gaussian function may be a wavelet. The Fourier transform of the nth order derivative of the Gaussian function is

$$H(\omega) = (j\omega)^n \exp\left(-\frac{\omega^2}{2}\right) \tag{10.2.31}$$

which is the Gaussian function multiplied by $(j\omega)^n$ so that $H(0) = 0$. The wavelet admissible condition is satisfied. Its nth derivative $H^{(n-1)}(0) = 0$. The wavelet is a regular wavelet of order n. Both $h(t)$ and $H(\omega)$ are indefinitely derivable. The wavelet transform coefficients decay with increasing of $1/s$ as fast as $(1/s)^{n-1/2}$.

The two-dimensional Mexican-hat wavelet is well known as the Laplacian operator, widely used for zero-crossing image edge detection.

10.3 Discrete and orthonormal wavelets

The continuous wavelet transform maps a one-dimensional signal to a two dimensional time-scale joint representation that is highly redundant. The time-bandwidth product of the continuous wavelet transform output is the square of that of the signal. For most applications the goal of signal processing is to represent the signal efficiently with fewer parameters. The use of discrete wavelets can reduce the time-bandwidth product of the wavelet transform output.

By the term discrete wavelets we mean, in fact, the continuous wavelets with discrete scale and translation factors. The discrete scale and translation factors can be expressed as $s = s_0^i$ and $\tau = k\tau_0 s_0^i$, where i and k are integers and $s_0 > 1$ is a fixed dilation step. The translation factor, $\tau = k\tau_0 s_0^i$, depends on the dilation step s_0^i. The corresponding discrete wavelets are written as

$$h_{i,k}(t) = s_0^{-i/2} h[s_0^{-i}(t - k\tau_0 s_0^i)]$$

$$= s_0^{-i/2} h(s_0^{-i} t - k\tau_0) \tag{10.3.1}$$

10.3.1 *Time-scale space lattices*

With the discrete wavelets, the wavelet transform of a continuous function is carried out at discrete times and frequencies that correspond to a sampling in the time-scale space. The time-scale joint representation of a wavelet transform with the discrete wavelet basis is a grid along the scale and time axes. To show this we consider localization points of the discrete wavelets in the time-scale space.

The sampling along the time axis has the interval $\tau_0 s_0^i$, which is proportional to the scale s_0^i. The time-sampling step is small for small-scale wavelet analysis and is large for large-scale wavelet analysis. With the varying scale the wavelet analysis will be able to "zoom in" on singularities of the signal using a more concentrated wavelet of very small scale. For this detailed analysis the time-sampling step is very small. Because only the signal detail is of interest, only a few small time steps would be needed. Therefore, the wavelet analysis provides a more efficient way to represent transient signals.

There is an analogy between the wavelet analysis and the microscope. The scale factor s_0^i corresponds to the magnification or the resolution of the microscope. The translation factor τ corresponds to the location where one makes observations with the microscope. If one looks at very small details, the magnification and the resolution must be large, which corresponds to a large and negative i. The wavelet is very concentrated. The step of translation is small, and that justifies the choice $\tau = k\tau_0 s_0^i$. For large and positive i, the wavelet is spread out, and the large translation steps $k\tau_0 s_0^i$ are adapted to this wide width of the wavelet analysis function. This is another interpretation of the constant-Q analysis property of the wavelet transform, discussed in Section 10.1.5.

The behavior of the discrete wavelets depends on the steps s_0 and τ_0. When s_0 is close to 1 and τ_0 is small, the discrete wavelets are close to the continuous wavelets. For a fixed scale s_0, the localization points of the discrete wavelets along the scale axis are logarithmic

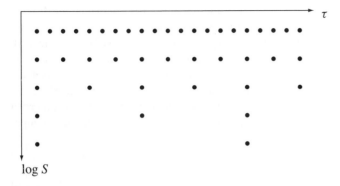

FIGURE 10.7
Localization of the discrete wavelets in the time-scale space.

as $\log s = i \log s_0$. We usually choose $s_0 = 2$ that corresponds to the dyadic sampling in frequency.

The frequency sampling interval 2^i is called one octave in music. One octave is the interval between two frequencies having a ratio of two. One-octave frequency band has a bandwidth equal to one octave.

The discrete time step is $\tau_0 s_0^i$. We usually choose $\tau_0 = 1$. Hence, the time-sampling step is a function of the scale and is equal to 2^i for the dyadic wavelets. Along the τ-axis the localization points of the discrete wavelets depend on the scale. The intervals between the localization points are equal within the same scale and are proportional to the scale s_0^i. The translation steps are small for small positive values of i with the small-scale wavelets, and are large for large positive values of i with large-scale wavelets. The localization of the discrete wavelets in the time-scale space is shown in Figure 10.7, which is logarithmic along the scale axis, $\log s = i \log 2$, and is uniform along the time axis τ with the time steps proportional to the scale factor $s = 2^i$.

10.3.2 *Wavelet series decomposition*

When the input function is continuous and the wavelets are continuous in scale and translation factors, one has the continuous wavelet transform. When the input function is continuous but the continuous wavelets are with discrete scale and translation factors, the wavelet transform results in a series of wavelet coefficients and is referred to as the wavelet series decomposition. There is analogy with the continuous Fourier transform and the Fourier series decomposition.

As we shall see in the following sections, implementation of the wavelet series expansion is essentially a discrete algorithm. The wavelet theory using the function space analysis shows that the wavelet series expansion and reconstruction of continuous functions can be computed with the multiresolution analysis algorithm using discrete iterated filters. An extra regularity condition is imposed on the orthonormal wavelet basis functions and on the discrete filters. When the wavelet transform is computed in the digital computer the input function and the iterated filters are all discrete. In this sense, the wavelet transform is the discrete wavelet transform.

One of the motivations for the wavelet series expansion and the discrete wavelet transform using the discrete wavelets is to reduce the time-bandwidth product of the wavelet transform output, that is, the number of the wavelet coefficients. The wavelet series expansion and the discrete wavelet transform represent signals with a sequence of wavelet coefficients, corresponding to sampling in the time-scale space. The sampling grid is determined by the discrete scale and translation factors. When $s_0 = 2$, the scale is sampled at a rate of 2^i, that is, by octave. As the scale factor increases by a factor of two one moves from one octave band to the next. The bandwidth of the signal is reduced by a factor of two. Thus, according to the Shannon sampling theorem, the time-sampling rate can also be reduced by two. As a consequence, the time-bandwidth product of the wavelet transform at the larger scale can be reduced by a factor of two. As we shall show in Section 10.8, the real time-bandwidth product of the orthonormal wavelet series expansion and the discrete wavelet transform can be equal to that of the input signal.

Another motivation for the wavelet series expansion and the discrete wavelet transform is that they can be implemented in the digital computer by a tree-structured algorithm in the multiresolution analysis framework. The wavelet iterative algorithm and the wavelet integrated circuits can compute the wavelet transform even faster than the FFT algorithm for computing the Fourier transform. In Section 10.8 we shall show that the number of operations of the wavelet transform is of the order of $O(L)$, where L is the size of the input data vector.

10.3.3   *Wavelet frame*

With the discrete wavelet basis a continuous function $f(t)$ is decomposed into a sequence of wavelet coefficients

$$W_f(i, k) = \int f(t) h^*_{i,k}(t) \, dt = \langle f, h_{i,k} \rangle \tag{10.3.2}$$

A rising question for the discrete wavelet transform is how well the function $f(t)$ can be reconstructed from the discrete wavelet coefficients:

$$f(t) = A \sum_i \sum_k W_f(i, k) h_{i,k}(t) \tag{10.3.3}$$

where A is a constant that does not depend on $f(t)$. Obviously, if s_0 is close enough to 1 and τ_0 is small enough, the wavelets approach as a continuum. Reconstruction (10.3.3) is then close to the inverse continuous wavelet transform. The signal reconstruction takes place without nonrestrictive conditions other than the admissible condition on the wavelet $h(t)$. On the other hand, if the sampling is sparse, $s_0 = 2$ and $\tau_0 = 1$, reconstruction (10.3.3) can be achieved only for very special choices of the wavelet $h(t)$.

The theory of wavelet frames provides a general framework that covers the above-mentioned two extreme situations. It permits one to balance between the redundancy, that is, the sampling density in the scale-time space, and the restriction on the wavelet $h(t)$ for reconstruction scheme (10.3.3) to work. If the redundancy is large with high oversampling, then only mild restrictions are put on the wavelet basis. If the redundancy is small with critical sampling, then the wavelet basis functions are very constrained.

Daubechies [8] has proven that the necessary and sufficient condition for the stable reconstruction of a function $f(t)$ from its wavelet coefficients $W_f(i, k)$ is that the energy, which is the sum of square moduli of $W_f(i, k)$, must lie between two positive bounds:

$$A\|f\|^2 \le \sum_{j,k} |\langle f, h_{i,k} \rangle|^2 \le B\|f\|^2 \tag{10.3.4}$$

where $\|f\|^2$ is the energy of $f(t)$, $A > 0$, $B < \infty$ and A, B are independent of $f(t)$. When $A = B$, the energy of the wavelet transform is proportional to the energy of the signal. This is similar to the energy conservation relation Eq. (10.2.7) of the continuous wavelet transform. When $A \ne B$ there is still some proportional relation between the two energies.

When Eq. (10.3.4) is satisfied, the family of basis functions $\{h_{i,k}(t)\}$ with $i, k \in \mathbf{Z}$ is referred to as a frame and A, B are termed frame bounds. Hence, when proportionality between the energy of the function and the energy of its discrete transform function is bounded between something greater than zero and less than infinity for all possible square integrable functions, then the transform is complete. No information is lost and the signal can be reconstructed from its decomposition.

Daubechies has shown that the accuracy of the reconstruction is governed by the frame bounds A and B. The frame bounds A and B can be computed from s_0, τ_0 and the basis function $h(t)$. The closer A and B, the more accurate the reconstruction. When $A = B$ the frame is tight and the discrete wavelets behave exactly like an orthonormal basis. When $A = B = 1$, Eq. (10.3.4) is simply the energy conservation equivalent to the Parseval relation of the Fourier transform. It is important to note that the same reconstruction works even when the wavelets are not orthogonal to each other.

When $A \neq B$ the reconstruction can still work exactly for the discrete wavelet transform if we use the synthesis function basis for reconstruction that is different from the decomposition function basis for analysis. The former constitute the dual frame of the later.

10.3.4 *Orthonormal wavelet transform*

In Section 10.2.1 we saw that the continuous wavelet transform with continuously varying scale and translation factors behaves just like an orthonormal transform, in the sense that the inverse wavelet transform permits us to reconstruct the signal by an integration of all the projections of the signal onto the wavelet basis. We call this the quasi-orthonormality of the continuous wavelet transform. Obviously a continuous wavelet basis is not orthonormal. The continuous wavelet transform is highly redundant.

The orthonormal wavelet decomposition has no redundancy in the signal representation. An orthonormal wavelet basis is possible only with the wavelets of discrete dilation and translation factors. The discrete wavelets defined in Eq. (10.3.1) can be made orthogonal to their own dilations and translations by special choices of the basis wavelet. This orthogonality means:

$$\int h_{i,k}(t)h_{m,n}^*(t)\,dt = \begin{cases} 1 & \text{if } i = m \text{ and } k = n \\ 0 & \text{otherwise} \end{cases} \tag{10.3.5}$$

for all the integers i and k. All the $h_{i,k}(t)$ are generated from the same basic $h(t)$ by means of discrete scaling and shift. With the same scale i, the $h_{i,k}(t)$ shifted with discrete steps k are orthogonal. With the same shift k, the $h_{i,k}(t)$ with discrete scales i are also orthogonal. For all the wavelets with different discrete scales or shifted to different discrete locations the inner products are equal to zero.

An arbitrary signal can be reconstructed as a sum of the orthogonal wavelet basis functions weighted by the wavelet transform coefficients:

$$f(t) = \sum_{i,k} W_f(i, k)h_{i,k}(t) \tag{10.3.6}$$

There exist well-behaved functions that can be used as the basic wavelet for generating the orthogonal wavelet basis.

The orthonormal wavelet basis permits the removal of redundancy in the signal representation. However, the orthogonality is not essential in the representation of signals. The wavelets need not be orthogonal. In some applications, the redundancy can help to reduce the sensitivity to noise.

10.4 Multiresolution analysis

Meyer and Mallat [9] found in 1986 that orthonormal wavelet decomposition and reconstruction can be implemented in the multiresolution signal analysis framework. The multiresolution analysis is now a standard way to construct orthonormal wavelet bases and to implement the orthonormal wavelets transforms. The multiresolution analysis is a technique that permits us to analyze signals in multiple frequency bands. There are two existing approaches of multiresolution analysis: the Laplacian pyramid and subband coding, which were developed independently in the late seventies and early eighties.

10.4.1 *Laplacian pyramid*

Multiresolution signal analysis was first proposed by Burt and Adelson in 1983 [10] for image decomposition, coding, and reconstruction. Multiresolution signal analysis is based on a weighting function also called the smoothing function. The original data, represented as a sequence of real numbers, $c_0(n)$, $n \in \mathbf{Z}$, is averaged in neighboring pixels by the weighting function. The weighting function can be a Gaussian function that is a low-pass filter. The correlation with the weighting function reduces the resolution of the signal data. Hence, after the averaging process, the data sequence is down-sampled by a factor of two, so that the resulting data sequence $c_1(n)$ is the averaged approximation of $c_0(n)$.

The averaging and down-sampling process can be iterated with the smoothing function dilated by a scale factor of 2 for averaging $c_1(n)$, and so on. In the iteration process the smoothing function is dilated with dyadic scales 2^i with $i \in \mathbf{Z}$ to average the signals at multiple resolutions. Hence, the original data is represented by a set of successive approximations. Each approximation corresponds to a smoothed version of the original data at a given resolution.

Given an original data of size 2^N, the smoothed sequence $c_1(n)$ has a reduced size 2^{N-1}. By iterating the process, the successive averaging and down-sampling result in a set of data sequences of exponentially decreasing size. If we imagine these sequences stacked on top of one another, they constitute a hierarchical pyramid structure with $\log_2 N$ pyramid levels. The original data $c_0(n)$ are the bottom or zero level of the pyramid. At ith pyramid level, the signal sequence is obtained from the data sequence in the $(i - 1)$th level by

$$c_i(n) = \sum_k p(k - 2n)c_{i-1}(k) \tag{10.4.1}$$

where $p(n)$ is the weighting function. The operation described in Eq. (10.4.1) is a correlation between $c_{i-1}(k)$ and $p(k)$ followed by a down-sampling by two. Because a shift by two in $c_{i-1}(k)$ results in a shift by one in $c_i(n)$, the sampling interval in level i is double that in the previous level $i - 1$. The size of the sequence $c_i(n)$ is half as long as its predecessor $c_{i-1}(n)$. When the weighting function is the Gaussian function, the pyramid of the smoothed sequences is referred to as the Gaussian pyramid. Figure 10.8 shows a part of the Gaussian pyramid.

By low-pass filtering with the weighting function $p(n)$, the high-frequency detail of the signal is lost. To compute the difference between two successive pyramid levels of different size, we have to expand first the data sequence $c_i(n)$. The expansion of $c_i(n)$ may be done in two steps: 1) inserting a zero between every samples on $c_i(n)$, that is up-sampling $c_i(n)$ by two; 2) interpolating the sequence with a filter whose impulse response is $p'(n)$. The expand process results in a sequence $c'_{i-1}(n)$ that has the same size as the size of $c_{i-1}(n)$. In general, $c'_{i-1}(n) \neq c_{i-1}(n)$. The difference is a sequence $d_{i-1}(n)$

$$d_{i-1}(n) = c_{i-1}(n) - c'_{i-1}(n) \tag{10.4.2}$$

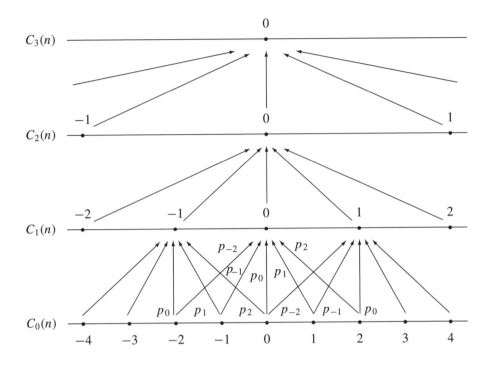

FIGURE 10.8
Multiresolution analysis Gaussian pyramid. The weighting function is $p(n)$ with $n = 0, \pm 1, \pm 2$. The even and add number nodes in $c_0(n)$ have different connections to the nodes in $c_1(n)$.

that contains the detail information of the signal. The differences between sequences of successive Gaussian pyramid levels are a set of sequences $d_i(n)$ that constitute another pyramid, called the Laplacian pyramid. The original signal can be reconstructed exactly by summing the Laplacian pyramid.

The Laplacian pyramid contains the compressed signal data in the sense that the pixel-to-pixel correlation of the signal is removed by averaging. If the original data is an image that is positively valued, the values on the Laplacian pyramid nodes are both positive and negative and are shifted toward zero, and can then be represented by fewer bits. The multiresolution analysis is then useful for image coding and compression.

The Laplacian pyramid signal representation is redundant. One stage of the pyramid decomposition leads to a half-size, low-resolution signal and a full-size, difference signal, resulting in an increase in the number of signal samples by 50%. Figure 10.9 shows the pyramid scheme.

10.4.2 *Subband coding*

Subband coding [11] is a multiresolution signal-processing approach that is different from the Laplacian pyramid. The basic objective of subband coding is to divide the signal spectrum into independent subbands in order to treat the signal subbands individually for different purposes. Subband coding is an efficient tool for multiresolution spectral analysis and has been successful in speech signal processing.

Given an original data sequence $c_0(n), n \in \mathbf{Z}$, the lower resolution approximation of the

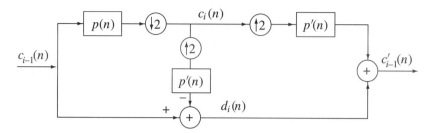

FIGURE 10.9
Schematic pyramid decomposition and reconstruction.

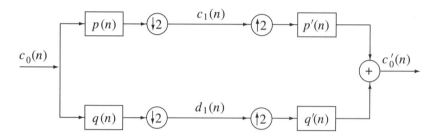

FIGURE 10.10
Schematic two-channel subband coding decomposition and reconstruction.

signal is derived by low-pass filtering with a filter having its impulse response $p(n)$

$$c_1(n) = \sum_k p(k - 2n)c_0(k) \tag{10.4.3}$$

that is the correlation between $c_0(k)$ and $p(k)$ down-sampled by a factor of two. The process is exactly the same as the averaging process in the pyramid decomposition. In order to compute the detail information that is lost by low-pass filtering with $p(n)$, a high-pass filter with the impulse response $q(n)$ is applied to the data sequence $c_0(n)$ as

$$d_1(n) = \sum_k q(k - 2n)c_0(k) \tag{10.4.4}$$

that is the correlation between $c_0(k)$ and $q(k)$ down-sampled by a factor of two. Hence, the subband decomposition leads to a half-size, low-resolution signal and a half-size, detail signal.

The subband analysis corresponds to a decomposition onto an orthonormal basis. Both $p(n)$ and $q(n)$ can be half-band filters. If they are ideal half-band filters, the low-pass filter $P(\omega)$ in the Fourier domain can be equal to 1 for $-\pi/2 \le \omega \le \pi/2$ and equal to 0 elsewhere, and the high-pass filter $Q(\omega)$ in the Fourier domain can be equal to 0 for $-\pi/2 \le \omega \le \pi/2$ and equal to 1 elsewhere. But in practice it is not necessary to use the ideal half-band filters.

To recover the signal $c_0(n)$ from the down-sampled approximation $c_1(n)$ and the downsampled detail $d_1(n)$, both $c_1(n)$ and $d_1(n)$ are up-sampled by a factor of two. The up-sampling is performed by first adding zeros between each node in $c_1(n)$ and $d_1(n)$ and then interpolating with the filters $p'(n)$ and $q'(n)$ respectively. Finally, adding together the two up-sampled sequences yields $c_0'(n)$. Figure 10.10 shows the scheme of the two-channel subband system.

The reconstructed signal $c_0'(n)$ usually is not identical to the original $c_0(n)$, unless the filters meet some specific constraints. The simplest case to analyze is that the subband analysis filters $p(n)$ and $q(n)$ and synthesis filters $p'(n)$ and $q'(n)$ are identical, $p(n) = p'(n)$ and $q(n) = q'(n)$. The perfect reconstruction is accomplished by:

$$c_0(n) = \sum_k [c_1(k)p(n - 2k) + d_1(k)q(n - 2k)] \tag{10.4.5}$$

that is the convolution between $c_1(k)$ and $d_1(k)$, which are up-sampled by a factor of two, and the synthesis filters $p(k)$ and $q(k)$, respectively. Note that the decomposition process described by Eqs. (10.4.3) and (10.4.4) is the correlation between $c_0(k)$ and the analysis filters $p(k)$ and $q(k)$. This is the standard expansion of a signal into an orthonormal basis and the reconstruction of the signal by weighted sum of the orthonormal projections. For the expansion and the reconstruction described in Eqs. (10.4.3), (10.4.4), and (10.4.5) to hold, the filters $p(n)$ and $q(n)$ should be orthogonal. When $p(n)$ and $q(n)$ are orthogonal, an application of the low-pass filter $p(n)$ to the second term in the right-hand side of Eq. (10.4.5), which is the detail information of $c_0(n)$ reconstructed from $d_1(k)$, should be equal to zero. Hence, we have

$$\sum_n p(n - 2m)q(n - 2k) = 0 \tag{10.4.6}$$

This is the cross-filter orthogonality condition between the low-pass and high-pass filters $p(n)$ and $q(n)$. On substituting for $c_1(n)$ and $d_1(n)$ of Eqs. (10.4.3) and (10.4.4) into Eq. (10.4.5) we obtain the condition for the perfect reconstruction

$$\sum_k [p(m - 2k)p(n - 2k) + q(m - 2k)q(n - 2k)] = \delta_{mn} \tag{10.4.7}$$

where δ_{mn} is the Kroneckev's function. The filters $p(n)$, $q(n)$ satisfying orthogonality condition (10.4.6) and the perfect reconstruction condition (10.4.7) constitute a two-band perfect reconstruction (PR) filter bank.

The two-band filter bank can be extended to the M-band filter bank by using a bank of M analysis filters followed by down-sampling and a bank of M up-samplers followed by M synthesis filters. The two-band filter bank can also be iterated: the filter bank divides the input spectrum into two equal subbands, yielding the low (L) and high (H) bands. The two-band filter bank can again be applied to these (L) and (H) half-bands to generate the quarter bands: (LL), (LH), (HL) and (HH). The scheme of this multiresolution analysis has a tree structure.

10.4.3 Subband filters

There is a close link between the orthonormal wavelet transform and the two-band paraunitary perfect reconstruction quadrature mirror filter (PR QMF) bank. Many concepts developed in the subband coding theory are used in the orthonormal wavelet transform theory. In this subsection, we define some special properties of the subband filters that are useful for the orthonormal wavelet transform.

The low-pass filters $p(n)$ in the subband coding are discrete and are, in general, real-valued sequences. The low-pass filter $p(n)$ is symmetric, $p(n) = p(-n)$. This is also true for some orthonormal wavelet bases. But in some other orthonormal wavelet bases such as the Daubechies bases, the low-pass filters $p(n)$ are not symmetric. The high-pass filters $q(n)$ are also discrete real-valued sequences and can be generated from the low-pass filters $p(n)$ with the cross-filter orthogonality condition and the perfect reconstruction condition.

The Fourier transforms of the filters $p(n)$ and $q(n)$ are

$$P(\omega) = \frac{1}{2} \sum_n p(n) \exp(-jn\omega)$$

$$Q(\omega) = \frac{1}{2} \sum_n q(n) \exp(-jn\omega) \tag{10.4.8}$$

that can be also expressed as the z-transform of the sequences $p(n)$ and $q(n)$:

$$P(z) = \frac{1}{2} \sum_n p(n) z^{-n}$$

$$Q(z) = \frac{1}{2} \sum_n q(n) z^{-n} \tag{10.4.9}$$

1. *Finite Impulse Response (FIR) filters*: The FIR filters have limited lengths and have compact supports: $p(n) = 0$ for $n < N_-$ and $n > N_+$. Only a finite number of $p(n)$ and $q(n)$ are not zero. The Fourier transforms $P(\omega)$ and $Q(\omega)$ of the FIR filters $p(n)$ and $q(n)$ with compact support would have no fast decay. There is a trade-off between the compactness and the regularity of the FIR filter.

2. *Scale and resolution*: In the time domain a large scale of the filter means global view and a small scale of the filter means detailed view, which is similar to the scale in maps. Scale change of continuous signals does not alter their resolution. The resolution of a continuous signal is related to its frequency content. The filtering with a half-band low-pass filter halves resolution of the signal, but keeps its scale.

   In digital signal analysis, however, changing the scale involves changing the clock rate of the time sequence and therefore changing the resolution. When a signal is transferred from scale 2^i to scale 2^{i+1}, its time clock rate is reduced by two. The time resolution is reduced by two. The size of the signal is also reduced by two.

3. *Down-sampling and up-sampling*: Down-sampling by a factor of two is a decimation operation, which can be achieved in two steps. First the signal $x(n)$ is sampled with the double sampling interval:

$$x'(n) = \begin{cases} x(n) & \text{for } n = 0, \pm 2, \pm 4, \ldots \\ 0 & \text{otherwise} \end{cases} \tag{10.4.10}$$

The intermediate signal $x'(n)$ has the same time clock rate as that of $x(n)$. Then, the time clock rate is reduced by two. The down-sampled signal $y(n)$ is obtained as

$$y(n) = x'(2n) \tag{10.4.11}$$

In the Fourier domain the spectrum $Y(\omega)$ is two times larger than $X(\omega)$ because $Y(\omega) = X(\omega/2)$. Thus, a low-pass filter should be used to avoid aliasing.

   Up-sampling by a factor of two is also implemented in two steps. First insert zeros between nodes of the signal $x(n)$ and then increase the time clock rate by two and let

$$y(n) = \begin{cases} x(n/2) & \text{when } n = 0, \pm 2, \ldots \\ 0 & \text{otherwise} \end{cases} \tag{10.4.12}$$

In the Fourier domain the spectrum $Y(\omega)$ is compressed by two with respect to $X(\omega)$, because $Y(\omega) = X(2\omega)$. Thus, a low-pass filter should be used for smoothing the up-sampled signal.

4. *Equal contribution constraint*: The low-pass filter $p(n)$ should satisfy the equal contribution constraint, stipulating that all the nodes in one resolution level contribute the same total amount to the next level and that the sum of all the weights for a given node n is independent of n. Note that the odd number nodes and the even number nodes in the data sequence $c_{i-1}(n)$ have two different connections with the weighting function $p(n)$, because of the down-sampling by 2 of $c_i(k)$. In an example shown in Figure 10.8, the even nodes in $c_0(n)$ are connected to $c_1(n)$ with the weighting factors $p(-2)$, $p(0)$, and $p(2)$, and the odd nodes are connected to $c_1(n)$ with the weighting factors $p(-1)$ and $p(1)$. Because the sum of all the weights for a given node is independent of n, the weighting function should satisfy:

$$\sum_n p(2n) = \sum_n p(2n+1) \tag{10.4.13}$$

5. *Quadrature Mirror Filters (QMF)*: Let $p(n)$ be a FIR filter with real coefficients; its conjugate mirror filter is defined as

$$q(n) = (-1)^n p(n) \tag{10.4.14}$$

or, equivalently, in the frequency domain $Q(\omega)$ is the conjugate mirror filter of $P(\omega)$ as

$$Q(\omega) = P(\omega - \pi) \tag{10.4.15}$$

The filter pair $P(\omega)$ and $Q(\omega)$ are called the mirror filters because on substituting for ω by $\pi/2 - \omega$ in Eq. (10.4.15) and noting that because $p(n)$ is real, $|P(\omega)|$ is an even function of ω, and we obtain

$$\left| Q\left(\frac{\pi}{2} - \omega\right) \right| = \left| P\left(\frac{\pi}{2} + \omega\right) \right| \tag{10.4.16}$$

This is the mirror image property of $|P(\omega)|$ and $|Q(\omega)|$ about $\omega = \pi/2$. Hence, the name quadrature mirror filter. Intuitively, if $P(\omega)$ is a half-band low-pass filter, and its conjugate mirror filter $Q(\omega)$ is a half-band high-pass filter, then the input spectrum in the full band $-\pi \leq \omega \leq \pi$ can be divided into two equal subbands by the analysis filters $P(\omega)$ and $Q(\omega)$.

6. *Power complementary filters*: The filter pair $\{P(\omega), Q(\omega)\}$ are called the power complementary filter if

$$|P(\omega)|^2 + |Q(\omega)|^2 = 1 \tag{10.4.17}$$

This relation shows the complementary property of the low-pass and high-pass filters.

7. *Paraunitary filter bank*: The subband filter bank theory emphasizes the frequency behavior of the filters $P(\omega)$ and $Q(\omega)$. It can be shown that cross-filter orthogonality condition (10.4.6) and perfect reconstruction condition (10.4.7), both in the time domain, can be satisfied when the filters $P(\omega)$ and $Q(\omega)$ in the frequency domain satisfy the paraunitary conditions, so that the matrix

$$\begin{vmatrix} P(\omega) & Q(\omega) \\ P(\omega + \pi) & Q(\omega + \pi) \end{vmatrix}$$

is paraunitary. The paraunitary filters must obey

$$\begin{vmatrix} P^*(\omega) & P^*(\omega + \pi) \\ Q^*(\omega) & Q^*(\omega + \pi) \end{vmatrix} \begin{vmatrix} P(\omega) & Q(\omega) \\ P(\omega + \pi) & Q(\omega + \pi) \end{vmatrix} = \begin{vmatrix} 1 & 0 \\ 0 & 1 \end{vmatrix} \tag{10.4.18}$$

Those properties will be discussed again in Section 10.6.1 for the orthonormality conditions of the multiresolution orthonormal wavelet transform.

10.5 Orthonormal wavelet transform

The orthonormal wavelet transform was developed independently from the multiresolution Laplacian pyramid and the subband coding. The first orthonormal wavelet basis was found by Meyer when he looked for the orthonormal wavelets that are localized in both time and frequency domains. The multiresolution Laplacian pyramid ideas of hierarchal averaging and computing the difference triggered Mallat and Meyer to view the orthonormal wavelet bases as a vehicle for multiresolution analysis [9]. Most known orthonormal wavelet bases are now constructed from the multiresolution analysis framework. In the multiresolution analysis framework, orthonormal wavelet decomposition and reconstruction use the tree algorithm that permits very fast computation of the orthonormal wavelet transform in the computer.

10.5.1 *Introduction*

Scaling functions

In the multiresolution analysis framework, the orthogonal wavelet transform is based on the scaling function. The scaling function is a continuous and, in general, real-valued function on the set of real numbers **R**. The scaling function is different from the wavelet. It does not satisfy the wavelet admissible condition: the mean value of the basic scaling function $\phi(t)$ is not equal to zero, but is usually normalized as

$$\int \phi(t)\,dt = 1 \qquad (10.5.1)$$

This is normalization with respect to the $\mathbf{L}^1(\mathbf{R})$ space. The scaling functions play the role of the average function. The correlation between the scaling function and a continuous function produces the averaged approximation of the function. The scaling function corresponds to the weighting function in the Laplacian pyramid and the low-pass filter in the subband coding. The difference is that the scaling function is continuous and the weighting function in the Laplacian pyramid and the low-pass filter in the subband coding are discrete filters.

The basic scaling function $\phi(t)$ is shifted by discrete translation factors to construct an orthonormal basis at the same resolution level. The scaling function $\phi(t)$ is also dilated by dyadic scale factors 2^{-i}

$$\phi_{i,k}(t) = 2^{-i/2}\phi(2^{-i}t - k) \qquad (10.5.2)$$

where the coefficient $2^{-i/2}$ is the normalization constant. The scaling function basis set is normalized in the $\mathbf{L}^2(\mathbf{R})$ norm. The dilated $\phi_{i,k}(t)$ forms an orthonormal basis at each resolution level i by the discrete translations. The scaling functions in the two adjacent resolution levels satisfy a relation, called the two-scale relation:

$$\phi(t) = \sum_k p(k)\phi(2t - k) \qquad (10.5.3)$$

which will be discussed in Section 10.5.3, where the elements in the discrete sequence $p(k)$ are called the interscale coefficients. The $p(k)$ corresponds to a discrete low-pass filter, which will be discussed in Section 10.5.4.

Wavelets

In the multiresolution analysis framework, the orthonormal wavelet bases are generated from the scaling function bases. The wavelet basis is orthogonal to the scaling function basis within the same scale. In order to emphasize the dependence of the wavelets to the scaling functions in the multiresolution analysis framework, from now on we change the notation of the wavelet and use $\psi(t)$ to denote the wavelets instead of $h(t)$ in the previous sections.

In the multiresolution analysis framework, the orthonormal wavelet transform is the decomposition of a function into approximations at lower and lower resolutions with less and less detail information by the projections of the function onto the orthonormal scaling function bases. The differences between each two successive approximations are computed with projections onto the orthonormal wavelet bases. Both the wavelet decomposition and reconstruction are computed with discrete iterated filters.

The scaling function and wavelet bases are orthonormal only with discrete translations and dilations. The decomposition of a continuous function onto the orthonormal scaling and wavelet function bases yields discrete sequences of expansion coefficients. Hence, there is an analogy of the orthonormal wavelet transform with the Fourier series decomposition.

The orthonormal wavelet transform is closely related to the subband coding. The discrete iterated low-pass and high-pass filters for the orthonormal wavelet transform are, in fact, the paraunitary PR QMF bank. However, an extra regularity condition is imposed on the scaling function and the wavelets. The orthonormal wavelet transform can be applied to continuous functions and therefore serves as a transform tool for analog signals. The multiresolution Laplacian pyramid and the subband coding are discrete. The multiresolution wavelet transform algorithm is also essentially discrete. But the algorithm leads to a wavelet series expansion that decomposes a continuous function into a series of continuous wavelet functions.

In the three following subsections we show first how the discrete translates of the scaling function and of the wavelet form the orthonormal bases at a given resolution, then how the scaling function generates the multiresolution analysis, and finally how the wavelet series decomposition and reconstruction are implemented by iterating the discrete filter bank.

10.5.2 *Orthonormal basis of the discrete translates*

The discrete translates of a basic scaling function $\phi(t)$ can form an orthonormal basis if $\phi(t)$ satisfies the orthonormality condition. An intuitive example is a scaling function that has a compact support: $\phi(t) = 1$ for $0 \leq t < 1$ and $\phi(t) = 0$ outside the interval $[0, 1)$. Obviously, the translations with integer steps of this rectangular function form an orthonormal set. This is the scaling function of Haar's wavelet basis.

Consider now a general basic scaling function $\phi(t)$ that is, in most cases, real valued. Its discrete translations form an orthonormal set $\{\phi(t - k)\}$:

$$\int \phi(t - k)\phi(t - k') \, dt = \delta_{k,k'} \qquad k, k' \in Z \qquad (10.5.4)$$

The orthonormality of the discrete translations of $\phi(t)$ is equivalent to the fact that the autocorrelation of $\phi(t)$ evaluated at discrete time steps $(k - k')$ must be zero everywhere except at the origin $k = k'$.

In the Fourier domain the orthonormality condition may be written as

$$\int |\Phi(\omega)|^2 \exp(-jn\omega) \, d\omega = \delta_{n,0} \qquad (10.5.5)$$

where $n = k - k'$ with $n \in \mathbf{Z}$, and $\Phi(\omega)$ is the Fourier transform of $\phi(t)$. Hence, the Fourier transform of $|\Phi(\omega)|^2$ evaluated at discrete steps n must be equal to zero except at the origin $n = 0$.

The Poisson summation formula establishes a relation between the Fourier series and Fourier transform:

$$\sum_n f(x + 2\pi n) = \frac{1}{2\pi} \sum_n F(n) \exp(jnx) \tag{10.5.6}$$

The conditions under which the Poisson summation formula holds is that both function $f(x)$ and its Fourier transform $F(n)$ satisfy

$$f(x) = O\left(\frac{1}{1 + |x|^r}\right) \quad \text{and} \quad F(n) = O\left(\frac{1}{1 + |n|^r}\right) \tag{10.5.7}$$

with $r > 1$. This condition means that both $f(x)$ and $F(n)$ are regular functions. Because $F(n)$ satisfies condition (10.5.7), $f(x)$ is necessarily continuous and m-times differentiable for $m \leq r$, and the series $\sum f(x + 2n\pi)$ in the left-hand side of Eq. (10.5.6) converges to a periodic function of period 2π. The Poisson summation formula then corresponds to the simple Fourier series decomposition of the periodic function $\sum f(x + 2n\pi)$.

Now assume that the Fourier spectrum $|\Phi(\omega)|^2$ of the basic scaling function $\phi(t)$ is regular, satisfying condition (10.5.7). Let $|\Phi(\omega)|^2$ be the $f(x)$ in the Poisson summation formula, Eq. (10.5.6); we obtain

$$\sum_n |\Phi(\omega + 2n\pi)|^2 = \frac{1}{2\pi} \sum_n R(n) \exp(jn\omega) \tag{10.5.8}$$

where $R(n)$ is the Fourier transform of $|\Phi(\omega)|^2$. If $\Phi(\omega)$ satisfies orthonormality condition (10.5.5), $R(n)$ would be equal to zero for $n \neq 0$ and equal to 2π for $n = 0$. Hence, the orthonormality condition becomes

$$\sum_n |\Phi(\omega + 2n\pi)|^2 = 1 \tag{10.5.9}$$

The sum of the series of the discretely translated Fourier spectrum $|\Phi(\omega)|^2$ must be equal to one.

To gain an insight of the orthonormality condition (10.5.9) let us expand a function $f(t)$ onto the orthonormal basis of the translates $\{\phi(t - k)\}$ with $n \in \mathbf{Z}$ that is

$$f(t) = \sum_k c(k)\phi(t - k) = \phi(t) * \sum_k c(k)\delta(t - k) \tag{10.5.10}$$

where $*$ denotes the convolution and $c(k)$ are the coefficients of expansion. In the Fourier domain this expansion becomes

$$F(\omega) = \Phi(\omega) \sum_k c(k) \exp(-jk\omega) = \Phi(\omega)M(\omega) \tag{10.5.11}$$

where $M(\omega)$ is defined as

$$M(\omega) = \sum_k c(k) \exp(-jk\omega) \tag{10.5.12}$$

which is a periodic with period 2π: $M(\omega) = M(\omega + 2n\pi)$. According to Parseval's relation of the Fourier transform

$$\frac{1}{2\pi} \int_0^{2\pi} |M(\omega)|^2 d\omega = \sum_n |c(n)|^2 \tag{10.5.13}$$

We can compute the energy of $f(t)$ by

$$\int_{-\infty}^{\infty} |f(t)|^2 \, dt = \frac{1}{2\pi} \int_{-\infty}^{\infty} |\Phi(\omega)|^2 |M(\omega)|^2 \, d\omega$$

$$= \frac{1}{2\pi} \sum_{n=-\infty}^{\infty} \int_{2\pi n}^{2\pi(n+1)} |\Phi(\omega)|^2 |M(\omega)|^2 \, d\omega$$

$$= \frac{1}{2\pi} \sum_{n=-\infty}^{\infty} \int_0^{2\pi} |\Phi(\omega + 2n\pi)|^2 |M(\omega + 2n\pi)|^2 \, d\omega$$

$$= \frac{1}{2\pi} \int_0^{2\pi} |M(\omega)|^2 \sum_{n=-\infty}^{\infty} |\Phi(\omega + 2n\pi)|^2 \, d\omega \qquad (10.5.14)$$

Because the orthonormality condition (10.5.9) is satisfied, $\sum |\Phi(\omega + 2n\pi)|^2 = 1$, we can write

$$\int |f(t)|^2 \, dt = \sum_n |c(n)|^2 \qquad (10.5.15)$$

Hence, the expansion of $f(t)$ onto the basis $\{\phi(t - k)\}$ is orthonormal.

When the Fourier transform $|\Phi(\omega)|^2$ is regular, satisfying condition (10.5.7) and the orthonormality condition (10.5.9), the integer translations of $\phi(t)$ form an orthonormal basis satisfying Eq. (10.5.4). Similarly, if a basic wavelet function $\psi(t)$ satisfies the wavelet admissible condition, and the regularity condition (10.5.7) and its Fourier transform of $\psi(t)$ satisfy the orthogonality condition (10.5.9):

$$\sum_n |\Psi(\omega + 2n\pi)|^2 = 1, \qquad (10.5.16)$$

the discrete translations of the wavelets $\{\psi(t - k)\}$ also form an orthonormal basis.

10.5.3   *Multiresolution orthonormal bases*

Scaling function bases

Assume that the basic scaling function $\phi(t)$ satisfies the orthogonality condition (10.5.9) so that its discrete translates $\{\phi(t - k)\}$ with integer translations $k \in \mathbf{Z}$ form an orthonormal set. The projection of a function $f(t) \in \mathbf{L}^2(\mathbf{R})$ on the orthonormal basis $\{\phi(t - k)\}$ is a correlation between the original function $f(t)$ and the scaling function $\phi(t)$ sampled at integer intervals.

The scaling function plays the role of a smoothing function in the multiresolution analysis. The projection of $f(t)$ on the scaling function basis results in a blurred approximation of $f(t)$. All the approximations of $f(t)$ form a subspace $\mathbf{V}_0 \in \mathbf{L}^2(\mathbf{R})$. The vector space $\mathbf{V}_0$ can be interpreted as the set of all possible approximations of functions in $\mathbf{L}^2(\mathbf{R})$ generated by the orthonormal set $\{\phi(t - k)\}$. The space $\mathbf{V}_0$ is spanned by the basis $\{\phi(t - k)\}$.

Now we consider the dilation of the scaling function. All the scaling functions are generated from the basic scaling function $\phi(t)$ by discrete dilations and translations as shown in Eq. (10.5.2):

$$\phi_{i,k}(t) = 2^{-i/2}\phi(2^{-i}t - k)$$

We shall restrict ourselves here to the dyadic scaling with $s = 2^i$ for $i \in \mathbf{Z}$, although other choices for s_0 for the scale $s = s_0^i$ are possible. The scaling functions of all scales 2^i with $i \in \mathbf{Z}$ generated from the same $\phi(t)$ are all similar in shape. Because the basic scaling function $\phi(t)$ generates the orthonormal basis $\{\phi(t - k)\}$ of $\mathbf{V}_0$ with an integer translation step, the dilated scaling function $\phi(t/2)$ will generate the orthonormal basis $\{\phi(2^{-1}t - k)\}$ of $\mathbf{V}_1$ with a translation step of 2, and $\phi(t/4)$ will generate the orthonormal basis $\{\phi(2^{-2}t - k)\}$ of $\mathbf{V}_2$ with a translation step of 4, and so on. There is then a set of orthogonal bases of scaling functions. Each scaling function basis is orthonormal within the same scale:

$$\langle \phi_{i,k}, \phi_{i,n} \rangle = \delta_{k,n} \tag{10.5.17}$$

for all k and $n \in \mathbf{Z}$.

The projections of functions in $\mathbf{L}^2(\mathbf{R})$ on the set of orthonormal scaling function bases form a set of subspaces $\mathbf{V}_i$. Each subspace $\mathbf{V}_i$ is the set of all possible approximations of functions in $\mathbf{L}^2(\mathbf{R})$ generated by the orthonormal scaling function basis $\{\phi(2^{-i}t - k)\}$. The subspace $\mathbf{V}_i$ is spanned by the orthonormal scaling function basis at the resolution level i. Hence, the scaling function $\phi(t)$ generates the subspaces of the multiresolution analysis.

The multiresolution analysis associated with the scaling function has some interesting properties.

The approximates of a function $f(t)$ at different resolutions must be similar, because they are all generated by the same scaling function with different scales. The approximation spaces $\mathbf{V}_i$ then may be deduced from each other by simple dilation.

$$f(t) \in V_i \Leftrightarrow f(2t) \in V_{i-1} \tag{10.5.18}$$

All the information useful to compute the approximate function at the coarser resolution level i is contained in the approximate function at the finer resolution level $(i - 1)$. Then $\mathbf{V}_i$ is a subspace in $\mathbf{V}_{i-1}$. This is a causality property. We have a fine-to-coarse sequence as

$$\cdots V_2 \subset V_1 \subset V_0 \subset V_{-1} \subset V_{-2} \cdots \subset L^2(R) \tag{10.5.19}$$

where $i \in \mathbf{Z}$. When the resolution increases with i tending to $-\infty$, the approximated function should converge to the original function. Any function in $\mathbf{L}^2(\mathbf{R})$ can be approximated as closely as desired by its project in $\mathbf{V}_i$ when i tends to $-\infty$. This property may be described as

$$\overline{\bigcup_i V_i} = L^2(R) \tag{10.5.20}$$

Conversely, when the resolution decreases to zero with i tending to $+\infty$, the approximations contain less and less information and converge to zero:

$$\bigcap_i V_i = \{0\} \tag{10.5.21}$$

In summary, the multiresolution analysis is generated by the scaling function $\phi(t)$. The scaling function $\phi(t)$ is scaled with the dyadic scaling factor 2^i. The discrete translates $\phi(2^{-i}t - k)$ form an orthonormal basis and span the subspace $\mathbf{V}_i$ at the resolution level i. All the dilates and translates are generated from a single basic scaling function and are not linearly independent.

Wavelet bases

The projection of a function on the orthonormal scaling function basis is the blurred approximation of the function at a particular resolution. Some information about the function is lost in the projection. We use the projections on the orthonormal wavelet bases to obtain the detail information of the function.

The wavelets are generated from the basic wavelet $\psi(t)$ by dyadic dilation and discrete translations:

$$\psi_{i,k} = 2^{-i/2}\psi(2^{-i}t - k) \tag{10.5.22}$$

The basic wavelet $\psi(t)$ satisfies the wavelet admissible condition (10.2.6). When the Fourier transform $\Psi(\omega)$ of the wavelet satisfies the orthogonality condition (10.5.9), the discrete translates of the wavelet $\{\psi(2^{-i}t - k)\}$ form an orthonormal basis at every scale 2^i. Furthermore, the set of the wavelet translates is orthogonal to the set of the scaling function translates within the same resolution

$$\langle \phi_{i,k}, \psi_{i,n} \rangle = 2^{-i} \int \phi_i(t - k)\psi_i(t - n)\,dt = 0 \qquad (10.5.23)$$

for all k and $n \in \mathbf{Z}$.

The projection of $f(t)$ on the orthonormal wavelet bases is a correlation between $f(t)$ and $\psi_i(t)$ sampled at discrete intervals. The projections of all functions in $\mathbf{L}^2(\mathbf{R})$ on the orthonormal wavelet basis $\{\psi(2^{-i}t - k)\}$ form a subspace $\mathbf{W}_i$. The subspace $\mathbf{W}_i$ is spanned by $\{\psi(2^{-i}t - k)\}$. As $\{\psi(2^{-i}t - k)\}$ is orthogonal to $\{\phi(2^{-i}t - k)\}$, the subspace $\mathbf{W}_i$ is an orthogonal complement of $\mathbf{V}_i$:

$$W_i \perp V_i \qquad (10.5.24)$$

Both $\mathbf{V}_i$ and $\mathbf{W}_i$ are the subspaces on $\mathbf{V}_{i-1}$: $\mathbf{V}_i, \mathbf{W}_i \in \mathbf{V}_{i-1}$. Because $\mathbf{W}_i$ is orthogonal to $\mathbf{V}_i$, the subspace $\mathbf{V}_{i-1}$ is the direct sum of $\mathbf{V}_i$ and $\mathbf{W}_i$.

$$V_{i-1} = V_i \oplus W_i \qquad (10.5.25)$$

Two-scale relations

The scaling functions and the wavelets form two orthonormal bases at every resolution level by their discrete translates. The scaling functions and the wavelets at multiple resolution levels are the dilated version of the basic scaling function and the basic wavelet, respectively.

Let $\phi(t)$ be the basic scaling function whose translates with an integer step span the subspace $\mathbf{V}_0$. In the next finer resolution the subspace $\mathbf{V}_{-1}$ is spanned by the set $\{\phi(2t - k)\}$, which is generated from the scaling function $\phi(2t)$ by a contraction with a factor of two and by translations with half integer steps. The set $\{\phi(2t - k)\}$ can also be considered as a sum of two sets of even and odd translates, $\{\phi(2t - 2k)\}$ and $\{\phi[2t - (2k + 1)]\}$; all are with integer steps $k \in \mathbf{Z}$. Hence, $\phi(t)$ can be expressed as a linear combination of the weighted sum of those two sets of $\phi(2t)$. We obtain therefore the two-scale relation Eq. (10.5.3):

$$\phi(t) = \sum_k p(k)\phi(2t - k)$$

that can also be considered as the projection of the basis function $\phi(t) \in \mathbf{V}_0$ onto the finer resolution subspace $\mathbf{V}_{-1}$. The two-scale relation, also called the two-scale difference equation (Eq. [10.5.3]), is the fundamental equation in the multiresolution analysis [12]. The basic ingredient in the multiresolution analysis is a scaling function such that the two-scale relation holds for some $p(k)$. The sequence $p(k)$ in the two-scale relation is the interscale coefficient that governs the structure of the scaling function $\phi(t)$. The orthonormality of the set of translates $\{\phi(t - n)\}$ is not as important as the two-scale relation.

Let $\psi(t) \in \mathbf{V}_0$ be the basic wavelet, which can also be expanded onto the orthonormal scaling function basis $\{\phi(2t - k)\}$ in $\mathbf{V}_{-1}$ as:

$$\psi(t) = \sum_k q(k)\phi(2t - k) \qquad (10.5.26)$$

where the sequence $q(k)$ is the interscale coefficient. Equation (10.5.26) is also a two scale relation useful for generating the wavelets from the scaling functions.

On the left-hand sides of the two scale relations, Eqs. (10.5.3) and (10.5.26), $\phi(t)$ and $\psi(t)$ are continuous. On the right-hand side of the two-scale relations, the interscale coefficients, $p(k)$

and $q(k)$, are discrete. The two-scale relations express the relations between the continuous scaling function $\phi(t)$ and wavelet $\psi(t)$ and the discrete sequences of the interscale coefficients $p(k)$ and $q(k)$.

Example 1   *Orthonormal Haar's bases*

The simplest orthonormal wavelet basis is the historical Haar wavelet. The Haar scaling function $\phi(t)$ is the simple rectangle function in the interval $[0, 1)$.

$$\phi(t) = \begin{cases} 1 & 0 \le t < 1 \\ 0 & \text{otherwise} \end{cases} \tag{10.5.27}$$

Obviously, the integer translates of $\phi(t)$ constitute an orthonormal basis, because $\phi(t-k)$ and $\phi(t-n)$ with $n, k \in \mathbf{Z}$ and $n \ne k$ do not overlap and

$$\int \phi(t-k)\phi(t-n)\, dt = \delta_{k,n}$$

The contracted Haar scaling function $\phi(2t)$ is a rectangular function in the interval $[0, 1/2)$. Its discrete translates with half integer steps form an orthonormal basis $\{\phi(2t-k)\}$. Automatically, within every fixed scale 2^{-i}, the Haar scaling functions form an orthonormal basis with the translation step of 2^{-i}.

As $\phi(2t-k)$ are of half integer width, the $\phi(t)$ in $\mathbf{V}_0$ can be expressed as a linear combination of the even and odd translates of $\phi(2t)$ in $\mathbf{V}_{-1}$:

$$\phi(t) = \frac{1}{\sqrt{2}}[\phi(2t) + \phi(2t-1)] \tag{10.5.28}$$

This is the two-scale relation described in Eq. (10.5.3) with the interscale coefficients

$$p(0) = \frac{1}{\sqrt{2}}, \qquad p(1) = \frac{1}{\sqrt{2}},$$

$$p(k) = 0 \qquad \text{for } k \ne 0, 1 \tag{10.5.29}$$

Haar's basic wavelet is defined as

$$\psi(t) = \begin{cases} 1 & 0 \le t \le \frac{1}{2} \\ -1 & \frac{1}{2} \le t \le 1 \\ 0 & \text{otherwise} \end{cases} \tag{10.5.30}$$

The integer translates of $\psi(t)$ constitute an orthonormal basis $\{\psi(t-k)\}$, because the wavelets $\psi(t-k)$ and $\psi(t-n)$ do not overlap for k and $n \in \mathbf{Z}$ and $k \ne n$:

$$\int \psi(t-k)\psi(t-n)\, dt = \delta_{k,n}$$

Also, within every fixed scale level i the discrete translates of the Haar wavelets form an orthonormal basis. The Haar wavelets are orthogonal to the Haar scaling function because

$$\int \phi(t-k)\psi(t-n)\, dt = 0$$

for all k and $n \in \mathbf{Z}$. The subspace $\mathbf{W}_0$ is spanned by the orthonormal basis $\{\psi(t-k)\}$, hence

$$W_0 \perp V_0$$

The Haar wavelet in $\mathbf{W}_0$ can also be expressed as a linear combination of the Haar scaling functions in $\mathbf{V}_{-1}$:

$$\psi(t) = \frac{1}{\sqrt{2}}[\phi(2t) - \phi(2t - 1)] \tag{10.5.31}$$

This is the two-scale relation described in Eq. (10.5.26) with the interscale coefficients

$$\begin{aligned} q(0) &= \tfrac{1}{\sqrt{2}} \qquad q(1) = -\tfrac{1}{\sqrt{2}} \\ q(k) &= 0 \qquad \text{for } k \neq 0, 1 \end{aligned} \tag{10.5.32}$$

The Haar wavelets are orthonormal at the same scale and also orthonormal across the scales. One can verify that

$$\int \psi_{i,k}(t)\psi_{m,n}(t)\, dt = \delta_{i,m}\delta_{k,n} \tag{10.5.33}$$

❚

10.5.4   *Multiresolution wavelet decomposition*

Let us consider now a function $f(t) \in \mathbf{V}_0$, which can be represented as a linear superposition of the translated scaling functions $\phi(t - k)$ in $\mathbf{V}_0$

$$f(t) = \sum_k c_0(k)\phi(t - k) \tag{10.5.34}$$

with the coefficients in the combination as

$$c_0(k) = \langle f, \phi_{0,k}\rangle = \int f(t)\phi(t - k)\, dt \tag{10.5.35}$$

Let P_i and Q_i denote the orthonormal projection operators on the subspaces $\mathbf{V}_i$ and $\mathbf{W}_i$ respectively. We now apply the entire multiresolution analysis to the function $f(t)$.

Orthonormal projections on the subspaces

The function to be analyzed is in $\mathbf{V}_0$. At the next coarser resolution $i = 1$, there are two mutually orthogonal subspaces $\mathbf{V}_1$ and $\mathbf{W}_1$, spanned by the orthonormal bases, $\{\phi_{1,k}(t)\}$ and $\{\psi_{1,k}(t)\}$ respectively. The subspace $\mathbf{W}_1$ is the orthogonal complement of $\mathbf{V}_1$. Because $\mathbf{V}_0$ is the direct sum of $\mathbf{V}_1$ and $\mathbf{W}_1$, $\mathbf{V}_0 = \mathbf{V}_1 + \mathbf{W}_1$, there is one and only one way to express a function $f(t) \in \mathbf{V}_0$ as the sum of two functions $v_1 + w_1$, where $v_1 \in \mathbf{V}_1$ and $w_1 \in \mathbf{W}_1$. In particular, $f(t) \in \mathbf{V}_0$, can be decomposed into its components along $\mathbf{V}_1$ and $\mathbf{W}_1$:

$$f = P_1 f + Q_1 f \tag{10.5.36}$$

The two components are the orthonormal projections of $f(t)$ on $\mathbf{V}_1$ and $\mathbf{W}_1$:

$$P_1 f = \sum_n c_1(n)\phi_{1,n}$$

$$Q_1 f = \sum_n d_1(n)\psi_{1,n} \tag{10.5.37}$$

where and from now on the explicit time dependence in f, $\phi_{i,k}$ and $\psi_{i,k}$ is not shown for ease of notation.

Because the scaling function set $\{\phi_{1,k}\}$ is orthonormal in $\mathbf{V}_1$ and the wavelet set $\{\psi_{1,k}\}$ is orthonormal in $\mathbf{W}_1$ and they are mutually orthogonal, multiplying the scaling function $\phi_{1,k}$

with both sides of the expansions (10.5.36) and computing the inner products yields

$$\langle \phi_{1,k}, f \rangle = \langle \phi_{1,k}, P_1 f \rangle \tag{10.5.38}$$

Multiplying $\phi_{1,n}$ with both sides of the expansions (10.5.37), computing the inner products and using Eq. (10.5.34) yields

$$c_1(k) = \langle \phi_{1,k}, P_1 f \rangle = \langle \phi_{1,k}, f \rangle$$

$$= \sum_n \langle \phi_{1,k}, \phi_{0,n} \rangle c_0(n) \tag{10.5.39}$$

where the inner product between the two scaling function sets $\{\phi_{1,k}\}$ and $\{\phi_{0,n}\}$ can be computed as

$$\langle \phi_{1,k}, \phi_{0,n} \rangle = 2^{-1/2} \int \phi\left(\frac{t}{2} - k\right) \phi(t - n)\, dt$$

$$= 2^{1/2} \int \phi(t)\phi[2t - (n - 2k)]\, dt \tag{10.5.40}$$

On substituting the two-scale relation for $\phi(t)$:

$$\phi(t) = \sum_n p(n)\phi(2t - n)$$

into Eq. (10.5.40) and using the orthonormality of the set $\{\phi(2t)\}$ we obtain

$$\langle \phi_{1,k}, \phi_{0,n} \rangle = 2^{-1/2} p(n - 2k) \tag{10.5.41}$$

Hence, Eq. (10.5.39) becomes

$$c_1(k) = 2^{-1/2} \sum_n p(n - 2k)c_0(n) \tag{10.5.42}$$

The sequence of $c_1(k)$ is the coefficient of the expansion of a continuous function $f(t)$ onto the continuous scaling function basis $\{\phi_{1,k}\}$ in $\mathbf{V}_1$. The sequence of $c_1(k)$ represents a smoothed version of the original data sequence $c_0(n)$ and is referred to as the discrete approximation of $c_0(n)$. According to Eq. (10.5.42), the discrete approximation $c_1(n)$ is the correlation between $c_0(n)$ and $p(n)$ down-sampled by a factor of two. The operation described in Eq. (10.5.42) is similar to the averaging followed by down-sampling by a factor of two described in Eqs. (10.4.1) and (10.4.3) in the Laplacian pyramid and the subband coding. The discrete interscale coefficients $p(n)$ correspond to the weighting function in the Laplacian pyramid and to the low-pass filter in the subband coding, and is called here the discrete low-pass filter.

Similarly, multiplying both sides of Eqs. (10.5.36) and (10.5.37) by the wavelet $\psi_{1,n}$ and computing the inner products yields

$$d_1(k) = \langle \psi_{1,k}, Q_1 f \rangle = \langle \psi_{1,k}, f \rangle$$

$$= \sum_n \langle \psi_{1,k}, \phi_{0,n} \rangle c_0(n) \tag{10.5.43}$$

Similar to Eq. (10.5.42), the inner product between wavelet set $\{\psi_{1,k}\}$ and scaling function set $\{\phi_{0,n}\}$ can be computed as

$$\langle \psi_{1,k}, \phi_{0,n} \rangle = 2^{1/2} \int \psi(t)\phi[2t - (n - 2k)]\, dt \tag{10.5.44}$$

Substituting the two-scale relation for $\psi(t)$:

$$\psi(t) = \sum_n q(n)\phi(2t - n)$$

into Eq. (10.5.44) and using the orthonormality of the set $\{\phi(2t)\}$ we have

$$\langle \psi_{1,k}, \phi_{0,n} \rangle = 2^{-1/2}q(n - 2k) \tag{10.5.45}$$

Hence, Eq. (10.5.43) becomes

$$d_1(k) = 2^{-1/2} \sum_n q(n - 2k)c_0(n) \tag{10.5.46}$$

According to Eq. (10.5.36) the orthonormal projection $Q_1 f$ on $\mathbf{W}_1$ is the detail information of $f(t)$. The sequence of $d_1(n)$ represents the difference between the original $f(t)$ and the approximation $P_1 f$ and is referred to as the discrete wavelet coefficients. The discrete sequence of interscale coefficients $q(n)$ corresponds to the high-pass filter in the subband coding and is called here the discrete high-pass filter.

Recursive projections

The orthonormal projections at one resolution level can continue to the next coarser resolution. The projection procedure can be iterated. At the next coarser resolution the subspace $\mathbf{V}_2$ and $\mathbf{W}_2$ are orthogonal complements, $\mathbf{V}_1 = \mathbf{V}_2 + \mathbf{W}_2$, and $\mathbf{V}_1$ is the direct sum of $\mathbf{V}_2$ and $\mathbf{W}_2$. We can decompose $P_1 f \in \mathbf{V}_1$ into two components along $\mathbf{V}_2$ and $\mathbf{W}_2$

$$P_1 f = P_2 f + Q_2 f \tag{10.5.47}$$

with

$$P_2 f = \sum_n c_2(n)\phi_{2,n}$$

$$Q_2 f = \sum_n d_2(n)\psi_{2,n} \tag{10.5.48}$$

Multiplying both sides of expansions (10.5.47) and (10.5.48) by $\phi_{2,k}$ and using the orthonormality of the set $\{\phi_{2,n}\}$ and the mutual orthonormality between $\phi_{2,n}$ and $\psi_{2,k}$ we obtain the discrete approximation $c_2(k)$ as

$$c_2(k) = \langle \phi_{2,k}, P_2 f \rangle = \langle \phi_{2,k}, f \rangle$$

$$= \sum_n \langle \phi_{2,k}, \phi_{1,n} \rangle c_1(n) \tag{10.5.49}$$

Similarly, multiplying $\psi_{2,k}$ with both sides of expansions (10.5.47) and (10.5.48), using the orthonormality of $\{\psi_{2,n}\}$ and the mutual orthogonality between $\phi_{2,k}$ and $\psi_{2,n}$, we obtain the discrete wavelet coefficients $d_2(k)$ as

$$d_2(k) = \langle \psi_{2,k}, Q_2 f \rangle = \langle \psi_{2,k}, f \rangle$$

$$= \sum_n \langle \psi_{2,k}, \phi_{1,n} \rangle c_1(n) \tag{10.5.50}$$

The decomposition into smoothed approximations and details at coarser resolutions can be continued as far as wanted. The procedure can be iterated as many times as wanted. The successive projections $P_i f$ correspond to a more and more blurred version of $f(t)$. The successive projections $Q_i f$ correspond to the differences in information between the two approximations of $f(t)$ at two successive resolution levels. At every step i one has the orthonormal projection of $P_{i-1}f$ along the subspaces $\mathbf{V}_i$ and $\mathbf{W}_i$

$$P_{i-1}f = P_i f + Q_i f$$

$$= \sum_k c_i(k)\phi_{i,k} + \sum_k d_i(k)\psi_{i,k} \tag{10.5.51}$$

It is easy to verify that similarly to Eqs. (10.5.41) and (10.5.45) we have

$$\langle\phi_{i,k}, \phi_{i-1,n}\rangle = 2^{-1/2}p(n-2k)$$

$$\langle\psi_{i,k}, \phi_{i-1,n}\rangle = 2^{-1/2}q(n-2k) \tag{10.5.52}$$

are independent of the resolution level i. It follows that

$$c_i(k) = 2^{-1/2}\sum_n p(n-2k)c_{i-1}(n)$$

$$d_i(k) = 2^{-1/2}\sum_n q(n-2k)c_{i-1}(n) \tag{10.5.53}$$

We define the low-pass and high-pass filtering operators L and H respectively such that the operations on a sequence $\alpha(n)$ are

$$(L\alpha)(k) = 2^{-1/2}\sum_n p(n-2k)\alpha(n)$$

$$(H\alpha)(k) = 2^{-1/2}\sum_n q(n-2k)\alpha(n) \tag{10.5.54}$$

Equations (10.5.53) can be shortened to

$$c_i = Lc_{i-1}$$

$$d_i = Hc_{i-1} \tag{10.5.55}$$

Wavelet series decomposition

The successive discrete approximation sequences $c_i(n)$ are lower and lower resolution versions of the original $c_0(n)$, each sampled twice as sparsely as its predecessor. The successive wavelet coefficient sequences $d_i(n)$ contain the difference in information between the two approximations at resolutions levels (i) and $(i-1)$. The sequences $c_i(n)$ and $d_i(n)$ can be computed from $c_{i-1}(n)$ by iterative filtering L and H, according to Eq. (10.5.55).

Continuing up to resolution M we can represent the original function $f(t)$ by a series of detail functions plus one smoothed approximation

$$f(t) = P_M L + Q_M f + Q_{M-1}f + \cdots + Q_1 f \tag{10.5.56}$$

and

$$f(t) = \sum_{k\in Z} 2^{-M/2}C_M(k)\phi(2^{-M}t - k)$$

$$+ \sum_{i=1}^M \sum_{k\in Z} 2^{-i/2}d_i(k)\psi(2^{-i}t - k) \tag{10.5.57}$$

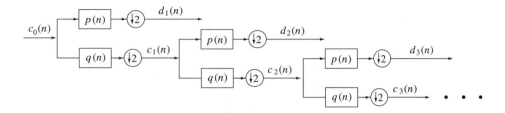

FIGURE 10.11
Schematic wavelet series decomposition in the tree algorithm.

Equation (10.5.57) is referred to as the wavelet series decomposition. The function $f(t)$ is represented as an approximation at resolution $i = M$ plus the sum of M detail components at dyadic scales. The first term in the right-hand side of Eq. (10.5.57) is the smoothed approximation of $f(t)$ with the scaling functions at very low resolution $i = M$. When M approaches to infinity, the projection of $f(t)$ smooths out any signal detail and converges to zero. The function $f(t)$ is then represented as a series of its orthonormal projections on the wavelet bases.

The wavelet series decomposition is a practical representation of the wavelet expansion and points out the complementary role of the scaling function in the wavelet decomposition. Note that in the wavelet series decomposition the function $f(t)$, the scaling function bases, and the wavelet bases are all continuous. The approximation coefficients $c_M(k)$ and the wavelet coefficients $d_i(k)$ with $i = 1, 2, \ldots, M$ are discrete. In this sense, the wavelet series decomposition is similar to the Fourier series decomposition.

The discrete approximations $c_i(n)$ and the discrete wavelet coefficients $d_i(n)$ can be computed with an iterative algorithm, described by Eq. (10.5.55). This is essentially a discrete algorithm implemented by recursive applications of the discrete low-pass and high-pass filter bank to the discrete approximations $c_i(n)$. The algorithm is called the tree algorithm. The first two stages of the tree algorithm for computing the wavelet decomposition are shown in Figure 10.11. The decomposition into coarser, smoothed approximations and details can be continued as far as wanted.

Example 2 *Decomposition with Haar wavelets*
A simple example for orthonormal wavelet decomposition is that with Haar's orthonormal bases. Let subspace $\mathbf{V}_0$ be spanned by the Haar scaling function basis $\{\phi(t - k)\}$, defined in Eq. (10.5.27). The projection of a function $f(t)$ on $\mathbf{V}_0$ is an approximation of $f(t)$ that is piecewise constant over the integer interval. The projection of $f(t)$ on $\mathbf{V}_{-1}$ with the orthonormal basis $\{\phi(2t - k)\}$ of the finer resolution is piecewise constant over the half-integer interval. Because $\mathbf{V}_{-1} = \mathbf{V}_0 + \mathbf{W}_0$, and

$$P_{-1}f = P_0 f + Q_0 f$$

The projection $Q_0 f$ represents the difference between the approximation in $\mathbf{V}_0$ and the approximation in $\mathbf{V}_{-1}$. The approximations $P_0 f$ in $\mathbf{V}_0$ and $P_{-1} f$ in $\mathbf{V}_{-1}$ and the detail $Q_0 f$ are shown in Figure 10.12. In Figure 10.12 the projection $Q_0 f$ is constant over half-integer intervals, which can be added to the approximation $P_0 f$ to provide the next finer approximation $P_{-1} f$.

When i approaches minus infinity with finer and finer resolution, the approximations $P_{-i} f$ will converge to the original function $f(t)$ as closely as desired.

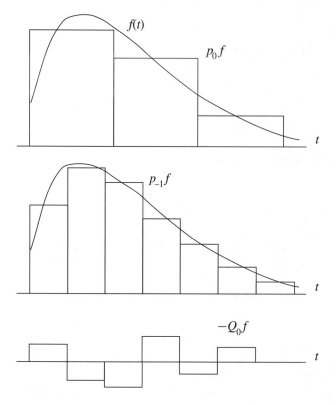

FIGURE 10.12
Orthogonal projections $P_0 f$ and $Q_0 f$ onto the Haar scaling function and wavelet bases. The projection at the next finer resolution, $P_{-1} f = P_0 f + Q_0 f$ [11].

The projection of $f(t)$ onto the subspace $\mathbf{V}_i$ spanned by the Haar scaling function basis at the resolution level i is

$$P_i f = \sum_k c_i(k)\phi_{i,k} \tag{10.5.58}$$

with the discrete approximation coefficients

$$c_i(k) = \langle f, \phi_{i,k}\rangle = 2^{-i/2} \int_{2^i k}^{2^i(k+1)} f(t)\, dt \tag{10.5.59}$$

Because $\phi_{i+1,k}$ is a rectangular function with width of 2^{i+1} and $\phi_{i,k}$ is with width of 2^i it is easy to verify that

$$\phi_{i+1,k} = 2^{-1/2}(\phi_{i,2k} + \phi_{i,2k+1})$$

$$\psi_{i+1,k} = 2^{-1/2}(\phi_{i,2k} - \phi_{i,2k+1}) \tag{10.5.60}$$

The discrete approximation $c_{i+1}(k)$ can be obtained directly by the orthonormal projection of $f(t)$ onto $\mathbf{V}_{i+1}$

$$c_{i+1}(k) = \langle f, \phi_{i+1,k}\rangle = 2^{-1/2}[c_i(2k) + c_i(2k+1)] \tag{10.5.61}$$

The difference between the two successive approximations is obtained using Eqs. (10.5.60) and (10.6.61)

$$P_i f - P_{i+1} f = \sum_k [c_i(k)\phi_{i,k} - c_{i+1}(k)\phi_{i+1,k}]$$

$$= \sum_k \{[c_i(2k)\phi_{i,2k} + c_i(2k+1)\phi_{i,2k+1}]$$

$$-2^{-1/2}[c_i(2k) + c_i(2k+1)]2^{-1/2}(\phi_{i,2k} + \phi_{i,2k+1})\}$$

$$= \frac{1}{2}\sum_k [c_i(2k) - c_i(2k+1)](\phi_{i,2k} - \phi_{i,2k+1})$$

$$= \frac{1}{2}\sum_k [c_i(2k) - c_i(2k+1)]\psi_{i+1,k} \tag{10.5.62}$$

Hence, the projection of $f(t)$ onto the subspace $\mathbf{W}_{i+1}$ is the difference between $P_i f$ and $P_{i+1} f$

$$Q_{i+1,k} = \sum_k d_{i+1}(k)\psi_{i+1,k} = P_i f - P_{i+1} f \tag{10.5.63}$$

provided that

$$d_{i+1}(k) = 2^{-1/2}[c_i(k) - c_i(2k+1)] \tag{10.5.64}$$

The interscale coefficients that are the discrete low-pass and high-pass filers $p(n)$ and $q(n)$ are given for Haar's bases by Eq. (10.5.29) and (10.5.32). Hence, the iterated filtering by L and H becomes

$$c_{i+1}(k) = \frac{1}{\sqrt{2}}\sum_n p(n-2k)c_i(n) = \frac{1}{\sqrt{2}}[c_i(2k) + c_i(2k+1)]$$

$$d_{i+1}(k) = \frac{1}{\sqrt{2}}\sum_n q(n-2k)c_i(n) = \frac{1}{\sqrt{2}}[c_i(2k) - c_i(2k+1)] \tag{10.5.65}$$

which agree with Eqs. (10.5.61) and (10.5.64).

10.5.5   *Reconstruction from the wavelet decomposition*

Recursive reconstruction

The original signal sequence $c_0(n)$ can be reconstructed from the sequences of the approximation coefficients $c_i(n)$ and of the wavelet coefficients $d_i(n)$ with $0 < i \leq M$, where $i = M$ is the lowest resolution in the decomposition. At each resolution level i we have the decomposition described by Eqs. (10.5.51) and (10.5.53). On multiplying both sides of Eq. (10.5.51) by $\phi_{i-1,n}$ and integrating both sides we obtain

$$c_{i-1}(n) = \langle P_{i-1}f, \phi_{i-1,n}\rangle$$

$$= \sum_k c_i(k)\langle\phi_{i,k}, \phi_{i-1,n}\rangle + \sum_k d_i(k)\langle\psi_{i,k}, \phi_{i-1,n}\rangle$$

$$= 2^{-1/2}\sum_k c_i(k)p(n-2k) + 2^{-1/2}\sum_k d_i(k)q(n-2k) \tag{10.5.66}$$

where the inner products $\langle\phi_{i,k}, \phi_{i-1,n}\rangle$ and $\langle\psi_{i,k}, \phi_{i-1,n}\rangle$ are obtained in Eq. (10.5.52) as the interscale coefficients $p(n-2k)$ and $q(n-2k)$. Hence, the discrete approximation $c_{i-1}(n)$ at

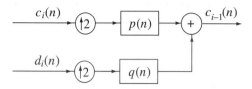

FIGURE 10.13
Schematic wavelet reconstruction.

the next finer resolution can be obtained as the sum of two convolutions between the discrete approximation $c_i(n)$ and the low-pass synthesis filter $p(n)$ and between the wavelet coefficients $d_i(n)$ and the high-pass synthesis filter $q(n)$.

The synthesis filters are identical to the analysis filters. But the filtering operations become the convolutions for synthesis instead of the correlations for analysis. To compute the convolution with the synthesis filters in Eq. (10.5.66) one must first put zeros between each sample of the sequences $c_i(n)$ and $d_i(n)$ before convolving the resulting sequences with the synthesis filters $p(n)$ and $q(n)$. The process is quite similar to the expand operation in the reconstruction algorithm of the multiresolution Laplacian pyramid and the subband coding. The reconstruction process can be repeated by iteration. We define the synthesis filtering operators L^* and H^* as

$$(L^*\alpha)(n) = \frac{1}{\sqrt{2}} \sum_k p(n - 2k)\alpha(k)$$

$$(H^*\alpha)(n) = \frac{1}{\sqrt{2}} \sum_k q(n - 2k)\alpha(k) \tag{10.5.67}$$

and rewrite Eq. (10.5.66) in a shortened form

$$c_{i-1} = L^*c_i + H^*d_i \tag{10.5.68}$$

The block diagram shown in Figure 10.13 illustrates the reconstruction algorithm, where the up-sampling by two means putting zeros between the sample of the sequences.

To reconstruct the original data $c_0(n)$ we start from the lowest resolution approximation c_M. According to Eq. (10.5.68) we have

$$c_{M-1} = H^*d_M + L^*c_M$$

$$c_{M-2} = H^*d_{M-1} + L^*(H^*d_M + L^*c_M)$$

$$= H^*d_{M-1} + L^*H^*d_M + (L^*)^2c_M \tag{10.5.69}$$

When the discrete approximation c_{i-1} is obtained from $c_i(n)$ and $d_i(n)$, the next finer approximation $c_{i-2}(n)$ can be obtained from the sequence $c_{i-1}(n)$ and the wavelet coefficients $d_{i-1}(n)$. The process can continue until the original sequence $c_0(n)$ is reconstructed. The reconstruction formula for the original sequence is

$$c_0 = \sum_{i=1}^{M}(L^*)^{i-1}H^*d_i + (L^*)^Mc_M \tag{10.5.70}$$

In this reconstruction procedure it is the low-pass filtering operator L^* that is iteratively applied to generate the finer resolution discrete approximations.

Convergence of the wavelet reconstruction

In this subsection we discuss a particular example of reconstruction from the wavelet series decomposition, where the original function to be decomposed is the scaling function itself. This reconstruction procedure is useful to compute the continuous scaling function and wavelets from the discrete low-pass and high-pass filters $p(n)$ and $q(n)$. The regularity condition of the scaling function and the wavelets will be introduced in Section 10.6.2 for ensuring the convergence of this reconstruction.

When the function $f(t)$ to be decomposed is itself the dilated scaling function $\phi(2^{-M}t - n)$ we have the simplest case of the wavelet series decomposition. In this case the wavelet series coefficients must be: $c_M(n) = \delta_{0,n}$ and $d_0 = \cdots = d_M = 0$ and the reconstruction formula becomes

$$c_0(n) = (L^*)^M c_M \tag{10.5.71}$$

It is therefore important to study the behavior of the iterated filtering operator $(L^*)^i e(n)$ for large i, where $e(n)$ is the sequence $\delta_{0,n}$ with only one non-zero entry for $n = 0$. Ideally we want $(L^*)^i e(n)$ to converge to a reasonably regular function when i tends to infinity.

With a graphic representation, we represent the sequence $c_M(n) = e(n)$ at the resolution level $i = M$ by a rectangular function $\eta_0(t)$:

$$\eta_0(t) = \begin{cases} 1 & -\frac{1}{2} \le t \le \frac{1}{2} \\ 0 & \text{otherwise} \end{cases} \tag{10.5.72}$$

as shown in Figure 10.14. Assume that the sequence $c_M(n)$ has the time clock rate of 1. At the next finer resolution level $i = M - 1$ the sequence $c_{M-1}(n)$ is

$$C_{M-1}(n) = L^* e(n) = \sum_k p(n - 2k)\delta_{0,k} = p(n) \tag{10.5.73}$$

In fact, to compute c_{M-1} we first increase the time clock rate such that $c_M(n)$ is with a time interval of length $1/2$. The $c_M(n)$ is then convolved with the discrete filter $p(n)$ that also has the time interval of $1/2$. The amplitude of $c_{M-1}(n)$ is equal exactly to $p(n)$, as shown in Figure 10.14. We represent $c_{M-1}(n)$ by a piecewise constant function $\eta_1(t)$, which is constant over the interval of $1/2$. It is easy to see that the $\eta_1(t)$ may be expressed as

$$\eta_1(t) = \sum_n p(n)\eta_0(2t - n) \tag{10.5.74}$$

Continuing with computing $c_{M-2}(n) = (L^*)^2 e(n)$, we put a zero between each node of the sequence $c_{M-1}(n)$ and increase the time clock rate. Thus, at this resolution level both the data sequences $c(n)$ and the filter $p(n)$ have the time interval of $1/4$. Their convolution yields the sequence $c_{M-2}(n)$. We represent $c_{M-2}(n)$ by the piecewise constant function $\eta_2(t)$ of a step of length $1/4$, as shown in Figure 10.14. It is easy to verify that

$$\eta_2(t) = \sum_n p(n)\eta_1(2t - n) \tag{10.5.75}$$

Similarly, $\eta_i(t) = (L^*)^i e(n)$ is a piecewise constant function with a step of length 2^{-i} and

$$\eta_i(t) = \sum_n p(n)\eta_{i-1}(2t - n) \tag{10.5.76}$$

When i approaches to infinity, $(L^*)^i e(n)$ can converge to a function $\eta_\infty(t)$ that can be continuous, or with finite discontinuities, even a fractal function. The sequence $(L^*)^i e(n)$ may not converge at all. Figure 10.14 shows an example where $\eta_\infty(t)$ is continuous. The condition for

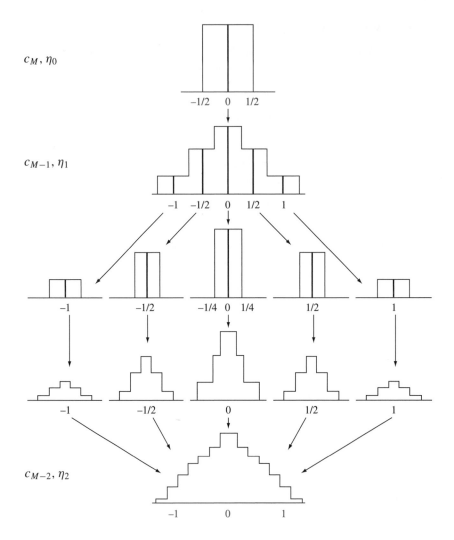

FIGURE 10.14
Reconstruction from $c_M(n) = 1$, for $n = 0$ and $c_M(n) = 0$ for $n \neq 0$ and the corresponding rectangle function $\eta_0(t)$. The time clock rate is equal to 1 for $c_M(n)$, $1/2$ for $c_{M-1}(n)$ and $1/4$ for $c_{M-2}(n)$ [13].

the reconstruction to converge to $\eta_\infty(t)$ is the regularity of the scaling function, which will be discussed in Section 10.6.2.

It can be shown that the Fourier transform of the piecewise constant function $\eta_\infty(t)$ is equal to the infinite product of the $P(\omega/2^i)$ with i tending to infinity, where $P(\omega)$ is the Fourier transform of the discrete low-pass filter $p(n)$. Let $\Phi(\omega)$ be the Fourier transform of $\phi(t)$. According to the relation between $P(\omega)$ and $\Phi(\omega)$, which will be discussed in Section 10.6.1, the $\eta_\infty(t)$ converges to the scaling function

$$\lim_{i \to \infty} \eta_\infty(t) = \phi(t) \qquad (10.5.77)$$

The above recursion process with the low-pass filter operator L^*, associated to the low-pass filter $p(n)$, is a reconstruction of the scaling function $\phi(t)$. Starting from the rectangular function η_0, the recursion of Eq. (10.5.76) gives the values of $\phi(t)$ of half-integers. Then, the recursion gives $\phi(t)$ at the quarter-integers, and ultimately, at all dyadic point $t = k/2^i$. Finer and finer detail of $\phi(t)$ is achieved by the recursion when the number of iterations i approaches to infinity. Therefore, the basic scaling function $\phi(t)$ is constructed from the discrete low-pass filter $p(n)$. This process is very useful to compute the continuous scaling and wavelet functions $\phi(t)$ and $\psi(t)$ from the discrete low-pass and high-pass filters $p(n)$ and $q(n)$.

Compact support of the wavelets

The wavelet transform is a local operation, so that the scaling functions and wavelets must have compact support. Assume that the discrete low-pass filter $p(n)$ has a compact support and is a FIR filter: $p(n) = 0$ for $n < N^-$, or $n > N^+$; then the corresponding basic scaling function and wavelet have compact support. This can be seen from the reconstruction of the scaling function $\phi(t)$ by recursion of η_i, discussed above. The recursion starts from the rectangular function $\eta_0(t)$ with $N_0^- = -1/2$ and $N_0^+ = 1/2$. At the next finer resolution level both $\eta_0(t)$ and $p(n)$ have the time interval of $1/2$. Their convolution yield $\eta_1(t)$ with the support of

$$N_1^- = \frac{1}{2}(N_0^- + N^-)$$

$$N_1^+ = \frac{1}{2}(N_0^+ + N^+) \tag{10.5.78}$$

With the recursion of Eq. (10.5.76), $\eta_i(t)$ will have the support as

$$N_i^- = \frac{1}{2}(N_{i-1}^- + N^-)$$

$$N_i^+ = \frac{1}{2}(N_{i-1}^+ + N^+) \tag{10.5.79}$$

Hence, when i approaches to infinity, N_i^- tends to N^- and N_i^+ tends to N^+, which implies the scaling function $\phi(t)$ with the compact support $[N^-, N^+]$. The discrete high-pass filter is generated from $p(n)$, as will be discussed in Section 10.6.1. It has the same support N as that of $p(n)$. Because only a finite number of $q(n)$ are nonzero, the wavelets $\psi(t)$ also have compact support.

In summary, in this section we introduced the orthonormal wavelet series decomposition and reconstruction. In the multiresolution analysis framework, the basic function of the orthonormal wavelet transform is the scaling function $\phi(t)$ that satisfies the two-scale relation. The discrete translates of the scaling functions form the orthonormal bases within each resolution level. The discrete translates of the wavelets also form the orthonormal bases within each resolution level. The scaling and the wavelet function bases are mutually orthogonal within the same resolution level. The recursive orthonormal projections on the multiresolution subspaces yield the wavelet series decomposition.

The wavelet series decomposition and reconstruction are computed by iterating the discrete low-pass filters $p(n)$ and the discrete high-pass filters $q(n)$, in the tree algorithms, in order to compute a set of discrete wavelet transform coefficients $d_i(n)$ and a set of discrete approximation coefficients $c_i(n)$.

10.6 Properties of the orthonormal wavelets

In this section we discuss the basic properties of the orthonormal scaling functions and wavelets. The orthonormality and regularity conditions are applied to the scaling function and the wavelets as well as to the discrete low-pass and high-pass filters. Most analysis on the filter properties will be done in the Fourier domain.

10.6.1   *Orthonormality conditions*

In Section 10.5.2 we showed that when the scaling function and the wavelets satisfy the orthonormality conditions the discrete translates of the continuous scaling functions and wavelets form the orthonormal bases. These orthonormality conditions are applied to the Fourier transforms $\Phi(\omega)$ and $\Psi(\omega)$ of the scaling function and the wavelet:

$$\sum_n |\Phi(\omega + 2n\pi)|^2 = 1$$

$$\sum_n |\Psi(\omega + 2n\pi)|^2 = 1 \tag{10.6.1}$$

In this section we shall study the implication of condition (10.6.1) on the discrete low-pass and high-pass filters and on their Fourier transforms. Here, the two-scale relation will play an important role, because the low-pass and high-pass filters $p(n)$ and $q(n)$ are the interscale coefficients in the two-scale relations.

Orthonormality of the discrete iterated filters

Let $\phi(t)$ and $\psi(t)$ be the basic scaling function and the wavelet, respectively, satisfying the orthonormality condition. The discrete translates of $\phi(t)$ and $\psi(t)$ with integer translation steps form two orthonormal bases of the subspace $\mathbf{V}_0$ and $\mathbf{W}_0$, respectively, and the two orthonormal bases are mutually orthogonal

$$\langle \phi_{0,k}, \phi_{0,n} \rangle = \delta_{k,n}$$

$$\langle \psi_{0,k}, \psi_{0,n} \rangle = \delta_{k,n}$$

$$\langle \psi_{0,k}, \phi_{0,n} \rangle = 0 \tag{10.6.2}$$

In Section 10.5.4 we obtained Eqs. (10.5.41) and (10.5.45) from the two-scale relations

$$\langle \phi_{1,k}, \phi_{0,n} \rangle = 2^{-1/2} p(n - 2k)$$

$$\langle \psi_{1,k}, \phi_{0,n} \rangle = 2^{-1/2} q(n - 2k)$$

Using these relations we can easily write the inner products at the resolution level $i = 1$ in terms of $p(n)$ and $q(n)$ as

$$\langle \phi_{1,k}, \phi_{1,k'} \rangle = \sum_{n,m} p(n - 2k) p(m - 2k') \langle \phi_{0,n}, \phi_{0,m} \rangle$$

$$= \sum_n p(n - 2k) p(n - 2k') = \delta_{k,k'}$$

$$\langle \psi_{1,k}, \psi_{1,k'} \rangle = \sum_{n,m} q(n - 2k) q(m - 2k') \langle \phi_{0,n}, \phi_{0,m} \rangle$$

$$= \sum_n q(n - 2k) q(n - 2k') = \delta_{k,k'}$$

$$\langle \psi_{1,k}, \phi_{1,k'} \rangle = \sum_{n,m} q(n - 2k)p(m - 2k')\langle \phi_{0,n}, \phi_{0,m} \rangle$$

$$= \sum_{n} q(n - 2k)p(n - 2k') = 0 \tag{10.6.3}$$

Note that the last equation in Eq. (10.6.3) is the time domain cross-filter orthonormality condition Eq. (10.4.6) obtained in Section 10.4.2 for the subband coding. According to Eq. (10.6.3) the two sets of discrete translates of low-pass filter and high-pass filter, $p(n)$ and $q(n)$, are orthonormal and mutually orthogonal within the same resolution level, just as their respective counterparts $\phi(t)$ and $\psi(t)$. Note that the translation steps of $p(n)$ and $q(n)$ in Eq. (10.6.3) are double integers. This remarque is very useful in the computing of the wavelet series decomposition.

Two scale relations in the Fourier domain

The two-scale relations in the multiresolution analysis represent the basic relations between the continuous scaling function $\phi(t)$ and wavelet $\psi(t)$ and the discrete low-pass and highpass filters $p(n)$ and $q(n)$:

$$\phi(t) = \sum_{k} p(k)\phi(2t - k)$$

$$\psi(t) = \sum_{k} q(k)\phi(2t - k)$$

In the multiresolution wavelet decomposition the sequence of interscale coefficients $\{p(n)\}$ plays the role of the weighting function and the sequence $\{q(n)\}$ is used to compute the detail information. The Fourier transform of the two-scale relations gives

$$\Phi(\omega) = \sum_{k} p(k) \int \phi(2t - k) \exp(-j\omega t) \, dt$$

$$= \frac{1}{2} \left[\sum_{k} p(k) \exp(-jk\omega/2) \right] \Phi\left(\frac{\omega}{2}\right)$$

$$= P\left(\frac{\omega}{2}\right) \Phi\left(\frac{\omega}{2}\right) \tag{10.6.4}$$

and similarly,

$$\Psi(\omega) = Q\left(\frac{\omega}{2}\right) \Phi\left(\frac{\omega}{2}\right) \tag{10.6.5}$$

where $P(\omega)$ and $Q(\omega)$ are the Fourier transforms of the sequences of the interscale coefficients $p(n)$ and $q(n)$.

$$P(\omega) = \frac{1}{2} \sum_{k} p(k) \exp(-jk\omega)$$

$$Q(\omega) = \frac{1}{2} \sum_{k} q(k) \exp(-jk\omega) \tag{10.6.6}$$

The $P(\omega)$ and $Q(\omega)$ can be expressed as the z-transforms of the sequences $p(n)$ and $q(n)$, respectively, with $z = \exp(jk\omega)$:

$$P(z) = \frac{1}{2} \sum_{k} p(k)z^{-k}$$

$$Q(z) = \frac{1}{2} \sum_k q(k) z^{-k} \qquad (10.6.7)$$

The $p(n)$ and $q(n)$ are the low-pass and high-pass filters for wavelet series decomposition and reconstruction. Their Fourier transforms $P(\omega)$ and $Q(\omega)$ are referred to as the quadrature mirror filters or the two-scale symbols. Both $P(\omega)$ and $Q(\omega)$ are periodic functions of period 2π.

According to the Fourier domain two-scale relation (10.6.4) the Fourier transform $\Phi(\omega)$ of the coarser resolution scaling function $\phi(t)$ can be computed from the twice wider Fourier transform $\Phi(\omega/2)$ of the finer resolution scaling function $\phi(2t)$ using the quadrature mirror filter $P(\omega/2)$. Equation (10.6.4) is a recursion equation. The recursion can be repeated m times to yield $\Phi(\omega/2)$, $\Phi(w/4), \ldots$, and so on, that gives

$$\Phi(\omega) = \prod_{i=1}^{m} P\left(\frac{\omega}{2^i}\right) \Phi\left(\frac{\omega}{2^m}\right) \qquad (10.6.8)$$

When m approaches to infinity we have the Fourier transform of the continuous scaling function expressed as

$$\Phi(\omega) = \prod_{i=1}^{\infty} P\left(\frac{\omega}{2^i}\right) \qquad (10.6.9)$$

because the scaling function $\phi(t)$ has a nonzero mean and is normalized with respect to the $\mathbf{L}^1(\mathbf{R})$ as

$$\int \phi(t)\, dt = \Phi(0) = 1 \qquad (10.6.10)$$

It can be proved that if for some $\varepsilon > 0$ the sequence of interscale coefficients $p(n)$ satisfies

$$\sum_n |p(n)||n|^\epsilon < \infty \qquad (10.6.11)$$

then the infinite product on the right-hand side of Eq. (10.6.9) converges pointwise and the convergence is uniform.

Similarly, we can replace the second term, $\Phi(\omega/2)$, in the right-hand side of Eq. (10.6.5) with the infinite product derived in Eq. (10.6.9):

$$\Psi(\omega) = Q\left(\frac{\omega}{2}\right) \prod_{i=2}^{\infty} P\left(\frac{\omega}{2^i}\right) \qquad (10.6.12)$$

Equations (10.6.9) and (10.6.12) relate the Fourier transform of the continuous scaling function and wavelet to the infinite product of the quadrature mirror filters that are the Fourier transforms of the sequences of interscale coefficients $p(n)$ and $q(n)$ in the two-scale relations.

Orthonormality condition on the conjugate quadrature mirror filters

We have shown in Section 10.5.2 that the orthonormality conditions for the basic scaling function and the basic wavelet are

$$\sum_n |\Phi(\omega + 2n\pi)|^2 = 1$$

$$\sum_n |\Psi(\omega + 2n\pi)|^2 = 1 \qquad (10.6.1)$$

When these conditions are satisfied, the discrete translates of $\phi(t)$ and $\psi(t)$ form the orthonormal bases at every given scale 2^i.

Substituting the Fourier domain two-scale relation (10.6.4) into the orthonormality condition (10.6.1) we write

$$\sum_n |\Phi(2\omega + 4n\pi)|^2 = |P(\omega)|^2 \sum_n |\Phi(\omega + 2n\pi)|^2 = |P(\omega)|^2$$

$$\sum_n |\Phi[2\omega + 2(2n+1)\pi]|^2 = |P(\omega + \pi)|^2 \sum_n |\Phi[\omega + (2n+1)\pi]|^2$$

$$= |P(\omega + \pi)|^2 \tag{10.6.13}$$

corresponding to the summation of the $\Phi(\omega)$ translated by $2n \bullet 2\pi$ and by $(2n+1) \bullet 2\pi$ respectively. Adding the two equations in (10.6.13) and using again the orthonormality condition (10.6.1) we have

$$|P(\omega)|^2 + |P(\omega + \pi)|^2 = 1 \tag{10.6.14}$$

This is the orthonormality condition for the square magnitude of the quadrature mirror filter $P(\omega)$ in the Fourier domain.

Similarly, the orthonormality condition for the conjugate quadrature mirror filter $Q(\omega)$ is

$$|Q(\omega)|^2 + |Q(\omega + \pi)|^2 = 1 \tag{10.6.15}$$

Cross-filter orthogonality

The scaling functions and the wavelets must be mutually orthogonal within the same scale:

$$\int \phi(t - n')\psi(t - k)\, dt = 0 \tag{10.6.16}$$

for all $n', k \in \mathbf{Z}$. In the Fourier domain the condition for the cross-filter orthogonality can be written as

$$\int \Phi(\omega)\Psi^*(\omega) \exp(-jn\omega)\, d\omega = 0 \tag{10.6.17}$$

where $n = n' - k$ and $n \in \mathbf{Z}$. We have used the Poisson summation formula, Eq. (10.5.8), in Section 10.5.2, which is

$$\sum_n f(x + 2n\pi) = \frac{1}{2\pi} \sum_n F(n) \exp(jnx) \tag{10.6.18}$$

Assume that the product $\Phi(\omega)\Psi^*(\omega)$ satisfies the regularity condition described in Eq. (10.5.7) and let $\Phi(\omega)\Psi^*(\omega)$ be the $f(x)$ in the Poisson summation formula; we have

$$\sum_n \Phi(\omega + 2n\pi)\Psi^*(\omega + 2n\pi) = 0 \tag{10.6.19}$$

because the Fourier transform of $\Phi(\omega)\Psi^*(\omega)$ is equal to zero according to the cross-filter orthogonality condition (10.6.17).

We separate the translations of $2k \bullet 2\pi$ and of $(2k+1) \bullet 2\pi$ of the product $\Phi(\omega)\Psi^*(\omega)$ and rewrite Eq. (10.6.19) as

$$\sum_n \Phi(2\omega + 4n\pi)\Psi^*(2\omega + 4n\pi) + \sum_n \Phi[2\omega + 2(2n+1)\pi]\Psi^*[2\omega + 2(2n+1)\pi] = 0$$

On substituting the Fourier domain two-scale relations (10.6.4) and (10.6.5) for $\Phi(\omega)$ and $\Psi(\omega)$ into the above expression and using the periodicity of period 2π of the quadrature mirror filters

$P(\omega)$ and $Q(\omega)$ we have

$$P(\omega)Q^*(\omega)\sum_n |\Phi(\omega + 2n\pi)|^2 + P(\omega + \pi)Q^*(\omega + \pi)\sum_n |\Phi[\omega + (2n+1)\pi]|^2 = 0$$

Using the orthonormality condition for $\Phi(\omega)$ described in Eq. (10.6.1) we have

$$P(\omega)Q^*(\omega) + P(\omega + \pi)Q^*(\omega + \pi) = 0$$

$$P^*(\omega)Q(\omega) + P^*(\omega + \pi)Q(\omega + \pi) = 0 \tag{10.6.20}$$

This is the cross-filter orthogonality condition for the conjugate quadrature mirror filters $P(\omega)$ and $Q(\omega)$ in the Fourier domain.

Using the z-transform with $z = \exp(i\omega)$ and noticing that $P(-z)$ and $Q(-z)$ correspond to $P(\omega+\pi)$ and $Q(\omega+\pi)$ and $P(z^{-1})$ and $Q(z^{-1})$ correspond to $P^*(\omega)$ and $Q^*(\omega)$, respectively, the Fourier domain cross-filter orthogonality conditions, Eq. (10.6.20), can be written in terms of the z-transform relations as

$$P(z)Q(z^{-1}) + P(-z)Q(-z^{-1}) = 0$$

$$P(z^{-1})Q(z) + P(-z^{-1})Q(-z) = 0 \tag{10.6.21}$$

High-pass filters $q(n)$

In particular, it is easy to verify that

$$Q(z) = z^{-(N-1)}P(-z^{-1}) \tag{10.6.22}$$

is a solution of the cross-filter orthonormality, Eq. (10.6.21), with an arbitrary even number N. The inverse z-transform of solution (10.6.22) gives the relation between the quadrature mirror low-pass filters $p(n)$ and high-pass filters $q(n)$ in the time domain. The high-pass filters $q(n)$ are obtained from the low-pass filters $p(n)$:

$$q(n) = (-1)^{n+1}p(N - 1 - n) \tag{10.6.23}$$

for an even N, so that $p(n)$ and $q(n)$ are mutually orthogonal. Note that according to Eq. (10.6.3) both the filters $p(n)$ and $q(n)$ are orthogonal to their own translates with double integer steps, so that an arbitrary even number N can be added in $p(N-1-n)$ of Eq. (10.6.23).

Paraunitary matrix

The three orthogonality conditions described in Eqs. (10.6.14), (10.6.15), and (10.6.20) are written in terms of the z-transform as

$$P(z)P(z^{-1}) + P(-z)P(-z^{-1}) = 1$$

$$Q(z)Q(z^{-1}) + Q(-z)Q(-z^{-1}) = 1$$

$$P(z)Q(z^{-1}) + P(-z)Q(-z^{-1}) = 0$$

$$P(z^{-1})Q(z) + P(-z^{-1})Q(-z) = 0 \tag{10.6.24}$$

Solution (10.6.22) implies that

$$Q(z^{-1}) = z^{N-1}P(-z) \tag{10.6.25}$$

Multiplying both sides of Eqs. (10.6.22) and (10.6.25) yields

$$|Q(z)|^2 = |P(-z)|^2 \tag{10.6.26}$$

or

$$|Q(\omega)|^2 = |P(\omega + \pi)|^2 \tag{10.6.27}$$

Comparing Eq. (10.6.27) with Eq. (10.4.16) for definition of the mirror filters, we say that the filters $P(z)$ and $Q(z)$ are the quadrature mirror filters.

From the orthonormality conditions described in Eq. (10.6.24) we have

$$|P(z)|^2 + |P(-z)|^2 = 1$$

$$|Q(z)|^2 + |Q(-z)|^2 = 1 \tag{10.6.28}$$

From Eqs. (10.6.26) and (10.6.28) we have

$$\begin{array}{ll} |P(z)|^2 + |Q(z)|^2 = 1 & |P(\omega)|^2 + |Q(\omega)|^2 = 1 \\ |P(-z)|^2 + |Q(-z)|^2 = 1 & |P(\omega + \pi)|^2 + |Q(\omega + \pi)|^2 = 1 \end{array} \tag{10.6.29}$$

According to Eq. (10.6.29) we say that the filter pair $\{P(z), Q(z)\}$ is power complementary. This property has an important implication. As $P(\omega)$ is a low-pass filter with

$$P(\omega)|_{\omega=0} = 1 \tag{10.6.30}$$

then $Q(\omega)$ must be a high-pass filter and

$$Q(\omega)|_{\omega=0} = 0 \tag{10.6.31}$$

The orthogonality conditions described in Eq. (10.6.24) and the power complementary properties described in Eq. (10.6.29) are equivalent to the requirement that the 2×2 matrix

$$\begin{vmatrix} p(z) & Q(z) \\ p(-z) & Q(-z) \end{vmatrix}$$

should be paraunitary:

$$\begin{vmatrix} p(z^{-1}) & P(-z^{-1}) \\ Q(z^{-1}) & Q(-z^{-1}) \end{vmatrix} \begin{vmatrix} p(z) & Q(z) \\ P(-z) & Q(-z) \end{vmatrix} = \begin{vmatrix} 1 & 0 \\ 0 & 1 \end{vmatrix} \tag{10.6.32}$$

for the conjugate quadrature mirror filters. The paraunitary properties are useful for designing the compactly supported orthonormal scaling function and the wavelet bases.

Example 3   *Orthonormality of Haar's bases*

Let us consider the orthonormality condition for Haar's bases, as an example. We know that Haar's bases are orthonormal at every scale. The Fourier transforms of the Haar scaling functions and wavelets are

$$\Phi(\omega) = e^{-j\omega/2} \frac{\sin(\omega/2)}{\omega/2}$$

$$\Psi(\omega) = e^{-j\omega/2} \frac{\sin^2(\omega/2)}{\omega/4} \tag{10.6.33}$$

We have for the scaling function $\Phi(\omega)|_{\omega=0} = 1$ and for the wavelet $\Psi(\omega)|_{\omega=0} = 0$. It can be verified that the orthonormality condition

$$\sum_n |\Phi(\omega + 2n\pi)|^2 = 1 \tag{10.6.34}$$

$$\sum_n |\Psi(\omega + 2n\pi)|^2 = 1 \tag{10.6.35}$$

is satisfied.

The interscale relations of the Haar scaling functions and the Haar wavelets are obtained in Section 10.5.3. On substituting the interscale coefficients: $p(n) = 1/\sqrt{2}$ for $n = 0, 1$ and $p(n) = 0$ otherwise, $q(n) = 1/\sqrt{2}$ for $n = 0$, $q(n) = -1/\sqrt{2}$ for $n = 1$, and $q(n) = 0$ otherwise of Haar's bases into the Fourier transform of $p(n)$ and $q(n)$, Eq. (10.6.6), we obtain the quadrature mirror filters of Haar's bases as

$$P(\omega) = 2^{-1/2} \cos\frac{\omega}{2} \exp(-j\omega/2)$$

$$Q(\omega) = j2^{-1/2} \sin\frac{\omega}{2} \exp(-j\omega/2) \tag{10.6.36}$$

It is easy to verify that Haar's quadrature mirror filters satisfy all the orthonormality conditions:

$$|P(\omega)|^2 + |P(\omega + \pi)|^2 = \frac{1}{2}\left[\cos^2\left(\frac{\omega}{2}\right) + \cos^2\left(\frac{\omega+\pi}{2}\right)\right] = 1$$

$$|Q(\omega)|^2 + |Q(\omega + \pi)|^2 = \frac{1}{2}\left[\sin^2\left(\frac{\omega}{2}\right) + \sin^2\left(\frac{\omega+\pi}{2}\right)\right] = 1 \tag{10.6.37}$$

and

$$P(\omega)Q^*(\omega) + P(\omega + \pi)Q^*(\omega + \pi) = -j\frac{1}{2}\cos\frac{\omega}{2}\sin\frac{\omega}{2}$$

$$-j\frac{1}{2}\cos\frac{\omega+\pi}{2}\sin\frac{\omega+\pi}{2}$$

$$= 0 \tag{10.6.38}$$

We have also that

$$|P(\omega)|^2 + |Q(\omega)|^2 = 1$$

$$|P(-\omega)|^2 + |Q(-\omega)|^2 = 1 \tag{10.6.39}$$

and that the matrix

$$\begin{vmatrix} p(\omega) & Q(\omega) \\ P(\omega + \pi) & Q(\omega + \pi) \end{vmatrix}$$

is paraunitary. ∎

10.6.2 *Regularity*

The regularity of the wavelet basis functions is an important property of the wavelet transform. Because of this property, the wavelet transform can be local in both time and frequency domains. In Section 10.2.2 we discussed the regularity condition for the continuous wavelet transform. For the wavelet transform coefficients to decay as fast as $s^{n+1/2}$ with an increase of

$(1/s)$, where s is the scale factor, the wavelet $\psi(t)$ must have the first $n + 1$ moments of the order $0, 1, \ldots, n$ equal to zero, and equivalently, the Fourier transform $\Psi(\omega)$ of the wavelets must have the first n derivatives of the order up to n equal to zero about zero frequency $\omega = 0$.

In this section we shall discuss the regularity condition on the orthonormal scaling functions and wavelets, and on the quadrature mirror filters $P(\omega)$ and $Q(\omega)$ in the multiresolution analysis framework. We shall discuss the regularity condition in a slightly different way from that in Section 10.2.2. The regularity conditions are applied for ensuring convergence of the reconstruction from the orthonormal wavelet decomposition. However, the regularity conditions obtained in both approaches are equivalent.

We now consider regularity as a measure of smoothness for scaling functions and wavelets. The regularity of the scaling function is determined by the decay of its Fourier transform $\Phi(\omega)$ and is defined as the maximum value of r such that

$$|\Phi(\omega)| \le \frac{c}{(1 + |\omega|)^r} \tag{10.6.40}$$

for $\omega \in \mathbf{R}$. Hence, the $|\Phi(\omega)|$ has exponential decay as ω^{-M}, where $M \le r$. This in turn implies that $\phi(t)$ is $(M - 1)$-times continuously differentiable, and both $\phi(t)$ and $\psi(t)$ are smooth functions.

Regularity condition

The reconstruction from the wavelet series decomposition is described by Eq. (10.5.70):

$$c_0 = \sum_{i=1}^{M} (L^*)^{i-1} H^* d_i + (L^*)^M c_M$$

where synthesis filtering operators L^* and H^* are defined as

$$(L^* \alpha)(n) = \frac{1}{\sqrt{2}} \sum_k p(n - 2k) \alpha(k)$$

$$(H^* \alpha)(n) = \frac{1}{\sqrt{2}} \sum_k q(n - 2k) \alpha(k)$$

In the reconstruction it is the low-pass filter L^* that is iterated. To ensure that this reconstruction will work, the discrete low-pass filter $p(n)$ has to satisfy the regularity conditions. In Section 10.5.5 we discussed the reconstruction of the scaling function from its wavelet series decomposition using the graphic representation. We defined a piecewise constant function $\eta_i(t) = (L^*)^i \eta_0(t)$, where $\eta_0(t)$ is a rectangular function. When the piecewise constant function $\eta_i(t)$ converges to a regular limit function as i indefinitely increases, we say that the reconstruction works.

The scaling function $\phi(t)$ is related to the quadrature mirror filter $p(n)$ in the Fourier domain by Eq. (10.6.9):

$$\Phi(\omega) = \prod_{i=1}^{\infty} P\left(\frac{\omega}{2^i}\right) \tag{10.6.9}$$

It can be proved that the infinite product in the right-hand side of Eq. (10.6.9) converges pointwise and uniformly for all $\omega \in \mathbf{R}$, if for some $\varepsilon > 0$,

$$\sum_n |p(n)| |n|^\epsilon < \infty$$

that is, $p(n)$ decays as fast as $n^{-\varepsilon}$. This is a very mild condition, because in most practical cases the low-pass filters $p(n)$ are the FIR filters with only a limited number of $p(n) \ne 0$.

Now, the regularity (10.6.40) of the scaling function should not only ensure that $\eta_\infty(t)$ converges, but also ensure that (1) $\eta_\infty(t)$ is sufficiently regular, or the Fourier transform of $\eta_\infty(t)$ has sufficient decay, and (2) $\eta_i(t)$ converges to $\eta_\infty(t)$ pointwise when i approaches to infinity. Daubechies [13] has proven that the above two conditions can be satisfied when the Fourier transform of the low-pass filter $p(n)$, the quadrature mirror filter $P(\omega)$, satisfies, for example,

$$P(\omega) = \left[\frac{1 + e^{-j\omega}}{2}\right]^M F(e^{-j\omega}) \tag{10.6.41}$$

or, written in terms of the z-transform, as

$$P(z^{-1}) = \left[\frac{1 + z^{-1}}{2}\right]^M F(z^{-1}) \tag{10.6.42}$$

where $M > 1$ and $F(z^{-1})$ is a polynomial in z^{-1}, or in $e^{-j\omega}$, of degree $N - 1 - M$ with real coefficients and satisfies some conditions so that the infinite product

$$\left|\prod_{k=0}^{\infty} F(z^{k/2})\right| \le 2^{k(N-M-1)} \tag{10.6.43}$$

converges and is bounded. Because the low-pass filter $p(n)$ is a FIR filter, $p(n) \ne 0$ only for $n = 0, 1, \ldots, N - 1$, and N is the length of $p(n)$, then the Fourier transform of $p(n)$:

$$P(\omega) = \frac{1}{2} \sum_n p(n) \exp(-jn\omega)$$

will be a polynomial in $e^{-j\omega}$, or in z^{-1}, of degree $N - 1$.

According to condition (10.6.41) and (10.6.42) the quadrature mirror filter $P(\omega)$ must have a sufficient number of zeros at $\omega = \pi$ or $z = -1$. We know that $P(\omega)$ has already at least one zero at $\omega = \pi$, because according to the paraunitary property discussed in Section 10.6.1, $P(\omega)$ and $Q(\omega)$ are the conjugate quadrature mirror filters as described in Eq. (10.6.27)

$$|P(\omega + \pi)|^2 = |Q(\omega)|^2$$

and Eq. (10.6.31)

$$Q(\omega)|_{\omega=0} = 0$$

Hence,

$$|P(\pi)|^2 = 0 \tag{10.6.44}$$

The quadrature mirror filter $P(\omega)$ must have at least one zero at $z = -1$ or at $\omega = \pi$, or it contains at least one term of $(1 + e^{-j\omega})$.

However, according to the regularity condition (10.6.41), $P(\omega)$ must have more zeros with the number of zeros $M > 1$ in order to insure convergence of the wavelet reconstruction. Regularity condition (10.6.41) implies that

$$|P(\omega)| = \left|\cos\frac{\omega}{2}\right|^M |F(e^{-j\omega})| \tag{10.6.45}$$

On substituting Eq. (10.6.45) into the infinite product form in Eq. (10.6.9) of the Fourier transform $\Phi(\omega)$ of the scaling function we obtain the regularity condition on $\Phi(\omega)$ as

$$|\Phi(\omega)| = \prod_{i=1}^{\infty} \left|\cos\frac{\omega}{2^{i+1}}\right|^{M} \cdot \left|\prod_{i=1}^{\infty} F(e^{-j\omega/2^{i}})\right| \tag{10.6.46}$$

but

$$\cos\frac{\omega}{2} = \frac{\sin\omega}{2\sin\frac{\omega}{2}}$$

The first infinite product term in Eq. (10.6.46) is therefore

$$\lim_{M\to\infty}\left[\prod_{i=1}^{M}\left|\frac{\sin(\omega/2^{i})}{2\sin(\omega/2^{i+1})}\right|\right]^{M} = \lim_{M\to\infty}\left[\left|\frac{\sin(\omega/2)}{2^{M}\sin(\omega/2^{M+1})}\right|\right]^{M} = \left[\frac{\sin(\omega/2)}{(\omega/2)}\right]^{M}$$

and

$$|\Phi(\omega)| = \left|\frac{\sin(\omega/2)}{\omega/2}\right|^{M}\left|\prod_{i=1}^{\infty} F(e^{-j\omega/2^{i}})\right| \tag{10.6.47}$$

The first term $|\operatorname{sinc}(\omega/2)|^{M}$ in the right-hand side of Eq. (10.6.47) contributes to the exponential decay of $\Phi(\omega)$ as ω^{-M}. The second term in the right-hand side of Eq. (10.6.47) is bounded as condition Eq. (10.6.43) shows.

The number M of zeros of the quadrature mirror filter $P(\omega)$ at $\omega = \pi$, or at $z = -1$ is a meaure of flatness of $P(\omega)$ at $\omega = \pi$. From the regularity conditions (10.6.41) and (10.6.47) we see that the exponential decay of the scaling function $\Phi(\omega)$ and the flatness of the quadrature mirror filter $P(\omega)$ are equivalent. The scaling function $\Phi(\omega)$ has the exponential decay as ω^{-M}. The quadrature mirror filter $P(\omega)$ has number M of zero at $\omega = \pi$. M is a measure of the regularity of the scaling function.

Smoothness of the quadrature mirror filters

The regularity condition also implies the smoothness of the iterated low-pass filter $p(n)$. We rewrite Eq. (10.6.41) as

$$P(\omega) = \exp(-jM\omega/2)\left(\cos\frac{\omega}{2}\right)^{M} F(\omega) \tag{10.6.48}$$

The rth derivative of Eq. (10.6.48) is

$$\frac{d^{r}P(\omega)}{d\omega^{r}} = \left(\cos\frac{\omega}{2}\right)^{M-r} g_{r}(\omega) \tag{10.6.49}$$

where $[\cos(\omega/2)]^{M-r}$ is the minimum power of $\cos(\omega/2)$ that the rth derivative contains, and $g_{r}(\omega)$ is the residual terms of the derivative. The term $[\cos(\omega/2)]^{M-r}$ makes the rth derivative of $P(\omega)$ equal to zero at $\omega = \pi$ for $r = 0, 1, \ldots, M-1$. On the other hand, using the Fourier transform relation between $P(\omega)$ and $p(n)$, Eq. (10.6.6), we have

$$\frac{d^{r}P(\omega)}{d\omega^{r}} = \sum_{n}(-jn)^{r} p(n)\exp(-jn\omega) \tag{10.6.50}$$

According to Eq. (10.6.49) the rth derivative of $P(\omega)$ is equal to zero at $\omega = \pi$; we then have from Eq. (10.6.50)

$$\left.\frac{d^{r}P(\omega)}{d\omega^{r}}\right|_{\omega=\pi} = (-j)^{r}\sum_{n} n^{r}(-1)^{n} p(n) = 0 \tag{10.6.51}$$

for $r = 0, 1, \ldots, M-1$. The iterated low-pass filter $p(n)$ is a smooth filter.

Substituting expression (10.6.22) for $Q(\omega)$

$$Q(z) = z^{-(N-1)} P(-z^{-1})$$

where N is any even number, into the regularity condition (10.6.42) we have

$$Q(z) = \left(\frac{1-z}{2}\right)^M z^{-(N-1)} F(-z^{-1}) \tag{10.6.52}$$

or in terms of the Fourier transform frequency ω as

$$Q(\omega) = \left(\sin \frac{\omega}{2}\right)^M g_r(\omega) \tag{10.6.53}$$

where $g_r(\omega)$ is the residual term. The $[\sin(\omega/2)]^M$ term insures that $Q(\omega)$ has the vanishing derivatives at $\omega = 0$

$$\frac{d^r Q(\omega)}{d\omega^r}\bigg|_{\omega=0} = 0 \tag{10.6.54}$$

Using the Fourier transform relation, Eq. (10.6.6), between $Q(\omega)$ and $q(n)$, the rth derivative of $Q(\omega)$ is equal to zero at $\omega = 0$ for $r = 0, 1, \ldots, M - 1$, and is equivalent to

$$\sum_n r^n q(n) = 0 \tag{10.6.55}$$

The quadrature mirror high-pass filter $q(n)$ has vanishing moment of order r.

The low-pass filter $p(n)$ is a FIR filter: $p(n) \neq 0$ only for $n = 0, 1, \ldots, N - 1$ and N is the length of $p(n)$. The compactness of $p(n)$ and $q(n)$, described by $(N - 1)$ is in contrast with the regularity of $P(\omega)$ and $Q(\omega)$ and of the scaling and wavelet functions, described by M, because according the regularity condition (10.6.41), $M < N - 1$. There is a tradeoff between the compactness and the regularity. We shall discuss this in Section 10.7.3.

10.7   Orthonormal scaling functions and wavelets

In this section we give some orthonormal wavelet bases. In general, the orthonormal wavelet bases generated in the multiresolution analysis framework are associated with the orthonormal scaling function bases. The different orthonormal scaling function and wavelet bases satisfy the orthonormality condition and the regularity condition in slightly different ways.

10.7.1   *Polynomial spline and wavelets*

One of the basic methods for constructing the orthonormal wavelet families involves the cardinal B-spline functions and the polynomial spline analysis, which are familiar in the approximation theory for interpolating a given sequence of data points. In this subsection we give a brief description of the multiresolution analysis with the polynomial cardinal B-spline scaling and wavelet function bases.

The B-splines of order m are generated by repeated convolutions of the first-order B-splines [14]. The first-order B-spline is the rectangular function as

$$\beta^1(t) = \begin{cases} 1 & \text{for } 0 \le t < 1 \\ 0 & \text{otherwise} \end{cases} \tag{10.7.1}$$

The mth order B-spline is then

$$\beta^m(t) = (\beta^{m-1} * \beta^1)(t)$$

$$= \int_{-\infty}^{\infty} \beta^{m-1}(t-x)\beta^1(x)\,dx = \int_0^1 \beta^{m-1}(t-x)\,dx \qquad (10.7.2)$$

where m is an arbitrary positive integer. Because of the repeated convolutions, the mth order B-spline has the support of $[0, m]$ in the time domain. This support increases with the order m. The B-spline is a symmetric function with respect to the center of its support $t = m/2$. The central B-splines are defined such that they are symmetric with respect to the origin.

The first-order B-spline $\beta^1(t)$ has been used as the Haar scaling function. It is not even continuous. The second-order B-spline, $\beta^2(t)$, is a triangle function, called the hat function. The high-order B-splines with $m > 2$ are smooth, bell-shaped functions. Let C^n denote the collection of all functions f such that the derivatives $f, f', \ldots, f^{(n)}$ are continuous everywhere, then the mth order B-splines have continuous derivatives of orders from 0 to $m-2$, $\beta^m \in C^{m-2}$.

According to definition (10.7.2), the Fourier transform of the mth order B-spline is

$$B^m(\omega) = \left(\frac{1 - e^{-j\omega}}{j\omega}\right)^m = e^{-jm\omega/2}\,\text{sinc}^m\left(\frac{\omega}{2}\right) \qquad (10.7.3)$$

Its modulus, $|B^m(\omega)|$, has the exponential decay as $1/\omega^m$. When m increases, the B-spline becomes more regular but its support $[0, m]$ becomes less compact. There is then the typical tradeoff between the regularity m and the compactness of the B-splines.

The polynomial splines are linear combinations of the translated B-splines. The polynomial splines $f^m(t)$ can be used to interpolate a given sequence of data point $\{s(k)\}$:

$$f^m(t) = \sum_k c(k)\beta^m(t-k) \qquad (10.7.4)$$

when the coefficients $c(k) = s(k)$, $f^m(t)$ is a cardinal spline interpolator, which has piecewise polynomial segments of m-degree. Hence, the mth order B-spline is the interpolation function. When $m = 1$, the sequence of point $\{s(k)\}$ with equal intervals is interpolated by $\beta^1(t)$, which is a staircase function, as shown in the case of the Haar scaling function approximation. When $m = 2$, the sequence of points are connected by the straight line segment in each interval $[k, k+1]$. When $m > 2$, the data points are interpolated by a function that is, in each interval, a polynomial in degree m.

The scaling function for the orthonormal wavelet transform is defined as a polynomial spline

$$\phi^m(t) = \sum_k p(k)\beta^m(t-k) \qquad (10.7.5)$$

that is a finite linear combination of translates of the B-spline. The B-splines themselves can also be used as the scaling functions. However, definition (10.7.5) provides some useful properties such as the orthonormality of the scaling functions. The set of functions $\{\phi^m(t-k)\}$ with integer translations $k \in \mathbf{Z}$ is the basis of a subspace S^m. The projection of a function $f(t)$ on S^m is

$$f(t) = \sum_k c(k)\phi^m(t-k) \qquad (10.7.6)$$

which is an approximation of $f(t)$, called the polynomial spline representation of $f(t)$. We want the minimum error polynomial spline representation.

When the coefficients $c(k)$ in the expansion (10.7.6) of $f(t)$ are equal to the sampled values $f(k)$ themselves, the expansion (10.7.6) is referred to as the cardinal spline representation. The corresponding basis functions $\phi^m(t)$ are referred to as the cardinal (or fundamental) splines. If we define a sequence b^n as the discrete mth order B-spline

$$b^m(t) = \beta^m(t)\big|_{t=n} \tag{10.7.7}$$

it can be shown that for the cardinal splines the coefficients $p(n)$ in the linear combination (10.75) must satisfy

$$p(n) = [b^n(n)]^{-1} \tag{10.7.8}$$

that corresponds to

$$(p * b^n)(k - n) = \delta_{k,n} \tag{10.7.9}$$

where $*$ denotes the convolution.

It can also be shown that when the order m of the spline approaches infinity, the cardinal spline $\phi^m(t)$ converges to sinc(t), which is the ideal interpolation function generating the minimum interpolation error, according to the Shannon sampling theorem.

The scaling functions can be dilated and translated to form the bases of subspaces S_i^m at different resolution levels (i). We note that the cardinal splines can be combinations of translated B-splines that have a support larger than $[0, m]$ and their integer translates do not necessarily form an orthonormal basis within the same resolution level.

The two-scale relation between the cardinal B-splines at two adjacent resolution levels is

$$\beta^m\left(\frac{t}{2}\right) = \sum_k u_2^m(k)\beta^m(t - k) \tag{10.7.10}$$

An admissible wavelet function $\psi(t)$ must satisfy two essential conditions. First, it must be included in the finer resolution approximation space. This means that the function $\psi(t/2)$ must be a polynomial spline of order m and can be represented as

$$\psi\left(\frac{t}{2}\right) = \sum_k q(k)\beta^m(t - k) \tag{10.7.11}$$

This is the two-scale relation for the spline wavelets. Second, at the same resolution level, the wavelet $\psi(t/2)$ must orthogonal to the polynomial spline subspace spanned by $\{\beta^m(t/2 - k)\}$

$$\left\langle \psi\left(\frac{t}{2}\right), \beta^m\left(\frac{t}{2} - k\right) \right\rangle = 0 \tag{10.7.12}$$

In the multiresolution analysis framework, the projection of a function $f(t)$ on the basic polynomial spline subspace S_0^m at the resolution $i = 0$ generates a polynomial spline representation and the projection of $f(t)$ on the subspace S_1^m at the resolution $i = 1$ yields a polynomial spline representation at a lower resolution. The projection of $f(t)$ on the subspace spanned by the wavelets $\{\psi(t/2 - k)\}$ corresponds to the difference or the detail between the two polynomial spline approximations.

10.7.2 *Lemarie and Battle wavelet bases*

The Lemarie and Battle wavelet bases are a family of orthonormal scaling functions and wavelets that are associated with the cardinal B-spline polynomials, with an additional condition that the Lemarie and Battle polynomial scaling functions and wavelets are orthonormal within the same resolution level [15]. The orthonormality is obtained by imposing the orthonormality constraints on the cardinal polynomial scaling functions and wavelets.

The Lemarie and Battle multiresolution bases are built from the polynomial cardinal B-spline of order $m = 2p + 1$. Hence, the approximations of a function in the Lemarie and

Battle multiresolution analysis are $(m-2)$-times continuously differentiable and equal to a polynomial of order $m = 2p + 1$ on each interval $[k, k+1]$ for $k \in \mathbf{Z}$.

When $p = 0$ and $m = 1$, the Lemarie and Battle scaling function is simply a rectangular function. Its Fourier transform is

$$B^1(\omega) = \left(\frac{1 - e^{-j\omega}}{j\omega}\right) \tag{10.7.13}$$

and we obtain the sum

$$\sum_k |B^1(\omega + 2\pi k)|^2 = 4\sin^2 \frac{\omega}{2} \sum_k \frac{1}{(\omega + 2\pi k)^2} \tag{10.7.14}$$

The condition for the integer translates of a scaling function forming an orthonormal basis is described in Eq. (10.5.9) as

$$\sum_k |\Phi(\omega + 2\pi k)|^2 = 1$$

When $B^1(\omega)$ satisfies the orthonormality condition we have

$$\sum_k \frac{1}{(\omega + 2\pi k)^2} = \frac{1}{4}\csc^2 \frac{\omega}{2} \tag{10.7.15}$$

Lemarie has found a scaling function that is associated with the nth order splines and its integer translates form an orthonormal basis within the same resolution level. The Lemarie-Battle scaling function is given by its Fourier transform as

$$\Phi(\omega) = \frac{1}{\omega^n} \left(\sum_k \frac{1}{(\omega + 2\pi k)^{2n}}\right)^{-1/2} \equiv \frac{1}{\omega^n \sqrt{\sum_{2n}(\omega)}} \tag{10.7.16}$$

where $n = 2p + 2$ and $\sum_{2n}(\omega) = \sum[1/(\omega + 2\pi k)^{2n}]$. We can differentiate identity (10.7.15) $2n - 2$ times to obtain

$$\sum_k \frac{1}{(\omega + 2\pi k)^{2n}} = \frac{1}{4(2n - 3)!} \frac{d^{2n-2}}{d\omega^{2n-2}} \csc^2 \frac{\omega}{2} \tag{10.7.17}$$

The Lemarie-Battle scaling function $\Phi(\omega)$ described in Eq. (10.7.16) can be computed with Eq. (10.7.17).

The low-pass quadrature mirror filter $P(\omega)$ can be obtained from the two-scale relation in the Fourier domain, described by Eq. (10.6.4)

$$\Phi(2\omega) = P(\omega)\Phi(\omega)$$

According to (10.7.16) we obtain

$$P(\omega) = \sqrt{\frac{\sum_{2n}(\omega)}{2^{2n} \sum_{2n}(2\omega)}} \tag{10.7.18}$$

The Fourier transform of the corresponding orthonormal wavelet can be derived from Eq. (10.6.5)

$$\Psi(\omega) = Q\left(\frac{\omega}{2}\right) \Phi\left(\frac{\omega}{2}\right)$$

where the conjugate quadrature mirror filter $Q(\omega)$ satisfying the orthonormal condition is

TABLE 10.1
The low-pass filter $p(n)$ of the Lemarie-Battle wavelet bases associated with the second order B-spline [16].

n	$p(n)$	n	$p(n)$
0	0.542	6	0.012
1	0.307	7	−0.013
2	−0.035	8	0.006
3	−0.078	9	0.006
4	0.023	10	−0.003
5	−0.030	11	−0.002

obtained from the paraunitary property as that described in Eq. (10.6.22)

$$Q(z) = z^{-(N-1)} P(-z^{-1})$$

$$Q(\omega) = e^{j(N-1)\omega} P^*(\omega + \pi)$$

Hence,

$$\Psi(\omega) = e^{j(N-1)\omega/2} P^* \left(\frac{\omega + \pi}{2} \right) \Phi \left(\frac{\omega}{2} \right) \tag{10.7.19}$$

When $p = 1$ and thus $m = 3, n = 4$, the Lemarie and Battle wavelet transform corresponds to a multiresolution approximation built from the cubic splines. Let

$$N_1(\omega) = 5 + 30 \cos^2 \frac{\omega}{2} + 30 \sin^2 \frac{\omega}{2} \cos^2 \frac{\omega}{2}$$

$$N_2(\omega) = 2 \sin^4 \frac{\omega}{2} \cos^2 \frac{\omega}{2} + 70 \cos^4 \frac{\omega}{2} + \frac{2}{3} \sin^6 \frac{\omega}{2}$$

The function $\sum_8(\omega)$ is given by

$$\sum_8(\omega) = \frac{N_1(\omega) + N_2(\omega)}{105 \sin^8 \frac{\omega}{2}} \tag{10.7.20}$$

From the expression for $\sum_{2n}(\omega)$, the Fourier transforms $\Phi(\omega)$ and $\Psi(\omega)$ of the scaling function and wavelet and the quadrature mirror filters $P(\omega)$ and $Q(\omega)$ can be calculated. Table 10.1 gives the first 12 coefficients of the discrete low-pass filter $p(n)$ that are the impulse response of $P(\omega)$, useful for the wavelet series decomposition. The high-pass filters $q(n)$ can be obtained from the low-pass filter $p(n)$ with Eq. (10.6.23). Figure 10.15 shows the Lemarie-Battle wavelet $\psi(t)$ and its Fourier transform $\Psi(\omega)$, which is given by Eq. (10.7.19). The wavelet is associated with the cardinal B-spline of order $m = 2$. Hence, the wavelet consists of the straight line segments between the discrete nodes. When m increases, the Lemarie-Battle scaling function and wavelet associated with high-order B-splines become more smooth. The Lemarie-Battle wavelet is symmetrical to $t = 1/2$ and has no compact support. The wavelet $\psi(t)$ decays slowly with time t.

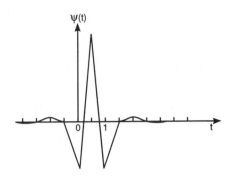

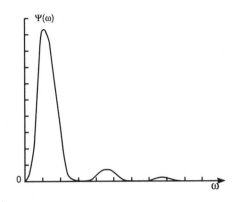

FIGURE 10.15
Lemarie-Battle wavelet and its Fourier transform associated with the second order B-spline [21].

10.7.3   *Biorthogonal wavelet bases*

The biorthogonal wavelet bases give more flexibility to the filter design. We define in the multiresolution framework two hierarchies of approximation subspaces:

$$\cdots V_2 \subset V_1 \subset V_0 \subset V_{-1} \subset V_{-2} \cdots$$
$$\cdots \bar{V}_2 \subset \bar{V}_1 \subset \bar{V}_0 \subset \bar{V}_{-1} \subset \bar{V}_{-2} \cdots \tag{10.7.21}$$

where the subspaces V_i are spanned by the translates of the scaling function $\phi(t)$ and $\bar{V}_i$ are spanned by the translates of the dual scaling function $\bar{\phi}(t)$. The wavelet subspace $\mathbf{W}_i$ is complementary to $\mathbf{V}_i$ in the finer-resolution subspace $\mathbf{V}_{i-1}$, but is not an orthogonal complement to W_i. Instead, $\mathbf{W}_i$ is the orthogonal complement to $\bar{\mathbf{V}}_i$. Similarly, the dual wavelet subspace $\bar{\mathbf{W}}_i$ is the orthogonal complement to $\mathbf{V}_i$. Thus,

$$W_i \perp \bar{V}_i \qquad \text{and} \qquad \bar{W}_i \perp V_i$$
$$\bar{V}_{i-1} = \bar{V}_i \oplus W_i \qquad \text{and} \qquad V_{i-1} = V_i \oplus \bar{W}_i \tag{10.7.22}$$

The orthogonality between the wavelet and the dual scaling function and between the scaling function and the dual wavelet can also be expressed as

$$\langle \bar{\phi}(t-k), \psi(t-n) \rangle = 0$$
$$\langle \bar{\psi}(t-k), \phi(t-n) \rangle = 0 \tag{10.7.23}$$

for any $n, k \in \mathbf{Z}$. We also expect the orthogonality between the scaling function and its dual and between the wavelet and its dual:

$$\langle \bar{\phi}(t-k), \phi(t-n) \rangle = \delta_{k,n}$$
$$\langle \bar{\psi}(t-k), \psi(t-n) \rangle = \delta_{k,n} \tag{10.7.24}$$

The orthogonality expressed in Eqs. (10.7.23) and (10.7.24) is referred to as the biorthogonality. Indeed, the biorthogonal scaling functions and wavelets can be found with the polynomial B-splines, scaling functions, and wavelets. The cross scale orthogonality of the wavelet and its dual can also be obtained

$$\langle \psi_{i,k}, \bar{\psi}_{m,n} \rangle = \delta_{i,m}\delta_{k,n} \tag{10.7.25}$$

Any function $f \in L^2(\mathbf{R})$ can be expanded onto the biorthogonal scaling function and wavelet bases as

$$f(t) = \sum_i \sum_k \langle f, \bar{\psi}_{i,k} \rangle \psi_{i,k}(t)$$

$$= \sum_i \sum_k \langle f, \psi_{i,k} \rangle \bar{\psi}_{i,k}(t) \tag{10.7.26}$$

and also

$$f(t) = \sum_i \sum_k \langle f, \bar{\phi}_{i,k} \rangle \phi_{i,k}(t)$$

$$= \sum_i \sum_k \langle f, \phi_{i,k} \rangle \bar{\phi}_{i,k}(t) \tag{10.7.27}$$

The implementation of the wavelet transform on the biorthogonal bases is also with the discrete low-pass and high-pass filters $p(n)$ and $q(n)$ in the multiresolution framework. In the reconstruction from the biorthogonal wavelet transform, however, the discrete synthesis filters $p_0(n)$ and $q_0(n)$ are not identical to the analysis filters $p(n)$ and $q(n)$. They can have no equal length. The discrete iterated filters are introduced with the two-scale relations as

$$\phi(t) = \sum_n p(n)\phi(2t - n)$$

$$\bar{\phi}(t) = \sum_n p_0(-n)\bar{\phi}(2t - n) \tag{10.7.28}$$

and

$$\psi(t) = \sum_n q(n)\phi(2t - n)$$

$$\bar{\psi}(t) = \sum_n q_0(-n)\bar{\phi}(2t - n) \tag{10.7.29}$$

One stage of the wavelet decomposition and reconstruction with the biorthogonal filter bank is shown in Figure 10.16. In Section 10.4.2 we discussed the condition for perfect reconstruction of the two-band filter bank, Eq. (10.4.7). The perfect reconstruction condition for the discrete biorthogonal filter bank can also be obtained from the subband coding theory. The perfect reconstruction conditions include orthogonality across the analysis and synthesis sections:

$$\sum_n p_0(2k - n)q(n - 2m) = 0$$

$$\sum_n q_0(2k - n)p(n - 2m) = 0 \tag{10.7.30}$$

and

$$\sum_n p_0(2k - n)p(n) = \delta_k$$

$$\sum_n q_0(2k - n)q(n) = \delta_k \tag{10.7.31}$$

for all $k, n \in \mathbf{Z}$.

Detailed design procedure for the biorthogonal wavelet bases can be found in the literature [14].

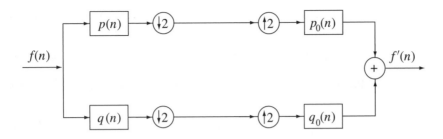

FIGURE 10.16
Schematic wavelet decomposition and reconstruction with the biorthogonal scaling function and
wavelet bases.

10.7.4   *Daubechies wavelets*

The Daubechies wavelet bases are a family of orthonormal, compactly supported scaling and
wavelet functions that have the maximum regularity for a given length of the support of the
quadrature mirror filters. The Daubechies scaling and wavelet functions are derived from the
regularity condition, Eq. (10.6.41), and the orthonormality, Eq. (10.6.14), on the quadrature
mirror filter $P(\omega)$ in the Fourier domain. For computing the orthonormal wavelet transform
in the computer, the Daubechies wavelets bases are given as a set of discrete low-pass and
high-pass filters $p(n)$ and $q(n)$. The orthonormal wavelet decomposition is implemented by
iterating those filters with the tree algorithm.

The quadrature mirror filter $P(\omega)$ in the Daubechies bases takes the form of Eq. (10.6.41):

$$P(\omega) = \left[\frac{1 + e^{-j\omega}}{2} \right]^M F(e^{-j\omega}) \tag{10.6.41}$$

which is the regularity condition applied on $P(\omega)$, where $M > 1$ is the regularity of the
scaling function $\Phi(\omega)$ in the sense of Eq. (10.6.40) and $F(e^{-j\omega})$ is a polynomial in $e^{-j\omega}$.
Equation (10.6.41) determines the structure of $P(\omega)$, and shows that $P(\omega)$ has M zeros at
$\omega = \pi$.

The length N of the discrete low-pass filter $p(n)$ is chosen first. The length of the discrete
high-pass filter $q(n)$ also is equal to N. If $p(n) \neq 0$ only for $n = 0, 1, \ldots, N-1$, $p(n)$ is said
to have compact support of length N and is called a FIR filter or an N-tap filter. Its Fourier
transform $P(\omega)$ is a polynomial in $e^{-j\omega}$ of degree $N - 1$:

$$P(\omega) = \frac{1}{\sqrt{2}} \sum_{n=0}^{N-1} p_m(n) \exp(-jn\omega)$$

Hence, the second term on the right side of Eq. (10.6.41), the polynomial $F(e^{-j\omega})$ in $e^{-j\omega}$ is
of degree $N - 1 - M$. The quadrature mirror filter $P(\omega)$ and its impulse response $p(n)$ is
determined by the choice of $F(e^{-j\omega})$.

Because all the coefficients of $F(e^{-j\omega})$ are real, $F^*(e^{-j\omega}) = F(e^{j\omega})$, $|F(e^{-j\omega})|^2$ is a
symmetric polynomial and can be rewritten as a polynomial in $\cos \omega$ or, equivalently, as a
polynomial in $\sin^2(\omega/2)$. Therefore, $|F(e^{-j\omega})|^2$ is some polynomial in $\sin^2(\omega/2)$ of degree
$N - 1 - M$ and is rewritten as $G[\sin^2(\omega/2)]$

$$|F(e^{-j\omega})|^2 = G\left(\sin^2 \frac{\omega}{2} \right) \tag{10.7.32}$$

Daubechies [13] considered the square modulus relation for $|P(\omega)|^2$. From Eq. (10.6.41) we have:

$$|P(\omega)|^2 = \left(\cos^2\frac{\omega}{2}\right)^M G\left(\sin^2\frac{\omega}{2}\right) \tag{10.7.33}$$

The orthonormality condition of the scaling function implies for the quadrature mirror filter $P(\omega)$:

$$|P(\omega)|^2 + |P(\omega + \pi)|^2 = 1 \tag{10.6.6}$$

Combining regularity condition (10.7.33) and orthonormality condition (10.6.6) and introducing a variable $y = \cos^2(\omega/2)$, we obtain

$$y^M G(1-y) + (1-y)^M G(y) = 1 \tag{10.7.34}$$

This equation has a solution of the form

$$G(y) = \sum_{k=0}^{M-1} \binom{M-1-k}{k} y^k + y^M R(1-2y) \tag{10.7.35}$$

that is a polynomial in y with the minimum degree $M - 1$, where $R(y)$ is an odd polynomial in y such that

$$R(y) = -R(1-y) \tag{10.7.36}$$

Different choices for $R(y)$ and M lead to different wavelet solutions. According to Eq. (10.7.35), the solution $G(y)$ is a polynomial in y of minimum degree $M - 1$. Daubechies chose the minimum degree $M - 1$ for $G(y)$ and let $R(y) = 0$. The solution $G(y)$ is then a polynomial $|F(e^{-j\omega})|^2$ in $e^{-j\omega}$ of the minimum degree $M - 1$.

$$|F(e^{-j\omega})|^2 = G\left(\sin^2\frac{\omega}{2}\right) = \sum_{k=0}^{M-1} \binom{M-1+k}{k} \sin^{2k}\frac{\omega}{2} \tag{10.7.37}$$

The right-hand side of Eq. (10.7.33) is a polynomial in $\cos^2(\omega/2)$ and $\sin^2(\omega/2)$ of degree $N - 1$. As the polynomial $G(y) = |F(e^{-j\omega})|^2$ in $\sin^2(\omega/2)$ is of degree $M - 1$, the term $[\cos^2(\omega/2)]^M$ would have the maximum value of $M = N/2$. Hence, the regularity M of the scaling function is the maximum. In Eq. (10.7.33) the term $[\cos(\omega/2)]^M$ insures $|P(\omega)|$ to have M vanishing derivatives at $\omega = 0$ and the polynomial $F(e^{-j\omega})$ insures $|P(\omega)|$ to have M vanishing derivatives at $\omega = \pi$. This corresponds to the unique maximally flat magnitude response. According to the relation for the conjugate quadrature mirror filters, Eq. (10.6.22), $Q(\omega)$ has M vanishing derivatives at $\omega = 0$.

Daubechies solved for $P(\omega)$ from $|P(\omega)|^2$ by spectral factorization. The Fourier transform $P(\omega)$ of the Daubechies scaling function is expressed as

$$P(\omega) = \left(\frac{1 + e^{-j\omega}}{2}\right)^M F(e^{-j\omega}) \tag{10.7.38}$$

with the polynomial $F(e^{-j\omega})$ expressed in Eq. (10.7.37).

The Daubechies basis functions are not given in analytical form. The decomposition and reconstruction are implemented by iterating the discrete low-pass and high-pass filters $p(n)$ and $q(n)$. The Daubechies scaling functions and wavelets constitute a family of orthonormal bases with compact support. For a given support, the Daubechies scaling functions and wavelets have the maximum regularity and the maximum number of vanishing moments. The regularity of the Daubechies scaling functions and wavelets increases linearly with the width of their support, that is, with the length N of the discrete filters $p(n)$ and $q(n)$, because $M = N/2$.

The values of the coefficients $P_M(n)$ for the cases $M = 2, 3, \ldots, 10$ are listed in Table 10.2. For the most compact support $M = 2$ and $N = 4$ the discrete low-pass filter is given analytically as

$$p(0) = \frac{1}{4}(1 + \sqrt{3})/\sqrt{2} = 0.483$$

$$p(1) = \frac{1}{4}(3 + \sqrt{3})/\sqrt{2} = 0.836$$

$$p(2) = \frac{1}{4}(3 - \sqrt{3})/\sqrt{2} = 0.224$$

$$p(3) = \frac{1}{4}(1 - \sqrt{3})/\sqrt{2} = -0.13 \tag{10.7.39}$$

The discrete high-pass filter $q(n)$ can be obtained from $p(n)$ by Eq. (10.6.23) where N is an even number.

It is easy to verify that the translates of $p(n)$ and $q(n)$ with double integer steps are orthonormal, respectively. Figure 10.17a and 10.17b show the Daubechies scaling function and wavelet with compact support $M = 2$ and $N = 4$ and their Fourier transforms. The scaling function is generated from the quadrature mirror filter $p(n)$ given in Eq. (10.7.39) with the method of reconstruction discussed in Section 10.5.5. Those functions have the most compact support and are neither very smooth nor regular. When the length of the filters $p(n)$ and $q(n)$ increases, the Daubechies scaling functions and wavelets become more smooth and regular, at the cost of a larger number of non-zero coefficients of $p(n)$ and $q(n)$ that results in large support widths for the scaling functions and wavelets.

Another important feature of Figure 10.17 is the lack of any symmetry or antisymmetry axis for the Daubechies scaling function and wavelet. Daubechies has shown that it is impossible to obtain an orthonormal and compactly supported wavelet that is either symmetric or antisymmetric around any axis, except for the trivial Haar wavelets.

10.8 Fast wavelet transform

The wavelet transform is not ready for explicit calculus. Only for a few simple functions does the wavelet transform have analytical solutions. For most functions the wavelet transforms must be computed in a digital computer. In the multiresolution analysis framework, the orthonormal wavelet transform is implemented by iterating the quadrature mirror filters in the tree algorithm. In the computer, the function to be transformed is discretized, and the iterated quadrature mirror filters are discrete. The orthonormal wavelet series decomposition and reconstruction are essentially discrete. The wavelet tree algorithm permits the fast wavelet transform. One of the main reasons for the recent success of the wavelet transform is the existence of this fast wavelet transform algorithm, which only requires a number $O(L)$ of the operations where L is the size of the initial data.

The discrete wavelet decomposition and reconstruction algorithms were discussed in Sections (10.5.4) and (10.5.5). In this section we need only to introduce the discrete wavelet matrix and discuss the number of operations and the time-bandwidth product of the wavelet transform output.

TABLE 10.2
The low-pass filter of the Daubechies wavelet bases with the support of the filter $N = 2M$ and
$M = 2, 3, \ldots, 10$ [13].

	n	$P_M(n)$		n	$P_M(n)$
$M = 2$	0	.482962913145	$M = 8$	0	.054415842243
	1	.836516303738		1	.312871590914
	2	.224143868042		2	.675630736297
	3	−.129409522551		3	.585354683654
$M = 3$	0	.332670552950		4	−.015829105256
	1	.806891509311		5	−.284015542962
	2	.459877502118		6	.000472484574
	3	−.135011020010		7	.128747426620
	4	−.085441273882		8	−.017369301002
	5	.035226291882		9	−.044088253931
$M = 4$	0	.230377813309		10	.013981027917
	1	.714846570553		11	.008746094047
	2	.630880767930		12	−.004870352993
	3	−.027983769417		13	−.000391740373
	4	−.187034811719		14	.000675449406
	5	.030841381836		15	−.000117476784
	6	.032883011667	$M = 9$	0	.038077947364
	7	−.010597401785		1	.243834674613
$M = 5$	0	.160102397974		2	.604823123690
	1	.603829269797		3	.657288078051
	2	.724308528438		4	.133197385825
	3	.138428145901		5	−.293273783279
	4	−.242294887066		6	−.096840783223
	5	−.032244869585		7	.148540749338
	6	.077571493840		8	.030725681479
	7	−.006241490213		9	−.067632829061
	8	−.012580751999		10	.000250947115
	9	.003335725285		11	.022361662124
$M = 6$	0	.111540743350		12	−.004723204758
	1	.494623890398		13	−.004281503682
	2	.751133908021		14	.001847646883
	3	.315250351709		15	.000230385764
	4	−.226264693965		16	−.000251963189
	5	−.129766867567		17	.000039347320
	6	.097501605587	$M = 10$	0	.026670057901
	7	.027522865530		1	.188176800078
	8	−.031582039318		2	.527201188932
	9	.000553842201		3	.688459039454
	10	.004777257511		4	.281172343661
	11	−.001077301085		5	−.249846424327
$M = 7$	0	.007852054085		6	−.195946274377
	1	.396539319482		7	.127369340336
	2	.729132090846		8	.093057364604
	3	.469782287405		9	−.071394147166
	4	−.143906003929		10	−.029457536822
	5	−.224036184994		11	.033212674059
	6	.071309219267		12	.003606553567
	7	.080612609151		13	−.010733175483
	8	−.038029936935		14	.001395351747
	9	−.016574541631		15	.001992405295
	10	.012550998556		16	−.000685856695
	11	.000429577973		17	−.000116466855
	12	−.001801640704		18	.000093588670
	13	.000353713800		19	−.000013264203

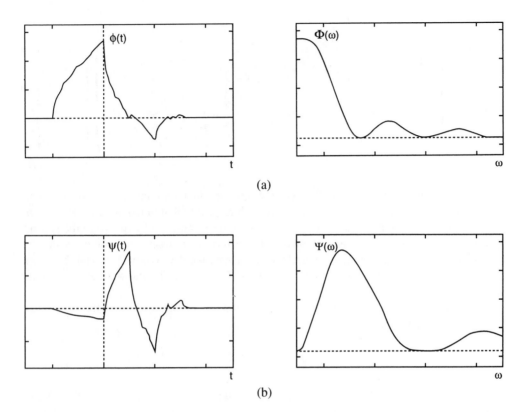

FIGURE 10.17
Daubechies scaling function (a) and wavelet (b) and their Fourier transforms with the compact support
$N = 4$ [13].

10.8.1   *Wavelet matrices*

The discrete orthonormal wavelet transform is a fast linear operation. Given a vector of data
that has a length of an integer power of two, the wavelet decomposition and reconstruction are
numerically computed by two recurring conjugate quadrature mirror filters $p(n)$ and $q(n)$ that
are the FIR filters compactly supported with a finite number N of nonzero coefficients. The
degree of the filters is $N - 1$. Hence, the two sets of coefficients form a $2 \times N$ matrix:

$$\begin{vmatrix} p(0) & p(1) & \dots & p(N-1) \\ q(0) & q(1) & \dots & q(N-1) \end{vmatrix} \tag{10.8.1}$$

The wavelet decomposition and reconstruction are computed with recursive applications of
the wavelet filter bank (10.8.1) in the tree algorithm. In the following, we show an example of
the wavelet decomposition and reconstruction with the Daubechies wavelets of $N = 4$, called
the DAUB4, in the matrix formalization.

Let $f(n)$ be the vector of initial data. We generate a wavelet transform matrix with the
translated discrete filters (see Eq. [10.8.2]) [17].

$$
\begin{pmatrix} c(1) \\ d(1) \\ c(2) \\ d(2) \\ \cdot \\ \cdot \\ \cdot \\ \cdot \\ \cdot \end{pmatrix} = \begin{pmatrix} p(0) & p(1) & p(2) & p(3) & & & \cdot & & \cdot & \cdot \cdot & & \cdot & & \cdot & & \cdot \\ p(3) & -p(2) & p(1) & -p(0) & & & \cdot & & & \cdot \cdot & & \cdot & & \cdot & & \cdot \\ \cdot & & \cdot & p(0) & p(1) & p(2) & p(3) & \cdot \cdot & \cdot & & \cdot & & \cdot & & \cdot \\ \cdot & & \cdot & p(3) & -p(2) & p(1) & -p(0) & \cdot \cdot & & \cdot & & \cdot & & \cdot \\ \cdot & & \cdot & \cdot & & \cdot & & \cdot & \cdot & \cdot & \cdot & & \cdot \\ \cdot & & \cdot & \cdot & & \cdot & & \cdot & \cdot & \cdot & \cdot & & \cdot \\ \cdot & & \cdot & \cdot & & \cdot & \cdot \cdot & p(0) & p(1) & p(2) & p(3) \\ \cdot & & \cdot & \cdot & & \cdot & \cdot \cdot & p(3) & -p(2) & p(1) & -p(0) \\ p(2) & p(3) & \cdot & & \cdot & & \cdot & \cdot \cdot \cdot & \cdot & p(0) & p(1) \\ p(1) & -p(0) & \cdot & & \cdot & & \cdot & \cdot \cdot \cdot & \cdot & p(3) & -p(2) \end{pmatrix} \begin{pmatrix} f(1) \\ f(2) \\ f(3) \\ f(4) \\ \cdot \\ \cdot \\ \cdot \\ \cdot \\ f(N) \end{pmatrix}
$$

$$(10.8.2)$$

In the wavelet transform matrix the odd rows are the low-pass filters $p(n)$. The low-pass filter $p(n)$ in the third row is translated by two with respect to that in the first row, and so on. The even rows are the high-pass filters $q(n)$ that are also translates by two from the second row, to the fourth row and so on. The wavelet transform matrix acts on a column vector of data, $f(n)$, resulting in two related correlations between the data vector $f(n)$ and the filters $p(n)$ and $q(n)$, that are the discrete approximation $c(n)$ and the discrete wavelet coefficients $d(n)$, respectively.

The high-pass filter $q(n)$ is obtained from the low-pass filter $p(n)$ from the cross-filter orthogonality Eq. (10.6.23)

$$
q(n) = (-1)^n p(N - 1 - n)
$$

Let $N = 4$ and $n = 0, 1, 2, 3$. We have $q(0) = p(3)$, $q(1) = -p(2)$, $q(2) = p(1)$ and $q(4) = -p(0)$.

It is easy to verify that the low-pass filter $p(n)$ is a smoothing filter, satisfying Eq. (10.6.51). With the coefficients of the Daubechies bases given in Eq. (10.7.39) we have

$$
p(0)^2 + p(1)^2 + p(2)^2 + p(3)^2 = 1 \tag{10.8.3}
$$

The high-pass filter $q(n)$ has the property that

$$
p(3) - p(2) + p(1) - p(0) = 0 \tag{10.8.4}
$$

corresponding to Eq. (10.6.31): $Q(0) = 0$, obtained from the orthonormality condition, and

$$
0p(3) - 1p(2) + 2p(1) - 3p(0) = 0 \tag{10.8.5}
$$

corresponding to Eq. (10.6.55) obtained for the regularity of the wavelets.

It is easy to see that in the wavelet transform matrix the orthonormality between the double integer translates of $p(n)$ and those of $q(n)$ and the cross-filter orthogonality between $p(n)$ and $q(n)$ because

$$
p(2)p(0) + p(3)p(1) = 0 \tag{10.8.6}
$$

It is also possible to reconstruct the original data $f(n)$ of length L from the approximation sequence $c(n)$ and the wavelet coefficients $d(n)$; both sequences are of the length $L/2$. According to Eqs. (10.8.3)–(10.8.6) we see that the wavelet transform matrix in Eq. (10.8.2) is orthonormal, so that its inverse is just the transposed matrix

$$
\begin{pmatrix}
p(0) & p(3) & \cdot & \cdot & \cdot\,\cdot & \cdot & \cdot & \cdot & \cdot & p(2) & p(1) \\
p(1) & -p(2) & \cdot & \cdot & \cdot\,\cdot & \cdot & \cdot & \cdot & \cdot & p(3) & -p(0) \\
p(2) & p(1) & p(0) & p(3) & \cdot\,\cdot & \cdot & \cdot & \cdot & \cdot & \cdot & \cdot \\
p(3) & -p(0) & p(1) & -p(2) & \cdot\,\cdot & \cdot & \cdot & \cdot & \cdot & \cdot & \cdot \\
\cdot & \cdot & \cdot & \cdot & \cdot\,\cdot & \cdot & \cdot & \cdot & \cdot & \cdot & \cdot \\
\cdot & \cdot & \cdot & \cdot & \cdot\,\cdot & p(2) & p(1) & p(0) & p(3) & \cdot & \cdot \\
\cdot & \cdot & \cdot & \cdot & \cdot\,\cdot & p(3) & -p(0) & p(1) & -p(2) & \cdot & \cdot \\
\cdot & \cdot & \cdot & \cdot & \cdot\,\cdot & \cdot & p(2) & p(1) & p(0) & p(3) & \cdot \\
\cdot & \cdot & \cdot & \cdot & \cdot\,\cdot & \cdot & p(3) & -p(0) & p(1) & -p(2) &
\end{pmatrix}
\tag{10.8.7}
$$

The discrete wavelet decomposition is computed by applying the wavelet transform matrix with operation (10.8.2) hierarchically with the down-sampling by a factor of two after each iteration. The down-sampling by two is implemented by a permutation of the output vector in the left-hand side of Eq. (10.8.2) as shown in diagram (10.8.8) where $N = 16$.

$$
\begin{pmatrix}
f(1)\\ f(2)\\ f(3)\\ f(4)\\ f(5)\\ f(6)\\ f(7)\\ f(8)\\ f(9)\\ f(10)\\ f(11)\\ f(12)\\ f(13)\\ f(14)\\ f(15)\\ f(16)
\end{pmatrix}
\xrightarrow[\;]{(10.8.2)}
\begin{pmatrix}
c(1)\\ d(1)\\ c(2)\\ d(2)\\ c(3)\\ d(3)\\ c(4)\\ d(4)\\ c(5)\\ d(5)\\ c(6)\\ d(6)\\ c(7)\\ d(7)\\ c(8)\\ d(8)
\end{pmatrix}
\xrightarrow[\;]{\text{permute}}
\begin{pmatrix}
c(1)\\ c(2)\\ c(3)\\ c(4)\\ c(5)\\ c(6)\\ c(7)\\ c(8)\\ d(1)\\ d(2)\\ d(3)\\ d(4)\\ d(5)\\ d(6)\\ d(7)\\ d(8)
\end{pmatrix}
\xrightarrow[\;]{(10.8.2)}
\begin{pmatrix}
C(1)\\ D(1)\\ C(2)\\ D(2)\\ C(3)\\ D(3)\\ C(4)\\ D(4)\\ d(1)\\ d(2)\\ d(3)\\ d(4)\\ d(5)\\ d(6)\\ d(7)\\ d(8)
\end{pmatrix}
\xrightarrow[\;]{\text{permute}}
\begin{pmatrix}
C(1)\\ C(2)\\ C(3)\\ C(4)\\ D(1)\\ D(2)\\ D(3)\\ D(4)\\ d(1)\\ d(2)\\ d(3)\\ d(4)\\ d(5)\\ d(6)\\ d(7)\\ d(8)
\end{pmatrix}
\xrightarrow[\;]{\text{etc.}}
\begin{pmatrix}
C(1)\\ C(2)\\ D'(1)\\ D'(2)\\ D(1)\\ D(2)\\ D(3)\\ D(4)\\ d(1)\\ d(2)\\ d(3)\\ d(4)\\ d(5)\\ d(6)\\ d(7)\\ d(8)
\end{pmatrix}
\tag{10.8.8}
$$

If the length of the data vector $N > 16$ there will be more stages of applying Eq. (10.8.2) and permuting. The final output vector will always be a vector with two approximation coefficients $C(1)$ and $C(2)$ at the lowest resolution, and a hierarchy of the wavelet coefficients $D'(1)$, $D'(2)$ for the lowest resolution, $D(1)$–$D(4)$ for higher resolution, and $d(1)$–$d(8)$ for still higher resolution, and so on. Notice that once the wavelet coefficients d' are generated they simply propagate through to all subsequent stages without further computation.

The discrete wavelet reconstruction can be computed by a simple reversed procedure, starting with the lowest resolution level in the hierarchy and working from right to left with diagram (10.8.8). The inverse wavelet transform matrix, shown in Eq. (10.8.7), is used instead of Eq. (10.8.2).

The above wavelet transform matrix method shows a clear figure of the discrete wavelet decomposition and reconstruction. The wavelet transform can also be computed with other methods iterating the discrete filters in the tree algorithms without using the wavelet transform matrix.

10.8.2   Number of operations

Let us now consider the number of operations required for the discrete orthonormal wavelet transform of a vector of data. Let L be the length of the data vector and N the length of the FIR filters $p(n)$ and $q(n)$. The wavelet transform is a local operation, usually $N \ll L$. At the highest frequency band the first stage of decomposition requires $2NL$ multiplications and additions.

In the tree algorithm at the next coarser frequency band the length of the vector of the discrete approximation $c(n)$ is reduced to $N/2$. Therefore the next stage of decomposition requires $2(NL/2)$ multiplications and additions. The total number of operations of the orthonormal wavelet decomposition is then

$$2\left(NL + \frac{NL}{2} + \frac{NL}{4} + \cdots\right) = 2NL\left(1 + \frac{1}{2} + \frac{1}{4} + \cdots\right) \sim 4NL \qquad (10.8.9)$$

The orthonormal wavelet transform requires only $O(L)$ computations. This is even faster than the FFT for the Fourier transform, which requires $O(L \log_2 L)$ multiplication and addition, due to its global nature.

10.8.3 *Time-bandwidth product*

The wavelet transform is a mapping of a function of time, in the one-dimensional case, to the two-dimensional time-scale joint representation. At first glance the time-bandwidth product of the wavelet transform output would be squared of that of the signal. In the multiresolution analysis framework, however, the size of the data vector is reduced by a factor of two in moving from one frequency band to the next coarser resolution frequency band. The time-bandwidth product also is reduced by a factor of two. If the original data vector $c_0(n)$ has L samples, in the tree algorithm for the wavelet decomposition shown in Figure 10.11 the first stage wavelet coefficients outputs $d_1(n)$ has $L/2$ samples, that of the second stage has $L/4$ samples, and so on. Let the length of the data vector $L = 2^K$; the total time-bandwidth product of the wavelet decomposition including all the wavelet coefficients $d_i(n)$ with $i = 1, 2, \ldots, K - 1$ and the lowest resolution approximation $c_{k-1}(n)$ is equal to

$$L\left(\frac{1}{2} + \frac{1}{4} + \cdots\right) \sim L \qquad (10.8.10)$$

There is even a slight decrease of the time-bandwidth product by the orthonormal wavelet transform.

10.9   Applications of the wavelet transform

In this section we present some basic applications of the wavelet transform for multiresolution transient signal analysis and detection, image edge detection, image compression, with some simple examples.

10.9.1 *Multiresolution signal analysis*

In this subsection we show an example of multiresolution analysis for a simple transient signal. Transient signals in the power system are nonstationary time-varying voltage and current signals that can occur as a result of changes in the electrical configuration, in industrial and residential loads, and of a variety of disturbances on transmission lines including capacitor switching, lightning strikes, and short-circuits. The waveform data of the transient signals are captured by digital transient recorders. Analysis and classification of the power system disturbance can help to provide more stability and efficiency in power delivery by switching transmission lines to supply additional current or switching capacitor banks to balance inductive loads, and help to prevent system failures.

The power system transient signals contain a range of frequencies from a few hertz to impulse components with microsecond rise times. The normal 60 Hz sinusoidal voltage and current waveforms are interrupted or superimposed with impulses, oscillations, and reflected waves. An experienced power engineer can visually analyze the waveform data in order to determine the type of system disturbance. However, the Fourier analysis with its global operation nature is not as appropriate for the transient signals as the time-scale joint representation provided by the wavelet transform.

Multiresolution wavelet decomposition of transient signals

The wavelet transform provides a decomposition of power system transient signals into meaningful components in multiple frequency bands, and the digital wavelet transform is computationally efficient [18]. Figure 10.1 in Section 10.1 shows the wavelet components in the multiple frequency bands. At the top is the input voltage transient signal. There is a disturbance of a capacitor bank switching on a three-phase transmission line. Below the first line are the wavelet components as a function of the scale and time shift. The scales of the discrete wavelets increase by a factor of two successively from SCALE 1 to SCALE 64, corresponding to the dyadic frequency bands. The vertical axis in each discrete scale is the normalized magnitude of the signal component in voltage. The three impulses in high-frequency band SCALE 1 correspond to the successive closing of each phase of the three-phase capacitor bank. SCALE 2 and SCALE 4 are the bands of system response frequencies. SCALE 4 contains the most energy from the resonant frequency caused by the addition of a capacitor bank to a primarily inductive circuit. The times of occurrence of all those components can be determined on the time axis. SCALE 64 contains the basic signal of continuous 60 Hz.

The wavelet analysis decomposes the power system transient into meaningful components, whose modulus maxima then can be used for further classification. The nonorthogonal multiresolution analysis wavelets with FIR quadratic spline wavelet filters were used in this example of the application.

Shift invariance

One problem in this application and many other applications with the orthogonal wavelet transform is the lack of shift invariance. The orthonormal wavelet transform is not shift invariant. When the input signal is shifted by one sampling interval distance the output of the orthonormal wavelet transform is not simply shifted by the same distance, but the values of the wavelet coefficients would be changed. Let the sampling rate of the original data be equal to one. Then, the low-pass and high-pass quadrature mirror filters have the translation steps of two, as described by Eq. (10.6.3). This is because the multiresolution wavelet decomposition is computed as correlations between the data and the filters down-sampled by a factor of two. Therefore, the shift of one pixel of the original data will change completely the results of the wavelet transform. This is a disadvantage of the orthonormal wavelet transform, because many applications such as real-time signal analysis and pattern recognition require shift invariant wavelet transforms. In the above application the orthonormal quadrature mirror filters have been found sensitive to translations of the input. Hence, nonorthonormal quadratic spline wavelets have been used.

10.9.2 *Signal detection*

The detection of weak signals embedded in a stronger stationary stochastic process, such as the detection of radar and sonar signals in zero-mean Gaussian white noise, is a well-studied problem. If the shape of the expected signal is known, the correlation and the matched filter provide an optimum solution in terms of the signal-to-noise ratio in the output correlation.

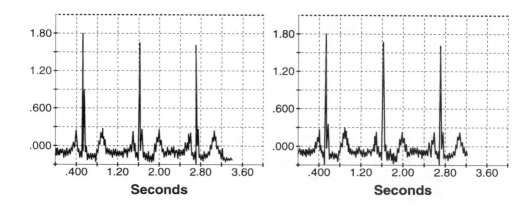

FIGURE 10.18
Left: Normal electrocardiogram; Right: Electrocardiogram with VLP abnormality [19].

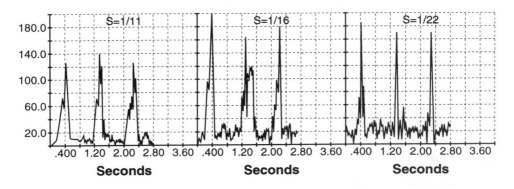

FIGURE 10.19
Wavelet transform of the abnormal electrocardiogram for scale factor $s = 11, 16, 22$. The bulge to the right of the second QRS peak for $s = 1/16$ indicates the presence of the VLP [19].

In the detection of speech or biomedical signals, the exact shape of the signal is unknown. The Fourier spectrum analysis could be effective for those applications, only when the expected signal has spectral features that clearly distinguish it from the noise. The effectiveness of the Fourier spectrum analysis is generally proportional to the ratio of the signal-to-noise energy. For short-time, low-energy transients, the change in the Fourier spectrum is not easily detected. Such transient signals can be detected by the wavelet transform. An example of electrocardiogram signal detection follows [19].

Figure 10.18 shows the clinical electrocardiogram with normal QRS peaks and an abnormality called ventricular late potentials (VLP) right after the second QRS peak. The amplitude of the VLP signal is about 5% of the QRS peaks. Its duration was about 0.1 second, or a little less than 10% of the pulse period. The VLPs are weak signals, swamped by noise, and they occur somewhat randomly.

Figure 10.19 shows the magnitude of continuous wavelet transform with the cos-Gaussian wavelets of scale $s = 1/11, 1/16$, and $1/22$. The peak after the second QRS spike observed for $s = 1/16$ is very noticeable and gives a clear indication of the presence of the VLP.

10.9.3   *Image edge detection*

Edges and boundaries, representing shapes of objects and intersections between surfaces and between textures, are among the most important features of images, useful for image segmentation and pattern recognition. An edge in an image is a set of locally connected pixels that are characterized by sharp intensity variation in their neighborhood. Edges are local features of an image.

The wavelet transform is a local operation. The wavelet transform of a constant is equal to zero and the wavelet transform of a function $f(t) = t^n$ is also equal to zero if the Fourier transform of the wavelet has the zero of order $n + 1$ about the frequency $\omega = 0$, as described in Section 10.2.3. Hence, the wavelet transform is useful for detecting singularities of functions and edges of images.

Edge detectors

The edge detectors first smooth an image at various scales and then detect sharp variation from the first- or second-order derivative of the smoothed images. The extrema of the first-order derivative correspond to the zero crossing of the second-order derivative and to the inflection points of the image.

A simple example of edge detectors is the first- or second-order derivatives of the Gaussian function $g(x, y)$. The Gaussian function $g_s(x, y)$ is scaled by a factor s. The first- and second-order derivatives, that is, the gradient and the Laplacian of $g_s(x, y)$ are the wavelets satisfying the wavelet admissible condition, as described in Section 10.2.4.

By definition the wavelet transform of an image $f(x, y)$ is the correlation between $f(x, y)$ and the scaled wavelets. We derive that

$$W_f(s; x, y) = f * (s\nabla g_s) = s\nabla(f * g_s)(x, y) \tag{10.9.1}$$

with the first-order derivative wavelet and that

$$W_f(x, y) = f * (s^2\nabla^2 g_s) = s^2\nabla^2(f * g_s)(x, y) \tag{10.9.2}$$

for the second-order derivative wavelet, which is the Mexican-hat wavelet. The wavelet transform is then the gradient or the Laplacian of the image smoothed by the Gaussian function $g_s(x, y)$ at the scale s.

The local extrema of the wavelet transform with the wavelet of the gradient of the Gaussian correspond to the zero-crossing of the wavelet transform with the wavelet of the Laplacian of the Gaussian and to the inflection points of the smoothed image $f * g_s(x, y)$. The first one is the Canny edge detector and the second one is the zero-crossing edge detector.

Two-dimensional wavelet transform

The continuous wavelet transform can be easily extended to a two-dimensional case for image processing applications. The wavelet transform of a two-dimensional image $f(x, y)$ is

$$W_f(s_x, s_y; u, v) = \frac{1}{\sqrt{s_x s_y}} \int \int f(x, y) \psi\left(\frac{x - u}{s_x}; \frac{y - v}{s_y}\right) dx\, dy \tag{10.9.3}$$

which is a four-dimensional function. It is reduced to a set of two-dimension functions of (u, v) with different scales, when the scale factors $s_x = s_y = s$. When $\psi(x, y) = \psi(r)$ with $r = (x^2 + y^2)^{1/2}$, the wavelets are isotropic and have no selectivity for spatial orientation. Otherwise, the wavelet can have particular orientation. The wavelet can also be a combination of the two-dimensional wavelets with different particular orientations, so that the two-dimensional wavelet transform has orientation selectivity.

The multiresolution orthogonal wavelet transform is computed by recursive projections on the orthonormal scaling function bases and wavelet bases. Let us consider a wavelet model that is based on a separable scaling function,

$$\phi(x, y) = \phi(x)\phi(y) \tag{10.9.4}$$

where $\phi(x)$ and $\phi(y)$ are one-dimensional scaling functions. The discrete translations of the dilated $\phi(x)$ and $\phi(y)$ generate the separable multiresolution approximation subspaces V_i as in the one-dimensional case. Extra importance is given to the horizontal and vertical directions in the image. The orthogonal projection of an image $f(x, y)$ on the set of scaling function at a resolution level i is therefore the inner products

$$c_i(x, y) = \langle f(x, y), \phi_i(x)\phi_i(y) \rangle \tag{10.9.5}$$

which is a coarse resolution approximation of $f(x, y)$.

The projection on the set of discrete translations of the scaling function is a correlation. To handle the problem of computing the correlation on the border of an image, we extend the original image such that the image is symmetric with respect to the horizontal and vertical borders of the original image.

As in the one-dimensional case, we generate the wavelet $\psi(x)$ and $\psi(y)$ from the scaling function $\phi(x)$ and $\phi(y)$ such that the set of discrete translations of $\psi(x)$ and of $\psi(y)$ are orthogonal to the set of discrete translations of $\phi(x)$ and $\phi(y)$, respectively. Then we define three two-dimensional wavelets as

$$\psi^1(x, y) = \phi(x)\psi(y)$$

$$\psi^2(x, y) = \psi(x)\phi(y)$$

$$\psi^3(x, y) = \psi(x)\psi(y) \tag{10.9.6}$$

The differences of information between the approximations $c_i(x, y)$ and $c_{i+1}(x, y)$ at two adjacent resolution levels are equal to the orthogonal projections of $f(x, y)$ on the three wavelet bases, which gives three detail images:

$$d_i^1(x, y) = \langle f, \psi^1 \rangle$$

$$d_i^2(x, y) = \langle f, \psi^2 \rangle$$

$$d_i^3(x, y) = \langle f, \psi^3 \rangle \tag{10.9.7}$$

In the two-dimensions, the wavelet decomposition with separable scaling functions and wavelets can be computed with the tree algorithm using the quadrature mirror filters $p(n)$ and $q(n)$, similar to the one-dimensional algorithm, described in Section 10.5.4. We first compute the correlation between rows of $c_{i-1}(x, y)$ with the one-dimensional filters $p(n)$ and $q(n)$ in the vertical direction. The two resulting images are then subsampled by a factor of two by retaining every other row. Then we compute the correlation between the resulting images and another $p(n)$ and $q(n)$ in the horizontal direction and retain every other column. The resulting images are then analyzed at the next lower frequency band by applying the filter bank recursively. The process is performed with a tree structure, as shown by Figure 10.20 [16].

Figure 10.21 shows a disposition of the three detail images and one approximation image at the resolution levels, 1, 2, 3. If the original image has L^2 pixels, each image $c_i(x, y)$, $d_i^1(x, y)$, $d_i^2(x, y)$ and $d_i^3(x, y)$ has $(N/2^i)^2$ pixels $(i > 0)$. The total number of pixels of an orthonormal wavelet representation is therefore still equal to L^2. The wavelet transform does not increase the volume of data, as discussed in Section 10.8. This is owing to the orthonormality of the

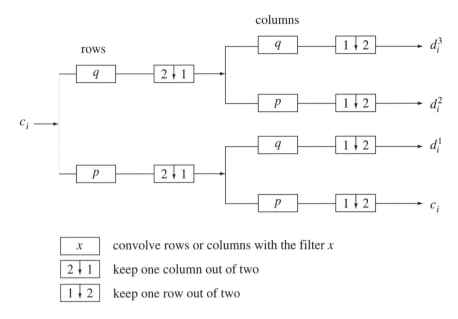

FIGURE 10.20
Schematic two-dimensional wavelet decomposition with quadrature mirror low-pass and high-pass
filters $p(n)$ and $q(n)$.

wavelet decomposition. More general nonseparable two-dimensional wavelet models are more
difficult to implement and are computationally more expensive.

Multiscale edges

The wavelet transform of a two-dimensional image for edge detection is performed at a set of
dyadic scales, generating a set of detail images. Similar to the reconstruction process described
in Section 10.5.5, the detail images from the wavelet decomposition can be used to reconstruct
the original image.

If the wavelet is the first-order derivative of the Gaussian smoothing function and we retain
the modulus maxima of the wavelet components then we obtain the edge images that correspond
to the maximum variation of the smoothed image at a scale s. The multiscale edge information
provided by the wavelet transform can also be used to analyze the regularity of the image
function by observing the propagation of the modulus maxima from one scale to the next
scale. A similar approach is used to reduce the noise in the image, because the noise has
different regularity than that of the image and image edges.

10.9.4 *Image compression*

Image compression uses fewer bits to represent the image information for different purposes,
such as image storage, image transmission, and feature extraction. The general idea behind it
is to remove the redundancy in an image to find more compact representation.

A popular method for image compression is so-called transform coding, which represents the
image in a different basis such that its coefficients are decorrelated. We saw in Section 10.5.4
that the multiresolution wavelet decomposition is projections onto subspaces spanned by the
scaling function basis and the wavelet basis. The projections on the scaling function basis
yield approximations of the signal and the projections on the wavelet basis yield the differences

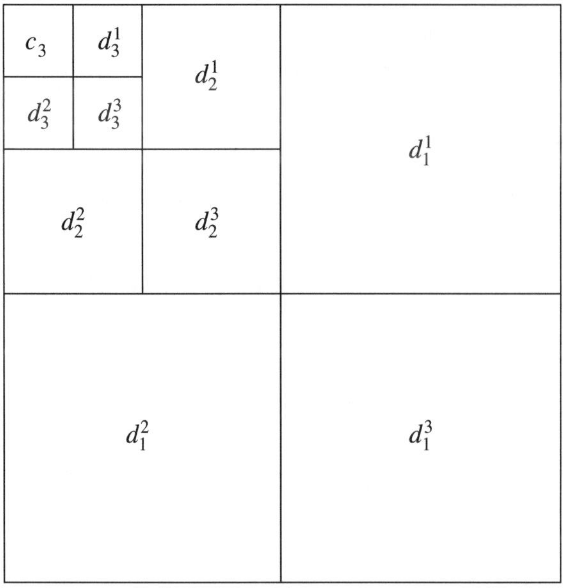

FIGURE 10.21
Presentation of the two-dimensional wavelet decomposition. The detail images d_i^n and the approximation c_i are defined by Eqs. (10.9.7) and (10.9.5).

between the approximations at two adjacent resolution levels. Therefore, the wavelet detail images are decorrelated and can be used for image compression. Indeed, the detail images obtained from the wavelet transform consist of edges in the image. There is little correlation among the values on pixels in the edge images.

One example of an image compression application is the compression of gray-scale fingerprint images using the wavelet transform [20]. The fingerprint images are captured as 500 pixels per inch and 256 gray levels. The wavelet subband decomposition is accomplished by the tree algorithm described by Figure 10.20. The dominant ridge frequency in fingerprint images is in roughly $\omega = \pi/8$ up to $\omega = \pi/4$ bands. Because the wavelet decomposition removes the correlation among image pixels, only the wavelet coefficients with large magnitude are retained. The wavelet decomposition uses pairs of symmetric biorthogonal wavelet filters with seven and nine taps.

The retained wavelet coefficients are subsequently coded according to a scalar quantizer and are mapped to a set of 254 symbols for Huffman encoding using the classical image coding technique. The wavelet filter values, the quantization rule, and the Huffman code table are included with the compressed images, so that a decoder can reconstruct approximations of the original images by performing the inverse wavelet transform. After compression at 20:1, the reconstructed images conserve the ridge features, ridge ending, or bifurcations that are definitive information useful for determination.

References

[1] Gabor, D. 1946. "Theory of communication." *J. Inst. Elec. Eng.*, 93, pp. 429–457.

[2] Wigner, E. 1932. "On the quantum correction for thermodynamic equilibrium." *Phys.*

Rev., 40, pp. 749–759.

[3] Rihaczek, A. W. 1969. *Principles of High Resolution Radar*, McGraw-Hill, New York.

[4] Gasquet, C., and Witomski, P. 1990. *Analyse de Fourier et Applications*, Chap. XII, Masson, Paris.

[5] Harr, A. 1910. Ph.D. dissertation, appendix.

[6] Martinet, R. K., Morlet, J., and Grossmann, A. 1987. "Analysis of sound patterns through wavelet transforms." *Intonational J. of Pattern Recogn. and Artificial Intellig.*, 1(2), pp. 273–302.

[7] Marr, D., and Hildreth, E. 1980. "Theory of edge detection." *Proc. Roy. Soc. London Ser. B*, 207, pp. 187–217.

[8] Daubechies, I. 1990. "The wavelet transform: time-frequency localization and signal analysis." *IEEE Trans. on Inf. Theory*, 36(5), 961–1005.

[9] Mallat, S. G. 1989. "Multifrequency channel decompositions of images and wavelet models." *Trans. IEEE on Acous. Speech and Sig. Proc. ASSP-37*, 12, pp. 2091–2110.

[10] Burt, P. J., and Adelson, E. H. 1983. "The Laplacian pyramid as a compact image code." *IEEE Trans. on Comm.*, 31(4), 532–540.

[11] Akansu, A. N., and Haddad, R. A. 1992. *Multiresolution Signal Decomposition*. Academic, Boston.

[12] Strang, G. 1989. "Wavelets and dilation equations: A brief introduction." *SIAM Review*, 31(4), 614–627.

[13] Daubechies, I. 1988. "Orthonormal bases of compactly supported wavelets." *Commun. on Pure and Appl. Math.*, XLI, pp. 909–996.

[14] Chui, C. K. (Ed.) 1992. *Wavelets: A Tutorial Theory and Applications*, Vol. 2, Academic, Boston.

[15] Lemarie, P. G., and Meyer, Y. 1986. "Ondelettes et bases Hibertiennes." *Revista Matematica Iberoamericana*, 1Y2, 2.

[16] Mallat, S. 1989. "A theory for multiresolution signal decomposition: The wavelet representation." *IEEE Trans. Pattern Anal. Machine Intell.*, PAMI-31, pp. 674–693.

[17] Press, W. H., et al. 1992. *Numerical Recipes*, 2nd edn., Chap. 13, Cambridge, London.

[18] Robertson, D. C., Camps,g O. I., and Mayer, J. 1994. "Wavelets and power system transients: Feature detection and classification." *Proc. SPIE*, 2242, pp. 474–487.

[19] Combes, J. M., Grosmann, A., and Tchamitchian, Ph. (Eds). 1990. *Wavelets*, 2nd edn., Springer-Verlag, Berlin.

[20] Hopper, T. 1994. "Compression of gray-scale fingerprint images." *Proc. SPIE*, 2242, pp. 180–187.

[21] Sheng, Y., Roberge, D., and Szu, H. 1992. "Optical wavelet transform." *Opt. Eng.*, 31, pp. 1840–1845.

[22] Szu, H., Sheng, Y., and Chen, J. 1992. "The wavelet transform as a bank of matched filters." *Applied Optics* 31(17), pp. 3267–3277.

[23] Freeman, M. O. 1995. "Wavelet signal representations with important advantages." *Photonics New*, August, pp. 8–14.

11

The Mellin Transform

Jacqueline Bertrand, Pierre Bertrand, and Jean-Philippe Ovarlez

CONTENTS

11.1 Introduction

In contrast to Fourier and Laplace transformations that were introduced to solve physical problems, Mellin's transformation arose in a mathematical context. In fact, the first occurrence of the transformation is found in a memoir by Riemann in which he used it to study the famous Zeta function. References concerning this work and its further extension by M. Cahen are given in [1]. However, it is the Finnish mathematician R. H. Mellin (1854–1933) who was the first to give a systematic formulation of the transformation and its inverse. Working in the theory of special functions, he developed applications to the solution of the hypergeometric differential equations and to the derivation of asymptotic expansions. The Mellin contribution gives a prominent place to the theory of analytic functions and relies essentially on Cauchy's theorem and the method of residues. A biography of R. H. Mellin including a sketch of his works can be found in [2]. Actually, the Mellin transformation can also be placed in another framework, which in some respects conforms more closely to the original ideas of Riemann. In this approach, the transformation is seen as a Fourier transformation on the multiplicative group of positive real numbers (i.e., group of dilations), and its development parallels the group-theoretical presentation of the usual Fourier transform [3, 4]. One of the merits of this alternative presentation is to emphasize the fact that the Mellin transformation corresponds to an isometry between Hilbert spaces of functions.

Besides its use in mathematics, Mellin's transformation has been applied in many different areas of physics and engineering. Maybe the most famous application is the computation of the solution of a potential problem in a wedge-shaped region where the unknown function (e.g., temperature or electrostatic potential) is supposed to satisfy Laplace's equation with given boundary conditions on the edges. Another domain where Mellin's transformation has proved useful is the resolution of linear differential equations in $x(d/dx)$ arising in electrical

0-8493-8342-0/96/$0.00 + $0.50
©1996 by CRC Press, Inc.

engineering by a procedure analogous to Laplace's. More recently, traditional applications have been enlarged and new ones have emerged. A new impulse has been given to the computation of certain types of integrals by O. I. Marichev [5], who has extended the Mellin method and devised a systematic procedure to make it practical. The alternative approach to Mellin's transformation involving the group of dilations has specific applications in signal analysis and imaging techniques. Used in place of Fourier's transform when scale invariance is more relevant than shift invariance, Mellin's transform suggests new formal treatments. Moreover, a discretized form can be set up and allows the fast numerical computation of general expressions in which dilated functions appear, such as wavelet coefficients and time-frequency transforms.

This chapter is divided into two parts that can be read independently. The first part (Section 2) deals with the introduction of the transformation as a holomorphic function in the complex plane, in a manner analogous to what is done with Laplace's transform. The definition of the transform is given in Section 11.2.1; its properties are described in detail and illustrated by examples. Emphasis is put in Section 11.2.1.6 on inversion procedures that are essential for a practical use of the transform. The applications considered in this first part (Section 11.2.2) are all well known: summation of series, computation of integrals depending on a parameter, solution of differential equations, and asymptotic expansions.

The second part (Section 11.3), which is especially oriented towards signal analysis and imaging, deals with the introduction of the Mellin transform from a systematic study of dilations. In Section 11.3.1, some notions of group theory are recalled in the special case of the group of positive numbers (dilation group) and Mellin's transformation is derived together with properties relevant to the present setting. The discretization of the transformation is performed in Section 11.3.2. A choice of practical applications is then presented in Section 11.3.3.

11.2   The classical approach and its developments

11.2.1   *Generalities on the transformation*

11.2.1.1   Definition and relation to other transformations

DEFINITION 11.2.1   *Let $f(t)$ be a function defined on the positive real axis $0 < t < \infty$. The Mellin transformation $\mathcal{M}$ is the operation mapping the function f into the function F defined on the complex plane by the relation*

$$\mathcal{M}[f; s] \equiv F(s) = \int_0^\infty f(t)\, t^{s-1}\, dt \tag{2.1}$$

The function $F(s)$ is called the Mellin transform of f. In general, the integral exists only for complex values of $s = a + jb$ such that $a_1 < a < a_2$, where a_1 and a_2 depend on the function $f(t)$ to transform. This introduces what is called the *strip of definition* of the Mellin transform that will be denoted by $S(a_1, a_2)$. In some cases, this strip may extend to a half-plane ($a_1 = -\infty$ or $a_2 = +\infty$) or to the whole complex s-plane ($a_1 = -\infty$ and $a_2 = +\infty$).

Example 11.2.1
Consider

$$f(t) = H(t - t_0)\, t^z \tag{2.2}$$

where H is Heaviside's step function, t_0 is a positive number, and z is complex. The Mellin transform of f is given by

$$\mathcal{M}[f; s] = \int_{t_0}^{\infty} t^{z+s-1} dt = -\frac{t_0^{z+s}}{z+s} \tag{2.3}$$

provided s is such that $\mathrm{Re}(s) < -\mathrm{Re}(z)$. In this case the function $F(s)$ is holomorphic in a half-plane.   ∎

Example 11.2.2
The Mellin transform of the function

$$f(t) = e^{-pt}, \qquad p > 0 \tag{2.4}$$

is equal, by definition, to

$$\mathcal{M}[f; s] = \int_0^{\infty} e^{-pt} t^{s-1} dt \tag{2.5}$$

Using the definition (see Appendix A) of the Gamma function, we obtain

$$\mathcal{M}[f; s] = p^{-s} \Gamma(s) \tag{2.6}$$

Recalling that the Gamma function is analytic in the region $\mathrm{Re}(s) > 0$, we conclude that the strip of holomorphy is a half-plane as in the first example.   ∎

Example 11.2.3
Consider the function

$$f(t) = (1+t)^{-1} \tag{2.7}$$

Its Mellin transform can be computed directly using the calculus of residues. But another method consists in changing variables in (2.1) from t to x defined by

$$t + 1 = \frac{1}{1-x}, \qquad x = \frac{t}{t+1}, \qquad dx = \frac{dt}{(t+1)^2} \tag{2.8}$$

The transform of (2.7) is then expressed by

$$\mathcal{M}[f; s] = \int_0^1 x^{s-1} (1-x)^{-s} dx \tag{2.9}$$

with the condition $0 < \mathrm{Re}(s) < 1$. This integral is known (Appendix A) to define the beta function $B(s, 1-s)$, which can also be written in terms of Gamma functions (see also Chapter 1, Section 2). The result is given by the expression

$$\mathcal{M}[f; s] = B(s, 1-s)$$

$$= \Gamma(s)\Gamma(1-s) \tag{2.10}$$

which is analytic in the strip of existence of (2.9). An equivalent formula is obtained using a property (Appendix A) of the Gamma function:

$$\mathcal{M}[f; s] = \frac{\pi}{\sin \pi s} \tag{2.11}$$

valid in the same strip.   ∎

Relation to Laplace and Fourier transformations   Mellin's transformation is closely related to an extended form of Laplace's. The change of variables defined by

$$t = e^{-x}, \qquad dt = -e^{-x} dx \tag{2.12}$$

transforms the integral (2.1) into

$$F(s) = \int_{-\infty}^{\infty} f(e^{-x}) e^{-sx} dx \tag{2.13}$$

After the change of function,

$$g(x) \equiv f(e^{-x}) \tag{2.14}$$

one recognizes in (2.13) the *two-sided* Laplace $\mathcal{L}$ transform of g usually defined by

$$\mathcal{L}[g; s] = \int_{-\infty}^{\infty} g(x) e^{-sx} dx \tag{2.15}$$

This can be written symbolically as

$$\mathcal{M}[f(t); s] = \mathcal{L}[f(e^{-x}); s] \tag{2.16}$$

The occurrence of a strip of holomorphy for Mellin's transform can be deduced directly from
this relation. The usual right-sided Laplace transform is analytic in a half-plane $\text{Re}(s) > \sigma_1$. In
the same way, one can define a left-sided Laplace transform analytic in the region $\text{Re}(s) < \sigma_2$.
If the two half-planes overlap, the region of holomorphy of the two-sided transform is thus the
strip $\sigma_1 < \text{Re}(s) < \sigma_2$ obtained at their intersection.

To obtain Fourier's transform, write $s = a + 2\pi j\beta$ in (2.13):

$$F(s) = \int_{-\infty}^{\infty} f(e^{-x}) e^{-ax} e^{-j2\pi\beta x} dx \tag{2.17}$$

The result is

$$\mathcal{M}[f(t); a + j2\pi\beta] = \mathcal{F}[f(e^{-x}) e^{-ax}; \beta] \tag{2.18}$$

where $\mathcal{F}$ represents the Fourier transformation defined by

$$\mathcal{F}[f; \beta] = \int_{-\infty}^{\infty} f(x) e^{-j2\pi\beta x} dx \tag{2.19}$$

Thus for a given value of $\text{Re}(s) = a$ belonging to the definition strip, the Mellin transform of
a function can be expressed as a Fourier transform.

11.2.1.2   Inversion formula

A direct way to invert Mellin's transformation (2.1) is to start from Fourier's inversion theorem.
As is well known, if $\hat{f} = \mathcal{F}[f; \beta]$ is the Fourier transform (2.19) of f, the original function is
recovered by

$$f(x) = \int_{-\infty}^{\infty} \hat{f}(\beta) e^{j2\pi\beta x} d\beta \tag{2.20}$$

Applying this formula to (2.17) with $s = a + j2\pi\beta$ yields

$$f(e^{-x}) e^{-ax} = \int_{-\infty}^{\infty} F(s) e^{j2\pi\beta x} d\beta \tag{2.21}$$

Hence, going back to variables t and s

$$f(t) = t^{-a} \int_{-\infty}^{\infty} F(s) \, t^{-j2\pi\beta} \, d\beta \qquad (2.22)$$

The inversion formula finally reads

$$f(t) = (1/2\pi j) \int_{a-j\infty}^{a+j\infty} F(s) \, t^{-s} \, ds \qquad (2.23)$$

where the integration is along a vertical line through $\text{Re}(s) = a$. Here a few questions arise. What value of a has to be put into the formula? What happens when a is changed? Is the inverse unique? In what case is f a function defined for all t's?

It is clear that if F is holomorphic in the strip $S(a_1, a_2)$ and vanishes sufficiently fast when $\text{Im}(s) \longrightarrow \pm\infty$, then by Cauchy's theorem, the path of integration can be translated sideways inside the strip without affecting the result of the integration. More precisely, the following theorem holds [6, 7]:

THEOREM 11.2.1
If, in the strip $S(a_1, a_2)$, $F(s)$ is holomorphic and satisfies the inequality

$$|F(s)| \leq K|s|^{-2} \qquad (2.24)$$

for some constant K, then the function $f(t)$ obtained by formula (2.23) is a continuous function of the variable $t \in (0, \infty)$ and its Mellin transform is $F(s)$.

Remark that this result gives only a sufficient condition for the inversion formula to yield a continuous function.

From a practical point of view, it is important to note that the inversion formula applies to a function F holomorphic in a given strip and that the uniqueness of the result holds only with respect to that strip. In fact, a Mellin transform consists of a pair: a function $F(s)$ and a strip of holomorphy $S(a_1, a_2)$. A unique function $F(s)$ with several disjoint strips of holomorphy will in general have several reciprocals, one for each strip. Some examples will illustrate this point.

Example 11.2.4
The Mellin transform of the function

$$f(t) = (H(t - t_0) - H(t)) \, t^z \qquad (2.25)$$

is given by

$$\mathcal{M}[f; s] = -\frac{t_0^{z+s}}{(z + s)} \qquad (2.26)$$

provided $\text{Re}(s) > -\text{Re}(z)$. Comparing (2.26) and (2.3), we see an example of two functions $F(s)$ having the same analytical expression but considered in two distinct regions of holomorphy: the inverse Mellin transforms, given respectively by (2.25) and (2.2), are indeed different (see Fig. 11.1).   ∎

Example 11.2.5   *Gamma function continuation*
From the result of example 11.2.2 considered for $p = 1$, the function $f(t) = e^{-t}$, $t > 0$ is known to be the inverse Mellin transform of $\Gamma(s)$, $\text{Re}(s) > 0$. Besides, it may be checked

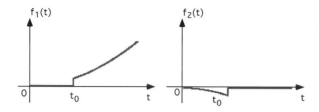

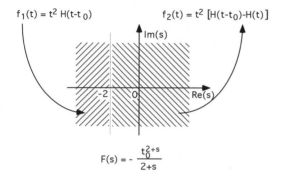

FIGURE 11.1
Examples of results when the regions of holomorphy are changed.

that $\Gamma(s)$ satisfies the hypotheses of Theorem 11.2.1; this is done by using Stirling's formula, which implies the following behavior of the Gamma function [5]:

$$|\Gamma(a + ib)| \sim \sqrt{2\pi}|b|^{a-1/2} e^{-|b|\pi/2}, \qquad |b| \to \infty \qquad (2.27)$$

Thus the inversion formula (2.23) can be applied here and gives an integral representation of e^{-t} as

$$e^{-t} \equiv (1/2\pi j) \int_{a-j\infty}^{a+j\infty} \Gamma(s)t^{-s}\,ds, \qquad a > 0 \qquad (2.28)$$

It is known that the Γ-function can be analytically continued in the left half-plane except for an infinite number of poles at the negative or zero integers. The inverse Mellin transform of the Gamma function for different strips of holomorphy will now be obtained by transforming the identity (2.28). The contour of integration can be shifted to the left and the integral will only pick up the values of the residues at each pole (Fig. 11.2). Explicitly, if $a > 0$ and $-N < a' < -N + 1$, N integer, we have

$$(1/2\pi j) \int_{a-j\infty}^{a+j\infty} \Gamma(s)t^{-s}\,ds = \sum_{n=0}^{N-1} \frac{(-1)^n}{n!}t^n + (1/2\pi j) \int_{a'-j\infty}^{a'+j\infty} \Gamma(s)t^{-s}\,ds \qquad (2.29)$$

Hence, the inversion formula of the Γ-function in the strip $S(-N, -N + 1)$ gives the result

$$(1/2\pi j) \int_{a'-j\infty}^{a'+j\infty} \Gamma(s)t^{-s}\,ds = e^{-t} - \sum_{n=0}^{N-1} \frac{(-1)^n}{n!}t^n, \qquad -N < a' < -N + 1 \qquad (2.30)$$

The integral term represents the remainder in the Taylor expansion of e^{-t} and can be shown to vanish in the limit $N \to \infty$ by applying Stirling's formula. ∎

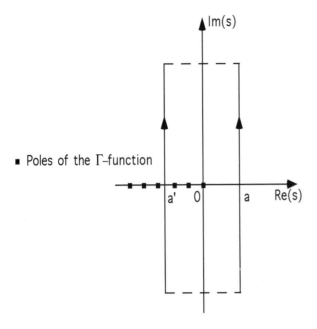

FIGURE 11.2
Different contours of integration for the inverse Mellin transform of the Gamma function. The
contributions from the horizontal parts go to zero as $\text{Im}(s)$ goes to infinity.

As a corollary of the inversion formula, a Parseval relation can be established for suitable
classes of functions.

COROLLARY *Let* $\mathcal{M}[f; s]$ *and* $\mathcal{M}[g; s]$ *be the Mellin transforms of functions* f *and* g *with
strips of holomorphy* S_f *and* S_g *respectively, and suppose that some real number* c *exists such
that* $c \in S_f$ *and* $1 - c \in S_g$. *Then Parseval's formula can be written as*

$$\int_0^\infty f(t)g(t)\,dt = \frac{1}{2\pi j} \int_{c-j\infty}^{c+j\infty} \mathcal{M}[f; s]\mathcal{M}[g; 1-s]\,ds \tag{2.31}$$

This formula may be established formally by computing the right-hand side of (2.31) using
(2.1):

$$\frac{1}{2\pi j} \int_{c-j\infty}^{c+j\infty} \mathcal{M}[f; s]\mathcal{M}[g; 1-s]\,ds = \frac{1}{2\pi j} \int_{c-j\infty}^{c+j\infty} \mathcal{M}[g; 1-s] \int_0^\infty f(t)\,t^{s-1}\,dt\,ds \tag{2.32}$$

Exchanging the two integrals

$$\frac{1}{2\pi j} \int_{c-j\infty}^{c+j\infty} \mathcal{M}[f; s]\mathcal{M}[g; 1-s]\,ds = \frac{1}{2\pi j} \int_0^\infty f(t) \int_{c-j\infty}^{c+j\infty} \mathcal{M}[g; 1-s]\,t^{s-1}\,dt\,ds \tag{2.33}$$

and using the inverse formula (2.23) for g leads to (2.31).

A different set of conditions ensuring the validity of this Parseval formula may be stated
(see, for example, [8, p. 108]). The crucial point is the interchange of integrals that cannot
always be justified.

11.2.1.3 Transformation of distributions

The extension of the correspondence (2.1) to distributions has to be considered to introduce a larger framework in which Dirac delta and other singular functions can be treated straightforwardly. The distributional setting of Mellin's transformation has been studied mainly by Fung Kang [9], A. H. Zemanian [6], and O. P. Misra and J. L. Lavoine [7]. We refer the interested reader to these works for a thorough treatment. As we will see, several approaches to the subject are possible, as was the case for Fourier's transformation.

It is possible to define the Mellin transform for all distributions belonging to the space $\mathcal{D}'_+$ of distributions on the half-line $(0, \infty)$. The procedure [9] is to start from the space $\mathcal{D}(0, \infty)$ of infinitely differentiable functions of compact support on $(0, \infty)$, and to consider the set Q of their Mellin transforms. It can be shown that it is a space of entire functions that is isomorphic, as a linear topological space, to the space $\mathcal{Z}$ of Gelfand and Shilov [10]. This space can be used as a space of test functions and the one-to-one correspondence thus defined between elements of spaces $\mathcal{D}(0, \infty)$ and Q can then be carried (i.e., transposed) to the dual spaces $\mathcal{D}'_+$ and Q'. In this operation, a Mellin transform is associated with any distribution in $\mathcal{D}'_+$ and the result belongs to a space Q' formed of analytic functionals (see Example 11.2.6 for an illustration). The situation is quite analogous to that encountered with the Fourier transformation where a correspondence between distributions spaces $\mathcal{D}'$ and $\mathcal{Z}'$ is established.

Actually, it may be efficient to restrict the class of distributions for which the Mellin transformation will be defined, as is usually done in Fourier analysis with the introduction of the space $\mathcal{S}'$ of tempered distributions [10]. In the present case, a similar approach can be based on the possibility to single out subspaces of $\mathcal{D}'_+$ whose elements are Mellin transformed into functions that are analytic in a given strip. This construction will now be sketched.

The most practical way to proceed is to give a new interpretation of formula (2.1) by considering it as the application of a distribution f to a test function t^{s-1}:

$$F(s) = \langle f, t^{s-1} \rangle \tag{2.34}$$

A suitable space of test functions $\mathcal{T}(a_1, a_2)$ containing all functions t^{s-1} for s in the region $a_1 < \mathrm{Re}(s) < a_2$ may be introduced as follows [7]. The space $\mathcal{T}(a_1, a_2)$ is composed of functions $\phi(t)$ defined on $(0, \infty)$ and with continuous derivatives of all orders going to zero as t approaches either zero or infinity. More precisely, there exists two positive numbers ζ_1, ζ_2, such that, for all integers k, the following conditions hold:

$$t^{k+1-a_1-\zeta_1} \phi^{(k)}(t) \longrightarrow 0, \qquad t \to 0 \tag{2.35}$$

$$t^{k+1-a_2-\zeta_2} \phi^{(k)}(t) \longrightarrow 0, \qquad t \to \infty \tag{2.36}$$

A topology on $\mathcal{T}$ is defined accordingly. It can be verified that all functions in $\mathcal{D}(0, \infty)$ belong to $\mathcal{T}(a_1, a_2)$.[1] The space of distributions $\mathcal{T}'(a_1, a_2)$ is then introduced as a linear space of continuous linear functionals on $\mathcal{T}(a_1, a_2)$. It may be noticed that if α_1, α_2 are two real numbers such that $a_1 < \alpha_1 < \alpha_2 < a_2$, then $\mathcal{T}(\alpha_1, \alpha_2)$ is included in $\mathcal{T}(a_1, a_2)$. One may so define a whole collection of ascending spaces $\mathcal{T}(a_1, a_2)$ with compatible topologies, thus ensuring the existence of limit spaces when $a_1 \to -\infty$ and/or $a_2 \to \infty$.[2] Hence the dual spaces of distributions are such that $\mathcal{T}'(a_1, a_2) \subset \mathcal{T}'(\alpha_1, \alpha_2)$ and $\mathcal{T}'(-\infty, +\infty)$ is included in all of them. Moreover, as a consequence of the status of $\mathcal{D}(0, \infty)$ relative to $\mathcal{T}(a_1, a_2)$, the space $\mathcal{T}'(a_1, a_2)$ is a subspace of distributions in $\mathcal{D}'_+$. The precise construction of these spaces is explained in reference [7]. A slightly different presentation is given in reference [6] and leads to these same spaces denoted by $\mathcal{M}'(a_1, a_2)$.

---

[1] More precisely, one can show that $\mathcal{D}(0, \infty)$ is dense in $\mathcal{T}(a_1, a_2)$.

[2] In fact, $\mathcal{T}(-\infty, a_2)$, $\mathcal{T}(a_1, +\infty)$, and $\mathcal{T}(-\infty, +\infty)$ are defined as inductive limits.

With the above definitions, the Mellin transform of an element $f \in \mathcal{T}'(a_1, a_2)$ is defined by

$$\mathcal{M}[f; s] \equiv F(s) = \langle f, t^{s-1} \rangle \tag{2.37}$$

The result is always a conventional function $F(s)$ holomorphic in the strip $a_1 < \mathrm{Re}(s) < a_2$.

In summary, every distribution in $\mathcal{D}'_+$ has a Mellin transform that, as a rule, is an analytic functional. Besides, it is possible to define subspaces $\mathcal{T}'(a_1, a_2)$ of $\mathcal{D}'_+$ whose elements f are Mellin transformed by formula (2.37) into functions $F(s)$ holomorphic in the strip $S(a_1, a_2)$. Any space $\mathcal{T}'$ contains, in particular, Dirac distributions and arbitrary distributions of bounded support. They are stable under derivation and multiplication by a smooth function. Their complete characterization is given by the following theorems.

THEOREM 11.2.2 *Uniqueness theorem [6]*
Let $\mathcal{M}[f; s] = F(s)$ and $\mathcal{M}[h; s] = H(s)$ be Mellin transforms with strips of holomorphy S_f and S_h respectively. If the strips overlap and if $F(s) \equiv H(s)$ for $s \in S_f \cap S_h$, then $f \equiv h$ as distributions in $\mathcal{T}'(a_1, a_2)$ where the interval (a_1, a_2) is given by the intersection of $S_f \cap S_h$ with the real axis.

THEOREM 11.2.3 *Characterization of the Mellin transform of a distribution in $\mathcal{T}'(a_1, a_2)$ [6, 7]*
A necessary and sufficient condition for a function $F(s)$ to be the Mellin transform of a distribution $f \in \mathcal{T}'(a_1, a_2)$ is

- *$F(s)$ is analytic in the strip $a_1 < \mathrm{Re}(s) < a_2$,*
- *for any closed substrip $\alpha_1 \leq \mathrm{Re}(s) \leq \alpha_2$ with $a_1 < \alpha_1 < \alpha_2 < a_2$ there exists a polynomial P such that $|F(s)| \leq P(|s|)$ for $\alpha_1 \leq \mathrm{Re}(s) \leq \alpha_2$.*

Example 11.2.6 *Example of analytic functional*
The function t^z, z complex, defines a distribution in $\mathcal{D}'_+$ according to

$$\langle t^z, \phi \rangle = \int_0^\infty t^z \phi(t)\, dt, \qquad \phi \in \mathcal{D}(0, \infty) \tag{2.38}$$

But it may be seen that this distribution does not belong to any of the spaces $\mathcal{T}'(a_1, a_2)$. Its Mellin transform may nevertheless be defined by the following formula:

$$\langle \mathcal{M}[t^z], \psi \rangle_M = \langle t^z, \phi \rangle \tag{2.39}$$

where $\langle \, , \, \rangle_M$ denotes duality in the space of Mellin transforms, $\phi \equiv \mathcal{M}^{-1}\psi$ is an element of $\mathcal{D}(0, \infty)$, and consequently, ψ is an entire function. According to (2.39) and definition (2.1), we obtain

$$\langle \mathcal{M}[t^z], \psi \rangle_M = \mathcal{M}[\phi; z + 1]$$

$$= \psi(z + 1) \tag{2.40}$$

Because distribution $\mathcal{M}[t^z]$ applied to ψ gives the value of ψ in a point of the complex plane, it can be symbolized by a delta function. To introduce the notation, we need the explicit form of duality $\langle \, , \, \rangle_M$, which comes out of Parseval's formula. According to (2.31), it is given for entire functions χ, ψ by

$$\langle \chi, \psi \rangle_M = \frac{1}{2\pi j} \int_{c-j\infty}^{c+j\infty} \chi(s)\psi(1-s)\, ds \tag{2.41}$$

where c is any real number. A more usual form is obtained by setting

$$\tilde{\psi}(s) \equiv \psi(1-s) \tag{2.42}$$

and

$$\langle \chi, \tilde{\psi} \rangle \equiv \langle \chi, \psi \rangle_M \tag{2.43}$$

$$= \frac{1}{2\pi j} \int_{c-j\infty}^{c+j\infty} \chi(s)\tilde{\psi}(s)\,ds \tag{2.44}$$

With these definitions, (2.40) can be written

$$\langle \mathcal{M}[t^z], \tilde{\psi} \rangle = \tilde{\psi}(-z) \tag{2.45}$$

and the notation

$$\mathcal{M}[t^z] = \delta(s+z) \tag{2.46}$$

can be proposed. Such Dirac distributions in the complex plane are defined in [10]. ∎

Example 11.2.7

The Mellin transform of the Dirac distribution $\delta(t - t_0)$ is found by applying the general rule (see also Chapter 1, Section 2.4)

$$\langle \delta(t - t_0), \phi \rangle = \phi(t_0) \tag{2.47}$$

to the family of functions $\phi(t) = t^{s-1}$. One obtains

$$\mathcal{M}[\delta(t - t_0); s] = \langle \delta(t - t_0), t^{s-1} \rangle$$

$$= t_0^{s-1} \tag{2.48}$$

for any value of the positive number t_0. Moreover, the result is holomorphic in the whole complex s-plane.

It is instructive to verify explicitly the inverse formula on this example. According to (2.23), the inverse Mellin transform $\mathcal{M}^{-1}[t_0^{s-1}; t]$ can be written as

$$\mathcal{M}^{-1}[t_0^{s-1}; t] = \frac{1}{2\pi j t_0} \int_{-j\infty}^{j\infty} \left(\frac{t}{t_0}\right)^{-s} ds \tag{2.49}$$

because the choice $a = 0$ is allowed by the holomorphy property of the integrand in the whole plane. Setting $s = j\beta$ in (2.49) and performing the integration leads to the equivalent expressions

$$\mathcal{M}^{-1}[t_0^{s-1}; t] = \frac{1}{2\pi t_0} \int_{-\infty}^{\infty} e^{-j\beta \ln(t/t_0)}\, d\beta$$

$$= t_0^{-1} \delta(\ln t - \ln t_0) \tag{2.50}$$

The expected result

$$\mathcal{M}^{-1}[t_0^{s-1}; t] = \delta(t - t_0) \tag{2.51}$$

comes out by using the classical formula (see also Table 2.4.1 of Chapter 1)

$$\delta(f(t)) = |f'(t_0)|^{-1} \delta(t - t_0) \tag{2.52}$$

in which $f(t)$ is a function having a simple zero in $t = t_0$. ∎

Example 11.2.8

Consider the distribution

$$f = \sum_{n=1}^{\infty} \delta(t - pn), \qquad p > 0 \tag{2.53}$$

Its Mellin transform is given by

$$\left\langle \sum_{n=1}^{\infty} \delta(t - pn), t^{s-1} \right\rangle = \sum_{n=1}^{\infty} (pn)^{s-1}$$

$$= p^{s-1} \sum_{n=1}^{\infty} n^{s-1} \tag{2.54}$$

The sum converges uniformly for $\mathrm{Re}(s) < 0$ and can be expressed in terms of Riemann's Zeta function [11] (see Appendix A). Explicitly, we have

$$\mathcal{M}\left[\sum_{n=1}^{\infty} \delta(t - pn); s \right] = p^{s-1} \zeta(1 - s), \qquad \mathrm{Re}(s) < 0 \tag{2.55}$$

∎

11.2.1.4 Some properties of the transformation.

This paragraph describes the effect on the Mellin transform $\mathcal{M}[f; s]$ of some special operations performed on f. The resulting formulas are very useful for deducing new correspondences from a given one.

Let $F(s) = \mathcal{M}[f; s]$ be the Mellin transform of a distribution that is supposed to belong to $\mathcal{J}'(\sigma_1, \sigma_2)$ and denote by $S_f = \{s : \sigma_1 < \mathrm{Re}(s) < \sigma_2\}$ its strip of holomorphy (σ_1 is either finite or $-\infty$, σ_2 is finite or ∞). Then the following formulas hold with the regions of holomorphy as indicated. The notation of functions will be used but this must not obscure the fact that f is a distribution and that all operations performed on f, especially differentiation, must be understood in the generalized sense of distributions.

- *Scaling of the original variable by a positive number*

$$\mathcal{M}[f(rt); s] = r^{-s} F(s), \qquad s \in S_f, \quad r > 0 \tag{2.56}$$

- *Raising of the original variable to a real power*

$$\mathcal{M}[f(t^r); s] = |r|^{-1} F(r^{-1} s), \qquad r^{-1} s \in S_f, \quad r \text{ real} \neq 0 \tag{2.57}$$

- *Multiplication of the original function by* $\ln t$

$$\mathcal{M}[(\ln t)^k f(t); s] = \frac{d^k}{ds^k} F(s), \qquad s \in S_f, \ k \text{ positive integer} \tag{2.58}$$

- *Multiplication of the original function by some power of t*

$$\mathcal{M}[(t)^z f(t); s] = F(s + z), \qquad s + z \in S_f, \quad z \text{ complex} \tag{2.59}$$

- *Derivation of the original function*

$$\mathcal{M}\left[\frac{d^k}{dt^k} f(t); s \right] = (-1)^k (s-k)_k F(s-k), \qquad s-k \in S_f, \ k \text{ positive integer} \tag{2.60}$$

where the symbol $(s - k)_k$ is defined for k integer by

$$(s - k)_k \equiv (s - k)(s - k + 1) \cdots (s - 1)$$

$$\equiv \frac{(s - 1)!}{(s - k - 1)!} \tag{2.61}$$

The coefficient (2.61) can also be written

$$(s - k)_k = \frac{\Gamma(s)}{\Gamma(s - k)}$$

so that formula (2.60) retains a meaning for non integer values of k [45]. On this point, the situation is similar to that encountered in the study of Fourier and Laplace transformations.

Formulas (2.59) and (2.60) can be used in various ways to find the effect of linear combinations of differential operators such that $t^k (d/dt)^m$, k, m integers. The most remarkable result is

$$\mathcal{M}\left[\left(t\frac{d}{dt}\right)^k f(t); s\right] = (-1)^k s^k F(s) \tag{2.62}$$

Other combinations can be computed. We have for example

$$\mathcal{M}\left[\frac{d^k}{dt^k} t^k f(t); s\right] = (-1)^k (s - k)_k F(s) \tag{2.63}$$

$$\mathcal{M}\left[t^k \frac{d^k}{dt^k} f(t); s\right] = (-1)^k (s)_k F(s) \tag{2.64}$$

where $s \in S_f$, k a positive integer and

$$(s)_k \equiv s(s + 1) \cdots (s + k - 1) \tag{2.65}$$

These relations are easily verified on infinitely differentiable functions. It is important to stress that they are essentially true for distributions. That implies in particular that all derivatives occurring in the formulas are to be taken according to the distribution rules. An example dealing with a discontinuous function will make this manifest.

Example 11.2.9

Consider the function

$$f(t) = H(t - t_0)t^z, \qquad z \text{ complex} \tag{2.66}$$

According to results of Example 11.2.1, the Mellin transform of f is given by

$$\mathcal{M}[f; s] \equiv F(s) = -\frac{t_0^{z+s}}{z + s} \tag{2.67}$$

Applying formula (2.60) for $k = 1$ yields

$$\mathcal{M}\left[\frac{df}{dt}; s\right] = -(s - 1)F(s - 1)$$

$$= (s - 1)\frac{t_0^{z+s-1}}{z + s - 1} \tag{2.68}$$

which can be rewritten as

$$\mathcal{M}\left[\frac{df}{dt}; s\right] = -z\frac{t_0^{z+s-1}}{z + s - 1} + t_0^{z+s-1} \tag{2.69}$$

or, recognizing the Mellin transforms obtained in Examples 11.2.1 and 11.2.7:

$$\mathcal{M}\left[\frac{df}{dt}; s\right] = \mathcal{M}\left[zH(t - t_0)t^{z+s-1}; s\right] + \mathcal{M}[t_0^z \delta(t - t_0); s] \tag{2.70}$$

This result shows explicitly that, in formula (2.60), f is differentiated as a distribution. The first term in (2.70) corresponds to the derivative of the function for $t \neq t_0$ and the second term is the Dirac distribution arising from the discontinuity at $t = t_0$. ∎

Additional results on the Mellin transforms of primitives can be established for particular classes of functions. Namely, if $x > 1$, integration by parts leads to the result

$$\mathcal{M}\left[\int_x^\infty f(t)\,dt; s\right] = \int_0^\infty s^{-1}x^s f(x)\,dx$$

$$= s^{-1}F(s+1) \tag{2.71}$$

provided the integrated part $s^{-1}x^s \int_x^\infty f(t)\,dt$ is equal to zero for $x = 0$ and $x = \infty$.

In the same way, but with different conditions on f, one establishes

$$\mathcal{M}\left[\int_0^x f(t)\,dt; s\right] = -s^{-1}F(s+1) \tag{2.72}$$

11.2.1.5   Relation to multiplicative convolution

The usual convolution has the property of being changed into multiplication by either a Laplace or a Fourier transformation. In the present case, a multiplicative convolution [9], also called Mellin-type convolution [6, 7], is defined that has a similar property with respect to Mellin's transformation. In the same way as the usual convolution of two distributions in $\mathcal{D}(\mathbb{R})$ does not necessarily exist, the multiplicative convolution of distributions in $\mathcal{D}'_+$ can fail to define a distribution. To avoid such problems, we shall restrict our considerations to spaces $\mathcal{J}'(a_1, a_2)$.

DEFINITION 11.2.2   *Let f, g be two distributions belonging to some space $\mathcal{J}'(a_1, a_2)$. The multiplicative convolution of f and g is a new functional $(f \vee g)$ whose action on test functions $\theta \in \mathcal{J}(a_1, a_2)$ is given by*

$$\langle f \vee g, \theta \rangle = \langle f(t), \langle g(\tau), \theta(t\tau) \rangle \rangle \tag{2.73}$$

It can be shown that $f \vee g$ is a distribution that belongs to the space $\mathcal{J}'(a_1, a_2)$.

If the distributions f and g are represented by locally integrable functions, definition (2.73) can be written explicitly as

$$\langle f \vee g, \theta \rangle = \int_0^\infty \int_0^\infty f(t)g(\tau)\theta(t\tau)\,dt\,d\tau \tag{2.74}$$

A change of variables then leads to the following expression for the multiplicative convolution of the functions f and g:

$$(f \vee g)(\tau) = \int_0^\infty f(t)g\left(\frac{\tau}{t}\right)\frac{dt}{t} \tag{2.75}$$

The so-called exchange formula for usual convolution has an analog for multiplicative convolution. It is expressed by the following theorem [6, 7].

THEOREM 11.2.4   Exchange Formula
The Mellin transform of the convolution product $f \vee g$ of two distributions belonging to $\mathcal{J}'(a_1, a_2)$ is given by the formula

$$\mathcal{M}[f \vee g; s] = F(s)G(s), \qquad a_1 < \text{Re}(s) < a_2 \tag{2.76}$$

where $F(s)$ and $G(s)$ are the Mellin transforms of distributions f and g respectively.

The proof is a simple application of the definitions. According to (2.37), the Mellin transform of distribution $f \vee g \in \mathcal{T}'(a_1, a_2)$ is given by

$$\mathcal{M}[f \vee g; s] = \langle f \vee g, t^{s-1} \rangle, \qquad a_1 < \mathrm{Re}(s) < a_2 \qquad (2.77)$$

or, using the definition (2.73) of convolution:

$$\mathcal{M}[f \vee g; s] = \langle f(t), \langle g(\tau), (t\tau)^{s-1} \rangle \rangle \qquad (2.78)$$

which can be rewritten as

$$\mathcal{M}[f \vee g; s] = \langle f(t), t^{s-1} \rangle \langle g(\tau), \tau^{s-1} \rangle \qquad (2.79)$$

Formula (2.73) allows us to consider the multiplicative convolution of general distributions not belonging to space $\mathcal{T}'(a_1, a_2)$. However, in that case, it is not ensured that the product exists as a distribution.

General properties of the multiplicative convolution   In this paragraph, f and g are distributions that belong to $\mathcal{T}'(a_1, a_2)$ and k is a positive integer. The following properties are easy to verify. In fact, some of them are a direct consequence of the exchange formula.

1. Commutativity

$$f \vee g = g \vee f \qquad (2.80)$$

2. Associativity

$$(f \vee g) \vee h = f \vee (g \vee h) \qquad (2.81)$$

3. Unit element

$$f \vee \delta(t - 1) = f \qquad (2.82)$$

4. Action of the operator $t(d/dt)$

$$\left(t\frac{d}{dt} \right)^k (f \vee g) = \left[\left(t\frac{d}{dt} \right)^k f \right] \vee g$$

$$= f \vee \left[\left(t\frac{d}{dt} \right)^k g \right] \qquad (2.83)$$

that is, it is sufficient to apply the operator to one of the factors.

5. Multiplication by $\ln t$

$$(\ln t)(f \vee g) = [(\ln t)f] \vee g + f \vee [(\ln t)g] \qquad (2.84)$$

6. Convolution with Dirac distributions and their derivatives

$$\delta(t - a) \vee f = a^{-1} f(a^{-1}t) \qquad (2.85)$$

$$\delta(t - p) \vee \delta(t - p') = \delta(t - pp'), \qquad p, p' > 0 \qquad (2.86)$$

$$\delta^{(k)}(t - 1) \vee f = (d/dt)^k(t^k f) \qquad (2.87)$$

PROOF OF RELATION (2.87)   According to the definition of the k-derivative of the Dirac distribution, the multiplicative convolution of f with $\delta^{(k)}(t - 1)$ is given by

$$\langle f \vee \delta^{(k)}(t - 1), \theta(t) \rangle = \langle f(t), \langle \delta^{(k)}(\tau - 1), \theta(t\tau) \rangle \rangle$$

$$= \langle f(t), \langle \delta(\tau - 1), (-1)^k \left(\frac{d}{d\tau} \right)^k \theta(t\tau) \rangle \rangle \qquad (2.88)$$

or, after performing an ordinary differentiation and applying the definition of δ:

$$\langle f \vee \delta^{(k)}(t-1), \theta(t) \rangle = \langle f(t), (-1)^k t^k \theta^{(k)}(t) \rangle \tag{2.89}$$

The usual rules of calculus with distributions and the commutativity of convolution yield

$$\langle \delta^{(k)}(t-1) \vee f, \theta(t) \rangle = \left\langle \left(\frac{d}{dt} \right)^k (t^k f(t)), \theta(t) \right\rangle \tag{2.90}$$

Finally, identity (2.87) follows from the fact that (2.90) holds for any function θ belonging to $\mathcal{T}(a_1, a_2)$. ∎

11.2.1.6   Hints for a practical inversion of the Mellin transformation

In many applications, it is essential to be able to perform explicitly the Mellin inversion. This is often the most difficult part of the computation and we now give some indications on different ways to proceed.

Compute the inversion integral   This direct approach is not always the simplest. In some cases, however, the integral (2.23) can be computed by the method of residues.

Use rules of Section 11.2.1.4 to exploit the inversion formula   Property (2.62) in particular can be used to extend the domain of practical utility of the inversion formula (2.23). Indeed, in the case where the Mellin transform $F(s)$, holomorphic in the strip $S(a_1, a_2)$ with (a_1, a_2) finite, does not satisfy condition (2.24), suppose that a positive integer k can be found such that

$$|s^{-k} F(s)| \leq K|s|^{-2} \tag{2.91}$$

The inversion formula (2.23) can now be used on the function $G(s)$ defined by

$$G(s) = (-1)^k s^{-k} F(s), \qquad a_1 < \mathrm{Re}(s) < a_2 \tag{2.92}$$

and yields a continuous function $g(t)$. Using rule (2.62) and the uniqueness of the Mellin transform, we conclude that the reciprocal of $F(s)$ is the distribution f defined by

$$f = \left(t \frac{d}{dt} \right)^k g(t) \tag{2.93}$$

In spite of the fact that the continuous function $g(t)$ is not necessarily differentiable everywhere, formula (2.93) remains meaningful because derivatives are taken in the sense of distributions. In fact, the above procedure corresponds to generalizing theorem 11.2.1 in the following form:

THEOREM 11.2.5   [7, 6]
Let $F(s)$ be a function holomorphic in the strip $S(a_1, a_2)$ with a_1, a_2 finite. If there exists an integer $k \geq 0$ such that $s^{2-k} F(s)$ is bounded as $|s|$ goes to infinity, then the inverse Mellin transform of $F(s)$ is the unique distribution f given by

$$f = \left(t \frac{d}{dt} \right)^k g(t) \tag{2.94}$$

where $g(t)$ is a continuous function obtained by the formula

$$g(t) = \frac{(-1)^k}{2\pi j} \int_{a-j\infty}^{a+j\infty} F(s) s^{-k} t^{-s} \, ds \tag{2.95}$$

with $a \in S(a_1, a_2)$.

Other inversion formulas may be obtained in the same way [7], by using rules (2.64) and (2.63) respectively. They are

$$f = t^k \left(\frac{d}{dt}\right)^k g(t), \qquad \text{where } g(t) = \frac{(-1)^k}{2\pi j} \int_{a-j\infty}^{a+j\infty} \frac{F(s)}{(s)_k} t^{-s}\, ds \qquad (2.96)$$

$$f = \left(\frac{d}{dt}\right)^k t^k g(t), \qquad \text{where } g(t) = \frac{(-1)^k}{2\pi j} \int_{a-j\infty}^{a+j\infty} \frac{F(s)}{(s-k)_k} t^{-s}\, ds \qquad (2.97)$$

Use the tables   In simple cases, exploitation of tables [12, 13, 14] and use of the rules of calculus exhibited in section 11.2.1.4 are sufficient to obtain the result.

In more difficult cases, it may be rewarding to use the systematic approach developed by Marichev [5] and applicable to a large number of functions. Suppose we are given a function $F(s)$, holomorphic in the strip $S(\sigma_1, \sigma_2)$, and we want to find its inverse Mellin transform. The first step is to try and cast F into the form of a fraction involving only products of Γ-functions, the variable s appearing only with the coefficient ± 1. This looks quite restrictive but in fact many simple functions can be so rewritten using the properties of Γ-functions recalled in Appendix A. Thus $F(s)$ is brought to the form

$$F(s) = C \prod_{i,j,k,l} \frac{\Gamma(a_i + s)\Gamma(b_j - s)}{\Gamma(c_k + s)\Gamma(d_l - s)} \qquad (2.98)$$

where C, a_i, b_j, c_k, d_l are constants and where $\text{Re}(s)$ is restricted to the strip $S(\sigma_1, \sigma_2)$ now defined in terms of these.

For such functions, the explicit computation of the inversion integral (2.23) can be performed by the theory of residues and yields a precise formula given in [5] as Slater's theorem. The result has the form of a function of hypergeometric type. The important point is that most special functions are included in this class. For a thorough description of the method, the reader is referred to Marichev's book [5], which contains simple explanations along with all the proofs and exhaustive tables.

Special forms related to the use of polar coordinates   [15] The analytical solution of some two-dimensional problems in polar coordinates (r, θ) is obtained by using a Mellin transformation with respect to the radial variable r. In this approach, one can be faced with the task of inverting expressions of the type $\cos(s\theta)\, F(s)$ or $\sin(s\theta)\, F(s)$. We will show that, for a large class of problems, the reciprocals of the products $\cos(s\theta)\, F(s)$ and $\sin(s\theta)\, F(s)$ can be obtained straightforwardly from the knowledge of the reciprocal of $F(s)$.

Let $f(r)$, f *real-valued*, be the inverse Mellin transform of $F(s)$ in strip $S(a_1, a_2)$, and suppose that f can be analytically continued into a function $f(z)$, $z \equiv re^{j\theta}$, in some sector $|\theta| < \beta$ of the complex plane. If the rule of scaling (2.56) can be extended to the complex numbers, we have

$$\mathcal{M}[f(re^{j\theta}); s] = e^{-j\theta s}\mathcal{M}[f(r); s] \qquad (2.99)$$

where the Mellin transforms are with respect to r.

In fact, this formula can be established by contour integration in a sector $|\arg z| < \beta$ where the function f is such that

$$z^s f(z) \longrightarrow 0 \qquad \text{as } |z| \to 0 \text{ or } \infty \qquad (2.100)$$

Remark that because f has a Mellin transform with strip of definition $S(a_1, a_2)$, this condition already holds on the real axis when $a_1 < \text{Re}(s) < a_2$.

Recalling that f is real valued, we can take the real and imaginary parts of (2.99) and obtain, for real s, the formulas

$$\mathcal{M}[\text{Re}(f(re^{j\theta})); s] = \cos(s\theta)F(s) \qquad (2.101)$$

$$\mathcal{M}[\text{Im}(f(re^{j\theta})); s] = -\sin(s\theta)F(s) \qquad (2.102)$$

These can be extended to complex s and yield the inverse Mellin transforms of $\cos(s\theta)F(s)$ and $\sin(s\theta)F(s)$ for $a_1 < \text{Re}(s) < a_2$ and $|\theta| < \beta$.

Example 11.2.10

To illustrate the use of the above rules, we shall perform explicitly the inversion of

$$F(s) = \frac{\cos s\theta}{s \cos s\alpha}, \qquad \alpha \text{ real} \qquad (2.103)$$

in the strip $0 < \text{Re}(s) < \pi/(2\alpha)$ [15].

Using the result of Example 11.2.3 and rule (2.57), we obtain

$$\mathcal{M}[(1 + r^2)^{-1}; s] = \frac{\pi}{2 \sin(\pi s/2)}, \qquad 0 < \text{Re}(s) < 1 \qquad (2.104)$$

Recalling that

$$\int_r^\infty \frac{dx}{1 + x^2} = \pi/2 - \tan^{-1} r \qquad (2.105)$$

and using rule (2.71) gives

$$\mathcal{M}[\pi/2 - \tan^{-1} r; s] = \frac{\pi}{2s \cos \pi s/2}, \qquad 0 < \text{Re}(s) < 1 \qquad (2.106)$$

Using again property (2.57) but with $\nu = \pi/2\alpha$ finally gives

$$\mathcal{M}[\pi/2 - \tan^{-1} r^\nu; s] = \frac{\pi}{2s \cos(\pi s/2\nu)}, \qquad 0 < \text{Re}(s) < \nu \qquad (2.107)$$

To find the domain in which function $f(z) = \pi/2 - \tan^{-1} z^\nu$, where $z = re^{j\theta}$, verifies the condition (2.100), we write it under the form

$$f(z) = (1/2j) \ln \left(\frac{z^\nu + j}{z^\nu - j} \right) \qquad (2.108)$$

subject to the choice of the determination for which $0 < \tan^{-1} r < \pi/2$. The result is $|\theta| \equiv |\arg(z)| < \pi/2$. Relation (2.101) yields the result

$$\mathcal{M}^{-1} \left[\frac{\cos(s\theta)}{s \cos(s\alpha)}; s \right] = \text{Re}(\pi/2 - \tan^{-1} z^\nu),$$

$$0 < \text{Re}(s) < \pi/(2\alpha), \qquad |\theta| < \pi/2 \qquad (2.109)$$

The real part of $f(re^{j\theta})$ is given explicitly by

$$\text{Re}(\pi/2 - \tan^{-1} z^\nu) = \begin{cases} 1 - \pi^{-1} \tan^{-1} \dfrac{2r^\nu \cos \nu\theta}{1 - r^{2\nu}} & 0 \leq r < 1 \\[3mm] \pi^{-1} \tan^{-1} \dfrac{2r^\nu \cos \nu\theta}{r^{2\nu} - 1} & r > 1 \end{cases}$$

∎

11.2.2 Standard applications

11.2.2.1 Summation of series

Even if a numerical computation is intended, Mellin's transformation may be used with profit to transform slowly convergent series either into integrals that can be computed exactly or into more rapidly convergent series.

Let S represent a series of the form

$$S = \sum_{n=1}^{\infty} f(n) \tag{2.110}$$

in which the terms are samples of a function $f(t)$ for integer values of the variable $t \in (0, \infty)$. If this function has a Mellin transform $F(s)$ with $S(a_1, a_2)$ as the strip of holomorphy, it can be written

$$f(t) = (2\pi j)^{-1} \int_{a-j\infty}^{a+j\infty} F(s)\, t^{-s}\, ds, \qquad a_1 < a < a_2 \tag{2.111}$$

Substituting this identity in (2.110) yields

$$S = (2\pi j)^{-1} \sum_{n=1}^{\infty} \int_{a-j\infty}^{a+j\infty} F(s) n^{-s}\, ds \tag{2.112}$$

Now, if $F(s)$ is such that sum and integral can be exchanged, an integral expression for S is obtained:

$$S = (2\pi j)^{-1} \int_{a-j\infty}^{a+j\infty} F(s)\zeta(s)\, ds \tag{2.113}$$

where $\zeta(s)$ is the Riemann zeta function defined by

$$\zeta(s) = \sum_{n=1}^{\infty} n^{-s} \tag{2.114}$$

The integral (2.113) is then evaluated by the methods of Section 2.1.5. The calculus of residues may give the result as an infinite sum that we hope will be more rapidly convergent than the original series.

Some care is necessary when going from (2.112) to (2.113). Actually, if the interchange of summation and integration is not justified, the expression (2.113) can fail to represent the original series (see [15, p. 216] for an example).

Example 11.2.11
Compute the sum

$$S(y) = \sum_{n=1}^{\infty} \frac{\cos ny}{n^2} \tag{2.115}$$

From Table 11.3 and properties (2.56) and (2.59), one finds

$$\mathcal{M}\left[\frac{\cos ty}{t^2}; s\right] = -y^{2-s}\, \Gamma(s-2)\, \cos(\pi s/2), \qquad 2 < \mathrm{Re}(s) < 3 \tag{2.116}$$

Hence the sum can be rewritten as

$$S = -(1/2\pi j) \int_{a-j\infty}^{a+j\infty} y^{2-s} \Gamma(s-2) \cos(\pi s/2)\zeta(s)\, ds \tag{2.117}$$

where the interchange of summation and integration is justified by absolute convergence. The integral can be rearranged by using Riemann's functional relationship (see for example [15])

$$\pi^s \zeta(1-s) = 2^{1-s} \Gamma(s) \cos(\pi s/2) \zeta(s) \tag{2.118}$$

Then (2.117) becomes

$$S = -(1/2\pi j) \int_{a-j\infty}^{a+j\infty} y^{2-s} 2^{s-1} \pi^s \frac{\zeta(1-s)}{(s-1)(s-2)} \, ds \tag{2.119}$$

The integral is easily computed by the method of residues, closing the contour to the left where the integrand goes to zero. The function $\zeta(s)$ is analytic everywhere except at $s = 1$ where it has a simple pole with residue equal to 1. The result is

$$S = \frac{y^2}{4} - \frac{\pi y}{2} + \frac{\pi^2}{6} \tag{2.120}$$

∎

11.2.2.2 Computation of integrals depending on a parameter

Essentially, the technique concerns integrals that can be brought to the form

$$K(x) = \int_0^\infty K_0(t) \, K_1(x/t) \, \frac{dt}{t}, \qquad x > 0 \tag{2.121}$$

One recognizes the expression of a multiplicative convolution. Such an integral can be computed by performing the following steps:

- Mellin transform functions K_0 and K_1 to obtain $\mathcal{M}[K_0; s]$ and $\mathcal{M}[K_1; s]$
- Multiply the transforms to obtain $\mathcal{M}[K; s] \equiv \mathcal{M}[K_0; s]\mathcal{M}[K_1; s]$
- Find the inverse Mellin transform of $\mathcal{M}[K; s]$ using the tables. The result will in general be expressed as a combination of generalized hypergeometric series.

For the last operation, the book by O. I. Marichev [5] can be of great help as previously mentioned in Section 11.2.1.6. The method can be extended to allow the computation of integrals of the form

$$K(x_1, \ldots, x_N) = \int_0^\infty K_0(t) \left[\prod_{i=1}^n K_i \left(\frac{x_i}{t} \right) \right] \frac{dt}{t} \tag{2.122}$$

where $x_1, \ldots, x_N$ are positive variables.

It can be verified that the multiple Mellin transform defined by

$$\mathcal{M}[K; s_1, \ldots, s_N] = \int_0^\infty \cdots \int_0^\infty K(x_1, \ldots, s_N) \, x_1^{s_1-1} \cdots x_N^{s_N-1} \, dx_1 \cdots dx_N \tag{2.123}$$

allows us to factorize the expression (2.122) as

$$\mathcal{M}[K; s_1, \ldots, s_N] = \mathcal{M}[K_0; s_1 + s_2 + \cdots + s_N] \prod_{i=1}^N \mathcal{M}[K_i; s_i] \tag{2.124}$$

Techniques of inversion for this expression are developed in a book by R. J. Sasiela [16] devoted to the propagation of electromagnetic waves in turbulent media.

11.2.2.3 Mellin's convolution equations

These are not always expressed with integrals of type (2.75), but also with differential operators that are polynomials in $(t(d/dt))$. Such equations are of the general form

$$Lu(t) \equiv \left(a_n \left(t\frac{d}{dt}\right)^n + a_{n-1}\left(t\frac{d}{dt}\right)^{n-1} + \cdots + a_0\right)u(t) = g(t) \qquad (2.125)$$

By using the identity

$$(t(d/dt))^k u(t) \equiv [(t(d/dt))^k \delta(t-1)] \vee u(t) \qquad (2.126)$$

they can be written as a convolution

$$\sum_{k=0}^{n} a_k(t(d/dt))^k \delta(t-1) \vee u(t) = g(t) \qquad (2.127)$$

The more usual Euler-Cauchy differential equation, which is written as

$$(b_n t^n (d/dt)^n + b_{n-1}t^{n-1}(d/dt)^{n-1} + \cdots + b_0)u(t) = g(t) \qquad (2.128)$$

can be brought to the form (2.125) by using relations such that

$$\left(t\frac{d}{dt}\right)^2 = t\frac{d}{dt} + t^2\frac{d^2}{dt^2} \qquad (2.129)$$

It can also be tranformed directly into a convolution that reads

$$\sum_{k=0}^{n} b_k t^k \delta^{(k)}(t-1) \vee u(t) = g(t) \qquad (2.130)$$

The Mellin treatment of convolution equations will be explained in the case of equation (2.127) because it is the most characteristic.

Suppose that the known function g has a Mellin transform $\mathcal{M}[g; s] = G(s)$ that is holomorphic in the strip $S(\sigma_l, \sigma_r)$. We shall seek solution u, which admits a Mellin transform $U(s)$ holomorphic in the same strip or in some substrip. The Mellin transform of equation (2.127) is obtained by using the convolution property and relation (2.62):

$$A(s)U(s) = G(s) \qquad (2.131)$$

where

$$A(s) \equiv \sum_{k=0}^{\infty} a_k(-1)^k s^k \qquad (2.132)$$

Two different situations may arise.

- $A(s)$ has no zeros in the strip $S(\sigma_l, \sigma_r)$. In that case, $U(s)$ given by $G(s)/A(s)$ can be inverted in the strip. According to Theorems 11.2.2 and 11.2.3, the unique solution is a distribution belonging to $\mathcal{T}'(\sigma_1, \sigma_2)$.

- $A(s)$ has m zeros in the strip. The main strip $S(\sigma_1, \sigma_2)$ can be decomposed into adjacent substrips

$$\sigma_l < \mathrm{Re}(s) < \sigma_1, \ \sigma_1 < \mathrm{Re}(s) < \sigma_2, \ldots, \sigma_m < \mathrm{Re}(s) < \sigma_r \qquad (2.133)$$

The solution in the k-substrip is given by the Mellin inverse formula:

$$u(t) = \frac{1}{2\pi j}\int_{c-j\infty}^{c+j\infty} \frac{G(s)t^{-s}}{A(s)}\, ds \qquad (2.134)$$

where $\sigma_k < c < \sigma_{k+1}$. There is a different solution in each strip, two solutions differing by a solution of the homogeneous equation.

11.2.2.4 Solution of a potential problem in a wedge [6, 7, 15]

The problem is to solve Laplace's equation in an infinite 2-dimensional wedge with Dirichlet boundary conditions. Polar coordinates with origin at the apex of the wedge are used and the sides are located at $\theta = \pm\alpha$. The unknown function $u(r, \theta)$ is supposed to verify

$$\Delta u = 0, \qquad 0 < r < \infty, \ -\alpha < \theta < \alpha \tag{2.135}$$

with the following boundary conditions:

- On the sides of the wedge, if R is a given positive number

$$u(r, \pm\alpha) = \begin{cases} 1 & \text{if } 0 < r < R \\ 0 & \text{if } r > R \end{cases} \tag{2.136}$$

  or, equivalently,

$$u(r, \pm\alpha) = H(R - r) \tag{2.137}$$

- When r is finite, $u(r, \theta)$ is bounded.
- When r tends to infinity, $u(r, \theta) \sim r^{-\beta}, \beta > 0$.

In polar coordinates, equation (2.135) multiplied by r^2 yields

$$r^2 \frac{\partial^2 u}{\partial r^2} + r \frac{\partial u}{\partial r} + \frac{\partial^2 u}{\partial \theta^2} = 0 \tag{2.138}$$

The above conditions on $u(r, \theta)$ ensure that its Mellin transform $U(s, \theta)$ with respect to r exists as a holomorphic function in some region $0 < \text{Re}(s) < \beta$. The equation satisfied by U is obtained from (2.138) by using property (2.59) of the Mellin transformation and reads

$$\frac{d^2 U}{d\theta^2}(s, \theta) + s^2 U(s, \theta) = 0 \tag{2.139}$$

The general solution of this equation can be written as

$$U(s, \theta) = A(s)e^{js\theta} + B(s)e^{-js\theta} \tag{2.140}$$

Functions A, B are to be determined by the boundary condition (2.137), which leads to the following requirement on U

$$U(s, \pm\alpha) = R^s \, s^{-1} \qquad \text{for } \text{Re}(s) > 0 \tag{2.141}$$

Explicitly, this is written as

$$A(s)e^{js\alpha} + B(s)e^{-js\alpha} = a^s s^{-1} \tag{2.142}$$

$$A(s)e^{-js\alpha} + B(s)e^{js\alpha} = a^s s^{-1} \tag{2.143}$$

and leads to the solution

$$A(s) = B(s) = \frac{R^s}{2s \cos(s\alpha)} \tag{2.144}$$

The solution of the form (2.140) that verifies (2.141) is given by

$$U(s, \theta) = \frac{R^s \cos(s\theta)}{s \cos(s\alpha)} \tag{2.145}$$

This function U is holomorphic in the strip $0 < \text{Re}(s) < \pi/(2\alpha)$. Its inverse Mellin transform is a function $u(r, \theta)$ that is obtained from the result of Example 11.2.10.

11.2.2.5 Asymptotic expansion of integrals

The Laplace transform $I[f; \lambda]$ defined by

$$I[f; \lambda] = \int_0^\infty e^{-\lambda t} f(t) \, dt \tag{2.146}$$

has an asymptotic expansion as λ goes to infinity that is characterized by the behavior of the function f when $t \to 0+$ [8, 12, 17]. With the help of Mellin's transformation, one can extend this type of study to other transforms of the form

$$I[f; \lambda] = \int_0^\infty h(\lambda t) f(t) \, dt \tag{2.147}$$

where h is a general kernel.

Examples of such h-transforms [8] are

- Fourier transform: $h(\lambda t) = e^{j\lambda t}$
- Cosine and Sine transforms: $h(\lambda t) = \cos(\lambda t)$ or $\sin(\lambda t)$
- Laplace transform: $h(\lambda t) = e^{-\lambda t}$
- Hankel transform: $h(\lambda t) = J_\nu(\lambda t)(\lambda t)^{1/2}$ where J_ν is the Bessel function of the first kind.
- Generalized Stieltjes transform: $h(\lambda t) = \lambda^\nu \int_0^\infty f(t)/(1 + \lambda t)^\nu \, dt$

A short formal overview of the procedure will be given below. The theory is exposed in full generality in [8]. It includes the study of asymptotic expansions when $\lambda \to 0+$ in relation to the behavior of f at infinity and the extension to complex values of λ. The case of oscillatory h-kernels is given special attention.

Suppose from now on that f and h are locally integrable functions such that the transform $I[f; \lambda]$ exists for large λ. The different steps leading to an asymptotic expansion of $I[f; \lambda]$ in the limit $\lambda \to +\infty$ are the following:

1. *Mellin transform the functions h and f and apply Parseval's formula.* The Mellin transforms $\mathcal{M}[f; s]$ and $\mathcal{M}[h; s]$ are supposed to be holomorphic in the strips $\eta_1 < \text{Re}(s) < \eta_2$ and $\alpha_1 < \text{Re}(s) < \alpha_2$ respectively. Assuming that Parseval's formula may be applied and using property (2.56), one can write (2.146) as

$$I[f; \lambda] = \frac{1}{2\pi j} \int_{r-j\infty}^{r+j\infty} \lambda^{-s} \mathcal{M}[h; s] \mathcal{M}[f; 1 - s] \, ds \tag{2.148}$$

where r is any real number in the strip of analyticity of the function G defined by

$$G(s) = \mathcal{M}[h; s] \mathcal{M}[f; 1 - s], \quad \max(\alpha_1, 1 - \eta_2) < \text{Re}(s) < \min(\alpha_2, 1 - \eta_1) \tag{2.149}$$

2. *Shift of the contour of integration to the right and use of Cauchy's formula.* Suppose $G(s)$ can be analytically continued in the right half-plane $\text{Re}(s) \geq \min(\alpha_2, 1 - \eta_1)$ as a meromorphic function. Remark that this assumption implies that $\mathcal{M}[f; s]$ may be continued to the right half-plane $\text{Re}(s) > \alpha_2$ and $\mathcal{M}[h; s]$ to the left $\text{Re}(s) < \eta_1$. Suppose, moreover, that the contour of integration in (2.148) can be displaced to the

right as far as the line $\mathrm{Re}(s) = R > r$. A sufficient condition ensuring this property is that

$$\lim_{|b|\to\infty} G(a + jb) = 0 \qquad (2.150)$$

for all a in the interval $[r, R]$.

Under these conditions, Cauchy's formula may be applied and yields

$$I[f; \lambda] = - \sum_{r<\mathrm{Re}(s)<R} \mathrm{Res}(\lambda^{-s} G(s)) + \frac{1}{2\pi j} \int_{R-j\infty}^{R+j\infty} \lambda^{-s} G(s)\, ds \qquad (2.151)$$

where the discrete summation involves the residues (denoted Res) of function $\lambda^{-s} G(s)$ at the poles lying inside the region $r < \mathrm{Re}(s) < R$.

3. *The asymptotic expansion.* The relation (2.151) is an asymptotic expansion provided the error bounds hold. A sufficient condition to ensure that the integral term is of order $O(\lambda^{-R})$ is that G satisfy

$$\int_{-\infty}^{\infty} |G(R + jb)|\, db < \infty \qquad (2.152)$$

The above operations can be justified step by step when treating a particular case. The general theory gives a precise description of the final form of the asymptotic expansion, when it exists, in terms of the asymptotic properties of h when $t \to +\infty$ and of f when $t \to 0+$.

The above procedure is easily adapted to give the asymptotic expansion of $I[f; \lambda]$ when $\lambda \to 0+$.

Example 11.2.12

Consider the case where the kernel h is given by

$$h(t) = \frac{1}{1+t} \qquad (2.153)$$

The integral under consideration is thus

$$I[f; \lambda] = \int_0^\infty \frac{f(t)}{1+\lambda t}\, dt \qquad (2.154)$$

The function f must be such that the integral exists. In addition, it is supposed to have a Mellin transform holomorphic in the strip $\sigma_1 < \mathrm{Re}(s) < \sigma_2$ and to have an asymptotic development as $t \to 0$ of the form

$$f \sim \sum_{m=0}^{\infty} t^{a_m} p_m \qquad (2.155)$$

where the numbers $\mathrm{Re}(a_m)$ increase monotonically to $+\infty$ as $m \to +\infty$, and the numbers p_m may be arbitrary.

To apply the method, we first compute the Mellin transform of h, which is given by

$$\mathcal{M}[(1 + t)^{-1}; s] = \frac{\pi}{\sin \pi s}, \qquad 0 < \mathrm{Re}(s) < 1 \qquad (2.156)$$

It can be continued in the half-plane $\mathrm{Re}(s) > 0$ where it has simple poles at $s = 1, 2, \ldots$ and decays along imaginary lines as follows:

$$\frac{\pi}{\sin \pi(a + jb)} = O(e^{-\pi|b|}) \qquad \text{for all } a \qquad (2.157)$$

As for function f, its behavior given by (2.155) ensures (see [8]) that the Mellin transform $\mathcal{M}[f; s]$ has an analytic continuation in the half-plane $\text{Re}(s) \leq \sigma_1 = -\text{Re}(a_0)$ that is a meromorphic function with poles at the points $s = -a_m$. Moreover, one finds the following behavior at infinity for the continued Mellin transform:

$$\lim_{|b| \to \infty} \mathcal{M}[f; a + jb] = 0, \qquad \text{for all } a < \sigma_2 \tag{2.158}$$

In this situation, the method will lead to an asymptotic expansion that can be written explicitly. For example, in the case where $a_m \neq 0, 1, 2, \ldots$, the poles of $\mathcal{M}[f; 1 - s]$, which occur at $1 - s = -a_m$, are distinct from those of $\mathcal{M}[h; s]$ at $s = m + 1$, and the expansion of I is given by

$$I[f; \lambda] \sim \sum_{m=0}^{\infty} \lambda^{-1-a_m} \frac{\pi}{\sin(\pi a_m)} \text{Res}_{s=1+a_m} \{\mathcal{M}[f; 1 - s]\}$$

$$+ \sum_{m=0}^{\infty} \lambda^{-1-m} \mathcal{M}[f; -m] \text{Res}_{s=m+1} \left\{ \frac{\pi}{\sin(\pi s)} \right\}$$

Hence,

$$I[f; \lambda] \sim \sum_{m=0}^{\infty} \lambda^{-1-a_m} p_m \frac{\pi}{\sin(\pi a_m)} + \sum_{m=0}^{\infty} (-1)^m \lambda^{-1-m} \mathcal{M}[f; -m] \tag{2.159}$$

In particular, if $f(t) = (1/t)e^{-(1/t)}$, all the p_m are equal to zero and the expansion is just

$$I[f; \lambda] \sim \sum_{n=0}^{\infty} (-1)^n (\lambda)^{-n-1} \Gamma(n + 1) \tag{2.160}$$

11.3 Alternative approach related to the dilation group and its representations

11.3.1 *Theoretical aspects*

11.3.1.1 Construction of the transformation

Rather than start directly by giving the explicit formula of the Mellin transformation, we will construct it as a tool especially devoted to the computation of functionals involving scalings of a variable. Such an introduction of the transform may be found, for example, in the book by N. Ya. Vilenkin [3] or, in a more applied context, in articles [18, 19, 20].

If $Z(v)$ is a function defined on the positive half-axis $(0, \infty)$, a scaling of the variable v by a positive number a leads to a new function $Z'(v)$, which is related to $Z(v)$ by the change of variable:

$$v \longrightarrow av \tag{3.1}$$

The set of such transformations forms a group that is isomorphic to the multiplicative group of positive real numbers. In practice, the scaled function is often defined by the transformation

$$Z(v) \longrightarrow a^{1/2} Z(av) \tag{3.2}$$

which does not change the value of the usual scalar product:

$$(Z_1, Z_2) \equiv \int_0^\infty Z_1(v) Z_2^*(v) \, dv \tag{3.3}$$

in which the symbol $*$ denotes complex conjugation. However, there are serious physical reasons to consider more general transformations of the form

$$\mathcal{D}_a : \quad Z(v) \longrightarrow (\mathcal{D}_a Z)(v) \equiv a^{r+1} Z(av) \tag{3.4}$$

where r is a given real number. The general correspondence $a \leftrightarrow \mathcal{D}_a$ is such that

$$\mathcal{D}_a \mathcal{D}_{a'} = \mathcal{D}_{aa'}$$

$$\mathcal{D}_1 \mathcal{D}_a = \mathcal{D}_a \mathcal{D}_1 = \mathcal{D}_a$$

$$(\mathcal{D}_a)^{-1} = \mathcal{D}_{a^{-1}} \tag{3.5}$$

Thus, for any value of r, the set of $\mathcal{D}_a$ operations constitutes a representation of a group. These transformations preserve the following scalar product:

$$(Z_1, Z_2) \equiv \int_0^\infty Z_1(v) Z_2^*(v) v^{2r+1} \, dv \tag{3.6}$$

that is, we have

$$(\mathcal{D}_a Z_1, \mathcal{D}_a Z_2) = (Z_1, Z_2) \tag{3.7}$$

The scalar product (3.6) defines a norm for the functions $Z(v)$ on the positive axis $\mathbb{R}^+$. We have

$$\|Z\|^2 \equiv \int_0^\infty |Z(v)|^2 \, v^{2r+1} \, dv \tag{3.8}$$

The corresponding Hilbert space will be denoted by $L^2(\mathbb{R}^+, v^{2r+1} \, dv)$. It is handled in the same manner as an ordinary L^2 space with the measure dv replaced by $v^{2r+1} \, dv$ in all formulas. In this space, the meaning of (3.7) is that the set of operations D_a constitutes a unitary representation of the multiplicative group of positive numbers.

The value of r will be determined by the specific applications to be dealt with. Examples of adjustments of this parameter are given in [21] where the occurrence of dilations in radar imaging is analyzed.

When confronted by expressions involving functions modified by dilations of the form (3.4), it may be advantageous to use a Hilbert-space basis in which the operators $\mathcal{D}_a$ have a diagonal expression. This leads to a decomposition of functions Z into simpler elements on which the scaling operation breaks down to a mere multiplication by a complex number. Such a procedure is familiar when considering the operation that translates a function $f(t), t \in \mathbb{R}$ according to

$$f(t) \longrightarrow f(t - t_0) \tag{3.9}$$

In that case, the exponentials $e^{\lambda t}$, λ complex, are functions that are multiplied by a number $e^{\lambda t_0}$ in a translation. If $\lambda = j\alpha$, α real, these functions are unitary representations of the translation group in $L^2(\mathbb{R})$ and provide a generalized[3] orthonormal basis for functions in this space. The coefficients of the development of function $f(t)$ on this basis are obtained by scalar product with the basis elements and make up the Fourier transform. In the present case,

---

[3]Such families of functions that do not belong to the Hilbert space under consideration but are treated like bases by physicists are called *improper bases*. Their use can be rigorously justified.

analogous developments will connect the Mellin transformation to the unitary representations of the dilation group in $L^2(\mathbb{R}^+, v^{2r+1} dv)$.

For simplicity and for future reference, the diagonalization of $\mathcal{D}_a$ will be performed on its infinitesimal form defined by the operator $\mathcal{B}$ whose action on function $Z(v)$ is given by

$$(\mathcal{B}Z)(v) \equiv -\frac{1}{2\pi j} \frac{d}{da}[(\mathcal{D}_a Z)(v)]_{(a=1)} \tag{3.10}$$

The computation yields

$$\mathcal{B} = -\frac{1}{2\pi j}\left(v\frac{d}{dv} + r + 1\right) \tag{3.11}$$

The operator $\mathcal{B}$ is a self-adjoint operator and the unitary representation $\mathcal{D}_a$ is recovered from $\mathcal{B}$ by exponentiation:[4]

$$\mathcal{D}_a = e^{-2\pi j a \mathcal{B}} \tag{3.12}$$

where the exponential of the operator is defined formally by the infinite series:

$$e^{-2\pi j a \mathcal{B}} = \sum_{n=0}^{\infty}(-1)^n \frac{(2\pi j a \mathcal{B})^n}{n!} \tag{3.13}$$

Here, we only need to find the eigenfunctions of $\mathcal{B}$, that is, the solutions of the differential equation

$$\mathcal{B}Z(v) = \beta Z(v) \tag{3.14}$$

with β real.

The solution is, up to an arbitrary factor,

$$E_\beta(v) = v^{-2\pi j\beta - r - 1} \tag{3.15}$$

As ensured by the construction, any member of this family of functions E_β is just multiplied by a phase when a scaling is performed

$$\mathcal{D}_a : E_\beta(v) \longrightarrow a^{-2\pi j\beta} E_\beta(v) \tag{3.16}$$

Moreover, the family $\{E_\beta\}$ is orthonormal and complete as will now be shown. The orthonormality is obtained by setting $v = e^{-x}$ in the expression

$$(E_\beta, E_{\beta'}) = \int_0^{\infty} v^{-2\pi j(\beta - \beta') - 1} dv \tag{3.17}$$

which becomes

$$(E_\beta, E_{\beta'}) = \int_{-\infty}^{\infty} e^{2\pi j(\beta - \beta')x} dx \tag{3.18}$$

The result is

$$(E_\beta, E_{\beta'}) = \delta(\beta - \beta') \tag{3.19}$$

To show completeness, we compute

$$\int_{-\infty}^{\infty} E_\beta(v)E_\beta^*(v') d\beta \equiv \int_{-\infty}^{\infty} e^{2\pi j\beta \ln(v/v')}(vv')^{-r-1} d\beta \tag{3.20}$$

$$= (vv')^{-r-1} \delta(\ln(v) - \ln(v')) \tag{3.21}$$

---

[4]This is known as Stone's theorem.

and, using the rule of calculus with delta functions recalled in (2.52), we obtain

$$\int_{-\infty}^{\infty} E_\beta(v) E_\beta^*(v')\, d\beta = v^{-2r-1}\delta(v - v') \tag{3.22}$$

Any function Z in $L^2(\mathbb{R}^+, v^{2r+1}\, dv)$ can thus be decomposed on the basis E_β with coefficients $\mathcal{M}[Z](\beta)$ given by

$$\mathcal{M}[Z](\beta) = (Z, E_\beta) \tag{3.23}$$

or explicitly

$$\mathcal{M}[Z](\beta) = \int_0^\infty Z(v) v^{2\pi j\beta + r}\, dv \tag{3.24}$$

The set of coefficients, considered as a function of β, constitutes what is called the Mellin transform of function Z. This definition coincides with the usual one (2.1) provided we set $s = r + 1 + 2\pi j\beta$. Thus,

$$\mathcal{M}[Z](\beta) \equiv \mathcal{M}[Z; r + 1 + 2\pi j\beta] \tag{3.25}$$

But the viewpoint here is different. The value of $\mathrm{Re}(s) = r + 1$ is fixed once and for all as it is forced upon us by the representation of dilations occurring in the physical problem under study. Thus, the situation is closer to the Fourier than to the Laplace case and an L^2 theory is developed naturally.

The property (3.22) of completeness for the basis implies that the Mellin transformation (3.23) from $Z(v)$ to $\mathcal{M}[Z](\beta)$ is norm preserving:

$$\int_{-\infty}^{\infty} |\mathcal{M}[Z](\beta)|^2\, d\beta = (Z, Z) \tag{3.26}$$

A Parseval formula (also called unitarity property) follows immediately:

$$\int_{-\infty}^{\infty} \mathcal{M}[Z_1](\beta)\mathcal{M}^*[Z_2](\beta)\, d\beta = (Z_1, Z_2) \tag{3.27}$$

where

$$\mathcal{M}^*[Z](\beta) \equiv [\mathcal{M}[Z](\beta)]^* \tag{3.28}$$

The decomposition formula of function $Z(v)$ on basis $\{E_\beta(v)\} \equiv v^{-2\pi j\beta - r - 1}$ that can be obtained from (3.24) and (3.19) constitutes the inversion formula for the Mellin transformation

$$Z(v) = \int_{-\infty}^{\infty} \mathcal{M}[Z](\beta) v^{-2\pi j\beta - r - 1}\, d\beta \tag{3.29}$$

By construction, the Mellin transformation performs the diagonalization of the operators $\mathcal{B}$ and $\mathcal{D}_a$. Indeed, by definition (3.23), the Mellin transform of the function $(\mathcal{B}Z)(v)$ is given by

$$\mathcal{M}[\mathcal{B}Z](\beta) = (\mathcal{B}Z, E_\beta) \tag{3.30}$$

or, using the fact that $\mathcal{B}$ is self-adjoint and that E_β is an eigenfunction of $\mathcal{B}$:

$$\mathcal{M}[\mathcal{B}Z](\beta) = \beta(Z, E_\beta) \tag{3.31}$$

Thus,

$$\mathcal{M}[\mathcal{B}Z](\beta) = \beta\mathcal{M}[Z](\beta) \tag{3.32}$$

In the same way, the Mellin transform of $\mathcal{D}_a Z$ is computed using the unitarity of $\mathcal{D}_a$:

$$\mathcal{M}[\mathcal{D}_a Z](\beta) = (\mathcal{D}_a Z, E_\beta)$$

$$= (Z, D_{a^{-1}} E_\beta) \tag{3.33}$$

Thus,

$$\mathcal{M}[\mathcal{D}_a Z](\beta) = a^{-2\pi j \beta} \mathcal{M}[Z](\beta) \tag{3.34}$$

All these results can be summed up in the following proposition.

THEOREM 11.3.1
Let $Z(v)$ be a function in $L^2(\mathbb{R}^+, \, v^{2r+1} \, dv)$. Its Mellin transform defined by

$$\mathcal{M}[Z](\beta) = \int_0^\infty Z(v) v^{2\pi j \beta + r} \, dv \tag{3.35}$$

belongs to $L^2(\mathbb{R})$.
  The inversion formula is given by

$$Z(v) = \int_{-\infty}^\infty \mathcal{M}[Z](\beta) v^{-2\pi j \beta - r - 1} \, d\beta \tag{3.36}$$

An analog of Parseval's formula (unitarity) holds as

$$\int_{-\infty}^\infty \mathcal{M}[Z_1](\beta) \mathcal{M}^*[Z_2](\beta) \, d\beta = \int_0^\infty Z_1(v) \, Z_2^*(v) \, v^{2r+1} \, dv \tag{3.37}$$

For any function Z, the Mellin transform of the dilated function

$$(\mathcal{D}_a Z)(v) \equiv a^{r+1} Z(av) \tag{3.38}$$

is given by

$$\mathcal{M}[\mathcal{D}_a Z](\beta) = a^{-2\pi j \beta} \mathcal{M}[Z](\beta) \tag{3.39}$$

11.3.1.2   Uncertainty relations

As in the case of the Fourier transformation, there is a relation between the spread of a function and the spread of its Mellin transform. To find this relation, we will consider the first two moments of the density functions $|Z(v)|^2$ and $|\mathcal{M}[Z](\beta)|^2$ connected by (3.37). The mean value of v with density $|Z(v)|^2$ is defined by the formula

$$\bar{v} \equiv \frac{(Z, vZ)}{(Z, Z)} \tag{3.40}$$

The mean value of v^2 is defined by an analogous formula. The mean square deviation σ_v^2 of variable v can then be computed according to

$$\sigma_v^2 \equiv \overline{(v - \bar{v})^2} \tag{3.41}$$

In the same way, in the space of Mellin transforms, the mean value of β is defined by

$$\bar{\beta} \equiv \frac{(\tilde{Z}, \beta \tilde{Z})}{(\tilde{Z}, \tilde{Z})} \tag{3.42}$$

where $\tilde{Z}$ denotes the Mellin transform of Z. Using Parseval formula (3.37) and property (3.32), one can also rewrite this mean value in terms of the original function $Z(v)$ as

$$\bar{\beta} = \frac{(Z, \mathcal{B}Z)}{(Z, Z)} \tag{3.43}$$

where the operator $\mathcal{B}$ has been defined by formula (3.11). At this point, it is convenient to introduce the following notation for any operator $\mathcal{O}$ acting on Z:

$$\langle \mathcal{O} \rangle \equiv \frac{(Z, \mathcal{O}Z)}{(Z, Z)} \tag{3.44}$$

and to rewrite (3.43) as

$$\bar{\beta} = \langle \mathcal{B} \rangle \tag{3.45}$$

The mean square deviation σ_β^2 of variable β can also be expressed in terms of the operator $\mathcal{B}$ according to

$$\sigma_\beta^2 = \langle (\mathcal{B} - \bar{\beta})^2 \rangle \tag{3.46}$$

A simple way to obtain a lower bound on the product $\sigma_\beta \sigma_v$ is to introduce the operator $\mathcal{X}$ defined by

$$\mathcal{X} \equiv \mathcal{B} - \bar{\beta} + j\lambda(v - \bar{v}) \tag{3.47}$$

where λ is a real parameter. The obvious requirement that the norm of $\mathcal{X}Z(v)$ must be positive or zero whatever the value of λ is expressed by the inequality

$$\|\mathcal{X}Z\|^2 = (Z, \mathcal{X}^*\mathcal{X}Z) \geq 0 \tag{3.48}$$

where $\mathcal{X}^*$ denotes the adjoint of $\mathcal{X}$. This constraint implies the positivity of the expression

$$\langle (\mathcal{B} - \bar{\beta} + j\lambda(v - \bar{v}))(\mathcal{B} - \bar{\beta} - j\lambda(v - \bar{v})) \rangle \tag{3.49}$$

Developing and using relations (3.41) and (3.46), we obtain

$$\lambda^2 \sigma_v^2 + j\lambda \langle v\mathcal{B} - \mathcal{B}v \rangle + \sigma_\beta^2 \geq 0 \tag{3.50}$$

The computation of $v\mathcal{B} - \mathcal{B}v$ yields

$$(v\mathcal{B} - \mathcal{B}v)Z(v) = \frac{-1}{2\pi j} \left(v^2 \frac{d}{dv} - v\frac{d}{dv}v \right) Z(v) \tag{3.51}$$

$$= \frac{1}{2\pi j} v Z(v) \tag{3.52}$$

With this result, condition (3.50) becomes

$$\lambda^2 \sigma_v^2 + (\lambda/2\pi)\bar{v} + \sigma_\beta^2 \geq 0 \tag{3.53}$$

The left member is a quadratic expression of the parameter λ. Its positivity whatever the value of λ means that the coefficients of the expression verify

$$\sigma_v \sigma_\beta \geq \bar{v}/4\pi \tag{3.54}$$

The functions for which this product is minimal are such that there is equality in (3.48). Hence, they are solutions of the equation

$$[\mathcal{B} - \bar{\beta} - j\lambda(v - \bar{v})]Z(v) = 0 \tag{3.55}$$

and are found to be

$$K(v) \equiv e^{-2\pi\lambda v} \, v^{2\pi\lambda\bar{v}-r-1-2\pi j\bar{\beta}} \tag{3.56}$$

These functions, first introduced by Klauder [22], are the analogs of Gaussians in Fourier theory.

11.3.1.3 Extension of the Mellin transformation to distributions

The definition of the transformation has to be extended to distributions to be able to treat generalized functions, such as Dirac's, which are currently used in electrical engineering. Section 11.2.1.3 provides a general view of the possible approaches. Here we only give a succinct definition that will generally be sufficient, and we show in explicit examples how computations can be performed.

First, a test function space $\mathcal{T}$ is constructed so as to contain the functions $v^{2\pi j\beta+r}$, $v > 0$, $\beta \in \mathbb{R}$ for a fixed value of r. Examples of such spaces are the spaces $\mathcal{T}(a_1, a_2)$ of Section 11.2.1.3, provided a_1, a_2 are chosen verifying the inequality $a_1 < r + 1 < a_2$ [6, 7]. Then the space of distributions $\mathcal{T}'$ is defined as usual as a linear space $\mathcal{T}'$ of continuous functionals on $\mathcal{T}$. It can be shown that the space $\mathcal{T}'$ contains the distributions of bounded support on the positive axis and, in particular, the Dirac distributions.

The Mellin transform of a distribution Z in a space $\mathcal{T}'$ can always be obtained as the result of the application of Z to the set of test functions $v^{2\pi j\beta+r}$, $\beta \in \mathbb{R}$, that is, as

$$\mathcal{M}[Z](\beta) \equiv \langle Z, v^{2\pi j\beta+r} \rangle \tag{3.57}$$

With this extended definition, it is easily verified that relations (3.32) and (3.39) still hold. One more property that will be useful, especially for discretization, is relative to the effect of translations on the Mellin variable. Computing $\mathcal{M}[Z]$ for the value $\beta + c$, c real, yields

$$\mathcal{M}[Z](\beta + c) = \langle Z, v^{2\pi j(\beta+c)+r} \rangle$$
$$= \langle Z v^{2\pi jc}, v^{2\pi j\beta+r} \rangle \tag{3.58}$$

and the result

$$\mathcal{M}[Z](\beta + c) = \mathcal{M}[Z v^{2\pi jc}](\beta) \tag{3.59}$$

Example 11.3.1
The above formula (3.57) allows us to compute the Mellin transform of $\delta(v - v_0)$ by applying the usual definition of the Dirac distribution:

$$\langle \delta(v - v_0), \phi \rangle = \phi(v_0) \tag{3.60}$$

to the function $\phi(v) = v^{2\pi j\beta+r}$, thus giving

$$\mathcal{M}[\delta(v - v_0)](\beta) \equiv \langle \delta(v - v_0), v^{2\pi j\beta+r} \rangle = v_0^{2\pi j\beta+r} \tag{3.61}$$

∎

FIGURE 11.3
Geometric Dirac comb in $\mathbb{R}^+$ space and arithmetic Dirac comb in the Mellin space.

Example 11.3.2   The Geometric Dirac Comb
In problems involving dilations, it is natural to introduce a special form of the Dirac comb defined by

$$\Delta_A^r(v) \equiv \sum_{n=-\infty}^{+\infty} A^{-nr}\, \delta(v - A^n)$$

$$\equiv \sum_{n=-\infty}^{+\infty} A^{nr}\, \delta(v - A^{-n}) \tag{3.62}$$

where A is a positive number. The values of v that are picked out by this distribution form a geometric progression of ratio A. Moreover, the comb Δ_a^r is invariant in a dilation by an integer power of A. Indeed,

$$D_A \Delta_A^r(v) \equiv A^{r+1} \delta_A^r(Av) \tag{3.63}$$

$$= \sum_{n=-\infty}^{\infty} A^{-(n-1)r}\, \delta(v - A^{n-1}) \tag{3.64}$$

$$= \Delta_A^r(v) \tag{3.65}$$

The distribution Δ_A^r will be referred to as the *geometric Dirac comb* and is represented in Figure 11.3.

Distribution Δ_A^r does not belong to $\mathcal{J}'$ and, hence, its Mellin transform cannot be obtained by formula (3.57). However, the property of linearity of the Mellin transformation and result (3.61) allow us to write

$$\mathcal{M}[\Delta_A^r](\beta) = \sum_{n=-\infty}^{+\infty} A^{2j\pi\beta n} \tag{3.66}$$

The right-hand side of (3.66) is a Fourier series that can be summed by Poisson's formula:

$$\ln A \sum_{n=-\infty}^{\infty} e^{2j\pi n\beta \ln A} = \sum_{n=-\infty}^{\infty} \delta\left(\beta - \frac{n}{\ln A}\right) \tag{3.67}$$

This leads to

$$\mathcal{M}[\Delta_A^r](\beta) = \frac{1}{\ln A} \sum_{n=-\infty}^{+\infty} \delta\left(\beta - \frac{n}{\ln A}\right) \tag{3.68}$$

Thus the Mellin transform of a geometric Dirac comb Δ_A^r on $\mathbb{R}^+$ is an arithmetic Dirac comb on $\mathbb{R}$ (cf. Fig. 11.3).   ∎

11.3.1.4   Transformations of products and convolutions

The relations between product and convolution that are established by a Fourier transformation have analogs here. Classical convolution and usual product in the space of Mellin transforms correspond respectively to a special invariant product and a multiplicative convolution in the original space. The latter operations can also be defined directly by their transformation properties under a dilation, as will now be explained.

Invariant product   The dilation-invariant product of the functions Z_1 and Z_2, which will be denoted by the symbol $\circ$, is defined as

$$(Z_1 \circ Z_2)(v) = v^{r+1} Z_1(v) Z_2(v) \tag{3.69}$$

The invariance means that a dilation D_a on each factor is equivalent to a dilation on the product, as shown in the following relation:

$$\mathcal{D}_a[Z_1] \circ \mathcal{D}_a[Z_2] = \mathcal{D}_a[Z_1 \circ Z_2] \tag{3.70}$$

Now, we shall compute the Mellin transform of the product $Z_1 \circ Z_2$. According to definition (3.35), this is given by

$$\mathcal{M}[Z_1 \circ Z_2](\beta) = \int_0^{+\infty} v^{r+1} Z_1(v) Z_2(v) v^{2j\pi\beta+r} \, dv \tag{3.71}$$

Replacing Z_1 and Z_2 by their inverse Mellin transforms given by (3.36) and using the orthogonality relation (3.19) to perform the v-integration, we obtain

$$\mathcal{M}[Z_1 \circ Z_2](\beta) = \int_{-\infty}^{+\infty} d\beta_1 \int_{-\infty}^{+\infty} \mathcal{M}[Z_1](\beta_1) \, \mathcal{M}[Z_2](\beta_2) \, d\beta_2 \, \delta(\beta - \beta_1 - \beta_2) \tag{3.72}$$

$$= \int_{-\infty}^{+\infty} \mathcal{M}[Z_1](\beta_1) \, \mathcal{M}[Z_2](\beta - \beta_1) \, d\beta_1 \tag{3.73}$$

where we recognize the classical convolution of the Mellin transforms.

THEOREM 11.3.2
The Mellin transform of the invariant product (3.69) of the two functions Z_1 and Z_2 is equal to the convolution of their Mellin transforms:

$$\mathcal{M}[Z_1 \circ Z_2](\beta) = (\mathcal{M}[Z_1] * \mathcal{M}[Z_2])(\beta) \tag{3.74}$$

Multiplicative convolution   For a given function Z_1 (resp. Z_2), the usual convolution $Z_1 * Z_2$ can be seen as the most general linear operation commuting with translations that can be performed on Z_1 (resp. Z_2). By analogy, the *multiplicative convolution* of Z_1 and Z_2 is defined as the most general linear operation on Z_1 (resp. Z_2) that commutes with dilations. More precisely, suppose that a linear operator $\mathcal{A}$ is defined in terms of a kernel function $A(v, v')$ according to

$$\mathcal{A}[Z_1](v) = \int_0^{+\infty} A(v, v') Z_1(v') \, dv' \tag{3.75}$$

Then the requirement that transformation $\mathcal{D}_a$ applied either on Z_1 or $\mathcal{A}[Z_1]$ yield the same results implies that

$$a^{r+1} \mathcal{A}[Z_1](av) = a^{r+1} \int_0^{+\infty} A(v, v') Z_1(av') \, dv' \qquad (3.76)$$

must be true for any function Z_1. Comparing (3.76) to (3.75), we thus obtain the following constraint on the kernel $A(v, v')$:

$$A(v, v') \equiv a \, A(av, av') \qquad (3.77)$$

valid for any a. For $a = v'^{-1}$, we obtain the identity

$$A(v, v') \equiv \frac{1}{v'} A\left(\frac{v}{v'}, 1\right) \qquad (3.78)$$

which shows that the operator $\mathcal{A}$ can be expressed by using a function of a single variable. Thus any linear transformation acting on function Z_1 and commuting with dilations can be written in the form

$$\int_0^{+\infty} Z_1(v') \, Z_2\left(\frac{v}{v'}\right) \frac{dv'}{v'} \qquad (3.79)$$

where $Z_2(v)$ is an arbitrary function.

It can be verified, by changing variables, that the above expression is symmetrical with respect to the two functions Z_1 and Z_2. It defines the multiplicative convolution of these functions, which is usually denoted by $Z_1 \vee Z_2$:

$$Z_1 \vee Z_2 \equiv \int_0^{+\infty} Z_1(v') \, Z_2\left(\frac{v}{v'}\right) \frac{dv'}{v'} \qquad (3.80)$$

On this definition, it can be observed that dilating one of the factors Z_1 or Z_2 of the multiplicative convolution is equivalent to dilating the result, that is,

$$D_a[(Z_1 \vee Z_2)(v)] \equiv [Z_1 \vee (D_a Z_2)](v) \qquad (3.81)$$

$$\equiv [(D_a Z_1) \vee Z_2](v) \qquad (3.82)$$

For applications, an essential property of the multiplicative convolution is that it is converted into a product when a Mellin transformation is performed:

$$\mathcal{M}[Z_1 \vee Z_2](\beta) = \mathcal{M}[Z_1](\beta) \, \mathcal{M}[Z_2](\beta) \qquad (3.83)$$

To prove this result, we write the definition of $\mathcal{M}[Z_1 \vee Z_2](\beta)$ which is, according to (3.35) and (3.83),

$$\mathcal{M}[Z_1 \vee Z_2](\beta) = \int_0^{\infty} v^{2\pi j\beta + r} Z_1(v') \, Z_2\left(\frac{v}{v'}\right) \frac{dv'}{v'} dv \qquad (3.84)$$

The change of variables from v to $x = v/v'$ yields the result.

THEOREM 11.3.3

The Mellin transform of the multiplicative convolution (3.80) of functions Z_1 and Z_2 is equal to the product of their Mellin transforms:

$$\mathcal{M}[Z_1 \vee Z_2](\beta) = \mathcal{M}[Z_1](\beta) \, \mathcal{M}[Z_2](\beta) \qquad (3.85)$$

REMARK It can be easily verified that the above theorems remain true if Z_1, Z_2 are distributions, provided the composition laws involved in the formulas may be applied. ∎

11.3.2 *Discretization and fast computation of the transform*

Discretization of the Mellin transform (3.35) is performed along the same lines as discretization of the Fourier transform. It concerns signals with support practically limited, both in v-space and in β-space. The result is a discrete formula giving a linear relation between N geometrically spaced samples of $Z(v)$ and N arithmetically spaced samples of $\mathcal{M}[Z](\beta)$ [19, 20, 23]. The fast computation of this discretized transform involves the same algorithms as used in the Fast Fourier Transformation (FFT).

Before proceeding to the discretization itself, we introduce the special notions of sampling and periodizing that will be applied to the function $Z(v)$.

11.3.2.1 Sampling in original and Mellin variables

Sampling and periodizing are operations that are well defined in the Mellin space of functions $\tilde{Z}(\beta)$ and can be expressed in terms of Dirac combs. We shall show that the corresponding operations in the space of original functions $Z(v)$ involve the geometrical Dirac combs introduced in Section 11.3.1.3.

Arithmetic sampling in Mellin space   Given a function $M(\beta) \equiv \mathcal{M}[Z](\beta)$, the arithmetically sampled function $M_S(\beta)$ with sample interval $1/\ln Q$, Q real, is usually defined by

$$M_S(\beta) \equiv \frac{1}{\ln Q} \sum_{n=-\infty}^{+\infty} \mathcal{M}[Z](\beta) \, \delta\left(\beta - \frac{n}{\ln Q}\right) \tag{3.86}$$

Remark that besides sampling, this definition contains a factor $1/\ln Q$ that is a matter of convenience.

To compute the inverse Mellin transform of this function $M_S(\beta)$, we remark that, due to relation (3.68), it can also be written as a product of Mellin transforms in the form

$$M_S(\beta) = \mathcal{M}[Z](\beta) \, \mathcal{M}[\Delta_Q^r](\beta) \tag{3.87}$$

where Δ_Q^r is defined in (3.62). Now applying property (3.83) on the Mellin transform of multiplicative convolution, we write M_S as

$$M_S(\beta) = \mathcal{M}[Z \vee \Delta_Q^r](\beta) \tag{3.88}$$

This relation implies that the inverse Mellin transform of the impulse function $M_S(\beta)$ is the function $Z^D(v)$ given by

$$Z^D(v) \equiv (Z \vee \Delta_Q^r)(v) \tag{3.89}$$

The definition of Z^D can be cast into a more explicit form by using the definition of the multiplicative convolution and the expression (3.62) of Δ_Q^r:

$$(Z \vee \Delta_Q^r)(v) = \int_0^{+\infty} Z\left(\frac{v}{v'}\right) \left[\sum_{n=-\infty}^{+\infty} Q^{-nr} \, \delta(v' - Q^n)\right] \frac{dv'}{v'} \tag{3.90}$$

The expression (3.89) finally becomes

$$Z^D(v) = \sum_{n=-\infty}^{+\infty} Q^{n(r+1)} \, Z(Q^n v) \tag{3.91}$$

As seen in Figure 11.4, function Z^D is constructed by juxtaposing dilated replicas of Z. This operation will be referred to as *dilatocycling* and the function Z^D itself as the *dilatocycled form* of Z with ratio Q. In the special case where the support of function Z is the interval

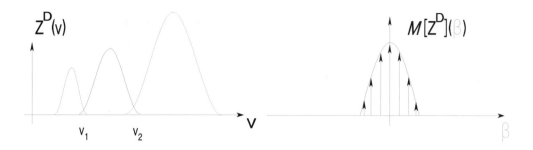

FIGURE 11.4
Correspondence between a dilatocycled form of a function and its Mellin transform.

$[v_1, v_2]$ and the ratio Q verifies $Q \geq v_2/v_1$, the restriction of Z^D to the support $[v_1, v_2]$ is equal to the original function Z.

RESULT   *The dilatocycled form Z^D of Z defined by multiplicative convolution with the geometric Dirac comb Δ_Q^r:*

$$Z^D(v) = (Z \vee \Delta_Q^r)(v) \tag{3.92}$$

where

$$\Delta_Q^r(v) \equiv \sum_{n=-\infty}^{+\infty} Q^{nr} \, \delta(v - Q^{-n}) \tag{3.93}$$

is connected by Mellin's correspondence to the sampled form of $\mathcal{M}[Z](\beta)$ given by

$$M_S(\beta) \equiv \frac{1}{\ln Q} \sum_{n=-\infty}^{+\infty} \mathcal{M}[Z](\beta) \, \delta\left(\beta - \frac{n}{\ln Q}\right) \tag{3.94}$$

▌

Geometric sampling in the original space   Given a function $Z(v)$, its geometrically sampled version is defined as the function Z_S equal to the invariant product of Z with the geometric Dirac comb Δ_q^r, q real, that is, as

$$Z_S \equiv Z \circ \Delta_q^r \tag{3.95}$$

or, using definition (3.69) of the invariant product:

$$Z_S(v) = Z(v) \sum_{n=-\infty}^{+\infty} q^{-nr} \, \delta(v - q^n) \, v^{r+1} \tag{3.96}$$

$$= \sum_{n=-\infty}^{+\infty} q^n \, Z(q^n) \, \delta(v - q^n) \tag{3.97}$$

This is a function made of impulses located at points forming a geometric progression (Fig. 11.5).

Let us compute its Mellin transform denoted by $M^P(\beta)$. Using definition (3.95) and property (3.74), we can write

$$M^P(\beta) \equiv \mathcal{M}[Z_S](\beta) \tag{3.98}$$

FIGURE 11.5
Correspondence between the geometrically sampled function and its Mellin transform.

$$= \mathcal{M}[Z \circ \Delta_q^r](\beta) \tag{3.99}$$

$$= (\mathcal{M}[Z] * \mathcal{M}[\Delta_q^r])(\beta) \tag{3.100}$$

Thus function $M^P(\beta)$ is equal to the convolution between $\mathcal{M}[Z]$ and the transform $\mathcal{M}[\Delta_q^r]$, which has been shown in (3.68) to be a classical Dirac comb. As a consequence, it is equal to the classical periodized form of $\mathcal{M}[Z](\beta)$, which is given explicitly by

$$M^P(\beta) \equiv \frac{1}{\ln q} \sum_{n=-\infty}^{+\infty} \mathcal{M}[Z]\left(\beta - \frac{n}{\ln q}\right) \tag{3.101}$$

If the function $M(\beta) \equiv \mathcal{M}[Z](\beta)$ is equal to zero outside the interval $[\beta_1, \beta_2]$, then to avoid aliasing, the period $1/\ln q$ must be chosen such that

$$\frac{1}{\ln q} \geq \beta_2 - \beta_1 \tag{3.102}$$

In that case, the functions $\mathcal{M}^P(\beta)$ and $(1/\ln q)\mathcal{M}(\beta)$ coincide on the interval $[\beta_1, \beta_2]$.

RESULT   *The geometrically sampled form of $Z(v)$ defined by*

$$Z_S(v) \equiv \sum_{n=-\infty}^{+\infty} q^n\, Z(q^n)\, \delta(v - q^n) \tag{3.103}$$

is connected by Mellin's correspondence to the periodized form of $\mathcal{M}[Z](\beta)$ given by

$$\mathcal{M}^P(\beta) = \frac{1}{\ln q} \sum_{n=-\infty}^{\infty} \mathcal{M}[Z]\left(\beta - \frac{n}{\ln q}\right) \tag{3.104}$$

∎

11.3.2.2   The discrete Mellin transform

Let $Z(v)$ be a function with Mellin transform $\mathcal{M}[Z](\beta)$ and suppose that these functions significantly differ from zero only on the intervals $[v_1, v_2]$ and $[\beta_1, \beta_2]$ respectively (see Fig. 11.6a). The operations leading to a discretized version of the transform are now detailed. As will be seen, the procedure is very similar to that used in the Fourier case.

Dilatocycle function $Z(v)$ with ratio Q. This operation leads to the function Z^D defined by (3.92). To avoid aliasing, the real number Q must be chosen such that

$$Q \geq \frac{v_2}{v_1} \tag{3.105}$$

The Mellin transform of Z^D is the sampled function $\mathcal{M}_S$ defined by (3.94) in terms of $\mathcal{M}[Z](\beta)$ (Fig. 11.6b).

Periodize $\mathcal{M}_S(\beta)$ with a period $1/\ln q$. This is performed by rule (3.104) and yields a function $\mathcal{M}_S^P(\beta)$ given by

$$\mathcal{M}_S^P(\beta) = \frac{1}{\ln q} \sum_{n=-\infty}^{\infty} \mathcal{M}_S\left(\beta - \frac{n}{\ln q}\right) \tag{3.106}$$

To avoid aliasing in β-space, the period must be chosen greater than the approximate support of $\mathcal{M}[Z](\beta)$ and this leads to the condition

$$\frac{1}{\ln q} \geq \beta_2 - \beta_1 \tag{3.107}$$

The inverse Mellin transform of $\mathcal{M}_S^P$ is the geometrically sampled form of Z^D (Fig. 11.6c). given, according to (3.103), by

$$Z_S^D(v) = \sum_{n=-\infty}^{\infty} q^n \, Z^D(q^n) \, \delta(v - q^n) \tag{3.108}$$

The use of (3.94) allows us to rewrite definition (3.106) as

$$\mathcal{M}_S^P(\beta) = \frac{1}{\ln q \, \ln Q} \sum_{n,p=-\infty}^{\infty} \mathcal{M}\left(\frac{p}{\ln Q}\right) \delta\left(\beta - \frac{n}{\ln q} - \frac{p}{\ln Q}\right) \tag{3.109}$$

We now impose that the real numbers q and Q be connected by the relation

$$Q = q^N, \qquad N \text{ positive integer} \tag{3.110}$$

This ensures that the function $\mathcal{M}_S^P$ defined by (3.106) is of a periodic impulse type that can be written as

$$\mathcal{M}_S^P(\beta) = \frac{1}{\ln q \, \ln Q} \sum_{n,p=-\infty}^{\infty} \mathcal{M}\left(\frac{p}{N \ln q}\right) \delta\left(\beta - \frac{nN + p}{N \ln q}\right) \tag{3.111}$$

or, changing the p-index to $k \equiv p + nN$,

$$\mathcal{M}_S^P(\beta) = \frac{1}{\ln q \, \ln Q} \sum_{n,k=-\infty}^{\infty} \mathcal{M}\left(\frac{k}{N \ln q} - \frac{n}{\ln q}\right) \delta\left(\beta - \frac{k}{N \ln q}\right) \tag{3.112}$$

Thus, recalling definition (3.104),

$$\mathcal{M}_S^P(\beta) = \frac{1}{\ln Q} \sum_{k=-\infty}^{\infty} \mathcal{M}^P\left(\frac{k}{\ln Q}\right) \delta\left(\beta - \frac{k}{\ln Q}\right) \tag{3.113}$$

Connect the v and β samples. This is done by writing explicitly that $\mathcal{M}_S^P$ as given by (3.113) is the Mellin transform (3.24) of Z_S^D defined by (3.108)

$$\mathcal{M}_S^P(\beta) = \sum_{n=-\infty}^{\infty} q^{n(r+1)} \, Z^D(q^n) \, e^{2j\pi n\beta \ln q} \tag{3.114}$$

This formula shows that expression $q^{n(r+1)} Z^D(q^n)$ for different values of n are the Fourier series coefficients of the periodic function $\mathcal{M}_S^P(\beta)$. They are computed as

$$Z^D(q^n) = q^{-n(r+1)} \ln q \int_0^{1/\ln q} \frac{1}{\ln Q} \sum_{k=-\infty}^{\infty} \delta\left(\beta - \frac{k}{\ln Q}\right) \mathcal{M}^P\left(\frac{k}{\ln Q}\right) e^{-j2\pi n\beta \ln q} \, d\beta$$

$$= \frac{q^{-n(r+1)}}{N} \sum_{k=K}^{K+N-1} \mathcal{M}\left(\frac{k}{\ln Q}\right) e^{-2j\pi kn/N} \tag{3.115}$$

where the summation is on those values of β lying inside the interval $[\beta_1, \beta_2]$. The integer K is thus given by the integer part of $\beta_1 \ln Q$.

Inversion of (3.115) is performed using the classical techniques of discrete Fourier transform. This leads to the discrete Mellin transform formula:

$$\mathcal{M}^P\left(\frac{m}{\ln Q}\right) = \sum_{n=M}^{M+N-1} q^{n(r+1)} Z^D(q^n) e^{2j\pi nm/N} \tag{3.116}$$

where the integer M is given by the integer part of $\ln v_1/\ln q$. In fact, because the definition of the periodized $\mathcal{M}^P$ contains a factor $N/\ln Q = 1/\ln q$, the true samples of $\mathcal{M}(\beta)$ are given by $(\ln Q/N)\mathcal{M}^P(m/\ln Q)$.

It is clear in formulas (3.115) and (3.116) that their implementation can be performed with an FFT algorithm.

Choose the number of samples to handle. The number of samples N is related to q and Q according to (3.110) by

$$N = \frac{\ln Q}{\ln q} \tag{3.117}$$

The conditions for nonaliasing given by (3.105) and (3.107) lead to the sampling condition

$$N \geq (\beta_2 - \beta_1) \ln\left(\frac{v_2}{v_1}\right) \tag{3.118}$$

which gives the minimum number of samples to consider in terms of the spreads of $Z(v)$ and $\mathcal{M}[Z](\beta)$. In practice, the spread of the Mellin transform of a function is seldom known. However, as we will see in the applications, there are methods to estimate it.

11.3.2.3   The reconstruction formula

In the same way as the Fourier transformation is used to reconstruct a band-limited function from its regularly spaced samples, Mellin's transformation allows us to recover a function $Z(v)$ with limited spread in the Mellin space from its samples spaced according to a geometric progression. If the Mellin transform $\mathcal{M}[Z]$ has a bounded support $[-\beta_0/2, \beta_0/2]$, it will be equal on this interval to its periodized form with period $1/\ln q = \beta_0$. Thus

$$\mathcal{M}[Z](\beta) = \sum_{n=-\infty}^{+\infty} \mathcal{M}[Z]\left(\beta - \frac{n}{\ln q}\right) g\left(\frac{\beta}{\beta_0}\right) \tag{3.119}$$

where the window function g is the characteristic function of the $[-1/2, 1/2]$ interval.

The inverse Mellin transform of this product is the multiplicative convolution of the two functions Z_1 and Z_2 defined as

$$Z_1(v) = \ln q \sum_{n=-\infty}^{+\infty} q^n Z(q^n) \delta(v - q^n) \tag{3.120}$$

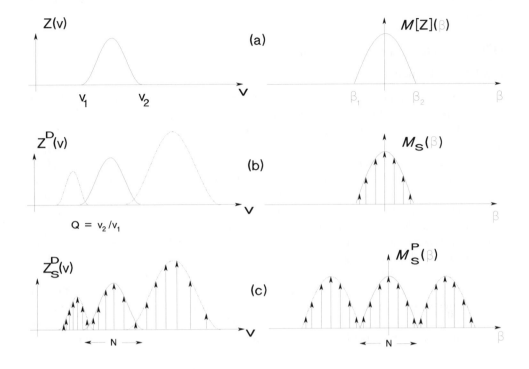

FIGURE 11.6
Construction scheme of the discrete Mellin transform. (a) Continuous form of the spectrum $Z(v)$ and its Mellin transform. (b) Dilatocycling of $Z(v)$ with the ratio $Q = v_2/v_1$ and the sampled form of its Mellin transform with rate $1/\ln Q$. (c) Geometric sampling with ratio q ($q^N = Q$) of the dilatocycled form and periodized form with period $1/\ln q$ of the sampled Mellin transform. Identification of the N samples in frequency and Mellin spaces leads to the discrete Mellin transform.

and

$$Z_2(v) = \int_{-\infty}^{+\infty} g\left(\frac{\beta}{\beta_0}\right) v^{-2j\pi\beta - r - 1}\, d\beta \tag{3.121}$$

$$= v^{-r-1}\, \frac{\sin(\pi\beta_0 \ln v)}{\pi \ln v} \tag{3.122}$$

The multiplicative convolution between Z_1 and Z_2 takes the following form

$$Z(v) = \int_0^{+\infty} \ln q \sum_{n=-\infty}^{+\infty} q^n Z(q^n)\delta(v' - q^n) \left(\frac{v}{v'}\right)^{-r-1} \frac{\sin\left(\pi\beta_0 \ln\left(\frac{v}{v'}\right)\right)}{\pi \ln\left(\frac{v}{v'}\right)} \frac{dv'}{v'} \tag{3.123}$$

and leads to the exact reconstruction formula

$$Z(v) = v^{-r-1} \sum_{n=-\infty}^{+\infty} q^{n(r+1)} Z(q^n) \frac{\sin \pi\left(\left(\frac{\ln v}{\ln q}\right) - n\right)}{\pi\left(\left(\frac{\ln v}{\ln q}\right) - n\right)} \tag{3.124}$$

This is the reconstruction formula of a function $Z(v)$ from its geometrically spaced samples $Z(q^n)$.

11.3.3 *Practical use in signal analysis*

11.3.3.1 Preliminaries

As seen above, Mellin's transformation is essential in problems involving dilations. Thus it is not surprising that it has come to play a dominant role in the development of analytical studies of wide-band signals. In fact, expressions involving dilations arise in signal theory any time the approximation of small relative bandwidth is not appropriate. Recent examples of the use of the Mellin transform in this context can be found in time-frequency analysis where it has contributed to the introduction of several classes of distributions [24–31]. This fast-growing field cannot be explored here but an illustration of the essential role played by Mellin's transformation in the analysis of wide-band signals will be given in Section 11.3.3.2 where the Cramer-Rao bound for velocity estimation is derived [32].

Numerical computation of Mellin's transform has been undertaken in various domains such as signal analysis [33, 34], optical image processing [35], or pattern recognition [36–38]. In the past, however, all these applications have been restricted by the difficulty of assessing the validity of the results, due to the lack of definite sampling rules. Such a limitation does not exist any more as we will show in Section 11.3.3.3 by deriving a sampling theorem and a practical way to use it. The technique will be applied in Sections 11.3.3.4 and 11.3.3.5 to the computation of a wavelet coefficient and of an affine time-frequency distribution [42–44].

11.3.3.2 Computation of Cramer-Rao Bounds for Velocity Estimation in radar theory [32]

In a classical radar or sonar experiment, a real signal is emitted and its echo is processed in order to find the position and velocity of the target. In simple situations, the received signal will differ from the original one only by a time shift and a Doppler compression. In fact, the signal will also undergo an attenuation and a phase shift; moreover, the received signal will be embedded in noise.

The usual procedure, which is adapted to narrow-band signals, is to represent the Doppler effect by a frequency shift [39]. This approximation will not be made here so that the results will be valid whatever the extent of the frequency band. Describing the relevant signals by their positive frequency parts (so-called analytic signals), we can write the expression of the received signal $x(t)$ in terms of the emitted signal $z(t)$ and noise $n(t)$ as

$$x_{\mathbf{a}'}(t) = a_1'^{-1/2} A_0 z(a_1'^{-1} t - a_2') e^{j\phi} + n(t) \tag{3.125}$$

where A_0 and ϕ characterize the unknown changes in amplitude and phase and the vector $\mathbf{a}' \equiv (a_1', a_2')$ represents the unknown parameters to be estimated. The parameter a_2' is the delay and a_1' is the Doppler compression given in terms of the target velocity v by

$$a_1' = \frac{c + v}{c - v}, \qquad c \text{ velocity of light} \tag{3.126}$$

The noise $n(t)$ is supposed to be a zero mean gaussian white noise with variance equal to σ^2. Relation (3.125) can be written in terms of the Fourier transforms Z, X, N of z, x, n (defined by (2.19)):

$$X_{\mathbf{a}'}(f) = a_1'^{1/2} A_0 e^{-2j\pi f a_1' a_2'} Z(a_1' f) e^{j\phi} + N(f) \tag{3.127}$$

The signal $Z(f)$ is supposed normalized so that

$$\|Z(f)\|^2 \equiv \int_0^\infty |Z(f)|^2 \, df = 1 \qquad (3.128)$$

Hence, the delayed and compressed signal will also be of norm equal to one. Remark that here we work in the space $L^2(\mathbb{R}^+, f^{2r+1} \, df)$ with $r = -1/2$ (cf. Section 11.3.1.1).

We will consider the maximum-likelihood estimates $\hat{a}_i$ of the parameters a_i'. They are obtained by maximizing the likelihood function $\Lambda(\mathbf{a}', \mathbf{a})$, which is given in the present context by

$$\Lambda(\mathbf{a}', \mathbf{a}) \equiv \frac{1}{2\sigma^2} |A(\mathbf{a}', \mathbf{a})|^2 \qquad (3.129)$$

where

$$A(\mathbf{a}', \mathbf{a}) \equiv \int_0^{+\infty} X_{\mathbf{a}'}(f) \, Z^*(a_1 f) \, e^{2j\pi a_1 a_2 f} \, df \qquad (3.130)$$

is the broad-band ambiguity function [40].

The efficiency of an estimator $\hat{a}_i$ is measured by its variance σ_{ij}^2 defined by

$$\sigma_{ij}^2 \equiv E[(\hat{a}_i - a_i)(\hat{a}_j - a_j)] \qquad (3.131)$$

where the mean value operation E includes an average on noise.

For an unbiased estimator ($E(\hat{a}_i) = a_i$), this variance satisfies the Cramer Rao inequality [41] given by

$$\sigma_{ij}^2 \geq (J^{-1})_{ij} \qquad (3.132)$$

where the matrix J, the so-called Fisher information matrix, is defined by

$$J_{ij} = \left(-E \left[\frac{\partial^2 \Lambda}{\partial a_i \partial a_j} \right] \right)_{ij} \qquad (3.133)$$

with the partial derivatives evaluated at the true values of the parameters. The minimum value of the variance given by

$$(\sigma_{ij}^0)^2 = (J^{-1})_{ij} \qquad (3.134)$$

is called the Cramer-Rao bound and is attained in the case of an efficient estimator such as the maximum-likelihood one.

The determination of the matrix (3.133) by classical methods is intricate and does not lead to an easily interpretable result. On the contrary, we shall see how the use of Mellin's transformation allows a direct computation and leads to a physical interpretation of the matrix coefficients.

The computation of J is done in the vicinity of the value $\mathbf{a} = \mathbf{a}'$, which maximizes the likelihood function Λ, and without loss of generality, all partial derivatives will be evaluated at the point $a_1 = 1, a_2 = 0$. Using Parseval's formula (3.37), we can write the ambiguity function $A(\mathbf{a}', \mathbf{a})$ as

$$A(\mathbf{a}', \mathbf{a}) = \int_{-\infty}^{+\infty} \mathcal{M}[X](\beta) \, \mathcal{M}^*[Z_{a_2}](\beta) \, a_1^{2j\pi\beta} \, d\beta \qquad (3.135)$$

with

$$Z_{a_2} \equiv Z(f) e^{-2j\pi a_2 f} \qquad (3.136)$$

In this form, the partial derivatives with respect to **a** are easily computed and the result is

$$\left\{\frac{\partial A}{\partial a_1}\right\} = 2j\pi \int_{-\infty}^{+\infty} \beta \, \mathcal{M}[X](\beta) \, \mathcal{M}^*[Z](\beta) \, d\beta \tag{3.137}$$

$$\left\{\frac{\partial A}{\partial a_2}\right\} = 2j\pi \int_{-\infty}^{+\infty} \mathcal{M}[X](\beta) \, \mathcal{M}^*[fZ(f)](\beta) \, d\beta \tag{3.138}$$

$$= 2j\pi \int_{0}^{+\infty} f \, X(f) \, Z^*(f) \, df \tag{3.139}$$

$$\left\{\frac{\partial^2 A}{\partial a_1 \partial a_2}\right\} = -4\pi^2 \int_{-\infty}^{+\infty} \beta \, \mathcal{M}[X](\beta) \, \mathcal{M}^*[fZ(f)](\beta) \, d\beta \tag{3.140}$$

$$\left\{\frac{\partial^2 A}{\partial a_1^2}\right\} = 2j\pi \int_{-\infty}^{+\infty} \beta(2j\pi\beta - 1) \, \mathcal{M}[X](\beta) \, \mathcal{M}^*[Z](\beta) \, d\beta \tag{3.141}$$

$$\left\{\frac{\partial^2 A}{\partial a_2^2}\right\} = -4\pi^2 \int_{0}^{+\infty} f^2 \, X(f) \, Z^*(f) \, df \tag{3.142}$$

where the curly brackets mean that the functions are evaluated for the values $a_1 = a_1' = 1$, $a_2 = a_2' = 0$.

The corresponding Fisher information matrix can now be computed. To obtain J_{11}, we substitute expression (3.129) in definition (3.133) and use (3.137) and (3.141)

$$J_{11} = -\frac{1}{\sigma^2} E\left[\text{Re}\left(A^* \left\{\frac{\partial^2 A}{\partial a_1^2}\right\} + \left\{\left|\frac{\partial A}{\partial a_1}\right|^2\right\}\right)\right] \tag{3.143}$$

$$= -\frac{1}{\sigma^2} \text{Re} \int_{-\infty}^{+\infty} \int_{-\infty}^{+\infty} E\left[\mathcal{M}[X](\beta_1)\mathcal{M}^*[X](\beta_2)\right] \mathcal{M}^*[Z](\beta_1)\mathcal{M}[Z](\beta_2)$$

$$\times \left[2j\pi\beta_1(2j\pi\beta_1 - 1) + 4\pi^2\beta_1\beta_2\right] d\beta_1 \, d\beta_2 \tag{3.144}$$

The properties of the zero mean white gaussian noise $n(t)$ lead to the following expression for the covariance of the Mellin transform of X:

$$E\left[\mathcal{M}[X](\beta_1)\mathcal{M}^*[X](\beta_2)\right] = A_0^2 \, \mathcal{M}[Z](\beta_1)\mathcal{M}^*[Z](\beta_2) + \sigma^2 \, \delta(\beta_1 - \beta_2) \tag{3.145}$$

Substituting this relation in (3.144), we obtain the expression of the J_{11} coefficient

$$J_{11} = \frac{4\pi^2 A_0^2}{\sigma^2} \sigma_\beta^2 \tag{3.146}$$

where the variance σ_β^2 of parameter β defined in (3.41) is given explicitly by

$$\sigma_\beta^2 = \int_{-\infty}^{+\infty} (\beta - \bar{\beta})^2 \, |\mathcal{M}[Z](\beta)|^2 \, d\beta, \qquad \bar{\beta} = \int_{-\infty}^{+\infty} \beta \, |\mathcal{M}[Z](\beta)|^2 \, d\beta \tag{3.147}$$

The computation of the J_{22} coefficient is performed in the same way and leads to

$$J_{22} = \frac{4\pi^2 A_0^2}{\sigma^2} \sigma_f^2 \tag{3.148}$$

where

$$\sigma_f^2 = \int_{-\infty}^{+\infty} (f - \bar{f})^2 \, |Z(f)|^2 \, df, \qquad \bar{f} = \int_{-\infty}^{+\infty} f \, |Z(f)|^2 \, df \qquad (3.149)$$

The computation of the symmetrical coefficient $J_{12} = J_{21}$ is a little more involved. Writing the definition in the form

$$J_{12} = -\frac{1}{\sigma^2} E\left[\mathrm{Re}\left(A^* \left\{\frac{\partial^2 A}{\partial a_1 \, \partial a_2}\right\} + \left\{\frac{\partial A^*}{\partial a_1}\frac{\partial A}{\partial a_2}\right\}\right)\right] \qquad (3.150)$$

and using relations (3.137)–(3.140) and (3.145), we get

$$J_{12} = \frac{4\pi^2 A_0^2}{\sigma^2} \, \mathrm{Re}\left[\int_{-\infty}^{+\infty} \beta_1 \, \mathcal{M}^*[f Z(f)](\beta_2) \, \mathcal{M}[Z](\beta_1) \, d\beta_1 \right.$$

$$\left. - \int_{-\infty}^{+\infty} \mathcal{M}^*[f Z(f)](\beta_1) \, \mathcal{M}[Z](\beta_1) \, d\beta_1 \int_{-\infty}^{+\infty} \beta_2 \, |\mathcal{M}[Z](\beta_2)|^2 \, d\beta_2\right] \qquad (3.151)$$

This expression is then transformed to the frequency domain using the Parseval formula (3.37) and the property (3.32) of the operator $\mathcal{B}$ defined by equation (3.11) (with $r = -1/2$). The result is

$$J_{12} = \frac{4\pi^2 A_0^2}{\sigma^2} \left[\mathrm{Re}\int_0^{+\infty} \mathcal{B}Z(f) \, f \, Z^*(f) \, df - \bar{\beta}\,\bar{f}\right]$$

$$= \frac{4\pi^2 A_0^2}{\sigma^2} \left[M - \bar{\beta}\,\bar{f}\right] \qquad (3.152)$$

where M is the broad-band modulation index defined by

$$M \equiv \frac{1}{2\pi} \mathrm{Im}\int_0^{+\infty} f^2 \, \frac{dZ^*}{df} \, Z(f) \, df \qquad (3.153)$$

The inversion of the matrix J just obtained leads, according to (3.132), to the explicit expression of the Cramer-Rao bound for the case of delay and velocity estimation with broad-band signals:

$$(\sigma_{ij}^0)^2 = \frac{\sigma^2}{4\pi^2 A_0^2 (\sigma_f^2 \sigma_\beta^2 - (M - \bar{\beta}\,\bar{f})^2)} \begin{pmatrix} \sigma_f^2 & \bar{\beta}\,\bar{f} - M \\ \bar{\beta}\,\bar{f} - M & \sigma_\beta^2 \end{pmatrix} \qquad (3.154)$$

Relation (3.126) allows us to deduce from this result the minimum variance of the velocity estimator:

$$E\left[(v - \hat{v})^2\right] = \frac{c^2}{4} E\left[(a_1 - \hat{a}_1)^2\right] \qquad (3.155)$$

$$= \frac{c^2}{4} (\sigma_{11}^0)^2 \qquad (3.156)$$

Comparing these results to the narrow-band case, we see that the delay resolution measured by σ_{22}^0 is still related to the spread of the signal in frequency

$$(\sigma_{22}^0)^2 \geq \frac{\sigma^2}{4\pi^2 A_0^2} \frac{1}{\sigma_f^2} \qquad (3.157)$$

while the velocity resolution now depends in an essential way on the spread in Mellin's space

$$E\left[(v - \hat{v})^2\right] \geq \frac{c^2 \sigma^2}{16\pi^2 A_0^2 \sigma_\beta^2} \qquad (3.158)$$

Thus, for wide-band signals, it is not the duration of the signal that determines the velocity resolution but the spread in the dual Mellin variable measured by the variance σ_β^2.

As an illustrative example, consider the hyperbolic signal defined by

$$Z(f) = f^{-2j\pi\beta_0 - 1/2} \tag{3.159}$$

Its Mellin transform that is equal to $\delta(\beta - \beta_0)$ can be considered to have zero spread in β. Hence, such a signal cannot be of any help if seeking a finite velocity resolution.

These remarks can be developed and applied to the construction of radar codes with given characteristics in the variables f and β [32].

The above results can be seen as a generalization to arbitrary signals of a classical procedure because, in the limit of the narrow band, the variance of the velocity estimator can be shown to tend toward its usual expression

$$E\left[(v - \hat{v})^2\right] = \frac{c^2\sigma^2}{16\pi^2 A_0^2 f_0^2} \frac{\sigma_f^2}{\sigma_t^2\sigma_f^2 - (m - f_0 t_0)^2} \tag{3.160}$$

where the modulation index m is given by

$$m = \frac{1}{2\pi} \operatorname{Im} \int_{-\infty}^{+\infty} t\, z^*(t)\, \frac{dz}{dt}\, dt = \frac{1}{2\pi} \operatorname{Im} \int_{-\infty}^{+\infty} t\, Z(f)\, \frac{dZ^*}{df}\, df \tag{3.161}$$

and the variance σ_t^2 by

$$\sigma_t^2 = \int_{-\infty}^{\infty} (t - \bar{t})^2 |z(t)|^2\, dt, \qquad \bar{t} = \int_{-\infty}^{\infty} t |z(t)|^2\, dt \tag{3.162}$$

11.3.3.3 Interpretation of the dual Mellin variable in relation to time and frequency

Consider a signal defined by a function of time $z(t)$ such that its Fourier transform $Z(f)$ has only positive frequencies (so-called analytic signal). In that case a Mellin transformation can be applied to $Z(f)$ and yields a function $\mathcal{M}[Z](\beta)$. But while variables t and f have a well-defined physical meaning as time and frequency, the interpretation of variable β and its relation to physical parameters of the signal has still to be worked out. This will be done in this section, thus allowing a formulation of the sampling condition (3.118) for the Mellin transform in terms of the time and frequency spreads of the signal.

As seen in Section 11.3.1.1, the Mellin transform $\mathcal{M}[Z](\beta)$ gives the coefficients of the decomposition of Z on the basis $\{E_\beta(f)\}$:

$$Z(f) = \int_{-\infty}^{\infty} \mathcal{M}[Z](\beta) E_\beta(f)\, d\beta \tag{3.163}$$

The elementary parts

$$E_\beta(f) = f^{-2\pi j\beta - r - 1} \equiv f^{-r-1} e^{j\phi(f)} \tag{3.164}$$

can be considered as filters with group delay given by

$$T(f) \equiv -\frac{1}{2\pi} \frac{d\phi(f)}{df}$$

$$= \frac{\beta}{f} \tag{3.165}$$

As seen in this expression, the variable β has no dimension and labels hyperbolas in a time-frequency half-plane $f > 0$. Hyperbolas displaced in time, corresponding to a group delay law $t = \xi + \beta/f$, are obtained by time shifting the filters E_β to $E_\beta^\xi(f)$ defined by

$$E_\beta^\xi(f) = e^{-2\pi j\xi f} f^{-2\pi j\beta - r - 1} \qquad (3.166)$$

A more precise characterization of signals (3.164) and hence of variable β is obtained from a study of a particular affine time-frequency distribution that is to dilations what Wigner-Ville's is to frequency translations. We give only the practical results of the study, referring the interested reader to the literature [24–26]. The explicit form of the distribution is

$$P_0(t, f) = f^{2r+2} \int_{-\infty}^{+\infty} (\lambda(u)\lambda(-u))^{r+1} \; Z(f\lambda(u)) \; Z^*(f\lambda(-u)) \, e^{2j\pi ftu} \, du \qquad (3.167)$$

where function λ is given by

$$\lambda(u) = \frac{ue^{u/2}}{2\sinh u/2} \qquad (3.168)$$

This distribution realizes an exact localization of hyperbolic signals defined by (3.166) on hyperbolas of the time-frequency half-plane as follows:

$$Z(f) = e^{-2\pi j\xi f} f^{-r-1} f^{-2j\pi\beta} \rightarrow P_0(t, f) = f^{-1} \delta(t - \xi - \beta/f) \qquad (3.169)$$

It can be shown that the affine time-frequency distribution (3.167) has the so-called tomographic property [24–26] that reads

$$\int_{-\infty}^{+\infty} dt \int_{0}^{+\infty} P_0(t, f) \, \delta(t - \xi - \beta/f) \, f^{-1} \, df = |\mathcal{M}[Z](\beta)|^2 \qquad (3.170)$$

Formulas (3.169) and (3.170) are basic for the interpretation of the β variable. It can be shown that for a signal $z(t) \leftrightarrow Z(f)$ having a duration $T = t_2 - t_1$ and a bandwidth $B = f_2 - f_1$, distribution P_0 has a support approximately localized in a bounded region of the half-plane $f > 0$ (see Fig. 11.7) around the time $\xi = (t_1 + t_2)/2$ and the mean frequency $f_0 = (f_1 + f_2)/2$. Writing that the hyperbolas at the limits of this region have the equation

$$t = \xi \pm \beta_0/f \qquad (3.171)$$

and pass through the points of coordinates $\xi \pm T/2$, $f_0 + B/2$, we find

$$\beta_0 = (f_0 + B/2)(T/2) \qquad (3.172)$$

The support $[\beta_1, \beta_2]$ of the Mellin transform $\mathcal{M}[Z](\beta)$ can thus be written in terms of B and T as

$$\beta_2 - \beta_1 = 2\beta_0 \qquad (3.173)$$

Condition (3.118), to avoid aliasing when performing a discrete Mellin transform, can now be written in terms of the time-bandwidth product BT and the relative bandwidth R defined by

$$R \equiv \frac{B}{f_0} \qquad (3.174)$$

The result giving the minimum number of samples to treat is

$$N \geq BT \left(\frac{1}{2} + \frac{1}{R}\right) \ln \frac{1 + R/2}{1 - R/2} \qquad (3.175)$$

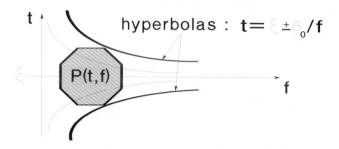

FIGURE 11.7
Time-frequency localization of a signal between hyperbolas with equations $t = \xi + \beta_0/f$ and
$t = \xi - \beta_0/f$.

11.3.3.4   The Mellin transform and the wavelet transform

The Mellin transform is well suited to the computation of expressions containing dilated
functions and, in particular, of scalar products such as

$$(Z_1, \mathcal{D}_a Z_2) = a^{r+1} \int_0^{+\infty} Z_1(f) \, Z_2^*(af) \, f^{2r+1} \, df \tag{3.176}$$

Because of the dilation parameter, a numerical computation of these functions of a by
standard techniques requires the use of oversampling and interpolation. By contrast, the
Mellin transform allows a direct and more efficient treatment. The method will be explained
in the example of the wavelet transform for one-dimensional signals. But it can also be used
in more general situations such as those encountered in radar imaging [42, 43].
   Let $s(t)$ be a real signal with Fourier transform $S(f)$ defined by

$$S(f) = \int_{-\infty}^{\infty} s(t) \, e^{-2j\pi tf} \, dt \tag{3.177}$$

The reality of s implies that

$$S(-f) = S^*(f) \tag{3.178}$$

Given a real function $\phi(t)$ (the so-called mother wavelet), one defines the continuous wavelet
transform of signal $s(t)$ as a function $C(a, b)$ of two variables $a > 0$, b real, given by
(Chapter 10)

$$C(a, b) = \frac{1}{\sqrt{a}} \int_{-\infty}^{+\infty} z(t) \, \phi^* \left(\frac{t - b}{a} \right) dt \tag{3.179}$$

   Transposed to the frequency domain by a Fourier transformation and the use of property
(3.178), the definition becomes

$$C(a, b) = 2 \operatorname{Re} \left\{ \sqrt{a} \int_0^{+\infty} Z(f) \, \Phi^*(af) \, e^{2j\pi fb} \, df \right\} \tag{3.180}$$

where Φ denotes the Fourier transform of ϕ.
   If we define the function $Z_b(f)$ by

$$Z_b(f) \equiv Z(f) \, e^{2j\pi bf} \tag{3.181}$$

the scale invariance property (3.39) of the Mellin transform with $r = -1/2$ and the unitarity property (3.37) allow us to write (3.180) in Mellin's space as

$$C(a, b) = \int_{-\infty}^{+\infty} \mathcal{M}[Z_b](\beta) \, \mathcal{M}^*[\Phi](\beta) \, a^{2j\pi\beta} \, d\beta \tag{3.182}$$

In this form, there are no more dilations and the computation of the wavelet coefficient reduces to Fourier and Mellin transforms that can all be performed using an FFT algorithm. First, the Mellin transform of the wavelet is computed once and for all. Then, for each value of b, one computes the Mellin transform of Z_b and the inverse Fourier transform with respect to β of the product $\mathcal{M}[Z_b](\beta) \, \mathcal{M}^*[\Phi](\beta)$. The complexity of this algorithm is given by $(2M+1)$ FFT with $2N$ points if the wavelet coefficients are discretized in (N, M) points on the (a, b) variables.

The same procedure can be applied to the computation of the broad band ambiguity function [40]. This function is used in problems of radar theory involving target detection and estimation of its characteristics (range, velocity, angle, ...). It is defined for an analytic signal $z(t)$ with Fourier transform $Z(f)$ by

$$X(a, b) = \frac{1}{\sqrt{a}} \int_{-\infty}^{+\infty} z(t) \, z^* \left(\frac{t}{a} - b \right) dt \tag{3.183}$$

$$= \sqrt{a} \int_0^{+\infty} Z(f) \, Z^*(af) \, e^{2j\pi abf} \, df \tag{3.184}$$

The parameter a and b are respectively called the Doppler compression factor and the time shift.

11.3.3.5 Numerical computation of affine time-frequency distributions [44]

In this section, the Mellin transformation is applied to the fast computation of the affine time-frequency distribution [24–26] given by

$$P_0(t, f) = f^{2r+2-q} \int_{-\infty}^{+\infty} (\lambda(u)\lambda(-u))^{r+1} \, Z(f\lambda(u)) \, Z^*(f\lambda(-u)) \, e^{2j\pi ftu} \, du \tag{3.185}$$

where the function λ is defined by

$$\lambda(u) = \frac{u e^{\frac{u}{2}}}{2 \sinh \left(\frac{u}{2} \right)} \tag{3.186}$$

and where r and q are real numbers.

Setting

$$\gamma = ft \quad \text{and} \quad \tilde{P}_0(\gamma, f) = P_0(t, f) \tag{3.187}$$

one can write (3.185) as

$$f^{-r-1+q} \, \tilde{P}_0(\gamma, f) = \int_{-\infty}^{+\infty} (\lambda(u)\lambda(-u))^{r+1}$$

$$\times \left[f^{r+1} \, Z(f\lambda(u)) \, Z^*(f\lambda(-u)) \right] e^{2j\pi\gamma u} du \tag{3.188}$$

To perform the Mellin transformation of this expression with respect to f, we notice that the term in brackets represents the invariant product of the two functions of f defined by $Z(f\lambda(u))$ and $Z^*(f\lambda(-u))$. By relation (3.74), we know that the Mellin transform of this product is equal to the convolution of the functions $\mathcal{M}[Z(f\lambda(u))]$ and $\mathcal{M}[Z^*(f\lambda(-u))]$. Besides, the scaling property (3.39) allows us to write

$$\mathcal{M}[Z(f\lambda(u))](\beta) = \lambda(u)^{-2j\pi\beta-r-1}\,\mathcal{M}[Z](\beta) \tag{3.189}$$

and

$$\mathcal{M}[Z^*(f\lambda(-u))](\beta) = \lambda(-u)^{-2j\pi\beta-r-1}\,\mathcal{M}^*[Z](-\beta) \tag{3.190}$$

where

$$\mathcal{M}^*[Z](-\beta) \equiv [\mathcal{M}[Z](-\beta)]^* \tag{3.191}$$

Introducing the notation

$$X(\beta, u) \equiv \lambda(u)^{-2j\pi\beta}\,\mathcal{M}[Z](\beta) \tag{3.192}$$

we can write the convolution between (3.189) and (3.190) as

$$(\lambda(u)\lambda(-u))^{-r-1} \int_{-\infty}^{+\infty} X(\beta_1, u)\, X^*(\beta_1 - \beta, -u)\, d\beta_1 \tag{3.193}$$

The Mellin transform of expression (3.188) is now written as

$$\mathcal{M}[f^{-r-1+q}\,\tilde{P}_0(\gamma, f)](\beta) = \int_{-\infty}^{+\infty} \left[\int_{-\infty}^{+\infty} X(\beta_1, u)\, X^*(\beta_1 - \beta, -u)\, d\beta_1 \right] e^{2j\pi\gamma u}\, du \tag{3.194}$$

The cross-correlation inside brackets is computed in terms of the Fourier transform of $X(\beta, u)$ defined by

$$F(\theta, u) = \int_{-\infty}^{+\infty} X(\beta, u)\, e^{-2j\pi\theta\beta}\, d\beta \tag{3.195}$$

and (3.194) becomes

$$\mathcal{M}[f^{-r-1+q}\,\tilde{P}_0(\gamma, f)](\beta) = \int_{-\infty}^{+\infty} \int_{-\infty}^{+\infty} F(\theta, u)\, F^*(\theta, -u)\, e^{2j\pi\theta\beta}\, e^{2j\pi\gamma u}\, d\theta\, du \tag{3.196}$$

Finally, inverting the Mellin transform by (3.36), recalling (3.187) and taking into account the property of the integrand in the change $u \to -u$, one obtains the following form of the affine Wigner function P_0

$$P_0(t, f) = 2\,\mathrm{Re}\left\{ f^{-q} \int_0^{+\infty} F(\ln f, u)\, F^*(\ln f, -u)\, e^{2j\pi t f u}\, du \right\} \tag{3.197}$$

where Re denotes the real-part operation. In this form, the numerical computation of P_0 has been reduced to a Fourier transform. The operations leading from Z to F are a Fourier and a Mellin transform, both of which are performed using the FFT algorithm. The approximate complexity of the whole algorithm for computing P_0 can be expressed in terms of the number of FFTs performed. If the time-frequency distribution $P_0(t, f)$ is characterized by (M, N) points respectively in time and frequency, we have to deal with $2M + 1$ FFT of $2N$ points and N FFT of M points. Figure 11.8 gives an example of affine distribution computed by this method.

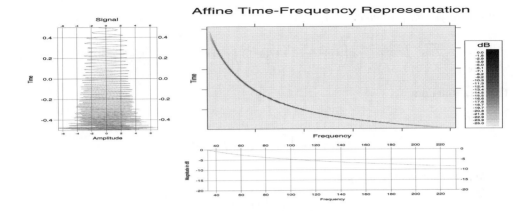

FIGURE 11.8
Affine time-frequency representation of a hyperbolic signal.

Appendix A. Some special functions frequently occurring as Mellin transforms

The gamma function (see also Chapter 1)

Definition. The gamma function $\Gamma(s)$ is defined on the complex half-plane $\text{Re}(s) > 0$ by the integral

$$\Gamma(s) = \int_0^\infty e^{-t}\, t^{s-1}\, dt \tag{3.198}$$

Analytic continuation. The analytically continued gamma function is holomorphic in the whole plane except at the points $s = -n$, $n = 0, 1, 2, \ldots$ where it has a simple pole.

Residues at the poles.

$$\text{Res}_{s=-n}(\Gamma(s)) = \frac{(-1)^n}{n!} \tag{3.199}$$

Relation to the factorial.

$$\Gamma(n+1) = n! \tag{3.200}$$

Functional relations.

$$\Gamma(s+1) = s\Gamma(s) \tag{3.201}$$

$$\Gamma(s)\Gamma(1-s) = \frac{\pi}{\sin(\pi s)} \tag{3.202}$$

$$\Gamma\left(\frac{1}{2}\right) = \sqrt{\pi} \tag{3.203}$$

$$\Gamma(2s) = \pi^{-1/2}\, 2^{2s-1}\, \Gamma(s)\Gamma(s+1/2)$$

(Legendre's duplication formula) $\tag{3.204}$

$$\Gamma(ms) = m^{ms-1/2}\, (2\pi)^{(1-m)/2} \prod_{k=0}^{m-1} \Gamma(s + k/m), \ \ m = 2, 3, \ldots$$

(Gauss-Legendre multiplication formula [5]) $\tag{3.205}$

Stirling asymptotic formula.

$$\Gamma(s) \sim \sqrt{2\pi}\, s^{s-1/2} \exp\left[-s\left(1 + \frac{1}{12s} + O(s^{-2})\right)\right] \quad s \to \infty, \quad |\arg(s)| < \pi \quad (3.206)$$

The beta function

Definition.

$$B(x, y) \equiv \int_0^1 t^{x-1} (1-t)^{y-1}\, dt \tag{3.207}$$

Relation to the gamma function.

$$B(x, y) = \frac{\Gamma(x)\Gamma(y)}{\Gamma(x+y)} \tag{3.208}$$

The psi function (logarithmic derivative of the gamma function)

Definition.

$$\psi(s) \equiv \frac{d}{ds} \ln \Gamma(s) \tag{3.209}$$

$$= -\gamma + \sum_{n=0}^{\infty} \left(\frac{1}{n+1} - \frac{1}{s+n}\right) \tag{3.210}$$

Euler's constant γ, also called C, is defined by

$$\gamma \equiv -\Gamma'(1)/\Gamma(1) \tag{3.211}$$

and has value $\gamma \cong 0.577\ldots$.

Riemann's zeta function

Definition.

$$\zeta(z, q) \equiv \sum_{n=0}^{\infty} \frac{1}{(q+n)^z}, \quad \mathrm{Re}(z) > 1, \ q \neq 0, -1, -2, \ldots \tag{3.212}$$

$$\zeta(z) \equiv \sum_{n=1}^{\infty} \frac{1}{n^z}, \quad \mathrm{Re}(z) > 1 \tag{3.213}$$

The function $\zeta(z)$ is analytic in the whole complex z-plane except in $z = 1$ where it has a simple pole with residue equal to $+1$.

Appendix B. Summary of properties of the Mellin transformation

Definition. The Mellin transformation of a function $f(t), 0 < t < \infty$ is defined by

$$\mathcal{M}[f; s] \equiv \int_0^{\infty} f(t) t^{s-1}\, dt$$

and the result is a function holomorphic in the strip S_f of the complex plane s.

When the real part $\mathrm{Re}(s) \equiv r + 1$ of s is held fixed, the Mellin transform is defined by

$$\mathcal{M}[f](\beta) \equiv \mathcal{M}[f; r + 1 + 2\pi j\beta]$$

In that case, it is an isomorphism between the space $L^2(\mathbb{R}^+, \, v^{2r+1} \, dv)$ of functions $f(t)$ on $(0, \infty)$ equipped with the scalar product

$$(f, g) \equiv \int_0^\infty f(t)g^*(t)t^{2r+1} \, dt$$

and the space $L^2(\mathbb{R})$ of functions $\mathcal{M}[f](\beta)$.

Moreover, the scaled function defined by

$$\mathcal{D}_a f(t) \equiv a^{r+1} \, f(at)$$

is transformed according to

$$\mathcal{M}[\mathcal{D}_a f](\beta) = a^{-j2\pi\beta} \mathcal{M}[f](\beta)$$

Inversion formulas.

$$f(t) = (1/2\pi j) \int_{a-j\infty}^{a+j\infty} \mathcal{M}[f; s] t^{-s} \, ds$$

$$f(t) = \int_{-\infty}^{+\infty} \mathcal{M}[f](\beta) \, t^{-2j\pi\beta - r - 1} \, d\beta$$

Parseval formulas.

$$\int_0^\infty f(t) \, g(t) \, dt = \frac{1}{2\pi j} \int_{c-j\infty}^{c+j\infty} \mathcal{M}[f; s]\mathcal{M}[g; 1 - s] \, ds$$

$$\int_0^\infty f(t) \, g^*(t)t^{2r+1} \, dt = \int_{-\infty}^\infty \mathcal{M}[f](\beta)\mathcal{M}^*[g](\beta) \, d\beta$$

Other basic formulas involving the Mellin transforms $\mathcal{M}[f; s]$ and $\mathcal{M}[f](\beta)$ are recalled in Tables 11.1 and 11.2.

Multiplicative convolution. It is defined by

$$(f \vee g)(t) \equiv \int_0^\infty f(\tau)f(t/\tau) \, (d\tau/\tau)$$

$$f \vee \delta(t - 1) = f$$

$$\left(t\frac{d}{dt}\right)^k (f \vee g) = \left[\left(t\frac{d}{dt}\right)^k f\right] \vee g$$

$$= f \vee \left[\left(t\frac{d}{dt}\right)^k g\right]$$

$$(\ln t)(f \vee g) = [(\ln t)f] \vee g + f \vee [(\ln t)g]$$

$$\delta(t - a) \vee f = a^{-1}f(a^{-1}t)$$

$$\delta(t - p) \vee \delta(t - p') = \delta(t - pp'), \qquad p, p' > 0$$

$$\delta^{(k)}(t - 1) \vee f = (d/dt)^k(t^k f)$$

TABLE 11.1
Properties of the Mellin transform in s variable (definition (2.1)). Here k is a positive integer.

Original Function	Mellin Transform	
$f(t), t > 0$	$\mathcal{M}[f; s] \equiv \int_0^\infty f(t)\, t^{s-1}\, dt$	Strip of holomorphy
$f(t)$	$F(s)$	S_f
$f(at), \quad a > 0$	$a^{-s} F(s)$	S_f
$f(t^a), a \text{ real} \neq 0$	$\lvert a \rvert^{-1} F(a^{-1} s)$	$a^{-1} s \in S_f$
$(\ln t)^k f(t)$	$\dfrac{d^k}{ds^k} F(s)$	$s \in S_f$
$(t)^z f(t), z \text{ complex}$	$F(s + z)$	$s + z \in S_f$
$\dfrac{d^k}{dt^k} f(t)$	$(-1)^k (s-k)_k F(s-k)$ $$(s-k)_k \equiv (s-k)(s-k+1)\cdots(s-1)$$	$s - k \in S_f$
$\left(t \dfrac{d}{dt}\right)^k f(t)$	$(-1)^k s^k F(s)$	$s \in S_f$
$\dfrac{d^k}{dt^k} t^k f(t)$	$(-1)^k (s-k)_k F(s)$	$s \in S_f$
$t^k \dfrac{d^k}{dt^k} f(t)$	$(-1)^k (s)_k F(s)$ $$(s)_k \equiv s(s+1)\cdots(s+k-1)$$	$s \in S_f$
$\displaystyle\int_t^\infty f(x)\, dx$	$s^{-1} F(s+1)$	
$\displaystyle\int_0^t f(x)\, dx$	$-s^{-1} F(s+1)$	
$\displaystyle\int_0^\infty f_1(\tau) f_2(t/\tau)\,(d\tau/\tau)$	$F_1(s) F_2(s)$	$s \in S_{f_1} \cap S_{f_2}$

TABLE 11.2
Some properties of the Mellin transform in β variable (definition (3.24)).

Original Function	Mellin Transform
$f(t), t > 0$	$\mathcal{M}[f](\beta) \equiv \int_0^\infty f(t) t^{2\pi j\beta + r}\, dt$
$f(t)$	$M(\beta)$
$\mathcal{D}_a f(t) \equiv a^{r+1} f(at), \quad a > 0$	$a^{-2\pi j\beta} M(\beta)$
$t^{2\pi jc} f(t), \; c \text{ real}$	$M(\beta + c)$
$\dfrac{-1}{2j\pi}\left(t\dfrac{d}{dt} + r + 1\right) f(t)$	$\beta M(\beta)$

TABLE 11.3
Some standard Mellin transform pairs.

Original Function	Mellin Transform	
$f(t), t > 0$	$\mathcal{M}[f; s] \equiv \displaystyle\int_0^\infty f(t)\, t^{s-1}\, dt$	Strip of holomorphy
$e^{-pt}, p > 0$	$p^{-s}\Gamma(s)$	$\mathrm{Re}(s) > 0$
$H(t-a)\, t^b, a > 0$	$-\dfrac{a^{b+s}}{b+s}$	$\mathrm{Re}(s) < -\mathrm{Re}(b)$
$(H(t-a) - H(t))\, t^b$	$-\dfrac{a^{b+s}}{b+s}$	$\mathrm{Re}(s) > -\mathrm{Re}(b)$
$(1+t)^{-1}$	$\dfrac{\pi}{\sin(\pi s)}$	$0 < \mathrm{Re}(s) < 1$
$(1+t)^{-a}$	$\dfrac{\Gamma(s)\Gamma(a-s)}{\Gamma(a)}$	$0 < \mathrm{Re}(s) < \mathrm{Re}(a)$
$(1-t)^{-1}$	$\pi \cot(\pi s)$	$0 < \mathrm{Re}(s) < 1$
$H(1-t)\,(1-t)^{b-1}, \mathrm{Re}(b) > 0$	$\dfrac{\Gamma(s)\Gamma(b)}{\Gamma(s+b)}$	$\mathrm{Re}(s) > 0$
$H(t-1)\,(t-1)^{-b}$	$\dfrac{\Gamma(b-s)\Gamma(1-b)}{\Gamma(1-s)}$	$\mathrm{Re}(s) < \mathrm{Re}(b) < 1$
$H(t-1)\sin(a \ln t)$	$\dfrac{a}{s^2 + a^2}$	$\mathrm{Re}(s) < -|\mathrm{Im}(a)|$
$H(1-t)\sin(-a \ln t)$	$\dfrac{a}{s^2 + a^2}$	$\mathrm{Re}(s) > |\mathrm{Im}(a)|$
$(H(t) - H(t-p))\ln(p/t), p > 0$	$\dfrac{p^s}{s^2}$	$\mathrm{Re}(s) > 0$
$\ln(1+t)$	$\dfrac{\pi}{s \sin(\pi s)}$	$-1 < \mathrm{Re}(s) < 0$
$H(p-t)\ln(p-t)$	$-p^s s^{-1}[\psi(s+1) + p^{-1}\ln\gamma]$	$\mathrm{Re}(s) > 0$
$t^{-1}\ln(1+t)$	$\dfrac{\pi}{(1-s)\sin(\pi s)}$	$0 < \mathrm{Re}(s) < 1$
$\ln\left|\dfrac{1+t}{1-t}\right|$	$(\pi/s)\tan(\pi s/2)$	$-1 < \mathrm{Re}(s) < 1$
$(e^t - 1)^{-1}$	$\Gamma(s)\zeta(s)$	$\mathrm{Re}(s) > 1$
$t^{-1} e^{-t^{-1}}$	$\Gamma(1-s)$	$-\infty < \mathrm{Re}(s) < 1$
e^{-x^2}	$(1/2)\Gamma(s/2)$	$0 < \mathrm{Re}(s) < +\infty$

TABLE 11.3
(continued)

Original function	Mellin transform	
$f(t), t > 0$	$\mathcal{M}[f; s] \equiv \int_0^\infty f(t)\, t^{s-1}\, dt$	Strip of holomorphy
e^{iat}	$a^{-s}\Gamma(s)e^{i\pi(s/2)}$	$0 < \mathrm{Re}(s) < 1$
$\tan^{-1}(t)$	$\dfrac{-\pi}{2s\cos(\pi s/2)}$	$-1 < \mathrm{Re}(s) < 0$
$\mathrm{cotan}^{-1}(t)$	$\dfrac{\pi}{2s\cos(\pi s/2)}$	$0 < \mathrm{Re}(s) < 1$
$\delta(t - p), p > 0$	p^{s-1}	whole plane
$\displaystyle\sum_{n=1}^{\infty} \delta(t - pn), p > 0$	$p^{s-1}\zeta(1 - s)$	$\mathrm{Re}(s) < 0$
$J_\nu(t)$	$\dfrac{2^{s-1}\Gamma[(s + \nu)/2]}{\Gamma[(1/2)(\nu - s) + 1]}$	$-\nu < \mathrm{Re}(s) < 3/2$
$\displaystyle\sum_{n=-\infty}^{+\infty} p^{-nr} \delta(t - p^n)$ $p > 0, r$ real	$\dfrac{1}{\ln p} \displaystyle\sum_{n=-\infty}^{+\infty} \delta\left(\beta - \dfrac{n}{\ln p}\right)$ $\beta = \mathrm{Im}(s)$	$s = r + j\beta$
t^b	$\delta(b + s)$	none (analytic functional)

Invariant product. It is defined by

$$(f \circ g)(t) \equiv t^{r+1} f(t)\, g(t)$$

$$D_a[f] \circ D_a[g] = D_a[f \circ g]$$

$$\mathcal{M}[f \circ g](\beta) = (\mathcal{M}[f] * \mathcal{M}[g])(\beta)$$

Useful formulas for discretization. In the following, the variable v goes from 0 to ∞ and $Z(v)$ is a (possibly generalized) function.

Geometric Dirac comb:

$$\Delta_Q^r(v) \equiv \sum_{n=-\infty}^{+\infty} Q^{-nr} \delta(v - Q^n), \qquad Q > 0$$

Dilatocycled form of function Z:

$$Z^D(v) \equiv \sum_{n=-\infty}^{+\infty} Q^{n(r+1)} Z(Q^n v), \qquad Q > 0$$

$$Z^D(v) = [\Delta_Q^r \vee Z](v)$$

$$\mathcal{M}[Z^D](\beta) = \frac{1}{\ln Q} \sum_{n=-\infty}^{+\infty} \mathcal{M}[Z](\beta)\, \delta\left(\beta - \frac{n}{\ln Q}\right)$$

Geometrically sampled form of function Z:

$$Z_S(v) \equiv \sum_{n=-\infty}^{+\infty} q^n \, Z(q^n) \, \delta(v - q^n)$$

$$Z_S(v) = (Z \circ \Delta_q^r)(v)$$

$$\mathcal{M}[Z_S](\beta) = \frac{1}{\ln q} \sum_{n=-\infty}^{+\infty} \mathcal{M}[Z] \left(\beta - \frac{n}{\ln q} \right)$$

Discrete Mellin transform pair.

$$\mathcal{M}^P \left(\frac{m}{N \ln q} \right) = \sum_{n=M}^{M+N-1} q^{n(r+1)} \, Z^D(q^n) \, e^{2j\pi nm/N}$$

$$Z^D(q^n) = \frac{q^{-n(r+1)}}{N} \sum_{k=K}^{K+N-1} \mathcal{M} \left(\frac{k}{N \ln q} \right) e^{-2j\pi kn/N}$$

where N is the number of samples.

References

[1] E. C. Titchmarsh, *Introduction to the Theory of Fourier Integrals*, Clarendon Press, Oxford, 1975.

[2] G. Elfving, *The History of Mathematics in Finland 1828–1918*, Frenckell, Helsinki, 1981 (ISBN 951-653-098-2).

[3] N. Ya. Vilenkin, *Special Functions and the Theory of Group Representations*, American Mathematical Society, Providence, RI, 1968.

[4] C. E. Reid and T. B. Passin, *Signal Processing in C*, Wiley, New York, 1992.

[5] O. I. Marichev, *Handbook of Integral Transforms of Higher Transcendental Functions: Theory and Algorithmic Tables*, Ellis Horwood Ltd., 1982.

[6] A. H. Zemanian, *Generalized Integral Transformations*, Dover Publications, New York, 1987.

[7] O. P. Misra and J. L. Lavoine, *Transform Analysis of Generalized Functions*, North-Holland Mathematics Studies, Vol. 119, Elsevier, Amsterdam, 1986.

[8] N. Bleistein and R. A. Handelsman, *Asymptotic Expansions of Integrals*, Dover Publications, New York, 1986.

[9] Fung Kang, Generalized Mellin transforms 1, *Sc. Sinica*, 7, pp. 582–605, 1958.

[10] I. M. Gelfand and G. E. Shilov, *Generalized Functions*, Vol. 1, Academic Press Inc., New York, 1964.

[11] H. M. Edwards, *Riemann's Zeta Function*, Academic Press, New York, 1974.

[12] V. A. Ditkin and A. P. Prudnikov, *Integral Transforms and Operational Calculus*, Pergamon Press, 1965; A. P. Prudnikov, Yu. A. Brychkov, and O. I. Marichev, *Integrals and Series*, Vol. 3: *More Special Functions*, Gordon and Breach, 1990.

[13] F. Oberhettinger, *Tables of Mellin Transforms*, Springer-Verlag, New York, 1974.

[14] I. S. Gradshteyn and I. M. Ryzhik, *Table of Integrals, Series and Products*, A. Jeffrey (Ed.), 5th edn., Academic Press, New York, 1994.

[15] B. Davies, *Integral Transforms and their Applications*, 2nd ed., Springer-Verlag, New York, 1984.

[16] R. J. Sasiela, *Electromagnetic Wave Propagation in Turbulence—Evaluation and Application of Mellin Transforms*, Springer Series on Wave Phenomena, Vol. 18, Springer-Verlag, 1994.

[17] G. Doetsch, *Laplace Transformation*, Dover Publications, New York, 1943.

[18] H. E. Moses and A. F. Quesada, The power spectrum of the Mellin transformation with aplications to scaling of physical quantities, *J. Math. Physics*, 15(6), pp. 748–752, June 1974.

[19] J. Bertrand, P. Bertrand, and J. P. Ovarlez, Discrete Mellin transform for signal analysis, *Proc. IEEE-ICASSP*, Albuquerque, NM, USA 1990.

[20] M. Bertero, Sampling theory, resolution limits and inversion methods, in *Inverse Problems in Scattering and Imaging*, M. Bertero and E. R. Pike (Eds.), Malvern Physics Series, Adam Hilger, 1992.

[21] J. Bertrand, P. Bertrand, and J. P. Ovarlez, Dimensionalized wavelet transform with application to radar imaging, *Proc. IEEE-ICASSP*, 1991.

[22] J. Klauder, in *Functional Integration: Theory and Applications*, J. P. Antoine and E. Tirapegui (Eds.), Plenum, New York, 1980.

[23] J. P. Ovarlez, *La Transformation de Mellin: un Outil pour l'Analyse des Signaux à Large-Bande*, Doctoral Thesis, University Paris 6, Paris, April 1992.

[24] J. Bertrand and P. Bertrand, Représentations temps-fréquence des signaux, *C. R. Acad. Sci. Paris*, 299, pp. 635–638, 1984.

[25] J. Bertrand and P. Bertrand, Affine time-frequency distributions, in *Time-Frequency Signal Analysis—Methods and Applications*, B. Boashash (Ed.), Chapter V, Longman-Cheshire, Australia, 1992.

[26] J. Bertrand and P. Bertrand, A class of affine Wigner functions with extended covariance properties, *J. Math. Phys.*, 33(7), pp. 2515–2527, 1992.

[27] R. A. Altes, Wide-band, proportional-bandwidth Wigner-Ville analysis, *IEEE Trans. Acoust., Speech, Signal Processing*, 38, pp. 1005–1012, 1990.

[28] G. Eichmann and N. M. Marinovich, Scale invariant Wigner distribution and ambiguity functions, *Proc. Int. Soc. Opt. Eng. SPIE*, 519, pp. 18–24, 1984.

[29] N. M. Marinovich, *The Wigner distribution and the ambiguity function: generalizations, enhancement, compression and some applications*, Ph.D. thesis, City University of New York, 1986.

[30] L. Cohen, The scale representation, *IEEE Trans. on Signal Processing*, 41(12), 1993.

[31] F. Hlawatsch, A. Papandreou, and F. G. Boudreaux-Bartels, Hyperbolic class, *IEEE Trans. on Signal Processing*, 41(12), 1993.

[32] J. P. Ovarlez, Cramer Rao bound computation for velocity estimation in the broad band case using the Mellin transform, *Proc. IEEE-ICASSP*, Minneapolis, MN, 1993.

[33] R. A. Altes, The Fourier-Mellin transform and mammalian hearing, *J. Acoust. Soc. Amer.*, 63, pp. 174–183, January 1978.

[34] P. E. Zwicke and I. Kiss, Jr., A new implementation of the Mellin transform and its application to radar classification of ships, *IEEE Trans. on Pattern Analysis and Machine*

Intelligence, 5(2), pp. 191–199, March 1983.

[35] D. Casasent and S. Psaltis, Position, rotation and scale invariant optical correlation, *Appl. Opt.*, 17, pp. 1559–1561, May 1978; Deformation invariant, space-variant optical pattern recognition, in *Progress in Optics XVI*, E. Wolf (Ed.), pp. 291–355, North-Holland, 1978.

[36] Y. Sheng and H. H. Arsenault, Experiments on pattern recognition using invariant Fourier-Mellin descriptors, *J. Opt. Soc. Am.*, 3(6), pp. 771–884, 1986.

[37] Y. Sheng and J. Duvernoy, Circular-Fourier-radial-Mellin transform descriptors for pattern recognition, *J. Opt. Soc. Am.*, 3(6), pp. 885–887, 1986.

[38] D. Casasent and S. Psaltis, New optical transforms for pattern recognition, *Proc. IEEE*, 5, pp. 77–84, January 1977.

[39] P. M. Woodward, *Probability and Information Theory with Applications to Radar*, Pergamon Press, New York, 1953.

[40] E. J. Kelly and R. P. Wishner, Matched filter theory for high velocity accelerating targets, *IEEE Trans. on Military Elect.*, Mil 9, 1965.

[41] H. L. Van Trees, *Detection, Estimation and Modulation Theory*, Parts I, II, and III, John Wiley and Sons, New York, 1971.

[42] J. Bertrand, P. Bertrand, and J. P. Ovarlez, The Wavelet Approach in Radar Imaging and its Physical Interpretation, in *Progress in Wavelet Analysis and Applications*, Y. Meyer and S. Roques (Eds.), Editions Frontières, 1993.

[43] J. Bertrand, P. Bertrand, and J. P. Ovarlez, Frequency directivity scanning in laboratory radar imaging, *International Journal of Imaging Systems and Technology*, 5, pp. 39–51, 1994.

[44] J. P. Ovarlez, J. Bertrand, and P. Bertrand, Computation of affine time-frequency distributions using the fast Mellin transform, *Proc. IEEE-ICASSP*, 1992.

[45] S. G. Samko, A. A. Kilbas, and O. I. Marichev, *Fractional Integrals and Derivatives, Theory and Applications*, Gordon and Breach, 1993.

12

Mixed Time-Frequency Signal Transformations

G. Faye Boudreaux-Bartels

CONTENTS

12.1   Introduction

Mixed time-frequency representations are transformations of time-varying signals that depict how the spectral content of a signal is changing with time. They are multidimensional, time-varying extensions to conventional Fourier transform spectral analysis of signals and systems. Most time-frequency representations (TFRs) transform a one-dimensional signal, $x(t)$, into a two-dimensional function of time and frequency, $T_x(t, f)$. These transformations represent a surface above the time-frequency "plane" which gives an indication as to which spectral components of the signal are present at a given time and their relative amplitude. They are conceptually similar to a musical score with time running along one axis and frequency along the other [50], [37]. Just as the location and the shape of the notes on a musical score represent the pitch, time of occurrence, and duration of each sound in a piece of music, so too does the location of the local maximum and the shape of the surface of the TFR give an indication as to the frequency content, onset, and duration of various dominant signal components. Such representations are useful for the analysis, modification, synthesis, and detection of a variety of nonstationary signals with time-varying spectral content [1], [4], [2], [3], [30], [91].

The purpose of this chapter is to give an overview of many of the linear and quadratic time-frequency representations that have been developed over the past 60 years. The chapter first reviews various one-dimensional spectral representations, such as the Fourier transform, the instantaneous frequency, and the group delay. A brief discussion follows of a few commonly used TFRs such as the short-time Fourier transform, the Wigner distribution, the Altes Q distribution, and the Bertrand unitary P_0 distribution. These TFRs will be used as examples in the subsequent section, which describes many useful properties that an ideal TFR should satisfy. Unfortunately, no one TFR exists which satisfies all of these desirable properties. The relative merits of these TFRs can be understood by grouping them into "classes" of TFRs that share two or more properties. The remainder of the chapter is devoted to defining and understanding these classes. Important insights into these classes of TFRs can be gained by examining a set of five two-dimensional kernel functions that are unique to each TFR. These kernels greatly simplify the analysis and application of TFRs.

Other tutorials that discuss properties TFRs satisfy or that give references to a variety of interesting applications can be found in [5], [6], [8], [10], [12], [13], [17], [16], [18], [20], [22], [23], [24], [25], [26], [30], [27], [29], [37], [38], [39], [41], [42], [43], [45], [47], [49], [61], [54], [56], [63], [60], [64], [65], [66], [67], [68], [71], [69], [73], [74], [72], [75], [81], [84], [86], [87], [88], [89], [90], [91], [92], [97], [99], [100], [103], [104], [109], [110], [111], [113], [114], [115], [116], [117], [118], [119], [121], [1], [4], [2], [3]. Although this chapter will deal primarily with the TFRs of continuous-time signals and systems, issues related to discrete-time implementation algorithms for TFRs are discussed in [12], [18], [21], [24], [34], [38], [45], [47], [48], [49], [60], [78], [85], [93], [95], [94], [102], [103], [104], [106], [107].

12.2   One-dimensional spectral representations

12.2.1   *Fourier transform*

Conventional spectral analysis of a signal has been based on Fourier transform techniques. The Fourier transform of a signal $x(t)$,[1]

$$X(f) = \int x(t)e^{-j2\pi ft}\, dt, \tag{1}$$

is a useful tool for analyzing the spectral content of a stationary signal and for transforming the difficult operations of convolution, differentiation, and integration into relatively simple algebraic operations in the Fourier dual domain [32], [33], [105], [101]. The inverse Fourier transform or synthesis equation,

$$x(t) = \int X(f)e^{j2\pi ft}\, df, \tag{2}$$

represents the signal $x(t)$ as a linear combination of infinite duration complex sinusoids, $e^{j2\pi ft}$. Throughout this chapter, the double arrow will be used to denote a Fourier transform pair:

$$x(t) \leftrightarrow X(f).$$

A variety of useful Fourier transform pairs and properties are summarized in Table 12.1. For example, the first entry states that $\delta(t - t_0)$, which is a Dirac impulse [101], [105] centered at time t_0, has a Fourier transform that is a complex exponential whose phase is proportional to t_0. The next entry is the "dual" Fourier transform relationship for a Dirac impulse centered at the frequency f_0. In fact, by making use of the Duality property of Fourier transforms, given in entry 13 in Table 1, almost all of the Fourier transforms can be written in dual pairs:

$$\text{(i)}\ x(t) \leftrightarrow X(f) \text{ and (ii)}\ y(t) = X(t) \leftrightarrow Y(f) = x(-f).$$

For example, in entry 2 in Table 12.1, since the Fourier transform of a rectangular function

---

[1]Unless otherwise noted, all limits of integration are assumed to be from $-\infty$ to ∞ and $j = \sqrt{-1}$. Also, throughout the text, lowercase letters denote a time-domain signal and uppercase letters denote its Fourier transform.

in the time domain is a sinc function in the frequency domain, then by the duality property, the Fourier transform of a sinc function in the time domain must be a rectangular function in frequency. In this example, the frequency reversal of the duality property has no effect as the rectangular function is an even function. The dilation property in entry 12 of Table 1 states that if a recorded segment of speech is played back on a tape recorder at five times the original recording speed ($a = 5$), then the bandwidth of the accelerated speech will be increased by a factor of 5. So compressing the signal in one domain has the inverse effect of dilating the signal in the Fourier dual domain. The effects of linear convolution, differentiation, dilation, and translation on a signal or its Fourier transform will be exploited throughout this chapter.

Traditional Fourier transform analysis techniques have several disadvantages. In the synthesis equation in (2), the value $X(f_0)$ of the Fourier transform at the frequency f_0 can be thought of as the weighting coefficient of the complex sinusoidal basis function, $e^{j2\pi f_0 t}$. Since these sinusoidal basis functions are infinite duration, Fourier analysis implicitly assumes that each sinusoidal component with nonzero weighting coefficient is always present, and hence that the spectral content of the signal under analysis is unchanging, i.e., stationary. However, many naturally occurring signals, such as speech, music, bio-sonar, etc., intrinsically have spectral characteristics that change with time. For example, from the Fourier transform of a piece of music, it would be relatively easy to discern the frequency or pitch of various notes that were played, but relatively difficult to extract when each note was played. Hence, Fourier transform spectral decomposition techniques are inadequate for the analysis of most real-world signals.

An example of the dichotomy that exists between time-domain analysis versus frequency-domain analysis accomplished via a Fourier transform is depicted in Figure 12.1. The time-domain signal, $x(t)$, which is plotted at the bottom of the figure, consists of three successive tone bursts. The exact value of the frequency of each tone is not immediately obvious. Its Fourier transform, whose squared magnitude $|X(f)|$ is plotted vertically on the left side of Figure 12.1, clearly shows the frequencies of the three tones, but obscures the time of their onset. In such a situation, it is desirable to have the ideal TFR, $T_x(t, f)$, such as the one plotted in the middle of Figure 12.1, which facilitates simultaneous analysis in both the time domain and the frequency domain.

12.2.2 *Instantaneous frequency and group delay*

The instantaneous frequency and the group delay of a signal are one-dimensional transformations which attempt to represent temporal and spectral signal characteristics simultaneously. They work best for phase modulated signals—i.e., $x(t) = e^{j2\pi\theta(t)}$ or $X(f) = e^{j2\pi\phi(f)}$, where $\theta(t), \phi(f)\epsilon\Re$, have only one frequency component present at any given time or one temporal component present at any given frequency, respectively. However, most signals that occur in nature are a rich mixture of spectral components. Further, even if the signal is mono-component, it is often corrupted by environmental or measurement noise.

An understanding of the definition of the instantaneous frequency, which has been frequently used in communication theory [101], [119], can be gained by examining a complex exponential. The (constant) frequency, f_0, of the complex exponential, $x(t) = e^{j2\pi f_0 t}$, is proportional to the derivative of the phase of the signal, i.e., $f_0 \propto \frac{d}{dt}(2\pi f_0 t)$. The instantaneous frequency of a time-varying signal is thus defined as the instantaneous change in the phase of that signal,

$$f_x(t) = \frac{1}{2\pi}\frac{d}{dt}\arg x(t). \tag{3}$$

For example, the signal $x(t) = e^{j\pi\alpha t^2}$ is called a linear FM chirp since its instantaneous frequency is the line $f_x(t) = \alpha t$. The slope or "sweep rate" α gives the change in frequency per unit time.

TABLE 12.1

Table of Fourier transform pairs and properties.

Here, $\text{rect}_a(t) = \begin{cases} 1, & |t| < |a| \\ 0, & |t| > |a| \end{cases}$, $\tilde{u}(t) = \begin{cases} 1, & t > 0 \\ 0, & t < 0 \end{cases}$, and $\text{sgn}(t) = \begin{cases} 1, & t > 0 \\ -1, & t < 0 \end{cases}$ are the

rectangular, unit step, and signum functions, respectively. $x(t) \leftrightarrow X(f)$ and $q(t) \leftrightarrow Q(f)$ are Fourier transform pairs.

	Signal or Property	$y(t)$	$Y(f) = \int y(t)e^{-j2\pi ft}\,dt$
1	Impulse $\leftrightarrow$ Exponential	$\delta(t - t_0)$	$e^{-j2\pi f t_0}$
		$e^{j2\pi f_0 t}$	$\delta(f - f_0)$
2	Box $\leftrightarrow$ Sinc	$\text{rect}_a(t)$	$\dfrac{\sin(2\pi af)}{\pi f}$
		$\dfrac{\sin(2\pi at)}{\pi t}$	$\text{rect}_a(f)$
3	One-sided Exponential	$e^{-\alpha t}\tilde{u}(t), \quad \alpha > 0$	$\dfrac{1}{\alpha + j2\pi f}$
		$\dfrac{1}{\alpha + j2\pi t}$	$e^{\alpha f}\tilde{u}(-f), \quad \alpha > 0$
4	Two-sided Exponential	$e^{-\alpha|t|}, \quad \alpha > 0$	$\dfrac{2\alpha}{\alpha^2 + 4\pi^2 f^2}$
		$\dfrac{2\alpha}{\alpha^2 + 4\pi^2 t^2}$	$e^{-\alpha|f|}, \quad \alpha > 0$
5	Unit Step	$\tilde{u}(t)$	$\frac{1}{2}\delta(f) - j\dfrac{1}{2\pi f}$
		$\frac{1}{2}\delta(t) - j\dfrac{1}{2\pi t}$	$\tilde{u}(-f)$
6	Signum Function	$\text{sgn}(t)$	$\dfrac{1}{j2\pi f}$
		$\dfrac{1}{j2\pi t}$	$\text{sgn}(-f) = -\text{sgn}(f)$
7	Gaussian $\leftrightarrow$ Gaussian	$\dfrac{1}{\sqrt{\sigma}}e^{-\pi(t/\sigma)^2}$	$\sqrt{\sigma}e^{-\pi(\sigma f)^2}$
8	Linear FM Chirp	$e^{j\pi\alpha t^2}$	$\dfrac{1}{\sqrt{-j\alpha}}e^{-j\pi f^2/\alpha}$
9	Axis Reversal	$x(-t)$	$X(-f)$
10	Convolution $\leftrightarrow$ Multiplication	$\displaystyle\int x(\tau)h(t - \tau)\,d\tau$	$X(f)H(f)$
		$x(t)h(t)$	$\int X(v)H(f - v)\,dv$
11	Differentiation	$\dfrac{d^n}{dt^n}x(t)$	$(j2\pi f)^n X(f)$
	Mult. by Fourier parameter	$(-j2\pi t)^n x(t)$	$\dfrac{d^n}{df^n}X(f)$

TABLE 12.1
(continued

	Signal or Property	$y(t)$	$Y(f) = \int y(t)e^{-j2\pi ft}\, dt$
12	Dilation $\leftrightarrow$ Compression	$x(at)$	$\dfrac{1}{|a|}X\left(\dfrac{f}{a}\right)$
13	Duality	$X(t)$	$x(-f)$
14	Linearity	$\alpha x(t) + \beta q(t)$	$\alpha X(f) + \beta Q(f)$
15	Translation $\leftrightarrow$ Phase Change	$x(t - t_0)$	$X(f)e^{-j2\pi ft_0}$
		$x(t)e^{j2\pi f_0 t}$	$X(f - f_0)$

A dual concept, known as group delay, is useful in filter analysis. If an ideal complex exponential of frequency f_0 is put into a linear time-invariant filter whose frequency response[2] is $H(f) = e^{-j2\pi f\tau}$, then the output of the filter is equal to the input delayed by τ [101]. Thus, in this simple case, temporal translation information can be obtained by looking at the derivative of the phase response of a filter, commonly known as its group delay,

$$\tau_h(f) = -\frac{1}{2\pi}\frac{d}{df}\arg H(f). \tag{4}$$

For linear phase systems, the group delay is nondispersive, i.e., constant for all frequencies.

These time-varying spectral representations give counterintuitive information for multi-component signals. For example, let

$$y(t) = e^{j2\pi f_1 t} + e^{j2\pi f_2 t}, \qquad a_1, a_2 \in \Re \tag{5}$$

$$= e^{j2\pi\theta(t)}, \qquad \text{where } \theta(t) = \arctan\left[\frac{\sin(2\pi f_1 t) + \sin(2\pi f_2 t)}{\cos(2\pi f_1 t) + \cos(2\pi f_2 t)}\right]$$

be the sum of two complex exponentials of frequency f_1 and f_2. The instantaneous frequency of $y(t)$ can be derived in closed form using the property $\frac{d}{da}\arctan b = \frac{1}{1+b^2}\frac{db}{da}$,

$$f_y(t) = \frac{1}{2\pi}\frac{d}{dt}\arctan\left(\frac{\sin(2\pi f_1 t) + \sin(2\pi f_2 t)}{\cos(2\pi f_1 t) + \cos(2\pi f_2 t)}\right) \tag{6}$$

$$= \frac{1}{2\pi}\frac{1}{1 + (\sin(2\pi f_1 t) + \sin(2\pi f_2 t))^2/(\cos(2\pi f_1 t) + \cos(2\pi f_2 t))^2}$$

$$\times 2\pi\left[\frac{f_1\cos(2\pi f_1 t) + f_2\cos(2\pi f_2 t)}{\cos(2\pi f_1 t) + \cos(2\pi f_2 t)}\right.$$

$$\left. - \frac{(\sin(2\pi f_1 t) + \sin(2\pi f_2 t))(-f_1\sin(2\pi f_1 t) - f_2\sin(2\pi f_2 t))}{(\cos(2\pi f_1 t) + \cos(2\pi f_2 t))^2}\right]$$

$$= \frac{f_1 + f_2}{2}.$$

---

[2] The frequency response of a linear, time-invariant filter is the Fourier transform of the filter's impulse response, $h(t)$ [101].

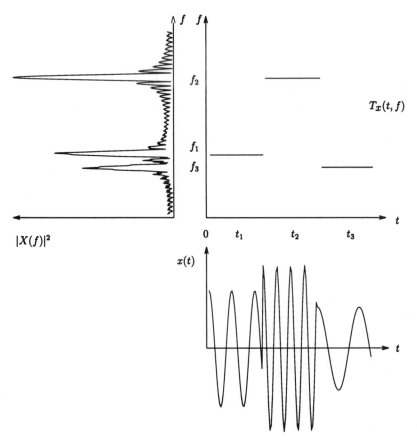

FIGURE 12.1
Ideal time-frequency representation for a signal that consists of three consecutive, short-duration tones of frequencies, f_1, f_2, and f_3, respectively. The corresponding time-domain signal is plotted on the bottom and its Fourier spectrum is plotted vertically along the left-hand side.

Thus, the instantaneous frequency of this signal is the numerical average of the frequency of its two sinusoidal components. This is not very useful information, in general, as there are an infinite number of pairs of frequencies, e.g., $f_1 + \epsilon$ and $f_2 - \epsilon$, which have the same numerical average for all ϵ. In addition, letting $f_2 = -f_1$ in (5), shows that the real signal $2\cos(2\pi f_1 t)$ has an instantaneous frequency equal to zero, which is clearly unintuitive.

12.3   Classical time-frequency representations

In this section, several commonly used linear and quadratic TFRs will be described. The first TFR is the Short-time Fourier transform (STFT), which is a linear transformation of the signal. The remaining TFRs that will be discussed in this section, such as the spectrogram, the Wigner distribution, the Altes Q distribution, and the unitary form of the Bertrand P_0 distribution, are quadratic functions of the signal.

12.3.1   *Short-time Fourier transform*

The most commonly used and easy to understand time-frequency representation is the short-time Fourier transform (STFT),

$$STFT_x(t, f; \Gamma) = \int x(\tau)\gamma^*(\tau - t)e^{-j2\pi f\tau}\, d\tau \tag{7}$$

$$= e^{-j2\pi tf} \int X(f')\Gamma^*(f' - f)e^{j2\pi tf'}\, df', \tag{8}$$

which is a linear function of the signal $x(t)$ [47], [92], [104], [103], [107], [112], [68], [69]. The first STFT equation in (7) indicates that the STFT can be thought of as the Fourier transform of a windowed segment of the data, $[x(\tau)\,\gamma^*(\tau - t)]$. Typically, the analysis window, $\gamma(t)$, is real and even, so that the STFT is equivalent to the Fourier transform of a segment of the signal centered at the output time, t. The second STFT equation in (8) illustrates that the STFT can also be thought of as a filtered version of the signal. If the analysis window is a lowpass function, then evaluating the STFT is equivalent to sending the signal through a bandpass filter $\Gamma^*(f' - f)$ centered at the output frequency, f. Thus, the STFT can be thought of as the frequency content of the signal near the output time t or the temporal fluctuations of the signal spectrum near the output frequency f. Analog STFT techniques known as spectrograms [83], [9] were originally used to analyze the local frequency content of speech signals. In the TFR literature, the squared magnitude of the STFT is sometimes referred to as the spectrogram:

$$SPEC_x(t, f; \Gamma) = |STFT_x(t, f; \Gamma)|^2$$

$$= \left| \int x(\tau)\gamma^*(\tau - t)e^{-j2\pi f\tau}\, d\tau \right|^2$$

$$= \left| \int X(f')\Gamma^*(f' - f)e^{j2\pi tf'}\, df' \right|^2. \tag{9}$$

The following two simple examples provide insight into the STFT. First, consider the case of the STFT of an impulsive signal, $x(t) = \delta(t - t_0)$. Using (7), it can be shown that the STFT of a Dirac impulse simplifies to the complex conjugate of the shifted STFT analysis window, $\gamma(t)$, modulated by a complex exponential:

$$STFT_x(t, f; \Gamma) = \int \delta(\tau - t_0)\gamma^*(\tau - t)e^{-j2\pi f\tau}\, d\tau \tag{10}$$

$$= \gamma^*(t_0 - t)e^{-j2\pi f t_0}.$$

Hence, if the window is real and even, then the nonzero support region of the STFT will be centered at $t = t_0$. The time duration of the STFT in this example is equal to the duration of the analysis window. The second example is that of a complex exponential whose Fourier transform, $Y(f) = \delta(f - f_0)$, is perfectly concentrated at the frequency f_0. Using (8) reveals that the magnitude of the STFT will be equal to the magnitude of the Fourier transform of the analysis window, shifted by an amount proportional to the frequency f_0 of the signal:

$$STFT_y(t, f; \Gamma) = e^{-j2\pi tf} \int \delta(f' - f_0)\Gamma^*(f' - f)e^{j2\pi f't}\, df' \tag{11}$$

$$= \Gamma^*(f_0 - f)e^{-j2\pi(f - f_0)t}.$$

The STFT's region of nonzero support about f_0 in this example is equal to the constant bandwidth of the analysis window for all output frequencies. Thus, the STFT gives a constant bandwidth or fixed resolution time-varying analysis.

There are several advantages and drawbacks to the STFT. The STFT is a linear signal transformation, i.e.,

$$y(t) = \alpha x_1(t) + \beta x_2(t) \quad \Rightarrow$$

$$STFT_y(t, f; \Gamma) = \alpha STFT_{x_1}(t, f; \Gamma) + \beta STFT_{x_2}(t, f; \Gamma). \tag{12}$$

It is easy to understand and compute and has been widely used in the analysis of quasi-stationary signals. In general, however, the STFT is complex-valued and its time duration and bandwidth are greater than that of the signal, producing a spreading of the support of the signal in the time-frequency plane. Second, there is a tradeoff between time resolution and frequency resolution. Consider the following multicomponent signal: $q(t) = \delta(t - t_0) + e^{j2\pi f_0 t}$. Using the linearity of the STFT in (12), it can be shown that the STFT of $q(t)$ is the sum of the two STFTs derived in equations (10–11). If the analysis window is the Gaussian function in Table 12.1,

$$\gamma(t) = \frac{1}{\sqrt{\sigma}} e^{-\pi(t/\sigma)^2} \leftrightarrow \Gamma(f) = \sqrt{\sigma} e^{-\pi(\sigma f)^2}, \tag{13}$$

then it can be shown using equations (10–12) that the STFT of $q(t)$ is as follows:

$$STFT_q(t, f; \Gamma) = \frac{1}{\sqrt{\sigma}} e^{-\pi(t-t_0)^2/\sigma^2} e^{-j2\pi f t_0} + \sqrt{\sigma} e^{-\pi\sigma^2(f-f_0)^2} e^{-j2\pi(f-f_0)t}. \tag{14}$$

This STFT corresponds to a two-dimensional Gaussian function centered at t_0 and f_0. Ideally, the STFT would be highly concentrated about the time $t = t_0$ and the frequency $f = f_0$, corresponding to the fact that the signal contains an impulse in time and an impulse in frequency at those locations. However, for the $STFT_q(t, f; \Gamma)$ to be highly concentrated in time, equation (14) reveals that the Gaussian analysis window must be very short duration, i.e., $\sigma \approx 0$. For the STFT to be highly concentrated in frequency about $f = f_0$, then the analysis window must be very narrowband, i.e., $\sigma \gg 0$. Clearly, both conditions on σ cannot be met simultaneously. The STFT can achieve either good time resolution or good frequency resolution but generally not both. Hence, the STFT works best on quasi-stationary signals, which are signals whose spectral content is changing slowly with time.

12.3.2 *Wigner distribution and Woodward ambiguity function*

The first time-frequency representation proposed was the Wigner distribution (WD) in the field of quantum mechanics [120], [37], [60], [88]. The WD of a signal $x(t)$,

$$WD_x(t, f) = \int x\left(t + \frac{\tau}{2}\right) x^*\left(t - \frac{\tau}{2}\right) e^{-j2\pi f \tau} d\tau \tag{15}$$

$$= \int X\left(f + \frac{\nu}{2}\right) X^*\left(f - \frac{\nu}{2}\right) e^{j2\pi t \nu} d\nu \tag{16}$$

$$= \iint AF_x(\tau, \nu) e^{j2\pi(t\nu - f\tau)} d\tau \, d\nu, \tag{17}$$

can be obtained using either the signal in (15) or its Fourier transform in (16). Equation (17) shows that the Wigner distribution is related to the Woodward narrowband ambiguity function (AF),

$$AF_x(\tau, \nu) = \int x \left(t + \frac{\tau}{2} \right) x^* \left(t - \frac{\tau}{2} \right) e^{-j2\pi\nu t} \, dt \tag{18}$$

$$= \int X \left(f + \frac{\nu}{2} \right) X^* \left(f - \frac{\nu}{2} \right) e^{j2\pi\tau f} \, df$$

$$= \int \int W D_x(t, f) e^{-j2\pi(\nu t - \tau f)} \, dt \, df, \tag{19}$$

via a two-dimensional Fourier transform. The AF is a two-dimensional auto-correlation function commonly used in radar and sonar to track the distance (range) and velocity (range rate) of a moving target [39], [121], [115], [109], [114].

For real, bandlimited signals, replacing the signal in (15–17) with its corresponding analytic signal,

$$z_x(t) = x(t) + j\hat{x}(t) \quad \leftrightarrow \quad Z_x(f) = \begin{cases} 2X(f), & f > 0 \\ X(f), & f = 0 \\ 0, & f < 0 \end{cases} \tag{20}$$

often simplifies the analysis. Here,

$$\hat{x}(t) = \frac{PV}{\pi} \int \frac{x(\tau)}{t - \tau} \, d\tau$$

is the Hilbert transform [105], [101] of $x(t)$ and PV indicates principal value. The spectrum of the analytic signal in (20) is proportional to that of the original signal for positive frequencies, but is equal to zero for negative frequencies. For real signals, this zeroing out of the negative frequency signal components results in no loss of information as the Fourier transform of a real signal is conjugate symmetric, i.e., $X(f) = X^*(-f)$. The TFR that results from this substitution,

$$W D_{z_x}(t, f) = \int z_x \left(t + \frac{\tau}{2} \right) z_x^* \left(t - \frac{\tau}{2} \right) e^{-j2\pi f\tau} \, d\tau, \tag{21}$$

is often referred to as the Wigner-Ville distribution [119], [24], [60], [61]. Using the Wigner-Ville distribution instead of the WD greatly simplifies the analysis of bandlimited signals, but can distort the analysis of lowpass signals.

The Wigner distribution can be computed in closed form for a variety of signals, as is indicated in Table 12.2. The first entry in Table 12.2 states that a Dirac impulse at $t = t_0$, $x(t) = \delta(t - t_0)$, has a WD that is also an impulse concentrated at the same time $t = t_0$.

PROOF Let $x(t) = \delta(t - t_0)$.

$$W D_x(t, f) = \int \delta(t + \tau/2 - t_0) \delta(t - \tau/2 - t_0) e^{-j2\pi f\tau} \, d\tau$$

$$= 2\delta(2(t - t_0)) e^{-j4\pi f(t - t_0)} = \delta(t - t_0). \qquad \blacksquare$$

In a dual manner, the second entry in Table 12.2 indicates that the WD of a complex exponential, i.e., $X(f) = \delta(f - f_0)$, is a Dirac function centered at the same frequency $f = f_0$. The next entry in Table 12.2 is that of WD of a linear FM signal, $x(t) = e^{j\pi\alpha t^2}$ with sweep rate α. Its Wigner distribution is a Dirac function perfectly concentrated along the signal's linear instantaneous frequency, $f_x(t) = \alpha t$.

PROOF Let $x(t) = e^{j\pi\alpha t^2}$. Using (15), one obtains the following proof:

$$WD_x(t, f) = \int e^{j\pi\alpha(t+\tau/2)^2} e^{-j\pi\alpha(t-\tau/2)^2} e^{-j2\pi f\tau} \, d\tau,$$

$$= \int e^{j\pi\alpha(t^2+t\tau+\tau^2/4-t^2+t\tau-\tau^2/4)} e^{-j2\pi f\tau} \, d\tau$$

$$= \int e^{-j2\pi(f-\alpha t)\tau} \, d\tau = \delta(f - \alpha t).$$ ∎

Another interesting example is that of the WD of a Gaussian signal. Recall from (13), the Fourier transform of a Gaussian signal is itself a Gaussian function, but one with inversely proportional variance. The WD of this Gaussian signal is the two-dimensional Gaussian function given in Table 12.2.

PROOF Let $g(t) = \frac{1}{\sqrt{\sigma}} e^{-\pi(t/\sigma)^2}$.

$$WD_g(t, f) = \frac{1}{\sigma} \int e^{-\pi[(t+\tau/2)/\sigma]^2} e^{-\pi[(t-\tau/2)/\sigma]^2} e^{-j2\pi f\tau} \, d\tau,$$

$$= \frac{1}{\sigma} \int e^{-\pi(t^2+t\tau+\tau^2/4+t^2-t\tau+\tau^2/4)/\sigma^2} e^{-j2\pi f\tau} \, d\tau$$

$$= \frac{1}{\sigma} e^{-2\pi(t/\sigma)^2} \int e^{-\pi\tau^2/(2\sigma^2)} e^{-j2\pi f\tau} \, d\tau$$

$$= \sqrt{2} e^{-2\pi[(t/\sigma)^2 + (\sigma f)^2]}. \tag{22}$$

∎

Note that if σ is very small, then the Gaussian signal $g(t)$ is very short duration, but its Fourier transform $G(f)$ is broadband in frequency; likewise, equation (22) reveals that the WD of $g(t)$ will be concentrated in time but broadband in frequency.

The ambiguity function of a variety of signals is given in the right-hand column of Table 12.2. Equations (17) and (19) indicate that the WD and the AF are two-dimensional Fourier transform pairs. Hence, we can exploit the Fourier transform properties in Table 12.1 to gain insight into the relationship between the AF and the WD of any signal. Some of these relationships are given in Table 12.3, which lists the effects of signal operations on the WD and AF. For example, the first entry in Table 12.3 shows that the WD is insensitive to the phase of proportionality constants, i.e.,

$$y(t) = e^{j\alpha} x(t) \quad \Rightarrow \quad WD_y(t, f) \equiv WD_x(t, f), \qquad \forall \, \alpha \in \Re. \tag{23}$$

The WD and the AF are not one-to-one signal transformations; rather, they are only unique up to a unit amplitude proportionality factor as indicated in (23). The next entry in Table 12.3 shows that reversing the time axis of the signal in turn reverses its frequency axis as well; this produces a corresponding reversal in both the time and frequency axes of the WD and the AF. The fifth entry in Table 12.3 states that if a signal is even or odd, then its WD and AF are scaled versions of one another. For example, if $x(t) = x(-t)$ is even, then its ambiguity function $AF_x(\tau, \nu) = \frac{1}{2} WD_x(\frac{\tau}{2}, \frac{\nu}{2})$.

TABLE 12.2

Signals with closed-form equations for their Wigner distribution and ambiguity function. Here, $\sigma > 0$, a, α, $c \epsilon \Re$, sgn(a), and rect$_a(t)$ are defined in Table 12.1; $u_n(t) = \frac{2^{1/4}}{\sqrt{n!}}e^{-\pi t^2}H_n(2\sqrt{\pi}t)$, $H_n(t) = (-1)^n e^{t^2/2}\frac{d^n}{dt^n}e^{-t^2/2}$ is the nth order Hermite polynomial; and $L_n(t) = \frac{1}{n!}e^t\frac{d^n}{dt^n}(t^n e^{-t}) = \sum_{k=0}^n \frac{n!}{k!(n-k)!}(-t)^k$ is the nth-order Laguerre polynomial.

Signal, $x(t)$	Fourier Transform, $X(f)$	Wigner Distribution, $WD_x(t,f)$	Ambiguity Function, $AF_x(\tau,\nu)$
$\delta(t - t_i)$	$e^{-j2\pi f t_i}$	$\delta(t - t_i)$	$e^{-j2\pi\nu t_i}\delta(\tau)$
$e^{j2\pi f_i t}$	$\delta(f - f_i)$	$\delta(f - f_i)$	$e^{j2\pi f_i \tau}\delta(\nu)$
$e^{+j\pi\alpha t^2}$	$\frac{1}{\sqrt{-j\alpha}}e^{-j\pi f^2/\alpha}$	$\delta(f - \alpha t)$	$\delta(\nu - \alpha\tau)$
$\frac{1}{\sqrt{j\alpha}}e^{j\pi t^2/\alpha}$	$e^{-j\pi\alpha f^2}$	$\delta(t - \alpha f)$	$\delta(\tau - \alpha\nu)$
$e^{j\pi(\alpha t^2 + 2f_i t + c)}$	$\frac{1}{\sqrt{-j\alpha}}e^{j\pi[c - (f-f_i)^2/\alpha]}$	$\delta(f - f_i - \alpha t)$	$\delta(\nu - \alpha\tau)e^{j2\pi f_i \tau}$
$\frac{1}{\sqrt{\sigma}}e^{-\pi(t/\sigma)^2}$	$\sqrt{\sigma}e^{-\pi(\sigma f)^2}$	$\sqrt{2}e^{-2\pi[(t/\sigma)^2 + (\sigma f)^2]}$	$\frac{1}{\sqrt{2}}e^{-(\pi/2)[(\tau/\sigma)^2 + (\sigma\nu)^2]}$
$\frac{1}{\sqrt{\sigma}}e^{-\pi(t/\sigma)^2}e^{j\pi\alpha t^2}$	$\frac{1}{\sqrt{\sigma[\sigma^{-2} - j\alpha]}}\exp\left[-\pi f^2\frac{\sigma^{-2}+j\alpha}{\sigma^{-4}+\alpha^2}\right]$	$\sqrt{2}e^{-2\pi[(t/\sigma)^2 + (\sigma)^2(f - \alpha t)^2]}$	$\frac{1}{\sqrt{2}}e^{-(\pi/2)[(\tau/\sigma)^2 + (\sigma)^2(\nu - \alpha\tau)^2]}$
$\frac{1}{\sqrt{\sigma}}e^{-\pi[(t-t_i)/\sigma]^2}e^{j2\pi f_i t}$	$\sqrt{\sigma}e^{-\pi\sigma^2(f - f_i)^2}e^{-j2\pi(f - f_i)t_i}$	$\sqrt{2}e^{-2\pi[[(t-t_i)/\sigma]^2 + \sigma^2(f - f_i)^2]}$	$\frac{1}{\sqrt{2}}e^{-(\pi/2)[(\tau/\sigma)^2 + (\sigma\nu)^2]}e^{j2\pi(f_i\tau - t_i\nu)}$
$\text{rect}_a(t)$	$\frac{\sin(2\pi af)}{\pi f}$	$\frac{\sin[4\pi(a - |t|)f]}{\pi f}\text{rect}_a(t)$	$\frac{\sin[\pi\nu(2a - |\tau|)]}{\pi\nu}\text{rect}_{2a}(\tau)$

TABLE 12.2
(continued)

Signal, $x(t)$	Fourier Transform, $X(f)$	Wigner Distribution, $WD_x(t,f)$	Ambiguity Function, $AF_x(\tau,\nu)$
$\dfrac{\sin(2\pi at)}{\pi t}$	$\text{rect}_a(f)$	$\dfrac{\sin[4\pi(a-|f|)t]}{\pi t}\text{rect}_a(f)$	$\dfrac{\sin[\pi\tau(2a-|\nu|)]}{\pi\tau}\text{rect}_{2a}(\nu)$
$e^{j\pi\alpha t^2}\text{rect}_a(t)$	$\dfrac{1}{\sqrt{-j\alpha}}\displaystyle\int e^{-j\frac{\pi}{\alpha}(f-\beta)^2}\dfrac{\sin 2\pi a\beta}{\pi\beta}\,d\beta$	$\dfrac{\sin[4\pi(a-|t|)(f-\alpha t)]}{\pi(f-\alpha t)}\text{rect}_a(t)$	$\dfrac{\sin[\pi(\nu-\alpha\tau)(2a-|\tau|)]}{\pi(\nu-\alpha\tau)}\text{rect}_{2a}(\tau)$
$\tilde{u}(t)=\begin{cases}1, & t>0\\0, & t<0\end{cases}$	$\dfrac{\delta(f)}{2}-\dfrac{j}{2\pi f}$	$\dfrac{\sin(4\pi ft)}{\pi f}\tilde{u}(t)$	$\left[\dfrac{\delta(\nu)}{2}-\dfrac{j}{2\pi\nu}\right]e^{-j\pi\nu|\tau|}$
$e^{-\sigma t}\tilde{u}(t)$	$\dfrac{1}{\sigma+j2\pi f}$	$e^{-2\sigma t}\dfrac{\sin 4\pi ft}{\pi f}\tilde{u}(t)$	$\dfrac{e^{-(\sigma+j\pi\nu)|\tau|}}{2\sigma+j2\pi\nu}$
$u_n(t),\ n=0,1,\ldots$	$(-j)^n u_n(f)$	$2e^{-2\pi(t^2+f^2)}L_n(4\pi(t^2+f^2))$	$e^{-\pi(\tau^2+\nu^2)/2}L_n(\pi(\tau^2+\nu^2))$
$\cos(2\pi f_i t)$	$[\delta(f+f_i)+\delta(f-f_i)]/2$	$[\delta(f+f_i)+\delta(f-f_i)+2\delta(f)\cos(4\pi f_i t)]/4$	$[\delta(\nu+2f_i)+\delta(\nu-2f_i)+2\delta(\nu)\cos(2\pi f_i\tau)]/4$
$\sin(2\pi f_i t)$	$j[\delta(f+f_i)-\delta(f-f_i)]/2$	$[\delta(f+f_i)+\delta(f-f_i)-2\delta(f)\cos(4\pi f_i t)]/4$	$-[\delta(\nu+2f_i)+\delta(\nu-2f_i)-2\delta(\nu)\cos(2\pi f_i\tau)]/4$
$\delta(t-t_i)+\delta(t-t_m)$	$e^{-j2\pi ft_i}+e^{-j2\pi ft_m}$	$\delta(t-t_i)+\delta(t-t_m)+$ $2\delta(t-\tfrac{t_i+t_m}{2})\cos(2\pi(t_i-t_m)f)$	$[e^{-j2\pi\nu t_i}+e^{-j2\pi\nu t_m}]\delta(\tau)+$ $e^{-j\pi(t_i+t_m)\nu}[\delta(\tau-(t_i-t_m))+\delta(\tau+(t_i-t_m))]$
$e^{j2\pi f_i t}+e^{j2\pi f_m t}$	$\delta(f-f_i)+\delta(f-f_m)$	$\delta(f-f_i)+\delta(f-f_m)+$ $2\delta(f-\tfrac{f_i+f_m}{2})\cos(2\pi(f_i-f_m)t)$	$[e^{j2\pi f_i\tau}+e^{j2\pi f_m\tau}]\delta(\nu)+$ $e^{j\pi(f_i+f_m)\tau}[\delta(\nu-(f_i-f_m))+\delta(\nu+(f_i-f_m))]$
$\displaystyle\sum_k c_k e^{j2\pi kf_0 t}$	$\displaystyle\sum_k c_k\delta(f-kf_0)$	$\displaystyle\sum_k|c_k|^2\delta(f-kf_0)+$ $\displaystyle\sum_k\sum_{m\neq k}c_k c_m^*\delta\left(f-\frac{k+m}{2}f_0\right)e^{j2\pi(k-m)f_0 t}$	$\displaystyle\sum_k|c_k|^2 e^{j2\pi kf_0\tau}\,\delta(\tau)+$ $\displaystyle\sum_k\sum_{m\neq k}c_k c_m^* e^{j\pi(k+m)f_0\tau}\delta(\nu-(k-m)f_0)$

PROOF   Let $x(t) = x(-t)$.

$$
\begin{aligned}
AF_x(\tau, \nu) &= \int x(t + \tau/2)x^*(t - \tau/2)e^{-j2\pi\nu t}\, dt \\
&= \int x(t + \tau/2)x^*(-(t - \tau/2))e^{-j2\pi\nu t}\, dt \\
&= \frac{1}{2}\int x\left(\frac{\tau}{2} + \frac{\gamma}{2}\right)x^*\left(\frac{\tau}{2} - \frac{\gamma}{2}\right)e^{-j2\pi(\nu/2)\gamma}\, d\gamma \\
&= \frac{1}{2}WD_x\left(\frac{\tau}{2}, \frac{\nu}{2}\right).
\end{aligned}
$$
∎

Just as Table 12.1 indicates that time shifts in a signal change the phase of its Fourier transform, so too does Table 12.3 indicate that time or frequency shifts in the WD cause phase changes in its Fourier transform, the AF.

PROOF   Let $WD_y(t, f) = WD_x(t - t_0, f - f_0)$ in (19).

$$
\begin{aligned}
AF_y(\tau, \nu) &= \int\int WD_x(t - t_0, f - f_0)e^{-j2\pi(\nu t - \tau f)}\, dt\, df \\
&= \int\int WD_x(t', f')e^{-j2\pi[\nu(t'+t_0)-\tau(f'+f_0)]}\, dt'\, df' \\
&= e^{-j2\pi(\nu t_0 - \tau f_0)}AF_x(\tau, \nu).
\end{aligned}
$$
∎

12.3.3   Altes Q or "wideband" Wigner distribution

The Altes Q distribution was originally proposed as a "wideband" version of the Wigner distribution [6], [97].

$$
Q_x(t, f) = f\int X(fe^{u/2})\, X^*(fe^{-u/2})\, e^{j2\pi t f u}\, du \quad , f > 0 \tag{24}
$$

$$
= f\int_0^\infty X(f\sqrt{\alpha})\, X^*(f/\sqrt{\alpha})\, \alpha^{j2\pi t f - 1}\, d\alpha
$$

$$
= \int\int HAF_X(\zeta, \beta)e^{j2\pi(tf\beta - \zeta\ln(f/f_r))}\, d\zeta\, d\beta. \tag{25}
$$

Here, f_r is a positive reference frequency. Marinovich proposed a dual formulation to (24), called the scale-invariant WD,

$$
MSIWD_x(t, c) = \int x(te^{\sigma/2})x^*(te^{-\sigma/2})e^{-j2\pi c\sigma}\, d\sigma,
$$

which used the time-domain version of the signal $x(t)$ instead of the signal spectrum used in (24) [51], [86], [57]. The Q distribution is related to the hyperbolic ambiguity function,

$$
HAF_X(\zeta, \beta) = \int_0^\infty X(fe^{\beta/2})\, X^*(fe^{-\beta/2})\, e^{j2\pi\zeta\ln(f/f_r)}\, df \tag{26}
$$

$$
= f_r\int e^b X(f_r e^{b + \beta/2})X^*(f_r e^{b - \beta/2})e^{j2\pi\zeta b}\, db
$$

TABLE 12.3

Effects of signal operations on the Wigner distribution and ambiguity function. Here, $\sigma > 0$, a, α, $c\in\Re$, and sgn(a) is the signum function defined in Table 12.1.

Signal, $y(t)$	Fourier Transform, $Y(f)$	Wigner Distribution, $WD_y(t, f)$	Ambiguity Function, $AF_y(\tau, \nu)$
$ax(t)$	$aX(f)$	$\|a\|^2 WD_x(t, f)$	$\|a\|^2 AF_x(\tau, \nu)$
$x(-t)$	$X(-f)$	$WD_x(-t, -f)$	$AF_x(-\tau, -\nu)$
$\sqrt{\|a\|}x(at)$	$\dfrac{1}{\sqrt{\|a\|}}X(f/a)$	$WD_x(at, f/a)$	$AF_x(a\tau, \nu/a)$
$\sqrt{\|a\|}X(at)$	$\dfrac{1}{\sqrt{\|a\|}}x(-f/a)$	$WD_x(-f/a, at)$	$AF_x(-\nu/a, a\tau)$
$x(t) = \pm x(\pm t)$	$X(f) = \pm X(\pm f)$	$\pm 2AF_x(2t, 2f)$	$\pm\dfrac{1}{2}WD_x(\tau/2, \nu/2)$
$x^*(t)$	$X^*(-f)$	$WD_x(t, -f)$	$AF_x^*(\tau, -\nu)$
$x(t-t_i)e^{j2\pi f_i t}$	$X(f-f_i)e^{-j2\pi(f-f_i)t_i}$	$WD_x(t-t_i, f-f_i)$	$AF_x(\tau, \nu)e^{j2\pi(f_i\tau - t_i\nu)}$
$x(t)h(t)$	$\displaystyle\int X(f')H(f-f')\,df'$	$\displaystyle\int WD_x(t, f')WD_h(t, f-f')\,df'$	$\displaystyle\int AF_x(\tau, \nu')AF_h(\tau, \nu-\nu')\,d\nu'$
$\displaystyle\int x(t')h(t-t')\,dt'$	$X(f)H(f)$	$\displaystyle\int WD_x(t', f)WD_h(t-t', f)\,dt'$	$\displaystyle\int AF_x(\tau', \nu)AF_h(\tau-\tau', \nu)\,d\tau'$
$x(t)e^{j\pi\alpha t^2}$	$\dfrac{1}{\sqrt{-j\alpha}}\displaystyle\int X(f-f')e^{-j\pi f'^2/\alpha}\,df'$	$WD_x(t, f-\alpha t)$	$AF_x(\tau, \nu-\alpha\tau)$

TABLE 12.3
(continued)

Signal, $y(t)$	Fourier Transform, $Y(f)$	Wigner Distribution, $WD_y(t,f)$	Ambiguity Function, $AF_y(\tau,\nu)$
$\int \sqrt{|\alpha|}\, e^{j\pi\alpha u^2} x(t-u)\,du$	$\sqrt{j\,\mathrm{sgn}(\alpha)}\, X(f)\, e^{-j\pi f^2/\alpha}$	$WD_x(t-f/\alpha,\, f)$	$AF_x(\tau-\nu/\alpha,\,\nu)$
$\displaystyle\sum_{i=0}^{N-1} x\!\left(t-(i-\tfrac{N-1}{2})T_r\right),$ $T_r>0$	$X(f)\,\dfrac{\sin(\pi T_r N f)}{\sin(\pi T_r f)}$	$\displaystyle\sum_{i=0}^{N-1} WD_x\!\left(t-(i-\tfrac{N-1}{2})T_r,\, f\right)$ $\displaystyle+2\sum_{i=0}^{N-2}\sum_{m=i+1}^{N-1} WD_x\!\left(t-\tfrac{(i+m)-(N-1)}{2}T_r,\, f\right)$ $\times\cos[2\pi T_r(i-m)f]$	$\displaystyle\sum_{n=-N+1}^{N-1} AF_x(\tau-nT_r,\,\nu)\,\dfrac{\sin\pi\nu T_r(N-|n|)}{\sin(\pi\nu T_r)}$
$\displaystyle\sum_{i=1}^{N} x(t-t_i)\,e^{j2\pi f_i t}$	$\displaystyle\sum_{i=1}^{N} X(f-f_i)\,e^{-j2\pi(f-f_i)t_i}$	$\displaystyle\sum_{i=1}^{N} WD_x(t-t_i,\, f-f_i)$ $\displaystyle+2\sum_{i=1}^{N-1}\sum_{m=i+1}^{N} WD_x\!\left(t-\tfrac{t_i+t_m}{2},\, f-\tfrac{f_i+f_m}{2}\right)$ $\times\cos 2\pi\left[(f_i-f_m)t-(t_i-t_m)f+\dfrac{f_i+f_m}{2}(t_i-t_m)\right]$	$\displaystyle AF_x(\tau,\nu)\sum_{i=1}^{N} e^{j2\pi(f_i\tau-\nu t_i)}$ $\displaystyle+\sum_{i=1}^{N}\sum_{\substack{m=1\\ m\neq i}}^{N} AF_x\big(\tau-(t_i-t_m),\,\nu-(f_i-f_m)\big)\exp\!\left[j2\pi\left(\dfrac{f_i+f_m}{2}\tau-\dfrac{t_i+t_m}{2}\nu+(f_i-f_m)\dfrac{t_i+t_m}{2}\right)\right]$

$$= \int_{-\infty}^{\infty} \int_{0}^{\infty} Q_X(t, f) e^{-j2\pi(tf\beta - [\ln(f/f_r)]\zeta)} \, dt \, df \qquad (27)$$

$$= \int \int Q_X\left(\frac{c}{f_r e^b}, f_r e^b\right) e^{-j2\pi(\beta c - \zeta b)} \, dc \, db, \qquad (28)$$

via Fourier and Mellin transformations. Like the Mellin transform [32], [94], [60],

$$MT_X(s) = \int_0^{\infty} X(u) e^{-j2\pi s \ln(u)} \frac{du}{u} \qquad (29)$$

$$= \int X(e^{\beta}) e^{-j2\pi s\beta} \, d\beta, \qquad (30)$$

the Altes Q distribution is useful for analyzing signals that have undergone scale changes, e.g., the compressions or dilations that occur in wideband Doppler analysis of a moving target. Both (27)–(28) and (29)–(30) demonstrate that the Mellin transform is equivalent to the Fourier transform if the argument of the function being transformed is first pre-warped in an exponential fashion. This allows the Mellin transform and the Altes Q distribution to be implemented efficiently using fast Fourier transform (FFT) techniques [95], [106].

The Altes Q distribution and the Hyperbolic AF of several signals are given in Table 12.4. This table shows that the Q distribution is well matched to signals with hyperbolic group delay. For example, the second entry in Table 12.4 states that the analytic signal $Y(f) = \frac{1}{\sqrt{f}} e^{-j2\pi c \ln f/f_r} \tilde{u}(f)$ has a Q distribution that is perfectly localized along the signal's hyperbolic group delay, $\tau_y(f) = c/f, f > 0$.

PROOF   Let $Y(f) = \frac{1}{\sqrt{f}} e^{-j2\pi c \ln f/f_r} \tilde{u}(f) = \frac{1}{\sqrt{f_r}} \left(\frac{f}{f_r}\right)^{-j2\pi c - \frac{1}{2}} \tilde{u}(f)$ where $\tilde{u}(f)$ is the unit step function defined in Table 12.1.

$$Q_Y(t, f) = f \int \frac{1}{\sqrt{f e^{u/2}}} e^{-j2\pi c \ln(f e^{u/2}/f_r)} \frac{1}{\sqrt{f e^{-u/2}}} e^{j2\pi c \ln(f e^{-u/2}/f_r)} e^{j2\pi tfu} \, du, \quad f > 0$$

$$= \int e^{-j2\pi c \ln e^u} e^{j2\pi tfu} \, du$$

$$= \int e^{j2\pi(tf - c)u} \, du = \delta(tf - c) = \frac{1}{|f|} \delta(t - c/f). \qquad (31)$$

∎

12.3.4   *TFR warping*

The Altes Q and the Wigner distributions are warped versions of each other [6], [97], [100].

$$Q_X(t, f) = W D_{\mathcal{W}X}\left(\frac{tf}{f_r}, f_r \ln \frac{f}{f_r}\right) \qquad (32)$$

$$W D_X(t, f) = Q_{\mathcal{W}^{-1}X}\left(t e^{-f/f_r}, f_r e^{f/f_r}\right) \qquad (33)$$

In (32)–(33), the signal is first pre-warped using the following unitary signal transformations,

$$(\mathcal{W}X)(f) = \sqrt{e^{f/f_r}} X\left(f_r e^{f/f_r}\right) \qquad (34)$$

$$(\mathcal{W}^{-1}X)(f) = \sqrt{\frac{f_r}{f}} X\left(f_r \ln \frac{f}{f_r}\right), \qquad f > 0, \qquad (35)$$

TABLE 12.4

Signals with closed form expressions for their Altes Q distribution and hyperbolic ambiguity function. Note that all signals in column one are assumed to be analytic, i.e., $Y(f) = 0$, $f < 0$. Here, $\tilde{u}(t)$ and $\text{rect}_a(t)$ are the unit step function and the rectangular function defined in Table 12.1. $u_n(t) = \dfrac{2^{1/4}}{\sqrt{n!}} e^{-\pi t^2} H_n(2\sqrt{\pi}\, t)$, $H_n(t) = (-1)^n e^{t^2/2} \dfrac{d^n}{dt^n} e^{-t^2/2}$ is the nth-order Hermite polynomial, and

$$L_n(t) = \frac{1}{n!} e^t \frac{d^n}{dt^n}\left(t^n e^{-t}\right) = \sum_{k=0}^{n} \frac{n!}{k!(n-k)!}\frac{(-t)^k}{k!} \text{ is the } n\text{th-order Laguerre polynomial.}$$

Analytic Signal, $Y(f)$, $f > 0$	Altes Distr., $Q_Y(t,f)$, $f > 0$	Hyperbolic AF, $HAF_Y(\zeta,\beta)$
$\sqrt{\dfrac{f_r}{f}}\, X(f_r \ln(f/f_r))$	$WD_X\left(\dfrac{tf}{f_r},\, f\ln\dfrac{f}{f_r}\right)$	$AF_X\left(\dfrac{\zeta}{f_r},\, f_r\beta\right)$
$\sqrt{\dfrac{f_r}{f}}\, e^{-j2\pi t_i f_r \ln(f/f_r)} = \left(\dfrac{f}{f_r}\right)^{-j2\pi t_i f_r - \frac{1}{2}}$	$\left\|\dfrac{f_r}{f}\right\|\delta\left(t - \dfrac{t_i f_r}{f}\right)$	$f_r e^{-j2\pi t_i f_r \beta}\delta(\zeta)$
$\sqrt{e^{f_i/f_r}}\,\delta\left(f - f_r e^{f_i/f_r}\right)$	$\left\|e^{f_i/f_r}\right\|\delta\left(f - f_r e^{f_i/f_r}\right)$	$\dfrac{1}{f_r} e^{j2\pi f_i \zeta/f_r}\delta(\beta)$
$\sqrt{\dfrac{f_r}{f}}\, e^{-j\pi\alpha f_r^2 (\ln(f/f_r))^2}$	$\left\|\dfrac{f_r}{f}\right\|\delta\left(t - \alpha f_r^2\dfrac{\ln(f/f_r)}{f}\right)$	$f_r\,\delta(\zeta - \alpha f_r^2\beta)$
$\sqrt{\dfrac{f_r}{f}}\, e^{-j\pi\alpha f_r^2(\ln(f/f_r))^2}\, e^{-j2\pi t_i f_r \ln(f/f_r)}$	$\left\|\dfrac{f_r}{f}\right\|\delta\left(t - \dfrac{f_r t_i}{f} - \alpha f_r^2\dfrac{\ln(f/f_r)}{f}\right)$	$f_r\,\delta(\zeta - \alpha f_r^2\beta)e^{-j2\pi t_i f_r\beta}$
$\sqrt{\dfrac{f_r}{-j\alpha f}}\, e^{j\pi[c-(f_r\ln(f/f_r)-f_i)^2/\alpha]}$	$\left\|\dfrac{f_r}{\alpha f}\right\|\delta\left(t - \left(\dfrac{f_r^2\ln(f/f_r)}{\alpha f} + \dfrac{f_i f_r}{\alpha f}\right)\right)$	$\dfrac{1}{f_r}\delta\left(\beta - \dfrac{\alpha}{f_r^2}\zeta\right)e^{j2\pi f_i\zeta/f_r}$
$\sqrt{e^{f_i/f_r}}\,\delta(f - f_r e^{f_i/f_r})$ $+ \sqrt{e^{-f_i/f_r}}\,\delta(f - f_r e^{-f_i/f_r})$	$\left\|e^{f_i/f_r}\right\|\delta(f - f_r e^{f_i/f_r})$ $+ \left\|e^{-f_i/f_r}\right\|\delta(f - f_r e^{-f_i/f_r})$ $+ 2\delta(f - f_r)\cos(4\pi f_i t)$	$\Big[\delta\left(\beta + \dfrac{2f_i}{f_r}\right) + \delta\left(\beta - \dfrac{2f_i}{f_r}\right)$ $+ 2\delta(\beta)\cos(2\pi f_i\zeta/f_r)\Big]/f_r$

TABLE 12.4
(continued)

Analytic Signal, $Y(f)$, $f > 0$	Altes Distr., $Q_Y(t, f)$, $f > 0$	Hyperbolic AF, $HAF_Y(\zeta, \beta)$
$\sqrt{\dfrac{f_r}{f}}\dfrac{\sin(2\pi a f_r \ln(f/f_r))}{\pi f_r \ln(f/f_r)}$	$\dfrac{\sin[4\pi(a f_r - \lvert tf\rvert)\ln(f/f_r)]}{\pi f_r \ln(f/f_r)}\text{rect}_a\left(\dfrac{tf}{f_r}\right)$	$f_r\dfrac{\sin[\pi\beta\,(2a f_r - \lvert\zeta\rvert)]}{\pi\zeta}\text{rect}_{2a}\left(\dfrac{\zeta}{f_r}\right)$
$\sqrt{\dfrac{f_r}{f}}\left[\dfrac{\delta(f - f_r)}{2} - \dfrac{j}{2\pi f_r \ln(f/f_r)}\right]$	$\dfrac{\sin 4\pi t f \ln(f/f_r)}{\pi f_r \ln(f/f_r)}\tilde{u}(tf/f_r)$	
$\sqrt{\dfrac{\sigma f_r}{f}}\,e^{-\pi(\sigma f_r \ln(f/f_r))^2}$	$\sqrt{2}\exp\left(-\pi\left[\left(\dfrac{tf}{\sigma f_r}\right)^2 + (\sigma f_r \ln(f/f_r))^2\right]\right)$	$\dfrac{1}{\sqrt{2}}\exp\left(-\dfrac{\pi}{2}\left[\left(\dfrac{\zeta}{f_r\sigma}\right)^2 + (\sigma f_r\beta)^2\right]\right)$
$(-j)^n\sqrt{\dfrac{f_r}{f}}\,u_n(f_r \ln(f/f_r))$	$2\exp\left(-2\pi\left[(tf/f_r)^2 + (f_r \ln(f/f_r))^2\right]\right)$ $\times L_n(4\pi((tf/f_r)^2 + (f_r \ln(f/f_r))^2))$	$\exp\left(-\dfrac{\pi}{2}\left[(\zeta/f_r)^2 + (f_r\beta)^2\right]\right)$ $\times L_n(\pi((\zeta/f_r)^2 + (f_r\beta)^2))$

respectively, followed by a warping of the time-frequency axis. Since $\mathcal{W}$ and $\mathcal{W}^{-1}$ in (34-35) are inverse operators, i.e., $(\mathcal{W}^{-1}\mathcal{W}X)(f) = X(f)$, then (32) can also be written as

$$Q_{\mathcal{W}^{-1}X}(t, f) = WD_X\left(\frac{tf}{f_r}, f_r \ln \frac{f}{f_r}\right), \tag{36}$$

which is the first entry in Table 12.4. Likewise, it can be shown that the Woodward AF in (18) and the hyperbolic AF in (26) are related by a simple axis scaling, once the signal has been pre-warped appropriately [97], [100]:

$$HAF_X(\zeta, \beta) = AF_{\mathcal{W}X}(\zeta/f_r, f_r\beta) \tag{37}$$

$$HAF_{\mathcal{W}^{-1}X}(\zeta, \beta) = AF_X(\zeta/f_r, f_r\beta). \tag{38}$$

Consequently, any table of WD and AF pairs, such as Table 12.2, can be transformed into an equivalent table of the Altes Q and HAF transform pairs by using (35) to warp the signal spectrum and (36) and (38) to warp the WD and the AF. For example, the second entry of Table 12.2 can be used to derive the third entry in Table 12.4.

PROOF   From the second entry in Table 12.2, let

$$X(f) = \delta(f - f_i) \Rightarrow WD_x(t, f) = \delta(f - f_i)$$
$$\Rightarrow AF_x(\tau, \nu) = e^{j2\pi f_i \tau}\delta(\nu).$$

By pre-warping $X(f)$ to form

$$Y(f) = (\mathcal{W}^{-1}X)(f) = \sqrt{\frac{f_r}{f}}\,\delta\left(f_r \ln \frac{f}{f_r} - f_i\right)$$
$$= \sqrt{e^{f_i/f_r}}\,\delta\left(f - f_r e^{f_i/f_r}\right),$$

one obtains using (36) and (38)

$$Q_Y(t, f) = WD_x\left(\frac{tf}{f_r}, f_r \ln \frac{f}{f_r}\right)$$
$$= \delta\left(f_r \ln \frac{f}{f_r} - f_i\right) = \left|e^{f_i/f_r}\right|\delta\left(f - f_r e^{f_i/f_r}\right)$$
$$HAF_Y(\zeta, \beta) = AF_x(\zeta/f_r, f_r\beta) = e^{j2\pi f_i\zeta/f_r}\delta(f_r\beta)$$

which corresponds to the third entry in Table 12.4.   ∎

12.3.5   Bertrand P_k distributions

The P_k distributions proposed by the Bertrands [19], [20], [95],

$$BP_k D_X(t, f; \mu) = f\int X(f\lambda_k(u))X^*(f\lambda_k(-u))\mu(u)e^{j2\pi tf(\lambda_k(u)-\lambda_k(-u))}\,du,$$

$$f > 0 \tag{39}$$

with

$$\lambda_0(u) = \frac{u/2e^{u/2}}{\sinh(u/2)} = \frac{u}{1 - e^{-u}},$$

$$\lambda_1(u) = \exp\left[1 + \frac{ue^{-u}}{e^{-u} - 1}\right],$$

$$\lambda_k(u) = \left[k\frac{e^{-u} - 1}{e^{-ku} - 1}\right]^{\frac{1}{k-1}}, \qquad k \neq 0, 1,$$

and $\mu(u) = \mu^*(-u)$, are affine TFRs that are covariant to all affine time transformations on analytic signals, i.e.,

$$Y(f) = \frac{1}{\sqrt{|a|}} X\left(\frac{f}{a}\right) e^{-j2\pi f t_0}, \qquad f > 0$$

$$\Rightarrow \quad BP_k D_Y(t, f; \mu) = BP_k D_X\left(a(t - t_0), \frac{f}{a}; \mu\right), \quad f > 0.$$

This affine covariance is useful in wideband Doppler applications. By properly selecting k, the versatile P_k distributions also have extended covariance to dispersive, i.e., nonconstant, time shifts on the signal including those proportional to power, hyperbolic, or logarithmic functions of frequency:

$$k \neq 0, 1; \; Y(f) = \frac{1}{\sqrt{|a|}} e^{-j2\pi(bf + cf^k)} X\left(\frac{f}{a}\right) \Rightarrow BP_k D_Y(t, f; \mu) = BP_k D_X\left(a(t - b - kcf^{k-1}), \frac{f}{a}; \mu\right) \quad (40)$$

$$k = 0; \quad Y(f) = \frac{1}{\sqrt{|a|}} e^{-j2\pi(bf + c\ln f)} X\left(\frac{f}{a}\right) \Rightarrow BP_0 D_Y(t, f; \mu) = BP_0 D_X\left(a(t - b - \frac{c}{f}), \frac{f}{a}; \mu\right) \quad (41)$$

$$k = 1; \; Y(f) = \frac{1}{\sqrt{|a|}} e^{-j2\pi(bf + cf\ln f)} X\left(\frac{f}{a}\right) \Rightarrow BP_1 D_Y(t, f; \mu) = BP_1 D_X\left(a(t - b - c[1 + \ln f]), \frac{f}{a}; \mu\right). \quad (42)$$

Note that the dispersive time shifts in the Bertrand P_k distributions in (40)–(42) are equivalent to the change in the group delay of $X(f)$ brought about by multiplication with the complex exponential terms used to form $Y(f)$. The most commonly used Bertrand distribution is the unitary form of the $k = 0$ or P_0 Bertrand distribution:

$$BP_0 D_X(t, f; \mu_0) = f \int X\left(f\frac{u/2e^{u/2}}{\sinh u/2}\right) X^*\left(f\frac{u/2e^{-u/2}}{\sinh u/2}\right) \frac{u/2}{\sinh u/2} e^{j2\pi t f u} \, du,$$

$$f > 0. \tag{43}$$

This special form of the P_0 distribution in (39) with $\mu_0(u) = \frac{u/2}{\sinh u/2}$ is called the unitary P_0 as the resulting TFR preserves inner products:

$$\int \int_0^\infty BP_0 D_X(t, f; \mu_0) BP_0 D_Y^*(t, f; \mu_0) \, dt \, df = \left|\int_0^\infty X(f) Y^*(f) \, df\right|^2.$$

12.3.6   *Cross terms of quadratic time-frequency representations*

The spectrogram, Wigner, Altes Q, and Bertrand P_k distributions, as well as several other "energetic" TFRs listed in later sections, are quadratic functions of the signal. These nonlinear functions produce "cross terms" which can make visual analysis of TFRs difficult. For example, the nonlinear operation,

$$|x(t) + y(t)|^2 = |x(t)|^2 + |y(t)|^2 + 2Real\{x(t)y^*(t)\},$$

is equal to the sum of the two "auto" terms $|x(t)|^2$, $|y(t)|^2$ plus the "cross term" $2Real\{x(t)y^*(t)\}$. The following discussion will focus on the characteristics of the cross terms of the WD [52], [70], [82].

The WD of a multicomponent signal,

$$y(t) = \sum_{i=1}^{N} x_i(t) \Rightarrow$$

$$WD_y(t, f) = \sum_{i=1}^{N} WD_{x_i}(t, f) + 2\sum_{i=1}^{N-1} \sum_{k=i+1}^{N} Real\{WD_{x_i x_k}(t, f)\}, \tag{44}$$

consists of N auto-terms, $WD_{x_i}(t, f)$, and $\frac{N(N-1)}{2}$ cross terms,

$$WD_{x_i x_k}(t, f) = \int x_i(t + \tau/2)x_k^*(t - \tau/2)e^{-j2\pi f\tau}\, d\tau = WD_{x_k x_i}^*(t, f). \tag{45}$$

An intuitive understanding of these cross terms can be obtained by analyzing the WD and the AF of the multicomponent signal $y(t)$ in (44) for the special case that each signal component, $x_i(t)$, is a shifted version of a basic envelope $x(t)$, i.e.,

$$x_i(t) = x(t - t_i)e^{j2\pi f_i t}. \tag{46}$$

The WD and the AF given below were taken from the last entry in Table 12.3:

$$WD_y(t, f) = \sum_{i=1}^{N} WD_x(t - t_i, f - f_i)$$

$$+ 2\sum_{i=1}^{N-1} \sum_{m=i+1}^{N} WD_x\left(t - \frac{t_i + t_m}{2}, f - \frac{f_i + f_m}{2}\right)$$

$$\times\ \cos 2\pi\left[(f_i - f_m)t - (t_i - t_m)f + \frac{f_i + f_m}{2}(t_i - t_m)\right] \tag{47}$$

$$AF_y(\tau, \nu) = AF_x(\tau, \nu) \sum_{i=1}^{N} e^{j2\pi(f_i \tau - \nu t_i)}$$

$$+ \sum_{i=1}^{N-1} \sum_{\substack{m=1 \\ m \neq i}}^{N} AF_x(\tau - (t_i - t_m), \nu - (f_i - f_m))$$

$$\times\ \exp\left[j2\pi\left(\frac{f_i + f_m}{2}\tau - \frac{t_i + t_m}{2}\nu + (f_i - f_m)\frac{t_i + t_m}{2}\right)\right] \tag{48}$$

The ith auto WD term in the first sum in (47) has been shifted by (t_i, f_i) in the same way that the basic signal component $x_i(t)$ in (46) was shifted. In the second summation, the cross WD term corresponding to the pair of auto terms $x_i(t)$ and $x_m(t)$ is equal to the WD of the envelope $x(t)$, shifted to midway in the time-frequency plane, i.e., $(\frac{t_i + t_m}{2}, \frac{f_i + f_m}{2})$, between the pair of auto signal components; the cross term oscillates with a spatial frequency proportional to the distance $(t_i - t_m, f_i - f_m)$ between the pair of auto terms. This is depicted in Figure 12.2. In (48), the AF of each auto term is equal to the AF of the envelope multiplied by a complex exponential; all auto AF terms map on top of one another at the origin of the AF plane. The cross terms in the second summation correspond to the AF of the basic envelope, $x(t)$, shifted

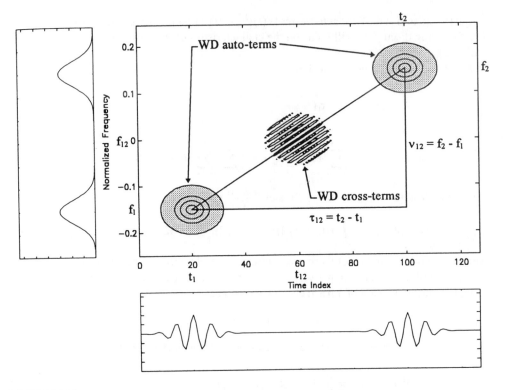

FIGURE 12.2

Interference geometry of the Wigner distribution (WD) of a two-component signal
$y(t) = \sum_{i=1}^{2} x(t - t_i)e^{j2\pi f_i t}$. The signal is plotted at the bottom of the figure and its Fourier spectrum
is plotted vertically along the left. The two auto terms are centered at (t_1, f_1) and (t_2, f_2). The
oscillatory WD cross term occurs midway between them at $(t_{12}, f_{12}) = (\frac{t_1+t_2}{2}, \frac{f_1+f_2}{2})$. (Figure taken
with permission from [45].)

away from the origin of the (τ, ν) plane by an amount equal to the distance between each pair
of signal terms, as indicated in Figure 12.3. Signal components which occur at the same time,
i.e., $t_i = t_m$, or at the same frequency, $f_i = f_m$, have cross terms which are shifted along
the axes in the AF plane. The greater the separation between any two signal components, the
more rapidly the corresponding cross term in (47) oscillates in the WD plane and the farther
away from the origin the cross term in (48) maps to in the AF plane. For example, Table 12.2
indicates that the WD of a cosine,

$$x(t) = \cos(2\pi f_i t) \quad \leftrightarrow \quad X(f) = [\delta(f + f_i) + \delta(f - f_i)]/2 \qquad (49)$$

$$\Rightarrow \quad WD_x(t, f) = [\delta(f + f_i) + \delta(f - f_i) + 2\delta(f)\cos(4\pi f_i t)]/4, \qquad (50)$$

consists of two auto terms, which are the impulses located at $f = \pm f_i$, plus the oscillatory
cross term located at mid-frequency $f = 0$. Increasing the sinusoidal frequency f_i increases
the oscillation rate of the cross term, but produces no change in its location. The AF of this
cosine,

$$x(t) = \cos(2\pi f_i t) \quad \Rightarrow \quad AF_x(\tau, \nu) = [\delta(\nu)\{e^{j2\pi f_i \tau} + e^{-j2\pi f_i \tau}\} + \delta(\nu+2f_i)+\delta(\nu-2f_i)]/4,$$

has two auto terms which always map to the origin of the AF plane and two cross terms which
occur at $\nu = \pm 2f_i$.

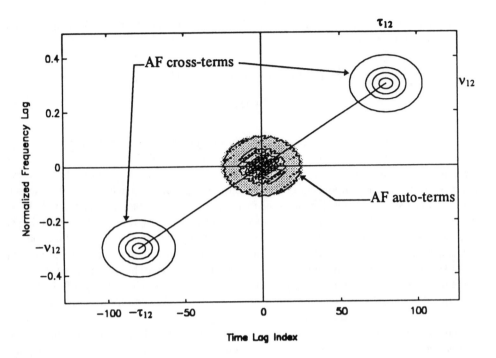

FIGURE 12.3

Interference geometry of the ambiguity function (AF) of a two-component signal
$y(t) = \sum_{i=1}^{2} x(t - t_i)e^{j2\pi f_i t}$. The two AF auto terms map to the origin, while the cross term maps
away from the origin to the locations corresponding to (τ_{12}, ν_{12}) and $(-\tau_{12}, -\nu_{12})$, where $\tau_{12} = t_2 - t_1$
is the temporal separation and $\nu_{12} = f_2 - f_1$ is the spectral separation between the two Wigner
distribution auto terms in Figure 12.2. (Figure taken with permission from [45].)

The characteristics of the cross terms of the spectrogram, Altes Q and the Bertrand P_k are
described in [76], [82], [12], [62], [46], [65]. The spectrogram contains undulating cross terms
that occur wherever the STFT of the auto signal components $x_i(t)$ overlap in the time-frequency
plane. Using (11) and the cosine example from (49), one can show that

$$x(t) = \cos(2\pi f_i t) \ \Rightarrow$$

$$SPEC_x(t, f; \Gamma) = \frac{1}{4}[|\Gamma(f_i - f)|^2 + |\Gamma(-f_i - f)|^2$$

$$+ 2 \, Real\{\Gamma^*(f_i - f)\Gamma(-f_i - f)e^{j2\pi(2f_i)t}\}]. \tag{51}$$

If the bandwidth of the analysis window, $\Gamma(f)$, is greater than the sinusoidal frequency f_i,
i.e., $|\Gamma(\pm f_i)| \neq 0$, then the cross term in (51) is nonzero and is modulated by a cosine whose
frequency, $2f_i$, equals the distance between the signal's spectral components at $f = \pm f_i$.
Likewise, the Bertrand P_k distributions produce an oscillatory cross term corresponding to
each pair of signal components; closed form expressions can be obtained for some of the
P_k distributions revealing that their cross terms occur at the location corresponding to the
generalized mean of the locations of the corresponding pair of auto terms.

The presence of cross terms can make visual analysis of nonlinear TFRs difficult. Two
basic approaches are used to minimize the effects of these cross terms. The first approach is to
use the analytic form of real, bandpass signals to zero out redundant negative frequency axis
components of the signal's Fourier transform. The advantage of replacing real signals with

their analytic counterpart can be demonstrated by comparing the Wigner-Ville distribution in (21) for a bandlimited cosine,

$$x(t) = \cos(2\pi f_i t) \Rightarrow z_x(t) = x(t) + j\hat{x}(t) = e^{j2\pi f_i t}$$

$$\Rightarrow WD_{z_x}(t, f) = \delta(f - f_i). \tag{52}$$

The analytic signal corresponding to the cosine is a single tone located at $Z_x(f) = \delta(f - f_i)$, and as such its WD is a simple impulse at $f = f_i$. Comparison of (50) with (52) demonstrates that using the Wigner-Ville distribution for bandlimited signals removes all cross WD terms that arise from signal components occurring on the negative frequency axis. The second approach to cross term removal exploits the fact that cross terms which oscillate rapidly in the time-frequency plane can be removed by smoothing or lowpass filtering the WD. Applying an ideal lowpass filter to the WD is equivalent to multiplying the signal's AF by a function that (i) is approximately equal to one in the region near the origin where the auto AF terms map to in (48) and (ii) is approximately zero in the region away from the origin of the AF plane where the AF cross terms map. Examples of TFRs that are a smoothed version of the WD are given in Table 12.5. The most commonly known smoothed WD is the spectrogram in (9). The pseudo WD and the smoothed pseudo WD use lowpass analysis windows $\gamma(\tau)$ and $s(t)$ to smooth out the oscillatory cross terms. Unfortunately, the cost to be paid for smoothing away of cross terms is a loss of resolution between signal components in the time-frequency plane [80], [55] and, as we shall see in the next section, a loss of desirable TFR properties [39].

12.4 Desirable properties of time-frequency representations

One way of selecting which TFR to use is to examine which one has the most desirable properties for the particular application at hand. This section is a summary of several such properties, listed in Table 12.6. This table can be broken up conceptually into the following categories of ideal TFR properties: covariance, statistical or energy distribution, signal analysis, localization, and inner products [39], [43], [60], [68], [69], [24], [10].

In the following section, each of the ideal TFR properties will be discussed individually and at least one example will be given demonstrating how to prove whether or not a certain TFR satisfies that property. Additional proofs can be found in [39], [60], [103], [10], [25], [81], [111].

12.4.1 *Covariance properties*

The covariance properties $P_1 - P_6$ in Table 12.6 state that certain operations on the signal, such as translation, dilation, or convolution, should be preserved in the TFR. That is, if the signal is changed in some way, then its TFR should change in exactly the same fashion.

P_1: **Frequency-shift covariance**

$$y(t) = x(t)e^{j2\pi f_0 t} \quad \Rightarrow \quad Y(f) = X(f - f_0) \quad \Rightarrow \quad T_y(t, f) = T_x(t, f - f_0)$$

TABLE 12.5
Many TFRs are equivalent to smoothed or warped Wigner distributions. Alternative formulations for these TFRs can be found in Tables 12.9, 12.11, 12.13, and 12.15. Here, $f_r > 0$ is a positive reference frequency,

$$(\mathcal{W}X)(f) = \sqrt{e f/f_r}\, X\left(f_r e^{f/f_r}\right),$$

$$(\mathcal{W}_\kappa X)(f) = \left|\kappa |f/f_r|^{(\kappa-1)/\kappa}\right|^{-1/2} X\left(f_r \mathrm{sgn}(f)|f/f_r|^{1/\kappa}\right), \quad \kappa \neq 0,$$

and $\mathrm{sgn}(f)$ is defined in Table 12.1. Also, $\psi_c(t, f) \leftrightarrow \Psi_C(\tau, \nu)$ and $\gamma(t) \leftrightarrow \Gamma(f)$ are Fourier transform pairs.

TFR Name	TFR Formulation
Cohen's class	$C_x(t, f; \Psi_C) = \int \int \psi_C(t-t', f-f') WD_x(t', f')\, dt'\, df'$
Pseudo Wigner	$PWD_x(t, f; \Gamma) = \int WD_\gamma(0, f-f') WD_x(t, f')\, df'$
Scalogram	$SCAL_x(t, f; \Gamma) = \int \int WD_\gamma\left(\frac{f}{f_r}(t'-t), f_r\frac{f'}{f}\right) WD_x(t', f')\, dt'\, df'$
Smoothed pseudo Wigner	$SPWD_x(t, f; \Gamma, s) = \int \int s(t-t') WD_\gamma(0, f-f') WD_x(t', f')\, dt'\, df'$
Spectrogram	$SPEC_x(t, f; \Gamma) = \int \int WD_\gamma(t'-t, f'-f) WD_x(t', f')\, dt'\, df'$
Altes Q	$Q_X(t, f) = WD_{\mathcal{W}X}\left(\frac{tf}{f_r}, f_r \ln\frac{f}{f_r}\right)$
κth power Wigner	$WD_X^{(\kappa)}(t, f) = WD_{\mathcal{W}_\kappa X}\left(\frac{t}{\kappa|f/f_r|^{\kappa-1}}, f_r \mathrm{sgn}(f)|f/f_r|^\kappa\right), \quad \kappa \neq 0$
Hyperbologram	$HYP_X(t, f; \Gamma) = \int \int_0^\infty WD_{\mathcal{W}\Gamma}\left(t'-\frac{tf}{f_r}, f'-f_r \ln\frac{f}{f_r}\right) WD_{\mathcal{W}X}(t', f')\, dt'\, df'$
Pseudo Altes Q	$PQ_X(t, f; \Gamma) = f_r \int_0^\infty WD_{\mathcal{W}\Gamma}\left(0, f_r \ln\frac{f}{f'}\right) WD_{\mathcal{W}X}\left(\frac{tf}{f_r}, f_r \ln\frac{f'}{f_r}\right)\frac{df'}{f'}$
Smoothed pseudo Altes Q	$SPQ_X(t, f; \Gamma, s) = f_r \int_{-\infty}^\infty \int_0^\infty s(tf-c) WD_{\mathcal{W}\Gamma}\left(0, f_r \ln\frac{f}{f'}\right)$ $\times WD_{\mathcal{W}X}\left(\frac{c}{f_r}, f_r \ln\frac{f'}{f_r}\right) dc\, \frac{df'}{f'}$

Property P_1 states that if a signal is modulated or shifted in frequency by an amount f_0, then the TFR of that signal should also be shifted by f_0. This property is very important for analyzing a variety of signals such as speech, music, or sonar. Both the WD and the spectrogram satisfy this property; the STFT does not.

TABLE 12.6
Ideal Time Frequency Representation (TFR) properties.

Property Name	TFR Property
P_1: Frequency-Shift Covariance	$T_y(t, f) = T_x(t, f - f_0)$ for $y(t) = x(t)e^{j2\pi f_0 t}$
P_2: Time-Shift Covariance	$T_y(t, f) = T_x(t - t_0, f)$ for $y(t) = x(t - t_0)$
P_3: Scale Covariance	$T_y(t, f) = T_x(at, f/a)$ for $y(t) = \sqrt{|a|}x(at)$
P_4: Hyperbolic Time Shift	$T_y(t, f) = T_x(t - c/f, f)$ if $Y(f) = \exp(-j2\pi c \ln\frac{f}{f_r})X(f)$
P_5: Convolution Covariance	$T_y(t, f) = \int T_h(t - \tau, f)T_x(\tau, f)\, d\tau$ for $y(t) = \int h(t - \tau)x(\tau)\, d\tau$
P_6: Modulation Covariance	$T_y(t, f) = \int T_h(t, f - f')T_x(t, f')\, df'$ for $y(t) = h(t)x(t)$
P_7: Real-Valued	$T_x^*(t, f) = T_x(t, f)$
P_8: Positivity	$T_x(t, f) \geq 0$
P_9: Time Marginal	$\int T_x(t, f)df = |x(t)|^2$
P_{10}: Frequency Marginal	$\int T_x(t, f)dt = |X(f)|^2$
P_{11}: Energy Distribution	$\int\int T_x(t, f)\, dt\, df = \int |X(f)|^2\, df$
P_{12}: Time Moments	$\int\int t^n T_x(t, f)\, dt\, df = \int t^n |x(t)|^2\, dt$
P_{13}: Frequency Moments	$\int\int f^n T_x(t, f)\, dt\, df = \int f^n |X(f)|^2\, df$
P_{14}: Finite Time Support	$T_x(t, f) = 0$ for $t \notin (t_1, t_2)$ if $x(t) = 0$ for $t \notin (t_1, t_2)$
P_{15}: Finite Freq. Support	$T_x(t, f) = 0$ for $f \notin (f_1, f_2)$ if $X(f) = 0$ for $f \notin (f_1, f_2)$
P_{16}: Instantaneous Freq.	$\dfrac{\int f T_x(t, f)\, df}{\int T_x(t, f)\, df} = \dfrac{1}{2\pi}\dfrac{d}{dt}arg\{x(t)\}$
P_{17}: Group Delay	$\dfrac{\int t T_x(t, f)\, dt}{\int T_x(t, f)\, dt} = -\dfrac{1}{2\pi}\dfrac{d}{df}arg\{X(f)\}$
P_{18}: Fourier Transform	$T_y(t, f) = T_x(-f, t)$ for $y(t) = X(t)$
P_{19}: Freq. Localization	$T_x(t, f) = \delta(f - f_0)$ for $X(f) = \delta(f - f_0)$
P_{20}: Time Localization	$T_x(t, f) = \delta(t - t_0)$ for $x(t) = \delta(t - t_0)$
P_{21}: Linear Chirp Localization	$T_x(t, f) = \delta(t - cf)$ for $X(f) = e^{-j\pi cf^2}$

TABLE 12.6
(continued)

Property Name	TFR Property
P_{22}: Hyperbolic Localization	$T_x(t, f) = \frac{1}{f}\delta(t - \frac{c}{f})$, $f > 0$ if $X_c(f) = \frac{1}{\sqrt{f}}e^{-j2\pi c \ln \frac{f}{f_r}}$, $f > 0$
P_{23}: Chirp Convolution	$T_y(t, f) = T_x(t - f/c, f)$ for $y(t) = \int x(t-\tau)\sqrt{|c|}e^{j\pi c\tau^2}d\tau$
P_{24}: Chirp Multiplication	$T_y(t, f) = T_x(t, f - ct)$ for $y(t) = x(t)e^{j\pi ct^2}$
P_{25}: Moyal's Formula	$\iint T_x(t, f)T_y^*(t, f)\, dt\, df = \left|\int x(t)y^*(t)\, dt\right|^2$

PROOF Let $Y(f) = X(f - f_0)$.

$$WD_y(t, f) = \int Y(f + \nu/2)Y^*(f - \nu/2)e^{j2\pi\nu t}\, d\nu$$

$$= \int X([f + \nu/2] - f_0)X^*([f - \nu/2] - f_0)e^{j2\pi\nu t}\, d\nu$$

$$= \int X([f - f_0] + \nu/2)X^*([f - f_0] - \nu/2)e^{j2\pi\nu t}\, d\nu$$

$$= W_x(t, f - f_0)$$

$$STFT_y(t, f; \Gamma) = e^{-j2\pi tf}\int Y(f')\,\Gamma^*(f' - f)e^{j2\pi tf'}\, df'$$

$$= e^{-j2\pi tf}\int X(f' - f_0)\,\Gamma^*(f' - f)e^{j2\pi tf'}\, df'$$

$$= e^{-j2\pi tf}\int X(\nu)\,\Gamma^*(\nu + f_0 - f)e^{j2\pi t[\nu + f_0]}\, d\nu$$

$$= e^{-j2\pi[f - f_0]t}\int X(\nu)\,\Gamma^*(\nu - [f - f_0])e^{j2\pi t\nu}\, d\nu$$

$$= e^{-j2\pi[f - f_0]t}STFT_x(t, f - f_0; \Gamma) \neq STFT_x(t, f - f_0; \Gamma)$$

$$SPEC_y(t, f; \Gamma) = |STFT_y(t, f; \Gamma)|^2 = |e^{-j2\pi[f - f_0]t}STFT_x(t, f - f_0; \Gamma)|^2$$

$$= SPEC_x(t, f - f_0; \Gamma) \qquad\blacksquare$$

P_2: Time-shift covariance

$$y(t) = x(t - t_0) \quad\Rightarrow\quad T_y(t, f) = T_x(t - t_0, f)$$

Property P_2 states that any time translations in the signal should be preserved in its TFR. Equivalently, P_2 states that an ideal TFR should be covariant to any constant shift in the signal's group delay. The WD and the spectrogram satisfy this property, but the STFT does not.

PROOF   Let $y(t) = x(t - t_0) \leftrightarrow Y(f) = e^{-j2\pi f t_0}X(f) \Rightarrow \tau_y(f) = \tau_x(f) - t_0.$

$$WD_y(t, f) = \int y\left(t + \frac{\tau}{2}\right) y^*\left(t - \frac{\tau}{2}\right) e^{-j2\pi f\tau} d\tau$$

$$= \int x\left(t + \frac{\tau}{2} - t_0\right) x^*\left(t - \frac{\tau}{2} - t_0\right) e^{-j2\pi f\tau} d\tau$$

$$= \int x\left([t - t_0] + \frac{\tau}{2}\right) x^*\left([t - t_0] - \frac{\tau}{2}\right) e^{-j2\pi f\tau} d\tau$$

$$= W_x(t - t_0, f)$$

$$STFT_y(t, f; \Gamma) = \int y(\tau)\gamma^*(\tau - t)e^{-j2\pi f\tau} d\tau$$

$$= \int x(\tau - t_0)\gamma^*(\tau - t)e^{-j2\pi f\tau} d\tau$$

$$= \int x(\beta)\gamma^*(\beta + t_0 - t)e^{-j2\pi f[\beta + t_0]} d\tau$$

$$= e^{-j2\pi f t_0} \int x(\beta)\gamma^*(\beta - [t - t_0])e^{-j2\pi f\beta} d\tau$$

$$= e^{-j2\pi f t_0} STFT_x(t - t_0, f; \Gamma) \neq STFT_x(t - t_0, f; \Gamma)$$

$$\left|STFT_y(t, f; \Gamma)\right|^2 = \left|e^{-j2\pi f t_0} STFT_x(t - t_0, f; \Gamma)\right|^2$$

$$= |STFT_x(t - t_0, f; \Gamma)|^2 \qquad \blacksquare$$

P_3: Scale covariance

$$y(t) = \sqrt{|a|}x(at) \quad \Rightarrow \quad T_y(t, f) = T_x\left(at, \frac{f}{a}\right)$$

To understand property P_3, recall that the dilation property in Table 12.1 indicates that if the time axis of a signal is compressed by a scalar factor a, then its Fourier transform is expanded by a factor of $1/a$. That is, if $y(t) = \sqrt{|a|}x(at)$, then $Y(f) = \frac{1}{\sqrt{|a|}}X\left(\frac{f}{a}\right)$. Hence, property P_3 states that if the signal's time axis is compressed by a scale factor a, then its TFR's time axis should also be compressed by a and its frequency axis expanded by the factor $1/a$. The Wigner distribution, the Altes Q distribution, and the Bertrand P_k distributions are scale covariant, but the STFT is not for any nontrivial scale factor $a \neq 1$.

PROOF   Let $y(t) = \sqrt{|a|}x(at) \leftrightarrow Y(f) = \frac{1}{\sqrt{|a|}}X\left(\frac{f}{a}\right).$

$$WD_y(t, f) = \int y\left(t + \frac{\tau}{2}\right) y^*\left(t - \frac{\tau}{2}\right) e^{-j2\pi f\tau} d\tau$$

$$= |a| \int x\left(\left[t + \frac{\tau}{2}\right]a\right) x^*\left(\left[t - \frac{\tau}{2}\right]a\right) e^{-j2\pi f\tau} d\tau$$

$$= \int x\left(at + \frac{u}{2}\right) x^*\left(at - \frac{u}{2}\right) e^{-j2\pi uf/a} du$$

$$= W_x \left(at, \frac{f}{a} \right)$$

$$Q_Y(t, f) = f \int Y(f e^{u/2}) Y^*(f e^{-u/2}) e^{j2\pi t f u} du \ , \ f > 0$$

$$= \frac{f}{a} \int X \left(\frac{f}{a} e^{u/2} \right) X^* \left(\frac{f}{a} e^{-u/2} \right) e^{j2\pi (at)(f/a)u} du, \qquad a > 0$$

$$= Q_X \left(at, \frac{f}{a} \right)$$

$$BP_k D_Y(t, f; \mu) = f \int Y(f \lambda_k(u)) Y^*(f \lambda_k(-u)) \mu(u) e^{j2\pi t f (\lambda_k(u) - \lambda_k(-u))} du, \quad f > 0$$

$$= \frac{f}{a} \int X \left(\frac{f}{a} \lambda_k(u) \right) X^* \left(\frac{f}{a} \lambda_k(-u) \right) \mu(u) e^{j2\pi (at)(f/a)(\lambda_k(u) - \lambda_k(-u))} du, \ a > 0$$

$$= BP_k D_Y \left(at, \frac{f}{a}; u \right)$$

$$STFT_y(t, f; \Gamma) = \int \sqrt{|a|} x(a\tau) \gamma^*(\tau - t) e^{-j2\pi f \tau} d\tau$$

$$= \frac{1}{\sqrt{a}} \int x(t') \gamma^* \left(\frac{t'}{a} - t \right) e^{-j2\pi f t'/a} dt', \quad a > 0$$

$$= \int x(t') \tilde{\gamma}^*(t' - at) e^{-j2\pi (f/a)t'} dt', \quad \text{where } \tilde{\gamma}(t) = \frac{1}{\sqrt{|a|}} \gamma \left(\frac{t}{a} \right)$$

$$= STFT_x \left(at, \frac{f}{a}; \tilde{\Gamma} \right) \neq STFT_x \left(at, \frac{f}{a}; \Gamma \right) \text{ unless } a = 1$$

P_4: **Hyperbolic time-shift covariance**

$$Y(f) = \exp \left(-j2\pi c \ln \frac{f}{f_r} \right) X(f) \quad \Rightarrow \quad T_y(t, f) = T_x \left(t - \frac{c}{f}, f \right)$$

Property P_4 states that an ideal TFR should be covariant to hyperbolic changes in the signal's group delay. If the signal's Fourier transform or spectrum undergoes a logarithmic phase change, then its group delay undergoes a hyperbolic shift, i.e., $\tau_y(f) = \tau_x(f) + \frac{c}{f}$. Property P_4 states that an ideal TFR of a logarithmic FM modulated signal should correspond to the TFR of the original signal, but with a dispersive temporal shift equal to the hyperbolic change in the group delay. The Altes Q distribution and the general form of the Bertrand P_0 distribution satisfy this property.

PROOF   Let $Y(f) = \exp(-j2\pi c \ln \frac{f}{f_r}) X(f) \Rightarrow \tau_y(f) = \tau_x(f) + \frac{c}{f}$.

$$Q_Y(t, f) = f \int Y(f e^{u/2}) Y^*(f e^{-u/2}) e^{j2\pi t f u} du, \quad f > 0$$

$$= f \int X(f e^{u/2}) e^{-j2\pi c \ln(f e^{u/2}/f_r)} X^*(f e^{-u/2}) e^{j2\pi c \ln(f e^{-u/2}/f_r)} e^{j2\pi t f u} du$$

$$= f \int X(fe^{u/2})X^*(fe^{-u/2})e^{-j2\pi c[\ln(f/f_r)+u/2-\ln(f/f_r)+u/2]}e^{j2\pi tfu}\,du$$

$$= f \int X(fe^{u/2})X^*(fe^{-u/2})e^{j2\pi[t-c/f]fu}\,du$$

$$= Q_X(t - c/f, f)$$

$$BP_0D_Y(t, f; \mu) = f \int X\left(f\frac{u/2e^{u/2}}{\sinh u/2}\right)e^{-j2\pi c\ln\left(\frac{f}{f_r}\frac{u/2e^{u/2}}{\sinh u/2}\right)}X^*\left(f\frac{u/2e^{-u/2}}{\sinh u/2}\right)$$

$$\times e^{j2\pi c\ln\left(\frac{f}{f_r}\frac{u/2e^{-u/2}}{\sinh u/2}\right)}\mu(u)e^{j2\pi tfu}\,du$$

$$= f \int X\left(f\frac{u/2e^{u/2}}{\sinh u/2}\right)X^*\left(f\frac{u/2e^{-u/2}}{\sinh u/2}\right)e^{-j2\pi c\ln e^u}\mu(u)e^{j2\pi tfu}\,du$$

$$= f \int X\left(f\frac{u/2e^{u/2}}{\sinh u/2}\right)X^*\left(f\frac{u/2e^{-u/2}}{\sinh u/2}\right)\mu(u)e^{j2\pi(t-c/f)fu}\,du$$

$$= BP_0D_X(t - c/f, f; \mu) \qquad \blacksquare$$

P_5: Convolution covariance

$$y(t) = \int h(t - \tau)x(\tau)\,d\tau \quad \Rightarrow \quad T_y(t, f) = \int T_h(t - \tau, f)T_x(\tau, f)\,d\tau$$

Property P_5 states that convolving two signals together in the time domain should produce the equivalent effect of convolving their corresponding TFRs together in the time domain. The WD is one of the few TFRs that satisfies this property.

PROOF   Let $y(t) = \int h(t - \tau)x(\tau)\,d\tau$.

$$WD_y(t, f) = \int y(t + t'/2)y^*(t - t'/2)e^{-j2\pi ft'}\,dt'$$

$$= \int \left[\int h(t+t'/2-\alpha)x(\alpha)\,d\alpha\right]\left[\int h^*(t-t'/2-\gamma)x^*(\gamma)\,d\gamma\right]e^{-j2\pi ft'}\,dt'$$

Substituting $\alpha = \tau + p/2$, $\gamma = \tau - p/2$, and $t' = q + p$ produces

$$WD_y(t, f)$$

$$= \int\int\int h\left((t-\tau)+\frac{q}{2}\right)h^*\left((t-\tau)-\frac{q}{2}\right)x\left(\tau+\frac{p}{2}\right)x^*\left(\tau-\frac{p}{2}\right)e^{-j2\pi f(q+p)}\,dq\,d\tau\,dp$$

$$= \int WD_h(t - \tau, f)WD_x(\tau, f)\,d\tau. \qquad \blacksquare$$

P_6: **Modulation covariance**

$$y(t) = h(t)x(t) \Rightarrow Y(f) = \int H(f - f')X(f') \, df'$$

$$\Rightarrow T_y(t, f) = \int T_h(t, f - f')T_x(t, f') \, df'$$

If two signals are modulated together in time, then Table 12.1 indicates that their Fourier transforms are convolved together in frequency. Similarly, property P_6 states that whenever the Fourier transforms of two signals are convolved together in the frequency domain, then the TFR of the resulting signal should be equal to the convolution in frequency of the two signals' respective TFRs. It can be shown using the dual to the proof in P_5 above that the WD satisfies this property.

12.4.2 *Statistical energy density distribution properties*

The second category of properties in Table 12.6 originates from the desire to generalize the concepts of the one-dimensional instantaneous signal energy, $|x(t_0)|^2$, and power spectral density, $|X(f_0)|^2$, into a two-dimensional statistical probability density function or an energy distribution, $T_x(t_0, f_0)$, which would ideally provide a measure of the local signal energy or the probability that a signal contains a sinusoidal component of frequency f_0 at the time t_0. Properties P_7–P_{13} state that such an energy distribution TFR should be real, nonnegative, and have its marginal distributions equal to the signal's temporal and spectral energy densities, $|x(t)|^2$ and $|X(f)|^2$ respectively. A TFR should also preserve the signal energy, mean, variance, and other higher order moments of the signal's temporal and spectral energy density. These ideal statistical or energy density TFR properties are described below.

P_7: **Real**

$$T_x(t, f) = T_x^*(t, f), \qquad \forall \, x(t)$$

For a TFR to be real, it must be equal to its own complex conjugate for all signals. The WD, the Altes Q distribution and the spectrogram are always real valued. Equations (10) and (11) are examples of the fact that the STFT is complex, which is why its squared magnitude, the spectrogram, is usually used for visual analysis.

PROOF

$$WD_x^*(t, f) = \left[\int x\left(t + \frac{\tau}{2}\right) x^*\left(t - \frac{\tau}{2}\right) e^{-j2\pi f \tau} \, d\tau \right]^*$$

$$= \int x^*\left(t + \frac{\tau}{2}\right) x\left(t - \frac{\tau}{2}\right) e^{j2\pi f \tau} \, d\tau$$

$$= \int x^*\left(t - \frac{t'}{2}\right) x\left(t + \frac{t'}{2}\right) e^{-j2\pi f t'} \, dt' = W_x(t, f)$$

$$Q_X^*(t, f) = f \int X^*(f e^{u/2})X(f e^{-u/2})e^{-j2\pi t f u} \, du$$

$$= f \int X^*(f e^{-\beta/2})X(f e^{\beta/2})e^{j2\pi t f \beta} \, d\beta = Q_X(t, f) \qquad \blacksquare$$

P_8: **Positivity**

$$T_x(t, f) \geq 0, \qquad \forall x(t)$$

If a TFR is to be interpreted as a two-dimensional distribution of signal energy, then it should be nonnegative. By definition, since the spectrogram is equal to the squared magnitude of the STFT, it is always nonnegative. However, Tables 12.2 and 12.4 reveal that the Wigner distribution and the Q distribution, respectively, have negative values for some signals. For example, Table 12.2 states that the WD of a rectangular box function is a variable width sinc function, which is frequently negative. The last five entries in Table 12.2 correspond to the WD of multicomponent signals. Each has a WD with cross terms that oscillate about zero.

P_9: **Time marginal preservation**

$$\int T_x(t, f) \, df = |x(t)|^2, \qquad \forall x(t)$$

If TFR is to be interpreted as a signal's two-dimensional energy distribution over the time-frequency plane, then integrating out the frequency variable should result in the signal's instantaneous energy in the time domain. The Wigner distribution satisfies this property; the spectrogram does not.

PROOF

$$\int W_x(t, f) \, df = \int \int x(t + \tau/2) x^*(t - \tau/2) e^{-j2\pi f \tau} \, d\tau \, df$$

$$= \int x(t + \tau/2) x^*(t - \tau/2) \delta(\tau) \, d\tau = |x(t)|^2$$

$$\int |STFT_x(t, f; \Gamma)|^2 \, df = \int \left[\int x(\tau) \gamma^*(\tau - t) e^{-j2\pi f \tau} \, d\tau \right]$$

$$\times \left[\int x^*(t') \gamma(t' - t) e^{j2\pi f t'} \, dt' \right] df$$

$$= \int \int x(\tau) x^*(t') \gamma^*(\tau - t) \gamma(t' - t) \delta(\tau - t') \, d\tau \, dt'$$

$$= \int |x(t')|^2 |\gamma(t' - t)|^2 \, dt'. \qquad \blacksquare$$

Hence, the marginal distribution corresponding to the two-dimensional spectrogram is equal to a weighted average of the signal's instantaneous signal energy, $|x(t')|^2$, in the neighborhood of the output time t. The weighting function is a shifted version of the analysis window. The spectrogram only satisfies P_9 if the analysis window is a Dirac impulse in time.

P_{10}: **Frequency marginal preservation**

$$\int T_x(t, f) \, dt = |X(f)|^2, \qquad \forall x(t)$$

If the TFR is the signal's two-dimensional energy distribution, then integrating out the time axis should result in the signal's spectral density function, $|X(f)|^2$. The Wigner and Altes Q distributions satisfy this property; the spectrogram does not.

PROOF

$$\int WD_x(t, f)\, dt = \int \int X(f + v/2)X^*(f - v/2)e^{j2\pi vt}\, dv\, dt$$

$$= \int X(f + v/2)X^*(f - v/2)\delta(v)\, dv = |X(f)|^2$$

$$\int Q_X(t, f)\, dt = \int f \int X(fe^{u/2})X^*(fe^{-u/2})e^{j2\pi tfu}\, du\, dt$$

$$= f \int X(fe^{u/2})X^*(fe^{-u/2})\delta(fu)\, du = |X(f)|^2$$

$$\int |STFT_x(t, f; \Gamma)|^2\, dt = \int \left[\int e^{-j2\pi tf}X(f')\Gamma^*(f'-f)e^{j2\pi tf'}\, df' \right]$$

$$\times \left[\int e^{j2\pi tf}X^*(v)\Gamma(v-f)e^{-j2\pi tv}\, dv \right]\, dt$$

$$= \int \int X(f')X^*(v)\Gamma^*(f'-f)\Gamma(v-f)\delta(v-f')\, df'\, dv$$

$$= \int |X(f')|^2|\Gamma(f'-f)|^2\, df'. \qquad\blacksquare$$

Hence, the spectrogram does not satisfy P_{10} unless the Fourier transform of the STFT analysis window is a Dirac function.

P_{11}: **Energy preservation**

$$\int \int T_x(t, f)\, dt\, df = \int |X(f)|^2\, df = E_x$$

If the TFR is a distribution of the signal's energy over the whole time-frequency plane, then integrating the TFR should give you back the total signal energy, E_x. The Wigner and Altes Q distributions satisfy this property. The proof is simplified by making use of the fact that they were shown to preserve the frequency-domain marginals in P_{10} above.

PROOF

$$\int \int WD_x(t, f)\, dt\, df = \int |X(f)|^2\, df = E_x$$

$$\int \int Q_X(t, f)\, dt\, df = \int |X(f)|^2\, df = E_x. \qquad\blacksquare$$

P_{12}: **Time moment preservation**

$$\int \int t^n T_x(t, f)\, dt\, df = \int t^n |x(t)|^2\, dt$$

The nth moment of a signal $g(t)$ is defined to be

$$m_g(n) = \int t^n g(t) \, dt.$$

Property P_{12} states that the value of the nth time moment of the signal's instantaneous energy, $|x(t)|^2$, and the nth time moment of the signal's TFR should be identical. The WD satisfies this property.

PROOF

$$\int \int t^n W D_x(t, f) \, dt \, df = \int \int \int t^n x(t + \tau/2) x^*(t - \tau/2) e^{-j2\pi f \tau} \, d\tau \, dt \, df$$

$$= \int \int t^n x(t + \tau/2) x^*(t - \tau/2) \delta(\tau) \, d\tau \, dt$$

$$= \int t^n |x(t)|^2 \, dt. \tag{53}$$

∎

P_{13}: **Frequency moment preservation**

$$\int \int f^n T_x(t, f) \, dt \, df = \int f^n |X(f)|^2 \, df$$

The nth moment of the power spectral density, $|X(f)|^2$, and the nth frequency moment of the signal's TFR should be identical. The Wigner and the Altes Q distributions satisfy this property.

PROOF

$$\int \int f^n W D_x(t, f) \, dt \, df = \int \int \int f^n X(f + v/2) X^*(f - v/2) e^{j2\pi v t} \, dv \, dt \, df$$

$$= \int \int f^n X(f + v/2) X^*(f - v/2) \delta(v) \, dv \, df$$

$$= \int f^n |X(f)|^2 \, df \tag{54}$$

$$\int \int f^n Q_X(t, f) \, dt \, df = \int \int \int f^{n+1} X(f e^{u/2}) X^*(f e^{-u/2}) e^{j2\pi t f u} \, du \, dt \, df$$

$$= \int \int f^{n+1} X(f e^{u/2}) X^*(f e^{-u/2}) \delta(f u) \, du \, df$$

$$= \int f^n |X(f)|^2 \, df. \tag{55}$$

∎

Note that an alternative way to prove that the WD and the Altes Q distribution satisfy the energy preservation property P_{11} is to evaluate (54)–(55) for the special case of $n = 0$.

12.4.3 *Signal analysis properties*

The next category of properties in Table 12.6 is P_{14}–P_{18} which arise from signal processing considerations. A TFR should have the same nonzero support, i.e., duration and bandwidth, as the signal under analysis. At any given time, t, the average or mean frequency should equal the instantaneous frequency of the signal, while the average or center of gravity of the TFR in the time direction should equal the group delay of the signal. These two properties have been used to analyze the distortion of audio systems and the complex FM sonar signals used by bats and whales for echolocation [75], [7], [5], [24], [69], [53]. Property P_{18} is the TFR equivalent of the Duality property of Fourier transforms in Table 12.1.

P_{14}: **Finite time support**

$$x(t) = 0 \text{ for } t \notin (t_1, t_2) \quad \Rightarrow \quad T_x(t, f) = 0 \text{ for } t \notin (t_1, t_2)$$

Property P_{14} states that if a signal starts at time t_1 and stops at time t_2, then an ideal TFR should also start and stop at the same time. This is a very intuitive property for the TFR to have if it is to be interpreted as a two-dimensional energy distribution; there should be no nonzero values of the TFR at any time before the signal starts up nor after the signal has stopped. However, this property is sometimes referred to as "weak" time support [43], as it simply guarantees that the TFR will have the same global time support as the signal under analysis. It does *not* guarantee that the TFR will be equal to zero whenever the signal or its spectrum are equal to zero. The WD satisfies the finite support property, but the STFT does not for any analysis window $\gamma(t) \neq k\delta(t)$.

PROOF Assume $x(t) = 0$ for $t \notin (t_1, t_2)$.

A sufficient condition for the WD of $x(t)$ in (15) to be equal to zero is that the signal product $[x(t + \tau/2)x^*(t - \tau/2)]$ inside the integral be zero everywhere, or equivalently, that the nonzero support region of these two shifted signals do not overlap. Since $x(t + \tau/2) = 0$ for $t \notin (2(t_1 - t), 2(t_2 - t))$ and $x^*(t - \tau/2) = 0$ for $t \notin (2(t - t_2), 2(t - t_1))$, then the WD in (15) will be equal to zero whenever the nonzero support region of $x(t + \tau/2)$ lies entirely to the right or entirely to the left of that of $x^*(t - \tau/2)$, i.e., $2(t - t_1) < 2(t_1 - t)$ or $2(t - t_2) > 2(t_2 - t)$, respectively. Simplifying these last two inequalities, we see that the WD of a finite duration signal is identically zero for $t < t_1$ and $t > t_2$. ∎

P_{15}: **Finite frequency support**

$$|X(f)| = 0 \text{ for } f \notin (f_1, f_2) \quad \Rightarrow \quad T_x(t, f) = 0 \text{ for } f \notin (f_1, f_2), \quad f_1 < f_2$$

Property P_{15} is the dual to the finite time support property above. It states that if the Fourier transform of the signal is bandlimited, then its TFR should also have the same nonzero support in the frequency domain. It is easy to show using the "frequency-domain" formulations of the WD and the STFT in (16) and (8), respectively, that the WD satisfies this property, but the STFT does not unless the Fourier transform of the analysis window is $\Gamma(f) = c\delta(f)$. Again, P_{15} is a 'weak' frequency support property, as it only guarantees that the global bandwidth of the TFR matches that of the signal spectrum.

P_{16}: **Instantaneous frequency**

$$\frac{\int f T_x(t, f) \, df}{\int T_x(t, f) \, df} = f_x(t) = \frac{1}{2\pi} \frac{d}{dt} \arg\{x(t)\} \tag{56}$$

Property P_{16} states that the first normalized moment in frequency of the TFR should be equal to the instantaneous frequency of the signal. Hence, this property asserts that the TFR's average value or center of gravity in the frequency direction should correspond to the signal's instantaneous frequency in (3). The WD satisfies this property.

PROOF   First, write the signal in terms of its polar form, $x(t) = A(t)e^{j2\pi\phi(t)}$, where $A(t) > 0$ is the real amplitude function and $\phi(t)$ is the phase of the signal. The proof will first evaluate the numerator of the expression in (56) and then the denominator. Let the following notation be used to represent the partial time derivative of a function, followed by that derivative being evaluated at the time t_0:

$$\dot{g}(t_0) = \left. \frac{\partial}{\partial \tau} g(\tau) \right|_{\tau = t_0} \tag{57}$$

$$\int f W D_x(t, f)\, df = \int f \int A(t+\tau/2)e^{j2\pi\phi(t+\tau/2)} A(t-\tau/2)e^{-j2\pi\phi(t-\tau/2)} e^{-j2\pi f\tau} d\tau\, df \tag{58}$$

$$= \int A(t+\tau/2)A(t-\tau/2)e^{-j2\pi[\phi(t-\tau/2)-\phi(t+\tau/2)]} \left[\int f e^{-j2\pi f\tau}\, df \right] d\tau$$

$$= \int A(t+\tau/2)A(t-\tau/2)e^{-j2\pi[\phi(t-\tau/2)-\phi(t+\tau/2)]} \frac{1}{j2\pi} \frac{\partial}{\partial\tau} \delta(\tau)\, d\tau$$

$$= \frac{1}{j2\pi} \frac{\partial}{\partial\tau} \left[A(t+\tau/2)A(t-\tau/2)e^{-j2\pi[\phi(t-\tau/2)-\phi(t+\tau/2)]} \right]|_{\tau=0}$$

$$= \frac{1}{j2\pi}[\dot{A}(t)A(t)/2 - A(t)\dot{A}(t)/2 + A^2(t)j2\pi\dot{\phi}(t)]$$

$$= A^2(t)\dot{\phi}(t)$$

$$\int W D_x(t, f)\, df = |x(t)|^2 = A^2(t) \quad \text{from } P_9. \tag{59}$$

Dividing (58) by (59), one obtains

$$\frac{\int f\, W D_x(t, f)\, df}{\int W D_x(t, f)\, df} = \frac{A^2(t)\dot{\phi}(t)}{A^2(t)} = \dot{\phi}(t) = f_x(t). \tag{60}$$

∎

This property is very useful for the analysis of FM signals. For example, we see from the third entry in Table 12.2 that the WD of a linear FM chirp in the time domain is a Dirac function centered along the chirp's linear instantaneous frequency. Since a Dirac function is symmetrical, this also shows that the average value or center of gravity of the chirp's WD in the frequency direction is also located along the signal's linear instantaneous frequency. Table 12.2 shows that all even signals modulated by a linear FM chirp have a WD that is symmetric with respect to the instantaneous frequency of the chirp; hence, the center of gravity of the WD in the frequency direction is equal to the signal's instantaneous frequency.

In general, the spectrogram does not satisfy this property [39]. Its center of gravity,

$$\frac{\int f\, SPEC_x(t, f; \Gamma)\, df}{\int SPEC_x(t, f; \Gamma)\, df} = \frac{\int \dot{\phi}(\tau)|x(\tau)|^2|\gamma(\tau - t)|^2\, d\tau}{\int |x(\tau)|^2|\gamma(\tau - t)|^2\, d\tau},$$

is only equal to the instantaneous frequency if the analysis window is a Dirac function.

P_{17}: **Group delay**

$$\frac{\int t T_x(t, f)\, dt}{\int T_x(t, f)\, dt} = -\frac{1}{2\pi} \frac{d}{df} \arg\{X(f)\}$$

Property P_{17} is the dual to property P_{16}. It states that the TFR's normalized average value or center of gravity in the time direction should be equal to the group delay of the signal in (4). The WD satisfies this property. The proof is similar to that of P_{16} above except that the signal's Fourier transform is expressed in polar form and the frequency-domain formulation of the WD in (16) is used. The spectrogram does not satisfy this property [39].

P_{18}: **Fourier transform**

$$y(t) = X(t) \;\Rightarrow\; Y(f) = x(-f) \;\Rightarrow\; T_y(t, f) = T_x(-f, t)$$

Property P_{18} is the TFR equivalent of the Duality property of the Fourier transform in Table 12.1, which states what happens to the Fourier transform if the time and frequency-domain forms of a signal $x(t)$ are switched. If $y(t)$ is set equal to the Fourier transform of $x(t)$, i.e., $y(t) = X(t)$, then the Fourier transform of $y(t)$ is equal to $x(t)$, but with its argument replaced by $t = -f$, i.e., $Y(f) = x(-f)$. Property P_{18} states that the TFR of the dual signal $y(t) = X(t)$ should likewise have the role of time and frequency interchanged, but with the frequency variable negated. The WD satisfies this property.

PROOF   Let $y(t) = X(t)$.

$$WD_y(t, f) = \int y(t + \tau/2) y^*(t - \tau/2) e^{-j2\pi f \tau}\, d\tau$$

$$= \int X(t + \tau/2) X^*(t - \tau/2) e^{-j2\pi f \tau}\, d\tau$$

$$= \int X(t + v/2) X^*(t - v/2) e^{j2\pi(-f)v}\, dv$$

$$= WD_x(-f, t). \qquad \blacksquare$$

This useful property can be used to simplify the derivation of many TFRs. For example, it has been shown in many articles that the WD of a rectangular function is a variable width sinc function [37], [103]. See Table 12.2. Since the rectangular and sinc functions are Fourier transform pairs in Table 12.1, then property P_{18} shows that the WD of a sinc function is equal to the WD of a rectangle, but with the time and frequency variables interchanged. That is, if

$$x(t) = \text{rect}_a(t) = \begin{cases} 1, & |t| < a \\ 0, & |t| > a \end{cases} \;\Rightarrow\; WD_x(t, f) = \frac{\sin[4\pi(a - |t|)f]}{\pi f} \text{rect}_a(t),$$

then

$$y(t) = X(t) = \frac{\sin 2\pi a t}{\pi t} \;\Rightarrow\; WD_y(t, f) = WD_x(-f, t) = \frac{\sin[4\pi(a - |-f|)t]}{\pi t} \text{rect}_a(-f)$$

$$= \frac{\sin[4\pi(a - |f|)t]}{\pi t} \text{rect}_a(f).$$

The negation of the frequency axis has no effect in this example since the magnitude and rectangular functions are even functions.

12.4.4 *Signal localization*

The group of properties P_{19}–P_{24} in Table 12.6 are ideal TFR localization properties that are desirable for high-resolution capabilities. These properties state that if a signal is perfectly concentrated in time or frequency (i.e., the signal is an impulse or a sinusoid) then its TFR should also be perfectly concentrated at the same time or frequency, respectively. Properties P_{21}–P_{22} state that the TFRs of a linear or hyperbolic spectral FM chirp signal should be perfectly concentrated along that signal's group delay. Property P_{24} states that a signal modulated by a linear FM chirp should have a TFR whose frequency axis has been sheared by an amount equal to the linear instantaneous frequency of the chirp. P_{23} is the dual to property P_{24}; it states that multiplication by a linear FM chirp in the frequency domain should shear the time axis of the TFR by an amount equal to the group delay of the chirp.

P_{19}: **Frequency localization**

$$X(f) = \delta(f - f_0) \quad \Rightarrow \quad T_x(t, f) = \delta(f - f_0)$$

This property states that if the signal is a complex sinusoid whose Fourier transform is perfectly concentrated about a certain frequency, f_0, then its TFR should also be perfectly concentrated about that same frequency. The WD and the Altes Q distribution satisfy this property. As equation (11) shows, the STFT does not, unless the Fourier transform of the analysis window is proportional to a Dirac function.

PROOF   Let $X(f) = \delta(f - f_0)$.

$$WD_x(t, f) = \int \delta(f + \nu/2 - f_0)\delta(f - \nu/2 - f_0)e^{j2\pi \nu t}\, d\nu$$

$$= 2\delta\left(f + \frac{2(f - f_0)}{2} - f_0\right)e^{j4\pi(f-f_0)t} = \delta(f - f_0)$$

$$Q_X(t, f) = f \int X(fe^{u/2})X^*(fe^{-u/2})e^{j2\pi tfu}\, du,$$

$$= f \int \delta(fe^{u/2} - f_0)\delta(fe^{-u/2} - f_0)e^{j2\pi tfu}\, du$$

$$= f \int \left[\frac{\delta(u - 2\ln(f_0/f))}{\left|\frac{1}{2}fe^{u/2}\right|}\right]\left[\frac{\delta(u + 2\ln(f_0/f))}{\left|-\frac{1}{2}fe^{-u/2}\right|}\right]e^{j2\pi tfu}\, du$$

$$= \frac{4}{|f|}\delta(4\ln(f_0/f))e^{j4\pi tf \ln(f_0/f)}$$

$$= \frac{4}{|f|}\frac{\delta(f - f_0)}{\left|4\frac{f}{f_0}[\frac{-f_0}{f^2}]\right|}e^{j4\pi tf \ln(f_0/f)} = \delta(f - f_0). \qquad \blacksquare$$

P_{20}: **Time localization**

$$x(t) = \delta(t - t_0) \quad \Rightarrow \quad T_x(t, f) = \delta(t - t_0)$$

This property states that if the signal is an impulse perfectly localized at time $t = t_0$, then its TFR should also be concentrated at time $t = t_0$. The WD satisfies this property. The proof is the dual to that used in property P_{19} above with the time-domain WD formulation in (15) being used in place of (16). The STFT does not satisfy this property in general; as (10) indicates, the STFT experiences spreading about $t = t_0$ equal to the duration of the analysis window.

P_{21}: **Linear chirp localization**

$$X(f) = e^{-j\pi cf^2} \quad \Rightarrow \quad T_x(t, f) = \delta(t - cf)$$

This property states that if a signal's Fourier transform is equal to a linear FM chirp, then its TFR should be perfectly concentrated along the chirp's linear group delay, $\tau_x(f) = cf$. The WD satisfes this property.

PROOF   Let $X(f) = e^{-j\pi cf^2}$.

$$WD_x(t, f) = \int e^{-j\pi c(f+v/2)^2} e^{j\pi c(f-v/2)^2} e^{j2\pi vt} \, dv$$

$$= \int e^{-j\pi c[f^2 + fv + v^2/4 - f^2 + fv - v^2/4]} e^{j2\pi vt} \, dv$$

$$= \int e^{j2\pi v(t-cf)} \, dv = \delta(t - cf). \tag{61}$$

∎

When property P_{18} is coupled with P_{21}, it can be shown that the WD also satisfies a dual property to Linear Chirp Localization. If the signal is a linear FM time-domain chirp, then its TFR should be concentrated along its instantaneous frequency. Thus, if $x(t) = e^{-j\pi ct^2}$, then $WD_x(t, f) = \delta(f + ct)$.

PROOF   Let $X(f) = e^{-j\pi cf^2}$ with WD, $WD_x(t, f) = \delta(t - cf)$ derived in (61). Coupling properties P_{18} and P_{21}, one can show that if $y(t) = X(t) = e^{-j\pi ct^2}$, then $WD_y(t, f) = WD_x(-f, t) = \delta(f + ct)$.   ∎

P_{22}: **Hyperbolic chirp localization**

$$X_c(f) = \frac{1}{\sqrt{f}} e^{-j2\pi c \ln \frac{f}{f_r}}, \quad f > 0 \quad \Rightarrow \quad T_{x_c}(t, f) = \frac{1}{f}\delta(t - \frac{c}{f}), \quad f > 0$$

This property is useful for analyzing FM signals whose group delay is hyperbolic. That is, signals with logarithmic phase spectra should have TFRs that are perfectly concentrated along its hyperbolic group delay, $\tau_x(f) = \frac{c}{f}\tilde{u}(f)$. Chirps with logarithmic phase are Doppler invariant; they have been used to model the biosonar signals used by bats [7], [53]. The Altes Q distribution and the Unitary Bertrand P_0 distribution are two of the few distributions that satisfy this property.

PROOF   Let $X(f) = \frac{1}{\sqrt{f}} e^{-j2\pi c \ln(f/f_r)} \tilde{u}(f)$.

$$Q_X(t, f) = f \int \frac{1}{\sqrt{fe^{u/2}}} e^{-j2\pi c \ln[fe^{u/2}/f_r]} \frac{1}{\sqrt{fe^{-u/2}}} e^{j2\pi c \ln[fe^{-u/2}/f_r]} e^{j2\pi tfu} \, du, \qquad f > 0$$

$$= \int e^{-j2\pi c[\ln(f/f_r) + u/2 - (\ln f/f_r) + u/2]} e^{j2\pi tfu} \, du$$

$$= \int e^{j2\pi(t - c/f)fu} \, du = \delta(t - c/f). \qquad ∎$$

P_{23}: **Chirp convolution**

$$y(t) = \int x(t-\tau)\sqrt{|c|}e^{j\pi c\tau^2}d\tau \quad \Rightarrow \quad T_y(t, f) = T_x\left(t - \frac{f}{c}, f\right)$$

To understand this property, recall from Table 12.1 that a Gaussian signal and its Fourier transform have inversely proportional variances. Thus, time-domain convolution of a signal $x(t)$ with a linear FM chirp with sweep rate c is equivalent to multiplying the Fourier transform of the signal with a linear FM chirp of sweep rate $-1/c$:

$$y(t) = \int x(t - \tau)\sqrt{|c|}e^{j\pi c\tau^2}d\tau \quad \leftrightarrow \quad Y(f) = X(f)\sqrt{j}e^{-j\pi f^2/c} \tag{62}$$

$$|Y(f)| = |X(f)| \tag{63}$$

$$\tau_y(f) = \tau_x(f) + f/c. \tag{64}$$

This multiplication in (62) of the signal spectrum with a linear FM chirp leaves the magnitude of $X(f)$ unchanged in (63), but changes the signal's group delay in (64) by the group delay of the chirp. Because changes in the group delay correspond to a temporal translation of each spectral component, property P_{23} states that the TFR of the convolution output $y(t)$ in (62) should therefore be equal to the TFR of $x(t)$, but with the time axis adjusted for the frequency-dependent change in the group delay brought about by the convolution. Note that this results in a shearing of the TFR. Such shearing has been exploited in RADAR analysis [109], [114]. By definition in (39), the Bertrand $k = 2$ distribution in (40) satisfies this property. The WD is one of the few other TFRs that satisfies this property.

PROOF   Let $Y(f) = X(f)\sqrt{j}e^{-j\pi f^2/c}$.

$$WD_y(t, f) = \int X(f + v/2)\sqrt{j}e^{-j\pi(f+v/2)^2/c}X^*(f - v/2)\sqrt{-j}e^{j\pi(f-v/2)^2/c}e^{j2\pi vt}\,dv$$

$$= \int X(f + v/2)X^*(f - v/2)e^{-j\pi[f^2+fv+v^2/4-f^2+vf-v^2/4]/c}e^{j2\pi vt}\,dv$$

$$= \int X(f + v/2)X^*(f - v/2)e^{j2\pi v(t-f/c)}\,dv = WD_x(t - f/c, f). \quad \blacksquare$$

P_{24}: **Chirp multiplication**

$$y(t) = x(t)e^{j\pi ct^2} \quad \Rightarrow \quad T_y(t, f) = T_x(t, f - ct)$$

This property is the dual to property P_{23} with multiplication by a linear FM chirp occurring in the time domain rather than in the frequency domain as before. If a signal is multiplied in the time domain by a linear FM chirp with sweep rate c, then its magnitude is unchanged, but its instantaneous frequency is changed by the linear instantaneous frequency of the linear FM chirp, i.e., $f_y(t) = f_x(t) - ct$. Property P_{24} states that the TFR of $y(t)$ should correspond to the TFR of $x(t)$ but with the frequency axis corrected to account for the time-dependent change in the signal's instantaneous frequency. The WD is one of the few TFRs to satisfy this property. The proof is similar to that of P_{23} except that the time-domain formulation of the WD in (15) is used. An alternative proof is to use the fact that the WD satisfies the convolution covariance property P_5 coupled with the fact that the WD of a linear FM chirp in Table 12.2 is a Dirac function centered around its instantaneous frequency:

$$y(t) = x(t)e^{j\pi ct^2} \quad \Rightarrow \quad WD_y(t, f) = \int WD_x(t, f')\delta((f - f') - ct)\, df'$$

$$= WD_x(t, f - ct).$$

Example

Table 12.2 states that the rectangular signal,

$$x(t) = \text{rect}_a(t) \Rightarrow WD_x(t, f) = \frac{\sin[4\pi(a - |t|)f]}{\pi f}\text{rect}_a(t),$$

has a WD that is equal to a sinc function whose mainlobe width varies with time. Near $t = 0$, the sinc function is a relatively narrow function of frequency centered near $f = 0$. For time values near the edges of the rectangle, the spectral width of the sinc's main lobe is very broad. This is an intuitive result as the middle of the rectangle is very smooth, so its WD should be narrowband and lowpass; however, near the edges of the rectangle, where a sharp discontinuity exists, the WD is broadband. Property P_{24} states that the WD of a chirp modulated rectangle,

$$y(t) = \text{rect}_a(t)e^{j\pi ct^2} \Rightarrow WD_y(t, f)$$

$$= WD_x(t, f - ct) = \frac{\sin[4\pi(a - |t|)(f - ct)]}{\pi(f - ct)}\text{rect}_a(t),$$

is a variable width sinc function as before, but centered along the chirp's linear instantaneous frequency, $f = ct$, in the time-frequency plane.   ∎

PROOF   Let $y(t) = \text{rect}_a(t)e^{j\pi ct^2}$.

$$WD_y(t, f) = \int e^{j\pi c(t+\tau/2)^2}\text{rect}_a(t + \tau/2)e^{-j\pi c(t-\tau/2)^2}\text{rect}_a(t - \tau/2)e^{-j2\pi f\tau}\, d\tau$$

$$= \int \text{rect}_a(t + \tau/2)\text{rect}_a(t - \tau/2)e^{-j2\pi(f-ct)\tau}\, d\tau$$

$$= 2\int \text{rect}_a(t')\text{rect}_a(2t - t')e^{-j4\pi(f-ct)(t'-t)}\, dt'$$

$$= \begin{cases} 0, & 2t + a < -a \\ 2\int_{-a}^{2t+a} e^{-j4\pi(f-ct)(t'-t)}\, dt' = \dfrac{\sin 4\pi(f - ct)(a + t)}{\pi(f - ct)}, & -a < 2t + a < a \\ 2\int_{2t-a}^{a} e^{-j4\pi(f-ct)(t'-t)}\, dt' = \dfrac{\sin 4\pi(f - ct)(a - t)}{\pi(f - ct)}, & -a < 2t - a < a \\ 0, & 2t - a > a \end{cases}$$

$$= \frac{\sin 4\pi(f - ct)(a - |t|)}{\pi(f - ct)}\text{rect}_a(t).$$ ∎

12.4.5 *Preserving inner products*

The last property in Table 12.6, known as Moyal's formula [39], [60] or the unitarity property, states that TFRs should preserve the signal projections, inner products, and norm metrics which are used frequently in signal detection, synthesis, approximation theory, and pattern recognition [56], [60], [96], [28], [69], [72], [77], [108], [69], [24]. It also states that if two basis functions are orthogonal, i.e., their inner product is equal to zero, then their respective TFRs should also be orthogonal. Hence, TFRs that satisfy Moyal's formula can be used to induce a set of two-dimensional, orthogonal basis functions, $T_{g_i}(t, f)$ from a set of one-dimensional orthogonal basis functions, $g_i(t), i = 1, \ldots, N$.

P_{25}: **Moyal's formula**

$$\left| \int x(t) y^*(t) \, dt \right|^2 = \int \int T_x(t, f) T_y^*(t, f) \, dt \, df$$

This property states that an ideal TFR should preserve inner products. It is analogous to Parseval's theorem for the Fourier transform [101], [105],

$$\int x(t) y^*(t) \, dt = \int X(f) Y^*(f) \, df,$$

which states that the inner product of two signals in the time domain should be equal to the inner product of their respective Fourier transforms in the frequency domain. The Wigner, Altes Q, and the Unitary Bertrand P_0 distributions satisfy this property.

PROOF

$$\int \int WD_x(t, f) WD_y^*(t, f) \, dt \, df$$

$$= \int \int \left[\int x(t+u/2) x^*(t-u/2) e^{-j2\pi f u} \, du \right]$$

$$\times \left[\int y^*(t+t'/2) y(t-t'/2) e^{j2\pi f t'} \, dt' \right] dt \, df$$

$$= \int \int \int x(t+u/2) y^*(t+t'/2) x^*(t-u/2) y(t-t'/2) \delta(t'-u) \, du \, dt \, dt'$$

$$= \int \int [x(t + u/2) y^*(t + u/2)][x(t - u/2) y^*(t - u/2)]^* \, du \, dt$$

$$= \left| \int x(t) y^*(t) \, dt \right|^2$$

$$\int\!\!\int Q_X(t, f) Q_Y^*(t, f) \, dt \, df$$

$$= \int\!\!\int \left[f \int X(f e^{u/2}) X^*(f e^{-u/2}) e^{j2\pi t f u} du \right]$$

$$\times \left[f \int Y^*(f e^{\beta/2}) Y(f e^{-\beta/2}) e^{-j2\pi t f \beta} d\beta \right] dt \, df$$

$$= \iiint f^2 X(f e^{u/2}) Y^*(f e^{\beta/2}) X^*(f e^{-u/2}) Y(f e^{-\beta/2}) \delta(f(u-\beta)) \, du \, d\beta \, df$$

$$= \iint f X(f e^{u/2}) Y^*(f e^{u/2}) X^*(f e^{-u/2}) Y(f e^{-u/2}) \, du \, df$$

$$= \left| \int X(a) Y^*(a) da \right|^2, \qquad f > 0 \qquad \blacksquare$$

12.5   Classes of TFRs with common properties

The list of ideal TFR properties given in the previous section is far from exhaustive; others have been proposed in [43], [60], [12], [122], [84], [1], [4]. Nonetheless, the checklist summary in Table 12.7 reveals that no known TFR satisfies all such properties. To understand the relative advantages of various TFRs, and to understand their inter-relationships, this section will group TFRs into classes. Each class is defined by the two or three ideal TFR properties that all member TFRs must satisfy. A review will be given of Cohen's class of shift covariant TFRs, the Affine class of affine covariant TFRs, the hyperbolic class, developed for signals with hyperbolic group delay, and the power class, which is useful for signals with power group delay [100], [97], [43], [60], [69], [98], [73].

The grouping of TFRs into classes sharing common properties has the following advantages. It provides very helpful insight as to which types of TFRs will work best in different situations. For example, many members of Cohen's class TFRs are well suited to constant bandwidth analysis whereas several affine class TFRs are best suited for multiresolution analysis. Within a given class, each TFR is completely characterized by a unique set of TFR-dependent kernels that can be compared against the list of class-dependent kernel constraints in Table 12.8 to quickly determine which ideal properties a given TFR satisfies. The shape of these kernels can be analyzed to ascertain which is best for auto term preservation or cross term removal in a particular application.

12.5.1   *Cohen's class of TFRs*

$$Cohen's \ class = \left\{ T_x(t, f) \mid y(t) = x(t - t_0) e^{j2\pi f_0 t} \Rightarrow T_y(t, f) = T_x(t - t_0, f - f_0) \right\}.$$

Cohen's class consists of all quadratic TFRs that satisfy the frequency-shift and time-shift covariance properties. Several TFRs in Cohen's class are listed in Table 12.9. Since each Cohen's-class TFR automatically satisfies properties P_1–P_2 in Table 12.6, then it must have a check mark in the first two property rows in Table 12.7 [40], [41], [43], [39], [69], [60]. Time and frequency-shift covariance are very useful properties in the analysis of speech, narrowband Doppler systems, and multipath environments.

Alternative formulations

Any TFR in Cohen's class can be written in one of the four equivalent "Normal Forms" [39], [60], [69],

$$C_x(t, f; \Psi_C) = \iint \varphi_C(t - t', \tau) \, x \left(t' + \frac{\tau}{2} \right) x^* \left(t' - \frac{\tau}{2} \right) e^{-j2\pi f \tau} \, dt' \, d\tau \qquad (65)$$

TABLE 12.7

Desirable properties satisfied by TFRs defined in Tables 12.9, 12.11, 12.13, and 12.15. In the second row, the letters c, a, h, and p indicate that the corresponding TFR is a member of the Cohen, affine, hyperbolic and κth power class, respectively. A $\sqrt{}$ indicates that the TFR can be shown to hold the given property. A number following the $\sqrt{}$ indicates that additional constraints are needed to satisfy the property. The constraints are as follows: (1) M=N; (2) $M > 1/2$; (3) $N > 1/2$; (4) $g(\tau)$ even; (5) $|\alpha| < 1/2$; (6) $M = 1/2$; (7) $N = 1/2$; (8) $\int_0^\infty |\Gamma(f)|^2\, df = 1$; (9) $\alpha = 1$; (10) $r = 0$, $\alpha = 1$, $\gamma = 1/4$; (11) $\alpha \neq 1$; (12) $|\rho_T(0)|^2 = \frac{1}{f_r}$; (13) $|\gamma(0)| = 1$; (14) $\gamma(0) = 1$; (15) $s(\beta)$ even; (16) $\int |\Gamma(b)|^2 \frac{db}{|b|} = 1$; (17) $s(c) \in$ Real; (18) $s(t) \in$ Real; (19) $S(0)|\gamma(0)|^2 =1$; (20) $\int |\gamma(t)|^2\, dt = 1$.

	TFR / ACK (c a)	BJD (a h)	BUD (c)	BCW (c a)	CKD (c)	CAS (c)	CAD (c)	FGQ (h)	GED (c)	GWD (c a)	HYP (h)	LMD (c a)	MTD (c)	NDD (c)	WDK (p)	PQD (h)	PWD (c)	RID (c a)	SCAL (a)	SPWD (c)	SPWD (h)	SPEC (c)	UAD (a)	WD (c a)
Property																								
1 Frequency Shift	√	√	√	√	√	√	√		√	√		√	√	√	√		√	√		√		√		√
2 Time Shift	√	√	√	√	√	√	√		√	√		√	√	√	√		√	√	√	√		√	√	√
3 Scale Covariance	√	√1	√1	√1			√	√	√1	√9	√	√	√9	√	√	√		√	√		√	√	√	√
4 Hyperbolic Time Shift	√	√						√			√					√								
5 Convolution									√6	√		√10	√10	√			√	√					√	√
6 Modulation									√7	√	√	√10	√10	√			√	√					√	√
7 Real-Valued	√	√	√	√	√4	√	√		√	√		√		√	√		√	√	√17	√18		√		√
8 Positivity					√	√		√			√							√	√					
9 Time Marginal	√	√	√	√	√		√		√	√		√	√11	√			√13	√						√
10 Frequency Marginal	√	√	√	√	√	√	√		√	√		√	√11	√				√						√
11 Energy Distribution	√	√	√	√	√	√	√		√	√	√8	√	√	√		√12	√13	√	√16	√12		√19	√20	√
12 Time Moments	√	√	√	√	√		√		√	√		√	√11	√			√13	√						√
13 Frequency Moments	√	√	√	√	√	√	√		√	√		√	√11	√				√						√
14 Finite Time Support		√		√	√					√5		√						√						√
15 Finite Frequency Support	√	√	√2	√		√5	√5		√5									√						√
16 Instantaneous Frequency	√	√	√2	√		√2	√2		√3			√		√			√	√		√	√		√	√
17 Group Delay	√	√	√3	√	√		√3		√3			√	√11	√	√		√14	√					√	√
18 Fourier Transform	√	√	√1	√		√	√1		√1			√	√9	√				√					√	√
19 Frequency Localization	√	√	√	√	√	√	√		√	√		√11	√11	√			√13	√		√11	√11		√	√
20 Time Localization	√	√	√	√	√	√	√	√	√	√		√11	√11	√	√15		√13	√		√11	√11		√	√
21 Linear Chirp Localization																								
22 Hyperbolic Localization	√	√																						
23 Chirp Convolution																							√	
24 Chirp Multiplication																							√	√
25 Moyal's Formula	√	√	√	√	√			√						√										√

TABLE 12.8

Kernel constraints needed to satisfy ideal TFR properties for the TFRs in Cohen's, affine, and κth power classes. Here, $\lambda_\kappa(u)$ is defined in (39) and $\mu(u) = \mu^*(-u)$.

Property Name	Kernel Constraints for Cohen's Class	Kernel Constraints for the Affine Class	Kernel Constraints for the Hyperbolic Class	Kernel Constraints for the κth Power Class
P_1: Frequency Shift Covariant	Always Satisfied	$\Psi_A^{(A)}(\zeta, \beta) = S_A(\zeta\beta)e^{-j2\pi\zeta}$		
P_2: Time Shift Covariant	Always Satisfied	Always Satisfied	$\Psi_H^{(H)}(\zeta, \beta) = S_{AH}(\beta)e^{-j2\pi\zeta\ln\mu_0(\beta)}$ with $\mu_0(\beta) = \frac{\beta/2}{\sinh(\beta/2)}$	$\Gamma_{PC}^{(\kappa)}(b_1, b_2) = \int \delta(b_1 - \lambda_\kappa(u)) \times \delta(b_2 - \lambda_\kappa(-u))\mu(u)du$
P_3: Scale Covariant	$\Psi_C(\tau, \nu) = S_C(\tau\nu)$	Always Satisfied	Always Satisfied	Always Satisfied
P_4: Hyperbolic Time Shift		$\Phi_A^{(A)}(b, \beta) = G_A^{(A)}(\beta)\delta\left(b + \frac{\beta}{2}\coth\frac{\beta}{2}\right)$		
P_5: Convolution Covariant	$\Psi_C(\tau_1 + \tau_2, \nu) = \Psi_C(\tau_1, \nu)\Psi_C(\tau_2, \nu)$	$\Psi_A^{(A)}(\zeta_1 + \zeta_2, \beta) = \Psi_A^{(A)}(\zeta_1, \beta)\Psi_A^{(A)}(\zeta_2, \beta)$	$\Phi_H^{(H)}(b_1, \beta)\Phi_H^{(H)}(b_2, \beta) = e^{b_1}\Phi_H(b_1, \beta)\delta(b_1 - b_2)$	
P_6: Modulation Covariant	$\Psi_C(\tau, \nu_1 + \nu_2) = \Psi_C(\tau, \nu_1)\Psi_C(\tau, \nu_2)$			
P_7: Real-Valued	$\Psi_C^*(-\tau, -\nu) = \Psi_C(\tau, \nu)$	$\Phi_A^{(A)}(b, \beta) = \Phi_A^{(A)*}(b, -\beta)$	$\Psi_H^{(H)*}(-\zeta, -\beta) = \Psi_H^{(H)}(\zeta, \beta)$	$\Phi_{PC}^{(A)}(b, \beta) = \Phi_{PC}^{(A)*}(b, -\beta)$
P_8: Positivity	$\Psi_C(\tau, \nu) = AF_\gamma(-\tau, -\nu)$		$\Psi_H^{(H)*}(\zeta, \beta) = HAF_\Gamma(-\zeta, -\beta)$	
P_9: Time Marginal	$\Psi_C(0, \nu) = 1$	$\int \Phi_A^{(A)}(b, 2b)\frac{db}{|b|} = 1$	$\Psi_H^{(H)}(c, \beta) = 0, \left|\frac{\zeta}{\zeta}\right| > \frac{1}{2}$	
P_{10}: Frequency Marginal	$\Psi_C(\tau, 0) = 1$	$\Phi_A^{(A)}(b, 0) = \delta(b + 1)$	$\Psi_H^{(H)}(\zeta, 0) = 1$	$\Phi_{PC}^{(A)}(b, 0) = \delta(b + 1)$
P_{11}: Energy Distribution	$\Psi_C(0, 0) = 1$	$\int \Phi_A^{(A)}(b, 0)\frac{db}{|b|} = 1$	$\Psi_H^{(H)}(0, 0) = 1$	$\int \Phi_{PC}^{(A)}(b, 0)\frac{db}{|b|} = 1$
P_{12}: Time Moments	$\Psi_C(0, \nu) = 1$			
P_{13}: Frequency Moments	$\Psi_C(\tau, 0) = 1$			
P_{14}: Finite Time Support	$\varphi_C(t, \tau) = 0, \left|\frac{t}{\tau}\right| > \frac{1}{2}$	$\varphi_A^{(A)}(b, \zeta) = 0, \left|\frac{\zeta}{\zeta}\right| > \frac{1}{2}$	$\Phi_H^{(H)}(c, \zeta) = 0, \left|\frac{\zeta}{\zeta}\right| > \frac{1}{2}$	$\phi_{PC}^{(A)}(b, \beta) = 0, \left|\frac{b+1}{\beta}\right| > \frac{1}{2}$
P_{15}: Finite Freq. Support	$\Phi_C(f, \nu) = 0, \left|\frac{f}{\nu}\right| > \frac{1}{2}$	$\Phi_A^{(A)}(b, \beta) = 0, \left|\frac{b+1}{\beta}\right| > \frac{1}{2}$		
P_{16}: Instantaneous Frequency	$\left.\frac{\partial}{\partial\tau}\Psi_C(\tau, \nu)\right|_{\tau=0} = 0$			
P_{17}: Group Delay	$\Psi_C(\tau, 0) = 1$ and $\left.\frac{\partial}{\partial\nu}\Psi_C(\tau, \nu)\right|_{\nu=0} = 0$	$\Phi_A^{(A)}(b, 0) = \delta(b + 1)$ and $\left.\frac{\partial}{\partial\beta}\Phi_A^{(A)}(b, \beta)\right|_{\beta=0} = 0$	$\Psi_H^{(H)}(\zeta, 0) = 1$ and $\left.\frac{\partial}{\partial\beta}\Psi_H^{(H)}(\zeta, \beta)\right|_{\beta=0} = 0$	$\Phi_{PC}^{(A)}(b, 0) = \delta(b + 1)$ and $\left.\frac{\partial}{\partial\beta}\Phi_{PC}^{(A)}(b, \beta)\right|_{\beta=0} = 0$
P_{18}: Fourier Transform	$\Psi_C(-\nu, \tau) = \Psi_C(\tau, \nu)$			
P_{19}: Freq. Localization	$\Psi_C(\tau, 0) = 1$	$\Phi_A^{(A)}(b, 0) = \delta(b + 1)$	$\Psi_H^{(H)}(\zeta, 0) = 1$	$\Phi_{PC}^{(A)}(b, 0) = \delta(b + 1)$
P_{20}: Time Localization	$\Psi_C(0, \nu) = 1$			
P_{21}: Linear Chirp Localization	$\Psi_C(\tau, \nu) = 1$			
P_{22}: Hyperbolic Localization	$\Psi_C(\tau - \frac{\nu}{c}, \nu) = \Psi_C(\tau, \nu)$		$\Psi_H^{(H)}(0, \beta) = 1$	
P_{23}: Chirp Convolution	$\Psi_C(\tau, \nu - c\tau) = \Psi_C(\tau, \nu)$			
P_{24}: Chirp Multiplication	$\Psi_C(\tau, \nu - c\tau) = \Psi_C(\tau, \nu)$			
P_{25}: Moyal's Formula	$|\Psi_C(\tau, \nu)| = 1$	$\int \Phi_A^{(A)*}(b\beta, \tilde{\eta}\beta)\Phi_A^{(A)}(\beta, \tilde{\eta}\beta)d\beta = \delta(b - 1), \forall\tilde{\eta}$	$|\Psi_H^{(H)}(\zeta, \beta)| = 1$	$\int \Phi_{PC}^{(A)*}(b\beta, \alpha\beta)\Phi_{PC}^{(A)}(\beta, \alpha\beta)d\beta = \delta(b - 1), \forall\alpha$

$$= \int \int \Phi_C(f - f', \nu)\, X\left(f' + \frac{\nu}{2}\right)\, X^*\left(f' - \frac{\nu}{2}\right)\, e^{j2\pi t\nu}\, df'\, d\nu \quad (66)$$

$$= \int \int \psi_C(t - t', f - f')\, WD_x(t', f')\, dt'\, df' \quad (67)$$

$$= \int \int \Psi_C(\tau, \nu)\, AF_x(\tau, \nu)\, e^{j2\pi(t\nu - f\tau)}\, d\tau\, d\nu, \quad (68)$$

or in the "bi-frequency" form

$$C_x(t, f; \Psi_C) = \int \int \Gamma_C(f - f_1, f - f_2) X(f_1) X^*(f_2) e^{j2\pi t(f_1 - f_2)}\, df_1\, df_2. \quad (69)$$

Each normal form is characterized by one of the four kernels $\varphi_C(t, \tau)$, $\Phi_C(f, \nu)$, $\psi_C(t, f)$, and $\Psi_C(\tau, \nu)$, which are interrelated by the following Fourier transforms,

$$\varphi_C(t, \tau) = \int \int \Phi_C(f, \nu) e^{j2\pi(f\tau + \nu t)}\, df\, d\nu = \int \Psi_C(\tau, \nu) e^{j2\pi \nu t}\, d\nu \quad (70)$$

$$\leftrightarrow \Phi_C(f, \nu)$$

$$\psi_C(t, f) = \int \int \Psi_C(\tau, \nu) e^{j2\pi(\nu t - f\tau)}\, d\tau\, d\nu = \int \Phi_C(f, \nu) e^{j2\pi \nu t}\, d\nu \quad (71)$$

$$\leftrightarrow \Psi_C(\tau, \nu)$$

$$\Gamma_C(f_1, f_2) = \Phi_C\left(\frac{f_1 + f_2}{2}, f_2 - f_1\right). \quad (72)$$

The kernels for the TFRs in Cohen's class are given in Table 12.10. They can be combined with equations (65)–(69) to provide alternative formulations for any of the Cohen's-class TFRs given in Table 12.9. These kernels can also be compared against the constraints in the second column of Table 12.8 to determine which properties a given Cohen's-class TFR satisfies. A checklist summary of the properties that each Cohen's-class TFRs satisfies is given in Table 12.7.

Implementation considerations

The formulations in (65)–(69) provide alternative ways of understanding and computing a given Cohen's-class TFR. For example, inserting the Choi-Williams or exponential distribution kernels in Table 12.10 into equations (65)–(69) yields the following five equivalent formulations for this popular TFR [35].

$$CWD_x(t, f; \sigma) = \sqrt{\frac{\sigma}{4\pi}} \int \int \frac{1}{|\tau|} \exp\left(\frac{-\sigma}{4}\left[\frac{t - t'}{\tau}\right]^2\right) x\left(t' + \frac{\tau}{2}\right) x^*\left(t' - \frac{\tau}{2}\right) e^{-j2\pi f\tau}\, dt'\, d\tau \quad (73)$$

$$= \sqrt{\frac{\sigma}{4\pi}} \int \int \frac{1}{|\nu|} \exp\left(\frac{-\sigma}{4}\left[\frac{f - f'}{\nu}\right]^2\right) X\left(f' + \frac{\nu}{2}\right) X^*\left(f' - \frac{\nu}{2}\right) e^{j2\pi t\nu}\, df'\, d\nu \quad (74)$$

$$= \sqrt{\frac{\sigma}{4\pi}} \int \int \int \frac{1}{|u|} \exp\left(\frac{-\sigma}{4}\left[\frac{t - t'}{u}\right]^2\right) e^{-j2\pi(f - f')u} WD_x(t', f')\, du\, dt'\, df' \quad (75)$$

$$= \int \int e^{-(2\pi \tau \nu)^2/\sigma}\, AF_x(\tau, \nu) e^{j2\pi(t\nu - f\tau)}\, d\tau\, d\nu \quad (76)$$

$$= \sqrt{\frac{\sigma}{4\pi}} \iint \frac{1}{|f_1 - f_2|} \exp\left(\frac{-\sigma}{4}\left[\frac{f - (f_1 + f_2)/2}{f_1 - f_2}\right]^2\right) \tag{77}$$

$$\times X(f_1) X^*(f_2) e^{j2\pi t(f_1 - f_2)} \, df_1 \, df_2.$$

The Choi-Williams exponential distribution (CWD) is often used as a compromise between the high-resolution but cluttered WD versus the smeared, but easy to interpret Spectrogram [35], [79], [24]. The most intuitive formulation for the CWD is (76), which states that the CWD is the two-dimensional Fourier transform of the product of the AF of the signal with a lowpass kernel. This kernel, $\Psi_{CWD}(\tau, \nu) = \exp[-(2\pi\tau\nu)^2/\sigma]$, is a Gaussian function when evaluated at the product of the kernel arguments $\tau\nu$. The disadvantage of equation (76) is that the signal must be known for all time before the AF can be computed. Fast algorithms for calculating the CWD are frequently implemented using equation (73), which can be computed directly from the signal $x(t)$. If σ is large, then the Gaussian kernel in (73) falls off quickly to zero, which means that the integration requires knowledge of only short, local segments of the signal, making real-time implementations a possibility. If the spectrum of the signal is very narrowband, then the frequency-domain formulations in (74) or (77) may prove more computationally efficient. The bi-frequency formulation in (77) states explicitly how signal components from different spectral bands interact and how they are weighted to produce the quadratic CWD.

Hence, the four normal forms offer various computational and analysis advantages. The first two normal forms in (65)–(66) can be computed directly from the signal, $x(t)$, or its Fourier transform, $X(f)$, via a one-dimensional convolution with $\varphi_C(t, \tau)$ or $\Phi_C(f, \nu)$, respectively. If the kernel $\varphi_C(t, \tau)$ is fairly short duration, then it may be possible to implement a discrete-time version of (65) on a digital computer in real time using only a small number of local signal samples. The third normal form in (67) indicates that any TFR in Cohen's-shift covariant class can be computed by convolving the TFR-dependent kernel, $\psi_C(t, f)$, with the Wigner distribution (WD) of the signal, defined in (15)–(17). Hence, the WD is one of the key members of Cohen's class and many TFRs correspond to smoothed WDs, as can be seen in the top of Table 12.5. Equation (71) and the fourth normal form in (68) indicate that the two-dimensional convolution in (67) transforms to multiplication of the Fourier transform of the kernel, $\psi_C(t, f)$, with the Fourier transform of the WD, which is the ambiguity function (AF). This last normal form provides an intuitive interpretation that the 'AF-domain' kernel, $\Psi_C(\tau, \nu)$, can be thought of as the frequency response of a two-dimensional filter. Equation (48) reveals that this AF plane kernel should be an ideal lowpass filter in order to retain AF auto terms which map to the origin and to reduce AF cross terms which map away from the origin in the AF plane.

Determination of TFR properties by kernel constraints

The kernels in (65)–(69) are signal-independent and provide valuable insight into the performance of each Cohen's-class TFR, regardless of the input signal. The kernels can be used to evaluate which ideal TFR property a given TFR satisfies and the relative merits of one TFR over another. Given on the left in each row in Table 12.8 is one of the ideal TFR properties discussed earlier; provided in the second column are the corresponding constraints that a TFR's kernel in (65)–(68) must satisfy for the ideal property to hold. For example, the second column in Table 12.8 reveals that the time-frequency marginal, moment, and localization properties are automatically satisfied by any Cohen's-class TFR whose AF plane kernel $\Psi_C(\tau, \nu)$ in Table 12.10 is equal to one along its axes, i.e., $\Psi_C(0, \nu) = 1 = \Psi_C(\tau, 0)$. The row for P_{25} in Table 12.8 indicates that Moyal's formula is satisfied by any Cohen's class TFR whose AF-domain kernel has unit modulus. Hence, it can be easily seen by examining the third column of Table 12.10 that the Wigner distribution with unit kernel $\Psi_{WD}(\tau, \nu) = 1$ trivially

TABLE 12.9

Cohen's-class TFRs. Here, $\text{rect}_a(t) = \begin{Bmatrix} 1, & |t| < |a| \\ 0, & |t| > |a| \end{Bmatrix}$, $AF_x(\tau, \nu)$ is defined in (18), and

$\tilde{\mu}(\tilde{\tau}, \tilde{\nu}; \alpha, r, \beta, \gamma) = \tilde{\tau}^2(\tilde{\nu}^2)^\alpha + (\tilde{\tau})^\alpha \tilde{\nu}^2 + 2r(((\tilde{\tau}\tilde{\nu})^\beta)^\gamma)^2$. Functions with lower- and uppercase letters, e.g., $\gamma(t)$ and $\Gamma(f)$, are Fourier transform pairs.

Cohen's-Class Distribution	Formula
Ackroyd	$ACK_x(t, f) = Re\{x^*(t)X(f)e^{j2\pi ft}\}$
Affine-Cohen Subclass	$AC_x(t, f; S_{AC}) = \int\int \frac{1}{|\tau|} S_{AC}\left(\frac{t-t'}{\tau}\right) x\left(t' + \frac{\tau}{2}\right) x^*\left(t' - \frac{\tau}{2}\right) e^{-j2\pi f\tau} dt'\, d\tau$
Born-Jordon	$BJD_x(t, f) = \int\int \frac{\sin(\pi\tau\nu)}{\pi\tau\nu} AF_x(\tau, \nu)e^{j2\pi(t\nu - f\tau)} d\tau\, d\nu$
	$= \int \frac{1}{\tau} \left[\int_{t-|\tau|/2}^{t+|\tau|/2} x\left(t' + \frac{\tau}{2}\right) x^*\left(t' - \frac{\tau}{2}\right) dt' \right] e^{-j2\pi f\tau} d\tau$
Butterworth	$BUD_x(t, f; M, N) = \int\int \left(1 + \left(\frac{\tau}{\tau_0}\right)^{2M}\left(\frac{\nu}{\nu_0}\right)^{2N}\right)^{-1} AF_x(\tau, \nu)e^{j2\pi(t\nu - f\tau)} d\tau\, d\nu$
Choi-Williams (Exponential)	$CWD_x(t, f; \sigma) = \int\int e^{-(2\pi\tau\nu)^2/\sigma} AF_x(\tau, \nu)e^{j2\pi(t\nu - f\tau)} d\tau\, d\nu$
	$= \int\int \sqrt{\frac{\sigma}{4\pi}} \frac{1}{|\tau|} \exp\left[-\frac{\sigma}{4}\left(\frac{t-t'}{\tau}\right)^2\right] x\left(t' + \frac{\tau}{2}\right) x^*\left(t' - \frac{\tau}{2}\right) e^{-j2\pi f\tau} dt'\, d\tau$
Cone Kernel	$CKD_x(t, f) = \int\int g(\tau)|\tau| \frac{\sin(\pi\tau\nu)}{\pi\tau\nu} AF_x(\tau, \nu)e^{j2\pi(t\nu - f\tau)} d\tau\, d\nu$
Cummulative Attack Spectrum	$CAS_x(t, f) = \left| \int_{-\infty}^{t} x(\tau)e^{-j2\pi f\tau} d\tau \right|^2$
Cummulative Decay Spectrum	$CDS_x(t, f) = \left| \int_{t}^{\infty} x(\tau)e^{-j2\pi f\tau} d\tau \right|^2$
Generalized Exponential	$GED_x(t, f) = \int\int \exp\left[-\left(\frac{\tau}{\tau_0}\right)^{2M}\left(\frac{\nu}{\nu_0}\right)^{2N}\right] AF_x(\tau, \nu)e^{j2\pi(t\nu - f\tau)} d\tau\, d\nu$
Generalized Rectangular	$GRD_x(t, f) = \int \text{rect}_1\left(|\tau|^{M/N}|\nu|/\sigma\right) AF_X(\tau, \nu)e^{j2\pi(t\nu - f\tau)} d\tau\, d\nu$
Generalized Wigner	$GWD_x(t, f; \tilde{\alpha}) = \int x\left(t + \left(\frac{1}{2} + \tilde{\alpha}\right)\tau\right) x^*\left(t - \left(\frac{1}{2} - \tilde{\alpha}\right)\tau\right) e^{-j2\pi f\tau} d\tau$
Levin	$LD_x(t, f) = -\frac{d}{dt} \left| \int_{t}^{\infty} x(\tau) e^{-j2\pi f\tau} d\tau \right|^2$
Margineau-Hill	$MH_x(t, f) = Re\left\{x(t)X^*(f) e^{-j2\pi ft}\right\}$
Multiform Tiltable Kernel	$MT_x(t, f; S) = \int\int S\left(\tilde{\mu}\left(\frac{\tau}{\tau_0}, \frac{\nu}{\nu_0}; \alpha, r, \beta, \gamma\right)^{2\lambda}\right) AF_x(\tau, \nu)e^{j2\pi(t\nu - f\tau)} d\tau\, d\nu$
	$S_{MTED}(\beta) = e^{-\pi\beta}, \quad S_{MTBUD}(\beta) = [1 + \beta]^{-1}$
Nutall	$ND_x(t, f) = \int\int \exp\left\{-\pi\left[\left(\frac{\tau}{\tau_0}\right)^2 + \left(\frac{\nu}{\nu_0}\right)^2 + 2r\left(\frac{\tau\nu}{\tau_0\nu_0}\right)\right]\right\} AF_x(\tau, \nu)e^{j2\pi(t\nu - f\tau)} d\tau\, d\nu$
Page	$PD_x(t, f) = 2 Re\left\{x^*(t)e^{j2\pi tf} \int_{-\infty}^{t} x(\tau)e^{-j2\pi f\tau} d\tau\right\}$

TABLE 12.9
(continued)

Cohen-Class Distribution	Formula
Pseudo Wigner	$PWD_x(t, f; \Gamma) = \int x \left(t + \frac{\tau}{2}\right) x^* \left(t - \frac{\tau}{2}\right) \gamma \left(\frac{\tau}{2}\right) \gamma^* \left(-\frac{\tau}{2}\right) e^{-j2\pi f \tau} \, d\tau$
	$\qquad = \int WD_\gamma(0, f - f') WD_x(t, f') \, df'$
Reduced Interference	$RID_x(t, f) = \int \int \frac{1}{|\tau|} s \left(\frac{t-t'}{\tau}\right) x \left(t' + \frac{\tau}{2}\right) x^* \left(t' - \frac{\tau}{2}\right) e^{-j2\pi f \tau} \, dt' \, d\tau$
	with $S(\beta) \epsilon \Re$, $S(0) = 1$, $\frac{d}{d\beta} S(\beta) \big|_{\beta=0} = 0$, $\left\{ S(\alpha) = 0 \text{ for } |\alpha| > \frac{1}{2} \right\}$
Rihaczek	$RD_x(t, f) = x(t) X^*(f) e^{-j2\pi t f}$
Smoothed Pseudo Wigner	$SPWD_x(t, f; \Gamma, s) = \int s(t - t') PWD_x(t', f; \Gamma) \, dt'$
	$\qquad = \int\int s(t - t') WD_\gamma(0, f - f') WD_x(t', f') \, dt' \, df'$
Spectrogram	$SPEC_x(t, f; \Gamma) = \left| \int x(\tau) \gamma^*(\tau - t) e^{-j2\pi f \tau} \, d\tau \right|^2 = \left| \int X(f') \Gamma^*(f' - f) e^{j2\pi t f'} \, df' \right|^2$
Wigner	$WD_x(t, f) = \int x \left(t + \frac{\tau}{2}\right) x^* \left(t - \frac{\tau}{2}\right) e^{-j2\pi f \tau} \, d\tau = \int X \left(f + \frac{\nu}{2}\right) X^* \left(f - \frac{\nu}{2}\right) e^{j2\pi t \nu} \, d\nu$

satisfies these properties. In fact, the checklist in Table 12.7 indicates that the WD satisifies P_1–P_7, P_9–P_{21}, and P_{23}–P_{25}. The product kernels, $\Psi_C(\tau, \nu) = S(\tau\nu)$, used in the reduced interference distributions (RID) and the Choi-Williams distributions (CWD) [79], [35] can easily be made to satisfy the time-frequency moment, marginal and localization properties simply by normalizing the kernel to be one at its origin, i.e., $S(0) = 1$. In addition, if $S(\beta)$ is an even function, then these distributions automatically satisfy the instantaneous frequency, group delay, and Fourier transform properties as well.

There is a tradeoff between Cohen's Class TFRs that have good cross-term reduction versus those TFRs that satisfy Moyal's formula or the marginal, moment, or localization properties. The WD kernel, $\Psi_{WD}(\tau, \nu) = 1$, satisfies the constraint for Moyal's formula, but it unfortu-nately acts as an allpass filter in the AF-domain formulation in (68), passing all cross terms in (48). For good cross-term reduction and little auto-term distortion, the kernel $\Psi_C(\tau, \nu)$ given in Table 12.8 should be as close as possible to an ideal lowpass filter. Kernels that are equal to one along the axes in the AF plane satisfy the time-frequency marginal, moment, and localization properties, but they cannot reduce the cross terms in (48) that occur along the AF plane axes, i.e., those cross terms which correspond to pairs of signal components that occur at the same time or at the same frequency. The Born-Jordon, Butterworth, Choi-Williams, generalized exponential, Auger generalized rectangular, reduced interference, and multiform tiltable kernel distributions [11], [43], [60], [45] all use one-dimensional prototype kernels in the third column of Table 12.10 that were designed to be lowpass and to be equal to one along the axes. For example, the Choi-Williams distribution has equal amplitude iso-contours along hyperbolas in the (τ, ν) plane; it is equal to one along the AF plane axes and decreases to zero away from the AF plane origin in a Gaussian fashion. It can be used to reduce those cross terms in (48) that map away from the axes. The CWD scaling factor σ determines the width of the passband region in the AF plane; small values for σ allow the user to select a narrow

TABLE 12.10

Kernels of Cohen's shift covariant class of time-frequency representations (TFRs) defined in Table 12.9. Here, $\bar{\mu}(\bar{\tau}, \bar{v}; \alpha, r, \beta, \gamma) = ((\bar{\tau})^2 ((\bar{v})^2) + ((\bar{\tau})^2)^\alpha (\bar{v})^2 + 2r((\bar{\tau}\bar{v})^\beta)^\gamma)$. Functions with lowercase and uppercase letters, e.g., $\gamma(t)$ and $\Gamma(f)$, indicate Fourier transform pairs.

TFR	$\psi_C(t,f)$	$\Psi_C(\tau,\nu)$	$\varphi_C(t,\tau)$	$\Phi_C(f,\nu)$
AC	$\displaystyle\int \frac{1}{|\tau|} s_{AC}\left(\frac{t}{\tau}\right) e^{-j2\pi f\tau}\, d\tau$	$S_{AC}(\tau\nu)$	$\dfrac{1}{|\tau|} s_{AC}\left(\dfrac{t}{\tau}\right)$	$\dfrac{1}{|\nu|} s_{AC}\left(-\dfrac{f}{\nu}\right)$
ACK	$2\cos(4\pi t f)$	$\cos(\pi\tau\nu)$	$\dfrac{\delta(t+\tau/2)+\delta(t-\tau/2)}{2}$	$\dfrac{\delta(f-\nu/2)+\delta(f+\nu/2)}{2}$
BJD		$\dfrac{\sin(\pi\tau\nu)}{\pi\tau\nu}$	$\begin{cases} \dfrac{1}{|\tau|}, & |t/\tau| < 1/2 \\ 0, & |t/\tau| > 1/2 \end{cases}$	$\begin{cases} \dfrac{1}{|\nu|}, & |f/\nu| < 1/2 \\ 0, & |f/\nu| > 1/2 \end{cases}$
BUD		$\left(1 + \left(\dfrac{\tau}{\tau_0}\right)^{2M}\left(\dfrac{\nu}{\nu_0}\right)^{2N}\right)^{-1}$		
CWD	$\sqrt{\dfrac{\sigma}{4\pi}}\displaystyle\int \frac{1}{|\beta|}\exp\left[-\frac{\sigma}{4}\left(\frac{t}{\beta}\right)^2\right] e^{-j2\pi f\beta}\, d\beta$	$e^{-(2\pi\tau\nu)^2/\sigma}$	$\sqrt{\dfrac{\sigma}{4\pi}}\dfrac{1}{|\tau|}\exp\left[-\dfrac{\sigma}{4}\left(\dfrac{t}{\tau}\right)^2\right]$	$\sqrt{\dfrac{\sigma}{4\pi}}\dfrac{1}{|\nu|}\exp\left[-\dfrac{\sigma}{4}\left(\dfrac{f}{\nu}\right)^2\right]$
CKD		$g(\tau)\,|\tau|\,\dfrac{\sin(\pi\tau\nu)}{\pi\tau\nu}$	$\begin{cases} g(\tau), & |t/\tau| < 1/2 \\ 0, & |t/\tau| > 1/2 \end{cases}$	
CAS		$\left[\dfrac{1}{2}\delta(\nu) + \dfrac{1}{j\nu}\right] e^{-j\pi|\tau|\nu}$		
CDS		$\left[\dfrac{1}{2}\delta(-\nu) - \dfrac{1}{j\nu}\right] e^{j\pi|\tau|\nu}$		
GED		$\exp\left[-\left(\dfrac{\tau}{\tau_0}\right)^{2M}\left(\dfrac{\nu}{\nu_0}\right)^{2N}\right]$	$\dfrac{\nu_0}{2\sqrt{\pi}}\left|\dfrac{\tau_0}{\tau}\right|^M \exp\left[\dfrac{-\nu_0^2\tau_0^{2M}t^2}{4\tau^{2M}}\right]$ $N = 1$ only	$\dfrac{\tau_0}{2\sqrt{\pi}}\left|\dfrac{\nu_0}{\nu}\right|^N \exp\left[\dfrac{-\tau_0^2\nu_0^{2N}f^2}{4\nu^{2N}}\right]$ $M = 1$ only
GRD		$\begin{cases} 1, & \left\||\tau|^{M/N}|\nu|/\sigma\right\| < 1 \\ 0, & \left\||\tau|^{M/N}|\nu|/\sigma\right\| > 1 \end{cases}$	$\dfrac{\sin\left(2\pi|\sigma||t|/|\tau|^{M/N}\right)}{\pi t}$	

TABLE 12.10
(continued)

TFR	$\psi_C(t,f)$	$\Psi_C(\tau,\nu)$	$\varphi_C(t,\tau)$	$\Phi_C(f,\nu)$
GWD	$\dfrac{1}{|\tilde\alpha|}e^{j2\pi tf/\tilde\alpha}$	$e^{j2\pi\tilde\alpha\tau\nu}$	$\delta(t+\tilde\alpha\tau)$	$\delta(f-\tilde\alpha\nu)$
LD		$e^{j\pi|\tau|\nu}$	$\delta(t+|\tau|/2)$	
MH	$2\cos(4\pi tf)$	$\cos(\pi\tau\nu)$	$\dfrac{\delta(t+\tau/2)+\delta(t-\tau/2)}{2}$	$\dfrac{\delta(f-\nu/2)+\delta(f+\nu/2)}{2}$
MT		$s\left(\tilde\mu\left(\dfrac{\tau}{\tau_0},\dfrac{\nu}{\nu_0};\alpha,r,\beta,\gamma\right)^{2\lambda}\right)$		
ND		$\exp\left[-\pi\tilde\mu\left(\dfrac{\tau}{\tau_0},\dfrac{\nu}{\nu_0};0,r,1,1\right)\right]$		
PD		$e^{-j\pi|\tau|\nu}$	$\delta(t-|\tau|/2)$	$\left[\delta\left(f+\dfrac{\nu}{2}\right)+\delta\left(f-\dfrac{\nu}{2}\right)+j\dfrac{\nu}{\pi(f^2-\nu^2/4)}\right]\bigg/2$
PWD	$\delta(t)WD_\gamma(0,f)$	$\gamma(\tau/2)\gamma^*(-\tau/2)$	$\delta(t)\gamma(\tau/2)\gamma^*(-\tau/2)$	$WD_\gamma(0,f)$
RGWD	$\dfrac{1}{|\tilde\alpha|}\cos(2\pi tf/\tilde\alpha)$	$\cos(2\pi\tilde\alpha\tau\nu)$	$\dfrac{\delta(t+\tilde\alpha\tau)+\delta(t-\tilde\alpha\tau)}{2}$	$\dfrac{\delta(f-\tilde\alpha\nu)+\delta(f+\tilde\alpha\nu)}{2}$
RID	$\displaystyle\int\dfrac{1}{|\beta|}s\left(\dfrac{t}{\beta}\right)e^{-j2\pi tf\beta}\,d\beta$	$S(\tau\nu)$, $S(\beta)\in\Re$, $S(0)=1$, $\dfrac{d}{d\beta}S(\beta)\big|_{\beta=0}=0$	$\dfrac{1}{|\tau|}s\left(\dfrac{t}{\tau}\right)$, $s(\alpha)=0,\ |\alpha|>\dfrac12$	$\dfrac{1}{|\nu|}s\left(-\dfrac{f}{\nu}\right)$, $s(\alpha)=0,\ |\alpha|>\dfrac12$
RD	$2e^{-j4\pi tf}$	$e^{-j\pi\tau\nu}$	$\delta(t-\tau/2)$	$\delta(f+\nu/2)$
SPWD	$s(t)WD_\gamma(0,f)$	$S(\nu)\gamma\left(\dfrac{\tau}{2}\right)\gamma^*\left(-\dfrac{\tau}{2}\right)$	$s(t)\gamma\left(\dfrac{\tau}{2}\right)\gamma^*\left(-\dfrac{\tau}{2}\right)$	$S(\nu)WD_\gamma(0,f)$
SPEC	$WD_\gamma(-t,-f)$	$AF_\gamma(-\tau,-\nu)$	$\gamma\left(-t-\dfrac{\tau}{2}\right)\gamma^*\left(-t+\dfrac{\tau}{2}\right)$	$\Gamma\left(-f-\dfrac{\nu}{2}\right)\Gamma^*\left(-f+\dfrac{\nu}{2}\right)$
WD	$\delta(t)\delta(f)$	1	$\delta(t)$	$\delta(f)$

passband for good cross-term reduction whereas large values for σ produce good auto-term preservation; unfortunately, both are not always possible with the CWD. The generalized exponential, Butterworth, and the multiform tiltable distributions are extensions to the CWD and the RID that use nonlinear axis mappings with several degrees of freedom that can be exploited to satisfy both passband and stopband constraints. Although the lowpass nature of product kernels is useful for cross-term reduction, it prevents any such TFR from satisfying Moyal's formula or the linear chirp localization property.

12.5.2   *Affine class of TFRs*

$$\textit{Affine class} = \left\{ T_x(t, f) \mid y(t) = \sqrt{|a|} x(a(t - t_0)) \Rightarrow T_y(t, f) = T_x \left(a(t - t_0), \frac{f}{a} \right) \right\}$$

TFRs that are covariant to scale changes and time translations, i.e., properties P_2–P_3 in Table 12.6, are members of the affine class [19], [20], [63], [111], [60]. Several such TFRs are given in Table 12.11. The scale covariance property, P_3, is useful for several applications, including wideband Doppler systems, signals with fractal structure [58], octave band systems like the cochlea of the inner ear, and short duration "transients" [2], [3], [36], [30], [91], [81], [90].

Alternative formulations

Any affine class TFR can be written in the four "normal forms" and "bi-frequency" form equations similar to those of Cohen's class [111], [60], [100]:

$$\mathcal{A}_x(t, f; \Psi_{\mathcal{A}}^{(A)}) = |f| \int \int \varphi_{\mathcal{A}}^{(A)}(f(t - t'), f\tau) x(t' + \tau/2) x^*(t' - \tau/2) \, dt' \, d\tau \tag{78}$$

$$= \frac{1}{|f|} \int \int \Phi_{\mathcal{A}}^{(A)}\left(-\frac{f'}{f}, \frac{\nu}{f}\right) X(f' + \nu/2) X^*(f' - \nu/2) e^{j2\pi t\nu} \, df' \, d\nu \tag{79}$$

$$= \int \int \psi_{\mathcal{A}}^{(A)}\left(f(t - t'), -\frac{f'}{f}\right) W D_x(t', f') \, dt' \, df' \tag{80}$$

$$= \int \int \Psi_{\mathcal{A}}^{(A)}\left(f\tau, \frac{\nu}{f}\right) A F_x(\tau, \nu) e^{j2\pi t\nu} \, d\tau \, d\nu \tag{81}$$

$$= \frac{1}{|f|} \int \int \Gamma_{\mathcal{A}}^{(A)}\left(\frac{f_1}{f}, \frac{f_2}{f}\right) X(f_1) X^*(f_2) e^{j2\pi t(f_1 - f_2)} \, df_1 \, df_2. \tag{82}$$

The affine-class kernels are interrelated by the same Fourier transforms given in (70)–(71), i.e.,

$$\varphi_{\mathcal{A}}^{(A)}(c, \zeta) \leftrightarrow \Phi_{\mathcal{A}}^{(A)}(b, \beta)$$

$$\psi_{\mathcal{A}}^{(A)}(c, b) \leftrightarrow \Psi_{\mathcal{A}}^{(A)}(\zeta, \beta).$$

The bi-frequency kernel is related to the normal form kernels by the following equation:

$$\Gamma_{\mathcal{A}}^{(A)}(b_1, b_2) = \Phi_{\mathcal{A}}^{(A)}\left(-\frac{b_1 + b_2}{2}, b_1 - b_2\right).$$

Note that the third normal form of the affine class in (80) involves an affine smoothing of

TABLE 12.11

TFRs in the affine class. The Bertrand P_0, generalized WD, Unterberger active and passive, and the Wigner distributions are special cases of the localized affine distributions with $G^{(A)}(\beta)$ and $F^{(A)}(\beta)$ defined accordingly; e.g., $F_{UAD}^{(A)}(\beta) = -\sqrt{1 + (\beta/2)^2}$ and $G_{UAD}^{(A)}(\beta) = 1$.

Affine Class Distr.	Formula
Ackroyd	$ACK_x(t, f) = Re\{x^*(t)X(f)e^{j2\pi ft}\}$
Affine-Cohen Subclass	$AC_x(t, f; S_{AC}) = \iint \dfrac{1}{\lvert\tau\rvert} s_{AC}\left(\dfrac{t-t'}{\tau}\right) x\left(t' + \dfrac{\tau}{2}\right) x^*\left(t' - \dfrac{\tau}{2}\right) e^{-j2\pi f\tau} \, dt' \, d\tau$
Affine-Hyp. Subclass	$AH_X(t, f; S_{AH}) = f \displaystyle\int s_{AH}(f(t-t')) BP_0 D_X(t', f; \mu_0) \, dt'$ $= f \displaystyle\int X\left(\dfrac{\beta/2 e^{\beta/2}}{\sinh\beta/2}\right) X^*\left(f\dfrac{\beta/2 e^{-\beta/2}}{\sinh\beta/2}\right) S_{AH}(\beta) \dfrac{\beta/2}{\sinh\beta/2} e^{j2\pi t f\beta} \, d\beta$
(Unitary) Bertrand P_0	$BP_0 D_X(t, f; \mu_0) = f \displaystyle\int X\left(f\dfrac{u/2 e^{u/2}}{\sinh(u/2)}\right) X^*\left(f\dfrac{u/2 e^{-u/2}}{\sinh(u/2)}\right) \dfrac{u/2}{\sinh(u/2)} e^{j2\pi t f u} \, du$
(General) Bertrand P_0	$BP_0 D_X(t, f; \mu) = \lvert f\rvert \displaystyle\int X\left(f\dfrac{\beta/2}{\sinh\beta/2} e^{\beta/2}\right) X^*\left(f\dfrac{\beta/2}{\sinh\beta/2} e^{-\beta/2}\right) \mu(\beta) e^{j2\pi t f\beta} \, d\beta$ $= \lvert f\rvert \displaystyle\int X\left(f\left[\dfrac{u}{2}\coth\dfrac{u}{2} + \dfrac{u}{2}\right]\right) X^*\left(f\left[\dfrac{u}{2}\coth\dfrac{u}{2} - \dfrac{u}{2}\right]\right) \mu(u) e^{j2\pi t f u} \, du$
Bertrand P_κ	$BP_\kappa D_X(t, f; \mu) = f \displaystyle\int X(f\lambda_k(u)) X^*(f\lambda_k(-u)) \mu(u) e^{j2\pi t f(\lambda_k(u) - \lambda_k(-u))} \, du,$ $\lambda_0(u) = \dfrac{u/2 e^{u/2}}{\sinh(u/2)}, \ \lambda_1(u) = \exp\left[1 + \dfrac{ue^{-u}}{e^{-u}-1}\right], \ \lambda_k(u) = \left[k\dfrac{e^{-u}-1}{e^{-ku}-1}\right]^{\frac{1}{k-1}}, k \neq 0, 1$ and $\mu(u) = \mu^*(-u)$
Born-Jordon	$BJD_x(t, f) = \displaystyle\iint \dfrac{\sin(\pi\tau v)}{\pi\tau v} AF_x(\tau, v) e^{j2\pi(tv - f\tau)} \, d\tau \, dv$
Choi-Williams Exp.	$CWD_x(t, f; \sigma) = \displaystyle\iint e^{-(2\pi\tau v)^2/\sigma} AF_x(\tau, v) e^{j2\pi(tv - f\tau)} \, d\tau \, dv$
Flandrin D	$FD_X(t, f) = f \displaystyle\int X\left(f\left[1 + \dfrac{u^2}{4}\right]\right) X^*\left(f\left[1 - \dfrac{u^2}{4}\right]\right) \left[1 - \left(\dfrac{u}{4}\right)^2\right] e^{j2\pi t f u} \, du$

TABLE 12.11
(continued)

Affine Class Distr.	Formula
Generalized Wigner	$GWD_x(t,f;\tilde\alpha) = \int x\left(t+\left(\frac{1}{2}+\tilde\alpha\right)\tau\right) x^*\left(t-\left(\frac{1}{2}-\tilde\alpha\right)\tau\right) e^{-j2\pi f\tau}\, d\tau$
Localized Affine	$LA_x(t,f;G^{(A)},F^{(A)}) = f\int\int X(f(-F^{(A)}(\beta)+\beta/2)) X^*(f(-F^{(A)}(\beta)-\beta/2)) G^{(A)}(\beta) e^{j2\pi tf\beta}\, d\beta$
Margineau-Hill	$MH_x(t,f) = Re\left\{x(t)X^*(f)e^{-j2\pi ft}\right\}$
Reduced Interference	$RID_x(t,f) = \int\int \frac{1}{|\tau|} s\left(\frac{t-t'}{\tau}\right) x\left(t'+\frac{\tau}{2}\right) x^*\left(t'-\frac{\tau}{2}\right) e^{-j2\pi f\tau}\, dt'\, d\tau$ with $S(\beta)\in\Re$, $S(0)=1$, $\frac{d}{d\beta}S(\beta)\big|_{\beta=0}=0$, $\left\{s(\alpha)=0 \text{ for } |\alpha|>\frac{1}{2}\right\}$
Rihaczek	$RD_x(t,f) = x(t)X^*(f)e^{-j2\pi tf}$
Scalogram	$SCAL_x(t,f;\Gamma) = \left\|\int x(\tau)\sqrt{\left\|\frac{f}{f_r}\right\|}\gamma^*\left(\frac{f}{f_r}(\tau-t)\right)d\tau\right\|^2 = \left\|\frac{f_r}{f}\right\|\left\|\int X(\hat f)\Gamma^*\left(f_r\frac{\hat f}{f}\right)e^{j2\pi tf}\hat f\, d\hat f\right\|^2$
Unterberger Active	$UAD_x(t,f) = f\int_0^\infty X(fu)X^*(f/u)[1+u^{-2}]e^{j2\pi tf(u-1/u)}\, du$ $= f\int\int X\left(f\left(\sqrt{1+(\beta/2)^2}+\beta/2\right)\right) X^*\left(f\left(\sqrt{1+(\beta/2)^2}-\beta/2\right)\right) e^{j2\pi tf\beta}\, d\beta$
Unterberger Passive	$UPD_x(t,f) = 2f\int_0^\infty X(fu)X^*(f/u)\left[\frac{1}{u}\right]e^{j2\pi tf(u-1/u)}\, du$ $= f\int X\left(f\left(\sqrt{1+\left(\frac{\beta}{2}\right)^2}+\frac{\beta}{2}\right)\right) X^*\left(f\left(\sqrt{1+\left(\frac{\beta}{2}\right)^2}-\frac{\beta}{2}\right)\right)\frac{1}{\sqrt{1+(\beta/2)^2}}e^{j2\pi tf\beta}\, d\beta$
Wigner	$WD_x(t,f) = \int x\left(t+\frac{\tau}{2}\right)x^*\left(t-\frac{\tau}{2}\right)e^{-j2\pi f\tau}\, d\tau = \int X\left(f+\frac{\nu}{2}\right)X^*\left(f-\frac{\nu}{2}\right)e^{j2\pi t\nu}\, d\nu$

the WD of the signal, i.e., the output frequency f is inversely proportional to the amount of time smoothing and proportional to the amount of frequency smothing [63]. In Table 12.11, various members of the affine class, such as the Bertrand P_0 distribution, the scalogram, and the Unterberger distributions, are defined.

The four normal form kernels used to formulate affine class TFRs in (78)–(81) are listed in Table 12.12. These kernels can be compared with the constraints listed in the third column of Table 12.8 to determine which of the ideal properties a given affine TFR satisfies. A checklist summary of the properties that the affine-class TFRs satisfy is provided in Table 12.7. Any affine-class TFR must necessarily have a check mark in the rows for P_2 and P_3. Because of the scale covariance property, many TFRs in the affine class exhibit constant-Q behavior, permitting multiresolution analysis [112], [60].

The scalogram [111], [60] in Table 12.11 is the squared magnitude of the recently introduced wavelet transform [49], [89], [112], [69], [36], [118], [60], [2], [3]:

$$
WT_x(t, f; \Gamma) = \int x(\tau) \sqrt{\left|\frac{f}{f_0}\right|} \gamma^* \left(\frac{f}{f_0}(\tau - t)\right) d\tau \tag{83}
$$

$$
= \int X(f') \sqrt{\left|\frac{f_0}{f}\right|} \Gamma^* \left(\frac{f_0}{f} f'\right) e^{j2\pi t f'} df'. \tag{84}
$$

It uses a special sliding analysis window, $\gamma(t)$, called the mother wavelet, to analyze local spectral information of the signal $x(t)$.

The mother wavelet is either compressed or dilated to give a multiresolution signal representation. Assume that $\gamma(t)$ is a bandpass filter centered near time $t = 0$ and frequency $f = f_0$ with (approximate) time duration D and one-sided spectral bandwidth B, i.e., $|\gamma(t)| \approx 0$, $|t| > D$ and $|\Gamma(f)| \approx 0$, $|f - f_0| > B$. Then the scaled mother wavelet in (83) is centered near the output time t and its time duration is $|\frac{f_0}{f}D|$. Likewise, $\Gamma^*(\frac{f_0}{f}f')$ in (84) is centered near the output frequency, f, and has spectral bandwidth $B|\frac{f}{f_0}|$. Hence, for $f > f_0$, the WT in (83) compresses the time duration of the mother wavelet while simultaneously expanding its spectral bandwidth in (84). In fact, the WT implements a "constant-Q" signal analysis, since the mother wavelet's quality factor [60],

$$
Q = \frac{\text{center frequency}}{\text{bandwidth}} = \frac{f}{B|f/f_0|} = \frac{f_0}{B}, \qquad f > 0,
$$

is constant for all output frequencies.

Thus, the scalogram can be thought of as the multiresolution output of a parallel bank of octave band filters [112], [111], [69]. High-frequency regions of the WT domain have very good time resolution whereas low-frequency regions of the WT domain have very good spectral resolution. The wavelet transform has been used to code images, to model the mid- to high-frequency operation of the cochlea, to track transients such as speech pitch and the onset of the QRS complex in ECG signals, and to analyze fractal and chaotic signals [118], [36], [2], [3], [1], [4], [91], [60], [30]. One drawback of the scalogram is its poor temporal resolution at low-frequency regions of the time-frequency plane and poor spectral resolution at high frequencies. The quadratic scalogram produces cross terms whenever signal components overlap [82]. Further, many discrete WT implementations do not preserve the important time-shift covariance property.

TABLE 12.12

Kernels of affine-class TFRs defined in Table 12.11. Members of the affine-Cohen (AC) subclass in the first row include the Ackroyd, Margineau-Hill, Born-Jordan, Choi-Williams, generalized Wigner, reduced interference, Rihaczek, and Wigner distributions. They have the following kernels: $S_{ACK}(b) = S_{MH}(b) = \cos\pi b$, $S_{BJD}(b) = \sin\pi b/\pi b$, $S_{CWD}(b) = e^{-(2\pi b)^2/\sigma}$, $S_{GWD}(b) = e^{j2\pi\tilde{\alpha}b}$, $S_{RID}(b) = S_{RID}(b) * (\sin\pi b/\pi b)$, $S_{RD}(b) = e^{-j\pi b}$, where $*$ denotes convolution and $S_{WD}(b) = 1$, respectively. Here, $u_\Gamma(c,\zeta) = \gamma(c + \zeta/2)\gamma^*(c - \zeta/2)$ and $U_\Gamma(b,\beta) = \Gamma(b + \beta/2)\Gamma^*(b - \beta/2)$, where $\gamma(t) \leftrightarrow \Gamma(f)$ are Fourier transform pairs.

TFR	$\psi_A^{(A)}(c,b)$	$\Psi_A^{(A)}(\zeta,\beta)$	$\varphi_A^{(A)}(c,\zeta)$	$\Phi_A^{(A)}(b,\beta)$
AC	$\int \frac{1}{\|\zeta\|} s_{AC}\left(\frac{1}{\zeta}\right) e^{-j2\pi c(1+b)\zeta}\, d\zeta$	$S_{AC}(\zeta\beta)e^{-j2\pi\zeta}$	$\frac{1}{\|\zeta\|} s_{AC}\left(\frac{c}{\zeta}\right) e^{-j2\pi\zeta}$	$\frac{1}{\|\beta\|} s_{AC}\left(-\frac{1+b}{\beta}\right)$
AH	$\int \frac{S_{AH}(\beta)\beta/2}{\sinh\beta/2}\delta\left(b + \frac{\beta}{2}\coth\frac{\beta}{2}\right) e^{j2\pi c\beta}\, d\beta$	$\frac{S_{AH}(\beta)\beta/2}{\sinh\beta/2} e^{-j2\pi\zeta\left[\frac{\beta}{2}\coth\frac{\beta}{2}\right]}$	$\int \frac{S_{AH}(\beta)\beta/2}{\sinh\beta/2} e^{-j2\pi\left(\zeta\left[\frac{\beta}{2}\coth\frac{\beta}{2}\right]-c\beta\right)} d\beta$	$\frac{S_{AH}(\beta)\beta/2}{\sinh\beta/2}\delta\left(b + \frac{\beta}{2}\coth\frac{\beta}{2}\right)$
Unitary BP_0D	$\int \frac{\beta/2}{\sinh\beta/2}\delta\left(b + \frac{\beta}{2}\coth\frac{\beta}{2}\right) e^{j2\pi c\beta}\, d\beta$	$\frac{\beta/2}{\sinh\beta/2} e^{-j2\pi\zeta\left[\frac{\beta}{2}\coth\frac{\beta}{2}\right]}$	$\int \frac{\beta/2}{\sinh\beta/2} e^{-j2\pi\left(\zeta\left[\frac{\beta}{2}\coth\frac{\beta}{2}\right]-c\beta\right)} d\beta$	$\frac{\beta/2}{\sinh\beta/2}\delta\left(b + \left[\frac{\beta}{2}\coth\frac{\beta}{2}\right]\right)$
General BP_0D_μ	$\int \mu(\beta)\delta\left(b + \frac{\beta}{2}\coth\frac{\beta}{2}\right) e^{j2\pi c\beta}\, d\beta$	$\mu(\beta) e^{-j2\pi\zeta\frac{\beta}{2}\coth\frac{\beta}{2}}$	$\int \mu(\beta) e^{-j2\pi\left(\zeta\frac{\beta}{2}\coth\frac{\beta}{2}-c\beta\right)} d\beta$	$\mu(\beta)\,\delta\left(b + \frac{\beta}{2}\coth\frac{\beta}{2}\right)$
FD	$\int\left[1-\left(\frac{\beta}{4}\right)^2\right]\delta\left(b+\left[1+\left(\frac{\beta}{4}\right)^2\right]\right) e^{j2\pi c\beta}\, d\beta$	$\left[1-\left(\frac{\beta}{4}\right)^2\right] e^{-j2\pi\zeta[1+(\beta/4)^2]}$	$\int\left[1-\left(\frac{\beta}{4}\right)^2\right] e^{-j2\pi(\zeta[1+(\beta/4)^2]-c\beta)} d\beta$	$\left[1-\left(\frac{\beta}{4}\right)^2\right]\delta\left(b+[1+(\beta/4)^2]\right)$
GWD	$\frac{1}{\|\tilde{\alpha}\|}e^{j2\pi c(1+b)/\tilde{\alpha}}$	$e^{-j2\pi\zeta[1-\tilde{\alpha}\beta]}$	$e^{-j2\pi\zeta}\delta(c+\tilde{\alpha}\zeta)$	$\delta(b+1-\tilde{\alpha}\beta)$
LA	$\int G(\beta)\delta(b-F(\beta))e^{j2\pi c\beta}\, d\beta$	$G(\beta)e^{j2\pi\zeta F(\beta)}$	$\int G(\beta)e^{j2\pi(\zeta F(\beta)+c\beta)} d\beta$	$G(\beta)\delta(b-F(\beta))$
SCAL	$WD_\gamma(-c/f_r, -f_r b)$	$AF_\gamma(-\zeta/f_r, -f_r\beta)$	$\frac{1}{f_r}\gamma\left(\frac{1}{f_r}(-c-\zeta/2)\right)\gamma^*\left(\frac{1}{f_r}(-c+\zeta/2)\right) = \frac{1}{f_r}u_\Gamma(-c/f_r, -\zeta/f_r)$	$f_r\Gamma\left(f_r(-b-\frac{\beta}{2})\right)\Gamma^*\left(f_r(-b+\frac{\beta}{2})\right) = f_r U_\Gamma(-f_r b, -f_r\beta)$
UAD	$\int \delta\left(b+\sqrt{1+\beta^2/4}\right)e^{j2\pi c\beta}\, d\beta$	$e^{-j2\pi\zeta\sqrt{1+\beta^2/4}}$	$\int e^{-j2\pi(\zeta\sqrt{1+\beta^2/4}-c\beta)} d\beta$	$\delta\left(b+\sqrt{1+\beta^2/4}\right)$
UPD	$\int\left[1+\frac{\beta^2}{4}\right]^{-\frac{1}{2}}\delta\left(b+\sqrt{1+\frac{\beta^2}{4}}\right)e^{j2\pi c\beta}\, d\beta$	$\left[1+\frac{\beta^2}{4}\right]^{-\frac{1}{2}} e^{-j2\pi\zeta\sqrt{1+\beta^2/4}}$	$\int\left[1+\frac{\beta^2}{4}\right]^{-\frac{1}{2}} e^{-j2\pi(\zeta\sqrt{1+\beta^2/4}-c\beta)} d\beta$	$\left[1+\frac{\beta^2}{4}\right]^{-\frac{1}{2}}\delta\left(b+\sqrt{1+\beta^2/4}\right)$
WD	$\delta(b+1)\delta(c)$	$e^{-j2\pi\zeta}$	$e^{-j2\pi\zeta}\delta(c)$	$\delta(b+1)$

12.5.3   Affine-Cohen subclass

$$\textit{Affine-Cohen subclass} = \left\{ T_y(t, f) \mid y(t) = \sqrt{|a|} x(a(t - t_0)) e^{j2\pi f_0 t} \right.$$

$$\left. \Rightarrow \ T_y(t, f) = T_x \left(a(t - t_0), \frac{f - f_0}{a} \right) \right\}$$

Those TFRs listed in both Table 12.9 and Table 12.11 are members of the intersection of Cohen's class of time-frequency shift-covariant TFRs with the affine class of affine-covariant TFRs. This intersection is depicted graphically in Figure 12.4. All shift-scale covariant TFRs must satisfy properties P_1–P_3 in Table 12.6. This subclass is characterized by those Cohen's-class TFRs whose AF-domain kernel, $\Psi_C(\tau, \nu) = fcn(\tau \nu)$, is a product kernel, that is, a one-dimensional function evaluated at the product $\tau \nu$. Inspection of the third column of Table 12.10 reveals that the affine-Cohen subclass includes the Ackroyd, Margineau-Hill, Born-Jordon, Choi-Williams, generalized Wigner, reduced interference, Rihaczek, and Wigner distributions. Further, the affine-class kernels used in (78)–(81) and the Cohen's-class kernels used in (65)–(68) of each shift-scale covariant TFR are related as follows:

$$\psi_{AC}^{(A)}(c, b) = \psi_{AC}(c, 1 + b) \quad \longleftrightarrow \quad \Psi_{AC}^{(A)}(\zeta, \beta) = \Psi_{AC}(\zeta, \beta) e^{-j2\pi\zeta} = S_{AC}(\zeta\beta) e^{-j2\pi\zeta}$$

$$\varphi_{AC}^{(A)}(c, \zeta) = \varphi_{AC}(c, \zeta) e^{-j2\pi\zeta} \quad \longleftrightarrow \quad \Phi_{AC}^{(A)}(b, \beta) = \Phi_{AC}(1 + b, \beta).$$

12.5.4   Hyperbolic class of TFRs

$$\textit{Hyperbolic class} = \left\{ T_X(t, f) \mid Y(f) = \frac{1}{\sqrt{|a|}} X\left(\frac{f}{a}\right) \frac{1}{\sqrt{f}} e^{-j2\pi c \ln(f/f_r)} \right.$$

$$\left. \Rightarrow \ T_Y(t, f) = T_X \left(a\left(t - \frac{c}{f}\right), \frac{f}{a} \right) \right\}$$

The hyperbolic class of TFRs consists of all TFRs that are covariant to scale changes and hyperbolic time shifts on analytic signals, i.e., properties P_3–P_4 in Table 12.6 [97], [98], [100]. They can be analyzed using the following alternative forms, valid for $f > 0$,

$$H_X(t, f; \Psi_H^{(H)}) = \int \int \varphi_H^{(H)}(tf - c, \zeta) \, v_X(c, \zeta) e^{-j2\pi [\ln(f/f_r)]\zeta} \, dc \, d\zeta \tag{85}$$

$$= \int \int \Phi_H^{(H)} \left(\ln \frac{f}{f_r} - b, \beta \right) f_r e^b X(f_r e^{b+\beta/2}) X^*(f_r e^{b-\beta/2}) e^{j2\pi tf\beta} \, db \, d\beta \tag{86}$$

$$= \int \int_0^\infty \psi_H^{(H)} \left(tf - t'f', \ln \frac{f}{f'} \right) Q_X(t', f') \, dt' \, df' \tag{87}$$

$$= \int \int \Psi_H^{(H)}(\zeta, \beta) \, HAF_X(\zeta, \beta) \, e^{j2\pi (tf\beta - [\ln(f/f_r)]\zeta)} \, d\zeta \, d\beta \tag{88}$$

$$= \frac{1}{f} \int_0^\infty \int_0^\infty \Gamma_H^{(H)} \left(\frac{f_1}{f}, \frac{f_2}{f} \right) X(f_1) X^*(f_2) e^{j2\pi tf \ln(f_1/f_2)} \, df_1 \, df_2, \tag{89}$$

where $Q_X(t, f)$ is the Altes Q distribution defined in (24), $HAF_X(\zeta, \beta)$ is the hyperbolic

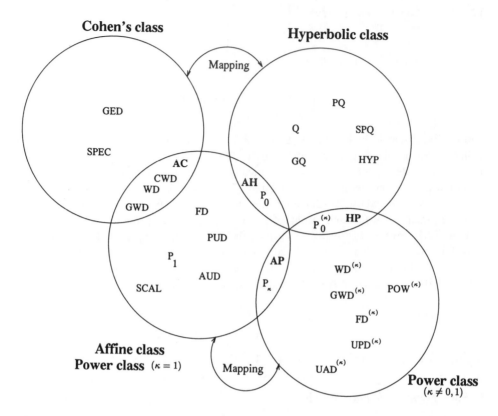

FIGURE 12.4
A pictorial summary of the different classes of QTFRs considered in this manuscript: Cohen's class, affine class, hyperbolic class, and κth power class ($\kappa \neq 0, 1$) together with their intersection subclasses and some important QTFR members. The spectrogram (SPEC), the Wigner distribution (WD), the generalized WD (GWD), the Choi-Williams exponential distribution (CWD), and the generalized exponential distribution (GED) are members of Cohen's class. The WD, GWD, CWD, scalogram (SCAL), Flandrin D (FD), passive Unterberger (UPD), active Unterberger (UAD), Bertrand P_0-distribution, Bertrand P_1-distribution, and Bertrand P_κ-distributions, $\kappa \neq 0, 1$, are members of the affine class. The affine-Cohen's intersection (AC) contains the WD, GWD, and CWD. The hyperbologram (HYP), the Altes-Marinovich Q-distribution (Q), the generalized Q (GQ), the pseudo Q (PQ), the smoothed pseudo Q (SPQ), the Bertrand P_0-distribution, and the κth power form of the Bertrand P_0-distribution ($P_0^{(\kappa)}$) are members of the hyperbolic class. The affine-hyperbolic intersection (AH) contains the Bertrand P_0-distribution. The powergram, POW$^{(\kappa)}$, the power Wigner distribution, WD$^{(\kappa)}$, the generalized WD$^{(\kappa)}$, GWD$^{(\kappa)}$, the Bertrand P_κ-distributions, the power Bertrand P_0-distribution, $P_0^{(\kappa)}$, the power FD, FD$^{(\kappa)}$, the power UPD, UPD$^{(\kappa)}$, and the power UAD, UAD$^{(\kappa)}$, are all P_κ-distributions. The hyperbolic-power intersection (HP) contains $P_0^{(\kappa)}$. The hyperbolic and Cohen's classes, and the affine and power classes are related through unitary mappings.

ambiguity function defined in (26), and the quadratic signal product,

$$v_X(c, \zeta) = \left(\int_0^\infty X(f) \left(\frac{f}{f_r} \right)^{j2\pi(c+\zeta/2)} \frac{df}{\sqrt{f}} \right) \left(\int_0^\infty X^*(v) \left(\frac{v}{f_r} \right)^{-j2\pi(c-\zeta/2)} \frac{dv}{\sqrt{v}} \right) \quad (90)$$

$$= \frac{1}{f_r} \rho_x([c + \zeta/2]/f_r) \rho_x^*([c - \zeta/2]/f_r), \quad (91)$$

is proportional to the product of modified Mellin transformations of the signal spectrum, where

$$\rho_x(t) = \int \sqrt{ef/f_r}\, X(f_r\, e^{f/f_r}) e^{j2\pi ft}\, df \leftrightarrow (\mathcal{W}X)(f) = \sqrt{ef/f_r}\, X(f_r\, e^{f/f_r}), \quad (92)$$

$$= \sqrt{f_r} \int_0^\infty X(v) \left(\frac{v}{f_r}\right)^{j2\pi f_r t} \frac{dv}{\sqrt{v}} \quad (93)$$

corresponds to the unitary warping of the frequency axis of the signal given in (34). The four normal form kernels in (85)–(88) are interrelated via the Fourier transforms in (70)–(71). The relationship of the hyperbolic bi-frequency kernel in (89) to the normal form II kernel in (86) is as follows:

$$\Gamma_H^{(H)}(b_1, b_2) = \frac{1}{\sqrt{|b_1 b_2|}} \Phi_H^{(H)}(-\ln\sqrt{|b_1 b_2|}, \ln(b_1/b_2)).$$

Table 12.13 reveals that the Altes Q distribution, the Bertrand P_0 distribution, and the hyperbologram are members of the hyperbolic class. Their corresponding hyperbolic kernels are given in Table 12.14. Listed in the fourth column of Table 12.8 are the constraints on these hyperbolic kernels needed for the corresponding TFR to satisfy the ideal properties listed in the first column. A summary of the ideal properties that the hyperbolic-class TFRs satisfy is provided in Table 12.7. Note that any member of the hyperbolic class must, by definition, have the rows for properties P_3 and P_4 checked off.

The hyperbolic class TFRs give highly concentrated TFR representations for signals with hyperbolic group delay. They are well suited for the analysis of self-similar random processes or of wideband Doppler-invariant signals similar to the biosonar signals used by bats and dolphins for echolocation [59], [7], [53], [69], [100].

Each hyperbolic-class TFR, kernel, and property corresponds to a warped version of a Cohen's-class TFR, kernel, and property, respectively [97], [100], [14], [15], [44]. That is, each hyperbolic-class TFR, H, corresponds to an axis-warped version of a corresponding Cohen's-class TFR, C_H,

$$H_X(t, f; \Psi_H^{(H)}) = C_{H_{\mathcal{W}X}} \left(\frac{tf}{f_r}, f_r \ln \frac{f}{f_r}; \Psi_{C_H}\right) \quad (94)$$

$$C_{H_X}(t, f; \Psi_{C_H}) = H_{\mathcal{W}^{-1}X}(te^{-f/f_r}, f_r e^{f/f_r}; \Psi_H^{(H)}), \quad (95)$$

provided that the signal is first pre-warped by $\mathcal{W}X$ or $\mathcal{W}^{-1}X$ defined in (34)–(35), respectively. Examples are given in the bottom of Table 12.5. Also, in (32), the Altes Q distribution is equal to the WD after warping both the signal and the time-frequency axes. In this case, $H = Q$ and $C_H = WD$ in (94). Likewise, the hyperbologram is a warped version of the spectrogram, provided that both the signal and the analysis window are pre-warped:

$$HYP_X(t, f; \Gamma) = SPEC_{\mathcal{W}X} \left(\frac{tf}{f_r}, f_r \ln \frac{f}{f_r}; \mathcal{W}\Gamma\right) \quad (96)$$

$$SPEC_X(t, f; \Gamma) = HYP_{\mathcal{W}^{-1}X} \left(te^{-f/f_r}, f_r e^{f/f_r}; \mathcal{W}^{-1}\Gamma\right).$$

The constant bandwidth analysis of many Cohen's-class TFRs, such as the spectrogram, is mapped into the multiresolution, constant Q analysis of many of the hyperbolic-class TFRs. Equation (96) demonstrates that the relatively new hyperbologram can be computed using standard spectrogram implementations. This one-to-one correspondence between TFRs in the Cohen and the hyperbolic classes greatly facilitates their analysis and gives alternative methods for calculating various TFRs [34], [106].

The hyperbolic and Cohen's-class kernels are related by a simple scaling factor, provided that any analysis windows are warped appropriately [100]:

$$\psi_H^{(H)}(c, b) = \psi_{C_H}\left(\frac{c}{f_r}, f_r b\right) \leftrightarrow \Psi_H^{(H)}(\zeta, \beta) = \Psi_{C_H}\left(\frac{\zeta}{f_r}, f_r \beta\right)$$

$$\varphi_H^{(H)}(c, \zeta) = \frac{1}{f_r}\varphi_{C_H}\left(\frac{c}{f_r}, \frac{\zeta}{f_r}\right) \leftrightarrow \Phi_H^{(H)}(b, \beta) = f_r \Phi_{C_H}(f_r b, f_r \beta)$$

$$\Gamma_H^{(H)}(b_1, b_2) = \frac{1}{\sqrt{|b_1 b_2|}}\Phi_{C_H}(-\ln\sqrt{b_1 b_2}, \ln(b_1/b_2)).$$

For example, comparing the WD kernels in Table 12.10 with the Altes Q kernels in Table 12.14 reveals that the Altes kernel

$$\psi_Q^{(H)}(c, b) = \psi_{WD}(c/f_r, f_r b) = \delta(c/f_r)\delta(f_r b) = \delta(c)\delta(b)$$

is identical to that of the WD; similarly, examining the spectrogram and hyperbologram kernels reveals that the kernels are related by an axis scaling and a warping of the spectrogram analysis window:

$$\psi_{HYP}^{(H)}(c, b) = \psi_{SPEC}(c/f_r, f_r b) = WD_{WT}(-c/f_r, -bf_r).$$

Properties of Cohen's-class TFRs also map in a one-to-one fashion to properties of the hyperbolic-class TFRs [100]. For example, the WD's perfect localization of linear FM chirps (P_{22}) warps to the Altes Q distribution's perfect localization for hyperbolic FM chirps (P_{23}).

Affine-hyperbolic subclass

$$Affine\text{-}hyperbolic\ subclass = \left\{T_Y(t, f) \mid Y(f) = \frac{1}{\sqrt{|a|}}X\left(\frac{f}{a}\right)e^{-j2\pi c \ln(f/f_r)}e^{j2\pi f t_0}\right.$$

$$\left.\Rightarrow T_Y(t, f) = T_X\left(a\left(t - t_0 - \frac{c}{f}\right), \frac{f}{a}\right)\right\}$$

TFRs that satisfy properties P_2–P_4 in Table 12.6 are members of the intersection of the affine class with the hyperbolic class, as depicted graphically in Figure 12.4. The most commonly known member of the affine-hyperbolic subclass is the unitary form of the Bertrand P_0 distribution. The first entry in Table 12.13 indicates that any member of the affine-hyperbolic subclass can be written as an affine smoothed version of the unitary Bertrand P_0 distribution. Evaluation of the first entry in Table 12.14 reveals that the two-dimensional kernels of the TFRs in this subclass simplify to a function of a one dimensional prototype $S_{AH}(\beta)$. This greatly simplifies the kernel constraints in Table 12.8.

12.5.5   κth power class

$$\kappa th\ power\ class = \left\{T_X(t, f) \mid Y(f) = \frac{1}{\sqrt{|a|}}X\left(\frac{f}{a}\right)\exp(-j2\pi c[\operatorname{sgn}(f)]|f/f_r|^\kappa)\right.$$

$$\left.\Rightarrow T_Y(t, f) = T_X\left(a\left(t - \frac{\kappa}{f_r}\left|\frac{f}{f_r}\right|^{\kappa-1}\right), \frac{f}{a}\right)\right\}$$

The power classes of TFRs consists of many classes of TFRs, each indexed by $-\infty < \kappa < \infty$

TABLE 12.13
Hyperbolic TFRs. Members of the localized hyperbolic (LH) subclass include the affine hyperbolic (AH) and the κth hyperbolic power subclasses as well as the Altes, Bertrand, P_0 and generalized Altes distributions. $(\mathcal{W}X)(f)$ is the unitary signal warping defined in (34), and $\mu(u) = \mu^*(-u)$ is a real and even function.

Hyperbolic TFR	Formula, $f > 0$
Affine-Hyperbolic Subclass	$AH_X(t, f; S_{AH}) = \|f\| \displaystyle\int \int s_{AH}(f(t - t'))\, BP_0D_X(t', f; \mu_0)\, dt'$
	$= \|f\| \displaystyle\int X\left(f\left(\frac{\beta}{2} \coth \frac{\beta}{2} + \frac{\beta}{2} \right) \right) X^*\left(f\left(\frac{\beta}{2} \coth \frac{\beta}{2} - \frac{\beta}{2} \right) \right) S_{AH}(\beta)\, \frac{\beta/2}{\sinh \beta/2} e^{j2\pi t f \beta}\, d\beta$
	$= \|f\| \displaystyle\int X\left(f\frac{\beta/2}{\sinh \beta/2} e^{\beta/2} \right) X^*\left(f\frac{\beta/2}{\sinh \beta/2} e^{-\beta/2} \right) S_{AH}(\beta)\, \frac{\beta/2}{\sinh \beta/2} e^{j2\pi t f \beta}\, d\beta$
Altes	$Q_X(t, f) = f \displaystyle\int \int X(fe^{u/2}) X^*(fe^{-u/2}) e^{j2\pi t f u}\, du$
Unitary Bertrand P_0	$BP_0D_X(t, f; \mu_0) = f \displaystyle\int \int X\left(f\frac{u/2}{\sinh(u/2)} e^{u/2} \right) X^*\left(f\frac{u/2}{\sinh(u/2)} e^{-u/2} \right) \frac{u/2}{\sinh(u/2)} e^{j2\pi t f u}\, du,$
General Bertrand P_0	$BP_0D_X(t, f; \mu) = f \displaystyle\int \int X\left(f\frac{\beta/2}{\sinh \beta/2} e^{\beta/2} \right) X^*\left(f\frac{\beta/2}{\sinh \beta/2} e^{-\beta/2} \right) \mu(\beta) e^{j2\pi t f \beta}\, d\beta$
	$= \displaystyle\int X\left(f\left(\frac{\beta}{2} \coth \left(\frac{\beta}{2} \right) + \frac{\beta}{2} \right) \right) X^*\left(f\left(\frac{\beta}{2} \coth \left(\frac{\beta}{2} \right) - \frac{\beta}{2} \right) \right) \mu(\beta) e^{j2\pi t f \beta}\, d\beta$
Generalized Altes	$GQ_X(t, f; \tilde{\alpha}) = f \displaystyle\int \int e^{-\tilde{\alpha}u} X\left(fe^{(\frac{1}{2} - \tilde{\alpha})u} \right) X^*\left(fe^{-(\frac{1}{2} + \tilde{\alpha})u} \right) e^{j2\pi t f u}\, du$
Hyp. Choi-Williams	$HCWD_X(t, f; \sigma) = \displaystyle\int \int e^{-(2\pi \tau v)^2/\sigma}\, AF_{\mathcal{W}X}(\tau, v) e^{j2\pi(\tau v f/f_r - \tau f_r \ln f/f_r)}\, d\tau\, dv$

TABLE 12.13
(continued)

Hyperbolic TFR	Formula, $f > 0$
κth Hyp. Power Subclass	$HP_X^{(\kappa)}(t,f;G^{(H)}) = f \int \int X\left(f\left(\frac{\kappa\beta/2}{\sinh(\kappa\beta/2)}\right)^{1/\kappa} e^{\beta/2}\right) X^*\left(f\left(\frac{\kappa\beta/2}{\sinh(\kappa\beta/2)}\right)^{1/\kappa} e^{-\beta/2}\right)\left(\frac{\kappa\beta/2}{\sinh(\kappa\beta/2)}\right)^{1/\kappa} G^{(H)}(\beta)e^{j2\pi t f \beta}\,d\beta$
	$= \left|\frac{f}{\kappa}\right| \int \int g^{(H)}\left(\frac{f}{\kappa}[t-t']\right) BP_0 D_X^{(\kappa)}(t',f;\mu_0)\,dt'$
Hyperbologram	$HYP_X(t,f;\Gamma) = \left|\frac{f_r}{f}\right| \left\| \int_{-\infty}^{\infty} \int_0^{\infty} X(\xi)\Gamma^*\left(\frac{f_r}{f}\xi\right) e^{j2\pi t f \ln(\xi/f_r)}\,d\xi \right\|^2$
	$= \int_{-\infty}^{\infty} \int_0^{\infty} Q_\Gamma\left(\frac{1}{f_r}\frac{f}{f'}(t'f'-tf),\, f_r\frac{f'}{f}\right) Q_X(t',f')\,dt'\,df'$
Localized Hyp. Subclass	$LH_X(t,f;G^{(H)},F^{(H)}) = f \int \int X\left(f e^{-F^{(H)}(\beta)+\beta/2}\right) X^*\left(f e^{-F^{(H)}(\beta)-\beta/2}\right) e^{-F^{(H)}(\beta)} G^{(H)}(\beta)e^{j2\pi t f \beta}\,d\beta$
Power-Warp Hyp. Subclass	$PWH_X(t,f;s^{(H)}) = \int s^{(H)}(-\eta)GQ_X(t,f;\eta)\,d\eta$
Pseudo Altes	$PQ_X(t,f;\Gamma) = f_r \int_0^{\infty} Q_\Gamma\left(0, f_r\frac{f}{f'}\right) Q_X\left(\frac{tf}{f'},f'\right)\frac{df'}{f'}$
Smoothed Pseudo Altes	$SPQ_X(t,f;\Gamma,s) = \int s(tf-c)\,PQ_X\left(\frac{c}{f},f;\Gamma\right)dc$
	$= f_r \int \int_0^{\infty} s(tf-c)\,Q_\Gamma\left(0, f_r\frac{f}{f'}\right) Q_X\left(\frac{c}{f'},f'\right)dc\,\frac{df'}{f'}$
Warped Cohen Class	$H_X(t,f;\Psi_H^{(H)}) = C_{H_WX}\left(\frac{tf}{f_r}, f_r \ln\frac{f}{f_r};\Psi_{C_H}\right)$

TABLE 12.14
Kernels of the hyperbolic class of TFRs defined in Table 12.13. Here, HAF is the hyperbolic AF, $\mu_0(\beta) = \frac{\beta/2}{\sinh(\beta/2)}$, $\mu(\beta) = S_{AH}(\beta)\mu_0(\beta)$,

$V_\Gamma(b,\beta) = f_r e^b e^{b+\beta/2}\Gamma^*\left(f_r e^{b-\beta/2}\right)$, $F^{(\kappa)}\left(f_r e^b\right) = \frac{1}{\kappa}\ln\frac{\sinh(\kappa\beta/2)}{\kappa\beta/2}$, $u_\Gamma(c,\zeta) = \gamma(c+\zeta/2)\gamma^*(c-\zeta/2)$, $U_\Gamma(b,\beta) = \Gamma(b+\beta/2)\Gamma^*(b-\beta/2)$, and $v_\Gamma(c,\zeta)$ and $(WT)(f)$ are defined in (90) and (34), respectively.

TFR	$\psi_H^{(H)}(c,b)$	$\Psi_H^{(H)}(\zeta,\beta)$	$\varphi_H^{(H)}(c,\zeta)$	$\Phi_H^{(H)}(b,\beta)$
AH	$\int S_{AH}(\beta)\delta(b-\ln\frac{\sinh\beta/2}{\beta/2})$ $\times e^{j2\pi c\beta}\,d\beta$	$S_{AH}(\beta)$ $\times e^{j2\pi\zeta[\ln\frac{\sinh\beta/2}{\beta/2}]}$	$\int S_{AH}(\beta)$ $\times e^{j2\pi(\zeta[\ln\frac{\sinh\beta/2}{\beta/2}]+c\beta)}\,d\beta$	$S_{AH}(\beta)$ $\times\delta(b-\ln(\frac{\sinh\beta/2}{\beta/2}))$
Unitary BP_0D	$\int\delta(b+\ln\mu_0(\beta))e^{j2\pi c\beta}\,d\beta$	$e^{-j2\pi\zeta\ln\mu_0(\beta)}$	$\int e^{j2\pi(c\beta-\zeta\ln\mu_0(\beta))}\,d\beta$	$\delta(b+\ln\mu_0(\beta))$
General BP_0D	$\int\frac{\mu(\beta)}{\mu_0(\beta)}\delta(b+\ln\mu_0(\beta))e^{j2\pi c\beta}\,d\beta$	$\frac{\mu(\beta)}{\mu_0(\beta)}e^{-j2\pi\zeta\ln\mu_0(\beta)}$	$\int\frac{\mu(\beta)}{\mu_0(\beta)}e^{j2\pi(c\beta-\zeta\ln\mu_0(\beta))}\,d\beta$	$\frac{\mu(\beta)}{\mu_0(\beta)}\delta(b+\ln\mu_0(\beta))$
Power $BP_0D^{(\kappa)}$	$\int\delta(b-F^{(\kappa)}(\beta))e^{j2\pi c\beta}\,d\beta$	$e^{+j2\pi\zeta F^{(\kappa)}(\beta)}$	$\int e^{j2\pi(c\beta+\zeta F^{(\kappa)}(\beta))}\,d\beta$	$\delta(b-F^{(\kappa)}(\beta))$
GQ	$\frac{1}{|\bar\alpha|}e^{j2\pi cb/\bar\alpha}$	$e^{j2\pi\bar\alpha\zeta\beta}$	$\delta(c+\bar\alpha\zeta)$	$\delta(b-\bar\alpha\beta)$
$HP^{(\kappa)}$	$\int G^{(H)}(\beta)\delta(b-F^{(\kappa)}(\beta))$ $\times e^{j2\pi c\beta}\,d\beta$	$G^{(H)}(\beta)e^{j2\pi\zeta F^{(\kappa)}(\beta)}$	$\int G^{(H)}(\beta)$ $\times e^{j2\pi[\zeta F^{(\kappa)}(\beta)+c\beta]}\,d\beta$	$G^{(H)}(\beta)$ $\times\delta(b-\frac{1}{\kappa}\ln(\frac{\sinh(\kappa\beta/2)}{(\kappa\beta/2)}))$
HYP	$Q_\Gamma(\frac{-c}{f_r e^{-b}},f_r e^{-b})$ $= WD_{WT}(\frac{-c}{f_r},-bf_r)$	$HAF_\Gamma(-\zeta,-\beta)$ $= AF_{WT}(\frac{-\zeta}{f_r},-f_r\beta)$	$v_\Gamma(-c,-\zeta)$ $= \frac{1}{f_r}u_{WT}(-c/f_r,-\zeta/f_r)$	$V_\Gamma(-b,-\beta)$ $= f_r U_{WT}(-f_r b,-f_r\beta)$
LH	$\int G^{(H)}(\beta)\delta(b-F^{(H)}(\beta))e^{j2\pi c\beta}\,d\beta$	$G^{(H)}(\beta)e^{j2\pi\zeta F^{(H)}(\beta)}$	$\int G^{(H)}(\beta)e^{j2\pi[\zeta F^{(H)}(\beta)+c\beta]}\,d\beta$	$G^{(H)}(\beta)\delta(b-F^{(H)}(\beta))$
PQ	$f_r\delta(c)Q_\Gamma(0,f_r e^b)$ $= \delta(c/f_r)WD_{WT}(0,f_r b)$	$f_r v_\Gamma(0,\zeta)$ $= u_{WT}(0,\zeta/f_r)$	$f_r\delta(c)v_\Gamma(0,\zeta)$ $= \frac{1}{f_r}\delta(\frac{c}{f_r})u_{WT}(0,\zeta/f_r)$	$f_r Q_\Gamma(0,f_r e^b)$ $= f_r WD_{WT}(0,f_r b)$
PWH	$\int s^{(H)}(\eta)\frac{1}{|\eta|}e^{-j2\pi\frac{cb}{\eta}}\,d\eta$	$S^{(H)}(\zeta\beta)$	$\frac{1}{|\zeta|}s^{(H)}(\frac{c}{\zeta})$	$\frac{1}{|\beta|}s^{(H)}(-\frac{b}{\beta})$
Q	$\delta(c)\delta(b)$	1	$\delta(c)$	$\delta(b)$
SPQ	$f_r s(c)Q_\Gamma(0,f_r e^b)$ $= s(c/f_r)WD_{WT}(0,f_r b)$	$f_r S(\beta)v_\Gamma(0,\zeta)$ $= S(f_r\beta)u_{WT}(0,\zeta/f_r)$	$f_r s(c)v_\Gamma(0,\zeta)$ $= \frac{1}{f_r}s(c/f_r)u_{WT}(0,\zeta/f_r)$	$f_r S(\beta)Q_\Gamma(0,f_r e^b)$ $= f_r S(f_r\beta)WD_{WT}(0,f_r b)$
Warped C_H	$\psi_{C_H}(c/f_r,f_r b)$	$\Psi_{C_H}(\zeta/f_r,f_r\beta)$	$\frac{1}{f_r}\varphi_{C_H}(c/f_r,\zeta/f_r)$	$f_r\Phi_{C_H}(f_r b,f_r\beta)$

TABLE 12.15

κth power class TFRs. $v_X^{(\kappa)}(c, \zeta)$ and $V_X^{(\kappa)}(b, \beta)$ are power signal products defined in (104)–(105), and the power functions $\xi_\kappa(b)$ and $\tau_\kappa(f)$ are given in (98) and (109), respectively.

κ^{th} PC TFR	Formula		
Affine Class	$\mathcal{A}_\kappa(t, f; \Psi_A^{(A)})$	$=$	$PC_X^{(1)}(t, f; \Psi_A^{(A)})$
Hyperbolic-Power	$HP_X^{(\kappa)}(t, f; G^{(H)})$	$=$	$\left\|\dfrac{f}{\kappa}\right\| \displaystyle\int g^{(H)}\left(\dfrac{f}{\kappa}(t - t')\right) BP_0 D_X^{(\kappa)}(t', f; \mu_0)\, dt', \quad f > 0$
Localized-Power	$LP_X^{(\kappa)}(t, f; G^{(A)}, F^{(A)})$	$=$	$\left\|\dfrac{f}{\kappa}\right\| \displaystyle\int \dfrac{G^{(A)}(\beta)}{\left\|-F^{(A)2}(\beta) - \beta^2/4\right\|^{\frac{\kappa-1}{2\kappa}}} X\left(f\xi_\kappa^{-1}(-F^{(A)}(\beta) + \beta/2)\right)$
			$\times X^*\left(f\xi_\kappa^{-1}(-F^{(A)}(\beta) - \beta/2)\right) e^{j2\pi t f\beta/\kappa}\, d\beta$
Pow. Unitary P_0	$BP_0 D_X^{(\kappa)}(t, f; \mu_0)$	$=$	$f \displaystyle\int X\left(\left(f\left(\dfrac{\kappa\beta/2}{\sinh\kappa\beta/2}\right)^{1/\kappa} e^{\beta/2}\right) X^*\left(f\left(\dfrac{\kappa\beta/2}{\sinh\kappa\beta/2}\right)^{1/\kappa} e^{-\beta/2}\right)\left(\dfrac{\kappa\beta/2}{\sinh\kappa\beta/2}\right)^{1/\kappa} e^{j2\pi t f\beta}\, d\beta\right.$
Bertrand P_k	$BP_k D_X(t, f; \mu)$	$=$	$f \displaystyle\int X(f\lambda_k(u)) X^*(f\lambda_k(-u))\mu(u) e^{j2\pi t f(\lambda_k(u) - \lambda_k(-u))}\, du, \quad f > 0$
			with $\lambda_k(u) = \left[k\dfrac{e^{-u} - 1}{e^{-ku} - 1}\right]^{\frac{1}{k-1}}, k \neq 0, 1, \lambda_1(u) = \exp\left[1 + \dfrac{ue^{-u}}{e^{-u} - 1}\right]$, and $\mu(u) = \mu^*(-u)$
Power Flandrin	$FD_X^{(\kappa)}(t, f)$	$=$	$\left\|\xi_\kappa\left(\dfrac{f}{f_r}\right)\right\| \displaystyle\int V_X^{(\kappa)}\left(\xi_\kappa\left(\dfrac{f}{f_r}\right)\left[1 + \left(\dfrac{\beta}{4}\right)^2\right], \xi_\kappa\left(\dfrac{f}{f_r}\right)\beta\right)\left[1 - \left(\dfrac{\beta}{4}\right)^2\right] e^{j2\pi t f\beta/\kappa}\, d\beta$

TABLE 12.15

κth power class TFRs. $v_X^{(\kappa)}(c,\zeta)$ and $V_X^{(\kappa)}(b,\beta)$ are power signal products defined in (104)–(105), and the power functions $\xi_\kappa(b)$ and $\tau_\kappa(f)$ are given in (98) and (109), respectively.

κ^{th} PC TFR	Formula	
Powergram	$POW_X^{(\kappa)}(t,f;\Gamma)$	$= \dfrac{f_r}{\|f\|}\left\|\int X(f')\Gamma^*(f_r f'/f)e^{j2\pi\frac{t}{\tau_\kappa(f)}\xi_\kappa(f'/f_r)}\,df'\right\|^2$
Pow. Unterberger Act.	$UAD_X^{(\kappa)}(t,f)$	$= \|\xi_\kappa(f/f_r)\|\int V_X^{(\kappa)}(\xi_\kappa(f/f_r)[\sqrt{1+(\beta/2)^2}\,,\,\xi_\kappa(f/f_r)\beta)e^{j2\pi t f\beta/\kappa}\,d\beta$
Pow. Unterberger Pass.	$UPD_X^{(\kappa)}(t,f)$	$= \|\xi_\kappa(f/f_r)\|\int V_X^{(\kappa)}(\xi_\kappa(f/f_r)\sqrt{1+(\beta/2)^2}\,,\,\xi_\kappa(f/f_r)\beta)\dfrac{1}{\sqrt{1+(\beta/2)^2}}e^{j2\pi t f\beta/\kappa}\,d\beta$
Power Generalized	$GWD_X^{(\kappa)}(t,f;\tilde\alpha)$	$= \dfrac{f_r}{\|\kappa\|}\int\dfrac{1}{\|(\xi_\kappa(f/f_r)-\tilde\alpha\beta)^2-\beta^2/4\|^{\frac{\kappa-1}{2\kappa}}}X(f_r\xi_\kappa^{-1}(\xi_\kappa(f/f_r)-\tilde\alpha\beta+\beta/2))$
		$\times X^*(f_r\xi_\kappa^{-1}(\xi_\kappa(f/f_r)-\tilde\alpha\beta-\beta/2))e^{j2\pi\frac{t}{\tau_\kappa(f)}\beta}\,d\beta$
Power Wigner	$WD_X^{(\kappa)}(t,f)$	$= \left\|\dfrac{f}{\kappa}\right\|\int\dfrac{1}{\|1-\beta^2/4\|^{\frac{\kappa-1}{2\kappa}}}X(f\xi_\kappa^{-1}(1+\beta/2))X^*(f\xi_\kappa^{-1}(1-\beta/2))e^{j2\pi t f\beta/\kappa}\,d\beta$

TABLE 12.16

Normal form kernels of the power-class TFRs defined in Table 12.15. Here, $AF_X^{(\kappa)}(\zeta, \beta)$, $v_\Gamma^{(\kappa)}(c, \zeta)$, $V_\Gamma^{(\kappa)}(b, \beta)$, and $(\mathcal{W}_\kappa X)(f)$ are defined in (107), (104), (105), and (110), respectively, and $U_\Gamma(b, \beta) = \Gamma(b + \beta/2)\Gamma^*(b - \beta/2)$.

PC TFR	$\psi_{PC}^{(A)}(\zeta, \beta)$	$\Psi_{PC}^{(A)}(\zeta, \beta)$	$\varphi_{PC}^{(A)}(c, \zeta)$	$\Phi_{PC}^{(A)}(b, \beta)$
Unitary $BP_0 D^{(\kappa)}$	$\int \frac{\beta/2}{\sinh \beta/2}\, \delta(b + \frac{\beta}{2}\coth(\beta/2))\, e^{j2\pi c\beta}\, d\beta$	$\frac{\beta/2}{\sinh \beta/2}\, e^{-j2\pi\zeta \frac{\beta}{2}\coth\frac{\beta}{2}}$	$\int \frac{\beta/2}{\sinh \beta/2}\, e^{j2\pi(c\beta - \zeta\frac{\beta}{2}\coth\beta/2)}\, d\beta$	$\frac{\beta/2}{\sinh \beta/2}\, \delta(b + \frac{\beta}{2}\coth\frac{\beta}{2})$
$LP^{(\kappa)}$	$\int G^{(A)}(\beta)\delta(b - F^{(A)}(\beta))e^{j2\pi c\beta}\, d\beta$	$G^{(A)}(\beta)e^{j2\pi \zeta F^{(A)}(\beta)}$	$\int G^{(A)}(\beta)e^{j2\pi(\zeta F^{(A)}(\beta)+c\beta)}\, d\beta$	$G^{(A)}(\beta)\delta(b - F^{(A)}(\beta))$
$FD^{(\kappa)}$	$\int (1 - (\frac{\beta}{4})^2)\delta(b + (1 + (\frac{\beta}{4})^2)e^{j2\pi c\beta}\, d\beta$	$[1 - (\frac{\beta}{4})^2]e^{-j2\pi\zeta(1+(\frac{\beta}{4})^2)}$	$\int (1 - (\frac{\beta}{4})^2)e^{j2\pi(c\beta - \zeta(1+(\frac{\beta}{4})^2))}\, d\beta$	$[1 - (\frac{\beta}{4})^2]\delta(b + (1 + (\frac{\beta}{4})^2))$
$GWD^{(\kappa)}$	$\frac{1}{|\tilde{a}|}\, e^{j2\pi c(b+1)/\tilde{a}}$	$e^{j2\pi c(\tilde{a}\beta - 1)}$	$e^{-j2\pi\zeta}\, \delta(c + \tilde{a}\zeta)$	$\delta(b + 1 - \tilde{a}\beta)$
$LP^{(\kappa)}$	$\int G^{(A)}(\beta)\, \delta(b - F^{(A)}(\beta))\, e^{j2\pi c\beta}\, d\beta$	$G^{(A)}(\beta)\, e^{j2\pi\zeta F^{(A)}(\beta)}$	$\int G^{(A)}(\beta)\, e^{j2\pi(c\beta + \zeta F^{(A)}(\beta))}\, d\beta$	$G^{(A)}(\beta)\delta(b - F^{(A)}(\beta))$
$POW^{(\kappa)}$	$WD_\Gamma^{(\kappa)}(-\tau_\kappa(f_r\xi_\kappa^{-1}(b))c, -f_r\xi_\kappa^{-1}(b))$ $= WD_{\mathcal{W}_\kappa\Gamma}(-c/f_r, -f_r b)$	$AF_\Gamma^{(\kappa)}(-\zeta, -\beta)$ $= AF_{\mathcal{W}_\kappa\Gamma}(-\zeta/f_r, -f_r\beta)$	$v_\Gamma^{(\kappa)}(-c, -\zeta)$	$V_\Gamma^{(\kappa)}(-b, -\beta)$ $= f_r U_{\mathcal{W}_\kappa\Gamma}(-bf_r, -f_r\beta)$
$UAD^{(\kappa)}$	$\int \delta\left(b + \sqrt{1 + (\frac{\beta}{2})^2}\right) e^{j2\pi c\beta}\, d\beta$	$e^{-j2\pi\zeta\sqrt{1+(\frac{\beta}{2})^2}}$	$\int e^{j2\pi\left(c\beta - \zeta\sqrt{1+(\frac{\beta}{2})^2}\right)}\, d\beta$	$\delta\left(b + \sqrt{1 + (\frac{\beta}{2})^2}\right)$
$UPD^{(\kappa)}$	$\int [1+(\frac{\beta}{2})^2]^{-1/2}\delta\left(b + \sqrt{1 + (\frac{\beta}{2})^2}\right)$ $\times e^{j2\pi c\beta}\, d\beta$	$[1 + (\frac{\beta}{2})^2]^{-1/2}$ $\times e^{-j2\pi\zeta\sqrt{1+(\frac{\beta}{2})^2}}$	$\int [1 + (\frac{\beta}{2})^2]^{-1/2}$ $\times e^{j2\pi\left(c\beta - \zeta\sqrt{1+\frac{\beta}{2}^2}\right)}\, d\beta$	$[1 + (\frac{\beta}{2})^2]^{-1/2}$ $\times\delta\left(b + \sqrt{1 + (\frac{\beta}{2})^2}\right)$
$WD^{(\kappa)}$	$\delta(c)\, \delta(b + 1)$	$e^{-j2\pi\zeta}$	$e^{-j2\pi\zeta}\, \delta(c)$	$\delta(b + 1)$

[73], [100]. These TFRs are scale covariant (property P_3) and power time-shift covariant, i.e.,

$$PC_Y^{(\kappa)}(t, f) = PC_X^{(\kappa)}\left(t - c\frac{d}{df}\xi_\kappa(f/f_r), f\right) \quad \text{for} \quad Y(f) = e^{-j2\pi c\xi_\kappa(f/f_r)}X(f) \tag{97}$$

$$= PC_X^{(\kappa)}\left(t - c\frac{\kappa}{f_r}\left|\frac{f}{f_r}\right|^{\kappa-1}, f\right)$$

where

$$\xi_\kappa(f) = \text{sgn}(f)|f|^\kappa \quad \text{for } \kappa \neq 0 \tag{98}$$

is a one-to-one phase function involving the κth power of frequency, f [100], [73], [98]. Consequently, the κth power class perfectly represents dispersive group delay changes in the signal that are proportional to powers of frequency. When $\kappa = 1$, the power class is equivalent to the affine class in (78). Any member of the power class can be written in the following four normal forms and the corresponding bi-frequency form:

$$PC_X^{(\kappa)}(t, f; \Psi_{PC}^{(A)}) = \left|\xi_\kappa\left(\frac{f}{f_r}\right)\right|\int\int \varphi_{PC}^{(A)}\left(\xi_\kappa\left(\frac{f}{f_r}\right)\left[\frac{t}{\tau_\kappa(f)} - c\right], \xi_\kappa\left(\frac{f}{f_r}\right)\zeta\right)v_X^{(\kappa)}(c, \zeta)\,dc\,d\zeta \tag{99}$$

$$= \frac{1}{\left|\xi_\kappa\left(\frac{f}{f_r}\right)\right|}\int\int \Phi_{PC}^{(A)}\left(\frac{-b}{\xi_\kappa\left(\frac{f}{f_r}\right)}, \frac{\beta}{\xi_\kappa\left(\frac{f}{f_r}\right)}\right)V_X^{(\kappa)}(b, \beta)e^{j2\pi\frac{t}{\tau_\kappa(f)}\beta}\,db\,d\beta \tag{100}$$

$$= \int\int \psi_{PC}^{(A)}\left(\xi_\kappa\left(\frac{f}{f_r}\right)\left[\frac{-t'}{\tau_\kappa(f')} + \frac{t}{\tau_\kappa(f)}\right], -\xi_\kappa\left(\frac{f'}{f}\right)\right)WD_X^{(\kappa)}(t', f')\,dt'\,df' \tag{101}$$

$$= \int\int \Psi_{PC}^{(A)}\left(\xi_\kappa\left(\frac{f}{f_r}\right)\zeta, \frac{\beta}{\xi_\kappa\left(\frac{f}{f_r}\right)}\right)AF_X^{(\kappa)}(\zeta, \beta)e^{j2\pi\frac{t}{\tau_\kappa(f)}\beta}\,d\zeta\,d\beta \tag{102}$$

$$= \frac{1}{|f|}\int\int \Gamma_{PC}^{(A)(\kappa)}\left(\frac{f_1}{f}, \frac{f_2}{f}\right)e^{j2\pi\frac{t}{\tau_\kappa(f)}\left[\xi_\kappa\left(\frac{f_1}{f_r}\right)-\xi_\kappa\left(\frac{f_2}{f_r}\right)\right]}X(f_1)X^*(f_2)\,df_1\,df_2. \tag{103}$$

Below are the definitions of the κth signal product,

$$v_X^{(\kappa)}(c, \zeta) = \rho_X^{(\kappa)}(c + \zeta/2)\rho_X^{(\kappa)*}(c - \zeta/2), \quad \text{with} \tag{104}$$

$$\rho_X^{(\kappa)}(c) = \int X(f)\sqrt{|\tau_\kappa(f)|}e^{j2\pi c\xi_\kappa(f/f_r)}\,df,$$

the κth power signal spectrum product,

$$V_X^{(\kappa)}(b, \beta) = \frac{f_r}{|\kappa||b^2 - \beta^2/4|^{\frac{\kappa-1}{2\kappa}}}X\left(f_r\xi_\kappa^{-1}(b + \beta/2)\right)X^*\left(f_r\xi_\kappa^{-1}(b - \beta/2)\right) \tag{105}$$

$$= f_r(\mathcal{W}_\kappa X)([b + \beta/2]f_r)(\mathcal{W}_\kappa X)^*([b - \beta/2]f_r)$$

the κth central member,

$$WD_X^{(\kappa)}(t, f) = WD_{\mathcal{W}_\kappa X}\left(\frac{t}{f_r\tau_\kappa(f)}, f_r\xi_\kappa\left(\frac{f}{f_r}\right)\right) \tag{106}$$

$$= \left|\frac{f}{\kappa}\right|\int \frac{1}{\left|1 - \frac{\beta^2}{4}\right|^{\frac{\kappa-1}{2\kappa}}}X\left(f\xi_\kappa^{-1}\left(1 + \frac{\beta}{2}\right)\right)$$

$$\times X^*\left(f\xi_\kappa^{-1}\left(1 - \frac{\beta}{2}\right)\right)e^{j2\pi\frac{tf}{\kappa}\beta}\,d\beta,$$

the κth power ambiguity function,

$$AF_X^{(\kappa)}(\zeta, \beta) = AF_{W_\kappa X}\left(\frac{\zeta}{f_r}, f_r\beta\right) \tag{107}$$

$$= \int \frac{f_r}{|\kappa| |b^2 - \frac{\beta^2}{4}|^{\frac{\kappa-1}{2\kappa}}} X\left(f_r\xi_\kappa^{-1}(b + \beta/2)\right)$$

$$\times X^*\left(f_r\xi_\kappa^{-1}\left(b - \frac{\beta}{2}\right)\right) e^{j2\pi\zeta b} \, db,$$

the kth power inverse phase function,

$$\xi_\kappa^{-1}(b) = \text{sgn}(b)|b|^{1/\kappa}, \qquad b \in \Re, \quad \kappa \neq 0, \tag{108}$$

the κth power group delay,

$$\tau_\kappa(f) = \frac{d}{df}\xi_\kappa\left(\frac{f}{f_r}\right) = \frac{\kappa}{f_r}\left|\frac{f}{f_r}\right|^{\kappa-1}, \qquad f \in \Re \text{ for } \kappa \neq 0, \tag{109}$$

and the kth power signal warping,

$$(W_\kappa X)(f) = \frac{1}{\sqrt{f_r\left|\tau_\kappa\left(f_r\xi_\kappa^{-1}\left(\frac{f}{f_r}\right)\right)\right|}} X\left(f_r\xi_\kappa^{-1}\left(\frac{f}{f_r}\right)\right)$$

$$= \frac{1}{\sqrt{\kappa}|f/f_r|^{\frac{\kappa-1}{2\kappa}}} X\left(f_r\text{sgn}(f)|f/f_r|^{1/\kappa}\right). \tag{110}$$

The bi-frequency kernel in (103) is related to the normal form kernels as follows:

$$\Gamma_{PC}^{(A)(\kappa)}(b_1, b_2) = \begin{cases} |\xi_\kappa'(\sqrt{|b_1 b_2|})| \Phi_{PC}^{(A)}\left(-\frac{\xi_\kappa(b_1) + \xi_\kappa(b_2)}{2}, \xi_\kappa(b_1) - \xi_\kappa(b_2)\right), & \kappa \neq 0 \\ \Gamma_{PC}^{(A)}(b_1, b_2), & \kappa = 1. \end{cases}$$

A list of power-class TFRs is given in Table 12.15. These PC TFRs are well matched to power chirps and dispersive power time shifts, e.g.,

$$X(f) = \sqrt{\tau_X(f)}e^{-j2\pi\xi_\kappa(f/f_r)} \Rightarrow WD_X^{(\kappa)}(t, f) = |\tau_\kappa(f)|\delta(t - c\tau_\kappa(f)).$$

The best known among them are the Bertrand P_k distributions. The PC TFR kernels for use in (99)–(102) are given in Table 12.16. The last column in Table 12.7 lists the kernel constraints for a given power class TFR to have ideal properties [100]. All TFRs, kernels, and properties of the power class correspond to a warped version of the TFRs, kernels, and properties, respectively, of the affine class. Every TFR, $\mathcal{A}_{PC}$, in the affine class that is covariant to scale changes and constant time shifts, maps to a corresponding power-class TFR

$$PC_X^{(\kappa)}(t, f; \Psi_{PC}^{(A)}) = \mathcal{A}_{W_\kappa X}\left(\frac{t}{f_r\tau_\kappa(f)}, f_r\xi_\kappa\left(\frac{f}{f_r}\right); \Psi_{\mathcal{A}_{PC}}^{(A)}\right) \tag{111}$$

that is covariant to scale changes and the dispersive time shifts in (97) that are proportional to powers of frequency. Just as in Cohen's-class to hyperbolic-class mapping in (94), the transformation in (111) involves a unitary warping of the signal and a scaling of the time-frequency axes. Likewise, affine-class TFR kernels and their corresponding PC TFR kernels are equivalent, provided any analysis windows are pre-warped appropriately. For example, the affine-class Unterberger active distribution, $UAD_X(t, f)$, in Table 12.11 maps to the power-class TFR, $UAD_X^{(\kappa)}(t, f)$, in Table 12.15. Their respective kernels in Tables 12.12 and 12.16 are identical. The scalogram in Table 12.11 maps to the powergram in Table 12.15; their respective kernels in Tables 12.12 and 12.16 differ only in the power warping in (110) of the analysis window spectrum $\Gamma(f)$.

12.6   Summary

This chapter has focused on linear and quadratic time-frequency representations and the different types of properties that they should satisfy. Properties of the Fourier transform, Mellin transform, and unitary operators were used to understand the relative merits of these time-varying spectral representations. No known TFR is ideal for all applications, so the search continues for new ones. There are many other TFRs in the literature that are highly nonlinear functions of the signal; they have been proposed to adapt automatically to changes in the signal, to be always nonnegative, to extend the concepts of higher order cummulants and spectra, or to solve time-varying constrained optimization problems [1], [4], [103], [116], [24]. Recent research has focused on the mapping of known TFRs to multidimensional representations of arbitrary variables [44].

Acknowledgements

The author would like to acknowledge the help given by Dr. A. Papandreou-Suppappola proof-reading this chapter, by Mr. S. Praveenkumar and Ms. R. Murray in formatting some of the tables and by Dr. Papandreow-Suppappola and Dr. A. Costa in generating the figures.

References

[1] *Proceedings IEEE-SP International Symposium on Time-frequency and Time-scale' Analysis*, Victoria, Canada, October 1992.

[2] Special issue on wavelet transforms and multiresolution signal analysis, in *IEEE Trans. Info. Th.*, 38, 1992.

[3] Special issue on wavelets and signal processing, in *IEEE Trans. Signal Proc.*, 1993.

[4] *Proceedings IEEE-SP International Symposium on Time-frequency and Time-scale Analysis*, Philadelphia, PA, October 1994.

[5] R. A. Altes, Detection, estimation, and classification with spectrograms, *Journal of the Acoustical Society of America*, 67, pp. 1232–1246, April 1980.

[6] R. A. Altes, Wide-band, proportional-bandwidth Wigner-Ville analysis, *IEEE Transactions on Acoustics, Speech, and Signal Processing*, 38(6), pp. 1005–1012, June 1990.

[7] R. A. Altes and E. L. Titlebaum, Bat signals as optimally Doppler tolerant waveforms, *Journal of the Acoustical Society of America*, 48(4), pp. 1014–1020, October 1970.

[8] M. G. Amin, Time-frequency spectrum analysis and estimation for non-stationary ran-

dom processes, in *Time-Frequency Signal Analysis*, B. Boashash (Ed.), pp. 208–232, Longman Cheshire, Melbourne, Australia, 1992.

[9] J. C. Anderson. Speech analysis/synthesis based on perception, Technical Report 707, MIT Lincoln Laboratory, Lexington, MAA, November 1984.

[10] F. Auger, *Représentations temps-fréquence des signaux non-stationnaires: Synthèse et contribution*, Ph.D. thesis, Universite de Nantes, Nantes, France, December 1991.

[11] F. Auger, Some simple parameter determination rules for the generalized Choi-Williams and Butterworth distributions, *IEEE Signal Proc. Letters*, 1994.

[12] F. Auger and C. Doncarli, Quelques commentaires sur des repreésentations temps-fréquence proposées récemment, *Traitement du Signal*, 9(1), pp. 3–25, 1992.

[13] F. Auger and P. Flandrin, The why and how of time-frequency reassignment, *IEEE-SP International Symposium on Time-Frequency and Time-Scale Analysis*, pp. 197–200, Philadelphia, Pennsylvania, October 1994.

[14] R. G. Baraniuk, Warped perspectives in time-frequency analysis, *IEEE-SP International Symposium on Time-Frequency and Time-Scale Analysis*, pages 528–531, Philadelphia, Pennsylvania, October 1994.

[15] R. G. Baraniuk and D. L. Jones, Unitary equivalence: A new twist on signal processing, *IEEE Transactions on Signal Processing*, to appear.

[16] R. G. Baraniuk and D. L. Jones, Shear madness: New orthonormal bases and frames using chirp functions, *IEEE Transactions on Signal Processing*, 41, pp. 3543–3549, December 1993.

[17] R. G. Baraniuk and D. L. Jones, A signal-dependent time-frequency representation: Optimal kernel design, *IEEE Transactions on Signal Processing*, 41, pp. 1589–1602, April 1993.

[18] M. Basseville, P. Flandrin, and N. Martin, Signaux non-stationnaires, analyse temps-fréquence et segmentation. fiches descriptives d'algorithmes, *Traitement du Signal*, 9(1 supplement), 1992.

[19] J. Bertrand and P. Bertrand, Affine time-frequency distributions, in *Time-Frequency Signal Analysis—Methods and Applications*, B. Boashash (Ed.), Chapter 5, pp. 118–140, Longman-Cheshire, Melbourne, Australia, 1992.

[20] J. Bertrand and P. Bertrand, A class of affine Wigner functions with extended covariance properties, *Journal of Mathematics and Physics*, 33, pp. 2515–2527, 1992.

[21] J. Bertrand, P. Bertrand, and J. P. Ovarlez, Discrete Mellin transform for signal analysis, *Proceedings IEEE International Conference on Acoustics, Speech, and Signal Processing*, pages 1603–1606, Albuquerque, New Mexico, April 1990.

[22] R. E. Blahut, W. Miller, and C. H. Wilcox, *Radar and Sonar, Part I*, Springer-Verlag, New York, NY, 1991.

[23] B. Boashash, Time-frequency signal analysis, in *Advances in Spectrum Estimation*, pp. 418–517, S. Haykin (Ed.), Prentice Hall, Inc., Englewood Cliffs, NJ, 1990.

[24] B. Boashash, editor, *Time-Frequency Signal Analysis—Methods and Applications*. Longman-Cheshire, Melbourne, Australia, 1992.

[25] G. F. Boudreaux-Bartels, *Time-frequency signal processing algorithms: Analysis and synthesis using Wigner distributions*, Ph.D. thesis, Rice University, Houston, Texas, December 1983.

[26] G. F. Boudreaux-Bartels, Time-varying signal processing using the Wigner-distribution time-frequency signal representation, in *Advances in Geophysical Data Processing*,

Vol 2. Two Dimensional Transforms, M. Simaan (Ed.), pp. 33–79, JAI Press Inc., Greenwich, CT, 1985.

[27] G. F. Boudreaux-Bartels, On the use of operators vs. warpings vs. axiomatic derivations of new time-frequency-scale (operator) representations, *Proceedings Twenty-Eighth Asilomar Conference on Signals, Systems and Computers*, Pacific Grove, CA, October/November 1994.

[28] G. F. Boudreaux-Bartels, Time-varying signal processing using Wigner distribution synthesis techniques, in *The Wigner Distribution—Theory and Applications in Signal Processing*, W. Mecklenbräuker (Ed.), North Holland Elsevier Science Publishers, to be published, 1995.

[29] G. F. Boudreaux-Bartels, Useful equations for time-frequency-scale representations, Technical Report 0895-0001, University of Rhode Island, Electrical Engg. Dept., Kingston, RI, 1995.

[30] G. F. Boudreaux-Bartels and F. Hlawatsch, References for the short-time Fourier transform, Gabor expansion, wavelet transform, Wigner distribution, ambiguity function, and other time-frequency and time-scale signal representations, Technical Report 0492-0001, University of Rhode Island, Electrical Engg. Dept., Kingston, RI, April 1992.

[31] G. F. Boudreaux-Bartels and R. L. Murray, Time-frequency signal representations for biomedical signals, in *Biomedical Engineering Handbook*, Chapter 57, pp. 866–885, CRC Press, Inc., Boca Raton, FL, 1995.

[32] R. N. Bracewell, *The Fourier Transform and its Applications*, 2nd ed., revised, McGraw-Hill Book Company, 1986.

[33] E. O. Brigham, *The Fast Fourier Transform and its Applications*, Prentice-Hall, Inc., Englewood Cliffs, NJ, 1988.

[34] K. G. Canfield and D. L. Jones, Implementing time-frequency representations for Non-Cohen Classes, *Proceedings Twenty-Seventh Asilomar Conference on Signals, Systems and Computers*, Pacific Grove, CA, November 1993.

[35] H. I. Choi and W. J. Williams, Improved time-frequency representation of multicomponent signals using exponential kernels, *IEEE Transactions on Acoustics, Speech, and Signal Processing*, 37, pp. 862–871, June 1989.

[36] C. K. Chui (Ed.), *Wavelets: A Tutorial in Theory and Applications*, Academic Press, Inc., Boston, MA, 1992.

[37] T. A. C. M. Claasen and W. F. G. Mecklenbräuker, The Wigner distribution—A tool for time-frequency signal analysis, Part I: Continuous-time signals, *Philips Journal of Research*, 35, pp. 217–250, 1980.

[38] T. A. C. M. Claasen and W. F. G. Mecklenbräuker, The Wigner distribution—A tool for time-frequency signal analysis, Part II: Discrete-time signals, *Philips Journal of Research*, 35, pp. 276–300, 1980.

[39] T. A. C. M. Claasen and W. F. G. Mecklenbräuker. The Wigner distribution—A tool for time-frequency signal analysis, Part III: Relations with other time-frequency signal transformations, *Philips Journal of Research*, 35, pp. 372–389, 1980.

[40] L. Cohen, Generalized phase-space distribution functions, *Journal of Mathematics and Physics*, 7, pp. 781–786, 1966.

[41] L. Cohen, Time-frequency distribution—A review, *Proceedings IEEE*, 77, pp. 941–981, July 1989.

[42] L. Cohen, A primer on time-frequency analysis, in *Time-Frequency Signal Analysis*,

B. Boashash (Ed.), pp. 3–42, John Wiley and Sons, Inc., New York, NY, 1992.

[43] L. Cohen, *Time-Frequency Analysis*, Prentice-Hall, Inc., Englewood Cliffs, NJ, 1995.

[44] L. Cohen and R. G. Baraniuk, Joint distributions for arbitrary variables, *IEEE-SP International Symposium on Time-Frequency and Time-Scale Analysis*, pp. 520–523, Philadelphia, Pennsylvania, October 1994.

[45] A. H. Costa, *Multiform, tiltable time-frequency representations and masked auto Wigner distribution synthesis*, Ph.D. thesis, University of Rhode Island, Kingston, RI, May 1994.

[46] G. Courbebaisse, Transformée bilinëaire temps-echelle des signaux asymptotiques d'énergie finie, in *14ème Coll. GRETSI*, Juan-les-Pins, France, 1993.

[47] R. E. Crochiere and L. R. Rabiner, *Multi-Rate Digital Signal Processing*, Prentice Hall, Inc., Englewood Cliffs, New Jersey, 1983.

[48] G. S. Cunningham and W. J. Williams, Fast implementation of generalized discrete time-frequency distributions, *IEEE Trans. on Signal Proc.*, 42, pp. 1496–1508, 1994.

[49] I. Daubechies, *Ten Lectures on Wavelets*, Society for Industrial and Applied Mathematics, Philadelphia, PA, 1992.

[50] N. G. DeBruijn, A theory of generalized functions with applications to Wigner distribution and Weyl correspondence, *Nieuw Archief voor Wiskunde*, 21(3), pp. 205–280, 1973.

[51] G. Eichmann and N. M. Marinovich, Scale-invariant Wigner distribution and ambiguity functions, *Proc. SPIE*, 519, pp. 18–24, 1985.

[52] P. Flandrin, Some features of time-frequency representations of multicomponent signals, *Proceedings IEEE International Conference on Acoustics, Speech, and Signal Processing*, pp. 41B.4.1–4, San Diego, CA, March 1984.

[53] P. Flandrin, Time-frequency processing of bat sonar signals, in *Animal Sonar Systems Symposium*, Hilsinger, Denmark, September 1986.

[54] P. Flandrin, *Représentations temps-fréquence des signaux non-stationnaires*, Ph.D. thesis, Institut National Polytechnique de Grenoble, France, 1987.

[55] P. Flandrin, Maximum signal energy concentration in a time-frequency domain, in *icassp*, pp. 2176–2179, 1988.

[56] P. Flandrin, A time-frequency formulation of optimum detection, *IEEE Transactions on Acoustics, Speech, and Signal Processing*, 36, pp. 1377–1384, September 1988.

[57] P. Flandrin, Scale-invariant Wigner spectra and self-similarity, *Proceedings European Signal Processing Conference, EUSIPCO–90*, pp. 149–152, Barcelona, Spain, September 1990.

[58] P. Flandrin, Fractional Brownian motion and wavelets, in *Wavelets, Fractals and Fourier transforms—New Developments and New Applications*, M. Farge, J. C. R. Hunt, and J. C. Vassilicos (Eds.), Oxford University Press, 1991.

[59] P. Flandrin, On the time-scale analysis of self similar processes, In *The Role of Wavelets in Signal Processing Applications*, B. W. Suter, M. E. Oxley, and G. T. Warhola (Eds.), number AFIT/EN-TR-92-3, pp. 137–150, 1992.

[60] P. Flandrin, *Temps-Fréquence*, Hermès, Paris, 1993.

[61] P. Flandrin and B. Escudié, Time and frequency representation of finite energy signals: a physical property as a result of a Hilbertian condition, *Signal Processing*, 2, pp. 93–100, 1980.

[62] P. Flandrin and P. Gonçalvès, Geometry of affine distributions, *IEEE-SP International*

Symposium on Time-Frequency and Time-Scale Analysis, pp. 80–83, Philadelphia, PA, October 1994.

[63] P. Flandrin and O. Rioul, Affine smoothing of the Wigner-Ville distribution, *Proceedings IEEE International Conference on Acoustics, Speech, and Signal Processing*, pp. 2455–2458, Albuquerque, NM, April 1990.

[64] D. Gabor, Theory of communication, *Journal of the IEE*, 93(III), pp. 429–457, November 1946.

[65] P. Gonçalvès, *Représentations temps-fréquence et temps-échelle bilinéaires: synthèse et contributions*, Ph.D. thesis, Institut National Polytechnique de Grenoble, France, November 1993.

[66] F. Hlawatsch, *A study of bilinear time-frequency signal representations with applications to time-frequency signal synthesis*, Ph.D. thesis, Technische Universität Wien, Vienna, Austria, 1988.

[67] F. Hlawatsch, Duality and classification of bilinear time-frequency signal representations, *IEEE Transactions on Signal Processing*, 39(7), pp. 1564–1574, July 1991.

[68] F. Hlawatsch, Time-frequency methods for signal processing, Technical Report 1291–0001, University of Rhode Island, Department of Electrical Engineering, Kingston, Rhode Island, December 1991.

[69] F. Hlawatsch and G. F. Boudreaux-Bartels, Linear and quadratic time-frequency signal representations, *IEEE Signal Processing Magazine*, 9(2), pp. 21–67, April 1992.

[70] F. Hlawatsch and P. Flandrin, The interference structure of the Wigner distribution and related time-frequency signal representations, in *The Wigner Distribution—Theory and Applications in Signal Processing*, W. Mecklenbräuker (Ed.), North Holland Elsevier Science Publishers, to be published, 1995.

[71] F. Hlawatsch and W. Krattenthaler, Bilinear signal synthesis, *IEEE Transactions on Signal Processing*, 40(2), pp. 352–363, February 1992.

[72] F. Hlawatsch and W. Krattenthaler, Signal synthesis algorithms for bilinear time-frequency signal representations, in *The Wigner Distribution—Theory and Applications in Signal Processing*, W. Mecklenbräuker (Ed.), North Holland Elsevier Science Publishers, to be published, 1995.

[73] F. Hlawatsch, A. Papandreou, and G. F. Boudreaux-Bartels, The Power Classes of quadratic time-frequency representations: A generalization of the Affine and Hyperbolic Classes, *Proceedings Twenty-Seventh Asilomar Conference on Signals, Systems and Computers*, pp. 1265–1270, Pacific Grove, CA, November 1993.

[74] F. Hlawatsch and R. L. Urbanke, Bilinear time-frequency representations of signals: The shift-scale invariant class, *IEEE Transactions on Signal Processing*, 42(2), pp. 357–366, February 1994.

[75] C. P. Janse and A. J. M. Kaizer, Time-frequency distributions of loudspeakers: The application of the Wigner distribution, *J. Audio Engineering Soc.*, 31, pp. 198–223, April 1983.

[76] J. Jeong and W. J. Williams, On the cross-terms in spectrograms, *Proc. IEEE Int. Symp. Ckts and Syst.*, pp. 1565–1568, 1990.

[77] J. Jeong and W. J. Williams, Time-varying filtering and signal synthesis using the extended discrete-time Wigner distribution. In B. Boashash and P. Boles, editors, *Proc. ISSPA90, Sig. Proc., Theories, Impl. and Appl.*, pp. 895–898, 1990.

[78] J. Jeong and W. J. Williams, Alias-free generalized discrete-time time-frequency dis-

tributions, *IEEE Transactions on Signal Processing*, 40, pp. 2757–2765, November 1992.

[79] J. Jeong and W. J. Williams, Kernel design for reduced interference distributions, *IEEE Transactions on Signal Processing*, 40(2), pp. 402–412, February 1992.

[80] D. L. Jones and T. W. Parks, A resolution comparison of several time-frequency representations, *IEEE Trans. on Signal Proc.*, 40, pp. 413–420, February 1992.

[81] S. Kadambe, *The Application of Time-Frequency and Time-Scale Representations in Speech Analysis*, Ph.D. thesis, University of Rhode Island, Kingston, RI, 1991.

[82] S. Kadambe and G. F. Boudreaux-Bartels, A comparison of the existence of 'cross terms' in the Wigner distribution and the squared magnitude of the wavelet transform and the short-time Fourier transform, *IEEE Trans. Signal Proc.*, 40(10), pp. 2498–2517, October 1992.

[83] R. Koenig, H. K. Dunn, and L. Y. Lacy, The sound spectrograph, *J. Acoust. Soc. Amer.*, 18, pp. 19–49, 1946.

[84] P. Loughlin, J. Pitton, and L. E. Atlas, Bilinear time-frequency representations: new insights and properties, *IEEE Trans. Signal Proc.*, 41, pp. 750–767, 1993.

[85] N. Marinovich and U. G. Oklobdzija, VLSI chip architecture for real time ambiguity function computation, *Proc. Asilomar Conf. Sig., Syst. and Comp.*, pp. 74–78, Pacific Grove, CA, November 1991.

[86] N. M. Marinovich, *The Wigner distribution and the ambiguity function: Generalizations, enhancement, compression and some applications*, Ph.D. thesis, The City University of New York, 1986.

[87] W. F. G. Mecklenbräuker, A tutorial on non-parametric bilinear time-frequency signal representations, in *Time and Frequency Representations of Signals and Systems*, G. Longo and B. Picinbono (Eds.), pp. 11–68, Springer-Verlag, New York, NY, 1989.

[88] W. F. G. Mecklenbräuker (Ed.), *The Wigner Distribution: Theory and Applications in Signal Processing*, North Holland Elsevier Science Publ., 1995.

[89] Y. Meyer, *Wavelets-Algorithms and Applications*, Society for Industrial and Applied Mathematics, 1993.

[90] R. L. Murray, *Dyadic wavelet transform based QRS detector*, Master's thesis, University of Rhode Island, Electrical Engg. Dept., Kingston, RI, 1993.

[91] R. L. Murray, Biomedical applications of time-frequency signal processing, Technical Report 0195-0001, University of Rhode Island, Electrical Engg. Dept., Kingston, RI, January 1995.

[92] S. H. Nawab and T. F. Quatieri, Short-time Fourier transform, In *Advanced Topics in Signal Processing*, J. S. Lim and A. V. Oppenheim (Eds.), Prentice Hall, Englewood Cliffs, NJ, 1988.

[93] A. H. Nuttall, Alias-free smoothed Wigner distribution function for discrete-time samples, Technical Report TR 8785, Naval Underwater Systems Center, New London, CT, October 1990.

[94] J. P. Ovarlez, *La Transformation de Mellin: Un Outil Pour L'Analyse des Signaux à Large Bande*, Ph.D. thesis, Thèse Univ. Paris 6, 1992.

[95] J. P. Ovarlez, J. Bertrand, and P. Bertrand, Computation of Affine time-frequency distributions using the fast Mellin transform, *Proc. IEEE Int. Conf. on Acoustics, Speech, and Signal Processing*, 5, pp. 117–120, San Francisco, CA, 1992.

[96] A. Papandreou, G. F. Boudreaux-Bartels, and S. M. Kay, Detection and estimation of

generalized chirps using time-frequency representations, *Proceedings Twenty-Eighth Asilomar Conference on Signals, Systems and Computers*, Pacific Grove, CA, October/November 1994.

[97] A. Papandreou, F. Hlawatsch, and G. F. Boudreaux-Bartels, The hyperbolic class of quadratic time-frequency representations Part I: Constant-Q warping, the hyperbolic paradigm, properties, and members, *IEEE Transactions on Signal Processing*, 41, pp. 3425–3444, December 1993.

[98] A. Papandreou, F. Hlawatsch, and G. F. Boudreaux-Bartels, A unified framework for the scale covariant affine, hyperbolic, and power class time-frequency representations using generalized time-shifts, *Proceedings 1995 International Conference on Acoustics, Speech and Signal Processing*, Detroit, MI, May 1995.

[99] A. Papandreou, S. M. Kay, and G. F. Boudreaux-Bartels. The use of hyperbolic time-frequency representations for optimum detection and parameter estimation of hyperbolic chirps, *IEEE-SP International Symposium on Time-Frequency and Time-Scale Analysis*, pp. 369–372, Philadelphia, PA, October 1994.

[100] A. Papandreou-Suppappola, *New classes of quadratic time-frequency representations with scale covariance and generalized time-shift covariance: Analysis, detection and estimation*, Ph.D. thesis, University of Rhode Island, Kingston, RI, May 1995.

[101] A. Papoulis, *Signal Analysis*, McGraw-Hill Book Company, New York, 1977.

[102] E. Peyrin and R. Prost, A unified definition for the discrete-time, discrete-frequency, and discrete-time/frequency Wigner distributions, *IEEE Trans. Acoust., Speech, Signal Proc.*, 34, pp. 858–867, 1986.

[103] B. Porat, *Digital Processing of Random Signals : Theory and Methods*, Prentice-Hall, Inc., Englewood Cliffs, NJ, 1993.

[104] M. R. Portnoff, Time-frequency representations of digital signals and systems based on short-time Fourier analysis, *IEEE Transactions on Acoustics, Speech, and Signal Processing*, 28, pp. 55–69, February 1980.

[105] A. Poularikas and S. Seeley, *Signals and Systems*, 2nd ed., PWD-Kent Publ., Boston, MA, 1991.

[106] V. S. Praveenkumar, *Implementation of Hyperbolic Class of time frequency distributions and removal of cross-terms*, Master's thesis, University of Rhode Island, Kingston, RI, May 1995.

[107] L. R. Rabiner and R. W. Schafer, *Digital Processing of Speech Signals*, Prentice Hall, Inc., Englewood Cliffs, NJ, 1978.

[108] S. Raz, Synthesis of signals from Wigner distributions : Representations in biorthogonal bases, *Signal Processing*, 20, pp. 303–314, 1990.

[109] A. W. Rihaczek, *Principles of High Resolution Radar*, McGraw Hill, New York, NY, 1969.

[110] M. D. Riley, *Speech Time-Frequency Representations*, Kluwer Academic Publ., Boston, MA, 1989.

[111] O. Rioul and P. Flandrin, Time-scale energy distributions: A general class extending wavelet transforms, *IEEE Transactions on Signal Processing*, 40(7), pp. 1746–1757, July 1992.

[112] O. Rioul and M. Vetterli, Wavelets and signal processing, *IEEE Signal Processing Magazine*, 8, pp. 14–38, October 1991.

[113] R. G. Shenoy and T. W. Parks, Affine Wigner distributions, *Proceedings IEEE Interna-

tional Conference on Acoustics, Speech, and Signal Processing, 5, pp. 185–188, San Francisco, CA, March 1992.

[114] M. I. Skolnik, *Introduction to Radar Systems*, McGraw Hill, New York, NY, 1962.

[115] J. M. Speiser, Wide-band ambiguity function, *IEEE Transactions on Information Theory*, 13, pp. 122–123, 1967.

[116] L. B. Stanković, S. B. Stanković, and Z. L. Usković, *Time-Frequency Signal Analysis*, EPSILON and MONTENEGROPUBLIC, Podgorica, Montenegro, 1994.

[117] E. F. Velez, *Transient Analysis of Speech Using the Wigner-Ville Distribution*, Ph.D. thesis, University of Vermont, Burlington, VT, 1989.

[118] M. Vetterli and J. Kovačević, *Wavelets and Subband Coding.*, Prentice-Hall, Inc., Englewood Cliffs, NJ, 1995.

[119] J. Ville, Théorie et applications de la notion de signal analytique, *Câbles et Transmission*, 2A, pp. 61–74, 1948. Translated into English by I. Selin, RAND Corporation Report T–92, Santa Monica, CA, August 1958.

[120] E. P. Wigner. On the quantum correction for thermo-dynamic equilibrium, *Physics Review*, 40, pp. 749–759, 1932.

[121] P. M. Woodward. Information theory and the design of radar receivers, *Proceedings of the Institute of Radio Engineers*, 39, pp. 1521–1524, 1951.

[122] Y. Zhao, L. E. Atlas, and R. J. Marks, The use of cone-shaped kernels for generalized time-frequency representations of nonstationary signals, *IEEE Transactions on Acoustics, Speech, and Signal Processing*, 38(7), pp. 1084–1091, July 1990.

Appendix 1: *Functions of a Complex Variable*[1]

Alexander D. Poularikas

1 Basic concepts

A complex variable z defined by

$$z = x + jy \tag{1.1}$$

assumes certain values over a region R_z of the complex plane. If a complex quantity $W(z)$ is so connected with z that each z in R_z corresponds with one value of $W(z)$ in R_w, then we say that $W(z)$ is a single-valued function of z

$$W(z) = u(x, y) + jv(x, y) \tag{1.2}$$

which has a **domain** R_z and a **range** R_w (see Figure 1.1). The function $W(z)$ can be **single valued** or **multiple valued**. Examples of single-valued functions include

$$W = a_0 + a_1 z + a_2 z^2 + \cdots + a_n z^n \qquad n \text{ integer}$$

$$W = e^z$$

Examples of multiple-valued functions are

$$W = z^n \qquad n \text{ not an integer}$$

$$W = \log z$$

$$W = \sin^{-1} z$$

DEFINITION 1.1 *A function* $W(z)$ *is* **continuous** *at a point* $z = \lambda$ *of* R_z *if, for each number* $\varepsilon > 0$, *however small, there exists another number* $\delta > 0$ *such that whenever*

$$|z - \lambda| < \delta \qquad then \qquad |W(z) - W(\lambda)| < \varepsilon \tag{1.3}$$

The geometric representation of this equation is shown in Figure 1.1.

---

[1] All contour integrals are taken counterclockwise, unless specifically indicated.

0-8493-8342-0/96/$0.00 + $0.50

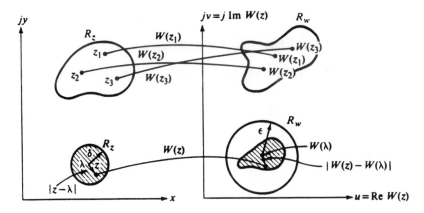

FIGURE 1.1
Illustration of the range and domain of complex functions.

DEFINITION 1.2   *A function $W(z)$ is* **analytic** *at a point z if, for each number $\varepsilon > 0$, however small, there exists another number $\delta > 0$ such that whenever*

$$|z - \lambda| < \delta \qquad then \qquad \left| \frac{W(z) - W(\lambda)}{z - \lambda} - \frac{dW(\lambda)}{dz} \right| < \varepsilon \qquad (1.4)$$

Example 1.1
Show that the function $W(z) = e^z$ satisfies (1.4).

Solution   From (1.4), we obtain

$$\lim_{z \to \lambda} \frac{W(z) - W(\lambda)}{z - \lambda} = \lim_{z \to \lambda} \frac{e^z - e^\lambda}{z - \lambda}$$

$$= \lim_{z \to \lambda} e^z \left[1 - \frac{(z - \lambda)}{2!} + \frac{(z - \lambda)^2}{3!} - \cdots \right]$$

$$= e^\lambda = \frac{de^z}{dz} \Big|_{z=\lambda}$$

which proves the assertion.   ▌

In this example, we did not mention the direction from which the z approaches λ. We might surmise from this that the derivative of our analytic function is independent of the path of z as it approaches the limiting point. However, this is not true in general. By setting $\lambda = z$ and $z = z + \Delta z$ in (1.4), we obtain an alternative form of that equation, namely,

$$\frac{dW}{dz} = \lim_{\Delta z \to 0} \left\{ \frac{W(z + \Delta z) - W(z)}{\Delta z} \right\} \qquad (1.5)$$

For a function to possess a unique derivative, it is required that

$$\frac{dW}{dz} = \lim_{\Delta z \to 0} \frac{\Delta W}{\Delta z} = \lim_{\substack{\Delta x \to 0 \\ \Delta v \to 0}} \frac{\Delta u + j \Delta v}{\Delta x + j \Delta y}$$

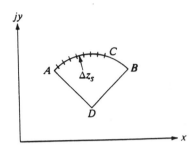

FIGURE 1.2
The path of integration in the complex plane.

But because

$$\Delta u = \frac{\partial u}{\partial x}\Delta x + \frac{\partial u}{\partial y}\Delta y$$

$$\Delta v = \frac{\partial v}{\partial x}\Delta x + \frac{\partial v}{\partial y}\Delta y$$

the unique derivative becomes

$$\frac{dW}{dz} = \lim_{\substack{\Delta x \to 0 \\ \Delta y \to 0}} \frac{\left(\frac{\partial u}{\partial x} + j\frac{\partial v}{\partial x}\right)\Delta x + j\left(\frac{\partial v}{\partial y} - j\frac{\partial u}{\partial y}\right)\Delta y}{\Delta x + j\Delta y}$$

For this to be independent of how Δx and Δy approach zero (that is, for the derivative to be unique), it is necessary and sufficient that $\Delta x + j\Delta y$ cancel in the numerator and denominator. This requires that

$$\frac{dW}{dz} = \frac{\partial u}{\partial x} + j\frac{\partial v}{\partial y} = \frac{\partial v}{\partial x} - j\frac{\partial u}{\partial y}$$

This condition can be met if

$$\frac{\partial u}{\partial x} = \frac{\partial v}{\partial y} \qquad \frac{\partial v}{\partial x} = -\frac{\partial u}{\partial y} \tag{1.6}$$

These are the **Cauchy-Riemann** conditions. If the function satisfies these equations, it possesses a unique derivative and it is analytic at that point. These conditions are necessary and sufficient.

Integration

Intergration of a complex function is defined in a manner like that for a real function, except for the important difference that the path of integration as well as the end points must be specified. A number of important theorems relate to integration, as we will discuss later.

Recall that the real integral $\int_a^b f(x)\,dx$ means that the x-axis is broken into tiny elements Δx from a to b, each element is multiplied by the mean value of $f(x)$ in the element, and then the sum of all such products from a to b is taken as $\Delta x \to 0$. The same general procedure is used to define the integral in the complex plane. Instead of being restricted to the x-axis, the path of integration can be anywhere in the z-plane, for example, the arc ABC in Figure 1.2. This arc is broken into n elements Δz, and the corresponding mean value of $W(z)$ over each element is written W_s. Now from the sum $\sum_{s=1}^{n} W_s \Delta z_s$, over all values of s from a to b, and

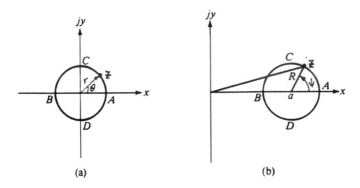

FIGURE 1.3
Integral of the function $W = 1/z$ over two paths.

take the limit $\Delta z_s \to 0, n \to \infty$. This limit, if it exists, is the integral

$$I = \int_a^b W(z)\, dz \tag{1.7}$$

The only innovation introduced here is that the path over which the integral is to be taken must be specified.

Example 1.2
Evaluate the integral in (1.7) for the function $W(z) = 1/z$ over the semicircles shown in Figure 1.3.

Solution   Refer first to Figure 1.3a, and introduce the polar coordinates

$$z = re^{j\theta} \qquad dz = jre^{j\theta}\, d\theta$$

Then

$$\int W(z)\, dz = \int \frac{dz}{z} = \int j\, d\theta$$

Over the path ACB, θ varies between 0 and π, and the integral equals $j\pi$. Over the path ADB, θ varies from 0 to $-\pi$, and the integral equals $-j\pi$. Thus, although the end points are the same, the integrals over the two paths are different. (The fact that one integral is numerically the negative of the other has no general significance.)

In evaluating the real integral by starting at A and integrating to B and then back to A, the result will be zero because the integral from A to B is the negative of the integral from B to A. The same result is not necessarily true for complex variables, unless the path from A to B coincides with the path from B to A. In the present complex integral, the integration from A to B via C and then back to A via D yields $j\pi - (-j\pi) = j2\pi$ and no zero.

Now consider the integration over the semicircle displaced from the origin, as shown in Figure 1.3b. Introduce the coordinates

$$z = a + Re^{j\psi} \qquad dz = j\, Re^{j\psi}\, d\psi$$

Then

$$\int_{ACBDA} \frac{dz}{z} = \int_0^{2\pi} \frac{j\,\mathrm{Re}^{j\psi}}{a + \mathrm{Re}^{j\psi}}\,d\psi = \ln(a + \mathrm{Re}^{j\psi})\Big|_0^{2\pi} = \ln z\big|_A^A = 0$$

The results of these calculations emphasize the fact that the two paths possess different features. The difference is that in Figure 1.3a, the path encloses a singularity (the function becomes infinite) at the origin, whereas the path in Figure 1.3b does not encloses the singularity and $W = 1/z$ is analytic everywhere in the region and on the boundary.

It is easily shown that the integrals of the function

$$W(z) = \frac{1}{z^2},\ W(z) = \frac{1}{z^3},\ \ldots,\ W(z) = \frac{1}{z^n}$$

around a contour encircling the origin of the coordinate axis are each equal to zero; that is,

$$\oint \frac{1}{z^2}\,dz = \oint \frac{1}{z^3}\,dz = \cdots = \oint \frac{1}{z^n}\,dz = 0 \tag{1.8}$$

where the contour is taken counterclockwise. ∎

Example 1.3
Find the value of the integral $\int_0^{z_0} z\,dz$ from the point $(0, 0)$ to $(2, j4)$.

Solution Because z is an analytic function along any path, then

$$\int_0^{z_0} z\,dz = \frac{z^2}{2}\bigg|_0^{2+j4} = -6 + j8$$

Equivalently, we could write

$$\int_0^{z_0} z\,dz = \int_0^2 x\,dx - \int_0^4 y\,dy + j\int_0^4 dy = \frac{x^2}{2}\bigg|_0^2 - \frac{y^2}{2}\bigg|_0^4 + jxy\big|_0^4$$

$$= 2 - \frac{16}{2} + j2 \times 4 = -6 + j8 \qquad\qquad ∎$$

We now state a very important theorem, and this is often referred to as the **principal theorem of complex variable theory**. This is the **Cauchy first integral theorem**.

THEOREM 1.1
Given a region of the complex plane within which $W(z)$ is analytic and any closed curve that lies entirely within this region, then

$$\oint_C W(z)\,dz = 0 \tag{1.9}$$

where the contour C is taken counterclockwise.

The integration over a closed path is called a contour integral. Also, by convention the positive direction of integration is taken so that when traversing the contour, the enclosed region is always to the left. The proof of this theorem depends on the fact that everywhere within C the Cauchy-Riemann equations are satisfied, $W(z)$ possesses a unique derivative at all points of the path.

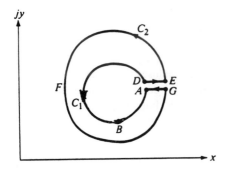

FIGURE 1.4
To prove the first corollary.

COROLLARY 1
If the contour C_2 completely encloses C_1, and if $W(z)$ is analytic in the region between C_1 and C_2 and also on C_1 and C_2, then

$$\oint_{C_1} W(z)\,dz = \oint_{C_2} W(z)\,dz \tag{1.10}$$

PROOF   Refer to Figure 1.4, which shows the two contours C_1 and C_2 and two connecting lines DE and GA. In the region closed by the contour $ABDEFGA$, the function $W(z)$ is analytic everywhere, and $\oint W\,dz = 0$ over the path. This means that

$$\int_{ABD} + \int_{DE} + \int_{EFG} + \int_{GA} = 0 \tag{1.11}$$

where $W(z)\,dz$ is to be understood after each integral sign. Now allow A to approach D, and G to approach E, so that DE coincides with AG. Then

$$\int_{DE} = \int_{AG} = -\int_{GA}$$

Also

$$\int_{ABD} = -\int_{C_1} \quad \text{and} \quad \int_{EFG} = \int_{C_2} \tag{1.12}$$

where strict attention has been paid to the convention given in the determination of the positive direction of integration around a contour. Combine (1.11) and (1.12) so that

$$-\int_{C_1} + \int_{C_2} = 0 \quad \text{or} \quad \int_{C_1} W(z)\,dz = \int_{C_2} W(z)\,dz$$

which was to be proved.   ∎

   This is an important theorem because it allows the evaluation around one contour by replacing that contour with a simpler one, the only restriction being that in the region between the two contours the integral must be regular. It does not require that the function $W(z)$ be analytic within C_1.

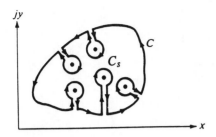

FIGURE 1.5
A contour enclosing n isolated singularities.

COROLLARY 2
If $W(z)$ has a finite number n of isolated singularities within a region G bounded by a curve C, then

$$\oint_C W(z)\, dz = \sum_{s=1}^{N} \oint_{C_s} W(z)\, dz \tag{1.13}$$

where C_s is any contour surrounding the sth singularity. The contours are taken in the counterclockwise direction.

PROOF Refer to Figure 1.5. The proof for this case is evident from the manner in which the first corollary was proved. ∎

COROLLARY 3
The integral $\int_A^B W(z)\, dz$ depends only upon the end points A and B (refer to Figure 1.2) and does not depend on the path of integration, provided that this path lies entirely within the region in which $W(z)$ is analytic.

PROOF Consider $ACBDA$ of Figure 1.2 as a contour that encloses no singularity of $W(z)$. Then

$$\oint_C = 0 = \int_{ADB} + \int_{BCA} \quad \text{or} \quad \int_{ABD} = \int_{ACB} \tag{1.14}$$

Hence, the integral is the same whether taken over path D or C, and thus is independent of the path and depends only on the end points A and B. ∎

THEOREM 1.2 *The Cauchy Second Integral Theorem*
If $W(z)$ is the function $W(z) = f(z)/(z - z_0)$ and the contour encloses the singularity at z_0, then

$$\oint_C \frac{f(z)}{z - z_0}\, d_z = j2\pi f(z_0) \tag{1.15}$$

or

$$f(z_0) = \frac{1}{2\pi j} \oint_C \frac{f(z)}{z - z_0}\, dz \tag{1.16}$$

PROOF Refer to Figure 1.6. Begin with the second corollary and draw a circle C_1 about the

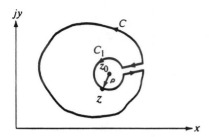

FIGURE 1.6
To prove the Cauchy second integral theorem.

point z_0. Then

$$\int_C \frac{f(z)}{z - z_0} \, dz = \int_{C_1} \frac{f(z)}{z - z_0} \, dz \tag{1.17}$$

Let $z' = z - z_0 = \rho e^{j\theta}$, which permits writing

$$\int_{C_1} \frac{f(z)}{z - z_0} \, dz = \int_0^{2\pi} \frac{f(z' + z_0)}{\rho e^{j\theta}} j\rho e^{j\theta} \, d\theta = j \int_0^{2\pi} f(z' + z_0) \, d\theta$$

In the limit as $\rho \to 0$, $z' \to 0$, and

$$j \int_0^{2\pi} f(z' + z_0) \, d\theta \bigg|_{\lim \rho \to 0} = 2\pi j f(z_0)$$

Combine with (1.17) to find

$$\int_C \frac{f(z)}{z - z_0} = 2\pi j f(z_0)$$

which proves the theorem. ∎

Derivative of an analytic function $W(z)$

The derivative of an analytic function is also analytic, and consequently itself possesses a derivative. Let C be a contour within and upon which $W(z)$ is analytic. Then if a is a point inside the contour (the prime indicates first-order derivative)

$$W'(a) = \lim_{|h| \to 0} \frac{W(a + h) - W(a)}{h} \tag{1.18}$$

and can be shown that

$$W'(a) = \frac{1}{2\pi j} \oint_C \frac{W(z) dz}{(z - a)^2} \tag{1.19}$$

where the contour C is taken in a counterclockwise direction. Proceeding, it can be shown that

$$W^{(n)}(a) = \frac{n!}{2\pi j} \oint_C \frac{W(z) dz}{(z - a)^{n+1}} \tag{1.20}$$

The exponent (n) indicates nth derivative and the contour is taken counterclockwise.

Taylor's theorem

Let $f(z)$ be analytic in the neighborhood of a point $z = a$. Let the contour C be a circle with center point a in the z-plane, and let the function $f(z)$ not have any singularity within and on the contour. Let $z = a + h$ be any point inside the contour; then by (1.15) we obtain

$$f(a + h) = \frac{1}{2\pi j} \oint_C \frac{f(z)dz}{z - a - h}$$

$$= \frac{1}{2\pi j} \oint_C f(z)\, dz \left\{ \frac{1}{z - a} + \frac{h}{(z - a)^2} \right.$$

$$\left. + \cdots + \frac{h^n}{(z - a)^{n+1}} + \frac{h^{n+1}}{(z - a)^{n+1}(z - a - h)} \right\}$$

$$= f(a) + hf^{(1)}(a) + \frac{h^2}{2!} f^{(2)}(a) + \cdots + \frac{h^n}{n!} f^{(n)}(a)$$

$$+ \frac{1}{2\pi j} \oint_C \frac{f(z)h^{n+1}dz}{(z - a)^{n+1}(z - a - h)}$$

But when z is on C the modulus $f(z)/(z - a - h)$ is continuous and therefore bounded. Its modulus will not exceed some finite number M. Hence, with $|z - a| = R$ for points on the circle, we obtain

$$\left| \frac{1}{2\pi j} \oint_C \frac{f(z)h^{n+1}dz}{(z - a)^{n+1}(z - a - h)} \right| \le \frac{M 2\pi R}{2\pi} \left(\frac{|h|}{R} \right)^{n+1}$$

where $|h|/R < 1$ and therefore tends to zero as n tends to infinity. Therefore, we have

$$f(a + h) = f(a) + hf^{(1)}(a) + \frac{h^2}{2!} f^{(2)}(a) + \cdots + \frac{h^n}{n!} f^{(n)}(a) + \cdots \tag{1.21}$$

or

$$f(z) = f(a) + (z - a)f^{(1)}(a) + \frac{(z - a)^2}{2!} f^{(2)}(a) + \cdots + \frac{(z - a)^n}{n!} f^{(n)}(a) + \cdots \tag{1.22}$$

where the numbers in the exponents indicate order of differentiation. The radius of convergence is such that it excludes from the interior of the circle that singularity of the function that is nearest to a.

Laurent's theorem

Let C_1 and C_2 be two concentric circles, as shown in Figure 1.4, with their center at a. The function $f(z)$ is analytic with the ring and $(a + h)$ is any point in it. From the figure and Cauchy's theorem we obtain

$$\frac{1}{2\pi j} \oint_{C_2} \frac{f(z)dz}{(z - a - h)} + \frac{1}{2\pi j} \oint_{C_1} \frac{f(z)dz}{(z - a - h)} + \frac{1}{2\pi j} \oint_{C_3} \frac{f(z)dz}{(z - a - h)} = 0$$

where the first contour is counterclockwise and the last two are clockwise. The above equation becomes

$$f(a + h) = \frac{1}{2\pi j} \oint_{C_2} \frac{f(z)dz}{(z - a - h)} - \frac{1}{2\pi j} \oint_{C_1} \frac{f(z)dz}{(z - a - h)} \tag{1.23}$$

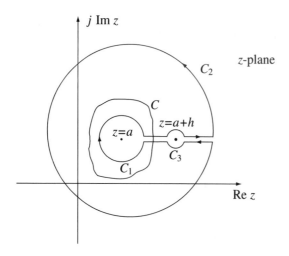

FIGURE 1.4
Explaining Laurent's theorem.

where both the contours are taken counterclockwise. For the C_2 contour $h < |(z-a)|$ and for the C_1 $h > |(z-a)|$. Hence we expand the above integral as follows:

$$f(a+h) = \frac{1}{2\pi j} \oint_{C_2} f(z) \left\{ \frac{1}{z-a} + \frac{h}{(z-a)^2} + \cdots + \frac{h^n}{(z-a)^{n+1}} + \frac{h^{n+1}}{(z-a)^{n+1}(z-a-h)} \right\} dz$$

$$+ \frac{1}{2\pi j} \oint_{C_1} f(z) \left\{ \frac{1}{h} + \frac{z-a}{h^2} + \cdots + \frac{(z-a)^n}{h^{n+1}} + \frac{(z-a)^{n+1}}{h^{n+1}(z-a-h)} \right\} dz$$

From Taylor's theorem it was shown that the integrals of the last term in the two brackets tend to zero as n tends to infinity. Therefore, we have

$$f(a+h) = a_0 + a_1 h + a_2 h^2 + \cdots + \frac{b_1}{h} + \frac{b_2}{h^2} + \cdots \tag{1.24}$$

where

$$a_n = \frac{1}{2\pi j} \oint_{C_2} \frac{f(z)dz}{(z-a)^{n+1}} \qquad b_n = \frac{1}{2\pi j} \oint_{C_1} (z-a)^{n-1} f(z)\, dz$$

The above expansion can be put in more convenient form by substituting $h = z - a$, which gives

$$f(z) = c_0 + c_1(z-a) + c_2(z-a)^2 + \cdots + \frac{d_1}{(z-a)} + \frac{d_2}{(z-a)^2} + \cdots + \frac{d_n}{(z-a)^n} + \cdots \tag{1.25}$$

Because $z = a + h$, it means that z now is any point within the ring-shaped space between C_1 and C_2 where $f(z)$ is analytic. Equation (1.25) is the Laurent's expansion of $f(z)$ at a point $z + h$ within the ring. The coefficients c_n and d_n are obtained from (1.24) by replacing a_n, b_n, z by c_n, d_n, ζ, respectively. Here ζ is the variable on the contours and z is inside the ring. When $f(z)$ has a simple pole at $z = a$, there is only one term, namely, $d_1/(z-a)$. If there exists an nth-order term, there are n terms of which the last is $d_n/(z-a)^n$; some of the d_n's may be zero.

If m is the highest index of the inverse power of $f(z)$ in (1.25) it is said that $f(z)$ has a pole of order m at $z = a$. Then

$$f(z) = \sum_{n=0}^{\infty} c_n(z-a)^n + \sum_{n=1}^{m} \frac{d_n}{(z-a)^n} \tag{1.26}$$

The coefficient d_1 is the **residue** at the pole.

If the series in inverse powers of $(z-a)$ in (1.25) does not terminate, the function $f(z)$ is said to have an **essential singularity** at $z = a$. Thus

$$f(z) = \sum_{n=0}^{\infty} c_n(z-a)^n + \sum_{n=1}^{\infty} \frac{d_n}{(z-a)^n} \tag{1.27}$$

The coefficient d_1 is the **residue** of the singularity.

Example 1.4

Find the Laurent expansion of $f(z) = 1/[(z-a)(z-b)^n]$ $(n \geq 1, a \neq b \neq 0)$ near each pole.

Solution First remove the origin to $z = a$ by the transformation $\zeta = (z-a)$. Hence we obtain

$$f(z) = \frac{1}{\zeta} \frac{1}{(\zeta+c)^n} = \frac{1}{c^n \zeta} \frac{1}{\left(1+\frac{\zeta}{c}\right)^n} \qquad c = a - b$$

If $|\zeta/c| < 1$ then we have

$$f(z) = \frac{1}{c^n \zeta} \left[1 - \frac{n\zeta}{c} + \frac{n(n+1)}{2!} \frac{\zeta^2}{c^2} - \cdots \right]$$

$$= \left[-\frac{n}{c^{n+1}} + \frac{n(n+1)\zeta}{2!c^{n+2}} - \cdots \right] + \frac{1}{c^n \zeta}$$

which is the Laurent series expansion near the pole at $z = a$. The residue is $1/c^n = 1/(a-b)^n$.

For the second pole set $\zeta = (z-b)$ and expand as above to find

$$f(z) = -\left(\frac{1}{c^{n+1}} + \frac{\zeta}{z^{n+2}} + \frac{\zeta^2}{c^{n+3}} + \cdots \right) - \left(\frac{1}{c^n \zeta} + \frac{1}{c^{n-1}\zeta^2} + \cdots + \frac{1}{c\zeta^n} \right)$$

The second part of the expansion is the principal expansion near $z = b$ and the residue is $-1/c^n = -1/(a-b)^n$. ∎

Example 1.5

Prove that

$$f(z) = \exp\left[\frac{x}{2}\left(z - \frac{1}{2} \right) \right] = J_0(x) + zJ_1(x) + z^2 J_2(x) + \cdots + z^n J_n(x) + \cdots$$

$$-\frac{1}{z}J_1(x) + \frac{1}{z^2}J_2(x) - \cdots + \frac{(-1)^n}{z^n}J_n(x) + \cdots$$

where

$$J_n(x) = \frac{1}{2\pi} \int_0^{2\pi} \cos(n\theta - x\sin\theta)\, d\theta$$

Solution The function $f(z)$ is analytic except the point $z = a$. Hence by the Laurent's theorem we obtain

$$f(z) = a_0 + a_1 z + a_2 z^2 + \cdots + \frac{b_1}{z} + \frac{b_2}{z^2} + \cdots$$

where

$$a_n = \frac{1}{2\pi j} \oint_{C_2} \exp\left[\frac{x}{2}\left(z - \frac{1}{z}\right)\right] \frac{dz}{z^{n+1}}, \qquad b_n = \frac{1}{2\pi j} \oint_{C_1} \exp\left[\frac{x}{2}\left(z - \frac{1}{z}\right)\right] z^{n-1}\, dz$$

where the contours are circles with center at the origin and are taken counterclockwise. Set C_2 equal to a circle of unit radius and write $z = \exp(j\theta)$. Then we have

$$a_n = \frac{1}{2\pi j} \int_0^{2\pi} e^{jx\sin\theta} e^{-jn\theta} j\, d\theta = \frac{1}{2\pi} \int_0^{2\pi} \cos(n\theta - x\sin\theta)\, d\theta$$

because the last integral vanishes, as can be seen by writing $2\pi - \varphi$ for θ. Thus $a_n = J_n(x)$, and $b_n = (-1)^n a_n$, because the function is unaltered if $-z^{-1}$ is substituted for z, so that $b_n = (-1)^n J_n(x)$. ∎

2 Sequences and series

Consider a sequence of numbers, such as those that arise in connection with the Z-transform. Suppose that the sequence of complex numbers is given as $z_0, z_1, z_2, \ldots$.

The sequence of complex numbers is said to **converge** to the limit L; that is,

$$\lim_{n\to\infty} z_n = L$$

if for every positive d there exists an integer N such that

$$|z_n - L| < \delta \qquad \text{for all } n > N$$

That is, a convergent sequence is one whose terms approach arbitrarily close to the limit L as n increases. If the series does not converge, it is said to **diverge**.

THEOREM 2.1
In order for a sequence $\{z_n\}$ of complex numbers to be convergent, it is necessary and sufficient that for all $\delta > 0$ there exists a number $N(\delta)$ such that for all $n > N$ and all $p = 1, 2, 3, \ldots$ the inequality $|z_{n+p} - z_n| < \delta$ is fulfilled.

The sum of an infinite sequence of complex numbers $z_0, z_1, \ldots$ is given by

$$S = z_0 + z_1 + z_2 + \cdots = \sum_{n=0}^{\infty} z_n \tag{2.1}$$

Consider the partial sum sequence of n terms, which is designated S_n. The infinite series converges to the sum S if the partial sum sequence S_n converges to S. That is, the series converges if for

$$S_n = \sum_{n=0}^{n} z_n \qquad \lim_{n\to\infty} S_n = S \tag{2.2}$$

When the partial sum S_n diverges, the series is said to diverge.

Comparison test

Let the terms of the numerical series (2.1) for all $n \geq n_0 \geq 1$ satisfy the condition $|z_n| \leq b_n$. Then the convergence of the series of positive terms $\sum_{n=1}^{\infty} b_n$ implies absolute convergence of the above series.

Limit comparison test

If the numerical series $\sum_{n=1}^{\infty} v_n$ converges absolutely and for the terms of the numerical series (2.1) there takes place the relationship

$$\lim_{n \to \infty} \left| \frac{z_n}{v_n} \right| = q = \text{const} < \infty$$

then series (2.1) converges absolutely.

D'Alembert's test

If for the terms of the numerical series (2.2) the finite limit

$$\lim_{n \to \infty} \left| \frac{z_{n+1}}{z_n} \right| = l$$

then for $0 \leq l < 1$ series (2.1) converges absolutely, for $l > 1$ series (2.1) diverges, and for $l = 1$ an additional test is required.

Root test

Consider the sequence

$$r_n = \sqrt[n]{|z_n|}$$

If this sequence converges to l as n approaches infinity, then the series (2.1) converges absolutely if $l < 1$ and diverges if $l > 1$.

3   Power series

A series of the form

$$W(z) = a_0 + a_1(z - z_0) + a_2(z - z_0)^2 + \cdots = \sum_{n=0}^{\infty} a_n(z - z_0)^n \tag{3.1}$$

where the coefficients a_n are given by

$$a_n = \frac{1}{n!} \frac{d^n W(z)}{dz^n} \bigg|_{z=z_0} \tag{3.2}$$

is a **Taylor** series that is expanded about the point $z = z_0$, where z_0 is a complex constant. That is, the Taylor series expands an analytic function as an infinite sum of component functions. More precisely, the Taylor series expands a function $W(z)$, which is analytic in the neighborhood of the point $z = z_0$, into an infinite series whose coefficients are the successive derivatives of the function at the given point. However, we know that the definition of a derivative of any order does not require more than the knowledge of the function in an arbitrarily small neighborhood of the point $z = z_0$. This means, therefore, that the Taylor series indicates that the shape of the function at a finite distance z_0 from z is determined by the behavior of the function in the infinitesimal vicinity of $z = z_0$. Thus, the Taylor's series implies that any analytic function has a very strong interconnected structure, and that by studying the function in a small vicinity of the point $z = z_0$, we can precisely predict what happens at the point $z = z_0 + \Delta z_0$, which is a finite distance from the point of study.

If $z_0 = 0$, the expansion is said to be about the origin and is called a **Maclaurin** series.

A power series of negative powers of $(z - z_0)$,

$$W(z) = a_0 + a_1(z - z_0)^{-1} + a_2(z - z_0)^{-2} + \cdots \tag{3.3}$$

is called a **negative power** series.

We first focus attention on the positive power series (3.1). Clearly, this series converges to a_0 when $z = z_0$. To ascertain whether it converges for other values of z, we write

THEOREM 3.1

A positive power series converges absolutely in a circle of radius R^+ centered at z_0 where $|z - z_0| < R^+$; it diverges outside of this circle where $|z - z_0| > R^+$. The value of R^+ may be zero, a positive number, or infinity. If $R^+ = $ infinity, the series converges everywhere, and if it is equal to zero the series converges only at $z = z_0$. The radius R^+ is found from the relation

$$R^+ = \lim_{n \to \infty} \left| \frac{a_n}{a_{n+1}} \right| \qquad \text{if the limit exists} \tag{3.4}$$

or by

$$R^+ = \lim_{n \to \infty} \frac{1}{\sqrt[n]{|a_n|}} \qquad \text{if the limit exists} \tag{3.5}$$

PROOF   For a fixed value z, apply the ratio test, where

$$z_n = a_n(z - z_0)^n$$

That is

$$\left| \frac{z_{n+1}}{z_n} \right| = \left| \frac{a_{n+1}(z - z_0)^{n+1}}{a_n(z - z_0)^n} \right| = \left| \frac{a_{n+1}}{a_n} \right| |z - z_0|$$

For the power series to converge, the ratio test requires that

$$\lim_{n \to \infty} \left| \frac{a_{n+1}}{a_n} \right| |z - z_0| < 1 \qquad \text{or} \qquad |z - z_0| < \lim_{n \to \infty} \left| \frac{a_n}{a_{n+1}} \right| = R^+$$

That is, the power series converges absolutely for all z that satisfy this inequality. It diverges for all z for which $|z - z_0| > R^+$. The value of R^+ specified by (3.5) is reduced by applying the root test.   ∎

Example 3.1
Determine the region of convergence for the power series

$$W(z) = \frac{1}{1 + z} = 1 - z + z^2 - z^3 + \cdots$$

Solution   We have $a_n = (-1)^n$, from which

$$R^+ = \lim_{n \to \infty} \left| \frac{(-1)^n}{(-1)^{n+1}} \right| = |1|$$

The series converges for all z for which $|z| < 1$. Hence, this expansion converges for any value of z within a circle of unit radius about the origin. Note that there will be at least one singular point of $W(z)$ on the circle of convergence. In the present case, the point $z = -1$ is a singular point. ∎

Example 3.2
Determine the region of convergence for the power series

$$W(z) = e^z = 1 + z + \frac{z^2}{2!} + \frac{z^3}{3!} + \cdots = \sum_{n=1}^{\infty} \frac{1}{n!} z^n$$

Solution　We have $a_n = 1/n!$ from which

$$R^+ = \lim_{n \to \infty} \left| \frac{(n+1)!}{n!} \right| = \lim_{n \to \infty} (n+1) = \infty$$

The circle of convergence is specified by $R^+ = $ infinity; hence $W(z) = e^z$ converges for all finite values of z. ∎

THEOREM 3.2
A negative power series (3.3) converges absolutely outside a circle of radius R^- centered at z_0, where $|z - z_0| > R^-$; it diverges inside of this circle where $|z - z_0| < R^-$. The radius of convergence is determined from

$$R^- = \lim_{n \to \infty} \left| \frac{a_{n+1}}{a_n} \right| \qquad \text{if this limit exists} \qquad (3.6)$$

or by

$$R^- = \lim_{n \to \infty} \sqrt[n]{|a_n|} \qquad \text{if the limit exists} \qquad (3.7)$$

PROOF　The proof of this theorem parallels that of Theorem 3.1.

If a function has a singularity at $z = z_0$, it cannot be expanded in a Taylor series about this point. However, if one deletes the neighborhood of z_0, it can be expressed in the form of a Laurent series. The Laurent series is written

$$W(z) = \cdots + \frac{a_{-2}}{(z - z_0)^2} + \frac{a_{-1}}{(z - z_0)} + a_0 + a_1(z - z_0) + a_2(z - z_0)^2 + \cdots$$

$$= \sum_{n=-\infty}^{\infty} a_n(z - z_0)^n \qquad (3.8)$$

If a circle is drawn about the point z_0 such that the nearest singularity of $W(z)$ lies on the circle, then (3.8) defines an analytic function everywhere within this circle except at its center. The portion $\sum_{n=0}^{\infty} a_n(z - z_0)^n$ is regular at $z = z_0$. The portion $\sum_{n=-1}^{-\infty} a_n(z - z_0)^n$ is not regular and is called the principal part of $W(z)$ at $z = z_0$.

The region of convergence for the positive series part of the Laurent series is of the form

$$|z - z_0| < R^+ \qquad (3.9)$$

while that for the principal part is given by

$$|z - z_0| > R^- \qquad (3.10)$$

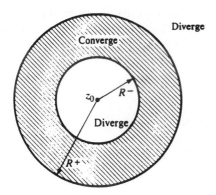

FIGURE 3.1

The evaluation of R^+ and R^- proceeds according to the methods already discussed. Hence, the region of convergence of the Laurent series is given by those points common to (3.9) and (3.10) or for

$$R^- < |z - z_0| < R^+ \qquad (3.11)$$

If $R^- > R^+$, the series converges nowhere. The annular region of convergence for a typical Laurent series is shown in Figure 3.1. ∎

Example 3.3
Consider the Laurent series $W(z) = \sum_n a_n z^n$ where

$$a_n = \begin{cases} \left(\frac{1}{3}\right)^n & \text{for } n = 0, 1, 2, \ldots \\ 2^n & \text{for } n = -1, -2, \ldots \end{cases}$$

Determine the region of convergence.

Solution   By (3.9) and (3.4) we have $R^+ = 3$. By (3.9) and (3.6) we have $R^- = 2$. Hence, the series converges for all z for which $2 < |z| < 3$. ∎

No convenient expression exists for obtaining the coefficients of the Laurent series. However, because there is only one Laurent expansion for a given function, the resulting series, however derived, will be the appropriate one. For example,

$$e^{1/z} = 1 + \frac{1}{z} + \frac{1}{2!z^2} + \frac{1}{3!z^3} + \cdots \qquad (3.12)$$

is obtained by replacing z by $1/z$ in the Maclaurin expansion of $\exp(z)$. Note that in this case the coefficients of all positive powers of z in the Laurent expansion are zero. As a second illustration, consider the function $W(z) = (\cos z)/z$. This is found by dividing the Maclaurin series for $\cos z$ by z, with the result

$$\frac{\cos z}{z} = \frac{1}{z}\left(1 - \frac{z^2}{2!} + \frac{z^4}{4!} - \cdots\right) = \frac{1}{z} - \frac{z}{2!} + \frac{z^3}{4!} - \cdots \qquad (3.13)$$

In this case, the Laurent expansion includes only one term $1/z$ in descending powers of z, but an infinite number of terms in ascending powers of z. That is, $a_{-1} = 1$ and $a_{-n} = 0$ if $n \neq 1$.

4 Analytic continuation

The Taylor theorem shows that if a function $f(z)$ is given by a power in z, it can also be represented as a power series in $z - z_0 = f[(z - z_0) + z_0]$ where z_0 is any point within the original circle of convergence, and this series will converge within any circle about z_0 that does not pass beyond the original circle of convergence. Actually, it may converge within a circle that does not pass beyond the original circle of convergence. Consider, for example, the function

$$f(z) = 1 + z + z^2 + \cdots = \frac{1}{1 - z} \qquad \text{for } |z| < 1$$

Choose $z_0 = j/2$, and the Taylor expansion of

$$f(z) = \frac{1}{1 - \left[\left(z - \frac{1}{2}j\right) + \frac{1}{2}j\right]} = \frac{1}{\left(1 - \frac{1}{2}j\right) - z'} \qquad z' = z - \frac{1}{2}j$$

in powers of z' is

$$f(z) = \frac{1}{1 - \frac{1}{2}j} + \frac{z'}{\left(1 - \frac{1}{2}j\right)^2} + \frac{z'^2}{\left(1 - \frac{1}{2}j\right)^3} + \cdots$$

This series must converge and be equal to the original function if $|z'| < 1/2$, because j is the point of the circle $|z| = 1$ nearest to $j/2$, a requirement of Taylor's theorem. Actually this series converges if $|z'| < |1 - \frac{1}{2}j| = \frac{1}{2}\sqrt{5}$.

Suppose that the considered series represented no previously known function. In this case, the new Taylor series would define values of an analytic function over a range of z where no function is defined by the original series. Then we can extend the range of definition by taking a new Taylor series about a point in the new region. This process is called **analytic continuation**. In practice, when continuation is required, the direct use of the Taylor series is laborious and is seldom is used. Of more convenience is the following theorem.

THEOREM 4.1
If two functions $f_1(z)$ and $f_2(z)$ are analytic in a region D and equal in a region D' within D, they are equal everywhere in D.

5 Singularities of complex functions

A singularity has already been defined as a point at which a function ceases to be analytic. Thus a discontinuous function has a singularity at the point of discontinuity, and multivalued functions have a singularity at a branch point. There are two important classes of singularities that a continuous, single-valued function may possess.

DEFINITION
A function has an **essential singularity** at $z = z_0$ if its Laurent expansion about the point z_0 contains an infinite number of terms in inverse powers of $(z - z_0)$.

DEFINITION

A function has a **nonessential singularity** or **pole of order** m if its Laurent expansion can be expressed in the form

$$W(z) = \sum_{n=-m}^{\infty} a_n(z - z_0)^n \tag{5.1}$$

Note that the summation extends from $-m$ to infinity and not from minus infinity to infinity; that is, the highest inverse power of $(z - z_0)$ is m.

An alternative definition that is equivalent to this but somewhat simpler to apply is the following: If $\lim_{z \to z_0}[(z - z_0)^m W(z)] = c$, a nonzero constant (here m is a positive number), then $W(z)$ is said to possess a pole of order m at z_0. The following examples illustrate these definitions:

1. $\exp(1/z)$ (see [3.12]) has an essential singularity at the origin.
2. $\cos z/z$ (see [3.13]) has a pole of order 1 at the origin.
3. Consider the function

$$W(z) = \frac{e^z}{(z - 4)^2(z^2 + 1)}$$

Note that functions of this general type exist frequently in the Laplace inversion integral. Because e^z is regular at all finite points of the z-plane, the singularities of $W(z)$ must occur at the points for which the denominator vanishes; that is, for

$$(z - 4)^2(z^2 + 1) = 0 \qquad \text{or} \qquad z = 4, +j, -j$$

By the second definition above, it is easily shown that $W(z)$ has a second-order pole at $z = 4$, and first-order poles at the two points $+j$ and $-j$. That is,

$$\lim_{z \to 4}(z - 4)^2 \left[\frac{e^z}{(z - 4)^2(z^2 + 1)} \right] = \frac{e^4}{17} \neq 0$$

$$\lim_{z \to j}(z - j) \left[\frac{e^z}{(z - 4)^2(z^2 + 1)} \right] = \frac{e^j}{(j - 4)^2 2j} \neq 0$$

4. An example of a function with an infinite number of singularities occurs in heat flow, wave motion, and similar problems. The function involved is

$$W(z) = 1/\sinh az$$

The singularities in this function occur when $\sinh az = 0$ or $az = js\pi$, where $s = 0, \pm 1, \pm 2, \ldots$. That each of these is a first-order pole follows from

$$\lim_{z \to j(s\pi/a)} \left(z - j\frac{s\pi}{a} \right) \frac{1}{\sinh az} = \frac{0}{0}$$

This can be evaluated in the usual manner by differentiating numerator and denominator (L'Hospital rule) to find

$$\lim_{z \to j(s\pi/a)} \frac{1}{a \cosh az} = \frac{1}{a \cosh js\pi} = \frac{1}{a \cos s\pi} \neq 0$$

6 Theory of residues

It has already been shown that the contour integral of any function that encloses no singularities of the integrand will vanish. (In this section all the contour integrals are taken counterclockwise unless it is indicated otherwise.) Now our purpose is to examine the integral, the path of which encloses one singularity, say at $z = z_0$. The Laurent expansion of such a function is

$$W(z) = \sum_{n=-\infty}^{\infty} a_n (z - z_0)^n$$

and so

$$\oint_C W(z)\, dz = \sum_{n=-\infty}^{\infty} a_n \oint_{C_n} (z - z_0)^n\, dz$$

But by (1.11), each term in the sum vanishes except for $n = -1$, with

$$\oint_C (z - z_0)^{-1}\, dz = 2\pi j$$

It then follows that

$$\oint_C W(z)\, dz = \sum_{n=-\infty}^{\infty} \oint_{C_n} (z - z_0)^n\, dz = 2\pi j a_{-1} \tag{6.1}$$

Because the integral $(1/2\pi j) \oint_C W(z)\, dz$ will appear frequently in subsequent applications, it is given a name; it is called the **residue of $W(z)$** at z_0 and is abbreviated Res(W).

From the second corollary (1.13), it follows that if $W(z)$ has n isolated singularities within C, then

$$\frac{1}{2\pi j} \oint_C W(z)\, dz = \sum_{s=1}^{n} \frac{1}{2\pi j} \oint_{C_s} W(z)\, dz = \sum_{s=1}^{n} \text{Res}_s(W) \tag{6.2}$$

or, in words, the value of the contour integral equals the sum of the residues within C. Observe that to evaluate integrals in the complex plane, it is only necessary to find the residues at the singularities of the integrand within the contour. One obvious way of doing this is (see [6.1]) to find the coefficient a_{-1} in the Laurent expansion about each singularity. However this is not always an easy task.

Several theorems exist that make evaluating residues relatively easy. We introduce these.

THEOREM 6.1
If the $\lim_{z \to z_0}[(z - z_0)W(z)]$ is finite, this limit is the residue of $W(z)$ at $z = z_0$. If the limit is not finite, then $W(z)$ has a pole of at least second order at $z = z_0$ (it may possess an essential singularity here). If the limit is zero, then $W(z)$ is regular at $z = z_0$.

PROOF Suppose that the function is expanded into the Laurent series

$$W(z) = \frac{a_{-1}}{z - z_0} + a_0 + a_1(z - z_0) + a_2(z - z_0)^2 + \cdots$$

Then the expression

$$\lim_{z \to z_0}[(z - z_0)W(z)] = \lim_{z \to z_0}[a_{-1} + a_0(z - z_0) + a_1(z - z_0)^2 + \cdots = a_{-1}$$

This proves the theorem. ∎

This process was previously used to ascertain whether or not a function had a first-order pole at $z = z_0$. Thus, referring back to the examples in Section 5, we have

$$\text{Res}\left(\frac{\cos z}{z}\right)_{z=0} = 1$$

$$\text{Res}\left[\frac{e^z}{(z-4)^2(z^2+1)}\right]_{z=j} = \frac{e^j}{(j-4)^2 2j}$$

$$\text{Res}\left(\frac{1}{\sinh az}\right)_{z=j(s\pi/a)} = \frac{1}{a\cos s\pi}$$

Many of the singularities that arise in system function studies are first-order poles. The evaluation of the integral is relatively direct.

Example 6.1
Evaluate the following integral

$$\frac{1}{2\pi j}\oint_C \frac{e^{zt}}{(z^2+\omega^2)}\,dz$$

when the contour C encloses both first-order poles at $z = \pm j\omega$. Note that this is precisely the Laplace inversion integral of the function $1/(z^2+\omega^2)$.

Solution   This involves finding the following residues

$$\text{Res}\left(\frac{e^{zt}}{z^2+\omega^2}\right)_{z=j\omega} = \frac{e^{j\omega t}}{2j\omega} \qquad \text{Res}\left(\frac{e^{zt}}{z^2+\omega^2}\right)\Big|_{z=-j\omega} = -\frac{e^{-j\omega t}}{2j\omega}$$

Hence,

$$\frac{1}{2\pi j}\oint_C \frac{e^{zt}}{z^2+\omega^2}\,dz = \sum \text{Res} = \left(\frac{e^{j\omega t}-e^{-j\omega t}}{2j\omega}\right) = \frac{\sin \omega t}{\omega}$$

A slight modification of the method for finding residues of simple poles

$$\text{Res } W(z_0) = \lim_{z\to z_0}[(z-z_0)W(z)] \tag{6.3}$$

makes the process even simpler. This is specified by the following theorem.   ∎

THEOREM 6.2
Suppose that $f(z)$ is analytic at $z = z_0$ and suppose that $g(z)$ is divisible by $z - z_0$ but not by $(z - z_0)^2$. Then

$$\text{Res}\left[\frac{f(z)}{g(z)}\right]_{z=z_0} = \frac{f(z_0)}{g'(z_0)} \qquad \text{where } g'(z) = \frac{dg(z)}{dz} \tag{6.4}$$

PROOF   We write the relation $(z - z_0)h(z) = g(z)$, then $g'(z) = (z - z_0)h'(z) + h(z)$ so that for $z = z_0$, $g'(z_0) = h(z_0)$. Then we have

$$\text{Res}\left[\frac{f(z)}{g(z)}\right]_{z=z_0} = \lim_{z\to z_0}\left[(z-z_0)\frac{f(z)}{g(z)}\right] = \lim_{z\to z_0}\left[\frac{f(z)}{h(z)}\right] = \frac{f(z_0)}{h(z_0)} = \frac{f(z_0)}{g'(z_0)}$$

which is the given result.   ∎

In reality, this theorem has already been used in the evaluation of $\text{Res}(1/\sinh az)_{z=j(s\pi/a)}$. Here $f(z) = 1$, $g(z) = \sinh az$, and $g'(z) = a \cosh az$.

As a second illustration, consider the previously used function

$$W(z) = \frac{e^z}{(z-4)^2(z^2+1)}$$

here we take

$$f(z) = \frac{e^z}{(z-4)^2}, \qquad g(z) = z^2 + 1$$

thus, $g'(z) = 2z$ and the previous result follows immediately with

$$\text{Res}\left[\frac{e^z}{(z-4)^2(z^2+1)}\right] = \frac{e^j}{(j-4)^2 2j}$$

Equation (6.3) permits a simple proof of the Cauchy second integral theorem (1.15). This involves choosing $g(z) = (z - z_0)$ in the integral

$$\frac{1}{2\pi j} \oint_C \frac{f(z)}{z - z_0} dz = \frac{f(z_0)}{1} = f(z_0) \tag{6.5}$$

Suppose that (6.5) is differentiated $n - 1$ times with respect to z_0. Then we write

$$\frac{d^{n-1} f(z_0)}{dz_0^{n-1}} \doteq f^{(n-1)}(z_0) = \frac{(n-1)!}{2\pi j} \oint_C \frac{f(z)}{(z - z_0)^n} dz \tag{6.6}$$

This specifies an any-order derivative of a complex function expressed as a contour integral.

Our discussion so far has concentrated on finding the residue of a first-order pole. However, (6.6) permits finding the residue of a pole of any order. If, for example, $W(z) = [f(z)/(z - z_0)^n]$, then evidently $W(z)$ has a pole of order n at $z = z_0$ because $f(z)$ is analytic at $z = z_0$. Then $f(z) = (z - z_0)^n W(z)$, and (6.6) becomes

$$\text{Res}(W(z))|_{z=z_0} = \frac{1}{2\pi j} \oint_C W(z)\, dz = \frac{1}{(n-1)!} \frac{d^{n-1}}{dz^{n-1}}[(z - z_0)^n W(z)]_{z=z_0} \tag{6.7}$$

Example 6.2
Evaluate the residue at the second-order pole at $z = 4$ of the previously considered function

$$W(z) = \frac{e^z}{(z-4)^2(z^2+1)}$$

Solution It follows from (6.7) that

$$\text{Res } W(z)|_{z=4} = \frac{1}{1!} \frac{d}{dz}\left[\frac{e^z}{z^2+1}\right]_{z=4} = \frac{9e^4}{289} \qquad\blacksquare$$

Example 6.3
Evaluate the residue at the third pole of the function

$$W(z) = \frac{e^{zt}}{(z+1)^3}$$

Solution A direct application of (6.7) yields

$$\text{Res } W(z)|_{z=-1} = \frac{1}{2!} \frac{d^2}{dz^2}(e^{zt})\bigg|_{z=-1} = \frac{1}{2} t^2 e^{-t} \qquad\blacksquare$$

There is no simple way of finding the residue at an essential singularity. The Laurent expansion must be found and the coefficient a_{-1} is thereby obtained. For example, from (3.12) it is seen that the residue of $\exp(1/z)$ at the origin is unity. Fortunately, an essential singularity seldom arises in practical applications.

Sometimes the function takes the form

$$W(z) = \frac{f(z)}{zg(z)} \tag{6.8}$$

where the numerator and denominator are prime to each other, $g(z)$ has no zero at $z = 0$ and cannot be factored readily. The residue due to the pole at zero is given by

$$\text{Res } W(z) = \left.\frac{f(z)}{g(z)}\right|_{z=0} = \frac{f(0)}{g(0)} \tag{6.9}$$

If $z = a$ is the zero of $g(z)$ then the residue at $z = a$ is given by

$$\text{Res } W(z) = \frac{f(a)}{ag'(a)} \tag{6.10}$$

If there are N poles of $g(z)$ then the residues at all simple poles of $W(z)$ are given by

$$\sum \text{Res} = \left.\frac{f(z)}{g(z)}\right|_{z=0} + \sum_{m=1}^{N}\left[f(z)/\left(z\frac{dg(z)}{dz}\right)\right]_{z=a_m} \tag{6.11}$$

If $W(z)$ takes the form $W(z) = f(z)/[h(z)g(z)]$ and the simple poles to the two functions are not common, then the residues at all simple poles are given by

$$\sum \text{Res} = \sum_{m=1}^{N}\frac{f(a_m)}{h(a_m)g'(a_m)} + \sum_{r=1}^{R}\frac{f(b_r)}{h'(b_r)g(b_r)} \tag{6.12}$$

Example 6.4

Find the sum of the residues $e^{2z}/\sin mz$ at the first $N + 1$ poles on the negative axis.

Solution   The simple poles occur at $z = -n\pi/m, n = 0, 1, 2, \ldots$. Thus

$$\sum \text{Res} = \sum_{n=0}^{N}\left[\frac{e^{2z}}{m\cos mz}\right]_{z=-n\pi/m} = \frac{1}{m}\sum_{n=0}^{N}(-1)^n e^{-2n\pi/m} \qquad \blacksquare$$

Example 6.5

Find the sum of the residues of $e^{2z}/(z\cosh mz)$ at the origin and at the first N poles on each side of it.

Solution   The zeros of $\cosh mz$ are $z = -j(n + 1/2)\pi/m$, n integral. Because $\cosh mz$ has no zero at $z = 0$, then (6.12) gives

$$\sum \text{Res} = 1 + \sum_{n=-N}^{N-1}\left[\frac{e^{2z}}{mz\sinh mz}\right]_{z=-(n+\frac{1}{2})\pi j/m} \qquad \blacksquare$$

Example 6.6

Find the residue of $ze^z/\sin mz$ at the origin.

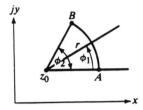

FIGURE 7.1

Solution Because near $z = 0$ $\sin mz \approx mz$ there is no pole at the origin and hence the integral $(1/2\pi j) \int_C z e^z \, dz / \sin mz$ is equal to zero for a contour encircling the origin with radius less than π/m. ∎

7 Aids to integration

The following three theorems will substantially simplify the evaluation of certain integrals in the complex plane. Examples will be found in later applications.

THEOREM 7.1

If AB is the arc of a circle of radius $|z| = R$ for which $\theta_1 \leq \theta \leq \theta_2$ and if $\lim_{R\to\infty}(zW(z)) = k$, a constant that may be zero, then

$$\lim_{R\to\infty} \int_{AB} W(z)\, dz = jk(\theta_2 - \theta_1) \tag{7.1}$$

PROOF Let $zW(z) = k + \varepsilon$, where $\varepsilon \to 0$ as R approaches infinity. Then

$$\int_{AB} W(z)\, dz = \int_{AB} \frac{k+\varepsilon}{z}\, dz = (k+\varepsilon)\int_{\theta_1}^{\theta_2} j\, d\theta = (k+\varepsilon) j(\theta_2 - \theta_1)$$

In carrying out this integration, the procedure employed in Example 1.2 is used. In the limit as R approaches infinity, (7.1) follows. ∎

This theorem can be shown to be valid even if there are a finite number of points on the arc AB for which the $\lim_{R\to\infty}(zW(z)) \neq k$, provided only that the limit remains finite for finite R at these points. This theorem can also be proved true if we choose $\lim_{R\to\infty}(z-a)W(z) = k$ when the integral is taken around the arc $\theta_1 \leq \arg(z-a) \leq \theta_2$ of the circle $|z-a| = r$.

THEOREM 7.2

If AB is the arc of a circle of radius $|z - z_0| = r$ for which $\varphi_1 \leq \varphi \leq \varphi_2$ (as shown in Figure 7.1) and if $\lim_{z\to z_0}[(z - z_0)W(z)] = k$, a constant that may be zero, then

$$\lim_{r\to 0} \int_{AB} W(z)\, dz = jk(\varphi_2 - \varphi_1) \tag{7.2}$$

where r and φ are introduced polar coordinates, with the point $z = z_0$ as origin.

PROOF The proof of this theorem follows along similar lines to that of Theorem 7.1. ∎

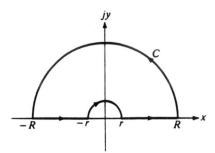

FIGURE 7.2

Note specifically that Theorem 7.1 will allow the evaluation of integrals over infinitely large arcs, whereas Theorem 7.2 will allow the evaluation over infinitely small arcs.

THEOREM 7.3

If the maximum value of $W(z)$ along a path C (not necessarily closed) is M, the maximum value of the integral of $W(z)$ along C is Ml, where l is the length of C. When expressed analytically, this specifies that

$$\left| \int_C W(z)\, dz \right| \le Ml \tag{7.3}$$

PROOF   The proof of this theorem is very simple if recourse is made to the definition of an integral. Thus, from Figure 7.2

$$\int_C W(z)\, dz = \lim_{\substack{\Delta z_s \to 0 \\ n \to \infty}} \sum_{s=1}^{n} W_s \Delta z_s \le M \lim_{\substack{\Delta z_s \to 0 \\ n \to \infty}} \sum_{s=1}^{n} \Delta z_s = Ml \qquad\qquad ∎$$

Jordan's Lemma 7.1

If $t < 0$ and

$$f(z) \to 0 \qquad as\ z \to \infty \tag{7.4}$$

then

$$\int_C e^{tz} f(z)\, dz \to 0 \qquad as\ r \to \infty \tag{7.5}$$

where C is the arc shown in Figure 7.3a.

PROOF   We must assume that the angle of C does not exceed π, $0 \le \arg z \le \pi$. This is not true if $c < 0$. However, the portion of C in the $\operatorname{Re} z < 0$ region will have length not exceeding $\pi |c|$. Hence, because of (7.4) the integration over this portion will tend to zero. From (7.4) it follows that, given $\varepsilon > 0$, we can find a constant r_0 such that

$$|f(z)| < \varepsilon \qquad for\ |z| > r_0$$

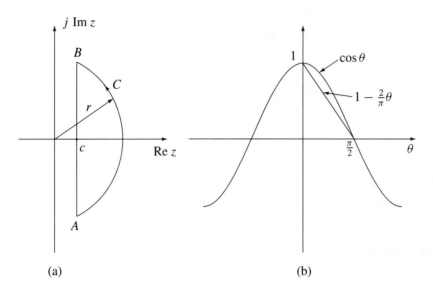

FIGURE 7.3

Hence, with $z = re^{j\theta}, r > r_0$ we obtain $(t < 0)$

$$\left| \int_C e^{tz} f(z) \, dz \right| = \left| \int_{-\pi/2}^{\pi/2} e^{tr(\cos\theta + j\sin\theta)} f(re^{j\theta}) jr e^{j\theta} \, d\theta \right|$$

$$< \varepsilon r \int_{-\pi/2}^{\pi/2} e^{tr\cos\theta} \, d\theta \le \varepsilon r 2 \int_0^{\pi/2} e^{tr(1-2\theta/\pi)} \, d\theta$$

$$= \frac{\varepsilon r \pi}{|t| r}(1 - e^{rt}) < \frac{\pi e}{|t|}$$

Because ε is arbitrarily small, the lemma is verified. ∎

From the above lemma we conclude that if $f(z)$ is analytic everywhere in the $\mathrm{Re}\, z \ge c$ region except at a number of poles, then

$$\int_{\mathrm{Br}} e^{tz} f(z) \, dz = -2\pi j \sum_{k=1}^n \mathrm{Res}_k \qquad t < 0 \tag{7.6}$$

where the Br stands for the Bromwich integration from $c - j\infty$ to $c + j\infty$, which is the line AB in Figure 7.3a. Res_k are the corresponding residues; the minus sign occurs because of the direction of integration along the Br line from B to A. The lemma can easily be extended for $t > 0$ and C be an arc lying on the $\mathrm{Re}\, z < c$ plane. The residues are given by

$$\int_{\mathrm{Br}} e^{tz} f(z) \, dz = 2\pi j \sum_{k=1}^n \mathrm{Res}_k \qquad t > 0 \tag{7.7}$$

THEOREM 7.4   Mellin 1
Let

a) $\phi(z)$ *be analytic in the strip* $\alpha < x < \beta$, *both alpha and beta being real;*

b) $\int_{x-j\infty}^{x+j\infty} |\phi(z)|\, dz = \int_{-\infty}^{\infty} |\phi(x+jy)|\, dy$ *converges*

c) $\phi(z) \to 0$ *uniformly as* $|y| \to \infty$ *in the strip* $\alpha < x < \beta$

d) θ = *real and positive: If*

$$f(\theta) = \frac{1}{2\pi j} \int_{c-j\infty}^{c+j\infty} \theta^{-z} \phi(z)\, dz \tag{7.8}$$

then

$$\phi(z) = \int_0^{\infty} \theta^{z-1} f(\theta)\, d\theta \tag{7.9}$$

THEOREM 7.5   Mellin 2
For θ real and positive, $\alpha < \mathrm{Re}\, z < \beta$, let $f(\theta)$ be continuous or piecewise continuous, and integral (7.9) be absolutely convergent. Then (7.8) follows from (7.9).

THEOREM 7.6   Mellin 3
If in (7.8) and (7.9) we write $\theta = e^{-t}$, t being real, and in (7.9) put p for z and $g(t)$ for $f(e^{-t})$, we get

$$g(t) = \frac{1}{2\pi j} \int_{c-j\infty}^{c+j\infty} e^{zt} \phi(z)\, dz \tag{7.10}$$

$$\phi(p) = \int_0^{\infty} e^{-pt} g(t)\, dt \tag{7.11}$$

Transformation of contour

To evaluate formally the integral

$$I = \int_0^a \cos xt\, dt \tag{7.12}$$

we set $v = xt$ that gives $dx = dv/t$ and, thus,

$$I = \frac{1}{t} \int_0^{at} \cos v\, dv = \frac{\sin at}{t} \tag{7.13}$$

Regarding this as a contour integral along the real axis for $x = 0$ to a, the change to $v = xt$ does not change the real axis. However, the contour is unaltered except in length.

Let t be real and positive. If we set $z = \zeta t$ or $\zeta = z/t$, the contour in the ζ-plane is identical in type with that in the z-plane. If it were a circle of radius r in the z-plane, the contour in the ζ-plane would be a circle of radius r/t. When t is complex $z = r_1 e^{j\theta_1}$, $t = r_2 e^{j\theta_2}$, so $\zeta = (r_1/r_2)e^{j(\theta_1-\theta_2)}$, r_1, θ_1 being variables while r_2 and θ_2 are fixed. If $z = jy = |z|e^{j\theta_1} = |z|e^{j\pi/2}$ and if the phase of t was $\theta_2 = \pi/4$ then the contour in the ζ-plane would be a straight line at 45 degrees with respect to the real axis. In effect, any figure in the z-plane transforms into a similar figure in the ζ-plane, whose orientation and dimensions are governed by the factor $1/t = e^{-j\theta_2}/r_2$.

Example 7.1
Make the transformation $z = \zeta t$ to the integral $I = \int_C e^{z/t}\, \frac{dz}{z}$, where C is a circle of radius r_0 around the origin.

Solution $dz/z = d\zeta/\zeta$ so $I = \int_{C'} e^{\zeta} \frac{d\zeta}{\zeta}$, where C' is a circle around the origin of radius r_0/r $(r = |t|)$.   ∎

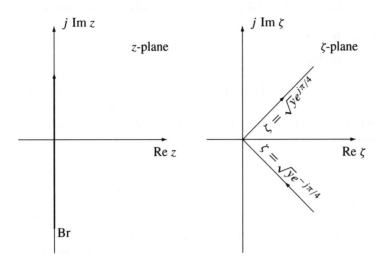

FIGURE 7.4

Example 7.2
Discuss the transformation $z = (\zeta - a)$, a being complex and finite.

Solution This is equivalent to a shift of the origin to point $z = -a$. Neither the contour nor the position of the singularities are affected in relation to each other, so the transformation can be made without any alteration in technique. ∎

Example 7.3
Find the new contour due to transformation $z = \zeta^2$ if the contour was the imaginary axis, $z = jy$.

Solution Choosing the positive square root we have $\zeta = (jy)^{1/2}$ above and $\zeta = (-jy)^{1/2}$ below the origin. Because

$$\sqrt{j} = (e^{j\pi/2})^{1/2} = e^{j\pi/4} \quad \text{and} \quad \sqrt{-j} = (e^{-j\pi/2})^{1/2} = e^{-j\pi/4}$$

the imaginary axis of the z-plane transforms to that in Figure 7.4. ∎

Example 7.4
Evaluate the integral $\int_C \frac{dz}{z}$, where C is a circle of radius 4 units around the origin, under the transformation $z = \zeta^2$.

Solution The integral has a pole at $z = 0$ and its value is $2\pi j$. If we apply the transformation $z = \zeta^2$ then $dz = 2\zeta \, d\zeta$. Also $\zeta = \sqrt{z} = \sqrt{r}e^{j\theta/2}$ if we choose the positive root. From this relation we observe that as the z traces a circle around the origin, the ζ traces a half-circle from 0 to π. Hence the integral becomes

$$2 \int_{C'} \frac{d\zeta}{\zeta} = 2 \int_0^\pi \frac{\rho j e^{j\theta}}{\rho e^{j\theta}} d\theta = 2\pi j$$

as was expected. ∎

8   The Bromwich contour

The Bromwich contour takes the form

$$f(t) = \frac{1}{2\pi j} \int_{c-j\infty}^{c+j\infty} e^{zt} F(z)\, dz \tag{8.1}$$

where $F(z)$ is a function of z, all of whose singularities lie on the left of the path, and t is the time, which is always real and positive, $t > 0$.

Finite number of poles

Let us assume that $F(z)$ has n poles at $p_1, p_2, \ldots, p_n$ and no other singularities; this case includes the important case of **rational transforms**. To utilize the Cauchy's integral theorem, we must express $f(t)$ as an integral along a closed contour. Figure 8.1 shows such a situation. We know from Jordan's lemma (see Section 7) that if $F(z) \to 0$ as $|z| \to \infty$ on the contour C then for $t > 0$

$$\lim_{R\to\infty} \int_C e^{tz} F(z)\, dz \to 0 \qquad t > 0 \tag{8.2}$$

and because

$$\int_{c-jy}^{c+jy} e^{tz} F(z)\, dz \to \int_{\mathrm{Br}} e^{tz} F(z)\, dz \qquad y \to \infty \tag{8.3}$$

we conclude that $f(t)$ can be written as a limit,

$$f(t) \xrightarrow[R\to\infty]{} \frac{1}{2\pi j} \int_C e^{zt} F(z)\, dz \tag{8.4}$$

of an integral along the closed path as shown in Figure 8.1. If we take R large enough to contain all the poles of $F(z)$ then the integral along C is independent of R. Therefore we write

$$f(t) = \frac{1}{2\pi j} \int_C e^{zt} F(z)\, dz \tag{8.5}$$

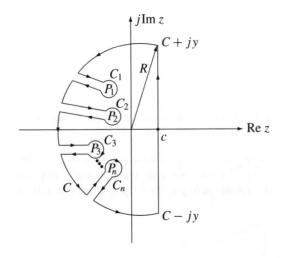

FIGURE 8.1

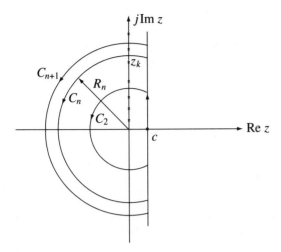

FIGURE 8.2

Using Cauchy's theorem it follows that

$$\int_C e^{zt} F(z)\, dz = \sum_{k=1}^{n} \int_{C_k} e^{zt} F(z)\, dz \tag{8.6}$$

where C_k's are the contours around each pole.

1. For simple poles we obtain

$$f(t) = \sum_{k=1}^{n} F_k(z_k) e^{z_k t} \qquad t > 0 \tag{8.7}$$

$$F_k(z_k) = F(z)(z - z_k)|_{z=z_k}$$

2. For a multiple pole of $m + 1$ multiplicity we obtain

$$\int_{C_k} e^{zt} F(z)\, dz = \int_{C_k} \frac{e^{zt} F_k(z)}{(z - z_k)^{m+1}}\, dz = \frac{2\pi j}{m!} \frac{d^m}{dz^m} \left[e^{zt} F_k(z) \right]\Big|_{z=z_k} \tag{8.8}$$

3. Infinitely many poles (see Figure 8.2)

If we can find circular arcs with radii tending to infinity such that

$$F(z) \to 0 \qquad \text{as } z \to \infty \qquad \text{on } C_n \tag{8.9}$$

Applying Jordan's lemma to the integral along those arcs, we obtain

$$\int_{C_n} e^{zt} F(z)\, dz \xrightarrow[n \to \infty]{} 0 \qquad t > 0 \tag{8.10}$$

and with C_n' the closed curve, consisting of C_n and the vertical line $\operatorname{Re} z = c$, we obtain

$$f(t) = \lim_{n \to \infty} \frac{1}{2\pi j} \int_{C_n'} e^{zt} F(z)\, dz \qquad t > 0 \tag{8.11}$$

Hence, for simple poles $z_1, z_2, \ldots, z_n$ of $F(z)$ we obtain

$$f(t) = \sum_{k=1}^{\infty} F_k(z_k)e^{z_k t} \tag{8.12}$$

where $F_k(z) = F(z)(z - z_k)$.

Example 8.1

Find $f(t)$ from its transformed value $F(z) = 1/(z \cosh az)$, $a > 0$.

Solution   The poles of the above function are

$$z_0 = 0, \qquad z_k = \pm j\frac{(2k-1)\pi}{2a} \quad k = 1, 2, 3, \ldots$$

We select the arcs C_n such that their radii are $R_n = jn\pi$. It can be shown that $1/\cosh az$ is bounded on C_n and, therefore, $1/(z \cosh az) \to 0$ as $z \to \infty$ on C_n. Hence

$$zF(z)|_{z=0} = 1, \qquad (z - z_k)F(z)|_{z=z_k} = \frac{(-1)^k 2}{(2k-1)\pi}$$

and from (8.12) we obtain

$$f(t) = 1 + \frac{2}{\pi}\sum_{k=1}^{\infty}\frac{(-1)^k}{2k-1}e^{z_k t} + \frac{2}{\pi}\sum_{k=1}^{\infty}\frac{(-1)^k}{2k-1}e^{-z_k t}$$

$$= 1 + \frac{4}{\pi}\sum_{k=1}^{\infty}\frac{(-1)^k}{2k-1}\cos\frac{(2k-1)\pi t}{2a} \qquad\qquad \blacksquare$$

Branch points and branch cuts

The singularities that have been considered are those points at which $|W(z)|$ ceases to be finite. At a branch point the absolute value of $W(z)$ may be finite but $W(z)$ is not single-valued, and hence is not regular. One of the simplest functions with these properties is

$$W_1(z) = z^{1/2} = \sqrt{r}e^{j\theta/2} \tag{8.13}$$

which takes on two values for each value of z, one the negative of the other depending on the choice of θ. This follows because we can write an equally valid form for $z^{1/2}$ as

$$W_2(z) = \sqrt{r}e^{j(\theta+2\pi)/2} = -\sqrt{r}e^{j\theta/2} = -W_1(z) \tag{8.14}$$

Clearly, $W_1(z)$ is not continuous at points on the positive real axis because

$$\lim_{\theta \to 2\pi}(\sqrt{r}e^{j\theta/2}) = -\sqrt{r} \qquad \text{while} \quad \lim_{\theta \to 0}(\sqrt{r}e^{j\theta/2}) = \sqrt{r}$$

Hence, $W'(z)$ does not exist when z is real and positive. However, the branch $W_1(z)$ is analytic in the region $0 \le \theta \le 2\pi$, $r \to 0$. The part of the real axis where $x \ge 0$ is called a **branch cut** for the branch $W_1(z)$ and the branch is analytic except at points on the cut. Hence, the cut is a boundary introduced so that the corresponding branch is single valued and analytic throughout the open region bounded by the cut.

Suppose that we consider the function $W(z) = z^{1/2}$ and contour C, as shown in Figure 8.3a, which encloses the origin. Clearly, after one complete circle in the positive direction enclosing the origin, θ is increased by 2π, given a value of $W(z)$ that changes from $W_1(z)$ to $W_2(z)$;

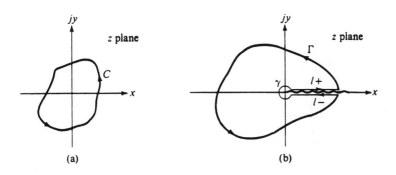

FIGURE 8.3

that is, the function has changed from one branch to the second. To avoid this and to make the function analytic, the contour C is replaced by a contour Γ, which consists of a small circle γ surrounding the branch point, a semi-infinite cut connecting the small circle and C, and C itself [as shown in Figure 8.3b]. Such a contour, which avoids crossing the branch cut, ensures that $W(z)$ is single valued. Because $W(z)$ is single valued and excludes the origin, we would write for this composite contour C

$$\int_C W(z)\,dz = \int_\Gamma + \int_{l-} + \int_\gamma + \int_{l+} = 2\pi j \sum \text{Res} \tag{8.15}$$

The evaluation of the function along the various segments of C proceeds as before.

Example 8.2
If $0 < a < 1$, show that

$$\int_0^\infty \frac{x^{a-1}}{1+x}\,dx = \frac{\pi}{\sin a\pi}$$

Solution Consider the integral

$$\oint_C \frac{z^{a-1}}{1+z}\,dz = \int_\Gamma + \int_{l-} + \int_\gamma + \int_{l+} = I_1 + I_2 + I_3 + I_4 = \sum \text{Res}$$

which we will evaluate using the contour shown in Figure 8.4. Under the conditions

$$\left| \frac{z^a}{1+z} \right| \to 0 \qquad \text{as } |z| \to 0 \qquad \text{if } a > 0$$
$$\left| \frac{z^a}{z+1} \right| \to 0 \qquad \text{as } |z| \to \infty \qquad \text{if } a < 1$$

the integral becomes by (7.1)

$$\int_\Gamma \to 0 \qquad \int_{l-} = -e^{2\pi ja} \int_0^\infty$$

by (7.2)

$$\int_\gamma \to 0 \qquad \int_{l+} = 1 \int_0^\infty$$

Thus

$$(1 - e^{2\pi ja}) \int_0^\infty \frac{x^{a-1}}{1+x}\,dx = 2\pi j \sum \text{Res}$$

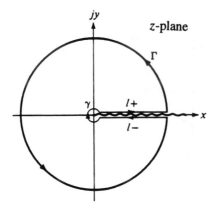

FIGURE 8.4

Further, the residue at the pole $z = -1$, which is enclosed, is

$$\lim_{z=e^{j\pi}} (1 + z) \frac{z^{a-1}}{1 + z} = e^{j\pi(a-1)} = -e^{j\pi a}$$

Therefore,

$$\int_0^\infty \frac{x^{a-1}}{x + 1} dx = 2\pi j \frac{e^{j\pi a}}{e^{j\pi a} - 1} = \frac{\pi}{\sin \pi a} \qquad \blacksquare$$

If, for example, we have the integral $(1/2\pi j) \int_{\mathrm{Br}_1} \frac{e^{zt} dz}{z^{v+1}}$ to evaluate with Re $v > -1$ and t real and positive, we observe that the integral has a branch point at the origin if v is a nonintegral constant. Because the integral vanishes along the arcs as $R \to \infty$, the equivalent contour can assume the form depicted in Figure 8.5a and marked Br_2. For the contour made up of Br_1, Br_2, the arc is closed and contains no singularities and, hence, the integral around the contour is zero. Because the arcs do not contribute any value, provided Re $v > -1$, the integral along Br_1 is equal to that along Br_2, both being described positively. The angle γ between the barrier and the positive real axis may have any value between $\pi/2$ and $3\pi/2$. When the only singularity is a branch point at the origin, the contour of Figure 8.5b is an approximate one.

Example 8.3
Evaluate the integral $I = \frac{1}{2\pi j} \int_{\mathrm{Br}_2} \frac{e^z dz}{\sqrt{z}}$, where Br_2 is the contour shown in Figure 8.4b.

Solution

1) Write $z = e^{j\theta}$ on the circle. Hence we get

$$I_1 = \frac{1}{2\pi j} \int_{-\pi}^{\pi} \frac{e^{re^{j\theta}} d(re^{j\theta})}{\sqrt{r}e^{j\theta/2}} = \frac{\sqrt{r}}{2\pi} \int_{-\pi}^{\pi} e^{r(\cos\theta + j\sin\theta) + j\theta/2} d\theta$$

2) On the line below the barrier $z = x \exp(-j\pi)$ where $x = |x|$. Hence the integral becomes

$$I_2 = \frac{1}{2\pi j} \int_{\infty}^{r} \frac{e^{xe^{-j\pi}} d(xe^{-j\pi})}{\sqrt{x}e^{-j\pi/2}} = \frac{1}{2\pi} \int_r^\infty e^{-x} x^{-1/2} dx$$

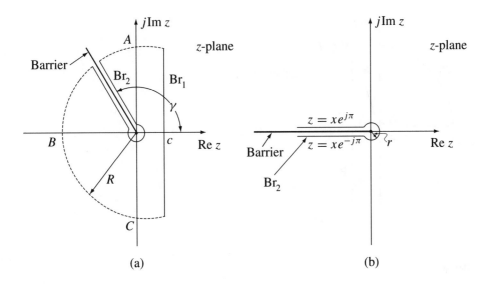

FIGURE 8.5

3) On the line above the barrier $z = x \exp(j\pi)$ and, hence,

$$I_3 = \frac{1}{2\pi j} \int_r^\infty \frac{e^{xe^{j\pi}} d(xe^{j\pi})}{\sqrt{xe^{j\pi/2}}} = \frac{1}{2\pi} \int_r^\infty e^{-x} x^{-1/2} \, dx$$

Hence we have

$$I_2 + I_3 = \frac{1}{\pi} \int_r^\infty e^{-x} x^{-1/2} \, dx$$

As $r \to 0$, $I_1 \to 0$ and, hence,

$$I = I_1 + I_2 + I_3 = \frac{1}{\pi} \int_0^\infty e^{-x} x^{-1/2} \, dx = \frac{\sqrt{\pi}}{\pi} = \frac{1}{\sqrt{\pi}} \qquad\blacksquare$$

Example 8.4
Evaluate the integral $f(t) = \int_{Br} \frac{e^{zt} e^{-a\sqrt{z}}}{\sqrt{z}} \, dz$, $a > 0$ (see Figure 8.6).

Solution   The origin is a point branch and we select the negative axis as the barrier. We select the positive value of $\sqrt{z}$ when z takes positive real values in order that the integral vanishes as z approaches infinity in the region $\text{Re } z > \gamma$, where γ indicates the region of convergence, $\gamma \leq c$. Hence we obtain

$$z = re^{j\theta} \qquad -\pi < \theta \leq \pi \qquad \sqrt{z} = \sqrt{r}e^{j\theta/2} \qquad (8.16)$$

The curve $C = Br + C_1 + C_2 + C_3$ encloses a region with no singularities and, therefore, Cauchy's theorem applies (the intergrand is analytic in the region). Hence

$$\int_C e^{zt} \frac{e^{-a\sqrt{z}}}{\sqrt{z}} \, dz = 0 \qquad (8.17)$$

It is easy to see that the given function converges to zero as R approaches infinity and therefore

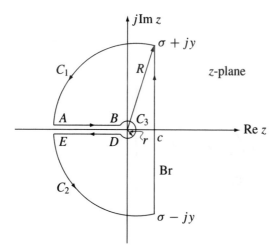

FIGURE 8.6

the integration over $C_1 + C_2$ does not contribute any value. For z on the circle we obtain

$$\left| \frac{e^{zt}e^{-a\sqrt{z}}}{\sqrt{z}} \right| \leq \frac{e^{rt}}{\sqrt{r}}$$

Therefore, for fixed $t > 0$ we obtain

$$\left| \int_{C_3} e^{zt} \frac{e^{-a\sqrt{z}}}{\sqrt{z}} \, dz \right| \leq 2\pi r \frac{e^{rt}}{\sqrt{r}} = \lim_{r \to 0} 2\pi r \frac{e^{rt}}{\sqrt{r}} = 0$$

because

$$\left| \int_C f(z) \, dz \right| \leq ML$$

where L is the length of the contour and $|f(z)| < M$ for z on C.

On AB, $z = -x$, $\sqrt{z} = j\sqrt{x}$, and on DE, $z = -x$, $\sqrt{z} = -j\sqrt{x}$. Therefore we obtain

$$\int_{AB+DC} e^{zt} \frac{e^{-a\sqrt{z}}}{\sqrt{z}} \, dz \xrightarrow[\substack{r \to 0 \\ R \to \infty}]{} -\int_\infty^0 e^{-xt} \frac{e^{ja\sqrt{x}}}{j\sqrt{x}} \, dx - \int_0^\infty e^{-xt} \frac{e^{-ja\sqrt{x}}}{-j\sqrt{x}} \, dx \qquad (8.18)$$

But from (8.1)

$$\int_{Br} e^{zt} \frac{e^{-a\sqrt{z}}}{\sqrt{z}} \, dz = 2\pi j f(t) \qquad (8.19)$$

and, hence, (8.17) and (8.19) reduce to

$$f(t) + \frac{1}{2\pi j} \int_0^\infty e^{-xt} \frac{e^{ja\sqrt{x}} + e^{-ja\sqrt{x}}}{j\sqrt{x}} \, dx = 0 \qquad (8.20)$$

If we set $x = y^2$ we have

$$\int_0^\infty e^{-xt} \frac{\cos a\sqrt{x}}{\sqrt{x}} \, dx = 2 \int_0^\infty e^{-y^2 t} \cos ay \, dy \qquad (8.21)$$

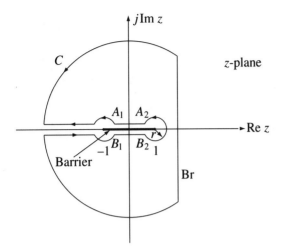

FIGURE 8.7

But (see Fourier transform of Gaussian function, Chapter 3)

$$2 \int_0^\infty e^{-y^2 t} \cos ay \, dy = \sqrt{\frac{x}{t}} e^{-a^2/4t} \tag{8.22}$$

and hence (8.21) becomes

$$f(t) = \frac{1}{\sqrt{\pi t}} e^{-a^2/4t} \tag{8.23}$$

∎

Example 8.5
Evaluate the integral $I = \frac{1}{2\pi j} \int_C \frac{e^{zt} dz}{\sqrt{z^2 - 1}}$ where C is the contour shown in Figure 8.7.

Solution   The Br contour is equivalent to the dumbbell-type contour shown in Figure 8.7, $B_1 B_2 A_2 A_1 B_1$. Set the phase along the line $A_2 A_1$ equal to zero (it can also be set equal to π). Then on $A_2 A_1$ $z = x$ from $+1$ to -1. Hence we have

$$I_1 = \frac{1}{2\pi j} \int_1^{-1} \frac{e^{xt} dx}{\sqrt{x^2 - 1}} = \frac{1}{2\pi} \int_{-1}^1 \frac{e^{xt} dx}{\sqrt{1 - x^2}} \qquad |x| < 1 \tag{8.24}$$

By passing around the $z = -1$ point the phase changes by π and hence on $B_1 B_2$ $z = x \exp(2\pi j)$. The change by 2π is due to the complete transversal of the contour that contains two branch points. Hence we obtain

$$I_2 = -\frac{1}{2\pi j} \int_{-1}^1 \frac{e^{xt} dx}{\sqrt{x^2 - 1}} = \frac{1}{2\pi} \int_{-1}^1 \frac{e^{xt} dx}{\sqrt{1 - x^2}} \tag{8.25}$$

Changing the origin to -1, we set $\zeta = z + 1$ or $z = \zeta - 1$, which gives

$$I_3 = \frac{e^{-t}}{2\pi j} \int \frac{e^{\zeta t} d\zeta}{\sqrt{[(\zeta - 2)\zeta]}} \tag{8.26}$$

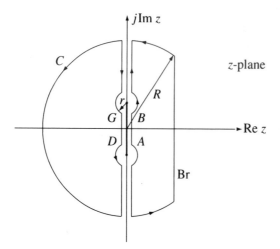

FIGURE 8.8

On the small circle with $z = -1$ as center, $\zeta = r \exp(j\theta)$ and we get

$$I_3 = \frac{e^{-t}}{2\pi} \int_{\pi}^{-\pi} \frac{e^{rt(\cos\theta + j\sin\theta) + (j\theta/2)}\sqrt{r}d\theta}{\sqrt{re^{j\theta} - 2}} \qquad (8.27)$$

When $\theta = 0$ the intergrand has the value $+\sqrt{r}e^{rt}/\sqrt{r - 2}$, and for $\theta = 2\pi$ the value is $-\sqrt{r}e^{rt}/\sqrt{r - 2}$. Therefore, the intergrand changes sign in rounding the branch point at $z = -1$. Similarly for the branch point at $z = 1$, where the change is from $-$ to $+$. As $r \to 0$, $I_3 \to 0$, and thus I_3 vanishes. The same is true for the branch point at $z = -1$. Therefore, by setting $x = \cos\theta$ we obtain

$$I = I_1 + I_2 = \frac{1}{\pi}\int_{-1}^{1}\frac{e^{xt}}{\sqrt{1 - x^2}}\,dx = \frac{1}{\pi}\int_{0}^{\pi}e^{t\cos\theta}\,d\theta = \frac{1}{\pi}\int_{0}^{\pi}\sum_{k=0}^{\infty}\frac{(t\cos\theta)^k}{k!}\,d\theta$$

$$= \frac{1}{\pi}\left[\pi + \pi\frac{1}{2}\frac{t^2}{2!} + \pi\frac{3}{4}\frac{1}{2}\frac{t^4}{4!} + \pi\frac{5}{6}\frac{3}{4}\frac{1}{2}\frac{t^6}{6!} + \cdots\right]$$

$$= 1 + \frac{t^2}{2^2} + \frac{t^4}{2^2 4^2} + \frac{t^6}{2^2 4^2 6^2} + \cdots = \sum_{k=0}^{\infty}\frac{\left(\frac{1}{2}t\right)^{2k}}{(k!)^2} = I_0(t)$$

where $I_0(t)$ is the modified Bessel function of the first kind and zero order. ∎

Example 8.6
Evaluate the integral $I = \int_C \frac{e^{zt}}{\sqrt{z^2+1}}\,dz$ where C is the closed contour shown in Figure 8.8.

Solution   The Br contour is equal to the dumbbell-type contour as shown in Figure 8.8, $ABGDA = C_1$. Hence we have

$$f(t) = \frac{1}{2\pi j}\int_{C_1}\frac{e^{zt}}{\sqrt{z^2 + 1}}\,dz \qquad (8.29)$$

But

$$\left| \frac{e^{zt}}{\sqrt{z^2 + 1}} \right| < \frac{e^{rt}}{\sqrt{r}\sqrt{2 - r}}$$

on the circle on the $+j$ branch point and therefore for $t > 0$

$$\left| \int \frac{e^{zt}}{\sqrt{z^2 + 1}} \, dz \right| < \frac{2\pi \sqrt{r} e^{rt}}{\sqrt{2 - r}} \to 0 \qquad \text{as } r \to 0$$

We obtain similar results for the contour around the $-j$ branch point. However

$$\text{On } AB, z = j\omega, \sqrt{1 + z^2} = \sqrt{1 - \omega^2}; \text{ on } GD, z = j\omega, \sqrt{1 + z^2} = -\sqrt{1 - \omega^2}$$

and, therefore, for $t > 0$ we obtain

$$f(t) = \frac{j}{2\pi j} \int_{-1}^{1} \frac{e^{j\omega t}}{\sqrt{1 - \omega^2}} \, d\omega + \frac{j}{2\pi j} \int_{1}^{-1} \frac{e^{j\omega t}}{-\sqrt{1 - \omega^2}} \, d\omega = \frac{1}{\pi} \int_{-1}^{1} \frac{\cos \omega t}{\sqrt{1 - \omega^2}} \, d\omega$$

If we set $\omega = \sin \theta$ (see also Chapter 1)

$$f(t) = \frac{1}{\pi} \int_{-\pi/2}^{\pi/2} \cos(t \sin \theta) \, d\theta = J_0(t)$$

where $J_0(t)$ is the Bessel function of the first kind. ∎

---

9 Evaluation of definite integrals

The principles discussed above find considerable applicability in the evaluation of certain definite real integrals. This is a common application of the developed theory, as it is often extremely difficult to evaluate some of these real integrals by other methods. We employ such methods in the evaluation of Fourier integrals. In practice the given integral is replaced by a complex function that yields the specified integrand in its appropriate limit. The integration is then carried out in the complex plane, with the real integral being extracted for the required result. The following several examples show this procedure.

Evaluation of the integrals of certain periodic functions (0 to 2π)

An integral of the form

$$I = \int_0^{2\pi} F(\cos \theta, \sin \theta) \, d\theta \tag{9.1}$$

where the integral is a **rational function** of $\cos \theta$ and $\sin \theta$ finite on the range of integration, and can be integrated by setting $z = \exp(j\theta)$,

$$\cos \theta = \frac{1}{2}(z + z^{-1}), \qquad \sin \theta = \frac{1}{2j}(z - z^{-1}) \tag{9.2}$$

The integral (9.1) takes the form

$$I = \int_C F(z) \, dz \tag{9.3}$$

where $F(z)$ is a rational function of z finite on C, which is a circle of radius unity with center at the origin.

Example 9.1

If $0 < a < 1$, find the value of the integral

$$I = \int_0^{2\pi} \frac{d\theta}{1 - 2a\cos\theta + a^2} \tag{9.4}$$

Solution   Introducing (9.2) in (9.4) we obtain

$$I = \int_C \frac{dz}{j(1 - az)(z - a)} \tag{9.5}$$

The only pole inside the unit circle is at a. Therefore, by residue theory we have

$$I = 2\pi j \lim_{z \to a} \frac{z - a}{j(1 - az)(z - a)} = \frac{2\pi}{1 - a^2} \qquad \blacksquare$$

Evaluation of integrals with limits $-\infty$ and $+\infty$

We can evaluate the integral $I = \int_{-\infty}^{\infty} F(x)\,dx$ provided that the function $F(z)$ satisfies the following properties:

1. It is analytic when the imaginary part of z is positive or zero (except at a finite number of poles).

2. It has no poles on the real axis.

3. As $|z| \to \infty$, $zF(z) \to 0$ uniformly for all value of $\arg z$ such that $0 \le \arg z \le \pi$, provided that

4. when x is real, $xF(x) \to 0$ as $x \to \pm\infty$, in such a way that $\int_0^{\infty} F(x)\,dx$ and $\int_{-\infty}^0 F(x)\,dx$ both converge.

The integral is given by

$$I = \int_C F(z)\,dz = 2\pi j \sum \text{Res} \tag{9.6}$$

where the contour is the real axis and a semicircle having its center in the origin and lying above the real axis.

Example 9.2

Evaluate the integral $I = \int_{-\infty}^{\infty} \frac{dx}{(x^2+1)^3}$.

Solution   The integral becomes

$$I = \int_C \frac{dz}{(z^2 + 1)^3} = \int_C \frac{dz}{(z + j)^3(z - j)^3}$$

which has one pole at j of order three (see [6.7]). Hence we obtain

$$I = \frac{1}{2!} \frac{d^2}{dz^2} \left[\frac{1}{(z + j)^3} \right]\bigg|_{z=j} = -j\frac{3}{16} \qquad \blacksquare$$

Example 9.3

Evaluate the integral $I = \int_0^{\infty} \frac{dx}{x^2+1}$.

Solution   The integral becomes

$$I = \int_C \frac{dz}{z^2 + 1}$$

where C is the contour of the real axis and the upper semicircle. From $z^2 + 1 = 0$ we obtain $z = \exp(j\pi/2)$ and $z = \exp(-j\pi/2)$. Only the pole $z = \exp(j\pi/2)$ exists inside the contour. Hence we obtain

$$2\pi j \lim_{z \to e^{j/\pi/2}} \left(\frac{z - e^{j\pi/2}}{(z - e^{j\pi/2})(z - e^{-j\pi/2})} \right) = \pi$$

Therefore, we have

$$\int_{-\infty}^{\infty} \frac{dx}{x^2 + 1} = 2 \int_0^{\infty} \frac{dx}{x^2 + 1} = \pi \qquad \text{or} \qquad I = \frac{\pi}{2} \qquad \blacksquare$$

Certain infinite integrals involving sines and cosines

If $F(z)$ satisfies conditions (1), (2), and (3), above, and if $m > 0$, then $F(z)e^{jmz}$ also satisfies the same conditions. Hence $\int_0^{\infty} [F(x)e^{jmx} + F(-x)e^{-jmx}] dx$ is equal to $2\pi j \sum \text{Res}$, where $\sum \text{Res}$ means the sum of the residues of $F(z)e^{jmz}$ at its poles in the upper half-plane. Therefore

   1. If $F(x)$ is an even function, that is, $F(x) = F(-x)$, then

$$\int_0^{\infty} F(x) \cos mx \, dx = j\pi \sum \text{Res} . \tag{9.7}$$

   2. If $F(x)$ is an odd function, that is, $F(x) = -F(-x)$, then

$$\int_0^{\infty} F(x) \sin mx \, dx = \pi \sum \text{Res} \tag{9.8}$$

Example 9.4

Evaluate the integral $I = \int_0^{\infty} \frac{\cos x}{x^2 + a^2} \, dx$, $a > 0$.

Solution   Consider the integral

$$I_1 = \int_C \frac{e^{jz}}{z^2 + a^2} \, dz$$

where the contour is the real axis and the infinite semicircle on the upper side with respect to the real axis. The contour encircles the pole ja. Hence

$$\int_C \frac{e^{jz}}{z^2 + a^2} \, dz = 2\pi j \frac{e^{jja}}{2ja} = \frac{\pi}{a} e^{-a}$$

However

$$\int_{-\infty}^{\infty} \frac{e^{jz}}{z^2 + a^2} \, dz = \int_{-\infty}^{\infty} \frac{\cos x}{x^2 + a^2} \, dx + j \int_{-\infty}^{\infty} \frac{\sin x}{x^2 + a^2} \, dx = \int_{-\infty}^{\infty} \frac{\cos x}{(x^2 + a^2)} \, dx$$

because the integrand of the third integral is odd and therefore is equal to zero. From the last two equations we find that

$$I = \int_0^{\infty} \frac{\cos x}{x^2 + a^2} \, dx = \frac{\pi}{2a} e^{-a}$$

because the integrand is an even function. $\blacksquare$

Example 9.5

Evaluate the integral $I = \int_0^\infty \frac{x \sin ax}{x^2 + b^2}\, dx$, $k > 0$ and $a > 0$.

Solution   Consider the integral

$$I_1 = \int_C \frac{z e^{jaz}}{z^2 + b^2}\, dz$$

where C is the same type of contour as in Example 9.4. Because there is only one pole at $z = jb$ in the upper half of the z-plane, then

$$I_1 = \int_{-\infty}^\infty \frac{z e^{jaz}}{z^2 + b^2}\, dz = 2\pi j \frac{jb e^{jajb}}{2jb} = j\pi e^{-ab}$$

Because the integrand $x \sin ax / (x^2 + b^2)$ is an even function, we obtain

$$I_1 = j \int_{-\infty}^\infty \frac{x \sin ax}{x^2 + b^2}\, dx = j\pi e^{-ab} \qquad \text{or} \qquad I = \frac{\pi}{2} e^{-ab} \qquad \blacksquare$$

Example 9.6

Show that $\int_{-\infty}^\infty \frac{x \sin \pi x}{x^2 + 2x + 5}\, dx = -\pi e^{-2\pi}$. $\blacksquare$

Integrals of the form $\int_0^\infty x^{\alpha-1} f(x)\, dx$, $0 < \alpha < 1$

It can be shown that the above integral has the value

$$I = \int_0^\infty x^{\alpha-1} f(x)\, dx = \frac{2\pi j}{1 - e^{j2\pi\alpha}} \sum_{k=1}^N \text{Res}\, [z^{\alpha-1} f(z)]\big|_{z=z_k} \qquad (9.9)$$

where $f(z)$ has N singularities and $z^{\alpha-1} f(z)$ has a branch point at the origin.

Example 9.7

Evaluate the integral $I = \int_0^\infty \frac{x^{-1/2}}{x+1}\, dx$.

Solution   Because $x^{-1/2} = x^{1/2-1}$ it is implied that $\alpha = 1/2$. From the integrand we observe that the origin is a branch point and the $f(x) = 1/(x + 1)$ has a pole at -1. Hence from (9.9) we obtain

$$I = \frac{2\pi j}{1 - e^{j2\pi/2}} \text{Res}\left[\frac{z^{-1/2}}{z+1}\right]\bigg|_{z=-1} = \frac{2\pi j}{j(1 - e^{j\pi})} = \pi$$

We can also proceed by considering the integral $I = \int_C \frac{z^{-1/2}}{z+1}\, dz$. Because $z = 0$ is a branch point we choose the contour C as shown in Figure 9.1. The integrand has a simple pole at $z = -1$ inside the contour C. Hence the residue at $z = -1 = \exp(j\pi)$ and is

$$\text{Res}|_{z=-1} = \lim_{z \to -1} (z + 1) \frac{z^{-1/2}}{z+1} = e^{-j\frac{\pi}{2}}$$

Therefore we write

$$\oint_C \frac{z^{-1/2}}{z+1}\, dz = \int_{AB} + \int_{BDEFG} + \int_{GH} + \int_{HJA} = e^{-j\pi/2}$$

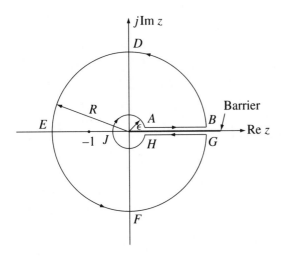

FIGURE 9.1

The above integrals take the following form:

$$\int_{\varepsilon}^{R} \frac{x^{-1/2}}{x+1}\, dx + \int_{0}^{2\pi} \frac{(Re^{j\theta})^{-1/2} j\, Re^{j\theta}\, d\theta}{1 + Re^{j\theta}}$$

$$+ \int_{R}^{\varepsilon} \frac{(xe^{j2\pi})^{-1/2}}{1 + xe^{j2\pi}}\, dx + \int_{2\pi}^{0} \frac{(\varepsilon e^{j\theta})^{-1/2} j\varepsilon e^{j\theta}\, d\theta}{1 + \varepsilon e^{j\theta}} = j2\pi e^{-j\pi/2}$$

where we have used $z = x\exp(j2\pi)$ for the integral along GH, because the argument of z is increased by 2π in going around the circle $BDEFG$.

Taking the limit as $\varepsilon \to 0$ and $R \to \infty$ and noting that the second and fourth integrals approach zero, we find

$$\int_{0}^{\infty} \frac{x^{-1/2}}{x+1}\, dx + \int_{\infty}^{0} \frac{e^{-j2\pi/2} x^{-1/2}}{x+1}\, dx = j2\pi e^{-j\pi/2}$$

or

$$(1 - e^{-j\pi}) \int_{0}^{\infty} \frac{x^{-1/2}}{x+1}\, dx = j2\pi e^{-j\pi/2} \qquad \text{or} \qquad \int_{0}^{\infty} \frac{x^{-1/2}}{x+1}\, dx = \frac{j2\pi(-j)}{2} = \pi \qquad \blacksquare$$

Miscellaneous definite integrals

The following examples will elucidate some of the approaches that have been used to find the values of definite integrals.

Example 9.8
Evaluate the integral $I = \int_{-\infty}^{\infty} \frac{1}{x^2 + a^2}\, dx, a > 0$.

Solution   We write (see Figure 9.2)

$$\int_{C} \frac{dz}{z^2 + a^2}\, dz = \int_{AB} + \int_{BDA} = 2\pi j \sum \text{Res}$$

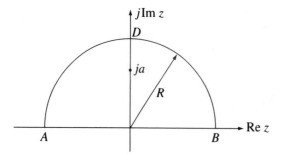

FIGURE 9.2

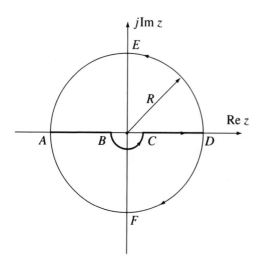

FIGURE 9.3

As $R \to \infty$

$$\int_{BDA} \frac{dz}{z^2 + a^2} = \int_0^\pi \frac{Rje^{j\theta}d\theta}{R^2 e^{j2\theta} + a^2} \xrightarrow[R \to \infty]{} 0$$

and, therefore, we have

$$\int_{AB} \frac{dx}{x^2 + a^2} = \int_{-\infty}^\infty \frac{dx}{x^2 + a^2} = 2\pi j \left. \frac{z - ja}{z^2 + a^2} \right|_{z=ja} = 2\pi j \frac{1}{2ja} = \frac{\pi}{a}$$

∎

Example 9.9
Evaluate the integral $I = \int_0^\infty \frac{\sin ax}{x} dx$.

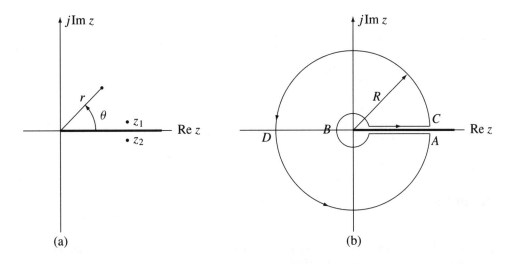

FIGURE 9.4

Solution   Because $\sin az/z$ is analytic near $z = 0$, we indent the contour around the origin as shown in Figure 9.3. With a positive we write

$$\int_0^\infty \frac{\sin ax}{x}\,dx = \frac{1}{2}\int_{ABCD}\frac{\sin az}{z}\,dz = \frac{1}{4j}\int_{ABCD}\left[\frac{e^{jaz}}{z} - \frac{e^{-jaz}}{z}\right]dz$$

$$= \frac{1}{4j}\int_{ABCDA}\frac{e^{jaz}}{z}\,dz - \frac{1}{4j}\int_{ABCDFA}\frac{e^{-jaz}}{z}\,dz = \frac{1}{4j}\left[2\pi j\frac{1}{1} - 0\right] = \frac{\pi}{2}$$

because the lower contour does not include any singularity. Because $\sin ax$ is an odd function of a and $\sin 0 = 0$, we obtain

$$\int_0^\infty \frac{\sin x}{x}\,dx = \begin{cases} \dfrac{\pi}{2} & a > 0 \\[2mm] 0 & a = 0 \\[2mm] -\dfrac{\pi}{2} & a < 0 \end{cases}$$

■

Example 9.10
Evaluate the integral $I = \int_0^\infty \frac{dx}{1+x^3}$.

Solution   Because the integrand $f(x)$ is odd, we introduce the $\ln z$. Taking a branch cut along the positive real axis we obtain

$$\ln z = \ln r + j\theta \qquad 0 \le \theta < 2\pi$$

The discontinuity of $\ln z$ across the cut is (see Figure 9.4a)

$$\ln z_1 - \ln z_2 = -2\pi j$$

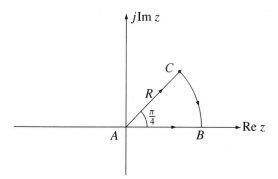

FIGURE 9.5

Therefore, if $f(z)$ is analytic along the real axis and the contribution around an infinitesimal circle at the origin is vanishing, we obtain

$$\int_0^\infty f(x)\,dx = -\frac{1}{2\pi j}\int_{ABC} f(x)\ln(z)\,dz$$

If further $f(z) \to 0$ as $|z| \to \infty$, the contour can be completed with CDA (see Figure 9.4b). If $f(z)$ has simple poles of order one at points z_k, with residues $\mathrm{Res}(f, z_k)$, we obtain

$$\int_0^\infty f(x)\,dx = -\sum_k \mathrm{Res}(f, z_k)\ln z_k$$

Hence, because $z^3 + 1 = 0$ has poles at $z_1 = e^{j\pi/3}$, $z_2 = e^{j\pi}$, $z_3 = e^{j5\pi/3}$, then the integral is given by

$$I = \int_0^\infty \frac{dx}{x^3+1} = -\left[\frac{j\pi/3}{3e^{2j\pi/3}} + \frac{j\pi}{3e^{j2\pi}} + \frac{j5\pi/3}{3e^{j10\pi/3}}\right] = \frac{2\pi\sqrt{3}}{9}$$ ∎

Example 9.11
Show that $\int_0^\infty \cos ax^2\,dx = \int_0^\infty \sin ax^2\,dx = \frac{1}{2}\sqrt{\frac{\pi}{2a}}$, $a > 0$.

Solution We first form the integral

$$F = \int_0^\infty \cos ax^2\,dx + j\int_0^\infty \sin ax^2\,dx = \int_0^\infty e^{jax^2}\,dx$$

Because $\exp(jaz^2)$ is analytic in the entire z-plane, we can use Cauchy's theorem and write (see Figure 9.5)

$$F = \int_{AB} a^{jaz^2}\,dz = \int_{AC} e^{jaz^2}\,dz + \int_{CB} e^{jaz^2}\,dz$$

Along the contour CB we obtain

$$\left|-\int_0^{\pi/4} e^{jR^2\cos 2\theta - R^2\sin 2\theta}\,j\,Re^{j\theta}\,d\theta\right| \le \int_0^{\pi/4} e^{-R^2\sin 2\theta}\,Rd\theta = \frac{R}{2}\int_0^{\pi/2} e^{-R^2\sin\phi}\,d\phi$$

$$\le \frac{R}{2}\int_0^{\pi/2} e^{-R^2\phi/\pi}\,d\phi = \frac{\pi}{4R}(1 - e^{-R^2})$$

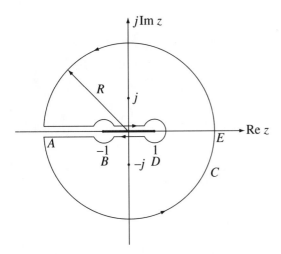

FIGURE 9.6

where the transformation $2\theta = \phi$ and the inequality $\sin \phi \geq 2\phi/\pi$ were used $(0 \leq \phi \leq \pi/2)$. Hence as R approaches infinity the contribution from CB contour vanishes. Hence

$$F = \int_{AB} e^{jaz^2}\, dz = e^{j\pi/4} \int_0^\infty e^{-ar^2}\, dr = \frac{1+j}{\sqrt{2}}\frac{1}{2}\sqrt{\frac{\pi}{a}}$$

from which we obtain the desired result. ∎

Example 9.12

Evaluate the integral $I = \int_{-1}^{1} \frac{dx}{\sqrt{1-x^2}(1+x^2)}$.

Solution   Consider the integral

$$\oint_C \frac{dz}{\sqrt{1 - z^2}(1 + z^2)}$$

whose contour C is that shown in Figure 9.6. On the top side of the branch cut we obtain I and from the bottom we also get I. The contribution of the integral on the outer circle as R approaches infinity vanishes. Hence, due to two poles we obtain

$$2I = 2\pi j \left[\frac{1}{2j\sqrt{2}} + \frac{1}{2j\sqrt{2}} \right] = \pi \sqrt{2} \quad \text{or} \quad I = \frac{\sqrt{2}}{2}\pi \qquad ∎$$

Example 9.13

Evaluate the integral $I = \int_{-\infty}^{\infty} \frac{e^{ax}}{e^{bx}+1}\, dx, \; a, b > 0$.

Solution   From Figure 9.7 we find

$$I = \int_C \frac{e^{az}}{e^{bz} + 1}\, dz = \int_C \frac{e^{az/b}}{e^z + 1}\, dz = 2\pi j \sum \text{Res}$$

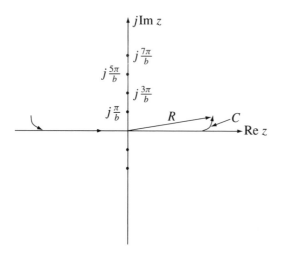

FIGURE 9.7

There is an infinite number of poles: at $z = j\pi/b$, residue is $-\exp(j\pi a/b)$; at $z = 3j\pi/b$, residue is $-\exp(3j\pi a/b)$, and so on. The sum of residue forms a geometric series and because we assume a small imaginary part of a, $|\exp(j2\pi a/b)| < 1$. Hence, by considering the common factor $\exp(j\pi a/b)$, we obtain

$$I = -\frac{2\pi}{b} j \frac{e^{j\pi a/b}}{1 - e^{j2\pi a/b}} = \frac{1}{b} \frac{\pi}{\sin(\pi a/b)}$$

The integral is of the form $\int e^{j\omega x} f(x)\, dx$ whose evaluation can be simplified by Jordan's lemma (see also [7.5])

$$\int_C e^{j\omega x} f(x)\, dx = 0$$

for the contour semicircle C at infinity for which $\mathrm{Im}(\omega x) > 0$, provided $|f(\mathrm{Re}^{j\theta})| < \varepsilon(R) \to 0$ as $R \to \infty$ (Note that the bound on $|f(x)|$ must be independent of θ). ∎

Example 9.14

A relaxed RL series circuit with an input voltage source $v(t)$ is described by the equation $L\, di/dt + Ri = v(t)$. Find the current in the circuit using the inverse Fourier transform when the input voltage is a delta function.

Solution The Fourier transform of the differential equation with delta input voltage function is

$$Lj\omega I(\omega) + RI(\omega) = 1 \qquad \text{or} \qquad I(\omega) = \frac{1}{R + j\omega L}$$

and hence

$$i(t) = \frac{1}{2\pi} \int_{-\infty}^{\infty} \frac{e^{j\omega t}}{R + j\omega L}\, d\omega$$

If $t < 0$, the integral is exponentially small for $\mathrm{Im}\,\omega \to -\infty$. If we complete the contour by a large semicircle in the lower ω-plane, the integral vanishes by Jordan's lemma. Because the contour does not include any singularities, $i(t) = 0$, $t < 0$. For $t > 0$, we complete the contour in the upper ω-plane. Similarly no contribution exists from the semicircle. Because there is only one pole at $\omega = jR/L$ inside the contour the value of the integral is

$$i(t) = 2\pi j \frac{1}{2\pi} \frac{1}{jL} e^{j(jR/L)t} = \frac{1}{L} e^{-\frac{R}{L}t}$$

which is known as the **impulse response of the system**.   ∎

10   Principal value of an integral

Refer to the limiting process employed in Example 9.9, which can be written in the form

$$\lim_{R \to \infty} \int_{-R}^{R} \frac{e^{jx}}{x} \, dx = j\pi$$

The limit is called the **Cauchy principal value** of the integral in the equation

$$\int_{-\infty}^{\infty} \frac{e^{jx}}{x} \, dx = j\pi$$

In general, if $f(x)$ becomes infinite at a point $x = c$ inside the range of integration, and if

$$\lim_{\varepsilon \to 0} \int_{-R}^{R} f(x) \, dx = \lim_{\varepsilon \to 0} \left[\int_{-R}^{c-\varepsilon} f(x) \, dx + \int_{c+\varepsilon}^{R} f(x) \, dx \right]$$

and if the separate limits on the right also exist, then the integral is convergent and the integral is written as $P \int$ where the P indicates the principal value. Whenever each of the integrals

$$\int_{-\infty}^{0} f(x) \, dx \qquad \int_{0}^{\infty} f(x) \, dx$$

has a value, here $R \to \infty$, the principal value is the same as the integral. For example, if $f(x) = x$, the principal value of the integral is zero, although the value of the integral itself does not exist.

As another example, consider the integral

$$\int_{a}^{b} \frac{dx}{x} = \log \frac{b}{a}$$

If a is negative and b is positive, the integral diverges at $x = 0$. However, we can still define

$$P \int_{a}^{b} \frac{dx}{x} = \lim_{\varepsilon \to 0} \left[\int_{a}^{-\varepsilon} \frac{dx}{x} + \int_{\varepsilon+}^{b} \frac{dx}{x} \right] = \lim_{\varepsilon \to 0} \left(\log \frac{\varepsilon}{-a} + \log \frac{b}{a} \right) = \log \frac{b}{|a|}$$

This principal value integral is unambiguous. The condition that the same value of ε must be used in both sides is essential; otherwise, the limit could be almost anything by taking the first integral from a to $-\varepsilon$ and the second from κ to b and making these two quantities tend to zero in a suitable ratio.

If the complex variables were used, we could complete the path by a semicircle from $-\varepsilon$ to $+\varepsilon$ about the origin, either above or below the real axis. If the upper semicircle were chosen, there would be a contribution $-j\pi$, whereas if the lower semicircle were chosen, the contribution to the integral would be $+j\pi$. Thus, according to the path permitted in the

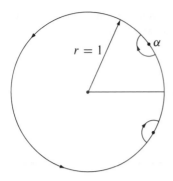

FIGURE 10.1

complex plane we should have

$$\int_a^b \frac{dz}{z} = \log \frac{b}{|a|} \pm j\pi$$

The principal value is the mean of these alternatives.

If a path in the complex plane passes through a simple pole at $z = a$, we can define a principal value of the integral along the path by using a hook of small radius ε about the point a and then making ε tend to zero, as already discussed. If we change the variable z to ζ, and $dz/d\zeta$ is finite and not equal to zero at the pole, this procedure will define an integral in the ζ-plane, but the values of the integrals will be the same. Suppose that the hook in the z-plane cuts the path at $a - \varepsilon$ and $a + \varepsilon'$, where $|\varepsilon| = |\varepsilon'|$, and in the ζ-plane the hook cuts the path at $\alpha - \kappa$ and $\alpha + \kappa'$. Then, if κ and κ' tend to zero so that $\varepsilon/\varepsilon' \to 1$, κ and κ' will tend to zero so that $\kappa/\kappa' \to 1$.

To illustrate this discussion, suppose we want to evaluate the integral

$$I = \int_0^\pi \frac{d\theta}{a - b\cos\theta}$$

where a and b are real and $a > b > 0$. A change of variable by writing $z = \exp(j\theta)$ transforms this integral to (where a new constant α is introduced)

$$I = \int_0^\pi \frac{2e^{j\theta}d\theta}{2ae^{j\theta} - b(e^{j2\theta} + 1)} = -\frac{1}{j} \int_C \frac{2dz}{bz^2 - 2az + b} = -\frac{1}{j} \int_C \frac{2dz}{b(z - \alpha)(z - \frac{1}{\alpha})}$$

where the path of integration is around the unit circle. Because the contour would pass through the poles, hooks are used to isolate the poles as shown in Figure 10.1. Because no singularities are closed by the path, the integral is zero. The contributions of the hooks are $-j\pi$ times the residue, where the residues are

$$-\frac{1}{j} \frac{\frac{2}{b}}{\alpha - \frac{1}{\alpha}} \qquad -\frac{1}{j} \frac{\frac{2}{b}}{\frac{1}{\alpha} - \alpha}$$

These are equal and opposite and cancel each other. Therefore, the principal value of the integral around the unit circle is zero. This approach for finding principal values succeeds only at simple poles.

11   Integral of the logarithmic derivative

Of importance in the study of mapping from z-plane to $W(z)$-plane is the integral of the logarithmic derivative. Consider, therefore, the function

$$F(z) = \log W(z) \tag{11.1}$$

Then

$$\frac{dF(z)}{dz} = \frac{1}{W(z)}\frac{dW(z)}{dz} = \frac{W'(z)}{W(z)}$$

The function to be examined is the following:

$$\int_C \frac{dF(z)}{dz}\,dz = \int_C \frac{W'(z)}{W(z)}\,dz \tag{11.2}$$

The integrand of this expression will be analytic within the contour C except for the points at which $W(z)$ is either zero or infinity.

Suppose that $W(z)$ has a pole of order n at z_0. This means that $W(z)$ can be written

$$W(z) = (z - z_0)^n g(z) \tag{11.3}$$

with n positive for a zero and n negative for a pole. We differentiate this expression to get

$$W'(z) = n(z - z_0)^{n-1}g(z) + (z - z_0)^n g'(z)$$

and so

$$\frac{W'(z)}{W(z)} = \frac{n}{z - z_0} + \frac{g'(z)}{g(z)} \tag{11.4}$$

For n positive, $W'(z)/W(z)$ will possess a pole of order one. Similarly, for n negative $W'(z)/W(z)$ will possess a pole of order one, but with a negative sign. Thus, for the case of n positive or negative, the contour integral in the positive sense yields

$$\int_C \frac{W'(z)}{W(z)}\,dz = \pm \int_C \frac{n}{z - z_0}\,dz + \int_z \frac{g'(z)}{g(z)}\,dz \tag{11.5}$$

But because $g(z)$ is analytic at the point z_0, then $\int_C [g'(z)/g(z)]\,dz = 0$, and by (6.1)

$$\int_C \frac{W'(z)}{W(z)}\,dz = \pm 2\pi j n \tag{11.6}$$

Thus the existence of a zero of $W(z)$ introduces a contribution $2\pi j n_z$ to the contour integral, where n_z is the multiplicity of the zero of $W(z)$ at z_0. Clearly, if a number of zeros of $W(z)$ exist, the total contribution to the contour integral is $2\pi j N$, where N is the weighted value of the zeros of $W(z)$ (weight 1 to a first-order zero, weight 2 to a second-order zero, and so on).

For the case where n is negative, which specifies that $W(z)$ has a pole of order n at z_0, then in (11.6) n is negative and the contribution to the contour integral is now $-2\pi n_p$ for each pole of $W(z)$; the total contribution is $-2\pi j P$, where P is the weighted number of poles. Clearly, because both zeros and poles of $F(z)$ cause poles of $W'(z)/W(z)$ with opposite signs, then the total value of the integral is

$$\int_C \frac{W'(z)}{W(z)}\,dz = \pm 2\pi j(N - P) \tag{11.7}$$

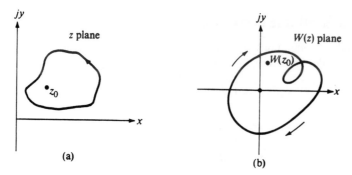

FIGURE 11.1

Note further that

$$\int_C W'(z)\, dz = \int_C \frac{dW(z)}{dz}\, dz = \int d[\log W(z)] = \int d[\log|W(z)| + j \arg W(z)]$$

$$= \log|W(z)|\Big|_0^{2\pi} + j[\arg W(2\pi) - \arg W(0)]$$

$$= 0 + j[\arg W(2\pi) - \arg W(0)]$$

so that

$$[\arg W(0) - \arg W(2\pi)] = 2\pi(N - P) \tag{11.8}$$

This relation can be given simple graphical interpretation. Suppose that the function $W(z)$ is represented by its pole and zero configuration on the z-plane. As z traverses the prescribed contour on the z-plane, $W(z)$ will move on the $W(z)$-plane according to its functional dependence on z. But the left-hand side of this equation denotes the total change in the phase angle of $W(z)$ as z transverses around the complete contour. Therefore, the number of times that the moving point representing $W(z)$ revolves around the origin in the $W(z)$-plane as z moves around the specified contour is given by $N - P$.

The foregoing is conveniently illustrated graphically. Figure 11.1a shows the prescribed contour in the z-plane, and Figure 11.1b shows a possible form for the variation of $W(z)$. For this particular case, the contour in the z-plane encloses one zero and no poles; hence, $W(z)$ encloses the origin once in the clockwise direction in the $W(z)$-plane.

Note that corresponding to a point z_0 within the contour in the z-plane, the point $W(z_0)$ is mapped inside the $W(z)$-plane. In fact, every point on the inside of the contour in the z-plane maps onto the inside of the $W(z)$-contour in the $W(z)$-plane (for single-valued functions). Clearly, there is one point in the z-plane that maps into $W(z) = 0$, the origin.

On the other hand, if the contour includes a pole but no zeros, it can be shown by a similar argument that any point in the interior of the z-contour must correspond to a corresponding point outside of the $W(z)$-contour in the $W(z)$-plane. This is manifested by the fact that the $W(z)$-contour is transversed in a counterclockwise direction. With both zeros and poles present, the situation depends on the value of N and P.

Of special interest is the locus of the network function that contains no poles in the right-hand plane or on the $j\omega$-axis. In this case the frequency locus is completely traced as z varies along the ω-axis form $-j\infty$ to $+j\infty$. To show this, because $W(z)$ is analytic along the this

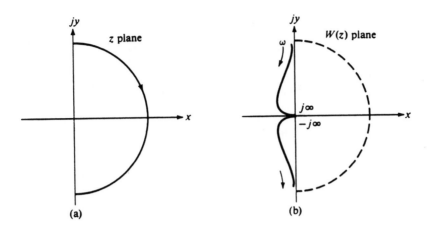

FIGURE 11.2

path, $W(z)$ can be written for the neighborhood of a point z_0 in a Taylor series

$$W(z) = \alpha_0 + \alpha_1(z - z_0) + \alpha_2(z - z_0)^2 + \cdots$$

For the neighborhood $z \to \infty$, we examine $W(z')$, where $z' = 1/z$. Because $W(z)$ does not have a pole at infinity, then $W(z')$ does not have a pole at zero. Therefore, we can expand $W(z')$ in a Maclaurin series

$$W(z') = \alpha_0 + \alpha_1 z' + \alpha_2(z')^2 + \cdots$$

which means that

$$W(z) = \alpha_0 + \frac{\alpha_1}{z} + \frac{a_2}{z^2} + \cdots$$

But as z approaches infinity, $W(\infty)$ approaches infinity. In a real network function when z^* is written for z, then $W(z^*) = W^*(z)$. This condition requires that $\alpha_0 = a_0 + j0$ be a real number irrespective of how z approaches infinity; that is, as z approaches infinity, $W(z)$ approaches a fixed point in the $W(z)$-plane. This shows that as z varies around the specified contour in the z-plane, $W(z)$ varies from $W(-j\infty)$ to $W(+j\infty)$ as z varies along the imaginary axis. However, $W(-j\infty) = W(+j\infty)$, from the above, which thereby shows that the locus is completely determined. This is illustrated in Figure 11.2.

Appendix 2: Series and Summations

Series

The expression in parentheses following certain of the series indicates the region of convergence. If not otherwise indicated it is to be understood that the series converges for all finite values of x.

Binomial

$$(x + y)^n = x^n + nx^{n-1}y + \frac{n(n-1)}{2!}x^{n-2}y^2$$

$$+ \frac{n(n-1)(n-2)}{3!}x^{n-3}y^3 + \cdots \qquad (y^2 < x^2)$$

$$(1 \pm x)^n = 1 \pm nx + \frac{n(n-1)x^2}{2!} \pm \frac{n(n-1)(n-2)x^3}{3!} + \cdots \text{ etc.} \qquad (x^2 < 1)$$

$$(1 \pm x)^{-n} = 1 \mp nx + \frac{n(n+1)x^2}{2!} \mp \frac{n(n+1)(n+2)x^3}{3!} + \cdots \text{ etc.} \qquad (x^2 < 1)$$

$$(1 \pm x)^{-1} = 1 \mp x + x^2 \mp x^3 + x^4 \mp x^5 + \cdots \qquad (x^2 < 1)$$

$$(1 \pm x)^{-2} = 1 \mp 2x + 3x^2 \mp 4x^3 + 5x^4 \mp 6x^5 + \cdots \qquad (x^2 < 1)$$

Reversion of series

Let a series be represented by

$$y = a_1 x + a_2 x^2 + a_3 x^3 + a_4 x^4 + a_5 x^5 + a_6 x^6 + \cdots \qquad (a_1 \neq 0)$$

to find the coefficients of the series

$$x = A_1 y + A_2 y^2 + A_3 y^3 + A_4 y^4 + \cdots$$

$$A_1 = \frac{1}{a_1} \quad A_2 = -\frac{a_2}{a_1^3} \quad A_3 = \frac{1}{a_1^5}(2a_2^2 - a_1 a_3)$$

$$A_4 = \frac{1}{a_1^7}(5a_1 a_2 a_3 - a_1^2 a_4 - 5a_2^3)$$

$$A_5 = \frac{1}{a_1^9}(6a_1^2 a_2 a_4 + 3a_1^2 a_3^2 + 14a_2^4 - a_1^3 a_5 - 21a_1 a_2^2 a_3)$$

$$A_6 = \frac{1}{a_1^{11}}(7a_1^3 a_2 a_5 + 7a_1^3 a_3 a_4 + 84a_1 a_2^3 a_3 - a_1^4 a_6 - 28a_1^2 a_2^2 a_4 - 28a_1^2 a_2 a_3^2 - 42a_2^5)$$

$$A_7 = \frac{1}{a_1^{13}}(8a_1^4 a_2 a_6 + 8a_1^4 a_3 a_5 + 4a_1^4 a_4^2 + 120a_1^2 a_2^3 a_4$$

$$+ 180a_1^2 a_2^2 a_3^2 + 132a_2^6 - a_1^5 a_7$$

$$- 36a_1^3 a_2^2 a_5 - 72a_1^3 a_2 a_3 a_4 - 12a_1^3 a_3^3 - 330a_1 a_2^4 a_3)$$

Taylor

1.

$$f(x) = f(a) + (x - a)f'(a) + \frac{(x-a)^2}{2!}f''(a) + \frac{(x-a)^3}{3!}f'''(a)$$

$$+ \cdots + \frac{(x-a)^n}{n!}f^{(n)}(a) + \cdots \qquad \text{(Taylor's Series)}$$

(Increment form)

2.

$$f(x + h) = f(x) + hf'(x) + \frac{h^2}{2!}f''(x) + \frac{h^3}{3!}f'''(x) + \cdots$$

$$= f(h) + xf'(h) + \frac{x^2}{2!}f''(h) + \frac{x^3}{3!}f'''(h) + \cdots$$

3. If $f(x)$ is a function possessing derivatives of all orders throughout the interval $a \leqq x \leqq b$, then there is a value X, with $a < X < b$, such that

$$f(b) = f(a) + (b - a)f'(a) + \frac{(b-a)^2}{2!}f''(a) + \cdots$$

$$+ \frac{(b-a)^{n-1}}{(n-1)!}f^{(n-1)}(a) + \frac{(b-a)^n}{n!}f^{(n)}(X)$$

$$f(a + h) = f(a) + hf'(a) + \frac{h^2}{2!}f''(a) + \cdots + \frac{h^{n-1}}{(n-1)!}f^{(n-1)}(a)$$

$$+ \frac{h^n}{n!}f^{(n)}(a + \theta h), \quad b = a + h, \quad 0 < \theta < 1.$$

or

$$f(x) = f(a) + (x - a)f'(a) + \frac{(x - a)^2}{2!} f''(a) + \cdots + (x - a)^{n-1} \frac{f^{(n-1)}(a)}{(n - 1)!} + R_n,$$

where

$$R_n = \frac{f^{(n)}[a + \theta \cdot (x - a)]}{n!}(x - a)^n, \quad 0 < \theta < 1.$$

The above forms are known as Taylor's series with the remainder term.

4. *Taylor's series for a function of two variables*
If $(h\frac{\partial}{\partial x} + k\frac{\partial}{\partial y})f(x, y) = h\frac{\partial f(x,y)}{\partial x} + k\frac{\partial f(x,y)}{\partial y}$;

$$\left(h\frac{\partial}{\partial x} + k\frac{\partial}{\partial y}\right)^2 f(x, y) = h^2 \frac{\partial^2 f(x, y)}{\partial x^2} + 2hk \frac{\partial^2 f(x, y)}{\partial x \partial y} + k^2 \frac{\partial^2 f(x, y)}{\partial y^2}$$

etc., and if $h(\frac{\partial}{\partial x} + k\frac{\partial}{\partial y})^n f(x, y)|_{\substack{x=a \\ y=b}}$ with the bar and subscripts means that after differentiation we are to replace x by a and y by b,

$$f(a + h, b + k) = f(a, b) + \left(h\frac{\partial}{\partial x} + k\frac{\partial}{\partial y}\right) f(x, y)\Bigg|_{\substack{x=a \\ y=b}} + \cdots$$

$$+ \frac{1}{n!} \left(h\frac{\partial}{\partial x} + k\frac{\partial}{\partial y}\right)^n f(x, y)\Bigg|_{\substack{x=a \\ y=b}} + \cdots$$

Maclaurin

$$f(x) = f(0) + xf'(0) + \frac{x^2}{2!} f''(0) + \frac{x^3}{3!} f'''(0) + \cdots + x^{n-1} \frac{f^{(n-1)}(0)}{(n - 1)!} + R_n,$$

where

$$R_n = \frac{x^n f^{(n)}(\theta x)}{n!}, \quad 0 < \theta < 1.$$

Exponential

$$e = 1 + \frac{1}{1!} + \frac{1}{2!} + \frac{1}{3!} + \frac{1}{4!} + \cdots$$

$$e^x = 1 + x + \frac{x^2}{2!} + \frac{x^3}{3!} + \frac{x^4}{4!} + \cdots \qquad \text{(all real values of } x)$$

$$a^x = 1 + x \log_e a + \frac{(x \log_e a)^2}{2!} + \frac{(x \log_e a)^3}{3!} + \cdots$$

$$e^x = e^a \left[1 + (x - a) + \frac{(x - a)^2}{2!} + \frac{(x - a)^3}{3!} + \cdots \right]$$

Logarithmic

$$\log_e x = \frac{x-1}{x} + \frac{1}{2}\left(\frac{x-1}{x}\right)^2 + \frac{1}{3}\left(\frac{x-1}{x}\right)^3 + \cdots \qquad \left(x > \tfrac{1}{2}\right)$$

$$\log_e x = (x-1) - \frac{1}{2}(x-1)^2 + \frac{1}{3}(x-1)^3 - \cdots \qquad (2 \geq x > 0)$$

$$\log_e x = 2\left[\frac{x-1}{x+1} + \frac{1}{3}\left(\frac{x-1}{x+1}\right)^3 + \frac{1}{5}\left(\frac{x-1}{x+1}\right)^5 + \cdots\right] \qquad (x > 0)$$

$$\log_e(1+x) = x - \frac{1}{2}x^2 + \frac{1}{3}x^3 - \frac{1}{4}x^4 + \cdots \qquad (-1 < x < 1)$$

$$\log_e(n+1) - \log_e(n-1) = 2\left[\frac{1}{n} + \frac{1}{3n^3} + \frac{1}{5n^5} + \cdots\right]$$

$$\log_e(a+x) = \log_e a + 2\left[\frac{x}{2a+x} + \frac{1}{3}\left(\frac{x}{2a+x}\right)^3 + \frac{1}{5}\left(\frac{x}{2a+x}\right)^5 + \cdots\right]$$

$$(a > 0, -a < x < +\infty)$$

$$\log_e \frac{1+x}{1-x} = 2\left[x + \frac{x^3}{3} + \frac{x^5}{5} + \cdots + \frac{x^{2n-1}}{2n-1} + \cdots\right], \qquad -1 < x < 1$$

$$\log_e x = \log_e a + \frac{(x-a)}{a} - \frac{(x-a)^2}{2a^2} + \frac{(x-a)^3}{3a^3} - \cdots, \qquad 0 < x \leqq 2a$$

Trigonometric

$$\sin x = x - \frac{x^3}{3!} + \frac{x^5}{5!} - \frac{x^7}{7!} + \cdots \qquad \text{(all real values of } x\text{)}$$

$$\cos x = 1 - \frac{x^2}{2!} + \frac{x^4}{4!} - \frac{x^6}{6!} + \cdots \qquad \text{(all real values of } x\text{)}$$

$$\tan x = x + \frac{x^3}{3} + \frac{2x^5}{15} + \frac{17x^7}{315} + \frac{62x^9}{2835} + \cdots + \frac{2^{2n}(2^{2n}-1)B_n}{(2n)!}x^{2n-1} + \cdots,$$

$$\left[x^2 < \tfrac{\pi^2}{4}, \text{ and } B_n \text{ represents the } n\text{th Bernoulli number.}\right]$$

$$\cot x = \frac{1}{x} - \frac{x}{3} - \frac{x^2}{45} - \frac{2x^5}{945} - \frac{x^7}{4725} - \cdots - \frac{2^{2n}B_n}{(2n)!}x^{2n-1} - \cdots,$$

$$[x^2 < \pi^2, \text{ and } B_n \text{ represents the } n\text{th Bernoulli number.}]$$

$$\sec x = 1 + \frac{x^2}{2} + \frac{5}{24}x^4 + \frac{61}{720}x^6 + \frac{277}{8064}x^8 + \cdots + \frac{E_n x^{2n}}{(2n)!} + \cdots,$$

$$\left[x^2 < \tfrac{\pi^2}{4}, \text{ and } E_n \text{ represents the } n\text{th Euler number.} \right]$$

$$\csc x = \frac{1}{x} + \frac{x}{6} + \frac{7}{360}x^3 + \frac{31}{15{,}120}x^5 + \frac{127}{604{,}800}x^7 + \cdots$$

$$+ \frac{2(2^{2n-1} - 1)}{(2n)!} B_n x^{2n-1} + \cdots,$$

$$[x^2 < \pi^2, \text{ and } B_n \text{ represents } n\text{th Bernoulli number.}]$$

$$\sin x = x \left(1 - \frac{x^2}{\pi^2}\right)\left(1 - \frac{x^2}{2^2\pi^2}\right)\left(1 - \frac{x^2}{3^2\pi^2}\right)\cdots \qquad (x^2 < \infty)$$

$$\cos x = \left(1 - \frac{4x^2}{\pi^2}\right)\left(1 - \frac{4x^2}{3^2\pi^2}\right)\left(1 - \frac{4x^2}{5^2\pi^2}\right)\cdots \qquad (x^2 < \infty)$$

$$\sin^{-1} x = x + \frac{x^3}{2\cdot 3} + \frac{1\cdot 3}{2\cdot 4\cdot 5}x^5 + \frac{1\cdot 3\cdot 5}{2\cdot 4\cdot 6\cdot 7}x^7 + \cdots \qquad \left(x^2 < 1, -\tfrac{\pi}{2} < \sin^{-1} x < \tfrac{\pi}{2}\right)$$

$$\cos^{-1} x = \frac{\pi}{2} - \left(x + \frac{x^3}{2\cdot 3} + \frac{1\cdot 3}{2\cdot 4\cdot 5}x^5 + \frac{1\cdot 3\cdot 5 x^7}{2\cdot 4\cdot 6\cdot 7} + \cdots\right)(x^2 < 1, 0 < \cos^{-1} x < \pi)$$

$$\tan^{-1} x = x - \frac{x^3}{3} + \frac{x^5}{5} - \frac{x^7}{7} + \cdots \qquad (x^2 < 1)$$

$$\tan^{-1} x = \frac{\pi}{2} - \frac{1}{x} + \frac{1}{3x^2} - \frac{1}{5x^5} + \frac{1}{7x^7} - \cdots \qquad (x > 1)$$

$$\tan^{-1} x = -\frac{\pi}{2} - \frac{1}{x} + \frac{1}{3x^2} - \frac{1}{5x^5} + \frac{1}{7x^7} - \cdots \qquad (x < -1)$$

$$\cot^{-1} x = \frac{\pi}{2} - x + \frac{x^3}{3} - \frac{x^5}{5} + \frac{x^7}{7} - \cdots \qquad (x^2 < 1)$$

$$\log_e \sin x = \log_e x - \frac{x^2}{6} - \frac{x^4}{180} - \frac{x^6}{2835} - \cdots \qquad (x^2 < \pi^2)$$

$$\log_e \cos x = -\frac{x^2}{2} - \frac{x^4}{12} - \frac{x^6}{45} - \frac{17x^8}{2520} - \cdots \qquad \left(x^2 < \tfrac{\pi^2}{4}\right)$$

$$\log_e \tan x = \log_e x + \frac{x^2}{3} + \frac{7x^4}{90} + \frac{62x^6}{2835} + \cdots \qquad \left(x^2 < \tfrac{\pi^2}{4}\right)$$

$$e^{\sin x} = 1 + x + \frac{x^2}{2!} - \frac{3x^4}{4!} - \frac{8x^5}{5!} - \frac{3x^6}{6!} + \frac{56x^7}{7!} + \cdots$$

$$e^{\cos x} = e \left(1 - \frac{x^2}{2!} + \frac{4x^4}{4!} - \frac{31x^6}{6!} + \cdots\right)$$

$$e^{\tan x} = 1 + x + \frac{x^2}{2!} + \frac{3x^3}{3!} + \frac{9x^4}{4!} + \frac{37x^5}{5!} + \cdots \qquad \left(x^2 < \tfrac{\pi^2}{4}\right)$$

$$\sin x = \sin a + (x - a) \cos a - \frac{(x - a)^2}{2!} \sin a$$

$$- \frac{(x - a)^3}{3!} \cos a + \frac{(x - a)^4}{4!} \sin a + \cdots$$

Hyperbolic and inverse hyperbolic

Table of expansion of certain functions into power series

$$\sinh x = x + \frac{x^3}{3!} + \frac{x^5}{5!} + \frac{x^7}{7!} + \cdots + \frac{x^{2n+1}}{(2n + 1)!} + \cdots \qquad |x| < \infty$$

$$\cosh x = 1 + \frac{x^2}{2!} + \frac{x^4}{4!} + \frac{x^6}{6!} + \cdots + \frac{x^{2n}}{(2n)!} + \cdots \qquad |x| < \infty$$

$$\tanh x = x - \frac{1}{3}x^3 + \frac{2}{15}x^5 - \frac{17}{315}x^7 + \frac{62}{2835}x^9 - \cdots$$

$$+ \frac{(-1)^{n+1}2^{2n}(2^{2n} - 1)}{(2n)!} B_n x^{2n-1} \pm \cdots^{(1)} \qquad |x| < \frac{\pi}{2}$$

$$\coth x = \frac{1}{x} + \frac{x}{3} - \frac{x^3}{45} + \frac{2x^5}{945} - \frac{x^7}{4725} + \cdots$$

$$+ \frac{(-1)^{n+1}2^{2n}}{(2n)!} B_n x^{2n-1} \pm \cdots^{(1)} \qquad 0 < |x| < \pi$$

$$\operatorname{sech} x = 1 - \frac{1}{2!}x^2 + \frac{5}{4!}x^4 - \frac{61}{6!}x^6 + \frac{1385}{8!}x^8 - \cdots + \frac{(-1)^n}{(2n)!} E_n x^{2n} \pm \cdots^{(2)}$$

$$|x| < \frac{\pi}{2}$$

$$\operatorname{cosech} x = \frac{1}{x} - \frac{x}{6} + \frac{7x^3}{360} - \frac{31x^5}{15,120} + \cdots + \frac{2(-1)^n(2^{2n-1} - 1)}{(2n)!} B_n x^{2n-1} + \cdots^{(1)}$$

$$0 < |x| < \pi$$

$$\operatorname{arg\,sinh} x = x - \frac{1}{2 \cdot 3}x^3 + \frac{1 \cdot 3}{2 \cdot 4 \cdot 5}x^5 - \frac{1 \cdot 3 \cdot 5}{2 \cdot 4 \cdot 6 \cdot 7}x^7 + \cdots$$

$$+ (-1)^n \cdot \frac{1 \cdot 3 \cdot 5(2n - 1)}{2 \cdot 4 \cdot 6 \ldots 2n(2n + 1)}x^{2n+1} \pm \cdots \qquad |x| < 1$$

$$\operatorname{arg\,cosh} x = \pm \left[\ln(2x) - \frac{1}{2 \cdot 2x^2} - \frac{1 \cdot 3}{2 \cdot 4 \cdot 4x^4} - \frac{1 \cdot 3 \cdot 5}{2 \cdot 4 \cdot 6 \cdot 6x^6} - \cdots \right] \qquad x > 1$$

$$\operatorname{arg\,tanh} x = x + \frac{x^3}{3} + \frac{x^5}{5} + \frac{x^7}{7} + \cdots + \frac{x^{2n+1}}{2n + 1} + \cdots \qquad |x| < 1$$

$$\operatorname{arg\,coth} x = \frac{1}{x} + \frac{1}{3x^3} + \frac{1}{5x^5} + \frac{1}{7x^7} + \cdots + \frac{1}{(2n + 1)x^{2n+1}} + \cdots \qquad |x| > 1$$

[1] B_n denotes Bernoulli's numbers.
[2] E_n denotes Euler's numbers.

Arithmetic Progression of the first order (first differences constant), to n terms,

$$a + (a + d) + (a + 2d) + (a + 3d) + \cdots + \{a + (n - 1)d\} \equiv na + \frac{1}{2}n(n - 1)d$$

$$\equiv \frac{n}{2}(\text{1st term} + n\text{th term}).$$

Geometric Progression, to n terms,

$$a + ar + ar^2 + ar^3 + \cdots + ar^{n-1} \equiv a(1 - r^n)/(1 - r)$$

$$\equiv a(r^n - 1)/(r - 1).$$

If $r^2 < 1$, the limit of the sum of an infinite number of terms is $a/(1 - r)$.

The reciprocals of the terms of a series in arithmetic progression of the first order are in **Harmonic Progression**. Thus

$$\frac{1}{a}, \qquad \frac{1}{a + d}, \qquad \frac{1}{a + 2d}, \qquad \cdots \qquad \frac{1}{a + (n - 1)d}$$

are in Harmonic Progression.

The Arithmetic Mean of n quantities is

$$\frac{1}{n}(a_1 + a_2 + a_3 + \cdots + a_n).$$

The **Geometric Mean** of n quantities is

$$(a_1 a_2 a_3 \cdots a_n)^{1/n}.$$

Let the **Harmonic Mean** of n quantities be H. Then

$$\frac{1}{H} = \frac{1}{n}\left(\frac{1}{a_1} + \frac{1}{a_2} + \frac{1}{a_3} + \cdots + \frac{1}{a_n}\right).$$

The arithmetic mean of a number of positive quantities is $\geqq$ their geometric mean, which in turn is $\geqq$ their harmonic mean.

$$1 + 2 + 3 + \cdots + n = \frac{n}{2}(n + 1) = \sum_{k=0}^{n} k$$

$$1^2 + 2^2 + 3^2 + \cdots + n^2 = \frac{n}{6}(n + 1)(2n + 1)$$

$$= \frac{n}{6}(2n^2 + 3n + 1) = \sum_{k=0}^{n} k^2$$

$$1^3 + 2^3 + 3^3 + \cdots + n^3 = \frac{n^2}{4}(n + 1)^2$$

$$= \frac{n^2}{4}(n^2 + 2n + 1) = \sum_{k=0}^{n} k^3$$

$$1 + 3 + 5 + 7 + 9 + \cdots + (2n - 1) = n^2 = \sum_{k=0}^{2n-1}(2k + 1)$$

$$1 + 8 + 16 + 24 + 32 + \cdots + 8(n - 1) = (2n - 1)^2$$

$$1 + 3x + 5x^2 + 7x^3 + \cdots = \frac{1 + x}{(1 - x)^2}$$

$$1 + ax + (a + b)x^2 + (a + 2b)x^3 + \cdots = 1 + \frac{ax + (b - a)x^2}{(1 - x)^2}$$

$$1 + 2^2 x + 3^2 x^2 + 4^2 x^3 + \cdots = \frac{1 + x}{(1 - x)^3}$$

$$1 + 3^2 x + 5^2 x^2 + 7^2 x^3 + \cdots = \frac{1 + 6x + x^2}{(1 - x)^3}$$

$$\frac{a[1 - (n + 1)a^n + na^{n+1}]}{(1 - a)^2} = \sum_{k=0}^{n} ka^k$$

$$\frac{a[(1 + a) - (n + 1)^2 a^n + (2n^2 + 2n - 1)a^{n+1} - n^2 a^{n+2}]}{(1 - a)^3} = \sum_{k=0}^{n} k^2 a^k$$

$$\frac{a}{(1 - a)^2} = \sum_{k=0}^{\infty} ka^k \qquad |a| < 1$$

$$\frac{a^2 + a}{(1 - a)^3} = \sum_{k=0}^{\infty} k^2 a^k \qquad |a| < 1$$

Appendix 3: Definite Integrals

$$\int_0^\infty x^{n-1} e^{-x}\, dx = \int_0^1 \left(\log \frac{1}{x}\right)^{n-1} dx = \frac{1}{n} \prod_{m=1}^{\infty} \frac{\left(1 + \frac{1}{m}\right)^n}{1 + \frac{n}{m}}$$

$$= \Gamma(n), n \neq 0, -1, -2, -3, \ldots \qquad \text{(Gamma Function)}$$

$$\int_0^\infty t^n p^{-t}\, dt = \frac{n!}{(\log p)^{n+1}} \qquad (n = 0, 1, 2, 3, \ldots \text{ and } p > 0)$$

$$\int_0^\infty t^{n-1} e^{-(a+1)t}\, dt = \frac{\Gamma(n)}{(a+1)^n} \qquad (n > 0, a > -1)$$

$$\int_0^1 x^m \left(\log \frac{1}{x}\right)^n dx = \frac{\Gamma(n+1)}{(m+1)^{n+1}} \qquad (m > -1, n > -1)$$

$\Gamma(n)$ is finite if $n > 0$, $\Gamma(n+1) = n\Gamma(n)$

$$\Gamma(n) \cdot \Gamma(1-n) = \frac{\pi}{\sin n\pi}$$

$\Gamma(n) = (n-1)!$ if $n = \text{integer} > 0$

$$\Gamma\left(\frac{1}{2}\right) = 2 \int_0^\infty e^{-t^2}\, dt = \sqrt{\pi} = 1.7724538509\ldots$$

$$\Gamma\left(n + \frac{1}{2}\right) = \frac{1 \cdot 3 \cdot 5 \cdot 7 \cdots (2n-1)}{2^n} \sqrt{\pi},$$

where n is an integer and > 0

$$\int_0^1 x^{m-1}(1-x)^{n-1}\, dx = B(m, n) \qquad \text{(Beta function)}$$

$$B(m, n) = B(n, m) = \frac{\Gamma(m)\Gamma(n)}{\Gamma(m+n)}$$

0-8493-8342-0/96/$0.00 + $0.50

where m and n are any positive real numbers

$$\int_0^1 x^{m-1}(1-x)^{n-1}\,dx = \int_0^\infty \frac{x^{m-1}dx}{(1+x)^{m+n}} = \frac{\Gamma(m)\Gamma(n)}{\Gamma(m+n)}$$

$$\int_a^b (x-a)^m(b-x)^n\,dx = (b-a)^{m+n+1}\frac{\Gamma(m+1)\cdot\Gamma(n+1)}{\Gamma(m+n+2)} \quad (m>-1,\ n>-1,\ b>a)$$

$$\int_1^\infty \frac{dx}{x^m} = \frac{1}{m-1} \qquad (m>1)$$

$$\int_0^\infty \frac{dx}{(1+x)x^p} = \pi\csc p\pi \qquad (p<1)$$

$$\int_0^\infty \frac{dx}{(1-x)x^p} = -\pi\cot p\pi \qquad (p<1)$$

$$\int_0^\infty \frac{x^{p-1}dx}{1+x} = \frac{\pi}{\sin p\pi}$$
$$= B(p,1-p) = \Gamma(p)\Gamma(1-p) \qquad (0<p<1)$$

$$\int_0^\infty \frac{x^{m-1}dx}{1+x^n} = \frac{\pi}{n\sin\frac{m\pi}{n}} \qquad (0<m<n)$$

$$\int_0^\infty \frac{x^a dx}{(m+x^b)^c} = m^{\frac{a+1}{b}-c}\left[\frac{\Gamma\left(\frac{a+1}{b}\right)\Gamma\left(c-\frac{a+1}{b}\right)}{\Gamma(c)}\right]$$
$$\left(a>-1, b>0, m>0, c>\frac{a+1}{b}\right)$$

$$\int_0^\infty \frac{dx}{(1+x)\sqrt{x}} = \pi$$

$$\int_0^\infty \frac{a\,dx}{a^2+x^2} = \frac{\pi}{2},\ \text{if } a>0; 0,\ \text{if } a=0; -\frac{\pi}{2},\ \text{if } a<0$$

$$\int_0^a x^m(a^2-x^2)^{\frac{n}{2}}\,dx = \begin{cases} \dfrac{1}{2}a^{m+n+1}B\left(\dfrac{m+1}{2},\dfrac{n+2}{2}\right) \\[2mm] \dfrac{1}{2}a^{m+n+1}\dfrac{\Gamma\left(\frac{m+1}{2}\right)\Gamma\left(\frac{n+2}{2}\right)}{\Gamma\left(\frac{m+n+3}{2}\right)} \end{cases}$$

$$\int_0^1 \frac{dx}{\sqrt{(1-x^n)}} = \frac{\sqrt{\pi}}{n}\frac{\Gamma\left(\frac{1}{n}\right)}{\Gamma\left(\frac{1}{n}+\frac{1}{2}\right)} \qquad (n>0)$$

$$\int_0^1 \frac{x^m\,dx}{\sqrt{(1-x^n)}} = \frac{\sqrt{\pi}}{n}\frac{\Gamma\left(\frac{m+1}{n}\right)}{\Gamma\left(\frac{m+1}{n}+\frac{1}{2}\right)} \qquad (m+1, n>0)$$

$$\int_0^1 x^m(1-x^2)^p\,dx = \frac{\Gamma(p+1)\Gamma\left(\frac{m+1}{2}\right)}{2\Gamma\left(p+\frac{m+3}{2}\right)} \qquad (p+1, m+1>0)$$

$$\int_0^1 x^m (1 - x^n)^p \, dx = \frac{\Gamma(p+1)\Gamma\left(\frac{m+1}{n}\right)}{n\Gamma\left(p + 1 + \frac{m+1}{n}\right)} \qquad (p+1, \, m+1, \, n > 0)$$

$$\int_0^1 \frac{x^m \, dx}{\sqrt{(1 - x^2)}} = \frac{2 \cdot 4 \cdot 6 \cdots (m-1)}{3 \cdot 5 \cdot 7 \cdots m} \qquad (m \text{ an odd integer } > 1)$$

$$= \frac{1 \cdot 3 \cdot 5 \cdots (m-1)}{2 \cdot 4 \cdot 6 \cdots m} \frac{\pi}{2} \qquad \begin{array}{c} (m \text{ an even,} \\ \text{positive integer)} \end{array}$$

$$= \frac{\sqrt{\pi}}{2} \frac{\Gamma\left(\frac{m+1}{2}\right)}{\Gamma\left(\frac{m}{2} + 1\right)} \qquad (m \text{ any value } > -1)$$

$$\int_0^\infty \frac{x^{p-1} \, dx}{1 + x} = \frac{\pi}{\sin(\pi - p\pi)} = \frac{\pi}{\sin p\pi} \qquad (0 < p < 1)$$

$$\int_0^\infty \frac{dx}{(1 + x)\sqrt{x}} = \pi$$

$$\int_0^\infty \frac{x^{p-1} \, dx}{a + x} = \frac{\pi a^{p-1}}{\sin p\pi} \qquad (0 < p < 1)$$

$$\int_0^\infty \frac{dx}{1 + x^p} = \frac{\pi}{p \sin \frac{\pi}{p}} \qquad (p > 1)$$

$$\int_0^\infty \frac{x^p \, dx}{(1 + ax)^2} = \frac{p\pi}{a^{p+1} \sin p\pi} \qquad (0 < p < 1)$$

$$\int_0^\infty \frac{x^p \, dx}{1 + x^2} = \frac{\pi}{2 \cos \frac{p\pi}{2}} \qquad (-1 < p < 1)$$

$$\int_0^\infty \frac{x^{p-1} \, dx}{1 + x^q} = \frac{\pi}{q \sin \frac{p\pi}{q}} \qquad (0 < p < q)$$

$$\int_0^\infty \frac{x^{m-1} \, dx}{(1 + x)^{m+n}} = \frac{\Gamma(m)\Gamma(n)}{\Gamma(m + n)} \qquad (m, n > 0)$$

$$\int_0^\infty \frac{x^{m-1} \, dx}{(a + bx)^{m+n}} = \frac{\Gamma(m)\Gamma(n)}{a^n b^m \Gamma(m + n)} \qquad (a, b, m, n > 0)$$

$$\int_0^\infty \frac{dx}{(a^2 + x^2)^n} = \frac{1 \cdot 3 \cdot 5 \cdots (2n - 3)}{2 \cdot 4 \cdot 6 \cdots (2n - 2)} \frac{\pi}{2a^{2n-1}} \qquad (a > 0; n = 2, 3, \ldots)$$

$$\int_0^\infty \frac{dx}{(a^2 + x^2)(b^2 + x^2)} = \frac{\pi}{2ab(a + b)} \qquad (a, b > 0)$$

$$\int_0^{\pi/2} (\sin^n x)\, dx = \begin{cases} \int_0^{\pi/2} (\cos^n x)\, dx \\[2mm] \dfrac{1\cdot 3\cdot 5\cdot 7\cdots (n-1)}{2\cdot 4\cdot 6\cdot 8\cdots (n)}\, \dfrac{\pi}{2}, & (n \text{ an even integer, } n \neq 0) \\[2mm] \dfrac{2\cdot 4\cdot 6\cdot 8\cdots (n-1)}{1\cdot 3\cdot 5\cdot 7\cdots (n)}, & (n \text{ an odd integer, } n \neq 1) \\[2mm] \dfrac{\sqrt{\pi}}{2}\, \dfrac{\Gamma(\frac{n+1}{2})}{\Gamma(\frac{n}{2}+1)}, & (n > -1) \end{cases}$$

$$\int_0^\infty \frac{\sin mx\, dx}{x} = \frac{\pi}{2}, \text{ if } m > 0; \text{ if } m = 0; -\frac{\pi}{2}, \text{ if } m < 0$$

$$\int_0^\infty \frac{\cos x\, dx}{x} = \infty$$

$$\int_0^\infty \frac{\tan x\, dx}{x} = \frac{\pi}{2}$$

$$\int_0^\pi \sin ax \cdot \sin bx\, dx = \int_0^\pi \cos ax \cdot \cos bx\, dx = 0, \qquad (a \neq b; a, b \text{ integers})$$

$$\int_0^{\pi/a} [\sin(ax)][\cos(ax)]\, dx = \int_0^\pi [\sin(ax)][\cos(ax)]\, dx = 0$$

$$\int_0^\pi [\sin(ax)][\cos(bx)]\, dx = \frac{2a}{a^2 - b^2}, \text{ if } a - b \text{ is odd, or zero if } a - b \text{ is even}$$

$$\int_0^\infty \frac{\sin x \cos mx\, dx}{x} = 0, \text{ if } m < -1 \text{ or } m > 1, = \frac{\pi}{4}, \text{ if } m = \pm 1; = \frac{\pi}{2}, \text{ if } m^2 < 1$$

$$\int_0^\infty \frac{\sin ax \sin bx}{x^2}\, dx = \frac{\pi a}{2} \qquad (a \leq b)$$

$$\int_0^\pi \sin^2 mx\, dx = \int_0^\pi \cos^2 mx\, dx = \frac{\pi}{2}$$

$$\int_0^\infty \frac{\sin^2 x\, dx}{x^2} = \frac{\pi}{2}$$

$$\int \frac{\cos mx}{1+x^2}\, dx = \frac{\pi}{2} e^{-|m|}$$

$$\int_0^\infty \cos(x^2)\, dx = \int_0^\infty \sin(x^2)\, dx = \frac{1}{2}\sqrt{\frac{\pi}{2}}$$

$$\int_0^\infty \frac{\sin x\, dx}{\sqrt{x}} = \int_0^\infty \frac{\cos x\, dx}{\sqrt{x}} = \sqrt{\frac{\pi}{2}}$$

$$\int_0^{\pi/2} \frac{dx}{1 + a\cos x} = \frac{\cos^{-1} a}{\sqrt{1 - a^2}} \qquad (a < 1)$$

$$\int_0^\infty \frac{dx}{a + b\cos x} = \frac{\pi}{\sqrt{a^2 - b^2}} \qquad (a > b \geq 0)$$

$$\int_0^{2\pi} \frac{dx}{1 + a \cos x} = \frac{2\pi}{\sqrt{1 - a^2}} \qquad (a^2 < 1)$$

$$\int_0^\infty \frac{\cos ax - \cos bx}{x} \, dx = \log \frac{b}{a}$$

$$\int_0^{\pi/2} \frac{dx}{a^2 \sin^2 x + b^2 \cos^2 x} = \frac{\pi}{2ab}$$

$$\int_0^{\pi/2} \frac{dx}{(a^2 \sin^2 x + b^2 \cos^2 x)^2} = \frac{\pi(a^2 + b^2)}{4a^3 b^3} \qquad (a, b > 0)$$

$$\int_0^{\pi/2} \sin^{n-1} x \cos^{m-1} x \, dx = \frac{1}{2} \mathrm{B}\left(\frac{n}{2}, \frac{m}{2}\right) \qquad (m \text{ and } n \text{ positive integers})$$

$$\int_0^{\pi/2} (\sin^{2n+1} \theta) \, d\theta = \frac{2 \cdot 4 \cdot 6 \cdots (2n)}{1 \cdot 3 \cdot 5 \cdots (2n+1)} \qquad (n = 1, 2, 3 \ldots)$$

$$\int_0^{\pi/2} (\sin^{2n} \theta) \, d\theta = \frac{1 \cdot 3 \cdot 5 \cdots (2n-1)}{2 \cdot 4 \cdots (2n)} \left(\frac{\pi}{2}\right) \qquad (n = 1, 2, 3 \ldots)$$

$$\int_0^{\pi/2} \sqrt{\cos \theta} \, d\theta = \frac{(2\pi)^{\frac{3}{2}}}{\left[\Gamma\left(\frac{1}{4}\right)\right]^2}$$

$$\int_0^{\pi/2} (\tan^h \theta) \, d\theta = \frac{\pi}{2 \cos\left(\frac{h\pi}{2}\right)} \qquad (0 < h < 1)$$

$$\int_0^\infty \frac{\tan^{-1}(ax) - \tan^{-1}(bx)}{x} \, dx = \frac{\pi}{2} \log \frac{a}{b} \qquad (a, b > 0)$$

The area enclosed by a curve defined through the equation $x^{\frac{b}{c}} + y^{\frac{b}{c}} = a^{\frac{b}{c}}$ where $a > 0$, c a positive odd integer and b a positive even integer is given by

$$\frac{\left[\Gamma\left(\frac{c}{b}\right)\right]^2}{\Gamma\left(\frac{2c}{b}\right)} \left(\frac{2ca^2}{b}\right)$$

$I = \iiint_R x^{h-1} y^{m-1} z^{n-1} \, dv$, where R denotes the region of space bounded by the co-ordinate planes and that portion of the surface $(\frac{x}{a})^p + (\frac{y}{b})^q + (\frac{z}{c})^k = 1$, which lies in the first octant and where $h, m, n, p, q, k, a, b, c$, denote positive real numbers is given by

$$\int_0^a x^{h-1} \, dx \int_0^{b\left[1-\left(\frac{x}{a}\right)^p\right]^{\frac{1}{q}}} y^m \, dy \int_0^{c\left[1-\left(\frac{x}{a}\right)^p - \left(\frac{y}{b}\right)^q\right]^{\frac{1}{k}}} z^{n-1} \, dz = \frac{a^h b^m c^n}{pqk} \frac{\Gamma\left(\frac{h}{p}\right) \Gamma\left(\frac{m}{q}\right) \Gamma\left(\frac{n}{k}\right)}{\Gamma\left(\frac{h}{p} + \frac{m}{q} + \frac{n}{k} + 1\right)}$$

$$\int_0^{\pi/2} \frac{dx}{a^2 \sin^2 x + b^2 \cos^2 x} = \frac{\pi}{2ab} \qquad (ab > 0)$$

$$\int_0^\pi \frac{dx}{a^2 \sin^2 x + b^2 \cos^2 x} = \frac{\pi}{ab} \qquad (ab > 0)$$

$$\int_0^{\pi/2} \frac{\sin^2 x \, dx}{a^2 \sin^2 x + b^2 \cos^2 x} = \int_0^{\pi/2} \frac{dx}{a^2 + b^2 \operatorname{ctn}^2 x} = \frac{\pi}{2a(a+b)} \qquad (a, b > 0)$$

$$\int_0^{\pi/2} \frac{\cos^2 x \, dx}{a^2 \sin^2 x + b^2 \cos^2 x} = \int_0^{\pi/2} \frac{dx}{b^2 + a^2 \tan^2 x} = \frac{\pi}{2b(a+b)} \qquad (a, b > 0)$$

$$\int_0^{\pi/2} \frac{dx}{(a^2 \sin^2 x + b^2 \cos^2 x)^2} = \frac{\pi}{4} \frac{(a^2 + b^2)}{a^3 b^3} \qquad (ab > 0)$$

$$\int_0^{\pi/2} \frac{\sin^2 x \, dx}{(a^2 \sin^2 x + b^2 \cos^2 x)^2} = \frac{\pi}{4a^3 b} \qquad (ab > 0)$$

$$\int_0^{\pi/2} \frac{\cos^2 x \, dx}{(a^2 \sin^2 x + b^2 \cos^2 x)^2} = \frac{\pi}{4ab^3} \qquad (ab > 0)$$

$$\int_0^\infty \sin(a^2 x^2) \, dx = \int_0^\infty \cos(a^2 x^2) \, dx = \frac{\sqrt{\pi}}{2a\sqrt{2}} \qquad (a > 0)$$

$$\int_0^\infty \sin \frac{\pi x^2}{2} \, dx = \int_0^\infty \cos \frac{\pi x^2}{2} \, dx = \frac{1}{2} \qquad \text{(Fresnel's integrals)}$$

$$\int_0^\infty \sin(x^p) \, dx = \Gamma\left(1 + \frac{1}{p}\right) \sin \frac{\pi}{2p} \qquad (p > 1)$$

$$\int_0^\infty \cos(x^p) \, dx = \Gamma\left(1 + \frac{1}{p}\right) \cos \frac{\pi}{2p} \qquad (p > 1)$$

$$\int_0^\infty \sin a^2 x^2 \cos mx \, dx = \frac{\sqrt{\pi}}{2a} \sin\left(\frac{\pi}{4} - \frac{m^2}{4a^2}\right) \qquad (a > 0)$$

$$\int_0^\infty \cos a^2 x^2 \cos mx \, dx = \frac{\sqrt{\pi}}{2a} \cos\left(\frac{\pi}{4} - \frac{m^2}{4a^2}\right) \qquad (a > 0)$$

$$\int_0^\infty \frac{\sin^{2p} mx}{x^2} \, dx = \frac{1 \cdot 3 \cdot 5 \cdots (2p-3)}{2 \cdot 4 \cdot 6 \cdots (2p-2)} \frac{|m|\pi}{2} \qquad (p = 2, 3, 4, \ldots)$$

$$\int_0^\infty \frac{\sin^3 mx}{x^3} \, dx = \frac{3}{8} m^2 \pi \qquad (m > 0)$$

$$\int_0^\infty \frac{\sin mx \cos nx}{x} \, dx = \pi/2 \qquad (m > n > 0)$$

$$= \pi/4 \qquad (m = n > 0)$$

$$= 0 \qquad (n > m > 0)$$

$$\int_0^\infty \frac{\sin mx \sin nx}{x} \, dx = \frac{1}{2} \log \frac{m+n}{m-n} \qquad (m > n > 0)$$

$$\int_0^\infty \frac{\cos mx \cos nx}{x} \, dx = \infty$$

$$\int_0^\infty \frac{\sin^2 ax \sin mx}{x} \, dx = \frac{\pi}{4} \qquad\qquad (2a > m > 0)$$

$$= \frac{\pi}{8} \qquad\qquad (2a = m > 0)$$

$$= 0 \qquad\qquad (m > 2a > 0)$$

$$\int_0^\infty \frac{\sin mx \sin nx}{x^2} \, dx = \frac{\pi m}{2} \qquad\qquad (n \geqq m > 0)$$

$$= \frac{\pi n}{2} \qquad\qquad (m \geqq n > 0)$$

$$\int_0^\infty \frac{\sin^2 ax \sin mx}{x^2} \, dx = \frac{m + 2a}{4} \log|m + 2a| + \frac{m - 2a}{4} \log|m - 2a| - \frac{m}{2} \log m$$

$$(m > 0)$$

$$\int_0^\infty \frac{\cos mx}{a^2 + x^2} \, dx = \frac{\pi}{2a} e^{-ma} \qquad\qquad (a > 0; m \geqq 0)$$

$$\int_0^\infty \frac{\sin^2 mx}{a^2 + x^2} \, dx = \frac{\pi}{4a}(1 - e^{-2ma}) \qquad\qquad (a > 0; m \geqq 0)$$

$$\int_0^\infty \frac{\cos^2 mx}{a^2 + x^2} \, dx = \frac{\pi}{4a}(1 + e^{-2ma}) \qquad\qquad (a > 0; m \geqq 0)$$

$$\int_0^\infty \frac{x \sin mx}{a^2 + x^2} \, dx = \frac{\pi}{2} e^{-ma} \qquad\qquad (a \geqq 0; m > 0)$$

$$\int_0^\infty \frac{\sin mx}{x(a^2 + x^2)} \, dx = \frac{\pi}{2a^2}(1 - e^{-ma}) \qquad\qquad (a > 0; m \geqq 0)$$

$$\int_0^\infty \frac{\sin mx \sin nx}{a^2 + x^2} \, dx = \frac{\pi}{2a} e^{-ma} \sinh na \qquad (a > 0; m \geqq n \geqq 0)$$

$$= \frac{\pi}{2a} e^{-na} \sinh ma \qquad (a > 0; n \geqq m \geqq 0)$$

$$\int_0^\infty \frac{\cos mx \cos nx}{a^2 + x^2} \, dx = \frac{\pi}{2a} e^{-ma} \cosh na \qquad (a > 0; m \geqq n \geqq 0)$$

$$= \frac{\pi}{2a} e^{-na} \cosh ma \qquad (a > 0; n \geqq m \geqq 0)$$

$$\int_0^\infty \frac{x \sin mx \cos nx}{a^2 + x^2} \, dx = \frac{\pi}{2} e^{-ma} \cosh na \qquad (a > 0; m > n > 0)$$

$$= -\frac{\pi}{2} e^{-na} \sinh ma \qquad (a > 0; n > m > 0)$$

$$\int_0^\infty \frac{\cos mx}{(a^2 + x^2)^2} \, dx = \frac{\pi}{4a^3}(1 + ma)e^{-ma} \qquad\qquad (a, m > 0)$$

$$\int_0^\infty \frac{x \sin mx}{(a^2 + x^2)^2} \, dx = \frac{\pi m}{4a} e^{-ma} \qquad\qquad (a, m > 0)$$

$$\int_0^\infty \frac{x^2 \cos mx}{(a^2 + x^2)^2} \, dx = \frac{\pi}{4a}(1 - ma)e^{-ma} \qquad (a, m > 0)$$

$$\int_0^\infty \frac{\sin^2 ax \cos mx}{x^2} \, dx = \frac{\pi}{2}\left(a - \frac{m}{2}\right) \qquad \left(a > \frac{m}{2} > 0\right)$$

$$= 0 \qquad \left(\frac{m}{2} \geqq a \geqq 0\right)$$

$$\int_0^\infty \frac{1 - \cos mx}{x^2} \, dx = \frac{\pi |m|}{2}$$

$$\int_0^\infty \frac{\sin^2 ax \sin mx}{x^3} \, dx = \frac{\pi am}{2} - \frac{\pi m^2}{8} \qquad \left(a \geqq \frac{m}{2} > 0\right)$$

$$= \frac{\pi a^2}{2} \qquad \left(\frac{m}{2} \geqq a > 0\right)$$

$$\int_0^\infty \frac{\sin mx}{\sqrt{x}} \, dx = \int_0^\infty \frac{\cos mx}{\sqrt{x}} \, dx = \frac{\sqrt{\pi}}{\sqrt{(2m)}} \qquad (m > 0)$$

$$\int_0^\infty \frac{\sin mx}{x\sqrt{x}} \, dx = \sqrt{(2\pi m)} \qquad (m > 0)$$

$$\int_0^\infty \frac{\sin mx}{x^p} \, dx = \frac{\pi m^{p-1}}{2 \sin\left(\frac{p\pi}{2}\right) \Gamma(p)} \qquad (0 < p < 2; m > 0)$$

$$\int_0^\infty e^{-ax} \, dx = \frac{1}{a} \qquad (a > 0)$$

$$\int_0^\infty \frac{e^{-ax} - e^{-bx}}{x} \, dx = \log \frac{b}{a} \qquad (a, b > 0)$$

$$\int_0^\infty x^n e^{-ax} \, dx = \frac{\Gamma(n + 1)}{a^{n+1}} \qquad (n > -1, \ a > 0)$$

$$= \frac{n!}{a^{n+1}} \qquad (n \text{ pos. integ.}, a > 0)$$

$$\int_0^\infty e^{-a^2 x^2} \, dx = \frac{1}{2a}\sqrt{\pi} = \frac{1}{2a}\Gamma\left(\frac{1}{2}\right) \qquad (a > 0)$$

$$\int_0^\infty x e^{-x^2} \, dx = \frac{1}{2}$$

$$\int_0^\infty x^2 e^{-x^2} \, dx = \frac{\sqrt{\pi}}{4}$$

$$\int_0^\infty x^{2n} e^{-ax^2} \, dx = \frac{1 \cdot 3 \cdot 5 \cdots (2n - 1)}{2^{n+1}a^n}\sqrt{\frac{\pi}{a}}$$

$$\int_0^1 x^m e^{-ax} \, dx = \frac{m!}{a^{m+1}}\left[1 - e^{-a}\sum_{r=0}^m \frac{a^r}{r!}\right]$$

$$\int_0^\infty e^{\left(-x^2 - \frac{a^2}{x^2}\right)} dx = \frac{e^{-2a}\sqrt{\pi}}{2} \qquad (a \geq 0)$$

$$\int_0^\infty e^{-nx}\sqrt{x}\,dx = \frac{1}{2n}\sqrt{\frac{\pi}{n}}$$

$$\int_0^\infty \frac{e^{-nx}}{\sqrt{x}}\,dx = \sqrt{\frac{\pi}{n}}$$

$$\int_0^\infty e^{-ax}\cos mx\,dx = \frac{a}{a^2 + m^2} \qquad (a > 0)$$

$$\int_0^\infty e^{-ax}\sin mx\,dx = \frac{m}{a^2 + m^2} \qquad (a > 0)$$

$$\int_0^\infty xe^{-ax}[\sin(bx)]\,dx = \frac{2ab}{(a^2 + b^2)^2} \qquad (a > 0)$$

$$\int_0^\infty xe^{-ax}[\cos(bx)]\,dx = \frac{a^2 - b^2}{(a^2 + b^2)^2} \qquad (a > 0)$$

$$\int_0^\infty x^n e^{-ax}[\sin(bx)]\,dx = \frac{n![(a - ib)^{n+1} - (a + ib)^{n+1}]}{2(a^2 + b^2)^{n+1}} \qquad (i^2 = -1,\ a > 0)$$

$$\int_0^\infty x^n e^{-ax}[\cos(bx)]\,dx = \frac{n![(a - ib)^{n+1} + (a + ib)^{n+1}]}{2(a^2 + b^2)^{n+1}} \qquad (i^2 = -1,\ a > 0)$$

$$\int_0^\infty \frac{e^{-ax}\sin x}{x}\,dx = \cot^{-1} a \qquad (a > 0)$$

$$\int_0^\infty e^{-a^2 x^2}\cos bx\,dx = \frac{\sqrt{\pi}}{2a}e^{\frac{-b^2}{4a^2}} \qquad (ab \neq 0)$$

$$\int_0^\infty e^{-t\cos\phi} t^{b-1}[\sin(t\sin\phi)]\,dt = [\Gamma(b)]\sin(b\phi) \qquad \left(b > 0,\ -\frac{\pi}{2} < \phi < \frac{\pi}{2}\right)$$

$$\int_0^\infty e^{-t\cos\phi} t^{b-1}[\cos(t\sin\phi)]\,dt = [\Gamma(b)]\cos(b\phi) \qquad \left(b > 0,\ -\frac{\pi}{2} < \phi < \frac{\pi}{2}\right)$$

$$\int_0^\infty \frac{e^{-ax^c} - e^{-bx^c}}{x}\,dx = \frac{1}{c}\log\frac{b}{a} \qquad (a, b, c > 0)$$

$$\int_0^\infty \frac{1 - e^{-ax^2}}{x^2}\,dx = \sqrt{(a\pi)} \qquad (a > 0)$$

$$\int_0^\infty \exp[-a^2 x^2 - \frac{b^2}{x^2}]\,dx = \frac{\sqrt{\pi}}{2a}e^{-2ab} \qquad (a, b > 0)$$

$$\int_0^\infty \frac{dx}{e^{ax} - 1} = \infty \qquad (a > 0)$$

$$\int_0^\infty \frac{x\,dx}{e^{ax}-1} = \frac{\pi^2}{6a^2} \qquad\qquad (a>0)$$

$$\int_0^\infty \frac{e^{-ax}-e^{-bx}}{x}\,dx = \log\frac{b}{a} \qquad\qquad (a,b>0)$$

$$\int_0^\infty \frac{dx}{e^{ax}+1} = \frac{\log 2}{a} \qquad\qquad (a>0)$$

$$\int_0^\infty \frac{x\,dx}{e^{ax}+1} = \frac{\pi^2}{12a^2} \qquad\qquad (a>0)$$

$$\int_0^\infty \frac{e^{-ax}}{x}\sin mx\,dx = \tan^{-1}\frac{m}{a} \qquad\qquad (a>0)$$

$$\int_0^\infty \frac{e^{-ax}}{x}\cos mx\,dx = \infty$$

$$\int_0^\infty \frac{e^{-ax}}{x}(1-\cos mx)\,dx = \frac{1}{2}\ln\left(1+\frac{m^2}{a^2}\right) \qquad\qquad (a>0)$$

$$\int_0^\infty \frac{e^{-ax}}{x}(\cos mx - \cos nx)\,dx = \frac{1}{2}\ln\frac{a^2+n^2}{a^2+m^2} \qquad\qquad (a>0)$$

$$\int_0^\infty \frac{e^{-ax}-e^{-bx}}{x}\cos mx\,dx = \frac{1}{2}\ln\frac{b^2+m^2}{a^2+m^2} \qquad\qquad (a,b>0)$$

$$\int_0^\infty e^{-ax}\cos^2 mx\,dx = \frac{a^2+2m^2}{a(a^2+4m^2)} \qquad\qquad (a>0)$$

$$\int_0^\infty e^{-ax}\sin^2 mx\,dx = \frac{2m^2}{a(a^2+4m^2)} \qquad\qquad (a>0)$$

$$\int_0^\infty \frac{e^{-ax}}{x}\sin^2 mx\,dx = \frac{1}{4}\ln\left(1+\frac{4m^2}{a^2}\right) \qquad\qquad (a>0)$$

$$\int_0^\infty \frac{e^{-ax}}{x^2}\sin^2 mx\,dx = m\tan^{-1}\frac{2m}{a} - \frac{a}{4}\ln\left(1+\frac{4m^2}{a^2}\right) \qquad\qquad (a>0)$$

$$\int_0^\infty e^{-ax}\sin mx\sin nx\,dx = \frac{2amn}{\{a^2+(m-n)^2\}\{a^2+(m+n)^2\}} \qquad\qquad (a>0)$$

$$\int_0^\infty e^{-ax}\sin mx\cos nx\,dx = \frac{m(a^2+m^2-n^2)}{\{a^2+(m-n)^2\}\{a^2+(m+n)^2\}} \qquad\qquad (a>0)$$

$$\int_0^\infty e^{-ax}\cos mx\cos nx\,dx = \frac{a(a^2+m^2+n^2)}{\{a^2+(m-n)^2\}\{a^2+(m+n)^2\}} \qquad\qquad (a>0)$$

$$\int_0^\infty \frac{e^{-ax}}{x}\sin mx\sin nx\,dx = \frac{1}{4}\log\frac{a^2+(m+n)^2}{a^2+(m-n)^2} \qquad\qquad (a>0)$$

$$\int_0^\infty e^{-a^2x^2} \cos mx \, dx = \frac{\sqrt{\pi}}{2a} e^{-m^2/(4a^2)} \qquad (a > 0)$$

$$\int_0^\infty x e^{-a^2x^2} \sin mx \, dx = \frac{m\sqrt{\pi}}{4a^3} e^{-m^2/(4a^2)} \qquad (a > 0)$$

$$\int_0^\infty \frac{e^{-a^2x^2}}{x} \sin mx \, dx = \frac{\pi}{2} \operatorname{erf}\left(\frac{m}{2a}\right) \qquad (a > 0)$$

$$\int_0^\infty \frac{e^{-ax}}{\sqrt{x}} \cos mx \, dx = \frac{\{a + \sqrt{(a^2 + m^2)}\}^{1/2}\sqrt{\pi}}{(a^2 + m^2)^{1/2}\sqrt{2}} \qquad (a > 0)$$

$$\int_0^\infty e^{-ax} \sin\sqrt{(mx)} \, dx = \frac{\sqrt{(\pi m)}}{2a\sqrt{a}} e^{-m/(4a)} \qquad (a, m > 0)$$

$$\int_0^\infty \frac{e^{-ax}}{\sqrt{x}} \cos\sqrt{(mx)} \, dx = \frac{\sqrt{\pi}}{\sqrt{a}} e^{-m/(4a)} \qquad (a, m > 0)$$

$$\int_0^\infty e^{-ax} \sin(px + q) \, dx = \frac{a\sin q + p\cos q}{a^2 + p^2} \qquad (a > 0)$$

$$\int_0^\infty e^{-ax} \cos(px + q) \, dx = \frac{a\cos q - p\sin q}{a^2 + p^2} \qquad (a > 0)$$

$$\int_0^\infty t^{b-1} \cos t \, dt = [\Gamma(b)] \cos\left(\frac{b\pi}{2}\right) \qquad (0 < b < 1)$$

$$\int_0^\infty t^{b-1} (\sin t) \, dt = [\Gamma(b)] \sin\left(\frac{b\pi}{2}\right) \qquad (0 < b < 1)$$

$$\int_0^1 (\ln x)^n \, dx = (-1)^n \cdot n!$$

$$\int_0^1 \left(\ln\frac{1}{x}\right)^{\frac{1}{2}} dx = \frac{\sqrt{\pi}}{2}$$

$$\int_0^1 \left(\ln\frac{1}{x}\right)^{-\frac{1}{2}} dx = \sqrt{\pi}$$

$$\int_0^1 \left(\ln\frac{1}{x}\right)^n dx = n!$$

$$\int_0^1 x \ln(1 - x) \, dx = -\frac{3}{4}$$

$$\int_0^1 x \ln(1 + x) \, dx = \frac{1}{4}$$

$$\int_0^1 \frac{\ln x}{1 + x} \, dx = -\frac{\pi^2}{12}$$

$$\int_0^1 \frac{\ln x}{1-x} dx = -\frac{\pi^2}{6}$$

$$\int_0^1 \frac{\ln x}{1-x^2} dx = -\frac{\pi^2}{8}$$

$$\int_0^1 \ln\left(\frac{1+x}{1-x}\right) \cdot \frac{dx}{x} = \frac{\pi^2}{4}$$

$$\int_0^1 \frac{\ln x \, dx}{\sqrt{1-x^2}} = -\frac{\pi}{2} \ln 2$$

$$\int_0^1 x^m \left[\ln\left(\frac{1}{x}\right)\right]^n dx = \frac{\Gamma(n+1)}{(m+1)^{n+1}}, \text{ if } m+1 > 0, \ n+1 > 0$$

$$\int_0^1 \frac{(x^p - x^q) dx}{\ln x} = \ln\left(\frac{p+1}{q+1}\right) \qquad\qquad (p+1 > 0, \ q+1 > 0)$$

$$\int_0^1 \frac{dx}{\sqrt{\ln\left(\frac{1}{x}\right)}} = \sqrt{\pi}$$

$$\int_0^\infty \ln\left(\frac{e^x + 1}{e^x - 1}\right) dx = \frac{\pi^2}{4}$$

$$\int_0^{\pi/2} \ln \sin x \, dx = \int_0^{\pi/2} \ln \cos x \, dx = -\frac{\pi}{2} \ln 2$$

$$\int_0^{\pi/2} \ln \sec x \, dx = \int_0^{\pi/2} \ln \csc x \, dx = \frac{\pi}{2} \ln 2$$

$$\int_0^\pi x \ln \sin x \, dx = -\frac{\pi^2}{2} \ln 2$$

$$\int_0^{\pi/2} \sin x \ln \sin x \, dx = \ln 2 - 1$$

$$\int_0^{\pi/2} \ln \tan x \, dx = 0$$

$$\int_0^\pi \ln(a \pm b \cos x) \, dx = \pi \log\left(\frac{a + \sqrt{a^2 - b^2}}{2}\right) \qquad\qquad (a \geqq b)$$

$$\int_0^\infty \frac{dx}{\cosh ax} = \frac{\pi}{2a}$$

$$\int_0^\infty \frac{x \, dx}{\sinh ax} = \frac{\pi^2}{4a^2}$$

$$\int_0^\infty e^{-ax} \cosh bx \, dx = \frac{a}{a^2 - b^2} \qquad (0 \le |b| < a)$$

$$\int_0^\infty e^{-ax} \sinh bx \, dx = \frac{b}{a^2 - b^2} \qquad (0 \le |b| < a)$$

$$\int_{+\infty}^1 \frac{e^{-xu}}{u} \, du = \gamma + \ln x - x + \frac{x^2}{2 \cdot 2!} - \frac{x^3}{3 \cdot 3!} + \frac{x^4}{4 \cdot 4!} - \cdots,$$

$$\text{where } \gamma = \lim_{z \to \infty} \left(1 + \frac{1}{2} + \frac{1}{3} + \cdots + \frac{1}{z} - \ln z \right)$$

$$= 0.5772157 \cdots \qquad (0 < x < \infty)$$

$$\int_0^{\pi/2} \frac{dx}{\sqrt{1 - k^2 \sin^2 x}} = \frac{\pi}{2} \left[1 + \left(\frac{1}{2} \right)^2 k^2 + \left(\frac{1 \cdot 3}{2 \cdot 4} \right)^2 k^4 \right.$$

$$\left. + \left(\frac{1 \cdot 3 \cdot 5}{2 \cdot 4 \cdot 6} \right)^2 k^6 + \cdots \right], \text{ if } k^2 < 1$$

$$\int_0^{\pi/2} \sqrt{1 - k^2 \sin^2 x} \, dx = \frac{\pi}{2} \left[1 - \left(\frac{1}{2} \right)^2 k^2 - \left(\frac{1 \cdot 3}{2 \cdot 4} \right)^2 \frac{k^4}{3} \right.$$

$$\left. - \left(\frac{1 \cdot 3 \cdot 5}{2 \cdot 4 \cdot 6} \right)^2 \frac{k^6}{5} - \cdots \right], \text{ if } k^2 < 1$$

$$\int_0^\infty e^{-x} \ln x \, dx = -\gamma = -0.5772157 \cdots$$

$$\int_0^\infty \left(\frac{1}{1 - e^{-x}} - \frac{1}{x} \right) e^{-x} \, dx = \gamma = 0.5772157 \cdots \qquad \text{(Euler's Constant)}$$

$$\int_0^\infty \frac{1}{x} \left(\frac{1}{1 + x} - e^{-x} \right) dx = \gamma = 0.5772157 \cdots$$

Appendix 4: *Matrices and Determinants*

1 General definitions

1.1. A matrix is an array of numbers consisting of m rows and n columns. It is usually denoted by a boldface capital letter, e.g.,

$$\mathbf{A} \quad \mathbf{\Sigma} \quad \mathbf{M}.$$

1.2. The (i, j) element of a matrix is the element occurring in row i and column j. It is usually denoted by a lowercase letter with subscripts, e.g.,

$$a_{ij} \quad \sigma_{ij} \quad m_{ij}.$$

Exceptions to this convention will be stated where required.

1.3. A matrix is called rectangular if m (number of rows) $\neq n$ (number of columns).

1.4. A matrix is called square if $m = n$.

1.5a. In the transpose of a matrix $\mathbf{A}$, denoted by $\mathbf{A}'$, the element in the jth row and ith column of $\mathbf{A}$ is equal to the element in the ith row and jth column of $\mathbf{A}'$. Formally, $(\mathbf{A}')_{ij} = (\mathbf{A})_{ji}$ where the symbol $(\mathbf{A}')_{ij}$ denotes the (i, j)th element of $\mathbf{A}'$.

1.5b. The Hermitian conjugate of a matrix $\mathbf{A}$, denoted by $\mathbf{A}^H$ or $\mathbf{A}^\dagger$, is obtained by transposing $\mathbf{A}$ and replacing each element by its conjugate complex. Hence if

$$a_{kl} = u_{kl} + i v_{kl},$$

then

$$(\mathbf{A}^H)_{kl} = u_{lk} - i v_{lk},$$

where typical elements have been denoted by (k, l) to avoid confusion with $i = \sqrt{-1}$.

1.6a. A square matrix is called **symmetric** if $\mathbf{A} = \mathbf{A}'$.

1.6b. A square matrix is called **Hermitian** if $\mathbf{A} = \mathbf{A}^H$.

1.7. A matrix with m rows and 1 column is called a column vector and is usually denoted by boldface, lowercase letters, e.g.,

$$\boldsymbol{\beta} \quad \mathbf{x} \quad \mathbf{a}.$$

0-8493-8342-0/96/$0.00 + $0.50
©1996 by CRC Press, Inc.

1.8. A matrix with one row and n columns is called a row vector and is usually denoted by a primed, boldface, lowercase letter, e.g.,

$$\mathbf{a}' \qquad \mathbf{c}' \qquad \mu'.$$

1.9. A matrix with one row and one column is called a scalar and is usually denoted by a lowercase letter, occasionally italicized.

1.10. The diagonal extending from upper left (NW) to lower right (SE) is called the principal diagonal of a square matrix.

1.11a. A matrix with all elements above the principal diagonal equal to zero is called a lower triangular matrix.

Example

$$\mathbf{T} = \begin{bmatrix} t_{11} & 0 & 0 \\ t_{21} & t_{22} & 0 \\ t_{31} & t_{32} & t_{33} \end{bmatrix} \text{ is lower triangular.} \qquad \blacksquare$$

1.11b. The transpose of a lower triangular matrix is called an upper triangular matrix.

1.12. A square matrix with all off-diagonal elements equal to zero is called a diagonal matrix, denoted by the letter $\mathbf{D}$ with subscript indicating the typical element in the principal diagonal.

Example

$$\mathbf{D}_a = \begin{bmatrix} a_1 & 0 & 0 \\ 0 & a_2 & 0 \\ 0 & 0 & a_3 \end{bmatrix} \text{ is diagonal.} \qquad \blacksquare$$

2   Addition, subtraction, and multiplication

2.1. Two matrices $\mathbf{A}$ and $\mathbf{B}$ can be added (subtracted) if the number of rows (columns) in $\mathbf{A}$ equals the number of rows (columns) in $\mathbf{B}$.

$$\mathbf{A} \pm \mathbf{B} = \mathbf{C}$$

implies

$$a_{ij} \pm b_{ij} = c_{ij}, \qquad i = 1, 2, \ldots m$$
$$j = 1, 2, \ldots n.$$

2.2. Multiplication of a matrix or vector by a scalar implies multiplication of each element by the scalar. If

$$\mathbf{B} = \gamma \mathbf{A},$$

then

$$b_{ij} = \gamma a_{ij}$$

for all elements.

2.3a. Two matrices $\mathbf{A}$ and $\mathbf{B}$ can be multiplied if the number of columns in $\mathbf{A}$ equals the number of rows in $\mathbf{B}$.

2.3b. Let $\mathbf{A}$ be of order $(m \times n)$ (have m rows and n columns) and $\mathbf{B}$ of order $(n \times p)$. Then the product of two matrices $\mathbf{C} = \mathbf{AB}$ is a matrix of order $(m \times p)$ with elements

$$c_{ij} = \sum_{k=1}^{n} a_{ik} b_{kj}.$$

This states that c_{ij} is the scalar product of the ith row vector of $\mathbf{A}$ and the jth column vector of $\mathbf{B}$.

Example

$$\begin{bmatrix} 3 & 4 & 2 \\ 2 & 3 & -1 \end{bmatrix} \begin{bmatrix} 1 & -2 & -4 \\ 0 & -1 & 2 \\ 6 & -3 & 9 \end{bmatrix} = \begin{bmatrix} 15 & -16 & 14 \\ -4 & -4 & -11 \end{bmatrix}$$

e.g.,

$$c_{23} = \begin{bmatrix} 2 & 3 & -1 \end{bmatrix} \begin{bmatrix} -4 \\ 2 \\ 9 \end{bmatrix}$$

$$= 2 \times (-4) + 3 \times 2 + (-1) \times 9 = -11 \qquad \blacksquare$$

2.3c. In general, matrix multiplication is not commutative:

$$\mathbf{AB} \neq \mathbf{BA}.$$

2.3d. Matrix multiplication is associative:

$$\mathbf{A}(\mathbf{BC}) = (\mathbf{AB})\mathbf{C}.$$

2.3e. The distributive law for multiplication and addition holds as in the case of scalars:

$$(\mathbf{A} + \mathbf{B})\mathbf{C} = \mathbf{AC} + \mathbf{BC}$$

$$\mathbf{C}(\mathbf{A} + \mathbf{B}) = \mathbf{CA} + \mathbf{CB}.$$

2.4. In some applications, the term-by-term product of two matrices $\mathbf{A}$ and $\mathbf{B}$ of identical order is defined as

$$\mathbf{C} = \mathbf{A} * \mathbf{B}$$

where

$$c_{ij} = a_{ij} b_{ij}.$$

2.5. $(\mathbf{ABC})' = \mathbf{C}'\mathbf{B}'\mathbf{A}'.$

2.6. $(\mathbf{ABC})^{\mathrm{H}} = \mathbf{C}^{\mathrm{H}}\mathbf{B}^{\mathrm{H}}\mathbf{A}^{\mathrm{H}}.$

2.7. If both $\mathbf{A}$ and $\mathbf{B}$ are symmetric, then $(\mathbf{AB})' = \mathbf{BA}$. Note that the product of two symmetric matrices is generally not symmetric.

3 Recognition rules and special forms

3.1. A column (row) vector with all elements equal to zero is called a null vector and is usually denoted by the symbol **0**.

3.2. A null matrix has all elements equal to zero.

3.3a. A diagonal matrix with all elements equal to one in the principal diagonal is called the identity matrix **I**.

3.3b. $\gamma \mathbf{I}$, i.e., a diagonal matrix with all diagonal elements equal to a constant γ, is called a scalar matrix.

3.4. A matrix that has only one element equal to one and all others equal to zero is called an elementary matrix $(\mathbf{EL})_{ij}$.

Example

$$(\mathbf{EL})_{23} = \begin{bmatrix} 0 & 0 & 0 & 0 & 0 \\ 0 & 0 & 1 & 0 & 0 \\ 0 & 0 & 0 & 0 & 0 \\ 0 & 0 & 0 & 0 & 0 \end{bmatrix}$$

The order of the matrix is usually implicit. ∎

3.5a. The symbol **j** is reserved for a column vector with all elements equal to 1.

3.5b. The symbol **j′** is reserved for a row vector with all elements equal to 1.

3.6. An expression ending with a column vector is a column vector.

Example

$$\mathbf{ABx} = \mathbf{y}$$

(It is assumed that rule 2.3a is satisfied, else matrix multiplication would not be defined.) ∎

3.7. An expression beginning with a row vector is a row vector:

Example

$$\mathbf{y}'(\mathbf{A} + \mathbf{BC}) = \mathbf{d}'.$$ ∎

3.8. An expression beginning with a row vector and ending with a column vector is a scalar:

Example

$$\mathbf{a}'\mathbf{Bc} = \gamma.$$ ∎

3.9a. If **Q** is a square matrix, the scalar $\mathbf{x}'\mathbf{Qx}$ is called a quadratic form. If **Q** is nonsymmetric, one can always find a symmetric matrix $\mathbf{Q}^*$ such that

$$\mathbf{x}'\mathbf{Qx} = \mathbf{x}'\mathbf{Q}^*\mathbf{x}$$

where

$$(\mathbf{Q}^*)_{ij} = \frac{1}{2}(q_{ij} + q_{ji}).$$

3.9b.   If $\mathbf{Q}$ is a square matrix, the scalar $\mathbf{x}^H\mathbf{Q}\mathbf{x}$ is called a Hermitian form.

3.10.   A scalar $\mathbf{x}'\mathbf{Q}\mathbf{y}$ is called a bilinear form.

3.11.   The scalar $\mathbf{x}'\mathbf{x} = \Sigma x_i^2$, i.e., the sum of squares of all elements of $\mathbf{x}$.

3.12.   The scalar $\mathbf{x}'\mathbf{y} = \Sigma x_i y_i$, i.e., the sum of products of elements in $\mathbf{x}$ by those in $\mathbf{y}$. $\mathbf{x}$ and $\mathbf{y}$ have the same number of elements.

3.13.   The scalar $\mathbf{x}'\mathbf{D}_w\mathbf{x} = \Sigma w_i x_i^2$ is called a weighted sum of squares.

3.14.   The scalar $\mathbf{x}'\mathbf{D}_w\mathbf{y} = \Sigma w_i x_i y_i$ is called a weighted sum of products.

3.15a.   The vector $\mathbf{Aj}$ is a column vector whose elements are the row sums of $\mathbf{A}$.

3.15b.   The vector $\mathbf{j}'\mathbf{A}$ is a row vector whose elements are the column sums of $\mathbf{A}$.

3.15c.   The scalar $\mathbf{j}'\mathbf{Aj}$ is the sum of all elements in $\mathbf{A}$. Schematically,

$\mathbf{A}$	$\mathbf{Aj}$
$\mathbf{j}'\mathbf{A}$	$\mathbf{j}'\mathbf{Aj}$

3.16a.   If $\mathbf{B} = \mathbf{D}_w\mathbf{A}$, then $b_{ij} = w_i a_{ij}$.

3.16b.   If $\mathbf{B} = \mathbf{A}\mathbf{D}_w$, then $b_{ij} = a_{ij} w_j$.

3.17.   Interchanging summation and matrix notation:

If

$$\mathbf{ABCD} = \mathbf{E},$$

then

$$e_{ij} = \sum_k \sum_l \sum_m a_{ik} b_{kl} c_{lm} d_{mj}.$$

The second subscript of an element must coincide with the first of the next one. Reordering and transposing may be required.

Example

If

$$e_{ij} = \sum_k \sum_l \sum_m a_{kl} b_{ki} c_{jm} d_{ml}$$

$$= \sum_k \sum_l \sum_m b_{ki} a_{kl} d_{ml} c_{jm},$$

then

$$\mathbf{E} = \mathbf{B}'\mathbf{A}\mathbf{D}'\mathbf{C}'. \qquad \blacksquare$$

3.18a. $\mathbf{A}'\mathbf{A}$ is a symmetric matrix whose (i, j) element is the scalar product of the ith column vector and the jth column vector of $\mathbf{A}$.

3.18b. $\mathbf{A}\mathbf{A}'$ is the symmetric matrix whose (i, j) element is the scalar product of the ith row vector and the jth row vector of $\mathbf{A}$.

4 Determinants

4.1a. A determinant $|\mathbf{A}|$ or $\det(\mathbf{A})$ is a scalar function of a square matrix defined in such a way that

$$|\mathbf{A}||\mathbf{B}| = |\mathbf{AB}|$$

and

$$\begin{vmatrix} a_{11} & a_{12} \\ a_{21} & a_{22} \end{vmatrix} = a_{11}a_{22} - a_{12}a_{21}.$$

4.1b. $|\mathbf{A}| = |\mathbf{A}'|$.

4.2.

$$\begin{vmatrix} a_{11} & a_{12} & a_{13} \\ a_{21} & a_{22} & a_{23} \\ a_{31} & a_{32} & a_{33} \end{vmatrix} = a_{11}a_{22}a_{33} + a_{12}a_{23}a_{31} + a_{13}a_{21}a_{32}$$

$$-a_{13}a_{22}a_{31} - a_{11}a_{23}a_{32} - a_{12}a_{21}a_{33}.$$

4.3.

$$\begin{vmatrix} a_{11} & a_{12} & \cdots & a_{1n} \\ a_{21} & a_{22} & \cdots & a_{2n} \\ & & \cdots & \\ a_{n1} & a_{n2} & \cdots & a_{nm} \end{vmatrix} = \sum (-1)^{\delta} a_{1i_1} a_{2i_2} \cdots a_{ni_n}$$

where the sum is over all permutations

$$i_1 \neq i_2 \neq \cdots i_n$$

and δ denotes the number of exchanges necessary to bring the sequence $(i_1, i_2, \ldots i_n)$ back into the natural order $(1, 2, \ldots n)$.

4.4. If two rows (columns) in a matrix are exchanged, the determinant will change its sign.

4.5. A determinant does not change its value if a linear combination of other rows (columns) is added to any given row (column).

Example

$$\begin{vmatrix} a_{11} & a_{12} & a_{13} & a_{14} \\ b_{21} & b_{22} & b_{23} & b_{24} \\ a_{31} & a_{32} & a_{33} & a_{34} \\ a_{41} & a_{42} & a_{43} & a_{44} \end{vmatrix} = \begin{vmatrix} a_{11} & a_{12} & a_{13} & a_{14} \\ a_{21} & a_{22} & a_{23} & a_{24} \\ a_{31} & a_{32} & a_{33} & a_{34} \\ a_{41} & a_{42} & a_{43} & a_{44} \end{vmatrix}$$

where

$$b_{2i} = a_{2i} + \gamma_1 a_{1i} + \gamma_3 a_{3i} + \gamma_4 a_{4i},$$

$$i = 1, 2, 3, 4,$$

$\gamma_1, \gamma_3, \gamma_4$ arbitrary. ∎

4.6. If the ith row (column) equals (a constant times) the jth row (column) of a matrix, its determinant is equal to zero ($i \neq j$).

4.7. If, in a matrix $\mathbf{A}$, each element of a row (column) is multiplied by a constant γ, the determinant is multiplied by γ.

4.8. $|\gamma \mathbf{A}| = \gamma^n |\mathbf{A}|$ assuming that $\mathbf{A}$ is of order ($n \times n$).

4.9. The cofactor of a square matrix $\mathbf{A}$, $\mathrm{cof}_{ij}(\mathbf{A})$, is the determinant of a matrix obtained by striking the ith row and jth column of $\mathbf{A}$ and choosing positive (negative) sign if $i + j$ is even (odd):

Example

$$\mathrm{cof}_{23} \begin{bmatrix} 2 & 4 & 3 \\ 6 & 1 & 5 \\ -2 & 1 & 3 \end{bmatrix} = - \begin{vmatrix} 2 & 4 \\ -2 & 1 \end{vmatrix}$$

$$= -(2 + 8) = -10.$$

4.10. (Laplace Development)

$$|\mathbf{A}| = a_{i1} \, \mathrm{cof}_{i1}(\mathbf{A}) + a_{i2} \, \mathrm{cof}_{i2}(\mathbf{A}) + \cdots + a_{in} \, \mathrm{cof}_{in}(\mathbf{A})$$

$$= a_{1j} \, \mathrm{cof}_{1j}(\mathbf{A}) + a_{2j} \, \mathrm{cof}_{2j}(\mathbf{A}) + \cdots + a_{nj} \, \mathrm{cof}_{nj}(\mathbf{A})$$

for any row i or any column j.

4.11. Numerical Evaluation of the determinant of a symmetric matrix.

Note: If $\mathbf{A}$ is nonsymmetric, form $\mathbf{A}'\mathbf{A}$ or $\mathbf{A}\mathbf{A}'$ by rule 3.18, obtain its determinant, and take the square root.

("Forward Doolittle Scheme," "left side")

Let

$$p_{11} = a_{11}, \qquad p_{12} = a_{12} = a_{21}, \ldots p_{1n} = a_{1n}$$

p_{11}	p_{12}	p_{13}	$\cdots$	p_{1n}
1	u_{12}	u_{13}	$\cdots$	u_{1n}
	a_{22}	a_{23}	$\cdots$	a_{2n}
	p_{22}	p_{23}	$\cdots$	p_{2n}
	1	u_{23}	$\cdots$	u_{2n}
		a_{33}	$\cdots$	a_{3n}
		p_{33}	$\cdots$	p_{3n}
		1	$\cdots$	u_{3n}
		$\cdot$	$\cdots$	$\cdot$
				a_{nn}
				p_{nn}
				1

$$u_{1i} = p_{1i}/p_{11} \qquad i = 1, 2, \ldots n$$

$$p_{2i} = a_{2i} - u_{12} p_{1i} \qquad i = 2, 3, \ldots n$$

$$u_{2i} = p_{2i}/p_{22}$$

$$p_{3i} = a_{3i} - u_{13}p_{1i} - u_{23}p_{2i} \qquad\qquad\qquad i = 3, 4, \ldots n$$

$$u_{3i} = p_{3i}/p_{33}$$

$$p_{ki} = a_{ki} - u_{1k}p_{1i} - u_{2k}p_{2i} - \cdots - u_{k-1,k}p_{k-1,i} \quad i = k, k+1, \ldots n$$

$$k = 2, 3, \ldots n$$

$$u_{ki} = p_{ki}/p_{kk}$$

If, at some stage, $p_{kk} = 0$, reordering of rows and columns may be required. If the matrix is positive-definite (see 8.16) (always true for $\mathbf{AA}'$ or $\mathbf{A}'\mathbf{A}$; see rule 10.24), none of the p_{kk} will be zero. The p_{ii} are called pivots. Then

$$|\mathbf{A}| = \prod_{i=1}^{n} p_{ii}.$$

Further, if $\mathbf{A}$ is partitioned

$$\mathbf{A} = \begin{bmatrix} \mathbf{A}_{11} & \mathbf{A}_{12} \\ \mathbf{A}_{12}' & \mathbf{A}_{22} \end{bmatrix}$$

where $\mathbf{A}_{11}$ is of order $(k \times k)$, then

$$|\mathbf{A}_{11}| = \prod_{i=1}^{k} p_{ii}.$$

(Numerical Examples: see 6.14.)

5 Singularity and rank

5.1. A matrix $\mathbf{A}$ is called **singular** if there exists a vector $\mathbf{x} \neq \mathbf{0}$ such that $\mathbf{Ax} = \mathbf{0}$ or $\mathbf{A}'\mathbf{x} = \mathbf{0}$. Note $\mathbf{x} \neq \mathbf{0}$ if a single element of $\mathbf{x}$ is unequal to 0. If a matrix is not singular, it is called **nonsingular**.

5.2. If a matrix $\mathbf{A}_1$ can be formed by selection of r rows and columns of $\mathbf{A}$ such that $\mathbf{A}_1\mathbf{x} \neq \mathbf{0}$ or $\mathbf{A}_1'\mathbf{x} \neq \mathbf{0}$ for every $\mathbf{x} \neq \mathbf{0}$, and if addition of an $(r+1)$st row and column would produce a singular matrix, r is called the **rank** of $\mathbf{A}$.

Example

$$\mathbf{A} = \begin{bmatrix} 2 & 4 & 6 \\ 1 & 3 & 7 \\ 3 & 7 & 13 \\ 1 & 1 & -1 \end{bmatrix}$$

Note that

$$\begin{bmatrix} 1, & 1, & -1 \end{bmatrix} \begin{bmatrix} 2 & 4 & 6 \\ 1 & 3 & 7 \\ 3 & 7 & 13 \end{bmatrix} = \begin{bmatrix} 0 & 0 & 0 \end{bmatrix}$$

and

$$[1, \quad -1, \quad -1] \begin{bmatrix} 2 & 4 & 6 \\ 1 & 3 & 7 \\ 1 & 1 & -1 \end{bmatrix} = [0 \quad 0 \quad 0]$$

but

$$\begin{bmatrix} 2 & 4 \\ 1 & 3 \end{bmatrix} \begin{bmatrix} x_1 \\ x_2 \end{bmatrix} \neq \begin{bmatrix} 0 \\ 0 \end{bmatrix}$$

or

$$[x_1 \quad x_2] \begin{bmatrix} 2 & 4 \\ 1 & 3 \end{bmatrix} \neq [0, \quad 0]$$

for any arbitrary

$$[x_1, \quad x_2] \neq [0, \quad 0].$$

Hence the matrix has rank 2. ∎

5.3. If $\mathbf{A}$ has rank r and if $\mathbf{A_1}$ is a nonsingular submatrix consisting of r rows and columns of $\mathbf{A}$, then $\mathbf{A_1}$ is called a basis of $\mathbf{A}$.

5.4a. The determinant of a square singular matrix is 0.

5.4b. The determinant of a nonsingular matrix is $\neq 0$.

5.5. $\text{rank}(\mathbf{AB}) \leq \min[\text{rank}(\mathbf{A}), \text{rank}(\mathbf{B})]$.

5.6. $\text{rank}(\mathbf{AA'}) = \text{rank}(\mathbf{A'A}) = \text{rank}(\mathbf{A})$.

5.7. $|\mathbf{A'A}| = |\mathbf{AA'}| = |\mathbf{A}|^2$ if $\mathbf{A}$ is square.

5.8. $|\mathbf{A'A}| = |\mathbf{AA'}| \geq 0$ for every $\mathbf{A}$ with real elements.

6 Inversion

Regular case, nonsingular matrices

6.1. If $\mathbf{A}$ is square and nonsingular ($|\mathbf{A}| \neq 0$), there exists a unique matrix $\mathbf{A}^{-1}$ such that $\mathbf{AA}^{-1} = \mathbf{A}^{-1}\mathbf{A} = \mathbf{I}$.

6.2. $(\mathbf{ABC})^{-1} = \mathbf{C}^{-1}\mathbf{B}^{-1}\mathbf{A}^{-1}$ (provided that all inverses exist).

6.3. $(\mathbf{A}^{-1})' = (\mathbf{A}')^{-1}$.

6.4. $\mathbf{Ax} = \mathbf{b}$ is a system of linear equations. If $\mathbf{A}$ is square and nonsingular, there exists a unique solution

$$\mathbf{x} = \mathbf{A}^{-1}\mathbf{b}.$$

6.5. $(\gamma\mathbf{A})^{-1} = (1/\gamma)\mathbf{A}^{-1}$.

6.6. $|\mathbf{A}^{-1}| = 1/|\mathbf{A}|$.

6.7. $\mathbf{D}_w^{-1} = \mathbf{D}_{1/w}$ where $\mathbf{D}$ is a diagonal matrix.

6.8. If

$$\mathbf{A} = \mathbf{B} + \mathbf{uv}',$$

then

$$\mathbf{A}^{-1} = \mathbf{B}^{-1} - \lambda\mathbf{yz}'$$

where

$$\mathbf{y} = \mathbf{B}^{-1}\mathbf{u}, \qquad \mathbf{z}' = \mathbf{v}'\mathbf{B}^{-1},$$

and

$$\lambda = 1/(1 + \mathbf{z}'\mathbf{u}).$$

Example 6.8.1

$$\mathbf{A} = \begin{bmatrix} 4 & 2 & 4 & 5 \\ 3 & 9 & 12 & 15 \\ 2 & 4 & 11 & 10 \\ 1 & 2 & 4 & 10 \end{bmatrix}$$

This matrix can be written as

$$\begin{bmatrix} 3 & 0 & 0 & 0 \\ 0 & 3 & 0 & 0 \\ 0 & 0 & 3 & 0 \\ 0 & 0 & 0 & 5 \end{bmatrix} + \begin{bmatrix} 1 \\ 3 \\ 2 \\ 1 \end{bmatrix} \begin{bmatrix} 1 & 2 & 4 & 5 \end{bmatrix} = \mathbf{B} + \mathbf{uv}'$$

$$\mathbf{B}^{-1} = \begin{bmatrix} 1/3 & 0 & 0 & 0 \\ 0 & 1/3 & 0 & 0 \\ 0 & 0 & 1/3 & 0 \\ 0 & 0 & 0 & 1/5 \end{bmatrix}$$

$$\mathbf{y} = \mathbf{B}^{-1}\mathbf{u} = \begin{bmatrix} 1/3 \\ 1 \\ 2/3 \\ 1/5 \end{bmatrix}$$

$$\mathbf{z}' = \mathbf{v}'\mathbf{B}^{-1} = \begin{bmatrix} 1/3 & 2/3 & 4/3 & 1 \end{bmatrix}$$

$$\mathbf{z}'\mathbf{u} = 1/3 \times 1 + 2/3 \times 3 + 4/3 \times 2 + 1 \times 1 = 6$$

$$\lambda = 1/7$$

$$\mathbf{A}^{-1} = \begin{bmatrix} 1/3 & 0 & 0 & 0 \\ 0 & 1/3 & 0 & 0 \\ 0 & 0 & 1/3 & 0 \\ 0 & 0 & 0 & 1/5 \end{bmatrix} - (1/7) \begin{bmatrix} 1/3 \\ 1 \\ 2/3 \\ 1/5 \end{bmatrix} \begin{bmatrix} 1/3 & 2/3 & 4/3 & 1 \end{bmatrix}$$

$$= (1/315) \begin{bmatrix} 100 & -10 & -20 & -15 \\ -15 & 75 & -60 & -45 \\ -10 & -20 & 65 & -30 \\ -3 & -6 & -12 & 54 \end{bmatrix}.$$

(This rule is especially useful if all off-diagonal elements are equal; then $\mathbf{u} = k\mathbf{j}$ and $\mathbf{v}' = \mathbf{j}'$ and $\mathbf{B}$ is diagonal.) ▮

6.9. Let $\mathbf{B}$ (elements b_{ij}) have a known inverse, $\mathbf{B}^{-1}$ (elements b^{ij}). Let $\mathbf{A} = \mathbf{B}$ except for one element $a_{rs} = b_{rs} + k$. Then the elements of $\mathbf{A}^{-1}$ are

$$a^{ij} = b^{ij} - \frac{kb^{ir}b^{sj}}{1 + kb^{sr}}.$$

6.10. (Partitioning)
Let

$$
\begin{array}{cc}
& (p) \quad (q) \\
\mathbf{A} = \begin{array}{c} (p) \\ (q) \end{array} & \left[\begin{array}{cc} \mathbf{B} & \mathbf{C} \\ \mathbf{D} & \mathbf{E} \end{array} \right]
\end{array}
\quad
\begin{array}{l} \text{(letters in parentheses} \\ \text{denote order of the submatrices).} \end{array}
$$

Let $\mathbf{B}^{-1}$ and $\mathbf{E}^{-1}$ exist. Then

$$\mathbf{A}^{-1} = \left[\begin{array}{cc} \mathbf{X} & \mathbf{Y} \\ \mathbf{Z} & \mathbf{U} \end{array} \right]$$

where

$$\mathbf{X} = (\mathbf{B} - \mathbf{CE}^{-1}\mathbf{D})^{-1}$$

$$\mathbf{U} = (\mathbf{E} - \mathbf{DB}^{-1}\mathbf{C})^{-1}$$

$$\mathbf{Y} = -\mathbf{B}^{-1}\mathbf{CU}$$

$$\mathbf{Z} = -\mathbf{E}^{-1}\mathbf{DX}.$$

6.11. (Partitioning of Determinants)
Let

$$|\mathbf{A}| = \left| \begin{array}{cc} \mathbf{B} & \mathbf{C} \\ \mathbf{D} & \mathbf{E} \end{array} \right| \quad \text{(same structure as in 6.10).}$$

Then

$$|\mathbf{A}| = |\mathbf{E}||(\mathbf{B} - \mathbf{CE}^{-1}\mathbf{D})| = |\mathbf{B}||(\mathbf{E} - \mathbf{DB}^{-1}\mathbf{C})|.$$

6.12. Let

$$\mathbf{A} = \mathbf{B} + \mathbf{UV}$$

where $\mathbf{B}(n \times n)$ has an inverse

$\mathbf{U}$ is of order $(n \times k)$, with k usually very small

$\mathbf{V}$ is of order $(k \times n)$

(the special case for $k = 1$ is treated in 6.8).
Then

$$\mathbf{A}^{-1} = \mathbf{B}^{-1} - \mathbf{Y}\Lambda\mathbf{Z}$$

where

$$\mathbf{Y} = \mathbf{B}^{-1}\mathbf{U}(n \times k)$$

$$\mathbf{Z} = \mathbf{VB}^{-1}(k \times n)$$

and

$$\Lambda(k \times k) = [\mathbf{I} + \mathbf{ZU}]^{-1}.$$

6.13. Let a_{ij} denote the elements of $\mathbf{A}$ and a^{ij} those of $\mathbf{A}^{-1}$. Then

$$a^{ij} = \mathrm{cof}_{ji}(\mathbf{A})/|\mathbf{A}|$$

where cof is the determinant defined in 4.9.

6.14. "Doolittle" Method of inverting symmetric matrices (see also 4.11). Let

$$p_{11} = a_{11}, \; p_{12} = a_{12} = a_{21}, \ldots p_{1n} = a_{1n} = a_{n1}.$$

Forward solution

p_{11}	p_{12}	p_{13}	$\cdots$	p_{1n}	1				
1	u_{12}	u_{13}	$\cdots$	u_{1n}	$u_{1\mathrm{I}}$				
	a_{22}	a_{23}	$\cdots$	a_{2n}	0	1			
	p_{22}	p_{23}	$\cdots$	p_{2n}	p_{21}	$p_{2\mathrm{II}}$			
	1	u_{23}	$\cdots$	u_{2n}	$u_{2\mathrm{I}}$	$u_{2\mathrm{II}}$			
		a_{33}	$\cdots$	a_{3n}	0	0	1		
		p_{33}	$\cdots$	p_{3n}	$p_{3\mathrm{I}}$	$p_{3\mathrm{II}}$	$p_{3\mathrm{III}}$		
		1	$\cdots$	u_{3n}	$u_{3\mathrm{I}}$	$u_{3\mathrm{II}}$	$u_{3\mathrm{III}}$		
		$\cdot$	$\cdots$	$\cdot$	$\cdot$	$\cdots$	$\cdot$		
				a_{nn}	0	0	0	$\cdots$	1
				p_{nn}	p_{n1}	$p_{n\mathrm{II}}$	$p_{n\mathrm{III}}$	$\cdots$	p_{nN}
				1	u_{n1}	$u_{n\mathrm{II}}$	$u_{n\mathrm{III}}$	$\cdots$	u_{nN}

$$u_{1i} = p_{1i}/p_{11} \qquad\qquad\qquad i = 1, 2, \ldots n, \mathrm{I}$$

$$p_{2i} = a_{2i} - u_{12}p_{1i} \qquad\qquad\quad i = 2, 3, \ldots n, \mathrm{I}, \mathrm{II}$$

$$u_{2i} = p_{2i}/p_{22}$$

$$p_{3i} = a_{3i} - u_{13}p_{1i} - u_{23}p_{2i} \qquad i = 3, 4, \ldots n, \mathrm{I}, \mathrm{II}, \mathrm{III}$$

$$u_{3i} = p_{3i}/p_{33}$$

$$p_{ki} = a_{ki} - u_{1k}p_{1i} - u_{2k}p_{2i} - \cdots - u_{k-1,k}p_{k-1,i} \quad i = k, k+1, \ldots n, \mathrm{I}, \mathrm{II}, \ldots K$$

$$\qquad\qquad\qquad\qquad\qquad\qquad\qquad\qquad k = 2, 3, \ldots n$$

$$u_{ki} = p_{ki}/p_{kk}$$

Backward solution

j **refers to Arabic,** *J* **refers to Roman numerals**

The elements of $\mathbf{A}^{-1}$ are a^{ij}

$$a^{nj} = u_{nJ} \qquad\qquad j = 1, 2, \ldots n;$$
$$J = \text{I}, \text{II}, \ldots N$$

$$a^{n-1\,j} = u_{n-1,J} - u_{n-1,n}a^{nj} \qquad j = 1, 2, \ldots (n-1);$$
$$J = \text{I}, \text{II}, \ldots (N-1)$$

$$a^{n-2,j} = u_{n-2,J} - u_{n-2,n}a^{nj} - u_{n-2,n-1}a^{n-1,j} \quad j = 1, 2, \ldots (n-2);$$
$$J = \text{I}, \text{II}, \ldots (N-2)$$

$$a^{n-k,j} = u_{n-k,J} - u_{n-k,n}a^{nj} - u_{n-k,n-1}a^{n-1,j}$$
$$- \cdots - u_{n-k,n-k+1}a^{n-k+1,j} \qquad j = 1, 2, \ldots (n-k);$$
$$J = \text{I}, \text{II}, \ldots (N-k);$$
$$k = 1, 2, \ldots (n-1),$$

and $a^{ji} = a^{ij}$.

Numerical example 6.14.1

Invert the matrix

$$\begin{bmatrix} 25 & 30 & -10 \\ 30 & 40 & -6 \\ -10 & -6 & 17 \end{bmatrix}$$

a_1	25	30	−10	1		
u_1	1	1.2	−0.4	0.04		
	a_2	40	−6	0	1	
	p_2	4	6	−1.2	1	
	u_2	1	1.5	−0.3	0.25	
		a_3	17	0	0	1
		p_3	4	2.2	−1.5	1
		u_3	1	0.55	−0.375	0.25

$$\begin{array}{ccc} 1.61 & -1.125 & 0.55 \\ -1.125 & 0.8125 & -0.375 \\ 0.55 & -0.375 & 0.25 \end{array}$$

Enter row a_1.
Elements in u_1 = elements in a_1 divided by $a_{11}(= 25)$.
Enter row a_2.

$$p_{22} = 40 - 1.2 \times 30 = 4$$

$$p_{23} = -6 - 1.2 \times (-10) = 6$$

$$p_{2\text{I}} = 0 - 1.2 \times 1 = -1.2$$

$$p_{2\text{II}} = 1$$

Elements in u_2 = elements in p_2 divided by $p_{22}(= 4)$.

Enter row a_3.

$$p_{33} = 17 - (-0.4) \times (-10) - 1.5 \times 6 = 4$$

$$p_{31} = 0 - (-0.4) \times 1 - 1.5 \times (-1.2) = 2.2$$

$$p_{3II} = 0 - 1.5 \times 1 = -1.5$$

$$p_{3III} = 1$$

Elements in u_3 = elements in p_3 divided by $p_{33}(= 4)$.
   Copy the right-hand side of the last (third) u-row as the last column below the double line.

$$a^{21} = -0.3 - 1.5 \times 0.55 = -1.125$$

$$a^{22} = 0.25 - 1.5 \times (-0.375) = 0.8125$$

$$a^{23} = 0 - 1.5 \times 0.25 = -0.375 \text{ (check against } a^{32}).$$

These are entered in the next to last (second) column below.

$$a^{11} = 0.04 - (-0.4) \times 0.55 - 1.2 \times (-1.125) = 1.61$$

$$a^{12} = 0 - (-0.4) \times (-0.375) - 1.2 \times 0.8125 = -1.125 \text{ (check against } a^{21})$$

$$a^{13} = 0 - (-0.4) \times (0.25) - 1.2 \times (-0.375) = 0.55 \text{ (check against } a^{31}).$$

6.15.   A matrix is called orthogonal if $\mathbf{A}' = \mathbf{A}^{-1}$ (or $\mathbf{AA}' = \mathbf{I}$).

7   Traces

7.1.   If $\mathbf{A}$ is a square matrix then the trace of $\mathbf{A}$ is $\text{tr}\,\mathbf{A} = \sum_i a_{ii}$, i.e., the sum of the diagonal elements.
   7.2.   If $\mathbf{A}$ is of order $(m \times k)$ and $\mathbf{B}$ of order $(k \times m)$ then $\text{tr}(\mathbf{AB}) = \text{tr}(\mathbf{BA})$.
   7.3.   If $\mathbf{A}$ is of order $(m \times k)$, $\mathbf{B}$ of order $(k \times r)$, and $\mathbf{C}$ of order $(r \times m)$, then

$$\text{tr}(\mathbf{ABC}) = \text{tr}(\mathbf{BCA}) = \text{tr}(\mathbf{CAB}).$$

7.3a.   If $\mathbf{b}$ is a column vector and $\mathbf{c}'$ a row vector, then

$$\text{tr}(\mathbf{Abc}') = \text{tr}(\mathbf{bc}'\mathbf{A}) = \mathbf{c}'\mathbf{Ab}$$

since the trace of a scalar is the scalar.
   7.4. $\text{tr}(\mathbf{A} + \gamma\mathbf{B}) = \text{tr}\,\mathbf{A} + \gamma\,\text{tr}\,\mathbf{B}$, where γ is a scalar.
   7.5. $\text{tr}(\mathbf{EL})_{ij}\mathbf{A} = \text{tr}\,\mathbf{A}(\mathbf{EL})_{ij} = a_{ji}$, where $(\mathbf{EL})_{ij}$ is an elementary matrix as defined in 3.4.
   7.6. $\text{tr}(\mathbf{EL})_{ij}\mathbf{A}(\mathbf{EL})_{rs}\mathbf{B} = a_{jr}b_{si}$.

(These rules are useful in matrix differentiation)

7.7. The trace of the second order of a square matrix $\mathbf{A}$ is the sum of the determinants of all $\binom{n}{2}$ matrices of order (2×2) that can be formed by intersecting rows i and j with columns i and j.

$$\text{tr}_2 \, \mathbf{A} = \begin{vmatrix} a_{11} & a_{12} \\ a_{21} & a_{22} \end{vmatrix} + \begin{vmatrix} a_{11} & a_{13} \\ a_{31} & a_{33} \end{vmatrix}$$

$$+ \cdots + \begin{vmatrix} a_{11} & a_{1n} \\ a_{n1} & a_{nn} \end{vmatrix} + \begin{vmatrix} a_{22} & a_{23} \\ a_{32} & a_{33} \end{vmatrix}$$

$$+ \cdots + \begin{vmatrix} a_{22} & a_{2n} \\ a_{n2} & a_{nn} \end{vmatrix} + \cdots + \begin{vmatrix} a_{n-1,n-1} & a_{n-1,n} \\ a_{n,n-1} & a_{nn} \end{vmatrix}.$$

7.8. The trace of the kth order of a square matrix is the sum of the determinants of all $\binom{n}{k}$ matrices of order $(k \times k)$ that can be formed by intersecting any k rows of $\mathbf{A}$ with the same k columns.

$$\text{tr}_k \, \mathbf{A} = \sum \begin{vmatrix} a_{i_1 i_1} & a_{i_1 i_2} & \cdots & a_{i_1 i_k} \\ a_{i_2 i_1} & a_{i_2 i_2} & \cdots & a_{i_2 i_k} \\ \cdot & \cdot & \cdots & \cdot \\ a_{i_k i_1} & a_{i_k i_2} & \cdots & a_{i_k i_k} \end{vmatrix}$$

where the sum extends over all combinations of n elements taken k at a time in order

$$i_1 < i_2 < \cdots < i_k.$$

7.9. Rules 7.2 and 7.3 (cyclic exchange) are valid for trace of kth order.

7.10. $\text{tr}_n \, \mathbf{A} = |\mathbf{A}|$ if $\mathbf{A}$ is of order $(n \times n)$.

8 Characteristic roots and vectors

8.1. If $\mathbf{A}$ is a square matrix of order $(n \times n)$, then $|\mathbf{A} - \lambda \mathbf{I}| = 0$ is called the characteristic equation of the matrix $\mathbf{A}$. It is a polynomial of the nth degree in λ.

8.2. The n roots of the characteristic equation (not necessarily distinct) are called the characteristic roots of $\mathbf{A}$

$$ch(\mathbf{A}) = \lambda_1, \lambda_2, \ldots \lambda_n.$$

8.3. The characteristic equation of $\mathbf{A}$ can be obtained by the relation

$$\lambda^n - (\text{tr} \, \mathbf{A})\lambda^{n-1} + (\text{tr}_2 \, \mathbf{A})\lambda^{n-2} - (\text{tr}_3 \, \mathbf{A})\lambda^{n-3} \cdots - (-1)^n (\text{tr}_{n-1} \, \mathbf{A})\lambda + (-1)^n |\mathbf{A}| = 0$$

where tr_k is defined in 7.8.

Example 8.3.1

$$\mathbf{A} = \begin{bmatrix} 25 & 30 & -10 \\ 30 & 40 & -6 \\ -10 & -6 & 17 \end{bmatrix}$$

$$\text{tr}\,\mathbf{A} = 25 + 40 + 17 = 82$$

$$\text{tr}_2\,\mathbf{A} = (25 \times 40 - 30 \times 30) + (25 \times 17 - 10 \times 10) + (40 \times 17 - 6 \times 6) = 1069$$

$$\text{tr}_3\,\mathbf{A} = |\mathbf{A}| = 25 \times 4 \times 4 = 400$$

(cf. 6.14 and procedure stated in 4.11)

Hence

$$\lambda^3 - 82\lambda^2 + 1069\lambda - 400 = 0.$$

The solutions (by Newton iteration) are

$$\lambda_1 = 65.86108$$

$$\lambda_2 = 15.75339$$

$$\lambda_3 = 0.38553.$$

These are the characteristic roots of $\mathbf{A}$.   ∎

8.4. $ch(\mathbf{A} + \gamma\mathbf{I}) = \gamma + ch(\mathbf{A})$.

8.5. $ch(\mathbf{AB}) = ch(\mathbf{BA})$

(except that $\mathbf{AB}$ or $\mathbf{BA}$ may have additional roots equal to zero).

8.6. $ch(\mathbf{A}^{-1}) = 1/ch(\mathbf{A})$.

8.7.   If $\lambda_1, \lambda_2, \ldots \lambda_n$ are the roots of $\mathbf{A}$ then

$$\sum_i \lambda_i = \text{tr}\,\mathbf{A}$$

$$\sum_{i<j} \lambda_i\lambda_j = \text{tr}_2\,\mathbf{A}$$

$$\sum_{i<j<k} \lambda_i\lambda_j\lambda_k = \text{tr}_3\,\mathbf{A}$$

$$\prod_i \lambda_i = |\mathbf{A}|.$$

8.8.   If $\mathbf{x}'$ denotes the radius vector (running coordinates $[x, y, z]$) and if a matrix $\mathbf{Q}$ is positive-definite, then

$$(\mathbf{x}' - \mathbf{x}_0')\mathbf{Q}^{-1}(\mathbf{x} - \mathbf{x}_0) = 1$$

is the equation of an ellipsoid with center at $[x_0, y_0, z_0] = \mathbf{x}_0'$ and semi-axes equal to the square roots of the characteristic roots of $\mathbf{Q}$.

8.9.   The characteristic roots of a triangular (or diagonal) matrix are the diagonal elements of the matrix.

8.10.   If $\mathbf{A}$ is a real matrix with positive roots, then

$$ch_{\min}(\mathbf{AA}') \le [ch_{\min}(\mathbf{A})]^2 \le [ch_{\max}(\mathbf{A})]^2 \le ch_{\max}(\mathbf{AA}')$$

where $ch_{\min}$ denotes the smallest and $ch_{\max}$ the largest root.

8.11.   The ratio of two quadratic forms ($\mathbf{B}$ nonsingular)

$$u = \frac{\mathbf{x}'\mathbf{Ax}}{\mathbf{x}'\mathbf{Bx}}$$

attains stationary values at the roots of $\mathbf{B}^{-1}\mathbf{A}$. In particular,

$$u_{\max} = ch_{\max}(\mathbf{B}^{-1}\mathbf{A}) \qquad \text{and} \qquad u_{\min} = ch_{\min}(\mathbf{B}^{-1}\mathbf{A}).$$

8.12. The equation system

$$\mathbf{Ax} = \lambda\mathbf{x}$$

permits nonzero solutions only if λ is one of the characteristic roots of $\mathbf{A}$. Such a solution $\mathbf{x}$ is called a characteristic vector.

8.13. If $\mathbf{x}$ is a solution to 8.12, so is $\gamma\mathbf{x}$ for an arbitrary scalar γ.

8.14. A solution $\mathbf{x}$ which has unit length ($\mathbf{x}'\mathbf{x} = 1$) is called the eigenvector associated with the characteristic root λ of $\mathbf{A}$. The vector is frequently denoted by $\mathbf{e}$.

8.15. A real symmetric matrix has real roots.

8.16. A matrix $\mathbf{A}$ is called positive-definite (abbreviated p.d.) if the quadratic form $\mathbf{x}'\mathbf{Ax} > 0$ for every $\mathbf{x} \neq \mathbf{0}$.

8.17. A matrix $\mathbf{A}$ is called positive-semidefinite (abbreviated p.s.d.) if the quadratic form $\mathbf{x}'\mathbf{Ax} > 0$ and/or $\mathbf{x}'\mathbf{Ax} = 0$ for some $\mathbf{x} \neq \mathbf{0}$.

8.18. A positive-definite real symmetric matrix has only positive characteristic roots.

8.19. If a real symmetric matrix is positive-semidefinite, it has no negative roots. The number of nonzero roots equals the rank of the matrix.

8.20. If all roots of a real symmetric matrix are distinct, the associated eigenvectors are distinct.

8.21. The matrix of eigenvectors

$$\mathbf{E} = [\mathbf{e}_1, \mathbf{e}_2, \ldots \mathbf{e}_n]$$

of a real symmetric matrix is (or can be chosen to be) orthogonal.

8.22. $\mathbf{AE} = \mathbf{ED}_\lambda$.

8.23. For a real symmetric matrix, $\mathbf{A} = \mathbf{ED}_\lambda\mathbf{E}'$ (decomposition into matrices of unit rank)

$$\mathbf{E}'\mathbf{AE} = \mathbf{D}_\lambda$$

where $\mathbf{D}_\lambda$ denotes the diagonal matrix of characteristic roots ordered in the same way as the eigenvector columns in $\mathbf{E}$.

8.24. If $f(\lambda)$ is a polynomial in λ, then

$$f(\mathbf{A}) = \mathbf{ED}_{f(\lambda)}\mathbf{E}^{-1}$$

where λ are the characteristic roots of $\mathbf{A}$ and $\mathbf{E}$ is the matrix of associated eigenvectors. If $\mathbf{A}$ is symmetric, $\mathbf{E}^{-1} = \mathbf{E}'$.

Example 8.24.1

Consider the matrix in 8.3 (and 6.14).

$$\mathbf{A} = \begin{bmatrix} 25 & 30 & -10 \\ 30 & 40 & -6 \\ -10 & -6 & 17 \end{bmatrix}.$$

The characteristic roots were found in Example 8.3.1,

$$\lambda_1 = 65.86108 \qquad \lambda_2 = 15.75339 \qquad \lambda_3 = 0.38553.$$

To find some $\mathbf{x}$ such that $\mathbf{Ax} = \lambda_1\mathbf{x}$, we arbitrarily set the first element of $\mathbf{x}$ equal to 1. Using only the first two rows of $\mathbf{A}$ we solve the equation system

$$25 + 30x_2 - 10x_3 = 65.86108$$

$$30 + 40x_2 - 6x_3 = 65.86108x_2$$

which yields $x_2 = 1.24294$ and $x_3 = -0.35729$. Substitution of these values into the third equation

$$-10 - 6x_2 + 17x_3 = 65.86108x_3$$

yields zero to five decimal places, indicating the accuracy of the first characteristic root. To reduce to unit length the characteristic vector

$$[\ 1 \quad 1.24294 \quad -0.35729 \ ]$$

we divide each element by

$$\sqrt{1 + 1.24294^2 + 0.35729^2}$$

and thus obtain the first eigenvector

$$[\ 0.61170 \quad 0.76030 \quad -0.21855 \ ].$$

This, written as a column vector, is $\mathbf{e}_1$. Repeating the same process for the second and third eigenvectors we obtain

$$\mathbf{e}_2 = \begin{bmatrix} -0.08659 \\ 0.33896 \\ 0.93681 \end{bmatrix} \qquad \mathbf{e}_3 = \begin{bmatrix} 0.78634 \\ -0.55412 \\ 0.27318 \end{bmatrix}.$$

The three vectors can be placed into the eigenvector matrix $\mathbf{E}$, which is easily seen to be orthogonal. ∎

9 Conditional inverses

9.1. Any matrix $\mathbf{A}$ (singular or nonsingular, rectangular or square) has some conditional or generalized inverse $\mathbf{A}^{(-1)}$ defined by the relation

$$\mathbf{AA}^{(-1)}\mathbf{A} = \mathbf{A}.$$

9.2. If (and only if) $\mathbf{A}$ is square and nonsingular, $\mathbf{A}^{(-1)}$ is unique and equals $\mathbf{A}^{-1}$. Otherwise there will be infinitely many matrices $\mathbf{A}^{(-1)}$ which satisfy the defining relation 9.1.

9.3a. If $\mathbf{A}$ is rectangular $(n \times m)$ of rank m, with $m < n$, then $\mathbf{A}^{(-1)}$ is of order $(m \times n)$ and $\mathbf{A}^{(-1)}\mathbf{A} = \mathbf{I}$ $(m \times m)$. Then $\mathbf{A}^{(-1)}$ is called an inverse from the left. $\mathbf{AA}^{(-1)} \neq \mathbf{I}$ in this case.

9.3b. If $\mathbf{A}$ is rectangular $(n \times m)$ of rank n, with $m > n$, then $\mathbf{A}^{(-1)}$ is of order $(m \times n)$ and $\mathbf{AA}^{(-1)} = \mathbf{I}$ $(n \times n)$. Then $\mathbf{A}^{(-1)}$ is called an inverse to the right. In this case,

$$\mathbf{A}^{(-1)}\mathbf{A} \neq \mathbf{I}.$$

9.3c. For a square, singular matrix, $\mathbf{AA}^{(-1)} \neq \mathbf{I}$ and $\mathbf{A}^{(-1)}\mathbf{A} \neq \mathbf{I}$.

Example 9.3.1

$$A = \begin{bmatrix} 3 \\ 2 \\ 1 \end{bmatrix}$$

The row vector $[\ 1/3 \quad 0 \quad 0\ ]$ is an inverse from the left. The row vector

$$[\ x \quad y \quad (1 - 3x - 2y)\ ]$$

is a conditional inverse of the above matrix A for any values of x and y. It is called the generalized inverse of A. ∎

Example 9.3.2

$$A = \begin{bmatrix} 1 & 2 & 3 \\ 2 & 5 & 6 \\ 3 & 7 & 9 \end{bmatrix}$$

A conditional inverse is

$$A^{(-1)} = \begin{bmatrix} 5 & -2 & 0 \\ -2 & 1 & 0 \\ 0 & 0 & 0 \end{bmatrix}$$

Here it was obtained by inversion of the basis (the 2×2 matrix in the upper left-hand corner) and replacement of the other elements by zeros. ∎

    9.4. A square matrix A is called idempotent if $AA = A^2 = A$.
    9.5. $AA^{(-1)}$ and $A^{(-1)}A$ are idempotent.
    9.6. All characteristic roots of idempotent matrices are either zero or one.
    9.7. A system of linear equations (m equations in n unknowns)

$$Ax = b$$

is called consistent if there exists some solution x that satisfies the equation system.

Example 9.7.1
The system

$$x + y = 2$$

$$2x + 2y = 4 \qquad \text{is consistent.} \qquad ∎$$

Example 9.7.2
The system

$$x + y = 2$$

$$2x + 2y = 5 \qquad \text{is inconsistent,}$$

for no pair of values (x, y) will satisfy this system. ∎

9.8. If, in a system of equations (rectangular or square)

$$\mathbf{Ax} = \mathbf{b}$$

$\mathbf{AA}^{(-1)}\mathbf{b} = \mathbf{b}$ for some conditional inverse $\mathbf{A}^{(-1)}$, then $\mathbf{AA}^{(-1)}\mathbf{b} = \mathbf{b}$ for every conditional inverse of $\mathbf{A}$, and $\mathbf{Ax} = \mathbf{b}$ is consistent. Conversely, if $\mathbf{AA}^{(-1)}\mathbf{b} \neq \mathbf{b}$ for some conditional inverse $\mathbf{A}^{(-1)}$, then $\mathbf{AA}^{(-1)}\mathbf{b} \neq \mathbf{b}$ for every conditional inverse of $\mathbf{A}$, and $\mathbf{Ax} = \mathbf{b}$ is inconsistent.

9.9. If $\mathbf{Ax} = \mathbf{b}$ is consistent, then $\mathbf{x} = \mathbf{A}^{(-1)}\mathbf{b}$ is a solution (generally a different one for each $\mathbf{A}^{(-1)}$).

9.10. Let $\mathbf{y}$ $(p \times 1)$ be a set of linear functions of the solutions $\mathbf{x}$ $(n \times 1)$ of a consistent system of equations $\mathbf{Ax} = \mathbf{b}$, given by the relation $\mathbf{y} = \mathbf{Cx}$. Then $\mathbf{y} = \mathbf{Cx}$ is called unique if the same values of $\mathbf{y}$ will result regardless of which solution $\mathbf{x}$ is used.

Example 9.10.1

$$3x + 4y + 5z = 22$$

$$x + y + z = 6$$

is a consistent system. One solution would be

$$x = 3 \qquad y = 2 \qquad z = 1.$$

Another solution is

$$x = 2 \qquad y = 4 \qquad z = 0.$$

The linear function

$$[\,7 \quad 9 \quad 11\,] \begin{bmatrix} x \\ y \\ z \end{bmatrix} = u$$

$(7x + 9y + 11z = u)$ will have the same value (50) regardless of which of the two (or any other) solutions is substituted. Thus u is unique. ∎

9.11. Let $\mathbf{Ax} = \mathbf{b}$ be a consistent system of equations. For $\mathbf{Cx} = \mathbf{y}$ to be a unique linear combination of the solution $\mathbf{x}$, it is necessary and sufficient that $\mathbf{CA}^{(-1)}\mathbf{A} = \mathbf{C}$. If this relation holds for some $\mathbf{A}^{(-1)}$, it will hold for every conditional inverse of $\mathbf{A}$. If it is violated for some $\mathbf{A}^{(-1)}$, it will be violated for every $\mathbf{A}^{(-1)}$, and $\mathbf{y}$ will be nonunique.

9.12. Let $\mathbf{A}$ be of rank r and select r rows and r columns that form a basis of $\mathbf{A}$. Then a conditional inverse of $\mathbf{A}$ can be obtained as follows: Invert the $(r \times r)$ matrix, place the inverse (without transposing) into the r rows corresponding to the column numbers and the r columns corresponding to the row numbers of the basis, and place zero into all remaining elements. Thus, if $\mathbf{A}$ is of order (5×4) and rank 3, and if rows 1, 2, 4 and columns 2, 3, 4 are selected as a basis, $\mathbf{A}^{(-1)}$, of order (4×5), will contain the inverse elements of the basis in rows 2, 3, 4 and columns 1, 2, 4, and zeros elsewhere. (See example 9.3.2.)

9.13. If $\mathbf{A}$ is a square, singular matrix of order $(n \times n)$ and rank r, let $\mathbf{M}$ be a matrix of order $[n \times (n - r)]$ and $\mathbf{K}$ another matrix of order $[(n - r) \times n]$ chosen in such a way that $\mathbf{A} + \mathbf{MK}$ is nonsingular. Then $(\mathbf{A} + \mathbf{MK})^{-1}$ is a conditional inverse of $\mathbf{A}$.

Example 9.13.1

$$\mathbf{A} = \begin{bmatrix} 3 & -1 & -1 & -1 \\ -1 & 3 & -1 & -1 \\ -1 & -1 & 3 & -1 \\ -1 & -1 & -1 & 3 \end{bmatrix}$$

is of order (4×4) and rank 3. Take $\mathbf{M} = \mathbf{j}$ (column vector of ones) and $\mathbf{K} = \mathbf{j}'$ (row vector of ones). Then $\mathbf{A} + \mathbf{MK} = \mathbf{A} + \mathbf{jj}' = 4\mathbf{I}$. Hence $(1/4)\,\mathbf{I}$ is a conditional inverse of $\mathbf{A}$.   ∎

9.14.   The "Doolittle" method (see 6.14) can be employed to obtain a conditional inverse of a symmetric matrix. If, at any stage, the leading element of the p-row is zero, that cycle is disregarded.

Example 9.14.1
Invert, conditionally, the matrix

$$\mathbf{A} = \begin{bmatrix} 4 & 2 & -2 & 4 \\ 2 & 17 & 11 & 6 \\ -2 & 11 & 10 & 1 \\ 4 & 6 & 1 & 30 \end{bmatrix}$$

4	2	−2	4	1				
1	.5	−.5	1	.25				
	17	11	6	0	1			
	16	12	4	−.5	1			
	1	.75	.25	−.03125	.0625			
		10	1	0	0	1		
		0						
			30	0	0	0	1	
			25	−.875	−.25	0	1	
			1	−.035	−.01	0	.04	

$$\begin{bmatrix} .29625 & -.0225 & 0 & -.035 \\ -.0225 & .065 & 0 & -.01 \\ 0 & 0 & 0 & 0 \\ -.035 & -.01 & 0 & .04 \end{bmatrix} = \mathbf{A}^{(-1)}$$ ∎

10   Matrix differentiation

10.1a.   If the elements of a matrix $\mathbf{Y}(m \times n)$ are functions of a scalar, x, the expression

$$\partial \mathbf{Y}/\partial x$$

denotes a matrix of order $(m \times n)$ with elements $\partial y_{ij}/\partial x$.

10.1b. If the elements of a column (row) vector $\mathbf{y}(\mathbf{y}')$ are functions of a scalar, x, the expression

$$\partial \mathbf{y}/\partial x \qquad (\partial \mathbf{y}'/\partial x)$$

denotes a column (row) vector with elements $\partial y_i/\partial x$.

10.2a. If y is a scalar function of $m \times n$ variables, x_{ij}, arranged into a matrix $\mathbf{X}$, the expression

$$\partial y/\partial \mathbf{X}$$

denotes a matrix with elements $\partial y/\partial x_{ij}$.

(Note: Partial differentiation is performed with respect to the element in row i and column j of $\mathbf{X}$. If the same x-variable occurs in another place as, e.g., in a symmetric matrix, differentiation with respect to the distinct (repeated) variable is performed in two stages.)

Example 10.2.1

If $y = \mathbf{j}'\mathbf{X}\mathbf{j}$ (sum of all elements of a square matrix), $\partial y/\partial \mathbf{X}$ is a matrix of ones. If $\mathbf{X}$ is symmetric, one can introduce a new notation $x_{ij} = x_{ji} = z_{ij}$. Then

$$\partial y/\partial z_{ij} = (\partial y/\partial x_{ij})(\partial x_{ij}/\partial z_{ij})$$

$$+ (\partial y/\partial x_{ji})(\partial x_{ji}/\partial z_{ij})$$

$$= 1 + 1 = 2 \quad (\text{if } i \neq j)$$

$$= 1 \qquad\qquad (\text{if } i = j). \qquad\blacksquare$$

10.2b. If y is a scalar function of n variables, x_i, arranged into a column (row) vector $\mathbf{x}(\mathbf{x}')$, the expression

$$\partial y/\partial \mathbf{x} \qquad (\partial y/\partial \mathbf{x}')$$

denotes a column (row) vector with elements $\partial y/\partial x_i$.

10.3. If $\mathbf{y}$ is a column vector with m elements, each a function of n variables, x_i, arranged into a row vector $\mathbf{x}'$, the expression $\partial \mathbf{y}/\partial \mathbf{x}'$ denotes a matrix with m rows and n columns, with elements $\partial y_i/\partial x_j$.

10.4. $\partial \mathbf{Y}/\partial y_{ij} = (\mathbf{EL})_{ij}$ (see definition of $(\mathbf{EL})$ in 3.4).

10.5. $\partial \mathbf{UV}/\partial x = (\partial \mathbf{U}/\partial x)\mathbf{V} + \mathbf{U}(\partial \mathbf{V}/\partial x)$.

10.6. $\partial \mathbf{AY}/\partial x = \mathbf{A}(\partial \mathbf{Y}/\partial x)$ (if elements of $\mathbf{A}$ are not functions of x).

10.7. $\partial \mathbf{Y}'/\partial y_{ij} = (\mathbf{EL})_{ji}$.

10.8. $\partial \mathbf{A}'\mathbf{YA}/\partial x = \mathbf{A}'(\partial \mathbf{Y}/\partial x)\mathbf{A}$.

10.9. $\partial \mathbf{Y}'\mathbf{AY}/\partial x = (\partial \mathbf{Y}'/\partial x)\mathbf{AY} + \mathbf{Y}'\mathbf{A}(\partial \mathbf{Y}/\partial x)$.

10.10. $\partial \mathbf{a}'\mathbf{x}/\partial \mathbf{x} = \mathbf{a}$.

10.11. $\partial \mathbf{x}'\mathbf{x}/\partial \mathbf{x} = 2\mathbf{x}$.

10.12. $\partial \mathbf{x}'\mathbf{Ax}/\partial \mathbf{x} = \mathbf{Ax} + \mathbf{A}'\mathbf{x}$.

10.13. (Chain Rule No. 1) $\partial \mathbf{y}/\partial \mathbf{x}' = (\partial \mathbf{y}/\partial \mathbf{z}')(\partial \mathbf{z}/\partial \mathbf{x}')$.

10.14. $\partial \mathbf{Ax}/\partial \mathbf{x}' = \mathbf{A}$.

10.15. $\partial \operatorname{tr}\mathbf{X}/\partial \mathbf{X} = \mathbf{I}$.

10.16. $\partial \operatorname{tr}\mathbf{AX}/\partial \mathbf{X} = \partial \operatorname{tr}\mathbf{XA}/\partial \mathbf{X} = \mathbf{A}'$.

10.17. $\partial \operatorname{tr}\mathbf{AXB}/\partial \mathbf{X} = \mathbf{A}'\mathbf{B}'$.

10.18. $\partial \operatorname{tr}\mathbf{X}'\mathbf{AX}/\partial \mathbf{X} = \mathbf{AX} + \mathbf{A}'\mathbf{X}$.

10.19. $\partial \log|\mathbf{X}|/\partial \mathbf{X} = (\mathbf{X}')^{-1}$ (log to base e).

10.20. $\partial \mathbf{Y}^{-1}/\partial x = -\mathbf{Y}^{-1}(\partial \mathbf{Y}/\partial x)\mathbf{Y}^{-1}$.

10.21.   (Chain Rule No. 2)

$$\partial y/\partial x = \text{tr}(\partial y/\partial \mathbf{Z})(\partial \mathbf{Z}'/\partial x)$$

where y and x are scalars. The scalar y is a function of $m \times n$ variables z_{ij}, and each of the z_{ij} is a function of x.

Example 10.21.1
Obtain $\log |\mathbf{R} - \mathbf{FF}'|/\partial \mathbf{F}$, where $\mathbf{R}$ is symmetric. By Chain Rule No. 2:

$\partial \log |\mathbf{R} - \mathbf{FF}'|/\partial f_{ij}$

$\quad = \text{tr}[\partial \log | \mathbf{R} - \mathbf{FF}'|/\partial(\mathbf{R} - \mathbf{FF}')][\partial(\mathbf{R} - \mathbf{FF}')/\partial f_{ij}]$ (since $\mathbf{R}$ and $\mathbf{FF}'$ are symmetric)

$\quad = \text{tr}(\mathbf{R} - \mathbf{FF}')^{-1}[\partial(\mathbf{R} - \mathbf{FF}')/\partial f_{ij}]$ (by 10.19)

$\quad = \text{tr}(\mathbf{R} - \mathbf{FF}')^{-1}[-(\partial \mathbf{F}/\partial f_{ij})\mathbf{F}' - \mathbf{F}(\partial \mathbf{F}'/\partial f_{ij})]$ (by 10.5)

$\quad = \text{tr}(\mathbf{R} - \mathbf{FF}')^{-1}[-(\mathbf{EL})_{ij}\mathbf{F}' - \mathbf{F}(\mathbf{EL})_{ji}]$ (by 10.4 and 10.7)

$\quad = -\text{tr}(\mathbf{R} - \mathbf{FF}')^{-1}(\mathbf{EL})_{ij}\mathbf{F}' - \text{tr}(\mathbf{R} - \mathbf{FF}')^{-1}\mathbf{F}(\mathbf{EL})_{ji}$

$\quad = -\text{tr}(\mathbf{EL})_{ij}\mathbf{F}'(\mathbf{R} - \mathbf{FF}')^{-1} - \text{tr}(\mathbf{EL})_{ji}(\mathbf{R} - \mathbf{FF}')^{-1}\mathbf{F}$ (by 7.3)

$\quad = -[\mathbf{F}'(\mathbf{R} - \mathbf{FF}')^{-1}]_{ji} - [(\mathbf{R} - \mathbf{FF}')^{-1}\mathbf{F}]_{ij},$

where $[\ ]_{ij}$ denotes the (i, j) element of the matrix in brackets (by 7.5),

$\quad\quad = -[(\mathbf{R} - \mathbf{FF}')^{-1}\mathbf{F}]_{ij} - [(\mathbf{R} - \mathbf{FF}')^{-1}\mathbf{F}]_{ij}$ (since $\mathbf{R} - \mathbf{FF}'$ is symmetric)

$\quad\quad = -2[(\mathbf{R} - \mathbf{FF}')^{-1}\mathbf{F}]_{ij}.$

Hence, by definition 10.2a,

$$\partial \log |\mathbf{R} - \mathbf{FF}'|/\partial \mathbf{F} = -2(\mathbf{R} - \mathbf{FF}')^{-1}\mathbf{F}. \quad\quad \blacksquare$$

10.22. $|\partial \mathbf{y}/\partial \mathbf{x}'| = J(\mathbf{y}; \mathbf{x})$ is called the Jacobian or **functional determinant** used in variable transformation of multiple integrals. Formally, if $\mathbf{y}$ is a column vector with m elements, each function of m variables x_i arranged into a row vector $\mathbf{x}'$,

$$dx_1 dx_2 \ldots dx_m = |\partial \mathbf{y}/\partial \mathbf{x}'|^{-1} dy_1 dy_2 \ldots dy_m.$$

10.23.   For a scalar y (a function of m variables x_i) to attain a stationary value, it is necessary that

$$\partial y/\partial \mathbf{x} = \mathbf{0}.$$

10.24.   For a stationary value to be a minimum (maximum) it is necessary that

$$\partial(\partial y/\partial \mathbf{x})/\partial \mathbf{x}' \quad\quad (-\partial(\partial y/\partial \mathbf{x})/\partial \mathbf{x}')$$

be a positive-definite matrix for the value of $\mathbf{x}$ satisfying 10.23.

Example 10.24.1

Find the values of β that minimize $u = \mathbf{x}'\mathbf{x}$ (the sum of squares of x_i) where $\mathbf{x} = \mathbf{y} - \mathbf{A}\beta$ (with $\mathbf{y}$ and $\mathbf{A}$ known and fixed).

$$\partial u/\partial \beta' = (\partial u/\partial \mathbf{x}')(\partial \mathbf{x}/\partial \beta') \quad \text{(by Chain Rule No. 1)}$$

$$= -2\mathbf{x}'\mathbf{A} \quad \text{(by 10.11 and 10.14)}$$

Hence

$$\partial u/\partial \beta = -2\mathbf{A}'\mathbf{x}$$

$$= -2\mathbf{A}'(\mathbf{y} - \mathbf{A}\beta).$$

Hence, for a stationary value, by 10.23, it is necessary that

$$\mathbf{A}'\mathbf{A}\hat{\beta} = \mathbf{A}'\mathbf{y}$$

where $\hat{\beta}$ denotes the values that make u stationary. Now,

$$\partial(\partial u/\partial \beta)/\partial \beta' = 2\partial(\mathbf{A}'\mathbf{A}\beta)/\partial \beta' = 2\mathbf{A}'\mathbf{A}.$$

If $\mathbf{A}$ has real elements, and if $\mathbf{A}'\mathbf{A}$ is nonsingular, then it is positive-definite (since, given an arbitrary real $\mathbf{x} \neq \mathbf{0}$, $\mathbf{x}'\mathbf{A}'\mathbf{A}\mathbf{x} = \mathbf{z}'\mathbf{z}$, with $\mathbf{z} = \mathbf{A}\mathbf{x}$; thus this is a sum of squares). Hence $\hat{\beta}$ minimizes u. ∎

10.25. (Generalized Newton Iteration)

Let $\mathbf{x}_0'$ be an initial estimate (m elements) of the roots of the m equations

$$\mathbf{f}(\mathbf{x}') = \mathbf{0}$$

where the m elements of the column vector $\mathbf{f}$ are each functions of $x_1, x_2, \ldots x_m$. Then an improved root is

$$\mathbf{x}_1 = \mathbf{x}_0 - \mathbf{Q}_0^{-1}\mathbf{f}(\mathbf{x}_0'),$$

where $\mathbf{Q}_0$ is the matrix of derivatives $\partial \mathbf{f}/\partial \mathbf{x}'$ evaluated at $\mathbf{x} = \mathbf{x}_0$. The usual procedure consists of evaluating $\mathbf{f}(\mathbf{x}_0')$, then solving $\mathbf{Q}_0\mathbf{u} = \mathbf{f}(\mathbf{x}_0')$ for $\mathbf{u}$. Then $\mathbf{x}_1 = \mathbf{x}_0 - \mathbf{u}$.

Example 10.25.1

Solve

$$f_1(x, y) = x^3 - x^2y + y^2 - 3.526 = 0$$

$$f_2(x, y) = x^3 + y^3 - 14.911 = 0$$

$$\mathbf{Q} = \begin{bmatrix} 3x^2 - 2xy & 2y - x^2 \\ 3x^2 & 3y^2 \end{bmatrix}.$$

Take $x_0 = 1$, $y_0 = 2$

$$f_1(x_0, y_0) = -0.526$$

$$f_2(x_0, y_0) = -5.911$$

$$\mathbf{Q}_0 = \begin{bmatrix} -1 & 3 \\ 3 & 12 \end{bmatrix}$$

$$-u + 3v = -0.526$$

$$3u + 12v = -5.911$$

yields $u = -0.55$, $v = -0.36$.

Then,

$$x_1 = x_0 - u = 1.55$$

$$y_1 = y_0 - v = 2.36$$

$$f_1(x_1, y_1) = 0.0976$$

$$f_2(x_1, y_1) = 1.9572$$

$$\mathbf{Q}_1 = \begin{bmatrix} -0.1085 & 2.3175 \\ 7.2075 & 16.7088 \end{bmatrix}$$

$$-0.1085u + 2.3175v = 0.0976$$

$$7.2075u + 16.7088v = 1.9572$$

yields $u = 0.157$, $v = 0.049$.

Then

$$x_2 = x_1 - u = 1.393$$

$$y_2 = y_1 - v = 2.311$$

$$f_1(x_2, y_2) = 0.03337$$

$$f_2(x_2, y_2) = 0.13443$$

$$\mathbf{Q}_2 = \begin{bmatrix} -0.61710 & 2.68155 \\ 5.82135 & 16.02216 \end{bmatrix}$$

$$-0.61710u + 2.68155v = 0.03337$$

$$5.82135u + 16.02216v = 0.13443$$

yields $u = -0.0068$, $v = 0.0109$.

Then,

$$x_3 = x_2 - u = 1.3998$$

$$y_3 = y_2 - v = 2.3001$$

(The exact roots are $x = 1.4$ and $y = 2.3$).

11   Statistical matrix forms

11.1.   Let E denote the expectation operator, and let $\mathbf{y}$ be a set of p random variables. Then

$$E(\mathbf{y}) = \boldsymbol{\mu}$$

states that $E(y_i) = \mu_i$ $(i = 1, 2, \ldots p)$.

11.2.   Let var denote variance. Then

$$\mathrm{var}(\mathbf{y}) = \boldsymbol{\Sigma}$$

denotes a $p \times p$ symmetric matrix whose elements are $\mathrm{cov}(y_i, y_j)$ and whose diagonal elements are $\mathrm{var}(y_i)$, where cov denotes covariance.

11.3. $E(\mathbf{Ay} + \mathbf{b}) = \mathbf{A}E(\mathbf{y}) + \mathbf{b} = \mathbf{A}\boldsymbol{\mu} + \mathbf{b}$.

11.4. $\mathrm{var}(\mathbf{Ay} + \mathbf{b}) = \mathbf{A}\,\mathrm{var}(\mathbf{y})\mathbf{A}' = \mathbf{A}\boldsymbol{\Sigma}\mathbf{A}'$.

11.5. $\mathrm{cov}(\mathbf{y}, \mathbf{z}')$ denotes a matrix with elements $\mathrm{cov}(y_i, z_j)$. $\mathrm{cov}(\mathbf{z}, \mathbf{y}') = [\mathrm{cov}(\mathbf{y}, \mathbf{z}')]'$.

11.6. $\mathrm{cov}(\mathbf{Ay} + \mathbf{b}, \mathbf{z}'\mathbf{C} + \mathbf{d}') = \mathbf{A}\,\mathrm{cov}(\mathbf{y}, \mathbf{z}')\mathbf{C}$.

11.7. $\mathrm{var}(\mathbf{y}) = E(\mathbf{yy}') - E(\mathbf{y})E(\mathbf{y}')$.

11.8. $\mathrm{cov}(\mathbf{y}, \mathbf{z}') = E(\mathbf{yz}') - E(\mathbf{y})E(\mathbf{z}')$.

11.9.   (Expected "sum of squares")

$$E(\mathbf{y}'\mathbf{Q}\mathbf{y}) = \mathrm{tr}[\mathbf{Q}\,\mathrm{var}(\mathbf{y})] + E(\mathbf{y}')\mathbf{Q}E(\mathbf{y}).$$

11.10.   If a matrix $\mathbf{Q}$ is symmetric and positive-definite, one can find a lower triangular matrix $\mathbf{T}$ (with positive diagonal terms, for uniqueness) such that $\mathbf{TT}' = \mathbf{Q}$. The matrices $\mathbf{T}$ and $\mathbf{T}^{-1}$ can be obtained from the Doolittle pattern (6.14) (Gauss elimination or square-root method) as follows: In each cycle, divide the p-row (left- and right-hand side) by $\sqrt{p_{ii}}$ (instead of p_{ii} for the u-row). Thus obtain rows designated as t-rows. The left-hand side (Arabic subscripts) is $\mathbf{T}'$, and the right-hand side (Roman subscripts) is $\mathbf{T}^{-1}$.

11.11.   If a coordinate system $\mathbf{x}$ is oblique, and if the cosines between reference vectors (scalar products of basis vectors of unit length) are stated in a symmetric matrix $\mathbf{Q}$, then $\mathbf{T}^{-1}\mathbf{x} = \mathbf{y}$ is an orthogonal system, where $\mathbf{T}$ is obtained from $\mathbf{Q}$ by 11.10.

11.12.   The likelihood function of a sample of size n from a multivariate normal distribution (p responses), with common variance-covariance matrix $\boldsymbol{\Sigma}(p \times p)$, and with means or main effects replaced by maximum-likelihood or least-squares estimates, can be written as

$$\log L = -\frac{np}{2} \log 2\pi - \frac{n}{2} \log |\boldsymbol{\Sigma}| - \frac{n}{2} \mathrm{tr}\,\boldsymbol{\Sigma}^{-1}\mathbf{S}$$

where $\boldsymbol{\Sigma}(p \times p)$ is the common variance-covariance matrix, and $\mathbf{S}$ is its maximum-likelihood estimate (matrix of sums of squares and products due to error, divided by sample size n) and log is base e.

11.13.   If $\boldsymbol{\Sigma}$ has a structure under a model or null hypothesis, and if elements of $\boldsymbol{\Sigma}$ are to be estimated, by maximum-likelihood, two cases can be distinguished:

(11.14) $\boldsymbol{\Sigma}^{-1}$ has the same structure (intraclass correlation, mixed model, compound symmetry, factor analysis).

(11.15) $\boldsymbol{\Sigma}^{-1}$ has a different structure (autocorrelation, Simplex structure).

11.14.   If the structure of $\boldsymbol{\Sigma}$ and $\boldsymbol{\Sigma}^{-1}$ are identical, and if u and v are elements (or functions of elements) of $\boldsymbol{\Sigma}^{-1}$, then estimates of $\boldsymbol{\Sigma}$ can be obtained from the relations (usually requiring Newton iteration; see 10.25):

$$\partial \log L / \partial u = \frac{n}{2} \mathrm{tr}\,\mathbf{A}(\boldsymbol{\Sigma} - \mathbf{S})$$

where $\mathbf{A} = \partial\mathbf{\Sigma}^{-1}/\partial u$ is frequently an elementary matrix (see 3.4, especially rules 7.5 and 7.6).

$$\partial^2 \log L/\partial u \partial v = \frac{n}{2}\, \mathrm{tr}(\partial\mathbf{A}/\partial v)(\mathbf{\Sigma} - \mathbf{S}) + \frac{n}{2}\, \mathrm{tr}\,\mathbf{A}\mathbf{\Sigma}^{-1}\mathbf{B}\mathbf{\Sigma}^{-1}$$

where $\mathbf{B} = \partial\mathbf{\Sigma}^{-1}/\partial v$. These rules are useful to obtain Newton iterations and asymptotic variance-covariance matrices of the estimates. The log is base e.

11.15.   If the structures of $\mathbf{\Sigma}$ and $\mathbf{\Sigma}^{-1}$ are different, then an estimate of $\mathbf{\Sigma}$ can be obtained from the relations

$$\partial \log L/\partial x = -\frac{n}{2}\, \mathrm{tr}\,\mathbf{A}(\mathbf{\Sigma}^{-1} - \mathbf{Q}),$$

where

$$\mathbf{Q} = \mathbf{\Sigma}^{-1}\mathbf{S}\mathbf{\Sigma}^{-1}$$

and

$$\mathbf{A} = \partial\mathbf{\Sigma}/\partial x \quad \text{(see comments in 11.14)}.$$

$$\partial^2 \log L/\partial x \partial y = -\frac{n}{2}\, \mathrm{tr}(\partial\mathbf{A}/\partial y)(\mathbf{\Sigma}^{-1} - \mathbf{Q}) + \frac{n}{2}\, \mathrm{tr}\,\mathbf{A}\mathbf{\Sigma}^{-1}\mathbf{B}(\mathbf{\Sigma}^{-1} - \mathbf{Q}) - \frac{n}{2}\, \mathrm{tr}\,\mathbf{A}\mathbf{Q}\mathbf{B}\mathbf{\Sigma}^{-1},$$

where

$$\mathbf{B} = \partial\mathbf{\Sigma}/\partial y$$

x and y are elements (or functions of elements) of $\mathbf{\Sigma}$. The comments of 11.14 apply, but the iterative procedure is considerably more complex. The log is base e.

Appendix 5: Vector Analysis

Definitions

Any quantity that is completely determined by its magnitude is called a **scalar**. Examples of such are mass, density, temperature, etc. Any quantity that is completely determined by its magnitude and direction is called a **vector**. Examples of such are velocity, acceleration, force, etc. A vector quantity is represented by a directed line segment, the length of which represents the magnitude of the vector. A vector quantity is usually represented by a boldface letter such as $\mathbf{V}$. Two vectors $\mathbf{V}_1$ and $\mathbf{V}_2$ are equal to one another if they have equal magnitudes and are acting in the same directions. A negative vector, written as $-\mathbf{V}$, is one that acts in the opposite direction to $\mathbf{V}$, but is of equal magnitude to it. If we represent the magnitude of $\mathbf{V}$ by v, we write $\mathbf{V} = v$. A vector parallel to $\mathbf{V}$, but equal to the reciprocal of its magnitude, is written as $\mathbf{V}^{-1}$ or $1/\mathbf{V}$.

The **unit vector** $\mathbf{V}/v$ ($v \neq 0$) is that vector which has the same direction as $\mathbf{V}$, but which has a magnitude of unity (sometimes represented as $\mathbf{V}_0$ or $\hat{\mathbf{v}}$).

Vector algebra

The vector sum of $\mathbf{V}_1$ and $\mathbf{V}_2$ is represented by $\mathbf{V}_1 + \mathbf{V}_2$. The vector sum of $\mathbf{V}_1$ and $-\mathbf{V}_2$, or the difference of the vector $\mathbf{V}_2$ from $\mathbf{V}_1$, is represented by $\mathbf{V}_1 - \mathbf{V}_2$.

If r is a scalar, then $r\mathbf{V} = \mathbf{V}r$ and represents a vector r times the magnitude of $\mathbf{V}$, in the same direction as $\mathbf{V}$ if r is positive, and in the opposite direction if r is negative. If r and s are scalars and $\mathbf{V}_1$, $\mathbf{V}_2$, $\mathbf{V}_3$ vectors, then the following rules of scalars and vectors hold:

$$\mathbf{V}_1 + \mathbf{V}_2 = \mathbf{V}_2 + \mathbf{V}_1$$

$$(r + s)\mathbf{V}_1 = r\mathbf{V}_1 + s\mathbf{V}_1; \qquad r(\mathbf{V}_1 + \mathbf{V}_2) = r\mathbf{V}_1 + r\mathbf{V}_2$$

$$\mathbf{V}_1 + (\mathbf{V}_2 + \mathbf{V}_3) = (\mathbf{V}_1 + \mathbf{V}_2) + \mathbf{V}_3 = \mathbf{V}_1 + \mathbf{V}_2 + \mathbf{V}_3.$$

0-8493-8342-0/96/$0.00 + $0.50
©1996 by CRC Press, Inc.

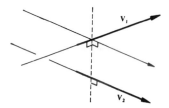

FIGURE 1

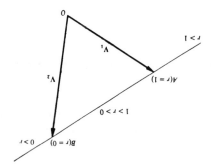

FIGURE 2

Vectors in space

A plane is described by two distinct vectors $\mathbf{V}_1$ and $\mathbf{V}_2$. Should these vectors not intersect one another, then one is displaced parallel to itself until they do (Fig. 1). Any other vector $\mathbf{V}$ lying in this plane is given by

$$\mathbf{V} = r\mathbf{V}_1 + s\mathbf{V}_2.$$

A **position vector** specifies the position in space of a point relative to a fixed origin. If, therefore, $\mathbf{V}_1$ and $\mathbf{V}_2$ are the position vectors of the points A and B, relative to the origin O, then any point P on the line AB has a position vector $\mathbf{V}$ given by

$$\mathbf{V} = r\mathbf{V}_1 + (1 - r)\mathbf{V}_2.$$

The scalar "r" can be taken as the parametric representation of P since $r = 0$ implies $P = B$ and $r = 1$ implies $P = A$ (Fig. 2). If P divides the line AB in the ratio $r : s$, then

$$\mathbf{V} = \left(\frac{r}{r + s}\right)\mathbf{V}_1 + \left(\frac{s}{r + s}\right)\mathbf{V}_2.$$

The vectors $\mathbf{V}_1, \mathbf{V}_2, \mathbf{V}_3, \ldots, \mathbf{V}_n$ are said to be **linearly dependent** if there exist scalars $r_1, r_2, r_3, \ldots, r_n$, not all zero, such that

$$r_1\mathbf{V}_1 + r_2\mathbf{V}_2 + \cdots + r_n\mathbf{V}_n = 0.$$

A vector $\mathbf{V}$ is linearly dependent on the set of vectors $\mathbf{V}_1, \mathbf{V}_2, \mathbf{V}_3, \ldots, \mathbf{V}_n$ if

$$\mathbf{V} = r_1\mathbf{V}_1 + r_2\mathbf{V}_2 + r_3\mathbf{V}_3 + \cdots + r_n\mathbf{V}_n.$$

Three vectors are linearly dependent if and only if they are coplanar.

All points in space can be uniquely determined by linear dependence on three **base vectors**, i.e., three vectors any one of which is linearly independent of the other two. The simplest set of base vectors are the unit vectors along the coordinate Ox, Oy, and Oz axes. These are usually designated by **i**, **j**, and **k**, respectively.

If **V** is a vector in space and a, b, and c are the respective magnitudes of the projections of the vector along the axes, then

$$\mathbf{V} = a\mathbf{i} + b\mathbf{j} + c\mathbf{k}$$

and

$$v = \sqrt{a^2 + b^2 + c^2}$$

and the direction cosines of **V** are

$$\cos\alpha = a/v, \qquad \cos\beta = b/v, \qquad \cos\gamma = c/v.$$

The law of addition yields

$$\mathbf{V}_1 + \mathbf{V}_2 = (a_1 + a_2)\mathbf{i} + (b_1 + b_2)\mathbf{j} + (c_1 + c_2)\mathbf{k}.$$

The scalar, dot, or inner product of two vectors $\mathbf{V}_1$ and $\mathbf{V}_2$

This product is represented as $\mathbf{V}_1 \cdot \mathbf{V}_2$ and is defined to be equal to $v_1 v_2 \cos\theta$, where θ is the angle from $\mathbf{V}_1$ to $\mathbf{V}_2$, i.e.,

$$\mathbf{V}_1 \cdot \mathbf{V}_2 = v_1 v_2 \cos\theta.$$

The following rules apply for this product:

$$\mathbf{V}_1 \cdot \mathbf{V}_2 = a_1 a_2 + b_1 b_2 + c_1 c_2 = \mathbf{V}_2 \cdot \mathbf{V}_1.$$

It should be noted that scalar multiplication is commutative:

$$(\mathbf{V}_1 + \mathbf{V}_2) \cdot \mathbf{V}_3 = \mathbf{V}_1 \cdot \mathbf{V}_3 + \mathbf{V}_2 \cdot \mathbf{V}_3$$

$$\mathbf{V}_1 \cdot (\mathbf{V}_2 + \mathbf{V}_3) = \mathbf{V}_1 \cdot \mathbf{V}_2 + \mathbf{V}_1 \cdot \mathbf{V}_3.$$

If $\mathbf{V}_1$ is perpendicular to $\mathbf{V}_2$, then $\mathbf{V}_1 \cdot \mathbf{V}_2 = 0$, and if $\mathbf{V}_1$ is parallel to $\mathbf{V}_2$, then $\mathbf{V}_1 \cdot \mathbf{V}_2 = v_1 v_2 = r w_1^2$.

In particular,

$$\mathbf{i} \cdot \mathbf{i} = \mathbf{j} \cdot \mathbf{j} = \mathbf{k} \cdot \mathbf{k} = 1,$$

and

$$\mathbf{i} \cdot \mathbf{j} = \mathbf{j} \cdot \mathbf{k} = \mathbf{k} \cdot \mathbf{i} = 0.$$

The vector or cross product of vectors $\mathbf{V}_1$ and $\mathbf{V}_2$

This product is represented as $\mathbf{V}_1 \times \mathbf{V}_2$ and is defined to be equal to $v_1 v_2 (\sin\theta)\mathbf{1}$, where θ is the angle from $\mathbf{V}_1$ to $\mathbf{V}_2$ and **1** is a unit vector perpendicular to the plane of $\mathbf{V}_1$ and $\mathbf{V}_2$ and so directed that a right-handed screw driven in the direction of **1** would carry $\mathbf{V}_1$ into $\mathbf{V}_2$, i.e.,

$$\mathbf{V}_1 \times \mathbf{V}_2 = v_1 v_2 (\sin\theta)\mathbf{1}$$

and

$$\tan\theta = \frac{|\mathbf{V}_1 \times \mathbf{V}_2|}{\mathbf{V}_1 \cdot \mathbf{V}_2}.$$

The following rules apply for vector products:

$$\mathbf{V}_1 \times \mathbf{V}_2 = -\mathbf{V}_2 \times \mathbf{V}_1$$

$$\mathbf{V}_1 \times (\mathbf{V}_2 + \mathbf{V}_3) = \mathbf{V}_1 \times \mathbf{V}_2 + \mathbf{V}_1 \times \mathbf{V}_3$$

$$(\mathbf{V}_1 + \mathbf{V}_2) \times \mathbf{V}_3 = \mathbf{V}_1 \times \mathbf{V}_3 + \mathbf{V}_2 \times \mathbf{V}_3$$

$$\mathbf{V}_1 \times (\mathbf{V}_2 \times \mathbf{V}_3) = \mathbf{V}_2(\mathbf{V}_3 \cdot \mathbf{V}_1 - \mathbf{V}_3(\mathbf{V}_1 \cdot \mathbf{V}_2)$$

$$\mathbf{i} \times \mathbf{i} = \mathbf{j} \times \mathbf{j} = \mathbf{k} \times \mathbf{k} = 0\mathbf{1} \text{ (zero vector)}$$

$$= \mathbf{0}$$

$$\mathbf{i} \times \mathbf{j} = \mathbf{k}, \qquad \mathbf{j} \times \mathbf{k} = \mathbf{i}, \qquad \mathbf{k} \times \mathbf{i} = \mathbf{j}.$$

If $\mathbf{V}_1 = a_1\mathbf{i} + b_1\mathbf{j} + c_1\mathbf{k}$, $\mathbf{V}_2 = a_2\mathbf{i} + b_2\mathbf{j} + c_2\mathbf{k}$, $\mathbf{V}_3 = a_3\mathbf{i} + b_3\mathbf{j} + c_3\mathbf{k}$, then

$$\mathbf{V}_1 \times \mathbf{V}_2 = \begin{vmatrix} \mathbf{i} & \mathbf{j} & \mathbf{k} \\ a_1 & b_1 & c_1 \\ a_2 & b_2 & c_2 \end{vmatrix} = (b_1 c_2 - b_2 c_1)\mathbf{i} + (c_1 a_2 - c_2 a_1)\mathbf{j} + (a_1 b_2 - a_2 b_1)\mathbf{k}.$$

It should be noted that since $\mathbf{V}_1 \times \mathbf{V}_2 = -\mathbf{V}_2 \times \mathbf{V}_1$, the vector product is not commutative.

Scalar triple product

There is only one possible interpretation of the expression $\mathbf{V}_1 \cdot \mathbf{V}_2 \times \mathbf{V}_3$ and that is $\mathbf{V}_1 \cdot (\mathbf{V}_2 \times \mathbf{V}_3)$, which is obviously a scalar.

Further, $\mathbf{V}_1 \cdot (\mathbf{V}_2 \times \mathbf{V}_3) = (\mathbf{V}_1 \times \mathbf{V}_2) \cdot \mathbf{V}_3 = \mathbf{V}_2 \cdot (\mathbf{V}_3 \times \mathbf{V}_1)$

$$= \begin{vmatrix} a_1 & b_1 & c_1 \\ a_2 & b_2 & c_2 \\ a_3 & b_3 & c_3 \end{vmatrix}$$

$$= v_1 v_2 v_3 \cos\phi \sin\theta$$

where θ is the angle between $\mathbf{V}_2$ and $\mathbf{V}_3$ and ϕ is the angle between $\mathbf{V}_1$ and the normal to the plane of $\mathbf{V}_2$ and $\mathbf{V}_3$.

This product is called the **scalar triple product** and is written as $[\mathbf{V}_1\mathbf{V}_2\mathbf{V}_3]$.

The determinant indicates that it can be considered as the volume of the parallelepiped whose three determining edges are $\mathbf{V}_1$, $\mathbf{V}_2$, and $\mathbf{V}_3$.

It also follows that cyclic permutation of the subscripts does not change the value of the scalar triple product so that

$$[\mathbf{V}_1\mathbf{V}_2\mathbf{V}_3] = [\mathbf{V}_2\mathbf{V}_3\mathbf{V}_1] = [\mathbf{V}_3\mathbf{V}_1\mathbf{V}_2]$$

but

$$[\mathbf{V}_1\mathbf{V}_2\mathbf{V}_3] = -[\mathbf{V}_2\mathbf{V}_1\mathbf{V}_3] \text{ etc.}$$

and

$$[\mathbf{V}_1\mathbf{V}_1\mathbf{V}_2] \equiv 0 \text{ etc.}$$

Given three non-coplanar reference vectors $\mathbf{V}_1$, $\mathbf{V}_2$, and $\mathbf{V}_3$, the **reciprocal system** is given by $\mathbf{V}_1^*$, $\mathbf{V}_2^*$, and $\mathbf{V}_3^*$, where

$$1 = v_1 v_1^* = v_2 v_2^* = v_3 v_3^*$$

$$0 = v_1 v_2^* = v_1 v_3^* = v_2 v_1^* \text{ etc.}$$

$$\mathbf{V}_1^* = \frac{\mathbf{V}_2 \times \mathbf{V}_3}{[\mathbf{V}_1\mathbf{V}_2\mathbf{V}_3]}, \qquad \mathbf{V}_2^* = \frac{\mathbf{V}_3 \times \mathbf{V}_1}{[\mathbf{V}_1\mathbf{V}_2\mathbf{V}_3]}, \qquad \mathbf{V}_3^* = \frac{\mathbf{V}_1 \times \mathbf{V}_2}{[\mathbf{V}_1\mathbf{V}_2\mathbf{V}_3]}.$$

The system $\mathbf{i}$, $\mathbf{j}$, $\mathbf{k}$ is its own reciprocal.

Vector triple product

The product $\mathbf{V}_1 \times (\mathbf{V}_2 \times \mathbf{V}_3)$ defines the **vector triple product**. Obviously, in this case, the brackets are vital to the definition:

$$\mathbf{V}_1 \times (\mathbf{V}_2 \times \mathbf{V}_3) = (\mathbf{V}_1 \cdot \mathbf{V}_3)\mathbf{V}_2 - (\mathbf{V}_1 \cdot \mathbf{V}_2)\mathbf{V}_3$$

$$= \begin{vmatrix} \mathbf{i} & \mathbf{j} & \mathbf{k} \\ a_1 & b_1 & c_1 \\ \begin{vmatrix} b_2 & c_2 \\ b_3 & c_3 \end{vmatrix} & \begin{vmatrix} c_2 & a_2 \\ c_3 & a_3 \end{vmatrix} & \begin{vmatrix} a_2 & b_2 \\ a_3 & b_3 \end{vmatrix} \end{vmatrix},$$

i.e., it is a vector, perpendicular to $\mathbf{V}_1$, lying in the plane of $\mathbf{V}_2$, $\mathbf{V}_3$.
Similarly,

$$(\mathbf{V}_1 \times \mathbf{V}_2) \times \mathbf{V}_3 = \begin{vmatrix} \mathbf{i} & \mathbf{j} & \mathbf{k} \\ \begin{vmatrix} b_1 & c_1 \\ b_2 & c_2 \end{vmatrix} & \begin{vmatrix} c_1 & a_1 \\ c_2 & a_2 \end{vmatrix} & \begin{vmatrix} a_1 & b_1 \\ a_2 & b_2 \end{vmatrix} \\ a_3 & b_3 & c_3 \end{vmatrix}$$

$$\mathbf{V}_1 \times (\mathbf{V}_2 \times \mathbf{V}_3) + \mathbf{V}_2 \times (\mathbf{V}_3 \times \mathbf{V}_1) + \mathbf{V}_3 \times (\mathbf{V}_1 + \mathbf{V}_2) \equiv 0.$$

If $\mathbf{V}_1 \times (\mathbf{V}_2 \times \mathbf{V}_3) = (\mathbf{V}_1 \times \mathbf{V}_2) \times \mathbf{V}_3$, then $\mathbf{V}_1$, $\mathbf{V}_2$, $\mathbf{V}_3$ form an **orthogonal set**. Thus $\mathbf{i}$, $\mathbf{j}$, $\mathbf{k}$ form an orthogonal set.

Geometry of the plane, straight line, and sphere

The position vectors of the fixed points A, B, C, D relative to O are $\mathbf{V}_1$, $\mathbf{V}_2$, $\mathbf{V}_3$, $\mathbf{V}_4$, and the position vector of the variable point P is $\mathbf{V}$.

The vector form of the equation of the straight line through A parallel to $\mathbf{V}_2$ is

$$\mathbf{V} = \mathbf{V}_1 + r\mathbf{V}_2$$
$$\text{or} \quad (\mathbf{V} - \mathbf{V}_1) = r\mathbf{V}_2$$
$$\text{or} \quad (\mathbf{V} - \mathbf{V}_1) \times \mathbf{V}_2 = 0,$$

while that of the plane through A perpendicular to $\mathbf{V}_2$ is

$$(\mathbf{V} - \mathbf{V}_1) \cdot \mathbf{V}_2 = 0.$$

The equation of the line AB is

$$\mathbf{V} = r\mathbf{V}_1 + (1 - r)\mathbf{V}_2$$

and those of the bisectors of the angles between $\mathbf{V}_1$ and $\mathbf{V}_2$ are

$$\mathbf{V} = r\left(\frac{\mathbf{V}_1}{v} \pm \frac{\mathbf{V}_2}{v_2}\right)$$
$$\text{or} \quad \mathbf{V} = r\left(\hat{\mathbf{v}}_1 \pm \hat{\mathbf{v}}_2\right).$$

The perpendicular from C to the line through A parallel to $\mathbf{V}_2$ has as its equation

$$\mathbf{V} = \mathbf{V}_1 - \mathbf{V}_3 - \hat{\mathbf{v}}_2 \cdot (\mathbf{V}_1 - \mathbf{V}_3)\hat{\mathbf{v}}_2.$$

The condition for the intersection of the two lines,

$$\mathbf{V} = \mathbf{V}_1 + r\mathbf{V}_3$$
$$\text{and} \quad \mathbf{V} = \mathbf{V}_2 + s\mathbf{V}_4$$
$$\text{is} \quad [(\mathbf{V}_1 - \mathbf{V}_2)\mathbf{V}_3\mathbf{V}_4] = 0.$$

The common perpendicular to the above two lines is the line of intersection of the two planes

$$[(\mathbf{V} - \mathbf{V}_1)\mathbf{V}_3(\mathbf{V}_3 \times \mathbf{V}_4)] = 0$$
$$\text{and} \quad [(\mathbf{V} - \mathbf{V}_2)\mathbf{V}_4(\mathbf{V}_3 \times \mathbf{V}_4)] = 0$$

and the length of this perpendicular is

$$\frac{[(\mathbf{V}_1 - \mathbf{V}_2)\mathbf{V}_3\mathbf{V}_4]}{|\mathbf{V}_3 \times \mathbf{V}_4|}.$$

The equation of the line perpendicular to the plane ABC is

$$\mathbf{V} = \mathbf{V}_1 \times \mathbf{V}_2 + \mathbf{V}_2 \times \mathbf{V}_3 + \mathbf{V}_3 \times \mathbf{V}_1,$$

and the distance of the plane from the origin is

$$\frac{[\mathbf{V}_1\mathbf{V}_2\mathbf{V}_3]}{|(\mathbf{V}_2 - \mathbf{V}_1) \times (\mathbf{V}_3 - \mathbf{V}_1)|}.$$

In general, the vector equation

$$\mathbf{V} \cdot \mathbf{V}_2 = r$$

defines the plane that is perpendicular to $\mathbf{V}_2$, and the perpendicular distance from A to this plane is

$$\frac{r - \mathbf{V}_1 \cdot \mathbf{V}_2}{v_2}.$$

The distance from A, measured along a line parallel to $\mathbf{V}_3$, is

$$\frac{r - \mathbf{V}_1 \cdot \mathbf{V}_2}{\mathbf{V}_2 \cdot \hat{\mathbf{v}}_3} \quad \text{or} \quad \frac{r - \mathbf{V}_1 \cdot \mathbf{V}_2}{v_2 \cos \theta}$$

where θ is the angle between $\mathbf{V}_2$ and $\mathbf{V}_3$.

(If this plane contains the point C, then $r = \mathbf{V}_3 \cdot \mathbf{V}_2$, and if it passes through the origin, then $r = 0$.)

Given two planes

$$\mathbf{V} \cdot \mathbf{V}_1 = r$$

$$\mathbf{V} \cdot \mathbf{V}_2 = s,$$

any plane through the line of intersection of these two planes is given by

$$\mathbf{V} \cdot (\mathbf{V}_1 + \lambda \mathbf{V}_2) = r + \lambda s$$

where λ is a scalar parameter. In particular, $\lambda = \pm v_1 / v_2$ yields the equation of the two planes bisecting the angle between the given planes.

The plane through A parallel to the plane of $\mathbf{V}_2$, $\mathbf{V}_3$ is

$$\mathbf{V} = \mathbf{V}_1 + r\mathbf{V}_2 + s\mathbf{V}_3$$

$$\text{or} \quad (\mathbf{V} - \mathbf{V}_1) \cdot \mathbf{V}_2 \times \mathbf{V}_3 = 0$$

$$\text{or} \quad [\mathbf{V}\mathbf{V}_2\mathbf{V}_3] - [\mathbf{V}_1\mathbf{V}_2\mathbf{V}_3] = 0$$

so that the expansion in rectangular Cartesian coordinates yields

$$\begin{vmatrix} (x - a_1) \cdot & (y - b_1) & (z - c_1) \\ a_2 & b_2 & c_2 \\ a_3 & b_3 & c_3 \end{vmatrix} = 0 \qquad (\mathbf{V} \equiv x\mathbf{i} + y\mathbf{j} + z\mathbf{k}),$$

which is obviously the usual linear equation in x, y, and z.

The plane through AB parallel to $\mathbf{V}_3$ is given by

$$[(\mathbf{V} - \mathbf{V}_1)(\mathbf{V}_1 - \mathbf{V}_2)\mathbf{V}_3] = 0$$

$$\text{or} \quad [\mathbf{V}\mathbf{V}_2\mathbf{V}_3] - [\mathbf{V}\mathbf{V}_1\mathbf{V}_3] - [\mathbf{V}_1\mathbf{V}_2\mathbf{V}_3] = 0.$$

The plane through the three points A, B, and C is

$$\mathbf{V} = \mathbf{V}_1 + s(\mathbf{V}_2 - \mathbf{V}_1) + t(\mathbf{V}_3 - \mathbf{V}_1)$$

$$\text{or} \quad \mathbf{V} = r\mathbf{V}_1 + s\mathbf{V}_2 + t\mathbf{V}_3 \qquad (r + s + t \equiv 1)$$

$$\text{or} \quad [(\mathbf{V} - \mathbf{V}_1)(\mathbf{V}_1 - \mathbf{V}_2)(\mathbf{V}_2 - \mathbf{V}_3)] = 0$$

$$\text{or} \quad [\mathbf{V}\mathbf{V}_1\mathbf{V}_2] + [\mathbf{V}\mathbf{V}_2\mathbf{V}_3] + [\mathbf{V}\mathbf{V}_3\mathbf{V}_1] - [\mathbf{V}_1\mathbf{V}_2\mathbf{V}_3] = 0.$$

For four points A, B, C, D to be coplanar, then

$$r\mathbf{V}_1 + s\mathbf{V}_2 + t\mathbf{V}_3 + u\mathbf{V}_4 \equiv 0 \equiv r + s + t + u.$$

The following formulae relate to a sphere when the vectors are taken to lie in three-dimensional space and to a circle when the space is two-dimensional. For a circle in three dimensions, take the intersection of the sphere with a plane.

The equation of a sphere with center O and radius OA is

$$\mathbf{V} \cdot \mathbf{V} = v_1^2 \qquad (\text{not } \mathbf{V} = \mathbf{V}_1)$$

$$\text{or} \qquad (\mathbf{V} - \mathbf{V}_1) \cdot (\mathbf{V} + \mathbf{V}_1) = 0,$$

while that of a sphere with center B radius v_1 is

$$(\mathbf{V} - \mathbf{V}_2) \cdot (\mathbf{V} - \mathbf{V}_2) = v_1^2$$

$$\text{or} \qquad \mathbf{V} \cdot (\mathbf{V} - 2\mathbf{V}_2) = v_1^2 - v_2^2.$$

If the above sphere passes through the origin, then

$$\mathbf{V} \cdot (\mathbf{V} - 2\mathbf{V}_2) = 0.$$

Note that in two-dimensional polar coordinates this is simply

$$r = 2a \cdot \cos\theta$$

while in three-dimensional Cartesian coordinates it is

$$x^2 + y^2 + z^2 - 2(a_2 x + b_2 y + c_2 x) = 0.$$

The equation of a sphere having the points A and B as the extremities of a diameter is

$$(\mathbf{V} - \mathbf{V}_1) \cdot (\mathbf{V} - \mathbf{V}_2) = 0.$$

The square of the length of the tangent from C to the sphere with center B and radius v_1 is given by

$$(\mathbf{V}_3 - \mathbf{V}_2) \cdot (\mathbf{V}_3 - \mathbf{V}_2) = v_1^2.$$

The condition that the plane $\mathbf{V} \cdot \mathbf{V}_3 = s$ is tangential to the sphere $(\mathbf{V} - \mathbf{V}_2) \cdot (\mathbf{V} - \mathbf{V}_2) = v_1^2$ is

$$(s - \mathbf{V}_3 \cdot \mathbf{V}_2) \cdot (s - \mathbf{V}_3 \cdot \mathbf{V}_2) = v_1^2 v_3^2.$$

The equation of the tangent plane at D, on the surface of sphere $(\mathbf{V} - \mathbf{V}_2) \cdot (\mathbf{V} - \mathbf{V}_2) = v_1^2$, is

$$(\mathbf{V} - \mathbf{V}_4) \cdot (\mathbf{V}_4 - \mathbf{V}_2) = 0$$

$$\text{or} \qquad \mathbf{V} \cdot \mathbf{V}_4 - \mathbf{V}_2 \cdot (\mathbf{V} + \mathbf{V}_4) = v_1^2 - v_2^2.$$

The condition that the two circles $(\mathbf{V} - \mathbf{V}_2) \cdot (\mathbf{V} - \mathbf{V}_2) = v_1^2$ and $(\mathbf{V} - \mathbf{V}_4) \cdot (\mathbf{V} - \mathbf{V}_4) = v_3^2$ intersect orthogonally is clearly

$$(\mathbf{V}_2 - \mathbf{V}_4) \cdot (\mathbf{V}_2 - \mathbf{V}_4) = v_1^2 + v_3^2.$$

The polar plane of D with respect to the circle

$$(\mathbf{V} - \mathbf{V}_2) \cdot (\mathbf{V} - \mathbf{V}_2) = v_1^2 \quad \text{is}$$

$$\mathbf{V} \cdot \mathbf{V}_4 - \mathbf{V}_2 \cdot (\mathbf{V} + \mathbf{V}_4) = v_1^2 - v_2^2.$$

Any sphere through the intersection of the two spheres $(\mathbf{V} - \mathbf{V}_2) \cdot (\mathbf{V} - \mathbf{V}_2) = v_1^2$ and $(\mathbf{V} - \mathbf{V}_4) \cdot (\mathbf{V} - \mathbf{V}_4) = v_3^2$ is given by

$$(\mathbf{V} - \mathbf{V}_2) \cdot (\mathbf{V} - \mathbf{V}_2) + \lambda(\mathbf{V} - \mathbf{V}_4) \cdot (\mathbf{V} - \mathbf{V}_4) = v_1^2 + \lambda v_3^2,$$

while the radical plane of two such spheres is

$$\mathbf{V} \cdot (\mathbf{V}_2 - \mathbf{V}_4) = -\frac{1}{2}(v_1^2 - v_2^2 - v_3^2 + v_4^2).$$

Differentiation of vectors

If $\mathbf{V}_1 = a_1\mathbf{i} + b_1\mathbf{j} + c_1\mathbf{k}$ and $\mathbf{V}_2 = a_2\mathbf{i} + b_2\mathbf{j} + c_2\mathbf{k}$, and if $\mathbf{V}_1$ and $\mathbf{V}_2$ are functions of the scalar t, then

$$\frac{d}{dt}(\mathbf{V}_1 + \mathbf{V}_2 + \cdots) = \frac{d\mathbf{V}_1}{dt} + \frac{d\mathbf{V}_2}{dt} + \cdots,$$

where

$$\frac{d\mathbf{V}_1}{dt} = \frac{da_1}{dt}\mathbf{i} + \frac{db_1}{dt}\mathbf{j} + \frac{dc_1}{dt}\mathbf{k}, \text{ etc.}$$

$$\frac{d}{dt}(\mathbf{V}_1 \cdot \mathbf{V}_2) = \frac{d\mathbf{V}_1}{dt} \cdot \mathbf{V}_2 + \mathbf{V}_1 \cdot \frac{d\mathbf{V}_2}{dt}$$

$$\frac{d}{dt}(\mathbf{V}_1 \times \mathbf{V}_2) = \frac{d\mathbf{V}_1}{dt} \times \mathbf{V}_2 + \mathbf{V}_1 \times \frac{d\mathbf{V}_2}{dt}$$

$$\mathbf{V} \cdot \frac{d\mathbf{V}}{dt} = v\frac{dv}{dt}.$$

In particular, if $\mathbf{V}$ is a vector of constant length then the right-hand side of the last equation is identically zero showing that $\mathbf{V}$ is perpendicular to its derivative.

The derivatives of the triple products are

$$\frac{d}{dt}[\mathbf{V}_1\mathbf{V}_2\mathbf{V}_3] = \left[\left(\frac{d\mathbf{V}_1}{dt}\right)\mathbf{V}_2\mathbf{V}_3\right] + \left[\mathbf{V}_1\left(\frac{d\mathbf{V}_2}{dt}\right)\mathbf{V}_3\right] + \left[\mathbf{V}_1\mathbf{V}_2\left(\frac{d\mathbf{V}_3}{dt}\right)\right]$$

and

$$\frac{d}{dt}\{\mathbf{V}_1 \times (\mathbf{V}_2 \times \mathbf{V}_3)\} = \left(\frac{d\mathbf{V}_1}{dt}\right) \times (\mathbf{V}_2 \times \mathbf{V}_3) + \mathbf{V}_1$$

$$\times \left(\left(\frac{d\mathbf{V}_2}{dt}\right) \times \mathbf{V}_3\right) + \mathbf{V}_1 \times \left(\mathbf{V}_2 \times \left(\frac{d\mathbf{V}_3}{dt}\right)\right).$$

Geometry of curves in space

s = the **length of arc**, measured from some fixed point on the curve (Fig. 3).

$\mathbf{V}_1$ = the position vector of the point A on the curve.

$\mathbf{V}_1 + \delta\mathbf{V}_1$ = the position vector of the point P in the neighborhood of A.

$\hat{\mathbf{t}}$ = the **unit tangent** to the curve at the point A, measured in the direction of s increasing.

The **normal plane** is that plane which is perpendicular to the unit tangent. The principal normal is defined as the intersection of the normal plane with the plane defined by $\mathbf{V}_1$ and $\mathbf{V}_1 + \delta\mathbf{V}_1$ in the limit as $\delta\mathbf{V}_1 - 0$.

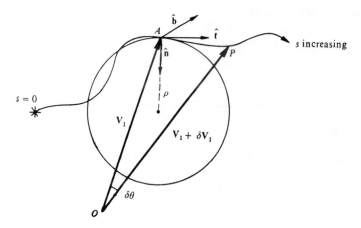

FIGURE 3

$\hat{\mathbf{n}}$ = the **unit normal** (principal) at the point A. The plane defined by $\hat{\mathbf{t}}$ and $\hat{\mathbf{n}}$ is called the **osculating plane** (alternatively, plane of curvature or local plane).

ρ = the radius of curvature at A.

$\delta\theta$ = the angle subtended at the origin by $\delta\mathbf{V}_1$.

$\kappa = \frac{d\theta}{ds} = \frac{1}{\rho}$.

$\hat{\mathbf{b}}$ = the **unit binormal**, i.e., the unit vector that is parallel to $\hat{\mathbf{t}} \times \hat{\mathbf{n}}$ at the point A.

λ = the **torsion** of the curve at A.

Frenet's Formulae:

$$\frac{d\hat{\mathbf{t}}}{ds} = \kappa\hat{\mathbf{n}}$$

$$\frac{d\hat{\mathbf{n}}}{ds} = -\kappa\hat{\mathbf{t}} + \lambda\hat{\mathbf{b}}$$

$$\frac{d\hat{\mathbf{b}}}{ds} = -\lambda\hat{\mathbf{n}}$$

The following formulae are also applicable:

Unit tangent:

$$\hat{\mathbf{t}} = \frac{d\mathbf{V}_1}{ds}.$$

Equation of the tangent:

$$(\mathbf{V} - \mathbf{V}_1) \times \hat{\mathbf{t}} = 0$$

$$\text{or} \quad \mathbf{V} = \mathbf{V}_1 + q\hat{\mathbf{t}}.$$

Unit normal:

$$\hat{\mathbf{n}} = \frac{1}{\kappa}\frac{d^2\mathbf{V}_1}{ds^2}.$$

Equation of the normal plane:

$$(\mathbf{V} - \mathbf{V}_1) \cdot \hat{\mathbf{t}} = 0.$$

Equation of the normal:

$$(\mathbf{V} - \mathbf{V}_1) \times \hat{\mathbf{n}} = 0$$

$$\text{or} \quad \mathbf{V} = \mathbf{V}_1 + r\hat{\mathbf{n}}.$$

Unit binormal:

$$\hat{\mathbf{b}} = \hat{\mathbf{t}} \times \hat{\mathbf{n}}.$$

Equation of the binormal:

$$(\mathbf{V} - \mathbf{V}_1) \times \hat{\mathbf{b}} = 0$$

$$\text{or} \quad \mathbf{V} = \mathbf{V}_1 + u\hat{\mathbf{b}}$$

$$\text{or} \quad \mathbf{V} = \mathbf{V}_1 + w\frac{d\mathbf{V}_1}{ds} \times \frac{d^2\mathbf{V}_1}{ds^2}.$$

Equation of the osculating plane:

$$[(\mathbf{V} - \mathbf{V}_1)\hat{\mathbf{t}}\hat{\mathbf{n}}] = 0$$

$$\text{or} \quad \left[(\mathbf{V} - \mathbf{V}_1)\left(\frac{d\mathbf{V}_1}{ds}\right)\left(\frac{d^2\mathbf{V}_1}{ds^2}\right)\right] = 0.$$

A **geodetic line** on a surface is a curve, the osculating plane of which is everywhere normal to the surface.

The differential equation of the geodetic is

$$[\hat{\mathbf{n}}d\mathbf{V}_1 d^2\mathbf{V}_1] = 0.$$

Differential operators—rectangular coordinates

$$dS = \frac{\partial S}{\partial x} \cdot dx + \frac{\partial S}{\partial y} \cdot dy + \frac{\partial S}{\partial z} \cdot dz.$$

By definition ,

$$\nabla \equiv \text{del} \equiv \mathbf{i}\frac{\partial}{\partial x} + \mathbf{j}\frac{\partial}{\partial y} + \mathbf{k}\frac{\partial}{\partial z}$$

$$\nabla^2 \equiv \text{Laplacian} \equiv \frac{\partial^2}{\partial x^2} + \frac{\partial^2}{\partial y^2} + \frac{\partial^2}{\partial z^2}.$$

If S is a scalar function, then

$$\nabla S \equiv \text{grad } S \equiv \frac{\partial S}{dx}\mathbf{i} + \frac{\partial S}{dy}\mathbf{j} + \frac{\partial S}{dz}\mathbf{k}.$$

Grad S defines both the direction and magnitude of the maximum rate of increase of S at any point. Hence the name **gradient** and also its vectorial nature. ∇S is independent of the choice of rectangular coordinates.

$$\nabla S = \frac{\partial S}{\partial n}\hat{\mathbf{n}}$$

where $\hat{\mathbf{n}}$ is the unit normal to the surface $S = $ constant, in the direction of S increasing. The

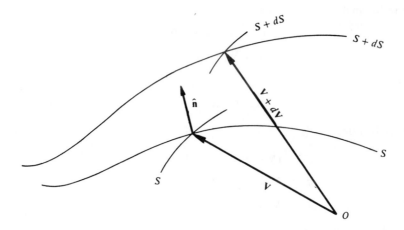

FIGURE 4

total derivative of S at a point having the position vector $\mathbf{V}$ is given by (Fig. 4)

$$dS = \frac{\partial S}{\partial n}\hat{\mathbf{n}} \cdot d\mathbf{V}$$

$$= d\mathbf{V} \cdot \nabla S$$

and the directional derivative of S in the direction of $\mathbf{U}$ is

$$\mathbf{U} \cdot \nabla S = \mathbf{U} \cdot (\nabla S) = (\mathbf{U} \cdot \nabla)S.$$

Similarly, the directional derivative of the vector $\mathbf{V}$ in the direction of $\mathbf{U}$ is

$$(\mathbf{U} \cdot \nabla)\mathbf{V}.$$

The **distributive** law holds for finding a gradient. Thus if S and T are scalar functions,

$$\nabla(S + T) = \nabla S + \nabla T.$$

The **associative** law becomes the rule for differentiating a product:

$$\nabla(ST) = S\nabla T + T\nabla S.$$

If $\mathbf{V}$ is a vector function with the magnitudes of the components parallel to the three coordinate axes V_x, V_y, V_z, then

$$\nabla \cdot \mathbf{V} \equiv \text{div } \mathbf{V} \equiv \frac{\partial V_x}{\partial x} + \frac{\partial V_y}{\partial y} + \frac{\partial V_z}{\partial z}.$$

The divergence obeys the distributive law. Thus, if $\mathbf{V}$ and $\mathbf{U}$ are vectors functions, then

$$\nabla \cdot (\mathbf{V} + \mathbf{U}) = \nabla \cdot \mathbf{V} + \nabla \cdot \mathbf{U}$$

$$\nabla \cdot (S\mathbf{V}) = (\nabla S) \cdot \mathbf{V} + S(\nabla \cdot \mathbf{V})$$

$$\nabla \cdot (\mathbf{U} \times \mathbf{V}) = \mathbf{V} \cdot (\nabla \times \mathbf{U}) - \mathbf{U} \cdot (\nabla \times \mathbf{V}).$$

As with the gradient of a scalar, the divergence of a vector is invariant under a transformation from one set of rectangular coordinates to another:

$$\nabla \times \mathbf{V} \equiv \text{curl } \mathbf{V} \quad (\text{sometimes } \nabla \wedge \mathbf{V} \text{ or rot } \mathbf{V})$$

$$\equiv \left(\frac{\partial V_z}{\partial y} - \frac{\partial V_y}{\partial z} \right) \mathbf{i} + \left(\frac{\partial V_x}{\partial z} - \frac{\partial V_z}{\partial x} \right) \mathbf{j} + \left(\frac{\partial V_y}{\partial x} - \frac{\partial V_x}{\partial y} \right) \mathbf{k}$$

$$= \begin{vmatrix} \mathbf{i} & \mathbf{j} & \mathbf{k} \\ \frac{\partial}{\partial x} & \frac{\partial}{\partial y} & \frac{\partial}{\partial z} \\ V_x & V_y & V_z \end{vmatrix}.$$

The **curl** (or **rotation**) of a vector is a vector that is invariant under a transformation from one set of rectangular coordinates to another:

$$\nabla \times (\mathbf{U} + \mathbf{V}) = \nabla \times \mathbf{U} + \nabla \times \mathbf{V}$$

$$\nabla \times (S\mathbf{V}) = (\nabla S) \times \mathbf{V} + S(\nabla \times \mathbf{V})$$

$$\nabla \times (\mathbf{U} \times \mathbf{V}) = (\mathbf{V} \cdot \nabla)\mathbf{U} - (\mathbf{U} \cdot \nabla)\mathbf{V} + \mathbf{U}(\nabla \cdot \mathbf{V}) - \mathbf{V}(\nabla \cdot \mathbf{U})$$

$$\text{grad}(\mathbf{U} \cdot \mathbf{V}) = \nabla(\mathbf{U} \cdot \mathbf{V})$$

$$= (\mathbf{V} \cdot \nabla)\mathbf{U} + (\mathbf{U} \cdot \nabla)\mathbf{V} + \mathbf{V} \times (\nabla \times \mathbf{U}) + \mathbf{U} \times (\nabla \times \mathbf{V}).$$

If

$$\mathbf{V} = V_x \mathbf{i} + V_y \mathbf{j} + V_z \mathbf{k}$$

$$\nabla \cdot \mathbf{V} = \nabla V_x \cdot \mathbf{i} + \nabla V_y \cdot \mathbf{j} + \nabla V_z \cdot \mathbf{k}$$

and

$$\nabla \times \mathbf{V} = \nabla V_x \times \mathbf{i} + \nabla V_y \times \mathbf{j} + \nabla V_z \times \mathbf{k}.$$

The operator ∇ can be used more than once. The number of possibilities where ∇ is used twice are

$$\nabla \cdot (\nabla \theta) \equiv \text{div grad } \theta$$

$$\nabla \times (\nabla \theta) \equiv \text{curl grad } \theta$$

$$\nabla(\nabla \cdot \mathbf{V}) \equiv \text{grad div } \mathbf{V}$$

$$\nabla \cdot (\nabla \times \mathbf{V}) \equiv \text{div curl } \mathbf{V}$$

$$\nabla \times (\nabla \times \mathbf{V}) \equiv \text{curl curl } \mathbf{V}.$$

The surface $PRS \equiv u = \text{const.}$, and the face of the curvilinear figure immediately opposite this is $u + du = \text{const.}$ etc.

In terms of these surface constants

$$P = P(u, v, w)$$

$$Q = Q(u + du, v, w) \quad \text{and} \quad PQ = h_1 du$$

$$R = R(u, v + dv, w) \qquad PR = h_2 dv$$

$$S = S(u, v, w + dw) \qquad PS = h_3 dw$$

where h_1, h_2, and h_3 are functions of u, v, and w.

In rectangular Cartesians **i**, **j**, **k**,

$$h_1 = 1, \qquad h_2 = 1, \qquad h_3 = 1.$$

$$\frac{\hat{\mathbf{a}}}{h_1}\frac{\partial}{\partial u} = \mathbf{i}\frac{\partial}{\partial x}, \qquad \frac{\hat{\mathbf{b}}}{h_2}\frac{\partial}{\partial v} = \mathbf{j}\frac{\partial}{\partial y}, \qquad \frac{\hat{\mathbf{c}}}{h_3}\frac{\partial}{\partial w} = \mathbf{k}\frac{\partial}{\partial z}.$$

In cylindrical coordinates $\hat{\mathbf{r}}$, $\hat{\boldsymbol{\phi}}$, $\hat{\mathbf{k}}$,

$$h_1 = 1, \qquad h_2 = r \qquad h_3 = 1.$$

$$\frac{\hat{\mathbf{a}}}{h_1}\frac{\partial}{\partial u} = \hat{\mathbf{r}}\frac{\partial}{\partial r}, \qquad \frac{\hat{\mathbf{b}}}{h_2}\frac{\partial}{\partial v} = \frac{\hat{\boldsymbol{\phi}}}{r}\frac{\partial}{\partial \phi} \qquad \frac{\hat{\mathbf{c}}}{h_3}\frac{\partial}{\partial w} = \hat{\mathbf{k}}\frac{\partial}{\partial z}.$$

In spherical coordinates $\hat{\mathbf{r}}$, $\hat{\boldsymbol{\theta}}$, $\hat{\boldsymbol{\phi}}$,

$$h_1 = 1, \qquad h_2 = r, \qquad h_3 = r\sin\theta$$

$$\frac{\hat{\mathbf{a}}}{h_1}\frac{\partial}{\partial u} = \hat{\mathbf{r}}\frac{\partial}{\partial r}, \qquad \frac{\mathbf{b}}{h_2}\frac{\partial}{\partial v} = \frac{\hat{\boldsymbol{\phi}}}{r}\frac{\partial}{\partial \theta}, \qquad \frac{\hat{\mathbf{c}}}{h_3}\frac{\partial}{\partial w} = \frac{\hat{\boldsymbol{\phi}}}{r\sin\theta}\frac{\partial}{\partial \phi}.$$

The general expressions for grad, div, and curl, together with those for ∇^2 and the directional derivative, are in orthogonal curvilinear coordinates given by

$$\nabla S = \frac{\hat{\mathbf{a}}}{h_1}\frac{\partial S}{\partial u} + \frac{\hat{\mathbf{b}}}{h_2}\frac{\partial S}{\partial v} + \frac{\hat{\mathbf{c}}}{h_3}\frac{\partial S}{\partial w}$$

$$(\mathbf{V}\cdot\nabla)S = \frac{V_1}{h_1}\frac{\partial S}{\partial u} + \frac{V_2}{h_2}\frac{\partial S}{\partial v} + \frac{V_3}{h_3}\frac{\partial S}{\partial w}$$

$$\nabla\cdot\mathbf{V} = \frac{1}{h_1 h_2 h_3}\left\{\frac{\partial}{\partial u}(h_2 h_3 V_1) + \frac{\partial}{\partial v}(h_3 h_1 V_2) + \frac{\partial}{\partial w}(h_1 h_2 V_3)\right\}$$

$$\nabla\times\mathbf{V} = \frac{\hat{\mathbf{a}}}{h_2 h_3}\left\{\frac{\partial}{\partial v}(h_3 V_3) - \frac{\partial}{\partial w}(h_2 V_2)\right\} + \frac{\hat{\mathbf{b}}}{h_3 h_1}\left\{\frac{\partial}{\partial w}(h_1 V_1) - \frac{\partial}{\partial u}(h_3 V_3)\right\}$$

$$+ \frac{\hat{\mathbf{c}}}{h_1 h_2}\left\{\frac{\partial}{\partial u}(h_2 V_2) - \frac{\partial}{\partial v}(h_1 V_1)\right\}$$

$$\nabla^2 S = \frac{1}{h_1 h_2 h_3}\left\{\frac{\partial}{\partial u}\left(\frac{h_2 h_3}{h_1}\frac{\partial S}{\partial u}\right) + \frac{\partial}{\partial v}\left(\frac{h_3 h_1}{h_2}\frac{\partial S}{\partial v}\right) + \frac{\partial}{\partial w}\left(\frac{h_1 h_2}{h_3}\frac{\partial S}{\partial w}\right)\right\}.$$

Transformation of integrals

s = the distance along some curve "C" in space and is measured from some fixed point

S = a surface area

V = a volume contained by a specified surface

TABLE 5.1
Formulas of vector analysis

	Rectangular coordinates	Cylindrical coordinates	Spherical coordinates
Conversion to rectangular coordinates		$x = r\cos\varphi \quad y = r\sin\varphi \quad z = z$	$x = r\cos\varphi\sin\theta \quad y = r\sin\varphi\sin\theta$ $z = r\cos\theta$
Gradient	$\nabla\phi = \dfrac{\partial\phi}{\partial x}\mathbf{i} + \dfrac{\partial\phi}{\partial y}\mathbf{j} + \dfrac{\partial\phi}{\partial z}\mathbf{k}$	$\nabla\phi = \dfrac{\partial\phi}{\partial r}\mathbf{r} + \dfrac{1}{r}\dfrac{\partial\phi}{\partial\varphi}\boldsymbol{\phi} + \dfrac{\partial\phi}{\partial z}\mathbf{k}$	$\nabla\phi = \dfrac{\partial\phi}{\partial r}\mathbf{r} + \dfrac{1}{r}\dfrac{\partial\phi}{\partial\theta}\boldsymbol{\theta} + \dfrac{1}{r\sin\theta}\dfrac{\partial\phi}{\partial\varphi}\boldsymbol{\phi}$
Divergence	$\nabla\cdot\mathbf{A} = \dfrac{\partial A_x}{\partial x} + \dfrac{\partial A_y}{\partial y} + \dfrac{\partial A_z}{\partial z}$	$\nabla\cdot\mathbf{A} = \dfrac{1}{r}\dfrac{\partial(rA_r)}{\partial r} + \dfrac{1}{r}\dfrac{\partial A_\varphi}{\partial\varphi}$ $+ \dfrac{\partial A_z}{\partial z}$	$\nabla\cdot\mathbf{A} = \dfrac{1}{r^2}\dfrac{\partial(r^2 A_r)}{\partial r} + \dfrac{1}{r\sin\theta}\dfrac{\partial(A_\theta\sin\theta)}{\partial\theta}$ $+ \dfrac{1}{r\sin\theta}\dfrac{\partial A_\varphi}{\partial\varphi}$
Curl	$\nabla\times\mathbf{A} = \begin{vmatrix} \mathbf{i} & \mathbf{j} & \mathbf{k} \\ \frac{\partial}{\partial x} & \frac{\partial}{\partial y} & \frac{\partial}{\partial z} \\ A_x & A_y & A_z \end{vmatrix}$	$\nabla\times\mathbf{A} = \begin{vmatrix} \frac{1}{r}\mathbf{r} & \boldsymbol{\phi} & \frac{1}{r}\mathbf{k} \\ \frac{\partial}{\partial r} & \frac{\partial}{\partial\varphi} & \frac{\partial}{\partial z} \\ A_r & rA_\varphi & A_z \end{vmatrix}$	$\nabla\times\mathbf{A} = \begin{vmatrix} \frac{\mathbf{r}}{r^2\sin\theta} & \frac{\boldsymbol{\theta}}{r\sin\theta} & \frac{\boldsymbol{\phi}}{r} \\ \frac{\partial}{\partial r} & \frac{\partial}{\partial\theta} & \frac{\partial}{\partial\varphi} \\ A_r & rA_\theta & rA_\varphi\sin\theta \end{vmatrix}$
Laplacian	$\nabla^2\phi = \dfrac{\partial^2\phi}{\partial x^2} + \dfrac{\partial^2\phi}{\partial y^2} + \dfrac{\partial^2\phi}{\partial z^2}$	$\nabla^2\phi = \dfrac{1}{r}\dfrac{\partial}{\partial r}\left(r\dfrac{\partial\phi}{\partial r}\right) + \dfrac{1}{r^2}\dfrac{\partial^2\phi}{\partial\varphi^2}$ $+ \dfrac{\partial^2\phi}{\partial z^2}$	$\nabla^2\phi = \dfrac{1}{r^2}\dfrac{\partial}{\partial r}\left(r^2\dfrac{\partial\phi}{\partial r}\right) + \dfrac{1}{r^2\sin\theta}\dfrac{\partial}{\partial\theta}\left(\sin\theta\dfrac{\partial\phi}{\partial\theta}\right)$ $+ \dfrac{1}{r^2\sin^2\phi}\dfrac{\partial^2\phi}{\partial\varphi^2}$

$\hat{\mathbf{t}} =$ the unit tangent to C at the point P

$\hat{\mathbf{n}} =$ the unit outward pointing normal

$F =$ some vector function

$ds =$ the vector element of curve ($= \hat{\mathbf{t}}\, ds$)

$dS =$ the vector element of surface ($= \hat{\mathbf{n}}\, dS$).

Then

$$\int_{(c)} \mathbf{F} \cdot \hat{\mathbf{t}}\, ds = \int_{(c)} \mathbf{F} \cdot ds$$

and when

$$\mathbf{F} = \nabla\phi,$$

$$\int_{(c)} (\nabla\phi) \cdot \hat{\mathbf{t}}\, ds = \int_{(c)} d\phi.$$

Gauss' Theorem (Green's Theorem)

When S defines a closed region having a volume V

$$\iiint_{(v)} (\nabla \cdot \mathbf{F})\, dV = \iint_{(s)} (\mathbf{F} \cdot \hat{\mathbf{n}})\, dS = \iint_{(s)} \mathbf{F} \cdot dS$$

also

$$\iiint_{(v)} (\nabla\phi)\, dV = \iint_{(s)} \phi\hat{\mathbf{n}}\, dS$$

and

$$\iiint_{(v)} (\nabla \times \mathbf{F})\, dV = \iint_{(s)} (\hat{\mathbf{n}} \times \mathbf{F})\, dS.$$

Stokes' Theorem

When C is closed and bounds the open surface S,

$$\iint_{(s)} \hat{\mathbf{n}} \cdot (\nabla \times \mathbf{F})\, dS = \int_{(c)} \mathbf{F} \cdot ds$$

also

$$\iint_{(s)} (\hat{\mathbf{n}} \times \nabla\phi)\, dS = \int_{(c)} \phi ds.$$

Green's Theorem

$$\iint_{(s)} (\nabla\phi \cdot \nabla\theta)\, dS = \iint_{(s)} \phi\hat{\mathbf{n}} \cdot (\nabla\theta)\, dS = \iiint_{(v)} \phi(\nabla^2\theta)\, dV$$

$$= \iint_{(a)} \theta \cdot \hat{\mathbf{n}}(\nabla\phi)\, dS = \iiint_{(v)} \theta(\nabla^2\phi)\, dV.$$

Appendix 6: Algebra Formulas and Coordinate Systems

Arithmetic progression[1]

An arithmetic progression is a sequence of numbers such that each number differs from the previous number by a constant amount, called the **common difference**.

If a_1 is the first term, a_n the nth term, d the common difference, n the number of terms, and s_n the sum of n terms,

$$a_n = a_1 + (n-1)d, \qquad s_n = \frac{n}{2}[a_1 + a_n], \qquad s_n = \frac{n}{2}[2a_1 + (n-1)d].$$

The arithmetic mean between a and b is given by $\frac{a+b}{2}$.

Geometric progression[1]

A geometric progression is a sequence of numbers such that each number bears a constant ratio, called the **common ratio**, to the previous number.

If a_1 is the first term, a_n the nth term, r the common ratio, n the number of terms, and s_n the sum of n terms

$$a_n = a_1 r^{n-1}; \; s_n = a_1 \frac{1 - r^n}{1 - r}$$

$$= a_1 \frac{r^n - 1}{r - 1} \qquad (r \neq 1)$$

$$= \frac{a_1 - r a_n}{1 - r}$$

$$= \frac{r a_n - a_1}{r - 1}$$

---

[1] It is customary to represent a_n by l in a finite progression and refer to it as the last term.

If $|r| < 1$, then the sum of an infinite geometrical progression converges to the limiting value

$$\frac{a_1}{1-r}, \qquad \left[s_\infty = \lim_{n \to \infty} \frac{a_1(1-r^n)}{1-r} = \frac{a_1}{1-r} \right]$$

The geometric mean between a and b is given by $\sqrt{ab}$.

Harmonic progression

A sequence of numbers whose reciprocals form an arithmetic progression is called an harmonic progression. Thus

$$\frac{1}{a_1}, \frac{1}{a_1+d}, \frac{1}{a_1+2d}, \ldots, \frac{1}{a_1+(n-1)d}, \ldots,$$

where

$$\frac{1}{a_n} = \frac{1}{a_1+(n-1)d}$$

forms an harmonic progression. The harmonic mean between a and b is given by $\frac{2ab}{a+b}$.

If A, G, H respectively represent the arithmetic mean, geometric mean, and harmonic mean between a and b, then $G^2 = AH$.

Factorials

$$\angle n = n! = e^{-n} n^n \sqrt{2\pi n}, \text{ approximately.}$$

Permutations

If $M = {}_n P_r = P_{n:r}$ denotes the number of permutations of n distinct things taken r at a time,

$$M = n(n-1)(n-2) \cdots (n-r+1) = \frac{n!}{(n-r)!}$$

Combinations

If $M = {}_n C_r = C_{n:r} = \binom{n}{r}$ denotes the number of combinations of n distinct things taken r at a time,

$$M = \frac{n(n-1)(n-2) \cdots (n-r+1)}{r!} = \frac{n!}{r!(n-r)!}$$

By definition $\binom{n}{0} = 1$.

Quadratic equations

Any quadratic equation may be reduced to the form,

$$ax^2 + bx + c = 0.$$

Then

$$x = \frac{-b \pm \sqrt{b^2 - 4ac}}{2a}.$$

If a, b, and c are real then:
If $b^2 - 4ac$ is positive, the roots are real and unequal;
If $b^2 - 4ac$ is zero, the roots are real and equal;
If $b^2 - 4ac$ is negative, the roots are imaginary and unequal.

Cubic equations

A cubic equation, $y^3 + py^2 + qy + r = 0$ may be reduced to the form,

$$x^3 + ax + b = 0$$

by substituting for y the value, $x - \frac{p}{3}$. Here

$$a = \frac{1}{3}(3q - p^2) \text{ and } b = \frac{1}{27}(2p^3 - 9pq + 27r).$$

For solution let,

$$A = \sqrt[3]{-\frac{b}{2} + \sqrt{\frac{b^2}{4} + \frac{a^3}{27}}}, \quad B = \sqrt[3]{-\frac{b}{2} - \sqrt{\frac{b^2}{4} + \frac{a^3}{27}}},$$

then the values of x will be given by,

$$x = A + B, \quad -\frac{A + B}{2} + \frac{A - B}{2}\sqrt{-3}, \quad -\frac{A + B}{2} - \frac{A - B}{2}\sqrt{-3}.$$

If p, q, r are real, then:

If $\frac{b^2}{4} + \frac{a^3}{27} > 0$, there will be one real root and two conjugate imaginary roots.

If $\frac{b^2}{4} + \frac{a^3}{27} = 0$, there will be three real roots of which at least two are equal.

If $\frac{b^2}{4} + \frac{a^3}{27} < 0$, there will be three real and unequal roots.

Trignometric solution of the cubic equation

The form $x^3 + ax + b = 0$ with $ab \neq 0$ can always be solved by transforming it to the trigonometric identity

$$4\cos^3 \theta - 3\cos \theta - \cos(3\theta) \equiv 0.$$

Let $x = m \cos \theta$, then

$$x^3 + ax + b \equiv m^3 \cos^3 \theta + am \cos \theta + b \equiv 4\cos^3 \theta - 3\cos \theta - \cos(3\theta) \equiv 0.$$

Hence

$$\frac{4}{m^3} = -\frac{3}{am} = \frac{-\cos(3\theta)}{b},$$

from which follows that

$$m = 2\sqrt{-\frac{a}{3}}, \quad \cos(3\theta) = \frac{3b}{am}.$$

Any solution θ_1 which satisfies $\cos(3\theta) = \frac{3b}{am}$, will also have the solutions

$$\theta_1 + \frac{2\pi}{3} \quad \text{and} \quad \theta_1 + \frac{4\pi}{3}.$$

The roots of the cubic $x^3 + ax + b = 0$ are

$$2\sqrt{-\frac{a}{3}}\cos\theta_1, \qquad 2\sqrt{-\frac{a}{3}}\cos\left(\theta_1 + \frac{2\pi}{3}\right), \qquad 2\sqrt{-\frac{a}{3}}\cos\left(\theta_1 + \frac{4\pi}{3}\right).$$

Example where hyperbolic functions are necessary for solution with latter procedure

The roots of the equation $x^3 - x + 2 = 0$ may be found as follows:
    Here

$$a = -1, \; b = 2, \; m = 2\sqrt{\frac{1}{3}} = 1.155$$

$$\cos(3\theta) = \frac{6}{-1.155} = -5.196$$

$$\cos(3\theta) = -\cos(3\theta - \pi) = -\cosh[i(3\theta - \pi)] = -5.196.$$

Using hyperbolic function tables for $\cosh[i(3\theta - \pi)] = 5.196$, it is found that

$$i(3\theta - \pi) = 2.332.$$

Thus

$$3\theta - \pi = -i(2.332).$$

$$3\theta = \pi - i(2.332)$$

$$\theta_1 = \frac{\pi}{3} - i(0.777)$$

$$\theta_1 + \frac{2\pi}{3} = \pi - i(0.777)$$

$$\theta_1 + \frac{4\pi}{3} = \frac{5\pi}{3} - i(0.777)$$

$$\cos\theta_1 = \cos\left[\frac{\pi}{3} - i(0.777)\right]$$

$$= \left(\cos\frac{\pi}{3}\right)[\cos i(0.777)] + \left(\sin\frac{\pi}{3}\right)[\sin i(0.777)]$$

$$= \left(\cos\frac{\pi}{3}\right)(\cosh 0.777) + i\left(\sin\frac{\pi}{3}\right)(\sinh 0.777)$$

$$= (0.5)(1.317) + i(0.866)(0.858) = 0.659 + i(0.743).$$

Note that

$$\cos \mu = \cosh(i\mu) \qquad \text{and} \qquad \sin \mu = -i \sinh(i\mu).$$

Similarly

$$\cos \left(\theta_1 + \frac{2\pi}{3} \right) = \cos[\pi - i(0.777)]$$

$$= (\cos \pi)(\cosh 0.777) + i(\sin \pi)(\sinh 0.777)$$

$$= -1.317,$$

and

$$\cos \left(\theta_1 + \frac{4\pi}{3} \right) = \cos \left[\frac{5\pi}{3} - i(0.777) \right]$$

$$= \left(\cos \frac{5\pi}{3} \right)(\cosh 0.777) + i \left(\sin \frac{5\pi}{3} \right)(\sinh 0.777)$$

$$= (0.5)(1.317) - i(0.866)(0.858) = 0.659 - i(0.743).$$

The required roots are

$$1.155[0.659 + i(0.743)] = 0.760 + i(0.858)$$

$$(1.155)(-1.317) = -1.520$$

$$(1.155)[0.659 - i(0.743)] = 0.760 - i(0.858).$$

Quartic equation

A quartic equation,

$$x^4 + ax^3 + bx^2 + cx + d = 0,$$

has the **resolvent cubic equation**

$$y^3 - by^2 + (ac - 4d)y - a^2d + 4bd - c^2 = 0.$$

Let y be any root of this equation, and

$$R = \sqrt{\frac{a^2}{4} - b + y}.$$

If $R \neq 0$, then let

$$D = \sqrt{\frac{3a^2}{4} - R^2 - 2b + \frac{4ab - 8c - a^3}{4R}}$$

and

$$E = \sqrt{\frac{3a^2}{4} - R^2 - 2b - \frac{4ab - 8c - a^3}{4R}}.$$

If $R = 0$, then let

$$D = \sqrt{\frac{3a^2}{4} - 2b + 2\sqrt{y^2 - 4d}}$$

and

$$E = \sqrt{\frac{3a^2}{4} - 2b - 2\sqrt{y^2 - 4d}}.$$

Then the four roots of the original equation are given by

$$x = -\frac{a}{4} + \frac{R}{2} \pm \frac{D}{2}$$

and

$$x = -\frac{a}{4} - \frac{R}{2} \pm \frac{E}{2}.$$

Partial fractions

This section applies only to rational algebraic fractions with numerator of lower degree than the denominator. Improper fractions can be reduced to proper fractions by long division.

Every fraction may be expressed as the sum of component fractions whose denominators are factors of the denominator of the original fraction.

Let $N(x)$ = numerator, a polynomial of the form

$$n_0 + n_1 x + n_2 x^2 + \cdots + n_i x^i$$

I. Non-repeated linear factors

$$\frac{N(x)}{(x-a)G(x)} = \frac{A}{x-a} + \frac{F(x)}{G(x)}$$

$$A = \left[\frac{N(x)}{G(x)}\right]_{x=a}$$

$F(x)$ is determined by methods discussed in the following sections.

Example

$$\frac{x^2 + 3}{x(x-2)(x^2 + 2x + 4)} = \frac{A}{x} + \frac{B}{x-2} + \frac{F(x)}{x^2 + 2x + 4}$$

$$A = \left[\frac{x^2 + 3}{(x-2)(x^2 + 2x + 4)}\right]_{x=0} = -\frac{3}{8}$$

$$B = \left[\frac{x^2 + 3}{x(x^2 + 2x + 4)}\right]_{x=2} = \frac{4+3}{2(4+4+4)} = \frac{7}{24}$$

II. Repeated linear factors

$$\frac{N(x)}{x^m G(x)} = \frac{A_0}{x^m} + \frac{A_1}{x^{m-1}} + \cdots + \frac{A_{m-1}}{x} + \frac{F(x)}{G(x)}$$

$$F(x) = f_0 + f_1 x + f_2 x^2 + \cdots, \qquad G(x) = g_0 + g_1 x + g_2 x^2 + \cdots$$

$$A_0 = \frac{n_0}{g_0}, \qquad A_1 = \frac{n_1 - A_0 g_1}{g_0}, \qquad A_2 = \frac{n_2 - A_0 g_2 - A_1 g_1}{g_0}$$

General term:[2]

$$A_k = \frac{1}{g_0}\left[n_k - \sum_{i=0}^{k-1} A_i g_{k-i} \right]$$

$$m = 1 \begin{cases} f_0 = n_1 - A_0 g_1 \\ f_1 = n_2 - A_0 g_2 \\ f_j = n_{j+1} - A_0 g_{j+1} \end{cases}$$

$$m = 2 \begin{cases} f_0 = n_2 - A_0 g_2 - A_1 g_1 \\ f_1 = n_3 - A_0 g_3 - A_1 g_2 \\ f_j = n_{j+2} - [A_0 g_{j+2} + A_1 g_{j+1}] \end{cases}$$

$$m = 3 \begin{cases} f_0 = n_3 - A_0 g_3 - A_1 g_2 - A_2 g_1 \\ f_1 = n_3 - A_0 g_4 - A_1 g_3 - A_2 g_2 \\ f_j = n_{j+3} - [A_0 g_{j+3} + A_1 g_{j+2} + A_2 g_{j+1}] \end{cases}$$

any m: $f_j = n_{m+j} - \sum_{i=0}^{m-1} A_i g_{m+j-i}$

Example

$$\frac{x^2 + 1}{x^3(x^2 - 3x + 6)} = \frac{A_0}{x^3} + \frac{A_1}{x^2} + \frac{A_2}{x} + \frac{f_1 x + f_0}{x^2 - 3x + 6}$$

$$A_0 = \frac{1}{6},$$

$$A_1 = \frac{0 - \left(\frac{1}{6}\right)(-3)}{6} = \frac{1}{12},$$

$$A_2 = \frac{1 - \left(\frac{1}{6}\right)(1) - \left(\frac{1}{12}\right)(-3)}{6} = \frac{13}{72},$$

$$m = 3 \begin{cases} f_0 = 0 - \frac{1}{6}(0) + \frac{1}{12}(1) - \frac{13}{72}(-3) = \frac{11}{24} \\ f_1 = 0 - \frac{1}{6}(0) - \frac{1}{12}(0) - \frac{13}{72}(1) = -\frac{13}{72} \end{cases}$$

---

[2]Note: If $G(x)$ contains linear factors, $F(x)$ may be determined by previous section I.

III. Repeated linear factors

$$\frac{N(x)}{(x-a)^m G(x)} = \frac{A_0}{(x-a)^m} + \frac{A_1}{(x-a)^{m-1}} + \cdots + \frac{A_{m-1}}{(x-a)} + \frac{F(x)}{G(x)}$$

Change to form $\frac{N'(y)}{y^m G'(y)}$ by substitution of $x = y + a$. Resolve into partial fractions in terms of y as described in Section II. Then express in terms of x by substitution $y = x - a$.

Example

$$\frac{x-3}{(x-2)^2(x^2+x+1)}.$$

Let $x - 2 = y$, $x = y + 2$

$$\frac{(y+2)-3}{y^2[(y+2)^2+(y+2)+1]} = \frac{y-1}{y^2(y^2+5y+7)} = \frac{A_0}{y^2} + \frac{A_1}{y} + \frac{f_1 y + f_0}{y^2+5y+7}$$

$$A_0 = -\frac{1}{7}, \qquad A_1 = \frac{1-\left(-\frac{1}{7}\right)(5)}{7} = \frac{12}{49},$$

$$m = 2 \begin{cases} f_0 = 0 - \left(-\frac{1}{7}\right)(1) - \left(\frac{12}{49}\right)(5) = -\frac{53}{49} \\[2mm] f_1 = 0 - \left(-\frac{1}{7}\right)(0) - \left(\frac{12}{49}\right)(1) = -\frac{12}{49} \end{cases}$$

$$\therefore \frac{y-1}{y^2(y^2+5y+7)} = \frac{-\frac{1}{7}}{y^2} + \frac{\frac{12}{49}}{y} + \frac{-\frac{12}{49}y - \frac{53}{49}}{y^2+5y+7}$$

Let $y = x - 2$, then

$$\frac{x-3}{(x-2)^2(x^2+x+1)} = \frac{-\frac{1}{7}}{(x-2)^2} + \frac{\frac{12}{35}}{(x-2)} + \frac{-\frac{12}{49}(x-2) - \frac{53}{49}}{x^2+x+1}$$

$$= -\frac{1}{7(x-2)^2} + \frac{12}{35(x-2)} + \frac{-12x-29}{49(x^2+x+1)} \qquad \blacksquare$$

IV. Repeated linear factors

Alternative method of determining coefficients:

$$\frac{N(x)}{(x-a)^m G(x)} = \frac{A_0}{(x-a)^m} + \cdots + \frac{A_k}{(x-a)^{m-k}} + \cdots + \frac{A_{m-1}}{x-a} + \frac{F(x)}{G(x)}$$

$$A_k = \frac{1}{k!}\left\{ D_x^k\left[\frac{N(x)}{G(x)}\right]\right\}_{x=a}$$

where D_x^k is the differentiating operator, and the derivative of zero order is defined as:

$$D_x^0 u = u.$$

V. Factors of higher degree

Factors of higher degree have the corresponding numerators indicated.

$$\frac{N(x)}{(x^2+h_1 x+h_0)G(x)} = \frac{a_1 x + a_0}{x^2+h_1 x+h_0} + \frac{F(x)}{G(x)}$$

$$\frac{N(x)}{(x^2 + h_1 x + h_0)^2 G(x)} = \frac{a_1 x + a_0}{(x^2 + h_1 x + h_0)^2} + \frac{b_1 x + b_0}{(x^2 + h_1 x + h_0)} + \frac{F(x)}{G(x)}$$

$$\frac{N(x)}{(x^3 + h_2 x^2 + h_1 x + h_0)G(x)} = \frac{a_2 x^2 + a_1 x + a_0}{x^3 + h_2 x^2 + h_1 x + h_0} + \frac{F(x)}{G(x)}$$

$$\vdots$$

Problems of this type are determined first by solving for the coefficients due to linear factors as shown above, and then determining the remaining coefficients by the general methods given below.

VI. General methods for evaluating coefficients

1.

$$\frac{N(x)}{D(x)} = \frac{N(x)}{G(x)H(x)L(x)} = \frac{A(x)}{G(x)} + \frac{B(x)}{H(x)} + \frac{C(x)}{L(x)} + \cdots$$

Multiply both sides of equation by $D(x)$ to clear fractions. Then collect terms, equate like powers of x, and solve the resulting simultaneous equations for the unknown coefficients.

2. Clear fractions as above. Then let x assume certain convenient values ($x = 1.0, -1, \ldots$). Solve the resulting equations for the unknown coefficients.

3.

$$\frac{N(x)}{G(x)H(x)} = \frac{A(x)}{G(x)} + \frac{B(x)}{H(x)}$$

Then

$$\frac{N(x)}{G(x)H(x)} - \frac{A(x)}{G(x)} = \frac{B(x)}{H(x)}$$

If $A(x)$ can be determined, such as by Method I, then $B(x)$ can be found as above.

Polar coordinates in a plane

Polar coordinates

In a plane, let OX (called the **initial line**) be a fixed ray radiating from point O (called the **pole** or **origin**). Then any point P, other than O, in the plane is located by angle θ (called the **vectorial angle**) measured from OX to the line determined by O and P and the distance r (called the **radius vector**) from O to P, where θ is taken as positive if measured counterclockwise and negative if measured clockwise, and r is taken as positive if measured along the terminal side of angle θ and negative if measured along the terminal side of θ produced through the pole. Such an ordered pair of numbers, (r, θ), are called **polar coordinates** of the point P. The polar coordinates of the pole O are taken as $(0, \theta)$, where θ is arbitrary. It follows that, for a given initial line and pole, each point of the plane has infinitely many polar coordinates.

Example

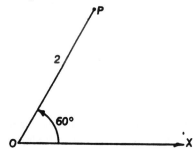

Some polar coordinates of P are: $(2, 60°)$, $(2, 420°)$, $(2, -300°)$, $(-2, 240°)$, $(-2, -120°)$. ∎

Points

Distance between P_1 and P_2:

$$\sqrt{r_1^2 + r_2^2 - 2r_1r_2\cos(\theta_1 - \theta_2)}$$

Points P_1, P_2, P_3 are collinear if and only if

$$r_2r_3\sin(\theta_3 - \theta_2) + r_3r_1\sin(\theta_1 - \theta_3) + r_1r_2\sin(\theta_2 - \theta_1) = 0.$$

Polygonal areas

Area of triangle $P_1P_2P_3$:

$$\frac{1}{2}[r_1r_2\sin(\theta_2 - \theta_1) + r_2r_3\sin(\theta_3 - \theta_2) + r_3r_1\sin(\theta_1 - \theta_3)]$$

Area of polygon $P_1P_2\cdots P_n$:

$$\frac{1}{2}[r_1r_2\sin(\theta_2 - \theta_1) + r_2r_3\sin(\theta_3 - \theta_2) + \cdots + r_{n-1}r_n\sin(\theta_n - \theta_{n-1}) + r_nr_1\sin(\theta_1 - \theta_n)]$$

The area is positive or negative according as $P_1P_2\cdots P_n$ is a counterclockwise or clockwise polygon.

Straight lines

Let p = distance of line from O, ω = counterclockwise angle from OX to the perpendicular through O to the line:

Normal form:　　　　$r\cos(\theta - \omega) = p$

Two-point form:　　　$r[r_1\sin(\theta - \theta_1) + r_2\sin(\theta - \theta_2)] = r_1r_2\sin(\theta_2 - \theta_1)$

Circles

Center at pole, radius a:　　　　　　　　　　　　　　　$r = a$

Center at $(a, 0)$ and passing through the pole:　　　$r = 2a\cos\theta$

Center at $(a, \frac{\pi}{2})$ and passing through the pole:　　　$r = 2a\sin\theta$

Center (h, α), radius a:　　　　　　　　　　　　$r^2 - 2hr\cos(\theta - \alpha) + h^2 - a^2 = 0$

Conics

Let $2p$ = distance from directrix to focus, e = eccentricity.

Focus at pole, directrix to left of pole: $\qquad r = \frac{2ep}{1-e\cos\theta}$

Focus at pole, directrix to right of pole: $\qquad r = \frac{2ep}{1+e\cos\theta}$

Focus at pole, directrix below pole: $\qquad r = \frac{2ep}{1-e\sin\theta}$

Focus at pole, directrix above pole: $\qquad r = \frac{2ep}{1+e\sin\theta}$

Parabola with vertex at pole, directrix to left of pole:

$$r = \frac{4p\cos\theta}{\sin^2\theta}$$

Ellipse with center at pole, semiaxes a and b horizontal and vertical, respectively:

$$r^2 = \frac{a^2 b^2}{a^2\sin^2\theta + b^2\cos^2\theta}$$

Hyperbola with center at pole, semiaxes a and b horizontal and vertical, respectively:

$$r^2 = \frac{a^2 b^2}{a^2\sin^2\theta - b^2\cos^2\theta}$$

Relations between rectangular polar coordinates

Let the positive x-axis coincide with the initial line and let r be nonnegative.

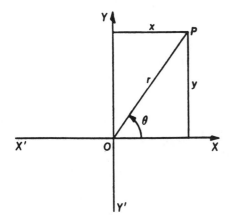

$$x = r\cos\theta, \quad y = r\sin\theta,$$

$$r = \sqrt{x^2 + y^2}, \quad \theta = \arctan\frac{y}{x},$$

$$\sin\theta = \frac{x}{\sqrt{x^2+y^2}}, \quad \cos\theta = \frac{y}{\sqrt{x^2+y^2}}$$

Rectangular coordinates in space

Rectangular (Cartesian) coordinates

Let $X'Y$, $Y'Y$, $Z'Z$ (called the y-**axis**, and the z-**axis**, respectively) be three mutually perpendicular lines in space intersecting in a point O (called the **origin**), forming in this way three mutually perpendicular planes XOY, XOZ, YOZ (called the xy-**plane**, the xz-**plane**, and the yz-**plane**, respectively). Then any point P of space is located by its signed distances x, y, z from the yz-plane, the xz-plane, and the xy-plane, respectively, where x and y are the rectangular coordinates with respect to the axes $X'X$ and $Y'Y$ of the orthogonal projection P' of P

on the xy-plane (here taken horizontally) and z is taken as positive above and negative below the xy-plane. The ordered triple of numbers, (x, y, z), are called **rectangular coordinates** of the point P.

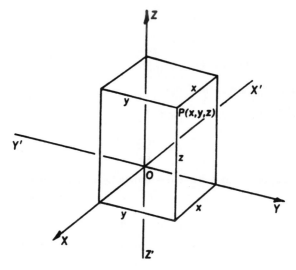

Points

Let $P_1(x_1, y_1, z_1)$ and $P_2(x_2, y_2, z_2)$ be any two points.

    Distance between P_1 and P_2: $\sqrt{(x_2 - x_1)^2 + (y_2 - y_1)^2 + (z_2 - z_1)^2}$

    Point dividing $P_1 P_2$ in ratio $\frac{r}{s}$: $\left(\frac{rx_2 + sx_1}{r+s}, \frac{ry_2 + sy_1}{r+s}, \frac{rz_2 + sz_1}{r+s} \right)$

    Midpoint of $P_1 P_2$: $\left(\frac{x_1 + x_2}{2}, \frac{y_1 + y_2}{2}, \frac{z_1 + z_2}{2} \right)$

    Points P_1, P_2, P_3 are collinear if and only if

$$x_2 - x_1 : y_2 - y_1 : z_2 - z_1 = x_3 - x_1 : y_3 - y_1 : z_3 - z_1.$$

Points P_1, P_2, P_3, P_4 are coplanar if and only if

$$\begin{vmatrix} x_1 & y_1 & z_1 & 1 \\ x_2 & y_2 & z_2 & 1 \\ x_3 & y_3 & z_3 & 1 \\ x_4 & y_4 & z_4 & 1 \end{vmatrix} = 0.$$

Area of triangle $P_1 P_2 P_3$:

$$\frac{1}{2} \sqrt{ \begin{vmatrix} y_1 & z_1 & 1 \\ y_2 & z_2 & 1 \\ y_3 & z_3 & 1 \end{vmatrix}^2 + \begin{vmatrix} z_1 & x_1 & 1 \\ z_2 & x_2 & 1 \\ z_3 & x_3 & 1 \end{vmatrix}^2 + \begin{vmatrix} x_1 & y_1 & 1 \\ x_2 & y_2 & 1 \\ x_3 & y_3 & 1 \end{vmatrix}^2 }$$

Volume of tetrahedron $P_1 P_2 P_3 P_4$:

$$\frac{1}{6} \begin{vmatrix} x_1 & y_1 & z_1 & 1 \\ x_2 & y_2 & z_2 & 1 \\ x_3 & y_3 & z_3 & 1 \\ x_4 & y_4 & z_4 & 1 \end{vmatrix}$$

Direction numbers and direction cosines

Let α, β, γ (called **direction angles**) be the angles that $P_1 P_2$, or any line parallel to $P_1 P_2$, makes with the x-, y-, and z-axis, respectively. Let $d = $ distance between P_1 and P_2.

Direction cosines of $P_1 P_2$:

$$\cos \alpha = \tfrac{x_2 - x_1}{d}, \quad \cos \beta = \tfrac{y_2 - y_1}{d}, \quad \cos \gamma = \tfrac{z_2 - z_1}{d}$$

$$\cos^2 \alpha + \cos^2 \beta + \cos^2 \gamma = 1$$

If a, b, c are direction numbers of $P_1 P_2$, then:

$$a : b : c = x_2 - x_1 : y_2 - y_1 : z_2 - z_1$$

$$= \cos \alpha : \cos \beta : \cos \gamma$$

$$\cos \alpha = \frac{a}{\pm\sqrt{a^2 + b^2 + c^2}}, \quad \cos \beta = \frac{b}{\pm\sqrt{a^2 + b^2 + c^2}},$$

$$\cos \gamma = \frac{c}{\pm\sqrt{a^2 + b^2 + c^2}}$$

Angle between two lines with direction angles $\alpha_1, \beta_1, \gamma_1$ and $\alpha_2, \beta_2, \gamma_2$:

$$\cos \theta = \cos \alpha_1 \cos \alpha_2 + \cos \beta_1 \cos \beta_2 + \cos \gamma_1 \cos \gamma_2$$

For parallel lines: $\alpha_1 = \alpha_2$, $\beta_1 = \beta_2$, $\gamma_1 = \gamma_2$
For perpendicular lines:

$$\cos \alpha_1 \cos \alpha_2 + \cos \beta_1 \cos \beta_2 + \cos \gamma_1 \cos \gamma_2 = 0$$

Angle between two lines with directions (a_1, b_1, c_1) and (a_2, b_2, c_2):

$$\cos \theta = \frac{a_1 a_2 + b_1 b_2 + c_1 c_2}{\sqrt{a_1^2 + b_1^2 + c_1^2}\sqrt{a_2^2 + b_2^2 + c_2^2}}$$

$$\sin \theta = \frac{\sqrt{(b_1 c_2 - c_1 b_2)^2 + (c_1 a_2 - a_1 c_2)^2 + (a_1 b_2 - b_1 a_2)^2}}{\sqrt{a_1^2 + b_1^2 + c_1^2}\sqrt{a_2^2 + b_2^2 + c_2^2}}$$

For parallel lines:

$$a_1 : b_1 : c_1 = a_2 : b_2 : c_2$$

For perpendicular lines:

$$a_1 a_2 + b_1 b_2 + c_1 c_2 = 0$$

The direction

$$(b_1 c_2 - c_1 b_2, c_1 a_2 - a_1 c_2, a_1 b_2 - b_1 a_2)$$

is perpendicular to both directions (a_1, b_1, c_1) and (a_2, b_2, c_2).

The directions (a_1, b_1, c_1), (a_2, b_2, c_2), (a_3, b_3, c_3) are parallel to a common plane if and only if

$$\begin{vmatrix} a_1 & b_1 & c_1 \\ a_2 & b_2 & c_2 \\ a_3 & b_3 & c_3 \end{vmatrix} = 0.$$

Straight lines

Point direction form: $\quad \frac{x-x_1}{a} = \frac{y-y_1}{b} = \frac{z-z_1}{c}$

Two-point form: $\quad \frac{x-x_1}{x_2-x_1} = \frac{y-y_1}{y_2-y_1} = \frac{z-z_1}{z_2-z_1}$

Parametric form: $\quad x = x_1 + ta, \ y = y_1 + tb, \ z = z_1 + tc$

General form:
$$\begin{cases} A_1x + B_1y + C_1z + D_1 = 0 \\ A_2x + B_2y + C_2z + D_2 = 0 \end{cases}$$

Direction of line: $\quad (B_1C_2 - C_1B_2, \ C_1A_2 - A_1C_2, \ A_1B_2 - B_1A_2)$

Projection of segment P_1P_2 on direction (a, b, c):

$$\frac{(x_2 - x_1)a + (y_2 - y_1)b + (z_2 - z_1)c}{\sqrt{a^2 + b^2 + c^2}}$$

Distance from point P_0 to line through P_1 in direction (a, b, c):

$$\sqrt{\frac{\begin{vmatrix} y_0 - y_1 & z_0 - z_1 \\ b & c \end{vmatrix}^2 + \begin{vmatrix} z_0 - z_1 & x_0 - x_1 \\ c & a \end{vmatrix}^2 + \begin{vmatrix} x_0 - x_1 & y_0 - y_1 \\ a & b \end{vmatrix}^2}{a^2 + b^2 + c^2}}$$

Distance between line through P_1 in direction (a_1, b_1, c_1) and line through P_2 in direction (a_2, b_2, c_2):

$$\pm \frac{\begin{vmatrix} x_2 - x_1 & y_2 - y_1 & z_2 - z_1 \\ a_1 & b_1 & c_1 \\ a_2 & b_2 & c_2 \end{vmatrix}}{\sqrt{\begin{vmatrix} b_1 & c_1 \\ b_2 & c_2 \end{vmatrix}^2 + \begin{vmatrix} c_1 & a_1 \\ c_2 & a_2 \end{vmatrix}^2 + \begin{vmatrix} a_1 & b_1 \\ a_2 & b_2 \end{vmatrix}^2}}$$

The line through P_1 in direction (a_1, b_1, c_1) and the line through P_2 in direction (a_2, b_2, c_2) intersect if and only if

$$\begin{vmatrix} x_2 - x_1 & y_2 - y_1 & z_2 - z_1 \\ a_1 & b_1 & c_1 \\ a_2 & b_2 & c_2 \end{vmatrix} = 0.$$

Planes

General form: $\quad Ax + By + Cz + D = 0$

   Direction to normal: $\quad (A, B, C)$

Perpendicular to yz-plane: $\quad By + Cz + D = 0$

Perpendicular to xz-plane: $\quad Ax + Cz + D = 0$

Perpendicular to xy-plane: $\quad Ax + By + D = 0$

Perpendicular to x-axis: $\quad Ax + D = 0$

Perpendicular to y-axis: $\quad By + D = 0$

Perpendicular to z-axis: $\quad Cz + D = 0$

Intercept form: $\quad \dfrac{x}{a} + \dfrac{y}{b} + \dfrac{z}{c} = 1$

Plane through point P_1 and perpendicular to direction (a, b, c):

$$a(x - x_1) + b(y - y_1) + c(z - z_1) = 0$$

Plane through point P_1 and parallel to directions (a_1, b_1, c_1) and (a_2, b_2, c_2):

$$\begin{vmatrix} x - x_1 & y - y_1 & z - z_1 \\ a_1 & b_1 & c_1 \\ a_2 & b_2 & c_2 \end{vmatrix} = 0$$

Plane through points P_1 and P_2 parallel to direction (a, b, c):

$$\begin{vmatrix} x - x_1 & y - y_1 & z - z_1 \\ x_2 - x_1 & y_2 - y_1 & z_2 - z_1 \\ a & b & c \end{vmatrix} = 0$$

Three-point form:

$$\begin{vmatrix} x & y & z & 1 \\ x_1 & y_1 & z_1 & 1 \\ x_2 & y_2 & z_2 & 1 \\ x_3 & y_3 & z_3 & 1 \end{vmatrix} = 0 \quad \text{or} \quad \begin{vmatrix} x - x_1 & y - y_1 & z - z_1 \\ x_2 - x_1 & y_2 - y_1 & z_2 - z_1 \\ x_3 - x_1 & y_3 - y_1 & z_3 - z_1 \end{vmatrix} = 0$$

Normal form ($p = $ distance from origin to plane: α, β, γ are direction angles of perpendicular to plane from origin):

$$x \cos \alpha + y \cos \beta + z \cos \gamma = p$$

To reduce $Ax + By + Cz + D = 0$ to normal form, divide by $\pm \sqrt{A^2 + B^2 + C^2}$, where the sign of the radical is chosen opposite to the sign of D when $D \neq 0$, the same as the sign of C when $D = 0$ and $C \neq 0$, the same as the sign of B when $C = D = 0$.

Distance from point P_1 to plane $Ax + By + Cz + D = 0$:

$$\frac{Ax_1 + By_1 + Cz_1 + D}{\pm \sqrt{A^2 + B^2 + C^2}}$$

Angle θ between planes $A_1x + B_1y + C_1z + D_1 = 0$ and $A_2x + B_2y + C_2z + D_2 = 0$:

$$\cos \theta = \frac{A_1 A_2 + B_1 B_2 + C_1 C_2}{\sqrt{A_1^2 + B_1^2 + C_1^2} \sqrt{A_2^2 + B_2^2 + C_2^2}}$$

Planes parallel: $A_1 : B_1 : C_1 = A_2 : B_2 : C_2$

Planes perpendicular: $A_1 A_2 + B_1 B_2 + C_1 C_2 = 0$

Spheres

Center at origin, radius r: $x^2 + y^2 + z^2 = r^2$

Center at (g, h, k), radius r: $(x - g)^2 + (y - h)^2 + (z - k)^2 = r^2$

General form:
$$\begin{cases} Ax^2 + Ay^2 + Az^2 + Dx + Ey + Fz + M = 0, \ A \neq 0 \\ x^2 + y^2 + z^2 + 2dx + 2ey + 2fz + m = 0 \end{cases}$$

Center: $(-d, -e, -f)$

Radius: $r = \sqrt{d^2 + e^2 + f^2 - m}$

Sphere on $P_1 P_2$ as diameter:

$$(x - x_1)(x - x_2) + (y - y_1)(y - y_2) + (z - z_1)(z - z_2) = 0$$

Four-point form:

$$\begin{vmatrix} x^2 + y^2 + z^2 & x & y & z & 1 \\ x_1^2 + y_1^2 + z_1^2 & x_1 & y_1 & z_1 & 1 \\ x_2^2 + y_2^2 + z_2^2 & x_2 & y_2 & z_2 & 1 \\ x_3^2 + y_3^2 + z_3^2 & x_3 & y_3 & z_3 & 1 \\ x_4^2 + y_4^2 + z_4^2 & x_4 & y_4 & z_4 & 1 \end{vmatrix} = 0$$

The seventeen quadric surfaces in standard form

1. Real ellipsoid: $\qquad\qquad\quad x^2/a^2 + y^2/b^2 + z^2/c^2 = 1$
2. Imaginary ellipsoid: $\qquad\quad\; x^2/a^2 + y^2/b^2 + z^2/c^2 = -1$
3. Hyperboloid of one sheet: $\qquad x^2/a^2 + y^2/b^2 - z^2/c^2 = 1$
4. Hyperboloid of two sheets: $\quad\; x^2/a^2 + y^2/b^2 - z^2/c^2 = -1$
5. Real quadratic cone: $\qquad\quad x^2/a^2 + y^2/b^2 - z^2/c^2 = 0$
6. Imaginary quadric cone: $\qquad x^2/a^2 + y^2/b^2 + z^2/c^2 = 0$
7. Elliptic paraboloid: $\qquad\qquad x^2/a^2 + y^2/b^2 + 2z = 0$
8. Hyperbolic paraboloid: $\qquad\; x^2/a^2 - y^2/b^2 + 2z = 0$
9. Real elliptic cylinder: $\qquad\quad x^2/a^2 + y^2/b^2 = 1$
10. Imaginary elliptic cylinder: $\quad\; x^2/a^2 + y^2/b^2 = -1$
11. Hyperbolic cylinder: $\qquad\quad\; x^2/a^2 - y^2/b^2 = -1$
12. Real intersecting planes: $\qquad x^2/a^2 - y^2/b^2 = 0$
13. Imaginary intersecting planes: $\; x^2/a^2 + y^2/b^2 = 0$
14. Parabolic cylinder: $\qquad\qquad x^2 + 2rz = 0$
15. Real parallel planes: $\qquad\qquad x^2 = a^2$
16. Imaginary parallel planes: $\qquad x^2 = -a^2$
17. Coincident planes: $\qquad\qquad\; x^2 = 0$

General equation of second degree

The nature of the graph of the general quadratic equation in x, y, z,

$$ax^2 + by^2 + cz^2 + 2fyz + 2gzx + 2hxy + 2px + 2qy + 2rz + d = 0,$$

is described in the following table in terms of ρ_3, ρ_4, Δ, k_1, k_2, k_3, where

$$e = \begin{bmatrix} a & h & g \\ h & b & f \\ g & f & c \end{bmatrix}, \quad E = \begin{bmatrix} a & h & g & p \\ h & b & f & q \\ g & f & c & r \\ p & q & r & d \end{bmatrix},$$

$$\rho_3 = \operatorname{rank} e, \quad \rho_4 = \operatorname{rank} E,$$

$$\Delta = \text{determinant of } E,$$

$$k_1, k_2, k_3 \text{ are the roots of } \begin{vmatrix} a-x & h & g \\ h & b-x & f \\ g & f & c-x \end{vmatrix} = 0.$$

Case	ρ_3	ρ_4	Sign of Δ	Nonzero k's same sign?	Quadric Surface
1	3	4	−	yes	Real ellipsoid
2	3	4	+	yes	Imaginary ellipsoid
3	3	4	+	no	Hyperboloid of one sheet
4	3	4	−	no	Hyperboloid of two sheets
5	3	3		no	Real quadratic cone
6	3	3		yes	Imaginary quadric cone
7	2	4	−	yes	Elliptic paraboloid
8	2	4	+	no	Hyperbolic paraboloid
9	2	3		yes	Real elliptic cylinder
10	2	3		yes	Imaginary elliptic cylinder
11	2	3		no	Hyperbolic cylinder
12	2	2		no	Real intersecting planes
13	2	2		yes	Imaginary intersecting planes
14	1	3			Parabolic cylinder
15	1	2			Real parallel planes
16	1	2			Imaginary parallel planes
17	1	1			Coincident planes

Cylindrical and conical surfaces

Any equation in just two of the variables x, y, z represents a **cylindrical surface** whose elements are parallel to the axis of the missing variable.

Any equation homogeneous in the variables x, y, z represents a **conical surface** whose vertex is at the origin.

Transformation of coordinates

To transform an equation of a surface from an old system of rectangular coordinates (x, y, z) to a new system of rectangular coordinates (x', y', z'), substitute for each old variable in the equation of the surface its expression in terms of the new variables.

Translation:

$$\begin{aligned} x &= x' + h \\ y &= y' + k \\ z &= z' + l \end{aligned}$$ The new axes are parallel to the old axes and the coordinates of the new origin in terms of the old system are (h, k, l).

Rotation about the origin:

$$x = \lambda_1 x' + \lambda_2 y' + \lambda_3 z'$$ The new origin is coincident with the old origin and

$$y = \mu_1 x' + \mu_2 y' + \mu_3 z'$$
$$z = \nu_1 x' + \nu_2 y' + \nu_3 z'$$

$$x' = \lambda_1 x + \mu_1 y + \nu_1 z$$
$$y' = \lambda_2 x + \mu_2 y + \nu_2 z$$
$$z' = \lambda_3 x + \mu_3 y + \nu_3 z$$

the x'-axis, y'-axis, z'-axis have direction cosines $(\lambda_1, \mu_1, \nu_1)$, $(\lambda_2, \mu_2, \nu_2)$, $(\lambda_3, \mu_3, \nu_3)$, respectively, with respect to the old system of axes.

Cylindrical coordinates

If (r, θ, z) are the cylindrical coordinates and (x, y, z) the rectangular coordinates of a point P, then

$$x = r\cos\theta, \quad r = \sqrt{x^2 + y^2},$$
$$y = r\sin\theta, \quad \theta = \arctan\frac{y}{x},$$
$$z = z, \qquad z = z.$$

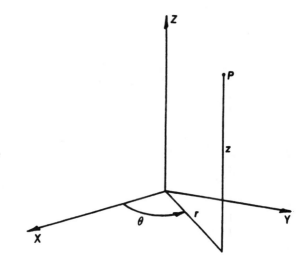

Spherical coordinates

If (r, θ, ϕ) are the spherical coordinates and (x, y, z) the rectangular coordinates of a point P, then

$$x = r\sin\theta\cos\phi,$$

$$y = r\sin\theta\sin\phi,$$

$$z = r\cos\theta,$$

$$r = \sqrt{x^2 + y^2 + z^2},$$

$$\theta = \arccos\frac{z}{\sqrt{x^2 + y^2 + z^2}},$$

$$\phi = \arctan\frac{y}{x}.$$

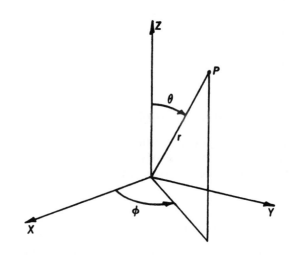

Index